Earth System Geophysics

Advanced Textbook Series

1. **Unconventional Hydrocarbon Resources: Techniques for Reservoir Engineering Analysis**
   Reza Barati and Mustafa M. Alhubail

2. **Geomorphology and Natural Hazards: Understanding Landscape Change for Disaster Mitigation**
   Tim R. Davies, Oliver Korup, and John J. Clague

3. **Remote Sensing Physics: An Introduction to Observing Earth from Space**
   Rick Chapman and Richard Gasparovic

4. **Geology and Mineralogy of Gemstones**
   David P. Turner and Lee A. Groat

5. **Data Analysis for the Geosciences: Essentials of Uncertainty, Comparison, and Visualization**
   Michael W. Liemohn

6. **Earth System Geophysics**
   Steven R. Dickman

7. **Earth's Natural Hazards and Disasters**
   Bethany D. Hinga

Advanced Textbook 6

Earth System Geophysics

Steven R. Dickman
Binghamton University, USA

This Work is a co-publication of the American Geophysical Union and John Wiley and Sons, Inc.

This edition first published 2025

© 2025 American Geophysical Union

Published under the aegis of the AGU Publications Committee

Matthew Giampoala, Vice President, Publications
Steven A. Hauck, II, Chair, Publications Committee
For details about the American Geophysical Union visit us at www.agu.org.

The right of Steven R. Dickman to be identified as the author of this work has been asserted in accordance with law.

Registered Office
John Wiley & Sons, Inc., 111 River Street, Hoboken, NJ 07030, USA

Editorial Office
111 River Street, Hoboken, NJ 07030, USA

For details of our global editorial offices, customer services, and more information about Wiley products visit us at www.wiley.com.

Wiley also publishes its books in a variety of electronic formats and by print-on-demand. Some content that appears in standard print versions of this book may not be available in other formats.

Library of Congress Cataloging-in-Publication Data Applied for:

Hardback 9781119627951

Cover Design: Wiley
Cover Image: © ANDRZEJ WOJCICKI/Getty Images

Set in 9.5/12.5pt STIXTwoText by Straive, Chennai, India

SKY10089345_102924

Dedication

—to Al Gore, for leading congressional support for the Internet, without which completing this book would have taken another decade.

—to Avraham Sternklar, for always reminding me to bring out the melody.

—to Bob Dickman, for early guidance on scientific writing (and thinking!).

—most of all, to Barb and Jennifer Dickman, for being Barb and Jennifer Dickman, and especially to Barb, for extreme patience as days and weeks stretched into months and years; this book would not have happened without her support.

Dedication

Contents

Preface   *xiii*
Acknowledgments   *xvii*
About the Companion Website   *xix*

Part I   An Earth System Science Framework

1   **The Birth of the Earth**.. *3*
   1.0   Motivation   *3*
   1.1   The Formation of the Solar System   *3*
       1.1.1   Overview: Contrasting Theories *Versus* Solar System Basics   *3*
       1.1.2   A Monistic Description of Solar System Formation   *7*
   1.2   Properties of the Solar System   *37*
       1.2.1   The Spacing of the Planetary Orbits   *37*
       1.2.2   Moment of Inertia: A Diagnostic Tool for Planetary Interiors   *42*
       1.2.3   A Brief Description of the Properties of Planets and Moons   *45*
   1.3   Life in the Solar System, and Beyond   *49*
       1.3.1   The Search for Planets   *49*
       1.3.2   Evidence for Life in the Universe   *54*
       1.3.3   Evidence for Life in Our Solar System   *57*

2   **The Evolution of Earth's Atmosphere**.. *71*
   2.0   Motivation   *71*
   2.1   The Differentiation of the Earth   *72*
       2.1.1   A Core by Condensation?   *72*
       2.1.2   An Act of Differentiation Created the Core   *74*
       2.1.3   Consequences of Core Formation   *75*
   2.2   The Faint Young Sun   *86*
       2.2.1   The Young Sun's Changing Luminosity Was Inevitable   *87*
       2.2.2   A Paradox, and Its Resolution   *88*
       2.2.3   The Urey Cycle   *90*
   2.3   Constraints on the Evolution of Atmospheric CO_2 *94*
       2.3.1   Levels of CO_2 Were (Relatively) Low During Ice House Climates   *94*
       2.3.2   Other Approaches, and a Synthesis   *97*

2.4   The Development of an Oxygen Atmosphere   99
        2.4.1   A Mostly Geology-Based Chronology of the Rise of Oxygen on Earth   100
        2.4.2   Oxygen and Evolution: An Overview   112
        2.4.3   Oxygen Chronology: A Synthesis   116

3   **The Climate System and the Future of Earth's Atmosphere**......................................119
    3.0   Motivation   119
    The Climate System   120
    3.1   The Circulation of the Atmosphere   120
        3.1.1   The Sun Is the Ultimate Driving Force   120
        3.1.2   Basic Concepts Underlying Atmospheric Circulation   122
        3.1.3   Global Atmospheric Circulation on a Nonrotating Earth   123
        3.1.4   Global Atmospheric Circulation on the Rotating Earth   124
        3.1.5   Complications of the Three-Cell Model   125
        3.1.6   Implications of the Three-Cell Model for Climate and Regional
                 Circulation   128
        3.1.7   A Brief Jovian Perspective   131
        3.1.8   Jet Streams in the Atmosphere   133
        3.1.9   Hurricanes   136
    3.2   The Circulation of the Oceans   136
        3.2.1   Thermohaline Convection   137
        3.2.2   Wind-Driven Circulation   143
        3.2.3   The Wind-Driven Oceans Move Heat, Too   147
    3.3   El Niño and the Southern Oscillation: A Coupled Atmosphere—Ocean
           Phenomenon   150
        3.3.1   El Niño   150
        3.3.2   Southern Oscillation   151
        3.3.3   The Mechanism of a Strong ENSO Event   155
        3.3.4   The Mechanism of a Weak ENSO Event   158
        3.3.5   The Return to Normalcy   159
        3.3.6   There's an Even Bigger Picture   160
    The Immediate Future of Our Atmosphere   163
    3.4   Preliminary Comments   163
    3.5   Solar Variability on Human Timescales   163
        3.5.1   Sunspot Cycles   163
        3.5.2   A Connection Between Sunspots and a Dramatic Change in Earth's
                 Climate?   165
        3.5.3   A Few Final Comments on Sunspots   169
    3.6   Anthropogenic Variations in Climate by the Emission of Greenhouse Gases   169
        3.6.1   Increases in Greenhouse Gas Abundances   170
        3.6.2   Direct *and* Indirect Impacts on Climate Expected From an Increase in
                 Greenhouse Gas Abundances   173
        3.6.3   Tempered Expectations: Complications in How These Consequences
                 Play Out   185
        3.6.4   Anthropogenic Variations in Climate: Evidence Concerning *Direct*
                 Consequences (—If You Insist)   193

3.6.5 Anthropogenic Variations in Climate: Evidence Concerning *Indirect* Consequences 201

3.6.6 Anthropogenic Climate Change: Some Final Thoughts 222

A Geophysical Perspective: *The Rest of This Textbook* 223

Part II A Planet Driven by Convection

4 Basics of Gravity and the Shape of the Earth..227

4.0 Motivation 227

4.1 The Nature of Gravity 228

4.1.1 Simple Expressions of the Law of Gravitation 228

4.2 Newton's Second Law and the Gravity Field 235

4.2.1 Cause and Effect, Mass and Weight 235

4.2.2 Earth's Gravity Field, and the Answer to a Really Fundamental Question 236

4.2.3 Weighing the Earth 239

4.3 The Gravity Field of a Three-Dimensional Earth 242

4.3.1 A Guiding Principle 242

4.3.2 More Consequences of Gravity Being an Inverse-Square Law Force 243

4.3.3 Revisiting Newton's Law of Gravitation, With Superposition 246

4.4 The Shape of the Earth, and Variations of Gravity With Latitude 248

4.4.1 A Motivation to Get Complicated 248

4.4.2 The Earth Is Not Spherical 248

4.4.3 Earth's Rotation Is the Cause 249

4.4.4 A Thorough Description of Centrifugal Force 252

4.4.5 Gravity *Versus* Centrifugal Force on a Rotating Earth 258

4.4.6 Indirect Effects of Centrifugal Force on Gravity, and the Idealized Earth 261

4.5 Kepler's Laws 265

5 Gravity and Isostasy in the Earth System..269

5.0 Motivation 269

5.1 Exploring the Earth System with Gravity 269

5.1.1 Scaling Down for Gravity Exploration 269

5.1.2 The Reduction of Gravity Data 274

5.1.3 An Application to the Earth System 281

5.1.4 Gravity Data Measured on a Moving Platform 285

5.2 Isostasy and the Earth System 287

5.2.1 Bouguer's Discovery and the Principle of Isostasy 287

5.2.2 Mechanisms to Achieve Isostatic Balance 289

5.2.3 The Moho and Other Evidence of Airy Isostasy 293

5.2.4 Airy Isostasy and the Oceanic Response to Atmospheric Pressure Fluctuations 294

5.2.5 A Third Mechanism for Achieving Isostatic Compensation 296

5.2.6 Isostatic Response to Surface Loads in the Earth System: Anomalous Regions 300

5.2.7 Isostatic Response to Surface Loads: Implications for Mantle Rheology 313

5.2.8 Global Constraints on Mantle Viscosity 319

6   Orbital Perspectives on Gravity *329*

  6.0   Motivation   329

  6.1   Tides   329

      6.1.1   Ebbs and Flows   329

      6.1.2   Tidal Forces   330

      6.1.3   The Response of the Oceans to Tidal Forces   332

      6.1.4   The Response of the Solid Earth to Tidal Forces   335

  6.2   Precession of the Equinoxes and Orbital Effects on Climate   338

      6.2.1   Precession   338

      6.2.2   Precession, the Core, and the Geomagnetic Field of the Earth   344

      6.2.3   Precession Can Affect the Earth's Climate   345

      6.2.4   Milankovitch and Mars   357

  6.3   Satellite Geodesy   359

      6.3.1   Satellite Orbital Precession   359

      6.3.2   The Geoid and Satellite Altimetry   368

      6.3.3   Geoid *Versus* Spheroid, and Geoidal Heights   379

      6.3.4   More Perspectives on Global Gravity and the Global Geoid   383

  6.4   Tidal Friction   394

      6.4.1   Another Way of Looking at Tides   394

      6.4.2   The Solid Earth Will End Up in the Middle of It All, and Suffer Greatly   396

      6.4.3   Tidal Friction Also Affects the Moon's Orbit   399

      6.4.4   Tidal Friction Has Consequences for the Earth System   401

      6.4.5   Tidal Friction Without Oceans, and Astronomical Implications   403

      6.4.6   Theories of the Origin of the Moon   405

7   Basics of Seismology *409*

  7.0   Motivation   409

  7.1   Stress and Strain   409

      7.1.1   Stress   410

      7.1.2   Stress: A Rigorous Description   412

      7.1.3   Strain   414

      7.1.4   Strain: A Rigorous Description   417

  7.2   Relations Between Stress and Strain in Elastic and Nonelastic Materials   419

      7.2.1   Ideal Models of Different Materials   420

      7.2.2   Material Properties of an Elastic Medium: *Elastic Parameters*   427

  7.3   Elastic Waves   431

      7.3.1   Descriptions of Waves   432

      7.3.2   Elastic Waves: The Wave Equation   433

      7.3.3   Elastic Waves: Reflection and Refraction   437

  7.4   Surface Waves and Free Oscillations   452

      7.4.1   Surface Waves   452

      7.4.2   Free Oscillations   457

  7.5   Seismic Waves and Exploration of the Shallow Earth   475

      7.5.1   Refraction Surveys; or, First Arrivals on a Flat Earth   475

	7.5.2	Refraction Surveys: An Illustration With Possible Hydrogeological Implications	*478*
	7.5.3	Refraction Surveys: Thoughts About Multilayered Situations	*481*
7.6	Seismic Waves and Exploration of the Whole Earth: Preliminaries		*482*
	7.6.1	Travel Times: Lateral Homogeneity Within the Earth	*482*
	7.6.2	Travel Times: Locating Earthquakes	*484*
	7.6.3	Travel Times: Identifying Phases on a Seismogram	*489*

8 Seismology and the Interior of the Earth *491*

8.0	Motivation		*491*
8.1	Seismology and the Dynamic Earth		*492*
	8.1.1	Defining Plate Tectonics	*492*
	8.1.2	Quantifying Plate Motions	*495*
	8.1.3	Plate Motions Through the Ages	*498*
	8.1.4	A Last Look at Plates and Plate Motions	*505*
	8.1.5	Travel Times and the Interior of the Earth	*510*
8.2	Seismology and the Large-Scale Structure of the Earth		*514*
	8.2.1	Travel Times: The Shadow Zone and the Core	*514*
	8.2.2	Travel Times: Determining Seismic Velocities Within the Earth	*516*
8.3	Seismic Velocities and the State of Earth's Interior		*521*
	8.3.1	Birch's Rule	*522*
	8.3.2	The Adams-Williamson Equations	*532*
	8.3.3	Seismic Tomography	*554*
8.4	Using Earth Models to Learn About the Composition of the Interior		*564*
	8.4.1	Equations of State	*564*
	8.4.2	High-Pressure Experiments	*568*

9 Heat From Earth's Interior *597*

9.0	Motivation		*597*
9.1	Measuring Heat Flow		*598*
	9.1.1	Basic Ideas and Practical Challenges	*598*
	9.1.2	Heat Flow Data	*600*
	9.1.3	Strengthening Our Theoretical Foundation of Heat Flow: An Introduction to Del	*606*
9.2	Heat Sources		*607*
	9.2.1	Radioactivity	*607*
	9.2.2	Gravitational Energy, Part One	*609*
	9.2.3	Heat of Compression	*609*
	9.2.4	Gravitational Energy, Part Two	*613*
	9.2.5	Moon-Forming Impact	*613*
	9.2.6	Tidal Friction	*614*
	9.2.7	Another Look at Radioactivity	*614*
	9.2.8	Growth of the Inner Core	*618*
	9.2.9	Some Reflections, and What Must Come Next	*619*
9.3	Transmission of Heat in Solids		*621*
	9.3.1	Conduction Plus Conservation Equals Diffusion	*621*

9.3.2   The Nature of Diffusion   *624*

9.3.3   Some Solutions to the Diffusion Equation   *627*

9.3.4   Learning From Failure: A Deeper Look Into Heat Flow by Conduction   *630*

9.4   Transmission of Heat in Fluids   *638*

9.4.1   Fluid Stability   *639*

9.4.2   How Convection Works in a Fluid   *640*

9.4.3   Heat Transmission in a Convecting Fluid   *649*

9.4.4   Horizontal Convection   *652*

9.4.5   Temperatures Within a Convecting Fluid; Fluid *Versus* Solid-State
          Convection   *658*

9.5   More on Surface Heat Flow in the Earth System   *666*

9.5.1   Geothermal Heat and the Thermohaline Circulation   *667*

9.5.2   Subsurface Temperature Variations and Climate Change   *669*

10 Geomagnetism and the Dynamics of the Core..*677*

10.0 Motivation   *677*

10.1 The Earth's Magnetic Field   *678*

10.1.1 Dipole Fields   *678*

10.1.2 An 'Elemental' Description of Magnetic Fields, with Reference
          to the Earth   *680*

10.1.3 Magnetic Fields: An Overview of the Earth System   *682*

10.2 Global Descriptions of the Internal Field   *707*

10.2.1 Satellite Missions Dedicated to Observing Earth's Magnetic Fields   *708*

10.2.2 Spherical Harmonics, Once Again   *711*

10.2.3 Back to the Surface: A Closer Look at the Crustal (and Geomagnetic)
          Fields   *718*

10.3 Snapshots in Time of the Geomagnetic Field   *726*

10.3.1 Current and Recent Snapshots   *726*

10.3.2 Snapshots Further Back in Time   *743*

10.4 Generation of the Geomagnetic Field   *755*

10.4.1 Preliminary Assessments   *755*

10.4.2 Fields Weaken, Fields Strengthen   *758*

10.4.3 Examples of Simple Dynamos   *765*

10.4.4 Inescapable Wisdom From Unavoidable Equations   *768*

10.4.5 Dynamo Flow in a Taylor-Proudman World   *776*

10.4.6 Some Final Thoughts   *788*

References   *791*

Index   *891*

Preface

An Earth System Science Approach to Geophysics, and a New Unifying Theme

Geophysics—the physics of the Earth—has always been a powerful way to understand the world. Connecting complex real-world phenomena to fundamental physical laws, and using those connections to deduce the nature of otherwise inaccessible regions of the Earth; framing natural processes and events in terms of cause and effect (which, after all, is what Newton's Second Law, usually written $\vec{F} = M\vec{a}$, embodies); constructing order-of-magnitude relationships to determine the relative importance of causative factors; and, of course, mathematically modeling and successfully predicting the future behavior of components of the Earth—all of these *clarify* the world around us wonderfully.

There was a time when geophysics was mainly devoted to the study of the 'solid earth'—the crust we stand on and the mantle and core below. That was a time when the theory of plate tectonics ruled as the unifying theme of Earth science; when the Cold War focused scientific research, demanding tools to (among other things) distinguish earthquakes from underground nuclear tests; and when our knowledge of the nature and behavior of the oceans and atmosphere was just beginning to blossom.

That time is past. We now see plate tectonics as part of a bigger process, whether on Earth or the other planets and moons. And, although seismology is crucial in defining the structure and properties of Earth's interior, it is no longer necessary to treat it as the central discipline of geophysical research. It can also be argued that our understanding of the atmosphere and oceans is now comparable to our understanding of the solid earth.

Over the past few decades, a newer and equally revolutionary paradigm—Earth System Science—has gained popularity in studies of the Earth. This paradigm recognizes the critical importance of interactions between the components of what is seen as an **Earth System**: the solid earth, oceans, and atmosphere, and even the biosphere (from a 'sphere' perspective, we could add the celestial sphere as well).

Geophysics is, at its essence, a multidisciplinary field. And even solid-earth geophysics provides invaluable tools to understand Earth System Science better, both conceptually and quantitatively. It makes sense to learn geophysics from an Earth System Science perspective.

In Earth System geophysics, it is not plate tectonics but rather the more fundamental theme of ***convection***—defined generally as a circulation of material driven by density differences, with lighter stuff rising and denser stuff sinking—that connects the components of the System. Mantle convection drives tectonics (with tangible consequences for the crust), but the core, oceans, and atmosphere also exhibit convective behavior; for that matter, within all stars, some planets, and even a few moons, convection of some type is the rule. Convection is indeed universal.

This textbook is a modest attempt to expand solid-earth geophysics within the framework of Earth System Science. Hopefully, such a perspective will increase the accessibility, and even popularity, of geophysics to Earth science students.

Who This Book Is For

Undergraduate and graduate students majoring in geophysics, physics, and engineering, as well as students working toward a master's in Earth science teaching, have all benefited from the courses this textbook is based on. However, this book's intended audience is primarily students of geology at the senior undergraduate or beginning graduate level, whose exposure to basic physical geology has been supplemented by at least one semester each of calculus and college physics but who may be somewhat unconfident about using math and physics to understand the Earth. This textbook *builds gradually* to advanced math and physics applications and should therefore be read in *sequential* order.

As an accompanying goal, then, this book can enhance the quantitative skills of the reader. After completing this textbook, the student will be familiar with such second-order partial differential equations governing Earth System processes as the diffusion equation and various fluid dynamic equations. Such familiarity will include an understanding of the origin, components, and uses of those equations; the ability to verify specific solutions mathematically; scaling techniques for interpreting and evaluating the equations; and a simple numerical technique for obtaining approximate solutions. Students with such familiarity should be well prepared for more advanced geophysics courses.

A Textbook for Different Courses

This textbook is designed for a full-year (two-semester) upper-level introductory course on the geophysics of the Earth System. But its breadth also allows for its use as the primary textbook in a variety of semester-long courses. Possibilities include the following:

- A traditional, solid-earth geophysics approach—but with applications to the Earth System—is possible by focusing (sequentially!) on Chapters 4 through 8 and portions of 10.

- Students with a major interest in global warming could start with Chapters 2 and 3, and then learn the geophysics behind some of their remote-sensing data (satellite geodesy) in Chapter 6 before finishing with some of the basics of fluid dynamics in Chapters 8 (hydrostatic equilibrium) and 9 (heat transmission, rigorously).

- Students with an interest in planetary geophysics would enjoy Chapter 1 on the origin and properties of the Solar System and the question of life beyond Earth; the section of Chapter 3 on solar variability; the orbital perspectives of Chapter 6 (the latter, including the Milankovitch hypothesis and tidal friction, would be preceded by Chapters 4 and 5 for background); and portions of the remaining chapters with comparisons to other planets and the Sun.

- For advanced students who have already taken an introductory geophysics course, the numerous references cited throughout this textbook (and available as a full bibliography in the Companion Website) could serve as guidance in an advanced geophysics seminar course, providing a 'starting point' for their journal research, while the book itself provides common background material for the rest of the class.

Whichever curricular path is followed, instructors may find this textbook's discussion of some topics unfairly brief, or even unrigorous. For such topics, this might have been a consequence of how I chose to avoid overwhelming the reader: simply put, on occasion it is better to present some concepts gradually and a bit loosely rather than all at once or with complete rigor. Instructors are encouraged to highlight their own selection of topics—those more 'limited' passages in the textbook then amounting to a 'first step' in their presentation.

As a traditional solid-earth geophysicist by training, I freely admit to a long-held bias in preferring internal to external causes wherever possible, for example climate change rather than asteroidal impact as the reason dinosaurs became extinct or outgassing rather than impact degassing as the main source of our second atmosphere. I encourage instructors using this textbook to provide their own balance as necessary. Instructors may also wish to point out to their students that, for some of the topics covered in this textbook, a passionate debate or a limited consensus might be the 'back story' underlying my seemingly factual presentation.

How to Read This Book

First of all, expect to encounter *a mixture of units* in this book, such as bars (or kbar) to describe pressure in the atmosphere and oceans but Pascals (or MPa or GPa) for pressure within the solid earth. This is unavoidable when dealing with the Earth System. Expect also to encounter, on occasion, somewhat 'improper' units for density (gm/cm^3 rather than kg/m^3) and temperature (°F rather than °C), in the hope of making those quantities more relatable.

Second, *notation* has very specific meaning and uses throughout this book.

- Important words, phrases, and sentences are *italicized*.
- A word or phrase being defined is in **bold** type.
- The symbol "~" is used to indicate similarity in magnitude, as opposed to equality or precision, as in "this textbook took ~10 years to write."
- Single quotes are used to reference an informal term or phrase with an everyday meaning; for example, "the planet Venus rotates

'backward' … " refers to our Earthly experience. It is not meant to imply either that some kind of value judgment is being placed on the state of that planet or that "forward" and "backward" are the terms formally used to describe a planet's sense of rotation.

- Textboxes are meant to provide a focus for the reader; their content should *not* be viewed as optional or random.
- Occasionally, the reader will be directly addressed with a question; these 'stop and think' questions are italicized and written in *brown* for emphasis. They are intended to keep the reader 'caught up' on the concepts being presented.
- In contrast to tradition, equations in this textbook are not numbered—an attempt to avoid 'number glaze,' as the level of mathematics is gently though inexorably increased. If prior equations are referred to without being restated, their original location can be identified simply by noting the earlier topic they relate to; this should be straightforward if the chapters are read sequentially. Additionally, although no equations the reader will encounter here are extraneous, *key* equations are highlighted in blue.
- Finally, the careful reader may note that I make frequent use of dashes and semicolons, with the result that the text includes many lengthy sentences; each such long sentence should be viewed as a complete thought. They may best be understood with the help of the sentence's punctuation, which serves to break up the thought into its components (for practice, review this sentence and the preceding one!).

Third, be warned: in this textbook, we will occasionally discuss both the currently accepted explanation of some phenomenon or observation and an alternative, no-longer-popular theory. For example, though I began teaching about the importance of global warming as far back as the 1980s, I still consider arguments against it to be worth teaching, and worth including in this book. Another example is the "nebula theory" describing how our Solar System formed, versus alternative, 'chance collision' type older theories. That is, some subjects are not presented as completely 'settled.' It will be up to the student to weigh both sides, judging their merits as well as defects, and understand why those alternatives are now less acceptable.

Fourth, be active! Students taking the courses upon which this textbook is based generally found my lectures to be well-organized and logical. But eventually I realized it was important to caution the class, at the outset, that finding that the material 'makes sense' is not the same as understanding it. What they needed to do was *digest* the material—put it into their own words—and *interrelate* it—connect concepts and facts from each day's lecture with those from previous lectures, gradually building a picture of how the Earth or Earth System works. Such a nonpassive approach is recommended for reading this textbook. And it should include confronting the material through homework assignments; problem sets based on those courses are available for each chapter at the Wiley website https://www.wiley.com/go/Dickman.

Steven R. Dickman
Binghamton University, USA
Summer 2006–2007
Winter/Spring/Summer 2013
2014–2024

Acknowledgments

During my early work on this textbook, occasional encouragement by Joe Preisig, Jon Aurnou, James Haddad, Al Tricomi, John Eidenier, Bennett Spevack, Joseph Newmark, Jon Burgman, Dave Yuen, and Roberto Sabadini, and logistical support by Joe Graney, played a significant role in keeping me from abandoning the project. As the project continued, tentative inquiries from numerous other friends and relatives were also appreciated. Additionally, I could not have completed this book without *huge* assistance from Elise Thornley and other staff members of the Binghamton University Library's Interlibrary Loan Office in obtaining pdfs of journal articles.

I also must acknowledge the many bright students in my classes at the State University of New York at Binghamton, SUNY-Binghamton, and Binghamton University; without their exceptional talents, I might not have been so motivated to develop the curriculum in those courses to the breadth and depth indicated by this book.

And, significant improvements in some chapters must be credited to the scientists (Jon Aurnou, Ben Chao, Richard Holme, Dominique Jault, James Kasting, Sébastien Merkel, Peter Olson, Roberto Sabadini, and several more who chose anonymity) who reviewed those chapters in later stages of this textbook's development. I am grateful for the time and thought they put into their reviews.

Finally, this book took shape thanks to the phalanx of editors, editorial assistants, and publication staff at AGU and Wiley who guided its development and brought it to life, starting with the initial encouragement by Brooks Hanson; continuing with advice and oversight by Ritu Bose, Noel McGlinchey, Summers Scholl, Neena Ganjoo, and others at Wiley; and culminating with editorial polishing by Jenny Lunn and the AGU Books Editorial Board, and its actual publication by Wiley.

About the Companion Website

This book is accompanied by a companion website:

www.wiley.com/go/Dickman

The website includes:

- References cited in all chapters
- Homework exercises for each chapter
- Table 3.3: Circumstantial evidence of global warming
- Brief guidance for instructors on mathematical challenges in each chapter

Part I

An Earth System Science Framework

1

The Birth of the Earth

(Did you read the Preface yet?)

1.0 Motivation

Earth is part of the Solar System, a collection of objects including eight officially designated planets, numerous moons orbiting those planets, and three major debris-filled regions, all orbiting a star we call the Sun. Knowledge of our astronomical setting is important for a number of reasons. The natural processes that take place on Earth's surface—whether the fluid dynamic processes within the oceans and atmosphere that create the climate of a region, or the geological processes of deposition and erosion that modify surface landforms—are driven to a large extent by heat energy from the Sun. The forces exerted by the Sun and Moon mobilize the oceans, deform the lithosphere, and influence the biosphere on a wide range of timescales. As distant as it is, even Jupiter forces measurable changes in our climate system. And, our surface has occasionally been altered by impacts of debris originating elsewhere in the Solar System.

Even the interior of our planet is connected to the rest of the Solar System. For example, convection within the Earth's mantle is fueled by a number of sources: energy derived from the decay of radioactive isotopes within the interior, catastrophic events that occurred early in the Earth's history, and even the highly compressed state of the interior—all of which Earth shares in common with the other major bodies of the Solar System. Thus, an appreciation of the 'basics' of the Solar System, including its overall composition, how it formed, and how it works today, will enhance our understanding of how the Earth operates.

Our Solar System is about $4\frac{1}{2}$ billion years (Byr) old; in contrast, the Universe began roughly 13–15 Byr ago. We are relative newcomers to the Universe. In all the eons that stars and galaxies have been around, are we unique in nature? That is, should we expect life, most importantly intelligent life, to exist (or to have existed) elsewhere? Civilizations are unlikely to flourish in extreme environments such as stars or the near-vacuum of outer space, so this question amounts to asking whether *planets* are a common or rare occurrence. Our understanding of how the Solar System came into existence will help answer that question.

1.1 The Formation of the Solar System

1.1.1 Overview: Contrasting Theories *Versus* Solar System Basics

The various theories proposed over the past four centuries to explain the formation of the Solar System fall into one of two categories: **monistic**, in which the Sun and planets all formed together from the gravitational collapse of a primitive nebula of gas, dust, and ice; or **dualistic**, in which a chance near-collision

Earth System Geophysics, Advanced Textbook 6, First Edition. Steven R. Dickman.
© 2025 American Geophysical Union. Published 2025 by John Wiley & Sons, Inc.
Companion Website: www.wiley.com/go/Dickman

between the Sun and a second massive object (such as another star or, as we might imagine today, a black hole, passing through our stellar neighborhood) drew a stream of gas out from the Sun that condensed into the planets.

Dualistic theories attempt to explain the origin of the planets, not the Sun. If planetary systems can only be created by chance encounters, though, they are not very likely to be found throughout the Universe. In contrast, because the processes underlying a nebula's collapse will, as a rule, produce planets as well as a star, monistic theories suggest at first glance that planets should be as numerous as the stars in the sky. Actually, because of what can happen during or after a nebula's collapse, as we will discuss in the last section of this chapter, it turns out that planets—including some very exotic types—might even be *more* abundant than stars!

Dualistic Theories. Dualistic theories were first proposed by Buffon in 1745, revived by Moulton and Chamberlin a century and a half later, and advocated by Jeans and Jeffreys (incidentally, Jeffreys, a seismologist, was possibly the most influential geophysicist of the twentieth century, due both to his longevity—with publications spanning at least eight decades—and his authorship of 'the' textbook in geophysics).

Hello, dear reader! I would like to take this opportunity to introduce you to one feature of this textbook: moments where you are addressed personally—what we might call 'stop and think' moments, occasions where you can come up with your own answers! Ready? Here's the first one:

It will quickly become apparent that dualistic theories, though based on valid physical principles, are problematic and not borne out by any observations. Can you think of any reasons—scientific or pedagogical—why we should devote time to discussing or even just presenting such theories?

In dualistic theories, the gas strung out between the Sun and the passing 'star' is envisioned to have been pulled out of the Sun by the gravitational (and tidal) forces exerted on the Sun's mass by the other object as it passed close to the Sun. One difficulty with such theories is that the material pulled out of the Sun would have been quite hot—the outer envelope of the Sun, called the **corona**, is of the order 10^6 degrees—so that any material not falling back to the Sun (by its own gravity) would be more likely to dissipate into space than to condense into planets. In fact, that envelope is so hot that even without the influence of any passing object, it is continually expanding away from the Sun (which replenishes what is lost). Such matter, composed of gases (mostly hydrogen) ionized by the high temperature and thus reduced to electrons, protons, and some heavier nuclei, expelled in all directions, is known as the **solar wind**.

Other difficulties faced by dualistic theories follow from our observations of the mechanical properties of the Solar System. *The orbits of the planets around the Sun are not random.* The **ecliptic** is defined to be the plane of the Earth's orbit around the Sun. Our first observation is that—with the exception of Mercury, the closest planet to the Sun, and Pluto, the farthest planet (if it is to be counted as a planet)—the orbital planes of all the planets lie within $3\frac{1}{2}°$ of the ecliptic; see Figure 1.1. Even the exceptions align roughly with it: Mercury's orbit tilts by $7°$, which is mild, considering Mercury is so close to the Sun that its movement is overwhelmingly dominated by the Sun's gravity; Pluto's orbital misalignment of $17°$ is somewhat larger, suggesting the need for further explanation, though even $17°$ is not a major tilt.

Second, all of the planets orbit the Sun in the same direction, a counterclockwise direction when viewed from above; *and* the Sun—containing by far the bulk of the Solar System's mass—spins on its axis in that counterclockwise direction as well. For that matter, with a few dramatic exceptions—Venus and Pluto, which rotate 'backward,' and Uranus, whose rotation axis is more or less *in the plane* of the ecliptic—most of the planets spin on their axes in that same counterclockwise sense.

Figure 1.1 The planets of the Solar System: Mercury, Venus, Earth, Mars, Jupiter, Saturn, Uranus, and Neptune (picture from http://nineplanets.org/overview.html). Below the photomosaic is a graph indicating the planets' orbital inclinations relative to the ecliptic (dashed horizontal line); the letters mark the locations of the planets (the inner planets are too close to be distinguished at this scale). Pluto (P) is also included for comparison. Distances are in units of astronomical units, with 1 AU = 93 million miles, the distance of Earth from the Sun. NASA.

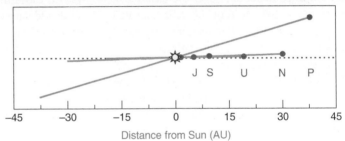

Distance from Sun (AU)

Those 'backward' spins, incidentally, may have developed well *after* Solar System formation was complete; Pluto's spin, for example, was likely a consequence of gradual tidal "locking" with its moon Charon. In the case of Venus, the situation is less clear, but it is thought that the opposition of tidal forces exerted by the Sun on the planet versus on its heavy atmosphere helped (along with very slight but persistent tugging by Earth's gravity) bring Venus to the extremely slow state of rotation it exhibits today (Gold & Soter, 1969; see also Bills, 2005 and references therein); under some conditions, those opposing tidal forces could have counteracted each other to the point where Venus' spin was reversed (Correia & Laskar, 2001). The kind of tidal effect, known as "tidal friction," which controlled Venus' rotation rate so much—and which also happens to cause tidal locking—is discussed in Chapter 6.

The orbital motions and spins of the Sun and planets indicate a *common origin*, not an independent and unrelated origin for the Sun as implied by dualistic theories. Moreover, with a dualistic cause, and depending on the passing star's trajectory, planetary systems would be almost as likely to revolve around polar orbits as they would with equatorial orbits (!); the observation that all of the planets in our Solar System orbit the Sun approximately about its equator requires, under a dualistic mechanism, that the passing star had to pass our Sun in (or very nearly in) its equatorial plane. Even worse, the passing star would have had to pass our Sun in the same counterclockwise sense as the Sun's own spin, or else the resulting planets would not orbit in that same sense. Altogether, the chance of the 'right' kind of Sun-star interaction required by dualistic theories to explain our Solar System seems remote.

Monistic Theories. The idea that the planets and Sun formed from one primordial nebula of gas and dust was pioneered by Descartes in 1644. Kant in 1755 and Laplace in 1796 were the first to provide a physical basis for the process of formation, invoking the principle of *angular momentum conservation* to propose that the nebula spun faster as it contracted (just as ice-skaters pulling in their arms will spin faster), accelerating to the point where it shed rings of matter which coalesced into the planets.

By itself, however, such coalescence is not guaranteed (e.g., Lowrie, 2007). Over the lifetime of the Solar System, for example, the Kuiper Belt, a ring of debris outside the orbit of Neptune, has not managed to coalesce into a unified mass; and, over the past 4 Byr, the Asteroid Belt, a zone of debris located between Mars and Jupiter, has similarly failed to aggregate into a planet. Evidently the mutual gravitational forces acting between the particles of such rings are too weak, and too dispersed, to overcome competing factors and fully coalesce (of course, there's more to that story—to be discussed later in this chapter).

In fact, the very orderly collapse envisioned by Laplace was overly simplistic. First of all, as the solar nebula contracted, it heated up, and a chaotic flow—turbulence—ensued, superimposed on the overall inward drift of the particles. Second, as we now understand, the contracting nebula evolved into a massive central proto-star surrounded by a thin disk of gas and dust; within that disk, the planets would grow, but as they did their orbits may have shifted, seemingly at random, as gravitational forces also evolved. Early on, this could have taken place under the influence of the disk at large (Goldreich & Tremaine, 1980; see also, e.g., Ward, 1997); later, it would have happened as multiple proto-planets fought to maintain their own orbital stability. Though it might seem counterintuitive, it was even possible for their gravitational forces to sling each other away from the Sun, *ejecting* some from the inner nebula (such *planet-planet scattering* is reviewed in, e.g., Armitage, 2014). Thus, the reality was not simply a stately, organized 'gathering in' of the nebula, occasionally punctuated by brief nebula-wide 'dumping' events to produce the planets.

Nebula theory has also been viewed as the basis for a geochemical 'orderliness' in our Solar System, with planetary composition considered to be a consequence of the overall temperature gradient within the disk as the planets formed, i.e., the disk was hottest near the developing Sun and coolest far away. Most fundamentally, away from the proto-Sun it would have been cold enough throughout the nebula's evolution for ices (such as water ice) to exist in abundance; indeed, we see proof of this in the icy moons, giant "ice" planets Uranus and Neptune, and numerous comets of the outer Solar System. Close in, just the opposite prevailed: ices would have vaporized, and the planets that developed in this environment—Earth and the other "terrestrial" planets—would have been more 'rocky' than icy.

However, the full range of planetary compositions in the Solar System, not just ice, has also been taken to be a consequence of the nebula's temperature gradient. Refractory substances like iron, for example, able to survive the high temperatures of the inner nebula without vaporizing, should be increasingly abundant (proportionally speaking) in planets closer to the developing Sun. And more generally, any given mineral (or metal or element) would, as some region of the nebula cooled below its *condensation temperature*, condense into solid particles, providing an abundance of rocky grains that could be incorporated into a nearby proto-planet. Any planet's composition, especially in the inner Solar System, would then reflect the availability and abundance of minerals that had condensed out of the nebula; and (reminiscent of the sequential orderliness of Laplace's ring-shedding), from planet to planet we would see trends—chemical gradients—ultimately reflecting the decrease in temperature with distance from the Sun, and how the decrease below condensation temperatures happened closer and closer to the Sun as the disk cooled.

Unfortunately, such clear geochemical patterns are not seen in the planets of the Solar System. Earth and Venus, for example, are believed to have iron cores very similar in mass (and similar in proportion to their silicate mantles), implying a negligible 'iron gradient' across at least a portion of the inner Solar System. A thorough analysis of numerous minerals and elements by Palme (2000) found no clear chemical gradient, no systematic variation in planetary composition, across the entire inner Solar System. He attributed this to a nearly zero temperature gradient (i.e., constant temperatures) within the nebula inside Earth's orbit, and just a slight gradient (modest cooling outward) through the rest of the inner region, according to the nebula model of Cassen (1994). Finally, if any planets (or their earlier proto-planet predecessors) have migrated away from their original orbits, moving closer to or farther away from the Sun, any original pattern of planet chemistry would have been disrupted (see Lissauer & de Pater, 2013); consider, for example, the quite exotic possibility (Hansen, 2009) that the four innermost planets all originated roughly in the vicinity of Earth's present-day orbit!

Speaking in general terms, we will see below that creating a star out of an expansive cloud of gas, dust, and ice requires concentrating both the nebula's mass *and* its developing heat within the central, would-be-a-star region of the nebula. The more successful those concentrations are, the greater the likelihood of a star emerging; conversely, the higher the temperature outside of the would-be star, the more difficult it would be to concentrate sufficient mass centrally—the heat would resist further contraction. As our nebula collapsed 'successfully,' on its way to producing and igniting a Sun, the planets would have formed in a comparatively cool environment. Observations discussed below (in Stage 6 of our scenario of Solar System evolution) confirm this prediction.

The thermal conditions of the primitive solar nebula certainly deserve attention in any monistic theory. But from our perspective,

perhaps a better question than which inner planets got more than a fair share of iron would be *what happened to their share of the gases, i.e., their atmospheres?* That question will also be addressed in the sections below.

1.1.2 A Monistic Description of Solar System Formation

Monistic theories have evolved significantly over the past several decades, with seminal work by Safronov, Cameron, Prentice, and others in the late 1960s and 1970s (see, e.g., Dermott, 1978), stimulated by the advent of the Space Age and aided by the use of computers to model the process of nebula collapse. More recently, abundant and transformative modern observations—not just from the Hubble Space Telescope but also from a variety of other satellite-based platforms and ground-based observatories using visible light, infrared, and radio wavelengths (see, e.g., Williams & Cieza, 2011)—have provided critical constraints on these theories, leading to an intense focus on the mechanisms of planetary growth and the evolution of planetary systems. Chemical analysis of both short- and long-lived isotopes and their distribution throughout the Solar System has also been invaluable. As a result of such modeling and observations, monistic theories now feature:

- an important role for the infant Sun, in a stage called the *T-Tauri* phase, in helping the nebula complete its evolution into a solar system;

- a sequence of gravitational and mechanical processes to grow planets efficiently from micron-sized particles to full and even giant planet size;

- limited accretion in the Asteroid Belt, now seen to have been the home of numerous planetesimals, and eventually some larger planetary 'embryos' and 'proto-planets' as well, but not ever full-sized planets (despite the classic, almost science-fiction-like view that planets did exist there but subsequently suffered catastrophic collision, leaving the fragments we would call asteroids);

- rapid accretion of planets and quick completion of the Solar System—over a timescale (less than ~10 million years (Myr)) for the Sun and giant planets that was at least an order of magnitude shorter than earlier models had postulated;

- early differentiation of the terrestrial planets into mantles and cores, likewise faster than in earlier models; and

- huge amounts of energy retained during planetary accretion, immediately rendering planetary surfaces and/or their interiors extremely hot.

One fundamental observation underpinning these refinements to monistic theory was the detection of flattened solar nebulas—called **circumstellar disks**, **accretionary disks**, **protoplanetary disks**, or 'proplyds'—surrounding proto-stars (Herbig-Haro and T-Tauri stars) similar to our early Sun, and the deduction that such disks have lifetimes of no more than about 10 Myr (Walter et al., 1988; Strom et al., 1989). In a way, the most striking discovery of the past few decades may be simply the realization, based on observations and models of these disks, that—when conditions are right—planetary growth is fast (e.g., ALMA Partnership et al., 2015; Chaussidon & Liu, 2015; Ballesteros-Paredes & Hartmann, 2007)!

A synthesis of various monistic models, including recent developments, is offered below, though we defer the discussion of one aspect of planetary evolution, namely their differentiation into mantle and core, until Chapter 2. We also note that the discovery of exotic planets in distant star systems is ongoing, and will continue to impact our understanding of the origin of our own. Such discoveries have so far revealed planets dramatically different from those in our Solar System—for example, gas giants (like Jupiter) closer to their star than Earth is to ours, in some cases closer even than Mercury! Whether those differences are merely a reflection of extreme planet migration or imply a novel mechanism of planet formation is still in question (e.g., Batygin et al., 2016).

Stage 0: The Setting. *Our* story begins more than $4\frac{1}{2}$ billion years ago, within a huge **molecular cloud** much like the thousands still dispersed throughout our Milky Way galaxy. These dense, dark clouds of extremely cold gas and dust—cold and dense enough for diatomic hydrogen and even more complex molecules to survive the interstellar medium—each stretch over distances of tens to thousands of light-years, and have the potential to incubate dozens to thousands of new stars.

We focus on one of the regions within this interstellar cloud where the gas and dust were particularly concentrated. Astronomers call this type of region a **molecular cloud core**; we will also call it a **primitive** (or **primordial**) **nebula**. This particular primitive solar nebula consisted of perhaps $\sim 3 \times 10^{30}$ kg of matter (equivalent to about one and a half times the current mass of the Sun) or more (e.g., Reipurth et al., 1993), and extended over a third of a light-year (1 light-year $\sim$ 10 trillion kilometers) in space or more; see, for example, Bergin and Tafalla (2007). Since the planets do not amount to much more than a tenth of 1% of the Sun's mass today, our premise is that the nebula originally contained significantly more than the current mass of the Solar System. As we will see, the reasons for such excess mass are quite consequential.

The composition of the primitive nebula would have included three types of substances: gases, ices (frozen gases), and mineral grains ('dust' particles). Given the composition of the Sun and other bodies of the Solar System, as well as present-day molecular cloud cores, we can be confident that the major constituent by far would have been hydrogen, followed by helium, with smaller amounts of neon and argon. At all temperatures of the nebula, even the low temperatures characterizing the nebula's earliest existence, these constituents (except possibly argon) would have remained in the gas phase.

The ices would be primarily water (H_2O) ice, frozen methane (CH_4), dry ice (frozen CO_2), and frozen ammonia (NH_3). Such compounds would remain in a frozen state as long as the nebula was cold, with initial temperatures of ~10–100K; but as the collapse proceeded and the inner nebula heated up, they would vaporize. The dust particles would include mainly iron & its oxides, and silicates of iron & magnesium, with lesser amounts of calcium and aluminum silicates. These particles could remain solid at temperatures up to ~1300–1400K (Lodders, 2003; note these are actually condensation temperatures), and would become the rocky material of the planets and moons.

A dominantly hydrogen (+ helium) nebula is consistent with our knowledge (from analysis of the light from stars throughout our galaxy and from even very distant galaxies, and analysis of the radio-wave emissions from gas dispersed throughout our galaxy) that the most abundant elements in the Universe are hydrogen and helium. These are the simplest elements, coming into existence after the "Big Bang" as the Universe expanded and began to cool, and were able to persist after the radiation permeating the once-hot Universe was finally too weak to annihilate hydrogen and helium nuclei. The early Universe would have been almost entirely H and He; as first-generation stars burned the H and then He, more complex elements, such as carbon and oxygen, would gradually be produced within them. In the later stages of their lives—and, in the case of supernovae, at the moments of their death—these stars would have dispersed a smattering of elements more complex than H and He throughout outer space. By the time our primordial nebula was ready to incubate a star, our galaxy had gone through two or more generations of such elemental syntheses.

Initially, when the nebula was very cool, the ices would have condensed onto the surfaces of the grains, trapping gases within; the resulting agglomerations would have a 'fluffy' structure and the consistency of dirty snowballs (Whipple, 1950, 1951). This is more than just a poetic description: such dirty snowballs could easily stick together upon impact, growing larger by **accretion**—an efficient alternative to gravity for getting large amounts of the tiny, disparately moving gas, ice, and dust particles of the early nebula to accumulate.

Stage 1: The Nebula Began to Collapse. Under gravitational forces acting mutually between all the particles of the primitive solar nebula, the nebula pulled in on itself and began to contract. If the nebula's mass was spread too thin—if there was too little mass, or it was distributed over too large a volume—it is possible that electromagnetic forces, or pressure forces generated within the nebula as it heated up, could effectively oppose further collapse; in fact, the percentage of nebulas succeeding in collapsing to the point of producing a star and planets is observed to be rather low (e.g., Dobbs et al., 2011; Williams & Cieza, 2011 and references therein). At some point in time, though, the gravitational forces acting within our nebula were sufficiently strong for the collapse to continue in earnest.

Interestingly, it is possible that the collapse of *our* primitive nebula was helped along, or even triggered, by the shock wave from a nearby supernova explosion (Cameron & Truran, 1977). The shock wave would have (temporarily) increased the nebula's density, and thus self-gravity, as it passed through.

Direct evidence for past supernovae can actually be held in our hands: uranium ores, for example, and minerals containing other heavy elements. Such elements are created within stellar furnaces only as the massive supernova star implodes, forcing lighter nuclei to fuse together in a process called **nucleosynthesis**; the subsequent supernova explosion would have scattered the star's insides, 'seeding' its stellar neighborhood with heavy elements (see Figure 1.2).

Some of the elements created by the nucleosynthesis were radioactive and have been decaying ever since then. For example, uranium isotopes U-235 and U-238, both created by supernova events, decay with half-lives of ~700 Myr and ~$4\frac{1}{2}$ Byr, respectively. To constrain the timing of a supernova that occurred just before our solar nebula's collapse, however, we focus on radioactive isotopes with much shorter half-lives; their abundance (decreasing

(a)

© Anglo-Australian Observatory

(b)

Figure 1.2 (a) Ground-based photographs of Supernova 1987A by David Malin at the Anglo-Australian Observatory (http://www.aao.gov.au/images/captions/aat050.html). The supernova, which occurred in the Large Magellanic Cloud (a satellite galaxy to the Milky Way), 170,000 light-years from Earth, was visible to the naked eye. The images on the left and right were taken, respectively, about 2 weeks after and before the day its 'nova light' first reached Earth in 1987 (differences in background picture quality are a result of atmospheric distortion). NASA / https://imagine.gsfc.nasa.gov/science/objects/supernovae2.html / last accessed September 09, 2023.
(b) Supernova remnant N132D, also from the Large Magellanic Cloud, a combination of satellite images from the Hubble Space Telescope and Chandra X-Ray Observatory. Light from the original explosion would have reached Earth about 3,000 years ago. The nearly spherical shockwave of expanding gases is 80 light-years in diameter (the stars appearing to fill it are in our galaxy and just happen to be in our line of sight). NASA / http://apod.nasa.gov/apod/ap051025.html / last accessed August 22, 2023.

by 50% over each successive half-life) would have dropped measurably during the interval between their creation and the collapse. Such isotopes include both heavy elements (such as hafnium-182 and iodine-129, which decay to tungsten-182 and xenon-129, respectively, with half-lives of 9 Myr and 16 Myr) and light elements (like aluminum-26, which decays to magnesium-26 with a half-life of 0.7 Myr); see Wadhwa et al. (2007). Thus, the fact (first pointed out by Reynolds, 1960) that we find the inert gas Xe-129 incorporated into meteorites, in iodine-bearing minerals, implies that the meteorites' parent body in the Asteroid Belt had incorporated those iodide minerals before the iodine decayed to xenon. It is hard to avoid the conclusion that the nucleosynthesis event occurred close to our primitive solar nebula in both space and time. Focusing on iron-60, which has a half-life of $1\frac{1}{2}$ Myr, Wadhwa et al. (2007; but see also Chaussidon & Liu, 2015 and references in Dones et al., 2015) tie the nucleosynthesis to a supernova occurring within a few Myr of the birth of the Solar System, less than ~10 light-years away (incredibly close, given the initial extent of the nebula!).

If other molecular cloud cores were close to our own, another way to trigger ours to collapse would be if one of them had already collapsed and formed a star; radiation pressure and the solar wind from that star could drive the intervening molecular cloud gas toward our nebula, compressing it and increasing its self-gravity.

Once the collapse was underway, it accelerated in the densest, central region in a manner often described as "inside out." In that region, gravity would be strongest between particles of gas and dust that were closest together, so particles nearest the center were pulled in first; but as they fell toward the center, they left a 'gap' behind. Their presence had helped support the weight of the layer above against the pull of gravity; without that support, that layer could no longer resist gravity's pull. Thus, as each layer fell toward the central mass, the lack of support it left for the layer above caused successively more distant layers to collapse inward.

Stage 2: The Collapse of the Nebula Created a Circumstellar Disk. For reasons too profound to be discussed here, *rotation* appears to be the natural state of the Universe. Clusters of galaxies rotate, as do the individual galaxies that comprise each cluster; molecular clouds within a galaxy rotate; star groups rotate; stars and planets rotate. Whether as a consequence of our galaxy's spin or the spin of our stellar neighborhood—or as a result of that supernova shock wave—the primitive solar nebula that would become our Solar System was characterized by an initial rotation.

Because of its rotation, the nebula possessed some angular momentum. As the nebula contracted, it would have had to rotate faster and faster *in order to conserve its angular momentum*—just as spinning ice-skaters, as they pull their arms in, conserve their own angular momentum by rotating faster.

As the particles of the nebula rotated faster, about whatever axis characterized its initial rotation, they would have experienced an increasing *centrifugal force* that opposed the collapse. Centrifugal force—sometimes described in physics classes as a "fictitious force"—is simply the kind of force you would feel while riding on a carousel in motion. The 'carousel force' acts outward, perpendicular to the spin axis of the carousel: it tries to fling you away from the carousel, not upward into the sky or down into the ground, but straight out. Gravity alone would have caused the nebula to assume an increasingly spherical shape as the collapse proceeded and each particle of the nebula was pulled toward the rest. But as the particles experienced an increasingly strong centrifugal force, they would resist the inward pull of gravity in the direction perpendicular to the nebula's axis (see Figure 1.3a).

Since centrifugal force did not oppose gravity at all in the direction parallel to the nebula's spin axis—the direction we would call 'up' or 'down' if we were on that carousel—the nebula would have *flattened* a lot as it collapsed. We can refer to this process as a flattening to the nebula's 'mid-plane.' In contrast, centrifugal

(a)

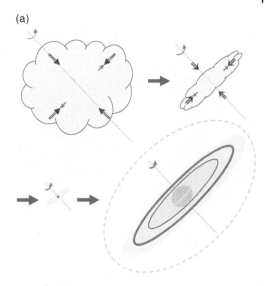

Figure 1.3 (a) The evolution (marked by heavy red arrows) of the primitive solar nebula. Mutual gravity (double arrows within the nebula) pulls all particles of the nebula in from all directions toward each other. The nebula possesses an overall spin, with the spin axis indicated by the light-blue-dotted straight line and an arrow at its top showing the sense of rotation. However, as the nebula contracts, it spins faster, generating a centrifugal force increasingly able to resist the collapse. Because the direction of the centrifugal force (single green arrows) is outward, perpendicular to the spin axis, such resistance occurs in that same direction. Gravity dominates, so the contraction is substantial, and the nebula shrinks to a fraction of its original size. Gravity has succeeded in flattening the nebula; but (in the final sketch, a close-up, not to scale) the resistance has produced a disk shape oriented perpendicular to the nebula's spin axis.
Studies of infrared emissions from some developing stars suggest that their disks actually flare, i.e., they are less flattened away from the central proto-star—a consequence of gravity being weaker, farther away (and reinforced by the star heating the gas and dust protruding above the disk); see, for example, D'Alessio et al. (1998); Dullemond and Dominik (2005); Williams and Cieza (2011); also, Dullemond and Dominik (2004); and Leinert et al. (2004).

force's opposition to gravity in the direction perpendicular to the nebula's axis made the collapse much less successful in that direction, leaving nebular matter in the mid-plane strung out away from the axis.

(b)

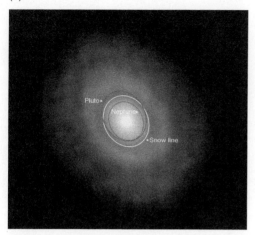

Figure 1.3 (b) An actual protoplanetary disk, around proto-star V883 Orionis in the constellation Orion; for scale, the orbits of Neptune and Pluto in our Solar System are also shown.
As discussed later in the text, the snow line marking the orbital distance beyond which ices could survive the heat of *our* proto-Sun was at ~3–4 AU; in the disk shown here, though, as a result of a flare-up by the proto-star, the snow line was observed to be at ~42 AU (Cieza et al., 2016). High-resolution image from the Atacama Large Millimeter/submillimeter Array (ALMA). European Southern Observatory / https://www.eso.org/public/images/eso1626e / last accessed August 22, 2023.

Very close to the center of the nebula, centrifugal force would always be ineffective—the carousel force is most disturbing at the carousel's outer rim, and negligible close to its spin axis, whereas gravity is strongest where the mass is closest together—so gravity would succeed in pulling the *central* portion of the dust cloud into a nearly spherical shape. Thus, a dichotomy was quickly established as the nebula contracted, between a massive spherical central body and the extensive but thin disk surrounding it (a face-on, radio telescope image of a protoplanetary disk in the Orion constellation is shown in Figure 1.3b). The dynamics were very different: gravity was opposed mainly by centrifugal force in the disk; in the central body, it was pressure forces that resisted gravity, especially as the center heated up. Central sphere and surrounding disk would

evolve very differently (one becoming a star, the other ending as an orbiting suite of planets), though they would continue to interact in key ways throughout their evolution.

The dichotomy between disk and sphere would be driven to extremes as the collapse continued, thanks to the effects of angular momentum. We can ease into that understanding by first exploring the size of the disk and the implications of angular momentum conservation; to complete our discussion of Stage 2, we then return to the question of how the collapse proceeded.

Comets Tell Us the Dimensions of That Circumstellar Disk. The disk surrounding the central proto-star was where 'dirty snowballs' would accrete to form larger and larger bodies that culminated in the planets and moons of the Solar System. But in the outer reaches of the disk, the nebula was too thin for accretion to proceed all the way, leaving an extensive, more or less flat region populated by large dirty snowballs. Occasionally, through random collisions, or responding to perturbations in the ambient gravity field, they would fall toward the Sun, their orbits becoming very elliptical (but remaining within the plane of the Solar System). In the inner Solar System—either approaching or receding from the Sun—their ice would melt to some extent, releasing the trapped gases. If this happened close to Earth, we could observe their disintegration: the freed gases (and dust particles) are pushed away from the Sun by the solar wind and the Sun's radiation pressure, creating one or more 'tails' (see Figure 1.4).

These disintegrating objects are, of course, known as *comets*. Flybys by space probes—beginning with those passing Comet Giacobini-Zinner and Comet Halley (1985 and 1986, respectively) and including more recently the *Stardust* mission to Comet Wild 2 and the *Deep Impact* mission to Comet Tempel 1 (2004 and 2005, respectively)—have confirmed the overall structure and composition of these large dirty snowballs, though minerals sampled from Wild 2 suggest that the outer nebula where the comet formed also acquired inner-nebula

Figure 1.4 (a) A picture of Halley's comet in 1910. 1911 Encyclopedia Britannica, taken from Edward Emerson Barnard / Wikimedia Commons / Public Domain. This comet, which visits the inner Solar System every ~76 years, has been recorded historically since at least 240 BCE after sightings by Chinese and Babylonian astronomers (Yeomans et al., 1986). It was seen before the start of the Battle of Hastings (1066 CE) and thought to portend defeat (Stein, 1910). And, it was first recognized as a repeat visitor around 1700 by Edmund Halley, whose orbital predictions represented a triumph of modern (Newtonian) physics (Hughes, 1987).
(b) The return of Halley's comet in 1986. NASA image as shown in Howell (2022) / NASA.

material flung outward by the same process envisioned in planet-planet scattering (see, e.g., Matzel et al., 2010; Brownlee, 2014).

The region where these comets originated is called the **Kuiper Belt** (a second, much more distant source region for comets will be discussed later in this chapter). Analyzing their orbits (including likely subsequent changes once they entered the main Solar System, due to interactions with giant planets), we can infer how far the Belt extends away from the Sun—thus, how far out the circumstellar disk stretched.

Such dimensions are perhaps best appreciated in terms of astronomical units, with **1 astronomical unit** (**1 AU**) defined as the distance from Earth to the Sun (see also the Figure 1.1 caption).

Here's a different way to think of these distances: if it takes light from the Sun 8⅓ minutes to reach the Earth, can you calculate how many AU add up to 1 light-year? So, how big was our primitive solar nebula at the start, in AU?

The Kuiper Belt begins just past the orbit of Neptune, which is a distance of 30 AU from the Sun; from observations of comets, we infer that the Belt reaches a radius of at least ~50 AU.

Our simple estimate of the disk size is a lower limit, in part because (as discussed below) its outer edge may have eroded away (cf. Adams et al., 2004). Though somewhat smaller than many disks observed throughout the galaxy (see, e.g., Figure 1.3b), it is nevertheless consistent with the range of observations (e.g., Strom & Edwards, 1993; Strom et al., 1993). Yorke and Bodenheimer (1999) point out that disk size can reflect the amount of angular momentum contained initially in the molecular cloud core.

To date, over a thousand Kuiper Belt Objects have been detected by telescope (Bannister et al., 2018). In 2006, NASA launched the *New Horizons* space probe with the goal of studying, up close, the Solar System beyond Neptune; that would include Pluto (and its five moons), located at the near edge of the Kuiper Belt. The probe reached the vicinity of Pluto in 2015; in 2019 it also flew by a small and unusually shaped Kuiper Belt Object, now known as "Arrokoth," which—at 44 AU—is the most distant body yet visited (see Tavares, 2020)! At such distances, of course, none of the objects of the Kuiper Belt are in 'comet mode.'

A Central Problem of Nebula Theory. As the collapse caused the great bulk of the nebula's material to be concentrated within a volume so much smaller, the nebula's spin—the spin of the central proto-star and the orbital velocities of the disk masses around the proto-star—must have increased appreciably, in accord with the ice-skater principle (conservation of angular momentum). And, with most of the nebula's mass ending up as part of the Sun, the *Sun* should now be rotating rapidly on its axis. In fact, the Sun does rotate impressively fast for an object so vast: with a diameter more than 100 times that of Earth's, it manages to rotate once on its axis every ~30 days; points on its equator circle its axis at speeds more than triple that on Earth!

The problem is that conservation of angular momentum should have left the Sun rotating even faster: in the early stages of collapse, as the radial extent of the bulk of the nebula contracted by a factor of ~100, the nebula's rotation rate should have amplified by a factor of ~10,000, to maintain the same angular momentum; and as masses fell further in, to within the orbit of the future Earth, their rotation should have accelerated by another factor of 40,000—reducing the period of spin of the central proto-Sun to less than about a day at that point.

Such a deficiency in the Sun's angular momentum is highlighted by the present-day disparity in how angular momentum is distributed between the Sun and the rest of the Solar System: the planets' angular momentum (which is almost entirely orbital) amounts to 99 1/2% of the total angular momentum in the Solar System; the Sun—despite having the great bulk of the system's mass—has only 1/2%.

We will return to this problem (also called *the angular momentum problem*) below, though it is worth mentioning here that the difficulty in explaining how the Sun lost most of its angular momentum caused dualistic theories to be preferred until well into the twentieth century (Seeds & Backman, 2013).

And Speaking of Angular Momentum… Particles of the contracting and flattening nebula that did not make it in as far as the proto-Sun would still have had to substantially increase their orbital velocity, in order to conserve their angular momentum. Thus, the masses populating the circumstellar disk would have been orbiting the proto-Sun quite rapidly, at least initially.

From conservation of angular momentum, those particles in the inner part of the disk, having traveled further in, would have been orbiting the proto-Sun with a faster angular velocity than those further out; in fact, we can envision a smooth increase in orbital angular velocity as the center is approached. Imagine the disk to be filled with particles, all of which are orbiting the center (with angular velocities reflecting that variation with distance). Now consider two adjacent orbits, one just slightly further out than the other and distinguished by particles moving with a slightly slower angular velocity. Friction from particles in that orbit will drag slightly on the faster-moving particles in the adjacent 'inside' orbit, slowing the latter's orbital velocity. That, in turn, reduces the centrifugal force they experience and allows the proto-Sun's gravity to pull them inward. At the same time, particles in that slightly further orbit are dragged ahead by friction with the faster particles just inside their orbit, slightly boosting the outer particles' orbital velocity, amplifying the centrifugal force acting on them and causing them to be flung out away from the proto-Sun, spreading the disk outward.

The motion of disk particles inward or outward is a consequence of changes in the balance of gravitational and centrifugal forces as friction operates. *The change in angular velocities caused by that friction also implies changes in the particles' angular momentum.* The net result, as pointed out by Lynden-Bell and Pringle (1974, based on remarkably prescient work), is that the inner portion of the disk, shedding its angular momentum, feeds the proto-Sun while angular momentum becomes concentrated in the disk's outer portion.

Feeding off the disk is evidently episodic rather than continual. The work of Lynden-Bell and Pringle also provided an explanation of why young stars occasionally flare, with the sudden brightness corresponding to "blobs of gas" from their disk falling into the stellar atmosphere (see Figure 1.3b). The end result (Chaisson & McMillan, 2005) was that the near part of the disk was incorporated into the proto-Sun, leaving the remainder (now a truly protoplanetary disk) for planet development.

Feeding the proto-Sun with lower angular momentum mass helps explain why the Sun has less angular momentum than it 'should,' but is not sufficient to resolve the central problem by itself. As we will see, full resolution awaits a later stage of the Sun's development.

Our protoplanetary disk would have been a chaotic environment, indeed, as particles shifted orbits and mass left the disk, changing the ambient gravity field and the way the remaining masses interacted. The turbulence characterizing the disk (Lissauer & de Pater, 2013) would have dissipated some of the particles' energy, adding significantly (Armitage, 2014) to the friction already invoked and perhaps also helping the masses accrete.

You may have noticed that our discussion of angular momentum changed from the implications of conserving it to the need to change it. We will discuss a similar conservation law (conservation of linear momentum, otherwise known as Newton's Second Law) in Chapters 4 and 6, and face the same distinctions. Conservation laws actually specify the conditions under which the property is either conserved or not. Here we have described conditions under which angular momentum can be reduced. Thinking once again of those spinning ice-skaters, can you explain why—even after they have pulled their arms in and sped up their rotation—their spin and angular momentum will, sooner or later, come to a halt?

Finally, with the above angular momentum considerations in mind, our dichotomous view of the collapsing nebula suggests the following scenario for our nebula's collapse. The nebular matter that was close in, to begin with, possessing low initial angular momentum, had none of the resistance to collapse exhibited by material falling in from great distances; it would have collapsed quickly into the central region, proceeding 'inside out' as noted earlier. As an illustration, the model developed by Laughlin and Bodenheimer (1994) begins

with a relatively small-sized molecular cloud core containing 1 solar mass. After just ~20 Kyr, the central region had already acquired 40% of the entire nebula mass, largely from the infall of that close-in, low-momentum matter; within ~50 Kyr (still only one-sixth of the time for the disk in their model to completely evolve), almost half of the nebula mass resided in the central core. At this point in their model, a similar amount of mass was distributed throughout a massive circumstellar disk, leaving just "remnants" of the original nebula surrounding the disk and core.

We can predict how the system evolved further. The disk would 'process' much of its mass, feeding most of it to the growing proto-star after having reduced the mass' angular momentum. At the same time, the disk would stretch out and 'spin up,' potentially even ejecting some of its outermost matter.

The disk was not, however, feeding the core or ejecting matter one molecule at a time. Almost from the start, other processes had been operating to accrete the nebula particles to larger size—which brings us to the next stage of development.

Stage 3: The Collapsing Nebula Became 'Clumpy.' Planets and moons would grow when the nebula particles, initially as small as micron-sized, accreted all the way to kilometer-sized and beyond. Most of this growth would take place within the disk (see below); but accretion was already underway before the disk had formed.

Early Accretion. Accretion had begun in the collapsing nebula through turbulence and through the process of disk formation itself.

- As discussed below, the nebula began to heat up as it contracted; however, the heating was not perfectly uniform. As a gas-solid mix, or in other words a lumpy fluid-like mass, the nebula would undergo fluid flow, as it continued its infall, in response to this uneven heating. But the nebula would have been a low-viscosity or 'thin' fluid, so that flow would have been **turbulent**: poorly ordered, even chaotic, flow in which the

Figure 1.5 Turbulence evident in a cold cloud core, revealed in this image from the Herschel infrared Space Telescope. NASA / http://www.nasa.gov/mission_pages/herschel/herschel-20091002a .html / last accessed August 22, 2023. As noted on the website, "The image reveals a cold and turbulent region where material is just beginning to condense into new stars. It is located in the plane of our Milky Way galaxy, 60 degrees from the center. Blue [coloring added] shows warmer material, red the coolest, while green represents intermediate temperatures. The red filaments are made up of the coldest material pictured here—material that is slightly warmer than the coldest temperature theoretically attainable in the universe."

trajectory of any small bunch of material would be nearly unconstrained by the trajectories of neighboring matter; see Figure 1.5 (also Cameron, 1975; Montmerle et al., 2006; Ballesteros-Paredes et al., 2007; and references in Dobbs et al., 2011).

Wait a minute, you say! Didn't we just learn that turbulence in the protoplanetary disk created a high-friction environment?

Turbulent flow is facilitated by the fluid having a low viscosity; but—whether in the nebula, disk, or even, as discussed in Chapter 6, in our oceans—turbulence has the effect *of amplifying friction because, in its 'one step forward, two steps to one side or the other' chaotic behavior, it inhibits the 'forward' progression of the flow.*

Turbulent fluids, behaving in such an unconstrained manner, exhibit whirlpools, or 'eddies,' on a small scale. In the nebula, eddies would have *accelerated* the dirty snowballs around and around, encouraging collisions between those sticky objects and thus fostering their growth by accretion (see also Johansen et al., 2014). This process would be a more efficient way to begin the growth of planets and moons than would random collisions of gas, ice, and dust particles within an orderly contracting nebula.

• The collapse and flattening of the nebula into a disk shape are sometimes described as a process of the nebular material **settling toward the mid-plane**. And as the dust particles, dirty snowballs, and eddy-generated clumps drifted 'down' through the nebula gas, *the larger ones*—less affected by gas drag (i.e., 'air' resistance)—*would settle faster*, overtaking and incorporating the smaller clumps and thus accreting into still larger masses. (This is reminiscent of how raindrops and hail grow larger as they fall through the lower atmosphere of Earth.)

"The rich got richer," it seems, on the way to forming a disk. Such a theme would recur later as the planets grew: by accretion, with larger clumps, having a broader cross-sectional area, experiencing more collisions and accreting faster; and then by gravity, with more massive clumps, having a stronger gravitational attraction, growing faster. At this point, however, this theme applied on very small scales: settling resulted in clumps up to millimeters and centimeters in size (Armitage, 2014).

Within the Protoplanetary Disk, Accretion Continued—and Then Some! Within the disk, clumps orbited the proto-Sun along with gas, dust, and ice yet to be accreted. We briefly mention four mechanisms (reviewed by Armitage, 2014) which successively build those clumps up to planet size. One has similarities to *settling*, and grows the clumps to meter size; two somewhat novel mechanisms appear capable of accelerating the clumps' growth, producing first **planetesimals**, that is, kilometer-sized masses, then even larger bodies; and a fourth, turbulent mechanism achieves similar growth but more gradually.

- In the protoplanetary disk, each clump's orbit would have been stable, with the outward centrifugal force it experienced balancing the inward pull of the proto-Sun's gravity. But at the distance of that clump, the gas would be orbiting at a slightly slower velocity: heat from the proto-Sun created an outward thermal pressure on the gas, counteracting the gravitational pull inward and decreasing the amount of centrifugal force needed to offset the net inward pull. The gas consequently dragged on the clump, slowing it down, reducing the centrifugal force it experienced, and allowing gravity to draw it inward toward the proto-Sun. This process is called **radial drift**.

  The 'good news' about radial drift is that it gave each clump an opportunity to interact with a lot of nebula particles as it spiraled inward, and grow to meter size, especially while it traveled through regions of the disk where ices were still plentiful. The 'bad news' is that radial drift was extremely rapid, and it was possible for the growing clumps to drift all the way into the proto-Sun in as short as a hundred years (Armitage, 2014)! Consequently, there had to be other processes that could quickly grow the clumps to a much larger (planetesimal) size, large enough that they could resist the gas drag.

- Thinking of the disk, in particular its dense mid-plane, as lumpy and fluid-like, the first novel mechanism envisions the mixture 'jostling' as it flowed around (or inward), with the gas, dust, and clumps pushing back on each other. Occasionally the 'fluid' would push together, temporarily forming dense clumps known as **streaming instabilities** (Youdin & Goodman, 2005); some could be large enough for their own gravity to pull in more mass, creating a number of planetesimals that would survive after the conditions leading to those instabilities had vanished.

- A less exotic alternative (or supplement) to streaming instabilities derives from the turbulence present in the protoplanetary disk. Barge and Sommeria (1995; see also Armitage, 2014) found that numerous smaller 'horizontal' eddies could coalesce into fewer but larger ones, quickly merging to form solitary 'regional' eddies. Those eddies could persist despite drag forces, accreting into large planetesimals closer to the proto-Sun and larger ones farther out. The model presented by Barge and Sommeria, with a simple variation in nebula density postulated versus distance from the proto-Sun, even predicted that the largest eddies would develop at about the distances of Jupiter and Saturn from the Sun.

- Our last mechanism is a somewhat novel combination of gas drag and gravity. It operated after planetesimals had been created but while the gaseous disk environment was still filled with numerous leftover centimeter-sized 'pebbles.' As the gas and pebbles flowed past a planetesimal, the planetesimal's gravity may have been strong enough to pull some of the pebbles to it, especially after gas drag had slowed the pebbles. Models show that such **pebble accretion** is actually a quite efficient way—orders of magnitude more efficient than collisions between planetesimals—to grow the planetesimal to proto-planet size or even planet size (e.g., Lambrechts & Johansen, 2012; Morbidelli & Nesvorny, 2012; Levison et al., 2015). As an added benefit, as discussed below, planets reaching a mass of several Earths could also serve as a 'seed' or 'growth core' around which to grow giant planets.

While planets were developing in the disk, the central region was evolving, too. We will briefly focus now on that central region, and save the final stage of turning proto-planets into planets and giant planets, and explaining how that planetary system became our Solar System, until after that.

Stage 4: The Collapse of the Nebula Caused It to Heat Up Dramatically. Pulled toward each other by their mutual gravity, the particles of the nebula were essentially falling toward the nebula's center. Their kinetic energy would be converted into heat upon impact with other particles—during the initial infall, on the circumstellar disk, or within the increasingly dense inner nebula.

The energy producing this heat is called **gravitational energy**. On a much smaller, human scale, we can think of gravitational energy as the kind of energy released when a box we were carrying falls to the ground; upon impact, most of that energy is manifest as heat, though some of it may go into breaking or deforming the box if that occurs. The magnitude of gravitational energy in this situation is small, both because the falling distance is short and because the amount of mass involved is small. In contrast, the amount of gravitational energy involved in assembling the Earth (discussed in Chapter 9), for example, was extremely large—so large that nearly all of it must have been radiated away into space, rather than being retained by the growing Earth.

Similarly, the gravitational energy released as heat from "intermolecular collisions" (Lissauer & de Pater, 2013) during the nebula's collapse was too great to have all been retained: it would have raised the temperature within the nebula so much that the resulting 'thermal' pressure would have prevented any further collapse. Instead, in the early stages of collapse, most of the gravitational energy would have been radiated away. The nebula stayed cool, and its collapse could proceed.

Initially, the ice, dust, and gas comprising the nebula would not have been evenly distributed. Gravitational forces driving the collapse would have varied within the nebula, depending on where matter was more concentrated, or less, so the collapse would not have been uniform, and neither would the production of heat. In the early stages of collapse, the resulting differences in nebula temperatures (accentuated by variations in radiative cooling) would have promoted turbulence and clumping, as discussed earlier. As the collapse proceeded, with most of the nebula's mass falling into the developing proto-Sun, most of the heat would simply be concentrated within the central region, and trapped by the increasing density there. On the other hand, because the circumstellar disk was thin, much of its heat could be radiated away.

As that central region collapsed, the interior layers of the proto-Sun would be increasingly compressed. Since—as a general rule of nature—things heat up when they are compressed, this amounted to an additional source of heat for the proto-Sun. We know from our everyday experience that compression generates heat; for example, when air is compressed within a bicycle pump, in the course of inflating the bicycle's tires, the pump housing becomes hot to the touch. The opposite situation, that expanding materials cool off, can easily be verified by feeling the air escaping (through the tire valve) from the tires into which it had previously been pumped. This general rule is also what enables the process of refrigeration to work: the cooling produced when a circulating refrigerant liquid, confined in narrow tubes behind the icebox, expands into larger tubing within, keeps the interior chilled.

Stage 5: A Proto-Sun Began to Shine in the Central Nebula. As the collapse continued, with the great bulk of the nebula falling all the way into the central region, huge amounts of heat were generated, and trapped there. The gravitational pull toward the center strengthened as more and more mass ended up there; mass infall accelerated, compression intensified, and temperatures shot up exponentially. The concentrated mass of the central region was on its way to becoming a star; and though thermonuclear reactions had *not* yet begun to occur within it (an activity that we will take to be the formal definition of a **star**), in as little as a few hundred thousand years it began to radiate huge amounts of heat and light.

We can describe two stages in the development of this "proto-star," based on astronomical observations and theory. In the **Herbig-Haro** (or **HH**) **phase**, enormous amounts of matter were expelled from the hot proto-star, at a rate as great as 2×10^{30} kg (1 solar mass) over ~1–10 Myr. This mass was ejected through the polar regions of the proto-star, and is thus called "bipolar jets" or a "collimated" solar wind (see Figure 1.6a, b, c).

During *and after* its HH phase, the hot proto-star would also have continually expelled matter from its entire surface, in all directions, like the solar wind of today only more intense. This solar wind, combined with the heat and light radiated by the proto-star, swept the inner nebula clean of gas and dust not yet incorporated into either the proto-star or the proto-planets and proto-moons of our Solar System. As the dust cleared, the proto-star (and its thinning circumstellar disk) would become visible to the rest of the galaxy. This second stage of the proto-Sun's development, in which the polar ejection has ended, leaving only the radial solar wind, and the proto-star has begun to be revealed, is called the **T-Tauri phase**.

The name for this phase comes from one such star in the constellation of Taurus the Bull. Other stars, such as some in the Orion Nebula (see Figure 1.7), also exhibit intense, T-Tauri-type behavior. Telescopes reveal a stellar nursery there: very brightly burning stars, with the nebular debris partially burned away—as if a curtain is being pulled back to reveal the infant stars. In the case of our solar nebula, we can similarly expect the heat and solar wind of the T-Tauri proto-Sun to have swept our Solar System clean of gas and small debris—probably within as little as 10 Myr.

Stage 6: The Protoplanetary Disk Began to Look Like Our Own Solar System. By the end of Stage 3, clumps in the protoplanetary disk had accreted to kilometer-sized or larger planetesimals. These planetesimals continued to grow through gravity and impacts, incorporating nearby material. Soon the larger ones began to interact with each other, their mutual gravitational attraction leading to jostling and

(a)

Figure 1.6 (a) (top) Sketch of Herbig-Haro bipolar jets expelled from the Sun in its late stage as a proto-Sun (http://www.nasa.gov/images/content/144594main_061303_DB2-nebula.jpg).
(bottom) Hubble Space Telescope image of actual bipolar jets streaming from the proto-stellar disk of Herbig-Haro 30, a proto-star in the constellation of Taurus. The jets extend roughly 30–40 AU vertically away from the disk; according to repeated images over a year, the jets are traveling at more than 200 km/sec (Koerner, 1997). NASA / https://esahubble.org/images/opo9524e / last accessed August 22, 2023.
The mid-plane, the densest, dustiest part of the disk—a 'dust lane' perpendicular to the jets—obscures the bright proto-star, but we can see its light reflected off the dust above and below the mid-plane.
NASA image downloaded from National Aeronautics and Space Administration, https://esahubble.org/images/opo9524e / last accessed August 22, 2023.

(b)

Figure 1.6 (b) Hubble image of the Carina Nebula, one of the largest stellar nurseries in our galaxy, about 7,500 light-years distant. The pillar of gas and dust rising toward the top of the image, 3 light-years in height, is topped by a Herbig-Haro star (HH901), with the HH jets clearly visible; the left-hand jet, more than half a light-year long, terminates in a shock-wave front impacting the surrounding gas and dust. Presumably the material closest to HH901 is in the process of forming a solar system. Image from NASA / http://www.spacetelescope.org/images/heic1007a / last accessed August 22, 2023.

(c)

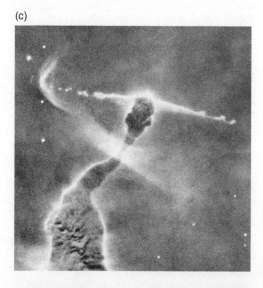

Figure 1.6 (c) Close-up of HH901. M. Livio, Hubble 20th anniversary team / https://stsci-opo.org/ STScI-01EVVDESMGAY5P97366ABGE28Z.pdf / last accessed August 22, 2023.

bumping, then successive collisions among them; and they grew in leaps and bounds. Even following near misses, if the bodies remained close for a long enough time, eventual collision and accretion were almost inevitable. And even if the colliding bodies disintegrated, the larger ones could recapture the fragmented material gravitationally. This period of efficient development has been described as "runaway growth."

The one weakness of this 'classic' model (Armitage, 2014) is the possibility of collisions being too energetic to sustain growth. An alternative, the pebble accretion mechanism, avoids the need for collisions and can yield efficient growth from planetesimal to proto-planet size, as long as the gas and pebbles surrounding the planetesimals are sufficiently abundant. This mechanism can also lead to exponential growth.

(a)

(b)

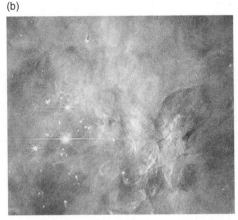

Figure 1.7 Orion Nebula as seen with the Hubble Space Telescope. (a) NASA, ESA, M. Robberto (Space Telescope Science Institute/ESA) and the Hubble Space Telescope Orion Treasury Project Team; the caption reads, "Stars are being born in the turbulent gas of the Orion Nebula." (b) in the center of Orion's "Trapezium," containing some of the nebula's stellar nurseries. J. Bally, D. Devine, & R. Sutherland, D. Johnson / http://antwrp.gsfc.nasa.gov/apod/ap971118.html / last accessed August 22, 2023.

The largest bodies thus produced were on their way to becoming planets or moons. But their ultimate fate depended strongly on their *location* within the disk, especially in relation to the developing Sun. In our Solar System, it turns out that the region between Mars and Jupiter (now occupied by the Asteroid Belt) marks a transition in planetary characteristics, a dividing line in planetary development that was simultaneously mechanical, thermal, and chemical. Beyond this region, even the heat of the T-Tauri Sun was not great enough to vaporize ice; here, the disk remained full of dirty snowballs, so that accretion was always efficient, and the proto-planets would start off fairly large. The dividing line or **snow line**, identified by a temperature of ~150K in the solar nebula, was at a distance of around 2–3 AU depending on how evolved the proto-Sun was (e.g., Garaud & Lin, 2007; Armitage, 2014; but see also Martin & Livio, 2012).

The Inner Planets. The proto-planets developing close to the central region would generally be on the losing end of fights with the proto-Sun's gravity for control of the disk's debris. These planets would have started out small and, with relatively weak gravitational fields, would tend to pull in mass only from their immediate neighborhoods. They would grow less efficiently, mainly accreting through random impacts.

In their earliest days, these planets might have had atmospheres; but because of their location those atmospheres could be eroded by the heat of the developing Sun. Even worse, the inner planets would experience the brunt of the proto-Sun's T-Tauri phase, and that intense heat and solar wind would decimate much of their remaining atmospheres. In short, these planets never had a chance—they would never grow to the enormous size of the outer planets!

Accretion by impact would bring energy as well as mass to the proto-planets. Depending on the characteristics of the impacts—how often they occurred and how fast they struck the Earth—as well as the material properties of the impacted proto-crust and the presence of a primitive atmosphere, it is possible that they produced widespread, perhaps even global, melting. The impacts could also vaporize and degas the crust; depending on how much silicate vapor, steam, and carbon dioxide were produced, the resulting 'fortified' atmosphere could blanket the surface and substantially prolong the melting. Thus, under the right conditions, a *magma 'ocean'* could develop. In some scenarios it would be shallow, in others, deep; and, it could be surficial or an interior layer. Some researchers (e.g., Abe, 1997; Elkins-Tanton, 2008) have found, however, that if one did develop, surface conditions would recover in $\sim 10^7$ yr, and the thermal state of the planetary interior might recover in $\sim 10^8$ yr or less, at least in the case of Earth.

With impactors (i.e., planetesimals) growing larger over time, we might expect a magma ocean to be more likely to develop during the later stages of planet development; but, inevitably, late-stage impacts would also be less frequent, so the net effect would be multiple, discrete magma ocean events (see also Tucker & Mukhopadhyay, 2014).

Some computer models predict that as the proto-planets populating the inner disk grew more massive, they would eventually distort each other's orbits gravitationally, making some orbits highly elliptical and causing orbits to cross. In this scenario it would be inevitable that some of the proto-planets would ultimately collide, in planet-altering giant impacts. At the tail end of Earth's accretionary period, it appears that Earth experienced just such an impact. As discussed in Chapter 6, some numerical models conclude that the impactor was a planet perhaps a tenth as massive as Earth and that the collision was a somewhat glancing blow, allowing Earth to survive and allowing our own Moon to form nearby from the leftover impactor particles. In other models, the impactor might even have been as massive as the Earth, and the collision more nearly head-on, merging the two planets into one mostly molten, partially vaporized, and rapidly spinning sphere, flinging off enough mass as it spun to coalesce into our Moon. Abundances of various radioactive isotopes and their decay products place the time of the giant impact at ~ 60–80 Myr after planetary accretion in the primitive solar nebula had begun (e.g., Touboul et al., 2007, Barboni et al., 2017; see also Tartèse et al., 2019).

The Outer Planets. The proto-planets far from the central region did not have to compete as strongly with the proto-Sun's gravity for control of the debris around them (for example, a planet like Earth would have a gravitational 'sphere of influence' five times as large if it was as far from the Sun as Jupiter is). Beyond the Asteroid Belt, the proto-planets could start out big, and easily grow bigger, gravitationally drawing in mass from an extensive surrounding volume, as well as accreting through random impacts of planetesimals. They would become the **giant planets** of the Solar System.

Since T-Tauri circumstellar disks appear to survive no more than ~ 10 Myr, these proto-planets must have pulled in the gas, dust, and ice around them very rapidly, and essentially grown to completion within 10 Myr. Theoretical models (discussed shortly) predict that such rapid, steady growth was possible there once those proto-planets had reached a mass about 10 times the current mass of Earth (e.g., Mizuno, 1980). To reach that critical size quickly, their gravitational growth could have been supplemented by pebble accretion, facilitated by the abundant gas and clumps present in this part of the disk.

Gravitational growth implies—as a consequence of the ice-skater principle—that as they pulled in more and more mass, these planets would spin faster and faster. We see the end result today: Jupiter, Saturn, Uranus, and Neptune, though huge, each take $\sim \frac{1}{2}$ day to rotate once about their axis. In contrast, the accretion-by-impact formation of the inner planets led to no such guarantee of a fast (or slow) rotation rate. Instead, whatever their initial spins, the nearness of those planets (and our Moon) to each other and to the Sun meant those spins were likely to be modified over time by tidal friction—for Venus (as mentioned earlier) and Mercury, solar tidal friction; and for the Earth, both solar and lunar tidal friction. The end result for the inner planets is that their rotation rates are slow, ranging from 1 day to 243 days for a single turn about their axis, and show no consistent pattern.

Location meant not only that the outer planets could grow large, but that they could stay large as well. The heat and solar wind that wiped out much of the inner planets' atmospheres spread out from the proto-Sun in all directions, and were much less intense by the time they reached the orbit of Jupiter. The outer planets would have retained much of their earliest atmospheres to the present day.

(Given our understanding of the composition of the primitive solar nebula, what would you expect the primary constituents of those atmospheres to be?)

Such atmospheric retention was not the fate of the inner planets.

How Hot Did the Evolving Disk Get? Molecular cloud interiors, like the cloud core that became our own primordial solar nebula, are typically quite cold to start with, ~10–20K. As the nebula contracted, it heated up, most of all in its central region; but observations of T-Tauri circumstellar disks suggest that those disks were probably not that hot. At the distance of Earth's orbit, the disk surfaces have temperatures in the range ~50–300 K, with most disks at ~75–175K (Beckwith et al., 1990); in the mid-planes of the disks, at that distance, temperatures might be ~200–750K (Woolum & Cassen, 1999). Only at distances closer than Mercury's orbit, or perhaps just beyond, would it be hot enough, with temperatures exceeding ~1350K, to vaporize silicates (Chambers, 2005; also Woolum & Cassen, 1999, who nevertheless stressed that the thermal regime could have been very different in earlier stages of nebula development).

At distances corresponding to the outer part of the Asteroid Belt and beyond—in other words, outside the snow line—it was cool enough at all times in the Solar System's development for ices to persist. We know this from the occurrence of hydrated minerals in some meteorites, which tells us that the outer Asteroid Belt (the source region of those meteorites) at the time the asteroid 'parents' were accreting was cold enough that dirty snowballs and ices were present and able to be directly incorporated into the asteroidal bodies (the resulting hydrated versus anhydrous or 'bimodal' distribution of meteorites, derived from asteroids outside versus inside the snow line, is summarized enthusiastically in Burbine and Greenwood, 2020). Beyond the Asteroid Belt, temperatures during planet formation never exceeded ~150K, so that, as previously discussed, this region was much more clumpy than the inner disk. Estimates (e.g., Cameron, 1975; Lecar et al., 2006; Garaud & Lin, 2007; Armitage, 2014) indicate that there could have been four times as much solid mass in the region of outer planet formation, most of it in the form of ices and dirty snowballs. It is no wonder that those planets were able to grow so much larger than the inner planets.

Constraints on the Growth of the Outer Planets. The gas giants Jupiter and Saturn are indeed composed primarily of gas—mainly hydrogen—though transforming to a liquid phase at depth due to increasing pressure (but see Lozovsky et al., 2017 and Wahl et al., 2017 for recent inferences of the giant planets' interior structure). However, their formation within this part of the accretionary disk—just beyond the snow line—guaranteed that they would also incorporate significant amounts of ice and rocky material. Jupiter, for example, which contains the mass of more than 300 Earths, is generally estimated to include roughly 10 Earth masses worth of rock and ice.

Older models of giant planet development envisioned rock-ice planetesimals growing to proto-planet size, at which point they would be able to create large atmospheres by attracting copious amounts of gas from the surrounding nebula. For conciseness (see Mizuno et al., 1978), those solid proto-planets were called "cores," but note they are quite distinct, in origin and function as well as composition, from the cores of the Earth and other terrestrial planets; we will distinguish them by referring to them as *growth cores*. More recent models emphasize pebble accretion as a quick way to grow planetesimals into growth cores with masses several times greater than Earth's.

But we also need a quick way to get huge amounts of gas around these growth cores before the accretionary disk is gone. In some scenarios, this was achieved by a process of "runaway gas accretion" (e.g., Mizuno et al., 1978; Mizuno, 1980; Pollack et al., 1996): if the ice-rock growth core was able to reach a critical mass, estimated to be ~10 Earth masses, the planet's gravitational pull would have been strong enough to compress the atmospheric gas, 'freeing up' higher levels of the atmosphere to accommodate more gas; the more the planet fed, the stronger the gravity, the more gas could be drawn in. This process is similar to the 'inside-out' collapse of the central nebula that accelerated the growth of the proto-Sun. Jupiter and Saturn could develop huge atmospheres because this process allowed them to directly pull in a lot of nebula gas.

The proportion of solids versus gas forming the growth cores would affect the fate of these planets. With a greater volume of solid debris, their gravity fields would strengthen faster, guaranteeing their maturity well before the nebula was swept clean, but potentially ending up with ice-rock growth cores that were a much bigger fraction of their total mass than we believe is reasonable. With a lower volume of debris, instead, such cores would not grow too large but neither would their gravity fields, raising the possibility that the nebula would be swept clean before they could reach maturity.

And no matter the proportion of debris to gas, at some point—as these developing planets efficiently pulled in nebula mass—they would have quickly exhausted the supply of gas and debris in their neighborhoods, terminating their growth.

Rice and Armitage (2003) have found that these issues can be resolved if Jupiter and Saturn existed within a turbulent region of the accretionary disk. In that case, small and random perturbations to their orbits driven by the turbulence would have accelerated their growth, yet allowed them to occupy enough changing neighborhoods to ensure modest-sized ice-rock growth cores. Evidently, then, the turbulent state of the disk continued through this stage of planetary development.

The Outer Solar System. The Solar System does not end at the orbit of the farthest planet. As we have discussed earlier, the Kuiper Belt begins just past the orbit of Neptune; populated with numerous large planetesimals, it is understood to be the source region for many comets that we observe to visit the inner Solar System.

But there are even more comets that have not originated in the Kuiper Belt; we know that because their orbits are far out of the plane of the ecliptic—they come into the inner Solar System from all 'vertical' as well as 'horizontal' directions. Additionally, their orbital paths imply that they would require hundreds of years or more (typically much more) to complete each orbit; in contrast, comets that started out in the Kuiper Belt all have shorter periods.

Such observations suggest the existence of a different kind of source region for those comets. Their long orbital periods suggest their source is located well beyond the Kuiper Belt. To explain the full 'vertical' range of their orbital inclinations, it must be a spherical region, not flat like the Kuiper Belt is (more or less); thus, this region is not associated with the protoplanetary disk. It is a part of our Solar System simply because every object in it orbits our Sun, under the influence of the Sun's gravity.

This distant spherical source region is known as the **Oort Cloud**. It surrounds the inner Solar System three-dimensionally at distances of ~10^4–10^5 AU (thus, distances as great as a light-year)! We can imagine it being filled with some of the original dirty snowballs of the primitive nebula, located too far out to have been incorporated into the protoplanetary disk and participate in the formation of the planets. However, it is also believed to contain numerous planetesimals that had existed in the vicinity of the giant planets: as those planets grew, sweeping up nearly everything in their neighborhood, their gravitational fields caused some planetesimals to be ejected from the protoplanetary disk; although most of those ejecta ended up beyond the nebula, it is estimated that perhaps ~10% (e.g., Dones et al., 2005 as cited in Morbidelli, 2008) ended up residing in the Oort Cloud (with their extremely elliptical orbits subsequently 'circularized' by galactic tides and, yes, passing stars, to keep them in the Cloud) (Shannon et al., 2015).

As an illustration of how populated the Oort Cloud is, we note that the same process of gravitational ejection that sent icy outer Solar System planetesimals into the Cloud would have acted, though to a much lesser extent, on rocky planetesimals in the inner Solar System. Shannon et al. (2015) modeled this process and found that only ~4% of the Cloud's mass should be rocky; but that would still amount (for typical-sized asteroids) to more 'asteroids' in the Oort Cloud than in the Asteroid Belt!

The orbital periods of Oort Cloud objects around the Sun range from tens of thousands to tens of millions of years; their orbital paths are mostly more or less circular, and they remain in the outer Solar System. Occasionally, however, chance collisions, near-collisions, or (more likely) perturbations in the gravity field through which they travel dramatically alter their orbits; with a fate similar to perturbed

Kuiper Belt Objects, they still orbit the Sun but with highly elliptical paths that bring them into the inner Solar System as comets (see Figure 1.8).

What might be an even rarer occurrence is taking place right now, though we will not experience it close-up: an object from the farthest reaches of the Oort Cloud—nearly 60,000 AU—was perturbed 300,000 yr ago into an orbit that has brought it to a distance of ~22 AU from the Sun (see Irving, 2021 and Deen, 2021). Provocatively, this Oort Cloud object appears to be quite large, with a diameter between 130 km and 370 km; but at its closest approach, it will remain at a safe distance, not quite reaching the orbit of Saturn.

Comets provide a window into the physical conditions and chemistry of the primitive solar nebula and early Solar System. But they are of interest to us for other reasons as well.

- Scientists (e.g., Oro, 1961; Hoyle & Wickramasinghe, 1980), and philosophers as well, have speculated that past cometary impacts or near-impacts have 'seeded' the Earth with water and organic compounds that may not (so it is postulated) otherwise have survived the heat and chaos of the inner Solar System in its early days. In this view, which has

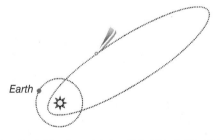

Figure 1.8 Comets originate in the Kuiper Belt or Oort Cloud, as debris—dirty snowballs—that usually remain far beyond even the outer planets of the Solar System. Perturbations to their orbits may cause them to follow highly elliptical trajectories that can bring them into the inner Solar System. As they move into the inner Solar System, heat from the Sun begins to vaporize them, producing 'jets' of material streaming away from the Sun that we see as their tail. Comet tails, which maintain their outward direction even as the comets swing around and move out from the Sun, generally include dust jets as well as gas (or "ion") jets. The latter stream directly away from the Sun under the influence of the solar wind, while the former are gradually left behind the incoming (e.g.) comet with the aid of solar "radiation pressure." Inset: photo of comet Hale-Bopp taken at the Kopernik Observatory on April 1, 1997 in Vestal, NY; the gas ion tail is bluish, while the dust tail is off-white. This long-period comet's orbit is perpendicular to the ecliptic! Kopernik Observatory & Science Center / http://www.kopernik.org/images/archive/hb0401-3.jpg / last accessed August 22, 2023.

come to be known as **panspermia**, comets have brought the chemical building blocks of life—or, in a more extreme version, actual life forms—to Earth again and again, establishing a means for life to develop here.

It might be noted that, in a cool circumstellar disk, there was no necessity for comets to bring water to Earth. In the region where Earth was forming, even after the disk had warmed to the point where dirty snowballs dissipated away, it probably remained cool enough that water vapor could adhere to nebular grains. As Earth accreted, it could have taken in 1–3 oceans worth of H_2O just by such adsorption (Drake & Campins, 2006).

An alternative to panspermia—that life arose naturally here, without assistance from outer space—is more widely accepted. Nevertheless, it should be noted that Kaiser et al. (2013) have verified by experimental simulations that simple hydrocarbons, such as those existing in the chilled 'soup' of a molecular cloud core, can in fact be transformed under the conditions prevailing in the early nebula into complex amino acids—the components of proteins. These organic compounds would then have been incorporated into the dirty snowballs (and future comets) of the outer Solar System. And, space probes of Comets Halley, Wild 2, Tempel 1, and others have shown the nuclei of those comets to be very dark, suggesting that they are indeed rich in hydrocarbons.

A 'far out' extension of the idea of panspermia, with the building blocks of life carried by wandering planets rather than comets, was suggested by Strigari et al. (2012). Such planets are discussed briefly later in this chapter. Finally, the basic idea underlying panspermia has been kept alive in the context of a meteorite from Mars, discussed near the end of this chapter.

- Massive impacts have the potential to devastate the Earth, by the heat and energy of impact or by the climate change—dubbed "nuclear winter"—forced by the impact as it sends dust billowing up into the stratosphere. More recently, this idea has been expanded to include situations where the climate is unstable—on the verge of changing for 'internal' reasons—and a smaller impact, despite producing only limited changes in the stratosphere, is nevertheless able to trigger the climate change.

For example, the massive impact ~65 Myr ago creating the Chicxulub Crater in the Yucatan Peninsula, Mexico, and sending a layer of iridium-filled dust around the globe, has been proposed as the cause of the major extinction marking the end of the Cretaceous period (the end of the Age of Dinosaurs, and the disappearance of many other species as well) (Alvarez et al., 1980; Renne et al., 2013). This widely popularized proposal illustrates the destructive power of massive impacts, though in this case, according to the geochemical evidence, the impactor was probably asteroidal rather than cometary (but see also Moore & Sharma, 2013).

Planets Did Not End Up Everywhere in the Inner Solar System. The Asteroid Belt, located between Mars and Jupiter, is filled with stony and metal 'chunks' (Figure 1.9), but no planets. Given the Asteroid Belt's distance from the Sun, and the fact that the snow line of the Solar System lies within the Belt, conditions *should* have been favorable for their growth. It appears, instead, that their development was a 'near miss': asteroids as large as Ceres, Pallas, and Vesta (the former nearly 1,000 km and the latter two each more than 500 km in diameter) are evidence that planetesimals had accreted there to the size of at least modest proto-planets. Together with the lack of full-sized planets, the existence of abundant debris during the time of planetary growth elsewhere implies that in the Asteroid Belt the process of planet formation was incomplete: interrupted, or interfered with.

As we will learn at the end of Chapter 4, the length of time it takes a body orbiting the Sun to complete each revolution depends on

Figure 1.9 Photograph of a somewhat potato-shaped Asteroid Ida taken by the *Galileo* space probe as it passed through the Asteroid Belt. Ida is about 52 km in length. To the right we can see a small 'moonlet' named "Dactyl"; Dactyl, the first confirmed satellite of an asteroid, orbits Ida every 1.5 days. NASA / http://solarsystem.nasa.gov/multimedia/gallery/Enhanced_Ida-browse.jpg / last accessed August 22, 2023.

its distance from the Sun. Within the bounds of the Asteroid Belt, there are several possible orbits whose period of revolution about the Sun happens to be in exact integer proportion to the period of Jupiter's orbit around the Sun; for example, for every five orbits completed by a body at that distance from the Sun, Jupiter completes exactly two orbits. Such **orbital resonances** are significant because planetesimals (or asteroids or meteorites) in a resonant orbit about the Sun are always closest to Jupiter at the same point in their orbit; they will be repeatedly and systematically subject to a slight tug from Jupiter's gravity at that point (more precisely, Jupiter's gravity will tug on them throughout their orbit, but that tug will be strongest at their point of nearness). Over time, these tugs will destabilize their orbit, and they will be ejected from that region of the Solar System. There are gaps in the Asteroid Belt today as a result of such resonances with Jupiter's orbit; the most dramatic gaps correspond to two asteroidal orbits and three asteroidal orbits per one Jovian orbit.

Within the evolving disk, proto-planets in the Asteroid Belt, as elsewhere, needed to accrete huge, 'planetary' amounts of mass from neighboring orbits to succeed in their development. Because Jupiter very likely grew rapidly, its influence on the Asteroid Belt would have been felt early on during this era of planetary development. Resonant gaps would have limited the ability of proto-planets there to acquire enough mass. Then, remaining as multiple proto-planets, and orbiting within a complex multibody gravity field (additionally perturbed by Jupiter's gravity), collisions were inevitable, with no fragment having sufficient gravity to draw them all in together as a planet.

Three additional implications of this orbital resonance mechanism are worth mentioning.

- Later in this chapter we will discuss the existence of planets and even planetary systems elsewhere in the galaxy. Many of the 'extrasolar' planets already detected are as massive as Jupiter (or more so). Those orbiting close to their star are known colloquially as **hot Jupiters**; for conciseness, we can refer to massive planets orbiting at distances more like Jupiter's as 'cold Jupiters.' Our explanation of how the Asteroid Belt was created suggests that extrasolar planetary systems should be expected to have *extrasolar asteroid belts* as well—at least, if they include a cold Jupiter. Cold Jupiters would have formed outside the snow line, where ice, dust, and gas were abundant, and they would have formed quickly via runaway gas accretion; meanwhile, around the snow line boundary, one or more proto-planets and lots of planetesimals, all accreting less mass due to their location, would have made slow progress as they struggled to grow. Under the gravitational influence of the nearby cold Jupiter, those boundary proto-planets and planetesimals would then have suffered a similar fate to the proto-planets in our own Asteroid Belt, with their accretion disrupted (cf. D'Angelo & Podolak, 2015).

In fact (e.g., Hines et al., 2006), observations of remnant protoplanetary disks around other stars, called *debris disks*, show that "a handful" of them are probably just such asteroid belts: their infrared emissions indicate they are relatively 'warm,' at temperatures (~100–300K) which are reasonable for the neighborhood of the snow line; most debris disks, in contrast, are cooler, with temperatures as low as ~50K, and are interpreted to correspond to farther out, Kuiper-type belts.

The other two implications refer to our own Asteroid Belt.

- We might imagine that, as the Herbig-Haro proto-Sun expelled a major fraction of its own mass—changing the proto-Sun's gravity field and thus modifying the orbits of every object in the Solar System (proto-planets and asteroids)—the resonant gaps would have varied as well, 'sweeping' through the width of the Asteroid Belt and depleting the Belt's mass even more. Similarly, as the T-Tauri heat and solar wind cleared the nebula gases (with a total mass at least an order of magnitude greater than that of all the planets; Ward, 1981) out of the Solar System, further modifying the ambient gravity field, a second sweep of the resonant gaps would have further depleted the Asteroid Belt (Heppenheimer, 1980; see also Ward, 1981).
- Resonant gaps, persisting to this day, provide a partial explanation of why the Asteroid

Belt is so depleted in mass: asteroidal fragments (meteorites) have continually been forced out by Jupiter's gravity. Random collisions among the asteroidal remnants and debris would have compounded the depletion (see, e.g., Minton & Malhotra, 2010). But what we observe today is striking: the total mass of the Asteroid Belt is estimated to be less than a thousandth of the mass contained in the Earth (Chambers, 2010)! Now, if the Belt had had the potential to produce a planet, in the absence of the factors just described, then (thinking of the range of actual planets in our Solar System) its original mass would have been more or less comparable to Earth's, or even greater. This suggests that still another mechanism contributed to the severity of the depletion. In Stage 8 below, we will discuss a Solar System–wide cataclysm—the final stage of Solar System evolution—and discover a way to produce that massive depletion.

Stage 7: The Sun Ignited. The T-Tauri proto-Sun continued to contract and heat up. At some point the temperatures within it reached millions of degrees, thermonuclear reactions commenced, and the star we call our Sun was born. The thermonuclear reactions were fusion reactions that cooked hydrogen into helium:

$$6H^+ \rightarrow \ldots \rightarrow He^{++} + 2H^+ + \text{heat},$$

with the intermediate reactions, unspecified here, resulting in two of the six protons (each a hydrogen nucleus) being converted into neutrons in the process of synthesizing each helium nucleus (each He nucleus has two protons and two neutrons). This fusion reaction leaves two protons free to participate in additional fusion reactions, leading to chain reactions and sustained solar burning if the central nebula has critical mass.

Why Did the Primitive Solar Nebula Have to Start Out with So Much Mass? At this point, we can understand the need for an initial nebula mass (~1½ solar masses) so much greater than the current mass of the Solar System.

- As the nebula began to contract, the heat generated within increased the gas pressure, resisting further compression. With a more massive nebula, the gravitational forces driving the collapse would be strong enough to overcome that early resistance.

- After the proto-Sun entered the Herbig-Haro phase, it would have lost a significant fraction of its mass from the bipolar outflow. Even in the less intense T-Tauri phase, it would have continued to lose mass via the T-Tauri solar wind.

- By shedding a significant amount of its mass, the HH/T-Tauri proto-Sun would lose much of its angular momentum.

The proto-Sun had attained a mass roughly equal to the Sun's present-day mass while the Herbig-Haro bipolar jets were still strong (Montmerle et al., 2006). It would maintain that mass, more or less, even during the remaining bipolar outflow and then strong solar wind because it continued to feed off the disk; it could maintain a now-reasonable rotation rate because that inner disk material had already shed most of its own angular momentum.

- After sufficient collapse, the density of material in the central portion of the massive nebula would be great enough to trap the heat generated by compression and gravitational infall, keeping enough of it from radiating away into space so that the central temperature could reach the critical value for ignition.

- A massive central region ensured that, following ignition, the Sun would have critical mass, allowing the thermonuclear reactions within it to continue.

The proto-Sun's loss of angular momentum by shedding mass helps resolve the central problem of nebula theory. Additional ways to resolve that problem derive from the fact that the proto-Sun was likely to possess a strong magnetic field, much stronger than the Sun's current field. Such a strong magnetic field would extend to the farther reaches of the protoplanetary disk, where everything takes much longer to orbit the Sun; any interactions between the field and dust grains there could *brake* the proto-Sun, decreasing its spin and angular momentum and also (by 'an equal and opposite reaction') transferring that angular momentum to the outer disk. Alternatively, interactions between the field and charged particles of the (T-Tauri) solar wind, the latter expelled radially outward from the hot proto-Sun, might also act to slow the proto-Sun's rotation (e.g., Montmerle et al., 2006).

Stage 8: A Heavy Bombardment Scarred the Inner Planets. The craters on the Moon, for example on the far side as shown in Figure 1.10, are evidence that meteorites, comets, and other space debris have hit the Moon—and, presumably, the planets, and other moons—throughout the life of the Solar System. Cratering would have been especially intense when the Moon was young, when there was still much debris around.

The near side of the Moon also shows evidence of cratering throughout its lifetime. Additionally, the near side features extensive dark areas, called lunar **maria** (plural for

Figure 1.10 Photograph of the far side of the Moon, taken from the Apollo Command Module. From Dr. N. Short's Remote sensing Tutorial / http:// rst.gsfc.nasa.gov / last accessed August 22, 2023.

Figure 1.11
Photograph of the near side of the Moon, the side visible from Earth, taken at Lick Observatory.
The bright-rayed crater near the bottom of the picture is *Tycho*. The *mare* enclosed by the red-dashed box is *Mare Tranquillitatis*, the "Sea of Tranquility," the site of the first human landing on the Moon. NASA / http://www.spacetelescope.org/images/html / last accessed August 22, 2023.

mare, which is Latin for "sea"); together, these constitute what is colloquially known as the Man in the Moon. The maria were the result of an exceptionally **heavy bombardment** of the lunar surface by massive bodies ending 3.8 Byr ago that created huge *impact basins*. The impacts were so energetic that they cracked the lunar lithosphere, allowing magma to seep up from below, filling the basins with dark basaltic lava; see Figure 1.11. The mare volcanism peaked ~3.8–3.2 Byr ago (e.g., Turner et al., 1973), though it started out ~4 Byr ago and may have continued on beyond its peak activity, declining significantly, for an additional 1 or even 2 Byr (e.g., Hiesinger et al., 2000, 2011).

The surfaces of Mercury (Figure 1.12) and Mars show evidence of similar huge impact basins, suggesting that the *entire* inner Solar System had been subjected to the Heavy Bombardment (e.g., Kring & Cohen, 2002). Any evidence for this on Earth, of course, would have been completely erased by subsequent geological and atmospheric processes. No atmosphere could persist on Mercury, due to its extreme proximity to the Sun and its relatively small size, and its small size may have limited its geological life, so—despite brief, contemporaneous volcanism—we can still see features on its surface created 4 Byr ago (Marchi et al., 2013). Mars may have had a thick atmosphere early on, as we will discuss later, and was geologically active, so impact basins from the Heavy Bombardment have been largely obscured—but they have not been entirely eliminated (e.g., Bottke & Andrews-Hanna, 2017).

Brief and Intense, or Drawn Out? As envisioned by Tera et al. (1973, 1974) and others, and based on the ages when moon rocks from these basins—brought back by Apollo astronauts and also Luna spacecraft—had been impacted, the barrage was understood to have happened beginning 4.1–4.0 Byr ago and ending 3.8 Byr ago; this period of time, occurring long after the Solar System formed, consequently became known as the **Late Heavy Bombardment**. Over the years, several

Figure 1.12 A photomosaic of the "mighty" multiringed Caloris Basin, the largest impact basin on Mercury, constructed from images obtained by MESSENGER, the first spacecraft to orbit that planet (https://messenger.jhuapl.edu/Explore/Science-Images-Database/pics/the_mighty_caloris_web-small.png; credit: NASA / Johns Hopkins / Carnegie Institution). The volcanism flooding into and surrounding the basin, filling it, is dated to about 3.8 Byr ago (Strom et al., 2011).

possible sources for the impactors have been considered, including the Asteroid Belt, the Kuiper Belt, a postulated belt of planetesimals located somewhere beyond the present-day orbit of Saturn, the Oort Clod, transient moons of Earth, remnants of planetary accretion, and remnants of the massive impact that formed our Moon. Evidence pointing to the Asteroid Belt as a likely source comes from geochemical similarities between the impactors, as inferred from the chemical and isotopic signatures they left in those moon rocks, and meteorites from the Asteroid Belt (specifically, meteorites that

had originated from differentiated asteroid parent bodies, bodies whose interiors had been heated enough for iron cores to have separated from rocky mantles) (Kring & Cohen, 2002). Those researchers also found evidence (consistent impact degassing times) from a range of meteorites implying that the asteroidal parent bodies themselves had been subject to that same barrage as the impactors were driven into the inner Solar System.

Whichever the source, though, they had all existed at the time of Solar System formation half a Byr earlier (or soon after, if they came from the Moon-forming impact); it is not obvious why the Solar System would suddenly experience a late 'spike' of impacts.

One intriguing mechanism involves migration of the outer planets' orbits. As envisioned by Gomes et al. (2005, summarized nicely in Morbidelli et al., 2007), after planet formation was complete, the outer planets were positioned differently than now: Jupiter was slightly farther out from the Sun, but Saturn, Uranus, and Neptune were less distant than they are now; the Kuiper Belt was also closer in, just half as far from the Sun, and perhaps two orders of magnitude more massive. Currently, Saturn's orbital period is about $2\frac{1}{2}$ times that of Jupiter, but those researchers speculated that back then, with Saturn closer, it was less than twice Jupiter's. Gravitational tugging from the massive Kuiper Belt gradually shifted the orbits of Jupiter and Saturn till those planets were in a 2:1 orbital resonance; at that time—possibly around 0.5–0.8 Byr after planet formation, depending on model parameters—their orbits changed rapidly, and they shifted toward their present configurations.

But the briefly endured 2:1 resonance wreaked havoc on the outermost planets' orbits; those orbits were so distorted that Neptune's even penetrated out into the Kuiper Belt. The net gravitational effect on the planetesimals of the Kuiper Belt was complete disruption, with a rain of planetesimals down toward the inner planets. A few would be captured along the way by the giant planets, adding to their collections of moons; as the majority of planetesimals fell through the Asteroid Belt, collisions and gravitational perturbations drained much of that Belt's supply of planetesimals (and asteroidal fragments), resulting in a Late Heavy Bombardment experienced by the inner Solar System.

According to Morbidelli et al. (2007), there are other orbital configurations of the outer planets (involving additional orbital resonances) which can also lead to the same results. A somewhat similar mechanism, postulated earlier by Levison et al. (2001), hypothesizes that, somehow, the formation of Uranus and

Neptune was delayed by a few hundred Myr; in that case, their subsequent accretion in a massive Kuiper Belt–type zone beyond the orbit of Saturn could have led to planetary migration, then disruption of the zone's remaining planetesimals, and, ultimately, depletion of the Asteroid Belt.

On the other hand, invoking mechanisms to create a spike of impact activity around 4 Byr ago may not be necessary: the Heavy Bombardment may have begun earlier, and either operated over an extended period of time (so that the spike was just the tail end of the accretionary period) or included other, discrete impact episodes. For example, it appears that one major impact site (called Nectaris) previously dated with an age of 3.9 Byr is more likely 4.2 Byr old (Fischer-Gödde & Becker, 2011, also Baldwin, 2006). And, there is at least one Moon rock, in the Descartes Highlands visited by Apollo 16 astronauts, which was derived from another basin-forming impact event dated 4.2 Byr (Norman & Nemchin, 2014).

It has also been argued that the apparent lateness of the Heavy Bombardment is an artifact of limited and perhaps biased sampling of impact basin rocks, at least in part a consequence of the dominance of the ~3.9-Byr-old Imbrium impact (e.g., Baldwin, 2006)—an 'overprint' by Imbrium's impact ejecta at the Apollo sampling sites (Norman & Nemchin, 2014). According to Chapman et al. (2007), the evidence and its interpretations are too equivocal to conclude whether the Heavy Bombardment was restricted to a narrow time period or not. Boehnke and Harrison (2016) have shown that it is even possible to produce an apparent spike in measured impact ages from an impact history that actually declines steadily from an initially high rate ~4.5 Byr ago.

The issue is further complicated by the long delay between the impact spike and the peak resulting volcanism. Nevertheless, the *end* of this period of impacts is well established at 3.8 Byr. Given the various uncertainties in its beginnings, a less presumptuous name for it might be the *terminal phase heavy bombardment* or the *terminated heavy bombardment*.

The Moon Shows the Power of the Bombardment. On the Moon, the Heavy Bombardment is believed to have created mass concentrations (nicknamed **mascons** by Muller & Sjogren, 1968) below some of the large impact basins (mostly the lava-filled ones); they have been detected by variations in the lunar gravity field experienced by satellites orbiting the Moon (Figure 1.13). Such mass excesses may represent the signature of the dense magma that filled the basins, carried upward by a delayed, broad regional uplift in response to the impact (Melosh et al., 2013). Mascons are associated with almost

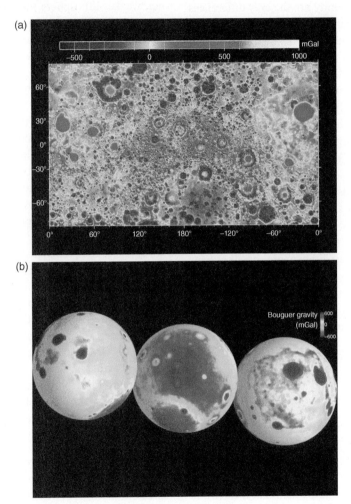

Figure 1.13 Lunar gravity reveals the existence of mascons.
(a) Variations in lunar gravity as measured by NASA's GRAIL mission. This map is centered on the far side of the Moon (longitude 180°); near side would center at 0° longitude.
Note that many craters are ringed by alternating gravity fields, so that (for example) an observer would measure a gravity high followed by a low followed by a strong high as they move from outside a crater into its center—in some cases over a very short distance. Such 'bumpiness' in the gravity field can be attributed to surficial effects including topography (Namiki et al., 2009).
(b) Removing the surficial effects (using techniques we will learn about in Chapter 4) allows the mascons to be seen more clearly. The center view focuses on the lunar far side; the left and right views are after and before the Moon has rotated to the far side view (so views are center → left → right, in order), showing different views mainly of the near side.
The units (see color scale at the tops of the images) are milliGals, defined as accelerations of 10^{-3}cm/sec². Images downloaded from NASA / http://www.nasa.gov/mission_pages/grail/multimedia/gallery/gallery-index.html / last accessed August 22, 2023.

a dozen and a half impact basins (see, e.g., Potts & von Frese, 2003), two-thirds of which are on the near side of the Moon. They help make the Moon's gravity field the "lumpiest" (A. Konopliv, cited in Bell, 2006) in the Solar System.

The Man in the Moon and the distribution of mascons are evidently not the only differences between the near and far sides of the Moon. Modeling of the lunar gravity field (Potts & von Frese, 2003, and references therein) finds the lunar crust to be about 13 km thicker on the far side than the near side—implying that the crust (and lithosphere) on the far side would be mechanically stronger. On the far side, the surface topography is rougher (Smith et al., 1997); Potts and von Frese find the depth of the crust-mantle boundary to be more variable also, leading to a less uniform crustal thickness. These characteristics suggest a colder as well as stronger crust (and lithosphere) on the far side and a warmer and weaker crust/lithosphere on the near side—just as, on Earth, hotter oceanic lithosphere is more pliable and has a smooth profile, and colder continental lithosphere is more rugged. Furthermore, impact basins on the far side have fewer 'rings' (i.e., ridges and rills) around them than those on the near side, suggesting that the far side lithosphere survived the massive impacts better, again leading to the conclusion that it is stronger and colder.

There are still other aspects of this asymmetry or dichotomy. The near and far sides of the Moon differ compositionally (e.g., Wasson & Warren, 1980), at least in part an effect of the composition of the impactors that created the large basins. And, at least partly as a consequence of the variations in surface topography and crustal thickness, the Moon's center is offset—by a couple of kilometers, toward the Earth but not exactly so (Smith et al., 1997; also Kaula, 1974)—from its center of mass.

In fact there has been a lot of research devoted to explaining the Moon's diverse asymmetry (various theories are referenced in Garrick-Bethel et al., 2012). Some of them invoke convection in the lunar interior as part of their explanation—perhaps not too surprising given the inferred thermal differences between the near and far sides; nevertheless, we will consider any discussion of the fundamental causes of asymmetry to be beyond the scope of this introductory chapter.

We can, though, appreciate why maria are more common on the Moon's near side: thinner, weaker lithosphere fractured more easily from those massive impacts; hotter interior magma, able to flow more easily and flow closer to the surface, oozed up to fill the impact basins.

Finally, a fact underlying our discussion of the Moon's near side and far side is that the Moon always shows Earth the same face—that is, the Moon spins on its axis at the same rate it orbits the Earth. We will learn in Chapter 6 how gravitational (actually tidal) forces exerted by a planet, repeatedly tugging on a moon as it both spins on its axis and orbits the planet, can end up locking the moon's spin to its orbit such that the same hemisphere always faces the planet, just as our own Moon does now. Because of this **tidal locking** or **spin-orbit coupling**, *'same-face moons' are actually somewhat commonplace in the Solar System* (one example of tidal locking, Pluto and its moon Charon, was noted early in this chapter). Given our Moon's origin—close to the Earth (so Earth's forces on it were strong) and hot from the impact that created it (so it was soft and 'compliant' in response to those forces)—tidal locking would have been achieved fairly soon after its formation.

The Bombardment May Have Had Life-Altering Effects on the Earth System. Given the larger size and stronger gravity of Earth compared to the Moon, the Heavy Bombardment would have been more intense here. Possible effects include creating the seasons as we know them, and significantly modifying the components of the Earth System—with obvious implications for life in both cases.

- With massive debris plunging in from odd directions under a heavy bombardment, their impacts could have tilted the rotation axis of the Earth away from any preexisting tilt; the latter would have represented a cumulative effect of the large impacts of late-stage accretion (Agnor et al., 1999) more than half a Byr earlier, which in turn had modified the 'primary tilt' acquired during turbulent nebula collapse and early planetary formation (see Barge & Sommeria, 1995). Today, rather than aligning perpendicular to the ecliptic (the plane of our orbit), the Earth's rotation axis is tilted by 23½°; equivalently, Earth's equator is tilted relative to the ecliptic by the same amount (see Figure 1.14).

The tilt of the Earth's axis is called its **obliquity**. Without an obliquity, the Earth would have nearly identical seasons—nearly, because Earth's orbit is not perfectly circular, so its varying distance from the Sun produces a slight seasonality in incoming solar heat. With an obliquity, there is a much bigger variation: more of the Sun's light shines directly on the hemisphere (northern or southern) that is tilted toward the Sun than on the hemisphere that is tilted away (see Figure 1.15). As the Earth proceeds along its orbit, the northern hemisphere receives the brunt of the Sun's heat for six months out of the year and the southern hemisphere for the other

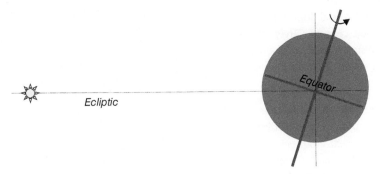

Figure 1.14 The orientation of the Earth in space. Relative to the ecliptic, the plane of Earth's orbit around the Sun, the Earth's equator is tilted by 23½°. With respect to the perpendicular to the ecliptic, the Earth's rotation axis is tilted by 23½°.

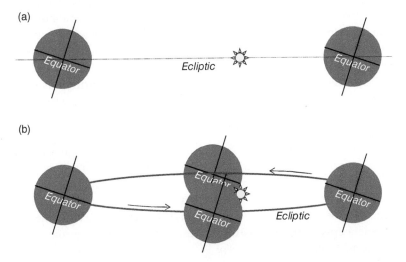

Figure 1.15 The tilt of the Earth is the main reason it has seasons. (a) A side view of the Earth's orbit, with Earth on the right during January and on the left during July. The Sun shines directly on the southern hemisphere during January and directly on the northern hemisphere during July, so northern hemisphere winter is in January and summer is in July. (b) A slanted view of the orbit, with Earth on the right during January and on the left during July. At the times of the *equinox*, about three months earlier or later, the Sun shines equally on both hemispheres.

six months, causing a significant and opposite seasonal variation in the solar heat received by each hemisphere.

- The evolution of Earth's atmosphere will be discussed in the next chapter. The role of the Heavy Bombardment in that evolution is unresolved, with some researchers predicting a loss or even complete 'blow-off' of the preexisting atmosphere due to the largest impacts, and others predicting a net gain of atmospheric gases. One factor affecting such predictions is the source region of the impactors: atmospheric erosion could dominate if the source is asteroidal (especially if the asteroids are inside the snow line); replenishment would dominate if the source is Kuiper Belt comets. Given the need to 'drain' the Asteroid Belt, as discussed above, we might consider an asteroidal source a bit more probable.

Furthermore, it is possible that the larger bombardments temporarily (for reasons also discussed in the next chapter) vaporized any watery oceans that had formed on Earth's surface (Zahnle & Sleep, 1997). And, the Heavy Bombardment may have produced a magma ocean, analogous to that hypothesized during accretion (though given the generally smaller size of asteroidal impactors, compared to late-accretion impactors, any magma ocean resulting from the Bombardment would more likely be surficial).

In terms of intensity and environmental consequences, the Heavy Bombardment dwarfed the ~65-Myr-old Chicxulub impact (Kring & Cohen, 2002). We imagine dire effects on life from the latter; but the former could have been orders of magnitude more extreme. Evidently, the best place for life to have developed—or survived—during this time might have been something like a hydrothermal vent or crustal fracture (see also Maher & Stevenson, 1988; Zahnle & Sleep, 1997)! Living on Earth's surface would have been tumultuous, even hellish; it is no surprise that this era (cf. Figure 1.16) has been named the Hadean eon!

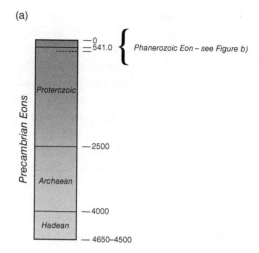

(a)

Figure 1.16 The International Commission on Stratigraphy (ICS) 2013 geologic timescale in millions of years ago. Adapted from Cohen et al. (2013).
(a) The major divisions of Earth's history, from Earth's birth sometime between 4,650 Myr and 4,500 Myr ago (or 4.65–4.5 Byr ago) to the present. The Phanerozoic Eon, marked by the appearance of visible fossils in the geologic record, was preceded by a period of time now called the Ediacaran (dashed line, from 635 to 541 Myr ago) now known to have supported a 'burst' of exotic life forms.

The threat to survival during the Heavy Bombardment was not just extreme events but also extreme changeability, as each major impact brought a much denser atmosphere or a much thinner one, loss of oceans, and extensive surficial melting—and each of these changes was transitory, but on its own timescale. For any life in existence during this era, adapting to such sudden and drastic changes in their environment would have been a huge challenge.

Over the eons, in the absence of external impacts, it turns out that *the Earth System* had a way to maintain an atmosphere and oceans and generally provide a more stable environment, with less extreme and more gradual variations. With the Bombardment over, life would develop and evolve *within* that System, connected to its geology and to the variations in its atmosphere and oceans—not just thriving in a much more compatible environment but able to interact with and even affect that environment. The role of the Earth

(b)

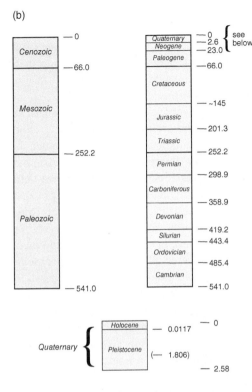

Figure 1.16 (b) The various eras (on the left) and periods (on the right) comprising the Phanerozoic. The recently named Paleogene and Neogene periods comprise what used to be known as the Tertiary period. The Quaternary period (see bottom also) is now defined by the start of glacial oscillations, 2.58 Myr ago (see Walker et al., 2018), accommodated by extending the beginning of the Pleistocene back to that time and terminating the Pliocene (a stage within the Neogene or Tertiary) at that time, rather than at 1.81 Myr. Adapted from Cohen et al. (2013).

System, and how it worked to produce a more life-friendly environment, are the subject of the next chapter.

1.2 Properties of the Solar System

1.2.1 The Spacing of the Planetary Orbits

As it turns out, the gap between neighboring planets doubles, moving out from the Sun! This profoundly simple property of our Solar System was empirically deduced in 1766 by a physics professor named Titius. But the simplicity of his observation stands in contrast to the tumult of the world's varied reactions to it.

For starters, Titius' observation is generally known as the Titius-Bode Law, and sometimes as Bode's Law, because of the plagiarism and rather successful publicizing of Titius' deduction as his own by an astronomer named Bode (Titius had inserted his theory into a treatise on natural philosophy he had been translating, as a translator's footnote; Bode incorporated it into the second edition of his own, very popular, astronomy textbook without attribution—at least, until subsequent editions (Nieto, 1972)). At the other end of its history is the striking—and possibly unique—editorial decision in recent years by a leading astronomical journal to reject the publication of articles concerned with observational reformulations of (in their words) Bode's Law (noted by Boss, 2006), a policy continued to the present day (www.elsevier.com/journals/icarus/0019-1035/guide-for-authors)! A quick survey of several recent astronomy textbooks unfortunately seems to support that policy, with the law mentioned not at all.

A First Look at the Titius-Bode Law, and Its Early Scientific and Popular Impacts. The law focused on the distance of each planet from the Sun, and not explicitly on the gap between adjacent planets. If r_n denotes the radial distance of the nth consecutively numbered planet from the Sun, measured in units of AU, then the law predicted that

$$r_n = 0.4 + 0.3 \times 2^n,$$

where $n = 0$ for Venus, $n = 1$ for Earth, $n = 2$ for Mars, and so on. For illustration, when $n = 1$, the distance r_1 would be $0.4 + (0.3 \times 2^1) = 0.4 + 0.6 = 1.0$ AU, verifying the distance for Earth.

You should verify that $r_0 = 0.7$ AU and $r_2 = 1.6$ AU, so the law predicts the gap between Earth and Mars to be exactly twice as large as the gap between Venus and Earth.

Mercury is included in this expression through the first term (the 0.4), that is, by excluding the 0.3×2^n term from the right-hand side (alternatively, this is sometimes described as evaluating the expression at $n = -\infty$); subsequent formulations (see below) have yielded a more natural inclusion of Mercury. Titius' essential idea, of the doubling of the gap between adjacent planets, is assured in this formulation by having an additional factor of 2 each time n increases by one.

As Figure 1.17 shows, the Titius-Bode Law works really well for most of the planets. That is impressive considering that, at the time of its formulation, the only planets known to exist beyond Earth were Mars, Jupiter, and Saturn; the Asteroid Belt was also unknown.

Titius' discovery—implying a regularity and order to the Solar System—had an immediate 'philosophical' attractiveness. But the law also amounted to a prediction of planets yet undiscovered, and its amazing success with such predictions guaranteed its credibility as well: in 1781 Uranus was discovered, confirming the rule for $n = 6$ and stimulating efforts to verify it for $n = 3$; and in 1801 the largest asteroid, Ceres, was discovered at the predicted distance.

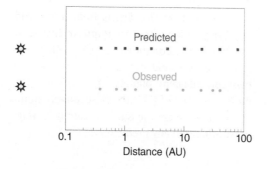

Figure 1.17 Titius-Bode Law predictions versus reality. Planetary distances from the Sun (☼, on the left) predicted by the law are plotted as brown squares; actual distances are shown as green circles. Distance is in units of AU (1 AU being the Earth-Sun distance) and is plotted on a log scale; for example, Mercury is at a distance of 0.4 AU, Earth ($n = 1$ in the law) is at 1 AU, and Saturn ($n = 5$ in the law) is at 10 AU. The asteroid Ceres, included as $n = 3$ in the law, may be viewed as representative of the Asteroid Belt; Pluto is included for comparative purposes for the case $n = 8$ but may also be viewed as representative of the near edge of the Kuiper Belt.

The law does not work well for Neptune or Pluto; on the other hand, Pluto's orbit has the curious property of intersecting Neptune's—at times Pluto is actually closer to the Sun than Neptune is—suggesting that other factors affected the orbits of those two bodies.

The early success of the Titius-Bode Law in predicting the orbital distance of an asteroid as well as unknown planets contributed to the popular interpretation of the Asteroid Belt, discussed earlier, as fragmented leftovers of an originally full-sized planet in that location, catastrophically destroyed (cf. Urey, 1951). And the law provided a reason for some to believe in the presence of additional, yet-undiscovered planets in the inner Solar System; such planets would correspond to $n = -1, -2$, etc., in the original formulation of the law. The most famous of these was a hypothetical planet named "Vulcan" orbiting the Sun inside Mercury's orbit, whose existence would also have explained observed peculiarities in the orbital path of Mercury over time (such variations would eventually be better explained by Einstein's theory of general relativity, which predicted distortions in the structure of space around the mass of the Sun amounting to modifications of the Sun's gravity; those modifications would cause additional perturbations to the orbit of Mercury, precisely as observed).

Subsequent Formulations of the Titius-Bode Law, Including an Angular Momentum Approach. Following Bode's publication of the law, several scientists determined that more precise relationships between planetary number n and radial distance r_n from the Sun were possible, by writing r_n as

$$r_n = a + b \times c^n$$

with somewhat different constants a, b, and c than those of Titius; or, even better,

$$r_n = b \times c^n,$$

with the possibility that b depends weakly on n as well. In these alternatives, a doubling of the planetary spacings would have corresponded roughly to a value of 2 for the parameter c

(but only approximately, because orbital spacing, not orbital radius, is what doubles); even without doubling, though, these formulations describe an orderly progression of the planets out from the Sun. In algebra, the latter type of formula is called a *geometric progression*.

Some scientists found that laws of this form *also* predicted accurately the distances of the moons of the giant planets from their parent planet; this added to the belief that such laws are based on fundamental laws of nature. Nieto (1972) provides details on such relations, and additional history on the impacts of Titius-Bode; see, e.g., Hayes and Tremaine (1998) for a summary of more recent developments.

Conservation of angular momentum has been a recurring theme in this chapter, in the collapse and evolution of the solar nebula. We find that it also plays a fundamental role in the Titius-Bode Law, as outlined in the steps below.

Filling in the details of this outline as we progress through Steps 1–4 will give you good practice manipulating basic formulas and equations, and is highly recommended!

Step 1. The **angular momentum** L of a point of mass m orbiting another is, by definition, $L \equiv rmv$ if the orbit is circular and of radius r; v is the speed of the orbiting mass (so the mass' *linear* momentum is mv). Note that if the mass takes T seconds to complete one orbit (so that T is the orbital **period**), covering a distance $2\pi r$ along the circular orbital path, then

$$v = \frac{2\pi r}{T},$$

allowing us to write the angular momentum of the mass as

$$L = \frac{2\pi r^2 m}{T}.$$

Step 2. In connecting L to the Titius-Bode Law, we might anticipate the planet's mass to be similarly irrelevant, so we graph in Figure 1.18 the relation within our Solar System between the *specific* angular momentum L/m, i.e., the planetary angular momentum *per unit mass*,

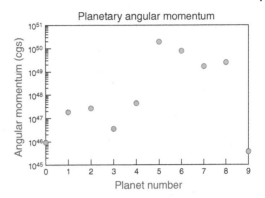

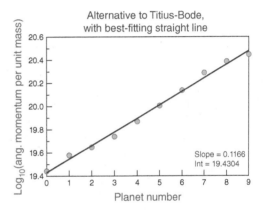

Figure 1.18 Alternative formulation of the Titius-Bode Law, relating log of the planetary angular momentum per unit mass versus planet number (lower graph); for illustration, the planetary angular momentum not per unit mass is plotted in the upper graph. This formulation was discovered during discussions with A. Whelsky. In this formulation, the planet number $n = 0$ corresponds to Mercury, $n = 1$ to Venus, etc. ($n = 4$ corresponds to Ceres); as in Figure 1.17, Pluto is also included here. Angular momenta (green circles) are computed using the actual distances of the planets; the best-fitting straight line is also shown, with its slope and y-intercept listed in the inset.

and planetary number. From Step 1, we can write L/m as $(2\pi r^2)/T$. Denoting this quantity for the nth planet by ℓ_n, Figure 1.18 reveals a strongly linear relationship between $\text{LOG}(\ell_n)$ and n. We can write such a relationship as

$$\text{LOG}\left(\ell_n\right) = A + Bn, n = 0, 1, 2, \ldots, 9,$$

where, with ℓ_n in cgs units, it turns out $A = 19.43$ and $B = 0.1166$.

A is a large number because the planets have so much angular momentum. But what does this relation tell us about how ℓ_n increases from planet to planet within the Solar System? Take the difference between this relation for the $(n+1)$th planet and for the nth planet; recalling that 'the difference of the LOGS is the LOG of the ratio,' you should find that the LOG of the ratio ℓ_{n+1}/ℓ_n depends fundamentally on B.

Step 3. In Chapter 4, we will discuss Kepler's Laws, which describe the orbit of any planet about the Sun. Kepler's Third Law, which relates the planet's orbital period (T) to its distance (r) from the Sun, says that T^2 is proportional to r^3,

$$T^2 \propto r^3.$$

It follows that $T \propto r^{3/2}$.

You should be able to show from this relation that L/m (i.e., ℓ) is proportional to $r^{1/2}$, or equivalently that ℓ equals a proportionality constant times $r^{1/2}$.

In terms of logarithms, then, $\text{LOG}(\ell)$ will differ from $\text{LOG}(r^{1/2})$ by a constant (the LOG of that proportionality constant). Finally, we can write $\text{LOG}(r^{1/2})$ as $\frac{1}{2} \text{LOG}(r)$. In short, thanks to Kepler's Third Law, the linear relation for $\text{LOG}(\ell_n)$ shown in Figure 1.18 implies a similarly linear relation for $\text{LOG}(r_n)$, say,

$$\text{LOG}(r_n) = A' + B'n, \; n = 0, 1, 2, \ldots, 9$$

Step 4. Finally, we can rewrite the preceding linear logarithmic relation as

$$r_n = 10^{(A'+B'n)}, \; n = 0, 1, 2, \ldots, 9$$

or

$$r_n = b \times c^n, \; n = 0, 1, 2, \ldots, 9$$

a geometric progression, with $b = 10^{A'}$ and $c = 10^{B'}$. This final result is actually the subsequent, more precise and popular geometric progression version of the Titius-Bode Law employed to characterize the orbits of the planets and also the giant planets' moons. The linear relation revealed in Figure 1.18 is then no different than the later version of

Titius-Bode; but its context serves to emphasize the underlying role of angular momentum in planetary spacings. If we began with that law and reversed the steps outlined here, we would end up showing that it implied a particular relationship between planetary number and specific angular momentum.

Our derivation also shows that the planet Mercury can be included without resort to mathematically ad hoc declarations (like specifying $n = -\infty$), and that in fact the actual numbering of the planets (e.g., whether we start at 0 or 1) is unimportant, as long as that numbering is consecutive (a result also shown by Prentice (1979), who numbered the planets consecutively—beginning with Neptune and moving inward).

What About That Doubling? As discussed in Nieto (1972), the fact that the later version of Titius-Bode employs a value of c different than 2 was used to conclude that the doubling of planetary spacings is not as significant as the orderliness of the geometric progressions, expressed by $r_n = b \times c^n$ no matter what the value of c is. There has even been speculation as to the physical significance of $c = 1.73$, the value which best fits $r_n = b \times c^n$ for both the planets and the moons of Jupiter. Incidentally, the value of B in our angular momentum analysis would lead to a value of B' such that $c = 1.71$.

Mathematically, however, such values of c are not necessarily inconsistent with a doubling of planetary spacings. That's because Titius' doubling expression is roughly equal to the geometric expression, as long as n is not too large and c is not 'too small.'

Here are the details. Within the range $0 \leq n \leq 9$ (or even $-2 \leq n \leq 9$, as considered by some researchers), the values of $\text{LOG}(0.4 + 0.3 \times 2^n)$ plotted versus n exhibit a linear dependence on n to fairly good approximation; for example, for $n \leq 7$,

$$\text{LOG}\left(0.4 + 0.3 \times 2^n\right) \cong -0.252 + 0.254n.$$

As we saw above in Steps 3 and 4, a linear relation like $\text{LOG}(r_n) \cong -0.252 + 0.254n$ can be rewritten in the form $r_n = b \times c^n$. That is,

$$(0.4 + 0.3 \times 2^n) = r_n \cong \underbrace{10^{(-0.252)}}_{b} \times \underbrace{10^{(0.254)n}}_{c^n}.$$

In this case, the value of c will be 1.79—not too far from our Solar System's value of 1.73.

Astrophysical Implications of the Titius-Bode Law. To this day, no consensus has been reached about the physical basis for the law; nevertheless, it can be used to better understand the Solar System. A comprehensive review of the history and science behind the Titius-Bode Law led Nieto (1972) to the conclusion that the law most likely reflects the importance of turbulence and/or electromagnetic forces in the primordial nebula, factors proposed over the years by a number of researchers. Because the law also evidently applies to the spacing of moons orbiting Uranus (as well as Jupiter and Saturn), however, we might want to discount electromagnetic causes, since Uranus' magnetic field probably did not exist until planet formation was complete (cf. Stanley & Bloxham, 2006).

Graner and Dubrulle (1994) argue that orbital laws characterized by geometric progressions, as in the later versions of Titius-Bode, are the natural outcome of planetary (or satellite) systems which are gravitationally dominated. Kepler's Third Law, for example, takes the same form (orbital period squared proportional to orbital radius cubed) whether it refers to planets orbiting the Sun or moons orbiting a planet; the 'scale' of the orbital distances depends only on the mass of the central body. Their arguments imply that *every* planetary system we observe (and any numerical model of a planetary system) should exhibit a Titius-Bode-type law. Any failure to obey such a law—for example, in our Solar System, the orbit of Neptune—would require explanation.

The "scale invariance" emphasized by Graner and Dubrulle (1994) includes the possibility of a turbulent protoplanetary disk; turbulent processes often feature similar behavior on both large and small length scales.

A common mechanism for both planetary and moon orbital spacings could then be some factor associated with the state of turbulence in the primitive nebula. In the analysis by Barge and Sommeria (1995) discussed earlier in this chapter, in fact, the eddies that coalesced into proto-planets within their turbulent accretionary disk model ended up at increasing distances from the Sun reminiscent of the Titius-Bode geometric progression.

Using an approach based instead on conservation laws, combined with specific models of the protoplanetary disk, Laskar (2000) showed that planetary accretion results in a variety of orbital laws depending on the initial mass distribution within the disk. He found that one—and only one—distribution, with a particular decrease in mass density with distance from the central body, yielded a Titius-Bode-type law.

Astronomical Confirmation of the Titius-Bode Law Is Mixed. So, is the Titius-Bode Law just a historically important curiosity? We first note that its prediction of a planet between Mars and Jupiter is not a defect: there would have been a planet there, had Jupiter not been so massive! Even the scientific discussions about the Asteroid Belt are phrased as if there 'should' have been a planet there: as discussed earlier, scientists worked to explain the depleted state of the Belt in terms of 'missing mass,' missing because the Belt *should* (it is argued) have had a planet's worth of mass.

In short, the only defect in the law's original predictions concerns one planet, Neptune (and one Kuiper Belt Object, Pluto). But how significant do you judge the law's successes—shown in Figure 1.17—to be? That is, how should the Titius-Bode Law be treated by scientists? As a step toward the answer, consider the following: early in this chapter, we noted that the observed prograde spins of the Sun and planets were consistent with a nebular origin of the Solar System; however, the spins of Venus and Uranus do not fit that model. Rather than dismiss the model, though, we found astrophysical reasons to explain the exceptions.

In this section, we have made much of the spacing exhibited by four orbital systems (one, planetary; three, satellite). But all of that is for just one star system, our own, with (e.g.) only cold giant planets and no hot Jupiters; perhaps we are, from an orbital spacing perspective, a statistical fluke in the Universe?

As discussed later in this chapter, thousands of planets orbiting other stars besides our Sun have been detected to date. By 2010, more than a dozen stars had been found which are orbited by at least three planets; by 2020, there were well over 200 (https://exoplanets .nasa.gov/discovery/exoplanet-catalog/). These 'exo-planetary systems' provide another opportunity to test the universality of the Titius-Bode Law, in at least one of the forms listed above. Such a test was conducted by Lovis et al. (2011), with their results shown graphically in Figure 1.19. Of the four planetary systems tested, three of them appear to obey the Titius-Bode Law reasonably well, judging by the goodness of fit by straight lines to their orbital distances. Lovis et al. take pains to emphasize that the significance of the tests illustrated in Figure 1.19 should not be overstated: they do not expect the law to be universal; instead, where it prevails it should merely be interpreted as signifying "a possible signature of formation processes."

To the results of Lovis et al., we can add that of Laskar (2000), who analyzed a system of three planets, and our own (unpublished) analysis of two of the most prolific star systems: Kepler-90, orbited by eight planets, and Trappist-1, orbited by seven. In all cases, the orbital radii of the planets were plotted versus planet number. The Kepler-90 system is also noteworthy in that all eight planets orbit at distances no greater than 1 AU; and in the Trappist-1 system, all seven planets orbit within a radius of 0.06 AU, each completing its orbit around the star in less than 3 weeks! The Trappist planetary system and the system examined by Laskar both exhibited well-defined Titius-Bode-type geometric progressions; but the Kepler-90 planetary system

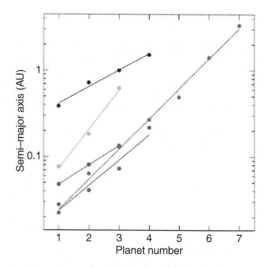

Figure 1.19 Tests of the Titius-Bode Law in four extrasolar planetary systems by Lovis et al. (2011) / EDP Science. Distances of planets (filled circles) from their star, plotted on a log scale in units of AU, versus planet number: green, star HD 69830; red, star HD 40307; magenta, star HD 10180; and blue, star GJ 581. Validity of the law is reflected in the goodness of fit to a straight line; on this basis, deviations from the law are significant only for star GJ 581.

These four systems are all categorized as low mass, i.e. having no Jupiter-type gas giant planets; for comparison, the fit to the *inner* planets of our Solar System is shown with the black circles and line. Note that most of the exo-planets shown have orbits much closer to their stars than even Mercury does in ours; evidently our inner Solar System is quite 'spacious.'

was found to exhibit only a very approximate geometric progression, in short failing to support the hypothesized universality of that law.

1.2.2 Moment of Inertia: A Diagnostic Tool for Planetary Interiors

When discussing the planets and moons of the Solar System, one obvious property of interest is their masses. Another important property is their *moment of inertia*, which measures how that mass is distributed throughout their volume. The distribution of mass reflects mostly how their density varies with depth, and ultimately implies whether or not they possess a metallic core—a key question, since, as we will see in subsequent chapters, the presence of a core can be related to the development of an atmosphere and even oceans, the vigorousness of plate tectonics, and the existence of a planetary magnetic field.

Basic Definitions. Moment of inertia can be defined for a continuum (a continuum is a three-dimensional mass, e.g., any 'macroscopic' material) or for a collection of discrete 'points' of mass (i.e., a system of particles). In either case the moment of inertia is defined with respect to a given *axis*, which we can think of as a reference axis, and which can be located either within or external to the mass system. Consider first a set of N particles, with masses m_i (m_1 through m_N); see Figure 1.20. Denote their perpendicular distances from a specified axis by $\mathscr{R}_i$ ($\mathscr{R}_1$ through $\mathscr{R}_N$). Note that $\mathscr{R}$ is not the same as a radius. The **moment of inertia** of each particle is defined to be the particle's mass multiplied by its squared perpendicular distance from the axis, thus $(\mathscr{R}_i)^2 m_i$; the moment of inertia $\mathscr{I}$ of the system of particles is simply the sum of those moments,

$$\mathscr{I} = (\mathscr{R}_1)^2 m_1 + \ldots + (\mathscr{R}_N)^2 m_N.$$

The latter can be concisely expressed as

$$\mathscr{I} = \sum_{i=1}^{N} (\mathscr{R}_i)^2 m_i.$$

From this definition, the moment of inertia can be interpreted as a weighted sum of the masses of the system of particles, with the weights given by the squared perpendicular distances of the particles from the axis. For a particular set of particles, $\mathscr{R}$ will be much larger if the particles—especially the more massive ones—are farther from the axis.

For a continuum, i.e., a three-dimensional mass, the moment of inertia is similarly defined, but with the summation (Σ) replaced with a continuous sum, i.e., an integral,

$$\mathscr{I} = \int \mathscr{R}^2 \, \mathrm{d}m$$

because we have a near-infinite collection of tiny (infinitesimally small) particles rather than a finite number of macroscopic, massive particles. Now $\mathscr{R}$ is the perpendicular distance of the infinitesimal particle of mass $\mathrm{d}m$ from the reference axis (see Figure 1.21), and the sum (integral) extends over all the particles comprising the continuum of interest.

Moments of Inertia of Homogeneous and Non-homogeneous Spheres. The *standard for comparison* in any geophysical discussion of the planets or moons of the Solar System is *the moment of inertia of a sphere of uniform density*. Consider a sphere of radius R, uniform density ρ, and total mass M; and take the reference axis to be a *diameter* of the sphere, i.e., the axis passes through the sphere's center. Henceforth, we will always use a diameter as the

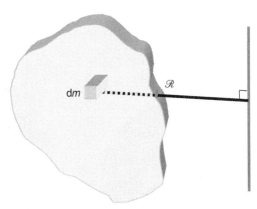

Figure 1.21 The moment of inertia of a continuous mass. In this figure, one tiny volume dVOL—of density ρ and containing an amount dm = ρ × dVOL of mass—has been selected to illustrate the integration. For this example, the moment of inertia is relative to the axis on the right.

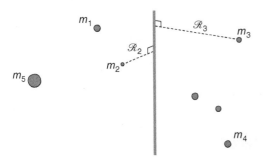

Figure 1.20 The moment of inertia of a set of discrete masses. The particles of mass $m_1, m_2, m_3, \ldots$ are positioned at perpendicular distances $\mathscr{R}_1, \mathscr{R}_2, \mathscr{R}_3, \ldots$ from the axis about which the moment of inertia is being measured.

reference axis; furthermore, in the remainder of this chapter, when the sphere in question is astronomical (e.g., the Earth, Moon, or Sun), that diameter will be taken to correspond to that body's rotation axis. By writing the elements of mass in terms of the density (which is mass per unit volume) using $\rho = dm/dv_{OL}$ and thus $dm = \rho\, dv_{OL}$, converting to spherical coordinates, and carrying out the integration (remembering that $\mathscr{R}$ is the perpendicular distance of dm from the reference axis, and not the radius), it can be shown that this sphere has moment of inertia

$$\mathscr{I} = (8/15)\pi R^5 \rho$$

with respect to the reference axis. Since the volume of the sphere is $(4/3)\pi R^3$ and its mass M equals ρ times its volume, we can write the moment of inertia of a homogeneous (uniform density) sphere as

$$\mathscr{I} = \frac{2}{5}MR^2.$$

Consider now (Figure 1.22) two spheres of the same size and total mass; one is homogeneous, and the other is not. Recall that the moment of inertia amounts to a weighted sum of the sphere's masses. If the density of the inhomogeneous sphere decreases with depth (so the sphere is 'top-heavy': Figure 1.22a), then the weights contributing to its inertia will be particularly large: some of the most massive particles are farthest from the axis. In sum, the top-heavy sphere will have a larger moment of inertia than the homogeneous sphere has.

On the other hand, if the density of the inhomogeneous sphere increases with depth

(a)

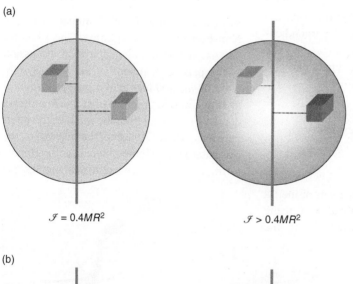

$\mathscr{I} = 0.4MR^2$ $\qquad$ $\mathscr{I} > 0.4MR^2$

(b)

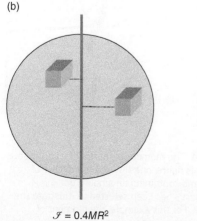
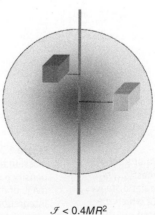

$\mathscr{I} = 0.4MR^2$ $\qquad$ $\mathscr{I} < 0.4MR^2$

Figure 1.22 The moment of inertia of a sphere with respect to its diameter. (a) Comparison of a homogeneous sphere and a top-heavy sphere, whose density decreases with depth toward the center. (b) Comparison of a homogeneous sphere and a sphere whose density increases with depth toward the center. All spheres are of radius R and contain the same total mass M. The moment of inertia of such a sphere is $0.4MR^2$ if the sphere is homogeneous.

(Figure 1.22b), then most of the massive particles are in toward the center, yielding smaller weights and thus a smaller moment of inertia than that of the homogeneous sphere.

Which situation more closely corresponds to the Earth, which is very nearly spherical and possesses a metallic core? Do you expect Earth's moment of inertia to be less than or greater than 0.4 times its mass times its squared radius?

Moment of Inertia of the Earth. Using satellite techniques we will briefly discuss in Chapter 6, it turns out that Earth's moment of inertia is approximately $0.33M_E R_E^2$ (if we denote the Earth's mass and mean radius by M_E and R_E).

How does that compare with your prediction above?

In fact, as we will learn, the Earth's density increases significantly toward the center, reaching ~13 gm/cm³; for comparison, Earth's surface density is typically ~2.7–2.9 gm/cm³. Evidently, an inertia coefficient (like 0.33) only a 'little' smaller than 0.4 indicates a very inhomogeneous sphere!

1.2.3 A Brief Description of the Properties of Planets and Moons

Our knowledge of planetary composition and structure has grown tremendously in recent decades through spectacularly successful data gathering by space probes such as the Mariner, Pioneer, and Voyager series, and more recently Magellan and Galileo. That knowledge is interesting in its own right but has also improved our understanding of the Earth's characteristics and behavior, through comparisons and contrasts with the other planets and moons.

Observations of those probes—or of natural satellites—orbiting a planet allow the mass and moment of inertia of the planet to be determined. Dividing the planet's mass by its volume yields the mean density of the planet. Tabulating these and other basic properties

(see Table 1.1) reveals that the planets group together fairly well depending on whether they correspond to *inner* or *outer* planets. The inner planets, including Mercury, Venus, Earth, and Mars, have a mean density of ~5 gm/cm³; a small mass; and a slow rotation (1 day or longer), but a fast revolution about the Sun (less than 2 years). If they have any atmosphere, it lacks appreciable hydrogen and helium. These terrestrial planets are rocky, being composed basically of ferromagnesian silicates and oxides, plus free iron. And, they are *chemically differentiated*—in short, they have iron cores.

In contrast, the outer planets, including Jupiter, Saturn, Uranus, and Neptune, have mean densities of ~1.2 gm/cm³; they are huge and massive; and they have fast rotation rates (typically ½ day) but revolve slowly about the Sun (periods of at least 12 years). They all have atmospheres, which are mainly hydrogen and helium, plus methane. Jupiter and Saturn are composed primarily of hydrogen, with helium plus smaller amounts of methane, ammonia, water, and carbon dioxide. Their interiors are *differentiated* essentially *by phase* rather than chemistry: as the pressure increases with depth, the gas compresses to a denser, liquid state then to a "liquid metallic" phase (H_2 gas → H_2 liquid → $2H^+$). Their overall composition must include the ice-rock *growth* cores—ten Earth masses worth, at least—that helped them pull in so much gas; but, thanks to that liquid metallic phase in their deep interior, which is able to dissolve water, rock, and iron (Wahl et al., 2013), there is likely no well-defined ice or rock core.

Pressures within Uranus and Neptune are not great enough to substantially increase their densities, or cause the phase change to metallic hydrogen; to explain their densities, we infer that the primary constituent of Uranus and Neptune is probably water. This is consistent with our earlier description of the growth of planets in the circumstellar disk: Uranus and Neptune developed well beyond the snow line, where icy planetesimals were abundant; but their ice-rock growth cores were too small to

Table 1.1 Planetary properties.

	Sun	Mercury	Venus	Earth	Moon	Mars	Jupiter	Saturn	Uranus	Neptune	Pluto
Mean radius (10^6 m)	696.0	2.440	6.052	6.371	1.738	3.397	71.49	60.27	25.56	24.76	1.195
Mass (kg)	1.99×10^{30}	3.30×10^{23}	4.87×10^{24}	5.97×10^{24}	7.35×10^{22}	6.42×10^{23}	1.90×10^{27}	5.68×10^{26}	8.68×10^{25}	1.02×10^{26}	1.25×10^{22}
Mean density (gm/cm^3)	1.41	5.43	5.24	5.51	3.34	3.93	1.33	0.69	1.27	1.64	1.75
Moment of inertia (kg-m^2)	5.68×10^{46}	6.49×10^{35}	5.88×10^{37}	8.02×10^{37}	8.73×10^{34}	2.71×10^{36}	2.36×10^{42}	4.05×10^{41}	1.26×10^{40}	$< 1.48 \times 10^{40}$	?
Inertia coeff. I/MR^2	0.059	0.330	0.330[e]	0.331	0.394	0.366	0.243	0.196	0.222	< 0.237	?
Atmosphere?	H, He	No	Heavy CO$_2$	N$_2$, O$_2$	No	Thin CO$_2$	H$_2$, He	H$_2$, He	H$_2$, He	H$_2$, He	Haze?
Fe core?	Yes	Yes	Yes	Yes	Yes (small)	Yes	Yes	Yes	Yes	Yes	?
Magnetic field[b]?	Yes	Yes (weak)	No	Yes	Crustal	Remanent	Yes	Yes	Yes	Yes	?
Primary source region for field	Outer, conv. zone		—	Fe core	—		H$^+$ 'core'	H$^+$ 'core'	Saline oceans	Saline oceans	
Rotational period (sidereal days)	25.1–34.4	58.8	−243.7	1.0	27.4	1.0	0.41	0.45¯	−0.72	0.67	−6.40

Length of year (Earth years)	—	0.241	0.615	1.000	0.0748	1.88	11.9	29.5$^-$	84.0	165$^-$	248
Obliquity[c]	7.25°	0.03°	177.4°	23.45°	6.68°	25.19°	3.12°	26.73°	97.86°	29.56°	122.5°
Ecliptic angle[d]	—	7.00°	3.39°	0°	5.1°	1.85°	1.31°	2.49°	0.77°	1.77°	17.15°
Other notable features				Life		Largest volcanoes in Solar System	Moon (Europa) most likely candidate for e.t. life	Extensive ring system			

Sources: Lide et al., 1996; Yoder, 1995; Williams, 2016; Schubert et al., 2001; Hester et al., 2002.

[a]See text for explanations.

[b]The Moon's crust is locally magnetized, and associated with impacts during the Heavy Bombardment (e.g., Schubert et al., 2001). Magnetization in Mars' crust is widespread and, as discussed at the end of this chapter, is probably a remanent field associated with ancient dynamo action in its core.

[c]Obliquity is defined here for each planet as the angle made by its spin axis relative to the perpendicular to the plane of *its own* orbit; for the Sun, it is the angle between its spin axis and the perpendicular to the *ecliptic*. The curious observation that the Sun's equator is slightly misaligned with respect to the ecliptic, i.e., to the orbits of all the planets, has been explained by Bailey et al. (2016) and Gomes et al. (2016) as a consequence of the gravitational effects of a massive "Planet 9" orbiting at the far edge of the Kuiper Belt.

[d]Ecliptic angle is defined here as the angle made by the plane of the body's orbit relative to the ecliptic.

[e]Venus' moment of inertia is not well determined; see Cook (1977) then Kaula (1979), or Schubert et al. (2001), for the basic reasons why.

enable runaway gas accretion and produce predominantly gaseous giant planets. Small rocky cores should remain within Uranus and Neptune.

In Chapters 7 and 8, we will learn about seismic waves and how observations of them can be used to infer a detailed picture of the Earth's interior structure and even composition. Seismic waves can propagate, one way or another, through just about any medium, and seismometers placed on the lunar and Martian surfaces have enabled us to detect the waves reverberating through those bodies. Analysis of such observations has led to a more robust picture of the interiors of the Moon and Mars.

- NASA's *InSight* mission landed on Mars in 2018, deploying a "complete geophysical observatory" (Lognonné et al., 2019) which included two sets of seismometers. The analysis by Stähler et al. (2021) of seismic data collected over the first 16 months following deployment concluded that Mars has a core of radius ~1,800 km (just over 50% of that planet's radius), composed primarily of liquid iron but with a significant fraction of a light alloy present. For comparison, Earth's core also has a radius slightly more than half our planet's radius; our liquid core also has a light alloy mixed with iron, but in a much smaller fraction.

  The moment of inertia inferred for Mars, 0.366, implies a less dramatic increase in Mars' density with depth than in the case of the Earth. The seismic analysis also found the densities in the Martian core to be only ~ twice that in the Martian mantle, whereas in Earth the typical core density is nearly three times that of our mantle, so the results are indeed consistent.

  The seismic results also generally confirm predictions (e.g., Khan et al., 2018; Rivoldini et al., 2011; also Konopliv et al., 2011) from theoretical models based on Mars' moment of inertia combined with its deformation by solar tidal forces, the latter inferred from observations of satellite orbital perturbations.

- Seismometers were deployed on the Moon back in the days of the Apollo Moon landings, and collected data for up to 8 years. Those data were reanalyzed by Weber et al. (2011) with the goal of refining our understanding of the deep lunar interior. They concluded that the Moon had a solid inner core of radius 240 km (which is about 14% of the total radius), overlain by a thin, liquid outer core with outer radius 330 km (19% of the total radius); the data are consistent with this outer core being composed of iron plus a small amount of light alloy. The core, in turn, is overlain by a partially molten transition zone ~150 km thick.

  The Moon's moment of inertia is nearly that of a homogeneous sphere. Weber et al. found that the density increase to the outer core was modest, whereas the density jumped substantially from outer to inner core; thus, the lunar moment of inertia reflects mainly the presence of a significantly denser but small inner core.

As we have learned in physics classes (and can verify with simple experiments), magnetic fields can be generated by electrical currents. One fundamental requirement for a planet to possess a magnetic field is thus that it contains an electrically conducting region; in the case of the Earth, the currents responsible for its magnetic field are produced mainly within its iron core. As their name implies, the liquid metallic interiors of Jupiter and Saturn are good electrical conductors; Uranus and Neptune do not contain that phase of hydrogen, but their saline oceans should be decent electrical conductors. All eight planets of our Solar System therefore have the capability of generating substantial electrical currents and should be expected to possess measurable magnetic fields. The lack of such a field emanating from our 'sister' planet Venus is striking—and will prove instructive (in Chapter 10) in understanding what it takes to generate planetary magnetic fields. The case of Mars is discussed at the end of this chapter.

Finally, the moons of the Solar System appear to be as varied in their properties as the planets. They can be predominantly rocky or contain abundant ice or water. Some moons (e.g., Jupiter's Io and Neptune's Triton) exhibit volcanic activity (Io, sulfur gas; Triton, nitrogen snow). Some moons, most importantly Jupiter's Europa, show evidence of tectonic activity (rifts, collision zones, etc.)—an "ice convection" analog to the silicate convection which occurs in the Earth's mantle. The existence of such geologic behavior in various moons helps us understand the Earth better; as we will see next, it also helps us identify where (besides Earth) life might exist in our Solar System.

1.3 Life in the Solar System, and Beyond

1.3.1 The Search for Planets

One of the motivations for studying the formation of our Solar System is that it can help us assess the likelihood of life existing throughout the Universe. The success of monistic theories in explaining our Solar System's formation suggests that planetary companions to stars might be plentiful. And in fact, astronomical evidence of extrasolar planets (**exoplanets**) has become increasingly abundant—including images of actual exoplanets (!), and evidence for the existence of planets so exotic that, were life to somehow develop on them, it would necessarily be unimaginably alien.

Such bizarre planets are possible because, starting with any nebula, its collapse and the formation of a central star are not guaranteed outcomes. Consider the bigger picture: as we discussed earlier in the context of our own primordial nebula, any nebula is part of a larger mass, a molecular cloud; and, that cloud would have contained many other densely concentrated nebulas or cores, each with the potential for star formation. But gravity cannot always overcome those forces resisting collapse, and during the molecular cloud's lifetime, typically several Myr or longer (e.g., Williams et al., 1999), only ~10% of those cores

will succeed in collapsing and igniting a star (Dobbs et al., 2011; also Lada & Lada, 2003). Cores that 'partially' collapse may nevertheless form planets—these have variously been named orphan, free-floating, unbound, rogue, and nomad planets. In that category we can also include planets that had been ejected from their original stellar environment, perhaps by the gravitational effects of massive planets that had migrated too close.

Furthermore, the partial collapse of still other cores, though insufficiently compressive to ignite a true star, may lead to a central mass—known as a *brown dwarf*—whose characteristics lie between those of a Jupiter-type planet and a small star.

The detection of actual nomad planets (and the least massive, most Jupiter-like brown dwarfs) has led some to infer that they are widespread, for example that there are at least as many Jupiter-sized nomads as there are stars (see, e.g., Sumi et al., 2011; Strigari et al., 2012; Dai & Guerras, 2018; and Bhatiani et al., 2019).

What do you think—could life develop, and persist, on a nomad planet? If it did, and some of its species evolved into thoughtful beings, what do you think their conception of time would be?

There are currently several different techniques for detecting exoplanets, varying widely in ease of detection and success rate.

- The most successful technique, called the *transit* method, looks for variations in the brightness of stars. Large planets orbiting their star in the plane of our view block some of the star's light periodically as they "transit" or pass in front of the star. The periodicity and extent of dimming would indicate both the planet's distance from the star and its size. The French ESA space telescope CoRoT (Convection, Rotation, and planetary Transits), launched at the end of 2006, was the first orbital telescope dedicated at least partially to the detection of extrasolar planets by this technique. NASA's *Kepler* space telescope, launched in early 2009 in an orbit around the Sun

(a)

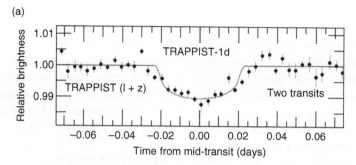

Figure 1.23 Techniques for detecting extrasolar planets.
(a) This graph (from Gillon et al., 2016) shows the dimming of light from a small, faint star, called an "ultracool dwarf" star, located in the constellation Aquarius only 39 light-years from Earth. The light dims periodically as several planets orbit the star; non-planetary causes for the dimming (such as 'sunspots' on the star's surface, a companion star, etc.) can be ruled out. Here the dimming from one of the planets, its effects isolated by extensive refinement of the observations, is shown by black dots (with error bars). The planet model that best fits these refined observations, indicated by the red curve, reveals a planet of near-Earth size located close to the "habitable zone" of this star (the distance from the star at which surface temperatures allow liquid water; but see Chapter 2!).
In honor of its discovery, the star system has been named TRAPPIST-1. Subsequent analysis, including observations from other ground-based telescopes and the Spitzer Space Telescope, has revealed a total of seven planets orbiting this ultracool dwarf, all similar to the Earth in size, mass, and potential habitability (Gillon et al., 2017).
Incidentally, using the orbital distances in Gillion et al. (2017), it turns out that (as noted earlier in this chapter) the seven planets in the TRAPPIST-1 system fit the later version of the Titius-Bode Law extremely well—even better than the planetary systems (shown in Figure 1.19) evaluated by Lovis et al. (2011).

similar to Earth's, possesses an ultrasensitive light-measuring capability and was designed with a wide field of view allowing simultaneous analysis of multiple stars, and is wholly dedicated to planet detection. As described in Figure 1.23a, ground-based telescopes such as TRAPPIST (the Transiting Planets and Planetesimals Small Telescope), which began operation in Chile in 2010, have also been successful in detecting exoplanets.

Half an orbit after transiting its star, the exoplanet will pass behind the star. Such a *"secondary eclipse,"* which can also be detected, provides additional and, in some cases, key information about the planet's orbit (e.g., Huber et al., 2017). The use of primary and secondary transit data, combined with analysis of the infrared emissions radiated from the exoplanet (especially if the exoplanet is a hot Jupiter), can tell us much about the state and composition of the exoplanet's atmosphere (e.g., Deming & Knutson, 2020).

- Our Sun and all other stars, in our galaxy for example, do not just sit still in space; they move with the galaxy as it rotates (taking about 225 Myr per revolution), and they move relative to each other. Hurtling through space, a star that has no planetary (or stellar) companions would follow a smooth path; but with planets—especially massive ones, especially close-in ones—circling it, the star would appear to follow a wavy trajectory as the star/planet system moved through space and the star and planets orbited around their common center of mass (Figure 1.23b, top). This wavy path (called a 'wobble' by some) could be visible if, as in the figure, the planet's orbital plane is perpendicular to our line of sight. If the orbit is *in* our line of sight—which, to some extent, is far more common—the wobble could be inferred from the periodic *Doppler-effect* shift in the colors (spectrum) of the star's light as the star alternately approached and receded from us (Figure 1.23b, bottom). The trajectory

(b)

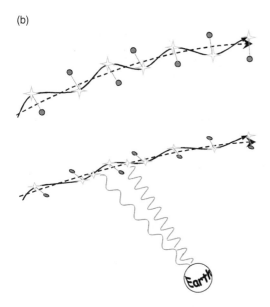

Figure 1.23 Techniques for detecting extrasolar planets.
(b) A wavy path taken by a star in space can result from a massive planet orbiting the star.
(Top) How the wavy path is created. The dashed line indicates the overall trajectory of the star-planet *system* through a region of the galaxy, as the planet (marked with a red-filled dot) continues to orbit the star.
As noted in the text, the manner in which this waviness is used to detect the planet depends on whether the planet's orbital plane (which is also the plane of the star's waviness)—the plane of this page—is perpendicular to our line of sight (e.g., as we look at this page) or in our line of sight (e.g., if our eyes were at the edge of this page).
If the planet's orbital plane is perpendicular to our line of sight, as shown here, we could conceivably measure the star's position over time to determine the wavy trajectory's amplitude and periodicity; however, this *astrometric* technique has not been very successful.
(Bottom) With the orbital plane oblique to us (i.e., somewhat edge-on), the star moves away from us then toward us in each cycle of waviness. The star's light is *Doppler-shifted* toward reddish then bluish frequencies during each cycle; from the shift in frequency, we can infer the velocity of the star's motion—its line-of-sight velocity or *radial velocity*—away from us or toward us.

wavelength (or periodicity) and waviness amplitude would indicate both the planet's orbital period—and thus, as a consequence of Kepler's Third Law, its distance from the star—and its mass. This Doppler effect, by

the way, is analogous to our experience of hearing an ambulance siren as it approaches (with a higher-pitch sound) then recedes (with a lower-pitch sound) from us; the analogy is imperfect because of differences in light versus sound waves (see Goldsmith, 1976), but the interpretation is the same.

- Still another technique, known as *microlensing*, depends on the ability of the gravitational field of a massive object to distort a passing beam of light—a key component of Einstein's general theory of relativity. As a 'near' star, moving through space, passes in front of a more distant star over a span of perhaps several weeks, its gravity can bend light from the distant source that would otherwise shoot past us, instead focusing it in our direction (see Figure 1.23c). If there are massive planets orbiting the near star, they can act as similar though weaker 'gravitational lenses,' further focusing the light that we receive from the distant star—but only temporarily (perhaps over a span of a day), making it appear to 'flash.' The brightness of the flash would be a function of the planet's mass, whereas the timing and duration of that flash would also depend on the planet's distance from its star and on that star's own motion.

- *Direct images* of exoplanets as they orbit their stars have, amazingly, been achieved in recent years (see Figure 1.23d), leading to 52 discoveries up to the present time, at distances ranging up to more than 600 light-years! Of course, such imaging requires that the light from the star be suppressed, either artificially (by digitally processing the image, or mechanically blocking that area the way we block sunlight to see the Sun's corona; Brennan, 2021) or naturally (if the star is still surrounded by a debris disk).

The first confirmed detection of an extrasolar planet was in 1992 (Wolszczan & Frail, 1992), though it involved planets orbiting an 'exotic' star (a pulsating neutron star); the detection was based on a variation of the second technique listed above, using the timing of the

(c)

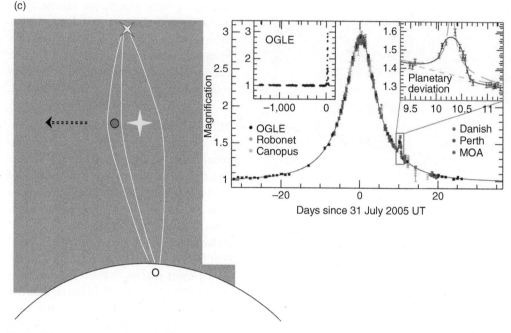

Figure 1.23 Techniques for detecting extrasolar planets.
(c) (Left) The light from a distant star can be deflected by a nearby star, which acts as a 'gravitational lens' to focus the distant star's light toward an Earth-based observatory (point O) as the nearby star (+ planets) passes in front of it. Gravitational microlensing by the star's planets would temporarily produce flashes of additional light. Sketch after Queloz (2006).
(Right) Observations by a multinational collaboration (various colored dots) showing the flash (see right inset graph) inferred to have been produced by a relatively small planet (only ~5 Earth masses) in orbit around a red dwarf star (the most common type of star in our galaxy) ~22,000 light-years from us. Figure from Beaulieu et al. (2006).

star's radio-wave pulses. The first confirmed detection of an extrasolar planet orbiting a Sun-type star was in 1995, also using the Doppler technique (Mayor & Queloz, 1995, who were awarded the Nobel Prize for their discovery). Other milestones, and a comprehensive discussion of detection techniques, may be found in Perryman (2018).

In the past 20 years or so, the pace of exoplanet discovery has accelerated and the number of confirmed detections has multiplied. As of January 2004, there were about 120 exoplanets detected; as of January 2006, ~160; by February 2007, more than 200; and by December 2010, more than 400 (Wright et al., 2011). According to NASA's Exoplanet Exploration website (Brennan, 2021), as of May 24, 2021 there were 4,389 confirmed discoveries.

The NASA website tallies the method of discovery as well, revealing that the great majority (76%) of those planets were found by the transit method; the Doppler technique (19%), micro-lensing (2%), and direct imaging (1%) were employed for most of the other detections. Three-quarters of the planets detected were part of multiplanet systems. And, where planet mass could be inferred, most—not surprisingly—were giant: almost 1,500 Neptune-like, and almost 1,400 Jupiter-like; less than 200 appeared to be terrestrial.

What do all these discoveries tell us about the possibility of life beyond our Solar System? The sheer number of planet detections suggests that planets may indeed be commonplace in our galaxy. More fundamentally, the fact of detection can be used to quantify the abundance of planets galaxy-wide. For example, the gravitational lensing technique has recorded relatively few planet detections; but a statistical analysis

(d)

2014-08-26   20 au   Jason Wang / Christian Marois

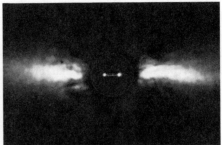

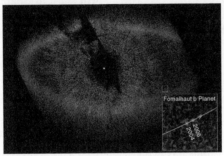

Figure 1.23 Techniques for detecting extrasolar planets.
(d) Direct imaging of exoplanets.
(Left) Exoplanets orbiting in a multiplanet system around star HR 8799 (yellow symbol at center of image), whose light is *imperfectly* blocked out. The initial discovery (Marois et al., 2008) was of three planets; the fourth, closest planet was detected two years later.
This snapshot is actually one of a sequence of images spanning 7 years, provided by C. Marois and compiled into a video by J. Wang, showing the orbital motions of these planets (Kaufman, 2017). Christian Marois / Wikimedia Commans / CC BY 3.0.
(Right) Superimposed images of a planet orbiting Beta Pictoris in 2003 and 2009. Direct light from the star (white dot at the center) has been blocked (black disc in the middle of the image), but with the remainder of the image enhanced we see the star's light brightly reflected off the debris disk.
This planet orbits its star (with the orbit drawn in as well) at roughly the distance Saturn is from our Sun. Image from A. Lagrange / https://www.eso.org/public/images/eso1024c / last accessed August 22, 2023.
(Bottom) Images can be deceptive. One of the first images of an exoplanet (Kalas et al., 2008), around the star Fomalhaut: light from the star is blacked out (and replaced with a white dot); the densest part of the circumstellar disk begins at the bright oval (beyond that, it is in shadow).
The planet is inside the tiny white square. The inset is a close-up of its locations in 2004 and 2006.
But its interpretation as a planet was problematic from the start (see, e.g., Janson et al., 2012 and Gaspar & Rieke, 2020). Numerical modeling by Gaspar and Rieke now suggests that the imaged planet was actually a cloud of dust (which has since dispersed)—the rarely captured aftermath of a collision between two 100-kilometer-sized planetesimals! Image from https://esahubble.org/images / credit to NASA, ESA, & P. Kalas (also Z. Levay) / last accessed August 22, 2023.

of those successes—combined with the fact that the chance of detection by that technique depends, for example, on the near-alignment of all three bodies with Earth, on the planet not being too close to the 'lens star,' and on the planet having sufficient mass (see Figure 1.23c, left)—has led some researchers to conclude that those detections actually imply that, on average, every star in our galaxy is likely to have one or two planets in orbit around them, at distances comparable to Saturn's orbit or less (Cassan et al., 2012)!

In contrast, a statistical analysis of the transiting technique based on a very large number of Kepler observations (successful and unsuccessful)—combined with the fact that the chance of detection by that technique depends, for example, on the orientation of the planet's orbit relative to our line of sight, on the planet not being too far from its star, and on the timing of the planet's transit relative to our observations—suggests that ~5% of Sun-like stars in our galaxy are orbited by Earth-sized planets at an Earth-like distance from them (Petigura et al., 2013).

1.3.2 Evidence for Life in the Universe

Is it sufficient for life that planets abound? That is, how easily can life, as we know it or otherwise, develop on planets (and moons)? As a partial answer to that question, we review the evidence for life beyond Earth. It should be noted that the evidence for *'intelligent'* life, by which we mean here life that has evolved both an awareness of outer space and the technology to probe it, will be very different than evidence merely indicating the presence of primitive life. Scientists have searched for both types of evidence, with quite different and provocative results.

Intelligent Life in the Universe? If life has appeared somewhere, can we expect that, at some point in its evolution, it will develop an awareness of the Universe—with an associated technology (e.g., some kind of telescopes, then satellites and rockets, and radio communication with the latter) that can enable our detection of it? With the help of a reasonable assumption, geophysical system science and anthropological science together suggest that, except under unusual circumstances, the answer may be affirmative.

If Intelligent Alien Life Is Diverse, Maybe Their Detection Is More Likely. We start with an assumption: that alien evolution, like evolution on Earth, would be governed by natural selection and the drive to survive

(without "selection," there would be no evolutionary path). A survival instinct ensures that, on planets and moons that rotate and derive needed energy (e.g., heat) from a sun, life will develop an awareness of day and night, and an appreciation of their sun. As that life becomes more advanced, self-conscious aliens may begin to associate their sun's brightness and position in the sky with the state of their surroundings, ultimately tying their survival instinct to the sun.

Geophysical system science tells us that, simply because planets and moons are round (!), alien life must be physically diverse. As we will learn in Chapter 3, the spherical shape of the Earth causes the amount of light and heat from the Sun received at the surface to vary with latitude, with the most intense incoming solar radiation in the tropics and the radiation decreasing toward the poles. To better adapt to these radiative conditions, humans evolved with, e.g., differences in skin color.

The variations in incoming heat also ultimately lead to different climate zones across the planet—on Earth, for example, tropical, temperate, and polar regions in each of the northern and southern hemispheres—which are further modified, to various extents, by such factors as the presence of mountains versus plains, and land versus oceans. On Earth, differences in their environment affected the lifestyle of people in different regions, contributing to different outlooks, philosophies, and religions (e.g., 'a harsh God in a harsh environment'). And differences in the regional abundance of natural resources had a major effect on the cultural and technological development of different societies (Diamond, 1997).

With these same natural factors, alien life—especially if land based—would evolve to be culturally and technologically as well as physically diverse.

Diamond (1997) has argued convincingly that, through the past few hundred years at least, technological developments by humans have been enhanced by the presence of multiple cultures; technology has not advanced by single 'heroic' acts of creation so much as by sequences of invention improved by individuals in first one society then another. The fastest development of technology resulted from "regions with large human populations, many potential inventors, and many competing societies" (Diamond, 1997; p. 261). One example of this process he discusses is the development of the internal combustion engine. We can cite as another example the "cold war," which took place during the twentieth century between some societies; it stimulated a "space race" between the primary antagonists that intensified our awareness of outer space and produced significant advancements in technology.

Cultural diversity in an alien world could thus be expected to lead to advances in technology, including some directed toward outer space.

It took life on Earth billions of years to evolve to the point where a species of intelligent life (as defined above) appeared, so our speculation regarding alien evolution is clearly extremely simplistic. Terrestrial evolution also reminds us that our ability to detect alien civilizations is time-constrained; that is, it depends on whether they advanced just at the time we would be detecting evidence of their existence. Furthermore, on Earth, physical and cultural diversity led to slavery and genocide, and some of the advances in technology led to destruction of our environment; among the time constraints, then, perhaps we should add a constraint that some of those civilizations may not exist for a long time. In short, it may be reasonable to search for evidence of intelligent alien life, but we should expect the chances of success to be low.

There has been an organized, systematic search to detect evidence of intelligent life in the Universe ever since 1960. Started by National Radio Astronomy Observatory (NRAO) astronomer Frank Drake, the Search for Extra-Terrestrial Intelligence (SETI) program was set up, using radio telescopes, to look for signals intentionally broadcast by an alien civilization. The idea is that such signals would have a distinct, universally recognizable character, allowing them to be identified above the background noise of outer space.

What signals could you devise that might fit those requirements?

Initially government-funded, SETI evolved into a fully privately funded foundation in 1995, and in 1999 added a 'populist' aspect (called "SETI@home") by employing the personal computers of thousands of volunteers, networked together, to analyze radio telescope data when their PCs were otherwise idle (in 2005, SETI@home transitioned to a more flexible infrastructure; see SETI, 2017). As of this time, the SETI program has not found signals that can be identified unequivocally as a deliberate alien broadcast. In 2003, however, SETI was reported to have detected an unusual signal on three occasions, from a region of the sky between the constellations Pisces and Aries, which has yet to be explained; interestingly, the signal was at a radio frequency associated with hydrogen—the most abundant molecule in the Universe and thus a potential signal in the 'universal' language of radio waves (Pickrell, 2006).

Requirements for Life as We Know It. If the life we are searching for is not sending out purposeful signals, then we are limited to searching for evidence of life as we know it—carbon-based, perhaps even within a DNA framework. Such life uses *organic chemicals* to produce amino acids, the building blocks of proteins; amino acids contain carbon, oxygen, hydrogen, and nitrogen. More complex organic compounds include nucleic acids, such as ribonucleic acid (RNA) and deoxyribonucleic acid (DNA), which are needed to regulate more advanced cell activities and reproduction, and might include phosphorus as well as C, O, H, and N.

Incidentally, our Solar System contained water, carbon dioxide, methane, and ammonia, as well as hydrogen, so most of those elements would have been readily available as the planets and moons formed. On Earth, the weathering of surface rocks puts phosphates into the soil and surface waters, so we can expect that under optimal conditions phosphorus would be available elsewhere in the Solar System, too.

It is worth noting that, by spectral analysis of the infrared radiation and radio waves emitted from molecular clouds, scientists have found evidence of organic compounds throughout the galaxy—perhaps 150 varieties, half discovered in the past 20 years, and including polycyclic aromatic hydrocarbons (possibly the oldest, most widespread, and abundant free organic molecules in space), aldehydes, alcohols, ethers, ketones, and amines, among others (e.g., Mattila et al., 1996; Draine, 2003; Woon, 2017; Müller et al., 2016; Ohishi, 2008; Bergin & Tafalla, 2007; Ehrenfreund & Charnley, 2000). For example, it has been estimated that there is enough ethyl alcohol in the Sagittarius B2 molecular cloud to produce ten thousand trillion trillion fifths (i.e., 10^{28} fifths) of 200-proof 'booze' (Zuckerman et al., 1975). The most complex organic molecules discovered to date include propyl cyanide (C_3H_7CN, both normal and isoforms) (Belloche et al., 2014); amino acetonitrile (NH_2CH_2CN), considered a possible precursor to the simplest amino acid (Belloche et al., 2008); and propylene oxide (CH_3CHCH_2O), in both its normal and mirror-image ('left-handed' and 'right-handed,' discussed below) forms (McGuire et al., 2016; see also Blue, 2016).

The Sagittarius B2 molecular cloud has been described as "the most massive star-forming region in our galaxy" (Belloche et al., 2014). Within that cloud, there are two portions that are warmer by ~100K or more than our own molecular cloud core was in its initial state; these two regions are molecular cloud cores which have already begun their collapse on the way to evolving into star systems. The analysis by Belloche et al. suggests that such temperatures are ideal for producing complex organic chemicals, promoting their formation on the dusty surfaces of dirty snowballs. It is easy to imagine that, eventually, those so-called "hot" molecular cloud cores (or, more precisely,

'slightly less frigid cores') will evolve into solar systems where life might well develop.

Of course, organic chemicals alone are not enough to establish that life exists on some planet or moon, or within some dust cloud. Life as we know it also requires *liquid water*; water is necessary because it allows complex organic compounds to form and react, and transports chemicals. On a macroscopic scale, it also serves to moderate temperatures in the environment—a bonus, at least for some types of life.

Still another requirement is that an *energy source* be available: life processes take energy.

Additionally, it would be helpful for the environment to *not* be *too extreme*. This is a relative issue: bacteria can survive in 'moderate' environments, such as within ice, or high up in our atmosphere. Some species of bacteria thrive in no-oxygen environments. And bacteria can survive in the solid earth, i.e., in the cracks and void spaces pervading surface rocks (Figure 1.24, left) (e.g., Gold, 1992 and references cited therein). In fact, rock-dwelling bacteria have been found in the crust kilometers deep and are considered likely to exist down to at least 4–5 km. In recent years, more complex life forms have also been discovered in the crustal subsurface. For example, Ciobanu et al. (2014) report the existence of fungi and primitive plants in the oceanic crust nearly 2 km below the sea floor, off the coast of New Zealand; Borgonie et al. (2015) have found flatworms, worms, and possibly a new species of crustacean (Figure 1.24, right) in deep fissures of some South African mines. It has been estimated that there is a greater total mass of carbon in rock-dwelling microbes than in all life on Earth's surface combined (Whitman et al., 1998 and Kallmeyer et al., 2012), though conservative estimates suggest the contribution is more like ~10% (McMahon & Parnell, 2014)—significantly less, but still impressive. One might even speculate that subsurface pore spaces provided a stable, protected environment for the earliest life on land.

Finally, life has been found in spectacular fashion near mid-ocean ridge vents at the

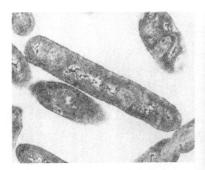
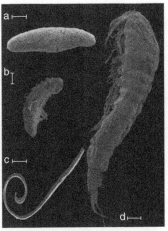

Figure 1.24 (Left) Rock-dwelling bacteria. Electron micrograph of bacillus strain TH-23 (from Thorn Hill, Virginia), found in drilling mud from a core drilled in 1992 sampling depths of ~2.7 km. Onstott et al. (1994) / John Wiley & Sons, Inc.
(Right) Scanning electron micrograph of complex organisms recovered from deep mines in South Africa: species of (a) Plathyelminthes (flatworms), (b) Annelida (segmented worms), (c) Nematoda (roundworms), and (d) Arthropoda (insects, crustaceans, etc.), the latter possibly a new species. Scale bars are in microns: (a) 50, (b) 50, (c) 100, (d) 20. Image from Borgonie et al. (2015) / Springer Nature.

sea floor, $2\frac{1}{2}$ km below sea level, beginning in 1977 at the Galapagos Rift and 1979 at the East Pacific Rise (where the first "black smokers" (Figure 1.25) were seen): thriving microbial colonies, living chemosynthetically off the metal-rich fluids and other chemicals pouring from the vents; and up the food chain, living nearby, communities of large clams and mussels—the former, ~1 foot in length—and even larger tubeworms, predator crabs, and novel species of fish (e.g., Corliss et al., 1979) (see Figure 1.25). At pressures of 250 atmospheres and temperatures ranging from –1°C away from the vent to up to 380°C adjacent to it, such life is robust indeed.

Many of these environments are extreme from *our* point of view; advanced forms of life, like us, evidently require more gentle environments!

With these requirements in mind, let's take a quick tour through our Solar System.

1.3.3 Evidence for Life in Our Solar System

Evidence from the Oort Cloud and Kuiper Belt. It is difficult to learn much about the composition of the bodies currently residing in the Kuiper Belt, because of their distance, faintness, and the fact that their surfaces tend to be covered by "volatile frosts" (Brown, 2012). Water ice, CH_4, CO, N_2, and probably silicates (like olivine, a ferromagnesian silicate abundant in our upper mantle) and ammonia (Brown, 2012) are among the constituents most easily and commonly detected; the presence of methanol (CH_3OH) and ethane (C_2H_6) is also indicated.

The reddish-brown color of the surfaces of some Kuiper Belt objects, including Pluto and Arrokoth, is generally interpreted to reflect a "tar-like" solid composed of complex organic quasi-polymer molecules named **tholins** by Sagan and Khare (1979). The tholins, which can be produced (e.g.) from irradiation of methane, other simple hydrocarbons, and nitrogen or ammonia, have been found to be a versatile food source for several types of Earthly bacteria—suggesting that they could serve as the base of a food chain in other environments where life might develop (Stoker et al., 1990).

Long-period Comet Hale-Bopp, with a 'nucleus' 30–40 km in diameter, was three times the size of Halley's Comet when it was

(a)

(b)

(c)

Figure 1.25 The sea-floor environment near mid-ocean rifts—for example, hydrothermal vents on the East Pacific Rise (left, right) and the Galapagos Rift (bottom), seen from the submersible vehicle *Alvin*—is exotic geologically and biologically. Photograph by W. Normark / https://pubs.usgs.gov/gip/dynamic/exploring .html / last accessed August 22, 2023.
In "black smokers" like the one shown here (a), so-called because of the iron sulfide minerals spewed out of it, the fluids can reach temperatures of 380°C, though temperatures drop rapidly away from the vent.
Photograph by W. Normark at the sea floor, the extreme darkness and pressures should prohibit life, yet it thrives, including tubeworms more than 2 m long (b) along with symbiotic bacteria, 'spaghetti worms' (or the tentacles thereof) (bottom), and even crabs (a). Photographs (b) by IFREMER and (c) by J. Childress / https:// www.whoi.edu/feature/history-hydrothermal-vents/discovery/1979.html / last accessed August 22, 2023.
Curiously, the temperatures at the head and tail of the tubeworms differ by a factor of nearly four (22°C versus 80°C, respectively).
Biological considerations suggest that the various vent areas are ecologically isolated from each other (Corliss et al., 1979).

detected in the inner Solar System in 1995. As it neared the Sun and began to ablate, it produced vapor jets composed of H_2O, CO, and CO_2, and dust jets rich in olivine, consistent with our model of the primitive solar nebula's dirty snowballs. Spectroscopic analysis also revealed numerous organic chemicals, including methanol (CH_3OH), formamide (NH_2CHO), methyl formate ($HCOOCH_3$), cyanoacetylene (HC_3N), methyl cyanide (CH_3CN), and dimethyl ether (CH_3OCH_3) (e.g., Biver et al.,

1997; Lis et al., 1997; Bockelee-Morvan et al., 1997; and the survey by Ehrenfreund & Charnley, 2000; but see also Rodgers & Charnley, 2001).

Even more significant are the results obtained from sampling short-period Comet Wild 2. Comet Wild 2, the target of NASA's *Stardust* mission, is considered a pristine comet: its orbit is believed to have changed during the past half century, thanks to the effect of Jupiter's gravity, for the first time bringing its closest point of approach to the Sun inside Jupiter's orbit; thus, it has been subject to a greater intensity of the Sun's heat only during the past several orbits. As a consequence, the dust and gas emanating from its surface are more representative of the primitive dirty snowballs of the early nebula than most other comets (Whalen & Baalke, 2003). Stardust was able to rendezvous with Wild 2 in 2004, sample the cloud of gas and dust surrounding the comet nucleus, and return the sample to Earth! It was confirmed by Elsila et al. (2009) that, in addition to organic chemicals like methylamine (NH_2CH_3) and ethylamine ($NH_2C_2H_5$), the sample included an abundance of non-terrestrial glycine (NH_2CH_2COOH), the simplest amino acid and, thus, as noted above, a building block of life.

More recent comet investigations, including NASA's *Deep Impact* mission to Comet Tempel 1 (e.g., A'Hearn et al., 2005) and the European Space Agency's *Rosetta* mission to Comet Churyumov-Gerasimenko (both comets being short period), also found a variety of organic compounds in their target comets; in particular, the *Rosetta* results showed the presence of glycine (McKay & Roth, 2021, who also provided an entertaining review of organic compounds found in various comets, as well as how they have been detected).

In short, observations of the outer Solar System suggest very strongly that *the potential for life in our Solar System has existed from its very earliest days*. From that perspective, it may be no surprise that life has managed to develop on Earth (though complications will be discussed in the next chapter); but what about elsewhere in the Solar System? Let's move closer in to find out.

Evidence from Jupiter's Moons. The four largest moons of Jupiter together provide the strongest evidence that conditions promoting life exist somewhere beyond Earth. When the *Galileo* spacecraft flew by the two largest moons, Callisto and Ganymede, in the late 1990s, it found their surfaces to be "sprinkled

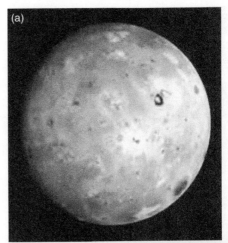

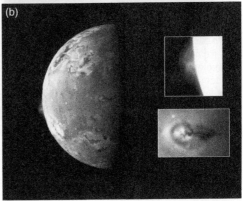

Figure 1.26 Photographs of Io taken by the *Galileo* space probe, NASA / https://solarsystem.nasa.gov/missions/galileo/overview / last accessed August 22, 2023. (a) The yellowish color of Io's surface is a result of ongoing sulfur eruptions.
(b) These images, taken on June 28, 1997, highlight just two sulfur eruptions. One eruption, visible on the edge of Io's image (also seen close-up in the upper inset photo) is 140 km high. We can also see a second eruption site, located in the middle of Io's image near the terminator (day/night boundary); the plume of this eruption, which is 75 km high, cast a reddish shadow, seen also in the lower inset photo. Throughout Galileo's mission, this latter eruption appeared in every image of this location—and in every such image taken by Voyager in 1979—suggesting that it has been a continuous eruption.

with organic chemicals" (McCord et al., 1997) (quote from *Press & Sun-Bulletin* October 10, 1997). Io (Figure 1.26), the closest of the four moons to Jupiter, exhibits ongoing, sulfurous eruptions from three dozen volcanoes; the

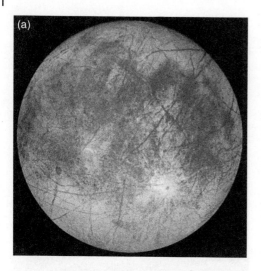 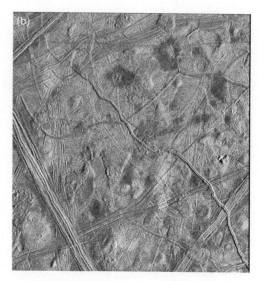

Figure 1.27 Photographs of Europa taken by the *Galileo* space probe (with false color added for clarity, (a) is from NASA / https://solarsystem.nasa.gov/missions/galileo/overview / last accessed August 22, 2023). The region displayed in (b) shows both a lack of cratering and a "chaotic" terrain with numerous ridges, pits, and domes, interpreted on the large scale as evidence of ice tectonics and therefore implying the existence of a subsurface ocean. Some of the smaller-scale features seen here, though, have also been interpreted as evidence of shallow intra-'plate' lakes or lenses (Schmidt et al., 2011). NASA / https://photojournal.jpl.nasa.gov/targetFamily/Jupiter?subselect=Target%3AEuropa%3A / last accessed August 22, 2023.

source of the volcanic activity, discussed in Chapter 6, is believed to be heat generated within the moon as it deforms periodically from the tidal forces of Jupiter.

None of those three moons appears to meet *all* of the requirements for life. However, sandwiched between them is another Jovian moon, Europa (Figure 1.27). When the Galileo probe took pictures of this icy moon's surface (on its nine flybys), scientists realized that the surface showed evidence of tectonic behavior. On Earth, the motion of lithospheric plates—driven by a hot, convecting mantle—results in rift zones, with new lithosphere created to fill in the gap between plates, and collision zones, and is accompanied by earthquakes and volcanoes. The surface of Europa exhibits ice-plate collisions, ice volcanoes, rifting, and new (i.e., fresh, relatively uncratered) ice plates. Such *ice tectonics* requires an energy source, just as Earthly tectonics does, and implies that the interior of Europa is convecting; that is, Europa's

interior is warm, and contains liquid water. The volume of water and ice on Europa is estimated to exceed all of that in Earth's oceans. Presumably it is the tidal forces of Jupiter that provide the heat for Europa's convection.

With tidal heat as an energy source, liquid water, and a mild environment, the only other ingredient required for life is organic chemicals. Because its neighboring moons possess them, it is *assumed* that Europa also contains organic compounds. With that assumption, Europa is generally viewed as the best candidate for extraterrestrial life at present.

Evidence from the Asteroid Belt. Anyone who has ever had the opportunity to touch a meteorite, or even to see one up close, knows the excitement of confronting something from 'out of this world.' From this perspective, it is not surprising that scientists have been analyzing meteorites for the presence of organic compounds for nearly 200 years. Studies by Berzelius in 1834, Wohler & Hornes in

1859, and Berthelot in 1868 (see Kvenvolden et al., 1970 for their references) all reported success in detecting organics. But for these and other studies, the main question has always been whether those compounds are indeed extraterrestrial, or are merely terrestrial contaminants.

Here, we focus on one carbonaceous chondrite: the Murchison meteorite, which fell to Earth in 1969, landing in Australia. This meteorite has been found to contain more than 80 amino acids and 8 nucleobases (some nucleobases make up the "base pairs" that hold the strands of an RNA or DNA double helix together) as well as a huge variety of other organic compounds (Kvenvolden et al., 1970; see also, e.g., Cronin et al., 1995; Sephton, 2002; Elsila et al., 2007; Schmitt-Kopplin et al., 2010; and Stoks & Schwartz, 1981; Martins et al., 2008; Ehrenfreund & Charnley, 2000). Callahan et al. (2011) have recently demonstrated that at least some of those nucleobases are "terrestrially rare" and thus likely extraterrestrial in origin.

Analyses of the amino acids in the Murchison meteorite by Kvenvolden et al. and subsequent researchers revealed some of them to be terrestrially rare as well—and others to be terrestrially 'familiar' but with a *structure* rarely seen on Earth, outside of laboratories. Organic molecules like amino acids can be constructed with what is termed right-handed or left-handed structure (the 'fingers' and 'thumb' of these molecules, determining their 'handedness' by their relative positions, are the C and H atoms and their functional groups (NH$_2$, COOH, etc.), linked in a chain structure with side chains); this property is known as their **chirality**. By pure chance (but see Gibney, 2014), possibly a mere consequence of the first life from which all subsequent life descended, *all* organisms on Earth (with the exception of bacteria; Radkov & Moe, 2014) are found with left-handed chirality. But about half of the Murchison amino acids possessed right-handed chirality! Furthermore, the peptides in Murchison (built up from its amino

acids) were not found with left-handed components either, unlike on Earth (Shimoyama & Ogasawara, 2002).

As additional evidence, the relative abundance of hydrogen isotopes (deuterium versus the more common hydrogen) in these organic compounds is consistent with the isotopic abundances inferred for cold interstellar molecular clouds rather than our present-day Solar System (e.g., Jefferts et al., 1973; Kolodny et al., 1980; Epstein et al., 1987; Pizzarello et al., 1991). Why this is so has been the subject of some speculation; for example, Pizzarello et al. (1991) proposed that cold asteroidal parent bodies in our primitive nebula initially contained ices and "deuterium-rich interstellar organics"; as they warmed, the ices melted, providing an environment for the more complex organic compounds to form from those simpler precursors. A discussion of such ideas is beyond the scope of this textbook; nevertheless, given the presence of rare, 'ambidextrous' and deuterium-rich amino acids, there appears to be little doubt that the Murchison amino acids had an extraterrestrial origin.

The Murchison meteorite has been described as "the astrobiology standard for detection of extraterrestrial organic matter" (El Amri et al., 2005; see also Sephton, 2002). But it does not stand alone; carbonaceous chondrite meteorites are generally found to contain measurable amounts of organic compounds, and even numerous amino acids (e.g., Cobb & Pudritz, 2014). As noted in the various references cited in this section (see also Glavin et al., 2011), several have been found to possess some of the key chemical features characterizing Murchison.

In sum, evidence from bodies originating in the Asteroid Belt, Kuiper Belt, and Oort Cloud, plus one Jovian moon (and, as it turns out, at least two other moons of the giant planets)—with additional supporting evidence from the galaxy at large—all suggest that the potential for life exists all around us. It is the *potential* for life rather than the *actuality*, though, because all the organic compounds detected are likely *abiotic* in origin.

On the other hand, "life finds a way," as mathematician-philosopher Ian Malcolm says in Michael Crichton's novel *Jurassic Park*. If life is indeed opportunistic, that is, if it is *driven* to develop (a trait some would consider to be the definition of life!), then it might not be unreasonable to conclude that life—perhaps even carbon-based life—can be expected to exist beyond Earth.

Evidence from Mars. Then there's Mars.... Mars has always held a fascination for people who believe in the possibility of alien life. In 1877 the Italian astronomer Schiaparelli turned his telescope to Mars and saw surface markings that looked like channels. Unfortunately, the Italian word for "channels" is "canali," so that reports of his discovery carried a connotation of civilization (purposeful construction) rather than nature. Observations of seasonal darkening across the surface of Mars, construed to be vegetation, and polar ice caps, assumed to be water ice, supported that connotation.

The cause of life on Mars was taken up by Lowell (perhaps more well known today for having organized the successful search for Pluto); his "brilliant public lectures," numerous articles, and three books on the subject stirred the creative imaginations—and interest in science—of people worldwide for decades (Zahnle, 2001). The popular culmination of that creativity included such tales of advanced Martian civilizations as *The War of the Worlds* and *The Martian Chronicles*.

The flyby of *Mariner IV* in 1965 revealed the Martian atmosphere to be thin and the Martian surface devoid of water and canals—though some channels were discovered (see below). The seasonal darkening could have been the result of dust storms; the polar ice was postulated to be frozen carbon dioxide (dry ice). Clearly, the primary evidence for civilization on Mars had failed; but in the popular imagination, it was still possible that Mars supported life.

In 1976, when the first *Mars Viking Lander* descended to the Martian surface, Ray Bradbury, the author of *The Martian Chronicles*, sat by the famous newscaster Walter Cronkite as the world waited to discover whether Mars was indeed populated by civilized beings, or microbes, or anything in-between. That *Lander* (see Figure 1.28), and a second *Lander* from the *Viking II* mission which reached Mars several weeks later, conducted four types of experiments (with a total of more than two dozen assays) to test for the existence of life (Klein, 1977). Three of those sets of experiments found no evidence of life. In the fourth type of experiment, organic nutrients and water were introduced into Martian soil samples; some of the outcomes from the experiments involving non-sterilized samples (sterilized samples were heated enough to presumably kill any Martian microbes) revealed that a small amount of the organic material had decomposed (been eaten?). This latter result was ambiguous, and was considered therefore inconclusive by some (e.g., Klein, 1977) but potentially positive by others (e.g., Levin & Straat, 1976, 2016; Navarro-González et al., 2010, 2011).

The Viking *Landers* also tested directly for the presence of organic compounds in Martian soil; their results were negative (Soffen, 1976; Biemann et al., 1977).

In recent years, those interested in the question of present-day life on Mars have focused on the possibility of water in the shallow Martian subsurface. It might be noted, incidentally, that the first set of experiments conducted by the *Viking Lander* demonstrated that the addition of water to the Martian soil samples did not 'revive' any dormant Martian life (Klein, 1977). There seems to be little doubt that there is currently no life on Mars.

In the mid-1990s, with a public still receptive to the idea of life on Mars, the controversy was revived by a few scientists and an intriguing 4-Byr-old (e.g., Halevy et al., 2011) meteorite. The meteorite was found in 1984 in Antarctica—actually an ideal location for collecting meteorites, because of the sharp contrast between those dark objects and the surrounding white ice; about two-thirds of all meteorites found are retrieved from the Antarctic (Grossman, 2017). Labeled ALH 84001 after

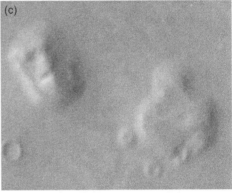

Figure 1.28 (a) Deep trenches dug on Mars by the *Viking Lander I* as part of its surface geology and biology explorations. The digging arm of the *Lander* is shown in the lower center of the image; meteorological instruments are to its left. Image by R. Van der Hoorn / Wikimedia Commons / Public Domain.
(b) Cartoon satirizing the failure of Viking and subsequent Rover missions to discover evidence of life on Mars; from http://www.cartoonaday.com/mars-lander-cartoon-curiosity-lands-on-mars / last accessed September 29, 2023.
(c) Photo from *Viking Orbiter* of region of Mars marked by scattered mesas showing a hill that looked to many like an artificially engineered face. NASA/JPL/Malin Space Science Systems / last accessed August 22, 2023.

the Allan Hills region where it was spotted, it was later said to have originated not in the Asteroid Belt but on the Martian surface. It would have been an energetic impact indeed that knocked this piece of old Martian crust into space, to orbit the Sun until some perturbation caused it to collide with the Earth. From measurements of cosmogenic isotopes of noble gases produced by cosmic rays during its travels, we infer that ALH 84001 was in space for ~17 Myr (Bogard, 1995); it fell to Earth about 13,000 years ago, judging from its setting in the ice in which it was found.

Fewer than 200 meteorites have been identified as Martian in origin—less than a third of a percent of all found meteorites; at the time of ALH 84001's identification, only about a dozen others were known (Grossman, 2017). Nearly all of these meteorites are classified as such because of the similarity of the gases trapped within them to the Martian atmosphere (as sampled by the *Viking Lander*) (see Treiman, 1995 for details); in the case of ALH 84001, however, the classification was based mainly on the similarity of its mineralogy to other Martian meteorites (Mittlefehldt, 1994).

Figure 1.29 The Martian meteorite ALH 84001. National Aeronautics and Space Administration / https://www.nasa.gov/mission_pages/mars/multimedia/pia00289.html / last accessed September 09, 2023. The inset photo, a scanning electron microscope image (NASA Public Domain), appears to reveal the existence of microscopic life forms; the length of the tube-like (worm-like?) structure is ~500 nm.

But what was most intriguing about this meteorite was that scientists analyzing it concluded (McKay et al., 1996) that ALH 84001 contained organic chemicals (polycyclic aromatic hydrocarbons); also, a suite of minerals more likely to exist, at least on Earth, through biological activity—and, tiny fossils of Martian microbes (see Figure 1.29)!

What McKay et al. discovered certainly was alien: those "nanofossils" were 100 times smaller than terrestrial bacteria. After word of their discovery leaked out, about a week before the journal article describing their work was to be published, NASA held a "hastily arranged" (David, 2016) press conference to present their discovery. Not surprisingly, the result was a 'media storm'; within hours, for example, President Clinton had released a congratulatory (though cautious) statement (Clinton, 1996). News of the discovery permeated the nation, capturing popular imagination. And politicians on both sides of the aisle even pledged

their support for bolstering NASA's mission to explore Mars—despite serious concerns about deficits in the U.S. budget (Lawler, 1996).

Even at the time of the announcement—perhaps in part *because of* the 'public relations' context of that announcement—reaction within the scientific community was mixed, with many skeptical (e.g., Kerr, 1996). Conceivably, the fossil-like features were simply inorganic carbonate globules, and the polycyclic aromatic hydrocarbons might just be terrestrial contamination from the surrounding Antarctic ice.

Still another possibility is that organic compounds were actually present on the Martian surface, but had been brought there from elsewhere in the Solar System (where, as we have seen, they are plentiful) by cometary impacts and an influx of interplanetary dust, and not produced in situ by Martian life. Consider, for example, the Martian meteorite named Tissint, which had impacted a desert location and been retrieved soon after falling to Earth in 2011; it contained a variety of organic compounds, and the analysis of Jaramillo et al. (2019) confirmed that they were unlikely to be from terrestrial contamination. However, we can also argue that those organic compounds were not likely to have derived from any Martian life, since Tissint formed during an era (much more recent than ALH 84001) when conditions on Mars had generally become even more inhospitable to life; this implies (as noted by Koike et al., 2020) an extra-planetary origin for those compounds.

In the case of ALH 84001, an analysis by Koike et al. (2020) similarly finds that terrestrial contamination was unlikely to be the source of its organic compounds. Those authors conclude the compounds were either brought to Mars by infall or synthesized (abiotically or possibly biotically) on Mars.

Of course, any conclusion regarding life on Mars based on the evidence from ALH 84001, given that meteorite's age and history, must relate to *ancient* rather than present-day life there. This limitation is also suggested by current surface conditions on Mars—including a dry, cold climate with a very thin atmosphere—and the mostly unequivocal results of the *Viking Lander* measurements. In fact, the evidence is strong that the ancient Martian atmosphere was probably thicker and warmer, and that Mars probably once had liquid water on its surface.

- Some evidence concerning Mars' ancient atmosphere is indirect but quite dramatic: the great Martian volcanoes. The greatest of these is Olympus Mons (Figure 1.30), the largest volcano in the Solar System, with three times the height of Mt. Everest—over 21 km. Olympus Mons is quite young, perhaps only ~25–100 Myr old, but other Martian volcanoes may date to as far back as 3 Byr ago, and magmatic activity in one form or another was widespread and prolific till at least 3 Byr ago (cf. Plescia & Saunders, 1979 and Carr & Head, 2010). Those volcanoes would have spewed prolific amounts of gas into the Martian atmosphere, especially carbon dioxide, of which the present atmosphere is a remnant. On Earth, volcanic eruptions release mainly water vapor and CO_2 into the air; correspondingly, the early, much thicker Martian atmosphere would have included abundant amounts of H_2O and CO_2. For example, Phillips et al. (2001, but see also Hirschmann & Withers, 2008) estimate that magmatic activity from the construction of just the Tharsis plateau and volcanoes could have produced an atmosphere 50% heavier than Earth's atmosphere is now, but composed entirely of CO_2, and also a global ocean 120 m deep.

Evidence that the ancient atmosphere of Mars could have been about as thick as Earth's is now *also* comes from the lack of small craters on the ancient Martian surface, with the would-be impactors having burned up before reaching the ground (Kite et al., 2014); from the implied long distances water could flow on the Martian surface (see below) without evaporating; and from the inferred ability (Wordsworth et al., 2013) of such an atmosphere to transport snow to the high elevations of Mars' Southern Highlands, successfully "recharging" the highland water sources.

Such an atmosphere would have warmed the surface, both by its denseness and by the greenhouse effects of its constituents. But those life-promoting conditions could not persist. Volcanic activity on Mars declined billions of years ago (Olympus Mons notwithstanding), and with the small planet's gravity being weak—little more than a third of Earth's—much of the gas not locked up in surface reactions or adsorption escaped the planet eons ago. We will return to this point below.

- The evidence for water is also striking; see Figure 1.31. We see Earth-like features that would typically be associated with fluvial processes, such as dendritic patterns, bars, and terracing, or with deposition in lakes; and there are more exotic

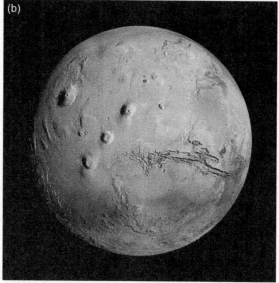

Figure 1.30 Martian volcanoes.
(a) Olympus Mons, the largest volcano in the Solar System. NASA / Public Domain.
(b) Hemispherical view of Mars; the 'diagonally' oriented Tharsis range, including Arsia, Pavonis, and Ascraeus Mons, is located between Olympus Mons and the 4000+-kilometer-long "grand canyon of Mars," Valles Marineris. Greg Shirah / NASA / Public Domain.
Incidentally, of all the sightings of 'canals' by Schiaparelli, Lowell, and others, through repeated observations, the most frequently seen—and one of the few non-imaginary channels—was Valles Marineris (Zahnle, 2001).

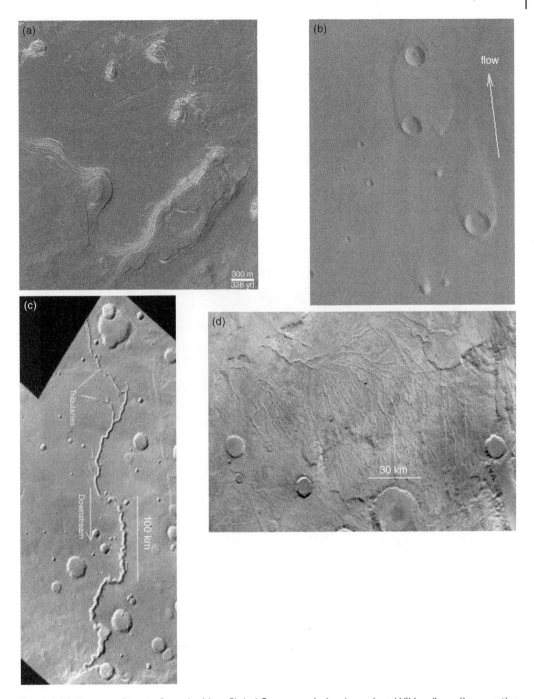

Figure 1.31 Features (images from the Mars Global Surveyor mission (upper) and Viking (lower)) suggesting the past presence of water on Mars.

(a) Colorized portion of Holden Crater, near the rim, showing "rounded slopes and buttes" and a layering suggestive of sedimentary lake deposits. National Aeronautics and Space Administration / https://science .nasa.gov/science-news/science-at-nasa/2000/ast04dec_2 / last accessed September 09, 2023.

(b) Eroded craters and streamlined islands (or perhaps leeward deposits) in Ares Vallis, with arrow indicating implied direction of water flow. NASA/JPL/Malin Space Science Systems.

(Lower) Outflow features (from NASA/JPL/Malin Space Science Systems). (c) not far from Valles Marineris.

(d) dendritic pattern in the Thaumasia region reminiscent of Earthly river deltas.

features, described as outflow channels, which are interpreted to be the result of catastrophic melting of subsurface ice or catastrophic release of groundwater under great pressure. Additionally, the geological explorations of more recent Mars orbiters equipped with advanced instrumentation, as well as close-up imaging and actual sampling by the Mars *Rovers*, have discovered rock types characteristic of evaporite basins (e.g., Squyres et al., 2004; Tosca & McLennan, 2006) and of long-distance river transport on Earth (Williams et al., 2013); see Figure 1.32.

Thus, there is little doubt that the essential requirements for life were met on ancient Mars, though to varying degrees: a moderate environment, an energy source (i.e., warmth from the Sun, trapped at the surface), and water, all thanks to the probable thick atmosphere early in Mars' history; and also organic chemicals. Interestingly, an analysis of meteorite ALH 84001 by Halevy et al. (2011)

Figure 1.32 Sedimentary rocks on Mars.
(a) The Mars rover *Opportunity* took this image near its landing site in Eagle Crater, an impact basin that is part of Meridiani Planum. These rocks were found to derive from basaltic weathering, and contain relatively high proportions of sulfate minerals. The spherules scattered about are hematite. The cross-bedding structure (red arrow) is ~6 cm in thickness. Image from Squyres et al. (2004) / American Association for the Advancement of Science.
(Lower, (b) to (d)) Color topographic map of the Aeolis Palus lowlands within Gale crater, an equatorial structure where the Mars rover *Curiosity* landed (white square). As it began to traverse the Bradbury Rise, *Curiosity* took this close-up image (c) showing pebbles and a conglomerate, quite similar in appearance to those seen in alluvial fan deposits like those of the Atacama Desert in Chile (d). Images from Williams et al. (2013) / American Association for the Advancement of Science.

supports this conclusion; they determined that the carbonates in that meteorite likely first precipitated at temperatures of ~20°C in a shallow (perhaps tens of meters below the surface) but evaporative aqueous environment (although they caution that the warmth could have been due to localized sources, e.g., 'geothermal' heat or impacts).

However, we have no way of knowing whether optimal conditions persisted long enough—and continuously enough (cf. Wordsworth, 2016)—for life to appear. At this point, the possibility cannot be ruled out. On Earth, as discussed in Chapter 2, the first life had developed by 3.5–3.8 Byr ago—the same era during which, on Mars, the fluvial features (e.g., in Meridiani Planum, Figure 1.32) were created (up to ~3.6—3.8 Byr ago, according to, e.g., Arvidson et al., 2003). If life began opportunistically on Earth, it may similarly have found a way during that era on Mars as well.

Hints of Earth System Geophysics. If life developed on ancient Mars, what happened to it—*what made it disappear?* One possible answer is consistent with the theme of Earth System Geophysics (—or in this case, Mars System Geophysics). As noted above, with Mars' relatively small size and low gravity, the gas molecules of its atmosphere would have gradually escaped into space; however, an important additional mechanism accelerating that loss—a mechanism that applied more to Mars than to Earth—may have been the disappearance of Mars' magnetic field. The magnetic field would have acted to deflect the charged particles of the solar wind away from the planet; without that shielding, those particles would have been much more effective in stripping the atmosphere (e.g., Jakosky et al., 2017).

Observations of widespread remanent crustal magnetization from the *Mars Global Surveyor* (Acuña et al., 1999) suggest that Mars had a strong global magnetic field in its youth. But the lack of magnetization in some impact basins—specifically, out of the 20 largest basins of estimated ages 4.22–3.81 Byr

ago, in the most recent 5 (e.g., Roberts et al., 2009), none of which are older than ~4.1 Byr or so—suggests that the field was absent when rocks there had recrystallized after impact, perhaps ~4.1 Byr ago (Lillis et al., 2008). (Note that aspects of Mars' magnetic field history are still disputed, e.g., Schubert et al., 2000; Kuang et al., 2008).

Mars' early magnetic field would have been maintained by 'dynamo' action in its core (a process we will discuss in Chapter 10 for the case of the Earth); the dynamo would have been driven by fluid convection in the Martian core. But the dynamo is only a side effect of the convection, whose main goal (at least, for thermal convection) would have been to transfer heat across the Martian core-mantle boundary, out of the hotter core into the cooler mantle. It has been postulated that something caused the *mantle*'s convection to slow or stop: perhaps volcanic degassing (and drying out) of the mantle, which would have increased its viscosity (Sandu & Kiefer, 2012); or, a transition from tectonic plates to a single, immobile rigid "lid" at the top of the mantle (if such a transition could have occurred quickly and early enough; cf. Nimmo & Stevenson, 2000). Heat subsequently building up at the base of the mantle would then have inhibited core convection. Alternatively, convective suppression could have resulted from heating of the lower mantle by the impacts which created the giant basins (Roberts et al., 2009).

Heat from the giant impacts may have also led to more uniform temperatures within the Martian core, stabilizing it directly against convection (Arkani-Hamed & Olson, 2010). Either way, the result was a termination of the Martian dynamo, baring the Martian atmosphere to the full ravages of the solar wind.

Earth or Mars System Geophysics is a size-dependent situation. As we will see in later chapters, the tendency for convection to occur is strongly size-dependent, and always more likely in a larger body. The factors that might have disrupted convection in the Martian mantle and core would have been less able to halt convection within the Earth; the survival of Earth's atmosphere was not so threatened. And Earth's greater mass (still another size dependence) also allowed it to retain more of its atmosphere than Mars did.

Conceivably, any life that arose on Mars simply could not survive the change to a much colder and drier environment when the atmosphere dissipated away. Earth's atmosphere has occasionally experienced transitions to colder and drier states (as we will see in the next chapter)—but nothing as extreme as a loss of most of its atmosphere.

2

The Evolution of Earth's Atmosphere

2.0 Motivation

In Chapter 1, we saw that, from the T-Tauri solar wind to accretional impacts—including the putative glancing collision with a planet-sized object that created our Moon—to the Heavy Bombardment, Earth was tormented episodically for more than half a billion years. These events wiped out much of Earth's earliest atmosphere, and supplied another; but, as we will see in this chapter, Earth was not just a passive witness to atmosphere-building.

The massive impacts Earth experienced also helped transform it, supplying some of the energy it would take to differentiate its interior into mantle and core. With this differentiation, which we discuss below, the planet came alive geologically: Earth began to experience systematic earthquakes, volcanic eruptions, and mountain building, and even began to replenish its atmosphere and oceans. In fact, the end result of differentiation was *the creation of the entire Earth System* as we know it: core, mantle, and crust as globally coherent layers, in addition to oceans and atmosphere, and—not long after conditions on, or near, Earth's surface had become less violent—the biosphere as well.

This Earth System defined a planet that was henceforth able to take a more active role in its own maintenance. For example, Earth's atmosphere is replenished by volcanic and magmatic *outgassing* associated with plate tectonics; plate tectonics, which began with the onset of mantle convection, reflects Earth's attempt to get heat (including that generated by differentiation) out of its interior. That convection is ongoing, allowing the replenishment to outlast any external threats, from those posed by the Late Heavy Bombardment to the Chicxulub impact to the present day. The continued existence of Earth's atmosphere is in large part the result of convection within Earth's interior.

The Earth is a convective machine. When the Earth came alive geologically, or soon after, the core would *also* have begun to convect, stirred by the heat of its formation and encouraged by the convection of the mantle above. The crust moves, as part of various lithospheric plates, because it is carried along with the convecting mantle. The atmosphere is driven to convect by heat from the Sun; and the oceans—created by the way, along with the atmosphere—are driven to circulate around by the atmosphere's convective motions, and also convect on their own.

Plate tectonics was once widely hailed as the "unifying theme" of Earth science. Clearly, though, *the process unifying the Earth* <u>*System*</u> *is—more fundamentally—convection*. And this process dominated the Earth System once the interior had differentiated.

The convective machine also protects Earth's atmosphere and oceans: the geomagnetic field generated by core convection shields the atmosphere from the solar wind (as mentioned in Chapter 1). This has helped the Earth System maintain its integrity over the eons.

Earth System Geophysics, Advanced Textbook 6, First Edition. Steven R. Dickman.
© 2025 American Geophysical Union. Published 2025 by John Wiley & Sons, Inc.
Companion Website: www.wiley.com/go/Dickman

The convective machine even enabled the Earth System to maintain an environment favorable for life for the billions of years *before* external factors alone should have allowed it. Astronomers talk about the **habitable zone**, the distance from a star at which water on a planet's surface can exist in a liquid state (and thus support life as we know it). But it wasn't until ~2 Byr ago that the Sun became bright enough for habitability at the distance of Earth's orbit. Nevertheless, as we will discuss in this chapter, the Earth System helped create its own habitable environment, and managed to keep it going during all that time.

In the next chapter, we will begin to understand and appreciate the convective paradigm of Earth System geophysics in the context of the atmosphere and oceans; later chapters will discuss mantle and core convection. A quantitative (fluid dynamic) investigation of convection will be delayed until the penultimate chapter; but that investigation will confirm that in large-scale systems, convection is essentially unavoidable—it really *is* universal.

For the Earth System, the universal theme of convection must be approached broadly. For example, viscosity is a key property of fluids which ultimately determines their ability to convect; but it will take a variety of gravitational, geodetic, and seismological observations to constrain that property in the core and especially throughout the mantle. The complex modes of mantle convection can only be revealed and appreciated with the help of seismological observations and tectonic models. And convection in the core is strongly affected by magnetic forces, which can be inferred by studying Earth's geomagnetic field observationally and theoretically. Such breadth will be reflected in the various chapters of this textbook.

This chapter will focus on the development and evolution of Earth's atmosphere. For billions of years, that atmosphere was not one in which today's humans could have survived; rather, it has naturally evolved into the advanced life-supporting atmosphere now enveloping us. Its evolution was driven mainly by interactions with the Earth's crust and (eventually) the biosphere, in the context of an ever-brightening Sun. As we will discuss below, these interactions produced both long-term *trends*, including decreases in the amount of carbon dioxide and increases in the amount of oxygen, and *extreme climate regimes* in the Earth System—"Ice House" and "Hot House" (or "Greenhouse") conditions persisting for many millions of years.

2.1 The Differentiation of the Earth

2.1.1 A Core by Condensation?

The gross structure of the solid earth today includes a thin rocky crust overlying a massive silicate mantle that, in turn, surrounds a metallic, mostly iron, core (see Figure 2.1a).

(a)

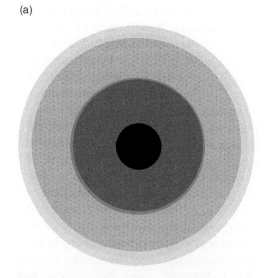

Figure 2.1 (a) The gross structure of Earth's interior, based on its compositional variations, as discussed later. The increasingly dark colors are meant to delineate the uppermost mantle, transition zone, lower mantle (shown as the dotted region, and containing the largest volume of any layer), and D″, all part of the mantle; and the core, including outer and inner portions. At this scale the crust (colored white) is too thin to see in detail. The interpretation of D″ as a compositionally distinct layer may be disputed.

Mid-twentieth-century theories such as those of Eucken (1944a, b as referenced in Urey 1951; also Turekian & Clark 1969 and others) held that the Earth formed *with* this layering. As Earth grew, its accretion was controlled by the volatility of the hot solar nebula gases: as the nebula cooled, the least volatile compounds—including, in particular, iron—were the first to condense directly out of the nebula, raining down on the proto-Earth to form its core; then slightly more volatile silicates condensed; and so on, ultimately to the most volatile ices and gases, building up the planet and its atmosphere sequentially in compositionally distinct layers. This 'volatility' model of planetary growth—envisioning that the Earth and other planets were layered (inhomogeneous) from the start—is summarized well in Jacobs (1987; see also Righter, 2003).

Proponents of this model also used it to explain trends in overall density from planet to planet. In the protoplanetary disk, dense, refractory iron would have settled through the nebula preferentially to the innermost regions, surviving comparatively well near the developing Sun and thereby accumulating larger cores for the inner proto-planets. This model thus predicted an increasing trend in planetary iron:silicate or core:mantle proportions with approach to the Sun.

But the Earth was not likely to have formed with an initial layering—even if the hot nebular environment envisioned by Eucken and others, in contrast to our version of the monistic theory highlighted in Chapter 1, was correct. Most fundamentally, Earth simply did not *have* to form that way! Natural geophysical processes, early in the development of the Earth, would automatically have caused an initially well-mixed interior to differentiate into mantle and core. And, as we know today, plate tectonic activity produces (as a by-product) oceanic, and eventually continental, crust (actually lithosphere); there is no need to invoke cosmic processes in order to create the outermost layer of the solid earth, either.

The same is true for Earth's envelope. The volatility theory posits that our atmosphere and oceans came from late-stage cometary (or, in general, nebular debris) infalls. But **outgassing**—venting of gases from Earth's interior—*also* replenishes the atmosphere and oceans, and has done so ever since the Earth came alive geologically.

If the volatility model were correct, we would expect Venus, which is significantly closer to the Sun than the Earth is, to have a proportionally larger core and less substantial atmosphere. But, based on its size, mass, and moment of inertia, we believe that Venus has a core and mantle similar in size to Earth's (see Chapter 1); we even think of Venus as Earth's 'sister' planet. And, for reasons (to be discussed later in this chapter) unrelated to the state of the solar nebula, Venus has an atmosphere almost two orders of magnitude heavier than Earth's, with only trace amounts of hydrogen and helium.

Furthermore, it turns out that forsterite, the magnesium silicate comprising much of the (upper) mantle, and iron are estimated to have condensation temperatures differing by perhaps only ~30°C (the uncertainty is a consequence of the dependence of condensation on nebula pressure and composition) (Grossman, 1977). And, within ~100°C of cooling from iron's condensation point, both substances are nearly completely condensed. It is therefore unlikely that the volatility model would have resulted in distinct cores and mantles (Stevenson, 1981).

Instead, the internal structure of the Earth and other inner planets was a consequence of *internal* processes that, as hinted at repeatedly in Chapter 1, produced a separation ("differentiation") of iron from rock.

Differentiation actually operated on both planetary and planetesimal scales. The solar nebula debris that accreted to create the inner planets was originally *chondritic*, containing both metal and silicate grains, along with some amount of volatiles. In the early stages of accretion within the nebula, as the planetesimals grew, some of them underwent their own differentiation, developing kilometer-sized iron cores surrounded by silicate mantles (Schersten et al., 2006) (for planetesimals accreting within or ending up within the Asteroid Belt, we would classify fragments of those differentiated bodies as *achondrite (stony) and iron meteorites*). But as the proto-planets grew, they accreted both primitive and differentiated planetesimals. The net effect was that those planets, including Earth, began in an intermixed state, without a distinct core or mantle.

2.1.2 An Act of Differentiation Created the Core

The subsequent differentiation of the Earth was a consequence of its interior heating up. As discussed quantitatively in Chapter 9, a number of heat sources may have played a role in this process, including the Earth's *accretion*, as the kinetic energy of infalling debris was converted to heat upon impact; *compression* of the interior, as the accreting mass piled up; and **radiogenic heat**, heat produced by the decay of radioactive isotopes incorporated into the Earth during its accretion. The accretional heat in particular was, potentially, devastatingly enormous; depending (among other factors) on the rate of accretion—slower accretion would have allowed more of the impact heat to radiate away into space—the Earth might have heated to the point where a surface or even subsurface layer melted completely, creating a magma ocean (see reviews by Solomatov, 2007 and Elkins-Tanton, 2012).

From these heat sources, even with only moderate rates of accretion, it would take no longer than a few tens of millions of years for the interior to become hot enough to begin chemically reducing the mineral mix, with free iron separating out as molten globs. Such globs, being denser than the surrounding silicate grains and primitive chondritic matrix, would sink down toward the center of the Earth.

Once begun, this process had the potential to accelerate considerably. The sinking iron would generate more heat, from friction as the globs squeezed between rock grains

and oozed down, and from their impact with deeper layers inside the Earth. The released heat would allow more iron to be reduced, producing more globs of molten metal that sunk down, releasing more heat, and so on. Such an 'avalanche' process could have completed core formation in dramatically short order, perhaps even within ∼50 Myr! This is an impressively brief time to redistribute all of the Earth's mass, with one third ending up as the core and the remaining two thirds comprising the mantle.

In fact, the hypothesis (described in Chapter 6) that our Moon was created by the collision of Earth with a planet-sized object is not viable *unless* core formation in general is rapid. The oldest zircon found within lunar rocks brought to Earth by Apollo astronauts is ∼4.42 Byr old (Nemchin et al., 2009), confirming that the Moon formed quite early in the history of the Solar System. And the Moon's bulk properties (mean density and moment of inertia, as well as surface composition) require it to be largely silicate. But if the impactor (and Earth) had not already differentiated prior to their collision, then the resulting moon—created from the remnants of the impactor's mantle in one version of the hypothesis, or from the spun-off mantle of the merged, impacted Earth in another—would not be as mantle-like and iron-poor in composition as our Moon actually is.

The contributions of the various heat sources to core formation depended on their timing (e.g., Rubie et al., 2015). Initially, accretional energy would have been mildly effective in heating Earth's surface layers, with much debris falling in but at slow speeds due to the proto-planet's small size and weak gravity; late-stage accretion, roughly 10–100 Myr later (e.g., Chambers, 2001; see also Rubie et al., 2015), would have been extremely effective, despite the scarcity of impactors, because those would have accreted to large size and the proto-Earth's stronger gravity would have accelerated their fall substantially. Radiogenic heat would have been quite significant during the first few million years of the solar nebula, as a result of the energetic decay of such short-lived isotopes as Al-26 and Fe-60; the early differentiation of small planetesimals, noted above, is attributed primarily to radioactive decay of those isotopes (Schersten et al., 2006). After those isotopes had mostly decayed away, Earth and other inner planets were left with the much less dramatic but geologically still consequential heat from long-lived isotopes. Finally, compressional heat would have intensified as the Earth grew, because of the increasing weight on interior layers.

In general, the process of planetary differentiation is more efficient if the heat sources are well distributed throughout the interior. For the Earth and other inner planets, early accretional heat would have been mainly surficial, whereas late-stage accretion, involving relatively large planetesimals, would have heated the deep interior more effectively. The limited number of large impacts, though, suggests that the differentiation itself might have proceeded quite unevenly. Eventually, the late-stage impacts plus the heat already released by differentiation may have made most of Earth's interior hot and soft enough for the iron globs to sink easily, releasing sufficient additional heat in turn for core formation to finish catastrophically. Or, conditions might have led to one or more late-stage magma oceans, which would provide a much quicker way than percolation for the iron to sink down through such layers of the mantle (see Rubie et al., 2015 for a review of 'sinking' mechanisms).

Either way, the timing and effectiveness of late impacts support the view that core formation was an inevitable part of Earth's accretion rather than a separate, later event (e.g., Wetherill, 1985).

The differentiation of the Earth did not sequester all of Earth's allotment of iron into the core; today, iron minerals exist, in relatively minor amounts, in both the crust and (as discussed in Chapter 8) the mantle. Thus, the differentiation was extensive but incomplete. This was probably the case also for Venus, Mars, and the Moon.

2.1.3 Consequences of Core Formation

The differentiation of the solid earth into core and mantle had major consequences for the entire Earth System.

Consequences for the Mantle. Only a small amount of the heat generated by core formation was needed to further the whole process, with the result that nearly all of that energy went into simply heating up the mantle and core, by more than 2,000°C (e.g., Birch, 1965; Flasar & Birch, 1973). The heat of core formation—added to the heat of late-stage accretion, which had already softened the interior—would have further weakened the solid mantle mechanically (see Figure 2.1b) and caused it to flow in a slow, "solid-state" convective circulation. This **thermal convection** would have been characterized by

(b)

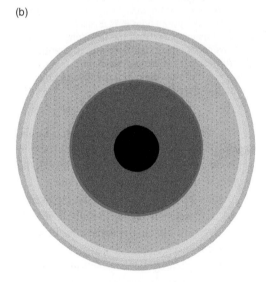

Figure 2.1 (b) The gross structure of Earth's interior, according to its mechanical properties, as discussed later in the text. Starting at Earth's surface and increasing with depth, we have the rigid lithosphere (in grey), upper mantle (tan), transition zone (yellow), lower mantle (dotted orange), and D″ (red); and the liquid outer core (brown) and solid inner core (black). At this scale, both the weak lower-crustal layer and the soft "low velocity zone" just below the lithosphere are too thin to see in detail. In this textbook we emphasize the role of D″ as a mechanically distinct layer—a convective "thermal boundary layer"; see Figure 2.2.

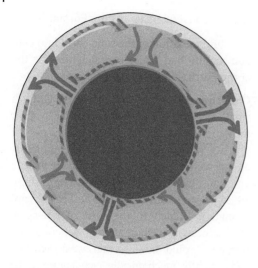

Figure 2.2 Convection in the Earth's mantle. In this idealized view, the entire mantle is involved and the convection is well organized into convection "cells" (as discussed in later chapters, the reality may be a lot less simple). Hot (red arrows) and therefore less-dense material rises from the depths of the mantle, spreading out as it reaches the base of the lithosphere, carrying the lithospheric plates (and any continents atop them) with it. Eventually the stuff cools (blue arrows), becoming dense enough to sink back deep into the mantle, thus completing the flow pattern.

relatively hot, light material rising up from the depths of the mantle, cooling substantially as it spread out and eventually becoming dense enough to sink back down through the mantle (see Figure 2.2). Ultimately, the whole mantle would have been involved: the whole mantle would overturn during the convective process.

With this convection, the Earth awoke geologically: carried by the convection 'currents' inching along at the top of the soft mantle, oceanic lithosphere would be pulled apart at rifts and mid-ocean ridges; magma would well up to fill the gaps, then cool and solidify as new oceanic lithosphere (Figures 2.3 and 2.4a). Moving away from the rift zones, the convection currents and the aging lithosphere they carried would continue to cool, eventually—in subduction zones—becoming dense enough to sink and return to the deeper mantle. As the cold lithosphere descended, sheet-like, it would reheat, generating magma

off its top surface that rose to create and feed a string of volcanoes—a volcanic island "arc" (Figures 2.3 and 2.4b). Collisions between lithospheric plates would build continents and build up mountains (Figure 2.4b). And all of these movements would generate earthquakes as well. In short, *plate tectonics began, as a consequence of core formation*. But plate activities are just the surface (and near-surface) manifestations of mantle convection.

Actually, we might even say that the tectonic *plates* themselves—and not just plate activities—are Earth's surface manifestation of mantle convection (e.g., Bercovici, 2003). As discussed in Chapter 8, the flow properties of mantle minerals are significantly temperature-dependent. As a consequence, the shallowest mantle—at the top of the convection cells—behaves much more rigidly and lithosphere-like than the bulk of the hotter and thus softer mantle found at depth; it becomes lithospheric. Furthermore, if this lithosphere is not too thick, or light, we can imagine that the forces of convection (including drag forces, and positive and negative buoyancy both from below and within the lithosphere) will cause it to move, creating in turn divergent and convergent boundaries that effectively separate the lithosphere into discrete *plates*.

However, even with a convecting mantle, the creation of distinct plates was not inevitable; depending on conditions, including not just lithospheric temperature, thickness, and composition but also more exotic material properties (see Bercovici, 2003), other tectonic modes are possible (see Brenner et al., 2020 and references therein). Some of these alternatives, which may prevail on other rocky planets and moons, are briefly discussed later in this textbook; one popular mode is known as the "rigid lid" or "stagnant lid." And, the very young Earth may have evolved from one mode to another (e.g., Sleep, 2000). But as far as we can tell, the plate tectonic mode was operating on Earth at least 3.2–3.5 Byr ago (Brenner et al.).

Consequences for the Core. Meanwhile, as suggested above, the core in a differentiated Earth would have formed hot, completely

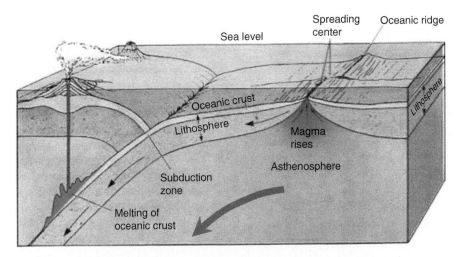

Figure 2.3 Basic concepts of plate tectonics. A tectonic "plate" is defined as a section of lithosphere that—driven by convection of the underlying mantle (whose flow is partially indicated here by a thick red arrow)—moves together, as a unit, relative to adjacent plates. Relative motion occurs at such plate boundaries as mid-ocean ridges (spreading centers) and subduction zones.
As shown here, oceanic lithosphere (which includes the oceanic crust and rigid top mantle) is created at mid-ocean ridges by rising magma as preexisting lithosphere pulls apart; as sea-floor spreading continues, the newly formed lithosphere solidifies onto and moves out with the older lithosphere. That plate motion is not on a flat plane or a horizontal cylindrical surface but along the Earth's spherical surface, requiring the mid-ocean ridge to be offset at various points (as shown here); the offsets are called transform faults, plate boundaries along which relative motion (between adjacent plates) is also unavoidable.
As the oceanic lithosphere cools, it becomes denser as well as thicker, and thus subsides, with the depth to the seafloor increasing.
Eventual collision of the plate with an adjacent lithospheric plate leads to its subduction. The descending plate heats up, because the interior is hotter and because of friction with the adjacent mantle, leading to melting; the resulting magma rises, creating volcanoes. The collision shown here is with another oceanic plate; in this case, the subduction is marked topographically by deep seafloor (an oceanic trench), and the surface volcanism appears as an "island arc." Collisions can also occur with continental plates (which include a thicker and lighter crust as well as top mantle); the continental lithosphere does not subduct but can thicken, producing mountain ranges.
Relative motion between adjacent plates is accomplished by both creep (steady or episodic) and earthquakes. Image from Skinner & Porter (1995) / John Wiley & Sons.

molten and predominantly iron. Once the outermost core had cooled sufficiently—a situation fostered by the convection of the overlying mantle (which brought heat from the interior up toward the surface)—the core would also have begun to convect, with hot, lighter fluid rising and cooler, denser fluid sinking. In ways that will be discussed in Chapter 10, *such motions of this electrically conducting fluid would end up creating and maintaining a planetary magnetic field.*

When igneous rocks containing iron or certain iron-bearing minerals cool sufficiently, they are able to record traces of the ambient magnetic field at that point in time. In ancient rocks, such *paleomagnetic* measurements document the continuous existence of the Earth's field (and thus ongoing core convection) extending back billions of years; some rock samples suggest the field operated as far back as 3.5 Byr ago (Yoshihara & Hamano, 2004; Tarduno et al., 2007). Longer ago than that, though, the situation is unclear. It is possible that an excessively hot lower mantle—proposed by Labrosse et al. (2007) as a relict of a dense deep magma ocean—initially blocked the core from cooling, delaying the onset of its convection (that same deep magma ocean might additionally have sequestered some iron

(a)

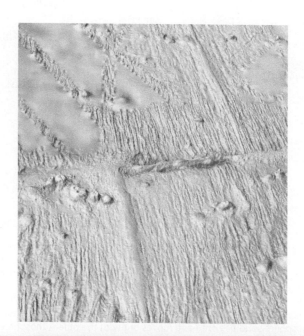

Figure 2.4 Examples of the plate tectonic features sketched in Figure 2.3.
(a) Divergent features.
(Upper) The East Pacific rise (a mid-ocean ridge separating the Pacific and Nazca plates in the Pacific, west of Central and South America). High-resolution bathymetry, between latitudes 9°N and 11°N, from the National Science Foundation's RIDGE 2000 program, downloaded from Bill Haxby / https://earthjay.com/?p=4019 / Last accessed September 14, 2023. Because plate spreading from this ridge is so rapid (e.g., roughly three times the spreading rate of the mid-Atlantic ridge), the central mountain chain is relatively small and lacks a central rift valley. Note the transform fault offsetting the rise at ~10° N. Numerous seamounts can also be seen.
For scale, note the offset is ~100 km wide.
(Lower) Divergence of the African and Arabian continental plates has left a rift that became the Red Sea. Photo from NASA / http://www.cosmosplus.com/user-content?page=27 / last accessed September 29, 2023. Photo from the International Space Station.

(b)

Figure 2.4 Examples of the plate tectonic features referred to in Figure 2.3.
(b) Convergent features.
(Upper) Collision between the northward-moving Pacific plate and the North American plate south of Alaska produced the Aleutian Islands, an island arc. Some of the islands are seen in this photo. Downloaded from https://www.nsf.gov/discoveries/disc_images.jsp?cntn_id=135851&org=NSF / last accessed August 22, 2023.
Actually, because of the great east-west breadth of the subducting Pacific plate, the volcanic arc stretches eastward beyond these islands, becoming the continental volcanic arc of the Alaskan Peninsula. To the west, the island arc extends through the width of the Pacific up to the Kamchatka Peninsula (at the edge of the Eurasian plate).
(Lower) Collision of the Nazca and South American plates—and subduction of the Nazca plate beneath South America—shortened and thickened the South American lithosphere, leading to the rise of the Andes mountain range. In this photo from the International Space Station, we are looking south, with the Chilean coast to the west of the mountains and Argentina to the east, both mostly cloud-covered. Photo downloaded from https://www.nasa.gov/image-article/andes-mountain-range / last accessed May 7, 2024.

minerals from differentiation, later redistributing them throughout the mantle (see Labrosse et al.) and thereby accounting for the minor amount of iron still present in the mantle). Alternatively, Zaranek and Parmentier (2004) and others envision a mantle that, after an initial overturn, achieved temporary stability and (briefly) stopped convecting, limiting cooling at the core boundary and thus inhibiting core convection for a time. Most likely, however, the lack of paleomagnetic evidence older than $3\frac{1}{2}$ Byr is simply a consequence of the scarcity and highly altered condition of such ancient rocks.

Differentiation may have also set the stage for *future* convection within the core. The density differences that make a fluid convect can be created by differences in temperature or in chemistry. Because the presence of a light alloy (such as sulfur) would depress the melting temperature of iron—allowing iron-silicate separation to proceed in an Earth not quite as hot as envisioned above—we might anticipate that differentiation ultimately produced a core that was primarily iron but contained some amount of light alloy as well. Possible candidates for that light alloy are evaluated geophysically in Chapter 8. Once the Earth's core had begun to solidify, producing a solid inner core, the light alloy—preferentially

excluded by the solidification and thus progressively enriching the surrounding fluid—would serve to promote chemical convection within the liquid outer core. With the Earth's interior also very hot after differentiation, we thus have the prospect of two effective driving mechanisms for core convection.

As we will discuss in Chapter 10, present-day observations of the geomagnetic field allow the speed of the core flow to be inferred. Estimates are typically hundredths of a cm/sec—not very fast, though actually orders of magnitude faster than the cm/yr speeds inferred (from plate velocities) for mantle convection.

Consequences for the Oceans and Atmosphere. Throughout its lifetime, the Earth has vented gases from its interior. A result of internal heating, chemical reactions, and even radioactive decay (decay of K-40, for example, produces argon gas), such venting occurs primarily through volcanoes (Figure 2.5a, b) and around mid-ocean ridges but also at geysers, hot springs, and hot faults. On a global scale, this outgassing is a by-product of ongoing mantle convection and plate tectonic activity. And when the core first formed, and the first upwellings of a convecting mantle occurred, there would have been a tremendous amount

(a)

Source: NASA

Figure 2.5 Earth's watery oceans and a predominantly carbon dioxide atmosphere resulted in part from outgassing in the early days of the Earth. (a) An example of modern-day outgassing: Cleveland Volcano, located in the Aleutian Islands, which has been very active for at least the past century. Satellite image Courtesy of NASA / http://www.volcanodiscovery.com/volcano-tours/278.html / last accessed August 22, 2023.

(b)

Figure 2.5 Earth's watery oceans and a predominantly carbon dioxide atmosphere resulted in part from outgassing in the early days of the Earth.
(b) Another example of modern-day outgassing: Augustine Volcano, shown here in August 2006 (http://www.avo.alaska.edu/volcanoes/) after a recent spurt of activity that began in mid-2005. This volcano, located at the entrance to Cook Inlet a few hundred kilometers from Anchorage, Alaska, has been very active during portions of the past few thousand years. The inset shows a MODIS satellite image of its plume, 90 miles long, during December 2005. Coombs, M. L. / https://www.avo.alaska.edu/images/image .php?id=11357 / last accessed September 14, 2023.

of degassing of the interior. This outgassing is familiarly known as "the *big burp*," and represents a surficial accompaniment to the differentiation of the interior.

We interrupt our descriptive narrative to bring you a basic quantitative exercise. The big burp was likely to be an extended event (—a better name for it thus being the *big bur-r-r-p*). For example, if mantle convection operated then with the same vigor as it does now, typical flow rates would probably not exceed ~10 cm/yr, the speed of the fastest plates driven by the convective flow now. In that case—imagining a convection cell as deep as the whole mantle (~2,900 km) and

hypothetically twice as wide as that (consider the distance on Earth today between ridges and trenches)—the *timescale* for one cycle of convection would have been ~175 Myr.

Can you verify this estimate? It is based on the rule that "distance equals rate times time."

If early mantle convection was more vigorous than now, with a hotter interior, the timescale would have been somewhat shorter. If the initial degassing required a few more cycles of convection, the big burp might even have outlasted the Heavy Bombardment.

Clues to the composition of the gases vented during and after the big burp come from analysis of present-day outgassing. The gases detected in volcanic eruptions include small amounts of H_2, HCl, CO, Cl_2, H_2S, SO_2, and notably N_2 and Ar, but by far the bulk of the gas—typically ~98%—is water vapor and also CO_2. The high percentage of water reflects to a large extent its presence in rehydrated minerals and in pore spaces within sediments and oceanic crust, recycled into the mantle as the sediment-covered oceanic lithosphere subducts, for example at island arcs, prior to eruption through overlying volcanoes (e.g., Sumino et al., 2010). And, as we will discuss later in this chapter, heating of surficial carbonate sediments and oceanic crust at subduction zones can similarly release carbon dioxide previously incorporated into those minerals. In the young Earth, then, the proportions of H_2O and CO_2 should have been lower; nevertheless, we can expect that the primary result of the big burp was the venting of large amounts of water vapor and carbon dioxide—thereby yielding a new atmosphere with an abundance of those gases. Then, or at some later time if Earth's surface was initially too hot, much of the water raining down onto the surface would have collected into the deepest basins, forming the earliest oceans. In sum, the ultimate consequence of the big burp was that it provided watery oceans and a CO_2 (plus water vapor) atmosphere.

A Timeline of Bumps, Bubbles, and Burps. Our model of Solar System formation discussed in Chapter 1 implies that outgassing would not be the only source of Earth's atmosphere and oceans during its infancy. The following scenario provides a speculative overview of how the various sources—and sinks—might have contributed to the early development of our atmosphere and oceans.

- In the beginning—as the Earth was accreting—it would have acquired nebula gases, mainly hydrogen, of course. During the early stages of accretion, this first atmosphere would have accumulated as more gas was pulled in. However, when the proto-Sun entered its T-Tauri phase, its intense solar wind began to sweep the solar nebula clean of gas and strip away Earth's first atmosphere. Thus, within less than ~10 Myr (as noted in Chapter 1), Earth's first atmosphere was being undermined.

  During this time, planetesimals were falling in, building the Earth; but they were too small and their impacts too low-energy to significantly affect the growth of the atmosphere (either positively or negatively).

- As accretion continued, the Earth grew larger, and collisions among the planetesimals led to *their* growth as well. Impacts with the Earth consequently became more energetic, leading to *impact degassing*. The planetesimals, and thus the Earth they were helping build, contained small amounts of carbonates and hydrated minerals, so the degassed volatiles would have included carbon dioxide and water vapor.

- By late-stage accretion, the impactors were even more massive, and their collisions with the developing Earth might have led to widespread surficial melting—a magma ocean. H_2O and CO_2 released from surface and subsurface rocks by that melting would have bubbled up rapidly through the magma ocean, adding profusely to the atmosphere (Elkins-Tanton, 2008).

  By the time roughly 50 or so million years had passed after the beginning of Earth's formation, and well into late-stage accretion, the Earth had differentiated into core and mantle, the mantle had started to convect, and tectonic processes were initiated. The mantle began to burp, and its outgassing vented prolific amounts of H_2O and CO_2 into the atmosphere.

  The magma ocean would have cooled quickly, its bubbling up dying away as solidification occurred within perhaps a few million years (Elkins-Tanton, 2008); after several more million years, the surface was cool enough to allow any water vapor that condensed and rained down to remain and pool, forming the first watery oceans and leaving a dense CO_2 atmosphere. As the burping continued, that carbon dioxide atmosphere would become denser.

Interestingly, as late-stage accretion progressed and the Earth grew close to its present size, any watery oceans that formed would never really go away. That is, even if a massive impact was able to completely vaporize the oceans, most of that water vapor would be retained in Earth's gravity field (Genda & Abe, 2005). Then, once surface temperatures had cooled sufficiently—a situation that could prevail because the remaining impactors, though very large, were few and likely far between—that water vapor would again condense and refill the ocean basins.

Those late-stage impactors would have added to Earth's atmosphere by impact degassing. At the same time, they could erode the atmosphere through a number of mechanisms (see, e.g., de Niem et al., 2012; Ahrens, 1993 for reviews). An impactor (or the shock-wave front preceding it) would have directly accelerated atmospheric gases to the sides as it plowed through on its way down to the surface, and the crustal debris thrown up on impact would have dragged some adjacent atmosphere back out toward space; however, such mechanisms are inefficient and involve only limited portions of the atmosphere. In contrast, the impact also would have vaporized part of the crust and much of the impactor; the vapor cloud, expanding upward and outward away from the point of impact, could accelerate atmospheric gases, 'blowing off' all of the atmosphere above the tangent plane at the impact point (Vickery & Melosh, 1990). The impact would, additionally, create a shock wave of deformation traveling through Earth's interior, 'pushing out' the solid surface on the other side of the Earth; if the impact is powerful enough, that opposite surface will rise sharply, creating a vapor cloud that can expand outward through a much greater volume of the atmosphere.

Earlier models (e.g., Ahrens, 1990, 1993) predicted that such a global blow-off from the vapor cloud would dominate over the degassing, so that the planet's atmosphere would be annihilated by very massive impacts.

Subsequent work (e.g., Genda & Abe, 2003; de Niem et al., 2012) has challenged that conclusion, finding that even with a giant impact, most of the atmosphere survives; the target planet can even acquire the atmosphere of the impactor (which had mostly survived as well).

The presence of oceans would have changed everything, though. A massive and very energetic impact could vaporize much of the oceans, and the rapidly expanding steam cloud would have augmented the silicate vapor cloud, intensifying the blow-off and driving out a substantial fraction of the atmosphere (Genda & Abe, 2005; see also Sleep et al., 1989; Zahnle, 2005). As noted above, most of the steam would not escape Earth's gravity, instead enduring to ultimately condense and replenish the oceans.

- Then, around 80 Myr after its accretion began, Earth was hit by a planet-sized body. Our Moon was a remnant of this collision, and, as noted before (and discussed in some detail in Chapter 6), if both the Earth and the planet-sized impactor had already differentiated, such a collision could explain the near-lack of a core in the Moon (e.g., Canup & Asphaug, 2001).

Models simulating this impact have generally focused on the fate of the planets' silicate and metal particles following the impact, and on the silicate vapor atmosphere it produced. But the Moon-forming impact can be expected to have shared key characteristics with other massive impacts in this era of late-stage accretion. In that case (e.g., Tucker & Mukhopadhyay, 2014) we can infer that the giant impact would indeed have blown off much of Earth's existing atmosphere.

As discussed in Chapter 6, the Moon-forming impact hypothesis envisions either a low-energy glancing blow, which might have incidentally created a limited magma ocean on Earth, or a direct hit, which completely melted and merged both Earth and the impactor. Either way, magma degassing would help replenish atmospheric CO_2 and

H_2O, until the magma solidified. In contrast to less massive late-accretion impacts, a heavy-enough "blanketing" atmosphere might have accumulated to delay surface cooling for millions more years (Sleep et al., 2014; see also Abe, 1997); that delay would, in turn, allow more CO_2 and H_2O to bubble up. Finally, despite modifications to the interior convection by the planet-altering impact, convective outgassing would continue to resupply the atmosphere with CO_2 and H_2O.

- For the next few hundred million years, external threats were minimal. Outgassing continued to refill the atmosphere, and the oceans deepened.

- That era of 'external peace' was followed by the Heavy Bombardment, and once again Earth was subjected to large impacts, with impact degassing; to possible magma oceans, with bubbles of volatile gases rising to the surface; and to expanding vapor clouds blowing off portions of any accumulating atmosphere.

But through it all, and after the Bombardment had terminated, mantle convection and outgassing continued, outlasting it all.

The preceding scenario—and it should be noted that it is only one of a myriad possible— relies heavily on the intensification of an impact-generated vapor cloud by vaporized oceans. It predicts a checkered history for our atmosphere and oceans, with very large impacts wiping out the former entirely and the latter (or portions thereof) temporarily. It would be a struggle, but thanks to outgassing an atmosphere would always, eventually, be reestablished.

If oceans were absent—or if their predicted effects on the vapor cloud are overstated—the atmosphere may have continued to accumulate through all of the late accretion and Heavy Bombardment. But even this conclusion is not certain, in part because of uncertainties regarding the nature of the impactor sources (the timing and energy of the impacts, and—for the Heavy Bombardment—whether their origin is asteroidal or cometary and thus water rich). For example, de Niem et al. (2012) conclude that, by the end of the Bombardment, Earth's atmosphere would have been augmented several-fold; in contrast, Pham et al. (2011) predict that, unless the impactors striking Earth had relatively high volatile content, Earth's atmosphere would have been substantially eroded by the end of the Bombardment.

The final outcome also depends on the assumed state of the Earth at the time of impact. Although de Niem et al. (2012) did not consider the effects of mantle outgassing, they point out that if Earth had a thicker initial atmosphere, the volatiles released by massive impacts would more likely be retained. An uninterrupted buildup of Earth's atmosphere for ~300 Myr prior to the start of the Bombardment would, consequently, have increased the chances that its final outcome 3.8 Byr ago was atmospheric gain rather than loss.

Between the time of core formation and the start of the Heavy Bombardment, our scenario envisions a thickening, predominantly carbon dioxide, atmosphere. By some time after the end of the Bombardment (if not sooner), it would have again been thickening and predominantly CO_2. If the effect of oceans on atmospheric loss from impacts is correctly predicted, outgassing had to have been a major contributor to that thickening atmosphere.

The Other Terrestrial Planets, Very Briefly. From the existence of iron cores within the other rocky planets, we know that they, too, had undergone differentiation, and thus outgassing. Like Earth, they also would have gained an atmosphere from accretion and impact degassing. With the exception of Mercury (which stood no chance of retaining *any* atmosphere, due to its small mass and extreme nearness to the Sun), all the rocky planets may well have experienced an early transition from a predominantly hydrogen atmosphere to one that was mainly CO_2 and H_2O.

But there would have been differences in their evolution. For example, given the small size of its core and the inference of widely dispersed iron minerals, Mars' differentiation was probably limited, with a relatively weak and brief 'burp.' Venus, though—Earth's 'sister' planet—would have at least started out like Earth, with the formation of a substantial core, major tectonic activity, and vigorous, ongoing burping.

Whatever atmospheres the inner planets acquired were protected to varying degrees from the solar wind by their core-generated magnetic fields: those fields could deflect solar wind particles, minimizing their erosion of the planets' upper atmospheres. Over the long term, the geomagnetic field has played a fundamental role in limiting the loss of Earth's atmosphere to outer space (e.g., Tarduno et al., 2010, 2015). In contrast, Mars' internal magnetic field is long gone; stripping of the Martian atmosphere (measured, for example, by NASA's MAVEN mission (Jakosky et al., 2015) beginning in 2014) is mostly unchecked, and in the past (as discussed in Chapter 1) may have contributed significantly to the loss of Mars' early, heavier atmosphere (LeBlanc et al., 2015; Jakosky et al., 2017, also Wei et al., 2012).

Venus has probably not (for reasons discussed in Chapter 10) had a core-generated magnetic field for at least the past $\frac{1}{2}$ Byr (Nimmo, 2002); given its relative proximity to the Sun, stripping of its atmosphere by the solar wind (measured indirectly by Grünwaldt et al., 1997) might be expected to be even more effective than what Mars is experiencing now. However, accounting for its greater planetary mass and other more exotic factors (specifically, induced external magnetic fields, which restore some protection) suggests about the same current rate of loss of atmospheric mass as for Mars (see Persson et al., 2020, 2021). On the other hand, outgassing on Venus has evidently continued to the present day—literally (Filiberto et al., 2020; Gülcher et al., 2020)!

In their 'youth,' both Venus and Mars would have lost most of their surface water (see, e.g., Way et al., 2016 and references therein for a variety of perspectives). For Venus, the Sun's heat alone (given its proximity to the Sun) could have caused it to lose an ocean's worth of water in only ~1 Byr (Kasting & Pollack, 1983) or less (Kulikov et al., 2007), with that water evaporated and the constituent hydrogen escaping into outer space. Mars' weak gravity contributed to its loss of surface water (and most of its atmosphere as well).

Our present-day observations of Mercury, which lacks an atmosphere, and Mars, which has a thin CO_2 atmosphere but shows evidence of a thicker one (and also the presence of surface water) in its distant past, are reasonably consistent with all these considerations. And, as predicted, the atmosphere of Venus is observed to contain only negligible amounts of water vapor, and is dominantly carbon dioxide. However, Venus' atmosphere turns out to be an extremely dense carbon dioxide atmosphere, with a mass estimated at ~90 times that of our entire atmosphere (Williams, 2013)—more massive even than might be explained by late-accretion impacts, an Earth-sized big burp, and heavy bombardments on an ocean-less planet. For that matter, Earth's present-day atmosphere is not at all consistent with our predictions of a dominantly CO_2 atmosphere after the Heavy Bombardment had ended. Clearly, we are missing a key aspect of atmospheric evolution. That will be addressed in the remainder of this chapter.

There is one last set of consequences of Earth's differentiation into mantle and core worth mentioning.

Consequences for the Biosphere. As life appeared on the surface, Earth's magnetic field would have helped protect it from solar wind particles; from cosmic rays (which are charged particles similar to those of the solar wind but from exotic sources like supernovae and black holes); and from the high-energy charged particles produced by interactions of the solar wind and cosmic ray particles with our atmosphere. The atmosphere itself provides even more protection. A striking illustration of the magnitude of such protection is provided by recent measurements revealing significant, even hazardous, levels of radiation experienced *outside* of our atmosphere *and* magnetic field, in the space between Earth and Mars (Zeitlin et al., 2013).

Throughout its past, as we will discuss in Chapter 10, the Earth's magnetic field has weakened, then strengthened, when the field polarity was in the process of reversing. We may speculate that, at such times, the weakened field might have allowed enough radiation through to Earth's surface to cause mutations in the DNA of some organisms—in the same way that alpha and beta particles produced during radioactive decay can lead to cancer—ultimately contributing to the evolution or extinction of those species. For example, Cooper et al. (2021a) found that carbon-14 levels in a well-preserved tree trunk, 42 Kyr old, recorded a cosmic ray "bombardment" (Voosen, 2021) of Earth's surface associated with just such a weakened field. Additionally, as Earth's field was weakening, the Sun was evidently experiencing its own weakened field (similar to the most extreme of the "solar minima" we will discuss in Chapter 3). According to Cooper et al. (2021a), the resulting ionization and erosion of the atmosphere, which could have included even the lower atmosphere (see Kirkby et al., 2016; Suter et al., 2014), led to climate change and, ultimately, extinction of some species—a conclusion that was controversial indeed, with objections quickly raised by Picin et al. (2021; see also Voosen) and countered by Cooper et al. (2021b). Their mechanism is somewhat more indirect than envisioned by Valet and Valladas (2010), who pointed to the depletion of the protective ozone layer at that time (resulting in more intense DNA-damaging ultraviolet light able to reach Earth's surface) as bringing about those extinctions.

We note also the proposal by Medvedev and Melott (2007; see also Erlykin et al., 2017 and references therein) that variations in cosmic ray intensity caused by changes in the position of our Solar System within the galaxy may have led to mass extinctions in Earth's past.

The existence of the Earth's magnetic field may be important for the biosphere in still other ways. As human civilization developed and advanced, the geomagnetic field has provided a reference system to aid navigation, by means of the compass. That reference system is useful only because the geomagnetic field produced in the core appears to have a simple 'bi-polar' structure (officially called a *dipole field*) with distinct 'north' and 'south' poles, as viewed from Earth's surface. Actually, as discussed in Chapter 10, the core field is quite complex; it is largely because of our distance from it, with the mantle intervening (—thus, a double benefit of differentiation!), that it appears primarily dipolar.

Other life forms—both primitive and advanced, from bacteria to bees, sharks, fish, turtles, lobsters, birds, dolphins, and cows—appear to instinctively use Earth's magnetic field for reference (see, e.g., Wiltschko & Wiltschko, 2005 and Gould, 2010 for reviews). For example, in trout (and other species as well) microscopic magnetic particles have been found attached to epithelial tissue, allowing the organisms to feel the force of Earth's field as they move through it (Eder et al., 2012). Those particles are micron-sized or smaller, iron-rich crystals, strongly magnetized for their tiny size (Eder et al., 2012), and are inferred (e.g., Blakemore, 1975 for the case of bacteria) to have been produced by the organisms (and not simply ingested from their surroundings). In contrast to such biogenic magnetite, sacs of electrically conducting fluid are thought to serve the same purpose in sharks and related species.

A few species of birds and turtles may actually be able to infer their location, by sensing variations in magnetic field intensity (Lohmann et al., 2007). More commonly, species use their sense of magnetic field directions: robins, homing pigeons, and other species of migratory birds, to orient their flight paths; underground mole rats (Marhold et al., 1997), compass termites, and honeybees, for example, to orient their nest- (or mound- or honeycomb-) building (Wiltschko & Wiltschko, 2005).

2.2 The Faint Young Sun

To some extent, the Sun's radiation determines the overall temperature of the Earth's surface.

It turns out that the characteristics of the young Sun—a fascinating topic on its own—imposed significant constraints on Earth's surface conditions, and thus on the state of our early atmosphere. These constraints will serve to broaden our understanding of that atmosphere beyond what was discussed in the timeline above and provide a framework for its subsequent evolution.

2.2.1 The Young Sun's Changing Luminosity Was Inevitable

The light and heat we observe radiating from the Sun are a by-product of the thermonuclear reactions in its interior that fuse hydrogen into helium. Over time, as more and more of the Sun's hydrogen is converted to helium, its internal density will rise, causing an increase in internal pressure. As the pressure builds, further compressing the interior,

temperatures within the Sun will rise. Those higher temperatures reflect a greater kinetic energy of the interior gas particles, including those ions that constitute the bulk of the interior plasma: hydrogen nuclei (protons). With greater energy, those protons can fuse together more efficiently; the net result is that, over time, the Sun's luminosity will increase.

As a consequence, for example, it is estimated that in the next 1 Byr the Sun will become 10% more radiant than it is now. The increased heat from the Sun will raise surface temperatures on Earth to the point where the oceans will evaporate, all water vapor will escape the planet, and Earth will dry up (Kump et al., 2010); the Earth will no longer be able to support life.

Correspondingly, in the past, the Sun must have been *less* radiant. Calculations suggest that 4.0–4.5 Byr ago the Sun was only 70% as luminous as it is today (see Figure 2.6a).

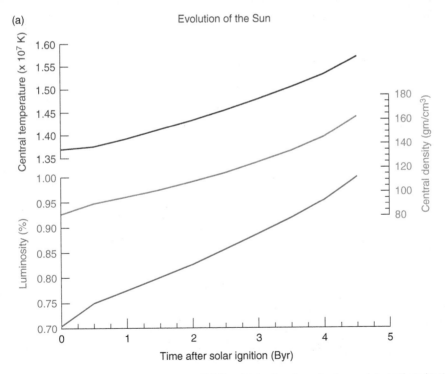

Figure 2.6 (a) Evolution of the Sun over the past 4.5 Byr. As the Sun fuses hydrogen into helium, its density increases; the middle curve shows the density at the Sun's center. The increased density raises the pressure, increasing the compression of the interior, which raises its temperature; the top curve shows the temperature increase at the center. With hotter temperatures, fusion reactions occur more efficiently, increasing the Sun's luminosity (lower curve). Data from Yoder (1995).

These predictions refer to the Sun after it had finished its intense T-Tauri stage and after it had ignited, that is, once it had begun to burn hydrogen normally as a "main-sequence" star.

2.2.2 A Paradox, and Its Resolution

The problem is that, with a solar luminosity of 70%, surface temperatures on Earth would have been about 30°C cooler than now, if our atmosphere then was like it is today. At such temperatures Earth's oceans would have been frozen; indeed, with the Sun's radiance increasing so gradually over the subsequent 4 Byr, there would have been no liquid water even as recently as ~2 Byr ago (see Figure 2.6b). This is problematic because there is abundant geological evidence, e.g. the very existence of ancient sedimentary rocks, which demonstrates that Earth's surface has supported running water as long ago as there is a rock record. In fact, the oldest rocks found on Earth are metamorphosed sedimentary rocks; see Figure 2.7.

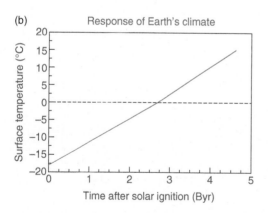

(b)

Figure 2.6 (b) Response of the Earth to the increased luminosity of the Sun (shown in Figure 2.6a) over the past 4.5 Byr. The curve shows the increase in Earth's average surface temperature assuming a present-day atmospheric composition, and including the greenhouse effect of increasing water vapor in the atmosphere as conditions moderate. The horizontal dashed line indicates the freezing temperature of water. Under these conditions—that is, without further warming from other effects such as additional greenhouse gases, as discussed in the text—Earth's surface would remain frozen until about 1.9 Byr ago. Data taken from Kump et al. (2010).

This inconsistency between astronomical and geological certainties (first pointed out by Sagan & Mullen, 1972) is called the **faint young Sun paradox**. Resolution of the paradox lies in the recognition (Owen et al., 1979) that the main constituent of Earth's post-accretion, post-Bombardment atmosphere was carbon dioxide. CO_2 is a greenhouse gas, meaning that it is able to trap some of the warmth of the Earth's surface—the latter, heated by incoming solar radiation—as that warmth tries to radiate away back into space; see Figure 2.8. With abundant CO_2, the Earth would have possessed a strong greenhouse atmosphere able to counteract the faintness of the Sun 4 Byr ago and keep surface water liquid.

How much CO_2 would it take? At a conference in Erice, Sicily in 1987, Jim Kasting (then at NASA Ames Research Center) told a captivated audience that anywhere from half an atmosphere to 10 atmospheres worth of CO_2 (that is, ½ bar to 10 bars of CO_2) would have kept Earth's oceans liquid 4 Byr ago, despite the Sun's faintness. Note that the greenhouse-warmed surface was not necessarily a pleasant environment back then. For example, with 10 atmospheres worth of CO_2, surface temperatures on Earth would have approached 100°C, even with a faint Sun; the oceans would not have boiled away only because of the weight of that heavy CO_2 atmosphere on the ocean surface.

Such a thick CO_2 atmosphere is not necessarily inconsistent with the timeline presented earlier: as noted above, magma oceans were capable of supplying several bars or more of CO_2, H_2O, and other volatiles to the surface following massive impacts; and, outgassing from an extended 'big burp,' prevailing between massive late-accretion impacts, during the subsequent ~300-Myr interim, and throughout the Heavy Bombardment that followed, was likely prolific.

In comparison to other rocky planets, is such a large amount of carbon dioxide believable? Mars' second atmosphere, primarily CO_2 to this day, was probably once heavy enough

Figure 2.7 The Jack Hills in Western Australia—including, in particular, Erawondoo Hill, shown here—are famous as the location of the oldest Earthly material yet found, zircon crystals ranging in age between 4 and 4.4 Byr old. Those zircons are embedded in metamorphosed conglomerates 3 Byr old, and by themselves have also been interpreted as evidence for liquid water on Earth (Mojzsis et al., 2001). Image downloaded from https://iugs-geoheritage.org/geoheritage_sites/archean-zircons-of-erawondoo-hill / with permission of International Union of Geological Sciences.

(a)   (b)

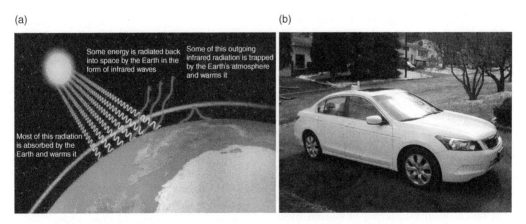

Figure 2.8 (a) Schematic illustration of the greenhouse effect. Radiation from the Sun (mostly at wavelengths of visible light) travels through the atmosphere and heats the Earth's surface. The Earth reradiates that heat—at lower frequencies and longer wavelengths, in the form of "black-body" infrared radiation—back into space; but at those wavelengths, some of the heat is prevented from escaping into space by water vapor, carbon dioxide, and other greenhouse gases in the atmosphere. Courtesy of EPA. (b) The author's car on a sunny late winter day. Despite a dusting of snow that morning, and air temperatures around freezing, the car's interior was about 30° F warmer—thanks to the greenhouse effect. *What do you think was causing the warmth within the car to be trapped? Hint: It was not anything in the air....*

(several bars, and thus three orders of magnitude greater than today, as implied by the evidence discussed in Chapter 1) to allow liquid water to exist at its surface, despite its great distance from the Sun—visible evidence, incidentally, of Mars' own ability to resolve the faint young Sun paradox! And, as noted above, Venus at present has an even heavier carbon dioxide atmosphere than we envision for the early Earth, with more than 90 bars of CO_2. Given such planetary realities, a better question might be *what happened to all that CO_2 on Earth?* Earth today has only trace amounts of CO_2. If Earth had such a heavy carbon dioxide atmosphere then, where has it all gone?

With the help of *water*, the answer turns out to be "into rocks," locked up as calcium carbonate (calcite) and other carbonate minerals, in what together is known as **the carbonate reservoir**. At present, there is enough carbon stored in continental sediments alone to produce 60–80 atmospheres of CO_2 (e.g., Kasting, 1993); the oceanic crust provides a comparable reservoir (—more than half as big; Sleep & Zahnle, 2001), and the mantle reservoir holds even more (Zhang & Zindler, 1993). Thus, the carbonate reservoir is easily big enough to have released into or taken from the atmosphere the 10 bars (at most) of CO_2 required to keep the early Earth warm.

2.2.3 The Urey Cycle

The geological processes by which carbon dioxide enters or leaves the carbonate reservoir constitute two sides of what is called the **Urey cycle** (Urey, 1952), also called the **carbonate-silicate cycle**. CO_2 is removed from the atmosphere and stored in the reservoir by *chemical weathering* of silicate rocks; heating of carbonate rocks by *tectonic activity*—metamorphism and magmatism, mainly at volcanoes and mid-ocean ridges—allows CO_2 to be vented back into the atmosphere. Chemical weathering of the land surface is possible because, in an atmosphere containing both water and carbon dioxide, some CO_2 will diffuse into water droplets and dissolve.

The dissolved carbon dioxide forms **carbonic acid**:

$$CO_2 + H_2O \rightarrow H_2CO_3,$$

making the raindrops acidic and promoting weathering. Even today, rain is naturally a bit acidic by this mechanism, though given the very low concentration of CO_2 in our atmosphere the resulting carbonic acid is weak; it is nevertheless possible for raindrops to gradually eat away and transform the crust as they strike the earth. In the early Earth, with atmospheric CO_2 abundant, rain would have been much more acidic and weathering by rainfall much more efficient.

In basic terms, the Urey cycle can be expressed as

$$\text{weathering} \rightleftharpoons \text{outgassing},$$

with CO_2 implicitly being taken from or released to the atmosphere by the left- or right-hand side. The symbol $\rightleftharpoons$ is *not* meant to imply that these opposing reactions balance each other. For a more explicit characterization of the Urey cycle, climatologists consider a representative continental crustal mineral known as wollastonite, whose chemical formula is $CaSiO_3$. In the presence of water, we have

$$CaSiO_3 + CO_2 + H_2O \underset{outgassing}{\overset{weathering}{\rightleftharpoons}} CaCO_3$$
$$+ SiO_2 + H_2O,$$

with $CO_2 + H_2O$ acting as carbonic acid to cause the weathering. As weathering of calcium silicates such as wollastonite proceeds, carbon dioxide is ultimately incorporated into calcium carbonate, $CaCO_3$, i.e., into the carbonate reservoir; the weathering also produces silica (quartz), SiO_2. Conversely, heating carbonate rocks frees carbon dioxide, which upon venting will end up in the atmosphere.

The Urey cycle can also operate when other types of minerals are weathered. For example, with $MgSiO_3$, a mineral known as enstatite, one possible reaction would be

$$MgSiO_3 + CO_2 + H_2O \underset{outgassing}{\overset{weathering}{\rightleftharpoons}} MgCO_3$$
$$+ SiO_2 + H_2O.$$

In the case of magnesium silicates, the mineral locking up carbonate by weathering may be $MgCO_3$, magnesite, as shown here, or—with more involved reactions based on more complex substances—minerals such as dolomite $(CaMg(CO_3)_2)$.

In general, the carbonate products of chemical weathering tend to remain in solution, carried by rivers or groundwater ultimately to the ocean where, under the right conditions, they can precipitate, forming sediments that will then settle onto the shallow seafloor. Through most of the Phanerozoic, a second, biological pathway into the worldwide carbonate reservoir has also existed: forams, coccoliths, and other organisms extract carbonate from seawater to build their shells, and those shells sink to the seafloor after the organisms die. Either way, abiotically or biologically, the "weathering" side of the Urey cycle thus actually includes both chemical weathering—for example, according to

$$CaSiO_3 + 2\,H_2CO_3 \longrightarrow Ca^{+2} + 2HCO_3^-$$
$$+ \, SiO_2 + H_2O,$$

showing the action of carbonic acid to disintegrate the silicate mineral and supply various cations and carbonates to the oceans—and carbonate precipitation, given by

$$Ca^{+2} + 2HCO_3^- \longrightarrow CaCO_3 + H_2CO_3$$

(see Kump et al., 2010). The net result (sum) of these two processes is the 'forward direction' weathering component of the Urey cycle equation written above.

With atmospheric carbon dioxide dissolved into the oceans as well, such weathering reactions also occur at the seafloor (Alt & Teagle, 1999). Additionally, the upper, more permeable layers of oceanic crust can become carbonatized through hydrothermal alteration: reactions with basalt (oceanic lithosphere is mainly basalt) produce calcite and other carbonate minerals (e.g., Nakamura & Kato, 2004). At present, the combination of weathering of the seafloor and alteration of oceanic crust is sufficient to lock up all of the carbon dioxide

degassed at mid-ocean ridges plus remove a fraction of the originally atmospheric CO_2.

Finally, as plate motions bring the oceanic lithosphere to a subduction zone, and the lithosphere begins its descent into the mantle, heating of the carbonatized crust and any carbonate sediments not scraped off onto the adjacent overriding plate will produce CO_2 that can be outgassed back into the atmosphere by the regional volcanism. However, some of the carbonate will continue down into Earth's deep interior, in effect removing that carbon dioxide from the Urey cycle for hundreds of millions of years (e.g., Sleep & Zahnle, 2001).

In the early Earth, when the mantle was actively convecting but before much continental crust had accreted, weathering and carbonatization of oceanic crust provided an effective way to lock up carbon dioxide dissolved in the seawater (Nakamura & Kato, 2004). And, during the Heavy Bombardment (which did not end until ~3.8 Byr ago), an additional sink for CO_2 would have been created by the pulverized and glassy products of the impacts: such *impact ejecta* are easily weathered, and could potentially have drawn down levels of atmospheric carbon dioxide significantly during the Hadean and earliest Archaean (Koster van Groos, 1988; see also Sleep & Zahnle, 2001).

It might be noted that weathering of continental crust (i.e., of its surface) was significantly enhanced beginning ~400 Myr ago by the development and proliferation of deep-rooted land plants, with "dramatic" effects (Berner, 1998) on CO_2 levels (see also, e.g., Taylor et al., 2009; Ibarra et al., 2019, and references therein).

Implications of the Urey Cycle. The Urey cycle—in a sense, the climatological equivalent of a law like "F = Ma" or Newton's Law of Gravitation in solid-earth geophysics—has profound implications for the long-term behavior and evolution of planetary atmospheres. For example, even in the era of the faint young Sun, Venus was too close to the Sun to be able to retain much water vapor in its atmosphere, and (as noted previously) would have dried out not long after the Heavy Bombardment had ended (Turbet et al., 2021). Without water, the weathering component of the Urey cycle could not operate, so from that point on, little of the CO_2 convectively outgassed from the interior of Venus could have been removed from its atmosphere. The result would be a *runaway greenhouse atmosphere*, with surface temperatures of hundreds of degrees centigrade; though the interior of Venus is similar enough to Earth's that we think of it as our sister planet, the surface of Venus early on began to follow a radically different history.

On Earth, the Urey cycle would not prevent extreme climates; but it could bring about the conditions for those extremes to, eventually, end. For example, eras dominated by icy climates and precipitation that was mostly frozen would be nearly as powerless as a waterless world to remove CO_2 from the atmosphere by weathering. During such eras, levels of carbon dioxide that had presumably been low (allowing glacial climates to develop in the first place) could eventually be boosted by tectonic outgassing, amplifying the greenhouse effect and warming the surface. That is, as long as Earth is tectonically active, the Urey cycle ensures that frigid climates cannot persist.

Throughout Earth's history, such eras have not lasted more than a couple of hundred million years. Do you think the similarity of this time span to the convective timescale noted earlier is coincidental or causal?

Astronomers searching for evidence of life, or of planets capable of supporting life, talk about the **habitable zone** (informally, the "Goldilocks zone"), which as stated earlier is the range of distances from a star within which temperatures on a planet's surface are moderate enough to allow liquid water to exist, and thus have the potential for life. The width of a planet's habitable zone depends on the specific composition assumed to characterize the planet's atmosphere. For terrestrial planets, the dependence on the Urey cycle

is so fundamental—continued habitability is unlikely without both outgassing and weathering—that predictions of their habitable zones generally start with a planetary atmosphere containing carbon dioxide, potentially in abundant amounts, as well as water. Investigations of these and other factors affecting the extent of a star's habitable zone have been carried out by, for example, Kasting et al. (1993) and Kopparapu et al. (2013) for "wet" planets like Earth and Abe et al. (2011) for 'mostly dry' planets.

Thanks to the Urey cycle, the early Earth was evidently able to create its own habitable environment, and life not only developed here but flourished as well—despite the Sun being so faint for the first 2.5 Byr following its formation that Earth would otherwise (e.g. with current levels of CO_2) have remained *outside* the faint Sun's habitable zone (see Figure 2.6b)! The story for Mars 3–4 Byr ago is less clear: for sustained habitability during that period, additional greenhouse gases (beyond a thick CO_2 atmosphere) would need to have been outgassed (e.g., Kasting, 1991; Ramirez et al., 2013); if habitability was episodic, though, meteorite impacts (generating a steam atmosphere), as well as intermittent volcanism, might have produced the transient warmth (cf. Wordsworth, 2016).

Given current conditions on Earth, weathering is very effective; for example, current rates of weathering could remove even 80 bars of carbon dioxide in only several hundred million years (Sleep & Zahnle, 2001). The 10 bars or less of CO_2 that kept the Earth warm 4 Byr ago could have disappeared in a geologic 'instant,' completely absorbed into the carbonate reservoir through weathering. But it took the Sun 4 Byr to reach its present luminosity; without ongoing outgassing, to replenish the atmosphere's CO_2, the oceans would have frozen—even as recently as 2 Byr ago. In short, biological life on Earth could never have developed without the Earth being *geologically* alive; and once life had developed, its survival has depended almost entirely (with one exception noted below) on such geological activity.

Weathering and plate tectonics (or mantle convection) depend on fundamentally different factors: the former is a function of atmospheric composition and temperature, whereas the latter reflects both the amount of heat trying to escape the interior and the ability of the mantle to flow (i.e., its rheology); so, it is not likely that weathering and outgassing activity ever balanced each other. According to the Urey cycle, then, we can anticipate that there would have been times in Earth's history when weathering prevailed, and little CO_2 remained in the atmosphere—resulting in **Ice House** climates, climates cold enough for ice to persist on Earth's surface; and there would have been times, for example when mantle convection was vigorous, during which outgassing dominated and CO_2 was abundant—times of **Hot House** climates.

So, how amazing is the Urey cycle?—a concept that applies 'at a point' on Earth's surface (where either weathering or outgassing is taking place, depending on the direction of the equation), but with profound global implications for the entire history of Earth's climate system!

The Urey Cycle Has Continued Influencing Climate to the Present. Earth's climate record during the past ~130 Myr provides dramatic illustrations of the climatological power of the Urey cycle.

The **mid-Cretaceous Hot House** lasted from ~130 Myr ago (especially ~ 120 Myr ago) till ~80 Myr ago and is widely believed to be one of the warmest climates of the past 500 Myr. Evidence includes the presence of turtles, dinosaurs, and other warm-adapted animals, as well as tropical vegetation, all north of the Arctic Circle or south of the Antarctic Circle (e.g., Barron, 1983; Vickers-Rich & Rich, 1993; Markwick, 1998; also Taylor & Taylor, 1990 and references in Rich et al., 1988). Temperatures in the tropics may have been ~10°C warmer than now in the late Cretaceous (see Pearson et al., 2001), and at least 20°C warmer in polar regions.

Such warmth is attributed mainly to increased carbon dioxide levels in the atmosphere from prolific outgassing. Mantle convection was quite vigorous, with faster plate velocities and massive volcanic activity, in large part associated with a *superplume* that overspread the Pacific seafloor (Larson, 1991; see also Courtillot & Olson, 2007) ~120 Myr ago. The volcanism created "large igneous provinces"—mostly continental flood basalts (nicknamed "traps"), but also including oceanic plateaus.

After the Cretaceous Hot House ended, there were still periods of extreme warmth, including a brief 'spike' of warmth 55 Myr ago known as the **Paleocene-Eocene Thermal Maximum** (**PETM**) and a more extended period of warmth, the **Early Eocene Climate Optimum** (**EECO**) ~52–50 Myr ago. The Cretaceous might have ended; but there was no indication that Earth's climate was about to change direction and begin its descent into the Ice House regime that would prevail to this day.

What initiated the climate reversal was the collision of a rapidly northward-moving Indo-Australian plate with the Eurasian plate, beginning ~50 Myr ago. The collision slowed plate motions globally and reduced convective outgassing. And, as the Tibetan Plateau and Himalayas began to be uplifted by the collision, a vast amount of 'fresh' continental crust would be exposed to weathering that could effectively lock up existing atmospheric CO_2; the weathering may have been enhanced by the uplift's changes in Asian wind patterns and resulting intensification of regional monsoons (Raymo et al., 1988; Ruddiman & Kutzbach, 1991).

The climate reversal was a major achievement of the Urey cycle. Significantly, however, though climatic cooling continued through the Cenozoic, Ice House levels were not reached until additional factors kicked in ~15 Myr later; those factors acted to change ocean circulation patterns such that the transport and thus distribution of heat within the climate system were measurably altered. The role of the oceans in climate will be discussed in Chapter 3.

Limitations of the Urey Cycle. The Urey cycle provides a very useful framework for understanding variations in climate on million-year timescales. However, as just noted, other factors besides outgassing and weathering are also capable of modifying Earth's climate. In addition to changes in atmospheric or oceanic circulation, which would affect the transport and distribution of heat within the climate system, changes in the reflectivity or "albedo" of Earth's surface, for example, can alter the amount of heat from the Sun that the climate system retains (either globally or regionally). These factors, which will be discussed in Chapter 3, mainly have had the effect of accelerating or delaying an impending Ice House; but they do not eliminate the fundamental weathering/outgassing imbalances that result in an Ice House or Hot House regime.

The Urey cycle does not account for other greenhouse gases that might have either maintained a Hot House climate even when carbon dioxide levels were low, or added to already high CO_2 levels. In particular, methane (CH_4)—a potent greenhouse gas—has been proposed as a critical supplement to CO_2 during the Archaean, when the Sun was still quite faint (e.g., Kiehl & Dickinson, 1987; Rye et al., 1995; Pavlov et al., 2000; Kasting & Siefert, 2002; and Haqq-Misra et al., 2008). More complex hydrocarbons such as ethane (C_2H_6), the product of photochemical reactions involving methane, may also have contributed to the overall greenhouse effect during the Archaean (Haqq-Misra et al., 2008). Modest levels of abiotic CH_4 would have been available from serpentinization of oceanic lithosphere (see, e.g., Schrenk et al., 2013); eventually, substantially greater amounts of biogenic CH_4 could have been provided by methanogenic bacteria, as life evolved in a low-oxygen atmosphere. Some researchers (e.g., Zachos et al., 2005 based on Dickens et al., 1995, 1997; see also Pagani et al., 2006) have also suggested an important role for methane as the cause of the PETM spike ~55 Myr ago. The question of CO_2 levels, which underlies any potential need for other greenhouse gases, will be explored in the next section.

2.3 Constraints on the Evolution of Atmospheric CO_2

2.3.1 Levels of CO_2 Were (Relatively) Low During Ice House Climates

The occurrence of an Ice House climate places a limit on the intensity of the greenhouse effect operating at that time, and thus implies an upper bound to the amount of greenhouse gas able to warm the Earth contemporaneously. If the greenhouse gases active during that climate regime include more than CO_2, the implied upper bound is still valid; the actual level of CO_2 in that case would be even less. In contrast, if more than one greenhouse gas is contributing significantly to a Hot House climate, the existence of the Hot House cannot by itself constrain the abundance of any one of those gases. Ice House climates thus provide a more useful, less ambiguous, constraint on CO_2 than do Hot House climates.

Earth has experienced several eras of Ice House climate that were severe enough to leave a sustained, widespread mark on the geologic record. The major Ice Ages that occurred within those periods left all the kinds of evidence of *glacial advance and retreat* that we are familiar with from the Pleistocene: striations and scouring, erratics, moraines, drumlins, and so on (see Figure 2.9).

An Episodic History. There is evidence that Earth experienced a period of glaciation—the Pongola Ice House—around 2.9 Byr ago (Young et al., 1998; also Kirschvink & Kopp, 2008); glaciation may even have occurred as far back as 3½ Byr ago (de Wit & Furnes, 2016). The most heavily studied of the early Ice Houses, called the *Huronian*, extended from ~2.4 Byr ago to 2.2 Byr ago and included three Ice Ages, whose glacial advances covered parts of what are now North America, Africa, and Australia. As we will see later in this chapter, the Huronian would not have come

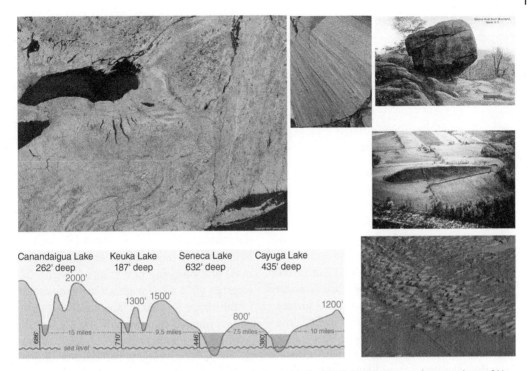

Figure 2.9 Examples of evidence showing that a region was glaciated, in this case various portions of New York State (satellite image, outlined in yellow, bounded to the northwest by Lake Ontario and the tip of Lake Erie; note also the Finger Lakes to the south of Lake Ontario).

Evidence (clockwise from top center) includes *striated rock surfaces* (in this case, the rock is known as Vroman's Nose, located in central New York State), with scratches produced by the overlying ice flow; *erratics*, rocks anomalous to their surroundings having been brought by the moving ice sheet (in this case, the gneiss known as Balance Rock is thought—see Hays (2020)—to have been carried ~25 miles south along the Hudson River Valley); *drumlins*, created beneath the ice sheets as they moved (in these images of an individual drumlin and a drumlin field south of Lake Ontario, north is on the left and ice sheets flowed to the south-southeast); and *U-shaped valleys*, valleys with steep sidewalls gouged out as the ice sheets flowed (shown by the west-east profile of the Finger Lakes, U-shaped valleys which became actual lakes when they were dammed by glacial debris to the south).

We also see the *terminal moraine* known as "Long Island," at the southern tip of the state, the debris (!) left by the southernmost edge of the ice sheet after its advance had terminated.

The examples shown here were all created during (or at the end of) the Pleistocene Ice Age, which began ~2½ Myr ago.

Image credits are Yarr65 / Adobe Stock Images; Walter Siegmund/Public domain; mindat.org / https://www.mindat.org/photo-370945.html / last accessed August 22, 2023; New York State Museum / http://www.nysm.nysed.gov/research-collections/geology/research/subglacial-landforms-and-processes / last accessed August 22, 2023; Cornell University / http://www.geo.cornell.edu/geology/faculty/RWA/photos/finger_lakes_topo_images / last accessed August 22, 2023.

about without a significant change in atmospheric chemistry that allowed high levels of methane—likely the product of bacterial life flourishing in the mid-Archaean—to have been eliminated.

More recent Ice House regimes include the Cryogenian (discussed below), which

occurred near the end of the Precambrian; a late Ordovician Ice House, which has been linked to a major extinction event (see Harper et al., 2014); an extended late Paleozoic (Carboniferous-Permian) Ice House spanning at least 80 Myr; and our current late Cenozoic Ice House, which began roughly 34 Myr ago with the appearance of ice in Antarctica, and which includes the Pleistocene Ice Age, beginning nearly 2.6 Myr ago. Specific dates and further aspects of the older Ice Houses are discussed by Catling and Kasting (2017).

It is hard to avoid seeing a connection between Ice Houses and the end of geologic eras. What reasons can you give for such a connection?

Constraints on CO_2 From Ice Houses. In general, the temperature of Earth's troposphere (the lowest layer of the atmosphere) is a function of both the amount of incoming solar radiation *and* the amount of outgoing heat trapped by greenhouse gases. At present—during our current Ice House—the Earth's average surface temperature is about 15°C. If we require that the Earth's surface temperature averaged globally during the Huronian should be roughly in the range 5°C–20°C—reflecting freezing temperatures from the poles down to mid-latitudes but more moderate conditions in the tropics—and if we also know the Sun's luminosity during those times (e.g., Figure 2.6), then we can calculate the amount of CO_2 consistent with those temperatures. At that 1987 conference, Jim Kasting presented calculations showing that during the Huronian glaciations, there must have been between 0.05 and 1 bar of CO_2 (but see below).

The Cryogenian (~720–635 Myr ago) featured major Ice Ages 720–700 and 640–635 Myr ago; a subsequent glaciation took place 585–580 Myr ago. Because Australia was among the land areas experiencing those Ice Ages, and Australia was located close to the equator during this period, some researchers have concluded that the entire globe must have been glaciated—a series of **Snowball Earth** episodes (Kirschvink, 1992; Hoffman et al., 1998)! In contrast, during the glacial maximum of the Pleistocene around 20,000 yr ago, ice sheets covered the Northern hemisphere only down to mid-latitudes. In a Snowball Earth, it would be difficult for life to survive anywhere (but see also Hoffman et al., 2017); given the apparent continuity of life through the Precambrian, later versions of this

hypothesis instead allowed for thin ice in the tropics, or even an ice-free band of tropical ocean (—a "waterbelt" Earth, including one version nicknamed the "Jormungand state" by Abbot et al., 2011). In some models of the Snowball Earth, outgassing ended each episode with a brief rebound into an extreme Hot House, which in turn lasted only until weathering could bring CO_2 levels back down to moderate amounts.

The amount of carbon dioxide in the atmosphere during this second major Ice House period can again be estimated by requiring that the combination of incoming solar radiation (with the Sun now almost at current brightness) and outgoing heat trapped by the CO_2 together yield an average surface temperature in the range 5°C–20°C. Kasting's 1987 calculations implied that, around 700 My ago, the level of CO_2 was between 0.005 and 0.03 bars. Given the possibility of Snowball conditions, the low end of these ranges—a mean surface temperature of ~5°C and a corresponding level of ~0.005 bars CO_2—may be preferred.

Why Has CO_2 Trended *Downward*? Over the long term, then, levels of carbon dioxide have dropped dramatically, by 1 to 3 orders of magnitude from 4 Byr ago to 700 Myr ago. Earth's atmosphere at present contains just 0.0004 bars of carbon dioxide—another order of magnitude lower from 700 Myr ago till now, and so low that, to map present-day variations in CO_2, we use units of *parts per million* (0.0004 bar out of 1 bar = 400 ppm). This overall trend toward lower levels of CO_2 in the Earth's atmosphere, revealed by successive Ice Houses, can be attributed to at least two factors. First, as the time of accretion and core formation receded into the distant past, with mantle convection succeeding in getting much of that early heat out of the Earth, mantle convection and plate tectonics would be expected to slow down, with a resulting decrease in outgassing. Second, as plate tectonics progressed, more continental crust would be created; unlike oceanic lithosphere, which is thin enough to bend and dense enough to (eventually) subduct along with the convecting mantle, continental lithosphere can only accumulate over time. The increase in land area would expose more crust to subaerial weathering, allowing more CO_2 to be locked up in the carbonate reservoir.

But those two factors are not the whole story. First of all, they are only overall trends, whereas the actual rate of outgassing and the accumulation and weathering of continental crust have varied through time. As noted earlier in this chapter, for instance, the spread of land plants beginning ~400 Myr ago accelerated the weathering of continental crust, significantly enhancing the trend toward lower atmospheric CO_2 for the next ~100 Myr. And examples of extreme variations that 'bucked the trend' were also noted earlier: extreme volcanism ~100 Myr ago in the mid-Cretaceous, which led to a major Hot House climate; and extreme weathering during the early Cenozoic when, following the Cretaceous Hot House, trends toward even hotter climates were turned around and transformed, beginning ~50 Myr ago, into a climatic cooling trend that brought us into our current Ice House ~35 Myr ago.

The variations in Earth's past climate, including all these extreme climate regimes, have happened within the context of a brightening Sun; but Earth's climate history has not been dominated by that increase in solar luminosity—consider, as proof, the present-day Ice House! Outgassing and weathering have competed through time—unevenly—to affect the abundance of CO_2; and Hot House and Ice House climates have resulted. But when they were over, the levels of CO_2 had drawn down or risen up to what, given the Sun's luminosity at that time, produced an intermediate, moderate climate; and the next extreme developed in that context. That is, the level of CO_2 has dropped over time, overall, *adjusting to an increasing solar luminosity,* so that variations in CO_2 *relative* to that drop are still able to produce either hot or cold climate extremes.

How is this possible? Earth's climate 'engine' may be driven, ultimately, by the Sun's heat energy, and the abundance of greenhouse gases may determine whether that engine operates in a hot or cold state; but *it is water—through its temperature-dependent weathering ability—that regulates the climate system and keeps those variations 'centered.'*

2.3.2 Other Approaches, and a Synthesis

Four data points may be enough to infer a trend in carbon dioxide abundance; but, spanning the past 4 Byr, that leaves a lot of gaps. We can use the results of other investigations following that conference in Erice (see, e.g., references in Kanzaki & Murakami, 2015) to fill in some of those gaps; the graph of CO_2 versus time shown in Figure 2.10 represents a synthesis of several such results.

For the Precambrian, Figure 2.10 includes a variety of estimates involving **paleosols**, which are defined as preserved ancient soils:

- a lower bound on CO_2 3.2 Byr ago (Hessler et al., 2004; see also Sheldon, 2006) based on the mineralogy of weathered paleosols;

- range estimates at 2.5 Byr (Sheldon, 2006) and 2.69 Byr (Driese et al., 2011) based on a mass balance involving those paleosols; and

- several range estimates between 2.77 Byr and 1.85 Byr (Kanzaki & Murakami, 2015) based on chemical interactions of pore waters with those paleosols.

Within the Phanerozoic, Figure 2.10 includes:

- the results of Berner (2006, his Fig. 18) based essentially on Urey-type modeling of CO_2 (see Berner & Kothavala, 2001), reproduced here for the past 530 Myr (with single estimates constructed from an average of his values including and excluding weathering of volcanic rocks, and some range estimates based on that inclusion or exclusion);

- highly precise estimates from Franks et al. (2014, their Fig. 4) over the past ~400 Myr based on models of gas exchange in fossil leaf samples;

(a)

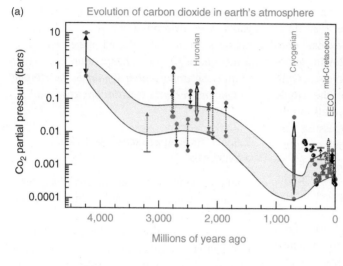

Evolution of carbon dioxide in earth's atmosphere

(b)

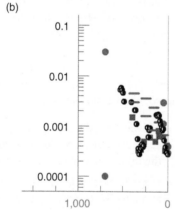

Figure 2.10 Selected estimates of the abundance of carbon dioxide in Earth's atmosphere. CO_2 partial pressure (pCO_2) is plotted on a log scale.
(a) Ranges of pressure are depicted with arrows: double-width arrows if based on known Ice House or Hot House climates at that time (and so labeled); single-width arrows if based on avoiding a frozen Earth; and dotted arrows if estimates (as described in the text) were made without reference to ambient climate. Range maximum is denoted by a red-filled circle or line symbol; range minimum is indicated in blue. The line symbol with a one-sided arrow 'sitting' on it represents a lower bound.
In the Phanerozoic, half-filled circles correspond to estimates by Berner (2006); dark brown squares correspond to estimates by Franks et al. (2014), as described in the text.
Superimposed on these estimates is the author's own completely subjective interpretation of the overall trend in CO_2 through time (gray band).
(b) Close-up of the data from the past 1 Byr, with arrows excluded for clarity.

- a lower bound on CO_2 during the Cretaceous Hot House based on global climate modeling (as discussed in Ruddiman, 2008);

- range estimates for 50 Myr ago (Lowenstein & Demicco, 2006) and, in effect, a strong lower bound on CO_2 at that time (Jagniecki et al., 2015), both results based on the stability field of a rare sodium carbonate mineral (note also the range estimates are consistent with Thrasher & Sloan, 2009); and

- present-day levels of CO_2, shown both as a minimum (preindustrial levels just under 300 ppm) and the value of 400 ppm achieved in the year 2013.

An alternative, systematic set of Phanerozoic estimates of CO_2 is provided by Beerling and Royer (2011).

Finally, Figure 2.10 also incorporates estimates (taken from Kasting, 1993; also 1987) of the range of possible CO_2 during Precambrian Ice House regimes; these are slightly lower bounds (roughly 0.02–0.30 bar for the Huronian and 0.0001–0.0300 bar for the Cryogenian) than those discussed above.

Despite the modeling and measurement uncertainties, this figure reinforces our earlier conclusion that the evolution of carbon dioxide was dominated by a trend toward decreasing levels of CO_2. Between the data points and projections, it also allows for the possibility of shorter-term variations, superimposed on that trend. The amplitude range of such variations may appear to remain high throughout Earth's past and to the present; but accounting for the log scale used in Figure 2.10 for pCO_2, all those superimposed variations probably trend toward decreasing amplitude, like the level of CO_2 itself.

2.4 The Development of an Oxygen Atmosphere

Earth's atmosphere today, excluding water vapor, is primarily nitrogen (78%) and oxygen (21%), with nearly 1% argon and trace amounts of everything else. Water vapor exists in variable amounts, typically up to ~4% in the tropics and less than 0.1% near the poles (and, e.g., ~1% in the mid-USA); but the actual amounts depend on regional factors and the season. Nitrogen and argon had built up over time through volcanic outgassing and other venting. But geologic processes are not likely to produce much free oxygen; instead, our oxygen atmosphere can be attributed to the biosphere: it is a by-product of photosynthesis! Photosynthetic activity by primitive bacteria—mainly cyanobacteria, commonly known as "blue-green algae," and their ancestors (see Fischer et al., 2016)—created an oxygen atmosphere, generation by generation, over a huge stretch of time; today, that atmosphere is maintained photosynthetically by all plant life, though cyanobacteria continue to play a significant role (e.g., Zimorski et al., 2019; Kasting & Siefert, 2002).

Photosynthesis can be summarized by the reaction

$$6CO_2 + 6H_2O + \text{sunlight} \rightarrow C_6H_{12}O_6 + 6O_2,$$

where the high-energy end product desired by the photosynthesizing organism is represented here by glucose, $C_6H_{12}O_6$. The actual, step-by-step reactions comprising photosynthesis in most cases take place with the help of chlorophyll. Sunlight provides the energy to power the reactions. Note, incidentally, the explicit importance of water to promote these reactions; life as we know it does require water.

An O_2-Rich Atmosphere Was *Not* Inevitable. The history of our oxygen atmosphere

begins *after*, rather than *when*, the first life evolved. After, because the earliest life may well have been chemosynthetic, relying on energy-rich chemicals rather than sunlight to create high-energy products. An example of chemosynthetic reactions might have been

$$12H_2S + 6CO_2 \rightarrow C_6H_{12}O_6 + 6H_2O + 12S$$

(Jørgensen et al., 1979, following Cohen et al., 1975, and Garlick et al., 1977; see also Pfennig & Widdel, 1982; Jannasch & Mottl, 1985; and Kelly & Wood, 2006; and note that such reactions are often written more generically in terms of $n[CH_2O]$ rather than $C_6H_{12}O_6$). Chemosynthetic organisms would have been able to survive harsh environments lacking sunlight, such as the dense, cloud-filled greenhouse atmosphere envisioned for the early Earth; hydrothermal vents at the seafloor near mid-ocean ridges (where thriving chemosynthetic microbial communities exist today); and even (for rock-dwelling bacteria) water-saturated pore spaces tens to hundreds of meters below the surface. But chemosynthesis did not create O_2 as a by-product.

The history of our oxygen atmosphere also begins *after*, rather than *when*, the first photosynthesizing bacteria evolved. After, because those bacteria may have been "anoxygenic," employing photosynthesis but not producing oxygen (e.g., Xiong, 2006); some species may even have consumed oxygen.

Furthermore, because oxygen is very reactive, it is likely that much of the earliest O_2 produced by oxygenic photosynthesis was consumed by chemical reactions with crustal minerals or atmospheric gases. At some point, the decay of organic matter by microbes would also become a significant "sink" of oxygen. In fact, one approach (e.g., Farquhar et al., 2011; Kasting, 2013) to the question of how atmospheric oxygen evolved, once oxygenic photosynthesis had begun, is to focus on the diverse sinks, and how they changed over time (e.g., geologically). The bottom line is, the development of an oxygen atmosphere was unavoidably a very slow, tentative process; for most of Earth's life, O_2 was at most a minor constituent.

Oxygenic photosynthesis did begin at some point in the Archaean, judging by the high organic content of Archaean shales (Lyons et al., 2014); based on an analysis of chromium isotopes by Crowe et al. (2013), discussed briefly below, and an analysis of molybdenum isotopes by Planavsky, Asael, et al. (2014), we can say that it was already underway by 3 Byr ago (see also Laakso & Schrag, 2017; Riding et al., 2014 and references therein).

In contrast to oxygen, the abundance of methane would have steadily increased in the Archaean atmosphere—from both abiotic and biogenic sources, as noted earlier. In a sense, the end of the Archaean was the battleground for dominance by methane versus oxygen, with the struggle embodied in the reaction

$$CH_4 + 2O_2 \rightarrow CO_2 + 2H_2O$$

(the destruction of methane would actually have been a more complex process, with multiple pathways, but this equation symbolizes some of its key aspects; Pavlov et al., 2001, Appendix A). As long as methanogenic bacteria proliferated and methane was abundant, any oxygen added to the atmosphere could be wiped out by reaction with methane; once oxygen had the upper hand, methane could no longer persist in abundance.

2.4.1 A Mostly Geology-Based Chronology of the Rise of Oxygen on Earth

Preliminaries. A rough chronology of the rise of an O_2 atmosphere can be deduced from various types of geologic evidence (e.g., Farquhar et al., 2011), as well as some biological considerations; it must be noted, however, that their interpretation is somewhat controversial at almost every turn (see, e.g., Lyons et al., 2014; Javaux & Lepot, 2018; and references therein). Indeed, the same geologic evidence has served as the basis for distinctly different scenarios than what is presented here. In the end, perhaps the best we can do is to use those

conflicting estimates to infer 'generous' upper and lower bounds on the abundance of oxygen in any given era.

With limited land on the early Earth, and the need for water, photosynthetic life might have developed in shallow pools but more likely would have flourished in the shallow ocean. Photosynthesis would have supplied oxygen first to the shallower regions of the oceans; as concentrations built up there, oxygen could then effuse into the atmosphere and, subject to ocean circulation patterns (discussed in the next chapter), penetrate into the deeper oceans. Thus, our chronology should consider three oxygen reservoirs—the shallow and deep oceans as well as the atmosphere. This three-component approach has also been taken for theoretical, biogeochemical models of oxygen's evolution (e.g., Laakso & Schrag, 2017). With this perspective in mind, the following scenario highlights four qualitatively distinct stages of development.

First Stage: The Earliest Life. Life appears to have developed on Earth *prior* to ~3.5 Byr ago, in the earliest part of the Archaean. Meta-sedimentary rocks about 3.8 Byr old from southwest Greenland that contain organic carbon 'smudges' are considered by some to represent the oldest evidence for life, though such a conclusion has been disputed; the controversy is nicely summarized in Wacey et al. (2008).

The oldest microbial *fossils* ever found include some from western Australia nearly 3.5 Byr old (Dunlop et al., 1978; Schopf & Packer, 1987); see Figure 2.11. Several of these (one is shown on the left of the figure, accompanied by an artistic reconstruction to suggest its interpretation) appear to be chains of photosynthesizing bacteria. Such organisms would be complex enough to imply that simpler forms of life existed at still earlier times; but their biological origin has also been questioned (Marshall et al., 2011).

More unequivocal are microfossils of chemosynthetic bacteria found not far away in the 3.4-Byr-old Strelley Pool Formation of western Australia (e.g., Wacey et al., 2010,

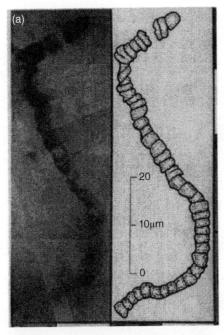
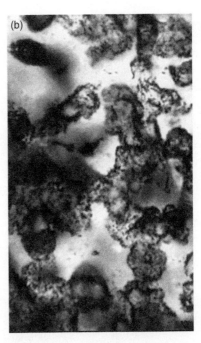

Figure 2.11 Some of the oldest microfossils ever found.
(a) Possible fossil from Apex Chert formation in western Australia, 3.485 Byr old; this photo, accompanied by an interpretive drawing, suggests a similarity to a chain of photosynthesizing cyanobacteria. Image from Schopf (1993) / American Association for the Advancement of Science. As noted in the text, however, more recent work casts doubt on the biological origin of this microstructure.
(b) Fossil chemosynthetic (sulfur-metabolizing) bacteria from Strelley Pool sandstone in western Australia, 3.4 Byr old. Wacey et al. (2011) / Springer Nature.

2011); see Figure 2.11b. These fossils reveal cell structures suggestive of colonies and chains, and include metabolic products (including iron pyrite, FeS_2) consistent with sulfate-reducing activity. Isotopic analysis of those products implies that the shallow-water coastal environment of these microbes was low in oxygen; if not an isolated occurrence, such low levels of oxygen would indicate that the atmosphere at that time was depleted in oxygen as well, at most perhaps a million times less than current levels (Wacey et al., 2010).

This region of western Australia, the Pilbara craton, also features a number of unusual mounds interpreted to be fossil stromatolite mats built by microorganisms (e.g., Walter et al., 1980; Allwood et al., 2006), probably sulfur-metabolizing (e.g., Bontognali et al., 2012); these are also 3.4–3.5 Byr old (see Figure 2.12). Although modern stromatolites are generally photosynthesizing, and

these Archaean mounds were shallow marine structures, it is unclear whether they were oxygenically photosynthesizing (see, e.g., Bosak et al., 2013).

Microbial fossils and stromatolites dating back almost 3.5 Byr are also found in the Kaapvaal craton of South Africa (in the section known as Barberton Mountain Land), with origins similarly inferred to be of a shallow-water environment (e.g., Muir & Hall, 1974; Knoll & Barghoorn, 1977; Walsh & Lowe, 1985; Tice & Lowe, 2004) or—for the slightly younger, 3.2-Byr-old sandstone there—even a tidal flats environment (Noffke et al., 2006).

The ancient environments in which these early organisms flourished were likely coastal or shallow-sea, not deep-ocean or deep within the crust. If life actually originated at the seafloor or within the crust, that origin would have had to take place even earlier, in order to leave enough time for organisms to evolve

Figure 2.12 Fossil structures suggesting the existence of microbes.
(a) 3.49-Byr-old "wrinkly mat" stromatolites (within the dashed green oval) from the Dresser Formation of East Pilbara, Australia. In modern stromatolites, the mat's top layer is home to cyanobacteria or other photosynthetic prokaryotes. Martin J. Van Kranendonk.
(b) 3.43-Byr-old conical stromatolites from the Strelley Pool Chert in western Australia. The white bar at the lower right is 10 cm in length. Awramik 2006 / Springer Nature.

and adapt to surface conditions. If life began as single cells, time would also be required for evolution into colonies. In short, life probably dawned on Earth soon after the Heavy Bombardment of the Hadean eon ended—in effect, almost as soon as conditions allowed!

Photosynthesis may or may not have begun by ~3½ Byr ago. In either case, however, at that time the level of oxygen in the oceans and atmosphere would have been quite low, as noted above. Higher levels have been inferred to prevail a half billion years later (Crowe et al., 2013)—but those levels were *still* significantly lower than present-day abundances. The inferences by Crowe et al. were based on an analysis of chromium isotopes in paleosols. In its oxidized state (Cr^{+6} rather than Cr^{+3}), chromium is soluble; subjected to weathering in an oxidative environment, those paleosols would become preferentially depleted (by runoff) in the heavier isotopes of chromium. Sedimentary deposits from such runoff would show the opposite, an enrichment of heavier chromium isotopes. Crowe et al. concluded from paleosols 3 Byr old, and contemporaneous sedimentary formations in the same region, that oxygen levels in the atmosphere probably reached more than ~0.6×10^{-4} bar at that time.

Over the eons, photosynthesizing bacteria would proliferate around the globe, and each microbial generation added incremental amounts of oxygen to their surroundings. But, given the high chemical reactivity of oxygen—so that once available, it would react with almost everything—the level of O_2 struggled to increase, and at times could even have fallen. At the same time, scattered "oxygen oases" (see Cloud, 1965; Olson et al., 2013) may have developed in the shallow ocean; in fact, there is evidence for one 3 Byr ago located in a coastal environment within South Africa's Pongola Basin (Planavsky et al., 2014a and Eickmann et al., 2018 following Crowe et al., 2013), and another at the edge of the Canadian Shield dating to ~2.8 Byr ago (Riding et al., 2014). In such oases, some small abundance of oxygen would be dissolved in the shallow waters, maintained by a corresponding level of O_2 in the overlying atmosphere (the way a half-consumed, capped bottle of soda maintains some carbonation with the help of the 'air' within the bottle, though more dynamically, as some exchange with the oceans and atmosphere outside the oasis will occur). Riding et al. estimated that oxygen above their oasis could have been up to ~0.012 bar, that is, up to 200 times more abundant than in the global Archaean atmosphere.

Even so, as we discuss next, it would take another half billion years for oxygen to pervade the seas and the atmosphere above on a *global* scale, and finally reach high enough levels for it to have a measurable—although not yet permanent, according to some—impact on the Earth System.

Second Stage: Global Oxygenation. By a couple of hundred million years into the Proterozoic, ~2.4–2.2 Byr ago, according to a wide range of evidence, oxygen had become a significant, though still minor, constituent of the atmosphere; peaking 2.3 Byr ago (Partin, Lalonde, et al., 2013), O_2 reached levels perhaps only two orders of magnitude less than its present-day abundance of 0.21 bar. This time, give or take that couple of hundred million years, is often called the **Great Oxidation Event** (Holland, 2002) or, alternatively, the **Great Oxygenation Event**, although—since it did not achieve the current ample level of oxygen—a more apt alternative name would be the **First Oxygenation Event**. Zimorski et al. (2019) describe its importance as marking "the appearance of continuous atmospheric O_2 in the geochemical record."

Evidence for the abundance of O_2 in the atmosphere comes from the presence or absence of oxidized minerals in the geologic record. For example, UO_2, the primary mineral comprising the uranium ore **uraninite**, is insoluble; but in its more oxidized state (as an ion with U^{+6} rather than U^{+4}), it is soluble. Geochemical modeling suggests that the transition to oxidation and solubility was achieved when ambient oxygen levels exceeded ~2×10^{-3} bar (Grandstaff, 1980). Below this critical level, physical erosion of the ore could leave small bits of rubble, called **detrital** uraninite, which were capable of withstanding dissolution throughout hundreds of kilometers (or more) of river transport to their point of deposition. Because detrital uraninite was present in sedimentary rocks until ~2.4–2.3 Byr ago (e.g., Johnson et al., 2014; Rye & Holland, 1998), we might infer that oxygen levels did not reach 2×10^{-3} bar until that time.

It should be noted that estimates of the critical level of oxygen for uraninite dissolution are somewhat uncertain, because the rate of dissolution also depends on the ambient level of carbon dioxide.

In a watery environment that is oxygenated, iron pyrite—which is otherwise insoluble—will decompose into various iron oxides and sulfate ions. Intact detrital pyrite is found in sedimentary rocks until ~2.4–2.2 Byr ago (e.g., Johnson et al., 2014; Rye & Holland, 1998), reinforcing our inference from detrital uraninite that oxygen levels were low until that time.

Most studies of detrital uraninite and pyrite evaluate the time during fluvial transport that it would take to dissolve the detrital grains. But physical abrasion of the grains must also occur; for the detritus to survive both processes until deposition, the ambient oxygen levels would have to be even lower than when only dissolution occurs. Modeling both processes, Johnson et al. (2014) find that oxygen levels prior to 2.4 Byr ago were probably no more than 3×10^{-5} bar.

Models of detrital weathering also show that, as oxygen levels rise above the minimal value, there are 'intermediate' oxygen levels which do away with uraninite but not pyrite, and then still higher levels where neither mineral can survive. The geologic record reveals that both minerals disappeared at roughly the same time, ~2.3 Byr (± 0.1 Byr) ago; thus, oxygen levels must have increased rapidly enough that the intermediate levels persisted only briefly (geologically speaking) (Johnson et al., 2014). From a geologic perspective, the rise in oxygen was indeed a single "event."

Upon closer inspection, though, and incorporating various other datasets, it has been argued that this First Oxygenation Event should more properly be viewed as a transition period during which atmospheric levels of oxygen may have increased, then faltered, then increased again. Based on the analyses presented by Partin et al. (Partin, Lalonde, et al., 2013; Partin, Bekker, et al., 2013), for example, we might infer that the transition lasted roughly a couple of hundred million years or so, during which time oxygen levels were reinvigorated (after faltering) thanks to a long episode of global carbon burial, known by its effects as the *Lomagundi excursion*. That burial, which amounted to the elimination of a major sink of oxygen from the atmosphere-ocean system for much of the transition period, has in turn been attributed to a prior intensification of plate tectonic activity, leading to more subduction of carbon sediments (Eguchi et al., 2020; see also, e.g., Kasting, 2013 and references therein, and Des Marais, 1997).

Strong bounds on the abundance of oxygen in the Archaean and earliest Proterozoic can be obtained from analyzing the proportion of different sulfur isotopes in pyrite, and in other Precambrian sulfide and sulfate minerals. It turns out that their **fractionation** (i.e., their abundance relative to the primary sulfur isotope, S-32) exhibits a striking change in pattern versus time, beginning just after the Archaean (Farquhar et al., 2000; see Johnston, 2011, for a comprehensive review of work on this subject). Most biological and other natural processes involving sulfur depend on the mass of the sulfur isotope, so the end products exhibit a mass-dependent fractionation. In contrast, photo-dissociation of sulfur gases and aerosols by ultraviolet light passing through the atmosphere acts equally on all sulfur isotopes, ultimately resulting in minerals with a more random, mass-independent sulfur fractionation. However, that process could have prevailed (and overwhelmed the mass-dependent fractionation) only while levels of ozone (O_3) remained low enough for ultraviolet light from the Sun to easily penetrate the lower atmosphere (e.g., Farquhar et al., 2001; Pavlov & Kasting, 2002). Farquhar et al. (2000) found that sulfur isotopes exhibited a mass-independent pattern throughout the Archaean, but since ~2.3 Byr ago the more typical mass-dependent fractionation has predominated. This change can be attributed to a noticeable increase in O_2 (and O_3) levels in the atmosphere, suppressing fractionation by ultraviolet photo-dissociation. Modeling the Archaean fractionation, Pavlov and Kasting (2002) determined an upper bound on O_2 to be 2×10^{-6} bar before the First Oxygenation Event. According to Johnston (2011) it is very likely that such low levels of oxygen characterized the entire Archaean.

Paleosols provide further evidence of oxygenation, thanks to the insolubility of ferric iron (Fe^{+3}). For example, paleosols can end up iron-poor or enriched in hematite (Fe_2O_3) after weathering, depending on the ambient level of oxygen. In a low-oxygen environment, iron (Fe^{+2}) leached out of the soil will remain dissolved and just wash away. In the presence of sufficient oxygen, though, iron leached from the soil will become more oxidized (Fe^{+3}) and precipitate out, enriching the soil with hematite. In the geologic record, paleosols formed prior to 2.44 Byr ago were iron-poor; the implied oxygen levels are estimated to have been less than ~5×10^{-4} bar (Rye & Holland, 1998).

Correspondingly, the appearance of hematite-rich "red beds" in sedimentary formations beginning around 2.3–2.2 Byr ago (see Roy et al., 1975; Rye & Holland, 1998) can also be attributed to increased oxygenation of the atmosphere since the Event.

All these differences in mineralogy and geochemical activity before versus after the Great Oxidation Event represent a fundamental qualitative change in the chemical state of the surficial and near-surficial Earth System. Though the levels of O_2 produced by the Event are inferred to have been low compared to present-day abundances (e.g., Holland, 2006), telling us that substantial increases would eventually follow, a globally oxidative environment could have had massive and "unprecedented" consequences for the biosphere (Hodgskiss et al., 2019).

Second Stage: Global Oxygenation—a BIF Perspective. The broadest picture of oxygen levels in the Earth System throughout the Precambrian may well be that derived from **banded iron formations** or **BIFs**. These sedimentary formations represent extensive iron deposits, amounting to 90% of Earth's mineable iron ore. They consist of alternating, millimeter- to centimeter-thick layers of chert (silica) and iron oxides (see Figure 2.13); the chert is often the reddish-orange mineral jasper, whereas the iron oxides can include hematite but are mainly magnetite and thus gray-black in color. One way BIFs are thought to have been produced was by the precipitation of iron as the oceans became increasingly oxygenated; the iron, which had remained dissolved in the ferrous state, became oxidized to ferric iron and precipitated out as various iron oxides.

In contrast to the geochemical evidence presented above, BIFs do not generally yield quantitative estimates of Archaean or Proterozoic oxygen levels; but their qualitative implications, combined with their occurrence over a span of 2 Byr, allow an oxygen chronology of the oceans to be constructed throughout that time span.

Deposition of the BIFs would have occurred preferentially in shallow-water rather than deep-ocean basins. Factors affecting deposition include the availability of dissolved iron—continental weathering and mid-ocean ridge hydrothermal vents have been advanced as sources of the iron—and the level of oxygen in the oceans; the latter factor in particular favors a shallow environment for the deposition. With the deep ocean anoxic through much (or all) of the Archaean, high concentrations of dissolved iron could build up there; transport by deep-ocean currents and upwelling would then bring large quantities of the iron to shallow environments, where oxidation, precipitation, and extensive deposition

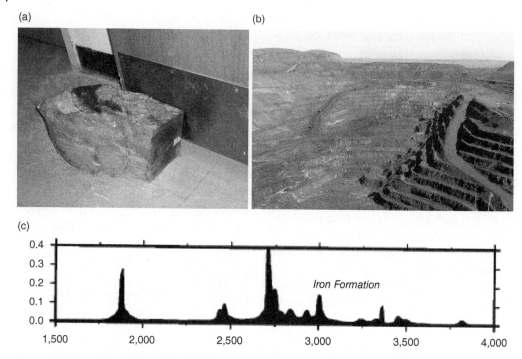

Figure 2.13 Banded iron formations (BIFs).
(a) A 2^+-Byr-old doorstop for the author's office: a chunk of BIF.
(b) The Mt. Whaleback open-pit iron mine, at the edge of the Hamersley Basin in western Australia. Image from NASA Earth Observatory / https://earthobservatory.nasa.gov/images/147416/mining-for-iron-at-mount-whaleback / Last accessed under September 14, 2023. Banded iron ore here is ~2.45–2.60 Byr old (Porter, 2013). More of the world's iron ore comes from western Australia than from any other province or country; and according to Porter (2013), the Mt. Whaleback ore body is "the largest known single, continuous iron ore deposit in Australia."
(c) History of banded iron formation production as inferred by Isley and Abbott (1999); the x-axis is years before present in million years. The y-axis can be interpreted essentially as the frequency of BIF occurrence; however, Isley and Abbott weight each occurrence by the uncertainty of the BIF's age, so the spikes shown may be partly a consequence of highly accurate age determinations as well as a relatively large number of BIF occurrences. Adapted from Isley & Abbott (1999) / John Wiley & Sons.

of BIFs could take place (Cloud, 1972; Holland, 1973).

A dependence on continental weathering, as well as the dependence on oxygen levels in the shallow ocean, might produce layering in the BIFs with a *seasonal* variation; if the layering is non-seasonal, other mechanisms would be required to produce it.

BIFs were created intermittently throughout the Precambrian (see Figure 2.13). According to the work of Isley and Abbott (1999), BIF production began in the earliest Archaean but was quite limited—consistent with our expectation that atmospheric and shallow-ocean oxygen levels were quite low throughout that

epoch. The main period of BIF deposition was ~3.0–2.4 Byr ago, peaking at ~2.7 Byr ago, evidence that oxygen continued to fill the shallow oceans (and eventually the atmosphere) in the mid-to-late Archaean.

The dramatic drop in BIF deposition after the 2.7 Byr peak was reached can conceivably be attributed either to a sudden drop in shallow-ocean oxygen levels—perhaps from widespread oxidation of organic carbon (see van Valen, 1971) and crustal minerals—or to a sudden drop in the abundance of dissolved iron. For example, the spike in BIF production itself, 2.7 Byr ago, might have consumed enough of the iron in the deep ocean to inhibit further production. For that matter, with ambient O_2 levels not very high to begin with, that spike in production may also have used up most of the ocean's oxygen (see Catling et al., 2001).

The analysis by Isley and Abbott (1999) shows strong correlations between BIF production and likely sources of iron (continental platforms, mid-ocean ridge hydrothermal fluids, etc.) created by mantle plume activity. These correlations are most evident (see Figure 2.14) during the periods 3.5–2.4 Byr ago and 1.9–1.8 Byr ago. Intriguingly, plume activity between 2.3 Byr ago and 2.0 Byr ago, with an implied availability of iron, somehow did not lead to BIF deposition.

Toward the end of the era of primary BIF deposition, a dramatic change in the climate system occurred: a shift into the Ice House regime known as the *Huronian*. The geologic record hints at cause and effect: it shows older, detrital pyrite and uraninite overlain directly ~2.4 Byr ago by Huronian deposits (Roscoe, 1969), including glacial ones, which are in turn overlain directly ~2.2 Byr ago by red beds and even some iron-rich paleosols (Rye & Holland, 1998). That is, oxygen levels were relatively low prior to the Huronian (Roscoe, 1969) but palpably higher afterward.

Could the level of oxygen reached ~2.4 Byr ago, which eliminated detrital uraninite from the geologic record and which was clearly on the rise, also have been critical for triggering the shift into an Ice House regime? One hypothesis is that increased production of O_2 by 2.4 Byr ago finally allowed atmospheric oxygen to gain the upper hand against methane (Kasting & Siefert, 2002). As discussed earlier, we expect that levels of biogenic methane would have built up during the Archaean, as primitive life proliferated. But the level of oxygen was also increasing, and large amounts of both gases could not coexist forever: once oxygen levels were high enough, methane would be effectively removed from the atmosphere (as noted earlier), leaving carbon dioxide as a by-product. Methane is a potent greenhouse gas. Whether levels of CH_4 (and CO_2) were sufficient to make the late Archaean climate a Hot House or barely temperate has been debated; but without question the loss of methane during the First Oxygenation Event left a much weaker overall greenhouse effect, and in the early Proterozoic the climate fell into

Figure 2.14 Correlations between production of banded iron formations (BIFs) and mantle plume activity as inferred by Isley & Abbott (1999). Isley & Abbott (1999) / John Wiley & Sons. Continental plume activity is inferred from continental flood basalts, also mafic dykes and layered intrusions; "global plumes" includes inferred oceanic plume activity as well. Cautions in the Figure 2.13 caption about interpreting the "iron formation" time series apply to all the time series here as well.

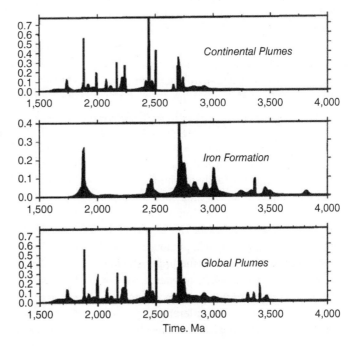

an Ice House (e.g., Haqq-Misra et al., 2008). (If ethane and other complex hydrocarbons were also present in the Archaean atmosphere, they would have added to the greenhouse effect before and the cooling effect after; Haqq-Misra et al., 2008; Lyons et al., 2014.)

Throughout and after the Huronian, photosynthesis by cyanobacteria continued to release oxygen into the shallow ocean and lower atmosphere. The oxidation of methane leading up to the Huronian would have consumed atmospheric O_2; but the decrease in that oxygen abundance must have been only temporary, given the contemporaneous evidence from paleosols and red beds. And the hiatus in BIF production between 2.4 and 2.0 Byr ago seen in Figure 2.14 may simply have been a side effect of the demise of methane: the oxidation of methane to carbon dioxide would have acidified the oceans, lowering the oceans' pH enough to slow the precipitation of iron oxides (see Vance, 1994). According to Pavlov et al. (2003), methane continued to have some influence on climate until the Cryogenian; overall, though, it would no longer dominate over oxygen. Oxygen had won the battle for the Earth System.

The interpretation of BIFs in terms of shallow-ocean oxygenation has been complicated in recent years by the recognition that oxidation of ferrous iron is also possible under *anoxic* conditions, by microbial activity. For example, "ferrous-oxidizing phototrophic bacteria" use light energy in the presence of CO_2 to convert Fe^{+2} to Fe^{+3} (e.g., Konhauser et al., 2002; Kappler et al., 2005; Hegler et al., 2008; see also Weber et al., 2006; Bekker et al., 2010). Not only is free oxygen not required for this oxidation, it would actually inhibit their functioning; Kappler et al. envision such microbes occupying a layer in the shallow ocean *beneath* the cyanobacteria, to avoid the oxygen produced by the latter's ongoing photosynthesis.

We can speculate that as the upper shallow ocean became increasingly oxygenated, oxygen would effuse upward into the atmosphere—in line with our previous inferences of oxygen levels from geological and geochemical evidence—but it would also diffuse downward, from its high concentration in the cyanobacterial layer into the oxygen-scarce ferrous-oxidizing microbial layer. As a consequence, microbial production of BIFs would likely have persisted only through the mid-Archaean, till perhaps ~3 Byr ago, while ambient oxygen levels were too low to inhibit their activity. However, a rigorous discussion of the implications of ferrous-oxidizing microbes for Earth's oxygen history is well beyond the scope of this textbook.

Third Stage: The Oxygenic Dark Ages. A consensus on the abundance of oxygen in the atmosphere over the nearly billion and a half years following the First Oxygenation Event seems to be lacking. There is clear evidence suggesting the abundance resulting from that Event was sustained, more or less, through the Proterozoic. Such evidence includes paleosols dating from ~1.9 to 2.2 Byr ago, whose iron enrichment implies oxygen levels in excess of ~0.03 bar (Rye & Holland, 1998); and the first appearance, at 1.9 Byr and continuing to the present time, of thick, extensive continental red beds (Cloud, 1972).

Consistent with this view, the absence of BIFs after 1.8 Byr ago (as illustrated in Figure 2.14) can be taken to mean that oxygen had now become plentiful enough in the shallow oceans and atmosphere that it even pervaded the deep oceans, terminating production of BIFs; any dissolved iron circulating out of mid-ocean ridge vents thereafter would become oxidized and precipitate out near the source (e.g., Holland, 1999). (However: more than a billion years later, the oceans of "Snowball Earth"—either ice-covered or nearly so—would be poorly enough ventilated that anoxic conditions would briefly prevail once again, allowing production of the very last BIFs.)

Finally, the lack of detritus in the sedimentary record after the First Oxygenation Event, particularly in deposits associated with *short* river transport systems, allowed Johnson et al. (2014; see also Holland, 2006) to conclude that the level of oxygen in the atmosphere has not dropped below 0.01 bar in the past ~2 Byr. In other words, atmospheric oxygen was roughly within an order of magnitude of present-day levels.

But there is also clear evidence indicating Proterozoic oxygen levels were drastically lower. The lack of fractionation exhibited by chromium isotopes in rocks from iron formations dated at ~1.9 Byr led Frei et al. (2009) to conclude that oxygen abundance in the atmosphere had decreased at that time, possibly even returning to Archaean levels. A decrease in O_2 at that time is also indicated by reduced uranium enrichment measured in iron formations (Partin, Lalonde, et al., 2013; see also other references therein); a similar reduction in uranium measured in shales implies that the decrease in atmospheric oxygen may have not only begun ~2 Byr ago but also extended through most of the Proterozoic (Partin, Bekker, et al., 2013). Conceivably, this drop in O_2 following the First Oxygenation Event was a rebound connected to the end of the Lomagundi excursion (Bekker & Holland, 2012; Partin, Bekker, et al., 2013).

If atmospheric levels of oxygen were low during the early to mid-Proterozoic, then it is unlikely that the deep ocean was highly oxygenated during that time span; in that case, another explanation for the disappearance of BIFs after ~1.8 Byr ago would be required. One popular alternative (Canfield, 1998) is that the deep ocean, still anoxic at this time, also became "sulfidic" with a relative abundance of H_2S that allowed dissolved iron to precipitate out as pyrite (FeS_2). Evidence for this includes a layer of pyrite seen in outcrops just above those last BIFs (Poulton et al., 2004). A more recent alternative (Slack & Cannon, 2009), recognizing that BIF termination was rather abrupt, attributes the end of BIF production to the massive Sudbury impact, which occurred at that time (and created the third largest impact crater seen on Earth today—of comparable size to the crater popularly associated with the extinction of the dinosaurs). This impact would have catastrophically mixed surface and deep waters worldwide, causing major changes in ocean chemistry and disrupting the process of BIF formation that had earlier prevailed.

Fourth Stage: The Earth System Becomes Oxygen-Rich. In the Phanerozoic, as life blossomed, the atmosphere finally achieved modern levels of oxygen. In the mid-Paleozoic, for example, we know such levels prevailed because of the appearance by ~420 Myr ago of fossil charcoal. Fossil charcoal is evidence of wildfires, which require the atmosphere to have had at least 0.13 bar of O_2 (Scott & Glasspool, 2006); that is, the atmosphere must have been at least 60% as oxygenated then as it is today (60% of the present atmospheric level or 60% **PAL**). Prior to that time, widespread vegetation able to fuel wildfires had not yet evolved, so this approach cannot rule out the possibility of similarly high levels of oxygen in earlier eras; on the other hand, wildfires have always been able to occur since that time, so we can view 60% PAL as a lower limit for the atmosphere over the rest of the Phanerozoic to the present day. But 420 Myr ago, such fire-ready vegetation was rare, and the evidence of wildfires led Scott & Glasspool to conclude oxygen levels then could even have exceeded 100% PAL.

Whether the rise to modern levels—this **Second Oxygenation Event**—began during the Phanerozoic or the Precambrian is unclear, however. *When* it began is important, because that may relate causally to the evolution of diverse and advanced life, potentially including the *Cambrian explosion* (which occurred in the first half of the Cambrian period) and possibly even the burst of exotic "Ediacaran fauna" that preceded it (635–541 Myr). One popular view is that the rise began earlier than the Cambrian—sometime in the late Proterozoic (in the Ediacaran or earlier), leading to its nickname as the *Neoproterozoic Oxygenation Event* (see Och & Shields-Zhou, 2012 for an extensive review; also, e.g., Ossa Ossa et al., 2022; Partin, Bekker, et al., 2013; and references therein).

As we saw in our discussion of earlier stages, most estimates of oxygen levels (or of the times of their occurrence) are indirect, derived from "proxy" measurements such as mineral enrichments and depletions or isotope fractionation; for this fourth stage, 'bioenergetic' models—to predict, for example, the oxygen requirements of some past species thought to be highly mobile—would be an especially relevant type of proxy. But one direct measurement worth noting was carried out by Blamey et al. (2016), who measured the oxygen trapped in halite crystals; the crystals had formed in a shallow marine environment ~800 Myr ago. The implied level of atmospheric oxygen, ~0.11 bar or slightly more than 50% PAL, is surprisingly high for that era.

As an example of an indirect estimate for this stage, Canfield and Teske (1996) analyzed sulfur isotopes in coastal sediments, concluding that atmospheric levels of O_2 had reached at least 5–18% PAL by 640 Myr ago. Both this and the preceding estimate support the view that the rise to modern levels of oxygen had begun in Neoproterozoic times.

A number of indirect estimates focused on the state of the deep ocean, specifically on whether it was fully oxygenated. Full oxygenation can help identify eras characterized by abundant O_2 in the atmosphere, and their timing implies when the rise preceding them—Neoproterozoic or otherwise—might have taken place. According to an analysis by Scott et al. (2008) of molybdenum enrichment observed in shales formed at the ocean bottom, deep-ocean oxygenation was achieved sometime between 663 and 551 Myr ago. A similar analysis by Sahoo et al. (2012), but including measurements of vanadium enrichment and also sulfur isotope fractionation, implied that the oceans were well oxygenated by ~632 Myr ago. Iron enrichment in seafloor sedimentary rocks led Canfield et al. (2007, but see also Canfield et al., 2008) to conclude that deep-ocean oxygenation was probably widespread almost 580 Myr ago. Based on these analyses, the rise in atmospheric oxygen must have taken place just before, or during, the Ediacaran.

From an analysis of molybdenum isotopes in marine shales, Dahl et al. (2010) concluded

that, following a smaller initial rise in oxygen ~550 Myr ago, oxygenation of the deep ocean was substantial by ~400 Myr ago; Sperling et al. (2015) obtained similar results from an examination of iron enrichment in shales, as constrained by other proxies. Dahl et al. described the mid-Paleozoic rise as possibly "the greatest oxygenation in Earth history." Stolper and Keller (2018) confirmed that deep-ocean oxygenation happened ~400 Myr ago; but their analysis of iron enrichment in seafloor basalts, which ranged in age from $3\frac{1}{2}$ Byr to 14 Myr, additionally led them to rule out a Neoproterozoic rise in oxygen.

Conceivably, the disparities among deep-ocean-based inferences of the rise in oxygen could stem (e.g., Stolper & Keller, 2018; Butterfield, 2009) from how well the various measurements have characterized the deep ocean, or whether the models used to connect dissolved oxygen in the deep ocean to the surface abundance of oxygen are sufficiently accurate. For some estimates, geographic variability may be a factor as well.

To some extent, though, the disparities in timing and intensity of the oxygen rise might simply reflect the fact that it was not a monotonic rise. Earlier, we acknowledged that the First Oxygenation Event might be better described as a 'transition period,' in which there were at least two upswings, with lower levels in-between. Considering similarities and differences between the two Oxygenation Events, including their durations as well as the conditions at the time they occurred, do you think it is reasonable to treat the Second Oxygenation Event as a transition period, too?

The rise in atmospheric oxygen peaked in the last quarter of the Paleozoic, at a level substantially greater than today. Beginning in the Devonian (as also implied by limited coal deposits), perhaps ~470–445 Myr ago (Lenton et al., 2016; also Dahl et al., 2010), and intensifying during the Carboniferous period (Berner & Canfield, 1989), plant life proliferated—especially land plants, whose spread was enhanced (thanks to continental drift) by the availability of extensive land areas situated at 'nurturing' tropical or temperate latitudes; such a proliferation added a lot of oxygen to the atmosphere through photosynthesis. Plants thrived in extensive swampy areas—even becoming the tropical rainforests sometimes called the *Coal Forests* of the Carboniferous, in the last third of that era (Sahney et al., 2010)—along continental margins (and inland seas), where they evolved into a variety of quite large plants and trees (Figure 2.15).

The normal 'cycle of life' should have balanced out any excessive production of oxygen, as the plants died, decayed, and were oxidized. However, these land plants were better able to resist bacterial decay (Berner & Canfield, 1989, and references therein). And, in those vast swamps and shallow seas, many of the plants that died became submerged without having been oxidized. The land plants also tended to be carbon rich. Some of all that organic material, gradually buried beneath sediments

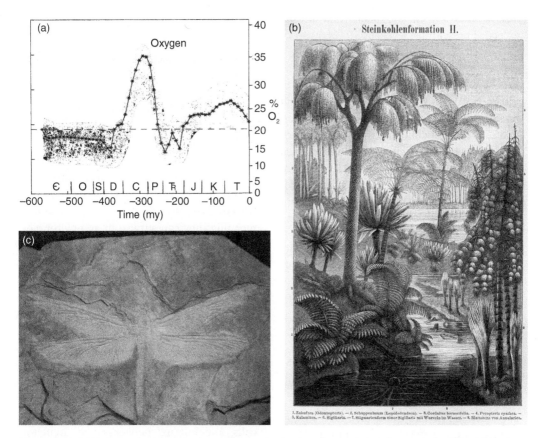

Figure 2.15 (a) Variation in atmospheric oxygen levels since the Precambrian. Berner & Canfield (1989) / American Journal of Science. Shading represents uncertainties based on model assumptions; the current level of O_2 (dashed line), 21% or 0.21 bar, is shown for comparison. Levels during the Carboniferous reached more than 30% (0.3 bar) but may have dropped to ~15% by the end of the Permian.
(b) Late nineteenth century etching of the Coal Forests of the Carboniferous by Bibliographisches Institut—Meyers Konversationslexikon. Meyers Konversationslexikon / https://en.wikipedia.org/wiki/Carboniferous#/media/File:Meyers_b15_s0272b.jpg / Public domain.
(c) Fossil of Meganeura, a type of giant dragonfly which flourished in the highly oxygenated atmosphere 300 Myr ago; the wingspan here is 28 inches. Ghedoghedo / https://commons.wikimedia.org/wiki/File:Meganeura_fossil_1JPG / CC-BY SA 3.0.

and slowly cooked, ultimately became the coal deposits of Pennsylvania and regions further south along the Appalachians. In any event, without the usual decay of the dead plants to remove O_2, a lot of oxygen was freed up and could remain in the atmosphere. It was estimated by Berner and Canfield (see also Berkner & Marshall, 1965 and Graham et al., 1995) that O_2 levels ~300 Myr ago reached almost 0.35 bar (167% PAL); see Figure 2.15. More recent estimates place the 'oxygen high' in the Permian, 270 Myr ago (Scott & Glasspool, 2006), or even the Triassic, up to 250 Myr ago (Krause et al., 2018), though in both studies oxygen abundance had already exceeded present-day levels by 300 Myr ago.

The rich oxygen of the Carboniferous did not persist, however. Under a glacial climate, the swampy Coal Forests collapsed by the end of the Carboniferous (see Sahney et al., 2010 and references therein); through most of the Permian, dry glacial climates continued, and sea level was lower. More of the crust was exposed to oxidation, and plant burial by submergence was no longer commonplace; the first major era of coal production was at an end. Additionally, a huge mass extinction event was taking place. Not surprisingly, then, oxygen levels declined—perhaps even to much less than the present abundance, though that would have been temporary (Berner & Canfield, 1989; see Figure 2.15).

2.4.2 Oxygen and Evolution: An Overview

As photosynthesizing life evolved beyond microbes, oxygen would have been produced at a much higher rate; depending on the effectiveness of the various oxygen sinks, levels of O_2 could have increased substantially. On the other hand, as we now acknowledge, more complex life has greater oxygen needs; it could therefore be argued that an increase in oxygen levels over time promoted their evolution. So which is the cause, and which is the effect? It turns out their relation to each other is far from straightforward, with evidence rarely definitive and often inconsistent. Lyons et al. (2014) caution about pitfalls in trying to assign cause and effect, and Erwin et al. (2011) and Butterfield (2009) argue that a more comprehensive approach, with many more factors being considered, is necessary to understand the true connections.

One of the most important advances in the evolution of life was the development of eukaryotes. In contrast to prokaryotes (like cyanobacteria), whose genetic material is distributed throughout their cell, all the more advanced kingdoms of life are **eukaryotic**, meaning their DNA strands are contained within a distinct cell nucleus (see Figure 2.16). From our perspective here, though, with a goal of relating evolution to the history of oxygen, *mitochondria* are a more significant evolutionary advance than enclosed DNA. **Mitochondria** are the structures powering each cell, using oxygen to metabolize carbs and fatty acids and generate energy for cellular activities. Prokaryotes do not have mitochondria, whereas almost all eukaryotes do (those eukaryotes without them are anaerobic, meaning they can survive in an environment lacking oxygen).

Clearly, possessing mitochondria increases the cell's (or organism's) variety and range of possible activities—a huge evolutionary advantage (e.g., Lane & Martin, 2010), but only if enough oxygen is present for them to function. Aerobic eukaryotes have high oxygen needs (e.g., Zimorski et al., 2019); this is especially true for more evolved organisms (Canfield et al., 2018), such as organisms with complex multicellularity—including differentiated cells, and internal networks like circulatory systems (e.g., Knoll, 2011)—and highly motile animals, which are able to intentionally move a part or all of their bodies and thus have a "high exercise metabolism." In short, advanced organisms could not proliferate until their environment (whether a regional oasis or the whole globe) was sufficiently oxygenated.

(a)

(b)

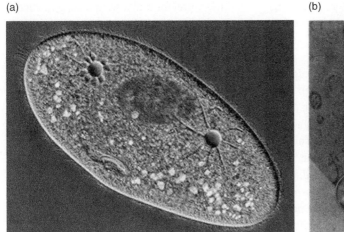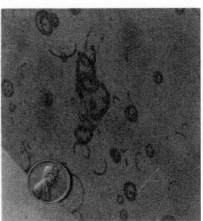

Figure 2.16 (a) Example of a one-celled eukaryotic organism, a Paramecium; the large dark nucleus is located between two "vacuoles." Nguyen Tan Tin / Flickr / CC BY-SA 2.0.
The length of this organism is ~½ cm; in contrast, mitochondria are perhaps 1 micron long (Korotaev et al., 2020).
(b) Fossil of the oldest known eukaryote, *Grypania Spiralis*, a coiled multicellular algae, found in a 1.9-Byr-old BIF in Michigan; image, with a penny shown for scale, is from Han & Runnegar (1992). Han & Runnegar (1992) / American Association for the Advancement of Science.

So, at this point in our discussion of oxygen and evolution, have you concluded which is the cause and which the effect? (Or is drawing a conclusion here premature?)

However, Zimorski et al. (2019) also note that many modern species of eukaryote are adapted to low-oxygen environments, implying that early eukaryotes would have been able to survive, if not proliferate, during oxygenically harsh times. The oldest known eukaryotic fossils (see Figure 2.16) date to the early Proterozoic, ~1.9 Byr ago (Han & Runnegar, 1992 but also Schneider et al., 2002), though their appearance became more frequent beginning ~1.7–1.4 Byr ago (e.g., Javaux & Lepot, 2018; also Porter & Knoll, 2000).

By ~500 Myr ago, if not earlier, eukaryotic photosynthesis had helped establish oxygen levels within an order of magnitude of present-day levels. Plant life gets most of the credit for finally achieving present-day levels of atmospheric oxygen, as explained above; but to clarify cause and effect, we will restate that explanation with a slightly different emphasis.

To recap: The rise in atmospheric oxygen to modern levels by ~400 Myr ago can be accounted for mainly by the existence, proliferation, and subsequent burial of land plants beginning 470 Myr ago. As they developed and evolved, O_2 was produced at a much higher rate; but they had greater oxygen needs as well, and their respiration consumed most of the available oxygen. Meanwhile, the burial of organic matter, especially land plants, before they could be oxidized, freed up a lot of oxygen to remain in the atmosphere; this burial helped atmospheric oxygen increase greatly into the mid-Paleozoic and culminate during (or following) the Carboniferous.

In other words, that major rise in O_2 was not a consequence of eukaryotes producing a lot of oxygen as much as it was the result of eukaryotes, particularly land plants, "producing" a lot of biomass, much of which would be buried. Note that today, land plants constitute about 80% of Earth's total biomass (Bar-On et al., 2018).

More Connections to Consider. Exploring the connections between oxygen and evolution a bit further, we note the following:

- In the present context, the most profound observation of all is that life began and evolved in anoxic environments (e.g., Fischer et al., 2016; Zimorski et al., 2019). This fundamentally constrains any cause-and-effect relationship between oxygen and evolution, at least in the far past. And, once oxygen began to pervade the atmosphere and shallow ocean, it would become necessary for organisms to develop an ability to "tolerate" oxygen (Fischer et al., 2016)—a necessity even today, as we are frequently reminded of the importance of taking "antioxidants" (whether in food or supplements) to help keep our cells healthy. Such adaptation would continue, with increasing sophistication, through the entire Archaean (Fischer et al. 2016, also David & Alm, 2011).

- Some potential connections—or *non-connections*—depend on the level of O_2 in the atmosphere or oceans at the time. According to Canfield et al. (2018), for example, oxygen in the atmosphere and oceans persisted at high-enough levels from ~1.4 Byr ago onward for organisms with complex multicellularity to prevail; but such life did not develop until ~800 Myr ago, more than half a billion years later! Oxygen might have 'opened the gates' to such evolutionary advances, but evolution worked in its own way and over its own timescales before proceeding through the gates.

  In contrast, Stolper and Keller (2018, also e.g. Sperling et al., 2015) found that the Ediacaran fauna had evolved in a world where oxygen levels would not be sufficient to meet their needs until much later, in the Cambrian.

  Some studies had concluded that two of the five mass extinctions which took place in the Phanerozoic were caused by low oxygen levels; but Belcher and McElwain (2008; see references within) find that O_2 levels at those times were not that low, and suggest such a conclusion should be reevaluated.

- In some instances, a direct, positive correlation between oxygen and evolution seems unavoidable. Dahl et al. (2010) pointed out that jawed fishes evolved to 'giant' size in the Paleozoic, growing from the Cambrian (where their typical length was a few centimeters) to the Ordovician (~30 cm) to the Devonian (several meters); according to Ferron et al. (2017), the maximum length of late Devonian jawed fish was 8.8 m. For these "high-energy" predators, as with modern fish, oxygen needs would have increased with size; correspondingly, oxygen levels must have either been increasing throughout this era in the atmosphere and oceans, or were already high.

  With a somewhat less direct approach, Bambach (1999) tied the appearance of those active Devonian jawed fish, hunters at the top of their food chain, to an increased productivity in the oceans able to meet their energy and metabolic needs.

  Chapelle and Peck (1999) provide a straightforward connection between O_2 and body size. Their study of numerous species of modern crustaceans (technically, amphipods) demonstrated a clear dependence of size (body length) on ambient oxygen levels; they concluded that more abundant oxygen would allow longer circulatory systems to successfully carry oxygen farther through the body.

- Basking in the mostly abundant oxygen of the Phanerozoic, species of fish and crustaceans were not the only animals that evolved to very large body size. A number of "gigantic" species, now extinct, come to mind, including dinosaurs in the late Jurassic and Cretaceous; giant ground sloths (see, e.g., Bargo, 2001) from the Eocene through the Pleistocene, evolving to a weight of ~4 tons by the late Pleistocene (Raj Pant et al., 2014); and giant snakes in the Paleocene ~60 Myr ago, ranging up to 15 m in length and 2 tons in weight (see Head et al., 2009). A serious evaluation of oxygen and "gigantism" as cause and effect requires that we look more broadly at all gigantic species.

We start with a noteworthy giant: the dragonfly of the Carboniferous, fossils of which have been found with wingspans exceeding 2 feet (Graham et al., 1995; Cannell, 2018; see Figure 2.15)—the largest insects ever to have lived (Vermeij, 2016). Like jawed fishes, the appearance of those giant insects took place around the time of the most abundant atmospheric oxygen in Earth's history, suggesting a very direct cause and effect. Kaiser et al. (2007) explain that, with such high oxygen levels, insects could overcome their normally limited ability to take in oxygen through "spiracle" holes in their exoskeletons and diffuse it throughout internal "tracheal" tubes, so they could successfully evolve to very large size (Butterfield, 2009, points out shortcomings of studies supporting this explanation). Flying insects would also benefit from the higher atmospheric density produced by the greater abundance of oxygen (Graham et al., 1995; Cannell, 2018).

By the end of the Paleozoic, giant dragonflies had become extinct. We can speculate on the cause: certainly, the end of superabundant O_2, which they had adapted to and depended on, would have been a factor; and perhaps also the sustained change to the cooler, drier climate of the Carboniferous-Permian Ice House.

Another impressive species from that era is also worth mentioning: giant millipedes, described as "the largest arthropod in Earth history" (Davies et al., 2022). Estimated to range up to more than 2½ m in length and weighing ~50 kg, they evolved during the Carboniferous and became extinct during the early Permian. Their gigantism and demise have been attributed to multiple causes, including mainly climate change but also oxygen levels and new predators (Davies et al., 2022; Schneider et al., 2010).

The evolution of gigantism throughout the Phanerozoic was studied comprehensively by Vermeij (2016; see also Butterfield, 2009). He examined different categories of giant species, such as marine predators and terrestrial herbivores, and considered various possible causes of their development. In general, there are several factors with the potential to promote or suppress the evolution of a gigantic species. Curiously, mass extinctions are *not* one of them; though they may 'clear the field' of competitive species to make way for the development of new and bigger species, mass extinctions occurring in the Phanerozoic were not followed by any gigantic species until, typically, another ~60–100 Myr had passed. Similarly, gigantism of any particular species developed long after the ancestral 'family' of related species had originated. As influential as those factors might have been, the long time-lag following their occurrence implies that an additional 'trigger' was needed to stimulate that gigantism.

For example, Head et al. (2009) considered high temperatures as the trigger for the evolution of those giant Paleocene snakes. In the case of the largest of modern whales, it appears to have been the great availability of their main food source (krill), triggered by favorable changes in climate and ocean upwelling (Slater et al., 2017), which led to their evolutionary survival (e.g., Goldbogen & Madsen, 2018; Goldbogen et al., 2019).

But the analysis by Vermeij did conclude there was a correlation between gigantism and atmospheric oxygen levels, at least in the Paleozoic (in agreement with Dahl et al., 2010). Because the largest species in all animal categories Vermeij studied were "aerobically active" species, with high metabolic demands, they would require food-rich and oxygen-rich environments. He found that the three non-marine giant species of the Paleozoic—which were in the categories of flying predators, terrestrial predators, and terrestrial herbivores—had evolved in size while oxygen levels were increasing, and finally appeared at roughly the time of the

Paleozoic oxygen 'high' (i.e., at or just after the Carboniferous). In short, the rise in oxygen was a contributing factor to the development of gigantism but evidently was the most important factor only during the Paleozoic.

All in all, evolutionary advances and changes in oxygen abundance through Earth's history are inconsistent with a unique cause-and-effect relation. Oxygen was a contributing cause for some evolutionary developments, and in some of those cases an additional factor (such as an increased food supply or a warming climate) subsequently triggered the actual development; at other times, the evolution of more advanced life contributed to a rise in oxygen levels, and in some of those cases an additional factor (e.g., large-scale plant burial) was critical.

In one situation, however, stronger conclusions may be reasonably justified: gigantism during the Paleozoic can be attributed mainly to superabundant oxygen levels; and, the extinction of those gigantic species was mainly a consequence of the subsequent drop in oxygen.

2.4.3 Oxygen Chronology: A Synthesis

A completely speculative graphical depiction of Earth's oxygen history is presented in Figure 2.17a (where it is shown as a lightly shaded region of the graph). The data points used to constrain this history, based on works cited below as well as work already discussed, are identified using different symbols to denote specific numerical values, ranges, and upper or lower bounds. Of the 53 data points included, 22 pertain to the Phanerozoic and 31 to the Precambrian. Almost half of the latter were based, at least in part, on isotope fractionation analyses; half of the former were mainly based on fossil charcoal and its implications for

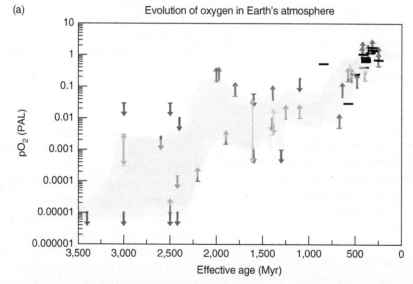

Figure 2.17 (a) Estimates of atmospheric oxygen abundance, pO_2, expressed in terms of present atmospheric level (PAL) using a logarithmic scale, versus time in millions of years before the present. The analyses yielding these estimates are discussed in the text. If the estimate from an analysis was intended to characterize pO_2 over a range of time (e.g., the mid-Archaean), then a single 'effective age' was picked here to represent that range.

Data points are symbolized according to the type of estimate: upper bound, red downward arrow originating at a red-filled circle; lower bound, blue upward arrow originating at a blue-filled circle; range, upward over downward pair of light brown arrows; single value, wide horizontal black line. Bounds originating at a small horizontal line of the same color are meant to indicate ≤ rather than <, or ≥ rather than >.

Light gray shading is an attempt to suggest a representative range of magnitudes for atmospheric oxygen through Earth's history (but see also Figure 2.18).

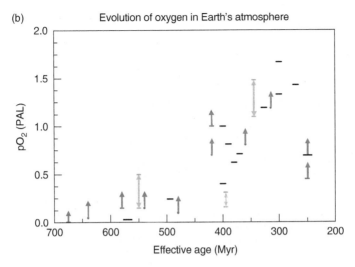

Figure 2.17 (b) Close-up over the past 700 Myr, with pO_2 relative to PAL using a linear scale.

wildfire activity (this type of data, as discussed earlier, has only been possible for the past 420 Myr or so), as constrained by Berner (2006).

References used to construct the data points, beyond those already cited above for their numerical results, include the following:

- for estimates employing isotope fractionation:

    Karhu and Holland (1996) (whose estimates, along with those of Rye & Holland, 1998, posit the greatest abundance of O_2 over the entire Precambrian), Farquhar et al. (2001), Pavlov et al. (2003), Planavsky, Reinhard, et al. (2014), Cole et al. (2016), Canfield et al. (2018), Crockford et al. (2018) (whose estimates were constrained by the level of photosynthetic biological activity—"primary productivity"—associated with different levels of O_2), and Krause et al. (2018);

- for estimates derived from fossil charcoal/ wildfire data:

    Belcher and McElwain (2008), Sperling et al. (2015), and Lenton et al. (2016), though principally Scott and Glasspool (2006) (as constrained by Berner, 2006);

- for estimates focusing on mineral enrichment or depletion in paleosols or sedimentary deposits:

Zbinden et al. (1988), Holland (1999), Anbar et al. (2007), Lyons et al. (2014), Sperling et al. (2015) (whose estimates were constrained by oxygen requirements, inferred either for the earliest animals or Cambrian-era animals), Zhang et al. (2016a) (with their controversial estimate subsequently debated by Planavsky et al., 2016 and Zhang et al., 2016b), and Stolper & Keller (2018) (with some of their estimates reinterpreted here using the methodology outlined in Riding et al., 2014, Supplementary Data); and

- for estimates based on other approaches:

    Han and Runnegar (1992) (whose estimates, using eukaryotic respiration, are as modified by Crockford et al., 2018 and whose age is revised according to Schneider et al., 2002), Canfield et al. (2007) (whose estimates used ocean models combined with constraints from oxygen requirements inferred for Ediacaran fauna), Dahl et al. (2010) (whose estimates employed a combination of the mineral enrichment and isotope fractionation approaches), and Zimorski et al. (2019) (whose estimates also were based on eukaryotic respiration).

In some cases, the estimate was intended by the authors to cover a large span of time; for

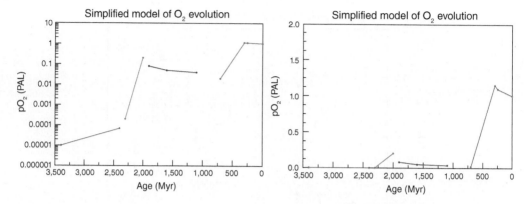

Figure 2.18 Simplified model of atmospheric O_2 evolution, based on data shown in Figure 2.17. The y-axis scale is logarithmic on the left and linear on the right. Eras depicted include the mid- and late Archean, First Oxygenation Event, mid-Proterozoic, and Neoproterozoic + Phanerozoic.

the sake of clarity (if not precision!), a single representative time was assigned for its display in Figure 2.17a, b.

Figure 2.17 includes a graph of oxygen chronology displayed using a log scale for oxygen abundance and (in the close-up, Figure 2.17b) one using a linear scale. What might be the advantages of either scale? What about the disadvantages?

Pursuing this a bit further, imagine depicting a rise from ~0 (or just over 0) to 10% and another rise from 10% to 100%; how different would those rises appear if a log scale was used versus if a linear scale was chosen? Do you think caution is warranted in describing a rise in oxygen as "great," based on one scale or the other?

Figure 2.17 implies that not all estimates of past oxygen levels are consistent with each other; this would still be the case if the estimates' uncertainties were included in the graph. But, as discussed earlier, it is also probable that oxygen did not remain at a constant level for geologically long periods of time—whether as a result of 'whiffs,' or an increase in sources or sinks of O_2; consequently, adjacent up and down arrows might merely reflect actual

fluctuations which have not been fully resolved in time.

Perhaps resigned to such poor resolution, a number of the preceding researchers have presented 'minimal' scenarios of Earth's oxygen history. In that same spirit, we reduce the data in Figure 2.17 to a similarly 'fundamental' variability; see Figure 2.18. At its most basic, then, the atmosphere can be characterized as being in a nearly anoxic state through the Archaean; oxygen levels rise sharply (as the First Oxygenation Event) in the early Proterozoic; oxygen remains within one or two orders of magnitude below current levels through the mid-Proterozoic; and oxygen rises fairly steadily, beginning at least 650 Myr ago, reaching current levels by the Carboniferous. This scenario is roughly consistent with the theoretical description by Laakso and Schrag (2017) of the atmosphere as having three fundamental states of oxygenation (with transitions from one state to another).

How does the rise from the first to the second fundamental state of our atmosphere compare with the rise from the second to the third? For a quantitative estimate, you can use either graph in Figure 2.18.

3

The Climate System and the Future of Earth's Atmosphere

3.0 Motivation

Chapter 2 demonstrated that *change has always been the rule* for Earth's climate system. Whether these changes were dramatic, as Earth entered or abandoned an Ice House or Hot House regime, or more subtle, as the composition of the atmosphere and oceans trended gradually, they occurred for the most part on geologic rather than human timescales. At present, and in the near future, though, as the threat of what is popularly called "global warming" materializes, due to a rise in atmospheric levels of greenhouse gases from human activity, additional changes in the climate system can be expected. Mostly, those changes will be only slight from a geologic perspective—for example, as global temperatures rise, our current Ice House climate regime will continue—but their societal impacts may well be significant, both in terms of their magnitude and with regard to the rapid rate at which the changes take place.

Thanks in large part to the increasingly comprehensive reports of the Intergovernmental Panel on Climate Change (**IPCC**), it has become clear to many scientists—climatologists and others—that human-caused global warming really is happening. But society's attempts to deal with global warming could lead to significant political and economic changes, so global warming has remained a contentious issue for some *outside* the scientific community. We will consequently find it important to evaluate arguments on both sides of the issue (—an approach that, as we saw in Chapter 1, can also strengthen our understanding of the subject).

This chapter will survey the near-term changes expected to our climate system in a globally warmed world. However, in order to understand how the climate system responds to the additional energy of global warming, we will first need to understand the energetics of the climate system—that is, how and why the atmosphere and oceans circulate. And we will also consider one of the alternatives to global warming: the possibility that the source of additional energy is a naturally variable Sun rather than a product of increased levels of CO_2 and other greenhouse gases in our atmosphere.

The anticipated near-term consequences of global warming turn out to be wide-ranging and substantial—so much so that, even if global warming were not yet a reality, societal action would still be called for, *now*, to minimize (or accommodate) those consequences; with global warming already underway, the need is even more urgent. We end this chapter by discussing some of the data and models used as the basis for that assertion, as well as some additional observations. Our final conclusion is that the evidence is not only consistent with the reality of global warming, it also argues against natural variability within the climate system as a viable alternative explanation. The climate system will always exhibit variations, on all timescales and regardless of

Earth System Geophysics, Advanced Textbook 6, First Edition. Steven R. Dickman.
© 2025 American Geophysical Union. Published 2025 by John Wiley & Sons, Inc.
Companion Website: www.wiley.com/go/Dickman

human interactions; but they will be super-imposed on rather than the ultimate cause of current and near-future global warming.

The Climate System

3.1 The Circulation of the Atmosphere

3.1.1 The Sun Is the Ultimate Driving Force

How Much Heat From the Sun Reaches the Earth? Earth's climate system is driven ultimately by heat from the Sun, but only a small fraction of that heat actually makes it down to the surface of the Earth. The energy generated by the Sun's thermonuclear reactions today, 3.8×10^{26} J/sec (that is, 3.8×10^{26} W), is radiated 'spherically,' i.e. in all directions. At the distance of Earth from the Sun, that radiation has spread out over a huge area, and Earth is a small planet; the fraction of it intercepted by the Earth (see Figure 3.1a) is only about half of one billionth of the total, or 1.7×10^{17} W.

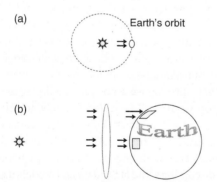

Figure 3.1 (a) An orbital view: the facing disk of the Earth receives a tiny fraction of the Sun's radiation. (b) A close-up, side view of Earth, with a hypothetical flat reference disk of the same diameter as the Earth positioned precisely in front of it. This disk and the Earth intercept the same *total* amount of solar radiation; but for the Earth, the radiation is spread out over a greater surface area away from the tropics, lowering the amount *per unit area* intercepted by the Earth.

As a 'standard' for reference, we imagine a flat disk positioned right in front of the Earth, and with the same diameter (Figure 3.1b); that disk will intercept all of the solar radiation that would otherwise reach the Earth, 1.7×10^{17} W. It is useful as a reference because the spherical curvature of the real Earth complicates the amount of radiation able to reach the surface. If the radius of the disk is a (the Earth's radius will also be a), then the disk will have an area equal to πa^2; given the value of a, this area will equal $\sim 1.3 \times 10^{14}$ m^2. It follows that the amount of energy per second striking each square meter of area of the intercepting disk is

$$1.7 \times 10^{17} \text{ W}/1.3 \times 10^{14} \text{ m}^2$$

or (with somewhat greater precision)

$$1{,}361 \text{ W/m}^2$$

(Kopp & Lean, 2011). In contrast, note (Figure 3.1b) that the amount of energy per second striking any square meter of the spherical Earth's surface will, in general, differ from that striking any other square meter, due to Earth's curvature—which is why our reference disk is so useful.

This $1{,}361$ W/m^2, sometimes expressed equivalently as 1.952 calories per minute per cm^2, is called the **solar constant** (although another term, **solar irradiance**, is also used—after all, the Sun's luminosity has not remained constant over time). The solar constant represents the intensity of light and heat theoretically available to drive Earth's climate. However, not all of the radiation intercepted by the disk actually makes it down to the surface of the Earth.

Fate of the Incoming Solar Radiation. There are three factors working to reduce the amount of solar energy that actually drives our climate system.

- The curved shape of the Earth. We note that the spherical Earth's surface area is $4\pi a^2$ if its radius is a—precisely four times the surface area of the intercepting disk! It follows that, at any instant, the incident radiation on the sunlit hemisphere (whose area is $2\pi a^2$) is spread over twice as much area as the intercepting disk; this reduces the amount per unit area by a factor of 2. Or, in the course of a day, the Sun's radiation is spread over Earth's entire surface. On average, then, the intensity of incoming radiation is one fourth of the solar constant or just over 340 W/m^2 (equivalently, just under 0.49 cal/minute per cm^2).

- Earth's atmosphere absorbs some of the incoming solar radiation before it reaches the ground. The typical amount is about 20%, including 3% from absorption by clouds and 16% from dispersed water vapor and dust.

- Earth's atmosphere and surface reflect some of the incoming solar radiation back into space. Clouds reflect 20% of the incoming radiation; atmospheric dust, 6%; and the surface, 4%.

Absorption and reflection together reduce the amount of incoming solar radiation that makes it down to the surface (and is retained) by a further 50%. The net effect of all three factors is that the incoming solar radiation—also known as the **insolation**—is one eighth of the solar constant, about 170 W/m^2 (or, just under 0.25 cal/minute per cm^2).

Incidentally, this magnitude of insolation implies that globally, the Earth's surface receives enough solar energy in *eight hours* to supply the world's needs for an entire year. Theoretically, at least, solar power has the potential to satisfy the energy demands of our society.

If the yearly energy needs of contemporary society amount to ~6 × 10^{20} J, can you verify the equivalence of 8 hours of sunlight? (It is probably easiest to start with that 1.7 × 10^{17} W, which, in more convenient units, is 1.7 × 10^{17} J/sec).

A key concept in climate studies is **albedo**, defined as how much a surface reflects light (expressed as a fraction or a percentage). For example, a pure white surface reflects 100% of incoming light, and has an albedo of 1 or 100%; a pure black surface absorbs all of the incoming

light, and has an albedo of 0. *The Earth's albedo is* 20% + 6% + 4% or 30%. With a greater or lesser albedo, Earth's 'equilibrium' surface temperature, representing a balance between incoming and outgoing radiation, would be lower or higher than the approximately 15°C it is now. During glacial periods, for example, the high albedo of the advancing ice sheets would have further cooled the planet. This cooling would also have initiated a *positive feedback loop* that would help the ice sheets grow even more: more cooling enabled larger ice sheets to exist, which increased the albedo and thus enhanced the cooling, leading to even more extensive ice, an even greater albedo and even more cooling; and so on. We describe this process as a **positive feedback**, because it acts to reinforce the initial cooling; a **negative feedback** would act to counteract the initial change. In contrast, once the climate began to warm and ice began to melt, the lower albedo would provide a positive feedback that promoted further warming and thus further melting of ice.

Such feedback loops can also act now, during our current interglacial period. Any global warming that begins to melt ice on Earth's surface will lower Earth's albedo, further warming the surface and the climate system, leading to additional ice melting—producing, in short, a positive feedback that enhances the initial global warming. And, since most of the ice to be melted lies in polar regions, that is where the albedo will drop the most and where the greatest amount of warming will be seen. That is, this positive feedback implies that global warming will not be globally uniform.

Of the two polar regions, which do you think will be warmed more, the Arctic or the Antarctic?

Finally, the same three limiting factors that reduce the magnitude of insolation also cause it to vary with latitude. The Earth's surface is curved, so that radiation striking the ground at high latitudes is spread out over a much greater area than that striking the tropics (see Figure 3.1b). Because of Earth's curvature, incoming radiation travels a longer distance through the atmosphere at high latitudes than near the equator, allowing more absorption (the Sun's rays always seem stronger in the tropics!). And because of Earth's curvature, incoming radiation at high latitudes strikes the surface at more of an angle, so more of it can bounce away. Thus, all three factors cause the insolation to be less at high latitudes—least at the poles—and more in the tropics. It turns out the insolation is about 68 W/m² (or ~0.1 cal/minute per cm²) at the poles and 275 W/m² (or ~0.4 cal/minute per cm²) at the equator, on average. That is, the tropical insolation is up to *four times* that in polar regions! As we will soon see, *this insolation difference—even more than the overall magnitude of the insolation itself—acts to drive the circulation of our atmosphere.*

3.1.2 Basic Concepts Underlying Atmospheric Circulation

Once those insolation differences set the atmosphere into motion, as we will describe shortly, that motion will be governed by a few basic processes. Heated air, being expanded and thus less dense than the surrounding air, will tend to rise; cooled air tends to sink. Humid air is less dense than dry air (perhaps in contrast to your intuition), and will tend to rise as well. Furthermore, when a 'parcel,' or small volume, of air rises, it encounters lower ambient atmospheric pressure (at greater altitude, there is less weight of the overlying atmosphere on it), so it decompresses and cools. As it rises, it is also continually giving off heat to its surroundings; eventually, it will cool enough to achieve a balance with the surrounding air, and its ascent will end. If it has risen enough, it will cool to the point where the moisture it contains *condenses*; clouds will form, and if the clouds are dense enough, it will rain!

But in order to condense, that is, go from the vapor state to a liquid state, the water vapor must shed some of its heat (—with less kinetic energy, the water molecules will be able to bond together loosely, as a liquid); this heat warms the rising and cooling air, counteracting the cooling to some extent and powering the rise to greater heights. For example, warm dry air rising from the surface to an altitude of 3 km will decompress and cool by about 29°C; if it is humid, it will end up (depending on how humid, and on the surface temperature) perhaps only 19°C cooler—a difference of ~10°C due to the heat of condensation!

The heat of condensation is why tropical storms like hurricanes can become so strong; ultimately, they are powered by the high humidity of their environment.

The circulation of the atmosphere—actually the **troposphere**, the lowest layer of the atmosphere, extending from the ground up to roughly 6 km at the poles or 16 km in the tropics—starts with the insolation differences at the equator and poles. With the insolation greatest at the equator, the air there is hotter than elsewhere, thus less dense, so it rises, to the top of the troposphere. The humidity in that rising air (which aided in its ascent) will have condensed, forming clouds and perhaps even producing a recurrent rain. At the same time, the insolation is a minimum at the poles; the air is cold and dry (cold air can only hold small amounts of water vapor at most), thus dense, and it sinks to the ground (see Figure 3.2a).

To fill the gap at the base of the equatorial troposphere, resulting from air there rising, ground-level winds begin to blow in toward the equator. As they blow in, they heat up and acquire humidity, rising as they approach the equator and thus drawing in more surrounding air. Eventually, even the air that had descended to the ground at the poles turns equatorward. Meanwhile, to fill the gap at the top of the polar troposphere, resulting from air there having sunk down, high-altitude winds blow in toward the pole. Lack of insolation and dryness ensure that the air newly arriving at the pole continues to sink, drawing in more surrounding air; ultimately, even the air that had risen up over the equator turns poleward.

(a)

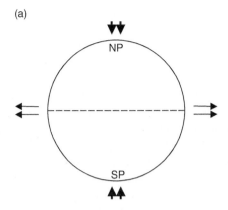

Figure 3.2 Idealized explication of the global "three-cell" (per hemisphere) circulation of the troposphere. The thickness of the troposphere (indicated by the length of the arrows) is exaggerated here for clarity.
(a) Side view of the vertical motions beginning the circulation: hot and humid (thus, very light) air rising over the equator (dashed line) and also cold and dry (thus, very dense) air sinking at the poles (NP = North Pole; SP = South Pole). For clarity, arrows indicating rising air are thinner; those indicating sinking air are thicker.

3.1.3 Global Atmospheric Circulation on a Nonrotating Earth

What happens next on the real Earth is fairly complicated, so we will 'ease into' that reality by first considering an imaginary Earth that does not rotate. The absence of rotation means, among other things, that the moving air will not experience the Coriolis force—a force that, on the real Earth, tends to deflect moving objects to the side. (Note: the Coriolis force will be formulated and explored quantitatively in Chapter 4.) On a nonrotating Earth, the air from the equator can continue moving poleward at the top of the troposphere, cooling and becoming denser as it flows all the way to the poles without deflection. By the time it reaches the poles, it will have long since shed its humidity as well, and this now cold and dry air will sink down to the surface. Meanwhile, the ground flow originating as sunken polar air can continue all the way to the equator, also undeflected, heating up as it moves toward lower latitudes where the insolation is greater,

and acquiring more humidity as it warms. Upon reaching the equator, this now warm and humid air, as light as can be, rises up. The end result for this nonrotating Earth is envisioned to be one 'cell' of tropospheric convection in each hemisphere, extending from equator to pole (see Figure 3.2b).

Although such a "one-cell" circulation pattern does not exist on the real Earth, our neighboring planet Venus rotates so slowly—once every 243 Earth days (and east to west, for that matter, unlike Earth)—that its Coriolis force is almost negligible; Venus has therefore been held up as an example of one-cell atmospheric circulation. However, the actual motion of Venus' troposphere is difficult to observe, because the atmosphere is so dense and because the top of the troposphere exhibits a high-speed "super-rotation" of its own (those winds, at speeds of ∼100 m/sec, circle the planet in only four Earth days!).

(b)

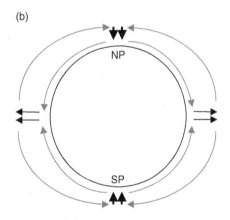

Figure 3.2 Idealized explication of the global "three-cell" (per hemisphere) circulation of the troposphere. The thickness of the troposphere (which extends from the surface up to the outer blue curved arrows) is exaggerated here for clarity.
(b) Side view of "one-cell" (per hemisphere) circulation resulting from insolation differences on an Earth that does not rotate. After rising above the equator, the air moves poleward, ending up dry and cold, and sinks to the ground at the poles. Meanwhile, the air that had previously descended over the poles moves equatorward, warming and becoming humid by the time it reaches the equator, where it rises, completing the pattern.

The super-rotation evidently creates an environment in the *upper* troposphere in which the Coriolis force is reasonably strong (with the high-speed flow making up for the planet's slow rotation; see Chapter 4), producing—according to recent observations (Svedhem et al., 2007)—pairs of giant hurricane-like vortices near the poles. Those observations also suggest that a one-cell-per-hemisphere circulation does operate on Venus but, curiously, with the downwelling at about ±60° latitude rather than at the poles. It remains unclear how the lower troposphere is circulating at high latitudes.

3.1.4 Global Atmospheric Circulation on the Rotating Earth

The Coriolis force, as noted earlier, acts on any moving mass to deflect it to one side. The amount of deflection, or equivalently the strength of the force, is proportional to how fast the planet is rotating (which is why it is generally negligible at the surface of Venus) and how fast the mass is moving (which is why it might be important in the super-rotating upper troposphere of Venus). This latter dependence explains why we do not experience the Coriolis force as we move around in everyday activities. Actually, the force is weak even for masses—like winds, or ocean currents—that are moving at a decent speed, so the deflection is only noticeable cumulatively, when those masses travel over a great distance.

The deflection by the Coriolis force acts relative to the direction of the mass' motion. In Earth's northern hemisphere, moving masses are deflected to *their* right; for example, winds blowing to the east will be deflected to their right, i.e. southward. 'Down under,' in the southern hemisphere, the deflection is reversed; for example, a wind blowing to the east will be deflected to its left, that is, northward. Since the deflection is opposite in both hemispheres, it makes sense that the Coriolis force must vanish (i.e., have zero magnitude) right at the boundary between the hemispheres, that is, at the equator.

The force increases away from the equator, and is strongest at the poles.

The air flowing between equator and pole shown in Figure 3.2b, whether at ground level or at the top of the troposphere, gets deflected by the Coriolis force. We can follow a parcel (small volume) of air through the circulation, beginning with its rise at the equator. At the top of the troposphere, it turns poleward, just as on a nonrotating Earth. We follow the parcel moving northward in the northern hemisphere. As it flows northward, away from the equator and toward the North Pole, the Coriolis force deflects it to its right—to the east. Throughout its travels away from the equator, it is giving off heat and humidity (assuming it has any left after its rise above the equator), so this parcel is becoming denser and denser. Eventually, having traveled a huge distance to the east as well as about a third of the way north to the pole, it sinks down—at ~30°N latitude. Upon hitting the ground, it can continue to move northward; or, it can return southward, back to the equator, to fill the ground-level gap in the troposphere there. If the parcel continues northward, it will again be deflected by the Coriolis force to the east (that is, to its right), traveling a long distance along Earth's surface. These ground-level winds are called the **mid-latitude westerlies** (meteorologists focus on the direction the winds come *from*, because that indicates whether they carry warmer or cooler air; technically, then, in the northern hemisphere these winds should be called mid-latitude *south*westerlies).

If the parcel that descended at ~30°N returns southward, it will be deflected by the Coriolis force to the west (that is, to its right), again traveling a long distance along Earth's surface, during which it warms up and acquires humidity. These surface winds (blowing from the east—technically the northeast here) are called the **tropical easterlies** or **Trade Winds**. By the time the parcel reaches the equator, completing the pattern, it has warmed appreciably and again contains a lot of moisture; its low density causes it to rise once more, and the flow pattern starts over again.

The convective flow pattern between the equator and 30°N is called the **Hadley cell** or **tropical convection cell**.

Meanwhile, the air descending over the North Pole, upon hitting the ground, had begun to move southward. Deflected by the Coriolis force to the west (i.e., its right) as it flowed, it became the surface wind known as a **polar easterly**. As it travels southwestward, it warms appreciably and gains humidity, and by the time it reaches ~55°N—60°N latitude, it is light enough to rise. At the top of the troposphere, it returns northward to the pole (deflected toward the east as it flows), completing the pattern we call the **polar convection cell**.

To complete the description of the atmosphere's circulation, we return to that parcel of air traveling along Earth's surface to the northeast as a mid-latitude westerly. Eventually (at ~55°N–60°N), it clashes with the southwestward-moving polar easterlies, and— despite having cooled significantly in its northward trek—is dragged upward with them; at the top of the troposphere, it may head toward the pole or, more likely, return back to 30°N, where it had previously descended (either way, it will be deflected to its right as it flows). At 30°N it will also descend, despite having warmed up in its southward travels, dragged downward with the descending air of the Hadley cell. The pattern thus created is called the **Ferrel cell** or **mid-latitude cell**. Because it is driven by the downwelling or upwelling of the adjacent cells rather than by its own intrinsic density differences, it is not a convection cell.

The circulation pattern we have inferred to be operating in this hemisphere, driven by the insolation differences with latitude and seriously modified by the Coriolis force, is called the **three-cell circulation**.

It will be instructive for you to repeat this analysis with a parcel rising at the equator but traveling southward toward the South Pole. Remember that the Coriolis deflection is opposite in the southern hemisphere. You should find that the result looks like the sketch in Figure 3.2c, which shows the global three-cell circulation.

(c)

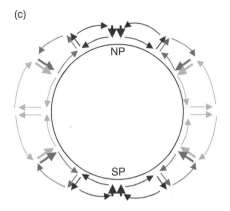

Figure 3.2 Idealized explication of the global "three-cell" (per hemisphere) circulation of the troposphere. The thickness of the troposphere (which extends from the surface up to the outer curved arrows) is exaggerated here for clarity. (c) Side view of circulation resulting from the insolation differences with latitude, as modified by the Coriolis force. After rising over the equator, the air moves poleward but gets deflected by the Coriolis force and ends up, dry and cooler, sinking to the ground at 30° latitude. Most of that air returns toward the equator, but some continues poleward at ground level. Meanwhile, the air that had descended over the poles moves equatorward but gets deflected by the Coriolis force. Warmer and somewhat humid by the time it reaches 55°–60° latitude, it ascends to the top of the troposphere and returns to the poles. The mid-latitude ground-level poleward-moving air is entrained upward with the rising air at 55°–60° latitude, then returns to its 30°-latitude starting point. As a side view, this sketch shows the vertical and north-south parts of the circulation, but not the east-west.

At the top of the troposphere, the temperature contrasts between the polar cell and mid-latitude cell, and between the mid-latitude cell and tropical cell, lead to the formation of strong narrow eastward-moving winds known as the jet streams.

The *surface winds* associated with this circulation in each hemisphere are shown in Figure 3.2d.

3.1.5 Complications of the Three-Cell Model

It is worth noting that the three-cell pattern we have deduced is an idealization of reality. First of all, as the seasons pass, the Sun will shine

(d)

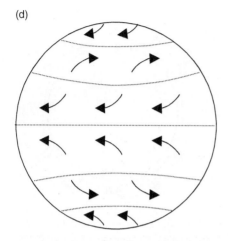

Figure 3.2 Idealized explication of the global "three-cell" (per hemisphere) circulation of the troposphere.
(d) Front view of the globe, showing the surface winds produced by the Coriolis deflection, which is toward the right of the wind direction in the northern hemisphere and toward its left in the southern hemisphere. The ground-level winds returning from 30° latitude to the equator are deflected westward in both hemispheres and are called the Trade Winds or tropical easterlies; a similar fate befalls the ground-level winds moving away from the poles toward 55° or 60° latitude, producing the polar easterlies. The ground-level winds moving from 30° latitude poleward toward 55°–60° latitude are called the mid-latitude westerlies.

more directly first on one hemisphere, then the other. But although the latitude of maximum insolation will travel up to $+23\frac{1}{2}°$ then down to 0°, then to $-23\frac{1}{2}°$ and back to 0°, the air and surface experiencing that insolation take time to heat up; as a result of such *thermal inertia*, what we might call the 'meteorological equator' will migrate north and south but with a more limited range (and with a time lag). Thus, in the hemisphere experiencing summer, the boundaries of the three cells will shift poleward, and the cells will contract poleward; in the hemisphere experiencing winter, the three cells will expand out, away from their pole, and the winter Hadley cell will even extend past 0° latitude (as its 'equatorial' boundary moves into the other hemisphere's tropics). The expansion or contraction of the Ferrel and

polar cells, though, is much more limited than that of the Hadley cell (see Figure 3.3).

It may be that the 'fight' between adjacent cells for domination as the seasons change is the origin of the popular northern hemisphere description of the month of March as "coming in like a lion but going out like a lamb." In any event, the three-cell circulation pattern we have constructed, shown in Figure 3.2a–d, is a *seasonal average*.

Additionally, with the southern hemisphere largely oceanic, its thermal inertia is greater than that of the northern hemisphere (water takes longer to heat up than land!), so the meteorological equator migrates over a more limited range south of 0° than north. The seasonally averaged meteorological equator is thus north of 0°. And, the polar cell in the southern hemisphere is relatively vigorous throughout the year, thanks to the chilling effect of the Antarctic ice sheet, pushing the other two cells out. The net result is that the average position of the meteorological equator (which we can also identify as the boundary between the two hemispheres' Hadley cells) is located a few degrees north of the equator!

There are other hemispherical asymmetries in the three-cell circulation. As suggested by Figure 3.3, the winter Hadley cell is much more energetic than the summer one. The intensity of the Hadley circulation is a function of the temperature difference between its equatorial and poleward boundaries. Because the summer Hadley cell is heated—the subsolar point, i.e. latitude of maximum insolation, is *within* the cell—that temperature difference is relatively small; in contrast, the winter cell is chilled at its poleward boundary, leading to a much larger difference.

Averaged over the seasons, the surface winds (Figure 3.2d) tend to be somewhat stronger in the southern hemisphere than the northern: with less land and more ocean south of the equator than north, those winds encounter fewer topographic barriers and lower surface friction.

Is the Three-Cell Model the Best Way to Describe the Large-Scale Circulation

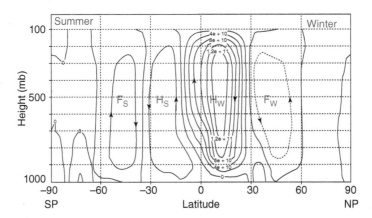

Figure 3.3 Observed north-south circulation of the troposphere, averaged over December, January, and February. Adapted from Vallis (2005). Cross-section (or, as described in Figure 3.2, side view) from South to North Pole extends from the surface of the Earth (where the atmospheric pressure is 1,000 mbar or 1 bar) up through the troposphere (atmospheric pressure at 16 km height above the equator is ∼100 mbar). We see a vigorous winter Hadley cell (H$_W$), extending from 30°N across the equator to more than 10°S. The boundary between winter and summer Hadley cells is the meteorological equator or ITCZ (see the text).

of Our Atmosphere? Despite its ability to explain several persistent features of the atmosphere—some of which are detailed in the next subsection—and despite its popularity at all levels of study, judging from its continued use in introductory textbooks and advanced research alike, the fundamental validity of this model has been questioned over the years.

- In mid-latitudes, some observations (Connolly et al., 2021; also alluded to in, for example, Aguado & Burt, 2007) show westerly winds at *all* levels of the troposphere, not just near the surface. At the higher levels, these westerlies would have a north-south (meridional) component, just like the surface westerly does; for example, in the northern hemisphere they would blow poleward from the southwest. This flow casts doubt on the existence of Ferrel cells, which feature the opposite, an equatorward 'return flow' in the upper troposphere, from the northeast to the southwest in the northern hemisphere. Such doubt has led to alternative models of the global circulation being proposed; see Figure 3.4.

Part of the problem is that the mid-latitude circulation is intrinsically *unsteady*, that is, variable over time, and highly asymmetric versus longitude (in other words, asymmetric in the east-west or "zonal" direction, thus a *zonal asymmetry*) (e.g., Vallis, 2005). Those factors could add much uncertainty to the time-averaged, zonally averaged meridional circulation—the pattern being depicted by the three-cell circulation. As stressed by Kump et al. (2010; also Ruddiman, 2014), even for the Hadley component the flow is unsteady: the winds do not blow continuously. The existence of mid-latitude asymmetries implies that the zonal averaging will partially cancel out, leaving the modeled Ferrel circulation as a "residual" flow (Peixoto & Oort, 1992; Kjellsson & Döös, 2012; Huang & McElroy, 2014).

Regardless of such observational challenges, however, those alternative global circulation models should not be preferred. First of all, as can be inferred from Figure 3.4 (top or middle profile), the meridional circulation implied when mid-latitude winds are westerly at all heights would deplete the Hadley cells, or (in more general terms) violate the principle of mass conservation. Second, theoretical solution of the equations governing atmospheric motion, forced appropriately, zonally averaged, and under

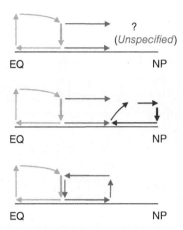

Figure 3.4 Simplified hemispheric profiles (equator to North Pole) of the meridional (North-South) global atmospheric circulation, from the surface of the Earth to the top of the troposphere, postulated by some as alternatives to the traditional three-cell model.

The first two profiles envision mid-latitude flow very different than a traditional Ferrel cell; the third profile is the traditional model modified by the absence of a polar cell.

Upon reflection, it should be evident that the first two profiles fail to conserve the system's mass, and would lead to depleted Hadley cells; the third profile, as noted in the text, treats frictional forces inconsistently from one side of the Ferrel cell to the other.

steady conditions, shows that an upper-level equatorward return flow is to be expected in mid-latitudes (Vallis, 2005). Finally, as it turns out, analysis of wind data spanning 1979–2018 (Liang & Gao, 2021; also Lachmy & Kaspi, 2020, Qian, Wu, & Liang, 2016; Qian, Wu, Leung, & Shi, 2016 and Wang et al., 2005 (their Fig. 4b) for somewhat shorter time spans of other datasets) has confirmed the existence of the traditional Ferrel cell circulation.

- Some models of the tropospheric circulation characterize the polar cells as intermittent, with weak flow at best, nonexistent at other times (as in Figure 3.3; see, e.g., Peixoto & Oort, 1992; Vallis, 2005; Huang & McElroy, 2014). Analysis of wind data confirms the relative weakness of the polar cell, typically four times weaker than the Ferrel cell

(Qian, Wu, Leung, & Shi, 2016). But the absence of polar cells would be problematic for the three-cell circulation, given their role in helping to drive the Ferrel cells' motion.

Additionally, a two-cell circulation (see Figure 3.4, bottom) may be inconsistent: it relies on frictional forces (exerted by the sinking air of the Hadley cell) being strong enough to drag down the adjacent Ferrel cell air; but somehow such forces are unable to entrain any polar air, and leave the polar air motionless (despite the rising air of the Ferrel cell at its poleward boundary).

- Over the past two decades, research has shown with increasing confidence that the northern hemisphere circulation can exhibit a *four*-cell pattern (!). Because this may be attributed to global warming, we defer discussion of it till later in this chapter.

3.1.6 Implications of the Three-Cell Model for Climate and Regional Circulation

Climate implications. The circulation of the Hadley and polar cells, as with any other convecting fluid, involves an exchange of heat between the warmer upwelling and colder downwelling cell boundaries. Ultimately, with the Ferrel cell as intermediary, the three-cell circulation brings tropical heat poleward and sends polar cold equatorward. But this process is not efficient, depending on mixing between adjacent cells to complete the transfer; and, of course, that north-south heat transfer is deflected to the east or west by the Coriolis force. In contrast, heat transfer in the atmosphere of a nonrotating planet, via its one-cell circulation, would be much more successful, and the overall temperature differences globally would be much moderated. On Earth, the mean temperature difference between the equator and the poles, roughly 50°C (e.g., Harvey et al., 2014), is more than twice what it would be if the Earth did not rotate (cf. Ingersoll, 2013).

The three-cell circulation determines more than just the overall thermal state of the atmosphere. At locations where the air is relatively warm and humid, thus less dense and rising, the weight of the atmosphere must be less than the surroundings; that is, the surface atmospheric pressure must be relatively *low*. At locations where the air is heavier and descending, the surface pressure is *high*. The three-cell circulation of the atmosphere implies that the **Polar Front**, the boundary between polar and Ferrel cells, will generally be characterized by low pressure. The boundary between the two hemispheres' Hadley cells, which we have nicknamed the 'meteorological equator' but is more formally known as the **Inter-tropical Convergence Zone** or **ITCZ** (and also called the **Doldrums**), will feature low pressures as well. In contrast, the boundary between Hadley and Ferrel cells—the **horse latitudes**—will generally be characterized by high pressure. These alternations are illustrated in Figure 3.5a.

The rising and sinking air at cell boundaries lead us to expect specific types of climate at those latitudes. As described earlier, rising air may cool enough for the moisture it contains to condense, producing clouds and perhaps rain; sinking air does the opposite, warming as it descends and compresses, evaporating any condensation. Consequently, the ITCZ and Polar Front should experience rainy, even stormy weather; the horse latitudes should feature fair, dry weather, ultimately even producing deserts at ~30° (N or S) latitude.

Regional Circulation. Till now we have envisioned tropospheric upwelling or down-welling as a 'latitudinal' process, in which all of the air along a cell boundary rises or sinks together at equal rates. But, more realistically, the rising or sinking is not likely to be uniform. For example, surface topography can hinder the ground-level winds that would feed the rising air or push out away from sinking air; or, an ocean—staying warmer than adjacent land areas during winter or cooler during the summer—can impart its own heat or chill to

(a)

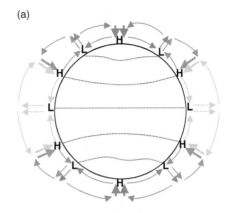

Figure 3.5 (a) Climate implications of the global "three-cell" (per hemisphere) circulation of the troposphere. The thickness of the troposphere (which extends from the surface up to the outer curved arrows) is exaggerated here for clarity. Side view of circulation reveals upwelling and downwelling regions, the former characterized by low surface pressure (marked as L) and the latter by high surface pressure (marked as H). Across the globe, the belts of high pressure, at roughly ±30° latitude, are known as the *horse latitudes*. As discussed in the text, these belts are typically the location of fair, dry weather, thus where deserts are found. The approximately equatorial zone of low pressures is the *Inter-tropical Convergence Zone* or *ITCZ*; this global 'trough' of low pressure is associated with recurrent rainy weather. The zone of low pressure at 55°–60° latitude in each hemisphere is called the *Polar Front*; regions within this zone tend to support persistent low-pressure, stormy weather systems.

the air above. Thanks to the Coriolis force, such nonuniformity means that the upwelling and downwelling will occur in spatially discrete, regional structures; the most common of these structures are called *cyclones* and *anticyclones*.

Consider, for example, the Polar Front in the northern hemisphere, and imagine a region within that zone of low pressure subjected nonuniformly to extra heat or humidity—a region, therefore, of especially low density. The surrounding atmosphere—under relatively higher pressure—will push into that lower-pressure area. As those winds blow in from all directions, they are deflected by the Coriolis force, to their right. If a balance is reached between the pressure 'gradient' pushing the winds in and the Coriolis force

(b)

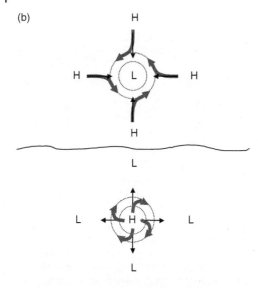

Figure 3.5 (b) Regional circulation of the troposphere in the northern hemisphere. Atmospheric pressures (L = low, H = high), initial wind directions (thin straight arrows), and developing wind patterns (thick curving arrows) are shown schematically looking down from above at the Earth's surface.
(Top) Low-pressure system, created when air pushes in from the surrounding higher pressures into the central low. In the northern hemisphere, those winds are deflected to their right, resulting ideally in counterclockwise or cyclonic flow around the low (along the dotted paths).
(Bottom) High-pressure system, formed when air pushes out to the surrounding lower pressures away from the central high. In the northern hemisphere, those winds are deflected to their right, resulting ideally in clockwise or anticyclonic flow around the high (along the dotted paths).
As discussed in the text, friction modifies the ideal flow, causing it to spiral in to the low or out from the high.

deflecting them to their right, the result will be *counterclockwise flow* (looking down from above) around the low (see Figure 3.5b, top). Such flow is called "*cyclonic*," and the resulting weather system is called a **low-pressure system** or **cyclone**. If the rising air within the low is humid enough and rises high enough, clouds will result, and perhaps rain—maybe even an intense storm.

In fluid dynamics, the balance between pressure forces and the Coriolis force is called **geostrophic flow** ("geo" for Earth and

"strophe" meaning "turning"). Typically, however, friction between the winds and their surroundings prevents a complete balance, slowing the winds and reducing the Coriolis force, so the pressure forcing prevails and the winds spiral in toward the central low. Such a *quasi-geostrophic* process feeds the low, allowing the rising air at the center to be replenished and keeping the low-pressure system alive.

Similarly, a relatively high pressure at the surface—resulting from overlying air that is particularly cool, perhaps, or dry—will push air out, away from it, toward surrounding lower-pressure areas; the Coriolis force will deflect those winds into *clockwise* (or *anticyclonic*) motion around the central high (see Figure 3.5b, bottom), and friction will allow a slight spiraling out of the flow. This is a **high-pressure system** or **anticyclone**, and with sinking air in its center it is characterized by fair weather.

Because the Coriolis force causes an opposite deflection and consequent sense of motion (clockwise versus counterclockwise) in the southern hemisphere, we avoid ambiguity by identifying low-pressure systems as cyclones and high-pressure systems as anticyclones, regardless of their sense of motion.

In short, the three-cell global circulation of the troposphere will be associated with discrete low-pressure systems along the Polar Front and discrete high-pressure systems along the horse latitudes. Given the humidifying influence of ocean versus land, low-pressure systems on the Polar Front tend to develop at specific oceanic locations; in the North Pacific and North Atlantic they are called the Aleutian low and the Icelandic low. Such lows, which are often called "persistent" or "semi-permanent" as a reminder of their recurrent connection to those locations, become intensified during winter, as the heat stored by the oceans during the preceding months warms the air above, in contrast to the chilled land adjacent to the oceans.

The discrete high-pressure systems that form along the horse latitudes are also identified with specific oceanic locations, one in each ocean basin. In the North Atlantic, the persistent high-pressure system, located about 30° south of the Icelandic low, is called the Bermuda/Azores high. This high-pressure system behaves reasonably in connection with the ocean it overlies, strengthening during summers—summer insolation heats up the surrounding land more, so the oceans act to chill the air above, relatively speaking—and weakening during winters, when the system splits and remnant high-pressure centers retreat to the adjacent land (Davis et al., 1997). However, in the other ocean basins, the connection of the persistent highs to the ocean is problematic: those high-pressure systems are evidently stronger in winter than summer (Fig. 7.7 of Rohli & Vega, 2015; also Salcedo-Castro et al., 2015).

As noted earlier, the horse latitudes are also the preferred zones around the world in which to find deserts; within those latitude bands, the actual desert locations are 'discretized' as well, a product of local conditions and the Coriolis force.

We might not expect to find any discrete low-pressure weather systems, persistent or otherwise, at the ITCZ; being essentially at the equator, the Coriolis force is too weak to produce any cyclonic motion. The ITCZ is sometimes consequently described as a low-pressure "trough" (with no identified centers). However, as we will discuss in the context of El Niño later in this chapter, the combination of land barriers and certain ocean currents leads to concentrations of extremely warm water in specific, more-or-less equatorial regions, resulting in a tendency for low-pressure systems to form at those locations.

A vivid illustration of the conversion of three-cell flow into discrete cyclonic structures is provided by the laboratory experiments of Nadiga and Aurnou (2008), which effectively simulated the process at the polar front boundary of a polar cell.

Finally, due to the presence of extensive land areas in the northern hemisphere, two additional regions stand out—incongruously, with respect to the three-cell circulation—as the location of persistent high-pressure systems. In both of these regions the land is covered by snow and ice through much of the winter, chilling the air above and causing that air to sink—creating the Siberian high (in northeastern Eurasia) and (in the northern plains of central Canada) the Canadian high.

3.1.7 A Brief Jovian Perspective

The concepts underlying our model of the regional and global circulation of Earth's atmosphere can be tested by application to the atmospheres of the giant planets. Jupiter, in particular, presents itself as the opposite extreme to Venus because of its rapid rotation, spinning once on its axis every 10 hours and thus creating a very strong Coriolis force.

Regionally, we would expect Jupiter to exhibit numerous cyclones and anticyclones. Globally, because the Coriolis deflection of the poleward and equatorward motion is so extreme, we would expect the global circulation to be more dominantly east-west than on Earth, leading to a greater number of cells—producing perhaps a five-cell or even seven-cell per hemisphere pattern. The multicolored, banded atmosphere of Jupiter (a reflection of the variety of gases and clouds present there) conveniently suggests the actual flow patterns, and, with a quick glance, confirms our expectations (see Figure 3.6).

In fact, the other giant planets—which also have spin rates faster than Earth's, Saturn rotating almost as rapidly as Jupiter, and Uranus and Neptune having ~17- or 16-hour 'days'—all exhibit banding (though, for compositional reasons, none as vivid as Jupiter's). As with Jupiter, they suggest global atmospheric circulation patterns like that on Earth but with more cells per hemisphere.

As viewed from space, of course, the Earth does not look banded; but on average, we can expect to see clouds in the vicinity of the ITCZ and the Polar Front, and the land or ocean *surface* (visible through clear skies) at the horse latitudes and poles. The banding of Jupiter consists of light zones, corresponding

Figure 3.6 Jupiter, as seen by the Hubble Space Telescope on May 5, 2006, exhibits cyclonic features and banding, reflecting the dominance of its Coriolis force. The anticyclonic Great Red Spot (seen below the equator, to the right in the image) has persisted for centuries; in contrast, a relatively new feature—sometimes called Red Spot, Jr—was created in 2000 after a gradual merger of several smaller (anti)cyclones; it acquired its reddish color in 2006. 'Junior,' with only half the diameter of the GRS, is seen here in the center of this image, a bit lower than the GRS. NASA / https://apod.nasa.gov/apod/ap060505 .html / last accessed August 22, 2023.

to high-altitude clouds, and dark belts, corresponding to lower-level clouds (Jupiter does not have a solid surface, but the absence of high-altitude clouds at these latitudes allows the deeper features to be seen). On both Earth and Jupiter, then, such alternations seem to imply latitudes of upwelling and downwelling, as expected in multicell per hemisphere atmospheric circulations.

Jupiter also exhibits a huge cyclonic area known as the Great Red Spot; this feature, which is evidently an anticyclone, is currently twice the size of Earth. Though its shape and circulation have changed (see, e.g., Rogers, 1995), it has been observed since the seventeenth century—possibly even by Galileo (who in 1610 was also the first person to see Jupiter's four large moons)—making it the longest-lived as well as largest weather system in the Solar System! The cause of the Spot has remained in question; conceivably, like our Siberian high,

it is produced, somehow, by more permanent features or conditions deeper within (or below) Jupiter's atmosphere. Such an unusual possibility (see Chapter 10) is supported by the observation that—although winds at its borders whip around at speeds of nearly 270 mph—the interior of the Great Red Spot is essentially stagnant, and isolated from the 'chaotic,' turbulent environment surrounding it (Mitchell et al., 1981).

As with our earlier application to Venus, however, the predicted global circulation patterns for Jupiter's troposphere mainly serve as a reminder that nature is complicated! Some of these complications stem from Jupiter's huge mass and fluid interior. Measurements by the Pioneer 10 and 11 spacecraft as they flew by Jupiter in 1973–1974 indicated that tropospheric temperatures were essentially uniform from equator to pole; but the variation in incoming solar radiation with latitude should, by itself, have left the polar regions 10°C–30°C colder (see, e.g., Ingersoll & Porco, 1978; Aurnou et al., 2008). It was already known that Jupiter radiates more heat than it receives from the Sun, much of it probably generated by the great compression of its interior, with additional contributions from left-over heat of formation or other sources. Some computer models (e.g., Aurnou et al., 2008) suggest that Jupiter's internal heat causes its fluid interior, extending deep below the troposphere, to convect in such a manner (discussed in Chapter 10 in the context of Earth's core) as to bring up more heat near the poles than the equator, compensating for the lack of polar heat from insolation.

The implications of such deep convective heat transport for the banding seen in Jupiter's troposphere are unclear. One possible conclusion is that the banding has nothing to do with insolation (e.g., Zhang & Schubert, 2000), undermining its interpretation as a consequence of multicell global circulation driven by equator-to-pole temperature differences. Alternatively, we might suppose that small but non-negligible equator-to-pole temperature differences persist (at least, within the uncertainty of the observations), so the compensation by internal convection is substantial but incomplete. In that case, Jupiter's deeply convecting fluid may, in its effects, simply be a more extreme version of what the fluid below *our* atmosphere does. As discussed later in this chapter, our own oceans—driven by the winds of the three-cell circulation—end up transporting a significant amount of heat poleward, about half as much as the atmosphere. This oceanic circulation helps reduce Earth's equator-to-pole temperature difference, though evidently not as fully as the convectively driven deep 'oceans' of Jupiter manage to achieve. The patterns of our ocean circulation reflect the three-cell tropospheric circulation and thus the insolation driving that circulation in the first place; it may be that the patterns of deep and tropospheric convection on Jupiter are coupled

and, similarly, ultimately connect to Jupiter's insolation (cf. Ingersoll et al., 2004).

Investigations into Jupiter's global circulation have a long history (see, e.g., Stone, 1973). Recent numerical models, as noted here, suggest a resolution of some issues (subject to future confirmation by observations of Jupiter's interior flow); however, the dynamics of Jupiter's atmosphere remain somewhat mysterious in other respects. Apparently (Vasavada & Showman, 2005), the great majority—something like 90%—of Jupiter's cyclonic features are anticyclones. And, although Jupiter's greater distance from the Sun significantly reduces the amount of sunlight it receives, and although Jupiter's equator-to-pole temperature difference would be less than Earth's even without the compensation from deep convection, the winds of Jupiter blow at speeds typically three to four times faster than on Earth (and even faster than that, at the edges of the bands) (e.g., Ingersoll et al., 2004). For that matter, Saturn and Neptune, despite being even colder than Jupiter, have still faster winds—the fastest in the Solar System (Hester et al., 2002)!

3.1.8 Jet Streams in the Atmosphere

The **jet streams** are narrow, concentrated, eastward-flowing winds near the top of the troposphere at the boundaries between warmer and colder air (see Figure 3.7): the **polar jet**,

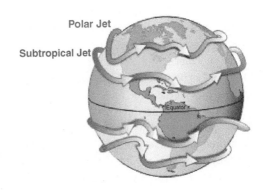

Figure 3.7 Cartoon depicting the jet streams. https://commons.wikimedia.org/w/index.php? curid=8697883) / Public Domain.

between the mid-latitude and polar cells, and the **subtropical jet**, between the tropical and mid-latitude cells. The polar jet is also called the **polar front jet** and the **eddy-driven jet**. The migration of the cells described earlier is reflected in corresponding migrations of the jets; at times when the subtropical jet—normally at ~30° latitude—is north of Binghamton, NY (42°N), for example, the tropical cell has pushed north of Binghamton, and the weather there will be decidedly tropical.

The existence of these jet streams derives from the contrast in air temperature at the boundary between adjacent cells. In the simplest terms, we can say that the temperature difference expands the warmer air mass and makes it taller; at any height, then, there is more air still higher in the warm air mass, thus more weight of overlying air, thus a greater pressure, compared to the same height in the cooler air mass. Weather systems, especially at the Polar Front, may add to or reduce those pressures somewhat. The lateral pressure differences induce a flow across the boundary, from the warmer to the colder air mass—a "thermal wind" initially blowing poleward, which is then deflected toward the east by the Coriolis force. The pressure difference increases with altitude, intensifying the thermal wind and resulting in 'jets' of concentrated eastward wind at heights near the top of the troposphere. The high-altitude intensification is further enhanced by the lack of friction experienced by the wind at such heights (—a different situation than experienced by ground-level winds). Because the temperature contrasts, like the Coriolis deflection, are mirror images in the northern and southern hemispheres (poleward in both cases), the jets in both hemispheres flow in the same direction, eastward.

Since the Coriolis force is stronger closer to the poles, the polar jets are more intense than the subtropical jets. The polar jets flow typically at heights of 8 to 10 km, with speeds of ~65 m/sec (~150 mph); but even the subtropical jets, at heights of 12 to 14 km (the troposphere is taller in the tropics), move fast enough, with speeds of ~40 m/sec (~90 mph),

that commercial airline pilots usually ascend to such heights in order to get a boost to their long-distance flights eastward.

Additional mechanisms to power the jet streams have also been proposed: weather systems at the Polar Front, to drive the polar jet (e.g., Vallis, 2005; Lee & Kim, 2003; see also Lachmy & Kaspi, 2020); and rotational effects, acting on the upper-level poleward-flowing winds of the Hadley cell, to drive the subtropical jet. Those rotational effects—which strengthen as the winds move away from the equator, culminating at the poleward edge of the cell—are taken to be either an increasing eastward deflection by the Coriolis force (cf. Peixoto & Oort, 1992), or an increase in the winds' west-to-east angular velocity as they conserve angular momentum (e.g., Lee & Kim, 2003; Aguado & Burt, 2007); the latter comes into play because the winds' distance from Earth's rotation axis is decreasing as the winds stream toward the poles. A full evaluation of these mechanisms is beyond the scope of this chapter; but it is worth noting that, like the thermal wind mechanism, they lead to jet streams which are linked to the Hadley-Ferrel and Ferrel-polar cell boundaries.

The jet streams naturally meander, swinging a bit north then south as they flow to the east (see Figure 3.7). Such meanders, like the jet streams' migrations, imply that the weather near cell boundaries will naturally be somewhat variable. The development of meanders can be related to the concept of *vorticity*, which will be discussed in Chapter 10; alternatively, and equivalently, the meanders can be viewed as a consequence of the Coriolis force.

Thus, consider a 'perturbation' to a jet stream in the northern hemisphere, perhaps caused by a high-altitude burst of wind that pushes the jet a bit northward, initiating a meander. The Coriolis force will deflect it to its right, so that—as it continues to flow to the east—it will gradually return to its initial latitude. But the Coriolis force is stronger to the north, so the jet is deflected a bit too much; it overshoots its initial latitude, gradually ending up a bit southward. Since the Coriolis force is weaker to the south, the forces originally producing the jet stream can dominate, and the jet is pushed northward once again. The combination of overshoot + restoring force (which of course is also the basis for the oscillations of a pendulum or spring) produces a periodic oscillation, called a **Rossby wave**, superimposed on the eastward flow of the jet stream. Because the Coriolis force provides a relatively weak restoring force, these meanders are truly *planetary*-scale *waves,* with wavelengths so long there might be only four or five cycles around the globe.

The jet streams are also able to steer storms that are large enough to penetrate the upper troposphere. For example, large low-pressure systems forming along the Polar Front near the Aleutian Islands can be pushed eastward by the polar jet. During extreme meanders, the jet can steer such storms as far south as the Gulf of Mexico, where the Gulf's warmth and humidity intensifies them; as the jet then pushes them up along the Atlantic coast, local residents will describe the winds as coming from the northeast (see Figure 3.8) and call the storms "nor'easters."

Very large meanders of the jet stream during the twentieth century were probably not all that common; but there is one we are all indirectly familiar with. By January 26 or 27, 1986, the northern hemisphere polar jet had meandered way south of its usual latitude and reached all the way to central Florida. The cold air accompanying it threatened the citrus

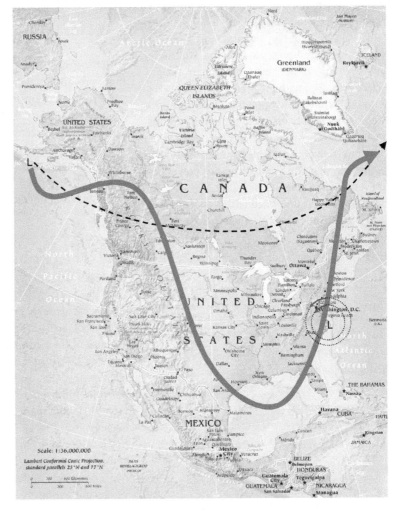

Figure 3.8 Evolution of a storm into a nor'easter. The storm ("L") may originate along the Polar Front near the Aleutians; steered by the polar jet, it will move eastward (dashed curve). However, if the jet is experiencing a large meander (heavy solid blue curve), it can push the storm over the Gulf of Mexico, where it will intensify. As the storm is steered up along the Atlantic coast, people just inland will experience winds (red arrows) coming from the northeast. Map from Geographic Guide.

crop, and news reports were full of pictures of orange growers deploying smoke pots and fans in their orchards to keep the oranges from freezing. Meanwhile, the Space Shuttle *Challenger* was on the launch pad at the Kennedy Space Center. During the night of January 27, temperatures dropped to 18°F, and they remained below freezing even up to launch time shortly before noon on the 28th. As it turned out (see Alley, 2003), the o-rings sealing the fuel tanks had cracked from the extreme cold; upon launch the fuel tanks leaked and the *Challenger*, just over a minute later, exploded, with tragic consequences.

3.1.9 Hurricanes

Hurricanes are intense low-pressure systems originating in the tropics and characterized by sustained winds of at least 74 mph (see Table 3.1). In order to generate such strong winds, the central pressure must be extremely low, drawing in the surrounding air (as in Figure 3.5b, top) with great energy. A very low central pressure is only possible if the air there is both appreciably warmed and quite humid, thus creating a very low central density. The very low density guarantees a rapid ascent of that air, and the great humidity powers it to even greater heights, thanks to the heat of condensation (as discussed earlier). Of course, such condensation will produce heavy clouds and rain as well.

Table 3.1 Categories of tropical storms, according to the National Hurricane Center.

Type of storm	Sustained winds at least
Tropical depression	Less than 39 mph
Tropical storm	39 mph
Category 1 hurricane	74 mph
Category 2 hurricane	96 mph
Category 3 hurricane	111 mph
Category 4 hurricane	130 mph
Category 5 hurricane	157 mph

Source: Adapted from http://www.nhc.noaa.gov/climo/.
Note: Hurricane categories are based on the Saffir-Simpson hurricane wind scale.

Hurricanes have their beginnings as low-pressure cyclonic systems in the tropics, that is, as tropical depressions; but that is problematic, since the global circulation typically produces rising air only at the Polar Front, far from the tropics, and at the equator (or ITCZ), where the Coriolis force is too weak to generate cyclonic motion. Ultimately, then, hurricanes form over the warmest of ocean waters outside the ITCZ; that warmth will both heat the air above and provide abundant water vapor. Such waters are called **hurricane breeding grounds**; as we see (Figure 3.9a, *and* 3.9b), they are found in the tropics, away from the equator.

Once established, the hurricane's fast winds cause more evaporation off the sea surface, strengthening it further. The massive storm will be pushed by the prevailing winds—the Trade Winds—westward; and the Coriolis force will deflect the storm poleward. If the storm travels to high-enough latitudes, the prevailing winds—now mid-latitude westerlies—will push it back toward the east. Though the path of the hurricane may be difficult to predict with precision, we know that eventually the storm will make landfall, or be deflected to colder latitudes, or both; at this point, the power of the storm will decline, and its life as a hurricane will end.

3.2 The Circulation of the Oceans

The oceans are a critically important component of Earth's climate system, not only because of their role in creating weather phenomena like hurricanes—or, more generally, because they can exchange heat and moisture with the atmosphere—but also because their large-scale circulation can supplement and even modify the global heat transfer carried out by the atmosphere. They do this *via* two completely different flow patterns that act on two completely different timescales: months versus millennia!

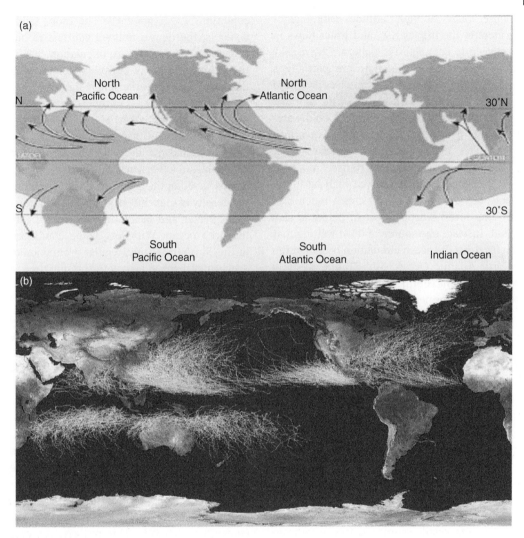

Figure 3.9 Hurricane breeding grounds.
(a) Dark gray-green areas denote regions of the world's oceans so warm—with temperatures exceeding 79°F (26°C)—they can potentially heat the overlying air and humidify it enough to create a central low pressure able to draw in winds at speeds of at least 74 mph. Arrows illustrate possible storm paths.
Image http://www.prh.noaa.gov/cphc/pages/climatology.php. / U.S. Department of Commerce / Public Domain.
(b) Actual tracks of tropical cyclones that occurred between 1985 and 2005. Colors indicate intensity: dark blue, tropical depression; light blue, tropical storm; hurricane categories 1 to 5 increasing from light tan to dark reddish brown. Image by Nilfanion / https://commons.wikimedia.org/wiki/File:Global_tropical_cyclone_tracks-edit2.jpg / Public Domain.

3.2.1 Thermohaline Convection

We consider first the slow, somewhat mysterious global-scale convective process known as the thermohaline circulation. In its original form, based on observations extending back at least to the 1925 *Meteor* oceanographic expedition (Schmitz, 1995, which also provides historical references), and as developed by such researchers as Stommel and Arons (1960), Gordon (1986) and Broecker (1987, 1991), dense waters at high latitudes sink, lighter waters in the tropics rise, and water flows in between to complete the pattern. The *density differences* driving this flow are partially a result of the temperature differences produced by the variation in insolation with latitude: warm surface water in the tropics, cold water near the poles; thus, for example, as warmth penetrates deeper into the tropical water column, that water will tend to rise.

The colder surface water at high latitudes, however, does not by itself create much of a driving force, since the ocean below is also cold. Instead, one source of downwelling that helps drive the ocean circulation occurs in the North Atlantic Ocean. Mid-latitude westerly winds blowing (north)eastward over the North American continent are fairly cold by the time they leave the land and cross the Atlantic, on their way to Western Europe. Over the ocean, they pick up heat from the North Atlantic surface water (and become warm enough to moderate the climate of Western Europe and beyond); the water they leave behind is now anomalously cold—just a few degrees above freezing (Broecker, 1991), thus relatively dense—causing it to sink deep into the ocean. This process acts continually; the water so created is called the **North Atlantic Deep Water** or **NADW**.

Of course, to fill the gap, surrounding waters—especially warmer surface water from the subtropics and even the tropics—must continually flow in toward the downwelling. That northward flow brings warmth to the region, increasing the surface temperature of the North Atlantic by several degrees (Manabe & Stouffer, 1988) and providing much heat that can be taken up by the westerly winds. The process of NADW formation is thus very efficient.

The density differences driving the convective flow are also caused by differences in *salinity*. The amount of salts (NaCl and others) in seawater is typically about $3\frac{1}{2}$ % (by weight), though it varies with both latitude and depth. With that salinity, the typical density of seawater is ~ 1.025 gm/cm^3 at the sea surface (the density of salt-free water under surface

conditions is ~ 1 gm/cm^3). The more saline water is, the denser it is—a contrast to our discussion of atmospheric circulation, where the presence of an 'impurity' (water vapor in that case) made the atmosphere lighter.

The northern reaches of the Atlantic, where NADW formation takes place, are a more saline environment than the northern Pacific Ocean—thanks mainly to the Gulf Stream and its northward extensions (such ocean currents are discussed in the next subsection)—making the density of that chilled North Atlantic water even greater and facilitating its descent into the deep ocean. With less saline conditions in the Northern Pacific, regular production of deep water there, at least at the present time, is not achievable (Warren, 1983; also Manabe & Stouffer, 1988; Broecker, 1991).

Meanwhile, at the coast of Antarctica, an amazing process occurs, transforming the very cold water there into very saline water, producing the densest water found anywhere in the oceans; this process is the freezing of coastal water into ice. As the freezing takes place, aided by cold winds blowing off the Antarctic ice sheet, the new ice accretes onto ice that is already there; but ice is mostly salt free, so the freezing water must expel its salts, and the surrounding water gains in salinity. The water so produced sinks rapidly to the seafloor, and is called the **Antarctic Bottom Water** or **ABW** (sometimes **AABW**). This process, centered mainly in the Weddell Sea and Ross Sea, is most effective during winter, producing ABW at the rate of up to a billion cubic feet per second!

The ABW has a counterpart in the Arctic, but with less impact on the flow. Arctic Bottom Water is produced during northern hemisphere winter, but in smaller amounts, and with most of that being trapped by seafloor topography within the Arctic Ocean basin.

In sum, then, oceanic convection is mainly driven by the downwelling accompanying the production of NADW and ABW (see Figure 3.10b). Such downwelling is much stronger than the tropical upwelling mentioned above, though the latter is also essential to the flow (if the flow is to conserve mass).

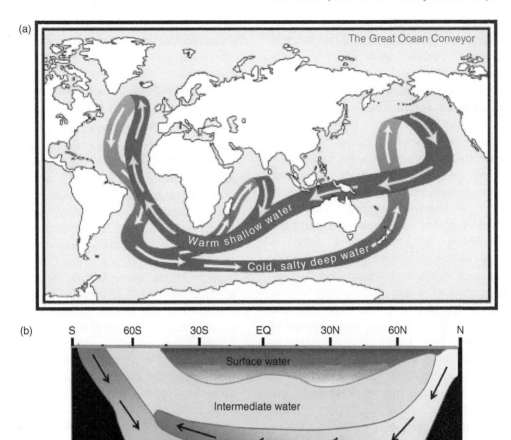

Figure 3.10 The "global conveyor belt" or thermohaline circulation of the world's oceans.
(a) An idealized depiction. Image from https://nsf.gov/news/mmg/media/images/image003.jpeg / National
Science Foundation / Public Domain.
(b) Simplified vertical cross-section illustrating the north-south thermohaline flow in the Atlantic Ocean.
Water masses—surface, intermediate, deep, and bottom waters—are distinguished by their temperatures,
salinities, dissolved oxygen, and other geochemical properties. Cross-sections of the Atlantic often include
an intermediate water mass outflowing at ~36°N from the Mediterranean Sea basin. Image adapted from
Wang et al. (2014) / U.S. Department of commerce / Public Domain.
Both of these images serve as popular representations of aspects of the thermohaline circulation; but, as drawn,
they are inconsistent with each other in one respect. Can you identify that disparity?

In recognition of the two sources of density differences that force it, this flow is called the **thermohaline circulation** (but not the more explicit "thermosaline" circulation). Given the pattern of flow inferred for it (see Figure 3.10a), and the fact that it is not confined to each hemisphere (unlike, for example,

the atmosphere's circulation), it is also known as the **global conveyor-belt circulation**.

It may seem unusual for this density-driven convection to involve *two* sources of density difference—a strangeness compounded by the possibility that those sources *may* act oppositely (e.g., the ocean near the horse latitudes will experience much evaporation, due to moderately high insolation and fair weather, leading to surface waters which are both warm and saline). But such "doubly driven" situations, more generally known as **double-diffusive convection**, actually occur throughout the Earth System.

The density differences driving the thermohaline circulation are slight, not exceeding a few percent globally and even less in any one region; consequently, the flow itself is quite slow. Upwelling in the tropics might occur at a rate of ~1 cm/day (see Stommel & Arons, 1960; also Munk, 1966). Applying a technique called "scaling," which we will learn about in Chapter 9, to the requirement of mass conservation, such a rate of upwelling implies that the lateral flow joining upwelling and downwelling regions may have speeds of ~17 m/day; with somewhat more sophisticated modeling, Stommel and Arons (1960) estimated north-south deep-ocean velocities of ~60 m/day. From these estimates, the timescale for a parcel of, say, NADW to reach the Southern Ocean or a parcel of ABW to reach the Aleutians (each of these traveling through just a portion of the entire global conveyor belt) is in the range of several hundred to more than two thousand years! Such long—millennial—timescales are consistent with measurements of the relative 'age' of bottom waters that had originated as ABW and subsequently crossed the equator to reach mid- and high-northern latitudes; those ages were inferred from the levels of oxygen or hydrogen isotope "tracers" (Garrison, 1993).

The Impact of the Thermohaline Circulation. Despite its sluggishness, the thermohaline circulation is not just a curiosity. As a global conveyor belt, it has the potential to communicate changes in climate experienced in one polar region to the other, on millennial timescales. With water's very high capacity to hold heat, it has the potential to force its own changes in climate, on those timescales. And as a flow pattern that extends down to the seafloor, it serves to ventilate the deep ocean,

providing dissolved oxygen and maintaining the near-freezing water temperatures there.

The most popular examples of the impact of this circulation on climate include one somewhat regional in focus, one specifically pole to pole, and one frighteningly global.

- **The Younger Dryas.** This was a cold period that temporarily reversed the warm-up taking place at the end of the Pleistocene around 12,000 yr ago; it began and ended rather suddenly, lasted about a millennium, and evidently affected mainly Europe. Broecker (1987, 1991 and references of earlier researchers therein) proposed that the Younger Dryas was caused by disruption of the conveyor belt; according to Meissner (2007), it was a complete shutdown of the Atlantic portion of that circulation. As the last glacial maximum wound down, extensive ice sheets melted away, leading to a massive addition of fresh water into the oceans—much of it, ultimately, into the North Atlantic. Impressively, this included a "pulse" of meltwater that itself lasted nearly three millennia (Ruddiman, 2014).

  Normally, as noted earlier, westerly winds blowing across the North Atlantic would have drawn warmth out of the surface waters, simultaneously creating the NADW and bringing warmth to northwest Europe. But that pulse of meltwater was too light to sink, even after passing winds had chilled it; the production of NADW ceased, the winds no longer carried any warmth, and glaciers could once again spread through Europe.

  Until, that is, the long-term 'memory' of the global conveyor belt—a memory carried in the momentum of its other components and in its thermal inertia—could (in our view) force a restoration of the circulation in the North Atlantic.

  A more comprehensive assessment of this hypothesis would include consideration of other North Atlantic cold periods (called *Heinrich events*; e.g., NOAA 2021), characterized by glacial surges and resulting in *ice-rafted debris* (carried by icebergs till they melted) being deposited on the seafloor; changes in the thermohaline circulation and NADW are thought to have played a key role in their occurrence (cf. McManus et al., 2004; Böhm et al., 201). A more comprehensive assessment should also consider the ability of salinity changes to affect the basic character of the circulation (e.g., Manabe & Stouffer, 1988; see also Maier-Reimer et al., 1990, whose model reveals a surprising sensitivity of the conveyor belt circulation to salinity).

And, there may well have been still other factors contributing to the onset or termination of the Younger Dryas. For example, the migration of humans from Asia to the Americas just prior to the start of the Younger Dryas was followed by the hunting of large North American mammals to extinction—an event called the "Pleistocene blitzkrieg" (Diamond, 1997). The absence of large mammals (which normally produce and emit methane during digestion) would have led to a decrease in methane levels in the atmosphere, perhaps significant enough to have cooled the climate and fostered the growth of glaciers. Also, during the Younger Dryas, high-latitude insolation continued to increase, due to orbital changes (see Chapter 6); perhaps that increase helped bring its glacial conditions to a close. On the other hand, such contributing factors are global, so a mechanism like that involving NADW production and the thermohaline circulation is still needed, to focus the effects on the North Atlantic.

- **A Climate Seesaw, From Pole to Pole.** In later chapters we will discuss how ice cores, drilled through mountain glaciers, ice caps, and ice sheets, can be used to shed light on regional and global climate over the past million years or so—most fundamentally, revealing a 'sawtooth' pattern of temperature variation that characterizes glacial-interglacial cycles. For now, we note that, within that overall sawtooth pattern, a number of shorter-term, millennial-scale

fluctuations can be seen. Greenland ice cores, in particular, reveal two dozen or so episodes of rapid and appreciable warming (by ~10°C!) during the most recent period of glaciation, between ~120 and ~15 Kyr ago; each lasted ~1/2 to 2 1/2 millennia and was generally followed by gradual cooling. Such warming episodes, which came to be known as *Dansgaard-Oeschger* events (see, e.g., Ruddiman, 2014; NOAA, 2021), may be attributed to the same thermohaline/NADW mechanism responsible for the Younger Dryas. As envisioned by Broecker (1991; see references in Buizert et al., 2015 for alternative mechanisms), the abundance of ice during the last glaciation might have caused the conveyor belt to switch off and on repeatedly: the warmth brought by westerly winds blowing across the Atlantic during normal thermohaline operation melted regional glaciers, whose meltwater eventually shut down NADW production, leading to cold westerly winds that chilled the climate till the ice sheets stopped melting, ultimately allowing NADW production to then resume, beginning the next cycle. This mechanism requires time, after NADW production has restarted, for the glaciers to rebuild their mass before the next round of melting begins; presumably, each time normal thermohaline operations resumed, the warm winds driving NADW production would have supplied enough humidity (and thus precipitation, as snow) to replenish the glacial mass.

Ice cores from Antarctica feature similar warming/cooling episodes (e.g., Brook et al., 2005), though their timing with respect to the northern hemisphere events was initially uncertain. The analysis by Buizert et al. (2015) established that the timing of Antarctic cooling (which, at ~1°C–2°C, was more subdued than in the North Atlantic) was correlated with North Atlantic Dansgaard-Oeschger warming, and Antarctic warming with Dansgaard-Oeschger cooling, in both cases with a lag of ~200 yr ($\pm$50 yr). This relation, whose underlying dynamics are vividly modeled by Rind et al. (2001) and others, has been called the *bipolar seesaw*. Presumably, disruption of the NADW, as discussed above, prevents the transport of heat toward the North Atlantic by the thermohaline circulation (see also Crowley, 1992; Stocker & Johnsen, 2003; Brook et al., 2005; Henry et al., 2019), causing it to build up farther south; the return of the NADW restores the coolness to the south – till the next cycle begins.

- **Future Climate Change.** Broecker's explanation of the Younger Dryas and Dansgaard-Oeschger oscillations turned out to resonate with the public (see Broecker, 1991) as well as with climate change researchers. Many interpreted his work as demonstrating that the addition of relatively small amounts of fresh water to the North Atlantic as a consequence of global warming could lead to drastic changes in climate in the near future—perhaps even in the coming decades (if not 'the day after tomorrow'...) (this hypothesis, and its converse, are reviewed in Raymo, 1994).

But in his work, that mechanism of climate change operated during or shortly after glacial periods, when large ice sheets still existed in northern mid- and high latitudes; upon melting, those ice sheets could effectively 'flood' the North Atlantic with great amounts of fresh water to disrupt the conveyor belt (cf. Rind et al., 2001). In contrast, the amount of ice currently available in the vicinity of the North Atlantic is limited.

Additionally, that flooding of the adjacent ocean would have to take place over a sustained period of time—like the "millennial pulse" that caused the Younger Dryas—in order to successfully fight the return flow of the thermohaline circulation.

Conveyor-Belt Complications. In recent years, attention has focused on the upwelling components of the thermohaline circulation, and it has become increasingly appreciated that the thermohaline forcing is too weak to drive even a sluggish upwelling in the tropics (see, e.g., Vallis, 2005). Indeed, one common way to induce a fluid to convect is by heating it from below; heating the oceans 'from above,' as initially suggested for the tropics, is simply inefficient. Worse still, that tropical heating is spread out diffusely over a broad area.

Vertical motions throughout the deep ocean can also be induced bathymetrically. Bottom topography can *direct* the horizontal currents produced by an initial downwelling *upward*. And, topographic features can *scatter* tidal currents, generating internal waves and buoyant instabilities (e.g., Munk, 1966; Munk & Wunsch, 1998). The preferred mechanism, however, is believed to be *surface winds*, 'stirring up' the oceans into a state of turbulence that would allow heat and mass to diffuse (upward) more energetically, leading to more effective upwelling over the entire heated area (e.g., Wunsch, 2002; Vallis, 2005). (Given the resulting chaotic state of the oceans, perhaps "shaken," and not "stirred," would be the more appropriate terminology....)

Such a turbulent state implies that depictions of the conveyor-belt circulation, rather than indicating the actual flow path, should be interpreted as an overall "mean flow" that comes to light when the actual, chaotic motions of the seawater are averaged. And, where upwelling takes place, the actual flow path should be over a much broader region than is depicted.

There are a number of features of the conveyor-belt circulation (Wunsch, 2002) that could serve as the basis for its name, if "thermohaline" is judged to be an incomplete (or even inaccurate) descriptor. In recent years, its most popular name has been the **meridional overturning circulation** (MOC), a recognition that the thermohaline overturning (upwelling and downwelling) is primarily north-south (i.e., meridional) in orientation (e.g., Figure 3.10). We note that such a name should more correctly be the **meridionally overturning circulation**, since the circulation is east-west as well as north-south, and it is only the overturning which is north-south.

In any case, we choose to continue here the use of a name that refers to the driving force of the circulation, highlighting its full name as the **wind-assisted thermohaline circulation** (in the spirit of oceanographic research, a more acronym-friendly name might be the wind-assisted overturning circulation or WAO circulation…).

Finally, nearly a century of observations of water properties (temperature, salinity, density, dissolved oxygen, and numerous ions) through the breadth and depth of the oceans has allowed the slow paths followed by NADW and ABW masses to be more accurately traced. A creative synthesis of these observations by Schmitz (1995) reveals that the wind-assisted thermohaline circulation is better described as having a four-layer structure (proposed earlier by Reid, 1981), so that it involves intermediate-depth waters as well as bottom and deep waters and (near-)surface water (see Figure 3.11). As a *multilevel global conveyor belt*, the upwelling can perhaps proceed more easily, locally, layer by layer (with appreciable horizontal transport between each step up), since extensive, uniform upwelling throughout the abyss of the oceans is no longer required (Schmitz, 1995). We might also surmise that such complexity would yield both greater natural variability in the circulation (Wunsch & Heimbach, 2013) and greater robustness; that is, with so many components working together, disrupting its operation and altering its impact on climate become more difficult.

3.2.2 Wind-Driven Circulation

The oceans exhibit a very different kind of circulation than the global conveyor belt. This second flow is driven directly by surface winds—the Trade Winds and mid-latitude westerlies—which blow in each hemisphere as part of the atmosphere's three-cell convection and drag the surface waters along with them. In some ways, this is the climate system's analog to plate tectonics; in the latter, of course, it is the surface layer of the solid earth being 'blown about' by the 'winds' of the mantle

as the mantle convects. And although the oceans do not act as a rigid lithospheric layer as they respond to the wind forcing, the flow that results has (as we will see shortly) very well-structured patterns.

The tropical easterlies and mid-latitude westerlies comprising the wind forcing are more widespread and more persistent than the NADW and ABW downwellings that drive the thermohaline convection. As a consequence, the wind-driven currents are more energetic; in most parts of the ocean, they typically have speeds of ~10 cm/sec—three to four orders of magnitude faster than the horizontal thermohaline flow—and in some areas the currents are even faster. On the other hand, despite acting over most of the oceans, the wind forcing is largely superficial, and is usually viewed as being able to induce flow only in the upper layers of the oceans.

There are four main components to the wind-driven circulation.

- The **Antarctic Circumpolar Current** or **ACC**, also known as the **West Wind Drift**. As its name implies, this current encircles Antarctica; driven by persistent mid-latitude westerlies, and—unlike other currents—unobstructed by land, it entrains water down to depths of at least a couple of kilometers and flows, west to east, at a speed of ~5 cm/sec. Its *volume transport*, that is, the volume of water passing any observation point per time, is ~150 million m^3/sec, making it the most massive 'river' system in the world. In units more commonly used by oceanographers, the ACC transports ~150 **Sverdrups (1 Sv ≡ 1 million m^3/sec)**; for comparison, the Mississippi River might transport roughly ½ Sv during a major flood.

- A cyclonic motion called a **gyre** in each of the five ocean basins: the North Atlantic, North Pacific, South Atlantic, South Pacific, and Indian Ocean (the bulk of which is in the southern hemisphere). Each gyre is created by the Trade Winds, which drag tropical water toward the west, and mid-latitude

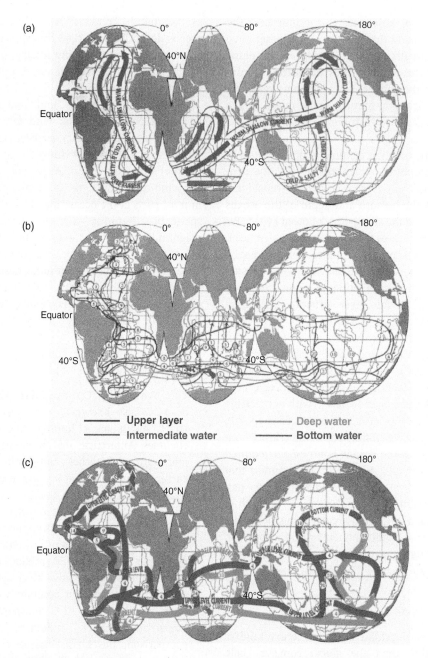

Figure 3.11 The multilevel global conveyor belt (wind-assisted thermohaline circulation) of the world's oceans, proposed by Schmitz (1995).

(a) The conveyor belt model of Broecker (see Figure 3.10a), redrawn from a basin perspective (the figure shows the Atlantic, Indian, and Pacific Ocean basins, from left to right). Schmitz (1995) / John Wiley & Sons.

(b) The four-layer global conveyor belt model of Schmitz (1995). Numbers in circles are estimated volume of water transported by the circulation, in units of 10^6 m^3/sec. Schmitz (1995) / John Wiley & Sons.

(c) A condensation of the four-layer model into three layers by combining upper and intermediate waters. Schmitz (1995) / John Wiley & Sons.

Subsequent work by Schmitz (1996), incorporating "a good bit of idealization," has led to further modification of these illustrations.

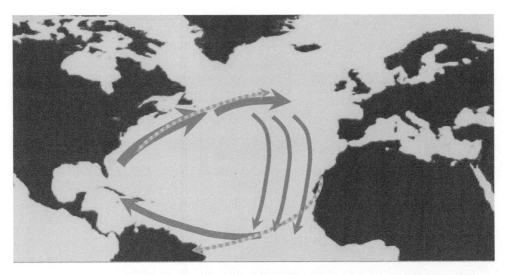

Figure 3.12 Creation of the gyre in the North Atlantic Ocean. The bottom edge of the figure is at the equator. Thin gray arrows depict the tropical easterly and mid-latitude westerly winds driving the flow; blue arrows indicate the currents making up the gyre. The Coriolis force, which strengthens with latitude, causes the western boundary current—here, the Gulf Stream—to intensify as it flows north and makes the return flow diffuse.

westerlies, which drag mid-latitude water toward the east (see Figure 3.12). The westward-moving tropical current will eventually encounter land, which blocks the flow and forces the current to shunt to the side—which, with the help of the Coriolis force, has already been happening. As it reaches the coast, then, it will flow *poleward*, up along the coast, as a **western boundary current** (western, from the perspective of the basin). At mid-latitude, the westerly winds drag the current back toward the east, where land is eventually encountered again and now, thanks once again to the Coriolis force, the current is deflected toward the equator; this "return flow" completes the cyclonic pattern.

Would you say the sense of motion in the North Atlantic and North Pacific resembles that of a cyclone or anticyclone? What about in the southern hemisphere basins?

Because the Coriolis force becomes stronger as the water flows to higher latitudes, the deflection experienced by the

western boundary current increases, and greater amounts of water are 'folded into' it, concentrating and intensifying it. We know some of these western boundary currents as the **Gulf Stream**, which flows northeastward along the east coast of the United States (see Figure 3.13); and the **Kuroshio**, with a similar flow along the east coast of Asia, including Japan. Typical velocities of these western boundary currents are ~2 m/sec—an order of magnitude faster than the rest of the gyre.

As they flow poleward, these surficial boundary currents increase in width and speed, so the volume of water they carry per second grows significantly; the Gulf Stream, for example, transports ~30 Sv along the Florida coast, ~85 Sv past the Carolinas, and ~150 Sv as it flows past Cape Cod. For comparison, the strongest components of the wind-assisted thermohaline circulation involve lateral flow of ~15–20 Sv (associated with the NADW; e.g., Schmitz, 1995; Gordon, 1986) and ~5–20 Sv (associated with the ABW; Schmitz, 1995).

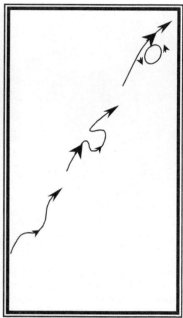

Figure 3.13 Satellite image showing the Gulf Stream as it flows northeastward along the western boundary of the North Atlantic basin. Colors added indicate temperature (red-orange = warm, blue-green = cold), revealing that the warm current meanders as it flows. Occasionally, more extreme meanders are 'pinched off' by a surge of water through the main line of the current (see sketch to the right); the resulting "rings"—which trap colder water (cold-core rings, with greenish centers) or warmer water (warm-core rings, with orange centers) depending on whether the original meander was open to the left (i.e., northwest) or to the right (i.e., southeast)—can persist for as long as a few years! Image from NASA / https:// earthobservatory.nasa.gov/features/Franklin/franklin_2.php / last accessed August 22, 2023.

In contrast, the current being dragged to the east, as the highest-latitude segment of the gyre, is experiencing the strongest Coriolis force it can. This strong force effectively deflects the flow equatorward all along the segment, producing a diffuse return flow (as in Figure 3.12). And, without a strengthening Coriolis force to intensify it as it flows toward the equator, the return current is weak as well, with typical speeds of ~10 cm/sec no stronger than anywhere else in the gyre. Despite their mundane character, however, for reasons discussed below, those currents are all named: for example, the **Canary Current** (the return flow of the North Atlantic gyre); the **California Current** (traveling southward past the west coast of North America, as the return flow of the North Pacific gyre); and the **Humboldt**
or **Peru Current** (traveling northward along the west coast of South America, as the South Pacific gyre's return flow).

Not surprisingly, the volume transport by these currents is lower than that carried by the western boundary currents, although their slower flow is compensated to some extent by their greater width. As an example, the work of a number of researchers (e.g., Stramma, 1984; Gould, 1985; Zhou et al., 2000) suggests that the southward transport of the Canary Current is ~10 Sv (we note, however, that, based on its dimensions—a width of ~10^3 km and a depth of ~½ km (Gyory et al., 2013)—we might estimate a typical value of ~50 Sv; though larger, this estimate is still smaller by half than the typical transport of the Gulf Stream).

- A narrow flow near the equator called the **equatorial countercurrent**. This flow results from the convergence of the Trade Winds at the ITCZ, which push water toward the western equatorial portions of the ocean basins; with no winds to hold the accumulation there (at this boundary of the Hadley cells, the air movement is upward), the piled-up water simply flows back to the east as an equatorial return flow.

- **Undercurrents**, which exist throughout the oceans and flow in the opposite direction to the overlying surface currents. Undercurrents are beyond the scope of this textbook.

The components of the wind-driven circulation (illustrated in Figure 3.14a, excluding the undercurrents) were first explained comprehensively by Munk (1950). These features can be recognized in Figure 3.14b, which displays ocean currents inferred from satellite altimetry observations (satellite altimetry and its use for such inferences will be discussed in Chapter 6).

3.2.3 The Wind-Driven Oceans Move Heat, Too

Like the three-cell circulation of the atmosphere and the thermohaline circulation of the oceans, the oceans' wind-driven circulation serves to transfer heat between high and low latitudes, taking warm waters poleward via its western boundary currents and bringing cool waters into the tropics with its return flow. Those currents are slower than the atmospheric circulation; on the other hand, water has a much higher heat capacity than air. It turns out that, in general, the wind-driven circulation is perhaps a quarter to half as effective as the atmosphere in warming the high latitudes and cooling the tropics (Trenberth & Caron, 2001; Vallis, 2005).

The coolness brought to the tropics by the return flow of the gyres is enhanced by the phenomenon of *coastal upwelling*. As those wind-driven ocean currents move equatorward along continental boundaries, the Coriolis force acts to deflect them, away from the coast; for example, the Humboldt (Peru)

(a)

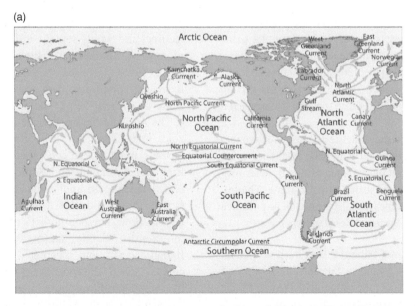

Figure 3.14 (a) Idealized illustration of surface currents in the world's oceans. Sketch, from https://www .ocean-pro.com/gulfstream/globaloceancurrents.jpg, shows the gyres in each ocean basin, the ACC, and the equatorial countercurrent. Additional currents, such as the various high-latitude extensions of the Gulf Stream, are beyond the scope of this textbook.

(b)

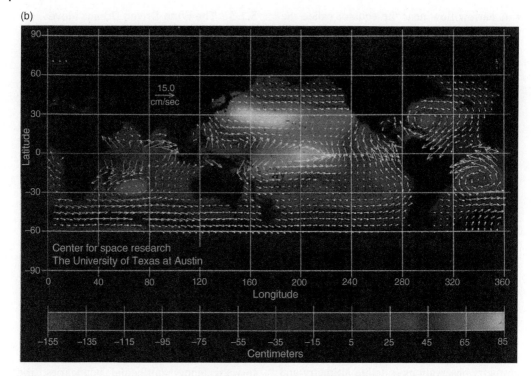

Figure 3.14 (b) Surface currents in the world's oceans, with velocities represented by white arrows, as inferred from observations of sea-surface topography (color contours) during 1987 by the altimetric satellite Geosat. In this Pacific-centered image, only the broad-scale features of these currents are shown. The University of Texas at Austin / Nerem et al. (1990) / https://www.csr.utexas.edu / Last accessed under September 22, 2023.

Current moves northward, toward the equator, with the west coast of South America to its right, and (being in the southern hemisphere) the Coriolis force deflects the current to its left. Deep water—cool water—rises to fill the gap, making those coastal waters even colder.

At times in the past, the oceans' effectiveness in transferring heat may have even been great enough that it played a major role in determining the state of Earth's global climate. The following examples also illustrate the importance of an Earth System Science perspective.

- We saw in Chapter 2 that carbon dioxide levels had begun to fall and the world had begun to cool following the PETM and EECO extreme warm events, thanks to the uplift and weathering of the Tibetan Plateau. But, despite the cooling, Antarctica—located at the South Pole, having been brought there

75 Myr earlier by continental drift—was ice-free, and would remain so for millions more years. Kennett (1977; see also Kennett et al., 1975) suggested that the transition to glacial conditions there was eventually achieved with the help of the ACC.

The ACC did not exist in the early Cenozoic, and could not exist until further tectonic motions had separated Antarctica from the Australian and South American plates, creating an unobstructed "Southern Ocean" that would allow a circumpolar current to flow. The Southern Ocean—where the South Atlantic, South Pacific, and Indian Oceans join together—was finally created when the northward motion of the Australian lithospheric plate separated Tasmania (and the South Tasman Rise in particular) from Antarctica, around 32 Myr ago, and the Drake Passage (between South America and the Antarctic Peninsula) had become open and deep, roughly 30 Myr ago (Lawver & Gahagan, 2003). It is possible the ACC could have bypassed the Drake Passage before separation from South America was complete: in the absence of an ice cover, West Antarctica (where the Antarctic Peninsula is located) is actually an archipelago, for the most part, allowing some through-flow; see the right-hand image of Figure 3.30a in this chapter. The ACC, then, would have been operating ~32 Myr ago, and was well established and at full strength by about 30 Myr ago.

In the absence of an Antarctic Circumpolar Current, the intense western boundary currents of the southern hemisphere (the predecessors of the Brazil current, East Australia current, and Agulhas) were able to bring their warmth all the way to the Antarctic coast. The development of the ACC, however, would have *isolated* the Antarctic land mass, *insulating* it from the warmth of those currents. As first proposed by Kennett, this insulation would have promoted the growth of ice throughout Antarctica, and helped take the world into its current Ice House climate regime.

In fact, various observations in and around Antarctica (with detailed analyses by Kennett 1977; Lawver & Gahagan, 2003) indicate that major continental glaciation had begun by about 33 Myr or so ago, with glaciation intensifying over the next few million years. This timing suggests ACC insulation was indeed a contributing factor to that glaciation.

One alternative to an ACC mechanism postulates that the continuing decline of atmospheric carbon dioxide (from weathering of the Tibetan Plateau, etc.) simply reached a critical point where the Antarctic was cold enough for glaciers to grow and begin to coalesce into a continental-scale ice sheet. Numerical modeling by DeConto and Pollard (2003) predicts that this mechanism is sufficient by itself, and that opening the Drake Passage gateway would enhance the glaciation only secondarily. This mechanism

does not explain why Antarctic glaciation developed around the time the ACC came into existence, however; if the effect of opening the Tasmanian gateway was also found to be secondary, an explanation for the timing of Antarctic glaciation would still be required.

As the climate cooled more and as the insulating effects of the ACC continued, the Antarctic ice sheet would grow further, reaching full size by perhaps 12 Myr ago. But there were glitches along the way: according to Lagabrielle et al. (2009), tectonic 'jostling' caused the Drake Passage to constrict somewhat between ~29 and 21 Myr ago, temporarily weakening the ACC and undermining its insulating effects; they speculate that this could account for short-lived warm-ups in climate that occurred globally ~26–25 Myr ago and 17–14 Myr ago.

- Plate tectonics and ocean circulation may also have collaborated more recently to affect Earth's climate. As the Americas drifted around over the past several million years, they eventually became a continuous land mass, with the closing of the Isthmus of Panama around 2.8 Myr ago (O'Dea et al., 2016). At that time, with currents no longer able to 'leak' from the North Atlantic into the North Pacific, the North Atlantic gyre would have intensified. A more intense western boundary current—that is, a stronger Gulf Stream—would have been able to push farther north, carrying its warmth (and associated humidity) with it. The result would have been increased precipitation in northwestern Europe, Greenland, and eastern North America, feeding the growth of northern hemisphere ice sheets in winter till they could survive and begin to advance year-round down to mid-latitudes. Thus, paradoxically, the additional warmth of a stronger Gulf Stream might have initiated the Pleistocene Ice Age (Keigwin, 1982; see also Kaneps, 1979).

According to Maier-Reimer et al. (1990) and Haug and Tiedemann (1998), closing of the Isthmus of Panama strengthened the

North Atlantic thermohaline circulation; with greater NADW production requiring a stronger return flow of warm surface waters to the north, Haug and Tiedemann concluded that this would have led to an increase in humidity above the North Atlantic, adding to that produced by the Gulf Stream intensification and setting the stage for the Pleistocene.

We can also speculate as to what factors might cause the wind-driven ocean circulation to change our climate in the near future. Obviously, over the short term, tectonic changes to the ocean basins and any resulting modification of ocean circulation would most likely be imperceptible. Another possibility that comes to mind is the addition of fresh water to the North Atlantic in the near future, anticipated by some as an impending consequence of global warming. But the mechanism by which North Atlantic meltwater affected thermohaline circulation during the Pleistocene or Younger Dryas does not apply to the Gulf Stream (or other components of the wind-driven circulation). That's because the production of NADW is not directly related to the processes generating the Gulf Stream—the Trade Winds dragging Atlantic water toward the southwest, and the Coriolis force deflecting that flow to its right and then intensifying it as it moves up the American east coast.

With the atmosphere as an intermediary, though, it is conceivable that the addition of water to the North Atlantic might somehow lead to changes in the North Atlantic gyre. That is, if the added water were sufficiently massive and warm, it might shift or intensify the Icelandic low or otherwise modify the Polar Front. The resulting changes in the westerly winds (and ultimately the Trade Winds) might affect the gyre and ultimately increase the intensity of the Gulf Stream—thereby boosting the equator-to-pole heat transport. We note that, in this scenario, such an effect would amount to a positive feedback, furthering the addition of meltwater into the North Atlantic.

In fact, in the next section we will study in detail a naturally occurring phenomenon—called El Niño and the Southern Oscillation—in which the presence of a pool of warm water (in this case, off the coast of South America) causes changes in the atmospheric three-cell circulation, which in turn lead to changes in the wind-driven circulation. There are several differences between this phenomenon and our speculative North Atlantic scenario; for one thing, the El Niño warm pool is not created by glacial meltwater. But there are enough similarities to suggest our scenario might be plausible.

That scenario is just one way that global warming might cause changes in atmospheric and oceanic circulation over the short term. In the second part of this chapter, we will see that climate models predict quite a variety of changes in wind and ocean patterns and intensity, as a consequence of global warming.

3.3 El Niño and the Southern Oscillation: A Coupled Atmosphere–Ocean Phenomenon

The oceans and atmosphere interact in any number of ways, for example mechanically as the global atmospheric circulation drives oceanic gyres, and thermally as the warmth and humidity of tropical waters fuel the growth of storms into hurricanes. El Niño and the Southern Oscillation are a complex set of ocean-atmosphere interactions that scientists at the National Oceanic and Atmospheric Administration (NOAA) have described as "second only to the … seasonal cycle in … [their] effects on climate around the globe." We begin with El Niño, an oceanic phenomenon.

3.3.1 El Niño

El Niño refers to the appearance of anomalously warm water—a *warm pool* covering an area that can significantly exceed the size of the continental United States—off the

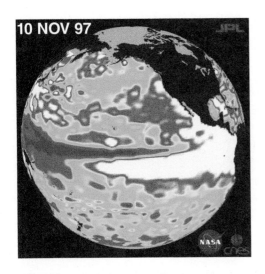

10 NOV 97

Figure 3.15 Sea surface temperatures inferred from satellite observations during the height of the 1997–1998 El Niño. Purple and blue colors correspond to the coolest temperatures, whereas white and red are the warmest. Note the massive warm pool off the coast of the Americas, extending entirely up the U.S. west coast and down the coast of Chile (at the edge of the image).
Actually, the satellite whose observations were used here, TOPEX/Poseidon, measured sea surface heights (SSH); the greatest SSH, in white, were at least 14 cm above average (corresponding to sea surface temperatures of ~29°C or more), and the least (in purple) were at least 18 cm below average. NASA / https://www.nasa.gov/centers/jpl/news/topexf-20050105.html / last accessed August 22, 2023.

Pacific coast of South America (Figure 3.15). Normally, the Humboldt (Peru) Current, the return flow of the South Pacific gyre, brings cool water to that area on its way north back to the tropics; together with the coastal upwelling it causes, as noted earlier, the net result might be coastal temperatures ~15°C–20°C. In contrast, the water of the El Niño warm pool may be ~20°C–23°C to begin with, and heat up to ~28°C or more (e.g., Rasmusson, 1985). The appearance of the warm pool heralds the start of an El Niño.

El Niños have been observed for well over a hundred years; their impacts have been inferred, from historical and environmental records, going back millennia. El Niños were reportedly named around the turn of the twentieth century by Peruvian fishermen who noticed that they appeared around Christmastime (the name "El Niño" refers to the Christ Child). However, they are not an annual event, instead occurring irregularly, perhaps every ~2–8 yr (see Table 3.2).

An El Niño warm pool has several 'local' consequences, including (for the most *extreme* El Niño events):

- disruption of the Peruvian fishing industry, as schools of anchovy—a major source of protein worldwide used mainly as fish-meal, to feed livestock—disappear from the surface waters, seeking a cooler and more nutrient-rich environment. In the years following the 1972–1973 El Niño, such disruptions combined with other factors (such as widespread droughts and the increasing success of soybeans as an alternative protein source) to cause the collapse of the Peruvian fishing industry (the largest in the world at the time; e.g., Ros-Tonen & van Boxel, 1999), leading the world to turn elsewhere for its protein needs (see Glantz, 2001);

- a rise in sea level, by as much as ~50 cm or more at the Peruvian coast (Chelton & Enfield, 1986)—a result of the additional mass of the warm pool; and

- intense coastal storms, with heavy rainfall and flooding—even in normally dry areas like the high desert of Chile (which is at the horse latitudes)—a result of the overlying air being warmed and humidified by the warm pool.

As our discussion of this phenomenon now broadens, though, we will see that the climate disruptions of an El Niño warm pool can extend around the world.

3.3.2 Southern Oscillation

The **Southern Oscillation** is a "seesaw" in atmospheric pressure over the tropical Pacific. There are two versions of this oscillation; in this section, we focus on the version that corresponds to *strong* El Niños. Recent categorizations of El Niños have distinguished them

Table 3.2 El Niño-Southern Oscillation (ENSO) events since the latter part of the nineteenth century.

Approx. date of ENSO event[a]	Significance
1876–1878	➤ Strongest event of the nineteenth century (Wolter, 2011)
1888–1891	➤ These four plus the previous:
1896–1897	combined with prevailing economic conditions, they are associated with the deaths of millions of people worldwide (Davis, 2000)
1899–1900	
1902–1903	
1905–1906	
1911–1912	
1914–1915	➤ Strong
1918–1919	➤ Strong
1923–1924	
1925–1926	➤ Strong
1929–1932	➤ Longest ENSO
1939–1942	➤ Longest ENSO; possibly included a strong ENSO 1940–1941
1951–1952	
1953–1954	
1957–1959	➤ First suggestion that El Niño and the Southern Oscillation were coupled (Bjerknes, 1969)
1963–1964	
1965–1966	
1968–1970	
1972–1973	➤ Strong; inspired scientific research (e.g., Wyrtki, 1975)
1976–1978	
1979–1980	
1982–1983	➤ Huge, with effects seen throughout the Pacific
	➤ Slowed Earth's rate of rotation
	➤ Term "teleconnections" popularized
1986–1988	➤ First to be detected first by satellite (Geosat)
1991–1995	➤ Longest-lasting (usually divided into 1991–1992 and 1994–1995)
1997–1998	➤ Most severe of the century
	➤ Global public awareness
2002–2003	➤ Mild
2004–2005	➤ Mild
2006–2007	➤ Strong
	➤ Heavy storms even in western North America (three major snowstorms over three weeks, centered in Colorado)

Table 3.2 (Continued)

Approx. date of ENSO event[a]	Significance
2009–2010	➤ Strong
	➤ Heavy storms even in eastern North America (blizzards in Southeast and mid-Atlantic states)
	➤ Strongest-measured "central Pacific"-type El Niño, where (in contrast to 'classical' El Niños), the warm pool was concentrated in the central, not eastern, equatorial Pacific (e.g., Ashok et al., 2007; Lee & McPhaden, 2010)
2015–2016	➤ As strong as the 1997–1998 ENSO? (McPhaden, 2016)
2018–2019	

Note: Historical inferences between 1871 and 2005 are based mainly on a combination of reconstructed atmospheric pressure and sea surface temperature data called the multivariate ENSO index—extended version (Wolter & Timlin, 2011; see also Wolter, 2011), an appropriate index for a coupled atmosphere-ocean phenomenon. See also the chronology in Glantz (2001).
[a] In general, each event started some months into the year given and ended similarly.

in other ways, such as according to whether the warm pool is mainly in the eastern versus central Pacific (see, e.g., Fedorov et al., 2015; also Table 3.2); using the different types of Southern Oscillation as a guide, however, our criterion here will be the size of their warm pool, with strong El Niños characterized by a massive warm pool.

Normally, as part of the three-cell atmospheric circulation in the southern hemisphere, atmospheric pressure at Earth's surface is low in the Doldrums, especially in the region of the western equatorial Pacific north of Australia (where, as Rasmusson (1985) phrases it, "the air rises in towering rainclouds"); and the pressure is high along the horse latitudes of ~30°S, for example in or near the South Pacific High, which is centered in the eastern portion of the south Pacific subtropics. Locations ranging from just off the coast of Chile to way out in the mid-Pacific at Easter Island (latitude 27°S) will generally experience higher pressures because of that subtropical high.

Occasionally, however, the pattern reverses, with relatively high pressure developing in the western equatorial Pacific and low pressures in the eastern subtropical (and tropical) Pacific, extending even to the Chilean coast. Such a reversal—a **Southern Oscillation**—amounts to a disruption of the three-cell circulation of the atmosphere.

This reversal is frequently measured using the **Southern Oscillation Index** (or **SOI**), originally defined (e.g., Quinn, 1974) as the difference in pressure between that at Easter Island and that at Darwin, Australia (the northernmost and most nearly equatorial city in Australia). A positive value of the SOI corresponds to the pressure at Easter Island exceeding that at Darwin—the norm for the three-cell circulation—whereas a negative value (higher pressure at Darwin) indicates that a seesaw has occurred.

With data from Easter Island prior to 1950 partially unavailable (see Quinn, 1974), the SOI has been redefined to involve the pressure at Tahiti versus Darwin (also, some researchers now define the SOI using a more elaborate formulation based on the pressure difference, rather than using just the difference itself). Given that the latitude of Tahiti is nearly 18°S and that, for atmospheric pressure over the Pacific Ocean, it is Easter Island that anti-correlates most strongly with locations near Darwin (demonstrated by Berlage as shown in Julian & Chervin, 1978), the redefined SOI may be a less-satisfactory

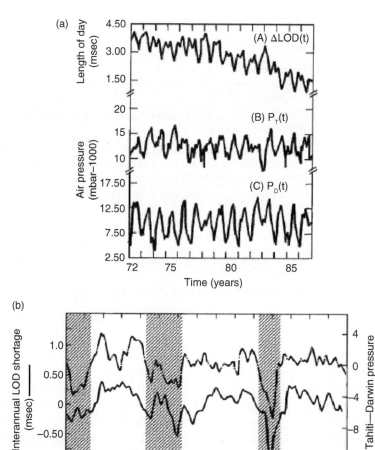

Figure 3.16 Observations of (a) surface atmospheric pressure at Tahiti and Darwin are used to construct (b) the Southern Oscillation Index (in this graph, defined simply as the pressure difference between those two sites). Shaded areas in (b) indicate times of El Niños, revealing the correlation between the two phenomena. Figures from Chao (1988). Chao's work also confirmed the ability of ENSO events to affect the rate of Earth's rotation (thus its length of day or LOD), a connection seen more clearly in (b) after the seasonal (annual) variation and a long-term trend in LOD, both evident in (a), have been removed. Note that defining the LOD curve in (b) 'negatively,' so that a negative value corresponds to a longer LOD, allows the correlation with ENSO to be more clearly seen.

measure of the pressure oscillation itself, but it is still useful. Figure 3.16a shows the atmospheric pressure measured at Tahiti and Darwin between 1972 and 1986, and Figure 3.16b uses those pressure differences (Tahiti minus Darwin pressure) to depict the SOI. The shaded areas in Figure 3.16b indicate times of El Niños. This analysis, by Chao (1988), is one of many examples demonstrating that reversals in pressure, i.e. negative SOI,

correspond to the appearance of warm pools off the South American coast, that is, times of El Niños. In short, the two phenomena are coupled! Such an idea was first suggested by Bjerknes (1969). In recognition of the coupling, the phenomena are generally referred to as El Niño and Southern Oscillation events or **ENSO** events (see Table 3.2).

This coupling means that the influence of an El Niño can extend well beyond the Pacific coast of South America, thanks to the atmospheric Southern Oscillation and its effects on the three-cell circulation. The ability of strong ENSO events to impact the whole world was made evident by the global devastation associated with the 1982–1983 El Niño. In fact, though, scientific analyses prior to that event had begun to document the connections, and the work of Rasmusson and Carpenter (1982, 1983) on earlier El Niños helped settle the issue. Perhaps it was the optimal timing of their publications, but one word they used (and it had been used at least as early as Bjerknes (1969) in this context) seemed especially apt as a description of the global reach of El Niño: its **teleconnections**, that is, its long-distance disruptions.

Examples of these teleconnections include droughts in tropical rainforest regions, even leading to record-setting, massive wildfires, such as in Indonesia and the Amazon during the 1997–1998 ENSO (discussed near the end of this chapter); heavy rainfall and floods in normally dry areas, such as Israel and other parts of the Mideast; heavy storms battering the U.S. west coast; and a decline in the frequency of hurricanes generated in the Atlantic Ocean and Caribbean.

3.3.3 The Mechanism of a Strong ENSO Event

A satisfactory explanation of how ENSO works has to account for the local and regional impacts of El Niño while at the same time recognizing the role of the atmosphere and elucidating the basis for any teleconnections.

ENSO events are believed to be triggered by a weakening of the Trade Winds over the Pacific (Wyrtki, 1975). As noted earlier, the Trade Winds not only drive the oceanic gyres, they also cause a "set-up" of warm water in the western equatorial regions of the ocean basins (some of which continues on as part of the gyres and some of which flows back east as the equatorial countercurrent). In the tropical Pacific, that warm water—the warmest temperatures anywhere, ocean-wide (McPhaden & Picaut, 1990)—is the warm pool that will eventually characterize an El Niño (see Figure 3.17a). When the Pacific Trades weaken, the bulk of that warm pool is no longer

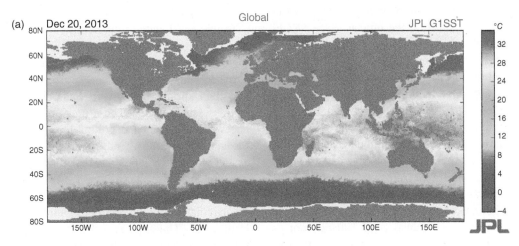

Figure 3.17 (a) High-resolution global contour map of sea surface temperature averaged over one day in 2013, produced by the Jet Propulsion Lab's *Regional Ocean Modeling System* group at Cal Tech. Temperatures are color-coded according to the scale at the right.
The warm pool in the western tropical Pacific, roughly centered on the ITCZ, is characterized by temperatures up to ~30°C or more over a wide area. As we see, temperatures in the eastern tropical Pacific are much cooler, typically ~20°C. A 'tongue' of that cold water (in green) is visible extending westward away from the coast of Ecuador.
Image downloaded from https://www.wikiwand.com/en/Sea_surface_temperature / NASA / Public Domain.

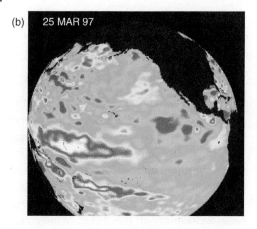

(b) 25 MAR 97

Figure 3.17 (b) Sea surface temperatures inferred from satellite observations, as in Figure 3.15, but preceding the climax of the 1997–1998 El Niño. We see here the warm pool to the east of its 'normal' location in the western tropical Pacific, beginning its eastward migration just after the El Niño was triggered. Downloaded from NASA / https://sealevel.jpl.nasa.gov/data/el-nino-la-nina-watch-and-pdo/historical-data/?page=0&per_page=40&order=publish_date+desc%2Ccreated_at+desc&search=&category=232 / last accessed August 22, 2023.

held to the western part of the basin, and it can flow eastward (see Figure 3.17b).

It was that kind of motion of the warm pool, in advance of the official 1986–1987 El Niño, that was detected by Geosat (Table 3.2) (Cheney & Miller, 1988), a satellite that was used to map sea surface topography. And that eastward motion of the warm pool during the strong 1982–1983 El Niño caused the solid earth to slow its rotation a bit (Table 3.2, Figure 3.16)—a compensation needed in order for the Earth System to conserve its total angular momentum. Wyrtki (1985) used sea-level observations to estimate that the eastward migration of that warm pool in 1982–1983 amounted to a transport of ~40 Sv—not as much as the most massive or strongest ocean currents, as we saw earlier in this chapter, but nevertheless quite respectable in comparison overall to the wind-driven gyres.

As the warm pool approaches the shores of South America (see Figures 3.15 and 3.18b, and compare with Figure 3.18a), the mass

of its water adds to the water already there, raising sea level. Furthermore, the presence now of warm seawater warms the air above and adds humidity to it, fueling the growth of low-pressure systems. With a large-enough warm pool, spreading along the coast, low-pressure systems may develop farther south, causing rainstorms and even flooding in a region more used to high pressures and dry weather. At the same time, the warm pool has left the western equatorial Pacific; the cooler waters taking its place chill the air above and provide less humidity, increasing the density of that air. Scientists tracking the changes in atmospheric pressure of the tropical Pacific note the reversal of the normal pressure pattern: a Southern Oscillation has taken place.

As the warm pool spreads out up and down along the South American coast, it interferes with the northward-moving Humboldt Current, and with the coastal upwelling. The cool, nutrient-rich waters vanish. The schools of anchovy disappear. Local fishermen announce the start of another El Niño.

The impact of the El Niño reaches beyond the coast of South America: with a weakened Humboldt Current, the South Pacific gyre has been disrupted.

The reversed state of the Southern Oscillation continues, with low-pressure storms in the east nearer to 30°S than to the equator—corresponding to the extensive warm pool—and high surface pressure in the Doldrums of the western Pacific. Those pressure differences can drive air in at ground level toward the storms, creating winds that oppose the Trade Winds; such winds, and the atmospheric upwelling at the storms' center, reinforce the weakening of the Trades, confounding the circulation of the southern hemisphere Hadley cell. The adjacent cells of the atmospheric circulation can be affected as well, and perhaps the cells adjacent to them. This in turn alters the wind-driven circulation of the rest of the oceans. In short, what began as a local perturbation to the tropical Pacific has now "teleconnected" to the global climate system.

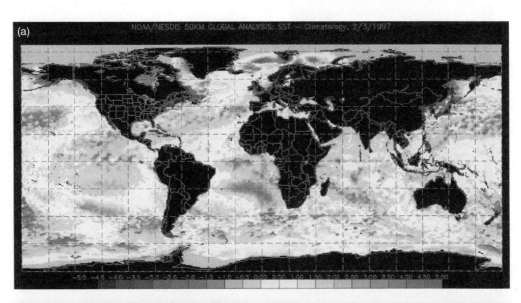

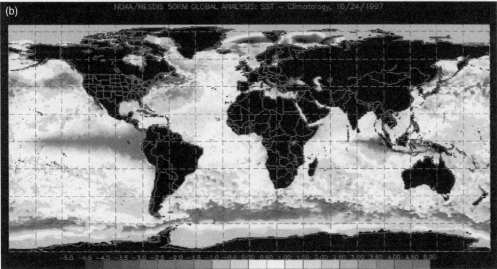

Figure 3.18 Sea surface temperature anomalies—that is, excess temperatures relative to the long-term mean (a map of those means will look like Figure 3.17a)—(a) before and (b) around the peak of the 1997–1998 El Niño. Warmest ocean temperatures are indicated by orange and red colors, and coolest by blue and purple; see color scales at the bottom of the figures.

In (a) February 1997, before the El Niño is fully underway, the warm pool still resides primarily in the western Pacific; temperatures are within ± ½°C of the long-term mean.

In (b), late in October 1997, a huge mass of warm water—with temperatures as much as 5°C greater than the average—had migrated eastward, already reaching the west coast of South America and spreading out northward and southward along the coast (this is also clearly visible in Figure 3.15).

Image downloaded from NOAA / https://www.ospo.noaa.gov/Products/ocean/sst/anomaly/1997.html / last accessed August 22, 2023.

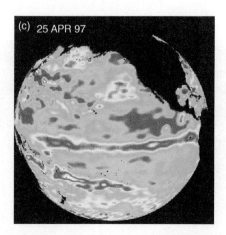
(c) 25 APR 97

Figure 3.18 (c) Sea surface temperatures inferred from satellite observations, as in Figures 3.15 and 3.17b, but corresponding to a month after the 1997–1998 El Niño was triggered. The eastward migration of the warm pool is concentrated along the equatorial "waveguide." Downloaded from NASA / https://sealevel.jpl.nasa.gov/elnino / last accessed August 22, 2023.

The southern hemisphere Hadley cell being weakened suggests that its boundaries, especially over the Pacific, will collapse inward slightly; the resulting shifts in all cell boundaries will be matched by shifts in the subtropical and polar jet streams, in turn causing changes in major-storm paths (cf. Hoerling & Ting, 1994; Graf & Zanchettin, 2012; Zhang et al., 2015; Zhang et al., 2019).

3.3.4 The Mechanism of a Weak ENSO Event

Characterized by a pool of warm water in the eastern Pacific that is both smaller and limited mainly to *equatorial* coastal areas of South America, there are some intriguing differences in the way a weak ENSO event operates.

First: In the Absence of ENSO. Normally, that is, in non-El Niño years, the equatorial waters off the west coast of South America are slightly cool, as a consequence of coastal upwelling. This naturally leads to the overlying air being chilled. The resulting 'dipolar' situation of heated, humid air rising over the warm pool in the western Pacific and slightly cooled, drier air descending over the equatorial eastern Pacific creates a narrow, mainly equatorial, circulation between the western low- and eastern high-pressure systems. After rising into the upper troposphere, some of the air from above the warm pool will flow directly eastward, supplying the air that will descend over the cooler waters in the east; to complete the pattern, the air at ground level returns westward to the warm-pool low.

As it travels westward, that ground-level air, essentially an equatorial trade wind, helps drag warm water to the western equatorial Pacific, intensifying the warm pool; back in the eastern Pacific, those equatorial trade winds have dragged water away from the Ecuadorian coast, producing an upwelling that strengthens the cool pool there (see Xie, 1998). Such positive feedback enhances the 'dipole' and thus reinforces the equatorial atmospheric circulation. Bjerknes (1969) named this flow pattern the **Walker circulation**, in honor of the researcher who first detected its effect on barometric pressure observations.

The narrow, essentially linear, flow characterizing the Walker circulation stands out in contrast to other regional and global flow in the oceans and atmosphere. Of course, large-scale linear flow can only be possible at the equator, where there is no Coriolis force deflection. But any flow in the oceans or atmosphere is subject to continual perturbations by its environment, so the question is, what keeps the Walker flow linear? For the westward ground flow, we might imagine that the southeasterly and northeasterly Trade Winds of the three-cell circulation, converging at the equator (roughly), help to keep any perturbations in check. For the eastward-flowing upper-troposphere portion of the circulation, the linearity is more strongly enforced: any north or south perturbation is deflected right back to 0° latitude by the Coriolis force.

Can you verify this constraint, as a consequence of the sense of deflection by the Coriolis force in each hemisphere?

In fact, the ability of the equator to act as a "wave guide," at least for eastward flow, is evident in the warming of the eastern Pacific Ocean during an El Niño: as seen in Figure 3.18b and c, the eastward motion of the warm pool is concentrated along the equator. Subsequently, much of its spread near the west coast of South America (as seen in Figure 3.18b) can probably be attributed to its being blocked at and reflected from that coast.

It is worth noting that the low pressure over the western Pacific warm pool, while continuing to be a part of the Walker circulation, is also participating in the three-cell circulation of the atmosphere. That is, the southern hemisphere Hadley cell is maintained in part as air rises through that low, and after flowing in the upper troposphere to the subtropical South Pacific High off the Chilean coast (centered near Easter Island), it descends and returns at ground level back to the low. Consequently, fluctuations in the Walker circulation can influence and be influenced by the broader-scale Southern Oscillation (Bjerknes, 1969).

Weak El Niños Are Associated With Localized Effects. The El Niño begins with a weakening of the equatorial trade winds comprising the return flow of the Walker circulation. With their ability to drag water to the western Pacific now diminished, water from the warm pool can make its way eastward; ultimately, that warm water (even if the bulk of it never makes it past the central Pacific) will counteract somewhat the cold waters off the Ecuadorian and possibly Peruvian coasts, leading scientists and fishermen to say an El Niño has arrived. This El Niño has reduced the extremes of the 'dipole' driving the Walker circulation, and interfered with the feedback that would have strengthened it. Atmospheric pressure over the western equatorial Pacific increases as pressure over the eastern equatorial Pacific falls; an equatorially based Southern Oscillation is taking place.

The warm waters brought to the west coast of South America are not sufficient in volume or energy to spread south along the coast as far as Chile. The rise in sea level from that warm pool is modest, as is its impact on the anchovy harvest. Storms associated with that warm pool do not reach 30°S or the Atacama Desert; any heavy rains and flooding are confined to Ecuador and northern Peru. Low-pressure systems do not develop near the southern horse latitudes, so the three-cell circulation is not much impacted (except for a slight weakening, in the Doldrums, of the western Pacific low). The Southern Oscillation of the atmosphere is confined mainly to the equatorial zone, limiting the teleconnections that would characterize a strong El Niño.

3.3.5 The Return to Normalcy

What Brings ENSO Conditions Back to Normal? Recognizing that El Niños are probably triggered by a weakening of the Trade Winds, it is often asserted that pre-El Niño conditions are restored when those winds regain their strength. But those winds will never strengthen, and might even remain in a reversed state, as long as the warm pool resides in the eastern or central Pacific—after all, winds must continue to blow toward the low-pressure system associated with the warm pool, wherever it is. A return to pre-El Niño conditions thus requires that the western Pacific warm pool first be replenished.

Then again, we can argue that the warm pool will only stay in the east as long as the Southern Oscillation remains in its negative phase (maintaining reversed or weakened winds that restrict any drift of the pool back to the west). And the Southern Oscillation can persist in its negative phase only while the Trade Winds are weakened or reversed.

Such circular reasoning is a consequence of ENSO being a coupled ocean-atmosphere process, with positive feedback operating in both states. An assertion that pre-El Niño conditions are restored when the Trades regain their strength is thus not much more enlightening than saying pre-El Niño conditions are restored when the El Niño (warm pool) ends.

ENSO does not operate within a vacuum, however, and that circular reasoning can be cracked open a bit with a more global perspective. Weakening or reversal of the Trade Winds and currents during a strong El Niño—that is, weakening or reversal of the "return flow" portions of the southern hemisphere Hadley cell and South Pacific gyre—amounts to a modification of the three-cell atmospheric and wind-driven oceanic circulations; teleconnections lead to even further modifications. But, as discussed earlier in this chapter, those global circulations are an unavoidable consequence of the variation of incoming solar radiation with latitude plus the Earth's prograde sense of rotation—and those factors have not changed since the ENSO began. Ultimately, every segment of that global circulation must return to its 'normal' state as well.

This suggests that the global circulation, in particular, some portion of it *outside* the tropical Pacific, may play a role in restoring the system to pre-El Niño conditions. Recognizing that the "chicken-or-egg" nature of an ENSO (Rasmusson & Wallace, 1983) may never be completely resolved, we nevertheless follow in the footsteps of El Niño and Southern Oscillation researchers and point to our own possible 'first chicken' (or is it 'first egg'?). As noted by the webmaster ("AES") who managed the NOAA website providing the images in Figure 3.18a and b, toward the end of the 1997–1998 ENSO (in February and March of 1998) the eastern warm pool appeared to be disappearing slowly around the edges. In those images (not shown here), the equatorial waters were still warm. Conceivably, it was the Humboldt Current, traveling northward from cooler waters near the Southern Ocean, that was eating away at the warm pool from the south, thus gradually reasserting itself. In those images, the western warm pool was still absent; but, of course, it would take time for the Humboldt Current to push through, and then for the return flow of the gyre to reach the western Pacific, allowing the western warm pool to be replenished.

Such a scenario is not inconsistent with comments by Wyrtki (1985).

For a weak El Niño, we can follow the same logic: the Walker circulation is embedded within the larger southern Hadley cell, and its reversal is a poor fit to the larger-scale circulation (e.g., the atmospheric ground flow from the southern horse latitudes to the western equatorial Pacific never stopped or reversed). In particular, since the South Pacific gyre was never overwhelmed by the Walker cell's reversal, we might imagine that the gyre continues to bring warm water to the western warm pool, eventually dominating the Walker cell reversal and reestablishing pre-El Niño conditions. But, as with strong El Niños, no matter our choice of 'chicken' or 'egg,' it is not a unique choice: we could alternatively argue, for example, that it is the southeasterly Trade Winds of the three-cell circulation which act to strengthen the equatorial trades, reasserting their normal direction of flow.

3.3.6 There's an Even Bigger Picture

The cause of the initial Trade Wind weakening is not agreed upon. Possibilities include the following:

- "(North)westerly wind bursts" or WWBs (e.g., Cheney & Miller, 1988; McCreary, 1976); these originate over east Asia and oppose the southeasterly Trade Winds or the equatorial trade winds in the western Pacific. The WWBs, in turn, are attributed to random fluctuations in the climate system.

- Pulses of undersea volcanic activity, from both erupted lava and intruded magma, at the East Pacific Rise (Shaw & Moore, 1988). The East Pacific Rise is an active mid-ocean ridge in the eastern half of the Pacific (we see the ridge's emergence from the ocean at Easter Island), and would provide frequent bursts of warmth up to the ocean surface (and thus into the atmosphere).

- Volcanic eruptions at Earth's surface (see, e.g., Kerr, 1983). The volcanic gas and dust can alter the insolation driving the three-cell circulation. Evidence for this possibility includes the timing of massive eruptions of El Chichón (in Mexico) in 1982 and Mt. Pinatubo (in the Philippines) in 1991.

WWBs have apparently become the most popular candidate as a trigger of ENSOs; but the other possibilities should not be disregarded. For example, studies have found correlations between historical eruptions and changes in equatorial sea surface temperatures associated with ENSO events, both observationally (e.g., Handler, 1984) and according to climate models (e.g., Liu et al., 2017, who, it might be noted, also conducted an observational analysis extending back over millennia, using 'proxies' for all variables). In the climate modeling, the resulting change in insolation was found to lead to wind bursts, a migrating ITCZ, and other anomalies in atmospheric circulation that—for volcanoes in the northern hemisphere or the tropics—affected the location and intensity of the tropical Pacific warm pool (for southern hemisphere volcanoes, the atmospheric cooling simply amplified the normal pre-ENSO conditions).

Beyond our categorization of ENSO events into those with a strong versus weak El Niño, ENSO events evidently differ according to their duration, teleconnections, the evolution and characteristics of the eastern and western warm pools, and so on. Some of that diversity must be expected given the fundamentally chaotic nature of the atmosphere and oceans. Some properties of ENSOs, such as their severity, may depend on the prior thermal state of the oceans (e.g., Harrison & Schopf, 1984; Cane & Zebiak, 1985; Fedorov et al., 2015). And some characteristics may simply result from the nature of the ultimate (volcanic, WWB, etc.) excitation—for example, how strong it was, where it occurred (as noted above), and whether it was a single pulse or a more extended excitation (e.g., Philander, 1981). But it is also possible that, at least for the more intense events, there is not just a single type of cause (this possibility was also noted by Wyrtki, 1975). For that matter, multiple mechanisms may even operate simultaneously. The irregularity of the ENSO recurrence

time (~2–8 yr) is also consistent with there being a variety of excitation mechanisms.

Or perhaps the specific excitation is not important. We have seen above that the natural or 'equilibrium' state of the ocean-atmosphere system is one that includes a warm pool in the western Pacific and cool waters in the eastern Pacific. According to Jin (1997; see also Cane & Zebiak, 1985; Vallis, 1986) and others, once something disturbs that equilibrium, the oceans and atmosphere will act literally like a coupled oscillator, swinging back and forth between warm-in-the-west/cool-in-the-east and cool-in-the-west/warm-in-the-east modes.

In an oscillator, an "overshoot" is needed to make the system 'snap back' toward its other extreme. In the Pacific, the overshoot is provided by positive feedback from the atmosphere, as we have already discussed; for example, the warm pool in the eastern Pacific (during an El Niño) creates an overlying low pressure, and the winds blowing toward it from the western Pacific drag more warm water eastward.

Jin (1997) finds that in the coupled ocean-atmosphere oscillator, *any* random perturbations to the system—we might specify WWBs or other random changes in tropical winds, also volcanic eruptions, pulses of heat at the ocean surface, and more—will perpetuate the alternations.

In this perspective, El Niño, where the warm pool is in the eastern (or perhaps central) Pacific and the western Pacific sustains cool waters, is viewed as one (extreme) state of the ocean-atmosphere oscillator; the opposite state, called **La Niña** for contrast, is a more extreme version—an overshoot—of what we have been calling normal or non-El Niño conditions: very cold temperatures in the eastern/central Pacific, and very warm temperatures in the western Pacific. An illustration of both oscillator states is shown in Figure 3.19.

The details of an ENSO oscillator are complex, and well beyond the scope of this textbook. We note, as an introduction to such complexity, that the warm-pool/cold-pool

Figure 3.19 The two states of the coupled ENSO oscillator, in connection with the 1997–1998 ENSO event. Sea surface temperatures as inferred from satellite observations of sea surface heights (see Figure 3.15 caption): white and red denote excessively warm waters, purple and blue excessively cool waters.
We see the warm pool off the (South) American west coast during the peak of the El Niño (upper image); less than two years later—almost a year after the El Niño had ended—the waters off the South American coast are abnormally cool, a condition some call La Niña (lower image).
These images focus on the central and eastern Pacific—the middle and the eastward side of the oscillating system; is what's happening in the western Pacific (seen at the left-hand edge of the images) consistent with what you'd expect for that side of the oscillator?
Image downloaded from NASA / https://sealevel.jpl.nasa.gov/data/el-nino-la-nina-watch-and-pdo/overview/#learn-more / last accessed August 22, 2023.

both rising higher and penetrating deeper than the denser oceanic crust). In the oceans, the boundary separating the warm surface layer from deeper and colder water is called the **thermocline**; the presence or absence of the warm pool corresponds to a deeper or shallower thermocline, and the oscillation between El Niño and La Niña conditions can be interpreted as a tilting back and forth of the tropical Pacific thermocline. The thermocline behaves dynamically like a mechanical spring, on a regional level at least, adding some support to the model of ENSO as a coupled oscillator.

Having introduced the concept of a thermocline, we note that there are actually two factors causing a warming of the eastern Pacific waters at the start of an El Niño (cf. Philander, 1981). When the Trade Winds are initially weakened, for example by a WWB, the disturbance also perturbs the local thermocline; the disturbance propagates as a special kind of wave, called a "Kelvin" wave, eastward along the equatorial waveguide. When the wave reaches the eastern Pacific, it depresses the thermocline there, opposing the coastal upwelling and thus reducing the amount of cold water comprising the surface waters (Harrison & Schopf, 1984). By itself, however (i.e., unless there is an additional source of warming, such as what could be provided by a mass of western warm-pool water migrating to the east), this mechanism might produce a relatively weak El Niño.

Perhaps splitting non-El Niño conditions— defined fundamentally by a warm pool in the western Pacific and a cool pool in the east—into 'routine' and extreme, with the extreme ones called La Niñas, bears some similarities to the view presented here of strong and weak El Niños (both of which are fundamentally defined by a warm pool in the eastern Pacific and a cool pool in the west, with only occasional extreme conditions—those being the strong El Niños).

Finally, however, we leave the world of esoteric climate dynamics by noting that, as an alternative to oscillator models, the extremes for non-El Niño conditions may simply be

oscillation we are envisioning can equivalently be described as an oscillation of energy (i.e., heat), mass, momentum, or angular momentum. For instance, the warm pool corresponds to a mass excess—originally, warm water piled up in the western Pacific; then, a large volume of it migrated to the east—but, although it is associated with a higher sea level, it is not simply a mass of warm water sitting on top of 'neutral' seawater. Instead, the mass of warm water both rises above and penetrates deeper into the oceans (in Chapter 5, we will see a similar situation in the solid earth, with the less-dense continental crust an artifact of the mathematics: in the eastern Pacific, for example, if El Niño represents above-average temperature conditions, there must be times of below-average temperatures there (hence La Niñas) as well as average times. Even in this case, La Niña is a useful designation, serving to warn countries worldwide of impending extremes.

The Immediate Future of Our Atmosphere

3.4 Preliminary Comments

There are three reasons to expect that Earth's climate will vary on human timescales. First, our climate is fundamentally driven by light and heat from the Sun, and the Sun's radiative output is not constant over time, even within our lifetime. Second, humans have been modifying the composition of the atmosphere, especially since the start of the Industrial Revolution more than 200 yr ago, in ways that *must* lead to changes in climate. We can be certain that change is inevitable because those modifications involve the addition of greenhouse gases and other substances that affect the transparency and heat capacity of the atmosphere; what is less certain is the *extent* of the consequent climate change (an uncertainty shared by solar forcing). Third, climate is naturally variable, certainly on geologically long timescales, as Chapter 2 made clear, but on human timescales also. The primary components of Earth's climate system, the atmosphere and oceans, are fluid, possessing (both individually and interactively) natural modes of variation on short, intermediate, and long timescales. A vivid example of this is ENSO, but other modes—with intriguing names like the quasi-biennial oscillation and Pacific decadal oscillation (popularly referred to as QBO and PDO), as well as the Heinrich and Dansgaard-Oeschger events noted earlier—have also been detected in the climate record, and have been the focus of much research.

In the remainder of this chapter, we will consider the first two of these possibilities, Sun-induced and human-induced (also called **anthropogenic**) changes in climate. The alternative to both—that we are just seeing transient, natural variability—will always loom in the background, regardless of even the most convincing evidence of change today. But, as time has passed, and the changes we see in climate have accumulated and become better defined as trends, rather than cycles or oscillations, the likelihood that we are in fact experiencing major anthropogenic change has become much clearer.

3.5 Solar Variability on Human Timescales

3.5.1 Sunspot Cycles

The Sun's radiance varies over a range of timescales. Many of these are associated with **sunspots**, dark regions on the surface of the Sun that can be as large as, or larger than, the Earth (see Figure 3.20a). They are dark because they are cold, typically ~1,500°C colder than normal; their resulting increased density

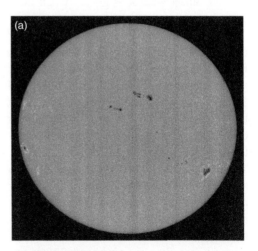

Figure 3.20 (a) Sunspots on the Sun, August 6, 2001. The large sunspot toward the lower right of this image could encompass four Earths. Courtesy of SOHO/Spaceweather.com / https://www .spaceweather.com / last accessed August 22, 2023.

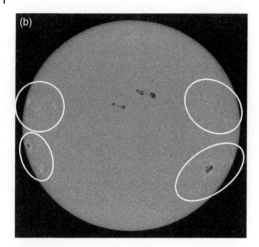

Figure 3.20 (b) Same image as in Figure 3.20a.
Bright areas, called faculae, are circled.

would cause them to sink deep into the solar interior were it not for locally strong magnetic fields that prop up the plasma. Sunspots radiate less light and heat than normal; however, sunspots are typically surrounded by extensive, brighter areas called **faculae** (also called **plages**, though formally there are slight differences between the two) (see Figure 3.20b), which more than compensate for the dark spots (e.g., Foukal & Lean, 1990). The net effect is that the Sun's radiance is greater when there are more sunspots than when there are fewer.

Every ~30 days or so, the Sun rotates once on its axis (actually, the Sun is a complex fluid body: the period of its rotation is 25 days at its equator, 28 days at mid-latitudes, and 36 days near the poles; as we will discuss in the somewhat analogous case of the Earth's core, in Chapter 10, such a "differential rotation" is tied internally to solar convection). The Sun's rotation carries the sunspots into and out of our view. Meanwhile, although the lifespan of most sunspots is much less than a month, a small number of large sunspots or sunspot "groups" may survive more than a month (see Bray & Loughhead, 1964); and the associated faculae can persist even longer (Willson, 1982). The net effect is that there is an approximately monthly variation in the Sun's radiance, amounting to roughly a tenth

of a percent, and a 'continuum' of variations on timescales ranging from days or less to several months (Willson, 1982; see also Kopp, 2016 for an even bigger picture).

It has long been recognized that the number of sunspots progresses periodically from a minimum (occasionally even zero) to a maximum (called the **solar maximum**) then back to a minimum every ~11 yr. This behavior is called the **sunspot cycle**. The periodicity in sunspot number is actually part of a 22-year "double sunspot cycle"; from the first to second 11-year portions, the global magnetic field of the Sun and the local magnetic field supporting the spots both reverse their polarity. Measurements of the Sun's radiance during recent cycles find it to be typically about a tenth of a percent greater during a solar maximum than a sunspot minimum (Fröhlich, 2009; Kopp, 2016), though the effect does vary from cycle to cycle.

Could such a modest variation in the heat from the Sun affect Earth's climate system? Over the last quarter of the twentieth century, R. Currie analyzed a variety of climate data—air temperature, rainfall, drought indices, tree-ring thickness, tropical storm activity, and so on—primarily using a technique of data analysis (called "maximum entropy" spectral analysis) capable of revealing faint signals buried in noisy or truncated data (e.g., Currie, 1991, 1993, 1995, 1996). He found that they all show an ~11-year periodicity. Other researchers have also detected that signal in global and regional surface temperatures (Lean & Rind, 2008), lower-troposphere temperature (Douglass & Clader, 2002), and sea surface temperature (Wilson et al., 2006) data. If those signals are all statistically significant, they would have to be a consequence of the sunspot cycle, because no phenomenon internal to the Earth varies with that timescale.

Sunspots have been observed for more than two thousand years; systematic observations began in 165 BC (e.g., Wittmann & Xu, 1987; Yau & Stephenson, 1988; but see also Ding et al., 1983) in China and 1610 in Europe (cf. Bicknell, 1968), with Galileo the

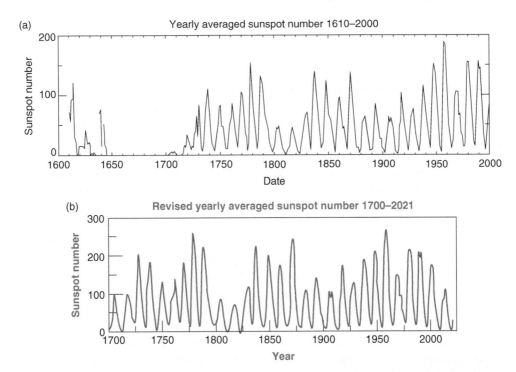

Figure 3.21 (a) Yearly averaged sunspot record from 1610 to 2000, showing the 11-year sunspot cycle but also revealing that the Sun is variable on many other timescales. http://www.spaceweather.com/glossary/sunspotnumber.html.
(b) Revised ("Version 2.0") yearly average of the daily total sunspot number, determined as a recalibration of previous observations 1700–2021; see Clette et al. (2014) for details. Data downloaded from the World Data Center–SILSO, Royal Observatory of Belgium, January 2022 at https://wwwbis.sidc.be/silso/datafiles. Note that data before/after 1849–1981 involved fewer/more than one observation per day. The revised time series does not show the slight increasing trend over the twentieth century discernable in the older compilation.

most renowned of the European observers. The European record of sunspot activity (see Figure 3.21) shows that during most of the past 400 yr, an 11-year sunspot cycle has been visible. But solar activity has evidently been influenced by other factors: the sunspot record shows "active" times, when sunspot numbers reached large values; "quiet" times, when solar maxima were muted; and even times when sunspot activity appeared to be minimal or nonexistent! There were 'stable' times, when sunspot activity looked similar from one cycle to the next, and 'unsteady' times, when successive cycles appeared to have little in common. After 1610 the most extreme period occurred between the years 1645 and 1715, a time called the **Maunder Minimum**. We would expect

that the Maunder Minimum corresponded to an overall decrease in the Sun's radiance, with the lack of dark spots offset by a lack of brighter faculae; the decrease in radiance might have been up to a few tenths of a percent.

3.5.2 A Connection Between Sunspots and a Dramatic Change in Earth's Climate?

It turns out that the Earth—or at least North America and Europe—experienced a relatively cold climate during the time of the Maunder Minimum, called the **Little Ice Age** (or **LIA**). Such synchronicity has been viewed as providing support for the claim that sunspot cycles can measurably affect our climate. But the facts are less supportive, and render the connection uncertain.

First of all, the chill in climate began around 1350–1450 AD (IPCC 2007, 2013) or even earlier—at least two to three hundred years before the apparent start of the Maunder Minimum. That cold marked the end of the previous so-called "medieval warm period" and may have caused flourishing Viking colonies (whose population had reached over 4000) at the Greenland coast—colonies used as a base to explore the New World—to be abandoned or wiped out (but see also Ruddiman, 2008; Young et al., 2015). Changes in weather patterns at this time may also have been responsible for droughts in the western United States that led to the Anasazi culture abandoning their cliff-dwelling settlements in the area around Mesa Verde.

During the Little Ice Age, mountain glaciers expanded, disrupting alpine villages (see Kump et al., 2010). Winters became snowier, summers cooler and wetter. Sea ice was more common in the North Atlantic (Ruddiman, 2008); winds strengthened, and the paths of eastward-moving mid-latitude storms ran 5° to 10° lower in latitude than normal (see Trouet et al., 2012 and references therein)—all of which turned the sea-going Age of Exploration, particularly between Europe and North America, into an even more challenging undertaking. Rivers and canals were typically frozen over for parts of the winters (see Figure 3.22); during one year, the western Baltic Sea was frozen, enabling people to ride carts directly from Germany to Sweden. At the start of the Revolutionary War, extremely severe winters made survival an issue for the American rebels at Valley Forge (and, 2 years later, even more so in Morristown, New Jersey; Raphael, 2021).

The end of the Little Ice Age, sometimes taken to be ~1850 AD, occurred surprisingly close to the present day. With its end, and climate beginning to warm up, crops could once again be grown easily in mid-latitudes, feeding a population that was beginning to burgeon

Figure 3.22 *Winter Landscape with Skaters and a Bird Trap* by Pieter Bruegel the Elder, painted in 1565, depicts life in what is now Belgium during the Little Ice Age. This painting is on display at the Royal Museum of Fine Arts in Brussels. Kunsthistorisches Museum, Vienna / https://commons.wikimedia.org/wiki/File:Circle_of_Pieter_Bruegel_the_Elder_-_Winter_Landscape_with_a_Bird_Trap.jpg / Public Domain.

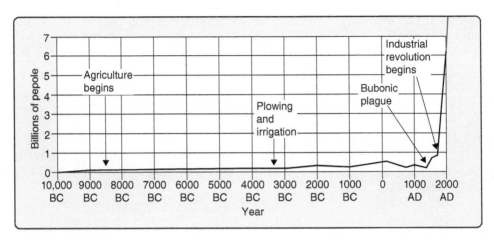

Figure 3.23 World population over the past 12,000 years, with some milestones noted. Image downloaded from http://images.slideplayer.com/35/10307621/slides/slide_23.jpg.
This image has been modified by adding a line (red) continuing the population growth to 8 billion in year 2022.

after the start of the Industrial Revolution (Figure 3.23).

In short, the time span of the Little Ice Age does not come close to matching that of the Maunder Minimum (though both the starting and ending dates of the LIA are not agreed upon). However, we note some complicating details. First, the coldest part of the Little Ice Age was reportedly during the 1600s and 1700s; it is conceivable that the Maunder Minimum acted to increase the severity of the LIA, though not initiating it (but see also Bradley & Jones, 1993). Second, other sunspot minima, though not as 'deep' as the Maunder Minimum, occurred during the centuries encompassing the Little Ice Age; these included the *Wolf Minimum* (ca. 1300–1350), the *Spörer Minimum* (ca. 1450–1540), and the *Dalton Minimum* (1790–1830). The Dalton Minimum is visible in Figure 3.21; the other two have been inferred from 'proxy' measures of solar activity, such as the solar wind's ability to modulate cosmic-ray production of C-14 from N-14 in Earth's atmosphere (with that carbon isotope subsequently incorporated into trees). These minima may have extended the span of cold climate both before and after the 1600s–1700s by a few hundred years.

Alternatively, volcanic eruptions might have 'jump-started' the Little Ice Age well before the

Maunder Minimum began. Work by Gao et al. (2008) found that volcanic eruptions throughout the 1200s resulted in massive additions of volcanic dust to the atmosphere, shielding Earth's surface from the Sun's warmth; Miller et al. (2012) demonstrated that their cooling effect could in fact have been the primary trigger to bring about the LIA. And, once the Little Ice Age was underway, it may have been maintained by the prolonged cooling from subsequent major volcanic eruptions. These eruptions included Lanzarote (Canary Islands) in 1730–1735, Laki (Iceland) in 1783, Vesuvius (Italy) in 1813, and—the most prolific of them all—Tambora (Indonesia) in 1815, which spewed so much dust into the atmosphere that 1816 became known as "the year without a summer" in North America and Europe (Figure 3.24).

But the cooling brought about by a single eruption, however intense, tends to be short-lived (as we can see in the effects on the upper atmosphere of major twentieth-century eruptions, discussed later in this chapter in the context of global warming and depicted in Figure 3.51); Tambora was a dramatic exception, and even its effects probably lasted no more than 2 yr. Miller et al. (2012) showed that volcanic activity throughout the entire

Figure 3.24 (a) The colors in *Chichester Canal*, painted by J. M. W. Turner in 1828, are said to have been inspired by the vivid sunsets resulting from atmospheric dust after Tambora's eruption. Painting on display at London's Tate Gallery. J. M. W. Turner / Wikimedia Commons / Public Domain.
(b) Crater left over after Tambora volcano, Indonesia, erupted in 1815; it is believed that eruption was so prolific it led to the "year without a summer" in 1816, which included snow falling in the summer in Europe (Italy) and Asia (China). Famines were widespread, and disease, ultimately resulting in the deaths of hundreds of thousands of people in Europe, the Middle East, and Asia (see, e.g., Kump et al., 2010). NASA / https://commons.wikimedia.org/wiki/File:STS026-38-056.jpg / Public Domain.

LIA could have maintained its cool temperatures and icy conditions with the help of a sea ice–ocean feedback (Zhong et al., 2011), involving changes in albedo and ocean circulation that prolonged the effects of each eruption.

Ironically, even in our discussion of the Sun as a cause of current and future climate change, it appears that the role of Earth-based natural variability cannot be ignored—and that in fact volcanic activity may have been the major driver of the Little Ice Age; nevertheless,

the effects of sunspot activity, especially the Maunder Minimum, should not be completely discounted. That is, there were probably multiple contributors to the Little Ice Age, and these included variations in sunspot activity.

In a further irony, natural variability will also figure in to our understanding of anthropogenic global warming below, but in a very different way. The "volcanic dust" referred to as a cause of climatic cooling (by its ability to reflect incoming sunlight) is mainly sulfate aerosols, produced ultimately from the sulfur gases in volcanic eruptions. But sulfate aerosols can also derive from pollution—from the sulfur gases generated as a by-product of our burning coal and oil. During the twentieth century, as we will see, such pollution actually had a significant cooling effect on global temperatures—which, had it been appreciated at the time, might have led to an earlier acceptance of the reality of global warming!

3.5.3 A Few Final Comments on Sunspots

The variation in sunspot number shown in Figure 3.21 appears to include a roughly 90-year periodicity superimposed on the 11-year cycle (combining, for example, to produce large solar max in ~1780, ~1870, and ~1960). That periodicity, known as the **Gleissberg Cycle** (see Hathaway, 2015 and references therein), is also seen in longer reconstructions of sunspot numbers and various proxy data sets; those longer time spans also reveal other quasi-periodic behavior on even longer timescales. Between 1981 and 1986, satellite observations directed toward the Sun measured a 0.02% per year drop in radiance; that decrease was interpreted by Reid (1987) to be part of the Gleissberg cycle. By implication, the Gleissberg cycle might lead to a nearly 1% peak-to-trough variation in solar output during each complete cycle.

Sunspot activity has, historically (and also politically, as a challenge to global warming), been the most commonly employed test of solar forcing of climate change. Solar variability certainly has the potential to modify Earth's climate. From the perspective of sunspot activity, it appears unlikely that the Sun has been more than a secondary contributor to climate change over the past several hundred years. However, there are other direct and proxy measures of solar irradiance; their implications for Earth's climate variations are beyond the scope of this textbook.

3.6 Anthropogenic Variations in Climate by the Emission of Greenhouse Gases

The greenhouse effect has always been important for life on Earth. We learned in Chapter 2 that without it, the inception of life would have been delayed for billions of years. We also saw that variations in the abundance of carbon dioxide and other greenhouse gases in our atmosphere, as regulated by the Urey cycle, produced major Hot House and Ice House climate regimes on Earth.

Over the eons, levels of carbon dioxide have decreased in order to 'balance' the increasing luminosity of the Sun; as the Sun got brighter, smaller fluctuations in the amount of CO_2 were able to cause those major shifts in climate regime. Today, such natural regulation of greenhouse gases has left only trace amounts of them in the atmosphere—levels small enough that human activities can actually modify them significantly. That is, the small changes in their abundance resulting from human activity can be expected to impact Earth's climate in measurable ways. For the foreseeable future, though, those anthropogenic impacts will take place *within* our current Ice House rather than bringing about its demise.

It has been proposed (Ruddiman, 2003) that the first human impact on the atmosphere began some 10,000 yr ago, with the onset of plant and animal domestication known as the Agricultural Revolution (Figure 3.23); clear-cutting the land would have increased the carbon dioxide content of the atmosphere, while producing crops like rice and raising livestock would have increased the methane content. Ruddiman's proposal sparked a lot of interest—and some controversy, with questions raised as to whether the relatively low human population at the time could have produced an appreciable impact, and whether the timing of these historical events is consistent with the measurements of carbon dioxide and methane over the past 10,000 yr; see also Ruddiman (2014).

There is no doubt, however, that human activity in the *Industrial Age*—primarily from the burning of fossil fuels to power various industries and transportation, and to provide heat—has significantly increased the amount of carbon dioxide, methane, and other greenhouse gases in the atmosphere. The birth of this era is traditionally taken to be 1776, marking the first commercial use of the steam engine (Lira, 2013), following the invention of an efficient version of it by Watt in the 1760s.

3.6.1 Increases in Greenhouse Gas Abundances

It is possible to measure CO_2 abundance in the past from bubbles of air found in glacial ice; these bubbles were trapped as the glaciers were covered, layer by layer, by more snow and ice each season. Measurements of the bubbles' composition show levels of carbon dioxide (Figure 3.25a) which do not exceed ~280 parts per million (ppm) for much of the past several thousand years (Figure 5.21b in Chapter 5 suggests a similar maximum extending back more than 400,000 yr). But since the official start of the Industrial Revolution, the abundance of CO_2 has increased steadily (see Figure 3.25a, b); by the year 2000 it was 370 ppm, and by the year 2014 it had reached almost 400 ppm—an increase by more than 40% during the past 230 yr.

The impact of such levels of atmospheric CO_2 on climate change will hopefully become clear by the end of this chapter. Anticipating that clarity, knowing the rate at which CO_2 is increasing with time will then help us determine whether severe actions are called for to mitigate those impacts. Calculate the number of ppm per year by which CO_2 increased between 1776 and 2000, and between 2000 and 2014; how do those rates of increase compare?

This trend has continued to the present day. By early August, 2021 (Figure 3.25a), for example, CO_2 had exceeded 415 ppm.

The input of carbon dioxide into the atmosphere between years 1800 and (e.g.) 2000, primarily from the burning of fossil fuels and also through deforestation, was probably more than 200 ppm, or at least a lot greater than the observed increase of ~100 ppm: an analysis of the global "carbon budget" implies that, of the CO_2 emitted into the atmosphere, anywhere from one third (Keeling et al., 2001) to one half (~52% for 2002–2011; IPCC, 2013) remains there while much of the rest diffuses down into the oceans (a small portion is incorporated into the biosphere).

Human activity has also led to increases in the abundance of other greenhouse gases in the atmosphere. Methane (CH_4) is a by-product of fossil fuel mining, biomass burning, and landfills; it is also produced by anaerobic bacteria in cultivated rice paddies (at the submerged roots) and in livestock intestines (resulting in a different kind of outgassing with the potential to affect climate!). Like carbon dioxide, CH_4 has been measured in glacial air bubbles; its pre–Industrial Age concentration in the atmosphere was about 0.7 ppm, but in more recent years its abundance has grown to more than 1.8 ppm (see Figure 3.26), reaching—and exceeding—1.9 ppm after August 2021 (e.g., https://gml.noaa.gov/ccgg/trends_ch4/).

Although methane is much less abundant than carbon dioxide, one molecule of CH_4 can trap as much warmth trying to exit the Earth as ~20 molecules of CO_2 (or more, depending on how its effects are calculated), so its impact on climate is not negligible.

An additional source of concern with methane is the existence of possibly vast reservoirs in the form of frozen methane-ice compounds, also called **methane hydrates** or **clathrates** (see Figure 3.27). These are located on land in permafrost regions and in the oceans buried beneath seafloor sediments at continental margins. The concern lies in the idea that once climate variations—such as global warming—have progressed to the point where high-latitude permafrost begins to melt, or the deep oceans begin to moderate in temperature, the methane thereby released will accelerate that climate change (see, e.g., Chadburn et al., 2017; Biskaborn et al., 2019; Lewkowicz & Way, 2019; Cory et al., 2013).

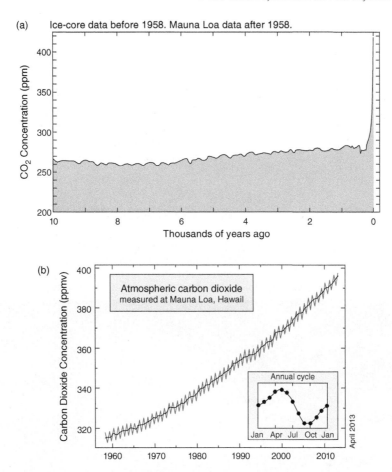

(a) Ice-core data before 1958. Mauna Loa data after 1958.

Figure 3.25 Historical rise of CO_2, quantified as parts per million (ppm).
(a) The rise over the past 10,000 years (ending in 2021), determined from bubbles of air trapped in ice, retrieved from various ice cores; and from Mauna Loa (see b). https://niklasrosenberg.com/blog/2021/8/10/how-bad-is-415-ppm-of-global-atmospheric-co2-concentration / Niklas Rosenberg.
(b) Atmospheric measurements atop Mauna Loa, Hawaii, since 1958, which as graphed is known as the *Keeling Curve* (shown here as monthly data 1958–2013). https://uwpressblog.com/2014/06/25/behind-the-covers-joshua-howes-behind-the-curve/ uwpressblog.
Superimposed on the trend, the small oscillations seen in (b), and enlarged in the inset as "Annual cycle," are the result of the biosphere 'breathing': during the summer, plants carry out a lot of photosynthesis, which removes CO_2 from the atmosphere; during winter, there is little photosynthesis but plants continue to respire, 'breathing out' CO_2 into the air. The atmosphere shows a net gain or loss of CO_2 according to the northern hemisphere seasons because more vegetation is located in the northern hemisphere.

Such a scenario reminds us of the importance of accounting for all the feedback mechanisms (such as this positive feedback) in modeling Earth's climate system.

Other greenhouse gases injected into Earth's atmosphere by human activity include tropospheric ozone (O_3), now the most widespread pollutant in the lower atmosphere (this ozone is distinguished from naturally occurring, stratospheric ozone); nitrous oxide (N_2O), whose abundance has grown significantly in the past two decades; and the family of human-made chemicals known as chlorofluorocarbons or CFCs. Although these latter exist only in tiny amounts, they are 10,000 times more effective than carbon dioxide is as a greenhouse gas. Banned by the Montreal Protocol and its seven subsequent 'Agreements'—because of their effects on the atmosphere's ozone layer—CFCs have been replaced by

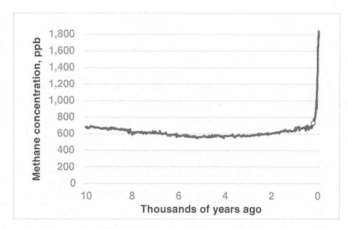

Figure 3.26 Historical rise of CH_4 over the past 10,000 years (ending at year 2018), determined from bubbles of air trapped in ice, retrieved from various ice cores; and, more recently, direct atmospheric measurements. CH_4 concentration in the year 1800 is marked with a circle. Image modified from https://www .darrinqualman.com/methane / darrinqualman. Units for concentration here are parts per billion (ppb).

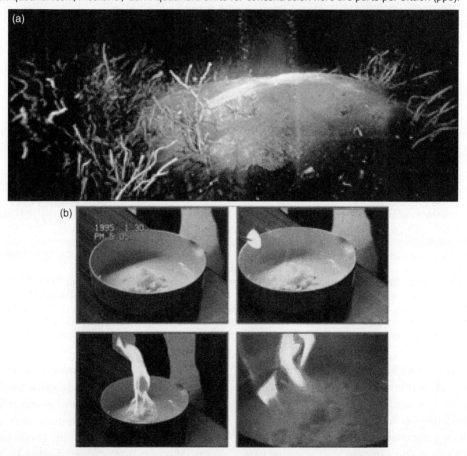

Figure 3.27 Pictures of clathrates. (a) At the bottom of the Gulf of Mexico, methane seeps up from below the seafloor; some of it creates icy mounds, as seen here, and can support chemosynthetic life. USGS / https://pl.wikipedia.org/wiki/Klatrat_metanu#/media/Plik:Seafloor_mounds.jpg / Public Domain. (b) Clathrates recovered from the seafloor. NOAA / https://oceanexplorer.noaa.gov/explorations/ deepeast01/background/fire/fire.html / last accessed August 22, 2023. These icy substances are able to burn because of the methane trapped within them.

other chemicals (HFCs and HCFCs) which are also strong greenhouse gases.

The ability of a greenhouse gas to modify climate depends both on its effectiveness 'per molecule' in trapping the Earth's outgoing warmth and on its abundance in the atmosphere. Overall, given current abundances, the contributions of these gases to an increased greenhouse effect—that is, their effective "radiative forcing" capabilities (expressed here as rounded percentages of the total)—are estimated to be as follows:

$$CO_2 : 56\% \quad CH_4 : 15\%$$

$$\text{Tropospheric ozone} : 12\%$$

$$N_2O : 5\% \quad \text{Tropospheric CFCs}$$

$$\text{(and successors)} : 11\%$$

(Tables 8.2 and 8.6 of IPCC 2013). Roughly, then, at present the total impact of greenhouse gas emissions is about twice that of carbon dioxide alone.

Note that these model percentages will change over time, as the abundances of the various greenhouse gases increase and also as atmospheric temperature and non-greenhouse gas composition change. Variations in such parameters as albedo (due to land use, black carbon settling on snow, etc.) and aerosol concentrations can further impact Earth's climate.

The emission of greenhouse gases into the atmosphere is fundamentally population-dependent. Population growth leads to increased fossil fuel consumption, thus more CO_2 (the ultimate product of burning any fossil fuel) and CH_4 (from mining of coal, oil, and natural gas). Population growth also leads to more food production (cattle *and* rice), thus more CH_4; and likely more fertilizer use, thus more N_2O. For developing countries, a rise in the standard of living leads to more energy consumption, thus (again) more CO_2 and CH_4; to more vehicle use, thus more tropospheric ozone and N_2O; and, for all their various uses, to more CFCs (and their successors). Given these population connections, it is worth noting that demographic predictions have the world's population increasing from 6 billion, reached near the end of 1999—rising to 7 billion, reached 12 yr later, and then 8 billion, reached 11 yr after that—to ~11 billion or more by the end of this century (Gerland et al., 2014). Most of that increase (see Figure 3.28) is projected to take place in developing countries, especially in Africa.

3.6.2 Direct *and* Indirect Impacts on Climate Expected From an Increase in Greenhouse Gas Abundances

The changes in climate caused by increased greenhouse gas concentrations in the atmosphere have the potential to be wide-ranging and severe, even in the near term. The most obvious impact would be increased temperatures in the troposphere—a global warming, due to the greenhouse effect. However, the warming at the latitudes where most people live would not be of a magnitude to send everyone indoors looking for air-conditioning—it would probably be at a rate of no more than a couple of degrees centigrade (e.g., IPCC 2007) or perhaps four degrees centigrade (IPCC 2013) increase per century. Instead, *it is the indirect consequences of global warming that may be the most threatening.*

In that case, it is clear that the climate impact of increased greenhouse gas concentrations would be *misrepresented* by the term "global warming." For that matter, given Earth's climate history, in which change has been the norm, the phrase "climate change" is not much better; specifying "anthropogenic climate change" or "accelerated climate change" would at least restrict the change in climate to the one of most concern today, but is still vague (is the change positive or negative?) and still misses the range of impacts. More explicit descriptions might include "global increase in atmospheric energy," which at least makes clear that the "change" involves an *increase* in energy (heat and otherwise); or, "multi-component climate system intensification;" or, simply, "climate disruption."

It is not surprising that more than one term can be appropriate here: global warming—or whatever you choose to call it—has societal as well

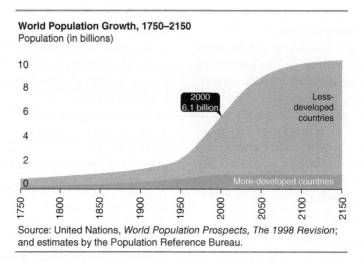

World Population Growth, 1750–2150
Population (in billions)

Source: United Nations, *World Population Prospects, The 1998 Revision*; and estimates by the Population Reference Bureau.

Figure 3.28 Population growth since 1750, predicted based on demographic analysis through the twenty-second century. By far, most of the increase since year 2000 has been and will continue to be from less-developed countries.
This image, from http://wiki.dickinson.edu/images/c/cb/World_population_growth.jpg and published in 2008, predicted a world population of ~10 billion by year 2100; more recent estimates (e.g., Gerland et al., 2014) predict a population likely exceeding 11 billion by then.
What do you think the effect on population growth will be as less-developed countries become more developed?

as scientific impacts. Furthermore, some terms might be associated with an intention *(educational, political, or other) on the part of the user; others are more neutral. Given the controversies, both scientific and societal, that have 'embroiled' the subject, which do you think would be the best term to use?*

In short, the result of increasing greenhouse gas abundances in the atmosphere will *include* a modern-day global warming, plus—potentially—a wide range of indirect consequences. We describe here three sets of indirect consequences we expect to result from global warming that have the potential to be particularly severe; in later subsections of this chapter, we will evaluate the evidence concerning their reality.

First Set of Indirect Consequences. The most straightforward of the indirect consequences of a warmed atmosphere is a rise in sea level. This would result from melting of glaciers and ice sheets, with the meltwater eventually filling the oceans, and from thermal expansion of seawater as the warmed air warms the oceans. Note that the former is a

mass effect, the latter, a volume effect, with the volume involved depending on how deep the warming penetrates into the sea.

Computer models predict that future global warming will be intensified at high latitudes (Figure 3.29a, b). As discussed earlier, positive feedback involving albedo can explain such intensification: as greenhouse gases in the troposphere trap outgoing warmth and cause some global warming, polar ice will begin to melt; the lower albedo (compared to that of ice) of the newly exposed land or ocean surface will result in more incoming sunlight being absorbed at those high latitudes than previously, multiplying the warming there. Since high latitudes are where most of the world's ice resides, even a mild global warming could affect ice sheets and sea level dramatically. Computer models also find the warming to preferentially occur during each hemisphere's winter (Figure 3.29a), a time when snow and ice would normally be accumulating on the ice sheets.

All in all, we would expect the impact of global warming on sea level to be substantial. For a 'ballpark' estimate, we note that at

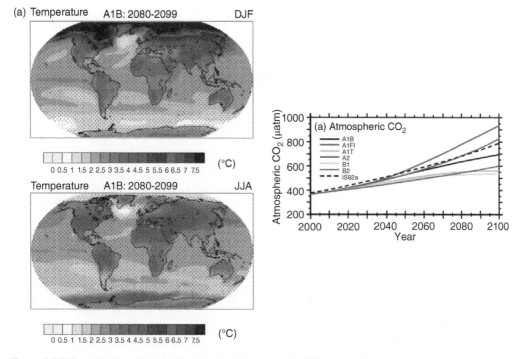

Figure 3.29 Two views on the warming predicted by the end of this century.
(a) Predicted seasonal temperature increases by 2080–2099 relative to the years 1980–1999 according to Scenario A1B in the IPCC 2007 report, for the months of December–February (upper left) and June–August (lower left).
The various IPCC 2007 scenarios for the atmospheric abundance of CO_2 (in units equivalent to ppm) are shown on the right. Scenario A1B posits an increase in atmospheric greenhouse gases to an equivalent 835 ppm of CO_2 in year 2100, more than doubling the greenhouse effect in that 100-year span. *For purposes of illustration but brevity, only the "middle-of-the-road" scenario A1B is considered in this textbook when discussing IPCC (2007).*
Images from IPCC (2007) / Intergovernmental Panel on Climate Change.

the end of the Pleistocene, as the ice sheets melted away, sea level rose (with more details provided later in this chapter) by ~120 m over ~13,000 yr—thus, a rate of nearly 1 m/century; with a temperature difference between glacial and interglacial periods of roughly 10°C (see Chapter 2), the response of the climate system to glacial → interglacial warming was evidently a rise in mean sea level of ~10 cm/century per degree of warming. With a similar response now, current global warming would lead to a sea-level increase of ~20–40 cm over this century. However, current warming is taking place within an already warmer climate (rather than starting out under glacial conditions), so glaciers and ice sheets are already

more vulnerable. Our 20–40 cm per century estimate, then, should be considered a lower bound.

For comparison, computer models referenced in IPCC 2007 predict rises of ~20–50 cm from 1980–1999 to 2090–2099 for the A1B scenario shown in Figure 3.29a. However, the most recent IPCC report (IPCC, 2013; Figure 3.29b) finds that, with improved sea-level models (newly developed models that better describe current and past sea-level observations), the rise in sea level might be double that in IPCC 2007, depending on the greenhouse gas scenario (see Rahmstorf, 2013). Thus, we should expect global sea level to rise

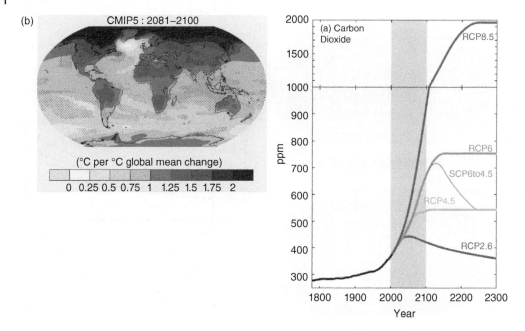

Figure 3.29 Two views on the warming predicted by the end of this century.
(b) Predicted annual temperature increases by 2081–2100 relative to the years 1986–2005, shown (on the left) as the amount of temperature increase for each degree that the globe warms. These predictions were based on numerous models, each run according to different CO_2 scenarios (right, in ppm), and averaged over all scenarios and models.
The scenarios are based on past levels of CO_2, predicted levels through the present century (gray shading), and more theoretical estimates through the next two centuries; some of the latter also considered socioeconomic development, energy use, agricultural factors, and land use (see IPCC, 2013). Images from IPCC (2013) / Intergovernmental Panel on Climate Change.
Do you think the approach taken here—which includes averaging over all scenarios—yields results that are less dependent on the choice of scenario?

by a meter, within the next two centuries at most.

A rise in sea level could have enormous societal consequences. Since 2003, more than half of all Americans live near (i.e., within 50 miles of) the coast; a significant rise in sea level would compel a large fraction of that populace to move away, tangibly impacting those in neighboring regions but with the cost of resettlement and other social services borne by all. Worldwide, more than 1 billion people live in very shallow coastal regions. Even a 1 m rise in sea level would impact them, through their land being inundated or increasingly vulnerable to storm damage; though such a sea-level increase may not be achieved for a

century, by then the number of people affected will be even greater (see, e.g., Strauss et al., 2021). The existence of a large number of "climate refugees" around the world is likely to heighten political and even military tensions. In some situations, however, and despite the inevitability of a rise in sea level, it is possible that adaptation, rather than emigration, could be a more desirable (McMichael et al., 2020) and successful strategy (e.g., Jamero et al., 2017; see also Kniveton, 2017), though it would still be costly.

A Looming Threat From West Antarctica. One complexity of the climate system is especially worth mentioning in the context of rising sea level. The West Antarctic Ice

Sheet (WAIS), in contrast to the much larger East Antarctic ice sheet located on the other side of the Transantarctic Mountains, rests on a crustal surface that is mostly below sea level (see Figure 3.30a). That is, if we could look beneath the ice covering the Antarctic continent, we would see a broad land surface in the East (though most of its coastal areas are below sea level), and then the mountain chain; but in the West we would find only an expanse of shallow seas containing a few large islands (and just a handful of small islands). The partly submerged nature of most West Antarctic ice makes it fundamentally more vulnerable to climate change than land-based ice sheets are. That vulnerability is compounded for many of

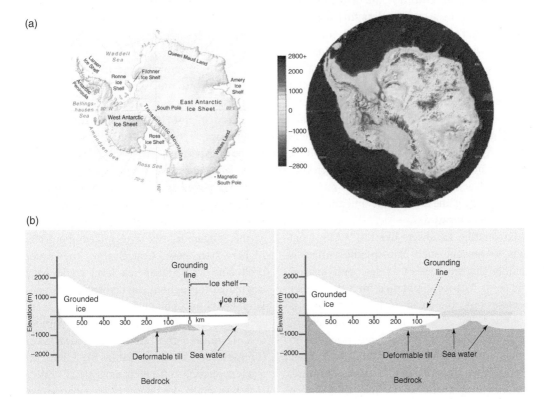

Figure 3.30 A mechanism for the collapse of the West Antarctic Ice Sheet.
(a) (Left) Map of Antarctica (http://www.grida.no/graphicslib/detail/antarctica-topographic-map_8716#). (Right) The bedrock surface of Antarctica: what Antarctica would look like without its ice cover. Color sc ale goes from 2800 m above sea level (dark reddish-brown) to 2800 m or more below sea level (very dark blue); all shades of blue are below sea level. Image from Fretwell et al. (2013) / Fretwell / CCBY 3.0 / Public Domain. Their analysis revealed a deeper bed and greater percentage of ice grounded below sea level for the *East* Antarctic ice sheet than previous investigations.
(b) Cross-section of a glacier in the WAIS with a "retrograde" bed—the bedrock and till upon which the ice is grounded slope *up* toward the coast. The ice extending beyond the grounding line is floating.
(Left) The ability of the glacier to flow toward the coast depends in part on friction with the bedrock and till (the thickness of the till layer—only a few meters—is exaggerated for illustration) and in part on the ability of the floating ice shelf to *buttress* the glacier. Oppenheimer (1998) / Springer Nature.
(Right) When the WAIS terminus recedes sufficiently, the retrograde bed leaves more inland portions of the remaining ice in contact with the ocean, and the grounded ice is now more vulnerable to further disintegration. And with the ice shelf gone, the glacier will flow more rapidly toward the coast, accelerating the disintegration. See the text for more details. Adapted from Oppenheimer (1998).

those glaciers by their geometry: beneath the ice, the sea bed slopes downward toward the glacier interior (see Figure 3.30b) rather than downward toward their outer edge (as with most continental glaciers).

For conciseness, we can define the outer edge of those glaciers by their **grounding line**, the point beyond which the ice is floating rather than grounded. It is the discharge of a glacier across its grounding line that will supply mass to the oceans, raising sea level. If the climate warms and a glacier's grounding line retreats, a 'reverse' or "retrograde" slope of the sea bed will lead to a thicker and deeper glacier at its edge (Figure 3.30b). Thicker glaciers tend to flow faster (due to their greater weight), leading to greater discharge, and deeper (more submerged) glaciers are more susceptible to melting by the encroaching oceans. Such positive feedback would accelerate the retreat of the grounding line at the same time the glacial mass surges forward.

For marine ice sheets like the WAIS, what acts to prevent such catastrophic surges is mainly the presence of an ice shelf just off the coast. The ice discharged from the glacier accretes onto the ice shelf; but if the shelf is confined—by adjacent ice, coastal geometry, or bottom topography—and has nowhere to grow, it will resist additional glacial discharges. However, with sufficient climatic warming, basal melting of the sea ice may thin the ice shelf, reducing its ability to buttress the glacier and leaving the glacier highly unstable. *The resulting glacial surges would cause an increase in mean sea level that is distinct from and in addition to the slow rise predicted from current sea-level trends and models.*

It should be noted (see, e.g., Oppenheimer, 1998; Huybrechts, 2009; Joughin & Alley, 2011) that both the possibility of large-scale glacial melting of West Antarctica—which was first envisioned by Mercer (1968; also 1978)—and its suddenness have been hotly debated. The debate has persisted in part because of the challenges involved in modeling all the different factors affecting glacial stability (see, e.g.,

Hughes, 1973; Weertman, 1974; Schoof, 2007; Thoma et al., 2008; Gudmundsson et al., 2012).

Combined with observational evidence, including open-ocean microfossils found beneath West Antarctic ice (e.g., Scherer et al., 1998), some recent computer models (e.g., Pollard & DeConto, 2009) have in fact supported the claim that the WAIS is intrinsically unstable and has exhibited extreme variations in its areal extent over the past few million years. At times, the ice sheet has expanded all the way to the edge of the continental shelf; at times, it was effectively nonexistent (leaving an open pathway to the ocean); and at times, like the present, it has persisted in an intermediate state. A "collapse" of the West Antarctic Ice Sheet at some point in the future, then, might not be unexpected.

Evidence for past WAIS collapses is also found in global sea-level data. For example, Mercer (1978) noted that 125,000 yr ago, during the previous interglacial period known as the Sangamon, sea level spiked about 6 m higher than now. Conceivably, the water producing that higher sea level could have originated throughout the cryosphere: Greenland contains enough ice to add ~7 m to mean sea level if it all melted; West Antarctica, 4.3 m, and the rest of Antarctica, 54 m (Fretwell et al., 2013, Table 8); mountain glaciers, though, ~0.5 m at most (IPCC 2013). However, given the relatively unstable nature of West Antarctica, the rise in sea level during the Sangamon is often attributed largely to the WAIS.

More precise considerations support that conclusion. First of all, if just the unstable West Antarctic ice is eliminated (i.e., just those glaciers whose bed slopes are retrograde), that meltwater would produce only a 3.4-m rise in global sea level on average (see Bamber et al., 2009; Fretwell et al., 2013). But in those distant locations used to infer the 6-m 'Sangamon spike,' the change in Earth's gravitational field caused by the loss of West Antarctic ice would have amplified the rise in sea level by about 10%–20% (Bamber et al., 2009; see Figure 3.31)

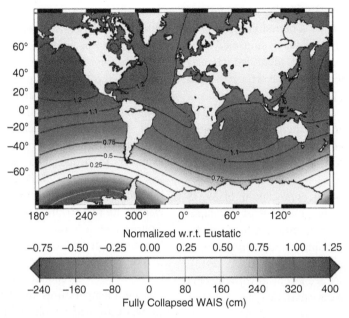

Figure 3.31 Prediction by Bamber et al. (2009) of the rise in mean sea level worldwide following a collapse of the WAIS. The rise is depicted in absolute terms, with the color scale in centimeters; and also relative to an idealized uniform (*eustatic*) rise in sea level, as a fraction. In their model, a full collapse of the WAIS would eustatically add 3.2 m to sea level (other models estimate that to be 3.4 m).

Their prediction accounts for the changes in gravity associated with the loss of WAIS ice and the gain of water in the world's oceans; the effects of the crust 'rebounding' where the weight of surface ice has disappeared, and subsiding under the additional weight of the added water; and secondary effects from a slight shift in Earth's rotation axis caused by the redistribution of mass in the Earth System.

Two factors are responsible for most of the departure we see in this map of the actual sea-level rise from an ideal uniform one. Prior to its collapse, the mass of ice in the WAIS had gravitationally pulled seawater up toward it, so that the sea surface was actually warped upward near the coast; after collapse, the lower gravity there, combined with the need to distribute the meltwater ocean-wide, results in a sea level which is much lower in that region than the eustatic amount. Meanwhile, ocean-wide, 'self-gravitation' of the oceans means that the additional gravity of the added water can pull more water toward itself, so that the sea-level rise away from the WAIS will be higher than eustatic. Image from Bamber et al. (2009) / American Association for the Advancement of Science.

or more (Mitrovica et al., 2009)—the geophysical reality is that an addition of meltwater to the oceans is never exactly globally uniform, because the gravity field changes in a nonuniform way (the hypothetical uniform rise, called **eustasy**, is really only an ideal)—so the net effect of melting unstable West Antarctic ice would be more like a 4-m rise in sea level at many locations. The contributions to the Sangamon spike from other cryospheric sources, such as the periphery and "Wilkes Land" portions of the East Antarctic ice sheet (which, as seen in Figure 3.30a, are evidently

marine based), the coastal portions of Greenland, and the remainder of the WAIS, therefore need only account for about one third of the total.

A connection between WAIS collapse and the Sangamon spike would have especially disturbing implications for our near future. Since the Sangamon at its warmest was only ~1°C–2°C higher globally than our temperature is at present, it is not hard to imagine that the world will face another collapse of the WAIS when global average temperatures rise by only another couple of degrees.

Higher global sea level during an interglacial about 400,000 yr ago—a period known not so much for its warmth as for its unusually long duration—has also been identified (Hearty et al., 1999; also Olson & Hearty, 2009; van Hengstum et al., 2009); corrected estimates of the rise (Raymo & Mitrovica, 2012), 6–13 m, suggest that at that time there was a substantial contribution from Greenland (and perhaps the EAIS periphery) in addition to a collapse of the WAIS.

Beyond a growing appreciation of WAIS instability, and of the possibility of a collapse of the WAIS, the key question is how rapidly future global warming would bring about its next collapse (if at all; see IPCC, 2013). That question has been kept alive by events taking place in West Antarctica (see our later discussion).

The 'other side of the coin' of cryospheric melting is also a serious threat to life. As mountain glaciers melt—for example, in the Tibetan plateau—populations dependent on them for their water needs will face shortages. Currently, it is estimated that about 40% of the world's population ultimately derive half their water supply from mountain glaciers (Gore, 2006).

Whether it is rising sea level or water shortages, these indirect physical consequences of global warming are certain to have a major societal impact. With enough of a rise in mean sea level, the resulting inundation of coastal cities might well, as mentioned earlier, lead to large-scale migrations away from such regions. And, according to some, we are already seeing the impact of water shortages in some of the recent and ongoing conflicts taking place around the globe: they are, at their core, water wars. As important as such global societal changes are, however, they are outside the scope of this textbook.

Second Set of Indirect Consequences. Another indirect consequence of global warming would be a *cooling* of the upper atmosphere—that's right, "global warming" refers to a warming of only the troposphere!

The temperature at any height in the atmosphere is a consequence of both the incoming solar radiation and the outgoing infrared warmth from Earth's surface. If greenhouse gases in the troposphere trap some of that outgoing heat, the troposphere will warm, but the atmosphere above will cool.

As the upper atmosphere cools, it will contract (this will be counteracted to some extent by the troposphere, which expands as it warms). The layer of the atmosphere above the troposphere is the **stratosphere**, which extends up to ~50 km altitude, and above the stratosphere are the mesosphere and then (beginning at ~80 km altitude) the thermosphere. The thermosphere is very tenuous, and the incoming solar radiation is able to heat it to very high temperatures; thanks to particularly energetic ultraviolet radiation, portions of it are even ionized. The strongest ionization occurs in a region more than ~250 km above Earth's surface—this is the main ionized layer, usually called simply the **ionosphere**. A weaker ionospheric layer exists at about 100 km altitude (and during daytime hours, solar radiation produces still other layers of ionization). As part of the upper atmosphere's contraction, the ionosphere(s) should sink closer to Earth's surface; this has been dubbed the *falling sky* effect.

As was first pointed out by Ramanathan (1988), 'global cooling' of the upper atmosphere, or not, could serve as a significant test of the cause of an observed tropospheric warming. That is, if the troposphere is warmed by increased incoming solar radiation, rather than by a greenhouse effect, that insolation would also heat the upper atmosphere—without limiting the outgoing warmth—resulting in a global warming of the entire atmosphere. Given the falling sky effect, the test of global warming *versus* increased insolation could also be carried out by measuring the altitude of the ionosphere, instead of upper atmosphere temperatures.

Another indirect consequence of global warming would be a growth of the **ozone holes**, that is, the springtime *depletion* of

(a)

$$\text{Production} \begin{cases} O_2 + uv \rightarrow O + O \\ O + O_2 \rightarrow O_3 \end{cases} \qquad \text{Destruction} \begin{cases} O_3 + uv \rightarrow O + O_2 \\ O + O \rightarrow O_2 \end{cases}$$

(b)

$$\text{Destruction by } \textit{chlorine} \begin{cases} O_3 + Cl = ClO + O_2 \\ ClO + O = Cl + O_2 \end{cases}$$

Figure 3.32 Schematic effects of chlorine (Cl) on the stratospheric ozone layer.
(a) The normal levels of ozone (O_3) in the stratosphere result from competing natural processes of production and destruction.
Production: incoming ultraviolet light (uv) causes photo-dissociation of oxygen; then, collision of single oxygen atoms with diatomic oxygen molecules creates ozone.
Destruction: uv also causes ozone to photodissociate. Because O_2 is so abundant, though, production of ozone dominates—until ozone levels have built up sufficiently to sustain a comparable rate of destruction. Ultimately a small residual (background) level is left in the stratosphere.
Note that when ozone has been photodissociated, the single oxygen atoms may combine with a variety of elements or compounds—with O, as shown here (on the right), producing more O_2; with O_2 (as on the left), leading once again to the production of more ozone; or with other elements, yielding compounds unrelated to oxygen that take them out of the picture permanently.
(b) Chlorine destroys ozone molecules by pulling off oxygen atoms from them; the resulting compound also prevents ozone formation by stealing single oxygen atoms to produce regular diatomic oxygen molecules. Because Cl is left over after these reactions, it is effectively a *catalyst*, available to act thousands of times to deplete ozone before eventually being consumed in some other way.

stratospheric ozone in polar regions. On the face of it, the ozone holes have little to do with global warming: chlorine (along with some other culprits)—produced by the ultraviolet photodissociation of chlorofluorocarbon molecules that had made their way up through the troposphere into the stratosphere—reacts with O_3 and O to destroy ozone *and* hinder its replenishment (see Figure 3.32). The destruction is most effective when the stratosphere is cold enough for *polar stratospheric clouds* (PSCs) to form: the ozone-depleting chemical reactions can take place an order of magnitude more efficiently on the PSC ice crystals—which act as 'platforms' to promote the reactions—than in thin air. The PSCs, which can include both water ice and nitric acid trihydrate ice (NAT ice, $HNO_3 \cdot 3H_2O$) crystals, form when the stratosphere is sufficiently cold (water ice forms at –83°C, NAT ice at –78°C) and sufficiently still.

So the primary connection between ozone depletion and global warming follows from the *cooling* effect of the latter on the upper atmosphere. A cooler stratosphere means the critical low temperatures for PSC formation will be reached farther away from the poles

(e.g., Figure 3.33) and will persist for a longer time; thus, more PSCs will form, the area they encompass will enlarge, and their lifetimes will lengthen. In short, the severity, area, and duration of the ozone holes will all rise.

Not surprisingly, there is also a positive feedback, in that the loss of ozone in the stratosphere represents the loss of a greenhouse gas able to absorb outgoing warmth from the Earth; thus, the loss of ozone cools the stratosphere more, leading to further depletion.

Third Set of Indirect Consequences. The *extreme weather effects* comprising this set of consequences are perhaps the hardest to model and require the most subtle explanations. We take a simple approach in this chapter, but by the end of this textbook, some of the analytical tools required for a more rigorous understanding of these phenomena will have been developed.

Global warming can be viewed as adding energy into the troposphere, both directly, as increasing amounts of heat are trapped there by greenhouse gases, and indirectly, as the troposphere's humidity increases. The latter takes place because warmer air can hold more

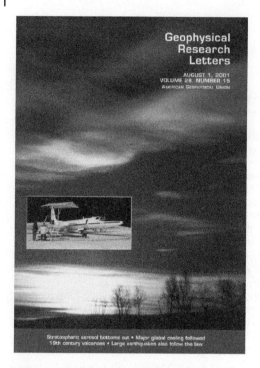

Geophysical
Research
Letters

AUGUST 1, 2001
VOLUME 28 NUMBER 15
AMERICAN GEOPHYSICAL UNION

Stratospheric aerosol bottoms out • Major global cooling followed
19th century volcanoes • Large earthquakes also follow the law

Figure 3.33 Polar stratospheric cloud (PSC) over Sweden, at latitude 68°N (cover picture from *Geophysical Research Letters*, August 1, 2001).

moisture; it adds energy indirectly because humid air releases heat of condensation as it rises. As a result of the greater humidity, we should expect to see more low-pressure weather systems with very low central pressures, thus more vigorously rising air drawing in stronger winds and producing more intense rain—in other words, more very strong storm systems.

Warmer oceans can be expected to directly power developing hurricanes by their heat as well as by the humidity they supply, producing even more intense hurricanes. With warmer oceans, the breeding grounds (Figure 3.9) that give birth to hurricanes will also be more extensive and last longer—resulting in a greater number of hurricanes each season, and a longer hurricane season.

The oceanic heat and humidity directly fueling hurricanes will also be a source of energy for every other low-pressure system passing over the oceans. With sufficient

global warming, one could imagine that such fuel would be sufficient to create hurricanes even outside of the hurricane breeding grounds. More frequent occurrence of such "extra-tropical" (or "subtropical") hurricanes might be still another sign that global warming has begun to dominate our climate system.

The greater intensity of storm systems (whether hurricanes or not) can be expected to result in more frequent and more severe flooding events.

As global warming increases the energy level of the atmosphere, it will become increasingly turbulent, with **wind shear** (variations in wind speed or direction, either from location to location or from one height to another) more prevalent. In very general terms, tornadoes are thought to develop—most often in certain environments (like "tornado alley" in the west-central United States)—from a combination of vertical wind shear, which can produce a long stream of 'rolling air,' and updrafts sufficiently strong to up-end the stream, producing a rotating column of air (Figure 3.34). A more energetic atmosphere should exhibit an increased number of tornadoes, with more of them severe. Additionally, tornadoes often form in association with severe thunderstorm systems called "supercells;" so, as global warming leads to more strong storms, we should expect to see a greater number of (severe) tornadoes.

Global warming can impact the three-cell circulation of the atmosphere in a variety of ways. We saw earlier in this chapter that the global circulation is driven by equator-to-pole temperature differences. Because of preferential warming of the Arctic (Figure 3.29a), that temperature difference in the northern hemisphere will be reduced, yielding a more sluggish circulation (see also references in Kossin, 2018). With the potential for WAIS collapse, the southern hemisphere circulation may well follow suit.

That preferential warming also means that the temperature contrast between adjacent cells will lessen: relative to the warming tropical cell, the mid-latitude cell will warm by

Figure 3.34 Cartoon suggesting very simply how tornadoes form. Wind shear produces a 'tube' of rolling air, which is then righted by an updraft. Image from https://www.accuweather.com/en/weather-news/watch-erupting-volcano-creates/23751780.

slightly more, and the polar cell by even more. A reduced temperature contrast implies a reduced density contrast, allowing the warmer cell to push into the cooler cell with less resistance from the cooler air mass. The cell boundaries will become 'wavier' as their 'rigidity' diminishes; north-south migration of the cells (seasonal or otherwise) will become easier, and variability of the weather—especially near the cell boundaries—should increase.

More mobile cell boundaries should lead more often to situations where tropical cells penetrate effectively poleward or polar cells penetrate effectively equatorward, resulting in more heat waves or cold snaps. (Imagine, for example, Alaska having mild temperatures or Florida having subfreezing temperatures....)

The cell variability should be reflected in a greater waviness—bigger meanders—of the jet streams, which track the cell boundaries from above. As with the explanation of the jet streams themselves, this variability can also be explained in terms of the force balance that creates them. With less temperature contrast between adjacent cells at the top of the troposphere, as global warming progresses, the pressure forces initiating the jet streams will be weaker, and the jet streams slower. A weaker Coriolis force—weaker, because it depends

on the velocity of the flowing air—will act to 'restore' the jet stream less effectively, allowing greater-amplitude meanders to happen. (Like rivers of water on the land below, the slower "rivers of air" constituting the jet streams will meander more; faster rivers would cut deeper into the 'streambed' and flow straighter.) See Francis and Vavrus (2012) and Cohen et al. (2014) for a more comprehensive discussion.

Deeper incursions of one cell into another—reflected in a larger amplitude, sharper meander of the jet stream—would result in more sudden changes in temperature experienced at the surface, as the meander progressed eastward as a Rossby wave and the surface experienced a change from, for example, a temperate to a polar air mass (see Figure 3.35). (Imagine, for example, a drop in temperature at some location of 30°F in the space of a few hours.)

To the extent that jet streams can steer large storms, with a weaker jet stream those storms will advance more slowly; this will result in greater amounts of rainfall in the regions experiencing the storms, thus more flooding episodes (see Kossin, 2018).

With more storms, more very large meanders of the jet stream (as in Figure 3.8), and the ability of the jet stream to steer those storms

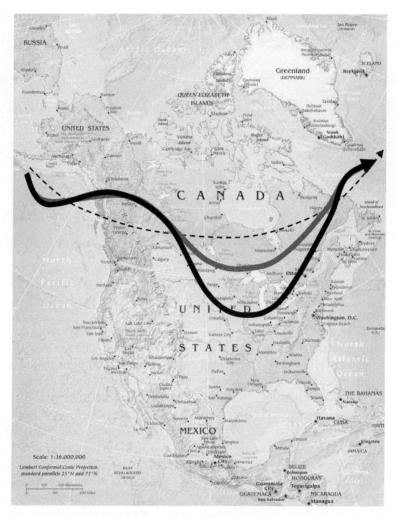

Figure 3.35 With cooler (warmer) air masses north (south) of the jet stream, larger-amplitude meanders of the jet stream (black curve versus brown curve) should be associated with more sudden changes in temperature—for example at the locations where the jet streams cross the dotted line—as the meanders progress to the east. Courtesy of Geographic Guide.

through those larger meanders, it follows that more nor'easters will be possible—and their intensity will be amplified by the greater warmth of the Gulf of Mexico or the Atlantic coastal waters, as the oceans continue to heat up.

The three-cell circulation will be further modified by the presence of a partially or completely ice-free Arctic Ocean, which through its warmth might foster low atmospheric pressure—an incongruous situation for polar latitudes (cf. Figure 3.5a). This situation has been studied broadly (e.g., Herman & Johnson,

1978; Cohen et al., 2014; also, the comprehensive essay by McSweeney, 2019 and references therein).

We might speculate that such a persistent low pressure could fundamentally disrupt the atmosphere's three-cell circulation. As we know, low atmospheric pressure is generally associated with rising air. That rising air would displace the northern boundary of the polar cell, normally characterized by air sinking at or near the North Pole, pushing that northern boundary slightly equatorward. In effect, it would create a new, 'Arctic' cell that would be driven, Hadley-like, by rising air near the Pole and by air sinking at the now-displaced polar-Arctic boundary. The resulting four-cell global circulation—Hadley, Ferrel, polar, and Arctic cells—is not what we would typically think of as (or as causing) extreme weather, but it is in itself extreme from a conceptual perspective!

We may also speculate about the effects of global warming on ENSO. A weaker wind-driven circulation in the northern hemisphere (and possibly in the southern hemisphere as well, eventually) suggests that the warm pool—set up during normal, non–El Niño conditions by converging trade winds and maintained by the North and South Equatorial Currents of the Pacific gyres, as shown in Figure 3.14a—might not be kept so effectively in the western equatorial Pacific. That is, migrations of the warm pool eastward might occur more frequently, i.e. El Niños might occur more often. Additionally, in a globally warmed world, with warmer oceans, we can expect the warm pool to be warmer and larger (both in the western Pacific *and* off the Peruvian coast), leading to an amplified Southern Oscillation and stronger teleconnections.

The disruption of the three-cell circulation by global warming, with associated changes in ocean circulation, should cause shifts in regional climate, changing weather patterns and thus rainfall, cloud formation, winds, and so on. These shifts are expected, according to IPCC researchers, to lead to longer-lasting and more extensive regional droughts (see Figure 3.36), which in turn would bring with them an increase in wildfires (see Figure 3.37). The latter could also amplify the global greenhouse warming, through the increase in CO_2 from the burned biomass and a decrease in albedo from the resulting soot particles as they settle to the ground.

As a result of regional climate changes, there will be shifts in natural vegetation (see Figure 3.38). A migration of agricultural crop belts could follow—for instance, a shift in the American wheat and corn belts northward toward Canada is often imagined—though their success would depend on how well those crops could adapt to changes in soil type, daylight hours, and other growing conditions. Such shifts in crop belts have the potential to affect world economies and perhaps even the balance of power between various countries. On the other hand, if a northward shift in "the world's bread-basket" fails to occur, the consequences—with continued global warming—could be dire.

Many of the predicted effects described in this section have the potential to impact the ability of various species of life to proliferate or, in some cases, to even survive. That assessment, which should also include consideration of past episodes of climate-related species extinction and radiation, is well beyond the scope of this chapter.

3.6.3 Tempered Expectations: Complications in How These Consequences Play Out

The anticipated consequences of global warming enumerated above follow from a basic appreciation of the way Earth's climate system works. But there are at least two factors complicating that vision of a globally warmed future.

Some Consequences Do Not Take Place Independently of Each Other. The climate system is complex, and some of the predicted effects can *interfere* with others. For example, increased 'background' winds—whether predicted from the higher energy level of the atmosphere or associated with increased storm activity—can result in stronger wind shear; and wind shear in the 'right' locations, such as the Caribbean or other parts of the tropical Atlantic, can shear off the tops of tropical storms, preventing them from developing further into hurricanes (see Kossin, 2017 and references therein). This may also occur during an ENSO event; a number of researchers (beginning with Gray, 1984, and others) have found that times of ENSO correlate with reduced hurricane activity in the North Atlantic, possibly a

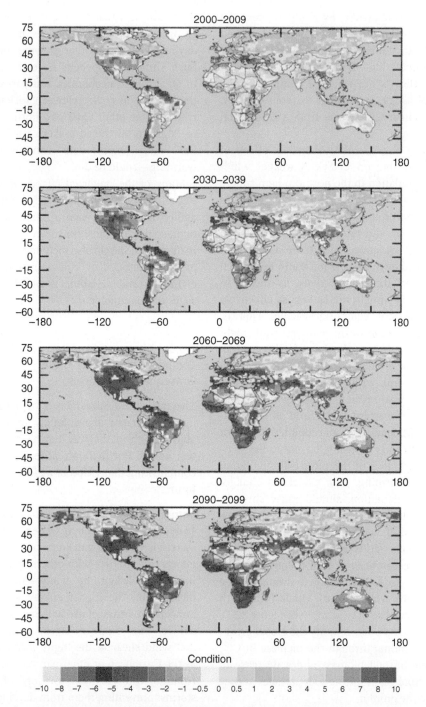

Figure 3.36 The potential for future drought, inferred from a variety of climate model predictions included in IPCC (2007), according to the analysis by Dai (2010, with corrections in 2012); image from http://www2 .ucar.edu/atmosnews/news/2904/climate-change-drought-may-threaten-much-globe-within-decades / University Corporation for Atmospheric Research. Courtesy Wiley Interdisciplinary Reviews. Areas colored red/purple/lilac are at the greatest risk for extreme drought; drought is unlikely in areas colored blue.

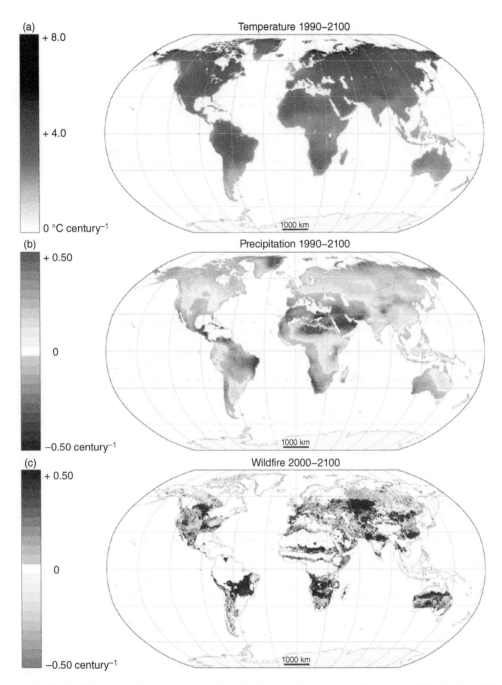

Figure 3.37 Predicted rates of change (per century) in temperature, precipitation, and wildfire for the A1B scenario described in Figure 3.29a, according to the analysis by Gonzalez et al. (2010) / John Wiley & Sons. Antarctica is excluded from the analysis. Precipitation and wildfire rates are fractional, relative to their twentieth-century averages.

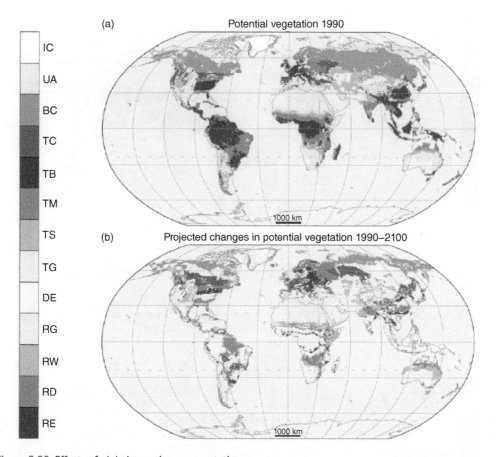

Figure 3.38 Effects of global warming on vegetation.
(a) Potential vegetation regimes near the end of the twentieth century. "Potential" vegetation refers to vegetation in the absence of farms and cities. Biomes (as color-coded on the left) include IC = ice; UA = tundra; BC = boreal conifer forest; TC, TB, TM, TS, and TG = temperate conifer forest, broadleaf forest, mixed forest, shrubland, and grassland; DE = desert; and RG, RW, RD, and RE = tropical grassland, woodland, deciduous broadleaf forest, and evergreen broadleaf forest (i.e., rainforest).
(b) Changes in regimes by the end of this century, compared to (a), predicted for the A1B warming scenario described in Figure 3.29a according to the analysis by Gonzalez et al. (2010). For all regions not grayed out, the color code on the left indicates which types of biomes are affected, with at least 50% confidence in the prediction.
Images from Gonzalez et al. (2010) / John Wiley & Sons, with reference to their Appendix S4.

consequence of the wind shear produced by El Niño–powered storms.

Such interactions, though, might not matter to strong hurricanes, which would be better able to survive wind shear. The end result may be that, in a globally warmed future, some ocean basins may experience fewer (or only marginally more) hurricanes, but the number of *intense* hurricanes will increase. Such conclusions are in line with predictions from recent climate models (e.g., Bengtsson

et al., 2007; Emanuel et al., 2008; Knutson et al., 2010).

Not All Consequences Happen Simultaneously. There are both temporal and spatial reasons for this. Our brief discussion here of those reasons will shed some light on how global warming works.

- Fundamentally, *global warming is a trend, not an event (or series of events)*! And by "trend," we mean a 'real-world' trend, not a mathematical trend. Mathematical trends—typically straight lines or quadratic curves, used to characterize the overall pattern in a data set versus time (for example) and thereby model changing phenomena—are smooth, well-defined functions that always increase or always decrease; real-world trends are irregular, suffering occasional and brief reversals as they move toward an overall increase or decrease.

Consider, for example, the trend of increasing global emissions of carbon dioxide over the twentieth century; can you suggest any factors that would have made the actual increase irregular?

What about the effect of that CO_2 increase on global temperatures—would you expect temperatures to trend upward with the same pattern of irregularity?

The differences between mathematical and real-world trends can be readily seen in many of the data sets and their analysis illustrated throughout this chapter.

Whatever components of global warming are trending—regional mean temperature of Greenland, Australian wildfires, global mean sea level, and so on—are also subject to *their own* irregularities, some of which will add to their trend while others undermine it. One year, for instance, the area of land burned by wildfires in Australia might reach a record level, but not (for whatever reasons) the next year; during those years, however, mean sea level might continue to rise. Thus, not all anticipated consequences of global warming will happen simultaneously.

Incidentally, the trending nature of each component of global warming means that we should not expect each year to top the previous. When we predict that global warming should lead to an increase in severe hurricane activity, for example, we do not mean that each year must have more strong hurricanes than the previous—that would be a mathematical trend—but only that the

overall (real-world) trend of strong hurricane occurrences should be an increase. In fact, successive record-breaking years of a trending component should be considered rare, and would be persuasive evidence indeed of the reality (and potential severity) of global warming. This interpretation applies to temperature as well: people who say that global warming must not be real if the average global temperature this year is not higher than last year's have failed to appreciate how real-world trends work.

- The response of most components of the climate system to global warming would probably be irregular even if those processes forcing the warming were perfectly steady. That's because *the climate system has multiple energy "sinks," or "reservoirs," to take up the energy of global warming.* An increase in climate energy from global warming can show up in one reservoir one year and another, the next. For example, additional heat retained by the atmosphere from new greenhouse gas emissions might be measured as warmer air temperatures one year, but over the next few years such heat could be absorbed by the oceans as they warm and by polar ice as it melts; and so on. As Trenberth and Fasullo (2013) point out, the energy—which begins as *radiant heat*, i.e. the incoming solar radiation—can end up as *internal heat* (increased temperatures), *latent heat* (causing melting or evaporation), *gravitational potential energy* (associated with upwelling), or *kinetic energy* (of winds and currents).

Furthermore, the effectiveness of any one reservoir in taking in that initial radiant heat might depend on the state of the other reservoirs; for example, polar ice is less likely to melt under strong surface winds, which can carry heat away from the exposed surface of the ice. And, these reservoirs are three-dimensional, so that energy transfer *within* them will not be instantaneous; vertical energy transfer within the oceans, for

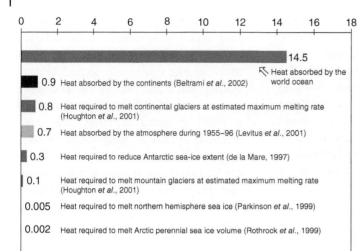

Figure 3.39 Observed increase in internal and latent heat within various components of the climate system between 1955 and 1998, in units of 10^{22} J. Levitus et al. (2005) / John Wiley & Sons. According to Levitus et al., these increases are most likely due mainly to global warming, with influences from sulfate pollution and natural variability.

example, may be quite slow. Figure 3.39 provides an illustration of the 'activity' of just *some* of these reservoirs during the second half of the twentieth century, quantifying the increase in internal and latent heat within the climate system during this time period.

Such behavior of the climate system should not come as a surprise to anyone, considering how the seasons develop or 'trend' each year. Each year, at most mid-latitude locations, the insolation steadily increases throughout springtime (as, each day, the Sun rises higher in the sky); and, overall, from the equinox to the solstice, the climate does warm. But the increasing warmth during springtime is not steady—*and not without occasional chilly reversals!* That is, the forcing (insolation) is smooth, but the response of the climate system is an irregular trend.

This occurs, at least in part, because the increasing heat is being taken up by different reservoirs, as radiant heat energy is converted to the various other types of energy. And, for the same reasons, as winter nears (with the noon Sun lower and lower in the sky each day), the forcing and climate both trend toward lower energy, but with the climate response

again irregular, and again with occasional warm reversals.

Analogously, even if the abundance of greenhouse gases happened to increase steadily from year to year, the resulting behavior of the climate system would be characterized by irregular, real-world trends.

- Lastly, even if those processes forcing global warming were perfectly steady, and even if the response of some component of the climate system to that forcing was (somehow) steady, our observations of that component might well show irregularity. That's because *some of the components we observe in order to reveal the effects of global warming also have contributions from sources unrelated to global warming*, and those sources might be intrinsically variable. For example, it is possible that the gradual disappearance of Arctic sea ice observed from year to year (discussed later in relation to Figure 3.46) is the result of a globally warming atmosphere and ocean, and from changes in climate conditions (cloudiness, winds, precipitation, warm ocean currents, and so on) associated with global warming; but the persistence of Arctic sea ice is *also* affected by such factors varying during the normal functioning of the climate system, independently of any global warming. The net effect may be that some years the ice survives better, or worse, than the overall trend predicts.

In sum, the multiplicity of sources and sinks more or less guarantees that the various consequences of global warming will exhibit irregular trends, making it unlikely that the consequences will all occur simultaneously.

A Case Study Involving ENSO. A provocative illustration of these complications (in relation to global warming) is provided in Figure 3.40, which shows Earth's global mean

surface temperature since 1970, relative to the twentieth-century average, and the timing of El Niño and La Niña events. Throughout most of this time span, the irregularities of the temperature trend evidently correspond to the swings in the ENSO cycle: times of 'positive' irregularity, when the mean surface temperature exceeds the overall trend, correlate with times of El Niños; and times of 'negative' irregularity, when the mean surface temperature falls short of the overall trend, correlate with La Niñas (see also Hansen et al., 2010 and references therein).

At first thought, a simple cause-and-effect explanation seems obvious—when there is a

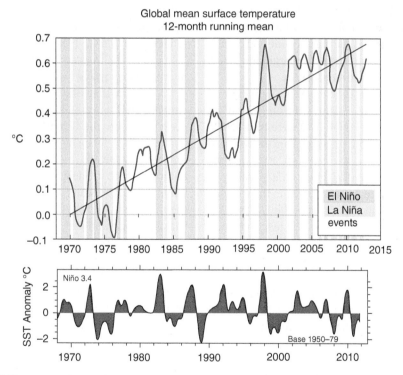

Figure 3.40 The top graph reveals correlations between irregularities in the global mean surface temperature (dark red curve, with each point an average over ±6 months, to smooth out shorter-term variations), relative to the overall trend of global warming (straight line), and the occurrence of an El Niño or La Niña. (The global mean surface temperature—or, more precisely, mean temperature "anomaly"—is discussed later in this chapter.)
Those occurrences are indicated by the sea surface temperature (SST) of the region that would be the El Niño warm pool during an El Niño. That region is defined in Trenberth and Fasullo (2013), where this figure is taken from, as the equatorial band within longitudes 120°W–170°W; its temperature relative to the 1950–1979 average is shown in the lower graph. As discussed earlier in the text, this region is anomalously warm during El Niño events and anomalously cold during La Niña events.

massive warm pool in the ocean, *of course* that will add to the global surface temperature!— but, in fact, such an interpretation is incorrect. As we saw when discussing ENSO earlier in this chapter, the sea surface warming that defines an El Niño is actually a redistribution of warm and cool waters across the Pacific; as warm waters accumulate in the eastern tropical Pacific, the seesaw observed in atmospheric pressure indicates that the warm pool is gone from the western Pacific, replaced by cooler waters. Such a redistribution does not affect Earth's mean surface temperature. On the other hand, the redistribution may not be entirely surficial, as the Pacific thermocline undergoes a tilt, shallowing to the west as cool waters dominate and deepening to the east as the warm pool expands there; but such opposing effects on the thermocline will cancel to some extent, and add only marginally to the global mean surface temperature.

Actually, even if the large temperature anomalies in the El Niño region shown in the lower part of Figure 3.40 were completely uncompensated rather than balanced by opposite anomalies in the western Pacific, that region (known as *El Niño 3.4*) is simply too small to contribute substantially to the correlations seen in the upper part of the Figure. The region in question covers an area just over 1% of the entire oceanic area. For the huge 1997–1998 ENSO event, the anomalous warmth shown in Figure 3.40 (lower graph) can contribute at most ~0.04°C to the spatially averaged, global temperature (4°C times 1%); but this is only ~20% of the variation that we see in 1997–1998 superimposed on the trend in Figure 3.40.

So, Pacific Ocean temperature variations during ENSO events are not directly responsible for the observed global warming, i.e. either the overall trend or the large irregularities superimposed on it. Instead, the strong correlation shown in Figure 3.40 implies that during an El Niño or La Niña, *teleconnections* play a significant role in affecting the global average temperature. The teleconnected atmosphere and oceans thus constitute a major reservoir for climate system energy.

What is the nature of that teleconnected reservoir? Over the past few decades, scientists have identified the Southern Ocean and Antarctic coast (particularly around the Weddell Sea and Antarctic Peninsula; see Figure 3.30a) as a region very responsive to ENSO fluctuations. The response there, described by McKee et al. (2011) as "representing the greatest temperature response to ENSO outside the equatorial Pacific," includes variations in high- and low-pressure systems in the atmosphere—changes in the Ferrel cell in the region—with surface winds occasionally spinning up a local ocean gyre and redistributing sea ice as well as impacting surface temperature (see Liu et al., 2002 and references in McKee et al., 2011).

Interestingly, as described by McKee et al., measurements in the near-bottom of the Weddell Sea show that ENSO variations—at least strong ENSO events—also impact the deep ocean, though the variations measured in bottom water temperature lag their surface forcing by typically more than a year. The authors ascribe the deep-ocean impact to variations in bottom water production caused by the changes in surface wind patterns; the delay between cause and effect is mainly because of the time it takes for those winds to clear sea ice away from the coast, creating a larger area to be subjected to freezing, and for the dense saline water thus left over to be transported off the ice shelf, allowing it to sink to abyssal depths. In effect, the deep Southern Ocean appears to act as a reservoir for climate energy, with characteristics distinct from the ocean surface layer reservoir.

Such a conclusion is reminiscent of the ocean modeling results of Meehl et al. (2011), Balmaseda et al. (2013), and Trenberth and Fasullo (2013), who all showed that the global ocean deeper than 700 m acts as an energy reservoir separate from the shallow oceans. Their deep ocean—which includes shallower depths in the oceans than are involved in the ENSO Southern Ocean reservoir—exhibits behavior that allows it to modify the progression of global warming, as it soaks up energy from the surface-layer and ultimately atmospheric reservoirs.

In other words, ENSO teleconnections help distribute global warming energy to deep and distant reservoirs, and in doing so cause the irregularities we see in the upward trend of Earth's mean annual surface temperature. Thus, it is the entire curve in Figure 3.40, irregularities and all, not just the straight line, which accurately describes the global warming signal in the latter part of the twentieth century.

3.6.4 Anthropogenic Variations in Climate: Evidence Concerning *Direct* Consequences (−If You Insist)

The expected consequences of global warming are wide-ranging and serious. To what extent are they already occurring?

We begin with the most direct consequence, an increase in tropospheric temperatures. Scientists initially focused on trends in the *globally averaged* surface temperature as a direct measure of global warming, because early climate models did not have the spatial resolution required for more regional temperature predictions. The general public shared that initial focus, which continues (among the public) till today.

Temperature Versus Temperature Anomaly. The global average temperature—a single number constructed by averaging temperatures at the poles, in the tropics, in desert regions, in the doldrums, and everywhere in between—is, however, not a very physically meaningful quantity, though it has been useful as a gross indicator of whether the Earth is in an Ice House or Hot House regime. Perhaps for this reason (NOAA 2018; also, e.g., Jones & Kelly, 1983 and Jones et al., 1986), climatologists have typically focused on **temperature anomalies**, departures in the temperature at a site from some reference value, rather than the 'absolute' or actual temperature values there. Classical data analysis techniques provide one simple way to achieve this "standardization": by subtracting the mean value of the data from each measured value at that site; the reduced data will represent departures from the mean.

It might be worth noting that this procedure is not the same as "normalization," in which standardized data is also rescaled such that the magnitude of the data is typically within ±1. Temperature data is often normalized as a way to identify outliers—extreme values and/or incorrect measurements.

With our interest in determining whether a trend of global warming exists at some location, a more useful version of the classic standardization would be to subtract from all measurements at that location the mean value calculated within a chosen reference epoch, such as the mid-twentieth century, rather than the mean value over the entire span of that site's data. With either version, standardization will produce a set of uniformly shifted data—shifted downward at a tropical site, upward at a polar site—but with the pattern of temperature variations there unaltered.

The temperature anomalies thus produced, now zero-mean within the chosen epoch, may more reasonably be compared site by site, and the anomalies from different sites combined in order to estimate regional, hemispheric, or global trends in temperature. Such goals suggest still another way to standardize the temperature data (see Figure 3.41): instead of the data at all sites being zero-mean (within a chosen epoch), we could equivalently designate one site as the reference location, and adjust the mean values of the other sites (within that epoch) to be the same as that of the reference. As Figure 3.41 implies, such adjustments would not affect any trends or other patterns of temperature change observed at a site.

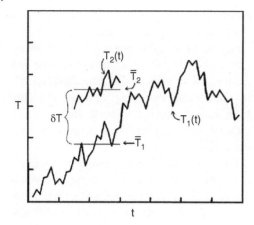

Figure 3.41 Hypothetical temperature data versus time at two sites. $\overline{T}_1$ and $\overline{T}_2$ are the mean values of these data sets within the epoch of interest (suggested by horizontal lines). Subtracting the difference δT in mean values from the second data set gives the two time series a common baseline and allows their trends to be better compared. Image from Hansen & Lebedeff (1987) / John Wiley & Sons.

Limitations of Temperature (Anomaly) Data; Are the Trends Accurate? To investigate global patterns of change, using globally averaged temperature anomalies is clearly an improvement over using globally averaged temperatures; but even so, their ability to reveal global warming trends is problematic.

- Because the various sources and sinks of global warming energy are not all 'in sync,' neither will be the changes in temperature they cause; spatially averaging those temperature changes may yield a mean global anomaly that is somewhat reduced in magnitude from the full effect, as some of those changes cancel each other.

  (Our case study above, based on temporal correlations between ENSO and changes in temperature, would have been affected only secondarily by reduced anomaly magnitudes.)

- We have seen that a decreased albedo in polar regions caused by global warming feeds back to amplify the temperature increase there. Global averaging will mask (to some extent) such amplifications as well.

- Even a relatively small global temperature anomaly might be associated with regional or indirect effects that are more substantial—and more indicative of the existence of global warming (remember: just because it is called "global warming" does not mean that atmospheric warming is its most important characteristic!).

Nevertheless, societal interest compels us to overlook these intrinsic shortcomings and evaluate the temperature evidence. Such an evaluation calls for a long data set, to better distinguish short-term patterns from long-term trends. Directly observed surface temperature anomalies extending back into the nineteenth century are shown in Figure 3.42a (top), graphed relative to a mid-late twentieth-century average; the observations comprising the anomaly in the year 2021 (—almost 1°C greater than that average) are illustrated in Figure 3.42a (bottom).

Unfortunately, such a long-term view forces us to confront some major practical deficiencies in global mean temperature data.

- Global averages were not possible before ~1860, due to severely limited spatial coverage (see IPCC 2007 for an enlightening discussion of temperature observations); but even by 1860 such temperature data covered only ~23% of the globe (with sites mainly in Europe, Russia, and North America) (Jones et al., 1986). The spatial distribution of land-based measurement sites improved with time (Figure 3.42b, left); prior to the International Geophysical Year (IGY) of 1957–1958, though, there was still no direct temperature data south of 67°S latitude (Hansen & Lebedeff, 1987)—that is, in *one quarter* of the southern hemisphere.

- Global averages require temperature data over the oceans as well as land areas. From 1880 till ~1980, this was achieved entirely

(a) **Global temperature anomaly** (relative to 1951–1980, °c)

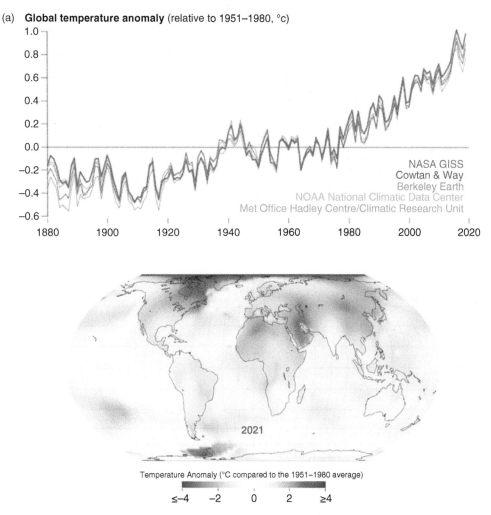

Figure 3.42 (a) (Top) Observed global annual temperature anomaly from 1880 to 2020, relative to the average global temperature during 1951–1980, according to various data sets (identified in the lower right and color-coded).
(Bottom) Observed temperature anomalies around the world during 2021. Both figures downloaded in January 2022 from https://earthobservatory.nasa.gov/world-of-change/global-temperatures.

from ship-based measurements, and until ~1920 those measurements yielded poor spatial coverage because they were confined to shipping lanes only. As shown in Figure 3.42b (right), coverage did not exceed 50% of the oceanic area until around the time of the IGY, and that was only possible using measurements sampled as infrequently as once per month. Data quality remained mixed until 1950, in large part because

of a change in shipboard measurement procedure in 1941, with a large correction required for earlier data (Folland et al., 2001). Buoys were used beginning in 1979, and satellite measurements from 1982 onward. But even today, problems remain in obtaining accurate temperatures in ice-covered seas by any method (see Hansen et al., 2010; also Rayner et al., 2003; Swanson, 2003).

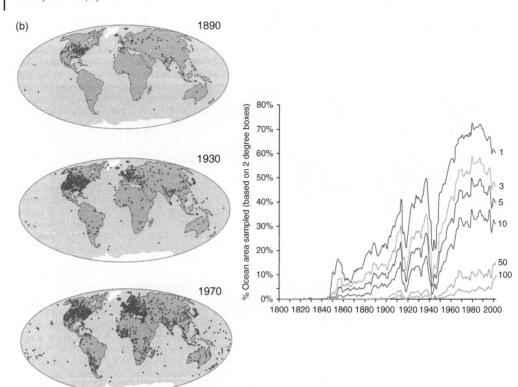

Figure 3.42 (b) (Left) Coverage by land-based stations that recorded surface temperature to determine the global mean temperature during three different years. Image from Ruddiman (2014) / W. H. Freeman and Company.
(Right) Coverage by ocean-based measurements of sea-surface temperature used to determine the global mean temperature over time. The number of measurements per month in each 2° × 2° area is shown to the right of each curve.
Thus, for example, the area of oceans sampled three times per month (light-blue curve) did not reach 30% until around 1910. Note also the impact of the First and Second World Wars on the percentage of oceanic areas sampled. Image by Worley et al. (2005) / John Wiley & Sons.

- Coverage is not the only deficiency in global mean temperature data. Over time, measurement sites have shifted—an unfortunate policy to follow, if the goal is to uncover *trends* in the data! To some extent, this has been in response to the development of **urban heat islands**, areas of enhanced heat associated both with the urban concentration of population and industry and with the absorbent (low-albedo) material covering building roofs, parking lots, and roadways (typically asphalt). The analysis by Peterson and Owen (2005), for example, suggested that temperatures may have been measurably distorted by urban heat island effects at nearly a third of the sites in their study.

  The potential for greenhouse-related temperature trends to be obscured by a shift in measurement sites is implied by the situation in the Greater Binghamton, NY area, where, in an attempt to eliminate the thermal distortions of urbanization, the official site for Binghamton temperatures was moved from downtown Binghamton to the county airport, 10 miles away and 200 meters higher in elevation—and typically several degrees cooler!

The overall urban impact on global mean temperature—whether a warming as expected from the concentration of population and industry or a cooling that results from a change in site location—can be assessed through models, or by comparisons between urban, suburban, and rural measurements. Brohan et al. (2006) proposed that urban effects might have added an uncertainty in the twentieth-century trend of global mean temperatures of up to ~0.05°C, thus (IPCC 2007) an order of magnitude less than the observed change. Hansen et al. (2010) found the net uncertainty to be even smaller, ~0.01°C, though they note that some cities and regions are known to have much larger effects.

Urban effects can be treated by excluding severely impacted sites from the mean temperature calculations, or by applying a correction, e.g. based on trends exhibited by rural stations not too distant. The land-based and marine temperature anomalies are typically processed with a variety of other corrections as well; see Jones et al. (1986) and Brohan et al. (2006) for examples.

The data in Figure 3.42a (top), the result of such corrections, do indeed show a trend toward higher temperatures. As evidence supporting the existence of global warming, however, that trend was often viewed skeptically, especially during the late 1980s and early 1990s and again in the early 2010s. For one thing, the rise has been erratic, faltering at times (e.g., ~2002–2012) and at other times even reversing (e.g., between 1940 and 1965).

Yes, we can now appreciate that this is a real-world trend, not a mathematical one. But without that 'surge' in warming beginning ~2015 (see Figure 3.42a), how confident would you be that the trend is significant? And what if the temperature starts trending down in, say, 2025?

Second, the magnitude of warming was small: overall, for example, the rise between 1900 and 2000 was only ~0.6°C. It's true that the rise since 2015 has been more dramatic; nevertheless, from a geologic perspective, even the entire increase from 1900 to 2020 amounts only to a minor warming (so far, at least)—much too small to be able to bring us out of our current Ice House climate, and not even comparable to the increase in temperature Earth experienced when shifting from glacial to interglacial periods during past ice ages.

Support From Computer Models? Despite the variety of other evidence beginning to accumulate in the late 1980s that global warming was a reality, the scientific community focused (unfortunately) on that approximately half-degree change in the globally averaged surface temperature. Computer models of the coupled, global circulation of the atmosphere and oceans predicted that the increase in greenhouse gas concentrations during the twentieth century should have raised the global average temperature by about a degree. Given the complexity of such models, being off by only half a degree might be considered a great success to some. But phrased as a result that the predicted increase in temperature was *twice* the observed increase, the disparity between prediction and observation was seen by many as a failure of the models to explain the observations. Many scientists therefore concluded that the trend must, instead, be reflecting some other, unknown process—or be nothing more than natural variability.

Because of the potential for global warming to cause widespread change to and disruption of society, it has often been assumed that countering it would be an expensive and all-consuming effort, requiring changes to the very foundations of society; this, in turn, has generated opposition to accepting its reality, especially among business-oriented conservative politicians in the United States (in other countries, political conservatives have shown no such reluctance). Such opposition—denying either that anthropogenic global warming has already begun or that any action should be taken to lessen its impact—found justification in the "failure" of those global warming models. We emphasize that even by the late 1980s, much evidence had already begun to accumulate demonstrating that *indirect* consequences of global warming were happening. But with global average temperature being viewed as the principal criterion for verifying global warming, the failure of numerical models to match temperature observations was critical.

The issue was resolved within the scientific community in 1996 when climate computer models were modified to include the effects of *sulfate aerosols* in the atmosphere. Those particles, produced when water droplets containing sulfuric acid (H_2SO_4) evaporate, act to reflect incoming solar radiation (both directly and from the clouds they create through a seeding effect), thus cooling the surface. The sulfate or sulfuric acid originates as sulfur dioxide emissions (e.g., from burning coal); earlier modelers had evidently not appreciated how widespread this air pollutant had become (in retrospect, the cooling during the mid-twentieth century seen in Figure 3.42a can be explained in part as a result of the ramping up of industrial activity, leading to more sulfur pollution, during and after the Second World War). Incorporating sulfate aerosols into the climate models, along with increased greenhouse gas concentrations, led to new predictions that the global average temperature would rise by only ~½°C—in agreement with observations.

Figure 3.43 illustrates the importance of sulfate aerosols more profoundly, showing what would happen to the global mean surface temperature by the year 2100 if those aerosols were entirely removed from the atmosphere in 2000. To avoid ambiguity, those predictions do not include the effect of increases in greenhouse gas abundances; their levels are instead assumed to remain unchanged throughout the twenty-first century. Of course, in reality, increases are likely unless a strong international climate agreement is fully implemented. But *if* the amount of greenhouse gases could remain at current levels from 2000 to 2100, Figure 3.43b shows that global surface temperatures would rise by century's end by an additional 1.0°C on average if the aerosols had been eliminated. Not surprisingly, these effects would vary regionally (Figure 3.43a), with some regions exhibiting substantially greater temperature increases by 2100.

One might say that the cooling effect of sulfate aerosols has been a 'silver lining' to the effects of global warming—although, considering the negative impacts of sulfur pollution, including acid rain and a variety of respiratory problems, perhaps 'gray-black lining' is more precise.

Figure 3.43b also illustrates the agreement possible between twentieth-century observations (black curve) and climate models, when the latter (blue curve prior to the year 2000) includes both greenhouse gas warming and the significantly counteractive effect of sulfate aerosols.

Temporal and Spatial Variations in the Temperature Trend. In recent decades, irregularities in the trend have continued. Ignoring for the moment the recent surge (~2015 to the present), the overall straight-line trend shown in Figure 3.42a (top) (and consistent with Figure 3.40 top) amounts to a warming of ~0.15°C per decade worldwide during 1970–2015, which, incidentally, is a greater rate of warming than any preceding ~45-year period since 1880. Within that 1970–2015 time span, however, there were substantial variations: over decadal time spans, for example, the observed increase was ~0.25°C/decade during 1994–2004, a rate 66% greater than the overall rate; but during 2002–2012 the rate was essentially zero or even marginally negative.

The "hiatus" during 2002–2012 was cited by some non-climatologists as proof that global warming was a misinterpretation of temperature data from the beginning—that we have just been seeing natural variations all along. As noted earlier, though, a "real-world" global warming trend does not guarantee a mathematically steady rise in global mean temperature, year after year, just as a cold spell during springtime does not mean summer will never happen: natural variations

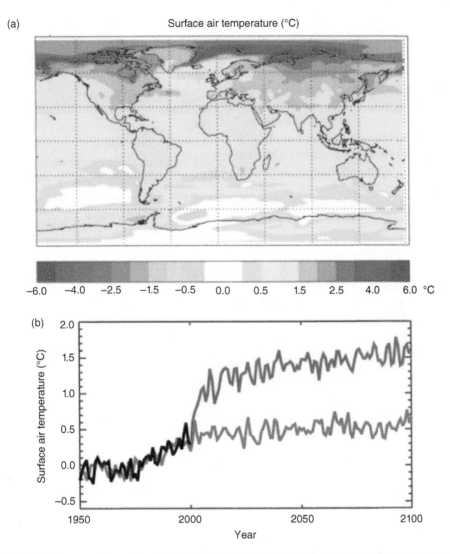

Figure 3.43 Effects on world's surface air temperature of removing all sulfate aerosols from the atmosphere in the year 2000. (a) Predicted changes by 2071–2100, and (b) average annual temperature versus time, relative to the mid-twentieth-century mean, predicted assuming greenhouse gas concentrations remain constant after year 2000, and aerosol concentrations are either unchanged (blue curve) or eliminated (red curve). Predictions (blue curve) prior to 2000 are based on observed greenhouse gas and aerosol concentrations. The observed change in mean annual temperature between 1950 and 2000 is also shown (black curve).
Images from IPCC (2007), Intergovernmental Panel on Climate Change.
Illustrations of the effects of sulfate and other aerosols on temperatures at locations around the globe at various times in the twentieth and twenty-first centuries (with the latter predicted according to IPCC 2013 scenarios) have been constructed by Shindell et al. (2013).

and global warming sources and sinks can all cause irregularities in the anthropogenic trend.

In other words, "hiatuses" should actually be expected to occur in the climate system as energy transfers between reservoirs. The ocean modelers cited earlier (Meehl et al., 2011; Balmaseda et al., 2013; Trenberth & Fasullo, 2013) found that the 2002–2012 hiatus could be attributed to a major transfer of global warming heat from the surface down to the deep ocean; the relatively long timescale required for that transfer can explain why the hiatus lasted so long.

In contrast, *indirect* evidence during the hiatus was quite clear. As described near the end of this chapter, the period 2002–2012 included 3 years in which strikingly extreme weather took place: 2003, which featured incredible tornado activity and a continent-wide deadly heat wave; 2005, memorable for extreme hurricane activity; and 2010, with a devastating heat wave and widespread flooding. Record after record was set, arguing compellingly for the existence, not absence, of global warming.

Regional- or even continental-scale observations, being less than global averages, may provide stronger tests of global warming. Recent climate models have in fact been able to achieve high spatial resolution. Those models (which also yielded the regional predictions of twenty-first-century climate factors illustrated in Figures 3.29 and 3.36–3.38) allow more meaningful comparison with observations over the twentieth century. Figure 3.44 (Figure 3.44a, providing an overview; and Figure 3.44b, more detailed and comprehensive) compares the observed surface temperatures of six continent-scale regions during the twentieth century with predictions that either include or exclude anthropogenic factors (greenhouse gas emissions *and* sulfate aerosols). In all cases, the predictions match observations much better—especially in the latter half of the twentieth century—if the anthropogenic factors are included.

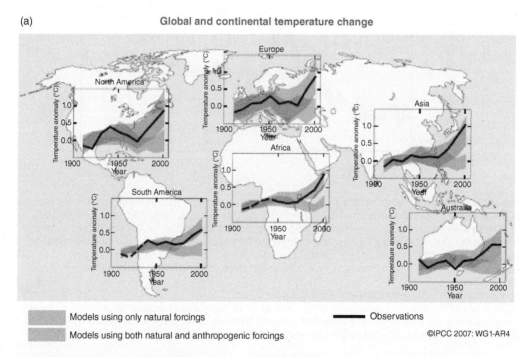

(a) **Global and continental temperature change**

Models using only natural forcings

Models using both natural and anthropogenic forcings

Observations

©IPCC 2007: WG1-AR4

Figure 3.44 Changes in surface temperatures on a continental scale.
(a) Decadal averages of the observed temperature change over the twentieth century (plotted relative to the 1901–1950 mean value: black curves), compared to the 5%–95% range of model predictions based on solar and volcanic activity (blue swath) or those plus anthropogenic factors (greenhouse gas emissions and sulfate aerosols) (pink swath). Dashed black curve indicates observed temperature estimates were based on limited (less than 50%) spatial coverage. Image from IPCC (2007) / Intergovernmental Panel on Climate Change.

(b)

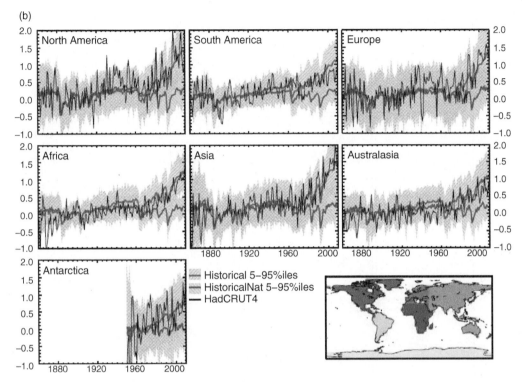

Figure 3.44 Changes in surface temperatures on a continental scale.
(b) Yearly values of the observed temperature change 1860–2010 over the twentieth century (plotted relative to the 1880–1919 mean value (1951–1980 for Antarctica): black curves), compared to the 5%–95% range of model predictions based on solar and volcanic activity (blue swath and curve) or those plus anthropogenic factors (tan swath and red curve). In this work, the anthropogenic factors include greenhouse gas emissions, sulfate aerosols, black carbon (soot), land use, and the effects of aerosols on cloud brightness and longevity. Analyzed regions are delineated in the map shown on the lower right. Image from Jones et al. (2013) / John Wiley & Sons.

3.6.5 Anthropogenic Variations in Climate: Evidence Concerning *Indirect* Consequences

It may be argued that more convincing evidence of ongoing global warming comes from its indirect consequences.

First Set of Indirect Consequences

Ocean Temperatures. As discussed earlier in this chapter, an increase in tropospheric temperatures should lead to warmer oceans as well. Figure 3.45 shows recent determinations by Levitus et al. (2009) of the heat content of the top 0.7 km of the oceans between 1955 and 2008, averaged over a great number of measurements worldwide. The observations show trends of increasing heat content over

time in all ocean basins; in most of the basins, those trends are the primary feature of the data. Earlier analysis of this data (Levitus et al., 2005) found that, over the second half of the twentieth century, the heat content of the upper 3 km of the ocean also increased.

Loss of Ice. The loss of ice is another expected consequence of tropospheric (and oceanic) warming, though we would do well to remember that any such warming takes place within an Ice House climate, and is far from the conditions that would create a world completely free of ice. That is, even with ongoing anthropogenic global warming, we should expect glaciers and especially ice sheets to exist on Earth's surface centuries into the future.

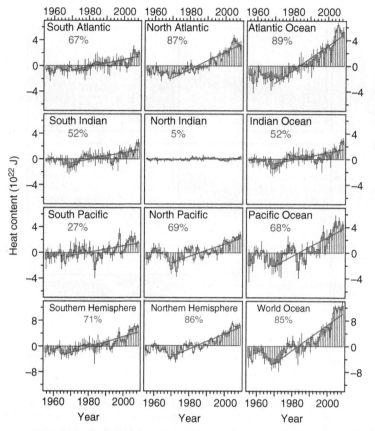

Figure 3.45 Change in the yearly heat content of the upper 700 m of the oceans between 1955 and 2008, from Levitus et al. (2009). Percentages in red indicate how much of the variability in the data can be described by a linear trend. However, like the global mean surface temperature, as discussed in the text, the global warming trends in these data sets are more complex than straight lines. Image from Supplemental Figure S11 of Levitus et al. (Note that the North Indian Ocean encompasses a relatively small area and thus has little heat content.)

Nevertheless, some glaciers, mountain glaciers, and sea ice have revealed a particular vulnerability over the past few decades; in some regions the loss of ice has been quite striking. For example, the decline in the area of the Arctic Ocean covered by sea ice (Figure 3.46a) has been appreciable; although this may be understood in terms of the positive feedback between albedo changes there and global warming, as discussed earlier, it is still alarming.

In some locations, like Mt. Kilimanjaro in Tanzania, Glacier National Park in Montana, and—on a larger scale—Alaska and parts of the Southern Andes (Figure 3.46b, top), glaciers may disappear entirely within a few decades; in other locations, including the Tibetan Plateau (Figure 3.46b, right), observations show a significant retreat of glaciers in some portions of it (Yao et al., 2012; Kumar et al., 2019) but glacial advances in others (see also Kargel et al., 2011).

Observations of the areal retreat of some mountain glaciers, combined with models of their thickness, have allowed estimates to be made of how much ice mass they have lost; by extrapolation to the thousands of other such glaciers, the net mass loss from the entire reservoir of mountain glaciers can be inferred. Estimates synthesized in the IPCC 2007 report yield a loss totaling ~2650 Gt ±750 Gt (Gt ≡ gigatonnes) during the 1993–2003 time span—implying a rate of loss 50% higher than in the preceding few decades (though observational error bars are too large to be certain; but see below).

Estimates of the total amount of ice in mountain glaciers are also highly uncertain (IPCC 2007), but such a loss probably represents ~1%–2% of the total, suggesting that mountains worldwide could be largely ice-free

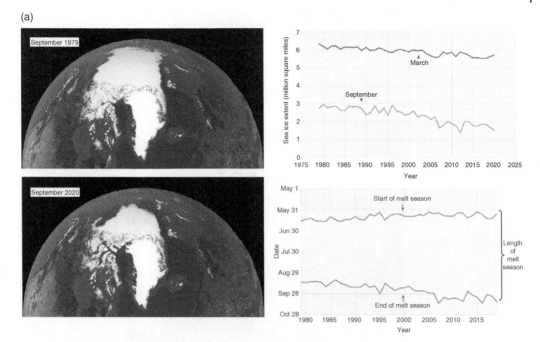

(a)

Figure 3.46 Examples of long-term ice melting around the world.
(a) Trend in the extent of Arctic sea ice, 1979–2020:
(Left) by summer's end, 1979 versus 2020. NASA Scientific Visualization Studio / https://svs.gsfc.nasa.gov / last accessed August 22, 2023.
(Right) annual variations in March extent and September extent (note for units that 10^6 mi.^{2} = 2.6×10^6 km^2) and length of melt season.
All images downloaded from https://www.epa.gov/climate-indicators/climate-change-indicators-arctic-sea-ice.
Looking at the history of summer melting, determine the rate at which sea ice is decreasing. If conditions remain unchanged, would you expect the Arctic to be ice-free in the summer by the end of this century? How about by 2050?

within several centuries (—a geologic instant!). However, it is possible that the 1993–2003 loss represented the melting of particularly vulnerable glaciers (smaller, lower-latitude or lower-elevation glaciers; glaciers in areas experiencing drought; etc.), and that more robust mountain glaciers might be expected to survive much longer—if conditions do not worsen.

Complicating any interpretation of glacial loss is the fact that the growth or even just stability of a glacier depends directly on continued snowfall in its "accumulation" zone, to balance a thinning elsewhere as the glacier flows downhill and its lower reaches ablate; thus, its growth can be hindered or fostered as wind patterns, cloudiness, and other local weather conditions change. Because all

of these changes may be connected to global warming, or may happen for unrelated reasons, citing the retreat or loss of mountain glaciers or sea ice as evidence of global warming may not always be justified. (By the same token, of course, localized glacial growth may not be evidence of the lack of global warming!)

Determining the mass budgets of the Greenland, West Antarctic, and East Antarctic ice sheets—which together hold 99% of all the world's ice—has proven difficult. Unlike glaciers, the areal extent of continental ice sheets like Greenland and East Antarctica is primarily controlled by the extent of the land they overlie, which has not changed significantly on the timescales we are considering here. Gain or loss of mass by those ice

(b)

Figure 3.46 Examples of long-term ice melting around the world.
(b) Glaciers.
(Top) Retreat of Glacier O'Higgins, located at high altitude in the south of Chile; 100 years ago, the spot where reporter S. Pelley ("The Age of Warming," *60 Minutes*) and glaciologist G. Cassasa sat was ice-covered (CBS Interactive Inc, http://www.cbsnews.com/2300-18560_162-2626628-7.html / last accessed August 22, 2023). Between 1945 and 2005, this glacier retreated 11½ km (Lopez et al., 2010).
(Middle) Retreat of Gangotri Glacier, the principal source of the Ganges River, from 1780 to 1971 (marked approximately by lighter-blue lines), and observed terminus in 2001 (dark-blue line), ~1 km further southeastward. Image by NASA / https://earthobservatory.nasa.gov/images/4594/retreat-of-the-gangotri-glacier / last accessed August 22, 2023.
(Lower) Further retreat of Gangotri, continuing through 2017. Image from Kumar et al. (2019) / with permission from ELSEVIER.

sheets would not, therefore, be visible as an advance or retreat of their terminus (as it was with, e.g., the Gangotri glacier in Figure 3.46); rather, as ice flows from the high interior to the lower-elevation coast, the mass balance of these ice sheets depends on whether their interior regions are accumulating enough mass (by snowfall) to compensate for the loss of mass at the coast.

We see the latter occurring at the coasts of Antarctica and Greenland, not just by the drainage of meltwater into the oceans, and the "calving" of coastal 'ice cliffs' to produce icebergs, but also more dramatically through the disintegration of West Antarctic floating ice (for example, the small Larsen A Ice Shelf in 1995, the much larger and older Larsen B in 2002, both seen in Figure 3.47a, and a massive chunk of the still larger Larsen C in 2017) and an expansion of ice streams, extending inland from the coast (e.g., on Pine Island, at the edge of the WAIS, and some in Greenland;

see Figure 3.47b). As noted earlier in this chapter, ice shelves and coastal glaciers can "buttress" inland glaciers, keeping them from flowing too rapidly, so their disappearance or instability might herald a greater loss of interior ice—facilitated by a proliferation of ice streams, bringing ice to the coast.

Yet we know that, even during the present interglacial period, the Antarctic and Greenland ice sheets have persisted; that is, precipitation (along with other processes) unrelated to global warming has managed to replenish their interior ice. So the question is, does that interior accumulation balance the loss at the margins currently taking place? Scientists have turned to satellite observations in order to shed light on the mass balance of the interiors of these already inaccessible regions. Satellites like GRACE (Gravity Recovery And Climate Experiment), launched in 2002, measure Earth's gravity field on a monthly basis; changes in gravity can

Figure 3.47 (a) Disintegration of the Larsen B Ice Shelf along the Antarctic Peninsula in January 2002. MODIS image courtesy of Ted Scambos, National Snow and Ice Data Center, University of Colorado, Boulder. In discussing the possibility of a collapse of the WAIS, Mercer (1978) wrote, "One of the warning signs that a dangerous warming trend is underway in Antarctica will be the breakup of ice shelves on both coasts of the Antarctic Peninsula. ... These ice shelves should be regularly monitored."
(b) Multisatellite inferences of ice movement in Greenland and Antarctica. Yellow/orange/red colors indicate location and speed of ice streams. The more active, plentiful, or rapid the ice streams are, the more efficiently they can drain interior ice. Image from Fricker (2010) / Springer Nature.

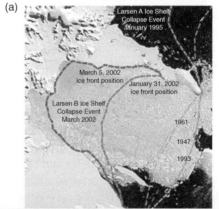

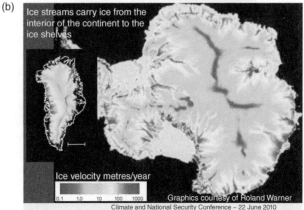

be interpreted—unfortunately, non-uniquely, as discussed in Chapter 6—in terms of gain and loss of surface mass (see, e.g., Tapley et al., 2019). Other satellites, like ICESat (Ice, Cloud, and land Elevation Satellite, launched in 2003), measure the elevation of the ice sheet's top surface as they orbit above; changes in that elevation are then modeled to infer changes in ice volume or mass. And, InSAR (Interferometric Synthetic Aperture Radar) instruments on a variety of satellites measure the speed at which surface ice deforms or flows, from which a significant portion of the ice mass budget can be inferred.

The ice flow revealed by InSAR, combined with measurements and modeling of surface variables (snowfall and ablation) and ice thickness, has provided a valuable perspective on the mass budgets of both ice sheets (Figure 3.48a). The analysis by Rignot et al. (2011) found that Greenland's mass balance (expressed as a *rate* of mass gain or loss) has tended to be negative—corresponding to a loss of mass—for most of the past two decades; has tended to become more negative over that time period; and has consistently been negative since around the year 2002. Much of this mass loss is attributed to enhanced melting of its surface ice. In contrast, the mass balance of Antarctica (Figure 3.48a, bottom) exhibits significant variability over timescales of a few years (a variability also found in GRACE data, as seen in the figure); nevertheless, the mass balance has tended to be negative for much (if not most) of the time period, especially after ~2006. It appears that, for Antarctica as well as Greenland, the first decade of this century has marked the transition to an era, continuing to the present, with mass loss, but *without* any annual mass gain.

Early results from GRACE indicated a definite loss of mass from both Antarctica and Greenland. Between 2002 and 2009, Antarctica lost an average of 190 Gt/yr ± 77 Gt/yr; not unexpectedly, most (two thirds) of that loss came from the West Antarctic Ice Sheet (Chen et al., 2009). The smaller Greenland ice sheet lost slightly more mass than Antarctica did,

on average ~210 Gt/yr during that time span (Velicogna, 2009). GRACE measurements also revealed the loss of mass to be accelerating; according to the analysis by Chen et al., this began around the end of 2005 or the beginning of 2006. The net result, according to Velicogna, was more than doubling of the rates from 2002 to 2009, from roughly 104 Gt/yr to 246 Gt/yr for Antarctica and from 137 Gt/yr to 286 Gt/yr for Greenland.

Such increases are not inconsistent with previous estimates, from a variety of measurement techniques, showing smaller amounts of ice loss—or even ice mass gain in the case of Antarctica—during earlier time spans. For example, the IPCC (2007, Table 4.6) report lists mass losses averaging ~75 ± 125 Gt/yr between 1993 and 2003 for Antarctica, and ~80 ± 25 Gt/yr for Greenland—yearly rates that are roughly 40% of those implied by the 2002–2009 GRACE results.

The GRACE mission ended in 2017 and was succeeded by the GRACE Follow-On nearly a year later. Figure 3.48b illustrates their results for 2002–2021 (presented in Chen et al., 2022; see also references within), for Greenland (top) and Antarctica (bottom). Over two decades or so (2002–2019), the loss of ice mass averaged 104 Gt ± 57 Gt per year in Antarctica and 261 Gt ± 45 Gt per year in Greenland (Velicogna et al., 2020), a higher rate for Greenland and lower rate for Antarctica than previously determined for 2002–2009.

The mass of each ice sheet, though characterized by a significant downward trend, also exhibits additional variability. One extreme example of this (Sasgen et al., 2020), visible in Figure 3.48b (top), occurred in 2019, when an ice loss of 600 Gt that summer led to a record loss of 532 Gt from Greenland for the entire year; this surpassed by ~15% the loss for the year 2012—which was a centuries-long record (see Mohajerani, 2020)! Interestingly, the preceding two summers had been cold, and the autumns and winters snowy, so that 2017 and 2018 (with models employed by Sasgen et al. to bridge the data gap) ended up with minimal ice loss; in short, the GRACE data of ice loss

Figure 3.48 (a) Mass balance in (top) Greenland and (bottom) Antarctica, quantified as *rate* of mass gain (positive) or loss (negative), and inferred from mass budget calculations that include surface measurements, modeling, and InSAR observations (black-filled circles and lines). Mass balance as inferred using GRACE data is also shown (red-filled circles and lines). Figure modified from Rignot et al. (2011) / John Wiley & Sons.

Despite variations over time, the mass balance in Greenland has consistently been negative (i.e., mass loss) since year 2000. Temporal variations are, mostly, significantly larger in Antarctica; but for much of the past two decades its mass balance has also been negative.

Zero gain, zero loss is indicated by the horizontal dashed light-brown line.

(a)
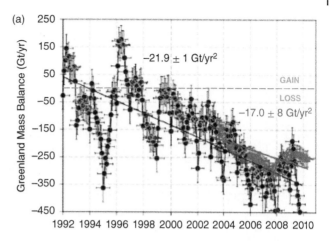
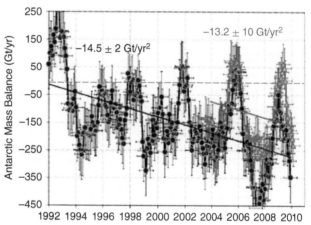

in Figure 3.48b displays a *real world*, not just mathematical, *trend*.

Results based on ICESat are similar to the preceding. Shepherd et al. (2012) combined ICESat and GRACE observations with other altimeter data (from ERS-1, ERS-2, and Envisat, all discussed in Chapter 6), and mass-balance models constrained by InSAR, to estimate changes in ice mass between 1992 and 2011 for Greenland, East and West Antarctica, and the Antarctic Peninsula. They inferred losses during 1992–2000 at a rate of 51±65 Gt/yr for Greenland and 48±65 Gt/yr for Antarctica (with only the WAIS ice loss statistically significant); and losses during 2000–2011 at a rate of 211±37 Gt/yr for Greenland and 87±43 Gt/yr for Antarctica (with the East Antarctic Ice Sheet gaining mass, but not

significantly, and ice loss everywhere else that is statistically significant).

Most recently, Smith et al. (2020; see references therein for studies from intervening years) estimated polar ice loss during 2003–2019 using data from the ICESat and ICESat-2 missions—the latter follow-on to ICESat launched in 2018, 9 yr after ICESat operation ended; they determined net losses averaging 200 Gt/yr ± 12 Gt/yr for Greenland and 103 (±69) Gt/yr for Antarctica, overshadowing small but increasing rates of accumulation throughout the interiors of those ice sheets. Furthermore, they were also able to distinguish changes in grounded ice mass from changes in floating ice mass (an important distinction since the latter would not contribute to changes in sea level); for Antarctica, grounded ice lost 118 Gt/yr ± 24 Gt/yr during this period

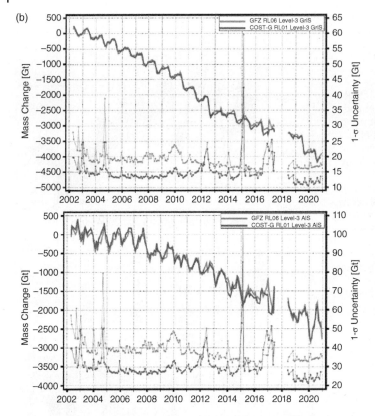

Figure 3.48 (b) Mass change of the Greenland (top) and Antarctic (bottom) ice sheets (in Gt), according to two analyses of GRACE data (red and blue solid curves); monthly values for 2002–2021. The gap corresponds to the time after GRACE ended and before GRACE Follow-On began. Uncertainties for each datum are shown as filled circles (red or blue), depending on the analysis, with the scale to the right.
Note the differences in both scales for Greenland versus Antarctica.
Images from Chen et al. (2022) after significant smoothing; typical year-to-year variability before smoothing is ~±137 Gt/yr for Greenland and ~±208 Gt/yr for Antarctica (Tapley et al. 2019).

while floating ice *gained* a highly uncertain 15 Gt/yr ± 65 Gt/yr. That gain evidently stemmed from a greater increase in floating ice around East Antarctica than the loss in floating ice in West Antarctica and the Antarctic Peninsula.

Finally, satellites have also been invaluable for revealing the 'big picture' when it comes to ice caps and glaciers, and that picture is unsettling. From GRACE data (Figure 3.48c) we see that, altogether, they have been losing ice mass at a rate of ~281 Gt ± 30 Gt per year on average during 2002–2019 (Ciraci et al., 2020)—essentially continuing the same rate of loss as estimated in IPCC (2007) for 1993–2003 (though with smaller uncertainties)—and implying a cumulative loss of well over 5,000 Gt for the first two decades of this century. Whether that is all attributable to global warming or not is uncertain, given the variety of factors mentioned earlier in this subsection; but it is clear that glaciers and ice caps represent the least stable part of the cryosphere. And that has

implications for the next indirect consequence of global warming, which we discuss now.

Increase in Mean Sea Level. As noted previously, a rise in sea level is expected from both melting ice (excluding sea ice, which is already floating, and ice whose meltwater has ponded on land rather than draining into the oceans) and from the thermal expansion of warmed oceans. But even without any current global warming, sea level would already be changing, as a consequence of Pleistocene deglaciation. At the end of the most recent glacial maximum, 21,000 yr ago, extensive continental ice sheets began to recede, and a large volume of meltwater began to be added to the oceans, ultimately producing a rise in sea level of about 120–130 m (Clark & Mix, 2002).

By 8,000 yr ago, the ice sheets had finished their recession, achieving roughly their present configuration, and were no longer contributing to the volume of seawater. But the

Figure 3.48 (c) Monthly variations (in Gt) in the combined mass of ice caps and glaciers worldwide, excluding those associated with the Greenland and Antarctic ice sheets, based on GRACE data for April 2002–September 2019. The "reference time" (arbitrarily assigned a mass of 0) is November 2010. The gray bar corresponds to the time after GRACE ended and before the GRACE Follow-On began. The best-fitting trend is shown in blue; the magnitudes of its linear component (in Gt/yr) and small but significant quadratic component (in Gt/yr^2) are shown in the box. Ciraci et al. (2020) / John Wiley & Sons.

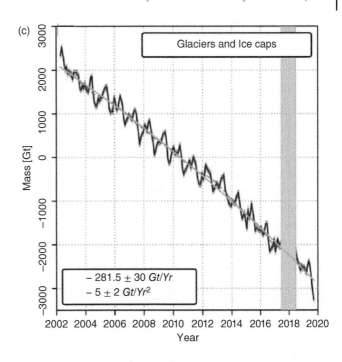

mass of northern hemisphere ice sheets during their prior glacial advance had weighed down Earth's crust; when those ice sheets disappeared, the crust (technically, the lithosphere) began to rebound vertically upward. Similarly, under the weight of the added meltwater, the ocean basins began to subside. Such rebound and subsidence, which will be discussed in detail in Chapter 5 (and which are called **glacial isostatic adjustments** or **post-glacial rebound** by some), are still taking place, at a rate of millimeters per year, and will continue for thousands of years more. Additionally, such mass redistributions are changing Earth's gravity field. In response to all of this, changes in sea level—measured as a relative difference, between the height of the sea surface and the height of the seafloor—will continue until the isostatic adjustments are complete.

Changes in sea level associated with global warming will be superimposed on the post-glacial effects. We can predict such superposed changes using the recently measured losses in ice mass described above. For our

predictions we note first, that the oceans cover about 70.8% of Earth's surface—thus, an area of 361×10^6 km^2—and second, that an increase in global mean sea level of 1 mm represents an additional mass of ~370 Gt or 370×10^{12} kg.

Can you verify this latter relation? Use the fact that seawater has a density of ~1.025 gm/cm^3 to convert the volume of that added layer to mass (we will be applying this relation neglecting the density difference between seawater and the fresh water draining into the oceans).

Loss of mass from mountain glaciers (2650 Gt), Greenland (1000 Gt), and Antarctica (600 Gt) during 1993–2003 (with the glaciers estimate from IPCC 2007 and the ice sheet estimates extrapolated from Shepherd et al., 2012) would then cause rises in sea level of 0.72 mm/yr, 0.27 mm/yr, and 0.16 mm/yr, respectively, for a total of 1.15 mm/yr; the total uncertainty would, very conservatively, not exceed ±0.62 mm/yr. During 2002–2019 the cumulative mass loss from glaciers was ~4,900 Gt (Ciraci et al., 2020), raising sea level by 0.76 mm/yr; the loss of ice from Greenland and Antarctica during 2003–2020 (see Chen et al., 2022) amounted to ~4,000 Gt and ~2,400 Gt, respectively, yielding sea-level rises of 0.60 mm/yr from Greenland and 0.36 mm/yr from Antarctica, and a total from all sources of 1.72 mm/yr ± 0.42 mm/yr.

Global warming will also cause a global expansion of the oceans, thus a further rise—called a **steric** (or **thermosteric**) rise—in sea level. To estimate this effect, we define the **thermal expansion coefficient α** to be the fractional increase in ocean volume, for each degree of warming,

$$\alpha \equiv \Delta V / V_0,$$

where V_0 is its original volume and ΔV is the volume increase per degree of warming; if the oceans extend over an area A, and their depth is originally h_0 (so that $V_0 = A \cdot h_0$, and $\Delta V = A \cdot \Delta h$), we can equivalently write

$$\alpha \equiv \Delta h / h_0.$$

It follows that the increase Δh in sea surface height per degree of warming is $\alpha \cdot h_0$, or a total of $(\alpha \cdot h_0) \cdot \Delta T$ if the oceans warm by ΔT degrees. For example, an increase in ocean temperature of 0.12°C—the warming from 1955 to 2003 measured in the top 700 m of the world's oceans, on average, according to Levitus et al. (2005, Table A1)—will cause the sea surface to rise by $\alpha \cdot (700 \text{ m}) \cdot (0.12°C)$. With the coefficient of thermal expansion of water measured at ~1.5 × 10⁻⁴/°C, the implied rise in sea level over that time period is 1.26 cm, corresponding to an average rate of rise during 1955–2003 (48 yr) of 0.26 mm/yr. Alternatively, using data from Levitus et al. (2009, Table S1), with a global ocean warming of 0.168°C over 39 yr (1969–2008), we might estimate an average rise from thermal expansion of 0.45 mm/yr.

There are complications attached to such estimates. The temperature of the oceans' surface waters is wide-ranging (exceeding 26°C in much of the tropics and barely reaching a few degrees above freezing at high latitudes), and the coefficient of thermal expansion of water is quite temperature dependent; the same increase in surface-water temperature therefore produces a different expansion in different regions of the oceans. Consequently, our predictions of the steric rise in sea level require a knowledge of the 'background' state of ocean temperatures, region by region; use of a correspondingly spatially dependent α; and global warming changes in temperature that are also specified region by region—not global averages such as the 0.12°C or 0.168°C we used above. To these complexities we can add two 'vertical' complications: near-freezing temperatures in the deepest ocean lead to significant variations in temperature with depth, particularly in tropical regions; and temperature increases imposed by global warming will take time to diffuse down through the water column (the process of diffusion will be discussed in great detail later in this textbook).

All of these complications can be addressed with the help of the *ARGO floats* (see Riser et al., 2016). These are freely drifting instrument packages deployed throughout the world's oceans, beginning in year 2000, which measure temperature, salinity, and pressure at their GPS-monitored locations, at various depths (their buoyancy is varied over time so that in each 10-day cycle, they will range between the surface and as much as 2,000–6,000 m depth). The ARGO program's target of 3,000 floats was reached in November 2007 (e.g., von Schuckmann & le Traon, 2011), though their spatial coverage was sufficient for accurate, detailed estimates of steric effects on sea level—and far greater than the coverage attained by previous generations of measurements (see Wijffels et al., 2008)—by 2004 (Cazenave et al., 2009) or 2005 (Dieng et al., 2015; Ablain et al., 2015; WCRP, 2018).

An example of the results obtained with ARGO data is shown in Figure 3.49a. For the period 2005–2016, the analysis by Tapley et al. (2019) determined the thermosteric rise in sea level to be 1.14 ± 0.15 mm/yr. Shorter-term variations in this trend are also visible, for example a faster increase from 2012 into 2015 and hints of a decrease in steric sea level from

(a)

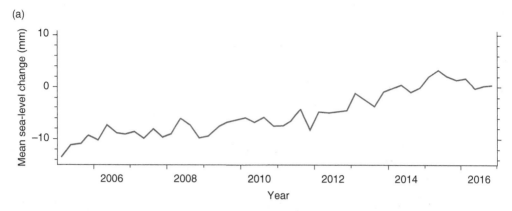

Figure 3.49 (a) Thermosteric changes in global sea level, in units of mm, based on ARGO data (depths less than 2,000 m) averaged in 3-month intervals, with annual and semiannual periodicities removed. Image adapted from Tapley et al. (2019), who estimated the trend in this data to be 1.14 ± 0.15 mm/yr; see also Figure 3.50b.

mid-2015 through 2016; these are conceivably the result of ENSO or other short-term variations in the climate system (see, e.g., Riser et al., 2016), though, of course, nonsteric contributions to sea level (e.g., glacial melting) can also be affected by such climate events.

This trend is comparable in magnitude to the trend of changes in sea level resulting from ice loss, and confirms that the steric effect is a significant contributor to the effects of global warming on sea level. The net effect of both is simply their sum; for 2005–2016, we have, approximately, 1.72 ± 0.42 mm/yr plus 1.14 ± 0.15 mm/yr, or 2.86 ± 0.45 mm/yr.

The variations in gravity monitored by GRACE revealed another process with the ability to affect global sea level: changes in how much (liquid) water there is on land, mainly as a consequence of changes in surface water and groundwater reservoirs. Quantifying such variations—whether produced by climate change (e.g., increased precipitation and flooding events) or societal activities (e.g., extracting groundwater or damming a river)—was largely a speculative effort until missions like GRACE. To some extent, those variations amount to a disruption of the natural *water cycle* (or *hydrologic cycle*), which describes the progression of water molecules from their evaporation off the ocean surface to their precipitation onto land, to their eventual

drainage back to the oceans (with perhaps a delay during temporary storage in a surface reservoir or subsurface aquifer). Because that global cycle involves an exchange of thousands of Gt of water between land and oceans (e.g., Reager et al., 2016; Jensen et al., 2013), even a modest disruption, involving hundreds of Gt of water, could affect sea level measurably.

The analysis of GRACE data by Reager et al. (2016) showed strong seasonal variations (implying a range of ~6,000 Gt/yr in the water exchange between oceans and land); subtracting that cycle revealed a trend of *increased* terrestrial water storage over time, at least during 2002–2014 (see Figure 3.49b). That increased storage would have delayed the return of water to the oceans, in effect lowering sea level by 0.33 mm/yr ± 0.16 mm/yr according to their analysis (see also Scanlon et al., 2018).

In short, our estimates are a rise in mean sea level (over 2003–2019) of 1.72 mm/yr ± 0.42 mm/yr resulting from ice loss; a rise (over 2005–2016) of 1.14 ± 0.15 mm/yr because of thermal expansion; and a drop (over 2002–2014) of 0.33 mm/yr ± 0.16 mm/yr from increased land storage. We thus predict the net effect to be roughly 2.5 mm/yr ± 0.5 mm/yr; considering that all three contributions vary in time, we might say that this prediction applies mainly to the period 2005–2014. So what do the observations say?

(b)

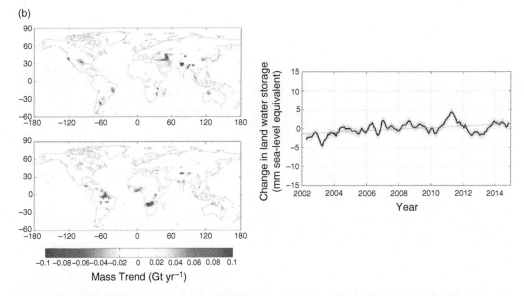

-0.1 -0.08 -0.06 -0.04 -0.02 0 0.02 0.04 0.06 0.08 0.1
Mass Trend (Gt yr⁻¹)

Figure 3.49 (b) Changes in terrestrial water storage between 2002 and 2014 from GRACE observations. Large seasonal signals were first removed to obtain the results shown here.
(Left) Regions that gained (blue) or lost (red) mass, according to the color scale below. The net global loss here is 350 Gt/yr; the net gain is 470 Gt/yr.
(Right) The net 'loss' of mean sea level, in units of mm, amounts to a slight gain in terrestrial water mass over time; the estimated linear trend is highlighted in orange. Image from Reager et al. (2016) / American Association for the Advancement of Science.

When sea-level observations at coastal and island tide-gauge stations worldwide are corrected for post-glacial rebound, an increase in global sea level is indeed observed. The corrected rate found by Church and White (2011)—similar to the results of other researchers, though there are differences due to the choice of ports and time span—is ~2.8 ± 0.8 mm/yr for 1993–2009. Their analysis also suggested that the rise in sea level has accelerated: going back to 1961, for example, the average rise (till 2009) was only 1.9 ± 0.4 mm/yr. Correspondingly, Dangendorf et al. (2017) found that three additional years of tide-gauge data (1993–2012) increased the trend to 3.1 ± 1.4 mm/yr.

Satellite altimetry, discussed in Chapter 6, provides a way to map the sea surface over time, allowing changes in its height to be measured ocean-wide (not just at a few scattered ports and island locations). The resulting global sea-level observations (see Figure 3.50a) show a trend of rising sea level, at an average rate of ~3.3 mm/yr ± 0.4 mm/yr during the time interval from 1993 through 2021.

Additionally, the upward trend in Figure 3.50a appears to steepen around ~2012—an acceleration of the sea-level rise, similar to what was inferred from analyses of GRACE data but several years later. If the trend is modeled as the simplest possible nonlinear trend, namely a quadratic curve, then the slope of the trend is found to increase by ~0.1 mm/yr per year (as noted in the figure); thus, the slope increases from ~1.9 mm/yr in 1993 to ~4.7 mm/yr in 2021. Such an acceleration, which is not unexpected in a warming world (though perhaps an irregular acceleration would be more realistic), has been found by several researchers (e.g., Dangendorf et al., 2017, 2019; Nerem et al., 2018; Frederikse et al., 2018).

Overall, the net trend based on our predictions is roughly in the same 'ballpark' as that based on altimetric observations; given the uncertainties, they are not even distinctly

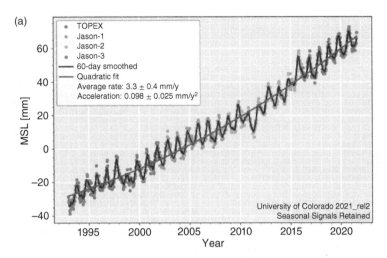

Figure 3.50 (a) Satellite altimetry determination of the global variation in sea level; data from TOPEX and Jason satellites—corrected for the response of the oceans to atmospheric pressure variations (using the "inverse barometer" model) and the effects of isostatic rebound (both discussed in Chapter 5)—was patched together (with offsets removed; see Nerem et al., 2010). The smoothed data features an annual variation superimposed on a slightly curving (quadratic) trend, discussed in the text. Image downloaded February 2022 from https://sealevel.colorado.edu/.

different, statistically speaking. But the data and trends are too variable over time for such a broad comparison to be meaningful; instead, a point-by-point comparison is called for. Figure 3.50b, from Tapley et al. (2019), shows (i) the thermosteric change in global sea level estimated from ARGO data; (ii) the total change in global sea level from mass exchange between oceans and land—whether from glacial melting or changes in water storage—as determined from GRACE measurements of the change in gravity over the oceans; (iii) the sum of mass and steric contributions; and, (iv) the altimetrically observed actual change in mean sea level, all at three-month intervals. For the time span of their analysis, the net predicted variations match the observed sea-level fluctuations almost point for point, and the rms differences are mostly quite small (and, as it turns out, the trends—listed in the figure caption—are also very consistent). Such strong agreement apparently extends for at least a few more years past the end of the analysis in the figure (see https://argo.ucsd.edu/science/global-phenomenon).

Second Set of Indirect Consequences

Stratospheric Cooling. Like much of the preceding evidence, the observations of stratospheric conditions are generally consistent with the existence of global warming, with the degree of consistency varying over time. Cooling of the stratosphere—unavoidable if global warming of the troposphere is the result of greenhouse gases rather than increased insolation, as discussed earlier in this chapter—occurred throughout the second half of the twentieth century, according to measurements from satellites and radiosondes (see Figure 3.51a). The cooling trend in the lower stratosphere, amounting to ~1°C from 1980 to 2000 but perhaps only half that rate prior to 1980, was superimposed on much small-scale variability—and interrupted by spikes of stratospheric warming, clearly associated with major volcanic eruptions.

More recently (Figure 3.51b), a comprehensive analysis presented by Randel et al. (2021) demonstrates that all layers of the stratosphere have continued to cool. After roughly 2005, though, temperatures in the lower stratosphere stopped paralleling those of the layers above,

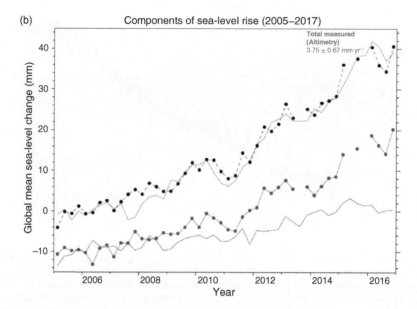

Figure 3.50 (b) Comparison of predicted contributions to the change in mean sea level between 2005 and 2016 versus that observed by satellite altimetry.
Red curve: thermosteric change in mean sea level based on ARGO data (as in Figure 3.49a). Blue curve and dots: change in mean sea level from changes in ocean mass, based on GRACE observations over the oceans. Black dashed line and dots: sum of the volume (red) and mass (blue) contributions to the change in sea level. Green: change in mean sea level observed by altimetric satellites (as in Figure 3.50a). Units are mm; each data set was processed by removing best-fitting annual and semiannual cycles and averaging the data at 3-month intervals. In this analysis by Tapley et al. (2019), the trends during 2005–2016 are estimated to be 1.14 ± 0.15 mm/yr (red), 2.46 ± 0.36 mm/yr (blue), 3.60 ± 0.39 mm/yr (black), and 3.75 ± 0.67 mm/yr (green). Image from Tapley et al. (2019) / Springer Nature.

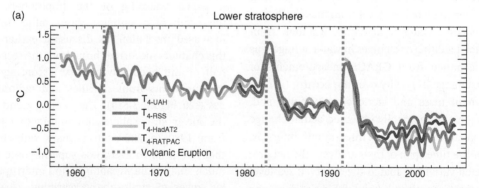

Figure 3.51 Measurements of anomalous stratospheric temperatures over time.
(a) In the lower stratosphere during 1958–2005, as processed by various research groups, from weather balloons (turquoise, pink curves) and microwave sounders on satellites (dark blue, green curves). Satellite data over different time spans is patched together; balloon data is limited to land areas. Times of major volcanic eruptions (Mt. Agung, El Chichón, Mt. Pinatubo) are indicated by dashed vertical lines. Image modified from IPCC (2007) / Intergovernmental Panel on Climate Change.

Figure 3.51 Measurements of anomalous stratospheric temperatures over time.
(b) Throughout the stratosphere during 1978–2020, as processed by various research groups, from satellite-based infrared sounders ("stratospheric sounding units" or SSU) and—for the lower stratosphere—microwave sounders.
The stratospheric temperatures are for the lower stratosphere (TLS), at altitudes ~13–22 km; and for ~20-km-thick middle- and upper-stratospheric layers (SSU1, SSU2, and SSU3) centered respectively near ~30 km, 38 km, and 45 km altitude, with anomalous temperatures offset by 1°C, 2°C, and 3°C.
As in Figure 3.51a, the effects of major eruptions—El Chichón in 1982 and Mt. Pinatubo in 1991—can be seen as 'warm' perturbations to the otherwise downward trends in stratospheric temperatures. Image from Randel et al. (2021) / American Meteorological Society.

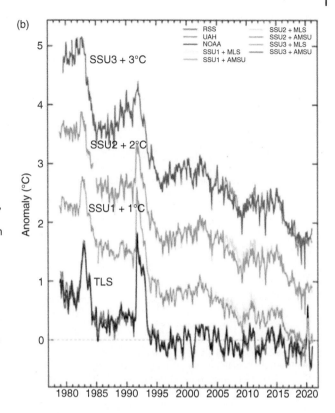

instead mostly leveling off then maintaining only a very mild cooling. Randel et al. attributed a brief spike in temperature at the end of this period to aerosols injected into the stratosphere by smoke from the extreme wildfires plaguing Australia in 2019–2020 (described below).

Further Ozone Depletion. The other indirect consequence of global warming predicted for the stratosphere, a worsening state of the ozone layer, emerged unambiguously over Antarctica in the 1980s and into the 1990s: an increasing area of increasingly severe ozone depletion—a worsening "hole" in the ozone layer—lasting longer each spring before natural processes replenished the ozone during the summer (see Figure 3.52 for the trends in maximum area and duration). The increase in duration was especially striking, with the hole covering a large area (say, at least 10×10^6 km^2) for 1 month in 1984 and increasing to 3 months in 1992. And each year after 1988 (except for

2019), the hole always attained a *far* greater area than that of the 14×10^6 km^2 Antarctic land mass below it.

After 1992, however, the peak area of the Antarctic hole seems to have leveled off, or even declined slightly, though with significant variability (as seen in Figure 3.52). Similarly, its duration apparently has 'saturated' at ~3 months; for example, up through 2006 (not shown in the figure), the duration was ~3 months every year but 2002 (when the duration fell below 3 months, and, curiously, the hole appeared to split and collapse).

Such changes in the behavior of the ozone hole have been attributed (e.g., Maycock et al., 2018) to changes in the abundance of chlorofluorocarbons and other ozone-destroying substances, probably beginning around 1998 or soon after (e.g., Harris et al., 2015). Restrictions on our use of ozone-destroying substances were possible thanks to global action—the 1987 Montreal Protocol and its subsequent revisions—stemming from global concern that

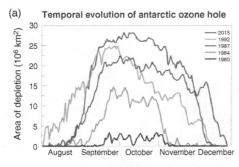

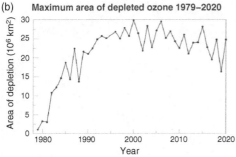

Figure 3.52 Characteristics of the Antarctic ozone hole. The hole is defined by ozone concentrations below 220 Dobson units (normal levels are ~320 D.u.). Data from NASA GSFC's *Ozonewatch* program (https://ozonewatch.gsfc.nasa.gov/meteorology/); data point for 1995 in the lower graph was unavailable. (a) Area of Antarctic ozone hole versus time for the years shown. Note that the duration of the hole increased from 1980 through at least 1992. (b) Yearly maximum area of Antarctic ozone hole between 1979 and 2020. Each year, that maximum always occurred in September or early to mid-October.

the Antarctic hole was worsening. Measured in terms of their ability to destroy ozone as effectively as chlorine (cf. Figure 3.32), the abundance of all those substances in the stratosphere declined, from ~1998–2000 (when levels peaked) through 2019, by an amount that brought us almost a quarter of the way back to the relatively low levels of 1980 (Vimont et al., 2021).

Another, more 'otherworldly,' influence on stratospheric temperatures can be seen in these data sets, especially for the upper stratosphere. Look at the drop in the SSU3 data in 2009–2010, for example, and again in 2020–2021. Earlier in this chapter, we discussed a phenomenon with a similar timescale of variation; can you identify it?

Third Set of Indirect Consequences
Extreme Weather: A Cautious Approach.
The indirect consequence of global warming with perhaps the most immediate impact on our daily lives is the predicted increase in extreme weather. In recent years, it has become apparent to many of us that extreme weather is on the rise. But weather events should be assessed cautiously before they are labeled "extreme." Weather is naturally variable, with both random components and periodicities on all timescales, including possibly longer timescales than the span of available data. And societal factors such as overall population growth, changing demographics including population shifts toward cities and coasts, and changing lifestyles can contribute to the impact of a moderate weather event, making it appear extreme. A conservative starting point might be to *consider only record-breaking events*. Thus, for example, an 'extreme' event in some region described as "the worst drought in a decade!" would have to be discounted, as would "the worst drought in a generation!!"; but "the worst drought ever recorded" there might be viewed as a valid extreme event.

An even more revealing diagnostic with this approach would be an *episode of successively broken records*. For example, record-high temperatures on a particular day occurring one year in some region might, or might not, be the result of natural variations; but if records are set and then broken day after day or year after year—as rare as that might be, given that we are looking at real-world trends—the cause is much less likely to be random natural factors.

Finally, from our earlier discussion on the spatially variable nature of global warming, it makes sense to search for local and regional extremes as well as global ones.

Extreme Weather: Some Scientific Evidence. The immediacy of extreme weather suggests that our discussion of its connections to global warming could follow an informal, *experiential* approach. We do indeed take that approach below, but do not mean to imply that rigorous scientific evaluation of the evidence concerning extreme weather is unnecessary

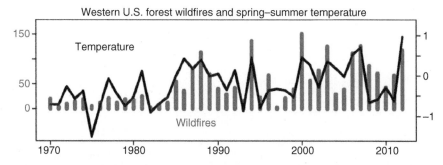

Figure 3.53 Number of large wildfires (defined as exceeding ~1,000 acres) per year between 1970 and 2012 in forested regions of the western United States (red histogram, with scale on the left), as determined by Westerling (2016). Such wildfires were much less frequent in the years before 1985.
Their frequency of occurrence correlates well with the average regional spring and summer temperature (black curve, with scale in °C on the right to indicate relative warmth or coolness of those seasons), especially after 1986. Figure from Westerling (2016) / The Royal Society.

or unfruitful; in fact, scientific studies over the years have convincingly documented a wide variety of supporting evidence. Examples include:

- an increase in large wildfires in the western United States from the 1980s onward (see Figure 3.53)—with a fourfold increase in their frequency of occurrence and a more than sixfold increase in the area burned (due also to longer times of burning) in the ~15 yr after 1985 compared to the preceding 15 yr (Westerling et al., 2006), and with those trends holding or even intensifying through 2012 (Westerling, 2016)—attributable primarily to contemporaneous warmer and drier conditions in the region;

- a time, 1997, when "more of the world was on fire ... than ever in recorded history" (Lovejoy, 1997), at least partly the result of droughts in the Indonesian and Amazon rainforests—that's *rain*forests!—burning up 20,000 km² in Indonesia and 40,000 km² in the Amazon according to Nepstad et al. (2004)—associated with the twentieth century's most intense ENSO, that of 1997–1998;

- a strong connection between record flooding that occurred in Great Britain in the Fall of 2000—during the wettest autumn on record (a record extending back more than 230 yr)—and climate change associated with increased greenhouse gas emissions (Pall et al., 2011); and

- a growing threat from hurricanes, measured by the increased proportion of major hurricanes (i.e., categories 3, 4, and 5) occurring worldwide (Figure 3.54a, from Kossin et al., 2020)—with the most intense hurricanes

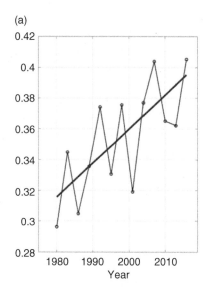

Figure 3.54 Intensification of hurricanes over time. (a) Proportion of major hurricanes (categories 3, 4, and 5) globally during 3-year periods from 1979 to 2017. The best-fitting linear trend highlights an overall increase in this fraction from less than 0.32 to almost 0.4 (that is, by the end of the most recent decade, roughly 40% of hurricanes around the world were category 3 or higher).
At that rate, what decade will see hurricanes more likely to be major than not?
Adapted from Kossin et al. (2020).

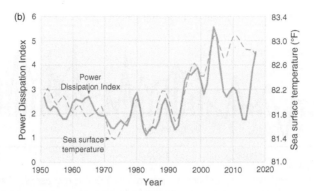

Figure 3.54 Intensification of hurricanes over time.
(b) The power dissipation index—a measure of storm (especially hurricane) destructiveness, based on its intensity and duration, defined by Emanuel (2005) —versus time in the North Atlantic (solid orange curve) for years 1949–2019, averaged in 5-year bins and plotted in the middle year (so, e.g., the datum for 2015–2019 is plotted at 2017). Units are 10^{11} m^3/sec^2.
The average September sea surface temperature within the Atlantic Ocean hurricane breeding grounds (dashed curve) is plotted over the same time span, but scaled for comparison. For the most part, the two data sets are closely correlated; the biggest exception appears around 2012.
Both curves exhibit fluctuations, reflecting natural modes of oscillation and the effects of ENSO. However, the PDI also shows a sharp increase from ~1990 to 2005, which disappears after ~2010 (since each point is a 5-year average) but returns 'with a vengeance' by 2017 (see also Murakami et al., 2018 and Klotzbach et al., 2022). Adapted from https://www.epa.gov/climate-indicators/climate-change-indicators-tropical-cyclone-activity#ref7, but see also Knutson et al. (2019).

(categories 4 and 5) *doubling* in number from the 1970s into the 2010s (Klotzbach & Landsea, 2015)—and also measured by their increased destructive potential, in the North Atlantic (Figure 3.54b) and possibly elsewhere (cf. Emanuel, 2005; Knutson et al., 2019), clearly connected to increasing ocean temperatures.

One additional example of scientifically documented extremes worth mentioning—disruption of the atmosphere's three-cell circulation—is more of a 'structural' than weather extreme. As noted earlier in this chapter, such a disruption would have been caused by rising polar air, resulting from low atmospheric pressure over the Arctic Ocean as the latter becomes nearly ice-free. Visible evidence of circulation there contrary to the normal three-cell pattern comes from the presence of clouds, whose formation requires rising air: during the latter half of the twentieth century, the North Pole has tended to be the cloudiest place on Earth during northern hemisphere summer (Peixoto & Oort, 1992)!

In fact, analysis of meteorological data from ~1980 by Qian et al. (2015; Qian, Wu, & Liang, 2016; Qian, Wu, Leung, & Shi, 2016) shows the existence of a fourth, "Arctic" cell (and somehow, in the southern hemisphere, perhaps a fourth "Antarctic" cell as well...); not surprisingly, it is weaker than the other cells, by one measure typically six times weaker than the adjacent polar cell. Liang and Gao (2021) find that the atmosphere has switched between three-cell and four-cell modes over the past 40 yr. However, the reality may be more complex, both in the structure of the cells (see Liang & Gao) and whether variations in Arctic sea ice are the cause of the anomalous regional atmospheric circulation or the effect (see Qian et al., 2015).

A very large number of studies dealing with extreme weather and its possible connection to global warming, beyond these examples, can be found in the scientific literature. Their synthesis, and a careful, quantified interpretation of the likelihood that each type of extreme weather has been or is the result of global warming, can be found in the IPCC reports

Table 3.3 Circumstantial evidence of global warming.

Date	Circumstantial evidence
September 1988	Hurricane Gilbert —most severe on record —central low = 26.13″; winds > 200 mph

(The above is just the first of hundreds of entries!)

>> **Table 3.3 may be found on the Companion Website** <<

(IPCC, 2007; IPCC, 2013; and the imminent IPCC, 2022).

Extreme Weather: Anecdotal Evidence. Table 3.3, which may be found online at the Companion Website (see the Preface for details) but the beginning of which is shown here, lists anecdotal evidence of various consequences of global warming, mainly involving extreme weather, that I compiled over the years from news reports as well as journal articles. Entries in the table include both global events and those occurring in the upstate NY region (centering on the Binghamton/Vestal locality). The entries are tersely written, but it is worth emphasizing that, despite their brevity, each one signifies a major event—in most cases, a record-setting event—that may have affected large numbers of people (e.g., see Figure 3.55).

That tabulation, though incomplete and based on my own subjectivities, perhaps conveys a sense of what everyday life in a globally warmed world will be like. In fact, it appears that all of the kinds of extreme weather anticipated from global warming, discussed earlier in this chapter, have already begun to appear over the past few decades.

Some events in Table 3.3—particularly dramatic examples of extreme weather during the past two decades—deserve elaboration.

- Extreme weather that occurred in 2003 included *tornado activity in the Midwest U.S.* and the *European heat wave of 2003*.

Normal springtime tornado activity in the United States Midwest took a dramatic turn in May of 2003 beginning with the appearance on May 4–5 of 80 tornadoes in five states overnight; by May 9, a record 276 tornadoes had been counted in 11 states—including three whose paths of destruction were more than 50 miles in length. The week of May 4–10 has been described as exhibiting more severe weather than any other week in U.S. history. From May 1–12, a record 412 tornadoes occurred; the number for the whole month of May, 562, was also a record (note: subsequent official numbers from NOAA (spc.noaa.gov) were 405 and 572, both still records).

High energy levels persisted in the atmosphere, and on June 24, the northern plains experienced an amazing outbreak of tornadoes: about 50 struck eastern South Dakota; normally, that state will face ~30 *per year*! The outbreak that day included six simultaneously emanating from one cloud. The news reports also showed numerous funnel clouds that did not touch down to the ground (so they were not counted).

Peaking in late July and August of that year (following an earlier heat wave in late May and June), the 2003 heat wave included the first time that temperatures in Britain ever officially reached 38°C (100°F); high temperatures of 40°C (104°F) in France and nearly 41°C (105°F) in Germany also set all-time records. Later reports described

Figure 3.55 Extreme weather strikes Binghamton, NY. Towns along the Susquehanna River in the Greater Binghamton area experienced a "hundred-year flood" in June 2006 following heavy rains along the East Coast. Five years later, in Sept. 2011, an extensive region of the northeast, including Binghamton, was flooded by the remnants of Tropical Storm Lee—less than 2 weeks after heavy rains from Hurricane Irene had saturated the ground (and contributed to that August becoming the wettest August on record in Binghamton). Experts called the damage experienced in the Binghamton region in 2011 the equivalent of a "five-hundred-year flood." 2006 Image from https://www.bayjournal.com/news/climate_change/floods-drive-binghamton-toward-sea-change-of-resiliency/article_e75c7f86-354d-11eb-b83f-ab0b9281bf63.html; more 'extreme' details about both flood events can be found at https://www.wunderground.com/blog/JeffMasters/flood-walls-hold-on-the-susquehanna.html. National Weather Service.

that summer as the hottest in Europe in the past 460 yr. The human consequences of this extended heat wave were disastrous: more than 27,000 heat-related deaths (some news sources declared up to 35,000 dead), including 15,000 in France alone; reportedly, these deaths resulted from a combination of the heat and so many caregivers abandoning their elderly patients for the traditional August vacation. (Note: a subsequent rigorous analysis, by Robine et al. (2008), placed the European heat-related death toll at more than 70,000.)

- The *2005 Atlantic hurricane season* was exceptional in many respects. The number of named storms, that is, tropical storms and hurricanes, in the Atlantic reached 27—so many that the U.S. weather service had to pick names from the Greek alphabet, having run out of English alphabet names! The season record was previously 21; in 2005 that number was reached as early as October 19—then that record was broken *six* times, leading up to tropical storm Zeta on December 30, 2005. The number of Atlantic hurricanes, 14, was also a record, twice breaking the previous record of 12.

The number of intense, category 4 and 5 hurricanes also set records in 2005. There were five such hurricanes in 2005, four of them category 5; in contrast, only one category 5 hurricane occurred in each of the years 1992, 1998, 2003, and 2004—and there were no category 5 events in the intervening years. One of those in 2005, Hurricane Rita, was not only one of the strongest Atlantic hurricanes ever recorded, it was *the* strongest ever recorded in the Gulf of Mexico. Another, Hurricane Wilma, grew from a tropical depression to a tropical storm to the strongest Atlantic hurricane on record—*all in less than 24 hours*!

- The year 2010 was particularly noteworthy for the *Russian heat wave of 2010* and the *Eurasian flooding of 2010.*

  Weeks after a record-setting heat wave struck the U.S. east coast and then the continental United States, Russia experienced a devastating heat wave, evidently associated with a large and persistent jet stream meander. First, there was record heat: it was the hottest July on record; in Moscow, temperatures on July 29 reached 38.2°C (101°F), the highest temperature ever recorded there for *any* day of the year. But then conditions worsened as wildfires began. By August 9 there were 550 wildfires, fed by severe drought and temperatures continuing near 100°F. Smog reached record-high levels, with carbon monoxide and other pollutants seven times the safe limit; mortality levels doubled; and about a fourth of the Russian wheat crop was destroyed (doubling the price of wheat worldwide). The total cost to the Russian economy was estimated to be $15 billion, with a total death toll of ~55,000.

  On July 30, 2010, flooding from monsoonal rains reached record levels on three major rivers in Pakistan. The flooding worsened over the next three weeks, ultimately leaving 1,700 people dead, 6 million people homeless, and 8 million others in need of aid; altogether, 20 million people were affected in some way by this flooding.

- In the past decade, all these types of extreme events continued to set records and even set successive records. For storm events, the 2020 Atlantic hurricane season stands out, especially in November of that year. Over a two-week period, Tropical Storm Eta—once again, the U.S. National Weather Service had run out of English alphabet names—formed, strengthened into Hurricane Eta, then further intensified, and made landfall four times, flooding regions from Nicaragua to North Carolina. Around that time, Theta (which formed in the *sub*tropical central Atlantic) became a record 29th named storm, and Iota became a record 30th named storm. In mid-November (!), Iota became a category 5 hurricane.

Despite extreme events such as those, however, these years may be best remembered as *the decade of the wildfire.* Record-setting wildfires hit one or more western U.S. states in every year of this decade. In 2013, Australia also suffered from wildfires. A massive wildfire in Alberta in 2016 caused the largest evacuation (88,000 people) in Canadian history; 2017 saw record wildfires in British Columbia. Wildfires also struck Europe, including areas above the Arctic Circle (!), in 2018 and 2020 and Siberia in 2020.

But record-setting and destructive wildfires took place most prominently and tragically in California (in 2017, 2018, and 2020) and, with even greater devastation, in parts of Australia (in 2019–2020). The most famous was probably the so-called Camp fire, which in November 2018 obliterated the town of Paradise in north-central California; it was the deadliest and most destructive wildfire in California history (thanks in part to a 7-year drought and strong Santa Ana winds). It also broke the record set a year earlier in California's wine country.

Records were also set and broken for the largest single wildfire (greatest area burned), in December 2017 and then August 2018; the latter event, named the Mendocino Complex fire, was notorious for the "fire-nado" it produced, with winds exceeding 150 mph equivalent to an EF3 tornado. The area-burned record was broken again in October 2020 by a wildfire twice as large. That wildfire 'season' (if such a term even applies anymore, wildfires by now springing up throughout the year) also set a record for the greatest total area burned: an area larger than the state of Connecticut.

In Australia the devastation from wildfires began in September 2019 and began to break records as a record drought and record temperatures baked the country. By February 2020, 33 people had died, and over 2,600 homes had been destroyed; an area four times the size of Massachusetts had been scorched, and within that area an estimated 1 billion animals had died.

The various sets of examples highlighted here, though they are all cases of extreme weather, also illustrate how global warming energy can manifest over time in one "reservoir," then another, and another.

For other examples of extreme weather, the reader should refer to Table 3.3.

3.6.6 Anthropogenic Climate Change: Some Final Thoughts

If we are looking for an answer to the question of whether anthropogenic global warming is underway, the preceding sections of this chapter provide dramatic supporting evidence indeed. We see trends in mean sea level, glacial melting, regional tropospheric and oceanic temperatures, stratospheric temperature, and the weather variables (rainfall, tornado activity, wildfires, etc.) comprising extreme weather events—everything expected as a consequence of global warming.

In fact, the expected consequences of global warming are so wide-ranging and serious that even without considering the underlying science, a reasonable assessment might be simply that—rather than doubting whether global warming has actually begun—its *risks* are so great that preventative or remedial action is necessary now.

But such a perspective, though accurate, misses the *biggest* picture. No carefully conducted, rigorous scientific analysis of global warming can be conclusive, thanks to the very nature of climate, *if* the observations being analyzed or modeled involve only individual events—a burst of tornado activity here or a

heat wave there, for example—or single types of evidence, e.g. trends only in sea level or only in temperature. Climate is naturally variable; it is possible that any one event or line of evidence we interpret as being a consequence of global warming is, instead, simply the product of natural variability.

Thus, those predicted consequences of global warming we actually observe may instead have a multitude of unrelated causes: some glaciers may have receded because of anomalous geothermal heat; the increase in the number of severe storms might be due to a random change in wind patterns affecting local humidity; the stratospheric cooling trend in the latter decades of the twentieth century may perhaps have been driven by random changes in stratospheric water vapor; the Russian heat wave could have been a result of 'chance' meanders of the polar and subtropical jet streams that became in phase with each other, amplifying their meanders; and so on.

But if *one* cause, such as global warming, can explain the *entire* suite of evidence, then it is most reasonable to prefer that one explanation (incidentally, such a logical "reduction" is known in philosophy as **Occam's Razor**). So it certainly is possible that natural variability within the climate system is responsible in some way for increased regional temperatures throughout the troposphere; in some other ways for dramatic loss of Arctic sea ice, and glacial ice; in still another way for a colder stratosphere; and in still other ways for more intense hurricanes, more frequent nor'easters and more frequent tornadoes, and for recent heat waves, floods, droughts, and wildfires. But all of these can also be explained as the result of just one process: ongoing global warming. *The ability of global warming alone to explain such diverse occurrences provides the most convincing evidence that we are indeed seeing global warming in action.*

A Geophysical Perspective:
The Rest of This Textbook

With the introduction to Earth System Science set out in Chapters 1, 2, and 3, we are now ready to explore the field of solid-earth geophysics. A geophysical approach will give us a better understanding of specific phenomena touched on in the first three chapters, including

- the collapse and evolution of the primitive solar nebula, which was guided by gravity and centrifugal force;

- the growth of the planets, promoted by turbulent fluid flow and constrained by the heat sources generated within the collapsing nebula and the need to conserve angular momentum;

- the big burp, a consequence of mantle convection, and the geomagnetic field, a consequence of core convection, both initiated by the differentiation of the young Earth's interior; and

- the behavior of Earth's climate system, i.e. the oceans and atmosphere, governed by the laws and equations of geophysical fluid dynamics.

Our study of geophysics will establish a framework for understanding the entire Earth System—especially the actions of and interactions among its various components—a framework for predicting their future behavior as well. This framework will incorporate basic physics and, as we progress from chapter to chapter, an increasing amount of mathematics and increasingly sophisticated equations to describe the behavior of the system. After our geophysical survey of the Earth System is complete, we will have the background and tools needed to study the entire Earth System, including Earth's climate system, quantitatively and predictively.

Part II

A Planet Driven by Convection

4

Basics of Gravity and the Shape of the Earth

4.0 Motivation

Gravity was instrumental in the evolution of the Solar System: it drove the contraction of the primitive solar nebula and the growth of planetesimals into planets and moons. Gravity provided two kinds of heat sources to help the interiors of the planets differentiate, and to ignite the Sun. And, gravitational tugs-of-war may have led to the final planetary rearrangements and collisions experienced within the young Solar System.

Gravity is also fundamentally responsible for the ways objects orbit each other, that is, Kepler's Laws. In Chapter 1, we hinted at the importance of Kepler's Third Law in allowing us to determine the distance at which an exoplanet orbits its star. The basic nature of gravity, as we will see in this chapter, means that Kepler's Third Law alternatively presents a way for us to calculate the mass of the star being orbited—or the mass of our Sun as a planet orbits it, or the mass of a planet, if the planet is orbited by a satellite.

Gravity is a key parameter in the convective processes operating throughout the Earth System. The convection within the atmosphere and oceans we discussed in Chapter 3 occurs because lateral density variations (created by heat or chemistry) are gravitationally pulled—more, or less, than the surrounding air or water—toward Earth's center: gravity gives those density variations **buoyancy**, negative or positive. The same situation takes place in mantle and core convection, though flow in those more massive realms is complicated by the fact that the moving fluid consequentially changes the gravity field causing the motion.

Gravity also helps determine the background state—the pressures and temperatures—within which the convection of any component of the Earth System takes place. The weight of a layer of air, seawater, rock, or metal exerts pressure on the layers below, leading to a substantial pressure increase with depth. Heat generated by the resulting compression contributes to an overall increase of temperature with depth (or decrease of temperature with height); as discussed later in this textbook, such a variation is called the *adiabatic gradient*. For convection to successfully occur, the buoyancy forces driving it must be sufficiently anomalous compared to those background temperature and pressure gradients; in thermal convection, for example, there must be additional heat sources which cause the medium to heat up with depth *more* than the adiabatic amount, in order to create a buoyancy that allows convection to prevail.

Finally, even if conditions are favorable, convection will not take place if it is too difficult for the medium to flow, that is, if its *viscosity* is too high. The viscosity of a medium is strongly influenced by the ambient pressure and especially temperature conditions; this connection between gravity and convection is particularly relevant to the mantle.

As we now begin our study of gravity—with the goal of building, from 'scratch,' a solid

Earth System Geophysics, Advanced Textbook 6, First Edition. Steven R. Dickman.
© 2025 American Geophysical Union. Published 2025 by John Wiley & Sons, Inc.
Companion Website: www.wiley.com/go/Dickman

quantitative foundation for understanding the Earth System geophysically—we will discover how the Earth's mass (and thus its mean density) and moment of inertia (and thus its gross structure) can be inferred. Gravity thereby gives us knowledge of some of Earth's most fundamental properties—the basis of any geophysical Earth model. Moreover, slight variations in gravity from place to place around the globe exist as a result of two factors: smoothly varying, north-south gravity variations, a consequence of Earth's daily rotation; and irregular variations, produced by mass anomalies scattered around the globe. By quantifying the smooth variations and including them in our Earth model, we can create a reference gravity field; using gravity measurements to survey the Earth will then help us pinpoint where the anomalies are located.

The Earth interacts gravitationally with all the objects of the Solar System—most significantly the Moon and Sun, and, to a lesser extent, Jupiter and Venus. We will learn in the next two chapters how those interactions are responsible for tides; for exotic changes to Earth's orbital properties, which are able to influence climate on long timescales; and for an amazing, ongoing process known as *tidal friction*.

4.1 The Nature of Gravity

4.1.1 Simple Expressions of the Law of Gravitation

In 1687, rigorously extending the ideas of others, Newton formulated and published his Law of Gravitation. From a verbal statement of this law, we can begin to understand the nature of gravity; stated quantitatively, it will allow us to predict the magnitude and consequences of the gravitational forces acting on different objects.

Verbal Formulation. Strictly speaking, the law applies to particles, or "points" that are vanishingly small in size but possess a finite amount of mass. We consider two such points,

separated by a specified distance. In words, Newton's Law states that

> *The gravitational force exerted by one point of mass on another has a _magnitude_ that is proportional to the product of the masses and inversely proportional to the square of the distance between them; it acts in the _direction_ straight from one mass toward the other, toward the mass exerting the force.*

Because the direction of the gravitational force is on the line between the two points, it is called a **central force**; because the mass exerting the force pulls the other mass toward it, gravity is an *attracting* force. Not all forces share these properties; for example, forces acting between electrical charges are central (and, for that matter, are inversely proportional to the square of the distance between the charges) but can be repelling or attracting depending on whether the charges are of the same or opposite sign. Magnetic forces, in contrast, are fundamentally different from electrical and gravitational forces, being neither proportional in magnitude to the inverse-squared distance between magnetic bodies nor central in direction (the force exerted by one magnet on another will, in most configurations, twist the other sideways as well as attract or repel it).

Our description of the Law of Gravitation contains two parts, one for its magnitude and the other for its direction, because—like all other forces—gravity is a **vector**, something that possesses both direction and magnitude (Figure 4.1). Though potentially quite abstract and mathematical, vectors are actually familiar to us in many contexts, whether to describe the movement of a car, or the flow of a wind or current; or, the action of a bat on a ball or a hammer on a nail. In all these cases, both the direction and the magnitude may be important to know. Pictorially, vectors are easily represented by arrows: the length of the arrow reflects the vector's magnitude, and the direction of the arrow indicates the direction of the vector. Quantitatively, the description of vectors will require us to employ symbolic

Point Mass

Point Mass

Figure 4.1 The gravitational force exerted by one particle on another is a vector. Arrows are an easy way to depict a vector, with their length corresponding to the *magnitude* of the vector and the direction in which they point being the *direction* of the vector.

vector notation, which will be introduced later in this chapter.

The fact that the gravitational force between two particles has a magnitude inversely proportional to the squared distance between the particles leads to some interesting consequences. The most basic consequence is that macroscopic balls or spheres of uniform density act gravitationally like points of mass; that is, *the net gravitational force exerted by a ball or sphere of uniform density is the same as if the sphere were replaced by a point of the same mass at the sphere's center.* This quite amazing

simplification of the law of gravity is discussed further below; for now, we note that it implies that Newton's Law can be tested materially in the laboratory, with three-dimensional objects, and is more than merely a statement of intangible interactions between idealized points of mass.

A second interesting consequence of gravity being precisely an inverse-square law force is that planetary orbits around the Sun (aside from small perturbations caused by the gravitational pull of the other planets, and an even smaller additional correction for Mercury required from considerations of special and general relativity) will be what we might call *stationary* orbits; see Marion (1980). A planet in a stationary orbit returns to the same location in space every cycle; with a 'migrating' orbit, though the planet is gravitationally bound to the Sun, the location it arrives at after each cycle gradually progresses around the Sun (see Figure 4.2).

Analysis of the orbits of Mercury and Mars, after accounting for the expected perturbations, allowed Talmadge et al. (1988) to conclude that any departure from precisely 2

looking down on the orbital plane...

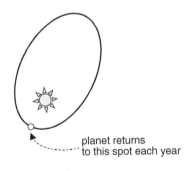

planet returns
to this spot each year

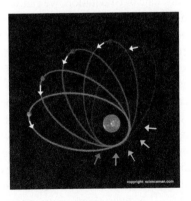

Figure 4.2 With gravity being an inverse-square law force and no other forces acting on it, the planet on the left follows a stationary orbital path, returning to the same location in space at the end of each cycle or 'year.' The point chosen for this illustration is the planet's "perihelion," the point in its orbit closest to the Sun.
This repeatability does not happen for the planet on the right, in a universe where the gravitational force is inversely proportional to distance raised to a power slightly different than 2. The orbit shifts continuously in a counterclockwise sense, but for clarity only five discrete orbits are shown, with increasing brightness; the small green arrows point (from a distance) to the perihelion for each orbit, thereby illustrating the orbital migration. Image on right is adapted from a blog by G. Martha, found at http://sciexplorer.blogspot.com/2011/04/mercury.html.

(in that inverse, squared exponent) was not significant out to at least nine decimal places.

This second consequence, more generally, applies to anything orbiting anything else, for example a moon or satellite orbiting a planet. Accurate tracking of the Moon by a technique called *lunar laser ranging* (to be discussed in Chapter 6), after accounting for perturbations to its orbit caused by the Earth, has similarly confirmed that any departures to its orbital velocity associated with a migrating orbit amount to less than a couple of billionths of a percent (Adelberger et al., 2003).

There are other consequences of the inverse-square nature of gravity worth mentioning, but, as they require us to consider explicitly the gravitational effects within a body, their discussion will be postponed until we learn how to deal with multiple masses. Still broader assessments of gravity's inverse-square dependence, including its connection to a hypothetical "fifth force" of nature, may be found in Adelberger et al. (2003) and Fischbach & Talmadge (1999); see also Ke et al. (2021) and references therein.

A Simple Gravity Formula and Big G. Before constructing a vectorial description of Newton's Law of Gravitation, we can write out a simpler, and occasionally useful, formula for the magnitude of the gravitational force. According to the verbal statement of the law, if one particle contains an amount of mass denoted M_1 and the other particle has mass M_2 within it, and if the particles are separated by a distance d, then the gravitational force is proportional to M_1 times M_2 and inversely proportional to d^2. Putting these together, the magnitude of the gravitational force exerted by either particle on the other, which we may denote F_{grav}, is determined according to

$$F_{\text{grav}} \propto \frac{M_1 M_2}{d^2}.$$

The verbal statement of Newton's Law expresses a proportionality, not an equality.

To make the law more useful, we introduce a proportionality constant, denoted G and called the **Universal Constant of Gravitation** (or,

simply, "Big G"). Thus,

$$F_{\text{grav}} = \frac{G M_1 M_2}{d^2}.$$

The magnitude of G determines how strong gravity is; as we will see later, it turns out that gravity is relatively weak, in comparison to other forces of nature. But some of those other forces are inversely proportional to distance cubed, or distance raised to even higher powers, which reduces their impact at increasingly large distances. The cosmic impact of electric forces, which are also inverse-square in nature, is reduced because they can be attractive or repulsive, whereas gravity can only be attractive (as far as we know!). In short, we find that gravity plays a dominant role on astronomical distance scales.

There was some degree of speculation throughout much of the twentieth century that the gravitational constant is not truly constant, for example that its magnitude has changed over the lifetime of the Universe. Informally this possibility is symbolized by saying that $\dot{G}$ ("G-dot") may not be zero, where the 'dot' is shorthand for the change of G with time (which we could write as $\Delta G / \Delta t$ or, in the symbology of calculus, as the time derivative dG/dt). Why should G change over time? Since Newton's Law explicitly involves only the two points of mass, we might think of Big G as a constant reflecting all the gravitational interactions with the rest of the Universe—a 'many-body' constant. That is, part of the force experienced by one point mass due to another may be the result of the latter's gravitational effect (albeit very slight) on that Universal matter, pulling it all in and changing a bit of what the first point experiences. Conceivably, then, G might be expected to diminish as the expansion of the Universe proceeds, with entire galaxies receding from one another and the mass of the Universe dispersing.

A decreasing strength of gravity would cause many changes throughout the Universe. One geophysical consequence would be that the Earth itself would expand as the gravitational force pulling all its mass together weakened; with its mass extending farther from its rotation axis, the Earth would slow its daily rotation, to conserve angular momentum. In fact the Earth's daily rotation *has* slowed over the eons, an observation that at least initially was invoked in support of the hypothesis. A second geophysical consequence would be the creation of surficial rifts around the globe, also a result of an expansion of the Earth; the existing worldwide mid-ocean ridge system might conceivably provide additional support for the idea of a weakening gravitational force.

The hypothesis that the gravitational constant is decreasing over time began with somewhat metaphysical speculation by Dirac in the late 1930s connecting G to the expanding Universe through 'dimensionless' ratios of physical parameters (Dirac, 1937). The possibility was quantified in the 1960s (Dicke, 1962) with a determination that the implied change in Big G should be a decrease of up to a few billionths of a percent per year; that is, the fractional change in G over time, which we could write as

$$\frac{\Delta G / G}{\Delta t},$$

thus,

$$\frac{1}{G}\frac{\Delta G}{\Delta t}, \text{ or } \frac{1}{G}\frac{dG}{dt}, \text{ or } \dot{G}/G,$$

should be $\sim\!-1\times10^{-11}$/yr to -3×10^{-11}/yr. The rate of expansion and thus the change in G is tied to the age of the Universe; with more modern estimates of that age, the fractional decrease in Big G would be roughly 2×10^{-11}/yr at most.

Because the Universe is billions of years old, such a rate of change in G would have had significant effects over the lifetime of the Universe—and even over the lifetime of the Earth! However, evidence has accumulated which provides bounds on possible changes in G (see also Gillies, 1997; Uzan, 2003, 2011; Will, 2006, and García-Berro et al., 2007, for a more comprehensive discussion of such estimates).

- Paleomagnetic data (discussed in Chapter 10) can be used to constrain changes in Earth's radius. Such data from sites of the same age and on the same continent must point to the same magnetic pole of that era, even when the sites are located far apart; in conjunction with the distance separating them, this allows Earth's radius to be estimated. As noted by Verhoogen et al. (1970), paleomagnetic data from the Permian era rule out any changes in radius by more than a few percent since that time, which implies (using the methodology of García-Berro et al., 2007) that $-\dot{G}/G$ cannot exceed $\sim\!1.2\times10^{-9}$/yr, on average, between the Permian and the present day.

- Changes in Earth's rate of rotation can be inferred from diverse kinds of observations, and used to constrain changes in Earth's moment of inertia and thus in G. Dicke (1966) inferred changes in the rate of rotation over the past several millennia from eclipse observations (discussed in Chapter 6); corrected for the effects of several geophysical processes in the Earth System, he estimated that $-\dot{G}/G$ should be $\sim\!3.8\times10^{-11}$/yr.

  Blake (1977) used inferences by Pannella (1972) of the decrease in Earth's rate of rotation from banding in mollusk shells from Ordovician to recent times (see also the analysis of Devonian coral data by R. Newton, 1968), discussed in Chapter 6, to conclude that $-\dot{G}/G$ did not exceed $(0.5 \pm 2)\times10^{-11}$/yr over the past 400 Myr.

- The lack of extensional features on the surface of a planet throughout its lifetime would argue against the possibility of a decrease in the Universal Constant of Gravitation. Crossley and Stevens (1976) used the lack of such features on the surface of Mercury—which, as pointed out in Chapter 1, has remained geologically pristine due to the lack of an atmosphere—to sharply constrain increases in that planet's radius; they concluded that $-\dot{G}/G$ was unlikely to be as large as $(8 \pm 5)\times10^{-11}$/yr, the value advocated by some cosmologists at that time.

  In a similar analysis for Mercury, McElhinny et al. (1978) determined upper bounds on $-\dot{G}/G$ to be either 2.5×10^{-11}/yr or 8×10^{-12}/yr, depending on the cosmological model for universal expansion considered.

These estimates refer to a change in Big G averaged over the lifetime of the Solar System.

- Changes in G cause changes in the Sun's gravitational field, leading to two types of consequences: changes in the motion of the planets and other bodies orbiting the Sun (e.g., Shapiro et al., 1971; Williams et al., 1996); and changes in the 'microlensing' or gravitational deflection of electromagnetic waves, such as radio waves and visible light, as they pass near the Sun (e.g., Will, 1981 and Ostro, 1993, also Shapiro et al., 1971 and Reasenberg et al., 1979). Tracking deep-space probes such as *Voyager* or *Cassini*, as they move out under the influence of the Sun's gravity, or tracking the motion of a planet as it progresses in its orbit around the Sun, can potentially constrain $\dot{G}$; alternatively, measuring the time delay in the radio waves sent to those probes and planets, as those signals reflect back to Earth, can (if they have passed close to the Sun) constrain G and thus (over time) $\dot{G}$.

Using mainly range data to the *Viking* Landers on the surface of Mars, as well as tracking data from other missions and for other planets (and also for the Moon, as discussed next), Hellings et al. (1983) got for $\dot{G}/G$ a value of $(2 \pm 4) \times 10^{-12}$/yr. Orbital determinations for Venus and Mercury (and Earth), improved through tracking of spacecraft either orbiting those planets or during flybys, allowed Anderson et al. (1992) to infer $\dot{G}/G = (0 \pm 2.0) \times 10^{-12}$/yr.

- A decrease in G over time would lead to an increase in the distance between the Earth and the Moon. In fact, as discussed in Chapter 6, that distance has been increasing, as a result of the phenomenon called tidal friction. Precise monitoring of the Earth-Moon distance over the past four or more decades, through lunar laser ranging, yields a value for $-\dot{G}/G$ of $(0.0 \pm$

$1.1) \times 10^{-12}$/yr (Williams & Dickey, 2002), $(0.4 \pm 0.9) \times 10^{-12}$/yr (Williams et al., 2004), or $(-0.2 \pm 0.7) \times 10^{-12}$/yr (Müller & Biskupek, 2007), after accounting for the effects of tidal friction.

The more precise estimates of $-\dot{G}/G$ are an order of magnitude or so smaller than called for by Dicke's prediction, suggesting that either G is, temporally, truly constant or its variations over the lifetime of the Solar System are too slight to make a measurable impact. Such a conclusion is reinforced by the strongest bound yet on $\dot{G}$,

$$\left| -\dot{G}/G \right| \leq 10^{-14}/\text{yr},$$

according to García-Berro et al. (2007), based on analysis (Bertotti et al., 2003) of an anomalous frequency shift in the signals received from the *Cassini* space probe as it was tracked; the shift was attributed to time delays as the signal passed by the Sun (on its way from Saturn to Earth).

A Vector Description of the Law of Gravity. Our formula above for Newton's Law of Gravitation is incomplete, because it says nothing about the *direction* of the gravitational force. The simplest way to remedy this omission would be to artificially introduce a vector (traditionally denoted $\hat{r}$, despite the fact that it is rarely meant to be a radius) whose direction is attracting one particle to the other, as required, and simply write

$$\vec{F}_{\text{grav}} = \frac{GM_1 M_2}{d^2} \hat{r}.$$

Unfortunately, such an expression is practically useless beyond its symbolic meaning; for example, $\hat{r}$ would have to be redefined every time one or both of the particles involved were varied. It does, however, serve to remind us that $\hat{r}$ must be a **unit vector** (as indicated by the $\hat{}$ symbol), i.e., have a magnitude of 1, otherwise it would multiply the other terms in the formula and produce a net magnitude for $\vec{F}_{\text{grav}}$ different than what is prescribed by Newton's Law.

The most useful way to express the Law of Gravitation fully is with vectors that are

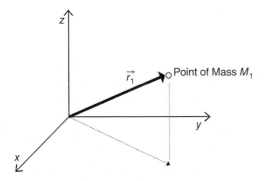

Figure 4.3 The vector $\vec{r}_1$ is shown in bold; the dotted arrow in the *x-y* plane, known as the "projection" of $\vec{r}_1$ onto the *x-y* plane, is shown for 3-D perspective.

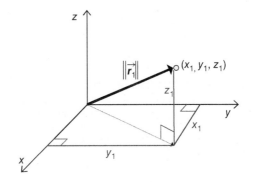

Figure 4.4 The components of a position vector are also the Cartesian coordinates of its tip. By applying the Pythagorean theorem to the horizontal then vertical triangles, you can show that $\|\vec{r}_1\|^2 = (x_1)^2 + (y_1)^2 + (z_1)^2$.

tied to a coordinate system. Referring to Figure 4.1 and those two particles of mass discussed earlier, we can set up a Cartesian (x-y-z) coordinate system within which the particles will be located. Let's say that one particle, containing an amount of mass M_1, is located at position $\vec{r}_1$ with respect to that system (see Figure 4.3). $\vec{r}_1$, called the "position vector" of the particle, is represented by an arrow that begins at the coordinate origin and stretches to the location of the particle; the direction of vector $\vec{r}_1$ points toward the particle, and the magnitude of $\vec{r}_1$—the length of the arrow—corresponds to the distance of the particle from the origin. $\vec{r}_1$ as a vector is symbolized here with a superscript arrow; the magnitude of the vector $\vec{r}_1$ is officially denoted with surrounding double bars, $\|\vec{r}_1\|$, but in a few situations can more easily be written as the vector without its arrow, r_1.

Within this coordinate system, the position of the particle can also be denoted by its x, y, and z coordinates, (x_1, y_1, z_1); it follows that these must also be the x, y, and z components of the position vector $\vec{r}_1$ (see Figure 4.4). To express this equivalence, we may write

$$\vec{r}_1 = (x_1, y_1, z_1).$$

From the Pythagorean theorem, as illustrated in Figure 4.4, we know that

$$\|\vec{r}_1\|^2 = (x_1)^2 + (y_1)^2 + (z_1)^2.$$

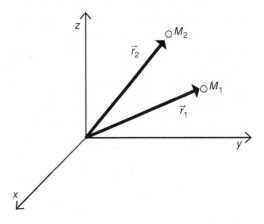

Figure 4.5 The two points of mass, M_1 and M_2, are located at the positions represented by vectors $\vec{r}_1$ and $\vec{r}_2$.

Now consider the second particle as well. Let M_2 denote the amount of mass it contains; represent its location within our Cartesian coordinate system by the position vector $\vec{r}_2$ (see Figure 4.5). From Newton's Law, we know that the magnitude of the gravitational force will involve the distance between the two particles; we use the rule for **vector addition** to determine that distance. According to that rule, two vectors are added by placing the tail of the second vector on the front tip of the first vector; in order to do this, as illustrated in Figure 4.6 for arbitrary vectors ($\vec{A}$ and $\vec{B}$), the second vector should

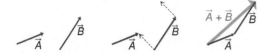

Figure 4.6 Procedure for determining the sum of two vectors $\vec{A}$ and $\vec{B}$. Move one vector toward the other, preserving its length and direction, until its tail is at the other's tip; the sum, as shown on the right, is the arrow extending from the tail of one to the tip of the other.

be moved parallel to itself so as to preserve its direction. The sum is a vector extending from the tail of the first vector to the front tip of the second vector. With some thought it should be apparent that this procedure is equivalent to adding the corresponding components of the vectors; if the vectors have components (x_A, y_A, z_A) and (x_B, y_B, z_B), then the vector representing the sum of these vectors equals $(x_A + x_B, y_A + y_B, z_A + z_B)$.

To determine the distance between particles M_1 and M_2 we denote the vector stretching from M_1 to M_2 by $\vec{?}$ (see Figure 4.7); the magnitude of $\vec{?}$ will equal the distance we seek. From the rule for vector addition, it follows that the sum of $\vec{r}_1$ and $\vec{?}$ must equal $\vec{r}_2$, that is,

$$\vec{r}_1 + \vec{?} = \vec{r}_2.$$

By subtraction,

$$\vec{?} = \vec{r}_2 - \vec{r}_1,$$

so that

$$\left\| \vec{?} \right\| = \left\| \vec{r}_2 - \vec{r}_1 \right\|.$$

Note, incidentally, that $\|\vec{r}_2 - \vec{r}_1\|$, the magnitude of vector $\vec{r}_2 - \vec{r}_1$, does not equal $\|\vec{r}_2\| - \|\vec{r}_1\|$ or $r_2 - r_1$ —a good thing, since otherwise we would be predicting that any two points an equal distance from the coordinate origin ($\|\vec{r}_2\| = \|\vec{r}_1\|$), no matter how far apart they were, would be separated by zero distance.

We are now able to express the magnitude of the gravitational force between particles M_1 and M_2:

$$F_{\text{grav}} = \frac{GM_1 M_2}{\left\| \vec{r}_2 - \vec{r}_1 \right\|^2}.$$

This is a more powerful expression than the earlier one with d in its denominator, since it can be used if we know the locations or coordinates of various particles but have not yet determined all of their relative distances (the "d" for each pair of particles).

The direction of the gravitational force can also be specified using our vectors. At this point, we need to employ more precise notation for the force vector. We define $\vec{F}_{12}$ to be the gravitational force exerted by mass M_1 on mass M_2. Because gravity is an attracting force, $\vec{F}_{12}$ must point from M_2 toward M_1. But $\vec{r}_2 - \vec{r}_1$ points oppositely, i.e., toward M_2; instead,

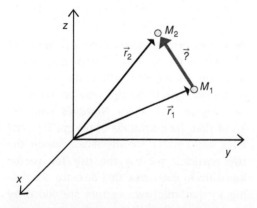

Figure 4.7 In order to determine the vector spanning the distance between two particles, apply the rule for vector addition to the position vectors of the particles.

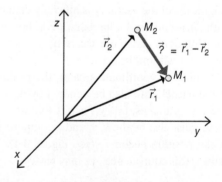

Figure 4.8 The gravitational force exerted by mass M_1 on mass M_2 attracts M_2 toward it, so the direction of that force should be the same as that of the vector $\vec{?}$ shown here. *Using the rule for vector addition, you should be able to show that* $\vec{?} = \vec{r}_1 - \vec{r}_2$.

the direction we seek is that of $\vec{r}_1 - \vec{r}_2$ (see Figure 4.8).

Putting it all together, finally, we write

$$\vec{F}_{12} = \underbrace{\frac{GM_1M_2}{\|\vec{r}_2 - \vec{r}_1\|^2}}_{\text{magnitude}} \underbrace{\frac{\vec{r}_1 - \vec{r}_2}{\|\vec{r}_1 - \vec{r}_2\|}}_{\text{direction}}.$$

In this expression, we have multiplied the magnitude of the gravitational force, written in accordance with Newton's Law, by a fraction labeled "direction." That fraction involves a vector, $\vec{r}_1 - \vec{r}_2$, which points in the required direction; and with $\vec{r}_1 - \vec{r}_2$ divided by its own magnitude, the "direction" fraction is a vector that still points from M_2 to M_1 but has a magnitude reduced to unity. Thus, we have created a unit vector to represent the direction of the gravitational force; as required, it is a quantity that can multiply the rest of the expression without affecting the overall magnitude of the force.

The two parts of this expression, the magnitude and direction of $\vec{F}_{12}$, manage to quite concisely incorporate everything in the original statement of Newton's Law of Gravitation.

We will see later that our vector approach is indispensable for dealing with the gravitational forces exerted by multiple particles, or by three-dimensional objects.

As preparation for that later theoretical development, can you construct the vector expression for $\vec{F}_{21}$, the gravitational force exerted by particle 2 on particle 1?

4.2 Newton's Second Law and the Gravity Field

4.2.1 Cause and Effect, Mass and Weight

Newton also formulated three laws of motion, known conveniently as his First, Second, and Third Laws. The **Second Law** relates force to acceleration, and will allow us to change our focus from gravitational forces to the motion associated with them. This law also applies only to point masses. It says that if a particle of mass M is acted on by any force $\vec{F}$ (gravitational or otherwise), its resulting acceleration is determined according to

$$\vec{F} = M\,\vec{a}.$$

The acceleration is written as a vector, $\vec{a}$, because it possesses both magnitude—the rate at which the particle's velocity is changing with time—and direction—the direction of velocity change. Newton's Second Law is a vector equation, saying both that the magnitudes of $\vec{F}$ and $M\,\vec{a}$ are equal ($F = Ma$) and that the directions of $\vec{F}$ and $\vec{a}$ are the same (the particle accelerates in the direction of the force).

This law may be viewed as the original definition of cause and effect (with $\vec{F}$ the cause and $\vec{a}$ the effect). The magnitude of the effect—the amount by which the particle accelerates—depends inversely on its mass: the greater the mass, the more resistance the particle has to being accelerated. In fact, M is sometimes called the **inertia** of the particle.

On the other hand, the equation can be inverted (to yield $\vec{a} = \vec{F}/M$), showing that acceleration may be viewed as equivalent to force, just rescaled.

In the case where a particle is experiencing a gravitational force, for example due to the Earth, the gravitational force is said to be the **weight**, $\vec{W}$, of the particle; and the particle's resulting acceleration, for example as it falls toward the ground when it is released, is called the **acceleration of gravity**—or, more familiarly, simply "the gravity"—and is denoted $\vec{g}$. In this case, for a particle of mass M, Newton's Second Law can be written

$$\vec{W} = M\,\vec{g}.$$

The **scalar** part of this equation (that is, the part not involving direction), $W = Mg$, can be used, for example, to convert mass to weight, or vice versa, or to calculate the weight of a person (treated as a point of mass, relatively speaking) standing on another planet, once the acceleration of gravity there is known.

4.2.2 Earth's Gravity Field, and the Answer to a Really Fundamental Question

Consider a massive body somewhere in the Universe and, near it, a particle of mass subjected to that body's gravity. From Newton's Law of Gravitation, we know that the force of gravity exerted on the particle will depend, in its strength and direction, on where that particle is located relative to the body, including its distance from the body and (in most cases) its angular position relative to the body. The acceleration of gravity experienced by the particle will similarly depend on its location. Thus, in general, *the gravity due to a body will vary spatially around it*, with diminishing effects farther away but also different effects in different directions. With this picture in mind, $\vec{g}$ is also called the **gravity field** of the body.

Earth System scientists deal with a variety of fields in their study of the Earth. These can be fields of vector quantities, such as the velocity of seawater throughout the oceans as tidal forces cause an ebb and flow; or the interior heat flow, which is generally directed outward as the Earth cools but also includes lateral flow away from hot structures like upwelling mantle plumes and volcanic magma chambers. They can also be scalar fields, such as Earth's surface topography, with its highs and lows; or the distribution of atmospheric pressure over the Earth's surface, resulting from the atmosphere's three-cell circulation. All of those fields exist within the physical context of Earth's vector gravity field.

Now consider that massive body again, imagining it to be the Earth, exerting gravitational forces on all particles within its gravity field. In order to measure the strength of the gravity field at some location, i.e., the magnitude of the gravitational acceleration g there, we can simply time the fall of a meter stick, determining its changing velocity as it falls and thus its acceleration. Measurements around the world show that, on average at the Earth's surface, $g \approx 9.80 \, \text{m/sec}^2$ (or, $g \approx 980 \, \text{cm/sec}^2$, or $g \approx 32 \, \text{ft./sec}^2$). *But why does it have that magnitude?*

To answer that question we will begin with the 'amazing fact' noted early in our discussion of the nature of gravity: uniform spheres act gravitationally as if they were point masses. That fact, it turns out, allows us to conclude that the Earth's gravity is the same, *to good approximation*, as if all of the Earth's mass is concentrated at a point at its center. But this is not because most of Earth's mass exists deep in the interior, near its center; in fact, only one third of its mass is even in the core (—despite its great density, the core has limited volume). Nor is the Earth even close to being a sphere of uniform density. However, the Earth is approximately *spherically symmetric*—that is, the Earth's interior approximates a set of concentric shells each of uniform density (though that density differs from shell to shell), and the gravitational force exerted by each of the shells is the same as if their mass were a point at the center.

Such a relation, which is explored later in this chapter in the context of something called "superposition," is worth restating as a general 'theorem':

> *The gravitational force exerted by a spherically symmetric body is exactly the same, in direction and magnitude, as if all its mass were concentrated in a point at its center.*

If we can treat the Earth as a point mass, then we can use Newton's Law of Gravitation to deduce the force it exerts on other point masses, such as a meter stick or person or other particle of mass M at the Earth's surface. Letting M_E denote the mass of the Earth, it then follows that M will experience a weight or gravitational force due to the Earth of magnitude

$$W \approx \frac{GMM_E}{\bar{r}^2},$$

if $\bar{r}$ is the distance of M from Earth's center, that is, if $\bar{r}$ is Earth's mean radius; see Figure 4.9.

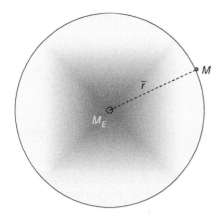

Figure 4.9 If all the mass of the Earth is concentrated in a point at its center, then we can use Newton's Law of Gravitation to calculate the gravitational force exerted by the Earth on a particle M, whether that particle is located at its surface (as shown here) or anywhere else outside the Earth.

We write approximately equal to remind us that Earth is only approximately spherically symmetric. At the same time, Newton's Second Law requires that $W = Mg$, so we can write

$$W \approx \frac{GMM_{\mathrm{E}}}{\bar{r}^2} \approx Mg.$$

This relation would be exact if the Earth were perfectly spherically symmetric. And, as implied by the 'theorem,' in the case of perfect spherical symmetry the direction of $\vec{g}$ would be straight toward the center of the Earth, where all of Earth's mass acts in effect as if it is located.

A Principle of Equivalence. The final step in our derivation would be to divide both sides of the equation by the particle mass M, yielding a convenient formula for the acceleration of gravity as measured at Earth's surface. But is M the same mass on both sides of the equation? The mass on the right, multiplying g, represents the particle's inertia, i.e., its resistance to being accelerated; the mass M in Newton's gravitational formula represents how strong a gravitational force is exerted by or experienced by the particle. There is no reason why inertial mass and gravitational mass should be the

same. The idea that they are identical is called the **Principle of Equivalence**.

Actually, sometimes this idea is called the *weak* Principle of Equivalence, to contrast it with a more stringent version employed in Einstein's general theory of relativity. If other forces had been acting on our particle, the right-hand side of the equation would have remained Ma, from Newton's Second Law, with a being the particle's net acceleration. A *strong* form of the Principle of Equivalence would be the assertion that the particle cannot distinguish a gravitational acceleration from any other acceleration of the same magnitude acting on it. Such an assertion includes but goes beyond the requirement that inertial and gravitational mass are the same. Finally, as we will discuss in Chapter 9, because Newton's Second Law implies that Ma has the units of force, Ma is sometimes referred to as the "inertial force" (in reality, of course, it is the net result of all the forces acting on the particle). Another 'strong' version of the Principle of Equivalence would be that, in the absence of other forces, the gravitational force and inertial force experienced by the particle are indistinguishable.

If the (weak) Principle of Equivalence did not hold—if we cannot divide out M from the equation—then the acceleration of a particle in Earth's gravitational field, for example when falling to the ground, would depend on its mass. Different materials, of different mass (like a feather and a hammer; watch Williams, 2008), would accelerate differently. Galileo's legendary experiments, revealing in one that different masses dropped from the Leaning Tower of Pisa fell at the same rate and in another that the period of swing of a simple pendulum (as it 'falls' back and forth through Earth's gravity field) is independent of the pendulum's mass, constituted, in retrospect, the first tests of the Principle of Equivalence. Other scientists measuring the degree of equivalence have included Newton, Eötvös, and Dicke. Modern tests of equivalence have focused on the fall of objects both in the laboratory (with objects falling short distances), finding the two types of mass identical to within a few parts in 10^7 on an atomic scale (Fray et al., 2004) and within a few parts in 10^{13} on a macroscopic scale (e.g., Schlamminger et al., 2007); and in the Solar System (with satellites 'falling' long distances around the Earth as they orbit, or the Earth and Moon as they orbit the Sun), finding the principle satisfied to within a few parts in 10^{13} (Will, 2006). In all cases, the uncertainties represent the limits of measurement capabilities at that time.

An Answer to That Question. We will assume that gravitational and inertial mass are exactly the same. Dividing both sides of the preceding 'weight' equation by M, we obtain

$$g = \frac{GM_E}{r^2}.$$

Most fundamentally, this equation tells us what the Earth's gravitational acceleration depends on: the Earth's mass and radius, as well as the gravitational constant. But does this combination of parameters end up having the observed value of $\sim 980 \, \text{cm/sec}^2$?

As we will learn later in this chapter, the value of GM_E can be deduced using Kepler's Third Law applied to anything orbiting the Earth; for high precision, we use artificial satellites, which can be carefully tracked, leading to (Dunn et al., 1999)

$$GM_E = 3.986004419 \times 10^{14} \, \text{m}^3/\text{sec}^2$$

(with an uncertainty of $\pm 0.000000002 \times 10^{14}$ m^3/sec^2).

The mean radius of the Earth can be estimated in a number of ways; for later purposes, we use the astronomical technique of Eratosthenes, an ancient Greek scholar who performed the calculation around 240 BCE using the Sun for reference. The analysis rephrased in terms of an arbitrary observed star is pictured in Figure 4.10.

At one site, the **zenith angle** of the star is measured; the zenith angle is the angle swept out as an observer points first to the star then to the zenith, i.e., to "straight up." For increased precision the zenith can be located by dropping a plumb bob, a weight attached to a line, which will point straight downward, then looking

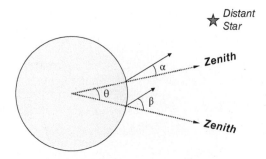

Figure 4.10 The geometry behind Eratosthenes' method of estimating Earth's radius. The zenith angles to the chosen star (whether that star is our Sun or not) are α at site one and β at site two. Because the star is far away, the direction to the star at both sites is the same (those arrows are parallel).

The central angle subtended by the radii to the two sites, θ, turns out to equal $\beta - \alpha$. To see this, imagine starting site two right next to site one (so $\theta = 0$ and $\beta = \alpha$); as site two is then moved farther and farther away, θ and β increase by the same amount.

back up along the plumb line. This process is repeated at a second site, north or south of the first, using the same star; because of the Earth's curvature, however, that star will appear higher or lower in the sky. If the second site is directly north or south of the first, the difference in zenith angles will equal the difference in latitude between the two sites.

For example, if one site is at the pole and the other is on the equator, then the star directly above the pole (the Pole Star) will lie on the horizon when seen from the equator; in that case, the zenith angle is $0°$ at the pole and $90°$ at the equatorial site, and the difference in angles does indeed equal the difference in latitude between equator and pole. In general, the difference in zenith angles will equal the central angle contained by the radii to the two sites (see Figure 4.10). If we denote the central angle by θ and the zenith angles by α and β, then

$$\theta = \beta - \alpha.$$

The reason for going into Eratosthenes' methodology in such detail is because of the way the central angle can be related to Earth's radius. As explained in Figure 4.11, if S denotes the "arc distance" S between the two sites, i.e., the distance along Earth's surface

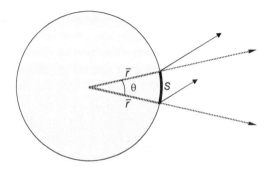

Figure 4.11 The circumference of a circle of radius $\bar{r}$ is $2\pi\bar{r}$. The entire circumference can be traced out by sweeping a radius around the center of the circle through a central angle of 360°, or 2π radians. The arc distance S, part of the circumference, is obtained by sweeping the radius only through a central angle of θ. In general, the portion of the circumference traced out should be in proportion to the amount of central angle swept out. Thus,

$$\frac{S}{2\pi\bar{r}} = \frac{\theta}{2\pi},$$

or

$$S = \bar{r}\,\theta.$$

between the sites, and the Earth's radius is $\bar{r}$, then

$$S = \bar{r}\,\theta.$$

It follows that

$$\bar{r} = S/\theta = S/(\beta - \alpha).$$

Eratosthenes knew the distance between his two sites, as they were in cities along a known caravan route. In our general analysis, we can perform a geodetic survey to calculate the distance S along Earth's surface; this will lead to

$$\bar{r} = 6{,}371 \text{ km.}$$

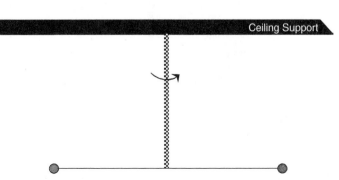

Figure 4.12 The horizontal pendulum used to 'weigh the Earth' consists of a beam suspended from a torsion fiber and able to swing horizontally. For our discussion, the beam is idealized as two spheres connected by a massless rigid wire.

Putting the values for $\bar{r}$ and GM_E into the formula for g, we indeed confirm that Earth's gravitational acceleration is about 980 cm/sec².

4.2.3 Weighing the Earth

As previously stated, Kepler's Third Law leads to a value for GM_E of 3.986×10^{14} m³/sec². This in turn implies a value for the mass of the Earth—once the value of Big G is known. The determination of G is often called "weighing the Earth." The first experiment to weigh the Earth was carried out by Cavendish in 1798. He used a 'horizontal pendulum,' a level beam able to swing in the horizontal plane while suspended from a 'torsion fiber' (Figure 4.12). The torsion fiber is so named because it has a *calibrated* resistance to twisting: that is, the fiber becomes twisted as the beam swings, resisting the motion; the magnitude of the twisting force ("torque") that slows and eventually stops the beam can be inferred (through its calibration) from the final angle of swing.

As a consequence of that resistance to twisting, the beam can swing periodically: initially forced out of its rest position, the beam swings back in as the twisted torsion fiber tries to untwist; but the beam's momentum carries it past its rest position, twisting the fiber in the opposite sense; the beam swings back out as the fiber untwists; and so on. The period of this swinging, which we will denote T_0, depends on how strongly the fiber resists twisting, with fibers of different composition resisting the twisting to different degrees. The material property describing the fiber's resistance is called its *torsion constant*.

To get the pendulum moving in the first place, of course, the beam's resistance to swinging—a consequence of its mass distribution (its *moment of inertia*)—must be overcome. The period of swing thus depends on both the torsion constant of the fiber and the moment of inertia of the beam; by measuring T_0, and also the beam's mass and dimensions (thereby determining its moment of inertia), the torsion constant can be deduced. This calibrates the torsion fiber—a necessary first step in Cavendish's experiment.

Cavendish placed two large spherical masses near the beam. Let M denote the mass of each of those spheres; the mass of the beam itself is approximated as being contained in two small spheres at the ends of the beam (Figure 4.12). Each of the small spheres, containing an amount of mass m, is located at a distance ℓ from the fiber, and the rest of the beam is massless (Figure 4.13).

When the large masses are brought near the beam, the beam will swing toward them. The motion of the beam indicates that the beam is being torqued by the gravitational forces acting between the large and small spheres. With **torque** defined as 'moment arm' times force, this gravitational torque will be the product of ℓ, the moment arm in this situation, and the gravitational force exerted by the large spheres on the small spheres. From Newton's Law of Gravitation, the force between each pair of 'near' masses after the beam has come to rest is GMm/d^2, where d is the final distance between the centers of the large and small spheres. Neglecting for simplicity the force and associated torque between each pair of 'far' masses m and M, and for that matter neglecting the fact that the beam is not truly massless (but note that there <u>are</u> techniques for including all these effects), the net torque is $2GMm\ell/d^2$. Our unknown in this formula is Big G.

The beam has continued to swing until the resisting torque on the fiber balanced it out (see Figure 4.14). Thanks to the torsion fiber's calibration, measuring the angle of swing allows the strength of the torque to be inferred independently of the gravity formula. Equating that torque to $2GMm\ell/d^2$ yields the magnitude of G.

Cavendish's achievement was groundbreaking, but even with modern instrumentation it is difficult to obtain great accuracy using his approach; this is at least partly because it is difficult to measure angles accurately. One variation on Cavendish's approach, developed by Heyl in the 1930s, found a very clever way to replace angular measurements with timing measurements; the latter are intrinsically easier to resolve with high precision. Heyl replaced the large spheres with massive, fixed cylinders—cylinders being easier to machine accurately than spheres—and the horizontal beam was placed into one of two possible rest positions (Figure 4.15a).

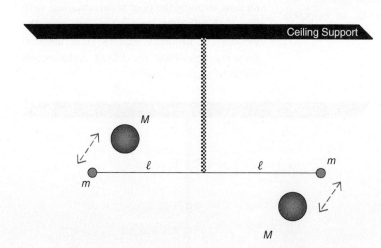

Figure 4.13 The large spheres, each of mass M, are placed in front of and behind the beam, i.e. in the same horizontal plane, not above or below it.

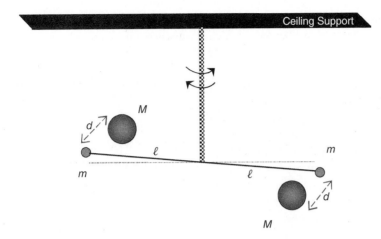

Figure 4.14 After the beam has come to rest, the separation between the large and small spheres is a distance *d*. For this experiment, ℓ, *d*, *m*, and *M* must be measured, as well as the angle the beam makes with its initial position.

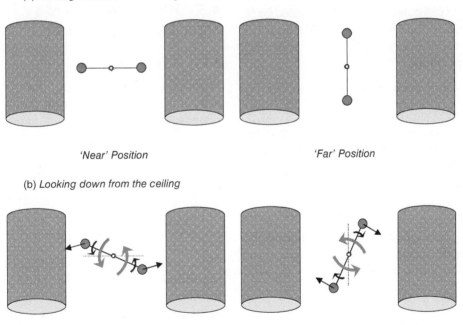

(a) *Looking down from the ceiling*

'Near' Position   'Far' Position

(b) *Looking down from the ceiling*

'Near' Position   'Far' Position

Figure 4.15 Heyl's modification of the Cavendish experiment.
(a) The horizontal beam is set up resting either 'near' to or 'far' from adjacent massive cylinders.
(b) The rest position of the beam is shown as a dotted line. When the beam is twisted around to the position shown, the torsion fiber tries to twist it back, with a restoring torque indicated by the curved bold orange arrows.
Straight black arrows denote the gravitational forces exerted on the near parts of the beam by the nearby massive cylinders (the 'far' gravitational forces are not shown here). Those gravitational forces have the effect of twisting the beam toward the cylinders. The resulting gravitational torque is indicated by the small curved black arrows.
In the near position, the gravitational torques *add* to the torsion fiber's restoring torque; in the far position, they *oppose* the restoring torque.

When the horizontal beam is twisted from its rest position, the torsion fiber from which it is suspended resists the twist and provides a restoring torque. In the 'near' position, the gravitational forces exerted by the massive cylinders pull on the beam and in effect <u>add</u> to the fiber's restoring torque (see Figure 4.15b); with a greater restoring torque, the beam will be forced back more quickly, and the resulting oscillation will have a shorter period than T_0, which was the unforced or natural period of swing of the beam. In the 'far' position, the gravitational pull of the cylinders acts to oppose the fiber's restoring torque, resulting in a net weaker torque and a period of oscillation greater than T_0. Heyl was able to relate the difference in those periods of oscillation to Big G, and with careful measurements of the periods determined a more precise value for G. From Heyl and Chrzanowski (1942), the result was $G = (6.673 \pm 0.003) \times 10^{-11}$ m^3/sec^{2}kg; more recently,

$$G = (6.67384 \pm 0.00080) \times 10^{-11} \text{ m}^3/\text{sec}^2\text{kg}$$

(CODATA, 2010).

The determination of a value for G yields key information about the Earth. With $GM_E = 3.986004419 \times 10^{14}$ m^3/sec^2 from Kepler's Third Law, we then find that

$$M_E = 5.9726 \times 10^{24} \text{ kg}.$$

Given the mean radius and thus volume of the Earth, this mass corresponds to an average density $\bar{\rho}$ for the Earth of

$$\bar{\rho} = 5.5138 \text{ gm/cm}^3,$$

where we use cgs (centimeter-gram-second) rather than mks (meter-kilogram-second) units here ($\bar{\rho} = 5.5138$ gm/cm$^3 = 5,513.8$ kg/m^3) for familiar comparisons; for example, the density of water is 1.0 gm/cm^3 under STP conditions. Since the density of crustal rocks is typically ~2.7–2.9 gm/cm^3, we have discovered a clue to Earth's interior: *the density in some portion of the interior must be significantly greater than 5.5 gm/cm^3, in order that the average density equal 5.5 gm/cm^3*. This result is consistent with Earth's moment of inertia, as discussed in Chapter 1.

[NOTE: still more recent determinations yield slightly different values, even if rounded off. Based on

$$G = (6.67430 \pm 0.00015) \times 10^{-11} \text{ m}^3/\text{sec}^2\text{kg}$$

from CODATA (2018; see also Li et al., 2018 and Xue et al., 2020), we have

$$M_E = 5.9722 \times 10^{24} \text{ kg}$$

and

$$\bar{\rho} = 5.5134 \text{ gm/cm}^3.$$

But most of the Earth models discussed in this book are not based on these latest values.]

4.3 The Gravity Field of a Three-Dimensional Earth

4.3.1 A Guiding Principle

Most objects in the Universe are not single particles of mass but rather—like the Earth—are composed of an arbitrarily large number of particles. To deal with the gravitational force or gravity field of such objects, we invoke a gravitational **superposition principle**:

The gravitational force on a body due to two masses equals the sum of the gravitational forces due to each mass.

Thus, the net force is not greater or less than the sum of the individual forces. For example, both the Sun and the Moon exert gravitational forces on the Earth. During an eclipse, the Moon does not 'screen out' the Sun's effect —blocking its gravity so that the Earth flies away out of the Solar System —or somehow greatly amplify the Sun's gravity so the Earth is pulled into the Sun. The net gravitational force experienced by the Earth will certainly vary over time, depending on the positions of the Sun and Moon, and even reach maximum and minimum values periodically over the course of a month or a year; but at no time is it amplified or reduced in anything more than an additive way.

The possibility of a screening effect (or its opposite) can be tested during eclipses. One such test was carried out during the 1997 total eclipse in northeast China (Yang & Wang, 2002). Within the uncertainties—a few parts in 10^8—associated with a very sensitive gravity meter that was used, observations of the strength of gravity showed no significant change over the tens of minutes preceding, during, and after totality. Such a conclusion, based on a more rigorous analysis, was also reached by Unnikrishnan et al. (2001).

Note that the gravitational superposition principle is a vector rule, because forces are vectors; that is, the net gravitational force due to the two masses is a *vector sum* of their individual forces.

The superposition principle can be logically extended from two objects to an arbitrarily large number of particles or, for that matter, to a three-dimensional continuum of mass. Viewing a collection of discrete particles as successive pairs of objects—the collection equals one particle plus the remainder; the remainder equals one particle plus the rest of the remainder; the rest of the remainder equals one particle plus the rest of the rest of the remainder; and so on—and invoking the superposition principle each time, it follows that the gravitational force due to a collection of many particles (or a three-dimensional continuum of mass, composed of an infinite number of infinitesimally small volumes of mass) equals the sum of the gravitational forces due to each.

4.3.2 More Consequences of Gravity Being an Inverse-Square Law Force

Early in this chapter, some consequences of gravity's inverse dependence on distance squared were explored, including that the gravitational force exerted by a homogeneous sphere is the same as if the sphere were replaced by a particle of equal mass at its center. Later on, we noted that such an amazing equivalence was simply a special case of a more general theorem involving the gravitational force exerted by a spherically symmetric body, i.e., a sphere whose density can vary radially (but not laterally). In fact, that more general equivalence follows fundamentally from the superposition principle, because any spherically symmetric body can be built up from a series of concentric spheres, each of the 'right' uniform density.

For example, let's imagine that the body's density increases systematically, layer by layer, from its surface down to its center. Our first concentric sphere of uniform density will be the same size as the body, with a density throughout it equal to that of the body's surface layer. Our second concentric sphere will be the size of the original body minus that top layer, and its density will be the difference between the densities of the body's top and second layers; when 'added' to the first uniform sphere, it will produce an object with the correct density of the original body's top *and* second layers. And so on, with deeper layers similarly accommodated using successively deeper concentric spheres.

Since the gravity due to each of these uniform spheres is the same as if their mass were concentrated at the center, *the net gravity due to the original spherically symmetric body will*—because of superposition—*equal the gravity due to a point at its center with the same total mass as that body.*

The gravity in these examples is of course the force exerted by the body on another mass, i.e., a mass *external* to the body. In later chapters we will consider the physical state (temperature, pressure, etc.) of Earth's interior, its effects on material properties, and the nature of any resulting flow. As mentioned at the beginning of this chapter, the interior pressure and temperature both depend on the strength of the gravitational field—but here we are talking about the *internal* field. For example, the weight of the minerals at some depth depends on how strongly they are pulled—by the rest of Earth's mass—toward Earth's center. This raises the interesting question of whether, and in what way, gravity varies *within* the Earth.

This question cannot be answered by invoking the 'amazing theorem,' even if the Earth were a sphere whose density is spherically symmetric or uniform: if its gravity within could be calculated by placing all its mass at the center, then as we went deeper and got closer to the center, we would feel a gravitational force (inversely proportional to distance squared) that became infinitely great.

Can you point to another factor whose variation with depth might keep internal gravity from blowing up?

As a first step toward investigating gravity within the Earth, we will use still another consequence of the inverse-square nature of gravity (combined, as you will see, with superposition): *in a hollow spherical shell of constant density, the gravitational force exerted by the shell on a particle located anywhere within its hollow must be zero.* Within that hollow, the particle can float, aimlessly, anywhere inside the hollow, and is pulled neither in toward the center nor out to the shell itself.

For a <u>thin</u> spherical shell of uniform density (Figure 4.16), this behavior is ultimately a result of the fact that the surface area of a sphere is proportional to the radius squared (just as its volume is proportional to the radius cubed). At whatever distance the particle is from the shell, the mass pulling it in one direction is, like the surface area, proportional to that distance squared (surface area $\propto r^2$, and mass = surface area × thickness × density, so mass $\propto r^2$ × thickness × density); thus, the mass divided by its distance squared—that is, the gravitational force it exerts—is the same for all portions of the shell. The particle is pulled equally in all directions, so (by superposition) the net force is zero.

A <u>thick</u> hollow shell which is either spherically symmetric or of uniform density can be viewed as a collection of thin shells, each of a constant density, as we considered before (though here without the need to extend the collection of shells any deeper than the base of that thick shell), so our argument applies in this case as well, and we obtain the same result, that the net gravity within the hollow must be zero.

Of course, the Earth is not hollow; but at any depth within it, it *can* be thought of as the 'sum' of the mass below that depth plus the mass above—that is, the sum of an approximately spherical mass below that depth plus a hollow spherical shell above. Consequently, the gravity at any depth within the Earth depends approximately (i.e., to the extent that Earth is spherically symmetric) only on the mass below!

Can you predict what the gravity at the center of the Earth should be in this case?

We will pursue this approach in later chapters as we quantify the physical state of Earth's interior.

A Digression on Divergence. A final consequence of the inverse-square nature of gravity provides us with an opportunity to introduce an advanced mathematical concept in a relatively simple way. Consider the empty space surrounding a point of mass; more specifically, imagine two concentric spherical surfaces surrounding a point mass at their center, as in Figure 4.17, and consider the empty space between those surfaces. If the inverse-square law is correct, the magnitude of that point mass's gravity field is stronger on the closer,

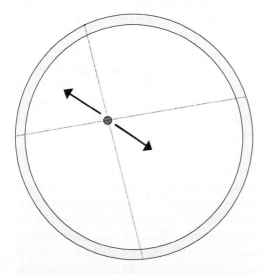

Figure 4.16 The lower right section of this thin shell contains more mass than the upper left section, in proportion to the square of its distance from the particle (●) within, so it exerts the same gravitational force on the particle as does the upper left section.

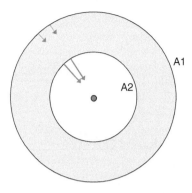

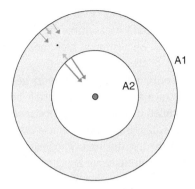

Figure 4.17 A1 and A2 are concentric spherical surfaces enclosing a particle at their center; the volume between A1 and A2 is shaded beige. The gravitational force (arrows) exerted by that particle and experienced on A1, the more distant surface, is smaller than that on A2 in inverse proportion to their squared distance from the particle. But the area of A1 is larger than that of A2 in direct proportion to those squared distances, so the product of the force times the area is the same on both surfaces.

Figure 4.18 Adding to the situation in Figure 4.17, the volume between A1 and A2 now contains a small mass, which exerts a gravitational force experienced on A1 and A2 as shown by small blue arrows. On A1, that force adds to the force exerted by the central particle (orange arrows), but on A2 it partially cancels the central particle's force (orange arrows). Consequently, the product of the force times the area will no longer be the same on both surfaces.

smaller surface and weaker on the further, larger surface in inverse proportion to the square of the radii. But the area of each of these surfaces is directly proportional to its radius squared. Thus, the product of the gravitational field and the surface area should be the same, and the difference between the outer *field × area* and the inner *field × area* should be 0.

This difference, which is calculated by considering how the magnitude of the gravitational field changes as we move from one surface to the other (a kind of gravity 'gradient'), relates to a mathematical concept we will explore in Chapter 9 called the *divergence*—in this case, the divergence of the gravity field.

Even with gravity being an inverse-square law force, however, zero divergence or zero difference is only possible if (as in Figure 4.17) there is no mass present within the volume between the surfaces. If even a little mass is present (see Figure 4.18), the gravity that mass exerts at points on the outer surface will add to that exerted by the central particle; but at points on the inner surface, its gravity will oppose that due to the central particle, since the masses are on opposite sides of that surface.

In this case, the difference or divergence will not be zero, and the value of the difference will depend on how much mass is enclosed within the volume.

Expressed in terms of calculus, the relation we are implying between divergence and enclosed mass is known as *Gauss' Law*. Because it only works if gravity is an inverse-square force, the law is considered an equivalent expression of the inverse-square nature of gravity. The product of gravitational field and surface area is called the gravity "flux." Not surprisingly—because electric forces depend inversely on the squared distance between electrical charges—there is also a version of Gauss' Law for electric forces, relating the electric flux to the net charge enclosed by a volume.

Our demonstration of Gauss' Law could have been presented more generally, in that the masses involved—the mass enclosed between those surfaces as well as the 'central' mass creating the ambient gravitational field—could have been three-dimensional spheres, of uniform or radially varying density, rather than point masses (as long as both surfaces A1 and A2 are outside the central mass, and fully enclose the second mass); but even so it would fall far short of a completely general proof of the law's validity. For example, the masses could be asymmetric 'blobs,' and the surfaces enclosing a blob need not be concentric or even spherical. In those cases we would need to first define 'flux' more generally, in terms of a type of multiplication between vectors called a "dot product" (a concept discussed in Chapter 6), and then use calculus to construct the net divergence, considering the surfaces one infinitesimal area at a time.

4.3.3 Revisiting Newton's Law of Gravitation, With Superposition

We can combine our earlier vector description of Newton's Law, expressing the gravitational force between two particles, with the superposition principle to determine the gravitational force acting between any two objects or collections of mass.

Consider first a set of N particles of mass; the ith particle (with i symbolically representing any integer from 1 to N), located at position $\vec{r}_i$, contains an amount of mass m_i (Figure 4.19a). These particles exert an overall gravitational force $\vec{F}$ on a particle of mass M located at position $\vec{R}$. The superposition principle says that $\vec{F}$ can be calculated by first determining the forces exerted by each of the particles on M, and then adding them. Let's denote by $\vec{F}_i$ the force exerted by m_i on M (see Figure 4.19b). According to Newton's Law of Gravitation,

$$\vec{F}_i = \frac{G m_i M}{\left\| \vec{r}_i - \vec{R} \right\|^2} \frac{\vec{r}_i - \vec{R}}{\left\| \vec{r}_i - \vec{R} \right\|};$$

this formula applies for $i = 1, 2, \ldots, N$. Invoking the superposition principle, we can write the total gravitational force on M as simply

$$\vec{F} = \vec{F}_1 + \vec{F}_2 + \ldots + \vec{F}_N \equiv \sum_{i=1}^{N} \vec{F}_i$$

$$= GM \sum_{i=1}^{N} \left(\frac{m_i}{\left\| \vec{r}_i - \vec{R} \right\|^2} \frac{\vec{r}_i - \vec{R}}{\left\| \vec{r}_i - \vec{R} \right\|} \right),$$

with the common term GM factored out of the sum for simplicity. The fact that $\vec{F}$ is a

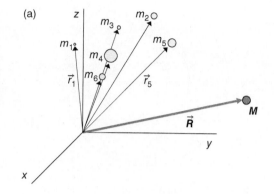

(a)

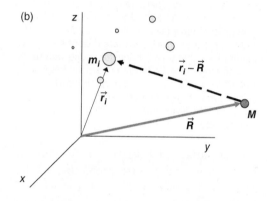

(b)

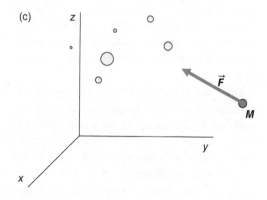

(c)

Figure 4.19 (a) A collection of particles of mass $m_1, m_2, \ldots$, located at positions $\vec{r}_1, \vec{r}_2, \ldots$, exerts gravitational forces on particle M, which is located at position $\vec{R}$.
(b) The gravitational force exerted by the *i*th particle attracts mass M toward it, and thus acts in the same direction as the vector $\vec{r}_i - \vec{R}$.
(c) The net gravitational force $\vec{F}$ experienced by mass M points generally toward the collection of particles.

Figure 4.20 To determine the net gravitational force exerted by a continuum on a point of mass M, we first quantify the force exerted by an infinitesimal portion of the continuum, containing an amount of mass dm; this mass attracts M toward it and thus acts in the same direction as the vector $\vec{r} - \vec{R}$.

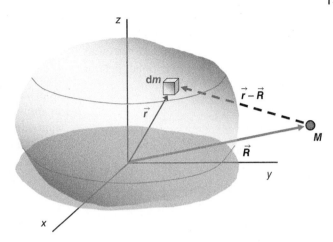

sum of vectors reminds us that the contributions to the net force experienced by M depend not only on the strengths of the individual forces (reflecting the magnitudes of the masses and their distances from M) but also on their orientations with respect to M (Figure 4.19c).

For a continuum of mass, Newton's Law of Gravitation can be applied to very small portions of mass within the continuum that approximate particles. In calculus terms, such 'infinitesimal' quantities are denoted dm, where "d" symbolizes a tiny or 'differential' amount. For the differential mass dm located at position $\vec{r}$ in Figure 4.20, Newton's Law tells us that the gravitational force $d\vec{F}$ it exerts on the particle of mass M located at position $\vec{R}$ can be written

$$d\vec{F} = \frac{GM\,dm}{\left\|\vec{r} - \vec{R}\right\|^2} \frac{\vec{r} - \vec{R}}{\left\|\vec{r} - \vec{R}\right\|}.$$

Note that this force is infinitesimal in magnitude because the mass exerting the force is infinitesimal. The net force due to the entire continuum of mass is finite rather than infinitesimal, so we denote it by $\vec{F}$. According to the superposition principle,

$$\vec{F} = \int d\vec{F},$$

where the sum is carried out in calculus terms with an integral. Substituting in, and again factoring out the common terms GM from the sum

(integral), we can write

$$\vec{F} = GM \int \left(\frac{dm}{\left\|\vec{r} - \vec{R}\right\|^2} \frac{\vec{r} - \vec{R}}{\left\|\vec{r} - \vec{R}\right\|} \right);$$

here the integral includes all the infinitesimal masses comprising the continuum, with the amount of mass dm and its position $\vec{r}$ varying as the integration is carried out over the continuum.

Note that this formula for $\vec{F}$ can also be obtained mathematically, by taking the "limit" of the preceding discrete formula as the number of particles in the set, N, approaches infinity; the particle masses each become infinitesimal, and the sum becomes an integral.

Finally, it is often more convenient to express the integral in terms of the volumes of the infinitesimal masses, since that can lead to very straightforward integration calculations (e.g., integration over x, y, and z). By definition, density equals mass per unit volume. If the volume of the differential mass dm is dv and its density is ρ, then the definition implies ρ equals dm/dv, so $dm = \rho\,dv$ and we have

$$\vec{F} = GM \int \left(\frac{\rho\,dv}{\left\|\vec{r} - \vec{R}\right\|^2} \frac{\vec{r} - \vec{R}}{\left\|\vec{r} - \vec{R}\right\|} \right).$$

This formula can easily be used to demonstrate that the gravitational force exerted on M by a sphere of uniform density (a sphere with $\rho = $ constant) is the same as if the mass of the sphere were concentrated in a point at its center. More generally, it can be used to determine the gravitational force exerted by an arbitrary mass. And, the gravitational field of an arbitrary mass can be found from

$$\vec{g} = G \int \left(\frac{\rho \, dv}{\left\| \vec{r} - \vec{R} \right\|^2} \frac{\vec{r} - \vec{R}}{\left\| \vec{r} - \vec{R} \right\|} \right).$$

4.4 The Shape of the Earth, and Variations of Gravity With Latitude

4.4.1 A Motivation to Get Complicated

The simple scalar formula that we previously derived for the Earth's gravitational acceleration was based on the assumption that Earth's internal density is spherically distributed, that is, that the Earth is spherically symmetric. In this case, we would (because of symmetry) measure the same magnitude of gravity everywhere around the globe, and (because the Earth's mass could equivalently be replaced by a point at Earth's center) gravity would everywhere point *downward*, toward the center.

Although spherical symmetry is a very good approximation for the Earth, it is not exactly true, either with respect to the Earth's density distribution or its outer shape. As a consequence, Earth's gravity field is not uniform, instead varying over its surface both in direction and magnitude. Additionally, as we will later see, the oceans must—because they are fluid—shift in response to even the slightest of those variations in gravity, compounding the asymmetry. All those gravity variations may be weak, but they are important, as they indicate where Earth's density may be anomalous, hinting at significant structures or behavior beneath the surface. Furthermore, when looking for the gravitational effects of climate change (including those associated with changes in ice sheet mass), we need to have an accurate picture of the 'background state' of Earth's gravity field upon which those effects are superimposed.

4.4.2 The Earth Is Not Spherical

One way that Earth is not spherically symmetric is that its overall shape is not perfectly round. Objects whose shape departs slightly from being spherical are often called *spheroidal* or *ellipsoidal* (see Figure 4.21a). Earth closely approximates an **oblate spheroid**, a spheroid whose largest dimension is its equator. If we denote its equatorial radius by a, and the radius to its pole by c (Figure 4.21b), then its sphericity or oblateness can be measured by its **flattening** (also called its geometric ellipticity), denoted f and defined as the fractional difference in radii:

$$f \equiv \frac{a - c}{a}.$$

Note that the flattening of a sphere (whose radii are all equal) is 0. The larger the flattening, up to a limit of 1, the more stretched out or ellipsoidal the oblate spheroid is. For the Earth, it turns out that

$$a = 6378.1366 \text{ km} \quad c = 6356.7518 \text{ km}$$

leading to

$$f = 1/298.2564$$

(IERS Conventions, 2010). The Earth's equatorial and polar radii differ by ~21 km, or about 1 part in 300—a third of a percent. Such a small

(a)

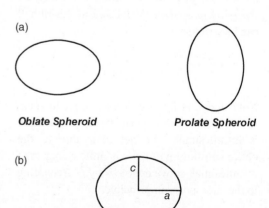

Oblate Spheroid   **Prolate Spheroid**

(b)

Figure 4.21 (a) Types of ellipsoidal shapes. (b) A side view of the Earth. *a* is a radius from the center to a point on the equator; *c* is the radius from the center to the North Pole.

flattening explains why treating the Earth's shape as spherical is quite a good approximation. Nevertheless, as we will see, Earth's flattening contributes measurably to its gravity.

4.4.3 Earth's Rotation Is the Cause

The Earth possesses a spheroidal rather than spherical shape because of its daily rotation. That rotation generates *centrifugal force*, which we first encountered in Chapter 1 and understood as the 'carousel' force that made the contracting primitive solar nebula disk-shaped; centrifugal force is also the primary cause of Earth's flattening.

Centrifugal force is related to another rotational force—the Coriolis force—which (as we saw repeatedly in Chapter 3) plays a central role in determining the character of ocean currents and especially atmospheric winds, and (as we will see in Chapter 10) is critically important in its effects on the motion of core fluid. The description of either of these forces will involve more sophisticated mathematical physics than we have employed so far. To ease the 'shock' of that description, and to better understand the nature of these forces, some background perspective here will be helpful.

The Truth About a Fiction. Many of us first hear about centrifugal force in introductory physics classes, where it is usually presented as "a fictitious force" (ironically, the Coriolis force, though its brother, is categorized as such only infrequently). To compound the murkiness of that terminology, this 'fictitious' force is generally discussed in the context of gravity, which is then labeled a "centripetal" force (centripetal means center-seeking, whereas centrifugal means center-fleeing). In intermediate physics courses, centrifugal force is still referred to as fictitious—despite the fact that a quantity to which it is intimately related is explicitly and unapologetically termed the "centrifugal" potential.

Newton's Laws of Motion are usually invoked as the basis for such a perspective. Newton's Second Law implies that an object not subjected to outside forces will not accelerate (zero F means zero a), which in turn means that the object will remain at rest or (if it was initially moving) continue on with uniform motion, i.e., in a straight line at constant speed; this consequence is actually known as **Newton's First Law**. Informally, we can think of this law as saying that a mass' *inertia* keeps it from changing its speed, or from moving if it was initially at rest, when there is no force acting on it. We measure the object's velocity and acceleration with respect to a reference frame. A reference frame in which Newton's First and Second Laws hold true is termed an **inertial reference frame**. The path taken by that unforced object in an inertial reference frame—a straight line at a constant speed—might be termed an **inertial path** (we can include the object being at rest as a special case of this).

If *two* reference frames are both inertial, then one is either at rest relative to the other or moving at a uniform velocity relative to it, in order that unforced particles maintain their inertial paths and not be accelerating in either frame. If a particle at rest in one frame appears to be accelerating in the other, then Newton's First Law is evidently being violated in at least one of those reference frames; accelerating reference frames are therefore noninertial.

Some reference frames are accelerating and thus noninertial because they are changing their direction rather than their speed. For example, in a rotating reference frame, a particle at rest will be changing direction (i.e., accelerating) as seen from a nonrotating frame; rotating reference frames, then, are just another kind of noninertial reference frame.

The Earth, as a rotating body, is a noninertial reference frame. An object at rest on Earth's surface will experience a centrifugal force, thus apparently violating Newton's Laws. Some scientists call centrifugal force fictitious because it results from a *state* (of rotation) rather than from a tangible *interaction* (with some other mass): according to **Newton's Third Law**, for every action there is an equal and opposite reaction; but what mass is there for the object to react to, what tangible interaction is there for it to have, when centrifugal force acts on it?

In this age of general relativity, however, where forces (like gravity) can be represented equivalently as changes in the *state* of space-time (i.e., its curvature; see Figure 4.22), such a distinction may simply be obsolete.

Some might view Newton's Laws as universal, and prefer to invalidate a force that seems inconsistent with them. But the reality is that *rotation is the norm in our Universe*: all the (major) bodies in our Solar System, including the planets and their moons and the Sun, spin on their axes; the planets and moons orbit the Sun; our Solar System orbits about the common center of our stellar neighborhood, and that set of stars (along with all the others in the Milky Way) orbits the center of our galaxy; and our galaxy orbits about the common center of our galactic neighborhood, together with other nearby galaxies. In fact, *it is inertial reference systems that are fictitious*—they are a simplification, a baseline, which allows us to gradually ease into the more realistic but mathematically and physically complex situations of rotating bodies and rotating reference frames.

Thus, it is more reasonable—and more in line with our daily experience—to simply distinguish inertial and noninertial reference frames, and accept that additional forces are experienced in the latter as a consequence of that reference frame's acceleration.

In a rotating system, up to three additional forces are created. They can be derived mathematically based on a principle of 'invariance,' a requirement that the position of a mass must be the same whether it is described using the coordinates and position vector of an inertial coordinate system or the coordinates and position vector of a rotating one; that is, the 'arrow' describing the position vector must (if they share a coordinate origin) have the same length and direction, no matter which reference frame is used. The details of their derivation are beyond the scope of this textbook (though the result is symbolized below). What the derivation reveals, however, is that, if the rotating system is spinning uniformly, a mass at rest within that system experiences a *centrifugal force* flinging it outward; and if the mass moves, it will *additionally* experience a *Coriolis force* deflecting it to the side. If the system's spin changes, it will experience still another force called the *Poincaré force* (sometimes also called the *Euler force*). These three forces are also quite different in magnitude. On the Earth, a typical mid-latitude magnitude for the centrifugal *acceleration* (i.e., the centrifugal force

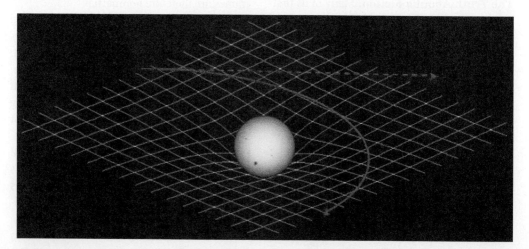

Figure 4.22 As interpreted by the general theory of relativity, the gravitational force exerted by a mass, here the Sun, acts to distort the surrounding 'metric' of space. Should gravity then be described as a "fictitious" force in this context, because a passing object is forced into an orbit by the curvature of space-time rather than by the mass itself? Adapted from http://pics9.this-pic.com/image/space%20time%20curvature.

per unit mass) is ~2 cm/sec²; the southward Coriolis acceleration experienced by a 10 m/sec westerly wind in the northern hemisphere is ~0.1 cm/sec²; and the Poincaré acceleration associated with a 1-sec seasonal variation in the length of Earth's day is ~10^{-8} cm/sec². All of them are much smaller than the gravitational acceleration experienced at Earth's surface; the Poincaré force is quite negligible unless the noninertial frame is rotating very unevenly.

The World Would Not Work Right Without Rotational Forces. Our willingness to accept that accelerating reference frames produce additional forces is not fickle; nor is it completely inconsistent with Newton's Laws. The three rotational forces are specific, and well defined (we will see the vector definitions for the centrifugal and Coriolis forces below); and they are purposeful. Their purpose stems from the fact (part of their derivation, actually) that *they combine so that, if the mass experiencing them is not otherwise accelerating (no other forces are acting on the mass), then their net result is to preserve the inertial path of that mass, as seen from an 'outside,' nonrotating (inertial) reference frame.*

This is most easily visualized in the kind of laboratory experiment, typically conducted in introductory physics classes, where a ball is sent across a rotating, frictionless turntable (e.g., Aharoni, 2007). From the point of view of the students, watching from a nonrotating reference frame (compared to the turntable, anyway), the ball follows a straight-line, inertial path, whereas—as viewed through a camera mounted on the turntable—the ball's trajectory as seen from the rotating frame is curved. The ball experiences rotational forces (as measured in the turntable's reference frame) that combine in just the right way so that its curved path is inertially straight.

We might say that what a mass in a rotating reference system experiences as centrifugal, Coriolis, and Poincaré forces is simply the mass' attempt to follow an inertial path. Things

are more complicated outside the lab, however: in the real world, other forces—gravity and friction, for example—generally foil that attempt and keep the mass as part of the rotating system.

Such complications can be appreciated by considering the situation of a driver and passenger in a car approaching an intersection (see Figure 4.23). The driver makes a sharp left turn, taking the turn at high speed, and both driver and passenger experience a strong centrifugal force trying to fling them to their right as they round the corner; it is a centrifugal force because, at least for the duration of the turn, they are in a rotating reference frame. And, were it not for frictional forces, seat belts, a locked door, and holding on like their lives depended on it, they might find themselves yielding to that center-fleeing force (plus the Coriolis force and perhaps even a Poincaré force), exiting the car onto the street and continuing on in a straight line—an inertial path, from the perspective of any pedestrian watching the event!

More generally, everyone and everything that is a part of the Earth System experiences the forces associated with the Earth's rotation; were it not for gravity holding us down to the Earth and pulling everything together, those 'fictitious' forces would eject us on a straight-line trajectory off the planet.

Our interpretation of rotational forces follows from the mathematical relation between the acceleration of a mass as measured with respect to the rotating reference frame and its acceleration as measured in an inertial reference frame. We denote these accelerations by $\vec{a}_{\text{ROT}}$ and $\vec{a}_{\text{IN}}$, and take a symbolic approach for simplicity. The derivation of our three rotational forces would have found that

$$\vec{a}_{\text{ROT}} = \vec{a}_{\text{IN}} + (\vec{a}_{\text{C}} + \vec{a}_{\text{COR}} + \vec{a}_{\text{POIN}}),$$

showing that masses in the rotating coordinate system experience three additional 'fictitious' accelerations (centrifugal, Coriolis, and Poincaré) not measured in an inertial reference system. Equivalently, that derivation

Figure 4.23 The white car speeds through the intersection in order to turn left before the traffic light turns red. As it makes the turn (bold dashed curve), it is, in effect, in a rotating coordinate system. Into the turn, the car will experience the centrifugal force indicated. If at that moment the passenger door flings open, the passenger—if not otherwise impeded—will fly out in response to that centrifugal force and end up following the inertial path shown. (For realism, we imagine that the door mechanism was strong enough to delay the door's opening prior to the moment shown.) pbk-pg/Shutterstock.com.

shows that the net acceleration as seen from an inertial point of view is

$$\vec{a}_{IN} = \vec{a}_{ROT} - (\vec{a}_C + \vec{a}_{COR} + \vec{a}_{POIN}).$$

It follows that, if no additional forces are acting on the mass, so that nothing impedes those rotational forces—not, for the case of the car passenger (for example), friction with the car seat, tension from a seat belt, or pressure from the car door (even as it flings open)—then those three rotational forces, experienced by everyone within the car, combine (with $\vec{a}_{ROT}$ precisely equal to $\vec{a}_C + \vec{a}_{COR} + \vec{a}_{POIN}$) to produce an inertial path for the passenger ($\vec{a}_{IN} = 0$).

Technically, of course, the students watching the turntable experiment and the bystander watching the car turning are themselves not exactly in an inertial reference frame, because Earth is spinning on its axis, thus they are not seeing a truly inertial path. To infer the true inertial path, we would have to account for the centrifugal, Coriolis, and Poincaré accelerations associated with Earth's spin—and the accelerations associated with Earth's orbit around the Sun, and the motion of our Solar System relative to the stellar neighborhood, and so on. In some situations, neglecting all these additional forces may be reasonable; in the Earth System, though, there are many situations where some of these forces are important (as we saw in Chapter 3 and will see in later chapters).

In this textbook, we will continue to treat rotational forces as real—and a necessary amendment to Newtonian physics. Describing them mathematically, therefore, is unavoidable.

4.4.4 A Thorough Description of Centrifugal Force

Writing out the expression for centrifugal force requires that we introduce two new concepts:

angular velocity and vector cross products. And, in order to be able to express the centrifugal force as generally as possible, we need to employ a *vector* definition of angular velocity.

Traditionally, angular velocity is represented by the Greek letter omega (Ω or ω). Consider now a point of mass M orbiting about an axis; we define its **angular velocity** $\vec{\Omega}$ to be a vector whose magnitude equals the angle swept out per time by the mass as it orbits (see Figure 4.24, left side). For example, as the Earth continues its daily spin, any particle on the surface (or in its interior) sweeps out an angle of 360° per day around its spin axis, equivalent to 2π radians per day.

By convention, the **direction of the angular velocity vector** is defined to be the right-hand thumb direction according to a *right-hand rule*: if the fingers of your right hand curve around with the particle in the direction of its orbit, then your thumb will point in the direction of the angular velocity vector (see Figure 4.24, right side). Or, looking down from above, if the particle orbit is counterclockwise, then the direction of $\vec{\Omega}$ is upward toward you, parallel to the orbital axis. With this convention, the direction of the angular velocity vector is in contrast—at right angles!—to the direction of the particle's

motion, i.e., the direction of the particle's velocity (which is sometimes called its "linear velocity"); the convention does, at least, allow the orbital axis to be easily identified. And, as we are about to see, though the direction of $\vec{\Omega}$ may seem counterintuitive, it predicts various quantities involving angular velocity to have intuitively reasonable magnitudes and directions.

Using the invariance of the position vector, as described earlier, it can be shown that the **centrifugal force** $\vec{F}_C$ experienced by a particle of mass M orbiting an axis with angular velocity $\vec{\Omega}$ must equal

$$\vec{F}_C = -M\,\vec{\Omega} \times (\vec{\Omega} \times \vec{R})$$

at the instant when the position of the particle is $\vec{R}$. In this formula, which we will take to be the definition of centrifugal force, the symbol "×" represents an operation between vectors known as the "vector cross product." Vector cross products are defined and explained in *Textbox A*. One property they have is that *the end result of a vector cross product is another vector*. As with any vector equation, the above definition of centrifugal force is actually a concise statement of two equations, one for the direction of $\vec{F}_C$ and the other for its magnitude.

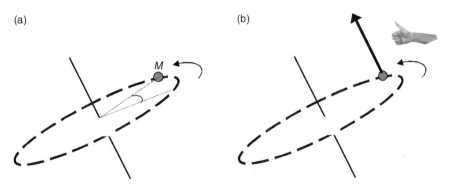

Figure 4.24 The angular velocity of a mass orbiting an axis is a vector, defined as follows. (a) The magnitude of this vector equals the angle swept out around the axis per time. (b) Using the right-hand rule described in the text, the direction of the vector in the situation shown here—the 'thumb' direction—is upward, parallel to the axis around which the mass orbits. In this example, the direction of the angular velocity vector remains the same at all points in the particle's orbit.

Textbox A: Vector Cross Products

Given vectors $\vec{A}$ and $\vec{B}$

their vector cross product is obtained by first placing the tails of these vectors together. The direction of the cross product is found by applying one of two right-hand rules: either (1) align the thumb and forefinger of your right hand with the first and second vectors, respectively, in which case your right-hand middle finger points in the direction of interest; or (2) align the edge of your right hand with the first vector such that you can close your hand until the edge aligns with the second vector, in which case your right-hand thumb points in the direction of interest.

It is always true that vector cross products are perpendicular to both of the original vectors. In this example, with both vectors $\vec{A}$ and $\vec{B}$ in the plane of this page, the cross product will be perpendicular to the page, pointing upward toward you.

The magnitude of the cross product is defined to be the product of the first vector's magnitude and the second vector's magnitude multiplied by the sine of the angle between the two vectors:

$$\left\| \vec{A} \times \vec{B} \right\| \equiv \left\| \vec{A} \right\| \left\| \vec{B} \right\| \sin\theta = AB \sin\theta,$$

if θ is the angle between $\vec{A}$ and $\vec{B}$.

One of the reasons the formula for $\vec{F}_C$ may seem so intimidating is because it is completely general: it is valid in any coordinate system, whether the particle is orbiting in the plane of the coordinate origin or not. Although the derivation is beyond the scope of this textbook, we can nevertheless convince ourselves that the formula works as we would expect. Applying the definition of vector cross product first to $(\vec{\Omega} \times \vec{R})$, we find—as illustrated in Figure 4.25—that $\vec{\Omega} \times \vec{R}$ is always *tangent* to the orbital path of M, pointing in the direction of its motion. By definition the magnitude of

this cross product, $\left\| \vec{\Omega} \times \vec{R} \right\|$, is

$$\left\| \vec{\Omega} \times \vec{R} \right\| = \left\| \vec{\Omega} \right\| \left\| \vec{R} \right\| \sin(\sphericalangle \, between)$$
$$= \Omega R \sin\theta.$$

Next, we apply the definition of vector cross product to $\vec{\Omega} \times (\vec{\Omega} \times \vec{R})$, that is, to the product of the vectors $\vec{\Omega}$ and $\vec{\Omega} \times \vec{R}$. From the right-hand rule, we find that the vector $\vec{\Omega} \times (\vec{\Omega} \times \vec{R})$ always points inward, directly toward the orbital axis, no matter where in its orbit the particle is (see Figure 4.26). And, the magnitude of

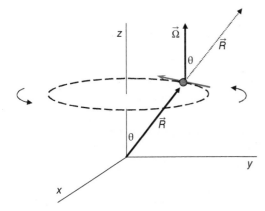

Figure 4.25 By displacing the position vector $\vec{R}$ so that its direction and magnitude are preserved (the length and direction of the arrow, and thus the vector itself, are unchanged), we can see that the angle between it and the angular velocity vector $\vec{\Omega}$ is θ. Applying the right-hand rule, beginning with the edge of the right hand aligned with $\vec{\Omega}$, then curling the hand in until the edge aligns with $\vec{R}$, we find that $\vec{\Omega} \times \vec{R}$ points tangent to the orbit, in the direction of the particle's motion.

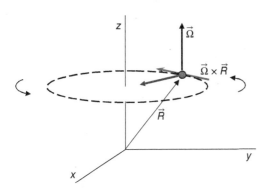

Figure 4.26 According to the right-hand rule, if $\vec{\Omega}$ is directed upward and $(\vec{\Omega} \times \vec{R})$ is tangent to the orbit, then $\vec{\Omega} \times (\vec{\Omega} \times \vec{R})$ points inward, toward the orbital axis.

$\vec{\Omega} \times (\vec{\Omega} \times \vec{R})$, $\left\| \vec{\Omega} \times (\vec{\Omega} \times \vec{R}) \right\|$, is by definition equal to

$$\left\| \vec{\Omega} \times (\vec{\Omega} \times \vec{R}) \right\|$$
$$= \left\| \vec{\Omega} \right\| \left\| \vec{\Omega} \times \vec{R} \right\| \sin(\sphericalangle \text{ between})$$
$$= \Omega \left\| \vec{\Omega} \times \vec{R} \right\| \sin(\sphericalangle \text{ between});$$

but because $\vec{\Omega}$ is always perpendicular to $(\vec{\Omega} \times \vec{R})$, the angle between them must be 90°,

so the magnitude of $\vec{\Omega} \times (\vec{\Omega} \times \vec{R})$ equals

$$\left\| \vec{\Omega} \times (\vec{\Omega} \times \vec{R}) \right\| = \Omega \left\| \vec{\Omega} \times \vec{R} \right\| \sin(90°)$$
$$= \Omega \left\| \vec{\Omega} \times \vec{R} \right\|$$
$$= \Omega^2 R \sin \theta,$$

using our previous result for $\left\| \vec{\Omega} \times \vec{R} \right\|$.

Our analysis implies that the direction of the centrifugal force, which is proportional to $-\vec{\Omega} \times (\vec{\Omega} \times \vec{R})$, is always directly *outward*, away from the orbital axis, no matter where the particle is located relative to the coordinate system or origin (Figure 4.27). This makes sense, and confirms our experience on a carousel: we are not flung upward, or downward, but straight outward away from the axis of the carousel.

The magnitude of the centrifugal force is also consistent with our experience on carousels. From our analysis, it follows that

$$\left\| \vec{F}_C \right\| = M\Omega^2 R \sin \theta,$$

or, in terms of the perpendicular distance $\mathcal{R}$ of the particle from the orbital axis (see Figure 4.27),

$$\left\| \vec{F}_C \right\| = M\Omega^2 \mathcal{R}.$$

Thus, the farther you stand on the carousel from its axis, the stronger the centrifugal force

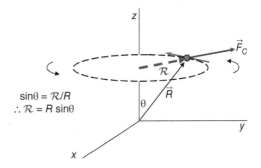

$\sin\theta = \mathcal{R}/R$
∴ $\mathcal{R} = R \sin\theta$

Figure 4.27 As derived in the text, the direction of $\vec{\Omega} \times (\vec{\Omega} \times \vec{R})$ is straight inward, pointing from the particle toward the orbital axis; thus, $-\vec{\Omega} \times (\vec{\Omega} \times \vec{R})$ and $\vec{F}_C$ point as shown, straight outward. The heavy dashed blue line beginning at the center of the orbit (not the coordinate origin) and pointing to the particle—in essence the 'projection' of the particle's position vector up onto the plane of its orbit—is perpendicular to the orbital axis of the particle, thus sinθ = length of dashed line / R.

you experience. And, the faster the carousel rotates (larger Ω), the greater the centrifugal force as well. It is interesting to note that the centrifugal force depends more on the rotational speed of the carousel than on your distance from the axis: double your distance, and the carousel force doubles; but double the rotation rate, and you are flung off with four times the force. To put it more familiarly: as you stand on the carousel, stepping a bit farther out is easy to handle, but sudden jerks of the carousel can cause you to lose your balance.

Also, according to our derivation, the expression for centrifugal force is independent of our choice for the coordinate origin—despite the explicit inclusion of the particle's position vector $\vec{R}$: thanks to the way vector cross products work, the force's magnitude is determined by the particle's distance from the orbital axis, not its distance from the origin. And, that dependence on distance explains why rotating objects bulge around their equator, and why the primordial nebula acquired a disk shape as it collapsed; in both cases, particles farthest from the axis experienced the strongest centrifugal force, and opposed gravity the most.

A few last comments from our analysis before returning to the Earth's rotation. We have focused on the centrifugal force on a particle as it orbits; so, dividing out the particle's mass yields a formula for the force per unit mass, that is, the *centrifugal acceleration*, experienced by the particle:

$$\vec{a}_C = -\vec{\Omega} \times (\vec{\Omega} \times \vec{R}).$$

The magnitude of this acceleration is

$$\|\vec{a}_C\| = \Omega^2 \mathcal{R}.$$

Earlier, we found that one part of our formula for centrifugal force, the vector cross product $(\vec{\Omega} \times \vec{R})$, had a direction tangent to the orbit and a magnitude $\Omega R \sin\theta$. We now see that we can write this magnitude as $\Omega \mathcal{R}$. When discussing how the radius of Earth could be measured, we found that the arc distance S along Earth's surface equaled that radius multiplied by the 'central angle' swept out by the arc. With some thought it should be clear that $\Omega \mathcal{R}$ equals the central angle swept out *per unit time* (e.g., per sec) times the radius of the orbit, thus it equals the arc distance along the orbit *per unit time*. But the distance covered by the particle per unit time, as it orbits, is its (linear) velocity. That is, $(\vec{\Omega} \times \vec{R})$ equals the particle's velocity in both direction and magnitude.

A homework exercise, in which the orbiting particle is replaced by a car driving around a racetrack, should help clarify these connections.

Finally, the same 'invariance' considerations that led to our vector formulation of centrifugal force also produce an expression for the Coriolis force acting on a mass as it moves over the Earth's surface. We will not pursue the Coriolis force quantitatively until Chapter 10 (with a nod to it in Chapter 6), but its vectorial aspects (vector cross products, again!) make it appropriate to introduce here. See *Textbox B* for the details.

Textbox B: The Coriolis Force

The Coriolis force acting on a particle of mass M, as it moves with linear velocity $\vec{v}$ with respect to the Earth, is defined to be

$$\vec{F}_{COR} = -2M\vec{\Omega} \times \vec{v},$$

where $\vec{\Omega}$ is the Earth's rotational angular velocity.

Just to clarify: linear velocity, with units like mph or m/sec, is just the 'regular' velocity we are used to dealing with every day. In general, M can be moving simply over the Earth's surface, or arbitrarily in three dimensions, so that $\vec{v}$ has a vertical as well as horizontal components; in all cases, though, v is still the distance along the path covered by M per time.

Textbox B: The Coriolis Force (*Continued*)

Take the *z*-axis to be aligned with Earth's rotation axis, and the coordinate system's origin to be at the Earth's center; that places the *x*- and *y*-axes within the plane of Earth's equator (dashed circle as shown below). One immediate consequence of the definition of $\vec{F}_{COR}$ is that, at the equator, any north-south motion of M (left side, red arrow) produces a zero Coriolis force: at the equator, such motion is parallel to the *z*-axis, making $\vec{v}$ and $\vec{\Omega}$ parallel; since the angle between those two vectors (when their tails are together) is zero, their cross product is also zero.

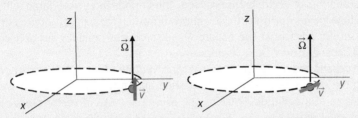

Furthermore, again applying the definition of vector cross product to $-2M\vec{\Omega}\times\vec{v}$, any east-west motion (right side, blue arrow) of the particle at the equator yields a vector pointing upward, that is, straight out away from the Earth; $\vec{F}_{COR}$ in this case produces no sideways deflection of the motion. In short, that portion of the Coriolis force important in climatology is effectively zero at the equator.

Away from the equator, a north-south motion of the particle is no longer strictly parallel to the rotation axis or $\vec{\Omega}$; and an east-west motion crossed with $\vec{\Omega}$ is a vector no longer purely upward. In general, the Coriolis force is fully three-dimensional, and does more to a moving particle than force it sideways; deflecting it around and moving it vertically at the same time, the net effect may be a spiral quality or 'helicity.' But for 'thin' fluids like the oceans and atmosphere, sideways deflection is the Coriolis force's main claim to fame.

With a bit of spherical geometry and repeated use of the cross product definition, we can obtain the surficial components of the Coriolis force.

Let's say that, at some instant while in motion, the particle is at position $\vec{R}$, which makes an angle θ with respect to the rotation axis. At this location, we 'break down' the vector $\vec{\Omega}$ into two other vectors: one, radially outward (in the same direction as $\vec{R}$), and the other, northward, tangential to Earth's surface.

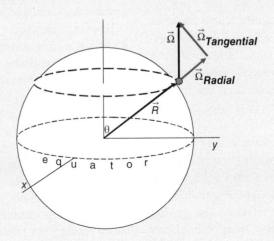

Textbox B: The Coriolis Force (*Concluded*)

Since

$$\vec{\Omega} = \vec{\Omega}_{radial} + \vec{\Omega}_{tangential},$$

it follows that the Coriolis force can be written

$$\vec{F}_{COR} = -2M\vec{\Omega}_{radial} \times \vec{v} - 2M\vec{\Omega}_{tangential} \times \vec{v}.$$

Using the properties of vector cross products, the Coriolis force associated with $\vec{\Omega}_{tangential}$ is zero if the particle moves north-south, and purely upward if the particle moves east-west—just like in the case of a particle at the equator—so it produces no sideways deflection. For the oceans and atmosphere, only $\vec{\Omega}_{radial}$ is important.

Thus, in the oceans and atmosphere, an eastward motion of velocity v in the northern hemisphere (along the path shown by the bold dashed line in the picture above) is deflected by a southward force of magnitude $2Mv\|\vec{\Omega}_{radial}\|$, using the rules of vector cross products; and a northward motion of velocity v is deflected eastward with a force of $2Mv\|\vec{\Omega}_{radial}\|$. Both of these amount to a deflection *to the right* of the particle's motion.

Can you verify these deflections? As a test of your understanding, repeat this analysis for motion in the southern hemisphere, and verify the deflections are opposite!

Finally, since $\vec{\Omega}_{radial}$ has a magnitude $\Omega\cos\theta$, we can define

$$f \equiv 2\Omega\cos\theta$$

and write the force magnitudes as fMv. f, which equals 0 at the equator, as it should, is often called the **Coriolis parameter**.

4.4.5 Gravity *Versus* Centrifugal Force on a Rotating Earth

The centrifugal force associated with Earth's daily rotation causes the Earth to bulge out at its equator, flattening it at the poles. As we will see, that ellipsoidal shape produces a comparatively large effect on gravity measurements around the globe. But even if Earth managed to remain perfectly spherical as it rotated, centrifugal force from its rotation would affect the gravitational force measured almost everywhere.

Conceptually, the reasons for this are straightforward. Consider measurements of gravity at a site located on Earth's surface at an arbitrary latitude (see Figure 4.28). If Earth is spherically symmetric, the gravitational force experienced there will always point downward, right toward the center of the Earth. But Earth's rotation exerts a centrifugal force at the site which is directed outward, away from the rotation axis. The two forces oppose each other (though centrifugal force is, of course, much weaker), so the net gravity experienced there is reduced.

However, although those forces oppose each other, their directions are in general not perfectly opposite. The effectiveness of centrifugal force in counteracting gravity thus *varies with latitude*, for two reasons: away from the equator, the magnitude of centrifugal force diminishes; and away from the equator, a smaller portion of centrifugal force directly opposes gravity. At the equator, the reduction is maximal; but closer to the poles, with centrifugal force at more nearly right angles to gravity, and the distance to the rotation axis much smaller, the reduction approaches zero (see Figure 4.29).

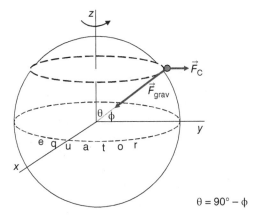

Figure 4.28 As a particle sits at rest on Earth's surface, the Earth's rotation carries it around Earth's rotation axis (here aligned with the z-axis). Earth's daily rotation in effect causes it to 'orbit' the axis. At the location of the particle (which could be described by latitude ϕ or colatitude θ), centrifugal force acts outward, perpendicular to the rotation axis; thus, it opposes gravity but not directly.

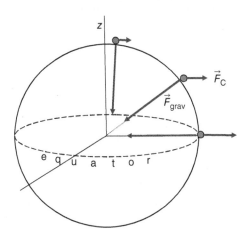

Figure 4.29 At different latitudes, gravity points at different angles (always toward Earth's center), but centrifugal force always acts outward, perpendicular to the rotation axis; thus, gravity is reduced most effectively at the equator and least effectively at the poles. The magnitude of centrifugal force opposing gravity is also greatest at the equator.

The Situation at the Equator. The maximum effect, at the equator, can be calculated easily. Since the two forces oppose each other directly there, centrifugal force simply subtracts from the gravitational force. We pose the calculation in terms of force per unit mass, i.e., acceleration. The centrifugal acceleration at the equator has magnitude $\Omega^2 \mathcal{R}$, and at the equator the distance $\mathcal{R}$ to the axis is simply the equatorial radius, a, 6.378×10^8 cm.

The value of the Earth's angular velocity, Ω, follows from the knowledge that Earth spins once on its axis every 23 hours 56 minutes and 4.10 sec, that is, every 86164.10 sec. We know this because we can observe any star that crosses the meridian overhead to cross again 86164.10 seconds later; for this reason, the length of Earth's day is called a **sidereal day**. Earth's angular velocity has a magnitude of 360° per sidereal day, or 2π radians per sidereal day, thus $2\pi/86164$ rad/sec; in short,

$$\Omega = 7.29 \times 10^{-5} \text{ rad/sec}.$$

Substituting this value into our formula for centrifugal acceleration, and recalling that radian units can be ignored, we find $\Omega^2 \mathcal{R} = 3.39$ cm/sec^2. This is the maximum possible centrifugal acceleration experienced on Earth due to Earth's daily rotation.

Incidentally, the sidereal day is not quite the same as the day that regulates our 'daily' lives. Human time is based on the **solar day**, the time it takes the Sun to appear overhead today after being overhead yesterday. The solar day is a bit longer than the sidereal day because the Earth moves incrementally in its orbit around the Sun each day; the Earth has to spin around a bit more for a point on its surface that faced the Sun yesterday to face it again today (see Figure 4.30). The amount of that additional spin differs depending on the time of year, because Earth's orbit is elliptical rather than circular, but on average the length of a solar day—the average being called a **mean solar day**—is precisely 24 hours.

The acceleration of gravity that we actually measure at the equator is about 978.03 cm/sec^2. This value of gravity represents the *net* effect of the gravitational force due to the Earth's mass, pulling a particle at the surface downward toward the center of the Earth, *and* the centrifugal force flinging it upward out into space

✷

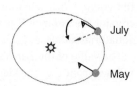

Figure 4.30 The difference between a sidereal day and a solar day, shown hypothetically over a 2-month span from May 16 to July 15 (i.e., one sixth of a year), looking down from above as the Earth orbits the Sun.

For simplicity, imagine that Earth's rotation axis is perpendicular to this page; that is, we will neglect the Earth's obliquity. The Earth must point to the same (distant) stars (upper left) at the end of each sidereal day, so a flag planted on the Earth—like that shown here—will point in the same direction in space after any number of sidereal days.

By pure coincidence, at the beginning of this time span, that flag also happened to point directly toward the Sun; thus, this time span began at local *noon*, with observers standing next to the flag seeing the Sun overhead.

The observers return to the flag location after two months, and the Earth has advanced in its orbit. But at the time the flag points directly to those distant stars (precisely 61 sidereal days after the experiment began), the Sun is nowhere near being overhead; in fact, it is only ~8 a.m. local time! The Earth must spin through an extra ~4 hours (the accumulated difference in time between sidereal and solar days, on average, over 2 months) for that to happen (dashed straight arrow).

Note that after a whole year, the accumulated difference amounts to a whole day: the year contains 366¼ sidereal days but only 365¼ solar days.

(Figure 4.29). At Earth's equator, then, since centrifugal force had fully opposed gravity, it must be that the gravity due to Earth's mass alone would be ~981 cm/sec² (that is, 978.03 + 3.39 cm/sec²).

The part of the total contributed by centrifugal acceleration even at the equator is reassuringly small—about a third of 1%. We can conclude that in most physically reasonable situations (i.e., except when rotation rates are extreme or planets are very light but very

large), gravity can be expected to dominate over centrifugal force.

The Situation at Other Latitudes. The effects of centrifugal force at sites away from the equator can be quantified somewhat. This situation is partially illustrated in Figure 4.28, but we need to express the site location in more familiar terms. Positions on Earth's surface are more typically expressed in terms of latitude ϕ, defined as the angle between the radius to the site and the equator, rather than colatitude (which we have already denoted θ), defined as the angle between the radius and the rotation axis; they are simply related according to $\theta = 90° - \phi$ (see Figure 4.31). So, if Earth's mean radius is $\bar{r}$ (i.e., approximating Earth, for now, as spherically symmetric with radius $\bar{r}$), the distance $\mathcal{R}$ of the site from Earth's rotation axis is

$$\mathcal{R} = \bar{r} \sin \theta = \bar{r} \cos \phi,$$

and the magnitude of the centrifugal acceleration experienced by the particle can be written

$$a_C = \Omega^2 \mathcal{R} = \Omega^2 \bar{r} \cos \phi.$$

The net gravity experienced at this site is the result of the gravitational attraction from Earth's mass combined with the effects of centrifugal acceleration at this latitude. Sometimes the net gravity, which is what we observe

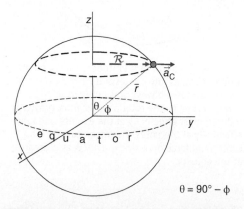

$$\theta = 90° - \phi$$

Figure 4.31 If we are located on Earth's surface at latitude ϕ or colatitude θ, and Earth is a sphere of radius $\bar{r}$, then our distance $\mathcal{R}$ from the rotation axis equals $\bar{r} \cos \phi$ or $\bar{r} \sin \theta$.

(after all, we can't stop the Earth from rotating!), is called the **observed gravity**, $\vec{g}_{obs}$, to distinguish it from the 'Newtonian' gravity $\vec{g}$ associated with Earth's mass. Technically, because the direction of $\vec{a}_C$ is not exactly opposite the direction of $\vec{g}$ (which is radially inward toward the Earth's center if Earth is spherically symmetric), the net effect should be calculated as a vector sum:

$$\vec{g}_{obs} = \vec{g} + \vec{a}_C.$$

One consequence of this vector sum is that the observed gravity will be in a slightly different direction than $\vec{g}$ due to centrifugal force: what we experience as 'downward' (the result of adding the vectors $\vec{g}$ and $\vec{a}_C$ in Figure 4.32) will be tilted slightly away from Earth's center. This deflection is small, though, because centrifugal force is so weak relative to gravity.

For the magnitude of the observed gravity, a simple approach might be to consider the net effect of gravity and only the *vertical* component of centrifugal acceleration, thus avoiding the need for a vector analysis. Such an approach would be reasonably accurate because the deflection of observed gravity away from vertical is so small. As explained in Figure 4.32, the vertical component of centrifugal acceleration has magnitude $a_C \cos\phi$. The net effect in this case is

$$g_{obs}(\phi) \cong g(\phi) - a_C \cos\phi,$$

thus

$$g_{obs}(\phi) \cong g(\phi) - \Omega^2 \mathcal{R} \cos\phi$$
$$= g(\phi) - \Omega^2 \bar{r} \cos^2\phi,$$

at latitude ϕ. These formulas make the dependence of the centrifugal effect on latitude explicit.

However, even within the accuracy of our approximation, this formula is *significantly incomplete*: centrifugal force also changes the shape of the Earth, and that further modifies its gravity; $g(\phi)$ is not simply the gravity of a spherically symmetric ball.

4.4.6 Indirect Effects of Centrifugal Force on Gravity, and the Idealized Earth

The spheroidal shape of the bulging Earth (caused by the centrifugal force associated with Earth's rotation) is responsible in turn for two effects on gravity. The *mass* of the equatorial bulge increases the gravitational force exerted on particles at near-equatorial latitudes; correspondingly, the lack of mass near the flattened poles decreases high-latitude gravity. On the other hand, particles near Earth's equator are a greater *distance* from the bulk of Earth's mass, leading to a weaker gravitational force; and particles near the poles are closer to most of Earth's mass, causing an increase in gravity.

Which effect do you think would be more important, mass or distance?

Based on Newton's Law of Gravitation, we expect that the *distance* effect should dominate, since the gravitational force is proportional to mass but also (inversely) to distance *squared*.

Altogether, then, there are three factors that cause gravity to vary from equator to pole:

- centrifugal force, which makes gravity at the equator < gravity at the poles

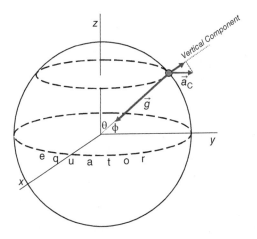

Figure 4.32 The angle between $\vec{a}_C$ and its vertical component is the latitude, ϕ, so the vertical component has magnitude $a_C \cos\phi$. If the accelerations were drawn to scale, g would be 300 times the size of $\vec{a}_C$.

- the mass of the bulge, which makes gravity at the equator > gravity at the poles
- the distance of the bulge, which makes gravity at the equator < gravity at the poles

The net effect is that *Earth's gravity is weakest at the equator and increases smoothly with increasing latitude*, becoming a maximum at the poles. In terms of gravitational acceleration, we measure

$$g \cong 978 \text{ cm/sec}^2 \text{ at the equator}$$
$$g \cong 983 \text{ cm/sec}^2 \text{ at the poles.}$$

This 5 cm/sec² difference is the largest variation in gravity anywhere around the globe. But it is only ~½%, and the portion associated with Earth's bulging shape is even less—a reflection both of the near-spherical symmetry of the Earth and the weakness of the gravitational force in nature.

It is possible to predict theoretically how gravity should vary with latitude as a result of those three factors—that is, predict theoretically what the gravity field of a spheroidal, rotating Earth should be. The parameters associated with those factors would include Earth's mass, M_E; its spin rate, Ω; its equatorial radius a and flattening f, to describe the 'distance' effect of its equatorial bulge; and some quantities related to Earth's moments of inertia, to describe the 'mass' effect of the bulge.

Recall from Chapter 1 that the moment of inertia of an object measures the mass distribution within the object relative to some axis; the mass of each particle constituting the object is 'weighted' (mathematically) by its squared distance from the axis, and the weighted masses are added to produce the overall moment of inertia. Chapter 1 focused on comparing the moments of inertia of spherically symmetric objects whose mass might be distributed uniformly *versus* those whose density might increase or decrease with depth. It should be noted that the moment of inertia of any of those spheres is unchanged by choosing a different axis through its center: *the moment of inertia of a spherically symmetric ball is the same about any of its diameters.*

The same cannot be said of ellipsoidally shaped objects such as an equatorially bulging Earth. By analogy with the notation used to describe Earth's shape (as in Figure 4.21), we will denote its moment of inertia through an equatorial diameter by A and through its polar axis by C (see Figure 4.33). The equatorial bulge of the Earth causes A and C to be unequal. Their difference, $C - A$, measures the difference in mass distribution caused by the bulge; accordingly, it reflects the mass effect of the bulge.

Finally, we define a parameter, denoted $\mathbf{J_2}$, to formally represent the mass effect of the bulge:

$$J_2 \equiv \frac{C - A}{M_E\, a^2}.$$

In this formula, the bulge term $C - A$ is 'scaled' by (divided by) a factor that represents the

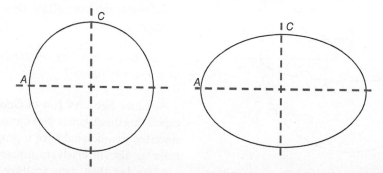

Figure 4.33 *A* and *C* denote the moments of inertia of both objects about axes through their equators and poles. For the sphere on the left, *A* and *C* will be equal; for the spheroid on the right, *A* and *C* will be unequal.
In that spheroid, which axis is the mass further from? Will that cause A to exceed C or C to exceed A?

order of magnitude of Earth's moment of inertia; if the Earth were homogeneous in density and only slightly ellipsoidal, for example, its moments of inertia would be approximately 0.4 times $M_E\,a^2$ (or times $M_E\,c^2$ or times $M_E\,\bar{r}^2$).

(Why the subscript 2 for J? Gravity theorists employ several "J" quantities related to Earth's mass and gravity field; for example, J_1 is used to describe an offset of Earth's center of mass. These J's are discussed in the context of spherical harmonics in Chapter 6.)

Satellite observations discussed in Chapter 6 yield

$$J_2 = 1.0826359 \times 10^{-3}$$

(Groten, 2004; IERS Conventions, 2010). This parameter is about a tenth of a percent, serving to remind us that the bulge is very small (and suggesting that it will produce only a small variation in gravity).

With these parameters, theoretical geophysicists have found that the shape as well as the surface gravity of the spheroidal, rotating Earth should vary with latitude in a simple manner. The radius of this Earth at latitude ϕ varies according to

$$r(\phi) = a(1 - f\sin^2\phi),$$

where

$$f = \frac{a - c}{a}$$

is the flattening of the ellipsoidal Earth; you can verify that this formula indeed produces an ellipsoid of revolution, as in Figure 4.21, with radius a at the equator (where $\phi = 0°$) and radius c at the poles (where $\phi = \pm90°$).

Furthermore, this Earth's net gravity at latitude ϕ is found to be

$$g_S(\phi) = g_{eq}\left(1 + \left[m + 2f - \tfrac{9}{2}J_2\right]\sin^2\phi\right)$$

—similar to the $a(1 - f\sin^2\phi)$ variation for $r(\phi)$, incidentally—where g_{eq} is the value of the Earth's gravitational acceleration at the equator, and $\boldsymbol{m}$ is defined to be the ratio of centrifugal to gravitational acceleration at the equator. For reasons explained below, we symbolize this gravity with a subscript S. From our

earlier estimates, $m \approx 3.39/978$; more precise observations yield

$$g_{eq} = 978.03278 \text{ cm/sec}^2$$

(Groten, 2004; IERS Conventions, 2010) and thus

$$m = 3.4677472 \times 10^{-3}.$$

The smallness of m reminds us that centrifugal force leads to only slight reductions in the strength of gravity. Our expression for gravity—known as **Clairaut's theorem** (1743) after the eighteenth-century French mathematical prodigy who first derived it—explicitly reveals the relative contributions of centrifugal force (m), the 'distance' effect of the bulge (f), and the 'mass' effect of the bulge (J_2) to the variation of gravity with latitude—all superimposed on the gravity due to the bulk of the Earth (the "1" in the equation).

The Earth whose gravity is given by this formula might be termed an "idealized" or **"Ideal Earth,"** in that it includes none of the complexities that make the real Earth's gravity so variable and difficult to predict—mass *excesses* such as buried ore bodies, descending lithospheric slabs, and bumps of core material protruding up into the mantle, and mass *deficits* like buried stream channels and ascending mantle plumes. The Ideal Earth's gravity variation derives from only the three factors considered here; and, as confirmed by Clairaut's theorem, two of them cause gravity to increase toward the poles (the $2f$ and m terms) while the third factor causes a decrease in gravity toward the poles (the $-\tfrac{9}{2}J_2$ term).

A few additional comments about Clairaut's theorem are in order. The gravitational acceleration due to any body, such as the Earth, depends, of course, on our distance from the body (or from the body's center of mass). Clairaut's theorem predicts the gravitational acceleration that would be experienced specifically *at sea level* of the Ideal Earth (other reference levels are possible, including some—see Molodensky, 1945; also Brovar & Yurkina, 2000—particularly useful for advanced applications, but those are beyond the scope of this textbook). Our ideal sea level is at a distance from the center of the Ideal Earth given by the above formula for $r(\phi)$; since, globally, $r(\phi)$ describes a slightly ellipsoidal shape, sea level of the Ideal Earth is known as the **Spheroid** (also, the **Reference Ellipsoid**). The value of gravity described by Clairaut's theorem is sometimes called **gravity on the Spheroid**, meaning gravity at sea level of the Ideal Earth. The Spheroid will be discussed more rigorously in Chapters 5 and 6.

Clairaut's theorem, derived theoretically by considering the work done against gravity when lifting a mass up above the Spheroid, has a higher-precision version:

$$g_S(\phi) = g_{eq} \left\{ 1 + \left[m + 2f - \tfrac{9}{2} J_2 - \tfrac{17}{14} mf \right] \right.$$
$$\left. \times \sin^2\phi - \left[\tfrac{5}{8} mf - \tfrac{1}{8} f^2 \right] \sin^2 (2\phi) \right\}$$

(Lambert, 1945). The additional terms in this version involve multiplication of small quantities (m and f, each much less than 1); the resulting products, mf and f^2, are each very small (*much much* less than 1) but, given the high precision of modern gravity measurements, are worth including.

Gravity on the Spheroid, $g_S(\phi)$, is an approximation to observed gravity ($g_{obs}(\phi)$) that includes the factors responsible for the biggest variations in gravity around the world. Measurements taken around the world will also include the effects of all the mass anomalies that distinguish the real world from our idealization. The values of m, f, J_2, and g_{eq} listed above yield a numerical form of Clairaut's theorem that *most closely* matches the global observations of gravity, in effect averaging out all those anomalies. From recent determinations of those parameter values, specified above, the resulting gravity formula is given by

$$g_S(\phi) = 978.03278\{1 + 5.2874069$$
$$\times 10^{-3} \sin^2 \phi - 0.0058615$$
$$\times 10^{-3} \sin^2(2\phi)\} \text{ cm/sec}^2$$

or, equivalently,

$$g_S(\phi) = 978.03278 + 5.171257 \sin^2 \phi$$
$$- 0.0057328 \sin^2 (2\phi) \text{ cm/sec}^2.$$

Note that this formula, which includes the effects of centrifugal force and the equatorial bulge—and thus describes the smooth variation of gravity with latitude that we had inferred earlier—correctly predicts a value of ~978 cm/sec^2 for gravity at the equator and ~983 cm/sec^2 for gravity at the poles. This formula is usually called **The international gravity formula**.

Alphabetic Closure. *Did you wonder why, when we described the spheroidal shape of the Ideal Earth, or its moments of inertia, we skipped the letter* b *(or* B*)?* The international gravity formula—which lacks any dependence on longitude—reminds us that the Ideal Earth is axially (rotationally) symmetric; thus, for example, the Ideal Earth's equator is perfectly circular (of radius a). For greater accuracy, we could represent the equator as an ellipse—traditionally, with semi-major radius a and semi-minor radius b, corresponding to moments of inertia A and B for this slightly-less-than-Ideal Earth; overall, this nearly Ideal Earth, characterized by radii a, b, and c (or moments A, B, and C, with $A < B < C$, and $A \neq B$ leading to variations in gravity with longitude), is called a *triaxial Earth*.

Do you think centrifugal force alone can be responsible for Earth's ellipsoidally shaped equator (or triaxiality)? The gravitational effects of triaxiality are found to be comparable in magnitude to the differences between $g_{obs}(\phi)$ and $g_S(\phi)$; does that suggest an alternative explanation?

The international gravity formula has two primary uses. First of all, it provides us with a reasonably accurate prediction of the value of gravity at sea level anywhere around the world. The mass excesses and deficits that were averaged out from our idealized model cause much smaller gravity perturbations than the ~5 cm/sec^2 variation from equator to pole that the international formula accounts for. As an illustration, using formulas that will be discussed in Chapter 5, it turns out that the additional gravity due to a mile-high plateau is no more than $\sim 0.2\,\text{cm/sec}^2$ (though a process called *isostasy*, which we will investigate in that chapter, may reduce the effect even further, to near zero!).

Secondly, in many situations it is not the bulge- or rotation-related variations in gravity that are of interest, but rather the local or regional mass excesses and deficits themselves; such mass anomalies may be of great geologic, tectonic, or geophysical significance. The international gravity formula accounts for the global-scale variations; by subtracting $g_S(\phi)$ from observations of gravity in a region, gravity anomalies corresponding to these smaller features may be revealed. Gravity observations in a region, being relatively easy and inexpensive to carry out, are often the first step in an exploration of the region (with potential economic or environmental benefits as a bonus).

Subtraction of the spheroid to reveal mass anomalies can also be carried out in a somewhat different manner. Because they are fluid, the oceans must respond to lateral variations in the ambient gravity field—piling up in regions where the mass below draws them in with a greater gravity, ebbing in places where gravity is deficient and the seawater is attracted elsewhere. Additionally, the sea surface exhibits peaks and troughs associated with ocean currents. As we will see in the next two chapters, all of these undulations of the sea surface are relatively small in height (generally tens of meters or less) and thus swamped by the overall spheroidal variation in sea level (which has the same 21-km variation in radius from equator to pole that describes the overall shape of the Earth). If we can eliminate that large-scale spheroidal variation, the remaining sea-level irregularities, on regional and smaller scales, may shed light on the ocean currents or the mass anomalies below the ocean surface. In this context, it is the spheroid itself rather than the international formula's values of gravity that must be effectively removed. This kind of analysis will be explored in the next two chapters.

4.5 Kepler's Laws

We end this chapter with a brief discussion of Kepler's Laws. These rules for planetary orbits were based on the highly precise observations of the astronomer Tycho Brahe (a very interesting character in the annals of astronomy, known more commonly as 'Tycho,' and honored by having a lunar crater, Martian crater, asteroid, and supernova named for him). The first two laws were deduced in 1609, the third law in 1619:

(I) Each planet orbits the Sun in a plane, and the orbit describes an ellipse, with the Sun at one focus.

(II) The line from the Sun to a planet sweeps out equal areas in equal times.

(III) The square of the orbital period of a planet is proportional to the cube of its mean distance from the Sun.

The foci (plural of focus) of an ellipse are defined as being the two points whose combined distance from any point on the ellipse is always the same (see Figure 4.34). The foci are

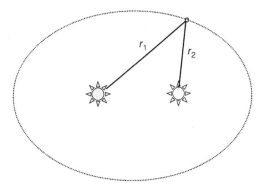

Figure 4.34 Ellipses are defined by the rule that the sum of the distances from every point on the ellipse to the two foci must be the same constant, for the whole ellipse. In the illustration, $r_1 + r_2 = constant$. The foci are shown here as suns. In fact, this rule provides a way to hand-draw an ellipse. Take a string and tack the two ends to your drawing board; place a pen against the string, stretch it taut, and trace out the path of the pen. The path will be elliptical, the tacks locate the foci, and the length of the string equals the value of the constant ($r_1 + r_2$ always equals the length of the string).

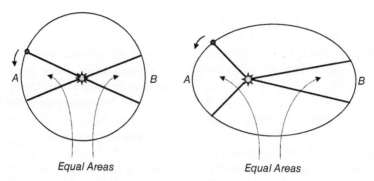

Equal Areas   *Equal Areas*

Figure 4.35 In both cases (circular and elliptical orbits), Kepler's Second Law requires that the area swept out by the planet during season "A" equals that swept out during season "B." Consequently, in the elliptical orbit, the planet must cover a longer distance along its orbit (i.e., must move faster) during season A than during season B.

to an ellipse what a center is to a circle, and as an ellipse becomes more circular, its foci collapse into the center.

In the special case where the orbit is circular, Kepler's Second Law tells us that the orbital speed of a planet remains constant as it circles around the Sun (see Figure 4.35). But in an elliptical orbit, the speed will vary; the Earth, for example, happens to be closer to the Sun during northern hemisphere winter, so Kepler's Second Law implies that the Earth must be traveling faster around the Sun during that season than during northern hemisphere summer.

In all three laws, the Sun plays a *central* role. In fact, *all three laws are a consequence of the nature of the gravitational force*—the gravitational force acting between the Sun and a planet. Orbits are elliptical, and planetary 'years' squared are proportional to their distances cubed, because the gravitational force depends inversely on the squared distance between Sun and planet. Equal areas are swept out as a planet orbits the Sun because the planet must conserve its orbital angular momentum; angular momentum is conserved because the Sun does not exert any torques on the planet—which in turn is true because gravity is a central force, acting straight between the Sun and the planet.

The fact that Kepler's Laws derive from properties of the gravitational force implies that those laws would hold true for the situation of *anything* orbiting *anything else*; for example, the orbits of satellites—natural or artificial—about a planet must also obey Kepler's Laws.

Not surprisingly, centrifugal force also figures in. For example, it makes sense that a moon following an elliptical orbit about a planet will orbit with faster angular velocity in the part of its orbit nearer to the planet (as required by Kepler's Second Law): when closer to the planet, the satellite experiences a stronger gravitational force pulling it in to the planet; a stronger orbital centrifugal force (flinging the satellite out) is needed to balance that gravity, keeping the satellite in orbit. And a faster angular velocity produces that increased center-fleeing force.

Our formula for centrifugal force allows us to 'prove' Kepler's Third Law in the simple case of a planet in a circular orbit. Let M_P denote the mass of the planet and M_S the mass of the Sun; take the radius of the orbit to be R_{PS} and the orbital angular velocity to be Ω_{PS} (see Figure 4.36). Note, incidentally, that in a circular orbit Ω_{PS} is constant throughout the orbit, and since the planet will sweep through 360° (2π radians) in the course of its 'year'—whose length we denote by T—it follows that

$$\Omega_{PS} = \frac{2\pi}{T}$$

(in units of rad/time).

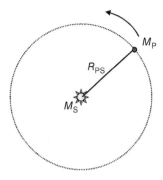

Figure 4.36 The planet, of mass M_P, is in a circular orbit at a distance R_{PS} from the Sun, whose mass is M_S. The angular velocity of the planet in its orbit around the Sun is Ω_{PS}.

Treating the planet and Sun as points of mass, Newton's Law of Gravitation implies that the magnitude of the gravitational force exerted on the planet by the Sun should equal

$$\frac{GM_P M_S}{R_{PS}^2}.$$

If the planet's mass is much less than the Sun's (which is an excellent approximation for all the planets in our Solar System except the two huge gas giants), then the planet's orbital axis passes nearly through the center of the Sun; in this case, the planet's distance R from the orbital axis is simply R_{PS}, and the centrifugal force experienced by the orbiting planet will be $M_P \Omega_{PS}^2 R$ or

$$M_P \Omega_{PS}^2 R_{PS}.$$

Finally, if the planet is in a stable orbit—that is, if it is neither falling into the Sun nor fleeing it—it must be true that the two forces acting on the planet, gravity and centrifugal force, precisely balance. Thus,

$$\frac{GM_P M_S}{R_{PS}^2} = M_P \Omega_{PS}^2 R_{PS}.$$

Dividing out the planet's mass and grouping terms, we find

$$GM_S = \Omega_{PS}^2 R_{PS}^3.$$

Since $\Omega_{PS} = 2\pi/T$, we can substitute in and solve for T^2, yielding

$$T^2 = \left\{ \frac{(2\pi)^2}{GM_S} \right\} R_{PS}^3.$$

In short, T squared is indeed proportional to R_{PS} cubed; the proportionality constant equals $(2\pi)^2/(GM_S)$, where M_S is the mass of the 'central' body, the Sun.

As some examples of the use of this relationship, consider first the orbit of the Earth around the Sun. With $T = 366\,\frac{1}{4}$ sidereal days $= 3.156 \times 10^7$ sec ($=$ the length of our year), $R_{PS} = 149.6 \times 10^6$ km, and $G = 6.67384 \times 10^{-11}$ m^3/sec^{2}kg, we find the Sun's mass to be 1.99×10^{30} kg. As a second example, consider a satellite orbiting Earth in a circular orbit of altitude 1,000 km (making its orbital radius 7,371 km); Kepler's Third Law implies that the satellite will take $\sim$6,300 sec $\sim$105 minutes $\sim$1¾ hours to complete each orbit.

Which mass did we use to deduce the satellite's orbital period—the Earth's, the Sun's, or the satellite's?

The details of our derivation must be modified if the planet's mass is great enough that both planet and Sun are orbiting about a distinct common center of mass, or if the orbit is not circular; but the applications and uses of Kepler's Third Law are unchanged.

5

Gravity and Isostasy in the Earth System

5.0 Motivation

In Chapter 4 it was suggested that observations of gravity can, in effect, provide a window into the Earth's interior: by comparing those observations with reference values of g, such as those prescribed by the International Gravity Formula, the locations and sizes of mass anomalies could be revealed. The detection and characterization of very *shallow* anomalies, which are often of interest in Earth System investigations, fall under the purview of *exploration* or *environmental geophysics*. In the first portion of this chapter, the preliminary data reduction techniques that reveal gravity anomalies, and make gravity measurements so useful for environmental geophysics, will be presented.

Some of the mass anomalies in mountainous regions, originating as deep as the crust-mantle boundary, turn out to have major, wide-ranging implications for Earth science and Earth System science: for the way mountains are built; for the ability of the 'solid' mantle to flow under stress; for the evolution of landforms on timescales of centuries to millennia; for the interactions between ice, oceans, and solid earth, e.g., during and after Pleistocene glaciation; and even for our ability to infer recent changes—associated with global warming—in the mass of ice sheets. All of these are connected by the subject known as *isostasy*, which constitutes the second portion of this chapter.

Earth is within reach of the gravity fields of the Moon, Sun, and other Solar System bodies, and our gravity field similarly extends beyond Earth's surface. A number of these interactions are of major importance to the Earth System; however, our discussion of them will be deferred to the next chapter.

5.1 Exploring the Earth System with Gravity

5.1.1 Scaling Down for Gravity Exploration

As we saw in the previous chapter, global variations in the gravity we observe over Earth's surface tend to be very small. Variations on regional and local scales are typically even smaller, as are variations in gravity over time. In recognition of Galileo's pioneering studies of gravity, the cgs unit for gravitational acceleration is called the **Gal**:

$$1 \text{ Gal} \equiv 1 \text{ cm/sec}^2;$$

thus, gravity at the equator is about 978 Gals, and the difference between equatorial and polar gravity is about 5 Gals. The traditional unit of gravity used in exploration is a milliGal or mGal:

$$1 \text{ mGal} \equiv 0.001 \text{ Gal} \equiv 0.001 \text{ cm/sec}^2.$$

In recent years, though, many applications of gravimetry to the Earth System have, more conveniently, employed units of microGals, where

$$1 \text{ μGal} \equiv 0.001 \text{ mGal} \equiv 10^{-6} \text{ cm/sec}^2.$$

Earth System Geophysics, Advanced Textbook 6, First Edition. Steven R. Dickman.
© 2025 American Geophysical Union. Published 2025 by John Wiley & Sons, Inc.
Companion Website: www.wiley.com/go/Dickman

As an example, consider a subsurface sand/gravel aquifer 20 m thick buried within a layer of shale (e.g., a relict stream channel). If the aquifer's density is 2.0 gm/cm³ and the surrounding shale has a density of 2.6 gm/cm³, then it turns out (using the "Bouguer" formula discussed below) that the gravity anomaly above that aquifer would be ∼−0.5 mGal or ∼−500 μGal. Traversing the area with a series of gravity measurements, each accurate to better than 0.5 mGal, might well allow the aquifer to be detected.

However, gravity measurements made at the Earth's surface generally require a number of corrections before an interpretation in terms of gravity anomalies is justified. Before getting to those corrections, some background material will provide an appropriate perspective.

The Capability of Gravity Meters, and the World of Microgravity. Gravity surveys can be performed on a local, regional, or global basis. Depending on the goals of the survey, the *actual* values of gravity may be recorded at all sites (yielding **absolute measurements** of g); or, the *differences* in gravity relative to a base station may be tabulated (yielding **relative measurements** of g). Absolute gravity meters or absolute *gravimeters* generally measure the acceleration of a 'test mass' falling within a larger automated apparatus; to improve precision, the test mass is dropped repeatedly—typically, thousands of times over the course of a day—and the measurements averaged. Gravimeters measuring relative gravity may contain a mass suspended by a spring or attached to a hinged horizontal beam; any *change* in gravity changes the weight of the mass, which can be detected through changes in the length of the spring or angle of the beam. Following pioneering work by Goodkind at UC San Diego beginning in the 1960s, *superconducting* (cryogenic) relative gravimeters have become the industry standard; these detect the change in gravity by the change in electrical current required to keep a small sphere magnetically suspended as its weight changes.

Crossley et al. (2013) have evaluated the usefulness of modern gravimeters for geophysical studies. Absolute gravimeters are typically not portable, and not inexpensive, but their accuracy is in the range 1–4 μGal. Relative gravimeters based on spring suspension can achieve a precision of 10 μGal or less; additionally, these instruments are relatively inexpensive and quite portable, thus great for fieldwork in which a region needs to be covered quickly with numerous measurements. However, they exhibit a "drift" in measurements over time—due to spring fatigue, static build-up, thermal expansion, and so on, of their internal components—that can amount to 5–300 μGals per day, limiting their effective accuracy for gravity surveys (after the drift has been modeled and partially eliminated) to the level of tens of μGal. Superconducting relative gravimeters exhibit almost no drift (less than 10 μGal per *year*) and are highly accurate—0.05 μGal or better!—though for field surveys, they are no more useful than are absolute gravimeters, due to their weight, expense, and power demands.

Relative measurements are, of course, less informative than absolute measurements; if actual values of gravity are needed from a survey, the relative changes in gravity must be *linked* to one or more absolute values. For example, if the difference in gravity measured between sites B and A is Δ_1, and the difference between C and A is Δ_2, A being our base station, then if the actual value at A is g_A, it follows that the actual values at the other two sites are $g_A + \Delta_1$ and $g_A + \Delta_2$.

Our ability to detect microgravity variations in the world around us opens a whole new frontier of Earth System geophysics. The spatial resolution implied by such measurements is impressive: a 1 μGal change in gravity at Earth's surface is the kind of variation seen in changing your elevation by ∼1/3 cm! The sensitivity of such measurements is equally astounding: the mass of the absolute gravimeter equipment itself actually exerts a detectable gravitational effect on its own measurements (∼0.2 μGal according to Liard et al., 2011)!

A more geophysical example of the sensitivity of modern gravimeters is the discovery (Nawa et al., 1998) of the Earth's background 'hum.' In response to a large earthquake or meteorite impact, the Earth will 'ring' like a bell; the different modes of deformation, each with a different 'pitch' or frequency, are called the **free oscillations** or **normal modes** of the Earth. These oscillations deform the entire Earth, including its shape; because they change the Earth's mass distribution, the deformations can be detected by gravimeters (as well as other kinds of instruments). The background hum, in contrast, is an ongoing but quiet ringing; in this case, the amplitudes of the normal modes are much reduced, producing very small changes in gravity detectable most reliably by continuously operating superconducting gravimeters. They are ongoing rather than episodic because they are not caused by individual, large earthquakes but are instead the result of continual deformation of the solid earth, most likely by atmospheric or oceanic processes. Earth's hum and free oscillations are discussed in some detail in Chapter 7.

Microgravity measurements have proven to be a valuable tool for exploring the Earth System. In many cases, surface or near-surface mass movements cause changes in gravity over *time* at the μGal or bigger level. Examples of interest include:

- the rise of the water table in an aquifer due to heavy rains, or its fall during a drought;

- seasonal variations in snowpack or soil moisture;

- subsurface groundwater flow forced by tectonic stresses at a subduction zone, or forced industrially by injection into a petroleum reservoir in order to recover more of the resource;

- the migration of the warm pool during ENSO;

- magma migration underneath a volcano;

- 'slow' and 'silent' earthquakes, in which the rock failure or slip taking place in the lithosphere or upper mantle occurs too slowly to generate seismic waves;

and so on. If the goal is not just detection but also quantification of the mass movement, gravity measurements will be particularly valuable. Some of these microGal applications are illustrated in Figure 5.1.

Of course, gravity observations at Earth's surface have demonstrated their usefulness for global-scale and deep-Earth processes as well. However, for reasons that will be discussed below involving gravity data corrections, as well as practical considerations, *satellite* determinations of the Earth's gravity field (independent of or combined with surface measurements) are preferable for global-scale or deep-source studies.

Even on a regional or local scale, microgravity studies have their challenges. Among other things, the accuracy attained by modern gravimeters is fine enough that they can *also* detect the changes in gravity generated by very unobtrusive but continual processes within the Earth System that may not be the main focus of the measurements—processes that serve instead to obscure the gravity signals of interest. Such processes include the periodic 'wobble' of Earth's rotation axis (which changes the centrifugal acceleration experienced by the gravimeter); variations in the local atmospheric pressure (which correspond to mass excesses, or deficits, in the atmospheric column); and coastal ocean tide 'loading' (which, like atmospheric pressure, includes the effects on gravity of both the mass of the tide and its depression of the underlying solid earth). All of these processes affect gravity at the level of 1–10 μGal, so in general they must be modeled and removed from the observations in order to reveal the gravity signals of interest.

Additionally, at the level of microGal precision, the approximations employed to obtain the International Gravity Formula (even the more precise version!) are insufficiently refined. Using a more advanced theoretical approach than Clairaut had taken to predict gravity of the Ideal Earth yields the highly precise expression

$$g_S(\phi) = g_{EQ} \frac{1 + k\sin^2\phi}{\sqrt{1 - \varepsilon^2\sin^2\phi}}$$

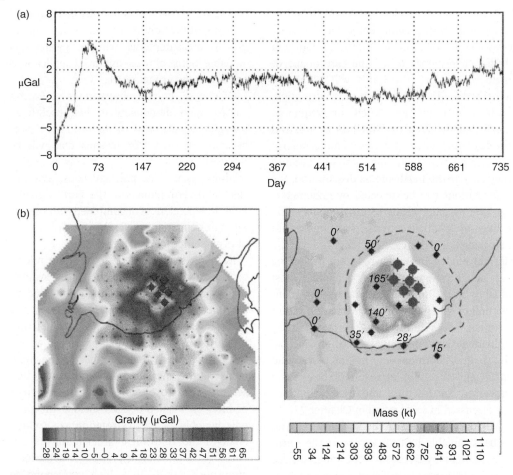

Figure 5.1 Examples of microgravity studies.

(a) Two years of superconducting gravity data at Table Mountain, Colorado, beginning April 14, 1995, recorded every few seconds; after corrections for tides, atmospheric pressure variations, the change in centrifugal force caused by 'wobbles' of the Earth's rotation axis, and drift, the largest remaining signal—an increase in gravity in the early portions of the data—is attributed to a rising water table due to heavy spring rains. From Crossley & Xu (1998) / Oxford University Press.

(b) Measurements around Prudhoe Bay, Alaska using field-portable absolute gravimeters, in 2007 compared to 2003, after similar corrections (except drift) to what was used in (a) above. The coast is the solid black curve with land to the south. It was hoped that decreasing reservoir pressures, and thus declining production, in this petroleum reservoir could be corrected by injecting water into the reservoir. Four episodes of injection (large blue dots = injection wells) took place from March 2003 through April 2007. The increase in gravity as water replaced gas in the pore spaces is seen in the left-hand plot; the inferred distribution of that water is shown in a close-up plot on the right. From Hare et al. (2008) / Society of Exploration Geophysicists.

for gravity on the Spheroid (e.g., Moritz, 2000). Here ε is the "eccentricity" of the Earth's shape ($\varepsilon = \sqrt{(a^2 - c^2)}/a$)—a measure of its ellipticity similar to the flattening f. k is the fractional difference in polar and equatorial 'gravity times radius' ($[c{\cdot}g_P - a{\cdot}g_{EQ}]/a{\cdot}g_{EQ}$), and

ϕ is ('geocentric') latitude. In recognition of the scientists that developed it and subsequently modified it, we call this expression the **Pizzetti-Somigliana gravity formula**. Using current values of the parameters g_{EQ}, ε, and k (Moritz, 2000; Ardalan & Grafarend,

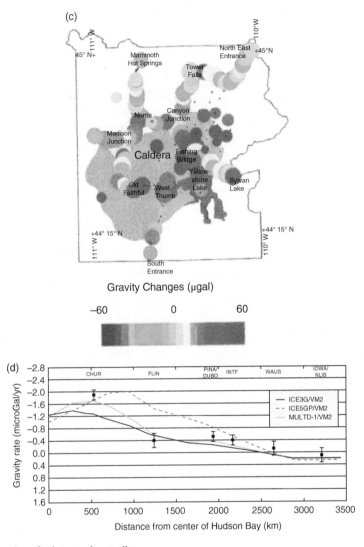

Figure 5.1 Examples of microgravity studies.
(c) Change in gravity 1987–1993 in northwestern Wyoming, including Yellowstone Park. Above the Yellowstone caldera, changes were primarily positive; but during this time, ground elevations (not shown) decreased by ~25 mm/yr, accounting for all of the gravity increase within the uncertainties. Despite the observed gravity increase, then, there was no substantial mass increase (e.g., by magma intrusion) but, instead, simply a depressurization as magma gases were carried out by hydrothermal circulation. Arnet et al. (1997) / John Wiley & Sons.
(d) Evaluation of late-Pleistocene ice-sheet distribution in North America using absolute gravimetry. The disappearance of the Laurentide ice sheet, centered over Hudson Bay, and the resulting rebound of the continental lithosphere (with the underlying mantle flowing in to support the rebound), has led to changes in gravity that are still occurring today. Long-term trends in gravity (filled squares) inferred between 1988 and 2005 at various sites (after correction for an unexpected 7-year variation observed at all sites) are compared with the predictions from three different ice models (curves). These results suggested that the ice sheet in Canada west of Hudson Bay might have been substantially thicker than previously thought. Lambert et al. (2006) / with permission of Elsevier.

2001; Groten, 2004), this highly precise latitude formula becomes

$$g_S(\phi) = 978032705$$

$$\cdot \frac{1 + 0.001931851353\sin^2\phi}{\sqrt{1 - 0.006694397902\sin^2\phi}} \, \mu\text{Gal.}$$

Despite its potential connections to Earth System geophysics, however, microGal topics are beyond the scope of this textbook. The formulas and corrections we develop here may not be sufficiently precise for microgravity studies; on the other hand, they provide an uncomplicated introduction to more advanced studies—illustrating, for example, the reasons why it is necessary to correct gravity observations and the techniques to carry out such data processing.

5.1.2 The Reduction of Gravity Data

We will consider any value of gravity at the Earth's surface to be the result of the gravitational pull of a reference Earth *plus* the effect of some anomalous mass whose character we would like to investigate. Two kinds of investigation can be envisioned.

For studies of global change, or other processes involving surficial or near-surficial mass, the reference Earth could be the Earth at some *time*, and the anomaly could refer to a subsequent regional change in Earth's mass distribution—for example, a higher water table resulting from an episode of heavy rain or the redistributions of mass associated with the melting of a glacier. In this case, a sequence of gravity measurements made over time in that region will reveal that an anomalous change in mass distribution has taken place; modeling the process (e.g., by specifying the aquifer's storativity or the rate at which the ground rebounds from the loss of ice) can yield estimates of the mass gain or loss.

More commonly, gravity measurements are used to identify and quantify *spatial* anomalies, i.e., locations on or at depth within the Earth where the density differs from that prescribed by the reference model. For example,

there might be a 'bump' on the core boundary, a feature not accounted for in an idealized reference model, protruding into the mantle and replacing silicate with denser iron. Such anomalous mass will add to the gravity measured up at the Earth's surface. At a measurement site on the surface, subtracting the gravity of the reference Earth—for example, using the International Gravity Formula, evaluated at the latitude of the site—can reveal the stronger gravity, thus the presence of a mass anomaly below. As we will see, the difference in gravity can also help us determine the magnitude of the mass anomaly.

The Elevation Effect. Quantifying the anomaly is not so easily achieved, however. The International Gravity Formula specifies the gravity of the Ideal Earth (our reference) *on* the Spheroid, i.e., *at sea level*; typically, though, land locations are appreciably above sea level (almost a kilometer, on average), and their height above the sea-level reference, placing them farther from the bulk of the Earth, leads to a smaller magnitude of gravity—regardless of the possible existence of a mass anomaly below. Simply comparing gravity on the Spheroid with the observed gravity, i.e., subtracting one gravity value from the other, may reveal more about the elevation of the observation site than it does about the subsurface mass distribution—unless the measurement is corrected for the site's elevation above sea level. (Complicating the picture, there are differences between sea level of the Ideal Earth and sea level of the *real* Earth; we will save those complications for the next chapter.)

The Elevation Correction. Our goal in making this correction is, formally, to reduce our gravity measurement to sea level, for comparison with gravity on the Spheroid. If we have a set of measurements from a gravity survey, however, we might be interested only in relative differences in gravity – i.e., in how the measurements compare to each other. As a convenient alternative, in that case, we can reduce all the measurements to the elevation of a *base station*, and compare them to the value of gravity at that base station; this approach would still highlight any mass anomaly. In short, sea level is the desired reference elevation in a global or regional survey; for local surveys, the reference level could be either sea level or the base station elevation.

To determine the elevation correction, we take advantage of the fact that, to good approximation in terms of its gravitational attraction, the Earth can be treated as a point of mass M_E at its center. The gravitational acceleration at a distance r from that point of mass has a magnitude

$$g = \frac{GM_E}{r^2}.$$

Using either algebra or calculus, this formula allows us to estimate how elevation has affected our measurements.

An Algebraic Approach. If an observer measured gravity at distance r and then moved to a distance $r + h$ from the center—that is, the observer rose by a height h—gravity would be different by the amount

$$\Delta g = \frac{GM_E}{(r+h)^2} - \frac{GM_E}{r^2}$$
$$= \frac{GM_E}{r^2}\frac{1}{(1+h/r)^2} - \frac{GM_E}{r^2}.$$

In most situations, h will be very small compared to r; for example, most mountains do not exceed a few kilometers in elevation, whereas r is typically ~6371 km. If h/r is a very small fraction, then to very good approximation

$$\frac{1}{(1+h/r)^2} \cong 1 - 2\left(\frac{h}{r}\right).$$

This rule, which actually follows from the binomial theorem, can be verified easily using actual numbers for h and r; for example, if $h/r = 0.001$ then $1/(1+h/r)^2$, which becomes $1/(1.001)^2$, does indeed very nearly equal 0.998. Using this rule, we then find

$$\Delta g \cong \frac{GM_E}{r^2}\left(1 - 2\frac{h}{r}\right) - \frac{GM_E}{r^2} = -2\frac{GM_E}{r^3}h.$$

A Calculus Alternative. Differentiating our starting formula for g tells us the 'rate' at which

g varies with radius or height:

$$\frac{dg}{dr} = -2\frac{GM_E}{r^3},$$

so the change Δg in gravity when our radius or elevation increases by an amount Δr or h is

$$\Delta g = \frac{dg}{dr}\Delta r = \frac{dg}{dr}h = -2\frac{GM_E}{r^3}h.$$

Again, this depends on h (i.e., Δr) being small compared to r. It is no surprise that we end up with the same result, since the calculus rule for differentiating powers is based on the same binomial theorem approximation.

With either approach, we could write the result as

$$\Delta g \approx -2g\frac{h}{r}$$

since $g \cong GM_E/r^2$. For $g \cong 982$ Gals and $r = 6.371 \times 10^6$ m,

$$\Delta g \approx -0.3083\,h\ \text{mGal}$$

if h is in meters; if h is in feet, the elevation effect is $-0.09396\,h$ mGal.

The units for g and r here are cgs and mks; using consistent units, can you redo this calculation and verify the effect is −0.3083 milliGals if h = 1 m?

The minus sign reflects the fact that gravity decreases, the farther from the Earth we get; as long as we are relatively near the surface (g ~982 Gals), it decreases by 0.3083 mGal per meter of elevation. The difference in gravity is also called the "free air" effect, that name serving as a reminder (as discussed below) that it does not include a mass effect.

If we want to compare two measurements of gravity made at different heights, we must recognize that the one made at the higher elevation will automatically be smaller, by 0.3083 times its additional height, than if it were at the same elevation. This effect must be *eliminated* before we compare the measurements, since it obscures any gravity differences which might be due instead to anomalous mass concentrations below ground.

Ultimately, we will usually want to compare our measurements to the predicted values given by the International Gravity Formula, i.e., to gravity on the Spheroid. In effect, that means we must correct each measurement to mean sea level: if a measurement was taken at elevation h (in meters) above sea level, we must add $0.3083h$ mGal to it—that is called the **free-air correction**, a correction to the observed gravity.

The *difference* between the corrected observation and the International Formula's value is called the **free-air anomaly**.

The Mass Effect. The free-air correction suggests the possibility of another correction. An observer at a higher elevation than sea level is most likely standing on solid ground, and the mass of the ground—that is, the mass between the observer and sea level—will also exert a gravitational force contributing to the observed value; but the free-air correction only corrects for the elevation difference between the observer and sea level.

For instance, gravity on top of a mountain, at elevation h meters above sea level, will be less (by $0.3083h$ mGal) than at sea level; but it is also larger than it would be if the mountain was not there, due to the additional downward gravitational pull of the mountain.

Remember that our goal is to compare observed gravity with *ideal* gravity, gravity on the Spheroid. The Ideal Earth is a smooth oblate Spheroid, devoid of mountains. If we are to have any hope of detecting subsurface anomalies, we must first eliminate the effect of the mass (as well as elevation) above sea level, before comparing observed and ideal gravity.

The Mass Correction. The mass correction is usually achieved in an *approximate* manner: we assume that the gravitational effect of the mountain's mass is the same as if the mountain had an infinite extent and no slope—as if the mountain was an infinitely long flat 'slab.' Then we use the fact (derived in *Textbox A*) that the gravitational attraction of an infinite flat slab is simply

$$2\pi G\rho h$$

if its density is ρ and its thickness is h. This approximation is not as drastic as it first appears, because of the nature of gravity: although the infinite slab brings in an infinite amount of mass, most of that mass is infinitely far away, and distance is the deciding factor. The difference in gravity between a mountain and an infinite flat slab of the same thickness and density is typically no more than a few milliGals, often less.

Thus, to eliminate the gravitational effect of some mass of density ρ that extends to a height h above mean sea level, $2\pi G\rho h$ should be subtracted from the measurement; this is called the **Bouguer correction**. The **Bouguer anomaly** is the *difference* between the observed gravity, corrected for elevation *and* mass, and gravity on the Spheroid at the same latitude.

The process of reducing gravity data can be illustrated as follows (see Figure 5.2). We imagine a volcano, rising to a height h above sea level; the typical density within the volume of the volcano is ρ. A group of observers is on the top of the volcano at point Q, thus at an elevation of h meters above sea level. For the latitude of point Q, the International Gravity

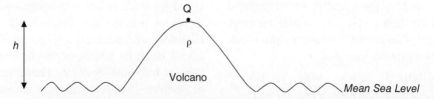

Figure 5.2 A conceptual example of gravity measurements and corrections. The initial gravity measurement is at point Q, which is at h meters elevation above sea level. The density of the volcanic mass is taken to be ρ.

Formula predicts the value of gravity on the Spheroid to be g_S. The observers at point Q measure gravity to be g_Q there. The following quantities may be considered:

$g_Q - g_S$;	such a difference is physically meaningless, since a number of unwanted factors (including mass and elevation) may be conspiring to make g_Q different from g_S
$(g_Q + 0.3083h)$;	this is the gravity that would be measured if the observers were at sea level, but with all the mass of the volcano scrunched up beneath them (that mass still tugs on the gravimeter)
$(g_Q + 0.3083h) - g_S$;	this difference is called the free-air anomaly
$(g_Q + 0.3083h - 2\pi G\rho h)$;	this is the gravity that would be measured if the observers were at sea level, and a volcano of density ρ never existed above sea level
$(g_Q + 0.3083h - 2\pi G\rho h) - g_S$;	this difference is called the Bouguer anomaly.

The reader should note the various signs used to apply the different corrections: the free-air correction is added to eliminate the elevation effect, whereas the Bouguer correction is subtracted to eliminate the mass effect.

The Bouguer anomaly sheds some light on the mass anomaly it helped identify. If the Bouguer anomaly is positive, then even after correcting for mass and elevation, the observed gravity must have been 'too large,' compared to the ideal gravity. This would indicate a *mass excess* below the observers. If, on the other hand, the Bouguer anomaly is negative, then the observed gravity is 'too low,' compared to the ideal, indicating a *mass deficit* below the observers. However, the gravitational anomaly does not require the mass anomaly to be below sea level. A deficit, for example, could indeed originate below sea level, so that g_Q is anomalously small to begin with, and the corrected observed gravity is less than ideal; *or*, the deficit could exist within the volcano above sea level, so the value chosen for density ρ in the Bouguer correction would have mistakenly been too high—thereby *also* yielding an anomalously small value for $(g_Q + 0.3083h - 2\pi G\rho h)$.

Textbox A: Mathematical Details of the Bouguer Approximation

This textbox is pretty much an exercise in integral calculus; if you are unsure about your calculus skills, a good start would be to go through the textbox, ignoring the equations for that first reading.

To show that the gravitational acceleration caused by an infinite flat slab of uniform density ρ and uniform thickness h is equal to $2\pi G\rho h$, we consider a particle located a distance L above the slab's top surface:

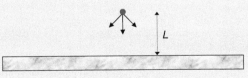

Textbox A: Mathematical Details of the Bouguer Approximation (*Concluded*)

By symmetry, being pulled equally to all sides, the particle experiences only a net *downward* gravitational force. Thus, the net gravity can be obtained by summing the vertical components of gravity due to all the elements of slab mass.

 We set up a Cartesian coordinate system at the top of the slab, beneath the particle; z is the vertical direction.

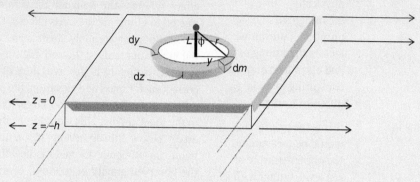

The gravitational field $d\vec{g}$ experienced by the particle due to a tiny element dm of slab mass is directed from the particle toward dm. The magnitude of the vertical component of that field is therefore $dg_z = \cos\phi \cdot dg$. Since by Newton's Law of Gravitation $dg = Gdm/r^2$, we can write

$$dg_z = \cos\phi \cdot \frac{Gdm}{r^2}.$$

The thin ring of mass shown in the illustration is composed of numerous elements like dm. They all exert gravitational forces on the particle, pulling it in a full '360-degree' range of directions; but if all of those mass elements have the same dimensions and density as dm, then the magnitudes of those forces are the same, and their vertical components are all the same. If the ring has thickness dz and width dy—so its volume (= length × width × thickness) equals $(2\pi y) \cdot dy \cdot dz$—then its mass equals $\rho(2\pi y)dy\,dz$. It follows that the net downward acceleration exerted by the ring on the particle is

$$dg_z = \cos\phi \cdot \frac{G\rho(2\pi y\,dy)\,dz}{r^2}.$$

Since $\cos\phi = L/r$ and $r^2 = L^2 + y^2$, this can be written

$$dg_z = 2\pi G\rho dz \cdot \frac{Ly\,dy}{(L^2 + y^2)^{3/2}}.$$

The net vertical gravitational pull from the entire slab can be calculated by summing the effects of the rings, as they expand out from the origin to infinity, summing also from the top to bottom of the slab:

$$net\ g_z = 2\pi G\rho \int_{z=-h}^{z=0} \int_{y=0}^{y=\infty} dz \frac{Ly\,dy}{(L^2 + y^2)^{3/2}} = 2\pi G\rho \int_{z=-h}^{z=0} dz = 2\pi G\rho h.$$

Note that this gravitational pull is independent of L: you can never really rise above the horizon of an infinite slab….

Topographic Effects. The premises of the Bouguer correction, namely that the mass beneath the measurement site is planar, infinite in extent, and uniform in density, suggest the possibility of still another correction, one that improves on those assumptions by accounting for variations in topography and density in the vicinity of the site. Technically, this will be a correction to a previous correction, not to the original measurement itself.

For example, consider that volcano shown in Figure 5.2. For the Bouguer correction, the mass of the volcano beneath point Q was approximated by an infinite flat slab of thickness *h*—an approximation that is increasingly inexact as we move away from point Q because it replaced the air above the downward-sloping surface of the volcano (all the way up to the height of point Q) with rock of density ρ. That 'false' mass is part of the whole Bouguer slab whose gravitational pull was found to equal $2\pi G\rho h$, so, to improve our correction for the volcano's mass we should *subtract* the gravity of the false mass from $2\pi G\rho h$.

Or, consider a mountain chain (not shown in Figure 5.2) with measurements potentially carried out both on the peaks and in the valleys between. For a measurement at a site in one of the valleys, say at height *h* above sea level, the Bouguer correction for the mass below will be based on a Bouguer slab of thickness (i.e., height) *h*; the adjacent peaks rise above the slab, and their 'extra' mass is not included in the correction. To include their effect on the site measurement—they pull upward on the gravimeter, opposing the downward pull of the slab represented by the Bouguer correction—we should *subtract* their gravitational force (i.e., acceleration) from $2\pi G\rho h$.

Alternatively, as noted earlier, our gravity measurements might be part of a local survey, with the reference level determined by the base station rather than sea level. In this case, the Ideal Earth is no longer our ultimate standard, and we will no longer be comparing our corrected measurements to gravity on the Spheroid; but we still have to account for the

elevation difference and mass between measurement site and base station—those factors will contribute to an apparent difference in gravity between the two locations, regardless of whether a true anomaly exists at the survey site or not. And we can still make the mass correction using a Bouguer slab as a reasonable first approximation; ultimately, then, even with a local survey, a further, topographic correction might be in order.

As it turns out, though, for a local survey, the application of a topographic correction is somewhat simpler. Any 'false' mass created within a Bouguer slab requires the same correction whether the site is higher than the base station or lower—either way, the gravitational effect of that mass must be subtracted from the slab's own $2\pi G\rho h$—just as we found for the regional survey. However, no topographic correction is required for any 'extra' mass protruding above the elevation of the slab, since that mass would affect both site and base station measurements the same way, leaving their difference in gravity unchanged. In fact, the most significant complication in a local survey concerns how to apply the elevation and Bouguer corrections themselves: if the survey site is above the base station, those corrections will be the opposite of when the site is lower than the base station.

For example, let's say the survey site happens to be at a lower elevation than the base station—so the goal of the elevation and mass corrections is, in effect, to bring the site measurement up to the level of the base station. In this case, will the elevation and Bouguer corrections (0.3083h and $2\pi G\rho h$, with h the elevation |difference| (treated as a positive quantity) between site and base station) be added to or subtracted from the site measurement?

The Topographic (or Terrain) Correction. To quantify the 'false' mass that must be removed from within Bouguer's infinite flat slab (and the 'extra' mass that must be added above it, if the survey is regional or global) in order to match the actual terrain, the elevation differences surrounding each measurement site must be known. If uniform density was a poor approximation for some parts of the slab, the actual density difference relative to that assumed for the slab must also be known, or at least estimated. For regional or global surveys, the density of any high terrain (i.e., extra mass) must of course be specified.

Traditionally—using an approach known as the "template method" (Hammer, 1939; see, e.g., Dobrin and Savit, 1988 for somewhat recent improvements)—the area around the measurement site is divided into a series of concentric rings, as in the *Textbox A* illustration; each ring is *further* subdivided into small sectors radially distributed around the ring. In each sector, the average elevation is calculated, yielding *h*, and a representative density within the sector is assigned, allowing the mass excess or deficit of that sector (compared to the Bouguer slab) to be determined. Finally, using formulas similar to those in *Textbox A*, the vertical component of the gravity corresponding to that 'corrective' mass can be calculated, yielding the desired topographic (terrain) correction.

Even with elevations available digitally and the template procedure automated, obtaining topographic corrections in a gravity survey is popularly described as "tedious." The corrections can be up to ~20 mGal where the terrain is very rugged (in the Andes, for example); but it is usually much less. According to Lowrie (2007), a good rule of thumb is that these corrections should be considered necessary if the elevation differences exceed 5% of the terrain's distance from the site—for example, if, 1 km away, the elevation differences exceed 50 m.

In some gravity surveys, the term "Bouguer anomaly" denotes a comparison in which each original measurement has been corrected for elevation, mass, and also terrain (e.g., Fowler, 2005; Lowrie, 2007). Alternatively, a comparison with corrections for elevation and mass might be referred to as a "simple Bouguer anomaly," and one with terrain corrections also as a "complete Bouguer anomaly" (USGS, 1997; Wynn et al., 1999; Forsberg, 2016).

Our Approach *Already* Included Latitude Effects, but … With a relative gravity survey, we can look forward to yet another correction, but in this case it is one that will make the data reduction more efficient. In the course of a regional survey, we would normally find ourselves computing gravity on the Spheroid (from the International Gravity Formula) at every site, in order to calculate free-air or Bouguer gravity anomalies. However, as an alternative to such repetitious computations, for example when survey sites differ in latitude from each other by much less than a degree, we can avoid calculating gravity on the Spheroid and simply correct all measurements for their difference in latitude relative to a designated base station. Such a basic correction is precisely what we would be interested in for a local survey as well.

The Latitude Correction. To determine this 'shortcut' we recall that it is differences in latitude that produce the smooth variation in gravity from equator to pole described by the International Gravity Formula. Let ϕ be the latitude of the reference site, i.e., the base station; from the International Formula, gravity on the Spheroid is given by

$$g_S(\phi) = 978.03278$$
$$+ 5.17126\sin^2\phi - 0.00573\sin^2(2\phi) \text{ Gal}$$

(rounded off slightly, given the typical precision of relative gravimeters noted earlier). If a second site is $\Delta\phi$ degrees north of the first site, the increase in gravity on the Spheroid from the base station to the second site equals the rate at which gravity increases with latitude $(dg_S/d\phi)$ times the increase in latitude $(\Delta\phi)$, or

$$\Delta g_{LAT} = \frac{dg_S}{d\phi}\Delta\phi$$
$$\approx 5.17126\frac{d}{d\phi}[\sin^2\phi]\Delta\phi \text{ Gal}$$

upon differentiating the International Formula; in writing this, we have ignored the $\sin^2 2\phi$ term since it is a thousand times smaller. Thus,

$$\Delta g_{LAT} \approx 5.17126\,[2\sin\phi\,\cos\phi]\,\Delta\phi \text{ Gal}.$$

For Δg_{LAT} to be in units of acceleration, of course, $\Delta\phi$ must be dimensionless, i.e. in radians. In that case, we can also write

$$\Delta g_{LAT} \approx 5.17126[2 \sin\phi \cos\phi] \quad \text{Gal}$$

per radian of latitude change. Since there are $360°$ in 2π radians, 1 radian equals $(360/[2\pi])$ degrees; thus,

$$\Delta g_{LAT} \approx 0.180511 \sin\phi \cos\phi \quad \text{Gal}$$

per degree of latitude change. And since there are 111 km per degree of latitude, we finally predict

$$\Delta g_{LAT} \approx 1.626 \sin\phi \cos\phi$$

mGal per km of N–S distance.

As before, this derivation could have proceeded without calculus, beginning instead with an expression for the sine of a sum of angles $(\phi + \Delta\phi)$ and then taking the difference between g_S at latitude $(\phi + \Delta\phi)$ and g_S at latitude ϕ. That sine expression is also the basis for the rule in calculus we just employed for taking derivatives of sinusoids $(d/d\phi\,[\sin^2\phi] = 2\sin\phi\cos\phi)$, so there should be no surprise once again that the results would be identical.

The formula we have derived says that gravity is larger (smaller) by $1.626\sin\phi\cos\phi$ mGal for each km the second site is north (south) of the first, in the northern hemisphere. For example, if the base station is at $\phi = 45°$ latitude, this latitude-induced change in gravity is 0.813 mGal per km north or south of the base station. The correction is *added* to gravity measured at the site if it is *south* of the base station, to compensate for the fact that gravity decreases toward the equator.

There Are Other Effects, too. In the course of a gravity survey, the ambient gravity field will change due to circumstances beyond the region—a consequence of the Earth wobbling, changes in the position of the Moon and Sun (which modify the tidal forces acting at any location), changes in the surrounding distribution of mass, and so on—and also as a consequence of instrumental drift. The geophysical effects can often be modeled and largely eliminated. For instrumental drift, the **drift correction** is usually determined by reoccupying the base station, and requiring that the gravity measured there be unchanged over the time span of the survey (after the geophysical variations have been removed); any change observed over time yields the *drift rate*, which can be applied to remove the drift from all measurements.

5.1.3 An Application to the Earth System

In upstate New York, the Susquehanna River cuts through glacially carved, till-covered, rolling hills as it meanders through the Greater Binghamton area, flowing briefly more or less east to west. From glacial times, as kilometer-thick ice sheets began to recede, through the present long-lived interglacial period, as the terrain morphed and water sources accumulated or evaporated away—in short, as the watershed evolved—the Susquehanna River channel would have migrated or jumped (avulsed). In geophysics classes at the local state university, gravity surveys have been carried out with the hope of revealing buried stream channels, artifacts of the Susquehanna's past.

But the region is hilly, and the variations in gravity apparent in profiles across the river valley have little connection with buried stream channels, being dominated instead by elevation and mass effects. We can easily estimate the order of magnitude of the gravity signal associated with a mass anomaly like the stream bed, versus that associated with topography. Let's say the density of the sandy-to-gravelly stream channel is ~1.8 gm/cm^3; the surrounding shale and till cover are somewhat denser, perhaps ~2.4 gm/cm^3. Thus, the *density deficiency* of the stream channel—corresponding to a deficit of mass per unit volume—is ~0.6 gm/cm^3. Using the Bouguer approximation, the gravity anomaly associated with a stream channel ~10 m thick would be roughly

$2\pi G\Delta\rho h = 2\pi$ (6.67384 × 10^{-8} cm^3/sec^{2}gm) × (0.6 gm/cm^3) × (1,000 cm) = ... = 0.25 mGal.

Meanwhile, topographic undulations of magnitude ~±100 m around the base station are producing gravity signals ~$2\pi G\rho h = 2\pi$ (6.67384 × 10^{-8} cm^3/sec^{2}gm) × (2.4 gm/cm^3) × (10,000 cm) = ... = 10.06 mGal for the mass between site and base. The elevation effects are even larger (~30.83 mGal) (as it turns out, for typical crustal densities, the free-air effect is always about three times the size of the Bouguer effect—again, distance trumps mass). Clearly, the measurements must be corrected for elevation and mass before we can hope to uncover gravitational evidence of the buried anomaly.

For the purpose of this survey—which was to simply illustrate the reduction of gravity measurements using those corrections—our classes focused on how to use the corrections themselves, and neglected more complex approaches that could increase their accuracy significantly. Such complexities include the incorporation of tidal models and frequent reoccupation of the base station, allowing more precise drift corrections to be inferred; spatially detailed topography, to construct a terrain correction; and repeat surveys (i.e., reoccupation of *all* sites), which, by averaging the measurements at each site, can substantially lower their error bars. An example of the great improvement possible is demonstrated by the work of Arnet et al. (1997) in their exploration of the Yellowstone caldera (see Figure 5.1c): the final precision of their gravity data was 0.012 mGal.

But, given the limitations of our analysis, as noted below, an anticipated gravity signal of 0.25 mGal is unlikely to be detected. As a classroom exercise, then, we revise our hypothesis to investigate whether a relict stream channel 40 m thick existed in our region; this corresponds to a gravity anomaly of ~1 mGal.

A sample of the 'raw' data is presented in Table 5.1 and plotted in Figure 5.3a; Table 5.1 also includes pertinent information regarding

Table 5.1 Raw data and site information for a gravity survey across the Susquehanna River valley, NY.[a]

Location	Base station	Site 1	Site 2	Site 3	Site 4	Site 5	Base station
Time of measurement	2:49 p.m.	3:01 p.m.	3:13 p.m.	3:24 p.m.	3:45 p.m.	4:02 p.m.	4:18 p.m.
Elevation, m	289.0	440.3	390.1	304.8	268.2	253.9	289.0
Northward distance from base station, km	—	−2.29	−0.76	+0.15	+0.75	+3.96	—
Local density, gm/cm^3	2.8	2.8	2.8	2.8	2.8	2.8	2.8
Raw gravity relative to base station, mGal	0	−31.10	−20.41	−3.33	+4.12	+8.29	−0.0449
Latitude correction, mGal	0	−1.85	−0.61	+0.12	+0.61		0
Free-air correction, mGal	−89.10	−135.74	−120.27	−93.97	−82.69		−89.10
Bouguer correction, mGal	+33.93	+51.70	+45.80	+35.79	+31.49		+33.93
Drift correction, mGal	0	−0.006	−0.012	−0.018	−0.028		−0.045
Corrected Δg, mGal	**55.17**	**54.81**	**54.68**	**54.75**	**54.74**	**53.59**	**55.17**

Note: Values here have been rounded after calculations were complete.

[a]"Raw gravity" is the difference in gravity at a site relative to the value measured at the base station at 2:49 p.m. Base station latitude is 42.0889°. Free-air and Bouguer corrections are, for simplicity, calculated relative to mean sea level; that approach leads to a nonzero final value for gravity at the base station, but does not alter the final differences in gravity relative to the base station.

the survey. Note that the second occupation of the base station yields a different measurement of gravity; for the sake of illustration, we will assume all of that difference is due to instrumental drift. (Because the survey lasted less than two hours, the positions of the Sun and Moon did not shift all that much, so tidal forces acting on the gravimeter probably did not change much, and our assumption may not be too drastic.) From Table 5.1, the drift of −0.0449 mGal over the 89 minutes of the survey implies a drift rate of ~−5.0449×10⁻⁴ mGal/min. Thus, for example, the gravimeter must have drifted by about [−5.0449 × 10⁻⁴ mGal/min.] ×

12 min. = −0.0061 mGal during the elapsed time between the base station measurement and the measurement at Site 1.

The reader should note carefully how raw gravity values are boosted or reduced to account for the various effects of elevation, mass, latitude, and drift—that is, note whether the various corrections are added or subtracted.

A useful exercise for you would be to fill in the missing values for Site 5 in Table 5.1, and verify the value of the corrected ∆g in the last row.

As shown in Figure 5.3a, b and Table 5.1, the corrections yield a greatly diminished variation

Figure 5.3 Relative gravity data across the Susquehanna River valley in Greater Binghamton, NY, plotted as a function of distance north of the base station. The blue dashed line indicates the current location of the Susquehanna River channel. (a) Raw data; no corrections have been applied to these data. *Do they imply that a buried stream channel might exist at the southernmost site?* (b) Reduced data, with corrections for elevation, mass, latitude, and drift as described in the text (see Table 5.1 also). The corrections are relative to mean sea level, thus the base station corrected value (red square) is no longer zero. *With these corrections to the data, where would you imagine the buried stream channel is located?*

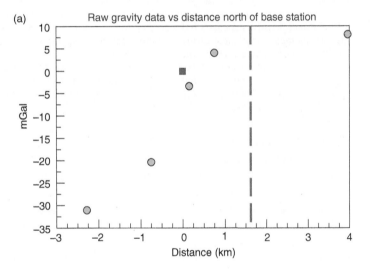

(a)

Raw gravity data vs distance north of base station

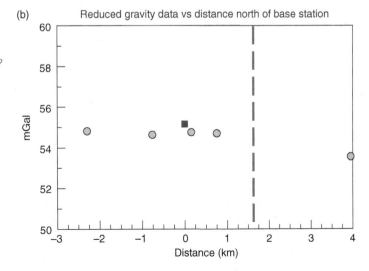

(b)

Reduced gravity data vs distance north of base station

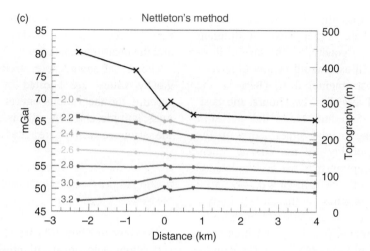

Figure 5.3 (c) Reduced gravity data based on different assumed densities for the Bouguer correction (labeled to left of curves, and ranging from 2.0 gm/cm³ for the light green curve to 3.2 gm/cm³ for the dark green curve), plotted along with the topography (black line, elevations in meters marked as "x"). The most likely *regional* density is chosen on the assumption that the corrected gravity, whether or not it reveals a possible anomaly, should bear little correlation with the topography (after the corrections). The optimal density in this case appears to be around 2.7–2.8 gm/cm³.
We called 2.7–2.8 gm/cm³ optimal because, from one end of the survey to the other, it results in the flattest gravity profile (unlike the observed topography). But, from a very local perspective, we could have chosen 2.4 gm/cm³ as optimal because it reflects none of the topography immediately surrounding the base station. Which perspective would you choose, and why?

of gravity. Such results certainly suggest that a relict stream channel is unlikely to exist south of the current river location (where Sites 1 and 2 are)—a conclusion bolstered by the fact that the actual topography there has a much higher elevation than the river, and that such elevation differences must have persisted from the Ice Age through to the present time. On the other hand, given the small magnitude of the hoped-for signal, ~1 mGal, as well as the lack of spatial detail and the absence of error bars, a buried stream channel cannot be ruled out to the north of the river (i.e., around Site 5), though such an inference would be highly speculative.

The analysis summarized in Table 5.1 carries some uncertainty because it assumed a particular density for the Bouguer correction. The density *initially* used, 2.4 gm/cm³, was based on the geology of the region, which, as noted before, is considered to be mainly shale, with a till cover. Such a justification is reasonable; but in many situations, a better estimate of the density to use in the Bouguer

corrections can be inferred from the data itself, using a technique called **Nettleton's method** (Nettleton, 1939).

Nettleton's method is based on the assumption that the reduced gravity values—corrected for elevation, latitude, and drift as well as mass—should exhibit no similarity to the surface topography. If too large a ρ is chosen, the Bouguer correction ($2\pi G\rho h$) could dominate the data reduction, yielding a corrected gravity profile that is a mirror image of the topography (since $2\pi G\rho h$ is subtracted); or, if too small a ρ is employed, the resulting gravity profile would parallel the topography. Testing a range of densities can reveal the one producing the least correlation with topography. In our case (see Figure 5.3c), the optimal ρ turns out to be ~2.7 gm/cm³ or ~2.8 gm/cm³, somewhat reasonable for shale but larger than we had posited for a shale-till mix (e.g., Manger, 1963)—perhaps caused by an unusually thin till cover in the survey region.

The final analysis summarized in Table 5.1 employs an optimal value of ρ, 2.8 gm/cm³.

We can use the discrepancy between initial and optimal values of density to suggest that our reduced gravity values are intrinsically uncertain by at least $2\pi G \Delta \rho h$, with $\Delta \rho \cong 2.8$ gm/cm^3 – 2.4 gm/cm^3 = 0.4 gm/cm^3, thus (with h relative to the base station) typically by ~0.71 mGal. But it is quite possible that, with a more spatially detailed survey, especially south of the base station (where the terrain is much more irregular), Nettleton's method would have constrained ρ more tightly, reducing the uncertainties and allowing us to have more confidence in the existence of a relict stream bed beneath Site 5.

Finally, a striking application of Nettleton's method may be seen in the work of Verdun et al. (2003) involving airborne gravity measurements over the French Alps; their analysis suggested that the interpretation of land-based gravity in one portion of that rugged area could be improved if the density of Alpine crust was 10% greater.

5.1.4 Gravity Data Measured on a Moving Platform

To derive a complete picture of how gravity varies over all of the Earth's surface, it is obviously necessary to take measurements on or over the *oceans*, and on land in regions of extreme terrain. This traditionally involved the use of ships and airplanes, raising the difficulty of conducting surveys with instruments that are moving. Even if the motion is at constant velocity (relative to the Earth), measurements of gravitational acceleration are still affected, as we can show.

Let's say that a particle of mass M—perhaps a boat (Figure 5.4)—is located (for simplicity) *on* Earth's surface, at latitude ϕ. If the particle is moving from west to east, covering a distance $\Delta \ell$ over a time interval Δt under its own power, then, in effect, it possesses an eastward velocity of magnitude $\Delta \ell / \Delta t$; we will denote its velocity by V. This eastward motion *adds* to the eastward motion of the particle as it is carried around the Earth's rotation axis every day by the Earth's spin, making its angular velocity greater than that of Earth. If we sat in space, watching the Earth rotate, the particle would take 1 sidereal day to return to its initial location if it were at rest; but if the particle is moving eastward (as in Figure 5.4), it will return to its initial location as seen from space in *less* than 1 sidereal day.

From our discussion in Chapter 4, we know that if the particle is at rest on Earth's surface, it experiences a centrifugal force of magnitude $M \Omega^2 R$, where Ω is the Earth's angular velocity and R is the perpendicular distance of the particle from Earth's rotation axis. The vertical component of this force—the portion

In space, looking down from above

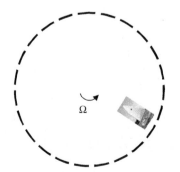
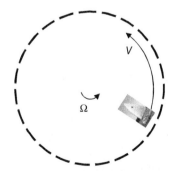

Figure 5.4 If the boat is at rest, Earth's daily rotation (with angular velocity Ω) carries it once around, returning it to its original location, in just under 24 hours. But if the boat is moving west to east, such motion brings it around as well, and it will take noticeably less time to circle around back to its original location. In this case, the boat's linear velocity (V) causes it to have an angular velocity greater than Ω.

that directly opposes gravity—has magnitude $M\,\Omega^2\mathcal{R}\,\cos\phi$.

The Eötvös Effect. While moving eastward, the particle does not change its latitude so that its distance from the axis is unchanged, but it experiences a stronger centrifugal force (and thus, incidentally, a more reduced gravity) because its effective angular velocity has increased. This change in centrifugal force is called the **Eötvös effect** (named after a Hungarian physicist equally famous for his experimental work, ending around 1909, on the (weak) Principle of Equivalence mentioned in Chapter 4).

If the 'particle' in question is a gravimeter, it will record the weaker-magnitude gravity resulting from that additional centrifugal force. Such a measurement will require correction for the Eötvös effect.

The Eötvös Correction. As illustrated in Figure 5.5, in the time Δt during which the particle moves through a distance $\Delta\ell$ along Earth's surface, it sweeps through a 'central angle' $\Delta\lambda$, where

$$\Delta\ell = \mathcal{R}\,\Delta\lambda;$$

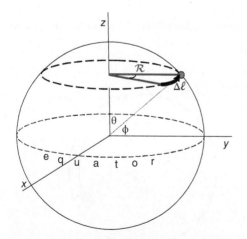

Figure 5.5 The particle at latitude ϕ moving west-to-east along Earth's surface covers a distance $\Delta\ell$ during a time interval Δt; in doing so, the perpendicular $\mathcal{R}$ from the rotation axis sweeps out a 'central' angle $\Delta\lambda$.

thus

$$\Delta\lambda = \Delta\ell/\mathcal{R}.$$

Such an increase in the central angle over time represents, in effect, an angular velocity

$$\Delta\lambda/\Delta t = \frac{\Delta\ell/\mathcal{R}}{\Delta t} = \frac{\Delta\ell}{\mathcal{R}\,\Delta t} = \frac{V}{\mathcal{R}}$$

produced by the particle's 'linear' velocity V along the surface.

Incidentally, does this formula, relating a particle's velocity to its angular velocity, look familiar to you? Check out the "earlier" final comments near the end of the "Thorough Description of Centrifugal Force" in Chapter 4.

In short, the moving particle experiences a net centrifugal force whose magnitude is given by

$$M\left(\Omega + \frac{V}{\mathcal{R}}\right)^2\mathcal{R}$$

instead of $M\,\Omega^2\mathcal{R}$. The *increase* in the particle's centrifugal force as a result of its motion is

$$M\left(\Omega + \frac{V}{\mathcal{R}}\right)^2\mathcal{R} - M\Omega^2\mathcal{R}.$$

Finally, the *vertical component* of the particle's increase in centrifugal force—this will be the additional amount by which the gravitational force is reduced—must be

$$M\left(\Omega + \frac{V}{\mathcal{R}}\right)^2\mathcal{R}\,\cos\phi - M\Omega^2\mathcal{R}\,\cos\phi$$

or

$$M\left(\Omega^2 + 2\Omega\frac{V}{\mathcal{R}} + \left(\frac{V}{\mathcal{R}}\right)^2\right)\mathcal{R}\,\cos\phi$$
$$- M\Omega^2\mathcal{R}\,\cos\phi,$$

and the first and last terms cancel. At reasonable speeds, $V/\mathcal{R}$ is considerably less than Ω so that $(V/\mathcal{R})^2$ is much less than $(V/\mathcal{R})$ times Ω; in this case, the vertical component simplifies to

$$2M\,\Omega V\,\cos\phi$$

to very good approximation. The reduction in gravitational acceleration from the Eötvös effect is therefore

$$2\Omega V\,\cos\phi.$$

This quantity is called the **Eötvös correction**, and allows measurements of gravity taken from boats or airplanes to be corrected for the effect of their motion.

The magnitude of the Eötvös correction is generally substantial. For example, a boat at 45° latitude moving eastward at a speed of only 1 mph with respect to the rotating Earth experiences a decrease in gravity of $2\Omega V \cos\phi = 2 \times (7.29 \times 10^{-5} \text{ rad/sec}) \times (1.61 \text{ km/hr}) \times \cos 45° = \ldots = 4.6 \times 10^{-3} \text{cm/sec}^2$, thus nearly 5 milliGals.

Consequently, each time gravity is measured, the speed of the boat (or plane) must be known to better than 1 mph for a survey accuracy of 5 mGal. In our derivation, however, the velocity V in the Eötvös correction represented the velocity of the boat or plane relative to the rotating Earth, i.e., relative to the terrestrial coordinate system. Before the use of GPS became widespread, ship and airplane speeds—though typically very well measured with respect to the surrounding water or air—were difficult to determine accurately with respect to reference points on the (distant) land surface of the Earth. One clever approach in that situation was to employ a 'repetitive' survey track, such as a N-S/E-W 'criss-cross' pattern, where the boat or plane loops around and passes through earlier measurement locations as it continues to measure gravity; at each of those *cross-over* points (i.e., repeat locations), gravity should be the same as it was earlier, aside from drift effects. Using such a constraint at the cross-over points helped those surveys achieve a final accuracy of perhaps ~0.1–5 mGal.

The need to apply Eötvös corrections to measurements of gravity on moving platforms points to the importance of *satellite* determinations of gravity; as we will discuss in the next chapter, such determinations do not employ gravimeters on the moving satellites, so do not involve an Eötvös effect. But satellite data is generally global or continental in scale, and ground-level measurements on stationary platforms usually have a local focus; without shipboard or airborne measurements, gravity

variations on more 'intermediate,' regional scales would remain unknown (e.g., Verdun et al., 2003; Bell et al., 1999). From the 1980s and especially the mid-1990s onward (e.g., Bell et al., 1999; Jekeli, 2016), GPS and other global navigation systems, through their ability to yield not only the position but also the velocity (and accelerations) of gravimeters in motion, have facilitated accurate Eötvös corrections for shipboard and airborne surveys, and enabled gravity determinations on those important intermediate scales (though, with other sources of measurement error (see, e.g., Bell et al., 1999), cross-over points and repetitive surveys are still important).

5.2 Isostasy and the Earth System

5.2.1 Bouguer's Discovery and the Principle of Isostasy

In 1749, as a controversy raged among scientists as to what the shape of the Earth should be (see, e.g., Watts, 2001 for an early history of isostasy), Bouguer published the results of his geodetic survey attempting to determine the flattening of the Earth. His determination involved directly measuring the Earth's radius at different locations, and identifying the smooth variation with latitude that is associated with a Spheroidal planet. At each site, the radius was estimated using star observations and the technique of Eratosthenes (see Figure 5.6) that we discussed in Chapter 4.

Bouguer expected that measurements he made in the Andes mountain region would be corrupted by the gravitational pull of the mountains: the plumb line used to determine the vertical direction would be deflected, because the plumb bob would be attracted toward the mass of the mountains; the zenith would be off target, and the zenith angles of the stars he measured would be in error, leading to incorrect estimates of the Earth's radius. Instead, however, his estimates of radius from

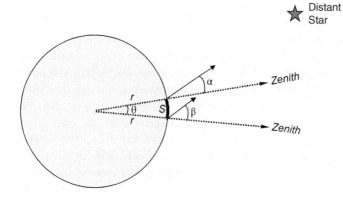

Figure 5.6 The zenith angles to the chosen star are α at one site and β at the other. As discussed in Chapter 4, the central angle θ subtended by the radii to the two sites equals $\beta - \alpha$; and if the arc distance S has been measured, the relation $S = r\theta$ allows the radius r to be determined.

the Andes region fit in nearly perfectly with the smooth variation versus latitude found elsewhere in his survey and around the world. It was almost as if the Andes Mountains, impressively massive as they appeared, were hollow, and actually contained no mass—or, more seriously, as if *the visible mass of the Andes Mountains above ground was* <u>*compensated*</u> *by a matching deficit of mass below ground.*

It is interesting to note that Bouguer's striking discovery had little impact in the scientific community—until, as proper in the world of science, it was confirmed by subsequent work: Everest's geodetic survey of the Himalayas, a century (!) later, yielded similar results to Bouguer's. And the science of **isostasy**, the name given to describe the state of mass balance or compensation of the Earth's crust, was born.

The essential principle of isostatic compensation is very straightforward. Consider any two regions of the Earth's surface, for example a mountainous region and a mountain-less 'plains' region of a continent; or, a plains region and a nearby oceanic region (see Figure 5.7). Draw *identical* vertical columns through these regions: the columns should have the same cross-sectional areas—denoted by A—and extend through the same height and depth, including air, crust, and even upper-mantle rock. Of course, depending on which region a column is located in, it may include very little air, much crust, and a little mantle (as in a mountainous column), or much air, water, and not very much crust and mantle (as in an oceanic column). The **principle of isostasy** asserts that, in a state of isostatic equilibrium,

Oceanic region · Continental 'plains' region · Mountainous or plateau region

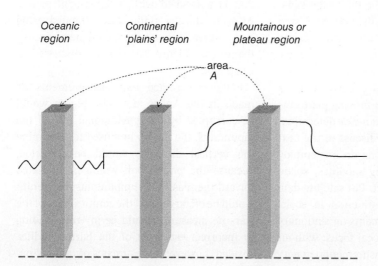

Figure 5.7 Setup for solving isostasy problems. To satisfy the principle of isostatic compensation, the three identical columns drawn through the crust in these regions must contain the same total amount of mass. The horizontal dashed line is the hypothetical *depth of compensation*, discussed in the text.

these *identical vertical columns must contain the same total amount of mass.*

It should be clear that this principle is consistent with Bouguer's experience in the Andes and Everest's in the Himalayas: if there is the same amount of mass below the surveyors as in the nearby mountains, there will be no deflection of the vertical, and no error in their determination of Earth's radius.

Upon reflection, it should also be apparent that it would be impossible to satisfy the principle of isostasy if the columns only penetrated down, say, to mean sea level, or to a few meters into the continental crust. The minimum depth that allows the principle to work is called the **depth of compensation**. As we will see shortly, that depth corresponds to the point below which there are no lateral variations in mass—the point below which there is no need for the columns to go: if the columns do extend below the depth of compensation, the same additional amounts of mass will be added to their columns, and their mass balance will not be altered.

5.2.2 Mechanisms to Achieve Isostatic Balance

Following Everest's confirmation of Bouguer's initial discovery, two theories were proposed to explain *how* the principle of isostasy might work. It is no surprise that these theories focused on the mass in different regions as being a function of surface topography—topography, in the form of mountains, was the reason isostasy had to be invoked in the first place. One such theory, postulated by Airy in 1855, held that the *thickness* of the crust must vary, depending somehow on the topography. Another theory, proposed by Pratt around the same time, envisioned the *density* of the crust to be varying in some way, as a function of the topography.

Pratt Isostasy. We explore the Pratt mechanism first, with the help of Figure 5.8. Identical columns are drawn in plains and mountainous regions of a continent. Both columns extend down to a hypothetical depth of compensation—hypothetical because the Pratt theory has no way of constraining the actual depth of compensation. Both columns have cross-sectional area A. In the plains region, the crustal density is ρ_1 and the depth of compensation is h_1 km below the surface; in the mountainous region, the density and depth to compensation are ρ_2 and h_2. In order for the principle of isostasy to be satisfied, the total mass in these columns must be the same.

Now, these columns include some amount of air, whose density is a thousand times smaller than that of rock. It is true that local or regional variations in atmospheric mass

Figure 5.8 Exploration of the Pratt mechanism of isostatic compensation. The horizontal dashed line is the hypothetical *depth of compensation*. The portions of the vertical columns containing air are drawn with light 'filling'; the rock portions have the densities and heights shown.

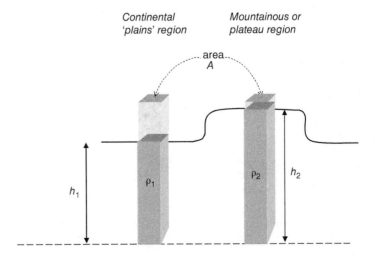

over time, 'loading' down the crust slightly more or less than on average, can actually cause the crust to depress or rebound a bit. Such changes can be detected and can influence measurements of gravity slightly. For isostatic analyses, the neglect of air mass and atmospheric loading of the crust constitutes a very minor approximation, one which we will henceforth employ in this chapter.

The volume of the plains column, excluding the air portion, is Ah_1, and its mass is $\rho_1(Ah_1)$; similarly, the mountainous column contains non-air mass $\rho_2(Ah_2)$. The principle of isostasy requires that

$$\rho_1(Ah_1) = \rho_2(Ah_2).$$

This mass balance can be written

$$\rho_1 h_1 = \rho_2 h_2;$$

indeed, this (and every other) isostasy calculation could be carried out 'per unit area' (i.e., with $A = 1$). The principle of isostasy thus implies

$$\rho_2 = \rho_1 \frac{h_1}{h_2},$$

so that $\rho_2 < \rho_1$ if $h_2 > h_1$: the crust within and beneath a mountain must be less dense, to compensate for the height of the mountain, according to the Pratt mechanism. This mechanism amounts to a statement of how crustal mountains are built up.

The Pratt mechanism is unsatisfying because it relies on a depth of compensation (for the principle of isostasy to be obeyed) but provides no basis for locating it. The mechanism has a more serious defect, though—a *dynamic*

failure. Observations of gravity around the world have tended to imply that the Earth acts to maintain a state of isostatic balance. In a region of the crust that maintains Pratt isostasy, we consider a mountain there which is eroding away over time. For simplicity, we will imagine that the eroded mass has been transported away from the whole region and is out of the picture. In order that the mass in mountainous columns remains the same as in neighboring columns, the mountain's density must increase as it erodes; that is, $\rho_2 = \rho_1 \, (h_1/h_2)$ requires that ρ_2 must increase as h_2 decreases from erosion. But the physical basis for such a density increase is unclear; if anything, as the overburden lessens, the rock density is more likely to decrease.

Airy Isostasy. The Airy mechanism postulates simply that lower-density crust *floats* upon greater-density mantle, in the same way that logs or icebergs float in water. For example (see Figure 5.9), logs of different sizes float 'proportionally,' with the biggest log floating the highest and penetrating the deepest into the water. Similarly, where the Earth's topography is higher, the crust must be thicker than elsewhere and must protrude more deeply into the mantle.

In Figure 5.10, we consider mountainous, plains, and oceanic regions, and draw identical vertical columns in those regions down into the mantle. The comparatively extra mass of the mountainous column above, say, sea level can be compensated precisely by a deficit below if the part of that column below sea level contains less of the more dense mantle and

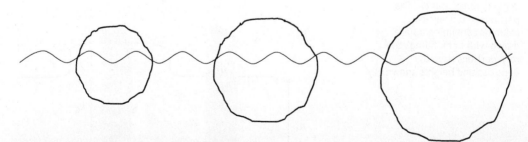

Figure 5.9 As a consequence of Archimedes' principle, hydrostatic equilibrium, and the principle of isostasy, the log that is thickest floats higher and deeper than the thinner logs.

Figure 5.10 Exploration of the Airy mechanism of isostatic compensation. The depth of compensation (horizontal dashed line) can be located at any depth at or below the deepest crust. The portions of the vertical columns containing air are drawn with light 'filling'; the crust is taken to have density ρ_C and the mantle, density ρ_M, with $\rho_C < \rho_M$.

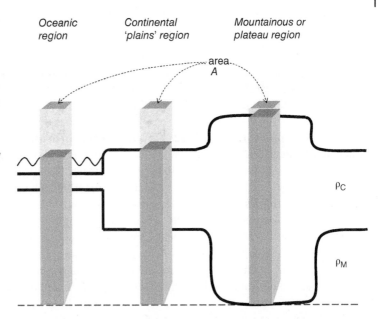

more of the less dense crust. That is, in order to satisfy the principle of isostasy, the crust has to penetrate deeper where it rises higher; the crust will be thickest beneath mountains, less thick in continental plains regions, and thinnest beneath the oceans.

For later discussion, we emphasize here that the compensation required by the principle of isostasy is achieved—according to the Airy mechanism—*directly* below the topography; wherever the topography rises, creating a mass excess above ground, the compensation by a deepening crust takes place below ground precisely at that location. Sometimes this mirror-image compensation mechanism is called **local Airy isostasy** or **local compensation**. The deficit of mass below the mountain is called the **mountain root**.

In contrast to the Pratt mechanism, with Airy isostasy the depth of compensation is well defined: it should be at the greatest depth to which the crust penetrates into the mantle in that region, e.g., in Figure 5.10 at the base of the mountain root.

In even greater contrast to the Pratt mechanism, the behavior of the crust when surface mass erodes away—the dynamic test that the Pratt mechanism failed—is perfectly reasonable under the Airy mechanism. If there is less mass excess above ground (as the eroded sediment is carried away out of the picture), the mass deficit below must diminish in order to regain an isostatic balance. The deficit is reduced by the crust 'popping up,' floating higher on the mantle, so that the crust does not penetrate as deeply and the mountain root shrinks. In fact, this is the same process by which logs float in water: if the top of a log is sliced away, the log will 'pop up' and float higher.

As the erosion of the mountain continues, the crust will continue to pop up, exposing it to further erosion. Ultimately, in order to completely level the crust down to the height of the surrounding plains, a thickness of crust equal to the mountain thickness *plus* the mountain root thickness must be eroded.

Local Airy isostasy is no different than what is known as *Archimedes' principle* in other contexts. Furthermore—as we will see in later chapters—the principle of isostasy implies that, because the mass of all three columns must be the same, the weight or pressure (due to overlying mass) must be the same all along the depth of compensation; local Airy isostasy is thus also identical to what is known as *hydrostatic equilibrium*. In short, it is not surprising that the Airy mechanism passes our dynamic test: it is based on established physical laws.

There is only one 'catch' about Airy isostasy: for the crust to be floating on the mantle, it requires that we treat the mantle as fluid! We will return to this assumption later in this chapter.

Airy isostasy also works in reverse. If mass is added to a region, for example if a river deposits sediment at the river delta, the crust will subside under the load, creating a crustal root which compensates the excess of sediment mass above with low-density crust below. Subsidence at the delta will create more space for more sediment to be deposited, and ultimately allows a very deep sedimentary column to be created—much taller, even, than the thickness of water there.

A last general comment. Formally, the Airy mechanism places no requirement on the crust being uniform density. In some applications, we might take the crustal density to be a constant ρ_C; and we will denote the mantle density by ρ_M. Airy isostasy can work as long as $\rho_M > \rho_C$. But even if the crustal composition varies laterally, the crust still floats on the mantle, and its thickness still depends on the surface topography; we just need to apply the principle of isostasy—with appropriate densities—to the various vertical columns. For example, oceanic crust tends to be distinctly denser than continental crust, but Airy isostasy still holds true in coastal regions, even though adjacent continental and oceanic columns will be of different densities. In some textbooks, the need to account for such differences in density leads the authors to conclude that Pratt isostasy applies, to some limited extent, around the world; but in reality, the underlying physical mechanism everywhere is always Airy.

Using Airy isostasy, we can estimate the thickness of mountain roots. Let's say that a mountain rises a height h above the surrounding plains, and its root extends through a depth H; and say that the crustal thickness away from the mountain is d (see Figure 5.11). As always, we draw identical vertical columns but neglect the mass of air above the crust. The volume of the mountainous column is $(h + d + H) \cdot A$, where A is the column's cross-sectional area; with ρ_C representing the crustal density, the mass contained in this column is $\rho_C \cdot (h + d + H) \cdot A$. The volume of the crustal portion of the plains column is $d \cdot A$, and that of the mantle portion is $H \cdot A$, leading to a mass for the plains column of $\rho_C \cdot d \cdot A + \rho_M \cdot H \cdot A$. If the region is in isostatic balance, the principle of

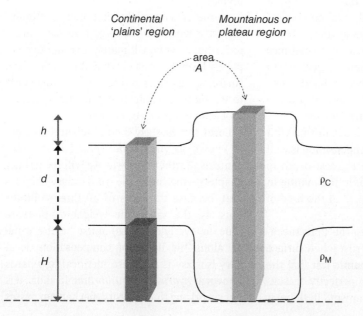

Figure 5.11 Further exploration of Airy isostasy. The depth of compensation is placed at the base of the mountain root, and the masses of air topping the vertical columns are neglected. In this analysis, the height of the mountain is *h*, the thickness of the mountain root is *H*, and the crustal thickness away from the mountain is *d*. The mantle portion of the plains column is drawn with a darker color.

isostasy tells us that the column masses must be equal, thus

$$\rho_C \cdot (h + d + H) \cdot A = \rho_C \cdot d \cdot A + \rho_M \cdot H \cdot A,$$

or

$$\rho_C \cdot (h + d + H) = \rho_C \cdot d + \rho_M \cdot H.$$

Subtracting out the $\rho_C \cdot d$ term from both sides (both columns contain the same crustal mass in that 'central' portion of their columns), and grouping terms involving H, we find

$$\rho_C \cdot h = (\rho_M - \rho_C) \cdot H.$$

This is the central algebraic statement of Airy isostasy. The term on the left represents the mass (per unit area, since we divided out the factor A) of the mountain above ground, i.e., the surficial mass excess; the right-hand term is the mass deficit (again, per unit area) represented by the mountain root, a deficit because the root is filled with rock of density ρ_C rather than ρ_M. And their equality represents an exact isostatic compensation.

Going one step further, our equation implies that

$$H = \frac{\rho_C}{\rho_M - \rho_C} h.$$

For typical, granitic continental material, $\rho_C \sim 2.7$ gm/cm^3; and we will later confirm from compositional studies that mantle material just below the crust is characterized by $\rho_M \approx 3.3$ gm/cm^3. Thus,

$$H \approx 4.2\, h.$$

That is, mountain roots extend downward four times the amount the mountains rise up above ground—4 km deeper for each kilometer in height! We can certainly expect mountain roots to be substantial, and crustal thickness to be significant in mountainous regions, if Airy isostasy prevails.

Like Pratt's theory, the Airy mechanism of isostatic compensation is a statement of how mountains build up: mountains are built by significant crustal accretion and thickening, with buoyancy associated with Archimedes'

principle operating simultaneously as the thickening crust continues to float on the underlying mantle fluid. And a substantial mountain root is created in the process.

Using vertical columns, the principle of isostasy, and the same type of analysis, we could also predict how much thicker the crust in a coastal plains area is than the adjacent oceanic crust—that is, envisioning the continental plains as rising up above the sea floor (perhaps suggested by Figure 5.10), we could predict using Airy isostasy how deep the 'continental root' goes, compared to oceanic crust. In this situation a couple of minor complications arise: oceanic columns include seawater as well as crustal and mantle rock; and oceanic crust, being basalt rather than granitic, is somewhat denser ($\sim$2.9 gm/cm^3) than continental crust.

Compared to our prediction of the depth of mountain roots being four times their height, do you think a greater density of oceanic crust would make continental roots deeper by more than four times their height, or less?

5.2.3 The Moho and Other Evidence of Airy Isostasy

As we have seen, a key variable in Airy isostasy is the depth of the boundary between the crust and the mantle. This boundary is called the **Mohorovičić discontinuity**, or **Moho** for short. Because the density contrast between crust and mantle is appreciable, it is possible for seismic waves, whether generated naturally by earthquakes or artificially by (e.g.) explosions, to trace out the Moho by reflecting off or traveling along the interface; from those waves, the depth to the Moho can be determined (techniques for doing this will be discussed in Chapter 7). The results of seismic surveys confirm the general picture of Airy isostasy: the crust is thickest beneath mountains and thinnest in oceanic regions. Typically, we find

oceanic crust $\sim$5 km thick,

placing the Moho in oceanic regions about 10 km below sea level (the seafloor is typically overlain by ~1 km thick sediments under ~4 km deep oceans),

continental crust ~35 km thick,

mountainous continental crust ~65 km thick or more.

Gravity observations provide additional evidence in support of Airy isostasy. Measurements taken in mountainous regions yield negative Bouguer anomalies: after the data has been corrected for the effects of elevation and mass of the mountains above sea level, the corrected gravity is less—by up to hundreds of mGal—than ideal gravity predicts at sea level. Furthermore, the anomalies are systematic rather than random: the higher the topography, the greater the magnitude of the anomaly. An example of this for the Andes is shown in Figure 5.12. Such large, trending, negative Bouguer anomalies are the consequence of mass deficits below ground, corresponding to mountain roots.

Additional gravitational confirmation that isostatic compensation prevails around the world comes from studies of the *Geoid*, to be discussed in the next chapter.

5.2.4 Airy Isostasy and the Oceanic Response to Atmospheric Pressure Fluctuations

An unusual application of Airy isostasy to purely *fluid* components of the Earth System may help reinforce the basic concepts we have been discussing. In this application, we consider columns of the atmosphere floating atop the oceans; our analysis will lead to an approximation commonly employed by researchers in some geophysical disciplines.

As a hurricane travels over the oceans, the low pressure in the hurricane's eye—pushing down less on the sea surface than the atmosphere does elsewhere—permits a higher sea level there, potentially adding significantly to the hurricane's storm surge as it approaches the coast. To quantify the hurricane eye's effect on sea level, we can imagine that the oceans respond to the low pressure of the eye 'statically' (i.e., without 'sloshing around'), by obeying hydrostatic equilibrium; in that case, the oceanic response will compensate the mass deficit of the eye's low pressure according to local Airy isostasy.

As in other situations, our approach will employ the principle of isostasy. Let H denote the rise in sea level beneath the hurricane's eye

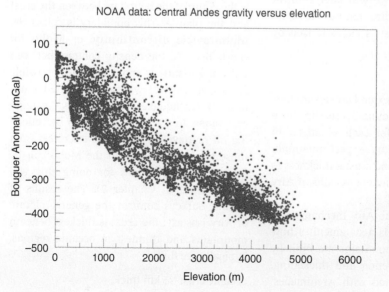

Figure 5.12 Bouguer gravity anomalies in the central Andes, plotted versus station elevation. Note the inverse correlation: the higher the elevation, the more negative the Bouguer anomaly; such a relationship is expected if the Andes mountains have roots in accord with Airy isostasy. Reduced data (6151 measurements) downloaded from http://www.ngdc.noaa.gov/mgg/gravity/1999/data/regional/andes97/.

(a)

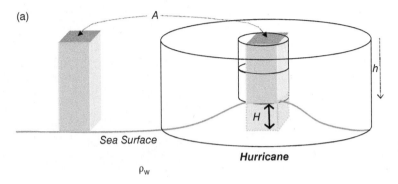

Figure 5.13 The oceanic response to atmospheric pressure variations in a hurricane.
(a) Pressure is lowest in the eye of the hurricane, and greatest away from the storm. The weight of the atmosphere, which determines the pressure it exerts on the sea surface, reflects the mass of the air column. We let *h* denote the height of the air column within the eye (but not draw it here to its full height). If the oceanic response is approximated as a static, isostatic one, we can use the principle of isostasy (as described in the text) to calculate the rise *H* in sea level within the eye of the hurricane, such that the two columns contain equal mass.

compared to elsewhere. We consider two identical vertical columns of cross-sectional area A (see Figure 5.13a); the hurricane column extends arbitrarily upward a distance h above the sea surface (e.g., to the stratosphere, or beyond), and the outside air column matches it, with a total height then of $(h + H)$. The seawater density is ρ_w.

The principle of isostasy tells us the total mass in those columns must be the same. We let m_1 denote the mass of air in the column within the hurricane's eye and m_0 the mass in the column outside the storm. The column within the eye also contains seawater, occupying a volume $(H \cdot A)$ and thus amounting to a water mass equal to $(H \cdot A)\rho_w$. If the columns are in isostatic balance, the principle of isostasy requires

$$m_0 = m_1 + (H \cdot A)\rho_w.$$

Ultimately, we want to relate H to the differences in atmospheric pressure exerted on the oceans. The mass of air contained in the columns determines their weight and thus the atmospheric pressure acting on the ocean surface. The total weight of air in the outside column is $m_0 \cdot g$, where g is the gravitational acceleration; that in the hurricane column is $m_1 \cdot g$.

We will learn in Chapter 7 that pressure, like any other type of stress, is defined as the force per unit area acting on a surface. In Figure 5.13a, the columns are acting on an area A of the sea surface, and the force is the weight of the air in the column. If the atmospheric pressure experienced at the ocean surface is denoted by P_1 within the hurricane's eye and by P_0 outside the storm, it follows that

$$P_0 = m_0 \cdot g/A \quad \text{and} \quad P_1 = m_1 \cdot g/A,$$

so

$$m_0 = P_0 \cdot A/g \quad \text{and} \quad m_1 = P_1 \cdot A/g.$$

Our statement of the principle of isostasy can then be written

$$P_0 \cdot A/g = P_1 \cdot A/g + (H \cdot A)\rho_w.$$

Dividing out the common factor of A, we solve for H, obtaining

$$H = \Delta P/(\rho_w g)$$

where $\Delta P \equiv P_0 - P_1$ denotes the drop in barometric pressure within the eye.

This formula for H is known as the **inverted barometer** approximation—'inverted' because the ocean surface is responding upside down to the pressure variation, rising beneath a low pressure and (in other situations) pushed down beneath a high pressure. For example, if the atmospheric pressure within the eye is 50 millibars less than outside the storm, putting ΔP, ρ_w, and g in consistent units yields $H \cong 50$ cm.

Our derivation of the inverted barometer nicely illustrates the principle of isostasy; but it also reminds us of its underlying assumptions. There are at least two worth questioning.

Is Isostasy the Whole Story Here? To the extent that the oceanic response to barometric pressure variations can be modeled as static, does an isostatic mass compensation mechanism fully explain the change in sea level? That is, under static conditions, how accurate is the inverted barometer?

Theoretical models tell us that the inverted barometer provides a reasonably accurate estimate of the static change in sea level *if* the breadth of the storm is not too great—in this case, it is the "local" aspect of local Airy isostasy that is key. It turns out that when the forcing by atmospheric pressure acts over larger scales, however, the oceans also respond to the change in gravity produced by the change in sea level, a phenomenon called "self-attraction" or "self-gravitation" whereby a higher sea level pulls more seawater toward it. Additionally, the weight of the higher sea level loads the underlying seafloor, deforming it and further changing the ambient gravity (as well as the height of the sea surface). Incorporating the effects of both loading and self-attraction leads to a more accurate model of the oceans' *static* response to atmospheric pressure fluctuations (Dickman, 1988; Boy et al., 1998).

Is a Static Description Sufficient? Static models of the oceanic response neglect the flow of the oceans—the currents—needed to produce the higher sea level in the first place. If the timescale over which the atmospheric pressure varies is sufficiently long, the currents will be slow and such neglect can be a reasonable approximation.

At the other end of the spectrum, on very short timescales, the oceans will certainly 'slosh around,' i.e. behave dynamically. We see this, for example (Figure 5.13b), in

Figure 5.13 The oceanic response to atmospheric pressure variations in a hurricane.
(b) Photograph showing the sea surface in the eye of Hurricane Katrina (August 2005). Winds in the eye of a storm should be minimal, suggesting that the 'whitecaps' in the picture were forced by pressure fluctuations. Those waves, reportedly 40–60 feet high, suggest that under some conditions the inverted barometer approximation may not be satisfactory. From NOAA storm chasers, National Oceanic and Atmospheric Administration (NOAA) / Public Domain.

photographs of the sea surface in the eye of Hurricane Katrina, which occurred in 2005. Winds in the eye of a hurricane should be minimal, suggesting that the 'whitecaps' in the picture were forced by pressure fluctuations. But those waves were reported to be 40–60 feet high! Evidently the oceans were responding dynamically (rather than statically) to the conditions at the time.

Both theoretical studies and observations suggest that the oceans behave dynamically even when the timescale of barometric pressure fluctuations is as long as a few weeks or more (see Dey & Dickman, 2010). On timescales of hours to weeks, then, an isostatic approach to storm surge—even just the part driven by atmospheric pressure—will be … incomplete. Nevertheless, the inverted barometer remains a popular approximation, and is widely used in some geophysical contexts.

5.2.5 A Third Mechanism for Achieving Isostatic Compensation

The Isostatic Response to *Changing* Surface Loads in the Earth System. The Earth System is mechanically active, with mass being redistributed within and even exchanged between its different components. As rivers transport sediment; as water moves through the hydrologic cycle; as ice sheets accumulate or ablate, advancing or retreating; as hot plumes of magma rise and cold lithospheric plates subduct and sink—on all timescales and throughout the breadth and depth of the Earth, the Earth's matter is in motion. When surface mass is redistributed, the weight on the earth changes; surface layers that might have originally been in equilibrium now experience mass excesses or deficits which are no longer compensated by the deficits or excesses below ground.

Yet observations (discussed in the next chapter) show that after broadscale mass shifts occurred in the past, compensation was eventually regained, or nearly so (and, recent and present-day mass excesses and deficits are in the process of achieving compensation). That is, *the Earth is able to respond to changes in surface mass distribution and restore isostatic balance*. This response includes a rising or subsiding Moho, as the crustal root thickness adjusts to the changing surface mass load; and motion of the mantle below the root, flowing in to buoy up the shrinking root or flowing away to allow the root room to grow.

But there is a problem. Airy isostasy was proposed well before Earth science was revolutionized by the unifying theory of continental drift or plate tectonics. We now recognize that the surface of the Earth is divided into a number of lithospheric 'plates' that behave nearly rigidly as they move with respect to each other, driven by mantle convection, generating earthquakes, colliding to form mountains, and rifting apart to form new lithosphere. Their nearly rigid behavior is what defines the lithosphere mechanically; the base of the lithosphere marks a sharp transition to the asthenosphere, a weak zone (probably extending down through the rest of the mantle) capable of slow, solid-state convective flow. Oceanic lithosphere is typically ~100 km thick; continental lithosphere, being generally older and colder, behaves rigidly down to

greater depths, perhaps ~150–200 km (e.g., Sleep, 2005) or more. In all cases, however, the lithosphere includes all of the crust and some amount of the uppermost mantle; thus, the lithosphere contains the Moho within it. And if the lithosphere is nearly rigid, *the Moho is essentially frozen in place*; it cannot migrate through the lithosphere, changing root thickness to compensate for changing surface mass loads. A rigid lithosphere also means that the accompanying mantle flow to support the growth or shrinking of a crustal root in response to changing surface loads cannot take place right at the depth of the root. The problem, then, is the need to reconcile traditional Airy isostasy with modern plate tectonics; that is, *how can isostatic compensation be achieved when a surface load has been placed on a lithospheric plate?*

It should be clarified that this problem concerns the addition of loads to a lithosphere that has already been formed. Traditional Airy isostasy works reasonably well for newly developing lithosphere: as a mountain is growing, and its mass accretes (for example, by a thickening lithosphere as plates collide), the physical law represented by Archimedes' principle ensures that a root will develop, i.e., that mountainous crust will sink down, with the mantle flowing as needed away from the root.

With older, colder lithosphere, the subsidence of the crust to form a root beneath a new load will instead be accomplished largely by a broad *flexure* of the lithosphere under the weight of the developing load. Given the internal strength of lithospheric material, very light surface loads—or even heavier loads concentrated in one spot—can be supported by the lithosphere without flexing it noticeably. For example, when the Empire State Building was erected in New York City—within an area encompassing less than one square city block, atop lithosphere much more than 100 km thick—it is not likely that the sub-Manhattan lithosphere flexed by a detectable amount. In contrast, loads such as the sediment deposited at the delta of a major river, a volcanic cone building up on the seafloor or on land, the advance or retreat of an ice sheet over land, and the mass of water filling an ocean basin after an ice sheet melts, can all be expected to engender a flexural response of the underlying lithosphere (with supporting flow by the *asthenospheric* mantle—the mantle below the lithosphere). Flexure is most successfully generated by loads that are both broadscale and massive.

Downward flexure of the lithosphere does indeed create a crustal root; see Figure 5.14a, b. As the lithosphere bends downward, the Moho is carried along with it, ending up deeper than in areas not flexing. Unlike traditional Airy isostasy, however, the Moho does not end up, mirror-like, reflecting the surface topography point by point, with the fourfold magnification we predicted from $H = [\rho_C/(\rho_M - \rho_C)] \cdot h$. Instead of such local compensation, lithospheric flexure produces a broad compensation over the entire region experiencing the flexure. We can call this mechanism **regional Airy isostasy**, to contrast it with *local* Airy isostasy, though it has also been called **flexural isostasy** and the **plate mechanism**.

As Figure 5.14b implies, a detailed gravity survey of the region might reveal the existence of that broad regional root: right beneath the seamount, for example, the mass deficiency of the root would be not quite sufficient to compensate the seamount's load; and, off the flanks of the seamount, there would be some mass deficiency (i.e., some root is present) below the seafloor but no overlying mass excess in need of compensating. This disparity is also illustrated in Figure 5.15. One of the first marine gravity surveys documenting these point-by-point imbalances was published by Vening-Meinesz in the 1930s. In honor of his discovery, regional Airy isostasy is also called the Vening-Meinesz mechanism of isostasy. Since Vening-Meinesz' work took place well before the introduction of plate tectonics, and because Turcotte (in the 1970s and 1980s) advanced the plate mechanism rigorously within a tectonics framework, we can also call flexural isostasy the **Vening-Meinesz–Turcotte mechanism**.

In the Vening-Meinesz–Turcotte mechanism, the regional density deficit $(\rho_M - \rho_C)$ between crust and mantle still ultimately drives the isostatic response to a surface load, but that response is modified by the lithosphere's *flexural rigidity*, a physical property that determines the lithosphere's ability to flex

(a) Before the undersea volcano forms

(b) After the undersea volcano has formed

Figure 5.14 Response of oceanic lithosphere to the formation of an undersea volcano. The Moho is shown as a bold line. For clarity, the oceans and crustal layer are vertically exaggerated (the oceans are typically ~5 km deep, and the Moho would typically be ~5 km below the seafloor, whereas the lithosphere is typically ~100 km thick).
(a) Before the undersea volcano forms.
(b) As the volcano builds up, loading the crust, the lithosphere flexes downward in response, creating a regional 'root' (shown as the brightly colored area).

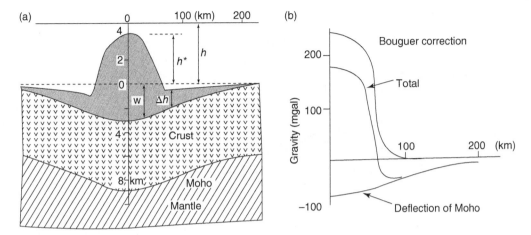

Figure 5.15 Gravity variations predicted around a seamount in a state of flexural isostatic equilibrium. From Lambeck (1988) / Cambridge University Press / Fig. 9.10.
(a) Model of seamount load on thin oceanic plate (vertical scale exaggerated for clarity).
(b) Calculated gravitational effect of the seamount, i.e. mass excess, labeled "Bouguer Correction," and of the flexural root, i.e. mass deficit, labeled "Deflection of Moho," both shown versus distance away from the load's center point. The imperfect compensation, point by point, is seen in their combined effect, labeled "Total," failing to equal zero over almost the entire region.

under the load. The flexural rigidity depends on material properties we will learn about in Chapter 7 related to how easily the lithospheric rock can deform, and also on how thick the plate is: it's a lot easier to flex a thin plate than a thick one.

The more flexurally rigid the lithosphere, the greater its ability to support a surface load; in this case, the plate will flex downward by a smaller amount but over a larger region. For weaker lithosphere, the surface load will depress it more but over a more limited area. In the limit of zero lithospheric rigidity, we regain local Airy isostasy: compensation exists precisely below the surface load.

Flexural Compensation Is Imperfect. Even considering the entire region, overall, the broad compensation between load and root that results from lithospheric flexure will never be perfect: the lithosphere's flexural rigidity allows it to support some weight without really flexing. Thus, lithospheric rigidity undercuts its ability to achieve mass compensation, whether point by point or regionally. As a consequence, the principle of isostasy (requiring equal mass in identical adjacent (e.g.) columns) cannot be invoked to determine

the magnitude of flexing—or the size of the root—ultimately achieved by the lithosphere.

The fact that there are forces involved in flexure—forces that can even dominate the buoyancy forces associated with the surface load and crustal root—suggests that a force approach would be a more accurate way to predict flexural subsidence (or rebound) and final root thickness than would a simple comparison of mass in adjacent columns. The resulting theoretical development (see, e.g., Turcotte & Schubert, 1982), as expected, shows that the predicted flexure depends on plate thickness and rigidity, as well as mantle and crustal densities.

The imperfections in flexural compensation also indicate the need for a more general definition of isostasy than "a state of mass balance"—if we are to include a compensation mechanism like regional flexure. Given the equivalence of Airy isostasy not only to Archimedes' principle but also to hydrostatic equilibrium (which is a force balance, as we will learn later in this textbook), a better definition of isostasy might be *a condition where, at some 'depth of compensation,' the region is in a state of hydrostatic equilibrium* (McNutt, 1980),

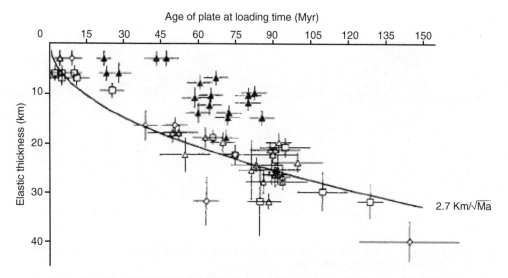

Figure 5.16 The flexural response of oceanic lithosphere to surface loads, as a function of the age (in Myr) of the lithosphere at the time of loading. In this graph, from Calmant et al. (1990) / John Wiley & Sons, and in most other works as well, the response is represented by a proxy nicknamed the "elastic thickness" of the lithosphere. The flexural rigidity of the lithosphere is proportional (according to theory) to the cube of this thickness.
The elastic thicknesses were inferred from observations of loading by seamounts and volcanoes in the Atlantic (open square symbols, with error bars), Indian (diamonds), and Pacific (triangles) Oceans; the filled triangles refer to observations in the south-central Pacific.
The analysis shows that the older the lithosphere is when loaded, the more rigid its flexural response is. The south-central Pacific shows this same relationship but with a lower rigidity (by a factor of $\sim 2^3 = 8$ at 80 Myr according to the graph).
Other examples may be found in Watts and Ribe (1984) and Watts (2001, discussed in Hillier, 2005).

or at least a hydrostatic equilibrium modified by flexural forces.

Flexure of the Oceanic Lithosphere Is Particularly Interesting. Regional isostasy provides a way to deduce important properties of the lithosphere. Oceanic lithosphere forms at mid-ocean ridges, and cools as mantle convection and seafloor spreading carry the young lithosphere away. The temperature of the lithosphere has a great impact on its strength, including its flexural rigidity: cooler lithosphere is stronger; and cooler lithosphere allows the asthenosphere immediately underneath to cool as well, ultimately resulting in a thicker lithosphere. And all of this depends on the lithosphere's age, i.e., on how much time has passed since it formed at the ridge; such dependence is illustrated in Figure 5.16. Of course, there may be complicating factors; for example, volcanism away from the ridge can

reheat the lithosphere, reducing the flexural rigidity and also resetting the lithosphere's cooling 'clock.'

Still another complication is what the lithosphere's *rheology* is, that is, whether it behaves like an elastic material—meaning that it responds instantaneously to the load—or nonelastically, with a response that develops over time (see, e.g., Watts & Zhong, 2000). This question is pursued in Chapters 7 and 8.

In the case of continental lithosphere, those complicating factors and others prevent any simple interpretation of its flexural rigidity (Burov & Diament, 1995).

5.2.6 Isostatic Response to Surface Loads in the Earth System: Anomalous Regions

When surface mass is redistributed, and various regions find themselves no longer in isostatic balance, it will take some time before the asthenospheric mantle has flowed in or out and the lithosphere has regained isostatic equilibrium. On a geologically or climatologically active Earth, some unbalanced regions can thus be expected to be present at any given moment. Gravity measurements can shed light on where those anomalous regions are, and on the degree of imbalance.

Isostasy Leads to Another Gravity Correction. In general, the existence of a crustal root—regardless of whether the region is in a state of local or regional isostatic balance, or neither—amounts to still another predictable source of gravity variations (like elevation) affecting all gravity measurements, one that can be accounted for and removed to search for more unusual anomalies. Referring to Figure 5.2 (see also Figure 5.11), a measurement made at point Q atop a mountain might be telling us that there is a real mass anomaly within or below an otherwise featureless mountain—or its value might simply be a consequence of the elevation of the site, or of the mass excess represented by the mountain above ground, *or* (as we now understand) of the mass deficit represented by the mountain root. We can express all of this, in the notation used earlier, as

$$g_Q = (g_S + \Delta g_{ANOM}) - 0.3083h + 2\pi G\rho_C h$$
$$- 2\pi G(\rho_M - \rho_C)H$$

where g_S is gravity on the Spheroid away from the mountain but at the same latitude, and Δg_{ANOM} is the contribution to g_Q from any truly anomalous mass. In this expression, we have also used the Bouguer approximation for the effect of the root, treated as an infinite flat slab of uniform density deficit $(\rho_M - \rho_C)$. And in this expression, note that the mass of the mountain above ground has added to the observed gravity, but the root deficit counters that.

This expression can be rearranged to solve for the anomalous gravity. In sum, correcting our observation g_Q for the effects of elevation, mass above ground, and crustal root mass deficiency, we define that anomalous gravity—the

isostatic anomaly—as

$$(g_Q + 0.3083h - 2\pi G\rho_C h + 2\pi G[\rho_M - \rho_C]H)$$
$$- g_S.$$

The approximate correction we have used here for the gravity of the root's mass deficit,

$$2\pi G(\rho_M - \rho_C)H,$$

is called the **isostatic correction**. This correction and the resulting gravity anomaly require that the root thickness H be known independently, e.g., from a seismic survey, or else it must be modeled in some way.

In general, mountain roots always exist, so this correction is always useful. We caution that its name is misleading, in that its application does not necessarily mean that the region is in a state of isostatic equilibrium, nor that the value of H it uses corresponds to a root of isostatic thickness. Interpretation of the isostatic anomaly also carries the same sort of ambiguity as the Bouguer anomaly: the source of the anomaly can be anywhere beneath point Q, including in the mountain root; in the latter case, one possible reason could be a nonisostatic root thickness.

Because crustal roots are many kilometers thick, the isostatic correction is typically $\sim 10-10^2$ mGal; removing the effect of the root may allow smaller anomalies of interest to be revealed. An example of this is given in Figure 5.17a, where isostatically corrected gravity measurements in the western United States—with the resulting isostatic anomalies of magnitude $\sim 1-10$ mGal—show correlations with mineral deposits. In contrast, Figure 5.17b shows an isostatic anomaly map of the continental United States, which reveals an intriguing large-magnitude (~ 50 mGal) structure called the "mid-continent gravity high"; this feature has been interpreted as an ancient failed rift.

Using Gravity to Assess Isostatically Anomalous Regions. Gravity data can also be used to locate probable regions of isostatic imbalance and to infer the degree of imbalance. We look first into the latter, because

(a)

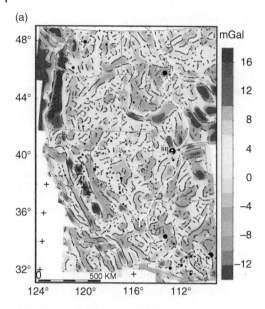

Figure 5.17 Uses for isostatic anomalies unrelated to isostasy.
(a) Isostatic anomaly map for the western United States, from Hildenbrand et al. (2000) / U.S. Department of the Interior. With complex, varying crustal root contributions to the observed gravity removed, the authors were able to show correlations between the residual gravity (the color scale, in mGal) and the locations of mineral deposits (black circles and squares), particularly in Idaho, Montana, and Utah. In the latter, the Bingham porphyry copper deposit, outside of Salt Lake City, is marked (right center on the map) as BH.

it is relatively straightforward. Our analysis will consider all types of gravity anomalies but focuses especially on Bouguer anomalies; recalling the fundamental ambiguity in all gravity data, noted above, we caution that this analysis may seem somewhat unsatisfying.

We have already seen that (negative) Bouguer anomalies can indicate the presence of a crustal root. With an Airy-type compensation mechanism, one approach to determining whether the root is the proper thickness to balance a surface load is to assume that the Bouguer anomaly we measure—let's denote it by Δg_B—is due primarily to the mass deficit of the root (rather than, say, some unrelated anomalous mass within the surface load or the lithosphere). In that case, to good

approximation, Δg_B must equal $2\pi G(\rho_M - \rho_C)H$, which is the gravitational pull of an infinite flat slab with the thickness and density contrast of the root deficit. Consequently, dividing Δg_B by $2\pi G$ allows us to estimate the deficit itself, $H \cdot (\rho_M - \rho_C)$. (Note: in this analysis, as with all isostatic calculations in which the column cross-sectional area A has been divided out, the deficits will be per unit area.)

- With the height h and density (say, ρ_C) of the surface load known, any disparity between the surface mass excess ($h \cdot \rho_C$) and our estimate of the root mass deficit ($\Delta g_B / 2\pi G$) can be interpreted as an isostatic imbalance.

- Alternatively, dividing our estimate ($\Delta g_B / 2\pi G$) of the root mass deficit by an estimate of the density contrast ($\rho_M - \rho_C$) yields a prediction of the root thickness H; again looking to quantify the isostatic imbalance, we can determine whether identical vertical columns through the mountain (or whatever the surface load is) versus outside of it contain equal amounts of mass.

- For that matter, once we have estimated H we can assess how close the region is to complete isostatic compensation simply by comparing our value of H with the root thickness expected from Airy isostasy when mass compensation is complete (as we have previously seen, that expected value is $\rho_C \cdot h / [\rho_M - \rho_C]$).

A related approach involving fewer calculations and assessments keeps us in the gravity domain. We know that, with a local Airy compensation mechanism, the root thickness H beneath a mountain of height h should be $\rho_C \cdot h / (\rho_M - \rho_C)$ if the region is in balance. So, if we apply to our gravity measurements an isostatic correction (which is of the form $2\pi G[\rho_M - \rho_C]H$) for the root using *that* value for H, and if the Bouguer anomaly, Δg_B, is thereby eliminated, we can conclude that the actual root thickness is indeed isostatic.

This conclusion would likely still hold if Δg_B is significantly reduced but not completely eliminated: a small residual could be the result of other mass anomalies, unrelated to isostatic processes, since (as noted earlier) such

(b)

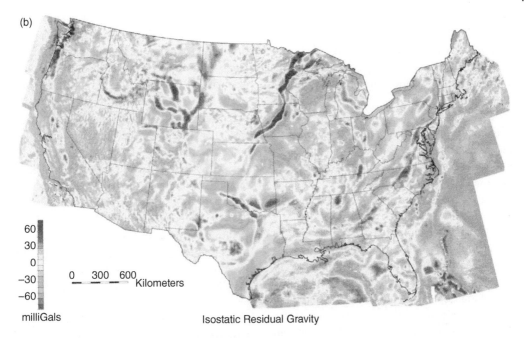

60
30
0
-30
-60

0   300   600 Kilometers

milliGals   Isostatic Residual Gravity

Figure 5.17 Uses for isostatic anomalies unrelated to isostasy.
(b) Isostatic anomaly map for the continental United States. As a first approximation, the root thicknesses needed for isostatic corrections were estimated from the surface topography and (offshore) the seafloor bathymetry. Many of the anomalous highs and lows are interpreted as what we might call tectonically produced 'scars.'
The "mid-continent gravity high" extending from northernmost Minnesota and Michigan down into Kansas can be interpreted as remnant dense oceanic crustal structures originating in failed mid-ocean rifting during the Precambrian (see also Stein et al., 2014).
Other notable features include the Amarillo-Wichita gravity high in northern Texas, with a similar rift-type origin (e.g., Robbins & Keller, 1992); structures associated with the Laramide orogeny in and around Wyoming; and a possible fossil plate boundary parallel to but to the east of the Appalachian Mountains.
Simpson et al. (1986) / John Wiley & Sons, Inc.

mass excesses and deficits often contribute only small amounts, perhaps several mGal, to observed gravity, whereas a root thickness off by hundreds of meters or more would have a gravity signature larger by an order of magnitude.

As for locating isostatically balanced regions, that order-of-magnitude difference in gravity suggests a quick though approximate method: look for where the Bouguer anomalies are large but the isostatic anomalies, calculated assuming an isostatic root thickness, are relatively negligible. Or, to identify isostatically uncompensated regions, we need only look for places where those anomalies have a large magnitude.

Mathematical Details Lead to a Procedural Shortcut. Actually, as long as we are assuming an isostatic root thickness, this quick approach can be made even quicker by phrasing it in mathematical terms. At a candidate location where the Bouguer anomaly is large, this approach, as just stated, directs us to: (1) specify the root thickness H, according to Airy isostasy, as $\rho_C \cdot h/(\rho_M - \rho_C)$; then (2) add the isostatic correction $2\pi G(\rho_M - \rho_C)H$ to the Bouguer anomaly,

$$(g_Q + 0.3083h - 2\pi G\rho_C h) - g_S,$$

in effect constructing the isostatic anomaly

$$(g_Q + 0.3083h - 2\pi G\rho_C h + 2\pi G[\rho_M - \rho_C]H)$$
$$- g_S,$$

and assess whether that anomaly is significantly smaller.

However, with our specified value of H from step (**1**), the isostatic correction simplifies to $2\pi G\rho_C h$ and the isostatic anomaly becomes

$$(g_Q + 0.3083h) - g_S,$$

which is the free-air anomaly! Of course, this makes sense, in that the Bouguer correction and the isostatic correction refer to the same amount of mass—one an excess, the other a deficit—*if* isostasy prevails, in which case their effects exactly cancel. But this suggests that the free-air anomaly can be used as a diagnostic to indicate regions of isostatic imbalance.

In short, a very quick (though still very approximate) way to identify isostatically anomalous regions would be to look for where the free-air anomalies have a large magnitude. Acknowledging that gravity anomalies can result from any number of anomalous mass distributions, only some of which bear on isostatic issues, this approach should be supplemented wherever possible with other types of observations.

The free-air approach is a common practice, though it has its pitfalls. For example, the Hudson Bay region has one of the largest (negative) observed free-air gravity anomalies worldwide, ranging up to ~-50 mGal. Based on this and on its location—at the center of the massive ice covering a good fraction of North America during the last glacial maximum—it was generally believed that this gravity anomaly represents the isostatic imbalance that remains today from the ice load's disappearance. However, analyzing the spatial patterns of these and other free-air gravity anomalies, Simons and Hager (1997) concluded that only about half of that maximum anomaly should be associated with incomplete rebound; they proposed that much of the remainder is a consequence of the deformation of the lithosphere (a 'down-warping') accompanying mantle convection (see also Peltier et al., 1992).

Finally, to assess the degree of isostatic balance in a region where flexure is likely to be important, we might consider using the same gravity anomalies as above. However, because of the 'incomplete' nature of this mechanism (incomplete amount of flexing and incomplete compensation, unless the load is extremely extensive or the lithospheric rigidity is nearly zero), the value of such interpretations as diagnostic of isostatic balance is less definitive. For example, the Bouguer anomalies associated with a crustal root will be smaller (less negative) when the root is achieved by flexing. And with regional compensation, imperfect as it is, the free-air anomaly will be nonzero even if there is isostatic balance. In short, for more precise and less ambiguous interpretations, modeling of the flexure and associated asthenospheric flow might be called for. Such an approach involves mechanical properties and dynamical equations we will learn about in Chapters 7–9, but one important property—viscosity—will be introduced below, in the next subsection.

Vertical Crustal Motions in Anomalous Regions. *Geodetic surveys* provide a direct and easy alternative to gravity measurements for locating isostatically anomalous regions, by detecting the *crustal subsidence or rebound* that must accompany the root's adjustment. Such surveys can use the modern tools of geodesy to quantify ongoing vertical crustal motion; past vertical movements can be inferred from geological proxies, such as the raised beaches cut by wave action in a region that had undergone isostatic rebound (e.g., Figure 5.18).

From these techniques, we can identify several processes acting throughout the Earth System that have produced isostatic imbalances on a regional scale. Those processes will prove to be invaluable for quantifying the mantle's viscosity structure.

- Erosion of extreme topography. In the case of the Himalayas, for example, a mountain range whose formation was complete ~40 Myr ago, uplift is observed at a rate of $\sim1.6\,\text{cm} \pm 0.8\,\text{cm}$ per year in that region (the southern Tibetan Plateau) (Xu et al., 2000). One explanation: erosion of those

Figure 5.18 Raised beaches in Bathurst Inlet, Nunavut (formerly part of the Northwest Territories), in the Canadian arctic. Models of Pleistocene deglaciation suggest that this region probably lost its ice load around 10,000–11,000 yr ago. Image by M. Beauregard / https://en.wikipedia.org/wiki/Post-glacial_rebound / CC BY 2.0.

highest peaks on Earth has led to an isostatic imbalance in which the mountain roots are compensating for more mass than now exists in the mountains; the lithosphere responds to this imbalance by flexing upward. Interestingly, in at least one part of this region, intense river incision into the bedrock (at a rate of ~1 cm/yr) and the resulting landsliding from mountain slopes keep pace with the uplift, such that the steep mountain slopes are maintained (Burbank et al., 1996).

- Accumulation of sediment in deltas of major river systems. An example of this is the Mississippi River, which currently deposits an estimated ~150 tons of sediment annually at its entrance into the Gulf of Mexico (Thorne et al., 2008); sedimentation was significantly greater prior to 1950 (Seed et al., 2006). Over the past several thousand years, such deposition has added ~10^2 m of sediment to the floor of the Gulf within the delta. The imbalance produced by this load requires the delta region to flex downward, ultimately lowering the seafloor more than the sedimentation is adding to it (if the sediment was as dense as the crust and the response was through a local Airy mechanism, the subsidence would have to create a root

four times as thick as the sediment layer). Relative to the sea surface (for reference), the seafloor is observed to subside by about 0.9 cm/yr more within the delta than outside it (see Penland & Ramsey, 1990). Mixed in with a variety of other factors—especially sediment compaction, fluid withdrawal, and the rise in global sea level—up to 25%–30% of this subsidence (Jurkowski et al., 1984; see also Yu et al., 2012 and Dokka, 2011) is attributed to a downward flexure of the lithosphere in response to the continuing sediment deposition.

Surprisingly, however, the entire region is not subsiding: north of the delta, we observe crustal uplift in southern Mississippi, with the center of the uplifting 'dome' rising at rates exceeding 3 mm/yr compared to the 'edges' (Jurkowski et al., 1984 and others). This has been interpreted as a secondary consequence of the sediment loading, an effect called the "forebulge" or "peripheral bulge," which will be discussed below in the context of a still different kind of surface loading.

There may be a human dimension to the subsidence of the Mississippi delta, as increasing portions of the city of New Orleans and its suburbs have found themselves below sea level. The subsidence has resulted in the loss of protective wetlands surrounding the city and in the lowering of floodwalls. When Hurricane Katrina hit the region in 2005, it is possible that these factors contributed to the death and destruction caused by the hurricane and by the subsequent failure to contain Lake Pontchartrain. But according to Seed et al. (2006), the immediate cause of the flooding was (preventable) erosion of the levees and floodwalls, and the failure to install floodgates in the drainage canals; and Dokka (2011) argues that, as far as subsidence is concerned, recent fluid withdrawal has been its primary cause.

- Mass redistributions between the cryosphere and hydrosphere caused by major shifts in climate. In the past, these have included the

(a)   (b)

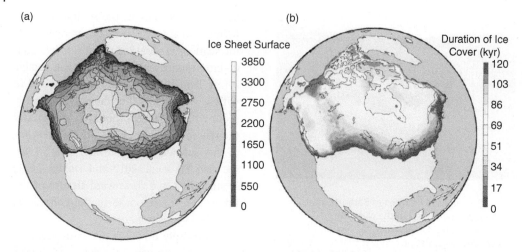

Figure 5.19 (a) One possible reconstruction of the North American ice load (the Laurentide (Hudson Bay), Innuitian (northern) and Cordilleran (western) ice sheets) at the last glacial maximum 21,000 yr ago; this reconstruction is based on paleoclimate, glaciological, and isostatic rebound models combined with evidence of the southern extent of the ice sheets. Elevation of ice sheet surface in meters. Marshall et al. (2002) / with permission of Elsevier.
(b) From the same analysis, one possible time history of these North American ice sheets, modeled over the past 120,000-year glacial cycle, shown as the number of thousands of years the ice persisted. Marshall et al. (2002).

disappearance of ice, and continental flooding, as Earth entered a Hot House climate; the growth of ice sheets and drop in sea level in the transition to an Ice House climate; and glacial advance and retreat within an Ice House climate, particularly during an ice age. Judging by their effect on sea level, the advance of northern hemisphere ice sheets down to mid-latitudes during glacial periods of the Pleistocene Ice Age (Figure 5.19 shows their maximum advance in North America during the last glacial period), and their disappearance during interglacial periods (especially near the end of the Pleistocene), led to the most significant, large-scale time-varying surface loads this planet has seen since the mid-Cretaceous Hot House!

These loads were able to generate major vertical movements of the lithosphere in part because of their massiveness (comprising a total volume of ice of perhaps 45 million km³ at their maximum) and their lateral extent (covering areas up to thousands of kilometers wide). But their effectiveness also depended on the fact that

the timescales over which the loads varied were *comparable* to the timescale of the mantle flow supporting the lithospheric flexure (for example, if cycles of glacial advance and retreat lasted only a few decades, there would have been too little time to produce much flow in the mantle).

During the glacial period of an ice age, the mass redistribution is manifested both in the growth of ice sheets on land and in the decrease in ocean volume (as water evaporated from the ocean surface ends up precipitating onto and thereby becoming incorporated into continental ice). The continental lithosphere subsides under the increased ice load, and the oceanic lithosphere flexes upward under a reduced seawater load. During the interglacial, as ice sheets melt away and the oceans are refilled, the situation reverses, with the continents rebounding and the seafloor subsiding.

At present, near formerly glaciated regions the rate of crustal rebound can be ~1 cm/yr or more (e.g., see Figure 5.20), and the free-air anomalies associated with a root that is still

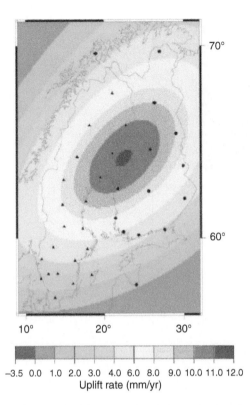

70°

60°

10°   20°   30°

-3.5 0.0 1.0 2.0 3.0 4.0 6.0 8.0 9.0 10.0 11.0 12.0
Uplift rate (mm/yr)

Figure 5.20 Crustal uplift observed in Finland and Scandinavia by the BIFROST (Baseline Inferences for Fennoscandian Rebound Observations, Sea-level, and Tectonics) program using GPS measurements between 1993 and 2000. Image from Johansson et al. (2002) / John Wiley & Sons.

too deep might be ~-20–-40 mGal (see, e.g., Walcott, 1973 and Peltier & Wu, 1982) though, as noted before, ~-25 mGal might better represent the largest of these. Of course, thousands of years ago, right after the ice sheets had disappeared, both the rate of rebound and the associated free-air anomalies would have been much greater.

Timescales of the Ice Age. Before pursuing isostatic connections further, however, we briefly focus on the *timescales* of these processes. Striking evidence for those timescales during the Pleistocene comes from variations in the abundance of oxygen and hydrogen isotopes versus depth (and thus time) in ice cores drawn from Antarctica and Greenland. These isotopes (which will henceforth be denoted symbolically by the more concise

^{18}O, for example, rather than O-18, just as we have increasingly employed other symbolic notation in the chapters on gravity, e.g., for calculus applications) yield a detailed record of climate variations over the past 800,000 yr (e.g., Jouzel et al., 2007). Heavier water molecules—containing ^{18}O, a heavier isotope of oxygen, rather than the more common ^{16}O, or deuterium (D, or formally ^{2}H) rather than the more common ^{1}H—cannot evaporate from the ocean surface, nor travel long distances through the atmosphere (to fall ultimately as snow onto the polar ice sheets), as easily as the more abundant and 'lighter' H_2O (see, e.g., Ruddiman, 2014; Appendix 1); but at higher temperatures, the percentage that succeeds increases. Changes in the proportion of D to ^{1}H or ^{18}O to ^{16}O measured in ice cores thus reveal how temperatures in the shallow ocean must have varied at those times.

In Antarctica, the longest records have been obtained from the East Antarctic Plateau: at Vostok Station, where the ice core goes back 420,000 yr (e.g., Petit et al., 1999); and at Dome C, which generated the 800,000-year time series. Vostok results are shown in Figure 5.21; they reveal that glacial times were characterized by steadily decreasing temperatures over spans of $\sim$100,000 yr—associated with a slow advance of the northern hemisphere ice sheets—alternating with a recovery to warm interglacial times (and a rapid melt-away of the ice sheets) for spans of $\sim$20,000 yr. The 'sawtooth' pattern of temperature variation on those timescales is matched in the variation of the volume of continental ice (discussed below) and—intriguingly—in the atmospheric abundance of greenhouse gases like carbon dioxide (and methane, not shown), inferred from the composition of air bubbles trapped in the ice cores.

That very distinctive sawtooth pattern is seen in ice cores from other sites as well. And further confirmation comes from the oceans. Scientists have analyzed the sediments blanketing the seafloor, focusing on the oxygen isotope proportions found in calcium carbonate ($CaCO_3$) shells of foraminifera; forams

(a)

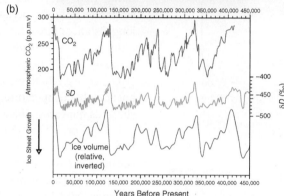

(b)

Figure 5.21 Climate study from the geochemistry of ice cores.
(a) "Pushing out the first ice core" in the West Antarctic Ice Sheet Divide Ice Core Project. National Science Foundation / https://waisdivide.unh.edu/ / last accessed August 22, 2023.
(b) Three measures of the state of our climate over the past 400,000$^+$ yr from the Vostok, Antarctica ice core—with older samples found deeper in the core—all exhibit the 'sawtooth' pattern characteristic of the late Pleistocene Ice Age. Figure from Sigman and Boyle (2000).
Top curve: variation in atmospheric abundance of carbon dioxide, as measured in gas bubbles trapped in the ice.
Middle curve: variation in the abundance of deuterium relative to common hydrogen; as described in the text, greater (less negative) proportions of the heavier isotope in the ice require a warmer climate.
Lower curve: variations in global ice volume, as inferred from changes in the proportion of ^{18}O, the heavier oxygen isotope, relative to common oxygen. Less ice worldwide (note inverted y-axis scale) requires warmer temperatures, which also results in a greater proportion of ^{18}O.
The symbolic quantities δD and $\delta^{18}O$ have complex definitions but in both cases depend on the ratio of heavier to lighter isotopes (D to ^1H or ^{18}O to ^{16}O).

are one-celled organisms with an impressive ability to extract dissolved calcium and carbonate from the surrounding seawater and build relatively large, generally globular or spiral, shells. The proportion of ^{18}O in the oceans relative to ^{16}O *increases* at colder temperatures because its evaporation from the sea surface is more difficult—leading to the *reduction* in ^{18}O we see in ice cores during glacial times; they are two sides of the same 'coin.' The isotopic composition of those carbonate shells reflects the proportions of ^{18}O and ^{16}O in seawater at

the times the shells were built; for example, if precipitated during colder times, a shell would have a slightly greater proportion of ^{18}O than in warmer times. Furthermore, such an increase is compounded biologically: at colder ambient temperatures, forams preferentially take in a greater fraction of ^{18}O into their shells. Ultimately, when the forams die, their shells rain down to the ocean bottom, accumulating over time. Deep-sea sediment cores, recording the variation in ^{18}O versus ^{16}O as a function of time, reveal the same kind of sawtooth pattern and the same dominant timescales over the past 700,000 yr as the ice cores.

Changes in the proportion of oxygen isotopes over time in deep-sea cores have also been used to infer changes in the volume of land-based ice sheets (e.g., Shackleton, 1967). As noted above, water evaporating off the sea surface is preferentially 'lighter'; and, as it travels through the atmosphere on the way to an ice sheet, the water vapor lightens further as the heavier molecules rain out. The end result of such *fractionation* is that ice sheets become huge reservoirs of ^{16}O—in effect taking ^{16}O out of the shallow ocean as they grow, leaving the oceans with a slightly increased proportion of ^{18}O. Conversely, when ice sheets melt and drain away back to the oceans, the proportion of ^{18}O in the oceans must decrease. It follows that the oxygen isotope ratio correlates with the volume of continental ice sheets.

Just so you get it straight: in terms of the proportion of ^{18}O to ^{16}O in sediment cores, is the correlation with ice volume a direct or inverse relation?

This correlation is not unexpected since the change in global temperature responsible for the variation in isotope abundances would also have driven the growth or shrinking of the ice sheets. Thus, for example, sections of the sediment cores containing proportionally more ^{18}O, corresponding to colder times, also represent times during the Ice Age when ice sheets could advance and cover more of the northern hemisphere continents. However, the actual correlation turns out to be complicated by time lags and nonlinearities—a consequence of the complexities of the fractionation, and of how ice sheets evolve as they grow and decay (see, e.g., Mix & Ruddiman, 1984).

The changes in ice-sheet volume also imply corresponding changes in global sea level; e.g., glacial times, when ice sheets were more expansive, correspond to times of lower sea level. A correlation between oxygen isotope abundances and ice-sheet volume thus requires that a correlation exists between those isotopes and sea level. Such a correlation is indeed found, using sea-level histories determined from the present-day height of ancient coral reefs above (or their depth of submergence below) sea level—combined with the fact that, while they were growing, those reefs would have remained more or less *at* sea level. That is, the changes in sea level implied by those reefs are consistent with the changes associated with our inferred glacial–interglacial ice volumes (e.g., Shackleton & Opdyke, 1973; Broecker & van Donk, 1970).

Sediment cores are also important because they encompass a time span more than two million years longer than the ice cores. And it turns out that, further back than 700,000 yr ago, the sawtooth pattern changed, in both the Dome C and marine sediment isotope records. The origin of the sawtooth-pattern timescales, and the nature of their changes nearly 3/4 Myr ago, will be discussed in the next chapter.

The loads of the Ice Age ebbed and flowed—slowly!—on these timescales. We pick up the story leading up to and following the most recent glacial maximum; first, the ice.

Northern hemisphere ice sheets had slowly been advancing southward since the end of the Sangamon (Eemian) interglacial ~120,000 yr ago; by 20,000 yr ago, the ice sheets averaged nearly 2 km in thickness (~2.5 km in North America; see Marshall et al., 2002) and covered an area 13 times the size of this hemisphere's only present-day ice sheet, in Greenland. According to one scenario (Clark et al., 2009), that areal extent was actually reached more than 26,000 yr ago, and remained largely stable for the next 6,000 yr.

Under these massive ice loads, the North American and Eurasian continental lithospheres flexed downward, attempting to achieve isostatic compensation through the Vening-Meinesz–Turcotte mechanism.

Then, around 20,000 yr ago and in a geologic instant, the climate warmed, the Ice Age ended (or, at least, the most recent glacial period ended), and the surface loads began

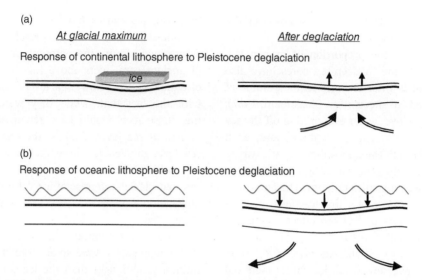

(a)

At glacial maximum   *After deglaciation*

Response of continental lithosphere to Pleistocene deglaciation

(b)

Response of oceanic lithosphere to Pleistocene deglaciation

Figure 5.22 Response of (a) continental and (b) oceanic lithosphere to the end of Pleistocene glaciation. Vertical scales are distorted to emphasize the movement of the Moho (bold line) and, in (b), the ocean surface.
(a) After the ice load has disappeared, the continental lithosphere flexes upward to eliminate a regional crustal root that is now too deep. Mantle material, represented by the double arrows, creeps in to fill the gap beneath the rising lithosphere and support the upward flexure.
(b) As continental ice sheets melt away and recede, the meltwater fills the oceans; the increased load causes the oceanic lithosphere to flex downward, producing a deeper regional root to compensate for the added surface load. Mantle material, represented by the double arrows, slowly flows out to allow the lithospheric subsidence to take place.

to melt away. The continental lithosphere has been rebounding ever since, to regain an isostatic balance (see Figure 5.22). In northwestern Europe, for example, it is estimated from ancient shorelines that over 275 m of uplift has already occurred (see Figure 5.23); from free-air gravity anomalies, perhaps another 120 m of rebound remains (e.g., Walcott, 1973; Steffen & Wu, 2011).

The ocean's story was mostly the opposite. As the ice sheets grew and advanced, following the end of the Sangamon, sea level dropped; from raised shorelines, coral reefs (whose growth surface should be close to sea level), submerged land vegetation, and other evidence, it is estimated that the oceans were as much as ~130 m below current sea level at the last glacial maximum (see Clark & Mix, 2002). To compensate isostatically for the diminished volume of seawater, the oceanic lithosphere would have responded generally by flexing upward and rebounding. But then, beginning ~20,000 yr ago, the oceans would have rapidly

filled with meltwater from the receding ice sheets, requiring the oceanic lithosphere to subside. It is believed that the oceans continued to fill until about 7,000 yr ago (with slight additions since then; Lambeck, 2002). Even after that, the seafloor has continued to subside isostatically, in order to balance the added mass of seawater with a deeper crust (see Figure 5.22).

As subsidence continued, with more effective flexure downward by the broader basins, seawater flowed in from the surroundings to fill those basins, causing a further rise in sea level, relative to the seafloor; and, in those basins, the water's gravitational attraction pulled still more water toward itself (—the 'self-attraction' discussed earlier in this chapter). With those enhancements—plus changes in the global gravitational field as the ice sheets melted away—the net result was a postglacial rise in sea level that was definitely *not* "eustatic" (globally uniform); see, e.g., Farrell and Clark (1976).

(a)

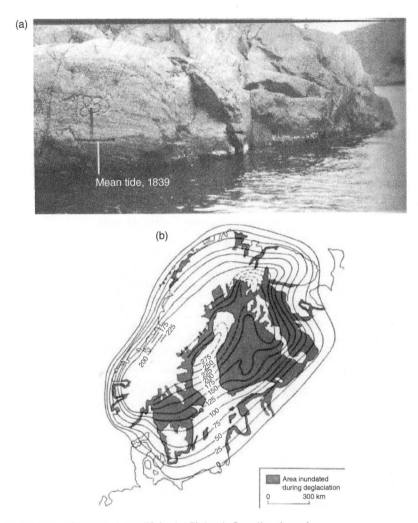

Mean tide, 1839

(b)

Area inundated during deglaciation

0   300 km

Figure 5.23 Evidence of postglacial uplift in the Finland–Scandinavia region.
(a) Rock cut (horizontal line) made at the coast near Oslo in the North Sea in 1839. This photo, taken in 1939, shows that the crust here rose ∼30 cm relative to the oceans over this time interval, as isostatic rebound continued. Image from Longwell et al. (1969) / John Wiley & Sons, Inc.
(b) Present-day contours (in meters) of former shorelines in Fennoscandia following the end of Pleistocene glaciation. More than 275 m of rebound has occurred to date in the center of uplift.
Image from John (2010) / taken from https://brian-mountainman.blogspot.com/2010/06/on-postglacial-rebound.html / Brian John.

One curious feature of the mass exchange between cryosphere and hydrosphere, first described by Farrell and Clark (1976), concerns sea level in the vicinity of the ice sheets. Although sea level was broadly lower during glacial times, the gravitational attraction of the ice sheets succeeded in pulling enough sea-water toward them that, near the Fennoscandian coast, sea level was actually higher than in the absence of the ice! And, when the ice melted away and the oceans filled, the decrease in local gravity led to a drop in sea level there. These regional effects amplified the isostatic subsidence associated with the ice load and the rebound associated with its disappearance, as well as contributing to the nonuniformity of the fall and rise in sea level.

The asthenospheric flow supporting the lithosphere's subsidence and rebound in its attempts to achieve isostasy led to further complications. As the lithosphere flexed downward beneath the massive load of a growing ice sheet, for example, it necessarily pushed mantle material below out to surrounding regions, with the net effect that the outlying lithosphere was pushed upward. On land, we would see this as a **peripheral bulge** (also called a **forebulge**), a ring of uplifted crust surrounding the subsiding ice (perhaps ~10^3 km away from the center of the ice cap); in adjacent oceanic regions, the peripheral bulge would add to the rebound already happening as sea level continued to lower.

If isostatic equilibrium was able to be achieved sometime after glacial maximum, the peripheral bulge would reach a maximum distance from the ice load; regardless, though, during the subsequent interglacial everything would reverse, including the bulge's outward migration: the bulge would 'collapse' back in as the asthenosphere flowed away from the periphery and back into the rebounding area. At the present time, for example, Laurentide rebound dominates much of North America—with the maximum uplift centered near southeastern Hudson Bay; but a collapsing forebulge has led to subsidence ringing the region of uplift. Along the Atlantic coast (where such vertical crustal movements are somewhat easier to deduce), the maximum observed subsidence extends from the New Jersey coast, with 13 m of subsidence in the past 6,000 yr ago (Stuiver & Daddario, 1963) and a current rate of ~$\frac{1}{2}$ cm/yr, southward to Chesapeake Bay; smaller but still significant amounts of subsidence are found outside that zone from Florida to Nova Scotia (Walcott 1972a,b, 1973).

It may seem surprising to be discussing the present-day subsidence of portions of a continent whose primary response to the end of the Pleistocene is crustal rebound. But, as Walcott (1972a, 1973) noted, subsidence at the periphery of the rebounding area "is as much a part of the pattern of [post-glacial] rebound as the uplift itself."

There were additional mass redistributions associated with Pleistocene glaciation and deglaciation. For example, during the Pleistocene, very large lakes existed in the American West, fed by the increased precipitation and cool temperatures south of the Cordilleran ice sheet; see Figure 5.24. One of the largest, Pleistocene Lake Bonneville, was, at its maximum, 50,000 km^2 in area and up to 300 m deep, with a load of 9×10^{15} kg (e.g., Bills et al., 2007). After the Pleistocene had ended, a catastrophic flood occurred (called the Bonneville Flood, it channeled a huge amount of water northward and then westward into the Pacific, at rates up to 10^6 m^3/sec (1 Sv) lasting for weeks; Jarrett & Malde, 1987); that plus climate changes taking place eliminated the lake's water supply, and the lake dried up nearly completely (leaving the Bonneville Salt Flats and salt-rich Great Salt Lake). With such a massive load of water gone, the lithosphere responded by flexing upward. In this case—for reasons discussed below related to the scale of the load—that rebound is now essentially complete; the lithosphere has rebounded 64 m, and the region has returned to a state of isostatic compensation (see Figure 5.24).

- Mass redistribution between the cryosphere and hydrosphere caused by modern-day shifts in climate. This is likely happening all around us—glaciers thinning or even disappearing, sea level rising—as a consequence of global warming. The isostatic response these mass shifts incur will be superimposed on the continuing rebound and subsidence associated with Pleistocene deglaciation (so, by the way, is the subsidence in the Mississippi River delta).

For reasons that will become evident in the next subsection, we consider instead an example (Larsen et al., 2005; see Figure 5.25) of the mass redistribution produced toward the end of the Little Ice Age. Our region of interest is centered on the Alaska Panhandle, which exhibits an ongoing rebound described as *the most rapid glacially caused uplift in the world*: three times the current rate of rebound in Hudson Bay or Fennoscandia, with peak rates exceeding ~3 cm/yr! One reason the uplift is so rapid

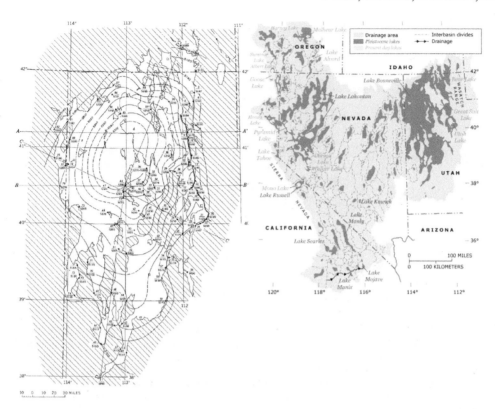

Figure 5.24 (Right) Pleistocene lakes (darker blue, with purple labels) and their present-day remnants (light blue, with light blue labels). The leading edge of the Cordilleran (western North American) ice sheet, and Pleistocene Lake Missoula, were hundreds of miles to the north, north of 46°N. Image from http://esp.cr.usgs.gov/projects/paleo_hyd/paleolakes.shtml / Geological Society of America.
(Left) Contours (at 20-foot intervals) of the elevation above mean sea level of former shorelines of Pleistocene Lake Bonneville; figure from Crittenden (1963b) / U.S. Department of the Interior. The center of uplift has rebounded by 210 ft. (64 m) relative to the outlying areas. With the analysis by Crittenden (1963a) but using the lake load estimated by Bills et al. (2007), we can conclude that the compensation for the reduced surface load is roughly complete.

is that it began so recently: in the year 1770 (± 20 yr), based on precise dating of former shorelines using tree ring counts. The load producing this uplift was a loss of ice estimated at ~2.7 × 10^15 kg, from the collapse of an ice field centered on Glacier Bay, where glaciers 1.5 km thick had dominated; most of the loss happened in about a century. So far, the Earth's surface has rebounded there by several meters.

This region was also the focus of repeated absolute gravity surveys, which allowed any variation in gravity over time to be determined. An analysis of that data by Sato et al. (2012) revealed decreases exceeding ~4 μGal/yr; their modeling experiments

demonstrated that such decreases were consistent with the observed rapid uplift in the Alaska Panhandle, and also resulted primarily from the Little Ice Age glacial loss described by Larsen et al. (2005). (Interestingly, their modeling results explained the absolute gravity data even better if present-day changes in ice mass, such as those described in Chapter 3, were also included.)

5.2.7 Isostatic Response to Surface Loads: Implications for Mantle Rheology

Till now we have emphasized the mass of the surface load, and the density contrast between

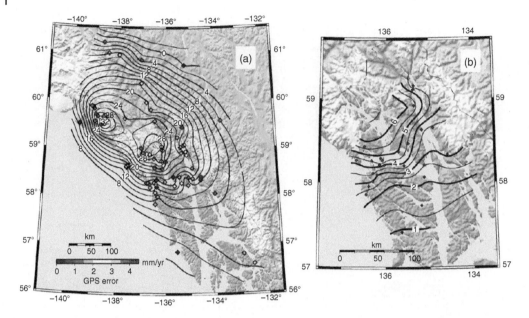

Figure 5.25 (a) Contours (mm/yr) of the current rate of crustal uplift in southeast Alaska (the Alaska Panhandle) and adjacent Canada based on GPS observations at 72 locations (diamond symbols, color fill indicating error bars using color scale shown), taken over a 5-year period (Larsen et al., 2005). The more centrally located center of rebound, with a rate of 30 mm/yr, is where the Glacier Bay Little Ice Age ice field had existed until the year 1770; the other center of uplift, to the northwest, has an even higher rate, 32 mm/yr, the result of that regional rebound from Glacier Bay and its surroundings compounded by subsequent severe thinning of the local Yakutat ice field.
(b) Contours of rebound (in meters) since uplift began in year 1770 ± 20 based on raised Little Ice Age shorelines documented at 27 sites. Larsen et al. (2005) / Elsevier.

crust and mantle, as the principal factors governing isostatic rebound. But the *rate* at which rebound takes place after the surface mass distribution has changed also depends on the ability of the mantle below the flexed lithosphere to flow away or back in. That flow rate in turn is a function of a material property of the mantle, its *viscosity*.

Both local and regional Airy isostasy are founded on the premise that the Earth's top layer is floating on the underlying mantle; our discussion of isostasy would be incomplete were the flow properties of the mantle not addressed. For simplicity, though, we will imagine here that the mantle flows as a true fluid, despite knowing (as later chapters will reinforce) that the mantle is solid; the reality, that the mantle responds to isostatic forcing with "solid-state" flow, is beyond the scope of this chapter.

Viscosity as a Property of Fluids. The **viscosity** of a fluid determines how much *resistance to relative motion* (resistance to 'shear') the fluid possesses as it flows. To illustrate this, we consider the 'bare-bones' situation of a shallow lake driven to move by a steady wind blowing along its top surface; we imagine that this happens far from the lake shore, so there are no lateral boundary constraints to worry about; and, we will neglect any forces caused by Earth's rotation—this is no wind-driven circulation of the oceans!

Because of its viscosity, the water must be immobile right at any fixed solid boundary, in this case the lake bed; at the same time, the water at the lake surface is forced to move, dragged by the wind stress acting on that surface. As a consequence of its molecular viscosity, each layer within the lake exerts a frictional force—a shear stress—on the layer

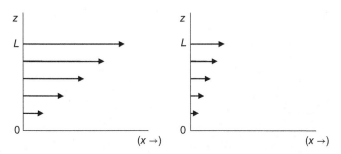

Figure 5.26 Velocity profiles—velocity as a function of height (z) within the fluid—representing the current produced within a shallow lake by a wind blowing steadily to the right; the lake bed is at $z = 0$ and the top surface is at $z = L$. Flow to the right is in the x direction. The length of each arrow indicates the velocity at that height. As a result of low molecular viscosity, the fluid on the left allows its velocity to increase significantly from layer to layer. With high molecular viscosity, the fluid on the right features a nearly uniform velocity (thus a very small gradient) throughout its height.

below, trying to accelerate it, and on the layer above, trying to slow it down. This inherent resistance to shearing, to one layer sliding over the next, produces a variation in velocity versus depth below the surface or versus height above the lake bed (see Figure 5.26); the variation is called a *velocity profile* or a *velocity gradient*. The viscosity determines whether the fluid will possess a steep velocity gradient as it flows, with the velocity changing significantly with height; or a modest gradient, in which case the velocity will hardly vary from bottom to top.

Which situation corresponds to the fluid having a low viscosity?

There are different types of viscosity, defined as they relate to different situations. Later in this textbook, we will encounter the *kinematic viscosity* as a parameter that expresses how easily the momentum of a fluid in motion can *diffuse* throughout the fluid. As a material property, though, the viscosity we have already introduced is the more relevant parameter; officially called the *molecular viscosity*, it expresses how 'sticky' a fluid is—how much friction its molecules generate as it flows. Given that our discussion above of molecular viscosity involves velocity gradients, it is not surprising that the two parameters are not completely distinct; their connections will be

discussed in Chapter 9. For now, we will continue to deal with molecular viscosity, which we will loosely call "the" viscosity; we will denote it by η (Greek letter eta).

To distinguish molecular from kinematic viscosity, the molecular viscosity is sometimes called the *dynamic viscosity*.

The fluid's resistance to shear can be expressed mathematically. As in Figure 5.26, we'll say that the fluid is flowing in the x direction, with velocity V, and that the height above the lake bed is the z direction. The velocity profiles in Figure 5.26 are distinguished by how much V changes with height, a quantity measured by the velocity gradient

$$\frac{\mathrm{d}V}{\mathrm{d}z}.$$

Stress will be explored thoroughly in Chapter 7; but for now, we will simply denote the shear stress exerted by each layer on the next due to viscous drag using the symbol σ. We can then define the **molecular viscosity η** according to

$$\sigma = \eta \frac{\mathrm{d}V}{\mathrm{d}z}.$$

With this definition, if η is bigger then $\mathrm{d}V/\mathrm{d}z$ must be smaller, for a given σ: greater viscosity forces the velocity to be more uniform with height, i.e. it enforces a smaller velocity gradient, in the context of a given shear

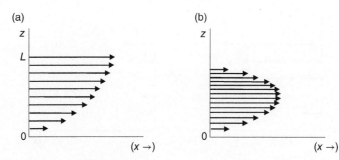

Figure 5.27 Velocity profiles for different situations, as discussed in the text.
(a) Velocity profile for a shallow lake or a river in the absence of wind stress or any other shear stress acting on the top surface. Presumably the motion occurred for other reasons, e.g. a slight topographic gradient (downhill to the right).
(b) Flow through a pipe. By symmetry, the shear stress in the middle of the flow (i.e., along the central axis) must be zero, making the flow in the top and bottom halves equivalent to the situation of the flow on the left.

stress. High viscosity, then, reflects a greater resistance to shearing flow.

Our definition of viscosity allows us to understand the physical basis for the profiles pictured in Figure 5.26 (it may be helpful now to glance at the profiles in Figure 5.27 for contrast). The linear variation in V versus height characterizing each of the two profiles in Figure 5.26 tells us that the velocity gradient dV/dz (pictorially, the slope of the profile) is constant through the full height of the lake. This in turn implies that the shear stress within the lake is the same at all heights. And that shear stress carries a particular significance: the stress in the very top layer of the lake must be the same as the stress acting on that top surface from right above the lake—with those layers separated by only an 'infinitesmal' distance, any discontinuous 'jump' in stress would be physically unrealistic. In short, the stress that is experienced within the lake (and the velocity profile) must correspond to a specified wind stress acting on the lake surface.

The condition of 'continuity of stress' across the lake's boundary surface implies that, in a situation where the water is flowing but there is no wind acting on the lake (with flow presumably caused by some other means, perhaps a drainage off to the side), the water velocity in the top layers of the lake would have to be practically uniform, yielding a nearly vertical profile in those layers, as in

Figure 5.27a, rather than linearly increasing as in Figure 5.26. That is, if $\sigma = 0$ at the surface, we must have $dV/dz = 0$ there. The 'continuity' constraint would also apply to river flow under windless conditions, and lead to the same kind of velocity profile.

Can you explain why even in windless situations the velocity gradient near the lake or river bed must be nonzero?

With brief reflection, we can easily predict the velocity profile in flow through a pipe. Symmetry requires that, in the center line or 'axis' of the pipe, dV/dz must be 0 (see Figure 5.27b); thus, the stress σ must also be 0 there (which makes sense since $dV/dz = 0$ means there is a lack of relative motion along the center line). Consequently, the velocity profile in either the top or bottom half of the pipe must be the same shape as in a zero wind stress situation, just upside down for the top half.

The units for molecular viscosity, as with all material properties, are a mix: from our defining equation, η must have units of stress divided by velocity gradient; since stress is given by force per unit area and the units for velocity gradient are velocity/distance or 1/time, η will have units of force × time/area. In cgs units, with force in dynes, this becomes dyne-sec/cm^2, which is called a **Poise**; in mks units, the molecular viscosity is measured as newton-sec/m² or **Pascal-sec**. Using Newton's Second Law to express force in more basic terms, we find

$$1 \text{ gm/cm-sec} \equiv 1 \text{ Poise} \quad \text{and}$$

$$1 \text{ kg/m-sec} \equiv 1 \text{ Pa-sec}.$$

As representative examples, water at STP has a viscosity of ~0.01 gm/cm-sec = 0.01 Poise, thus 0.001 Pa-sec; the viscosity of tar (pitch) at STP is ~2×10^9 Poise. Viscosity is one of those material properties whose magnitude encompasses a huge range in nature; as we will now learn, the Earth's mantle can be characterized by a viscosity many orders of magnitude greater than these examples (but the upper limit for viscosity in the Earth System appears to be even greater still!).

Inferences of Mantle Viscosity From Isostasy. Observations of isostatic rebound and subsidence can be used to infer the viscosity of the mantle. The approach is straightforward, though the mathematical and modeling details may be intimidating. The analysis requires that *the current rate of uplift* and *the remaining uplift*, each as a function of distance from the center of the rebounding area, be known. The rate of uplift can be measured by geodetic surveys or estimated from the rebound implied by (e.g.) raised beaches; the remaining uplift can be inferred from free-air gravity anomalies, or estimated by modeling the amount of mass still uncompensated.

Applying these to the equation defining viscosity allows that viscosity to be deduced. In the simplest terms, the approach works along these lines.

- Focusing on a volume of the lithosphere and mantle beneath the rebounding area, conservation of mass requires that the rate of rebounding mass-flow upward, out of the volume, must equal the rate of flow into the volume from the adjacent mantle beneath the lithosphere. The rate of uplift thus constrains the velocity of the asthenospheric inflow. If the inflow is modeled simply, for example as flow within a layer (which can be deep or shallow, thick or thin)—similar

to the traditional fluid dynamic situation of flow through a pipe—then the velocity gradient, dV/dz, can be inferred.

- The shear stress, σ, can be inferred from the amount of remaining uplift—that is, from how much more of the root remains to be adjusted in order to achieve mass compensation. That mass anomaly (the root excess) determines the strength of the upward pressure driving the rebound to buoy or flex upward. But the anomaly, in general, diminishes with distance from the center of rebound, implying that the upward pressure will similarly vary: it will be greatest at the rebound's center and decrease to zero at the 'nodes' of the rebound. The increase in pressure, moving inward toward the center, serves to create a shear stress on the lithosphere from the mantle.

Substituting these inferences of shear stress and velocity gradient into the equation defining viscosity yields an estimate of its magnitude.

Such an approach has been employed in numerous regions, including the following, to estimate the mantle's viscosity:

- *Hudson Bay region of North America*, which had been the center of the Laurentide ice sheet till the ice sheet's final recession and disappearance 20,000–7,000 yr ago;

- *Baltic Sea region of northwestern Europe*, the center of the Fennoscandian (Finland–Scandinavian) ice sheet, which began its retreat around 18,000 yr ago and had disappeared by 9,000 yr ago;

- *Lake Bonneville area of western North America*, with the drying up of Pleistocene Lake Bonneville by 9,000 yr ago;

- *Mississippi River delta*, as ongoing sediment deposition and subsidence take place; and

- *Alaska Panhandle area*, following the disappearance of the Glacier Bay Little Ice Age ice field and surrounding ice 2½ centuries ago.

Table 5.2 illustrates some of the viscosity estimates that have been obtained for

Table 5.2 A selection[a] of mantle viscosity estimates derived from observations of crustal rebound or subsidence.

Region	Inferred viscosity (Poise)[b]	Scale (km)[c]
Hudson Bay	1.6×10^{21}	3,300
Fennoscandia	3×10^{21}	1,800
Lake Bonneville	1.8×10^{20}, 1.2×10^{21}	450
Lake Bonneville	$3 \times 10^{19} - 3 \times 10^{20}$	450
Mississippi River delta	3×10^{20}	300
Alaska Panhandle	4×10^{19}	300

[a]These estimates assumed an elastic plate overlying a viscous or viscoelastic mantle; the authors (and others) also considered viscoelastic plates and/or multiple layers.

[b]Values would be 10 times smaller in units of Pa-sec.

Hudson Bay result from Cianetti et al. (2002) assumes an elastic lithosphere 120 km thick; they also find the lower-mantle viscosity is almost 15 times greater.

Fennoscandia result from Lambeck et al. (1998) assumes an elastic lithosphere 75 km thick; they also find the lower-mantle viscosity is more than 15 times greater.

First set of Bonneville results are from Bills et al. (1994) assuming an elastic lithosphere 25 km thick, and from Bills and May (1987) assuming an elastic lithosphere 23 km thick; the latter also find a higher-mantle viscosity below a depth of 300 km.

Second set of Bonneville results are from Nakiboglu and Lambeck (1983) assuming an elastic lithosphere 30 km thick.

Mississippi delta result from Jurkowski et al. (1984) assumes an elastic lithosphere 30 km thick.

Alaska result from Larsen et al. (2005) is for an asthenosphere assumed to be 110 km thick under an elastic lithosphere 60 km thick; they also find the viscosity of the mantle below the asthenosphere is 100 times greater.

[c]Estimate of typical size (lateral dimension) of the region.

these regions. Such estimates depend to some extent on the type of Earth model the analysis employed; the results in Table 5.2 are restricted to elastic lithospheric plates (rather than, e.g., viscoelastic) overlying a viscous or viscoelastic mantle (elastic and viscoelastic materials will be discussed in Chapter 7). Our first conclusion from these viscosity estimates is that they are all quite large; the mantle behaves as a very 'thick,' slow-moving fluid! Such a high viscosity explains why we can think of the mantle as solid: on everyday timescales there is essentially no detectable flow; it is only over spans of decades to centuries that significant flow will occur, and it will take millennia for the flow accompanying a particular isostatic adjustment to be complete.

Because of such viscosities, convective flow in the mantle, which drives plate tectonics, must be similarly slow, with timescales of millennia to eons to achieve significant amounts of continental drift. In fact, it is easy to imagine

that a viscosity higher by only a few orders of magnitude would have rendered the mantle so 'stiff' that measurable flow would not be possible, even over the lifetime of the Earth. That is, the mantle viscosity we determine from isostatic investigations has the power to place bounds on just how universal the convection paradigm is; what we find, as it turns out, is a low-enough viscosity to grant the mantle membership in the family of convecting Earth System components. However, that conclusion is subject to further discussion (in the next subsection, as well as in later chapters) because, as we will see shortly, these estimates may not apply to the deeper mantle.

The slow speed of the mantle as it flows to support isostatic rebound or subsidence implies that some fraction of the uplift being observed at present is a continuing response to past loading events, going back millennia. That is, the *loading history* of the region must have affected the current rate and cumulative amount of rebound or subsidence observed at any time. Consequently, matching observations of uplift rate or remaining uplift in order to infer viscosity requires that we know not only the size of the load that disappeared or was emplaced, but also the past 'comings and goings' of the load over many thousands of years. The analyses in Table 5.2 all employed some model of their load history.

The viscosity estimates in Table 5.2 are all high; but they are clearly not the same—not even the same order of magnitude. There are several possible reasons for these differences. First of all, regional variations in the thermal regime of the upper mantle can affect its ability to flow. For example, the Basin and Range province, which includes the Lake Bonneville region, features high heat flow and the Yellowstone 'superplume;' a greater warmth of the lithosphere and upper mantle there would be expected to soften the rock and enhance its ability to 'flow,' thus lowering its viscosity. In contrast, Hudson Bay lies within the Canadian Shield, whose great age and tectonic inactivity imply that it might be relatively cold even down well into the mantle, resulting in a higher viscosity.

Second, areas like southeast Alaska and the Basin and Range are tectonically active (e.g., the Alaska Panhandle is located along the boundary between the North American and Pacific plates, with the Fairweather and Denali Faults running along the western and eastern boundaries of Glacier Bay; and the Wasatch Fault runs along the eastern border of the former Lake Bonneville). Fault movements will compound the observations of isostatic uplift.

Beyond these factors, Table 5.2 hints at a consistent pattern: the smaller areas of rebound or subsidence yield lower estimated viscosities. Earlier studies, which focused primarily on Fennoscandia and Lake Bonneville, revealed such a dependence more explicitly. This was first interpreted by Kuo (as noted in the Lake Bonneville study by Crittenden, 1963a) as a consequence of the mantle's *viscosity increasing with depth*—or, putting a slightly different spin on his reasoning, a *low-viscosity channel*

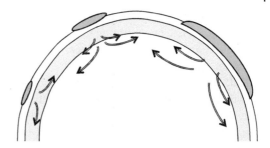

Figure 5.28 Dependence of mantle flow accompanying lithospheric flexure on the scale of the surface load. The lithosphere is shown here as the outer shell; the postulated low-viscosity channel is indicated by the shaded layer below the lithosphere; and the mantle below that channel has higher viscosity but can still flow. In this illustration, the surface loads (blue-filled lenses) are mass excesses, and the mantle flows away from them. The larger the scale of the load, the greater is the volume of the mantle participating, so the deeper is the flow (indicated by arrows), and the more nearly the inferred viscosity represents the whole mantle (or the mantle below the channel).

just below the lithosphere. This interpretation is illustrated in Figure 5.28. For a surface load encompassing a small region, the mantle flow accompanying rebound/subsidence will take place primarily within that low-viscosity zone; but the broader the (lateral) scale of rebound, the greater the percentage of mantle flow deeper than that channel, and the less the flow will be influenced by the low-viscosity channel. In short, Table 5.2 suggests that the mantle below the lithosphere might consist of a low-viscosity layer, characterized by $\eta \sim 3 \times 10^{19} - 3 \times 10^{20}$ Poise, and a stiffer layer below that with η at least $\sim 10^{21}$ Poise.

5.2.8 Global Constraints on Mantle Viscosity

The idea that the mantle's viscosity might increase with depth raises a concern: might the deep mantle be so viscous as to suppress convection in the lower mantle? In fact, for several decades the belief that convection of the solid earth was restricted to the upper mantle (defined, unofficially for now, as that portion of the mantle above ~ 670 km depth) was held to be reasonable, even favored. After all, earthquakes, one of the main products of a convecting mantle, had never been detected with an origin as deep as 700 km (e.g., Rees & Okal, 1987). (On the other hand, though, think of the friction at the boundary between the highly viscous, convecting upper mantle and an immobile lower mantle: wouldn't the shear stresses be too great? And, how would heat from the convecting core get through an immobile lower mantle?)

(Which is more convincing to you: the observational evidence supporting the belief, or some theoretical arguments raising doubt against it?)

A viscosity of $\sim 10^{24}$ Poise was often invoked as the critical value for the lower mantle: a viscosity exceeding that magnitude would incapacitate lower-mantle convection. Using the principle illustrated in Figure 5.28 suggests that we could resolve the question of lower-mantle (or whole-mantle) convection by analyzing isostatic-type situations where the forcing is *global* in scale; with only a small fraction of the flow involving the mantle right below the lithosphere, the viscosity so determined would reflect mainly the properties of the lower mantle.

In fact, several global-scale processes have been used to quantify the viscosity structure of the whole mantle.

- Ocean loading. The growth or disappearance of ice loads on land may be local or regional, but the corresponding lowering or raising of sea level—with all the complications discussed earlier—of course takes place ocean-wide. Figure 5.29 shows an example of the rise in sea level experienced in the tropics due to Pleistocene deglaciation, from the time of the last glacial maximum to the present.

- Laurentide rebound. The Laurentide ice sheet was continental in scale, and the rebound of Hudson Bay is extensive enough that the mantle flow it generated has penetrated all the way to the bottom of the mantle. Not surprisingly, models of that flow must be more complex than those representing Fennoscandia or Bonneville.

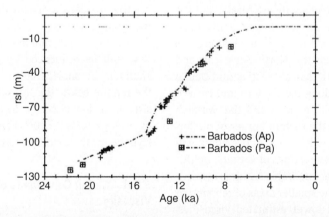

Figure 5.29 Example from Peltier (1998) / John Wiley & Sons of the 'far-field' effects of the last glacial maximum and subsequent deglaciation: relative sea level (rsl) history off the coast of Barbados, an island in the southern Caribbean. Sea level was determined using corals, mostly of a species (*Acropora Palmata*, + symbols) that usually lives within ~5 m of the sea surface; the other species plotted here (*Porites Asteroides*, boxed + symbols) lives in the deeper water. Coral depths include a slight correction for tectonic movements associated with the subduction of the South American plate beneath the island.
The coral data was originally collected by Fairbanks (1989) and Bard et al. (1990).
Curve shown is a best fit to the data based on Peltier's ICE-4G deglaciation model and Earth model with a 120 km lithosphere, an upper-mantle viscosity of $\sim 4 \times 10^{20}$ Poise, and a viscosity through the lower mantle that starts at $\sim 1.6 \times 10^{21}$ Poise and levels off around 3×10^{21} Poise.

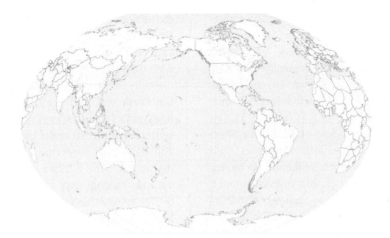

Figure 5.30 Map of the world, centered on the Pacific Ocean. Subsidence of the seafloor reduces the distance of seawater from the rotation axis; this happens in the tropics and equatorial regions more than at mid- and high latitudes. With most rebounding land following Pleistocene deglaciation existing at northern hemisphere mid- and high latitudes, the rebound puts that rock at increased distance from the axis. Both types of mass redistribution thus contribute to making the Earth more round, reducing Earth's equatorial bulge. Image by TUBS (2012) downloaded from http://commons.wikimedia.org/wiki/File:World_location_ map_(W3_Western_Pacific).svg.

- The change in Earth's equatorial bulge, and consequent change in Earth's rate of rotation, from Pleistocene deglaciation. The mass exchange between cryosphere and hydrosphere as glaciers melted, and the resulting lithospheric flexure and astheno-spheric flow that continue to this day, have been redistributing mass relative to Earth's polar axis; that is, Earth's moment of inertia has been changing.

Given the distribution of continents versus oceans (Figure 5.30), it should not be surprising that the axial moment of inertia has been *decreasing*, ever since the melting of ice was complete and water stopped filling the oceans: seafloor subsidence has decreased the amount of mass farthest from the axis, in tropical regions, which are largely oceanic (—think of the vast central Pacific!). At the same time, continental rebound, with support from asthenospheric inflow, has carried mass upward, that is, farther from the axis, in mid-latitudes. The net effect has been an increase in the sphericity of the Earth—a reduction of Earth's equatorial bulge.

The reduction of the bulge is a global action, so the mantle flow that supports this readjustment must involve the entire mantle, down to the core boundary. The *rate* at which the bulge shrinks depends on the viscosity of the mantle—especially the lower mantle, which (with depths exceeding $\sim$670 km) contains most of the mantle's volume.

In Chapter 4, we defined a quantity J_2, which was a dimensionless measure of the size of the equatorial bulge. If the bulge is decreasing over time as a result of deglaciation, then so is J_2. As discussed in the next chapter, J_2 can be determined from satellite observations, so the rate of change in J_2 over time, which is commonly denoted $\dot{J}_2$ and pronounced "J_2-dot," provides a numerical constraint on models of postglacial rebound—particularly on the value of the lower-mantle viscosity.

Conservation of angular momentum requires the Earth to rotate more quickly if its axial moment of inertia gets smaller, just as skaters pulling in their arms reduce their own moments of inertia and must spin faster. As a consequence, the Earth's *length of day*—known among specialists as the **LOD**—will shorten increasingly; this provides an alternative 'observable' for constraining Earth's viscosity structure. However, many processes in the Earth System—stratospheric winds, ocean currents, and El Niño (e.g., Cheng & Tapley, 2004), to name a few—are able to affect Earth's LOD on various timescales; and there is one special phenomenon (to be discussed in the next chapter) which, though related to tides, is able to change LOD particularly effectively on timescales of millennia. When corrections are made for the latter, observations of long-term changes in LOD do reveal an acceleration of Earth's spin (sometimes this postglacial acceleration is called the **nontidal acceleration**). The nontidal change in LOD is estimated to be ~-0.6 milliseconds per century (Stephenson, 2003), which ultimately (as derived in a homework problem for this chapter) implies $\dot{J}_2 \sim -3.5 \times 10^{-11}$/yr or a decrease in J_2 of $\sim 0.03\%$ over 10,000 yr.

- The movement of Earth's rotation pole. Postglacial rebound and seafloor subsidence change more than Earth's polar moment of inertia: the oceans, though concentrated at low latitudes, are not distributed symmetrically with respect to Earth's axis; and continental rebound, though taking place mainly at midlatitudes, is not occurring symmetrically around the world, either. The mass exchanges following Pleistocene deglaciation have thus been asymmetric, and have had the effect of gradually tilting Earth's 'symmetry' axis, the geographic axis that divides Earth's mass equally and whose endpoints are the South Pole and North Pole; the North Pole has consequently been drifting, toward—no surprise here—Hudson Bay, the center of the largest rebounding region.

The natural tendency of the Earth is to rotate about its symmetry axis. Let's imagine that something occurs to change the distribution of mass in the Earth System—for example, a subducting lithospheric plate suddenly descends a few meters deeper during a major earthquake, displacing mantle material; such a mass redistribution may cause the symmetry axis to lose its alignment with the rotation axis (such a misalignment formally defines a "**wobble**"). In that case, the Earth will find itself rotating about a nonsymmetry axis, which reduces its angular momentum. To make up the difference and continue conserving its total angular momentum, the rotation axis itself 'winds around' the symmetry axis: the rotation pole circles around the North Pole. That gives the Earth the appearance of 'wobbling' about on its axis. (This type of wobble, which is unforced ('free') once the earthquake is over, takes about 14 months to circle around once, given the properties of the Earth.) However, the energy required to maintain the misalignment will eventually dissipate, the wobble will decay, and the two axes ultimately coincide again.

Of course, earthquakes are far from the only way mass is exchanged within the Earth System; varying weather systems, changing sea level, winds, and ocean currents (e.g., Nastula & Ponte, 1999; Aoyama & Naito, 2001; Gross et al., 2003; Naghibi et al., 2017 and references therein), ENSO (e.g., Kolaczek et al., 2000), seasonal and random changes in surface water and groundwater storage (e.g., Hinnov & Wilson, 1987; Chao & O'Connor, 1988; Chen & Wilson, 2005), tidal variations (e.g., Dickman, 1993; Dickman & Gross, 2010), and other surficial and internal processes (e.g., Dickman, 1983), force additional misalignments between the rotation axis and the symmetry axis, so that wobbles are being 'excited' all the time.

The tilt of the Earth during a wobble means that, from our point of view, all the stars in the sky appear to shift (e.g., northward, if Earth's axis tilts toward the south), allowing the amplitude of wobble to be measured at any moment in time; for more precision, we can use artificial satellites instead of stars as our reference. Such observations have revealed a perhaps unexpected result: small-amplitude wobble

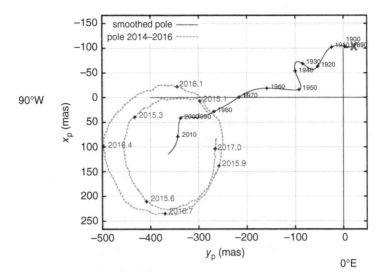

Figure 5.31 Current wobble and (approximate) true polar wander depicted using a coordinate system set up at the top of the world, with *x*- and *y*-axes as shown. Units are milliarcseconds (mas); note 100 mas $\cong$ 3.1 meters $\cong$ 10 ft. along the Earth's surface. The data is presented here with the coordinate origin offset slightly from the actual North Pole location for ~1900 (marked with a red X, between the labels for 1890 and 1910).

The Earth's wobble, based on a combination of optical telescope observations and the space geodetic techniques discussed in Chapter 6, is shown from the end of 2014 to the beginning of 2017 (dotted red curve).

As discussed in the text, the wobbling rotation pole circles around the symmetry pole, with a period we see here to be just over a year; by averaging or 'smoothing' the rotation pole positions over one such cycle, we can approximate that symmetry pole as the center of the circle. Using wobble observations from ~1900 CE (not shown here) to the present, one cycle at a time, that technique yields the approximate path of the symmetry pole (solid black curve, marked every 10 yr from 1900 to 2010). The Earth has been wobbling around the symmetry axis during all this time, following the symmetry axis as the latter has tilted in the general direction of Hudson Bay.

Because the Earth possesses a small, long-period (~30-year) wobble not averaged out over each 12-month or 14-month cycle, however, the smoothing technique used here results in a slightly distorted symmetry pole location; accounting for its effect on the determination of the symmetry pole path, Dickman (1981) found an overall rate of true polar wander of about 3.5 mas/yr—a rate of nearly a degree of latitude per million years if it will continue for that long—toward ~80°W longitude and Hudson Bay, at least during 1900–1978. Image adapted from http://hpiers.obspm.fr/eop-pc/index.php?index=pm#mean-pole.

is a big-picture phenomenon! That is, it apparently takes a lot of energy (—a lot of redistributed mass) to cause even a small wobble: during more than a century of observations, the total misalignment has not amounted to more than ~10 m of offset between the rotation pole and North Pole.

The asymmetric mass redistributions following Pleistocene deglaciation have caused Earth's symmetry axis to tilt progressively—very gradually but very measurably. And Earth's rotation axis has dutifully been following, unavoidably wobbling around the symmetry axis (in response to the many 'excitations' within the Earth System), but with the center of the wobble—the symmetry pole—heading as well toward North America. Figure 5.31 shows deductions of the position of Earth's symmetry pole (North Pole), based on observations of the rotation pole during the past century or so. Despite the drift of the symmetry pole, which we can presume has been ongoing since the last glacial maximum, the two poles have always managed to stay within several meters of each other.

In the language of plate tectonics, the slow long-term drift or 'secular' motion of the poles, and tilt of the Earth's axes, would be called a **true polar wander** (see also Dickman, 1977 and Argus & Gross, 2004).

The wandering rotation pole brings some complications with it. As we saw in Chapter 4, the Earth's equatorial bulge is a consequence of the centrifugal force associated with Earth's spin, reflecting in particular the property that centrifugal force is strongest at locations on Earth farthest from the spin axis—that is, at points on the equator! But when the rotation axis tilts, the bulge is no longer aligned perpendicular to it, and points on the equator are no longer farthest from the axis. As a consequence, the lithosphere flexes globally—and flow throughout the mantle is induced—as Earth attempts to shift its equatorial bulge to stay aligned perpendicular to the rotation axis (see Figure 5.32). The readjustment of the equatorial bulge, involving the entire mass of the mantle, further shifts the Earth's symmetry axis, which amplifies the drift. The more the bulge adjusts, the more the symmetry axis shifts, so the more the rotation axis shifts, so the more the bulge adjusts, and so on. Such feedback is somewhat analogous to the donkey that continues to lunge forward to reach a carrot attached to a stick held out in front of it by the rider: despite each lunge, the stick and thus the carrot keep their distance from the donkey, motivating another lunge (since the donkey never reaches the carrot, perhaps "feedback" is the wrong term!).

Like the decrease in J_2 and the size of the bulge, the rate at which the bulge migrates as it follows the drifting rotation axis depends on the viscosity of the mantle. And because of the truly global scales involved, flow in this case also takes place through the entire depth of the mantle. The rate of true polar wander, like the nontidal acceleration, is thus especially useful for constraining the viscosity of the lower mantle.

In this discussion, we have maintained that the observed twentieth-century drift of the poles is a consequence of Pleistocene deglaciation; but any process that redistributes mass and shifts the symmetry axis (exciting wobble, as a by-product) clearly has the potential to produce that drift, if it acts effectively enough over the long term. For example, changes in terrestrial water storage (Chao, 1995; see also Paulson et al., 2007), the cumulative effect of earthquakes (e.g., Chao & Gross, 1987; Alfonsi & Spada, 1998, but see also Cambiotti et al., 2016; Chao & Ding, 2016), and mantle convection (e.g., Steinberger & O'Connell, 2002; Richards et al., 1999)—or various components of it, including sub-duction of lithospheric slabs (e.g., Spada et al., 1992) or subduction plus lithospheric accretion at mid-ocean ridges (Alfonsi & Spada, 1998), nonisostatic mountain-building (Vermeersen et al., 1994), lithospheric plate displacements (e.g., Dickman, 1979), and convectively pro-duced gravity anomalies and topographic highs and lows (e.g., Mitrovica et al., 2015 based on Mitrovica & Forte, 2004)—have all been studied for their ability to generate true polar wander. The contributions of all of these sources to the recent polar drift are generally found to be small, of order 10% at most. Given also the obvious link between deglaciation and the direction of the observed drift, which was toward Hudson Bay, we may feel confident that this drift was primarily a consequence of Pleistocene deglaciation.

In this context, we also mention the effect of mantle convection on our estimation of the true polar wander induced by any of the preceding processes. Convective upwellings and downwellings can modify the Earth's bulge (C–A); if the bulge is enlarged, that will stabilize Earth's rotation and resist any polar drift, especially on long timescales and if the lower mantle is not too viscous (Mitrovica et al., 2005).

To understand the stabilizing effect of the bulge, imagine the Earth had no bulge and was instead perfectly round; in that case, can you locate the symmetry axis the Earth would naturally want to rotate about?

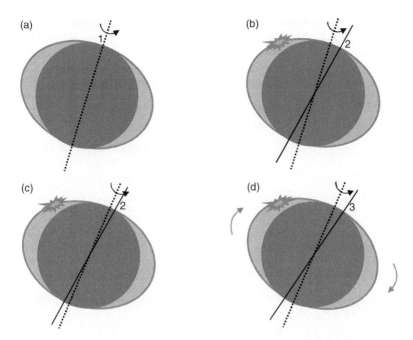

Figure 5.32 Deglaciation causes a drift of the rotation pole. (a) The undisturbed initial state of the Earth, rotating about its symmetry axis (position 1). The equatorial bulge (lightly shaded) is aligned perpendicularly to both axes. (b) Crustal rebound at a mid- or high northern latitude (gray 'starburst') elevates some mass; as a result, Earth's mass is now symmetric about a slightly tilted axis (solid line, position 2, tilted *away* in order to 'balance' the additional mass). (c) The rotation axis (dotted line) follows the shifted symmetry axis, attempting to realign with it; but when the rotation axis tilts, that (d) causes the equatorial bulge to readjust (orange arrows), trying to remain perpendicular to that axis. That causes a further shift in the symmetry axis (position 3), and the poles continue to drift.
How would this scenario change if the excitation in (b) was 'negative,' such as the loss of ice that preceded the rebound? Now imagine both the ice loss and the rebound occurring—how does the rotation pole drift, and is your prediction consistent with observations such as those in Figure 5.31?

With ongoing mantle convection enhancing Earth's bulge, the true polar wander produced in response to deglaciation could be significantly less than what is observed, if the viscosity of the lower mantle is sufficiently low. In this situation, deglaciation might no longer be a major contributor to the observed drift (leaving the direction of the drift toward Hudson Bay a curious coincidence). For studying the true polar wander, and its implications for mantle viscosity, this points to the importance of a comprehensive analysis, in which all likely sources of drift are considered together, and the net effect evaluated. The final portion of this chapter will set the stage for just such an analysis.

Results from a selection of globally constrained analyses are listed in Table 5.3. The inferred viscosities differ in part because of differences in the Earth models: the number of layers considered for the mantle; whether the mantle material is incompressible or not (a material property we will learn about in a later chapter); the deglaciation load history employed; and so on. These results confirm the overall high viscosity of the mantle, and additionally reveal the likelihood of a somewhat greater viscosity for the lower mantle than for the upper mantle. But the implied lower-mantle viscosity is not too large to prevent whole-mantle or lower-mantle convection.

Still another type of global observation has been used in recent years to constrain the viscosity structure of the deep mantle. Based on gravity anomalies interpreted to be associated

Table 5.3 A selection of mantle viscosity estimates derived from global constraints: observations of changes in Earth's rotation rate (or $\dot{J}_2$), the secular drift of the rotation pole, and relative sea level around the world.

Authors	Inferred upper-mantle viscosity (Poise)[a]	Inferred lower-mantle viscosity (Poise)[a]
Yuen et al. (1982)	8×10^{21} [b]	$1 \times 10^{22} - 4 \times 10^{22}$
Peltier & Jiang (1996)	1×10^{22} (assumed)	1.8×10^{22}
		1.3×10^{23} below 1,400 km (if lower mantle above 1,400 km has 1×10^{22} viscosity)
Johnston & Lambeck (1999)[c]	3.2×10^{21} (assumed)	$4.7 \times 10^{22} - 9 \times 10^{22}$
Nakada & Okuno (2003)[d]	5×10^{21}	$1 \times 10^{22} - 2 \times 10^{22}$

Note: See also the discussion for "a changing world."
[a] Values would be 10 times smaller in units of Pa-sec.
[b] Their value of the upper-mantle viscosity is the average of a viscosity of 2.5×10^{21} Poise in an asthenosphere 150 km thick and a viscosity of 1×10^{22} Poise in the rest of the upper mantle.
[c] Their analysis includes modeling of current (global warming–produced), as well as Pleistocene, ice loss and sea-level rise.
[d] Values based on my interpretation of their Figs. 1–3, with an elastic lithosphere of thickness ~50–100 km.

with the mass excesses and deficits of mantle convection, this global constraint is the *Geoid*, which will be discussed in the next chapter.

A Changing World. Global warming will cause large portions of the Antarctic and Greenland ice sheets, and mountain glaciers worldwide, to wane, leading to a global mass redistribution that modifies Earth's moments of inertia and symmetry axis, thus affecting the nontidal acceleration and the secular drift of the pole. Eventually, decreased crustal loads from the loss of ice will also produce a rebound of those areas, and increased sea level will lead to seafloor subsidence, each further modifying the acceleration and the drift.

Recent estimates of these rotational parameters indicate that global warming is already having a measurable impact on them.

It should be noted that the value of $\dot{J}_2$ stated earlier, -3.5×10^{-11}/yr, was based on records of the timing of ancient eclipses and represents an average over the past 2,500 yr (Stephenson, 2003). Modern *space geodetic* techniques, including observations of satellites and quasars, as discussed in the next chapter, have been used by numerous researchers to estimate $\dot{J}_2$ beginning in the last quarter of the twentieth century. The analysis by Cheng

et al. (2013; see also references therein) was based on a relatively long span of satellite data (1976–2011) and was able to rigorously separate out shorter timescales of variation in J_2 from the long-term trend of interest. For $\dot{J}_2$ the various studies found general agreement with the ancient eclipse value until the 1990s, at which time an increase in $\dot{J}_2$ (i.e., to a less negative value) began to manifest itself; according to Cheng et al. (2013) that increase has resulted in a value of $\dot{J}_2 \sim -0.6 \times 10^{-11}$/yr (say, for the year 2000) (see also, e.g., Roy & Peltier, 2011). Cheng et al. additionally showed that various climatic processes, particularly an accelerating mass loss from mountain glaciers and ice caps and the resulting rise in global sea level, were likely responsible for the increase in $\dot{J}_2$.

Similarly, recent observations of polar drift suggest that its rate and direction may have changed, though the extent of those changes is not yet settled. The polar drift rate quoted earlier—3.5 mas/yr (milliarcseconds/yr) directed toward Hudson Bay (see Figure 5.31 caption)—was inferred from star observations during 1900–1978. Space geodetic techniques, in some cases combined with traditional star observations, have also yielded estimates of Earth's axial tilt and polar wander, and (e.g.) for the time span 1976–1994 implied either the same rate (~3.4 mas/yr; McCarthy & Luzum, 1996) or a slightly higher rate (~4.1 mas/yr; Gross & Vondrak, 1999) of drift, in both analyses directed toward Hudson Bay.

However, Roy and Peltier (2011) found that the secular polar drift changed after ~1992, with the average rate during the next 17 yr dropping by half, and the direction slightly more eastward, toward Newfoundland. Chen et al. (2013) found an even sharper change in the polar drift: from 1982 to 2004, as measured using yearly average rotation pole positions (*their* Fig. 2), the rate was ~3 mas/yr, directed slightly eastward of Hudson Bay; but during 2005 conditions changed dramatically, resulting in a drift rate of ~6 mas/yr over the next 7 yr directed much more to the east. They also demonstrated that this newer, much more rapid drift could be completely explained as a consequence of mass redistributions inferred from GRACE data, especially ice melting (data we saw in Chapter 3) and the associated rise in sea level, and changes in land water storage.

The rough agreement among these estimates should improve as a longer time span of observations becomes available and the trends can be better distinguished from other kinds of variations. Nevertheless, some of the analyses convincingly demonstrate that the mass redistributions associated with modern-day global warming are capable of explaining the magnitudes of the changes in $\dot{J}_2$ and polar drift. At this point, the observations also appear to suggest that global warming's global effects were 'kicking in' during the mid-1990s and then especially around 2005.

Recognition of the likely impact of global warming on polar motion has led the rotational (geodetic) community (IERS, 2010) to adopt a cubic rather than linear description of the current polar drift as their conventional model (Cheng et al., 2013).

The ability of global warming to alter $\dot{J}_2$ and the secular polar drift provides an opportunity to simultaneously determine the mantle's viscosity *and* quantify the net warming-induced melting. As illustrations of the potential for this approach, we note early analyses carried out by James and Ivins (1997) and Sabadini et al. (2002; see also Tosi et al., 2005); both groups combined an ice history for Pleistocene deglaciation with various possible scenarios for present-day melting, and determined the viscosity of the lower mantle (and upper mantle, in the latter work) which best fits the observed $\dot{J}_2$ and rate of polar drift for 1900–1990 or 1994.

The preferred solution of James and Ivins (1997) (with the upper-mantle viscosity assumed to be 10^{22} Poise) included a lower-mantle viscosity of ~2×10^{23} Poise, a mass loss by Antarctica of ~275 Gt/yr, and a mass loss by Greenland of ~145 Gt/yr. In the work by Sabadini et al. (2002), the best of their best fits was for an upper-mantle viscosity of ~1.3×10^{21} Poise and a lower-mantle viscosity of ~3.2×10^{23} Poise, and required a mass loss of ~250 Gt/yr from Antarctica and ~144 Gt/yr from Greenland. Based on the time spans corresponding to the rate of polar drift and the value of $\dot{J}_2$ used in these studies ($\dot{J}_2$ was taken to be ~-2.7×10^{-11}/yr), the inferred mass losses would correspond roughly to year ~1992 (or perhaps the time span 1986–1996). Such mass loss represents the loss of grounded ice (and the corresponding addition of meltwater to the oceans); the melting of floating ice (or the nearby ponding of meltwater) would not change Earth's mass distribution or affect its rotation. The inferred mass losses are large—roughly comparable to those derived from GRACE satellite data for the subsequent decade, but distinctly greater than those listed in the IPCC (2007) report, as discussed in Chapter 3 of this textbook.

According to these early globally constrained analyses, modeling present-day melting of continental ice sheets yields viscosities ~2×10^{23}–3×10^{23} Poise for the lower mantle, perhaps an order of magnitude or so higher than the estimates in Table 5.3—but, again, *not* too large to limit mantle convection.

The broadest investigation to date of the causes of the secular polar motion is the study by Adhikari et al. (2018). They estimated the polar drift caused by Pleistocene deglaciation to be at a rate roughly only one third of what has been observed (e.g., Figure 5.31) over most of the twentieth century. The main part of their investigation concerned additional sources of polar drift excitation: the effects of surface mass variations—including groundwater storage (Wada et al., 2012), water impoundment in dams and reservoirs from human activity (Chao et al., 2008), modeled density changes within the oceans (Landerer et al., 2009), and mass exchange between the oceans and glaciers (e.g., Marzeion et al., 2015), Greenland (Kjeldsen et al., 2015), and Antarctica (Smith et al., 2017; Miles et al., 2016; Christ et al., 2014; and others)—combined with the effects of mantle convection, which they estimated from different gravity, tectonic, and seismic models.

With some realizations of their convective models, Adhikari et al. (2018) found that the total predicted true polar wander from all sources matched the observed polar drift fairly well or even quite well, implying that convective effects might be contributing to the drift even more than Pleistocene deglaciation or present-day surficial excitations are. However, the recent analyses of $\dot{J}_2$ and the polar drift discussed above (by Roy & Peltier, 2011; Cheng et al., 2013; and Chen et al., 2013) showed that the end of the twentieth century and start of the twenty-first century was a transition period for those rotational parameters; the surface mass excitations, thanks to global warming (and human activity), were also undergoing transitions. The analysis by Adhikari et al. (2018) employed averages to represent each parameter during the twentieth century and into the twenty-first, so it is not certain how well the rotational changes are matched. We can, at least, conclude that the recent variability of those parameters over time, with a resulting complexity that challenges our combined understanding of rotational phenomena, Earth system processes, and interior Earth properties, is still another consequence of global warming.

6

Orbital Perspectives on Gravity

6.0 Motivation

The gravitational interactions between the Earth and external bodies—especially the Moon, Sun, and artificial satellites—are of great significance in the Earth System. Some of those interactions produce tidal forces, which affect not only the oceans but also the solid earth (and even the atmosphere). Other gravitational interactions cause the Earth's precession, the so-called *Precession of the Equinoxes*, and change Earth's orbit; these phenomena may in turn have helped to regulate Earth's climate during the Pleistocene.

The concept of precessional torques also leads us to understand how Earth's gravity causes perturbations in satellite orbits, thus how precise tracking of satellites can reveal the global variations in Earth's gravity field.

The discussion of satellite orbit perturbations, or satellite geodesy in general, will compel us to revisit the subject of sea level, at which time we will focus explicitly on sea level as a special gravitational surface known as the *Geoid*. We will find the Geoid—the real-world analog to the Ideal Earth's Spheroid—to be a very useful way to explore both shallow (seafloor) and deep mass anomalies. And we will learn how analysis of sea-level fluctuations *superimposed* on the Geoid reveals the large-scale circulation of the oceans.

The final portion of this chapter is devoted to an investigation of an amazing interaction between the oceans, solid earth, Moon, and Sun known as *tidal friction*; it comes closer than any other single topic in this textbook to epitomizing the fusion of geophysics with the various sciences devoted to Earth System components. It will also provide constraints on the origin of the Moon, and help us better understand how the Solar System works.

6.1 Tides

6.1.1 Ebbs and Flows

Loosely speaking, the term **tides** refers to *periodic effects of the Sun and Moon*, and thus encompasses a broad range of phenomena. Most of these periodic effects are created by tidal forces, which derive from the gravitational attraction of the Sun and Moon. Acting on the oceans, they produce the ocean tides we are all familiar with; acting elsewhere in the hydrosphere, for example in rivers, lakes, and coastal aquifers, they can produce other, more unusual periodic effects or tides. The lunar and solar tidal forces stirring the oceans also act on the solid earth, which, as we have already seen, can be deformed; the resulting periodic deformations of the solid earth are called **body tides**, **earth tides**, or **solid-earth tides**. Periodic variations in the gravitational acceleration itself, as the Moon and Sun move through the sky, are called **gravitational tides**, as are the periodic variations in gravity caused by Earth's tidal deformations.

Earth System Geophysics, Advanced Textbook 6, First Edition. Steven R. Dickman.
© 2025 American Geophysical Union. Published 2025 by John Wiley & Sons, Inc.
Companion Website: www.wiley.com/go/Dickman

In contrast to gravitationally based tides, periodic effects of the Sun's *heat*—called **radiational tides**—also exist, with variations from day to night and season to season. Unlike tides of the oceans or solid earth, tides of the atmosphere are dominantly (though not entirely) radiational; as a result, the largest of the atmospheric tides is by far the daily variation. Within the ionosphere, which participates in the upper atmosphere's daily tide, the streaming of electrically charged particles in response to tidal heating creates a daily **magnetic tide**.

Interactions between the components of the Earth System produce what might be called 'secondary' tides. The periodically varying weight of the atmosphere or oceans on the underlying solid earth, during high versus low tide, deforms the solid earth, creating what is known as the **load** tide. The periodically varying weight of the atmosphere *on* the oceans produces what might be called a pressure-driven or 'barometric' ocean tide. Then, too, seasonal variations within the hydrosphere—in runoff, precipitation, and groundwater storage—can fill and drain the oceans, slightly, over the course of a year, producing a periodic, *mass*-driven component of the annual ocean tide (the annual tide also has gravitational and thermal components, associated with the Earth's changing distance from the Sun as it orbits the Sun in an elliptical path).

In the discussion below, we will focus on direct rather than secondary tides and specifically on those driven ultimately by gravity rather than solar heat.

6.1.2 Tidal Forces

The simplest view of direct, gravity-driven tides is that they result *only* from the gravitational pull of the Sun (in the case of the solar tide) or the Moon (for the lunar tide); thus, for example, when the Sun is overhead, its pull (on the land or ocean below) is the greatest it can be, so the solar tide is highest. Yet if this were the case, there would be only *one* high tide and *one* low tide per day due to the Sun. In fact, the Sun causes *two* high and *two* low tides per day. This fact implies that there must be *another* force acting on the Earth, to produce the second solar high tide.

Reconsidering Kepler's Third Law allows us to identify this second force. As discussed in Chapter 4 (for the simple case of a circular orbit), this law depended on the counterbalancing effects of gravity and orbital centrifugal force. We might restate the law as follows:

If one mass is orbiting around another, and is neither falling into nor flying away from it, then the gravitational pull of the second mass on the first must be balanced by *the centrifugal force that the first mass experiences in its orbit about the second mass.*

In the case of the Earth orbiting the Sun, for example, the balance can be expressed quantitatively as

$$\frac{G M_S M_E}{R_S{}^2} = M_E \Omega_S{}^2 R_S,$$

where the term on the right is the orbital centrifugal force experienced by the Earth; R_S is the distance of the Earth from the Sun, the Earth's orbital angular velocity around the Sun is Ω_S, and M_E and M_S are the masses of the Earth and Sun.

But Kepler's Third Law, like the laws upon which it is based (Newton's Laws), technically applies only to point masses; the balance implied by the law will hold exactly true only at the Earth's center (or center of mass). Figure 6.1 illustrates the problem, for an Earth imagined to possess a highly exaggerated, cylindrical shape. Everywhere outside of that center point, the two forces fail to balance. Points on the planet closer to the Sun experience a stronger gravitational force due to the Sun, in accordance with Newton's Law of Gravitation; points farther from the Sun experience a weaker gravity. At the same time, given the dependence of centrifugal force on distance from the orbital axis, closer (farther) points experience a slightly weaker (stronger) centrifugal force.

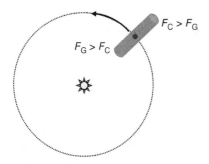

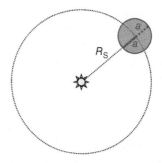

Figure 6.1 Tidal imbalance for a planet in a circular orbit about the Sun. To illustrate the implications of Kepler's Third Law, the planet is modeled as being a long cylindrical rod directed, for the moment, toward the Sun. At the center of mass of the rod (represented by a filled circle), Kepler's law requires that the Sun's gravitational pull on the planet (F_G) is balanced by the planet's orbital centrifugal force (F_C); at all other points on or within the rod, either the Sun's gravity exceeds the orbital centrifugal force ($F_G > F_C$) or vice versa. The two forces are in maximum imbalance at the ends of the rod.

The imbalance between these forces—also called a *differential force* or **the tidal force**—is responsible for the rise and fall of the sea surface that we call the ocean tides. The magnitude of the maximum tidal acceleration (maximum tidal force per unit mass) can be calculated relatively easily. We treat the Earth as spherical, for simplicity, and denote its radius by a; R_S is the distance from the Sun to the center of the Earth (see Figure 6.2). Consider first a point on the day side of the Earth, on the surface and at 'noon,' i.e. when the Sun is overhead. Since the distance of this point from the Sun is $(R_S - a)$, a unit mass there will experience a gravitational force given by

$$\frac{GM_S}{(R_S - a)^2} = \frac{GM_S}{R_S^2} \frac{1}{\left(1 - a/R_S\right)^2}$$
$$\approx \frac{GM_S}{R_S^2}\left(1 + 2\frac{a}{R_S}\right),$$

using the same type of approximation made in the previous chapter for the free-air effect and noting that $a \ll R_S$. The centrifugal force experienced by the unit mass is

$$\Omega_S^2(R_S - a) \approx \Omega_S^2 R_S,$$

Figure 6.2 Tidal forces acting on the Earth. The Earth's relative size is exaggerated here for clarity. In this illustration, Earth is in a circular orbit of radius R_S, so the distance of its center from the Sun is R_S; Earth's mean radius is a.

to a similar approximation. The imbalance, or tidal acceleration, is therefore well-approximated as

$$\frac{GM_S}{R_S^2}\left(1 + 2\frac{a}{R_S}\right) - \Omega_S^2 R_S \equiv +\frac{2a\,GM_S}{R_S^3},$$

where we have obtained the expression on the right using Kepler's Third Law (which, as written above, implies that $GM_S/R_S^2 = \Omega_S^2 R_S$). The plus sign indicates that the imbalance favors gravity, and this tidal force pulls the oceans (for instance) *in* toward the Sun.

The dependence of the imbalance inversely on distance *cubed* confirms that tidal forces are not the same as gravitational forces.

If the unit mass is on the surface but on the night side of Earth, i.e. at the 'midnight' location, then its distance from the Sun is $R_S + a$, so it experiences a weaker gravitational force per unit mass, $GM_S/(R_S + a)^2$, due to the Sun, and a slightly stronger centrifugal acceleration, $\Omega_S^2(R_S + a)$. With the same kind of derivation, we find the tidal imbalance here to be

$$-\frac{2a\,GM_S}{R_S^3},$$

with the negative sign reminding us that centrifugal force dominates, and the oceans and other masses at this location are flung *out* away from the Sun.

Pure Newton Works, Too. It is possible to quantify the tidal force without recourse to centrifugal force. When we subtracted $\Omega_S^2 R_S$, in order to measure the force imbalance, we were in effect (because $\Omega_S{}^2 R_S = GM_S/R_S{}^2$, by Kepler's Third Law) subtracting the gravity due to the Sun, experienced at the Earth's center, from the gravity experienced at our noon and midnight locations. That is, tidal forces can be viewed as a *differential gravity* force—the effect of the Sun's gravity pulling more at the noon location than on the 'rest' of the Earth, and pulling less at the midnight location than the rest of the Earth. In other words, the high tide closest to the Sun exists because that mass of ocean is pulled more toward the Sun than is the solid earth; and a second high tide, on the other side of the Earth from the Sun, exists because that mass of ocean is 'left behind,' pulled toward the Sun less than is the solid earth. And, as a spatially differential gravity force, it makes sense that the tidal force depends inversely on distance cubed, since the derivative of $1/r^2$ is proportional to $1/r^3$.

As the Moon and Earth orbit each other, the same types of force imbalance exist. Repeating the analysis, we find the maximum force imbalance per unit mass is

$$\pm \frac{2a\,GM_M}{R_M{}^3},$$

where M_M is the mass of the Moon, and R_M is its mean distance from the Earth.

The solar and lunar force imbalances produce periodic variations in the acceleration of gravity measured on Earth's surface—gravitational tides—as the Earth rotates and points on the surface move from being 'noon' to 'midnight' locations and back. The magnitude of the maximum variation is

$$\Delta g_M = \pm \frac{2a\,GM_M}{R_M{}^3} \cong \pm 0.1099 \text{ mGal}$$

$$\Delta g_S = \pm \frac{2a\,GM_S}{R_S{}^3} \cong \pm 0.05045 \text{ mGal}$$

so the range of the lunar tidal force is twice that of the solar tidal force. It can easily be verified that the solar *gravitational* pull experienced on Earth is almost 200 times greater than the lunar pull (the tidal force is not gravity!); the lunar *tidal* force exceeds the solar, even with the mass of the Moon so much less

than the Sun's, because tidal forces depend on the *cube* of the distance, and the Moon is much closer to Earth than the Sun is.

6.1.3 The Response of the Oceans to Tidal Forces

The Shape of the Oceans: The Tidal Bulge. With the Moon's tidal force greatest in magnitude—both on the near and far sides of the Earth—right below it, the preceding analysis suggests that the oceans should respond by rising up at those locations, straight toward or away from the Moon. The tidal force exerted by the Sun should cause a similar response of the oceans relative to it, though with an amplitude smaller by half. The oceans flow toward these locations, supplying seawater to the developing high tides, drawing down sea level everywhere else. At the 'sides' of the Earth (dawn and dusk locations or moonrise and moonset locations), the situation is the most dire: though the magnitudes of the gravitational and centrifugal forces are equal, they are not precisely opposite in direction; the difference is a small tidal force directed 'downward,' toward Earth's center (see Figure 6.3), compounding the loss of seawater to the growing high tides.

As viewed from space, the net effect is that the oceans assume a bulging shape; this ellipsoidal *tidal bulge* is aligned with the Moon (for the lunar tide) or Sun (for the solar tide), with maximum ocean mass (high tides) beneath the Moon or Sun, and a relative lack of ocean mass (low tides) at right angles to that. As the Earth spins, an observer sitting on the Earth's surface will be carried through two high tides and two low tides each day (see Figure 6.4a). For the solar tide, the observer will encounter high tides every 12 hours; for the lunar tide, because the Moon moves in its orbit around the Earth a noticeable bit every day (so the Moon is overhead 50 minutes later each day), the observer will see high tides every 12 hours 25 minutes.

For a quick estimate of the size of the ocean tidal bulge, we can try comparing it to another bulge of the Earth: the equatorial bulge. Such

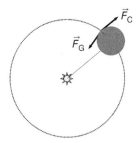

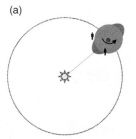

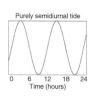

Figure 6.3 Tidal forces experienced on Earth's surface. For simplicity, the Earth's obliquity is neglected here, that is, the Earth is here spinning about an axis perpendicular to its orbit. The Earth's relative size is exaggerated for clarity, as are the gravitational ($\vec{F}_G$) and centrifugal ($\vec{F}_C$) forces acting at a 'dawn' location. The net result of both forces acting at this location, the vector sum $\vec{F}_G + \vec{F}_C$ (shown as the small red arrow), is incomplete cancellation (i.e., an imbalance) because $\vec{F}_G$ and $\vec{F}_C$ point in slightly non-opposite directions.

Figure 6.4 The tidal bulge created in the Earth's oceans by the Sun. For simplicity, the bulge is drawn as if the oceans exist everywhere and can respond to tidal forces without obstruction by any continents, i.e. the Earth is ocean-covered. The relative sizes of the Earth and especially of the tidal bulge are exaggerated here for clarity. The Earth's rotation pole is marked by a filled circle. (a) Looking down on the Earth's orbit from above. The solar tidal bulge is aligned directly toward the Sun. For a first approximation here, the tilt of the Earth in its orbit (i.e., its obliquity) is taken to be zero. The same observer is shown a few hours before and a few hours after noon. As the Earth spins, the observer is carried around the Earth, and sees two equally high tides and two equally low tides per day.

The graph on the right, which covers a 26-hour period of time for clarity, illustrates the twice-a-day rise and fall of the ocean surface as seen by the observer.

a comparison may not be that far-fetched—as we saw in Chapter 5, the Earth's equatorial bulge can migrate as the pole wanders, adjusting to changing conditions (e.g., changing centrifugal force) over time, so it is reasonable to view that bulge as a (highly viscous) *fluid* response of the solid earth. However, equatorial bulge adjustment takes centuries to millennia, at least—clearly a very nontidal timescale—so for our comparison, we will just consider the end result, an equatorial bulge of magnitude 21.4 km, in response to a combination of (rotational) centrifugal acceleration and gravity deficiency totaling 5.17 Gal (according to the International Gravity Formula). For the lunar high tide, caused by a force per unit mass of 0.1099 mGal, the proportional bulge amplitude would be ~21.4 km × (0.1099 mGal/5.17 Gal) ~45 cm; for the solar high tide, it would be ~21 cm.

These numbers are estimates of what is termed the **static** or **equilibrium ocean tide**, a tide in which the change in sea level exactly and instantaneously balances the forcing, leaving no energy for currents. Later in this chapter, we will learn the basis for a more accurate way to estimate such tidal

amplitudes; but our estimates here are acceptable as order-of-magnitude calculations—in part because static tides are a fiction, anyway (though a popular one, as they avoid the real complexities of *dynamic* tides). Static tides cannot rise and fall, which we know real tides do; and in reality, it takes some time to achieve any balance with the forces responsible for the flow. Furthermore, tidal currents are an important aspect of tides, no matter how slow or weak they might be. Even in the adjustment of the equatorial bulge, which we just imagined for simplicity to be an equilibrium response to centrifugal force and gravity, the viscous 'currents' in the mantle accommodating the adjustment may be of great interest. The tides we experience at coastlines are even more dynamic, with some of the tidal energy partitioned into strong currents that hinder the static response, plus resonances

(b)

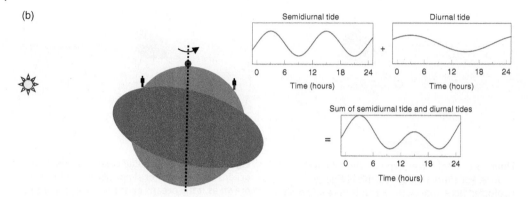

Figure 6.4 The tidal bulge created in the Earth's oceans by the Sun. For simplicity, the bulge is drawn as if the oceans exist everywhere and can respond to tidal forces without obstruction by any continents, i.e. the Earth is ocean-covered. The relative sizes of the Earth and especially of the tidal bulge are exaggerated here for clarity. The Earth's rotation pole is marked by a filled circle.
(b) Side view of the Earth, here acknowledging that the tilt of the Earth is nonzero, so the plane of the ecliptic is tilted with respect to the plane of Earth's equator. Even in this case, the tidal bulge points directly toward the Sun (or, for lunar tides, the Moon). The tilt of the tidal bulge relative to circles of latitude causes the diurnal inequality: the observer, shown here at noon and midnight but carried around with the Earth every day, sees two high and two low tides per day; however, being closer to the peak of the bulge at noon and farther from it at midnight, the observer sees one higher high tide and one lower high tide each day. The low tides, at dawn and dusk, are affected much less by the tilt of the tidal bulge.
The graphs on the right illustrate how a semidiurnal tide and diurnal tide can combine to produce the diurnal inequality. The resulting high tides, at hours 3 and 15, are very different in amplitude; low tides, at hours 9 and 21, are not.

and 'funneling' effects in bays and estuaries that can significantly amplify the tide height and currents. The mid-ocean tide can also experience delays in reaching equilibrium, and resonances, but these are much less extreme, making the static tide model a somewhat better approximation for the tide globally than it is coastally.

Static Complications. The static tidal bulge remains aligned with the Moon (in the case of the lunar tide) or Sun (for the solar tide) at all times. But the Moon and Sun lie essentially in the plane of the ecliptic, and thus—because of the Earth's obliquity—are rarely in the plane of Earth's equator. The bulge will therefore be *tilted* with respect to the equator throughout most of the Moon's orbit around the Earth or the Earth's orbit around the Sun. As shown in Figure 6.4b, an observer at rest on Earth's surface, carried around the Earth's axis each day, will experience one very-high high tide

and one not-so-high high tide each day; this is known as the **diurnal inequality**. (At some latitudes, the inequality is so strong it may even appear as if there is only one high tide per day.) It is often convenient to think of the diurnal inequality as the superposition of the same amplitude tide repeating every 12 hours (or 12 hr 25 min) *plus* a 24-hour (or 24 hr 50 min) tide—that is, the superposition of a **semidiurnal tide** plus a **diurnal tide** (Figure 6.4b, right-hand side); the superposition adds to one of the semidiurnal highs and subtracts from the other.

Still another complication is that both the lunar orbit around the Earth and the Earth's orbit around the Sun are elliptical; with the distance from Earth to Moon or Earth to Sun varying over time, the tidal force—and thus the ocean tide amplitudes—will also vary over time. The variation in lunar distance every month will produce a semimonthly or "fortnightly" tidal variation (because of symmetry, the tidal bulge is the same whether the Moon is on one side of the Earth or the other; cf. Figure 6.4b), and also a monthly variation (because of the diurnal inequality); similarly, the solar variations in distance will lead to semiannual and annual variations. In the same way that the diurnal *variation* could be viewed as the sum of the main semidiurnal tide plus an additional diurnal tide, all of these distance-related *variations* can be thought of as **long-period tides**: a *fortnightly lunar tide, monthly lunar tide, semiannual solar tide,* and *annual solar tide.*

In this discussion, we have treated the lunar and solar tides as independent, which they are. But there is a periodic *situation* that results from their superposition and may be fairly described as a tide as well. Twice a 'month' as the Moon orbits the Earth, the Moon will either be right between the Sun and Earth or right on the other side of the Earth from the Sun. Twice a month, then, the lunar tidal bulge will be aligned with the Sun, i.e. with the solar tidal bulge. At those times, the bulges will add, and the net high tide will be the highest possible; this is called a **spring tide**. Twice a month, but a week later or earlier, the bulges will be at right angles to each other, counteracting each other, producing the smallest possible net tide; this is called the **neap tide**. The spring-neap tidal variation is another fortnightly tide, though its period is slightly different from the orbital, purely lunar fortnightly tide because the alignment of lunar and solar tidal bulges is delayed each time by the slight progression of the Earth and Moon in their orbit around the Sun (—in the same way that the solar day is a bit longer than the sidereal day).

In our discussion of tides, we have managed to maintain a space-based view, for the most part. But there are even more complications when a consistent Earth-based view is adopted; the diurnally rotating Earth imparts complex time variations to the positions and orientations of the Moon and Sun. The net result is that, instead of there being a single semidiurnal tide (from the Moon or from the Sun), a single diurnal tide (from the Moon or from the Sun), and a limited number of long-period tides, there are a nearly infinite *suite* of tides (most of diminishing importance), occurring within a 'band' of frequencies that are all semidiurnal, plus another suite occurring within a diurnal 'band,' plus a third suite all long-period.

6.1.4 The Response of the Solid Earth to Tidal Forces

We have already seen evidence that the solid earth can deform in response to applied forces. In response to surface loads, the lithosphere flexes, and the underlying mantle flows to achieve isostatic compensation. In response to $4\frac{1}{2}$ Byr of centrifugal force, the Earth developed an equatorial bulge; and that bulge has continued to readjust to the changing speed and tilt of the rotation axis following Pleistocene deglaciation. And, as we will study in the next chapter, waves of deformation radiate throughout the Earth every time an earthquake occurs.

The solid earth experiences the same differential forces that cause tides in the ocean. We should thus expect the solid earth to respond to tidal forces, by bulging toward and away from the Moon, or the Sun. These tidal bulges should maintain their alignments as the Earth continues to rotate, that is, as particles of the rotating solid earth are carried around through the bulges. Thus, there should be tides of the solid earth, essentially semidiurnal—the solid earth should bulge up and flatten down, following the Moon or Sun, twice every day. And, superimposed on this semidiurnal rise and fall of the Earth's solid surface, it should also exhibit the same types of diurnal and long-period variations displayed by the oceans.

Such tides do indeed exist! They are the earth tides mentioned earlier, and they can be measured by the tilt or uplift of the ground. The peak amplitude of uplift is found to be ~30 cm—not all that large compared to the dynamic ocean tide in coastal regions, but involving a huge amount of mass nevertheless (enough to amplify the tidal variation in gravity by an additional ~30%, though the net gravity tide is also affected by other secondary factors).

The Ramifications of Solid-Earth Tides Are Astronomical. Tidal forces can act on any body. The Earth exerts tidal forces on the Moon, as the Moon orbits the Earth, and the lunar material deforms—with a fixed, permanent tidal bulge, since the Moon keeps the same face to the Earth throughout its orbit. Artificial satellites in orbit around the Earth also experience tidal forces, though greatly reduced in magnitude since their radii (the "a" in the expressions for tidal acceleration like $2aGM_S/R_S^3$) are comparatively tiny and the difference in gravity across the satellite is so small.

You might want to verify this for yourself—after all, distance is the most important factor in tidal forces, and the satellite's orbital radius is much smaller than Earth's distance to the Sun (or Moon).

Because the tidal force depends inversely on the *cube* of the orbiting object's distance from the orbited mass, the amount of deformation can increase significantly if that distance is small. If the orbiting material is closer to the central mass than a certain critical distance, called the **Roche limit**, then the tidal bulge created may become so large—the satellite or moon may become so stretched out—that it will, theoretically at least, break apart (see Figure 6.5). The most well-known example is the rings of Saturn, which lie approximately just within Saturn's Roche limit. It was demonstrated a century and a half ago in a prize-winning essay by Maxwell that the tidal forces exerted by Saturn on its rings would be enough to break them apart, if they had ever existed as solid rings. The rings discovered in recent years to be encircling the other giant planets of our Solar System are also pretty much within those planets' Roche limits and,

in line with Maxwell's analysis, must also be independently orbiting chunks of debris.

In practice, for a moon or artificial satellite orbiting a planet, the internal cohesive strength of the orbiting material—and the nuts and bolts holding it together, in the case of an artificial satellite—will usually help it to withstand the tidal forces successfully. The Roche limit is thus more realistically discussed in the context of moons that are in the process of accreting, rather than already formed moons. For Saturn's rings—and, incidentally, though they lie just within Saturn's Roche limit, so do two tiny '*shepherding*' *moons*, one inside the closest ring and the other near the outermost—it is not that they were the result of a moon disintegrating after wandering too close to Saturn; rather, the ring material was never able to accrete into a moon. In the prevailing view (but see also Iess et al., 2019 and references therein), that material is assumed to be primordial, solar nebula debris that began to coalesce in the early days of the Solar System, only to be broken apart by Saturn's tidal forces.

Even if the tidal forces of a planet do not succeed in shattering its moon, or if the moon is outside the Roche limit, those forces continually 'work' the moon, stretching and flattening it again and again to produce its tidal bulge. This generates a small amount of heat from internal friction, throughout the moon's interior, during each tidal cycle. Moons in closer orbits will experience much stronger tidal forces and be forced to deform into much larger tidal bulges, significantly increasing the degree of tidal heating. Add to this some amount of water in the moon's interior (see, e.g., Hirth & Kohlstedt, 1996), or an internal rheology that is particularly frictional (Renaud & Henning, 2018), and the result could be intense, even leading to visible volcanic activity.

Many of the moons in our Solar System, like our own Moon, show the same face to the planet they orbit; that is, they spin on their axes at the same rate they orbit around their planet. This phenomenon, called

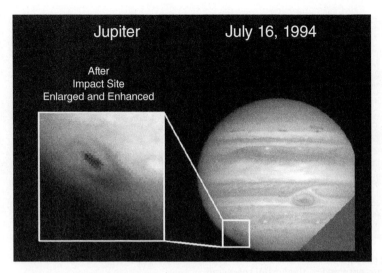

Figure 6.5   A comet pays the price for crossing the Roche limit. Comet Shoemaker-Levy 9 (the ninth comet of its type discovered, in 1993, by astronomers Carolyn and Gene Shoemaker and David Levy) had, according to subsequent analysis, been captured by and was orbiting Jupiter. According to calculations by Nakano and Marsden (Marsden, 1993), in 1992 its highly elliptical orbit brought it well within the planet's Roche limit, and it broke up into numerous fragments.

In July 1994, as their orbit evolved further, those fragments collided successively with Jupiter over a period of a week. This Hubble Space Telescope image was taken after the collision of Fragment A with Jupiter; the impact left a scar in its atmosphere, thousands of kilometers long, which lasted for months. NASA / http://www2.jpl.nasa.gov/sl9/image18.html / last accessed August 22, 2023.

synchronous rotation, **tidal locking**, and **spin-orbit coupling** (as noted in Chapter 1), is another consequence of tidal bulges and the internal friction that produces tidal heating and will be discussed at the end of this chapter. If the orbit of a tidally locked moon is perfectly circular, then the particles of the moon are carried through a tidal bulge at the same rate the bulge shifts to stay aligned with the planet; the tidal bulge will be permanently fixed within the body of the moon. In that case, there will be no continual flexing (just a permanent deformation) and no tidal heating. However, if the moon's orbit is elliptical, then an 'along-axis' flexing will still occur—a periodic increase and then decrease in the stretch of the bulge without an angular shift—because the moon's distance from its planet will decrease and then increase through its orbit. And, from Kepler's Second Law, the moon's speed around its planet will vary slightly through an elliptical orbit, so its orbital motion is slightly ahead of and then slightly behind its spin during

each orbit (in the case of our Moon, this back-and-forth **libration** of its face allows us to see a bit more than 50% of the lunar surface). With an elliptical orbit, then, especially one close to its planet, a tidally locked moon will still experience some degree of tidal heating.

In the case of Io, the closest of the large moons of Jupiter, tidal heating is intense, as we infer from its ongoing volcanic eruptions (check out Figure 1.26 in Chapter 1), despite its tidally locked orbit around Jupiter. This is possible (Peale et al., 1979) because its orbit is highly elliptical, a condition that in turn is maintained by resonances with nearby moons (Europa's orbital period is approximately twice Io's, and Ganymede's is approximately four times Io's, allowing those moons to tug regularly on Io). A similar situation with Enceladus as it orbits closely around Saturn accounts for the plumes of water vapor spewed from geysers (see Figure 6.6) and large, 'tiger stripe' fractures dotting its surface.

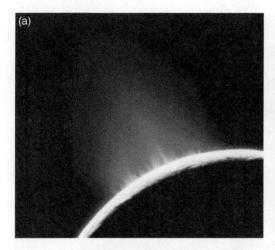
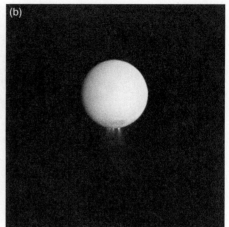

Figure 6.6 An example of tidal heating: water vapor and ice crystals erupted from the surface of Saturn's moon Enceladus, observed by the *Cassini* spacecraft.
(a) False-color image taken in November 2005. NASA / https://apod.nasa.gov/apod/ap071013.html / last accessed August 22, 2023.
(b) Visible-light image taken in December 2009 from a distance of about 600,000 km, showing four distinct plumes. NASA / http://saturn.jpl.nasa.gov/photos/imagedetails/index.cfm?imageId=4152 / last accessed August 22, 2023.

6.2 Precession of the Equinoxes and Orbital Effects on Climate

6.2.1 Precession

What do a spinning top, the spinning Earth, and the advance and retreat of ice sheets during an ice age have in common? The answer appears to be a phenomenon called *precession*. For the Earth, that phenomenon originates—like the tides—in the gravitational forces of the Sun and Moon.

Precessional Torques. We consider the Sun's effect on the Earth first. For simplicity, we imagine that the oblate Spheroid characterizing the shape of the Earth consists of three parts: a spherical part, comprising the bulk of the Earth; and the two portions of the bulges, one nearer to and the other farther from the Sun (see Figure 6.7). The Earth's bulge is tilted with respect to the ecliptic, the plane of Earth's orbit around the Sun, by 23½°, the same amount (of course) by which Earth's rotation axis tilts in space. This obliquity results in part

of the bulge standing out above the ecliptic, and part below.

The Sun exerts a gravitational pull on every particle of the Earth, including its equatorial bulge. The Earth's spherical bulk, nearer bulge, and farther bulge are each pulled straight toward the Sun, as Newton's Law of Gravitation demands (see Figure 6.7). For the spherical bulk of the Earth, this force is clearly in the plane of the ecliptic, directed from the center of the Earth toward the Sun. The Sun's pull on the two bulge components also acts primarily to pull those bulges in the general direction of the Sun (i.e., parallel to the ecliptic)—after all, the bulges only protrude a few thousand kilometers or so out of the ecliptic, and they are 150 million km from the Sun. But even though those protrusions are relatively small, the Sun's pull on them *also* acts slightly downward, in the case of the bulge above the ecliptic, and slightly upward, for the bulge below the ecliptic.

The net result is that all three portions of the Earth are pulled in the general direction of the Sun—an effect that is precisely

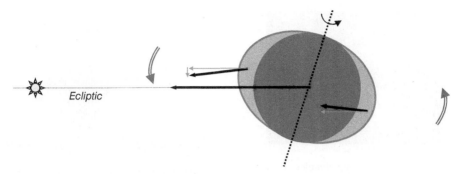

Figure 6.7 Earth experiences a precessional torque because its bulge is tilted with respect to the ecliptic, the plane of Earth's orbit around the Sun.
A side view of the Earth in its orbit is shown here. Conceptually, the Earth can be treated as a spherical portion, the bulk of the Earth, plus two bulges, one above the ecliptic and one below. The Sun's gravitational forces acting on these three portions are shown as bold vectors.
The forces on the bulge portions are resolved into components parallel to and perpendicular to the ecliptic (light arrows). The parallel components, along with the force on the bulk of the Earth, are together balanced by the orbital centrifugal force on the whole Earth, and the Earth continues to orbit the Sun. The perpendicular components amount to a torque on the Earth, acting to twist the Earth (double arrows) in an attempt to straighten it and align its bulge with the ecliptic.

counterbalanced by the outward centrifugal force generated as the Earth orbits the Sun—leaving a slight downward/upward pull that attempts to twist the Earth until its bulges line up with the ecliptic. This remaining, unbalanced twisting force or *torque* causes the Earth to precess (to be described momentarily) as it rotates.

The Moon's gravity also pulls on the entire Earth, and in this case there is also an unbalanced force on the Earth's bulge that acts to twist the Earth until its bulge is aligned. The lunar orbit is nearly in the plane of the ecliptic, so the precessional torque exerted by the Moon adds nearly directly to the Sun's torque. As in the situation of tidal forces, the Moon's precessional torque is larger than the Sun's, and for the same reason: the closeness of the Moon more than compensates for its small mass. Because the unbalanced force producing the precessional torque results from the *difference* between the total gravitational force on the bulge and the 'counterbalanced' component of it, it ends up being inversely proportional to the *cube* of the distance from the Moon (or Sun)—just like the tidal force, another differential force. And, like the tidal force, the lunar precessional torque is approximately

twice as strong as the solar precessional torque.

Precession in Everyday Life. If the Earth were not rotating, its response to the Moon's and Sun's precessional torques would be simply to twist as directed, straighten up, and align its bulge with the ecliptic, at which point the torque would vanish. Because the Earth rotates, however, and is consequently required to obey the law of angular momentum conservation, it turns out that its bulge will never be brought into alignment with the ecliptic! To demonstrate this, we consider two examples. First, imagine a bicycle or motorcycle in motion. Perhaps anticipating a turn around an approaching corner, the rider inclines the bike toward one side (see Figure 6.8a). At all times, the Earth's gravity pulls down on the bike and rider, while the roadway pushes up on the bike at the points of contact with its wheels; after the bike is tilted, however, their center of mass is no longer over the wheels, so those forces are not directly opposed. The net result is a (counterclockwise) torque in which the Earth's gravity tries to make the bike flop over. If the bike is moving extremely slowly, the torque may well succeed;

(a)

(b)

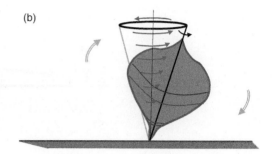

Figure 6.8 (a) Stability of a moving bicycle. As discussed in the text, a tilt of the bike results in a clockwise torque which tries to make the bike flop over; the spinning wheels resist gyroscopically, and keep the torque from succeeding. Image from Pixabay/Pexels.

It appears in this photo that the rider and bike are about to enter a curve. As they round the curve, the resulting outward centrifugal force—which is resisted by the wheels at their points of contact with the road surface, due to friction—will act to produce an additional (and potentially very strong) counterclockwise torque also opposing the flopover. In the end, though, and regardless of any twists or turns in the road, factors relating to bicycle design may impart the greatest amount of stability to the moving bike (see Meijaard et al., 2007; also Thiagarajan, 2015; Hunt, 2016).

Figure 6.8 (b) Precession of a spinning top. The top (brown shape) experiences a precessional torque (pale double arrows) because the Earth's gravity pulls it down from its center of mass while the table pushes up at the point of contact. Whether or not that point of contact is fixed, the spin axis of the top (black line) ends up describing a conical path about the vertical (red arrows) as the top precesses to conserve angular momentum. The 'north pole' of the top will describe a more or less circular path.

if the bike is moving rapidly, though, it can even be inclined nearly to the ground and will still manage to avoid flopping over: its rapidly spinning wheels will not succumb to the torque.

Second, consider a gyroscope or top (—*take a break from reading, and try this for yourself!*). If the top is placed on a table in a nearly upright but tilted position, then it experiences a torque: gravity tries to pull it down, and the table pushes back, pushing it up at its point of contact. If the top is not spinning, it will simply flop over; but if it is spinning, its response to the torque is to add a lateral rotation to its spin. That is, its spin axis will gradually rotate around, describing a conical path around the vertical as it continues to spin smoothly (see Figure 6.8b). This motion—a rotation of the rotation axis—is called a **precession**. Such an unusual motion may seem contrary to our intuition of how the top should respond to the torque; but it is no more contrary than other rotational phenomena, like centrifugal force and the Coriolis force, not to mention other

quantities involving vector cross products (see *Textbox A*).

The Earth's Axial Precession. The Earth responds to the torques of the Moon and Sun in similar fashion to the spinning top, by exhibiting a precessional motion (Figure 6.9). As a consequence of its precession, its orientation in space varies, periodically. Like the precessing top, the angle of tilt does not change as the Earth precesses—the Earth maintains its obliquity, and does not straighten up—but its rotation axis points differently in space at different times through each precessional cycle.

The Earth's precession turns out to be very slow: the Earth spins on its axis several million times for each cycle of precession! In fact, very little precessional motion takes place even over the span of a year, as the Earth orbits the Sun.

In addition to the slowness of its precessional motion, the Earth's precession differs from that of a top because of the sense of its precession. You can verify for yourself that tops precess in the same sense as their spin: if they rotate in a counterclockwise sense, as viewed from above, for example, then their axes will precess

Textbox A: Torques and Angular Momentum

Ultimately, we must turn to equations describing physical laws to understand how torques affect spinning objects like tops and the Earth. We begin by recalling Newton's Second Law, which relates the acceleration of a particle to the net force acting on it:

$$\vec{F} = M\vec{a},$$

as discussed in Chapter 4. Since acceleration is the rate of change with time of the velocity, that is,

$$\vec{a} = \frac{d\vec{v}}{dt}$$

where $\vec{v}$ is the particle's velocity, we can write the Law as

$$\vec{F} = M\frac{d\vec{v}}{dt} = \frac{dM\vec{v}}{dt}.$$

In this form, Newton's Second Law is often called "conservation of momentum," since in the absence of an applied force ($\vec{F} = 0$), the particle's (linear) momentum $M\vec{v}$ is conserved (constant).

For a single particle, the extension to 'angular' quantities is straightforward. Vector-multiplying both sides of the above equation by the position, $\vec{r}$, of the particle yields

$$\vec{r} \times \vec{F} = \vec{r} \times \frac{dM\vec{v}}{dt}.$$

Such vector multiplication may seem like a questionable 'trick'—after all, vector cross products involve factors relating to both magnitude and direction, as we saw in Chapter 4—but you should be reassured that, were we to deal separately with those factors, we would end up with the same result. Since

$$\frac{d(\vec{r} \times M\vec{v})}{dt} = \frac{d\vec{r}}{dt} \times M\vec{v} + \vec{r} \times \frac{dM\vec{v}}{dt}$$

by the product rule of calculus, and since $d\vec{r}/dt \equiv \vec{v}$, it follows that the first term on the right is zero (—do you see why?); so, we can write

$$\vec{r} \times \vec{F} = \vec{r} \times \frac{dM\vec{v}}{dt} = \frac{d(\vec{r} \times M\vec{v})}{dt}.$$

With a few symbols, we can make this equation understandable and believable. The term in parentheses, $\vec{r} \times M\vec{v}$, is the 'moment arm' of the particle (relative to the coordinate origin) times the particle's linear momentum—that is, it is the particle's angular momentum! We will denote that by $\vec{L}$, as we did in Chapter 1. And the left-hand side, the moment arm times the applied force, is the angular force or torque, which we might symbolize as $\vec{\tau}$. Our equation, then, is simply "torque equals the rate of change with time of the angular momentum"—the angular version of Newton's Second Law:

$$\vec{\tau} = \frac{d\vec{L}}{dt}.$$

Like Newton's Law, it says that momentum—angular, in this case—is conserved, that is, constant, in the absence of applied torques. It also says that nonzero torques change the particle's angular momentum.

For systems of particles like the Earth (which we would more rigorously call a continuum), we can add the equations for every particle: on the left-hand side, we would have the sum of

Textbox A: Torques and Angular Momentum (*Concluded*)

all the torques experienced by the particles; on the right-hand side, we would have the sum of the rates of change of every particle's angular momentum, thus the rate of change of the sum of all the particles' angular momentums. We can write this as

$$\vec{\tau} = \frac{d\vec{L}}{dt},$$

where now $\vec{L}$ is the *total* angular momentum of the system and $\vec{\tau}$ is the net torque experienced by the system. $\vec{\tau}$ includes torques the particles exert on each other (due to molecular bonds, friction, gravity, and so on) as well as external torques associated with gravitational forces exerted by other bodies (like the Moon and Sun).

Finally, making the assumption that the torque exerted by any particle on any other is equal and opposite to the torque exerted by the other particle on it, the internal torques cancel out. This allows us to write

$$\vec{\tau}_{E} = \frac{d\vec{L}}{dt},$$

where the subscript reminds us that only external torques can modify a system's angular momentum.

For the spinning top pictured in Figure 6.8b, can you describe how the torque it experiences modifies its angular momentum?

In our discussion of tidal friction later in this chapter, we will be exploring a 'tidal' torque exerted by the Moon on the Earth, and the equal and opposite torque the Earth exerts on the Moon. Because those torques are equal and opposite, the net torque of the "Earth-Moon system" vanishes—implying that the angular momentum of the Earth-Moon system remains constant.

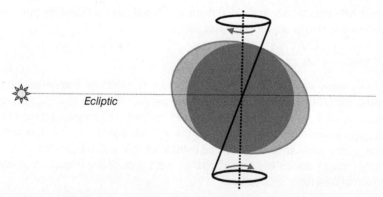

Ecliptic

Figure 6.9 The Earth's axial precession is characterized by its rotation axis (solid black line) describing a conical motion around the perpendicular to the ecliptic (the perpendicular is shown by a dotted line). The angle of tilt between Earth's axis and the perpendicular remains at 23½° throughout the precessional cycle.

in a counterclockwise sense, too (unless they possess a very anomalous shape). This is called a *prograde* precession. But the precessional torque experienced by the Earth tries to twist it to an upright position (relative to the ecliptic), which is opposite to the torque acting on the top (that torque tries to make the top flop over; compare Figures 6.7 and 6.8b). As a result, Earth's precession is *retrograde*, that is, in the opposite sense from its spin.

Quantifying the Precession Observationally. We can actually measure the rate of precession exhibited by the Earth. The **celestial pole** is the point in the sky above Earth's rotation axis. We can locate it by observing the stars at night: each night, as the Earth's daily rotation carries us around the axis, it appears to us instead as if all the stars are following circular paths around that axis (see Figure 6.10). The point around which all the stars appear to circle is the celestial pole. Currently, the celestial pole is very near to a nondescript star in the "Little Dipper" or tail portion of the constellation Ursa Minor (the Little Bear), called *Polaris* or the *Pole Star*. But as a consequence of the Earth's precession, the Earth's rotation axis will gradually shift its direction in space, and the celestial pole will change. For example, in ancient Egyptian times, the celestial pole was observed to be in the constellation of Draco the Dragon.

Each cycle of precession is very long, but the path traversed by the celestial pole in the sky is huge: over the next half cycle of precession, the rotation axis will move from its present orientation, $23\frac{1}{2}°$ to one side of the perpendicular to the ecliptic, to a mirror image located $23\frac{1}{2}°$ to the other side of that perpendicular. Over each cycle, then, the celestial pole travels along the circumference of a circular area $47°$ in diameter; the precession, though slow, is therefore detectable on human timescales. The rate of precession, which can be measured as an angular velocity (precession is, after all, a rotation [of the rotation axis] around an axis), is found to be

$$\Omega_{\mathrm{P}} = 7.741 \times 10^{-12} \ \mathrm{rad/sec}$$

(Capitaine et al., 2003), implying that one complete cycle of precession would take 25,722 yr.

Precession occurs slowly because the torques exerted by the Moon and Sun on Earth's

Figure 6.10 To an observer sitting on Earth's surface, Earth's rotation appears to carry all the stars around each night. The point in the sky encircled by all the stars is the celestial pole. The photo shown here, of star trails over Mauna Kea, Hawaii, illustrates the northern celestial pole. Image thanks to Peter Michaud / https://noirlab.edu/public/images / last accessed August 22, 2023.
Can you explain why the north celestial pole is so low in the sky, in this photograph?

equatorial bulge are weak; they are weak, in turn, because the Earth's bulge is very slight. From the perspective of solid-earth geophysics, *one of the most important reasons to study the precession of the Earth is that it allows us to estimate the size of the equatorial bulge, from observations of the rate of precession.* In other contexts, we might describe the size of the bulge geometrically, in terms of the amount Earth's equatorial radius exceeds its polar radius; but because gravity is involved, i.e. the gravitational force acting on the bulge, it is more relevant to quantify the bulge by its mass distribution as well as its shape. Thus, we measure the bulge by $C - A$, the amount Earth's polar moment of inertia C exceeds its equatorial moment of inertia A, following the notation of Chapter 4.

Quantifying the Precession Theoretically is Also Possible. For simplicity, the Moon and Sun can be taken to be points of mass at some distance from the Earth. To calculate the precessional torque they exert on the Earth, we could adapt the theory used to determine the gravitational field of a bulging Earth (which led in Chapter 4 to Clairaut's theorem and the International Gravity Formula, but for this application would be generalized to specify the direction as well as magnitude of the field). That theory would then yield the gravitational force that the Earth and its bulge exert on the Sun or Moon. The force exerted by the Sun or Moon on the Earth and its bulge will be equal and opposite. In this way, we could find how the strength of the torque ($\vec{r} \times$ force) acting on the bulge depends on the size of the bulge, $C - A$.

Meanwhile, the law of conservation of angular momentum (*Textbox A*) allows us to relate torques to the resulting changes in Earth's angular momentum. In the case of precession, the change in Earth's angular momentum is proportional to the rate of precession, Ω_P (see Figure 6.9). Connecting all these relationships, we would find—not unexpectedly—that Ω_P is proportional to

$$\frac{C - A}{C}$$

(so the bigger the bulge, the faster the rate of precession). The ratio $(C - A)/C$ is reminiscent of another measure of the bulge, the parameter J_2. In that case the bulge $(C - A)$ was scaled by $M_E a^2$, but here the relevant scale is Earth's moment of inertia, C, about its rotation axis, since Earth's rotational angular momentum is proportional to C.

In analogy with the quantity used to describe the geometric flattening of the spheroidal Earth, $(a - c)/a$, the ratio $(C - A)/C$ is called the **dynamical flattening** or **dynamic ellipticity** of the Earth and traditionally denoted **H**. Its theoretical connection to the rate of precession, combined with the observed value for Ω_P, yields

$$H \equiv \frac{C - A}{C} = 1/305.456$$

$$= 3.273795 \times 10^{-3}$$

(IERS Conventions, 2010). The dynamical flattening is thus about a third of a percent— reminding us again that the bulge is very small (and almost equal to the geometric flattening; if the Earth's interior had uniform density, it turns out they would be exactly equal).

6.2.2 Precession, the Core, and the Geomagnetic Field of the Earth

The equatorial bulge, which plays a central role in Earth's precession, is created by the centrifugal force of Earth's daily rotation. As discussed in Chapter 4, the magnitude of centrifugal force is proportional to the distance from the rotation axis. It follows that interior layers of the Earth should bulge *less* than the surface, and the bulge should diminish progressively (to zero) as the Earth's center is approached. The Earth's core, for example, is estimated to have a dynamical flattening of ~1/393 (Wahr, 1981) on the theoretical assumption that the Earth's interior is in a state of 'balance' (hydrostatic equilibrium, to be discussed in Chapter 8); observations of some rotational fluctuations of the Earth have been interpreted as requiring that the core's dynamical flattening be ~1/373 (Gwinn et al., 1986) or ~1/379 (Mathews et al., 2002).

Either way, the core's H is significantly less than that of the whole Earth—roughly 20% less. In contrast, the mantle (or mantle plus crust) contains most of the Earth, and most of the equatorial bulge, so it turns out to have nearly the same dynamical flattening as the whole Earth. Thus, we are faced with the curious situation in which the core and mantle have significantly different dynamic ellipticities. For the sake of 'round numbers,' let's say the core's H is three fourths of the mantle's. If the core and mantle were uncoupled—that is, if they were able to rotate independently—then the lunisolar precessional torques would cause them to precess at very different rates: the core would precess at only three fourths the rate of the mantle's precession; in 26,000 yr, the mantle would complete a cycle of precession but the core would still have a quarter cycle to go, and in 52,000 yr the core would be half a cycle behind. That means that, if at some time the core and mantle had been aligned, with their rotation axes pointing to the same spot in space, then 52,000 yr later their axes would be on opposite sides of the precessional cone, 47° apart.

That misalignment would be catastrophic for the Earth. We will learn in a later chapter how relative motion within the fluid core, and between the core and lower mantle, acts to create components of the geomagnetic field. The huge misalignment we have envisioned between core and mantle would create extremely large relative velocities between them, as each body continues to rotate once per day around those misaligned axes; the end result would be magnetic fields experienced even at Earth's surface with a devastatingly large amplitude. Since there is no evidence for such geomagnetic field fluctuations, we conclude that *the mantle and core cannot rotate independently; they precess together*.

Such a conclusion implies that there must be another force acting at the core-mantle boundary which supplements the lunisolar precessional torque on the core by the additional 20% or so required for it to precess along with the mantle. So, what is the source of the additional torque acting on the core during precession? Although we can expect a variety of forces to act between a core and mantle in relative motion—viscous drag, induced electromagnetic forces, and pressure forces from a topographically uneven core boundary (known as 'bumps on the core'), for example—all of those turn out to be relatively ineffective, for reasons having to do with the state of Earth's interior that will be discussed in later chapters.

There is a curious history to this subject, which was ultimately resolved with the help of very complex mathematics to show what our intuition tells us anyway: that *the ellipsoidal shape of the core-mantle boundary*, i.e. of the base of the mantle, acts on the core when it begins to be misaligned, pressuring it to remain aligned. In other words, the top of the core and the base of the mantle share a certain shape (with a flattening of just over ~1/400), and there is no way to fit the core ellipsoid inside the mantle cavity ellipsoid unless the two ellipsoids remain aligned. This forcing has been called *inertial coupling* or *Poincaré coupling*.

6.2.3 Precession Can Affect the Earth's Climate

Till now our reference frame for measuring the shifting of Earth's orientation during precession has been inertial space, i.e. the fixed stars. This is appropriate given that the motion of the celestial pole is the primary manifestation of this precession. But as the Earth precesses, its orientation is also changing with respect to the *Sun*, and that can impact our climate. This is a more complicated subject than the axial precession, in part because Earth's orientation relative to the Sun is continually changing anyway, even without considering precessional effects, simply because it is orbiting around the Sun with a tilted axis (that change in relative orientation is of course what produces the seasons).

A useful way to measure the effects of precession relative to the Sun is to mark the changes in key orbital positions—what we might call

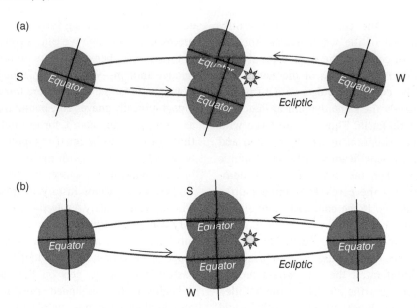

Figure 6.11 The effect of precession on the seasons. The Earth's orbit is shown here with exaggerated eccentricity, and the Sun appears correspondingly much farther from the center of the orbital ellipse than in reality.
(a) A slanted view of the orbit at the present time, illustrating 'seasonal markers': winter solstice (W) on the right, summer solstice (S) on the left, and vernal and autumnal equinoxes in-between.
(b) A slanted view of the orbit a quarter cycle from now. The Earth's axis has precessed such that it is directed out of the page, toward the viewer (23½° toward us). Winter solstice (W) now occurs much earlier in the orbit.

seasonal markers (see Figure 6.11). In the present era (Figure 6.11a), late in December, the Sun is directly shining the farthest south it can on the southern hemisphere; that is the first day of southern hemisphere summer and the first day of northern hemisphere winter, and we (with a northern hemisphere bias) call it the **winter solstice**. Six months earlier or later, near the end of June, the Sun is directly shining the farthest north it can on the northern hemisphere: our **summer solstice**, the first day of northern hemisphere summer, and the first day of southern hemisphere winter. Finally, at times about halfway in-between the solstices, the Earth reaches a point in its orbit where the Sun is directly over the equator, shining equally on both hemispheres; those are the **vernal** and **autumnal equinoxes**.

Now imagine the Earth just a couple of thousand years hence. The Earth's axis has precessed a little, in a retrograde sense, which means Earth will experience its winter solstice (with the Sun shining most directly on the southern hemisphere) a little *sooner* in its orbit than indicated on the right-hand side of Figure 6.11a.

To visualize this in 3-D, mark a ball with an 'equator' and then shine a flashlight on it under solstice conditions. Now 'precess' the ball a little; if you have 'precessed' it correctly, you will have to back up the ball in its orbit, as you keep shining the flashlight on it, to regain the solstice.

In fact the other seasonal markers will also be reached a little sooner than at present.

These changes in the timing of the seasons might be more apparent a quarter of the precessional cycle from now (shown in Figure 6.11b). Because the times of equinoxes (as well as solstices) occur progressively earlier in the Earth's orbit, the overall phenomenon is sometimes called the **Precession of the Equinoxes**.

The Earth's distance from the Sun varies over the course of its elliptical orbit. At present (Figure 6.11a), the Earth is closest to the Sun in early January, just two weeks after the winter solstice, and it is at its greatest distance from the Sun in early July, two weeks after summer solstice. The Sun's radiation spreads out in all directions; our closeness to the Sun during northern hemisphere winter allows us to capture a bit more of that radiation, making northern hemisphere winter a bit milder—but also making southern hemisphere summer a bit more intense. The converse holds true during northern hemisphere summer, with Earth more distant from the Sun: a milder summer for us, and a stronger winter 'down under.' That is, at present, the elliptical orbit of the Earth leads to milder seasons in the northern hemisphere and more extreme seasons in the southern hemisphere.

Half a precessional cycle from now, though, the solstices have 'backed up' in the orbit half way around, so the winter solstice occurs when the Earth is just about at its farthest distance from the Sun and the summer solstice when Earth is nearly at its closest. At that time, it will be the northern hemisphere with the more extreme seasons. During each cycle of precession, then, the *intensity* of the seasons in each hemisphere varies.

It should be mentioned that, despite Earth's orbital 'backsliding,' our calendar will *always* say that the winter solstice occurs in December and the summer solstice in June. The reason is that our calendar is based on the **tropical year**, defined as how long it takes the Earth to go from vernal equinox to vernal equinox—from the first day of spring to the first day of spring. This is conceptually different from the time it takes the Earth to complete one orbit, which can be called a **sidereal year** (i.e., a year as measured relative to the stars—just like a sidereal day is measured, in contrast to a solar day). Our seasonal markers are defined only by their relation to the Sun, not to the orbit or the stars. As we have just seen, Earth's axial precession causes the seasonal markers to 'back up' in the orbit, a bit each year—so the tropical year

is completed a bit sooner than the sidereal year—but the calendar maintains its connection to the markers. Thus, for example, at the present time (Figure 6.11a) the first day of winter ("W"), in December, occurs at a point in Earth's orbit nearly closest to the Sun; six or seven thousand years from now (Figure 6.11b), the first day of winter ("W") will occur at a point in the orbit a quarter of the way around earlier—but the calendar will still say that it is late December. One might say that the axial precession is matched by a procession of the calendar, backward, through the orbit.

Though our calendars preserve the timing of the seasonal markers, we can always track the precession by changes in the position of the celestial pole relative to the stars, that is, relative to constellations—or, by the changes in the constellations we see at the time of seasonal markers. Each cycle of the Precession of the Equinoxes brings us through the zodiac.

Other Perturbations to Earth in Its Orbit. The Earth experiences gravitational forces exerted by all the objects in the Solar System, although most of them are too small and/or too far away to affect our planet measurably. The most important exception to this is Jupiter, which—though its orbit is an additional 630 million km out from the Sun—has a tenth of a percent of the Sun's mass. Jupiter's gravity perturbs the Earth in three novel ways:

- it modifies Earth's *obliquity*, causing the tilt to vary periodically between 22° and 24½° (thus, a range of about ±1° around its current value) on a timescale of ~41,000 yr;

- it causes the *ellipticity* of Earth's orbit to vary quasi-periodically, with the shape of the orbit ranging from somewhat stretched out to very nearly circular, on hundred-thousand-year (and longer) timescales; and

- it causes the elliptical shape of Earth's *orbit* to advance progressively over time.

The timescales involved in these perturbations are long because Jupiter's forcing is weak, and because it is acting on an Earth of wildly varying distance and orientation: at times, as Earth and Jupiter both revolve around the Sun, the Earth may be on the same side of the Sun as Jupiter, or on the far side away from Jupiter; and Earth's rotation axis may point toward or away from Jupiter.

Basics of Orbital Geometry. The orbital ellipticity involved in these perturbations requires some explanation. As per Kepler's Laws, the Earth's orbit is, in general, elliptical in shape rather than circular. We measure how stretched out a planetary orbit is using parameters analogous to those employed to quantify the flattening of the Spheroid. The **ellipticity** e of the orbit is defined as

$$e \equiv \frac{a-b}{a},$$

where for this discussion a and b are the semimajor and semiminor axes of the orbital ellipse (see Figure 6.12). If the orbit is circular, then $a = b$, and $e = 0$; the more stretched out the orbit is, the closer e will be to 1. For the Earth, whose mean distance from the Sun is

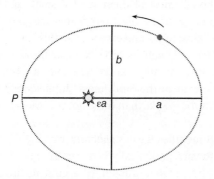

Figure 6.12 Geometry of the Earth's orbit around the Sun, looking down from above. The largest and smallest diameters of the ellipse, called the major and minor axes, are bold lines; their half magnitudes are denoted a and b. The Sun is at one focus (defined in Chapter 4) of the elliptical orbit. Analysis of the orbital geometry implies that the distance of either focus from the center of the ellipse is equal to εa, where the eccentricity ε is defined in the text; for most ellipses, that distance is much greater than $a - b$ (though smaller than is shown in this highly exaggerated sketch). The perihelion P is also defined and discussed in the text.

~93 million miles or nearly ~150 million km, observations show that, currently,

$$a \cong 149{,}598{,}261 \text{ km} \qquad \text{and}$$

$$b \approx 149{,}577{,}000 \text{ km},$$

yielding

$$e \cong 0.0001421;$$

clearly, Earth's elliptical orbit is nearly circular.

The semimajor axis of Earth's orbit, a, roughly our average distance to the Sun, provides a 'yardstick' (as we saw in Chapter 1) to measure distances within our Solar System, for example the sizes of planetary orbits. Astronomers call the magnitude of a an **astronomical unit** or **AU**. For example, the semimajor axis of Jupiter's orbit is just over 5 AU.

The *eccentricity* of an orbit is another measure of its shape. We met such a parameter in Chapter 5 in the context of the highly precise Pizzetti-Somigliana gravity formula for gravity. As before, we denote it by ε:

$$\varepsilon \equiv \frac{\sqrt{a^2 - b^2}}{a},$$

where, again, a and b refer in this discussion to orbital distances. The eccentricity is often employed rather than the ellipticity in discussions of Earth's orbital variations because it appears in the orbital mechanics theory underlying Kepler's Laws. ε is still within the limits of 0 and 1, though generally larger than e; for the Earth's current orbit, $\varepsilon \sim 0.0169$. One benefit of using ε rather than e to describe the orbit is that it provides a simple formula for determining the distance of the Sun, located at one focus of the orbital ellipse, from the center of the ellipse; it turns out to simply equal εa. For the Earth's orbit at present, that distance is ~2.5 million km—a significant distance, climatologically, but still a small fraction of the orbital dimensions.

For the Earth, how does that distance compare to a − b, the amount of 'stretch' in its orbit?

The definition of ε, combined with a little algebra, implies that for any elliptical orbit that is not circular, εa is always much bigger than $a - b$; that is, the Sun's distance from the center of the ellipse is much bigger than the amount the ellipse is 'squashed.' Thus, out of all the points on the ellipse, the point at the near end of the ellipse's major axis (marked P in Figure 6.12) is the one closest to the Sun; the minor axis of the ellipse defines the shortest 'diameter' but not the shortest distance to the Sun. (Mathematically, if $\varepsilon a > a - b$ then, switching terms, $b > a - \varepsilon a$.) The closest point to the Sun in Earth's orbit is called the **peri-helion**. And, the far end of the major axis is the point in the orbit farthest from the Sun: the **aphelion**. From Figure 6.12, it follows that the distance to the Sun is $a - \varepsilon a$ at perihelion and $a + \varepsilon a$ at aphelion; these distances amount to ~147 million km and ~152 million km,

respectively—quite different than a and b in magnitude!

Jupiter's modulation of Earth's orbital eccentricity leads to variations in the size of Earth's orbital axes, the perihelion and aphelion distances, and the distance of the Sun from the orbital center. ε varies between the extremes of very nearly 0 (a nearly circular orbit) and 0.06 (during which time the major and minor axes differ by an order of magnitude more than they do now). With a very nearly circular orbit, the Sun is nearly equidistant from the Earth year-round; in the most eccentric orbit, the Sun is 18 million km closer at perihelion than at aphelion. The variations in ε are not exactly regular in time but are characterized by a period of ~413,000 yr and several periods close to ~100,000 yr (see Figure 6.13a). The climatic

Figure 6.13 (a) Eccentricity of the Earth's orbit around the Sun over the past million years, using the results of calculations by Berger and Loutre (1999). Counting the number of peaks in this time span implies that the eccentricity varies most basically on a timescale of ~100,000 yr; but to produce the variability we see in the amplitudes of those 100,000-year peaks requires several sinusoids with similar periods (to add to or subtract from each other with slightly different regularity). It turns out there are four sinusoids, with periods in the range 95,000–131,000 yr. Additionally, the time series shows a modulation of these 100,000-year peaks with a period (high to low to high) of roughly 400,000 yr.
(b) Insolation variations over the past million years at 65°N during mid-July, in units of W/m². These variations are predicted from the changes in orbital eccentricity and obliquity, and the effects of precession, as discussed in the text.
Data from Berger and Loutre (1999).

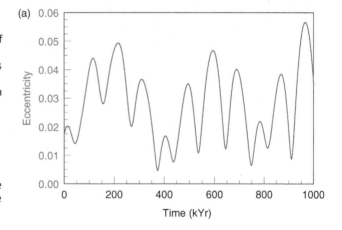

(a)

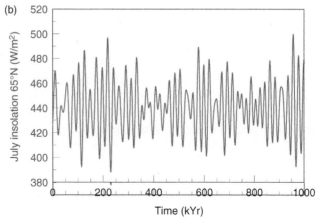

(b)

implications of these variations in ε will be pursued shortly.

The advance in Earth's orbit caused by Jupiter's gravitational tugging amounts to successive changes in the *location* of the orbit's major and minor axes around the Sun. In effect, this is a rotation of the whole orbit, and is described as a precession of the orbit. Since the nearer end of the major axis is the perihelion, this behavior is also called the **precession of the perihelion**. Mercury also exhibits such a phenomenon (as do other planets), though in the case of Mercury, as alluded to in Chapters 1 and 4, it is driven by additional factors beyond the gravity of other planets.

Orbital precession complicates the situation for Earth. For example, we can now define a third type of year: the time it takes for the Earth to travel from perihelion to perihelion, called the **anomalistic year**. Because Earth's orbit advances over time (the precession of the perihelion is prograde), it takes the Earth a bit longer—more than 7 minutes, as it turns out, thanks to Jupiter's tugging—to reach the perihelion again after completing an orbit (i.e., after completing a sidereal year). For comparison, the tropical year is about 20 minutes shorter than the sidereal year; thus, the precession of the perihelion is a slower phenomenon than the axial precession.

We have seen that the axial precession changes the orbital timing of the seasons. The precession of the perihelion *also* changes the timing of the seasons because of the Earth's tilt. Earth takes ~26,000 yr to complete each cycle of axial precession, that is, 26,000 yr for the rotation axis to point to the same celestial pole it had at the start of the cycle; meanwhile, its orbit has advanced somewhat. The axial precession is retrograde, whereas the orbital precession is prograde; the net result is that the Earth returns *earlier* to the point in its orbit with its 'original' orientation relative to the Sun.

Our discussion of angular velocity in Chapter 4 may be helpful here to illustrate the effect. For the moment, let's denote the period of axial precession by T_{AX} and the period of orbital precession by T_{OR}; from the 7-minute advance in the perihelion per year, we can infer that $T_{OR} \sim 72,700$ yr. We'll assume Earth's orbit is circular, with radius R_S, and for simplicity we'll choose the present-day summer solstice (the point marked S in Figure 6.11) as the 'starting point' in the orbit from which to track things; we want to determine how long it takes before the solstice returns to that point in its orbit. The angular velocity of the summer solstice around Earth's orbit is $-2\pi/T_{AX}$, with the minus sign accounting for its clockwise sense of motion; the angular velocity of the perihelion is $2\pi/T_{OR}$. Thus, the linear velocities of the solstice and perihelion around the orbit are (from the formula $\vec{V} = \vec{\Omega} \times \vec{R}$ in Chapter 4) $-2\pi R_S/T_{AX}$ and $2\pi R_S/T_{OR}$. In an amount of time Δt, the solstice travels a distance $(2\pi R_S/T_{AX})\Delta t$ clockwise around the orbit, while the perihelion covers a distance $(2\pi R_S/T_{OR})\Delta t$ counterclockwise. If, at that time, the two points meet again, then the total distance they have covered, coming from opposite directions, must be the circumference of the orbit, $2\pi R_S$. It follows that $(2\pi R_S/T_{AX})\Delta t + (2\pi R_S/T_{OR})\Delta t = 2\pi R_S$. Solving this relation for Δt yields the time to complete a 'solstice cycle': we find that the orbital advance brings the solstice back to the same point in its orbit after only $\Delta t \sim 19,000$ yr—about 7,000 yr sooner than without the gravitational effect of Jupiter.

We could say that 26,000 yr is the period of axial precession relative to the fixed stars but—because of this advance of the orbit—the period is only 19,000 yr relative to the Earth's orbit around the Sun.

There are additional complications. The other planets, primarily Venus (which, though much less massive than Jupiter, is also much closer), also modify Earth's orbit. Venus' gravity, continually tugging on Earth as the two planets' relative distances and orientations vary, manages to produce a *second* orbital precession! This precession of the perihelion, like that due to Jupiter, is prograde; but in this case the net result, i.e. when combined with axial precession, is a period of 23,000 yr for the

Earth to regain its 'original' orientation with respect to the Sun.

Each of these orbital progressions produces a periodic variation in the timing of the solstices and equinoxes relative to the Sun, and are called **climatic precession** or **the general precession**. They are superimposed on each other (they happen simultaneously), but they have different periodicities, tied to the orbits of different planets. Sometimes, for simplicity, they are combined into a single, conceptual, climatic precession: on average, Earth's orbit advances by almost 5 minutes every year, so the solstice regains its match with the perihelion every 21,000 yr. (On the subject of simplicity we note also the popular use of multiple terms to describe these phenomena—for example, axial precession is also called the precession of the equator, and orbital precession is also called the precession of the ecliptic—and by the frequent interchange of various terms; for example, "Precession of the Equinoxes" is used to refer to the axial precession, the overall climatic precession, or individual components of the climatic precession.)

Implications for Climate, and the Milanko-vitch Hypothesis. At this point, such details may seem to be of diminishing importance (!); but these phenomena (precession, as well as the obliquity and eccentricity variations) have climatic consequences. We have already seen that the Precession of the Equinoxes periodically changes the intensity of the seasons. The periodic changes in Earth's obliquity and orbital eccentricity caused by other planets can also affect the intensity of the seasons. At times when the tilt is greater, for instance, high latitudes on Earth will experience the Sun's rays a bit more directly during their summers; and insolation will be somewhat weaker during their winters. Greater obliquity, then, makes all the seasons more intense. When the orbital eccentricity is greater, the Sun's distance (εa) from the center of the elliptical orbit increases, making it closer to Earth around the time of perihelion and farther from Earth six months later—exacerbating the differences in

northern and southern hemisphere seasons we already experience due to Earth's elliptical orbit now (discussed earlier with reference to Figure 6.11a).

The net effect of these rotational and orbital phenomena on insolation is illustrated in Figure 6.13b for a northern hemisphere latitude in midsummer. At that latitude, the net effect, up to ~10%, seems fairly substantial. *Is it possible that such changes in insolation produce significant long-term changes in Earth's climate (and if so, what pattern of climate change, versus time, do you think would result)?*

Such a possibility was the basis for a hypothesis proposed by the Serbian 'renaissance' scientist *Milankovitch* through a series of papers in 1914–1924 (much of his work was actually completed in 1914–1918 during World War I, when he was under house arrest—a political prisoner, due to his ethnicity). Milankovitch proposed that *the ice ages were a consequence of changes in insolation due to the* combined *effects of variations in eccentricity, variations in obliquity, and precession.* (For conciseness, we will henceforth refer to all three processes as 'orbital' factors.) Since changes in obliquity impact high latitudes—where ice sheets start out—most of all, we might expect that orbital factor to be the most important of the three.

The history of Milankovitch's hypothesis presents a striking example of the curious interplay between theory and data in the advance of science. His hypothesis was a theoretically much more rigorous version of ideas proposed by other scientists over the preceding hundred years (see, e.g., the review in Berger, 1988); yet it was considered just one of a variety of possible explanations, and was met initially with no particular excitement. One reason for the lack of enthusiasm was the lack of geological evidence to support the theory's implication that there should have been numerous glacial-interglacial cycles during the Pleistocene—perhaps dozens, given the timescales of the orbital variations; the lack of evidence was unavoidable, since each glacial advance would have been obscured by the next.

Then, in 1976, dramatic supportive evidence was found within deep-sea sediment cores, in the record of oxygen isotope variations they contained, and the pendulum of scientific opinion swung the other way—as far as it could. Analysis of those cores revealed that, over the past 450,000 yr, world temperatures and glaciation had varied in a 'sawtooth' pattern (see Figure 6.14), with times of gradually cooling temperatures and intensifying glacial conditions alternating with shorter spans of rapidly warming temperatures and retreating ice sheets (Hays et al., 1976). From those cores, the periods of glaciation lasted ~100,000 yr, and interglacial periods lasted ~15,000–20,000 yr. That sawtooth pattern—with those same timescales—would subsequently be found in the isotope variations measured in other sediment cores and in ice cores, as we discussed in Chapter 5 (see, e.g., Figure 5.21 in that chapter).

The similarity of those timescales to at least some of the timescales of orbital variation implied that a cause-and-effect relation exists between the orbital phenomena and ice-age climate conditions, i.e. whether glacial or interglacial conditions prevail.

In hindsight, we can understand qualitatively how the orbital factors have to vary in order to promote continental glaciation. The sawtooth pattern tells us that it is a lot easier to cause ice sheets to recede than to advance (see also Birchfield, 1977). On a year-by-year basis, then, mild northern hemisphere summers are a prerequisite for the growth and advancement of ice sheets (hence the focus on 65°N summer insolation in Figure 6.13b). In terms of precession, obliquity variations, and eccentricity variations, such conditions would be more likely to occur when June is at aphelion, the Earth's obliquity is less, and the orbital eccentricity is low, respectively. The timing that determines how these three effects combine (as seen in the varying amplitudes in Figure 6.13b) will be important as well.

But the orbital variations, and the changes in insolation they produce, are sinusoidal oscillations—not exactly a sawtooth pattern! Ruddiman (2014; see our Figure 6.15) shows how periodic signals with orbital timescales might underlie the most recent 'tooth' of the sawtooth pattern. Unfortunately, that pattern does not repeat precisely from tooth to tooth (see, e.g., Figure 6.14). Rigorously establishing the connection between climatic and orbital timescales has, instead, generally been achieved using a data analysis technique called "spectral analysis," in which data that is a function of time (a *time series*) is decomposed into its "spectrum," the set of sinusoids

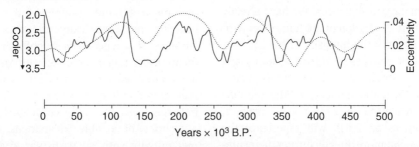

Figure 6.14 Oxygen isotope analysis in marine sediment cores from the southern Indian Ocean. Variations in the enrichment of ^{18}O are expressed as a quantity denoted $\delta^{18}O$ (solid curve). Higher numbers in its scale (downward, on left-hand side) correspond to increased retention of ^{18}O by the carbonate sediment and seawater, and should be interpreted as indicating cooler temperatures and greater amounts of continental glaciation. Cycles of glaciation and deglaciation are best understood by having time go from past to present, i.e. from right to left (showing gradual cooling through glacial periods followed by rapid warming during the shorter interglacials).

Dotted curve shows the variation in orbital eccentricity (scale on right) predicted over the past 500,000 yr, based on earlier theory than in our Figure 6.13.

Figure adapted from Hays et al. (1976).

Figure 6.15 Inference by Ruddiman (2014 / Macmillan Learning; his Fig. 10–14) of how sinusoidal signals contribute to one ^{18}O sawtooth pattern, spanning the end of the Sangamon interglacial ~120,000 yr ago to the beginning of the present interglacial. The periods of those sinusoidal signals, indicated in the figure, correspond to the 'Milankovitch' orbital timescales of eccentricity, obliquity, and precession, with the eccentricity variation dominating.
As noted in the text, the relative contributions of sinusoids of all periods to the sawtooth pattern are best quantified using the technique of spectral analysis.

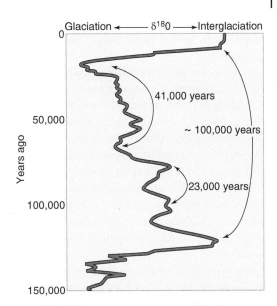

or periodicities best representing it. That is, the technique determines the amplitudes of sinusoids of those frequencies which, when added together, produce the time series of interest; the largest-amplitude sinusoids in the spectrum are construed to be the periodicities most responsible for the data. Spectral analysis demonstrates that the sawtooth patterns in sediment cores and ice cores are dominated by sinusoids with periods of ~100,000 yr, with ~40,000-year and 20,000-year periodicities contributing to lesser degrees (e.g., Hays et al., 1976).

Spectral analysis is an important technique used to explore a wide range of time-varying phenomena encountered in the Earth System, including tides, wobble, seismic waves, and the Earth's magnetic field, though it can also be used to reveal the characteristics of any nongeophysical time series (consider the behavior of the stock market...!). It is discussed in Chapter 7 in the context of global oscillations of the Earth ("normal modes") produced by earthquakes. That discussion includes an example of the spectrum of paleoclimatic temperatures calculated from the Vostok ice core by Petit et al. (1999), revealing the Milankovitch periodicities.

The authors of the discovery in 1976 were circumspect in their conclusions, claiming only that the orbital factors influenced the *timing* of the glacial advances and retreats: they called the orbital factors "the pacemaker of the ice ages" (Hays et al., 1976). And that conclusion still stands today. But the Earth Science community was not so cautious, and for perhaps the next two decades innumerable journal articles and presentations at conferences claimed (or even just assumed) that the Milankovitch orbital factors were the *cause* of the ice ages, and of a wide host of other features seen to vary (on orbital timescales or not) in nature. Students new to Earth Science during these decades would be hard-pressed to believe anything other than that all variations in the geologic record were Milankovitch-type in origin and scale. In reality, though, causality had never been established theoretically.

Milankovitch Can Only Be Part of the Answer. Most obviously, if orbital effects were the cause of the Pleistocene Ice Age, why didn't that ice age extend back in time earlier than $2\frac{1}{2}$ Myr ago?—certainly the orbital factors continued to operate before then! For that matter, why wasn't most of Earth's climate history a perpetual Ice Age within a never-ending Ice House regime? Clearly other factors, especially internal factors associated with the Earth's climate system, must play a role in determining whether an ice age starts or ends—and how much each orbital perturbation is able to impact climate.

In fact, the sawtooth pattern itself is evidence that Earth's climate system modifies the orbital effects: though spectral analysis demonstrates that the sawtooth is characterized by Milankovitch periodicities, its nonsinusoidal shape (in contrast to the eccentricity, obliquity, and precessional oscillations) shows that the climate system has not responded passively to orbital forcing but has selectively amplified or suppressed various components (frequencies) of the forcing—with a sawtooth as the net end result.

The insolation variations in Figure 6.13b represent the net forcing from all three types of orbital factors, with periodicities ranging from ~20 Kyr to 400 Kyr; but inspection of that figure fails to show a strong 100-Kyr signal: even though an eccentric orbit is crucial for the ability of precession to affect insolation (as discussed earlier), the change in insolation caused by a change in the eccentricity of Earth's orbit ends up being relatively small. Yet the Earth's climate response, as seen in Figures 6.14 and 6.15, is dominated by the 100,000-year timescale. We might conclude that the climate system has somehow—surprisingly—preferentially amplified the 100,000-year insolation signal.

It turns out (according to spectral analysis by, e.g., Petit et al., 1999) that the sawtooth pattern of CO_2, as measured in trapped air bubbles in ice cores (see Figure 5.21 in Chapter 5), is even more dominated by eccentricity, and less influenced by precession and obliquity variations, than the isotopically determined global temperature is.

Much research has been devoted to accounting for the dominance of the 100,000-year signal in climate data. Some of those studies characterize the climate system as possessing a natural resonance with a 100,000-year timescale, enabling the system to amplify weak 100,000-year variations in insolation and take the Earth from the beginnings of an ice age into cycles of glacial advance and retreat. Using insolation variations over the past 125,000 yr, Marsiat et al. (1988) found that feedback involving ice-sheet albedo (i.e., an albedo–temperature feedback) was able to lengthen the timescale for ice-sheet growth such that large ice sheets could persist for 100,000 yr.

Alternatively, Ridgwell et al. (1999; also Raymo, 1997) suggested that ice-sheet growth might be particularly effective at times when the maximum summer insolation is relatively low; focusing on how eccentricity modulates precessional effects on insolation (the variation in seasonal intensity due to Earth's elliptical orbit becomes more severe, then less, as ε changes), we would expect to see relatively low insolation every ~100,000 yr. Arguing also that ice sheets would tend to collapse if they became too massive—thus, positing a threshold for ice-sheet instability—they concluded that the result would be a strong 100,000-year signal in glaciation.

The threshold model of ice-sheet growth developed by Huybers and Wunsch (2005), in contrast, was best able to recreate the observed sawtooth climate pattern if obliquity was the key orbital factor; they postulated that the impact of obliquity on high-latitude climate could depend on ice-sheet thickness, and argued that this dependence could delay the impact of that orbital factor, allowing substantial ice-sheet growth until high-obliquity insolation managed to undermine the ice sheet, one or two obliquity cycles later—and resulting in an apparent 10^5-year timescale.

Still other models of ice-sheet dynamics (Birchfield & Weertman, 1978; Imbrie & Imbrie, 1980) produced amplification of the 100,000-year signal by specifying not only that the growth of an ice sheet depends differently on insolation than its retreat does, but also that the size of an ice sheet is sensitive to the *cumulative* (or "integrated") insolation rather than to the insolation itself. We can explain the significance of this latter requirement mathematically: adding the effects of insolation day by day tends to bring out its long-term trends, including those influenced by eccentricity. For a physical explanation, it might be argued that cumulative insolation reflects the duration of each season, which is governed to some extent by Kepler's Second Law (Huybers, 2006) and thus relates to Earth's orbital radius, so it contains an eccentricity-dependent variation.

With the discovery that greenhouse gas concentrations from ice cores exhibit a sawtooth variation and a strong 100,000-year spectral peak, it has been argued (e.g., Shackleton, 2000) that the 100,000-year timescale of ice-sheet advance and retreat is driven by a similar timescale in Earth's global carbon cycle, rather than by ice-sheet dynamics. Explanations as to why CO_2 might exhibit that periodicity in the first place are beyond the scope of this chapter.

Into all these theoretical possibilities, data threw another, irrefutable complication: as dominant as the 100,000-year periodicity is during the late Pleistocene, it was essentially absent from the early and mid-Pleistocene! In sediment cores going back to the beginning of the Pleistocene (Figure 6.16), we see a transformation in the sawtooth pattern: prior to ~1.2–1.3 Myr ago, glacial periods exhibited dominant timescales of ~40,000 yr rather than 100,000 yr! As demonstrated more clearly

in Figure 6.16b, the change in the dominant period is not sudden but actually occurs as a transition, and is more or less complete at around ~900,000–700,000 yr ago.

There is also a change in the amplitude of the pattern, which is noticeably smaller in the early and mid-Pleistocene; from Figure 6.16b we see that what makes the pattern larger during the late Pleistocene is the addition of the 100,000-year signal.

Because of the strength of the obliquity effect on insolation, it is not surprising that we see a 40,000-year signal throughout the Pleistocene. The gradual inclusion and dominance of the 100,000-year signal have been explained as a threshold effect (e.g., Raymo, 1997): as temperatures continued to cool through the Pleistocene, perhaps aided by a trend of lower CO_2 due to weathering, a decrease in volcanism, or other processes, conditions gradually reached a point where ice sheets could grow larger and better survive times of high insolation; the end result would be lengthened glacial periods.

Alternatively (or together with the cooling), other geologic factors may have led to more massive ice sheets and thus longer periods of glaciation. After earlier glacial advances

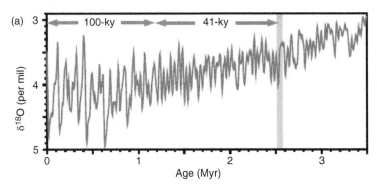

Figure 6.16 Oxygen isotope variations (increasing downward) from a sediment core in the equatorial Pacific near the East Pacific Rise obtained by the Deep Sea Drilling Project.
(a) Note the cooling (increase in ^{18}O) prior to 2½ Myr ago as Earth's climate descended into the Pleistocene Ice Age. Once within the Pleistocene, the character of the variations in temperature (or ice volume) changed, with the peaks and troughs initially closer together then more spread out; this is reflected in the time series' spectrum, where the dominant periodicity was ~40,000 yr but changed to 100,000 yr. This was evidently accompanied by an increase in the amplitude of the fluctuations.
According to Clark et al. (1999), the spectral change happened ~1.2 Myr ago. Alternatively, however, we might infer a transition period from ~1.2 Myr ago to ~0.9 Myr or even ~0.7 Myr ago. Image from Clark et al. (1999) / American Association for the Advancement of Science.

(b)

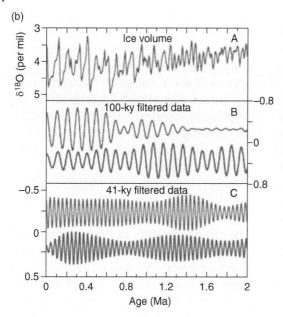

Figure 6.16 (b) On closer inspection, the transition period between ~1.3 Myr ago and 0.9 Myr – 0.7 Myr ago becomes clear. Part A of this figure shows the same data as (a) above. Once the spectrum of that data is computed, we can determine the sinusoids comprising those portions of the data with periods near 100,000 yr (Part B) and near 40,000 yr (Part C) (blue curves; red curves are similarly filtered components of the corresponding astronomical time series). We see that the 40,000-year signal continues more or less unchanged throughout the mid- and late Pleistocene, whereas the 100,000-year signal starts up around 1.3 or 1.4 Myr ago and reaches full strength after 0.7 Myr ago. Image also from Clark et al. (1999) / American Association for the Advancement of Science.

had repeatedly gouged and plowed the land, the thick continental soils of the Pliocene would have gradually been worn away, exposing bedrock; later glaciers, advancing over bedrock instead of soil, would have found it more difficult to flow, slowing their thinning and advance, and causing them to build up in size (Clark et al., 1999).

Paillard (1998) has developed an abstract 'combination' model of Earth's climate system, with surprising results. It was constructed with two climate state thresholds—marking the transitions from interglacial to weak glacial and from weak glacial to strong glacial conditions (thus, a three-state climate system)—and an ice-volume threshold which trended over time rather than staying constant; and, the model was forced by cumulative insolation, which (as mentioned above) serves to amplify the eccentricity signal. Impressively, the model's output reproduces all the main features of sediment core data over the past 2 Myr: it exhibits a sawtooth pattern; a spectrum featuring orbital periods; and a transition, around 1 Myr ago, from the dominant ~40,000-year signal in mid-Pleistocene to a strong ~100,000-year signal in the late Pleistocene.

The existence of a very long-term isotope record in marine sediment cores allows us to search for the *other* signal associated with eccentricity variations: the 400,000-year periodicity. As noted by Muller and MacDonald (1997; see also our Figure 6.16a), such a variation is weak at best. We can interpret that outcome as further evidence that the dominance of the 100,000-year signal is the result of an active rather than passive climate system responding selectively and 'nonlinearly' to the external, astronomical forcing; we would not see much evidence of the 100,000-year periodicity either if it weren't for the climate system.

In sum, the combination of data and theory has brought the pendulum back to a reasonable, middle ground. The Milankovitch mechanism accounts to some extent for the timing of glacial advances and retreats, but does not constitute the fundamental cause of ice ages. Changes in ocean circulation caused by continental drift—as mentioned in Chapter 3—provide a reasonable explanation for the onset of our current Ice House climate regime *and* for the start of Pleistocene glaciation. And, changes within the Earth System are likely to be responsible for the transition from an obliquity-dominated glacial cycle to an eccentricity-dominated one.

6.2.4 Milankovitch and Mars

If civilizations had ever existed on Mars, Milankovitch theory would have been the foundation of every cultural tradition there; Mars' orbital variations are simply too extreme to ignore, thanks to its relative proximity to Jupiter. Mars' obliquity, currently ~25.2°, can vary by up to ±13° (depending on the epoch)—a range that is an order of magnitude larger than our own—over a timescale of ~120,000 yr; its orbital eccentricity can vary by up to ±0.02 on timescales of ~95,000 yr but over ~2 Myr will reach extremes of ~0 and 0.12—with the latter equal to twice our own maximum eccentricity (Ward, 1979).

Meanwhile, the axial precession of Mars contributes another 10^5-year timescale. With moons of negligible mass, the only precessional torque Mars experiences is from the Sun, thus 2½ times weaker than our solar torque because of its greater distance from the Sun; on the other hand, its dynamical flattening is about ⅔ greater than ours (Williams, 2013). Altogether (recalling that for Earth the solar torque is one third of the total), it turns out that the axial precessional torque experienced by Mars is about six times weaker than Earth's, leading to a much longer period of axial precession, ~171,000 yr (Konopliv et al., 2006). Finally, with a strong precession of the perihelion due to Jupiter, Mars experiences a general (climatic) precession with a period of 102,000 yr (Carr, 1990).

Of course, for these major orbital variations to have a correspondingly more extreme impact on the climate of Mars than do the orbital perturbations that affect Earth, Mars would need to have an Earth-like climate system: highly energetic and highly variable, with exchanges of heat, mass, and momentum between its atmosphere, water, and ice reservoirs (and even the solid earth). In its first billion years, its atmosphere thick with CO_2 and containing appreciable amounts of H_2O, and liquid water on its surface, Mars' climate system may have been much more Earth-like, and perhaps some climate-related

processes exhibited orbitally forced 10^5-year cycles of large magnitude. Rhythmically layered terrains within the extensive "Arabia Terra" region of Mars, for example, suggest that sedimentary processes may have been influenced by orbital effects in that era (e.g., Lewis et al., 2008). But there is no evidence that Earth-type cycles of continental-scale advances and retreats of massive glaciers and ice sheets were ever a component of the Martian climate system.

At present, under the dry, thin-atmosphere conditions that prevail now, climate on Mars refers mainly to variations (e.g., seasonal) in atmospheric temperature and composition, with a small mass exchange involving the polar ice caps. One might imagine, however, a Martian 'parallel' to ice ages: *dust ages*, in which times of incessant dust storms of continental and even global scale alternate with times in which large-scale dust storms are infrequent. The dust storms we see on Mars today (e.g., Figure 6.17) tend to occur during the southern hemisphere's spring and early summer, when more intense (or, less feeble) solar heating produces strong winds, capable of lifting dust up in the thin atmosphere; warmer conditions could also release CO_2 from the soil into the atmosphere, producing a 'thicker' atmosphere that could support the transport of dust grains more efficiently even when the winds are weak (Pollack, 1990; see also NASA Science News, 2001). Both mechanisms suggest that optimal conditions for generating dust storms would occur more often when Mars is near perihelion *and* when its obliquity is high; thus, in-phase versus out-of-phase combinations of those orbital factors should lead to cycles of extreme versus limited dust storm activity.

A provocative idea by Head et al. (2003) goes one step further, speculating that at times of the highest obliquity, such as prevailed around 2 Myr ago, insolation would have heated the polar regions of Mars sufficiently to vaporize water ice, which then could have spread equatorward through the atmosphere almost

Figure 6.17 Hubble space telescope images showing stages of a dust storm on Mars. The storm began on June 26, 2001 (left) in the Hellas Basin, and became global a day later (NASA Science News, 2001). It lasted three months before diminishing; the right-hand image shows the planet 71 days after the storm began. Image from https://mars.nasa.gov/resources/7891/hubble-sees-a-perfect-dust-storm-on-mars/.

to the Martian tropics, covering all but the tropics of the planet's surface with a thin layer of dusty snow and ice. This scenario has been termed a *Martian ice age*, though it differs from Earth's recent ice ages in a number of respects, including that the spread of ice is stimulated by polar heating, and that no massive ice sheets are actually involved.

Evidence of present-day orbital influences on small-scale geological features is seen in laminated dust/ice deposits visible near Mars' south polar region (e.g., Toon et al., 1980) (see Figure 6.18); and in gullies at midlatitudes, interpreted to be created by runoff from subsurface water ice that melted at times of extreme obliquity (e.g., Costard et al., 2002).

We close this section by noting the impact Milankovitch had on popular culture regarding Mars. Milankovitch had conducted some early research on the climate of Mars, establishing that its average surface temperature was (at present) much too cold for liquid water. At the time, public imagination was still enflamed by the notion of advanced civilizations on Mars—civilizations that were capable of creating an elaborate system of canals across its surface. Milankovitch's research threw

Figure 6.18 Laminated terrain in the south polar region of Mars, thought to be the result of orbital effects. The terraces seen in the upper portion of the image extend for hundreds of kilometers. Area pictured is ~50 km wide. Toon et al. (1980) / with permission from ELSEVIER.

cold water, as it were, on those passions (Macdougall, 2004), though (given our unending preoccupation with the possibility of life on Mars, as discussed in Chapter 1) perhaps not sufficiently.

6.3 Satellite Geodesy

6.3.1 Satellite Orbital Precession

Our discussion of gravity measurements in the preceding chapter ended with the recognition that a complete picture of Earth's gravity field is not really possible without satellites, thanks to both the global coverage they provide and the substantial improvements they bring to shipborne and airborne measurements (by enabling accurate corrections for the Eötvös effect). The geophysical benefits to monitoring satellites became evident soon after the Space Age began (and the 'Space Race,' between the United States and the Soviet Union, its rival superpower) following the launch of Sputnik in 1957. Most fundamentally, tracking satellite orbital paths provided a novel way to deduce patterns in Earth's gravity field; for example, a 1959 journal article (O'Keefe et al., 1959) headlined the discovery—based on the way satellite orbits were affected by the global gravity field—that the Earth is "pear-shaped" (with that shape superimposed on Earth's much larger equatorial bulge). Satellites can tell us about gravity because, regardless of the type of satellite, or the method of tracking it, all satellites are subject to Earth's gravitational forces. And the end result is that all satellite orbits are *perturbed*: they undergo precession, changes in obliquity, and changes in eccentricity.

We have seen that the gravitational force of the Sun and Moon on Earth's equatorial bulge creates a precessional torque on the Earth. The circumstance of a satellite orbiting the Earth at an inclination to the equator shares many similarities with that situation. As before, we can imagine the Earth to consist of a spherical mass containing the bulk of the planet, plus the equatorial bulge in two portions. The mass of the 'near' bulge pulls the satellite down toward it, or up toward it (see Figure 6.19), in effect exerting a torque on the satellite that tries to align the plane of the satellite's orbit with the plane of the bulge (the equator). Since the satellite has angular momentum, from its revolution about the Earth, the torque will never achieve that realignment; rather, the *axis* about which the satellite orbits—the satellite's angular momentum axis—will precess, about an axis perpendicular to the equatorial plane.

In either case—Earth precession or satellite precession—the rate of precession depends on the torque's strength, which depends in turn on how much the Earth is bulging. By measuring the rate of precession, the amount of bulge can be determined. As discussed earlier, the amount of equatorial bulge can be described by $C-A$, where C and A are the polar and equatorial moments of inertia of the Earth. However, though Earth and satellite precession both depend on $C-A$, the actual quantities measured (in the case of Earth's precession it was H, the dynamical flattening) differ because of the different roles of the Earth. Satellite precession derives from the gravitational force exerted by the Earth, calling for $C-A$ to be scaled by M_E or, for proper units, $M_E\,a^2$—ultimately leading to the parameter J_2 whose variations over time we found so useful for constraining the viscosity of the mantle.

Figure 6.19 Torques (orange arrows) caused by the gravitational attraction of the Earth's bulges on an orbiting satellite cause the satellite orbit to precess; the orbital precession is characterized by the axis of the orbit describing a conical motion around the perpendicular to the equator.

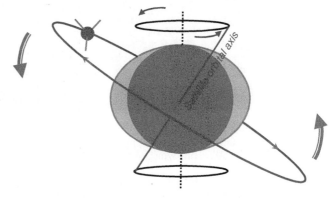

The combination of results from satellite precession and the Precession of the Equinoxes, by the way, yields fundamental information about the distribution of mass within the Earth:

$$\frac{J_2}{H} = \frac{(C - A)/M_E a^2}{(C - A)/C} = \frac{C}{M_E a^2};$$

with the values of J_2 and H known from precessional observations, we find $C = 0.33070 M_E \cdot a^2$ (where a is Earth's equatorial radius; or, in terms of Earth's mean radius $\bar{r}$, $C = 0.32996 M_E \bar{r}^2$). The importance of the 0.33 coefficient was already discussed in Chapter 1.

If the Earth were ideal, only its equatorial bulge would be able to exert torques on a satellite. In the real Earth, any anomalous mass (i.e., anomalous as compared to the Ideal Earth) can cause *additional* torques beyond that exerted by the bulge. For example, the extensive iron ore deposits southwest of Moscow (Russia) will exert a gravitational force that tugs repeatedly on any satellite orbiting the Earth, especially when the satellite passes close by; if the ore body is not in the plane of the satellite orbit, those tugs will torque the orbit. These result in further orbital perturbations, complicating the basic precession of the satellite.

Such torques are generally weaker than that due to the bulge, especially for satellites at high altitudes; but there are special cases in which the bulge's torque vanishes—if the satellite orbit is equatorial or, if the bulge is axially symmetric, polar—so in those cases the anomalous torque will dominate. Depending on orbital altitude and inclination, then, the satellite perturbations altogether could have a range of periods. By closely tracking the satellite's orbit, and by tracking satellites with different altitudes and orbital inclinations, the perturbations—and thus the mass anomalies responsible for them—can be deduced. Ultimately the Earth's gravity field can be determined, on a global scale.

There are Different Ways to Track Satellites Accurately. One sequence of satellites used to infer Earth's gravity field is LAGEOS (LAser GEOdynamics Satellite); LAGEOS I was launched in 1976, LAGEOS II in 1992. These satellites can be tracked precisely by bouncing laser beams, sent from a ground station, off their surfaces; those surfaces are covered with numerous "corner-cube" retroreflectors, allowing light from any angle to be reflected back in the same direction (and giving them the appearance of a half-meter-size golf ball; see Figure 6.20). A similar satellite, called Starlette and launched by the French government in 1975, was actually the first dedicated laser-tracked satellite.

The reflected laser beams allow the satellite **range**, which is the distance to the satellite from the tracking station, to be determined as the satellite moves through its orbit. Measuring the time it takes the laser pulse to travel from station to satellite (which is half the time it takes to send and then receive the reflected signal back at the station), the range—using

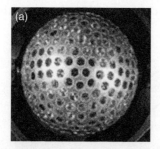 

Figure 6.20 Satellite laser ranging.
(a) Photograph of LAGEOS I. This satellite measures 60 cm in diameter and is covered by 426 'corner-cube' retroreflectors. From NASA / http://msl.jpl.nasa.gov/QuickLooks/lageosQL.html / last accessed August 22, 2023.
(b) A pulse of laser light for the purpose of satellite tracking is captured in this photograph of the Royal Greenwich Observatory's 'old solar dome' at Herstmonceux. RGB Ventures/Alamy Stock Photo.

"distance equals rate × time"—will equal the velocity of light multiplied by the one-way travel time. Accurate range determination depends on knowing the speed of light in the atmosphere (which differs from that in a vacuum because of the air itself), and also requires correction for the motion of the station (carried along with the rotating Earth) and the motion of the satellite during the laser signal transmission.

Whether observing Starlette, LAGEOS, or any other of the several laser satellites now orbiting the Earth, the precision of **satellite laser ranging** (**SLR**) is high enough that

- the set of range observations must account for the wobbles and changing rotation rate of the Earth—such changes in the Earth's angular velocity vector alter the positions of the tracking stations, relative to the satellite;
- the inferred gravity field must include components produced by the solid earth and ocean tides (the mass redistributions created by those tides detectably perturb the satellite orbit); and
- the displacements of the tracking stations over a set of observations must be modeled or corrected for continental drift (plate motions are on the order of 1–10 cm per year).

At present, SLR—with the range from a single pulse determined with a precision better than 1 cm—routinely provides high-precision models of Earth's gravity field, variations in Earth's rotation, the ocean tides, and tectonic plate motions.

It is worth noting that SLR also yields the most fundamental gravitational information of all about the Earth: GM_E. This follows directly from Kepler's Third Law, with the orbital radius and period deduced precisely from the laser-tracked satellites.

One popular alternative to laser ranging is **Doppler tracking**, in which radio signals broadcast by the satellite are received at a ground station, or vice versa; the Doppler effect on the signal's frequency due to the relative motion between satellite and station is inferred, and used to calculate the satellite's relative velocity or **range rate**. Observations from repeated broadcasts and from stations throughout the tracking network allow the satellite's orbit, and orbital perturbations, to be determined. In contrast to SLR, clear skies are not needed for Doppler tracking to work.

Accurate range determination from Doppler tracking depends on knowing the speed of light (i.e., the speed of the radio waves) in the atmosphere, which differs from that in a vacuum because of the air itself; additionally, the variable effects on the speed of light and on the path of the radio transmission caused by the electrically conducting ionosphere and by water vapor in the troposphere must be accounted for. At present, Doppler tracking is capable of achieving a precision of ~10 cm in the satellite orbit after a few hours of observation (AVISO, 2013). Like SLR, Doppler tracking contributes to models of Earth's gravity field, Earth rotation, and tides.

The use of orbital perturbations to determine the suite of mass anomalies responsible for torqueing the satellite (or to quantify the corresponding gravitational field) is technically phrased as an "inversion" of the (orbital) data: using the *effects* of the torqueing to infer the *causes* rather than the converse. The inversion process is a 'bootstrap' one, beginning, of course, with the J_2 determination; as more mass anomalies are identified (or as the description of the gravity field is refined, building on earlier gravity models), the satellite orbit can be more accurately deduced, and the range observations will yield increasingly subtle perturbations.

Spherical Harmonics Help Us See the Big Picture. The mass anomalies and gravity field determined by the analysis are typically described within a mathematical framework known as *spherical harmonics* (see *Textbox B*). Spherical harmonics originate as the solution to an equation (called Laplace's equation) that all inverse-square-type forces, like gravity, must obey in outer space. Their derivation and mathematical specifications are beyond the scope of this textbook; but even a qualitative description of them is worth the effort, as they will prove useful in several contexts. In general, the simplest kind of "harmonic" functions are the sines and cosines we are all familiar with; among other things, those functions are the solution to equations governing the behavior of harmonic oscillators. Sines and cosines actually provide a fairly powerful means of describing any conceivable one-dimensional pattern, using an appropriate combination of sinusoids of different wavelengths (such combinations, called *Fourier Series*, are discussed in Chapter 7). *Spherical* harmonics are just harmonic functions—combinations of sines and cosines—used to describe patterns on the surface of a sphere. The harmonic 'wavelengths' in north-south and west-east directions are referenced by two parameters called the harmonic *degree* and *order*; they are explained in more detail in *Textbox B*.

Textbox B: Spherical Harmonics

Harmonic functions like sines and cosines can be used to reconstruct any one-dimensional pattern (i.e., any function of a single variable). We illustrate that here by reproducing a triangular pattern, reconstructing it to good approximation (as shown on the right) with just a few sinusoids (in this case, $0.86\cos(\pi X/2) + 0.0955\cos(\pi X/(2/3)) + 0.0344\cos(\pi X/(2/5))$, shown on the left but offset vertically for clarity). Because the original pattern has a long 'wavelength' (2, in this case, if we define wavelength here as the distance between zero-crossings), the wavelengths of the sinusoids to build it are also mostly long (mostly 2, a small amount of 2/3, and a little bit of 2/5).

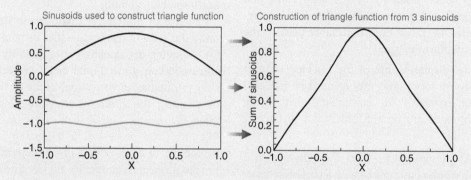

Spherical harmonics are two-dimensional harmonic functions—specific combinations of sine and cosine functions of latitude and longitude—that can be used to reconstruct any pattern on the surface of a sphere.

What global geophysical quantities can you envision being well described using spherical harmonics?

The basic types of spherical harmonic patterns, with a color scheme of orange for positive values and blue for negative in the range −1 to +1, are shown here (Hoover et al., 2009):

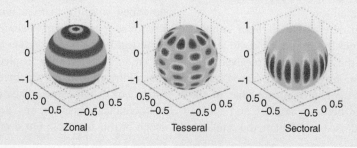

Textbox B: Spherical Harmonics (*Continued*)

Zonal harmonics are constant (either positive or negative) along a given latitude, so they only vary as you change your latitude, e.g. as you go from pole to pole. Zonal harmonics can change sign only with latitude; sectorial (or sectoral) harmonics can change sign only as you change your longitude, i.e. as you move east ↔ west; and tesseral harmonics can vary in sign both with latitude and longitude.

There are a huge variety of spherical harmonics even of one type of pattern, distinguished by the 'wavelengths' in their patterns, i.e. the distance from peak to peak or trough to trough in the north-south and/or west-east direction. To reproduce an arbitrary pattern on Earth's surface, we are likely to need many zonal, tesseral, and sectorial harmonics of all different wavelengths.

Mathematically, though, all these harmonics are usually differentiated by quantities more akin to 'wave numbers' rather than wavelengths: the *degree* and *order* of each specific harmonic, denoted ℓ and m, respectively; these parameters suggest how many cycles of peaks and troughs the pattern has in the north-south and west-east directions (so they are related to wavelength). In general, an order m harmonic is a function like cos(m·longitude), so those harmonic patterns have m full wavelengths going west to east around the Earth. The north-south parts of the functions, which relate to the harmonic degree, are more obscure, but the result is that the degree ℓ order m pattern changes sign (from positive to negative or vice versa) $\ell - m$ times going from pole to pole.

For example, consider the degree 16 order 9 ($\ell = 16, m = 9$) tesseral harmonic (Barthelmes, 2013), shown below with magenta areas indicating positive values; blue, negative.

If you chose a specific circle of latitude and went around the Earth you would experience $m = 9$ peaks and 9 troughs; if you moved down a meridian from pole to pole, you would experience $\ell - m = 16 - 9 = 7$ changes from peak to trough or trough to peak.

Zonal spherical harmonic functions are defined by $m = 0$, which means they are axially symmetric (west to east they have zero wavelengths, and go from peak to trough zero times); they vary only in the north-south direction. When $m = \ell$, the pattern is sectorial, and there is no sign change ($\ell - m = 0$) in the north-south direction. All other patterns ($0 < m < \ell$, as the rules of spherical harmonics do not allow $m > \ell$) are tesseral or 'tile-like.'

The lowest-degree zonal harmonics vary *only* on global scales, as shown on the next page in this side view (north pole at the top) of the degree 2, 3, and 4 zonal harmonics superimposed on a sphere; positive regions are outside the sphere, negative regions are within.

Textbox B: Spherical Harmonics (*Continued*)

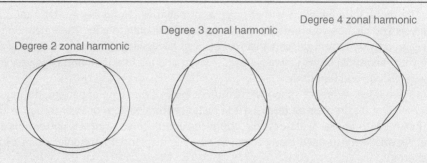

The shape of the degree 2 zonal harmonic evidently makes it ideal for describing an equatorial bulge, or a tidal bulge. The degree 3 harmonic resembles a pear shape, and that of degree 4 is diamond-like.

Because spherical harmonics are fundamentally sinusoidal, one might expect them to be useful mainly for describing sinusoidal-type patterns on the globe. The illustration shown below demonstrates that, to the contrary, spherical harmonics can recreate reasonably well *any* pattern on the surface of a sphere. In this case, the pattern is the distribution of continents (whose locations are assigned a value of 0) versus oceans (whose locations are assigned a value of 1), and a combination of harmonics through degree and order 180 was used (from Dickman, 2010). (This pattern is recreated only "reasonably well" because, as indicated by the color scale on the right, it does not achieve values of essentially just 0 or 1.)

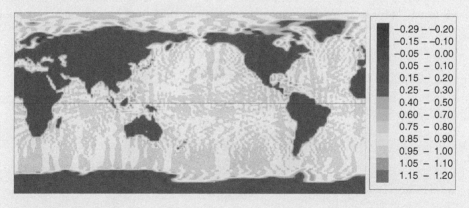

How do we know what combination of harmonics will yield the pattern we would like to recreate—not just which harmonic types, and which degrees and orders, but in what proportions they should be added? Mathematically, the way to answer this question is to start by allowing for all types, degrees, and orders of harmonics, and then determine how much of each, proportionally, to include in the combination; those whose proportional contributions are very small would be the harmonics not needed for that pattern.

So, let's denote the latitude of a point on the Earth's surface by ϕ and the longitude by λ; and let's say the pattern we would like to reproduce is described mathematically by a function of latitude and longitude denoted $P(\phi, \lambda)$. We will denote the spherical harmonic function of degree ℓ, order m by $Y_{\ell m}(\phi, \lambda)$. To express the pattern of interest as a combination of spherical

Textbox B: Spherical Harmonics (*Continued*)

harmonics, we write

$$P(\phi, \lambda) = C_{00} \cdot Y_{00}(\phi, \lambda) + C_{10} \cdot Y_{10}(\phi, \lambda) + C_{11} \cdot Y_{11}(\phi, \lambda) + C_{20} \cdot Y_{20}(\phi, \lambda)$$
$$+ C_{21} \cdot Y_{21}(\phi, \lambda) + \dots$$

or (more concisely)

$$P(\phi, \lambda) = \sum_{\ell, m} \{ C_{\ell m} \cdot Y_{\ell m}(\phi, \lambda) \},$$

where the coefficient $C_{\ell m}$ represents the proportion of the $Y_{\ell m}$ harmonic included in the combination. Just as it was necessary to deduce the coefficients of $\cos(\pi X/2)$ and other sinusoids in order to recreate the triangle function pictured near the top of this textbox, the problem now is to determine the magnitudes of all the $C_{\ell m}$.

Spherical harmonics, like sines and cosines, have a mathematical property called *orthogonality* which can help us figure out the $C_{\ell m}$. Orthogonality means that, if you multiply two harmonics and integrate the product in the right way, the integral will be zero unless the harmonics are the same.

As a relatively simple illustration, consider the longitude part of a spherical harmonic, $\sim \cos(m\lambda)$, where λ can range from 0 to 360°, i.e. from 0 to 2π; from first-year calculus and trig. identities, it can be shown that

$$\int \cos(m\lambda) \cdot \cos(n\lambda) d\lambda = 0 \quad \text{unless } m = n,$$

that is, unless the harmonics are of the same order.

Thanks to orthogonality, then, we can 'pick out' each coefficient from the full combination. Let's say we are interested in the degree 2 order 1 coefficient. If we multiply our equation by $Y_{21}(\phi, \lambda)$, then

$$Y_{21}(\phi, \lambda) \cdot P(\phi, \lambda) = C_{00} \cdot Y_{00}(\phi, \lambda) \cdot Y_{21}(\phi, \lambda) + C_{10} \cdot Y_{10}(\phi, \lambda) \cdot Y_{21}(\phi, \lambda)$$
$$+ C_{11} \cdot Y_{11}(\phi, \lambda) \cdot Y_{21}(\phi, \lambda) + C_{20} \cdot Y_{20}(\phi, \lambda) \cdot Y_{21}(\phi, \lambda)$$
$$+ C_{21} \cdot Y_{21}(\phi, \lambda) \cdot Y_{21}(\phi, \lambda) + \dots$$

or,

$$\int Y_{21}(\phi, \lambda) \cdot P(\phi, \lambda) = \int C_{00} \cdot Y_{00}(\phi, \lambda) \cdot Y_{21}(\phi, \lambda) + \int C_{10} \cdot Y_{10}(\phi, \lambda) \cdot Y_{21}(\phi, \lambda)$$
$$+ \int C_{11} \cdot Y_{11}(\phi, \lambda) \cdot Y_{21}(\phi, \lambda) + \int C_{20} \cdot Y_{20}(\phi, \lambda) \cdot Y_{21}(\phi, \lambda)$$
$$+ \int C_{21} \cdot Y_{21}(\phi, \lambda) \cdot Y_{21}(\phi, \lambda) + \dots$$

where $\int$ symbolizes (without any details) an integration over all the latitudes and longitudes of the sphere. The orthogonality condition tells us that all the integrals will be zero except the one involving C_{21}:

$$\int Y_{21}(\phi, \lambda) \cdot P(\phi, \lambda) = 0 + 0 + 0 + 0 + C_{21} \int Y_{21}(\phi, \lambda) \cdot Y_{21}(\phi, \lambda) + 0 + 0 \quad + \dots.$$

By performing some straightforward integrations, we can solve for the value of C_{21}—or any other $C_{\ell m}$—thereby revealing the correct proportion of each harmonic to the pattern of interest.

Textbox B: Spherical Harmonics (*Concluded*)

The solutions to Laplace's equation (as noted earlier) yield specific formulas for all the harmonic functions. Examples include the zonal harmonics

$$Y_{20}(\phi, \lambda) = {}^3\!/_2 \sin^2 \phi - {}^1\!/_2 \qquad\qquad (\ell = 2, m = 0)$$

$$Y_{30}(\phi, \lambda) = {}^5\!/_2 \sin^3 \phi - {}^3\!/_2 \sin \phi \qquad (\ell = 3, m = 0)$$

$$Y_{40}(\phi, \lambda) = {}^{35}\!/_8 \sin^4 \phi - {}^{15}\!/_4 \sin^2 \phi + {}^3\!/_8 \quad (\ell = 4, m = 0)$$

(their odd-seeming coefficients are designed such that every zonal $Y_{\ell m}$ equals 1 at the North Pole, where ϕ is 90°); these functions were illustrated above in a side view relative to a sphere (but $-Y_{20}$, which has the shape of an oblate rather than prolate spheroid, was actually shown there). An example of a tesseral harmonic is

$$Y_{42}(\phi, \lambda) = \left\{ \left[{}^{105}\!/_2 \sin^2 \phi - {}^{15}\!/_2 \right] \cos^2\phi \right\} \sin 2\lambda \quad (\ell = 4, m = 2).$$

According to the 'pattern rules' noted earlier in this textbox, this harmonic should have 2 full wavelengths going west-east around the Earth and should change sign twice going from pole to pole; can you identify the terms in this formula that cause such variations?

Our main application of spherical harmonics in this chapter will be to depict Earth's gravity field. The coefficients that yield the pattern of the Earth's gravity field (as seen, e.g., in Figure 6.21) are often called *Stokes coefficients*. The magnitudes of those coefficients tell us which harmonics contribute most strongly to the gravity field, thus which harmonic patterns characterize the field.

The Gravity Field Revealed. The usefulness of spherical harmonics to represent Earth's gravity field goes beyond their ability to portray spatial patterns; in the case of gravity, those spherical harmonics also have physical connections. Clairaut's Theorem, which was discussed in Chapter 4 (as the basis for the International Gravity Formula), showed that the gravity field of the Ideal Earth could be expressed in terms of physical parameters like $J_2 \equiv (C-A)/M_E \cdot a^2$, which symbolizes the equatorial bulge; its effect on gravity was to contribute a term of the form $J_2 \sin^2$(latitude) to the overall variation. But, as noted in *Textbox B*, the degree 2 zonal spherical harmonic function can also be described using a $\sin^2$(latitude) term. And, if we wrote the gravity field as a sum of 'coefficients times spherical harmonic functions' (a sum of $C_{\ell m} \cdot Y_{\ell m}(\phi, \lambda)$, in the notation of the *textbox*), orthogonality would require us to identify the coefficient of Y_{20}—representing how much of the gravity field can be described

by a bulge-type pattern—as J_2, or at least (depending on mathematical details related to how Y_{20} is defined) as a quantity that involves J_2. (The connection to J_2 should come as no surprise since we already noted in the *textbox* that the variation of Y_{20} with latitude gives it the shape of an equatorial bulge. It is perhaps more of a surprise that, as it turns out, way back in Chapter 4 we were already expressing things in terms of spherical harmonics!)

The connection between Y_{20} and J_2 still holds when we expand the gravity field of the *real* Earth (which is replete with mass anomalies in addition to the bulge) in a series of spherical harmonics: the degree 2 zonal harmonic relates directly to the Earth's equatorial bulge. And the magnitude of its coefficient, J_2, can be determined by tracking satellites and observing the main precession of their orbits.

Traditionally, the coefficients of the zonal harmonics in an expansion of the real Earth's gravity field, reflecting the relative contributions of different zonal patterns to the overall field, are denoted by J_ℓ, where ℓ is the degree of the harmonic, rather than $C_{\ell 0}$.

Given the shape of the degree 3 zonal harmonic (see *Textbox B*), we can similarly identify the coefficient J_3 with the gravitational effect of the Earth's pear shape; thus, tracking some of the 'lesser' perturbations exhibited by a satellite orbit will allow the magnitude of J_3 to be determined. That pear shape, by the way, is a mere 'blip' on the equatorial bulge, gravitationally speaking: satellite data confirms that J_3 is three orders of magnitude smaller than J_2. Earth can be quite well approximated as an oblate spheroid.

As noted earlier, the variety of mass anomalies representing the real Earth can produce a wide range of orbital perturbations. Conceivably, some of those masses might be described by individual spherical harmonics; but in general, numerous harmonics will be required to recreate the pattern of any one mass anomaly (actually, even precisely describing an equatorial bulge requires more than a degree 2 zonal harmonic). If we are interested in describing Earth's gravity field as completely as possible, our spherical harmonic expansion should include zonal, tesseral, and sectorial components of all degrees and orders. Observations of the various orbital perturbations will allow the magnitudes of those spherical harmonic coefficients to be estimated.

As discussed in *Textbox B*, the degree and order of any spherical harmonic function relate to its 'wavelengths;' for example, an order m harmonic will have a wavelength (peak-to-peak) of $\sim 360°/m$ in longitude. Consequently, in order to identify regional and local mass anomalies, and their contributions to Earth's gravity, we must determine the gravity field to a high degree and order. Low-degree harmonics relate mainly to global aspects of the field.

An example of a satellite-based gravity model is shown in Figure 6.21. One of a series produced at NASA's Goddard Space Flight Center, this particular Goddard Earth Model,

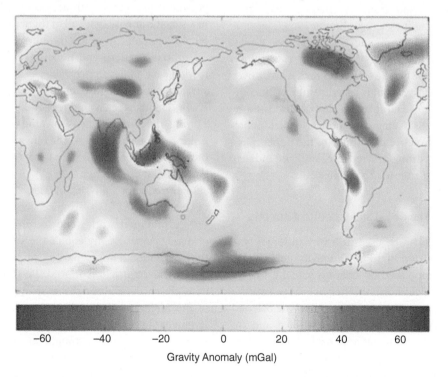

Gravity Anomaly (mGal)

Figure 6.21 Map of gravity anomalies relative to the Spheroid, based on tracking of numerous satellites (mostly by SLR, also Doppler plus some optical tracking; Lemoine et al., 2002) over a span of decades. Image from https://www2.csr.utexas.edu/grace/gallery/gravity/03_07_GRACE.html / NASA / Public domain.

designated GEM-T1 because it was designed to be used in the reduction of data from the TOPEX satellite (discussed below), includes spherical harmonic coefficients of the gravity field complete to degree and order 36.

The global gravity map shown in Figure 6.21 reveals peaks and troughs that bear little connection to whether their location is oceanic or continental. That is, the most broad-scale mass excesses and deficits seen around the globe—continents and oceans—show no common patterns of gravity highs and lows; for example, we do not see highs prevailing on continents and lows throughout the ocean basins. *On such global scales, mass compensation—isostasy—seems indeed to be a good approximation.*

That is not to say that the patterns in Figure 6.21 are random, however. For example, from the gravity highs ringing the Pacific—in the Andes and at the various trench zones of the Australian/Pacific and Australian/Eurasian plate boundaries—we can infer a significant contribution to the gravity field from plate tectonics, and from the mantle convection underlying it (see, e.g., Kaula, 1969; Hager & Richards, 1989; Zhong & Davies, 1999; Čadek & Fleitout, 1999, 2003; Steinberger & Holme, 2002, 2008). The strong gravity low over Hudson Bay could be a result of incomplete postglacial rebound, but as discussed in Chapter 5, it may also include contributions from convective distortions of the upper mantle. The strong gravity low south of India is discussed later in this chapter.

6.3.2 The Geoid and Satellite Altimetry

Satellites can be used in a very different way to help determine Earth's gravity field. Radar altimeters placed on board a satellite can be used to determine the height of the satellite above the sea surface: radar pulses are beamed downward, reflect off the sea surface, and are received by radar antennas back at the satellite; half their travel time down and back to the altimeter yields the satellite's height. If the satellite is also tracked, with laser or Doppler tracking (or both), that is, if the satellite can be located with respect to the solid earth, then that height relative to the sea surface can be converted into the distance of the sea surface above the center of the Earth, thus the elevation of the sea surface relative to some reference level of the Earth (see Figure 6.22). As the satellite moves through its orbit, beaming thousands of pulses downward per second, the undulations of the sea surface can be mapped.

The Physics of Level Surfaces. The connection of sea-surface undulations or 'topography' to Earth's gravity field is based on the concept of equipotential surfaces. An **equipotential surface**—also called a **level surface** or a **horizontal surface**—is defined to be *any surface that is everywhere perpendicular to the direction of gravity*. An example of an equipotential surface is the floor in a well-made building; crossing the floor should not have to require an uphill climb, for instance. Since there are many such floors in any building, it follows that there are really an infinite number of possible equipotential surfaces corresponding to a specific gravity field, all of them extending out horizontally from the building (or from any initial location) and none of them intersecting each other. Conceivably, each of these surfaces could extend completely around the Earth. Their shapes would reflect the spatially varying directions of gravity around the globe. Equipotential surfaces are thus invaluable for providing a global picture of one part of Earth's gravity field: its direction.

In introductory physics classes, "work" is defined as force times distance; for example, the gravitational work done moving a point of mass against gravity equals the strength of the gravitational force multiplied times the distance the particle is moved. But the work done clearly depends on the relative directions of the force and the particle's displacement. If gravity in some location acts downward, say, and we lift a box straight up, our exertion testifies to the work we are doing; but if (after lifting the box) we move it perfectly horizontally (imagine hugging the box close, and then walking so that the box does not bob

(a)

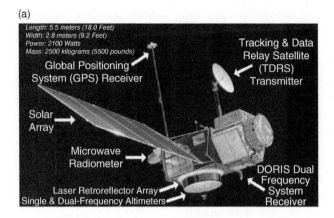

(b)

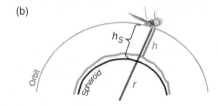

Figure 6.22 Satellite altimetry.
(a) Schematic of the altimeter satellite TOPEX-Poseidon; the altimeter sends and receives pulses of radar reflected off the sea surface to determine the satellite height. For precise orbit determination, TOPEX is tracked using laser, Doppler (DORIS), and GPS methods. Satellite operation is controlled by instructions through the TDRS transmitter, and the microwave radiometer determines moisture content in the atmosphere below. NASA / https://ilrs.gsfc.nasa.gov/missions/satellite_missions/past_missions/topx_general.html / last accessed August 22, 2023.
(b) Mapping the sea surface (in blue). The height *h* of the satellite above the ocean is determined by timing the return of the radar pulse. Satellite tracking determines the radius *r* of the satellite orbit, thus the height h_S of the satellite relative to the Spheroid. Subtraction yields ($h_S - h$), the height of the sea surface with respect to the Spheroid. From http://topex-www.jpl.nasa.gov/technology/technology.html / NASA / Public Domain.

up and down), there is essentially no work done against gravity, and our only exertion of energy goes into overcoming the inertia of our enhanced mass and accelerating as we begin to walk. And, if we take a previously lifted box and let go, the falling box is working with rather than against gravity; energy is released, rather than needing to be supplied.

A better definition of work, then, would be vector-based. If a particle is moved through a distance $\vec{d}$ against a force $\vec{F}$, the **work done** equals

$$-\vec{F} \cdot \vec{d};$$

here the symbol "•" denotes the vector operation called a "*dot product.*" Unlike the vector cross product operation we encountered in

Chapter 4, the result of a dot product is a scalar (nonvector) quantity, a magnitude, not another vector. The **magnitude of the dot product** equals the magnitude of the first vector times the magnitude of the second vector times the *cosine* of the angle between the two vectors:

$$\vec{A} \cdot \vec{B} \equiv A\,B\;\cos(\sphericalangle\;between)$$

for any two vectors $\vec{A}$ and $\vec{B}$. Because the cosine of 90° is zero, this definition implies that no work is done when the particle is displaced laterally, i.e. at right angles to gravity, as expected. The minus sign in the definition is required so that when a particle is falling (the $\sphericalangle$ *between* is 0°, and $\cos(0°) = +1$), the work done to move the particle is negative (gravitational energy is released); and when the particle is raised against gravity (the ∢ *between* is 180°, and $\cos(180°) = -1$), the work done on the particle is positive.

Our vector definition implies that no work is done when moving a particle along an equipotential surface—any displacement along the surface is always perpendicular to the direction of gravity. Theoretically, any equipotential surface could be traced out by starting at a point on the surface and always moving so that no work (positive or negative) is incurred.

The formal name for these surfaces stems from their relation to energy. With **energy** defined as the ability to do work, a particle is said to increase its gravitational energy—its potential energy—if it is lifted against gravity. From our vector definition of work, that increase in potential energy can be found as force 'dotted into' displacement:

$$\Delta P.E. = -\vec{F} \cdot \vec{d};$$

for example, if a point mass M is lifted vertically through a height h, its gravitational potential energy increases by an amount Mgh, if the acceleration of gravity there is g. **Gravitational potential** is defined as gravitational potential energy per unit mass. For example, when the point mass M is lifted vertically through a height h, its gravitational potential increases by an amount gh. If the change in potential energy equals force dotted into displacement, then the change in potential must equal force per unit mass—in other words, acceleration—dotted into displacement:

$$\Delta \text{potential} \equiv \Delta P.E./M = -\vec{F} \cdot \vec{d}/M$$

$$= -(\vec{F}/M) \cdot \vec{d} \equiv -\vec{g} \cdot \vec{d}.$$

Here's the punchline: if moving along an equipotential surface involves no work, that surface must have constant potential energy, thus *constant potential*—hence the name.

Given two level (equipotential) surfaces, the work done in moving from any location on one to any location on the other must always be the same, since the difference in potential is always the same. In a well-made building, moving from one floor to the next takes the

same amount of energy whether you use stairs at the front of the building or the back. Consequently, once one equipotential surface has been delineated, any other one within a given gravity field could be found by determining all the points that are a set amount of "work done" away.

At the risk of 'physics overload,' one additional technical detail must be noted. On Earth, the gravitational force we experience—called the *observed* gravity in Chapter 4—is the net result of both the gravitational force exerted by Earth's mass and the centrifugal force due to its rotation. The work we do raising a unit mass against observed gravity is a bit different than just the gravitational potential; in recognition of this, geophysicists call that work done the **geopotential**. And, with reference to the Earth, **equipotential surfaces** are defined as *surfaces of constant geopotential.*

In general, equipotential surfaces never intersect, as noted above; but they need not always maintain a fixed distance from each other. At locations where gravity is stronger, for example, the same amount of work will be completed by moving up through a smaller height than elsewhere: greater F (or g) allows a smaller d. As an example, given the International Gravity Formula, level surfaces surrounding the (Ideal) Earth cannot be spherical, or even maintain the same polar flattening: because gravity is strongest at the poles, the equipotential surfaces there must bunch together, in order to maintain the same energy difference. And, because gravity is weakest at the equator, the equipotential surfaces of the rotating, spheroidal Earth must bulge out increasingly there, to maintain the same energy difference.

It follows that the *spacing* of adjacent equipotential surfaces indicates where gravitational forces are stronger or weaker. If, in some region, then, we map the elevations of the equipotential surfaces with contours—just like the contours on a topographic map—we can see where the equipotentials bunch together or spread apart. Places where the potential has *steeper gradients* (changing,

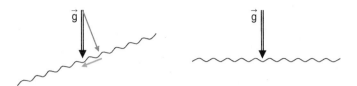

Figure 6.23 Every fluid surface at rest must be perpendicular to the local gravity field (right-hand illustration); if not (left-hand illustration), although most of the gravitational force might still act to hold the water down (blue arrow), there would be a component of gravity acting along the surface (green arrow), and the fluid would flow freely, not resting until it had become perpendicular to gravity.

from one equipotential level to another, over shorter distances) will mark the locations of stronger gravitational forces. Places where the equipotentials are more spread out (where the potential changes more gradually) mark the locations of weaker gravitational force. Incidentally, the connection to gradients actually has a mathematical basis (—calculus, again!), but is beyond the scope of this chapter.

To bring us back to Earth (!): the most important example of an equipotential surface is the oceans' **mean sea level**, that is, the ocean surface with all disturbances (such as tides) averaged out. In fact, every fluid surface at rest must be an equipotential surface. The reason for this, as illustrated in Figure 6.23, is simply that, if the surface was not perpendicular to the direction of gravity, it would *flow* until it was; if it is not flowing, and not impeded, it must already have come to rest at the required 90° angle to $\vec{g}$.

Mapping the Undulations of the Geoid by Satellite Altimetry. The Earth may have an infinite number of equipotential surfaces, but there is one of particular interest, called the **Geoid**. Our initial definition is as follows:

> *the Geoid is that equipotential surface of the real Earth which coincides with mean sea-level over the oceans.*

Identifying the Geoid with mean sea-level only over the oceans is a necessary restriction because sea level exists only where there are oceans, but the Geoid, defined with respect to Earth's gravity field, encircles the globe. There is no guarantee that the Geoid would

remain at sea level in continental regions—no guarantee that if the continents could be made permeable, the infiltrating seawater would conform to the same level as it trickled into the continental interior. Indeed, given the fact that mass is distributed irregularly around the world, and given the way gravity varies globally, it would be surprising if the infiltrated sea surface did not rise or fall by several meters from its coastal level.

A more precise definition of the Geoid—one that better accounts for the complexities of the Earth System—will be discussed later in this chapter.

Since the Geoid coincides with mean sea level over the oceans, altimetric satellites essentially trace out the oceanic portion of the Geoid with their radar, as they circle the globe. The world turns beneath the satellites, so (even without the effects of satellite orbit precession) each orbit around the Earth traces out a continually varying path which, after several days, can lead to dense, fairly complete sampling of the ocean surface (see Figure 6.24). Altimetry yields an oceanic Geoid with unprecedented accuracy and spatial resolution. The technique can pinpoint the mean sea surface to within a few centimeters or better in height, over the area—called the 'footprint'—illuminated by the satellite's radar beam; after the satellite's orbit has been determined, the end result at present is Geoid elevations with an accuracy of ~10 cm (Rosmorduc et al., 2011). With a footprint effectively ~2–10 km in size and the sea surface 'sampled' at intervals of several kilometers, the spatial resolution along the ground track is typically much less than a degree of latitude or longitude; with

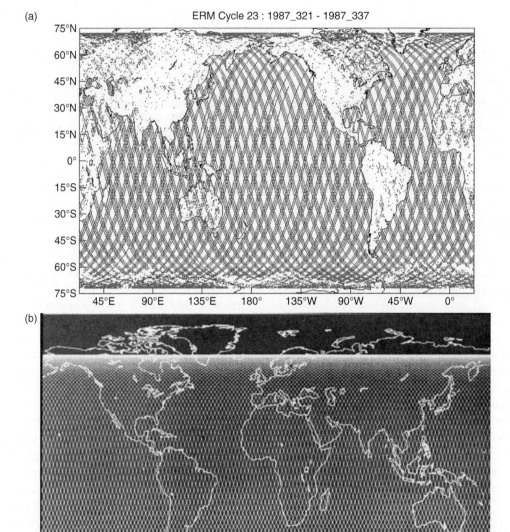

Figure 6.24 Altimetry satellite ground tracks. As the world turns beneath the orbiting satellite, the altimeter's radar illuminates a gradually changing pathway along the ocean surface.
(a) Some ground tracks of Geosat during 1987 for a period of 17 days. During this cycle, the spatial resolution was atypically coarse; this cycle was chosen to illustrate the tracks more clearly. Given the satellite's inclination, the ocean surface is illuminated only within ±72° latitude.
(b) Ground tracks for TOPEX-Poseidon, between ±66° latitude. The satellite orbit was designed such that it repeats the same ground tracks (to within ~1 km) every 9.9 days. Image from Le Traon (2007).

(a)

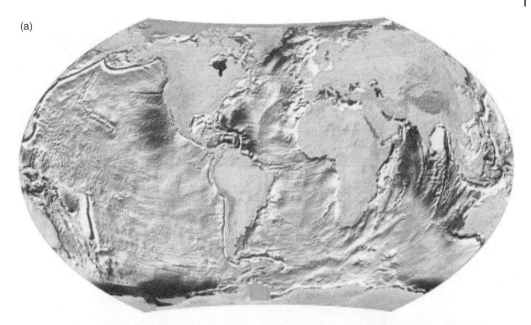

Figure 6.25 (a) The mean sea surface (altimetric Geoid), based mainly on altimetric measurements (7 years of TOPEX-Poseidon data, 5 years of ERS-1 and ERS-2 data, and 2 years of Geosat data); taken from Rosmorduc et al. (2011) / ESA and CNES.

tracks separated by $\sim10^2$ km, the cross-track resolution is perhaps a degree or so, although the drift in tracks over a long time span can increase that resolution substantially (e.g., Sandwell et al., 2014). An example of an altimetric Geoid is shown in Figure 6.25a.

The information harvested from satellite altimetry is rich beyond our wildest dreams. The high resolution achieved in altimetric Geoids reveals that the sea surface is *distorted* by the gravitational effects of ocean-bottom topography—seamounts, fracture zones, trenches, ridges, etc. These are short-wavelength features of the gravity field—high harmonics, if you will—which are not completely eliminated by the broad-scale mechanisms of regional isostatic compensation; and, as a bonus for us, they provide a detailed picture of the seafloor never before seen. In fact, more than one third of all seamounts in existence, overlooked by shipboard surveys, remained undiscovered and uncharted until the first altimetric Geoid maps were produced.

With short-wavelength features in the altimetric Geoid so evidently resembling seafloor topographic structures, it is easy to conclude that those structures are the *cause* of such geoidal features; that is, *components of the gravity field described by high spherical harmonic degree and order can be attributed to shallow sources* (shallow mass excesses and deficits). Such a conclusion is not surprising given the nature of gravity and its strong dependence on distance. The inverse statement is also generally (though not exclusively) true: *components of the gravity field described by spherical harmonics of low degree and order can be attributed to deeper sources*; thus, the broad-scale gravity field depicted by Figure 6.21 tells us more about the deeper Earth.

Based on a formula derived later in this chapter, it is possible to estimate the magnitude of the gravity anomaly corresponding to each altimetrically observed geoidal fluctuation high or low; as seen in Figure 6.25b, the spatial detail of the resulting maps is astonishing. The seafloor laid bare by these and by the high-resolution altimetric Geoids themselves amounts to a combination of tectonic and gravity information about the

(b)

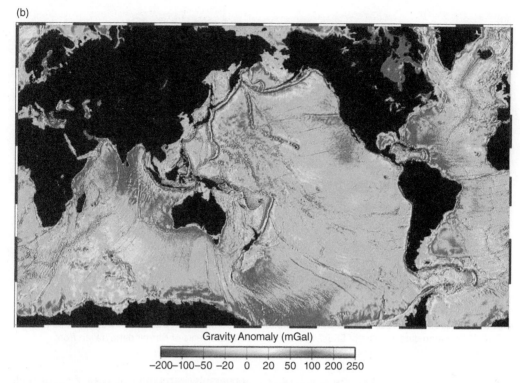

Gravity Anomaly (mGal)

−200−100−50 −20 0   20 50 100 200 250

Figure 6.25 (b) Marine gravity anomalies based on satellite altimetry data, notably CryoSat-2 and Jason-1. See, e.g., Sandwell et al. (2014); NOAA / Wikimedia Commons / Public Domain and https://topex.ucsd.edu/grav_outreach/index.html#grav_map. Color scale shown here is approximate.

oceanic lithosphere that allows us to study its structural, thermal, rheological, and dynamical behavior in ways never before envisioned. It defines plate boundaries; constrains lithospheric cooling rates through their associated bounds on subsidence rates; provides a variety of surface (i.e., seafloor) loads whose partial or complete isostatic compensation constrains the sublithospheric mantle viscosity or its rheological nonlinearity; and reveals details of mantle convection's upper boundary layer.

Even with just such information (let alone what it can tell us about other aspects of the Earth System, as described below), it is not surprising that the value of satellite altimetry was recognized early in the technique's development. After some preliminary experiments to establish the capability of spaceborne altimeters (see, e.g., Rosmorduc et al., 2011), SeaSat—the first satellite dedicated comprehensively to remotely sensing the ocean with

an altimeter (and other instruments)—was launched in 1978 by NASA. For 3½ months it collected a huge amount of data, data that would take years to fully analyze; but then its electrical system completely short-circuited, and its mission ended. From those few months of data, however, its value was evident, and with improvements in methodology and technology, the next generation of altimetric satellites was born with the launch of the very next altimetric satellite, Geosat, by the U.S. Navy in 1985. The first phase of that mission, nearly 1½ yr long, was high resolution and classified; despite the official end of the Cold War in 1989 with a declaration by Soviet and American leaders, following the fall of the Berlin Wall, it was not until 1995 that scientists were able to obtain that data for scientific analysis (Sandwell & Smith, 1997). The second, 3-year, unclassified phase of Geosat (now in a different orbit) also spawned a wealth of

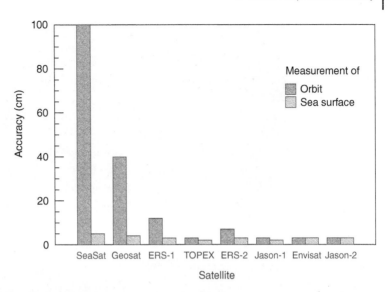

Figure 6.26 Evolution of satellite altimetry: improvement in orbit determination (radial component) and measurement of sea-surface height. Data (except for Jason-2) taken from Rosmorduc et al. (2011) / ESA and CNES.

scientific studies over the next several years focusing on the oceanic lithosphere and plate tectonics, ocean currents, ocean-atmosphere interactions, and more.

The accuracy of the altimetric Geoid is limited by the precision of the altimeter instrument itself as well as our ability to accurately determine the satellite's orbit. Improvements over time have been dramatic (see Figure 6.26), facilitated by 'next-generation' instrumentation and by 'bootstrap' improvements in the gravity field from satellite tracking.

Third-generation altimetric satellites include the European Space Agency's ERS-1 and ERS-2 (ERS stands for European Remote Sensing Satellite), launched in 1991 and 1995; TOPEX-Poseidon (TOPEX stands for topography of the oceans experiment), a joint project of France and the United States, launched in 1992, and its follow-on missions, Jason-1, launched in 2001, and Jason-2, launched in 2008 (with the same orbits as TOPEX); and Envisat, launched in 2002 by the European Space Agency (ESA). What might be called "fourth-generation" altimetric satellites have also been developed: specialized altimeters, some laser based rather than radar based, to measure the topography of the land or, in polar regions, the ice surface. These include ICESat (the Ice, Cloud, and land Elevation Satellite), launched in 2003 by NASA, with

further goals of quantifying cloud properties, including those of polar stratospheric clouds, and determining the height of the vegetative canopy on land; CryoSat(-2), dedicated to measuring polar ice topography and inferring the thickness of floating ice, launched by the ESA in 2010; and SARAL, a cooperative venture between India and France launched in 2013.

Satellite altimetry has yielded a wealth of data concerning the atmosphere and oceans. For example, the radiometers used to measure the effect of humidity in the troposphere on the speed of the radar pulse (and thus its effect on the height determination) allow us to construct global maps of atmospheric water vapor abundance (Figure 6.27a). And, as noted below, the character of the radar reflections provides information on the surface winds blowing over the ocean. But as we now discuss, the greatest impact satellite altimetry has had on climatology has been on what it reveals about the oceans.

Satellite Oceanography. The sea surface rises and falls periodically, but also episodically, on global spatial scales and on regional and local scales, all as a result of various processes acting simultaneously and all potentially detectable

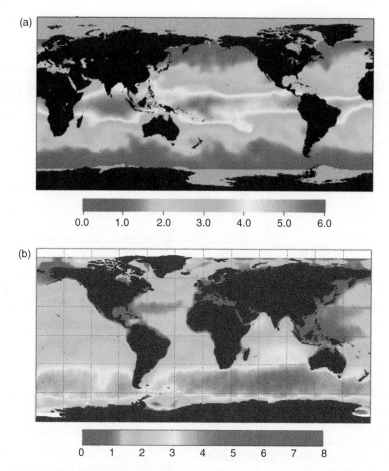

Figure 6.27 (a) Tropospheric water vapor in Spring 1997 measured by the TOPEX radiometer. Colors denote the mass of water vapor and liquid water in a unit-area column of atmosphere below the satellite; units are gm/cm^2. Note the high concentration of humidity in the InterTropical Convergence Zone—as expected since upwelling there drives much of the atmosphere's three-cell circulation. Figure adapted from Le Traon (2007). (b) Significant wave height, in meters, during summer 1995 determined from ERS-2 altimetry. Evidently the Atlantic was almost as calm as the Pacific at this time. Meanwhile, the perpetual storminess of the Southern Ocean (e.g., Babanin et al., 2019) produces very rough seas.
A recent analysis by Young and Ribal (2019) has found significant wave heights to be increasing worldwide but especially in the Southern Ocean, since at least the 1980s. We might reasonably attribute these increases to global warming. Figure adapted from Bosch (2010).

by altimetry. These processes include tides, acting mainly globally and with known periodicities; variations in the weight of the atmosphere (i.e., variations in atmospheric pressure), which occur on all possible timescales, and on regional and local as well as broadly global spatial scales; and surface winds, whose unsteadiness over time generates small-scale disturbances—waves—of the sea surface. In all of these cases, the rise and fall of the sea surface is supported by currents. Wind forcing over the long term also produces a distinctly different oceanic response, one characterized predominantly by currents; these currents—the response to steady (persistent) surface winds dragging the oceans along as they blow—amount to a well-defined and global, slowly fluctuating current system whose idealization (the *wind-driven circulation*) we described in

Chapter 3. But even in this case, there is a sea-surface component: adjustments of the *sea surface* that accompany these currents, called the oceans' **dynamic topography**.

All these variations in sea-surface height are part of what satellite altimeters measure. Altimetry thus provides an opportunity to improve our understanding of these processes. Altimetric determinations of tides, for example, can be combined with tide-gauge measurements (or ocean-bottom sensors recording the increase in pressure during high tide) and with tidal perturbations in satellite orbits to produce highly accurate tide models (many of these models also include theoretical constraints on the tide's dynamics, for example on the amount of friction experienced by the currents, in mid-ocean or near the coasts). Altimetric inferences of wind- or pressure-forced effects on sea level can similarly be used to develop or refine models of such variations in sea level.

In addition to mapping the sea surface by measuring the travel times of radar pulses, altimeters also record the strength of the reflected pulse; because roughness and ripples of the ocean surface produced by surface winds cause the reflected pulse to scatter and weaken, the strength of those surface winds can be inferred. This is key information gathered far from the land-based weather stations we used to rely on for surface wind measurements! Satellite altimeters also record the shape of the reflected pulse *waveform*; distortions in that waveform, resulting from later reflection off wave crests 'farther out' in the illuminated area, provide information on the typical wave height (called the *significant wave height*) within the footprint (Figure 6.27b).

From all the fluctuations in sea level, it is clear that the time-averaged sea level—which we call **mean sea level** or the **mean sea surface**—does not necessarily average all these processes out. For tides, as an illustration, we would have to average sea level over decades in order to average out all of the long-period tides (one of the tidal periodicities, called the *nodal tide* and associated with a

precession of the lunar orbit, has a period of almost 19 yr). Even if we were willing to take a really long time average, though, at least one of the processes we have mentioned would not average to zero: the dynamic topography associated with steady wind-driven circulation of the oceans.

Earlier, we had envisioned mean sea level as a 'zero-disturbance' sea surface—sea level with all disturbances removed. In that case, mean sea level was an equipotential surface, and our subsequent definition of the Geoid was correct. However, mean sea level is traditionally defined as we have just stated, as the sea surface averaged over time; this requires that we adjust our definition of the **Geoid**:

> *the Geoid is that equipotential surface of the real Earth which coincides over the oceans with the zero-disturbance sea surface.*

The difference between mean sea level and the zero-disturbance sea level is that the former includes the mean dynamic topography:

> *mean sea level*
>
> = *zero-disturbance sea level*
>
> + *mean dynamic topography.*

We will discover that the oceans' dynamic topography is at most a couple of meters in amplitude, so that identifying the Geoid with mean sea level is not a bad first approximation. The mean sea surface can be calculated by time-averaging the altimetric sea surface heights. Thus, if we are willing to neglect the mean dynamic topography, satellite altimetry alone can be used to infer an approximate Geoid over the oceans. Alternatively, either in combination with other observational techniques (SLR, ground-based measurements, etc.) or through modeling, altimetric measurements allow both the oceanic Geoid and the oceans' mean dynamic topography to be resolved.

Though small, the dynamic topography of the oceans is important because of its ability to shed light on the underlying currents. We can illustrate how this works using concepts we have already discussed, without sacrificing too much rigor. Let's say we are interested in studying the Gulf Stream. Although this current is part of the North Atlantic gyre (as discussed in Chapter 3), along most of its length it is far removed from the tropical easterly and midlatitude westerly winds powering the whole gyre; its dynamics are therefore dominated by the Coriolis force. As the current first began to move northward, that force deflected it to its right, building up sea level—creating positive dynamic topography—to its right as it continued to flow northward. As that topography grew, it acted increasingly as a weight on the water below it—a 'high-pressure' center—and this weight exerted an outward pressure on the periphery of the gyre. Eventually, a balance developed between the Coriolis force deflection to the east and the outward pressure force pushing the water to the west (back toward the North American east coast), and, like the high-pressure cyclone illustrated in Figure 3.5b of Chapter 3, the water has continued to flow up the coast, following a clockwise path around the center of the gyre.

As the water streams northward, the Coriolis force intensifies, requiring a stronger pressure force to balance it. Thus, the Gulf Stream should exhibit a higher dynamic topography as it travels northward.

Imagine (Figure 6.28) that the ocean surface is a height h_{HP} above the Geoid, as measured within the 'high pressure' to the right of

the Gulf Stream, but at a height h_{GS} above the Geoid as measured at the location of the Stream itself. Those heights are dynamic topography associated with the Gulf Stream. Take the density of seawater to be ρ_w everywhere. At the depth of the Geoid, the weight of the overlying ocean in a column with cross-sectional area A cm^2 equals its mass times gravity, or its density times its volume times gravity, thus $\rho_w \cdot h_{HP} \cdot A \cdot g$ beneath the high pressure and $\rho_w \cdot h_{GS} \cdot A \cdot g$ beneath the current. The additional weight of the dynamic topography in the 'high pressure' is therefore $\rho_w \cdot [h_{HP} - h_{GS}] \cdot A \cdot g$, or $\rho_w \cdot \Delta \cdot A \cdot g$ if we denote the difference in height by Δ. The additional pressure (force per unit area) counteracting the Coriolis force is then $\rho_w \cdot \Delta \cdot g$.

Later in this textbook, we will learn how to express quantitatively the forces associated with pressure differences. For now, Figure 6.28 should at least suggest to you that the pressure difference $\rho_w \cdot \Delta \cdot g$ is more effective if it acts over a shorter distance, i.e. if L is small, than if it acts over a much broader distance; that is, the pressure force here should be proportional to $(\rho_w \cdot \Delta \cdot g)/L$. It turns out that, if the mass of Gulf Stream water experiencing the pressure force is M, the magnitude of the pressure force will equal $M \cdot \Delta \cdot g/L$.

At the same time (using *Textbox B* in Chapter 4), the Coriolis force experienced by that Gulf Stream water has a magnitude $f \cdot M \cdot v$ if the current is moving at velocity v; f is the Coriolis parameter for the latitude

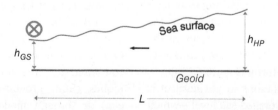

Figure 6.28 Determination of the velocity of an ocean current (e.g., the Gulf Stream) from the dynamic topography associated with it. The current is on the left, moving into the page as indicated by the blue-green "×" marking the 'tail' of the vector arrow.
Even with variations in sea level over time averaged out, the mean sea surface does not coincide with the Geoid here because of this steady current, as explained in the text. In the close-up sketch shown here, the disparity—the dynamic topography—equals h_{GS} beneath the current and h_{HP} to the right, at a location a distance L away from the current. The additional weight of the dynamic topography exerts a pressure force to the left, as indicated by the bold arrow.

where this is taking place. If the forces balance, we have

$$f \cdot M \cdot v = M \cdot g \cdot \Delta / L$$

or

$$v = g \cdot \Delta / (L \cdot f),$$

allowing the current velocity to be deduced from the *slope* (Δ/L) of the dynamic topography.

Because these currents are calculated assuming a balance between the Coriolis and pressure forces, they can be called *geostrophic currents*. We have encountered geostrophic situations before, in Chapter 3, as a way of describing the cyclones and anticyclones that epitomize regional atmospheric circulation. In the simplest of those situations, a geostrophic balance led to cyclonic winds circling continually around a central high or low pressure; in the more realistic *quasi-geostrophic* situations where friction is present, the weather systems were cyclonic, overall, but the winds gradually spiral out from or in toward the center.

An example of geostrophic ocean currents as determined altimetrically from dynamic topography is shown in Figure 3.14b of Chapter 3. A more precise example from one of the newest techniques in satellite geodesy, discussed below, may be seen in Figure 6.33.

6.3.3 Geoid *Versus* Spheroid, and Geoidal Heights

In Chapter 4, the Spheroid was defined as *sea level of the Ideal Earth*. In retrospect, it should be clear that we were actually defining the Spheroid as an *equipotential* surface, one that related to the gravity field of our 'ideal' approximation to the real Earth. Theoretically, this is perfectly reasonable, in that equipotential surfaces can be defined with respect to any gravitational field. And, to the extent that the Ideal Earth is a meaningful concept, defining the Spheroid in terms of sea level is not ambiguous: on the Ideal Earth, the oceans are subjected to no disturbances of any kind, so "sea level," "mean sea level," and "zero-disturbance sea level" are all the same.

Furthermore, on the Ideal Earth the oceans extend over the whole globe, avoiding the challenges of trying to pin down the Geoid on land.

Nevertheless, the Spheroid is only a useful (ideal) reference if it can be connected to the real Earth. Our formal definition of the **Spheroid** makes this explicit:

> *The Spheroid is that equipotential surface of the Ideal Earth which coincides as closely as possible with the Geoid.*

The shape of the Spheroid is, well, spheroidal, of course—unavoidably, since the net gravitational force that the Spheroid must remain perpendicular to varies around the globe because of the equatorial bulge and centrifugal force. The Geoid surface must remain perpendicular to a very different gravitational force—that of the real Earth—but even the real Earth has a bulge and rotates, and even on the real Earth those two factors are the primary cause of variations in gravity worldwide. The 21-kilometer ($= a-c$) variation in radius of the Ideal Earth from pole to equator is shared by the real Earth, and by the real oceans; the rise and fall of sea level caused by any disturbance is merely a small perturbation superimposed upon a 21-kilometer bulge of the global sea surface.

Since the Spheroid and the Geoid both possess equatorial bulges, they already coincide closely (though not perfectly); the Ideal Earth *is* a good approximation to the real Earth, in terms of equipotential surfaces. What interests scientists *more*, however, are all the mismatches—the variations in the shape of the Geoid *superimposed* on the bulge shape. With none of the complexities of the real Earth, and consequently a smoothly varying, well-defined ellipsoidal shape, the Spheroid makes a very useful reference surface to measure satellite orbits or heights—or the sea surface—with respect to. Contrasts between the Geoid and Spheroid then allow the complexities of the real Earth to be revealed.

We have already seen some of these complexities in altimetric maps of the sea surface, which warps accordingly over mid-ocean

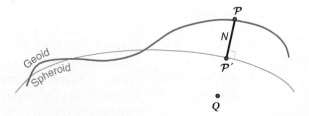

Figure 6.29 Profiles of the Spheroid and Geoid in some region of the Earth; the Spheroid curve is of course much smoother than the Geoid. If the entire Earth were shown, both profiles would look ellipsoidal; but the Geoid, though ellipsoidal to a first approximation, does not *match* the Spheroid. At point P on the Geoid, a perpendicular line is dropped down to point P' on the Spheroid; the geoidal height there, N, is the length of that line.

ridges, seamounts, scarps, and trenches. These altimetric maps are not maps of the Geoid itself (even after removing the smaller-scale, dynamic topography associated with steady ocean currents) but rather are maps of the difference between the Geoid and Spheroid; had they been maps of the full Geoid, they would have been overwhelmed by the bulge shape, and little else could have been discerned.

The differences between Geoid and Spheroid can be measured by the distance between these equipotential surfaces. We define the **geoidal height** at some location to be the height of the Geoid above the Spheroid (i.e., the vertical distance of the Geoid above the Spheroid; see Figure 6.29, also Figure 6.22b). The altimetric Geoids shown in this chapter are maps of *geoidal heights* over the oceans.

Calculating Geoidal Heights, for a Global Picture. Other techniques, including tracking satellite orbit perturbations and incorporating surface gravity measurements, allow the Geoid to be resolved on land as well as over the oceans. In these situations, the geoidal heights can be calculated purely from gravity or related data. We can illustrate how this might work using concepts we have already developed; our approach has its subtleties—we'll be replacing, in part, the cleverness of a mathematical derivation with verbal arguments—but the resulting formula is attractively simple. We will focus on quantifying the energy level that characterizes the Geoid and Spheroid: they are surfaces of constant geopotential, or (equivalently)

surfaces of constant geopotential energy; but what is that constant?

Our derivation is a tale of two worlds. An observer at point P in Figure 6.29 is simultaneously standing on the Geoid in the real world, and at a height N above the Spheroid in the ideal world. How is gravity experienced in these worlds?

- The Geoid is at a particular geopotential level, in that it takes a certain amount of *work* to move against Earth's gravity from point Q, say, to point P; and we know we can express that work in terms of the real Earth's gravity. We also know (as concluded in *Textbox B*) that it is possible to describe Earth's gravity in terms of spherical harmonics, with the appropriate combination of harmonics determined by the values of the Stokes coefficients. In short, we can express the geopotential level of the Geoid as a sum of spherical harmonics, in some particular combination.

- Similarly, the Spheroid is well defined theoretically, and we can express its geopotential level—based on the work done in moving from point Q to point P' on the Spheroid—in terms of its own spherical harmonic sum. That sum will be much more limited than the sum that produces the Geoid's geopotential: since the Spheroid is completely defined by Earth's mass, bulge, and rotation rate (the latter determining the resulting centrifugal force), the sum will include, to very good approximation (Stacey & Davis, 2008), only a couple of harmonics (degree 0 for the mass, and the zonal degree 2 for the other two factors).

But the two worlds intersect. To see this, let's step back for a moment and consider the Spheroid and Geoid more fundamentally. These surfaces are each defined as an equipotential surface, so that the amount of work done moving a mass *along* either surface is zero; but the work done moving a mass *to* that surface from a location at another 'height' will, of course, be nonzero. It takes work, fighting the real Earth's gravity, to move the mass from point Q to point P on the Geoid; and it takes work, fighting the Ideal Earth's gravity, to move the mass from point Q to point P' on the Spheroid. If you think about it the right way—here's where some subtlety must be appreciated—you should be able to convince yourself that the amount of work in these two cases must be equal! That is, placing one equipotential surface "as close as possible" to the other equipotential surface—required as part of the formal definition of the Spheroid—means that, overall, you do as much work moving from some location onto *one* surface as onto the *other*. (And prescribing the 'closeness' of those surfaces in terms of energy or work rather than distance makes sense for equipotential surfaces.) *In short, the Geoid's geopotential level and the Spheroid's geopotential level must be the same.*

The rest of the derivation, short of the final step, is just a matter of notation, to symbolize the preceding concepts.

- On the Ideal Earth, the geopotential level at height N is greater (if $N > 0$) than the geopotential level on the Spheroid; we know that because it takes work, fighting the Ideal Earth's gravity, to move a mass from position P' to position P. That is, the geopotential of the Ideal Earth is a function of position. Symbolically, we might represent the value it attains at position p by $U_S(p)$ ("$_S$" stands for "spheroidal," and U is the traditional geophysical symbol for geopotential), and

the amount of work it takes to go from position P' to position P is $U_S(P) - U_S(P')$.

- On the Ideal Earth, equipotential surfaces are defined by $U_S(p) =$ constant, with the value of the constant depending on the particular surface. We will denote the constant corresponding to the Spheroid by c_S, thus (with P' on the Spheroid) $U_S(P') = c_S$.

- Similarly, we could represent the real Earth's geopotential by a function $U(p)$ of position p. As discussed above, U can be expressed in terms of a spherical harmonic sum involving Stokes coefficients, with the values of the spherical harmonic functions (those combinations of sines and cosines) determined once p is specified. And, we can denote the constant corresponding to the Geoid by c_G, that is, $U(P) = c_G$.

With this notation, our 'subtle' conclusion above can be concisely expressed as $U(P) = U_S(P')$, or $c_G = c_S$. Consequently, the amount of energy expended on the Ideal Earth going from position P' to position P, that is, $U_S(P) - U_S(P')$, could be subtly rewritten as $U_S(P) - U(P)$. This latter quantity can be calculated easily using the spherical harmonic expansions for U and U_S, evaluated at the real-world location P.

Incidentally, since the real world and ideal world share the same mass, equatorial bulge, and rotation rate, the subtraction that yields this difference ends up producing the same spherical harmonic sum previously written for just the Geoid, with the same Stokes coefficients—but with the degree 0 and zonal degree 2 harmonics eliminated by that subtraction.

The final step of our derivation is easily achieved. The amount of work it takes to move a unit mass from position P' to position P on the Ideal Earth is also, by definition, the force per unit mass times the displacement,

$$g_S(\phi)\, N,$$

where the gravitational force per unit mass here is $g_S(\phi)$, gravity on the Spheroid at the latitude of position P. That work must equal the difference in Ideal Earth's geopotential at $\mathcal{P}$ and $\mathcal{P}'$, $U_S(\mathcal{P}) - U_S(\mathcal{P}')$, thus

$$g_S(\phi)\, N = U_S(\mathcal{P}) - U(\mathcal{P}),$$

yielding at last

$$N = [U_S(\mathcal{P}) - U(\mathcal{P})]/g_S(\phi).$$

This simple formula is known as **Bruns formula**. At any location, the value of g_S, calculated from the International Gravity Formula for that latitude, and the difference in geoidal and spheroidal geopotentials, computed from their spherical harmonic expansions, yields the geoidal height. Since the geopotentials' spherical harmonic expansions can be applied at any location around the world (the surface patterns they represent are global, as are the harmonic functions), Bruns formula can yield geoidal heights over land as well as the ocean.

Illuminating Terminology. There is a complexity underlying the preceding that should be mentioned. Throughout our derivation, we have asserted that the difference in geopotential between two equipotential levels equals the work done when moving a unit mass from the lower level to the higher (in our case, from $\mathcal{P}'$ to $\mathcal{P}$). The basis for making such an assertion is a widely used mathematical relation between gravity and gravitational potential; this relation involves a generalization of the 'gradient' operation we saw in Chapter 5 (we will finally write out the generalized relation explicitly in Chapter 10, though in various other contexts it will be hinted at before that point). However, an equally useful alternative formulation is found—in some texts and, in particular, in a classic monograph popular in geodetic research (Heiskanen & Moritz, 1967)—in which the (minus) sign of that relation is reversed; with that approach to deriving Bruns formula, then, the change in geopotential when moving from $\mathcal{P}'$ to $\mathcal{P}$ is not interpreted physically in terms of work done or energy expended but rather is expressed purely mathematically.

With signs switched, the end result would be Bruns formula with the geopotential difference reversed. Since our own result is correct, and physically consistent with definitions and mathematical relations used throughout this textbook, the best we can do—with the purpose of motivating some enlightening terminology—is to preserve the above result but write it in the Heiskanen-Moritz style:

$$N = -[U(\mathcal{P}) - U_S(\mathcal{P})]/g_S(\phi).$$

The quantity in brackets, the difference between real-Earth and Ideal-Earth geopotential functions is essentially the gravitational potential of the Earth with its bulge and overall mass subtracted out, i.e. the geopotential that results from all the mass anomalies of the real Earth besides its equatorial bulge. We could call that the geopotential "anomaly" or the **anomalous potential**, but it is more commonly known as the **disturbing potential**. Note that at locations where the disturbing potential equals zero, there is no anomaly, and the Geoid and Spheroid will coincide ($N = 0$).

Finally, in altimetric situations, where geoidal heights are measured directly, we note that Bruns formula can be 'inverted' in order to estimate gravity anomalies from geoidal heights. This would be particularly useful in otherwise inaccessible regions, and particularly revealing when the altimetry data has high spatial resolution. Very simply, the approach to take is suggested by the fact that a change in geopotential versus height relates to the work done, i.e. gravity·distance; that is, 'differencing' the Bruns formula equation (i.e. differentiating it with respect to height) will produce a quantity on the right-hand side of the equation which involves gravity. More specifically, differentiating U will lead to a real-world gravity factor; differentiating U_S will lead to an Ideal-Earth gravity factor; and the subtraction of one from the other will yield the gravity anomaly.

This anomaly actually corresponds to a free-air anomaly; can you explain why?

Meanwhile, differencing or differentiating the left-hand side of the equation produces a term that relates to how geoidal height varies along the Geoid, that is, the *slope* of the Geoid. The details of this relation, and how the inversion to obtain a map of gravity anomalies can be achieved from those slopes, are beyond the scope of this chapter but can be found in Sandwell and Smith (1997).

6.3.4 More Perspectives on Global Gravity and the Global Geoid

Other relatively new techniques for determining the Earth's gravity field and Geoid deserve mention. First of all, the establishment of a *Global Positioning System* (*GPS*) framework of satellites—at present, a 'constellation' of 31 satellites (a 'nominal' set of 24 are nearly always operational), all in orbits at 55° inclination with respect to the equator, arranged so that at least four will be in sight from any point on Earth (see Figure 6.30a)—has allowed precision tracking of the orbital perturbations of other, non-GPS satellites with respect to the GPS reference frame, and ultimately with respect to the Earth.

Perhaps even more powerful is the 'tandem' approach of the *GRACE* (Gravity Recovery And Climate Experiment) satellite, launched in 2002 at an inclination of 89° under the auspices of the American and German governments, with subsequent participation by the ESA; the mission ended in 2017 and was succeeded by the GRACE Follow-On (both discussed in Chapter 3). As indicated in Figure 6.30b, GRACE was actually twin satellites whose separation distance (typically ~200 km) was precisely monitored; as they orbited the Earth, passing over regions of stronger or weaker gravity, first one then the other would accelerate, as they each passed closest to the anomaly. At an altitude of ~450 km, the change in separation distance was capable of yielding accurate values of gravity at a spatial resolution of ~150 km for the gravity field (Tapley et al., 2007), 300 km for its changes over time (Chen et al., 2022); and with an 89° orbital inclination, its coverage was indeed global.

Early analysis of GRACE data (Tapley et al., 2005) led to geoidal height maps reportedly accurate to 1 cm even for spherical harmonic components of degree and order up to 70; a more recent gravity model based on GRACE data alone (Figure 6.31) resolves spherical harmonic components through degree and order 180—a substantial improvement in resolution over earlier degree and order 36 models like that shown in Figure 6.21. In Figure 6.31, we can easily see the dichotomy in scale characterizing the Earth's gravity field. Around the Pacific Rim, there are sharp highs and lows extending hundreds of kilometers or more along plate boundaries but only ~100 km or

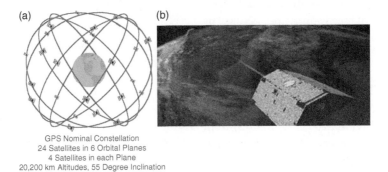

(a) (b)

GPS Nominal Constellation
24 Satellites in 6 Orbital Planes
4 Satellites in each Plane
20,200 km Altitudes, 55 Degree Inclination

Figure 6.30 Newer approaches to measuring Earth's gravity field.
(a) Nominal constellation of 24 GPS satellites, distributed around the globe but all of which are inclined 55° with respect to Earth's equator. http://www.colorado.edu/geography/gcraft/notes/gps/gps_f.html / P. Dana.
(b) Artist's rendition of the GRACE mission, with the two identical satellites orbiting 220 km apart. The satellites were tracked within the GPS system and also by laser ranging. Changes in the distance separating them as they approached and then receded from mass anomalies below were monitored, allowing the gravitational effects of those anomalies to be inferred. NASA/JPL-Caltech / Public Domain.

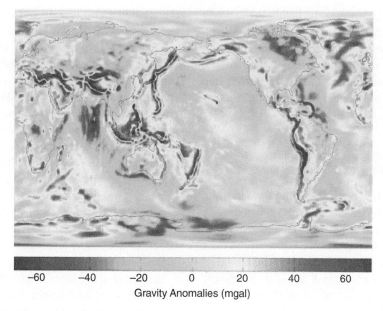

Gravity Anomalies (mgal)

Figure 6.31 Gravity model GGM03S, based on ~4 years of GRACE data. Note that many of the subduction zones around the Pacific Rim show both gravity highs and gravity lows, as discussed in the text; and that those sharp features, and others, are superimposed on broader, regional highs and lows. Tapley et al. (2007) / NASA / Public Domain.

so across them; the lows are, of course, associated with the mass deficits of seafloor trenches. These high-resolution features—high harmonics of the gravity field—are superimposed on a *broader* (lower-degree harmonic) 'ring of high gravity' around the Pacific Rim presumably associated with the *deeper* mass excesses of the subducting lithospheric slabs.

To achieve spatial resolution on land comparable to that seen over the oceans by altimetry, however, even GRACE satellite data must be combined with additional data. Figure 6.32a shows the Earth Gravitational Model 2008 (EGM2008), derived by scientists at the U.S. National Geospatial-Intelligence Agency (Pavlis et al., 2008, 2012, 2013) from a combination of near-surface (airborne and submarine), surface (land and ocean), and GRACE gravity measurements; altimetric data (Geosat, TOPEX and their follow-on missions, ERS-1 and -2, Envisat and ICE-Sat); and high-resolution topographic data (to correct the gravity data and to model it in 'unobserved' locations). One indication

of its resolving capability is that it was constructed in spherical harmonic terms using harmonics to degree and order 2159—an unprecedented undertaking requiring the determination of more than $4\frac{1}{2}$ million harmonic coefficients!

Combinations of all the techniques should also yield the truest rendition of the global Geoid. Figure 6.32b shows the high-resolution Geoid produced from EGM2008. Comparison with the TOPEX altimetric geoid shows that EGM2008 can reproduce it to within ~5 cm ocean-wide (Pavlis et al., 2008).

With a spherical harmonic expansion through degree and order 2159, can you calculate the spatial resolution of this Geoid in kilometers? For simplicity, focus just on the east-west resolution at the equator.

Finally, the newest (and quite novel) technique for measuring Earth's gravity field and determining the Geoid was launched in 2009 by the ESA. Called *GOCE* (Gravity field and steady-state Ocean Circulation Explorer), it was a single satellite with dual roles: tracked

(a)

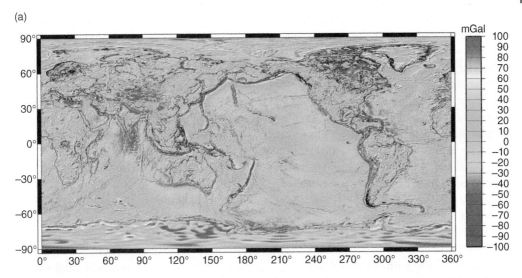

Figure 6.32 (a) Free-air gravity anomalies with respect to the Spheroid (approximate sea level), based on the EGM2008 combination gravity model described in the text. Downloaded from http://earth-info.nga.mil/GandG/wgs84/gravitymod/egm2008/anomalies_dov.html) / National Geospatial-Intelligence Agency.

(b)

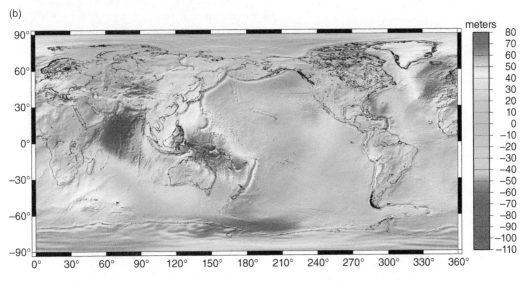

Figure 6.32 (b) High-resolution Geoid (2.5″ × 2.5″) based on the combination gravity model EGM2008. Geoidal heights range from almost − 110 m to +80 m. N. Pavlis, pers. comm. (2013); see Pavlis et al. (2012).

by SLR and GPS, it provided an(other) estimate of the harmonics of the gravity field and the long-wavelength Geoid; as it circled the Earth, however, it also acted as an orbiting gravimeter, with an onboard gravity 'gradiometer' continually measuring the differences in gravity from one side of the satellite to the other, in each of three mutually perpendicular directions. To achieve its goal

of high resolution, mGal precision for the gravity field and centimeter-level precision for the Geoid, it was in a strikingly low-altitude orbit—barely more than 250 km above Earth's surface—which meant it was also subject to strong drag forces by the atmosphere; to maintain its orbit, those forces required continual compensation from onboard thrusters. Additionally (—think "Eötvös effect"!), the satellite's own angular velocity had to be continually monitored, so all three sets of gravity differences could be corrected.

Combined with high-resolution altimetric mean sea-surface maps, GOCE was able to produce detailed maps of dynamic ocean topography; an example of the geostrophic currents inferred from such topography is shown in Figure 6.33. GOCE data has also been combined with LAGEOS and GRACE observations and with gravity anomalies over land areas derived from the EGM2008 model to produce another ultra-high resolution global gravity field model—up to degree and order 2190—known as EIGEN-6C4 (Förste et al., 2014). Using both the broad-scale and

high-order harmonics of EIGEN-6C4, Marotta et al. (2020) were able to distinguish the physical properties and processes characterizing different types of subduction zones (with implications even for the likelihood of large subduction zone earthquakes). The GOCE mission ended in 2013 (lasting twice as long as planned) when the satellite ran out of fuel.

A Tentative Interpretation of the Geoid, With Cautions. At this point we finally take the opportunity to examine the Geoid quantitatively. First of all, as Figure 6.32b reveals, the typical size of geoidal heights worldwide is tens of meters; compared to the typical radius to the Geoid from Earth's

(a)

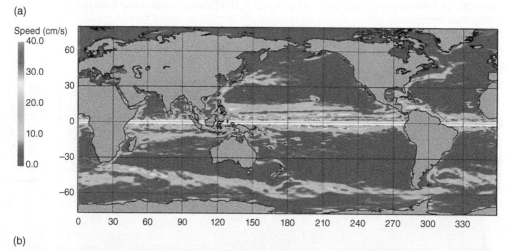

(b)

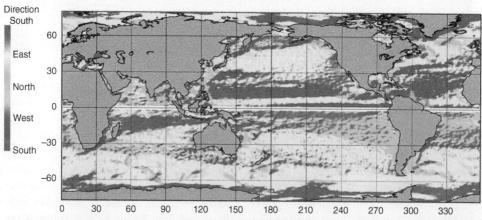

Figure 6.33 The geostrophic currents constituting the steady circulation of the oceans, as inferred from a preliminary release of GOCE gravity data (analyzed by the 'direct method'), using spherical harmonics through degree and order 240, combined with a 12-year altimetric mean sea surface known as DNSC08 (compiled by the Danish National Space Center) similar to that depicted in Figure 6.25a. (a) Current speeds; (b) Current directions. Analysis and images by Knudsen et al. (2011) / European Space Agency.

center, ~6,371 km—or even compared to the size of the equatorial bulge, ~21 km—these variations are small. That is, *the Geoid is well approximated by the Spheroid.*

As shown in Figure 6.32b, the biggest geoidal anomaly in the world is the geoidal low south of India, where the Geoid dips more than 100 m below the Spheroid. In this region, a ship passing from west to east south of India would gradually sink deeper and deeper toward Earth's center (and then find itself shallower and shallower as it gradually emerged from the low), all the while remaining on the ocean surface and 'feeling' like it was moving strictly horizontally—which it was, always moving perpendicularly to Earth's gravity!

The cause of that geoidal low has never been agreed upon. Finding a geophysically sound explanation is challenging in part because the low is broad enough in scale to suggest that its source is somewhat deep, i.e. in the sublithospheric mantle, yet that source also amounts to the most severe mass deficit, in terms of its impact on Earth's gravity, anywhere on the planet. One intriguing hypothesis posits that the origin of the low is somehow related to the Deccan Traps, an extensive province of continental flood basalts in the eastern central portion of the Indian subcontinent produced by extraordinarily prolific volcanism between 68 Myr and 60 Myr ago. The Traps would later pass right over the low's present location, as the Indian plate moved northward on its way to colliding with Eurasia, hinting at a connection between the two features. The vaguely north-south orientation of the region's gravity low seen in recent gravity models (e.g., Figures 6.31, 6.32a) also suggests the low might be related to the Indian plate's northward displacement. And, the north-south trending Chagos-Laccadive ridge, a bathymetric feature in the Indian Ocean (above sea level, we see this as the island chain called the Maldives) stretching from near the currently active Réunion hotspot northward toward the Deccan Plateau, adds to the evidence; it also implies that the mantle plume now associated with the Réunion hotspot was the source of the Deccan flood basalts (e.g., Richards

et al., 1989), though this has been disputed (Sheth, 2006).

As it moved northward, the Indian plate did not reach the location of the low until ~45–50 Myr ago, so the low was not directly caused by the formation of the Traps. If the source of the geoidal low is indeed below the lithosphere, however, we can speculate that that source was related to a *middle* phase of the Deccan/Réunion mantle plume; that is, perhaps the low derives from a 'fossilized' plume conduit left over after the 'Maldive' stage of plume activity had ended and before the current Réunion stage began. That middle phase could mark a gap in volcanic activity (see, e.g., Bredow et al., 2017) as the plume structure caused it to shift from its prolific but relatively short-term "plume head" phase (which produced the Traps) to the milder but ongoing "plume tail" phase (which produced the island chains).

With the exception of that low, the major geoidal high in the southwest Pacific, and a North Atlantic high (see Figure 6.32b), the variations in geoidal height are fairly muted.

The long-wavelength variations in geoidal heights around the world are often called the **undulations of the Geoid**. As already noted with reference to the gravity field in Figure 6.21, the undulations (see Figure 6.34a) do not correspond well to land versus ocean location; such a lack of correlation establishes the global validity of the principle of isostasy, and is consistent with the sources of the undulations not being surficial. *They are a glimpse into Earth's interior!*

A variety of possible sources for the undulations have been proposed. These range from the warping of key 'transitional' layers within the mantle (Stacey, 1977) (to be discussed in Chapter 8), perhaps caused by mantle convection; to large-scale convection patterns and viscosity structure throughout the mantle (e.g., Hager & Clayton, 1989; Moucha et al., 2007; Tosi et al., 2009; and many others); to 'bumps' on the core surface, which could also perturb convection in the outer core, thereby affecting the geomagnetic field (Hide & Malin, 1970; see also Lambeck et al., 1982).

(a)

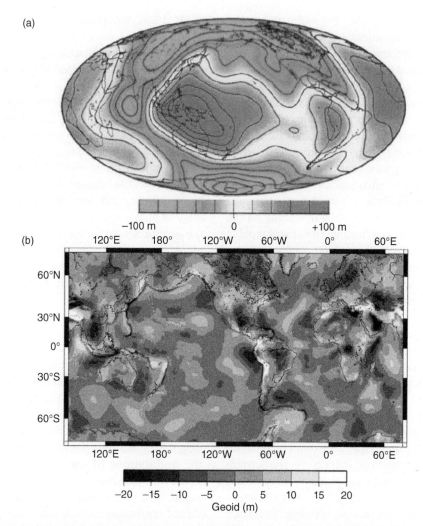

(b)

Figure 6.34 The long-wavelength variations in geoidal heights, also called the undulations of the Geoid.
(a) Geoid constructed from degrees 2–12 spherical harmonics by Čadek & Fleitout (1999) / John Wiley &
Sons, centered over 180°E longitude, with surficial (lithospheric) effects removed (using Doin et al., 1996).
Note that the Spheroid employed by Čadek and Fleitout to construct the undulations is defined slightly
differently than our own definition.
(b) Geoid 'filtered' to show only wavelengths shorter than ~5,000 km, equivalent to spherical harmonics of
degrees exceeding 9—thus, more of an 'intermediate' wavelength Geoid. King (2002) / John Wiley & Sons.

To the extent we hope to glean knowledge about mantle convection from the Geoid, it might be noted (Richards & Hager, 1984 and references therein, also Hager et al., 1985; Hager & Richards, 1989) that broad-scale deformations of the mantle, including mainly its top surface but also internal boundaries and the core-mantle boundary, caused by convection can make a greater contribution to the undulations of the Geoid than can the density anomalies responsible for the convection in the first place. Those deformations, called the *dynamic topography* associated with mantle convection by Hager et al. (1985; also, e.g., Lemoine et al., 1998; Čadek & Fleitout, 1999, 2003; King, 2002), do indeed share some similarities with the dynamic topography of the sea surface discussed earlier in this chapter, most fundamentally that both originate with currents of some kind (convective or wind driven). But, just to be clear, there are differences, too. Their timescales are hugely dissimilar; the forces governing the currents that produce them are only marginally related (buoyancy forces for one, Coriolis and pressure forces for the other); and one kind of dynamic topography is manifested well below the sea surface, whereas the other is physically superimposed on the Geoid.

Gravity Is Only Part of the Story. We have previously noted one major limitation of using gravity as a tool to investigate the interior of the Earth, namely, the weakness of the gravitational force. It is with the long-wavelength Geoid that we confront a *second* such major limitation: the *nonuniqueness* of gravity. We know that gravity is nonunique because multiple mass distributions within an object can produce the same gravity field outside it. The simplest, and perhaps the most striking, example of this (a consequence of the 'amazing fact' introduced in Chapter 4) is that all spheres of the same total mass *M*—but *regardless* of their internal mass distribution (as long as it is spherically symmetric)—act gravitationally like the same point mass, and exert the same gravitational acceleration on another body (GM/d^2 if d is our distance from the sphere's center). Such mass distributions include a top-heavy sphere (with most of its mass in its outer layers), a sphere of uniform density, and a sphere that is hollow except for its outer skin and a single massive particle at its center (but with the same total mass *M*).

Those examples suggest that in any situation, it might be possible to eliminate some of the postulated mass distributions because they are physically unrealistic. In the Earth System, however, where core, mantle, crust, hydrosphere, and atmosphere all have complexities of structure and behavior, it is easy to envision any number of possible mass distributions that are physically realistic; this compounds the challenge of dealing with the nonuniqueness of gravity interpretation.

The variety of physically reasonable mass distributions can usually be narrowed by supplementing the gravity data with additional geophysical information. That could include, for example, the density, structure, or temperature at some depth; evidence that certain types of processes are taking place within some region of the Earth (e.g., the geomagnetic field is evidence of convection within the outer core); and models describing how material properties such as rigidity or viscosity might vary through the interior. These subjects will be discussed in subsequent chapters. An illustration of the value of supplemental data is presented below in the context of time variations of gravity, where the interpretation of gravity data in terms of seasonal hydrological processes is substantiated by independent hydrologic models (after seasonal effects caused by the atmosphere and oceans had been removed from the gravity data).

Finally, one property of gravity that helps in its interpretation is a general relation between the 'wavelength' or harmonic degree of a gravity feature and the depth of the source that produced it: *the narrower the spatial scale of the gravity variation (i.e., the higher the harmonic degree or order describing it), the shallower the source is likely to be*. We applied such a rule implicitly when making a quick interpretation of subduction-area gravity anomalies in Figure 6.31.

The other side of the 'coin'—that deeper sources produce broader-scale (lower harmonic) gravity features—is often more casually (and less rigorously) stated as a claim that broader-scale features are the result of deeper sources (in reality, though, broad-scale features can *also* be produced by broad-scale shallow sources). With reference to Figures 6.21 and 6.34, however, we are able to conclude that the broad undulations of the Geoid result from deeper sources by *first* using the lack of correlation between the gravity field or Geoid and land versus ocean locations to assert that isostasy prevails, more or less, around the world; and *then* recognizing that isostatic compensation significantly reduces the contributions of shallow sources to the gravity field.

The relation between depth of source and the spatial scale over which a field varies will be exploited in Chapter 10 to understand the various sources contributing to Earth's geomagnetic field.

Time-Dependent Variations in Earth's Gravity Field. For years, as more and more satellites were tracked and the sea surface was repeatedly mapped, determinations of the Earth's gravity field and the Geoid have been refined, made more precise and detailed by 'bootstrap,' leading up to the impressive maps highlighted in this chapter. But there is more gravity can teach us about the Earth. The Earth System is not immobile; even the solid earth deforms under stress. Mass moves within each component of the Earth System, and can even be exchanged between some of the components, on a variety of timescales. Those movements and mass exchanges will be reflected, to some extent, in temporal changes of the global gravity field.

We have already (in Chapter 5) discussed changes in the broadest scale of the gravity field, that is, in J_2. A linearly decreasing trend in that harmonic was first detected from changes in the orbit of LAGEOS (Yoder et al., 1983), and ascribed to the effects of postglacial rebound. Subsequent SLR analyses expanded the choice of satellites—Cheng et al. (2013), for example, analyzed eight laser-tracked satellites over a span of more than three decades—and confirmed the trend in J_2; they also detected a change in it (as noted in Chapter 5), beginning in the 1990s (see Figure 6.35), which Cheng et al. (2013) convincingly argued was likely the result of modern-day global warming.

On human timescales, the largest variations in mass within the Earth's climate system may well be seasonal variations. Cheng et al. (2013) found a seasonal variation in J_2, with a typical amplitude of ~3 × 10^{-10} (see Figure 6.35). We can explain the variation as follows. In the northern hemisphere winter, much water is locked up as ice and snow on land; that water is released beginning in springtime and ultimately flows back to the oceans. Because the tropics have more oceans and the northern midlatitudes more land, a part of this annual mass redistribution in effect modifies the equatorial bulge, albeit only slightly. This is by no means the largest portion of the seasonal exchange—so we can expect to see variations in other Stokes coefficients as well—but it does involve an exchange of mass between latitudes.

We can even estimate how much mass is exchanged. At a latitude of 45°, the perpendicular distance from Earth's rotation axis to the surface is about 4,500 km; the perpendicular distance at the equator is, of course, 6,378 km. As an upper limit, if we idealize the mass involved as a single point of mass M_{seas}, its moment of inertia at the equator is

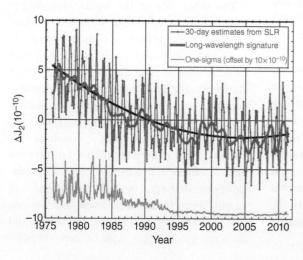

Figure 6.35 Variation in Earth's oblateness as measured by changes in J_2 between 1976 and 2011 from laser-tracked satellites. Each data point (blue dot) is an estimate from 30 days of tracking. The changes in J_2 (blue curve) are dominated on the short term by a seasonal variation, whose amplitude is seen to vary from year to year; but there is also a long-term trend (red curve, created by averaging the data over a progressive or 'sliding' time span), which appears to be mainly quadratic (best-fitting black curve). As noted in the text, the quadratic likely results from postglacial changes in J_2 compounded by the effects of modern-day global warming. Cheng et al. (2013) / John Wiley & Sons.

M_{seas} times $(6,378 \text{ km})^2$, its moment of inertia at midlatitude is M_{seas} times $(4,500 \text{ km})^2$, and the change in inertia is the difference, or M_{seas} times 2.04×10^{12} kg m^2 if M_{seas} is in units of kg. Recalling the definition of J_2 from Chapter 4, if we then think of the change in J_2 as being this change in inertia divided by $M_E a^2$, the results of Cheng et al. (2013) imply that the net amount of mass exchanged seasonally between the tropics and midlatitudes is up to $\sim 4 \times 10^{16}$ kg. For comparison, this amount of mass is several times the surface load of Pleistocene Lake Bonneville at its maximum (see Chapter 5) and is exchanged on a seasonal basis.

Which do you think would be greater—this seasonal mass exchange or the present-day net mass loss from Antarctic and Greenland ice sheets? Look at Figure 6.35 for a clue; then check out the numbers in Chapter 3, from GRACE data (section 3.6.5), to determine the answer (note that you will have to be able to convert Gt to kg!).

Satellite missions dedicated to measuring temporal variations in global gravity with high spatial resolution (i.e. high degree and order spherical harmonics) include CHAMP and GRACE. CHAMP (<u>Ch</u>allenging <u>M</u>inisatellite <u>P</u>ayload) was a German satellite (with participation by the United States and France), tracked by SLR and GPS, and in orbit during the years 2000–2010. It was designed to monitor spatial and temporal variations in Earth's gravity and magnetic fields at high spatial resolution from low orbit (altitude $\sim$450–350 km). Recent reprocessing of its tracking data has yielded trends and seasonal variations in gravity both globally and regionally (Weigelt et al., 2013). The impact of CHAMP on our knowledge of the magnetic field has been especially significant, and will be discussed in Chapter 10.

The GRACE mission has been prominent and widely successful, yielding monthly estimates of changes in the global gravity field, generally with resolution corresponding to harmonic degree and order 130 or higher, with a precision in geoidal heights of at least 2 cm (Chen et al., 2022) to 4 cm (Dahle et al., 2019) for degree and order up to 60 or higher (with much greater precision for the low-degree, broad-scale harmonics; e.g., Tapley et al., 2004b).

Figure 6.36 shows seasonal variations in geoid heights around the world based on

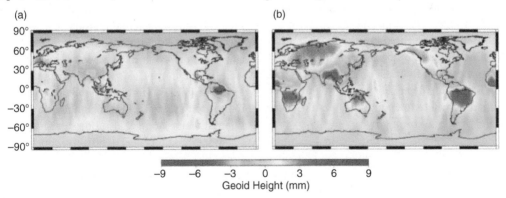

Figure 6.36 Residual annual variation in the Geoid based on early GRACE data; the effects of solid earth and ocean tides, atmospheric pressure variations, and the oceanic responses to winds and atmospheric pressure have all been removed. The J_2 portion of this variation has also been removed, as have harmonics of degree exceeding $\sim$100. Analysis and images by Tapley et al. (2004a) / American Association for the Advancement of Science.
(a) The winter-to-summer part of the annual variation, also called the "cosine" component of the variation, during 14 of the months between April 2002 and December 2003.
(b) The spring-to-fall or "sine" component of the variation.
Note that some portion of the variations shown in these images, and the lack of dramatic variations between winter and summer (a), may be the result of incomplete data for January and June 2003.

GRACE data. This analysis, by Tapley et al. (2004a), excludes the J_2 or degree 2 zonal component's contribution; a number of modeled effects on the Geoid—tides, mass variations in the atmosphere, and oceanic mass variations forced by winds and atmospheric pressure fluctuations—have also been removed. The largest remaining contribution to this variation is likely to be hydrological, that is, seasonal changes in surface water, groundwater, and ice mass, and such an expectation is confirmed by comparison with independent hydrological models. The greatest magnitude annual variation is centered on the Amazon rainforest; as subsequently estimated, its 2 cm range (+1.4 cm to −0.8 cm April to October 2003) suggests a fluctuation in the mass balance of that watershed comparable to the discharge of the entire Mississippi River (Tapley et al., 2004a; see also Wagner & McAdoo, 2004), thus a mass exchange of perhaps $\sim 2 \times 10^{14}$ kg over a 6-month span.

Nonuniqueness, Again. At this point, we briefly note that time-dependent gravity brings with it an added dimension of nonuniqueness. One way this can happen is when gravity is used to infer a change in glacial mass in a region (such as one full of mountain glaciers) where there has been glacial melting, but the meltwater is prevented by the terrain from draining away into the oceans and instead ponds locally. In that case, gravity surveys and satellites would detect no change in gravity! In a situation where the ponding is partial, there would be variations in gravity, but they would underestimate the true change in ice mass. Alternative techniques to infer the changing mass of glaciers would be needed to allow the extent of ponded versus drained meltwater to be estimated and the nonuniqueness minimized.

Finally, variations in Earth's gravity over time can of course result from episodic events rather than periodic or trending processes. In 2004, one of the largest earthquakes in recorded history struck off the west coast of Sumatra, in a portion of the subduction zone where the Indian/Australian plate is descending shallowly, toward the northeast, beneath the Eurasian plate (in this region, the 'Eurasian' plate actually includes several subplates and a microplate; Lay et al., 2005). The quake lasted for 10 minutes at the source, as the sudden slip of one plate relative to the other propagated along the fault for a record 1,300 km (cf. Ammon et al., 2005). The rapid movement of the seafloor displaced a large amount of seawater, causing a major tsunami that spread through the Pacific with devastating results. The earthquake also changed the region's distribution of stresses—the stress 'field'—which led to ongoing 'postseismic' deformation within both plates as they adjusted to a post-earthquake state.

The mass redistributions associated with the earthquake and the postearthquake deformation affected the gravity field detectably; but the changes in gravity were small—of the order of µGal. de Linage et al. (2009) used GRACE data to determine first the 10-day changes in gravity, then monthly averages, for the Sumatran region (see, e.g., Cambiotti & Sabadini, 2013 and references therein for other GRACE-based studies of this and other really large earthquakes, and Cambiotti et al., 2020 for further references and for assessment of a future satellite mission that will be especially useful for 'gravitational seismology' applications). Their data set was long enough (spanning 2 yr prior, to 2 yr following the earthquake) to allow seasonal signals (associated mainly with regional monsoon effects on continental water storage, also tides and regional ocean circulation) to be reliably eliminated from those monthly averages. Using their corrected data set, both the coseismic change in gravity around the time of the event and a post-earthquake gravity signal were inferred; see Figure 6.37.

The coseismic deformation of the Sumatra earthquake is understood to have involved, in the west of the region, a 'rebound' upward of the Eurasian plate's leading edge (at the trench); that rebound, which also precipitated the tsunami, occurred when the plate

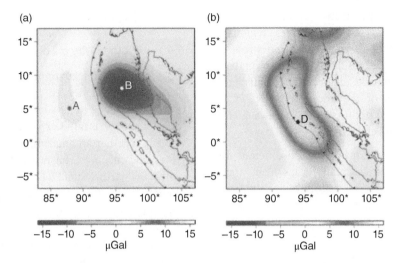

Figure 6.37 Changes in gravity associated with the December 2004 Sumatra earthquake as inferred from GRACE measurements by de Linage et al. (2009) / Oxford University Press. The data, spanning 4½ years of monthly averaged values, was processed so that coseismic (a) and postseismic (b) effects on gravity, 26 months after the earthquake, could be discerned. Maximum coseismic changes in gravity were +4 µGal (at point A) and − 16 µGal (at point B); maximum postseismic change was +12 µGal (at point D), near the epicenter of the earthquake. Background curves indicate the coast (solid curves) and the trench (sawtooth curve).

'unlocked' from the ongoing downward drag of the subducting Indian/Australian plate beneath it. The rebound would have been accompanied by a compensating downward tilt of the plate over a broad area off to the east. Changes in gravity corresponding to both coseismic displacements are seen in Figure 6.37a.

Comparisons with predicted changes in gravity, from models of the various processes (solid earth, oceanographic, and climatological) acting in the region around the time of the event, demonstrated that the inferred coseismic changes in gravity were mainly the result of the solid-earth deformations associated with the earthquake itself *and* the static mass redistribution of the oceans in response to the earthquake's change in gravity (de Linage et al., 2009).

Even more surprising, the GRACE data revealed that the effect on gravity of the postseismic deformation was almost as large as the coseismic changes (12 µGal versus 16 µGal, in absolute value). Ogawa and Heki

(2007) reached the same conclusion by considering the corresponding changes to the Geoid implied by GRACE. The shape of the postseismic effect on gravity (see Figure 6.37) suggests that the deformation was related to processes occurring at the subduction zone. Proposed mechanisms that could have led to continued deformation over the 2 yr following the earthquake include

- "afterslip," that is, ongoing, *aseismic* slip (meaning slip produced along the fault in the absence of additional earthquakes)—which might more simply be called post-earthquake "fault creep"—generated because the earthquake did not relieve enough of the tectonic stress previously built up to allow the fault to 'stick' (see, e.g., Marone et al., 1991; also Vigny et al., 2005);

- viscous relaxation of the whole mantle, in order to relieve the shear stresses generated by the earthquake outside and below the coseismic region (see, e.g., Pollitz et al., 2006); and

- "poroelastic rebound," in which water or another pore fluid diffuses through the crust or upper mantle from higher-stress (compressed) to lower-stress regions of the subduction zone (with the stress differences the result of the earthquake or post-earthquake deformation), causing an expansion of the lower-stress region and thus a local uplift (see, e.g., Ogawa & Heki, 2007; also Peltzer et al., 1998).

These processes may work on different time and spatial scales, and it has been argued that more than one must have been responsible for the Sumatra postseismic observations (e.g., Grijalva et al., 2011; de Linage et al., 2009).

It is ironic that our big-picture description of how Earth's gravity varies over time may depend significantly on the small-scale behavior of the Earth: how frictional forces can keep a fault 'locked' against slip; how a tectonic plate bends when its edge eventually is unlocked; how the mantle can flow viscously under stress; how well pore fluids can diffuse through the crust (and possibly the upper mantle). We will see the same small-scale influence on a (really) big-picture phenomenon as we discuss the final topic of this chapter. And all of this amounts to a further motivation for the next chapter, where we will discover that the entire field of seismology depends on understanding the various ways that arbitrarily small volumes of Earth material can respond to applied forces.

At the same time, our exploration of time variations in Earth's gravity, such as those recorded at the time of the Sumatra earthquake, reminds us that, in our effort to understand how the Earth works, the answer may require looking at *all* components of the Earth System.

A Tidal Segue. Speaking of material properties—and viscosity, in particular—the viscosity of the oceans should be ~0.01 Poise (or 0.001 Pa-sec), the same as that of water, right? It turns out that, according to the way ocean tides behave, the answer is both yes and no. And the difference is by orders of magnitude!

6.4 Tidal Friction

6.4.1 Another Way of Looking at Tides

Tidal friction is an extraordinary gravitational phenomenon experienced by our planet that owes its existence to the Sun and Moon. As a topic for study it is mind-expanding, managing to bring together an amazing variety of disciplines, and encompassing nearly every component of the Earth System! It is a great way to end our discussion of gravity and the Earth System, requiring us to use concepts we have already learned, and serving as a fine review of much of what has been presented along the way.

We have seen that the Sun and Moon act to raise tides in the oceans. As viewed from space, the tidal bulge remains aligned with the Sun or Moon, and the Earth rotates 'underneath' it, carrying an observer standing on the Earth's surface past two high and two low tides every day. Seen from Earth, the tidal bulges or 'high tide waves' travel around the world, following the Sun and Moon as they move across the sky.

In actuality, as the high tide travels around the globe, it encounters *frictional resistance* from the seafloor. There are two sources of friction. The first type of friction stems from the inherent stickiness, or molecular viscosity, of water; as the water flows over the seafloor, its viscous nature slows it down, trying to cancel the relative motion between the ocean and the seafloor. However, the viscosity of water is low, and that viscous drag is not very effective.

The second type of friction originates in the dynamical behavior of the oceans, and the viscosity associated with it is called *turbulent viscosity* or *eddy viscosity*. Consider ocean waves, perhaps generated offshore by winds, heading toward a coast. They move in more slowly as they near the coast, causing the waves behind to plow into them; they continue to slow, and the water piles up more, and finally the waves crest and then break. *That the waves slow at all is a result of the increasing shallowness of the water* (see Figure 6.38).

As the high tide waves generated by the Sun or Moon travel through shallow seas, the

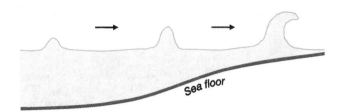

Sea floor

Figure 6.38 Cartoon illustration of waves breaking as they approach the coast. The speed of the waves depends on the depth of the ocean: waves travel faster in deeper water. As the waves move in toward the shore, they slow down, so that waves behind them plow into them; this causes the waves to build up and ultimately break.

shallowness would tend to slow them down; but the tidal force that creates them requires them to travel always at the same speed: they must traverse the globe in 24 hours (24 hours 50 minutes in the case of the Moon, because the Moon moves that much farther ahead in its orbit around the Earth). The *conflict* between this forcing and their natural tendency to slow down impedes their passage, causing the tidewater to flow chaotically—turbulently; such turbulence amounts to a large resistive force acting between the oceans and the seafloor.

The passage of the high tide wave around the globe bears some similarity to what is known in elementary physics classes as a *damped harmonic oscillator*. One traditional example of a damped oscillator is a mass connected to a spring, in turn connected to a viscous 'dashpot' fixed to a wall; if we are considering *forced* oscillations, the system may be driven by a flywheel (see Figure 6.39). When the flywheel is operating, the mass is forced in and out periodically. The spring can respond instantaneously by compressing or stretching as needed, but it takes time for the dashpot

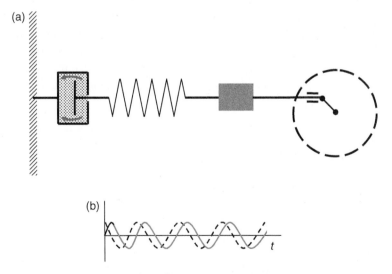

Figure 6.39 Forcing of a damped harmonic oscillator.
(a) The dashpot on the left, fixed to a wall, is filled with a highly viscous fluid; in order for the plate within it to move to the left or right, it must displace that fluid around it. The dashpot plate is connected to a spring that can stretch or compress. The mass connected to the spring and dashpot is forced in and out periodically by the flywheel.
(b) Schematic response of the mass, showing its lag. The forcing (dashed curve) and resulting displacement of the mass (solid green line) are plotted versus time.

Looking down from above the Earth in its orbit …

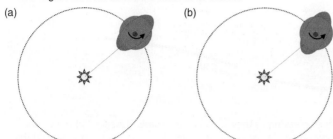

(a)   (b)

Figure 6.40 (a) Frictionless oceans would respond instantaneously to the Sun's tidal forces, rising without delay to produce a tidal bulge (high tide wave). (b) In viscous oceans, there is a lag in the oceans' response; by the time the high tide forms, Earth's rotation has carried the tidal bulge ahead of its 'frictionless' position.

plate to push through the pot's fluid, and this causes a *delay* or **phase lag** in the oscillation of the mass. If the flywheel pushes the mass in and out once every second, as an example, the system will eventually—after it gets "organized," and any transient oscillations have died down—move in and out every second; but it will do so a little *after* the forcing peaks!

In the same way, friction causes the ocean tidal wave to lag behind in responding to the tidal force. That force requires the high tide to completely circuit the globe in 24 hours, in the case of the Sun, but the friction causes it to peak some minutes later than it should. The daily spin of the Earth on its axis (being much faster than the Earth's orbital motion about the Sun) carries that sluggish tidal bulge past the Sun; relative to the Sun, the tidal peak is ahead of it (see Figure 6.40). Without friction, the tidal bulge would have risen instantaneously in response to the solar tidal force. Although it may appear that the cause has been preceded by the effect, as viewed from space, causality is not violated: a person sitting on Earth's surface and carried with it will observe the Sun overhead, i.e. it will be noon time, *before* reaching the high tide.

Such delays in the high tide can be observed at coastal locations. Satellite determinations of the ocean tides imply that globally, the misalignment of the high-tide-to-high-tide axis relative to the Earth-Sun axis (see Figure 6.40), and the misalignment relative to the Earth-Moon axis, is typically a few degrees (Christodoulidis et al., 1988; Stacey & Davis, 2008), which would take an observer an

additional ~12–24 minutes to cover. That magnitude delay implies a turbulent 'tidal' viscosity orders of magnitude larger than the molecular viscosity of seawater; relevant eddy viscosity estimates presented in Lambeck (1980) range from 10^7 to 10^{11} cm²/sec (equivalent to 10^7 to 10^{11} Poise) with our preference for Zahel's (1978) estimate of 5×10^9 cm²/sec.

6.4.2 The Solid Earth Will End Up in the Middle of It All, and Suffer Greatly

In much the same way that the Sun and Moon pulled gravitationally on the Earth's equatorial bulge, causing a torque that made the Earth precess (Figures 6.7 and 6.9), the Sun and Moon gravitationally pull on the ocean tidal bulge they each created. This torque tries to make the tidal bulge realign with the Sun or Moon (see Figure 6.41). *The lunar torque is*

Looking down from above the Earth in its orbit …

Figure 6.41 The Sun's gravitational pull on the tidal bulge it raised in the Earth's oceans causes a torque on the bulge, and thus on the solid earth beneath, which attempts to realign the bulge (and the Earth) with the Sun.

stronger, both because its tidal bulge is larger and because the Moon is closer. The lunar tidal bulge is larger because, as we have seen, the tidal force depends inversely on distance cubed. The torque, reflecting a "differential" gravity force analogous to that causing the Precession of the Equinoxes, is also inversely proportional to distance cubed. The total effect thus depends inversely on distance raised to the *sixth* power, making distance the major consideration and the Moon's role unquestionably dominant.

Friction between the oceans and solid earth causes the torque to act against the Earth's rotation, which had carried the bulge ahead (see Figure 6.41). In short, the torque acts to slow the Earth down; this is called **tidal friction**. Tidal friction is mostly a lunar effect, given the nearness of the Moon to the Earth.

Geological and Other Evidence of Tidal Friction. There is paleontological evidence that in fact the Earth's rotation has slowed by a significant amount in its history. Coral skeletons exhibit *growth ridges* or rings which were deposited daily (see Figure 6.42a); because their growth is affected by temperature conditions and tide heights, growth rings are accentuated seasonally and monthly.

(a)

cm

Figure 6.42 (a) (Left) Rugose coral from the middle Devonian Bell Shale, found in Michigan.
(Right) Close-up of marked section. Daily and monthly growth ridges are evident. House (1995) / The Geological Society, photos from C. Scrutton.

Counting the rings and dating the coral enables the number of days and months per year millions of years ago to be inferred. Wells (1963; also Scrutton, 1964) found that in mid-Devonian times, ~385 Myr ago, there were about 400 days per year.

On the other hand, observations of the positions of the Sun, Mercury, and Venus over the last few hundred years have established that our year remains constant in duration. Thus, the analysis of coral data implies that the day used to be shorter—only ~22 hours long in the Devonian! As the length of day has increased over the eons, Earth's spin rate has decreased. The average rate of deceleration of the Earth on its axis ($d\Omega/dt$) between then and now is about -5.7×10^{-22} rad/sec^2; this corresponds to an increase in the length of day of about 2.1 msec per century.

Starting with Wells' inference of 400 days per year in the Devonian, can you verify the numbers it implies (a 22-hour day back then, and an increase in LOD of 2 msec/century)?

Since Wells' initial work, a number of similar analyses have appeared, by him and others, based on corals, mollusks, and stromatolites of various ages (see Lambeck, 1980 for an extensive discussion). The interpretation of growth ridges and stromatolite laminations is somewhat controversial (as even Wells, 1963 acknowledged), although all of the analyses imply that Earth has slowed, and many of the analyses yield deceleration rates or numbers of days per year consistent with the above. As an example, using bivalve mollusks, Pannella (1972) determined that about 70 Myr ago there were ~375 days per year, thus ~23 1/3 hours in the day. Stromatolites provide the oldest paleontological estimates of Earth's past rotation rate, more than 1.7 Byr ago (Pannella, 1972); both stromatolites and layered inorganic sedimentary structures called *tidal rhythmites* (Figure 6.42b)—the latter based on daily deposition of sediment influenced by tidal and seasonal periodicities (see, e.g., Williams, 1997, 2000)—have the potential to

(b)

1 mm

Figure 6.42 (b) (Left) Stromatolite from upper Cambrian Conococheague in Maryland, showing daily and monthly laminations. Pannella et al. (1968) / American Association for the Advancement of Science (AAAS). (Right) Tidal rhythmite from late Proterozoic Reynella siltstone in south Australia. Its layering features diurnal and smaller semidiurnal laminae consisting of light, fine-grained sandstone and siltstone alternating with dark (more opaque) clayey material; during sluggish neap tides, increased deposition of the latter produces the thick and very dark "mud drapes" seen near the top and bottom of this image. Scale in lower right corner is 1 cm. Thanks to G. Williams for providing this high-resolution image. Williams (2000) / American Geophysical Union.

quantify the amount of tidal friction going back 2.5–3 Byr.

For the more recent past, historical records of eclipses from ancient Babylon, ancient China, medieval Europe and China, and the medieval Arab world have confirmed the existence of tidal friction (e.g., Stephenson, 2003). Total eclipses are dramatic events, and mention of them is often found in written records. We can accurately predict the *astronomical* times of those past eclipses; but a difference in the length of day of a couple of milliseconds, accumulating daily over the centuries, can make quite a difference as to which portion of the Earth's surface will experience the eclipse's shadow. Stephenson (2003) cites, as

an example, a total eclipse in 136 BCE recorded in ancient Babylon which would have been seen instead more than 9,000 km to the west if the length of day had been constant over the past two millennia.

Eclipse records extend back to 1223 BCE (de Jong & van Soldt, 1989), providing valuable estimates of changes in the length of day in the geologically recent past. However, such estimates are postglacial; the changes in the length of day actually observed include the effects of postglacial rebound—the decrease in Earth's equatorial bulge (or in J_2) and consequent "nontidal" acceleration associated with the end of Pleistocene glaciation—discussed in Chapter 5. The deceleration in Earth's spin rate due to tidal friction can be inferred from eclipse analyses only by correcting for the post-glacial effects. Modern-day estimates of tidal deceleration, for example from satellite (e.g., Christodoulidis et al., 1988) or astronomical observations, also require such correction; the paleontological and sedimentary estimates do not.

6.4.3 Tidal Friction Also Affects the Moon's Orbit

By symmetry, if the Moon is tugging on the Earth's tidal bulge, the Earth must be exerting a complementary torque on the Moon. The gravitational pull of the tidal bulges attracts the Moon toward the Earth, with the near bulge causing a slight net attraction toward *it*, in effect acting to gravitationally pull the Moon ahead (forward), out of its orbit (see Figure 6.43). This torque thus has the effect of throwing the Moon outward, *away* from the Earth!

More rigorously, the expulsion can be modeled in terms of the angular momenta of

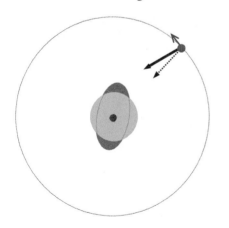

Figure 6.43 Tidal friction and the lunar orbit. The Earth as a whole pulls the Moon toward it; orbital centrifugal force also acts on the Moon, and the balance between the two forces allows the Moon to stay in orbit around the Earth (Kepler's Third Law). But with the ocean tidal bulge misaligned, the pull (especially by the 'near' bulge, shown with a bold arrow resolved into components) adds a small forward component to the forces acting on the Moon which very gradually draws the Moon ahead, out of its current orbit. All quantities are exaggerated for clarity.

the Earth and Moon, together with Kepler's Third Law. Since the angular momentum of the Earth-Moon system must be conserved, such an approach allows the problem to be phrased in terms of the slowing of Earth's rotation: the decrease in Earth's angular momentum from its decreasing spin must be compensated by an increase in the Moon's angular momentum, achieved by an increase in the Moon's orbital distance ("moment arm").

As we saw in Chapter 1, in the course of discussing the Titius-Bode Law, the angular momentum of an orbiting point of mass m equals

$$mrv$$

if its orbit is circular with radius r and it is moving through its orbit at velocity v. In that discussion, we found that the angular momentum could also be written

$$\frac{2\pi r^2 m}{T},$$

where T is the orbital period. But, as we now know, $2\pi/T$ is simply the mass' orbital angular velocity. Altogether, then, the Moon's orbital angular momentum is approximately

$$M_M \, \Omega_M \, (r_M)^2$$

if its mass is M_M and it orbits the Earth at a distance r_M with angular velocity Ω_M. This expression implies that the Moon's orbital angular momentum will increase as the Moon recedes from the Earth.

The situation is actually a bit more complicated. We know from Kepler's Third Law that the period of one object's orbit around another relates to the orbital radius; thus, the outer planets, for example, take longer to orbit the Sun than do the inner planets. Analogously, the Moon's orbital period increases as it recedes from the Earth. But that means its angular velocity decreases correspondingly.

Which do you think affects the Moon's angular momentum more: its increasing distance or decreasing angular velocity?

According to Kepler's Third Law, angular velocity is inversely proportional to distance raised to the 3/2 power; this follows from the square of the orbital period T_M being proportional to distance cubed:

$$T_M^2 \propto (r_M)^3, \quad \text{so } T_M \propto (r_M)^{3/2}, \quad \text{so}$$
$$\Omega_M \propto (r_M)^{-3/2}.$$

It follows that the effect of the Moon's angular velocity on its angular momentum is somewhat smaller than the direct, distance-squared effect of its increasing radius.

The resulting increase in lunar angular momentum is able to compensate the Earth-Moon system for the Earth's loss of angular momentum. From the inferred rate of deceleration ($d\Omega/dt$) of the Earth, one can theoretically predict how much the Moon's orbital radius and thus angular momentum must increase in order to achieve that compensation. We expect that at present the Moon should be receding from the Earth by about 3–5 cm/yr.

Reflectors placed on the lunar surface by the Apollo astronauts (plus another reflector from an unmanned Soviet mission, Lunokhod 2, in 1973) have allowed precise tracking of the Moon's orbital motions since 1969. The technique, known as **lunar laser ranging (LLR)**, is analogous to SLR and simply requires timing pulses of laser light as they travel from a terrestrial observatory to a lunar reflector and back to the observatory (but photoreceptors capable of detecting only a few photons per return are required!). Such measurements over the past few decades have verified that the Moon's distance from us is increasing—and at a rate of 3.79 ± 0.05 cm/yr (Williams & Dickey, 2002). *Such measurements provide the strongest confirmation of tidal friction.*

If the Moon is receding from the Earth, it must in the past have been much closer. Extrapolation from the present rate of recession yields ~1.6 Byr ago as the time, often called the *Gerstenkorn event*, when the Moon was so close to Earth it was at the Roche limit (Stacey & Davis, 2008; see also MacDonald, 1964). Such an extrapolation is quite provocative because it implies—going back a bit more in time, bringing the Moon even closer to Earth—that the Moon has existed only for less than a couple of billion years; or, that the Moon was captured by the Earth within the past ~1½ Byr.

Yet we know from 'Moon rocks,' brought back from the lunar highlands by the Apollo astronauts and dated at more than 4 Byr in age, that the Moon is much older than a couple of billion years, and was in all likelihood created around the same time as Earth. Geological evidence of tidal deposition in sedimentary rocks 3.2 Byr ago (Eriksson & Simpson, 2000), plus the existence of nearly 3-Byr-old stromatolites and tidal rhythmites, also require the Moon to have been orbiting Earth, intact, well before a Gerstenkorn event. Furthermore, an event involving the Moon's creation or even its extreme nearness to Earth would be cataclysmic: with the Moon inside the Roche limit, the solid-body tides *on Earth* would be kilometers high, twice a day! But there is no evidence in the geological record of such a cataclysm.

The most reasonable conclusion is simply that the magnitude of tidal friction has varied in the past, invalidating the extrapolation. If most of the dissipation of tidal energy occurs in shallow seas, where the 'conflict' between free and forced oceanic behavior is most severe, then it is to be expected that—as plate motions have continued, causing changes in the configuration and volume of the oceans, and thus the extent of shallow seas—there have been times of greater effectiveness and times of lesser effectiveness of tidal friction. Such variability is evident in the results of Pannella et al. (1968), who analyzed paleontological evidence of tidal friction over the past 500 Myr (see Figure 6.44). Pannella et al. also attributed that variability in tidal friction to changes in the distribution of shallow seas from continental drift.

Our expectation of variable tidal friction is also supported by the work of Williams (1997), who estimated the length of day 620 Myr ago, 900 Myr ago, and 2.5 Byr ago from tidal rhythmites; he found the length of day 620 Myr ago to be the same as in the mid-Devonian, thus implying a hiatus in tidal friction and lunar recession between ~400 Myr and ~600 Myr ago.

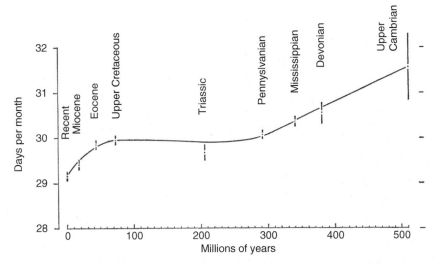

Figure 6.44 Effect of tidal friction on the Earth's spin, measured relative to the synodic month (also called the lunar month: the time from full moon to full moon or spring tide to spring tide). Calculations (dots, with error bars) by Pannella et al. (1968), based mainly on growth lines in mollusks. As we move along the curve toward the present, tidal friction slows both the Earth's spin and, as discussed in the text, the Moon's orbital velocity. But the tidal bulge is always ahead of the Moon, so we always measure a decrease in the number of days per month. However, the rate of slowing, indicated by the slope of the curve, has often varied during this time. Image from Pannella et al. (1968) / American Association for the Advancement of Science.

The timescale of tidal friction variability implied by both studies, $\sim 10^2$ Myr, is consistent with a mechanism driven by plate tectonics. Such a timescale would also explain why inferences of tidal friction over the past 3,000 yr, based on eclipse data, are not (de Jong & van Soldt, 1989) significantly different than present-day estimates.

To avoid having the Moon too close to the Earth throughout the past 4 Byr, then, we require only that the effectiveness of shallow seas was much less at times in the past—perhaps because they encompassed a smaller area, or were located in remote or poorly connected regions—resulting in a rate of lunar recession at those times which was much slower.

As the Moon recedes from the Earth and the Earth's daily rotation slows down, there may come a time in the far future when the lagged tidal bulge will be carried forward by the Earth slowly enough that the Moon, as it orbits the Earth ever more distantly, will keep pace with the bulge. At that time, the torque between the bulge and the Moon will vanish,

and lunar tidal friction will cease. But the story doesn't stop there! The solar tidal friction torque, weak as it is, will still be operating, and will slow the Earth's rotation further. Darwin (that's Sir George Darwin, one of the foremost geophysicists of the nineteenth century and a son of Charles), in his popular science classic *The Tides* (Darwin, 1899), explains with fascinating style but complete rigor how the solar effect will cause lunar tidal friction to be reversed, bringing the Moon in toward the Earth until they finally collide! Of course, the distribution of shallow seas will continue to vary, so when such a reversal occurs cannot be predicted; but at the present rate of lunar recession, the reversal would occur billions of years from now—by which time the Sun would have already evolved into a Red Giant star, perhaps even 'swallowing' the Earth and Moon as it expanded to a huge size....

6.4.4 Tidal Friction Has Consequences for the Earth System

In Chapter 3, we saw that the global pattern of atmospheric circulation is a direct consequence of the Coriolis force, whose 'sideways' deflection of winds driven ultimately by the pole-to-equator temperature gradient produces a three-cell circulation pattern in each hemisphere. The strength of the Coriolis force in turn depends directly on the magnitude of Earth's angular velocity. In the past, with the Earth having been less tidally decelerated than it is now, the Coriolis force would have been stronger. There is tentative evidence from tidal rhythmites that the length of day might have been as short as ~17 hours around 2½ Byr ago (Williams, 1997). If Earth's length of day was ever half its present value (which presumably would have been prior to 2½ Byr ago), we can speculate that the Earth—rotating so fast its day was nearly as short as Jupiter's and Saturn's—would have been meteorologically similar to those gas giants: their atmospheric circulation is dominated by the Coriolis force, resulting in a seven-cell (or more) per hemisphere circulation pattern (e.g., Figure 3.6 of Chapter 3) plus widespread cyclonic activity. On Earth, the pole-to-equator temperature gradient would be greater than on those two planets, both because Earth is a smaller planet and because it is closer to the Sun, so the Coriolis domination would not be so extreme, and perhaps the net result would be 'only' a five-cell circulation. Even on such an Earth, the surface winds from such a circulation pattern would have produced a very different wind-driven circulation in the oceans than we see today, even if the ocean basins happened to be geographically similar.

Whether or not the Coriolis force had strengthened to such an extreme as to impose a vastly different structure on the global atmospheric and oceanic circulation is highly speculative; but even a moderately stronger Coriolis force would have impacted the climate system in diverse ways. A Coriolis force that increases more with increasing latitude would have produced more intense western boundary currents and would have produced stronger jet streams (with fewer meanders). And with a greater 'vorticity' or 'twistiness' generated

by stronger Coriolis forces, we might expect storms to have been more widespread and tornadoes to have occurred more frequently.

Later in this textbook, we will learn about the geomagnetic field and the processes in the core that create and maintain it. The existence of the geomagnetic field—and its strength, and possibly even its ability to reverse polarity—depends directly on the Coriolis force. A stronger Coriolis force in the past would have led to a stronger magnetic field, one that conceivably might have 'saturated' (i.e., reached its maximum possible strength given the conditions in the core) earlier in Earth's history; had a stronger 'high-harmonic' (shorter wavelength, also called 'non-dipole') component; and, speculatively, a lesser tendency to reverse.

In contrast, in the far future, when the Moon has receded greatly from the Earth, Earth's rotation rate may be slow enough to render the Coriolis force negligibly weak. At that time, the geomagnetic field would be feeble. And, our atmospheric circulation might more resemble that of slow-rotating Venus, characterized (very roughly, as noted in Chapter 3) by a one-cell-per-hemisphere circulation pattern, and leading in turn to nearly stagnant oceans (if, by that time, the oceans had not already boiled away from an ever-brightening Sun). In that future scenario, the only relevance of the oceans to climate would be through their heat-storing ability. Tides would be smaller—only solar tides would be non-negligible—and would occur with much longer periodicities; future geological and paleontological 'clocks' like rhythmites and corals, if they exist at all, might be of little use to scientists of the even farther future.

Finally, if the Earth was spinning faster in the past, its equatorial bulge would have been greater; in the future, tidal friction will cause the Earth to bulge less. Combined with a different spin rate, such an Earth will respond differently than it does now to the torques exerted by the Sun and Moon on its bulge, resulting in a different period for its axial precession. The changing distance of the Earth from the Moon also will alter the precessional torque and precessional period. Additionally, the Earth will respond differently to the torques exerted by Jupiter and Venus, yielding different periods than we observe at present for its orbital precession, the variation in its obliquity, and the variations in its orbital eccentricity. In short, the Milankovitch timescales so influential in determining the timing of glacial advances and retreats during the Pleistocene would have been different during ancient ice ages and will be different in the far future.

Unfortunately, Earth's orbit is mathematically "chaotic" (e.g., Laskar, 1989), meaning that the effects of numerous ongoing but mostly quite minor orbital perturbations *accumulate*, making prediction of Earth's orbit into the far distant past or future somewhat unreliable. Such perturbations include physical changes in Earth's orbit, due to the gravitational forces exerted by all other planets, their moons, even asteroids; drag by the solar wind; and so on. The perturbations can also be numerical: tiny disparities in the properties of the Solar System, like the masses of the planets or Sun; or precision errors in how calculated numbers are rounded off, as the equations are solved from one time to the next. In the case of the Earth, extrapolations extending much more than several million years (Varadi et al., 2003) become highly uncertain.

Nevertheless, even treating such calculations as quite approximate, it appears that the Milankovitch periodicities may have undergone significant changes over time. Calculations by Berger et al. (1989) suggest that the farther into the past we go, the shorter those timescales will be. For example, the obliquity variation, a 41,000-year periodicity now, might conceivably have been ~32,000 yr in the mid-Devonian (Berger, et al., 1989; House, 1995). Thus, a periodicity that is roughly ~20,000–25,000 yr as seen in some glacially modified formations from the Precambrian might actually have derived from obliquity variations rather than Earth's precession.

6.4.5 Tidal Friction Without Oceans, and Astronomical Implications

If Earth had no oceans, there would still be tidal friction. Like the oceans, the solid earth responds to tidal forces, rising up beneath and opposite the Sun and Moon with high earth tides, creating a tidal bulge. Like the oceans, the solid earth possesses internal friction—generally called *anelasticity*—which causes a delay in its response, a phase lag, which in turn allows the solid earth's tidal bulge to be carried ahead by Earth's daily rotation; like the oceans, that solid-earth tidal bulge is pulled back by the gravity of the Moon (and Sun), causing Earth's rotation to slow. And, like the oceans, that solid-earth tidal bulge pulls the Moon ahead in its orbit, causing the Moon to recede over time from the Earth.

However, solid-earth tidal friction differs from oceanic tidal friction in two respects: the amplitude of the solid-earth tidal bulge is smaller than the oceanic one; and the internal friction in the solid earth is much less than the turbulent friction in shallow seas. The phenomenon of tidal friction is, as a consequence, dominantly oceanic.

Anelasticity is actually not very well determined at tidal periods, being masked by the ocean tides; and models of how anelasticity varies with frequency (used to infer internal friction at tidal periods from observations at nontidal frequencies) tend to be quite simplistic (as argued by, e.g., Dickman & Nam, 1998). Estimates generally agree that the tidal phase lag associated with anelasticity is very small, perhaps ~0.21° (e.g., Munk, 1997) or even ~0.10° (see Dickman & Nam, 1998; then Sailor & Dziewonski, 1978).

Planets not possessing oceans but composed at least partially of silicates like our mantle can therefore be expected to exhibit tidal friction, just not to the degree exhibited by our nearly ocean-covered Earth. For moons, however, the situation is much more interesting. Our own Moon, for example, experiences body tides due to the Earth (as well as the Sun); based on our earlier analysis of the lunar tide, we could even write an expression for the maximum tidal force experienced by the Moon due to Earth:

$$\pm \frac{2a_{\mathrm{M}}GM_{\mathrm{E}}}{R_{\mathrm{M}}^{3}},$$

where R_{M} is still the Earth-Moon distance, but now the numerator contains M_{E} rather than M_{M} because it is the Earth's gravity which is ultimately responsible for this tidal force. Since the Earth is so much more massive than the Moon (by a factor of ~80, though the lunar radius, a_{M}, smaller by a factor of ~4, counteracts its impact, a little), the bulge of the Moon's body tide is much bigger than even Earth's ocean tidal bulge. As a consequence, if the phase lag were the same, the tidal friction torque exerted by Earth on the Moon's tidal bulge would be much stronger than the Moon's torque on Earth's bulge. The torque on the Moon is made larger still by the fact that it is Earth's gravity—and another factor of M_{E} rather than M_{M}—exerting the torque.

Subject to possible differences in the phase lag, then, the net result is, potentially, greater body-tide friction experienced by the oceanless Moon than ocean-tide friction experienced by the Earth! This result is something we can expect for most of the moons in our Solar System, as long as the mass of the planet is significantly greater than the mass of the moon orbiting it or the moon is very close by.

In the most extreme cases—for example, moons close to the most massive of planets (Io, the closest moon of Jupiter, is an obvious choice)—internal friction should also generate significant heat, as the moon is tidally flexed almost beyond endurance.

In the past, when the Moon was much closer to Earth—given the dependence of the tide and the torque on distance—the dominance of body-tide friction would have been greater still. The tidal friction experienced by the Moon back then would have acted to slow its own spin drastically until, eventually, as the Moon continued to recede from Earth and its orbital angular velocity slowed, its bulge was no longer carried ahead of the Earth. At that time, its spin rate and orbital angular velocity matched, and the Moon became **tidally locked** to the Earth, always showing the same face to Earth (tidal locking is also called **synchronous rotation** or **spin-orbit coupling**). Tidal locking, causing the Moon's tidal bulge to remain aligned with the Earth ever since, represented the end of significant tidal friction on the Moon. (Technically, a minor amount of tidal friction would still be required, to maintain the alignment with Earth as the Moon receded and its orbital velocity continued to slow.)

Our expectation is that most moons in the Solar System should by now be tidally locked to the planets they circle. For the larger moons, this is generally true. (Charon, a moon of Pluto just an order of magnitude less massive than the 'planet' it orbits, has *also* tidally locked Pluto to *it*—creating a mutual, 'binary system' locking!) But tidal locking means that the tidal bulge of a moon due to its planet is not only aligned with the planet but is also essentially permanent: if some astronauts had placed a flag where the high tide occurred, at some moment, on the surface of such a moon, they would find that the high tide never moved away from that location, never progressed around the moon. For our Moon, the Galilean moons of Jupiter, and other moons, then, the only periodic flexing they experience is from minor body tides, rising and falling on longer timescales, associated with the Sun or with the slight eccentricities of their own orbits. The extreme heat generated internally during early, pre-locked stages of tidal friction is, for those moons, by now a thing of the past. In the case of Io, the tidal heat producing the sulfurous volcanism we see on its surface today is thought to be possible because nearby moons continually force Io's orbit to be elliptical, allowing Jupiter's tidal forces to periodically stretch (but not misalign) Io's tidally locked tidal bulge (Lainey et al., 2009; Schubert, 2009). In the case of Europa, Jupiter's second closest moon, the internal heat producing liquid oceans beneath its icy surface (and perhaps supporting life there) may also result from a similar mechanism.

6.4.6 Theories of the Origin of the Moon

Our determination that the Moon was closer to Earth the farther back in time we go—with a potential Gerstenkorn event looming—suggests the possibility, if not the necessity, of a special origin for the Moon, beyond simple accretion within the evolving primitive solar nebula. A special origin is demanded, however, by the Moon's composition. There are two key aspects to this constraint, the primary one being that the Moon is significantly depleted in iron, compared to Earth (and the rest of the inner Solar System); a second compositional constraint is introduced later. The inference of iron depletion follows from the Moon's moment of inertia, which implies that it has at most only a small core; from its mean density, which is comparable to that of Earth's upper mantle silicates; and from Moon rocks obtained by the Apollo astronauts. These imply that the iron fraction in the Moon is no more than ~10% of its mass, whereas the Earth is more than 30% Fe; more generally, they indicate that the Moon is very much like the Earth's *mantle* in composition.

Simple accretion also fails a different kind of test for the Moon. As we saw in Chapter 1, the evolution of the primitive solar nebula led to terrestrial planets and giant planets; the giant planets, growing primarily through their own gravitational attraction, spun rapidly on their axes because of the requirement to conserve angular momentum, whereas the inner planets grew by random collision and ended up with slow rotation.

We understand now, however, that the Earth must have rotated faster in its youth, and has slowed significantly because of tidal friction. But even while tidal friction was acting, the total angular momentum of the Earth-Moon system would have remained constant at all times (as pointed out in *Textbox A*). If we extrapolate back to the time of accretion, and imagine all that angular momentum concentrated into an Earth-Moon 'embryo,' we would find that the embryo spun at a rate a factor of ~5 times Earth's current spin rate (e.g., Asphaug, 2014). That would give the newly accreted Earth (or Earth + Moon) a daily spin twice as fast as the gas giants, which is unrealistic for a body so much less massive than the giants. In short, accretion does not work here; something more than accretion must have given the Earth-Moon a rotational 'boost' at some time in the past.

There is always the possibility that the Moon is depleted in iron for purely fortuitous reasons—chance variations in the composition of the primordial solar nebula in the region where the Moon originated. And the angular momentum of the Earth-Moon system may have been amplified in the past by additional events (such as the Heavy Bombardment). Nevertheless, a number of alternative hypotheses have been proposed to explain the origin of the Moon (see, e.g., Boss & Peale, 1986). We briefly note three: the *capture* hypothesis, the *fission* hypothesis, and the *giant impact* hypothesis.

- According to the capture hypothesis, the Moon originated in a region of the solar nebula presumably distant enough from the Earth to explain its different composition. As a result of its orbit being perturbed, the Moon was subsequently captured by the Earth gravitationally; given the evidence for a lunar tidal influence on Earth extending at least 3 Byr into the past, as discussed above, the time of capture would have to have been in the first billion years or so of the Earth's life.

  But to avoid the problem of a Gerstenkorn event, the captured Moon must always have been outside the Earth's Roche limit—even when it was first captured, even if that was more than 3 Byr ago. This reinforces the need we saw earlier for a variable rate of tidal friction through much of the Moon's history.

- There are two types of fission hypotheses. We know that in the past, when the Moon was much closer to the Earth, conservation of angular momentum in the Earth-Moon system requires that the Earth rotated much faster on its axis than it does now. The older version of the fission hypothesis envisions a developing, fluid-like proto-Earth, largely differentiated, that was spinning so fast it became unstable, and a portion of its (iron-poor) mantle split off to become the Moon.

An alternative fission hypothesis (Andrews, 1982) views core formation within the Earth as a potentially unstable process. While differentiation was in progress, there might have been molten iron as well as a silicate mantle in the outer half of the Earth, with the iron sinking down through the outer half to settle onto and surround an as-yet undifferentiated interior. Because the undifferentiated central region was less dense than the iron above it (it would have a density corresponding to Earth's average composition), it would exist in an unstable equilibrium at the Earth's center; any perturbation could dislodge it to one side, at which time it would 'pop up' toward the surface. Its collision with the material above it would then dislodge a portion of the silicate mantle, sending that into orbit as the Moon.

- With elements of the preceding two hypotheses, the giant impact hypothesis (see, e.g., references in Asphaug, 2014) is currently the most popular explanation of the Moon's origin, thanks in part to rigorous and convincing computer modeling by Canup and Asphaug (2001) and others. The leading version of this hypothesis presupposes an Earth whose accretion was otherwise complete, with a mass $\gtrsim 90\%$ of its current mass, struck by another body (described as Mars-sized) with a mass of 10%–15% of Earth's mass. Both the Earth and its planet-sized impactor had already differentiated into mantle and core; with a glancing blow, a portion of the impactor's mantle would survive its fragmentation and vaporization, able to re-form beyond the Roche limit as an iron-depleted Moon (the rest of the impactor would have been absorbed by the Earth). A ~45° angle of impact (relative to head-on) is optimal for minimizing the amount of core retained by the surviving Moon and also maximizing the leftover mass. Based on various isotopic analyses, this giant impact is estimated to have occurred ~60 Myr after accretion in the solar nebula began (Touboul et al., 2007; Barboni et al., 2017).

The giant impact hypothesis leaves us with an implicit problem: if the impactor, envisioned to be the current size of Mars, had had time to differentiate into mantle and core, why did Mars itself never differentiate to that extent?

More recently, the giant impact hypothesis has become somewhat uncertain (see Canup, 2014 and Asphaug, 2014 for critical reviews). The discovery that lunar rocks have essentially the same oxygen isotope signature as Earth (Wiechert et al., 2001) has led to the conclusion that the Moon is not just generally mantle-like in composition, but is essentially identical to Earth's mantle. Such a conclusion follows automatically in fission models but is problematic for giant impact models because, as described above, they envision the Moon to have formed from a portion of the impactor's mantle, following that glancing blow, not the Earth's.

It might appear that this second compositional constraint could be satisfied by requiring the impactor to have accreted under identical conditions as Earth had in the primitive solar nebula, for example in nearly the same orbit as Earth. But, as pointed out by Pahlevan and Stevenson (2007), a Mars-sized planet would have accreted from a very limited neighborhood of the nebula, whereas larger planets also grew larger by impacts originating throughout the inner solar system; unless the inner nebula was chemically homogeneous, it would thus have been unlikely for the impactor and Earth to have identical compositions.

Alternatively, the impactor could have been Earth-sized—that is, both objects would have been planets, already differentiated, each with about half of Earth's present-day mass. Computer simulations show that, in this case, a Moon satisfying *both* chemical constraints can form from a collision between them that is somewhat more nearly head-on and less of a glancing blow (Canup, 2012). Such a high-energy collision would evolve differently than what we found to result from a glancing blow. The impact would have vaporized part of Earth's mantle, as well as that of the impactor, and melted the rest of both planets; the planets would have recoiled, recollided, and merged into a single differentiated molten sphere, and then spun around rapidly till mantle material was thrown off into space to subsequently coalesce as the Moon. In this case the Moon would be iron-poor and (isotopically) identical to the mantle it came from, as required. However, such an impact would have left the Earth and Moon with *too much* angular momentum; in order to bring it down to reasonable values, an unusual type of orbital resonance (see Ćuk & Stewart, 2012), acting between the Moon and the Sun, must be invoked.

A variation on this theme (Ćuk & Stewart, 2012, whose work conceptually preceded Canup, 2012) considers that Earth might already have been spinning at a very fast rate (something we might infer anyway, given that it has been progressively slowing down from tidal friction), e.g. as the result of late-accretion large impacts, and proposes that the Moon-forming giant impact 'spiked' Earth's spin rate even more, facilitating the ejection of a portion of Earth's mantle. This hypothesis is like an impact-driven fissioning mechanism. Computer simulations show that it succeeds if the impact is somewhat head-on, as above, and retrograde

(i.e., opposing the Earth's sense of rotation). But it also yields an Earth with too much angular momentum, again requiring an unusual Moon-Sun resonance to decimate the momentum to acceptable levels.

Incidentally, the isotopic match between the Earth and the Moon implies that the Moon-forming impact was the final giant impact, since any additional massive impacts striking Earth would have altered Earth's isotopic signature (Ćuk & Stewart, 2012). For the same reason, if the Moon was created by fissioning from the Earth, that event would have had to take place at the end of the accretion era to avoid any subsequent massive late-stage impacts. Problematically, models of the late stage of accretion (Agnor et al., 1999) show that in some cases, the last impact was large but not the largest (implying the Moon-forming impact had already occurred); and in some cases, the last impact might have reduced the spin of the Earth (so fissioning would have occurred prior to it).

The giant impact hypothesis is clearly a fortuitous way for Earth to acquire its moon—not only was it a chance event, but it also had to occur at just the proper angle and at just the right time. But the other two mechanisms are also fortuitous: capture is only commonplace for massive planets (due to their strong gravity fields) and for planets near debris belts (Mars, near the Asteroid Belt, and Pluto (if we count it as a planet), near *or in* the Kuiper Belt); and fission, which if not an extraordinary occurrence would have also produced a moon for our sister planet Venus.

But, then, maybe "fortuitous" is the reality, after all: outside of the Pluto-Charon pair, our Moon is the largest and most massive by far, in proportion to its planet, of the more than 160 moons in the Solar System . . . .

7

Basics of Seismology

7.0 Motivation

Seismology is concerned with the phenomenon of earthquakes, and with the waves radiated worldwide when an earthquake occurs. Both the earthquake event—which represents the sudden failure, by fracture or by slip, of rock under stress—and the resulting waves depend on the kinds of material experiencing the failure and transmitting the waves. By analyzing seismic waves, we will ultimately find it possible to infer some of the material properties, including density and elastic properties, which characterize the depths of the Earth.

Like gravity, seismology is a tool for geophysical exploration of the deep interior of the Earth as well as the shallow subsurface. And, like gravity, seismic waves are 'integrative': just as the gravity signal measured by a gravimeter represents the integrated effects of all masses, the seismic 'signal' recorded by a seismometer represents the accumulated effects on the seismic wave of all the materials it encountered in its travels from source to receiver. However, unlike gravity, the focus of seismology is typically as much on the propagation of the signal as it is on the end result—the journey, not just the destination; with sufficiently detailed global coverage, the integrated data we measure can be "inverted" or unraveled to produce models of subsurface or interior properties that are strikingly high in spatial resolution.

The seismic waves that probe Earth's interior, revealing its layered structure, also propagate to some extent through the oceans and atmosphere, allowing seismic techniques to be used to investigate 'remotely' the layering and material properties of those components of the Earth System as well. But it's a two-way street: the oceans and atmosphere also generate some of the seismic energy we measure propagating through the solid earth.

All of this compels a global perspective—perhaps even a cosmic one, since the techniques for analyzing seismic wave propagation also turn out to be important for probing the interior of the Sun, and potentially other stars. In the next chapter, we will learn how the material properties inferred from seismic waves propagating within the Earth shed light on the physical state of the interior and imply what its behavior must be. But we begin this chapter on a small scale, as we first learn the basic concepts of seismology.

7.1 Stress and Strain

What makes one material different than others? To a seismologist, material properties are revealed by *the way the material responds to applied forces*. Within a section of oceanic lithosphere, for example, the lower crust and uppermost mantle differ not simply because the latter is 14% denser but because it is stiffer, that is, 75% harder to compress and 50% harder

Earth System Geophysics, Advanced Textbook 6, First Edition. Steven R. Dickman.
© 2025 American Geophysical Union. Published 2025 by John Wiley & Sons, Inc.
Companion Website: www.wiley.com/go/Dickman

to flex than the former. Such material properties come into play because of the type of applied forces seismologists are concerned with, namely, stresses; and they are measured by the kind of response that is the standard in seismology, namely, strain.

7.1.1 Stress

A **stress** is created from *forces that act on surfaces. The magnitude of the stress is equal to the force per unit area.* Since forces are vector quantities, stresses are as well; *the direction of a stress vector is the same as that of the force that created it.*

For any object, the surface experiencing the force and resulting stress can be an internal surface or the outside surface of the object itself. When the surface is within the object, it may be useful to think of the stress as acting on the material on one side of the surface *due to* the material on the other side; see Figure 7.1.

Since the forces creating the stress can have any orientation with respect to the surface of interest, the stress can also have any orientation. For reasons that will become apparent, two specific orientations are singled out: stresses that are *perpendicular* to the surface are called **normal stresses**; stresses that act *along* the surface are called **shear stresses**.

The most common type of surface force is *pressure*. Pressure is a normal stress: it always acts perpendicular to surfaces—as anyone swimming underwater can attest to, recognizing that the pressure on their eardrums persists no matter what their orientation. In situations where nothing is moving, pressure arises from weight, the weight of overlying material; in situations involving fluids in motion, there can be dynamic pressures as well. Thus, on and within the Earth (except under vacuum conditions in a laboratory), pressure will always be operating.

As some illustrations of pressure, consider first a 1 kg cube lying on a table at Earth's surface. With the acceleration of gravity being ~9.8 m/sec^2, the cube has a weight ($W = Mg$) of 9.8 kg-m/sec$^2 \equiv 9.8$ **Newtons (N)**. If the cube is 1 m on a side, the area of the cube in contact with the table is 1 m^2, implying that the pressure is 9.8 N/m^2; this is also called 9.8 **Pascals (Pa)**. In contrast, if the cube is only 1 cm on a side (but still contains 1 kg of mass), its weight is the same as before, but now the stress it exerts on the table is 9.8 N/(0.01 m)^2 = 98,000 Pa—a much greater pressure because the stress is 'concentrated' over a much smaller surface area.

Cgs or meteorological units for pressure may be easier to relate to; see Table 7.1. In this textbook, with its focus on the Earth System, such units are preferred for describing surface and near-surface situations. The weight of the 1 kg cube is (1,000 gm) × (980 cm/sec^2), that is, 9.8 × 10^5 gm-cm/sec$^2 \equiv 9.8 × 10^5$ **dynes**. If the cube is 1 m on a side, the area in contact with the table is (100 cm)^2, yielding a pressure of 98 dynes/cm^2 on the table. The Earth's atmosphere exerts a pressure at ground level of typically 10^6 dynes/cm^2, so 10^6 dynes/cm^2 is called 1 **bar**. The pressure exerted by the large cube is, therefore, just under 10^{-4} bar or 0.1 mbar.

Can you figure out the pressure due to the small cube? In this case, the weight of 9.8 × 10^5 dynes acts over an area of just 1 cm^2. What is your answer, approximately, in bar? (Does that seem like a lot? Imagine that weight of 2.2 pounds concentrated into a cube 1 cm on a side!)

Note in comparison to the preceding paragraph, by the way, that 1 bar = 10^5 Pa.

As another illustration, we can estimate the pressure within the Earth, halfway down to the Earth's center.

- Crudely taking the density of the outer half of the Earth to be approximately equal to Earth's mean density, 5.5 gm/cm^3 (clearly an overestimate), the mass of the outer half would be $\frac{4}{3}\pi[(6.371 × 10^8$ cm)^3—(3.186 × 10^8 cm)^3], which is its volume, times 5.5 gm/cm^3, thus 5.2 × 10^{27} gm.

- Even more crudely, taking the gravitational acceleration *within* the Earth to be about

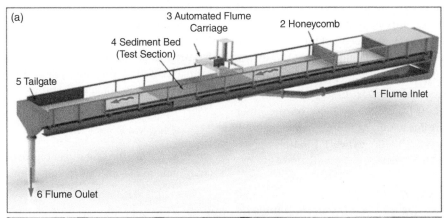

Figure 7.1 Stress in a sedimentological setting.
(a) Illustration of the kind of laboratory equipment—a flume—used for such studies. The direction of the controlled water flow is illustrated by blue arrows. In this case, the erosion around bridge abutments in a river channel was being modeled. Downloaded from https://www.fhwa.dot.gov/publications/research/infrastructure/structures/bridge/17054/003.cfm.
(b) Stress in action. Close-up cross section of a dune created during a flume experiment simulating sediment transport in a river channel; the direction of flow is indicated by a large blue arrow. The cross-section exhibits a cross-stratification, seen as alternating dark and light (fine- and coarse-grained) dipping layers, which results from the migration of dunes (along with superimposed ripples and other smaller bedforms) as points of deposition and erosion vary during the water flow. The surface of interest (the top of a fine-grained layer) is marked with a bold dashed brown line. The stresses experienced by the sediment below that surface are caused by (1) the weight of the sediment above, (2) the weight of the water above the sediment, and (3) forces generated by the flow of the water and transmitted down through the sediment to that surface. Photo courtesy of J. Bridge; taken from Reesink & Bridge (2007) / with permission from ELSEVIER.

Table 7.1 Units for pressure, in the Earth System.

	Meteorological units	cgs units	mks units	
Typical surface pressure	1 bar	10^6 dyne/cm^2	10^5 N/m^2	0.1 MPa
Typical interior pressure	1 Mbar	10^{12} dyne/cm^2	10^{11} N/m^2	100 GPa*

*GPa = gigapascal = 10^9 Pascals.

the same as at Earth's surface, $\sim10^3$ cm/sec^2 (more on this approximation later in the next chapter), the weight of the outer half of the Earth would be roughly 5.2×10^{30} dynes.

- Finally, with the 'inner sphere' having a surface area of $\sim4\pi(3.186 \times 10^8$ cm$)^2$ = 1.3×10^{18} cm^2, the pressure on the inner sphere due to this overburden will be approximately 5.2×10^{30} dynes/1.3×10^{18} cm^2 = 4.1×10^{12} dynes/cm^2 = 4.1 Mbar.

To gain greater familiarity with all the units we will be dealing with in this chapter, you should repeat these calculations (for mass, weight, and pressure) entirely in mks units.

Primarily because of our choice for the typical density of the outer half of the Earth, our deduction of the pressure halfway down into the interior is indeed an overestimate—but only, as it turns out, by a factor of ~2.

One of the most common types of shear stress is *friction*. For example, if you rub your hands together, friction acts along the surface of each hand, *opposing* the rubbing motion. The magnitude of the frictional stress would equal the magnitude of the rubbing force applied, divided by the area of hands in contact with each other.

Note that not all forces act on surfaces. For example, the 1 kg cube considered above experiences the gravitational force of the Earth—giving the cube its weight, 9.8 N—regardless of the size of the cube or the area of cube in contact with the table. Those gravitational forces act on the particles within the cube; the surface area of the cube faces is irrelevant to those forces. Forces such as gravity are called *volume forces* or *body forces*, in contrast to forces that can act on surfaces. Only for surface forces does it make sense to talk about "force per unit area."

7.1.2 Stress: A Rigorous Description

In general, any stress is not likely to be a purely normal stress or a pure shear stress. And even if the stress vector acting on some surface is

resolved into its three (x, y, and z) components, none of those components is likely to be purely normal or shear, either—unless the surface happens to be aligned with or perpendicular to an axis of the coordinate system. That is, if a stress vector is denoted by $\vec{s}$, and we write $\vec{s}$ in terms of its components (s_x, s_y, s_z) as

$$\vec{s} = s_x\,\hat{x} + s_y\,\hat{y} + s_z\hat{z},$$

where $\hat{x}$, $\hat{y}$, $\hat{z}$ are the unit directions of the x-y-z coordinate axes (see Figure 7.2), then neither

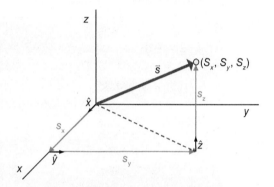

Figure 7.2 The components of a stress vector, and hints of a bigger picture.
If the coordinates of the stress vector $\vec{s}$ are (s_x, s_y, s_z), we can write

$$\vec{s} = (s_x, s_y, s_z).$$

But, with unit vectors $\hat{x}$, $\hat{y}$, and $\hat{z}$ in the x-, y-, and z-directions, we can treat those coordinates as component vectors:

$$s_x\,\hat{x},\ s_y\,\hat{y},\ \text{and}\ s_z\hat{z}.$$

Then, recalling how vectors add (see, e.g., Chapter 4), we know that $s_x\hat{x} + s_y\hat{y}$ will equal the dashed vector; and the dashed vector plus $s_z\hat{z}$ will equal $\vec{s}$. In sum,

$$s_x\,\hat{x} + s_y\,\hat{y} + s_z\hat{z} = \vec{s}.$$

Thus, the x-, y-, and z-components of a vector added together vectorially yield the total vector. This description can apply to *any* vector. But for a stress, whether any of the component vectors (e.g., $s_x\hat{x}$) is a normal or shear stress depends on the *surface* experiencing the stress; in general, we can imagine that surface (with any particular orientation) is located at the point of contact (o). As discussed in the text and illustrated in Figure 7.3, isolating the normal and shear stresses will require a more fundamental decomposition, one relating to that surface.

$s_x \hat{x}$, $s_y \hat{y}$, or $s_z \hat{z}$ is, in general, a normal or shear stress.

Nevertheless, it is possible to decompose any stress vector into 'components' that are *all* either normal or shear! The key is to perform the decomposition, not relative to fundamental directions (such as $\hat{x}$, $\hat{y}$, and $\hat{z}$), but rather with respect to fundamental *surfaces*—perhaps not an unreasonable approach since stresses derive from surface forces. For such a decomposition, though, we will first need to describe the orientation of any given surface. The traditional approach is to define the **orientation of a surface** as the direction of the perpendicular to the surface; see Figure 7.3a. Since the only property we are interested in concerning that perpendicular is its direction, we use vectors of unit magnitude; as in the case of unit vectors like $\hat{x}$, $\hat{y}$, and $\hat{z}$, we denote the orientation of a surface, in general, by the unit-magnitude normal vector $\hat{n}$.

We focus on a very small region of the material under stress, and—as indicated in Figure 7.3b—imagine the stress vector $\vec{s}$ to be acting on a surface (bounded by bold lines in the figure) with some particular orientation $\hat{n}$. To decompose our stress vector, we envision surfaces that are perpendicular to the $\hat{x}$-, $\hat{y}$-, and $\hat{z}$-directions, and create a right-angled tetrahedron based on those three surfaces, with the surface experiencing our stress vector constituting the fourth, front face of the tetrahedron. That front face is squeezed and twisted by $\vec{s}$; but *those effects can also be produced by a combination of stresses acting on the three side faces of the tetrahedron*—perhaps, as in Figure 7.3b, where $\vec{s}$ is coming from above and to the side of $\hat{n}$, a mostly normal,

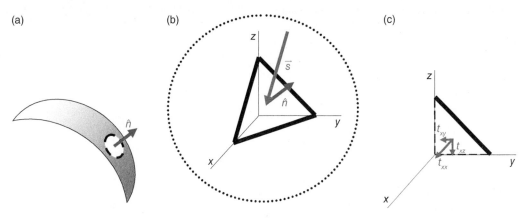

Figure 7.3 General analysis of stress.
(a) The orientation of any surface, for example the circular area within a larger curving wedge-shaped region, can be represented by the perpendicular direction to that surface. By convention, we describe that perpendicular using a vector of unit magnitude, denoted $\hat{n}$, directed perpendicularly *outward*, away from the surface experiencing the stress.
(b) Now focusing in on a small volume within a solid material, we identify three fundamental surfaces, defined by $\hat{x}$, $\hat{y}$, and $\hat{z}$ (i.e., these surfaces are perpendicular to the $\hat{x}$, $\hat{y}$, and $\hat{z}$ directions), which comprise the side faces of a right-angled tetrahedron; and we consider the stress vector $\vec{s}$ acting on the fourth (tilted) surface of the tetrahedron, a surface whose orientation is given by $\hat{n}$. In the situation shown here, $\vec{s}$ and $\hat{n}$ are not parallel (or antiparallel), suggesting that $\vec{s}$ will be exerting both normal and shear stresses on this surface.
In practice, we may be given this stressed surface to begin with; in that case, $\hat{n}$ will determine the sizes of the side faces.
(c) Finally, we focus on one side face of the small volume in (b), the face perpendicular to the $\hat{x}$ direction. The stress acting on that face, ultimately a result of $\vec{s}$, was denoted $\vec{t}_x$; its $\hat{x}$, $\hat{y}$, and $\hat{z}$ components are denoted t_{xx}, t_{xy}, and t_{xz}, respectively. With respect to this side face, those components act as one normal and two shear stresses.
Analogous decompositions apply to the other side faces.

compressive stress acting downward on the $\hat{z}$ face, also a shear stress in the $-\hat{y}$-direction, plus smaller stresses on the $\hat{y}$ face and even smaller stresses on the $\hat{x}$ face. If we denote those three stresses by $\vec{t}_x$, $\vec{t}_y$, and $\vec{t}_z$ (with "t" for traction)—for example, $\vec{t}_x$ will be the stress acting on the face perpendicular to the $\hat{x}$-direction—then such an equivalence is expressed as

$$\vec{s} = \vec{t}_x + \vec{t}_y + \vec{t}_z,$$

amounting to a decomposition by surface rather than by direction.

There are two reasons that such a decomposition is useful. First, when each of those t-vectors is resolved into x-, y-, and z-components, the geometry of the situation implies that one of those components will be a normal stress, and the other two will be shear stresses; as shown in Figure 7.3c, for example, whatever the direction of $\vec{t}_x$, its x-component $(t_x)_x$—or, simply, t_{xx}—will be in the direction that is perpendicular to the $\hat{x}$-face, and its y- and z-components t_{xy} and t_{xz} will be in directions that are *along* the $\hat{x}$-face.

The collection of all nine components of the three t-vectors,

$$\begin{pmatrix} t_{xx} & t_{xy} & t_{xz} \\ t_{yx} & t_{yy} & t_{yz} \\ t_{zx} & t_{zy} & t_{zz} \end{pmatrix}$$

traditionally denoted $\underline{\underline{t}}$, thus contains three normal stresses (t_{xx}, t_{yy}, and t_{zz}) and six shear stresses. This 3×3 collection of stress components is called the **stress tensor**. Tensors can be thought of as 'higher-order' vectors (for example, stress vectors are only 1×3 collections of stress components).

As stated earlier, seismology is concerned with the responses of materials to applied stresses. Materials respond differently to normal stresses than to shear stresses; so it is important to be able to decompose the applied stresses into those two types.

The second reason decomposing the stress vector into a stress tensor is useful is that the elements of the tensor can be recombined to produce the stress vector acting on *any*

surface, not just the tetrahedral front or side faces of Figures 7.3b and 7.3c. This is achieved employing a simple formula: if the surface of interest has an orientation given by a unit perpendicular $\hat{n}$, then the x-, y-, and z-components of the associated stress vector are

$$s_x = t_{xx}\, n_x + t_{xy}\, n_y + t_{xz}\, n_z$$
$$s_y = t_{yx}\, n_x + t_{yy}\, n_y + t_{yz}\, n_z$$
$$s_z = t_{zx}\, n_x + t_{zy}\, n_y + t_{zz}\, n_z$$

where the x-, y-, and z-components of $\hat{n}$ are (n_x, n_y, n_z) and the elements of the stress tensor t_{xx}, t_{xy}, ... are unchanged. That is, at any location within the Earth, once the stress vectors (and thus the stress tensor) associated with the right-angle tetrahedron have been specified, the state of stress there is completely known—the stress vector acting on a surface with any arbitrary orientation can be determined. At each location, $\vec{s}$ may be different depending on the surface's orientation; but $\underline{\underline{t}}$ is unchanged.

(If you are familiar with linear algebra, it will be easy to show that the preceding formula can be written as

$$\begin{pmatrix} s_x \\ s_y \\ s_z \end{pmatrix} = \begin{pmatrix} t_{xx} & t_{xy} & t_{xz} \\ t_{yx} & t_{yy} & t_{yz} \\ t_{zx} & t_{zy} & t_{zz} \end{pmatrix} \cdot \begin{pmatrix} n_x \\ n_y \\ n_z \end{pmatrix}$$

or, more compactly, as

$$\vec{s} = \underline{\underline{t}} \cdot \hat{n},$$

where the 'dot-product' between the matrix $\underline{\underline{t}}$ and the unit vector $\hat{n}$ follows the rules of matrix multiplication.)

7.1.3 Strain

Strain is the internal deformation or distortion of an object as the result of applied forces. Strain may take the form of an expansion or a contraction, a twisting or a bending. Strain is *dimensionless*, i.e. it is measured without units: **strain** *is the fractional change in dimensions within an object.*

For example, if a bar of taffy originally 20 cm long is pulled until it has become 30 cm long, the amount of strain in the direction of its length is

$$\frac{30 \text{ cm} - 20 \text{ cm}}{20 \text{ cm}} = \frac{10}{20} = 0.5 \text{ or } 50\%.$$

As the taffy is stretched, it gets thinner; if it was originally 9 cm wide and thins to 6 cm, the strain in the direction of its width is

$$\frac{6 \text{ cm} - 9 \text{ cm}}{9 \text{ cm}} = -\frac{3}{9} = -0.33 \text{ or } -33\%.$$

Two special types of strain are worth noting: *longitudinal strain* (also called *axial strain*) and *shear strain*—with obvious connections to the stresses that most directly cause them. These are idealized in Figure 7.4 with a cross-sectional view of a bar of some material under various stresses. In all cases, the strain is measured as the fractional change in dimensions of the bar. One type of longitudinal strain, an extension (Figure 7.4a, also called an expansion or dilation; the opposite would be a contraction), can be produced by applying a tensional stress to the $\hat{x}$-face of the bar. With the length taken to be in the x direction, the

original length x_0, and the increase in length Δx, the strain is

$$\frac{\Delta x}{x_0};$$

often this strain is denoted by e_{xx}. Both of the examples of taffy deformation mentioned above are longitudinal strains (perhaps e_{xx} and e_{yy}).

Shear strain can be created in the bar, as shown in Figure 7.4b (left-hand picture), by applying a shear stress to its top face (the $\hat{z}$-face). With the points on the top face shearing to the right by an amount Δx, and z_0 the original length in the z-direction, the shear strain is traditionally defined as

$$\frac{1}{2}\frac{\Delta x}{z_0};$$

tentatively, we might denote this strain by e_{xz}. The notation and the factor of $\frac{1}{2}$ will be discussed shortly.

Shear strain changes the angular configuration of the object. In Figure 7.4b (left), for example, the adjacent faces of the bar make

(a) Longitudinal Strain

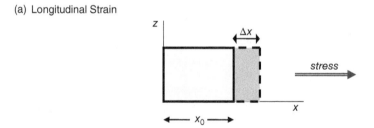

(b) Shear Strain

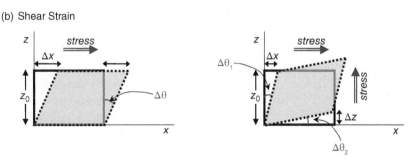

Figure 7.4 Idealized types of strain, applied to a bar of material (extending into and out of the page) whose cross-section is shown here. (a) In response to a tensional normal stress acting on its right-hand face, the bar expands to the right. (b) In response to a shear stress acting along its top (left-hand picture) or top and right-side (right-hand picture) faces, the bar shears to the right or up and to the right. See text for details.

a 90° angle before the shear stress is applied; afterward, adjacent faces intersecting at the origin make a smaller angle—smaller by an amount $\Delta\theta$. Because $\tan(\Delta\theta) = \Delta x/z_0$, we can write that shear strain as

$$e_{xz} \equiv \frac{1}{2}\frac{\Delta x}{z_0} = \frac{1}{2}\tan(\Delta\theta) \cong \frac{1}{2}\Delta\theta;$$

the approximation that $\tan(\Delta\theta)$ equals $\Delta\theta$ (with the angles measured in radians) is generally an excellent one because in most real-world situations, the strain experienced by rocks is quite small.

From our analysis so far, we can already discover some differences between longitudinal and shear strain. Shear strain reflects angular distortions, whereas longitudinal strain measures linear deformation. And, as inspection of Figure 7.4 will confirm, the material changes volume when there is longitudinal strain but not when there is shear strain. We can verify these differences via everyday experiences. If the bar of material is replaced with a deck of cards, for example, application of a stress *along* the top of the deck (cf. Figure 7.4b, left-hand side) would shear the cards out but leave them intact; in order to decrease their volume (e.g., crush them), we would have to apply a pressure acting downward on the top of the deck, and that would create a longitudinal strain.

Even under idealized conditions, shear strain need not be as simple as we have considered so far. As shown in Figure 7.4b (right-hand sketch), a 'double stress,' i.e., a shear stress applied to both the top and side faces of the bar, with the bottom not held fixed, can lead to shearing in both the x- and z-directions. If the right-hand top and bottom corners of the bar cross section are displaced upward by an amount Δz as a result of the shearing, the *net* shear strain—still denoted e_{xz}, by the way, though in this case it just as clearly might be denoted e_{zx}—will be

$$e_{xz} \equiv \frac{1}{2}\left(\frac{\Delta x}{z_0} + \frac{\Delta z}{x_0}\right).$$

One explanation for the factor of $\frac{1}{2}$ included in the traditional definition of shear strain is that it acts to *average* the x-z strain and the z-x strain.

The angular strain is the same in the doubly sheared case as in the partially sheared situation. Denoting (see Figure 7.4b) the change in the vertical orientation of the bar by $\Delta\theta_1$ and the change in its horizontal orientation by $\Delta\theta_2$, we have for very small strain

$$\Delta\theta_1 \cong \frac{\Delta x}{z_0} \quad \text{and} \quad \Delta\theta_2 \cong \frac{\Delta z}{x_0},$$

leading to

$$e_{xz} = \frac{1}{2}(\Delta\theta_1 + \Delta\theta_2) \equiv \frac{1}{2}\Delta\theta,$$

where $\Delta\theta$ is the total angular distortion of the bar (the total decrease in the corner angle). Note that, even in this more complicated deformation, there is no change in the volume of the bar.

How good an approximation is it to say that the strain in real-life situations will be small? The strain associated with the ground motion produced by earthquakes can be measured or inferred; based on the displacements ('slip') of the largest earthquakes, the strain near the fault is estimated to be typically $\sim 10^{-4}$, that is, 0.01% (e.g., Stacey & Davis, 2008; Turcotte & Schubert, 2002), though conditions may be significantly nonuniform along the fault (cf. Holdahl & Sauber, 1994). In near-source environments, then, our small-strain approximation might not be very accurate. But far from the source the story would be different: the waves of deformation, called **teleseisms**, radiated worldwide by an earthquake generate longitudinal strains in the range 10^{-6}–10^{-12}. And, as a different type of measure, the ongoing rate of shear strain (i.e., not associated with specific earthquake events) along California's San Andreas fault has been estimated to be $\sim 10^{-6}$/year (though it varies both spatially and temporally; e.g., Savage et al. 1981).

The solid earth also strains in a variety of ways not related to seismic activity. When an aquifer dewaters, for example, the pore pressure available to support the weight of the overlying rock is reduced, allowing that portion of the aquifer to further compress and the surface to subside. Or, strain is produced when an ice sheet retreats and the underlying lithosphere flexes upward.

Figure 7.5 Tidal strain. In response to tidal forces, the solid earth assumes a bulging shape. Here the height of the body tide at the peak of the bulge is Δr, and Earth's radius is r.

Earth tides are still another example of solid earth deformation. As we discussed in Chapter 6, the solid earth takes on an ellipsoidal shape as a consequence of lunisolar tidal forces (see Figure 7.5), with the high tide having an amplitude of ~30 cm. That is, the Earth's radius increases by 30 cm at the peaks of the tidal bulge. It follows that the radial (longitudinal) strain caused by earth tides is $\Delta r/r = (30 \text{ cm}) / (6.371 \times 10^8 \text{ cm}) = 4.7 \times 10^{-8}$.

It should be noted that this tidal deformation also generates other types of strain. As the bulge peaks at any location, adjacent locations experience shear strain and the surface tilts. And, the load of the ocean's high tide compresses the solid earth, leading to additional axial strains, shear strains, and surface tilting, especially at coastal locations. Such load strains are made more complex because the ocean tide, as noted near the end of Chapter 6, is likely to be 'out of sync' with the body tide, due to tidal friction and perhaps additional dynamic effects.

7.1.4 Strain: A Rigorous Description

Because strain is a measure of the differential motion of particles within a material, it should not be surprising to learn that *calculus* provides a natural way to quantify strain. We begin by considering two particles in an object that is initially undeformed, and set up a coordinate system that will allow us to measure the particles' movements. One of the particles is initially located at position $\vec{R}$ relative to the origin, and the second particle is initially a distance L from the first (see Figure 7.6a). As a result of applied forces, the object may be displaced and/or deformed; we denote by $\vec{u}$ the displacement of the first particle. Following its displacement, that particle will be located at position $\vec{r}$ with respect to the origin, where

$$\vec{r} = \vec{R} + \vec{u}$$

(see Figure 7.6b). For later use, we note in particular that the vectors $\vec{u}$ and $\vec{r}$ could be resolved into x-, y-, and z-components (u_x, u_y, and u_z) and (r_x, r_y, and r_z).

The second particle will also have been displaced by the applied forces, but not necessarily by the same amount or in precisely the same direction as the first particle. A simple example of such differential displacement would be when a kitchen sponge is squeezed or a spongey football is kicked. And most fluid flow, including the flow through a pipe described in Chapter 5, involves shearing and thus differential displacement. So, we will allow for the likelihood that the second particle is displaced by an amount $(\vec{u} + d\vec{u})$ instead of $\vec{u}$; see Figure 7.6b. Here we write $d\vec{u}$

(a) Before forces are applied

(b) After forces have been applied

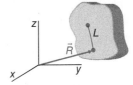

Figure 7.6 Rigorous description of strain.
(a) Before forces are applied to an object, one of its particles is located at position $\vec{R}$ with respect to the coordinate origin. A second particle is a distance L from the first.
(b) After the application of forces, the object may have been displaced and/or deformed. If the displacement of the first particle is $\vec{u}$, then it is now at position $\vec{r}$, where $\vec{r} = \vec{R} + \vec{u}$. As illustrated, the second particle is now a distance ℓ from the first. See the text for remaining definitions.

rather than $\Delta\vec{u}$ in recognition of the situation commonly treated in seismology where the strain is very small—"infinitesimal" rather than "finite" strain. As a result of its additional displacement, the second particle may now be a distance ℓ from the first particle. Following our original discussion of strain, we could define the strain within this object to be

$$\frac{\ell - L}{L},$$

the fractional change in its dimensions.

But such a definition is limited, telling us only how much longitudinal strain the object suffered, and only with reference to those two particles, at the instant their separation equals ℓ. A more informative quantity might be $d\vec{u}/d\vec{r}$, at least symbolically (there is no way to divide one vector by another, because directions cannot be divided). The denominator, $d\vec{r}$, could be interpreted—depending on the viewpoint of the user—as measuring the changing position of the first particle as the deformation continues, or a change in the selection of the first particle (any particle, at any position $\vec{r}$, can be chosen as the first); the ratio $d\vec{u}/d\vec{r}$ gives us the additional displacement of the second particle relative to that change—thus, some measure of the interior strain.

Even better, we can consider

$$\frac{du_i}{dr_j},$$

where the subscripts i and j refer to x-, y-, or z-components of the vectors $d\vec{u}$ and $d\vec{r}$. This set of 'fractions' is known as the **displacement gradient**, and provides a way to evaluate the full three-dimensional variety of strains in the object's interior. For example, if $i = y$ (if we want to choose the y-component of $d\vec{u}$) and $j = x$ (for the x-component of $d\vec{r}$), this fraction would be du_y/dr_x, and would represent (in some way) the shear strain associated with the additional displacement of the second particle in the y-direction relative to the changing x-position of the first particle. Clearly, the displacement gradient includes much more

information than the simple, 'linear' fractional change $(\ell - L)/L$.

(—Perhaps *too* much information! du_y/dr_x, du_y/dr_y, and du_y/dr_z, for example, are all meaningful quantities; that is, we should expect that u_y—or, for that matter, $\vec{u}$ itself—is a function of all three variables (r_x, r_y, and r_z). But that means we should be writing "partial" derivatives for the displacement gradient. We will reserve such sophistication for later chapters in this textbook.)

If strain is the response of a material to being stressed, we might expect the stress and strain to be proportional to each other in some way. And if a stress *tensor* is the most rigorous way to describe stress, it would be reasonable to imagine that there is a tensor way to describe strain as well. The displacement gradient is a 3×3 collection of terms (since each of i and j can have three values); and it turns out that it is a true tensor as well. But the stress tensor is symmetric, in that $t_{ij} = t_{ji}$; for example, $t_{yx} = t_{xy}$ (if it wasn't, then 'tractions' on adjacent side faces of the tetrahedron in Figure 7.3b would not balance each other, and the tetrahedron would spin chaotically under stress—violating the law of angular momentum conservation). We can construct a **symmetric strain tensor**, which will now officially be denoted $\underline{\underline{e}}$, from the displacement gradient by simply defining its ijth element as

$$e_{ij} = \frac{1}{2}\left(\frac{du_i}{dr_j} + \frac{du_j}{dr_i}\right),$$

where i and j can each be selected from x, y, and z.

We will explore the relationships between stress and strain in the next section. Here it is merely noted that e_{xx}, e_{yy}, and e_{zz}, the diagonal elements of the strain tensor, correspond to the simple longitudinal strains described and depicted above (for example, e_{xx}, which equals du_x/dr_x, is similar to $\Delta x/x$). They can most easily be produced by the application of normal stresses. And, the 'off-diagonal' elements of $\underline{\underline{e}}$, such as e_{zx}, correspond to the shear strains discussed above—and can be produced directly by shear stresses.

7.2 Relations Between Stress and Strain in Elastic and Nonelastic Materials

We are now ready to discover what makes one material different than another. Very simply, materials are defined ideally by their response to applied stress.

- **Elastic materials** have the ability to return immediately to their initial, unstrained state after the stress is removed. A common example of elastic materials is a rubber band; if you stretch out a rubber band, it will 'snap back' once you let go. Most solid materials exhibit elastic behavior *if* the applied stresses are small enough—an assumption that is frequently made.

- **Elastic-plastic** or **ductile materials** behave elastically when a small stress is applied, but as the stress is increased beyond a critical point, they will deform permanently and continuously, exhibiting a plastic 'flow.' A familiar example of plastic materials is taffy candy, which appears to be a normal (elastic) solid unless you try to pull it apart; if you pull hard enough, it will begin to stretch out and thin. The minimum stress large enough to produce plasticity (ductility) is called the **finite strength** of the material.

- **Elastic-brittle materials** behave elastically in response to small stresses, but break once they have been deformed to a critical point; that critical point of strain is called the **elastic limit**. Interestingly, at cold temperatures taffy will behave brittlely.

The behavior of these types of materials is shown schematically in Figure 7.7.

Rocks that we find on or near Earth's surface behave as elastic-brittle solids; under sufficiently great regional tectonic stresses, for example, the crust may fail at *fault zones*. Deeper within the Earth, however, rocks exhibit the ability to flow.

Why do you think this should be?

Although a number of factors, such as composition and especially water content (discussed in the next chapter), may play a role, geophysicists also point to the physical conditions

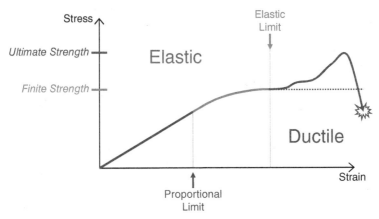

Figure 7.7 Schematic illustration of stress versus strain in some materials, all of which behave elastically for small stress. The reason stress is plotted on the *y*-axis is discussed later in the text.
For small enough stress levels, these materials will be *linearly elastic*, i.e., the stress is linearly proportional to the strain; when the resulting strain exceeds the *proportional limit*, the relation between stress and strain may become nonlinear, so that (e.g.) increasing the stress a bit more causes a lot more strain. When the strain reaches the *elastic limit*—which occurs when the stress is high enough to exceed the material's *finite strength*—plastic flow begins for ductile materials; for brittle substances (not shown), material failure occurs at this point. For many elastic-plastic materials, failure will occur later, when the stress exceeds the material's *ultimate strength*; for perfectly plastic materials (ideal ductility, shown by the dotted line), failure does not occur.

experienced by the rocks: the higher temperatures, which can soften the material and make it easier for it to flow; and the greater pressures, which confine the rock and discourage any increase in its volume (when rocks fracture, their net volume—rock plus new void space—is greater than before).

Figure 7.7 illustrates the behavior of materials that are elastic, at least under conditions of low stress. But some materials do not behave elastically even when the applied stress is tiny.

- *Fluids* will never return to their unstrained state when the applied stress is relaxed, if that stress is a *shear* stress. The shear stress created by a wind blowing over a lake, for example, will cause the surface waters of the lake to begin to flow, producing a shear strain in those waters. As long as the wind stress continues, the flow will continue—at a rate that depends on the lake water's resistance to shearing, that is, on its viscosity—and the shear strain within the surface layers will build up. But when the wind dies away and the stress vanishes, the flow will not reverse, even after it has stopped, and the accumulated strain will not be undone.

- *Viscoelastic materials* display a combination of elastic and fluid behavior. They may exhibit an instantaneous response to the applied stress, as elastic materials do, and a response that, fluid-like, grows as the stress is applied for a longer span of time. In general, some permanent deformation remains in viscoelastic materials once the applied stresses have been removed.

7.2.1 Ideal Models of Different Materials

Simple physical models of various types of materials can help us understand how material responses to stress differ, and also suggest how the stress-strain relations can be described mathematically. In all of these models, illustrated in Figure 7.8, we consider the applied stress to be an impulse that begins at time t_0, continues at constant amplitude for a duration

Δt, and then drops instantaneously back to zero. Symbolically, but without focusing on details such as the types of stress and strain involved in these models, we will denote the stress by S and the strain by e.

- **Elastic Materials.** The traditional model of an elastic material is a spring, for example with one end connected to a wall and the other end subject to the applied stress S (caused in this model by tensile or compressive forces). The resulting strain e is measured by the (fractional) amount the spring is stretched or compressed. The behavior of the spring in general is described by some kind of proportionality between stress and strain:

$$S \propto e^n$$

where the exponent n allows for some degree of nonlinearity (e.g., if $n = 2$); or, more simply,

$$S \propto e$$

for linear elasticity. With either formulation, the strain vanishes when the stress is removed, so there is no permanent deformation, as required for elastic materials.

What do these formulas imply about the strain if the stress is held constant? Is your answer consistent with the graph in Figure 7.8b?

We can make these relationships more useful by inserting proportionality constants; the behavior of linearly elastic materials, for example, would be described by

$$S = Ce,$$

where C is an **elastic 'constant,'** more commonly called an **elastic 'modulus'** or **'parameter.'** For the spring model of an elastic material, C would represent the *stiffness* of the spring: the larger C is, the stiffer the spring, and the more stress it would take to achieve a given strain.

In physics classes, we learn that the behavior of springs is governed by **Hooke's Law**, which says that when the spring is pulled out, the force with which it pulls back, attempting to restore its original state, is proportional to the amount by which the spring had been stretched (in fact, the proportionality constant between that force and the stretch is officially defined as the spring's stiffness). Clearly, the amount of stretching relates to the strain experienced by the spring, and the restoring force corresponds to the applied stress which caused the deformation in the first place. So the linear relation written above is equivalent to Hooke's Law (and C relates directly to the stiffness). Finally, following Hooke's lead, we have written the relation as $S \propto e$ rather than $e \propto S$.

We have seen in this chapter that there are various types of stresses that can be applied to a material, which in turn can deform in various ways. Hooke's Law characterizes all elastic materials, but it is too simple, referring only to a tensile normal stress applied to a material that can deform in only one 'longitudinal' way. Because of the complexity of real, three-dimensional masses, in geophysics we are compelled to invoke a **generalized Hooke's Law**: *every component of stress is proportional to every component of strain*; in terms of the tensors described earlier, we could write

$$t_{ij} \propto e_{kl},$$

where each of the indices $i, j, k,$ and l can correspond to the x-, y-, or z-direction. Furthermore, for each combination of indices there can be a proportionality constant, so the generalized Hooke's Law can be expressed as

$$t_{ij} = C_{ijkl}\, e_{kl}.$$

Since there are $3 \times 3 = 9$ possible elements of each of the two tensors, there are conceivably $(3 \times 3) \times (3 \times 3) = 81$ different proportionality constants; because those tensors are symmetric, though, there will actually be fewer distinct proportionality constants than 81! Later in this chapter, we will identify specific proportionality constants which relate certain kinds of stress and strain.

- **Viscous Materials.** We model the stress-strain behavior of a fluid using a 'dashpot,' which contains the fluid, and a plate within the dashpot to generate the stress—a shear stress, as the fluid is forced to flow around the plate when the plate is displaced. The plate is pulled for a time interval Δt, to create the stress impulse shown in Figure 7.8a; the strain is measured by the amount by which the plate displaces. The plate must be pulled more and more to maintain a fixed level of stress—the condition illustrated in Figure 7.8a—so the deformation will continue to grow: the fluid within the dashpot will continue to flow around the displaced plate, allowing the plate to displace further. This behavior is in contrast to that of a spring, where a constant level of stress maintains the spring's initial extension, but no more—the spring does not stretch beyond that.

The greater the applied stress (i.e., the more strongly the plate is pulled), the faster the plate will move through the fluid, and the faster the strain will build up. It follows that for viscous materials, *the stress is proportional*, not to the strain produced, but *to the rate of strain*.

This is also suggested by the graph in Figure 7.8b for a viscous material, in which the slope of the strain 'curve' (i.e., the slope de/dt of the linear segment beginning at time t_0) would be greater—the strain would increase faster—if the stress were greater. We can express this dependence as

$$S \propto \frac{de}{dt} \quad \text{or} \quad S \propto \dot{e},$$

where t denotes time and the superscript dot is shorthand for a time derivative.

This proportionality applies to all fluids, but the proportionality constant depends on which fluid is being stressed: the actual amount of strain built up (or fluid displaced) depends on the ease with which the fluid can shear around the edges of the plate; that is, it depends on the fluid's *viscosity*.

(a)

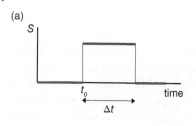

(b)

Elastic

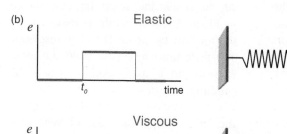

Viscous

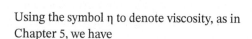

Figure 7.8 Ideal models of different materials, characterized by their strain response, e, to an applied impulse of stress, S.
(a) The applied tensile stress is graphed versus time as an impulse beginning at time t_0 and lasting for time Δt.
(b) The resulting strain versus time is graphed on the left, and the ideal spring or dashpot model of the material type is shown to the right of the graph, for elastic and viscous fluid classifications.

Using the symbol η to denote viscosity, as in Chapter 5, we have

$$S = \eta \frac{de}{dt} \quad \text{or} \quad S = \eta \dot{e}.$$

This relation, which implies that in higher viscosity fluids, a greater amount of stress will be required to achieve a given rate of strain, is actually equivalent to the formula presented in Chapter 5; that formula employed the velocity gradient—which is analogous to strain rate—to define viscosity as a resistance to shearing.

From the dashpot model or the graph depicted in Figure 7.8, we see that there is always permanent deformation in a fluid after the applied stress is removed: when the stress moving the plate is relaxed, the plate will stop, but not return to its initial position.

- **Maxwell Viscoelastic Materials.** Modeled as a spring and a dashpot connected "in series," that is, in a direct line with each other, it is easy to imagine how this type of material—called a "Maxwell solid" for short—deforms: when the tensile stress is first applied, the dashpot remains rigid, but the spring immediately stretches elastically;

then, as the stress is maintained, the plate gradually pulls outward. When the stress is removed, the viscous deformation simply stops; but there is an instantaneous relaxation by the spring. The net effect, as indicated on the graph in Figure 7.8c, is a permanent deformation of the system.

Mathematically, deriving the governing equation is straightforward once it is realized that both the spring and the dashpot must be experiencing the same stress S. For the spring, we had found $S = Ce$, which can be written

$$\dot{S} = C\dot{e}$$

by differentiating both sides of the equation (as before, "˙" signifies time derivative); the spring's rate of strain is thus $\dot{S}/C$. Under the same stress S, the strain rate for the dashpot is S/η. The total strain rate—which we will now denote by $\dot{e}$—is the sum of the spring's strain rate and the dashpot's; that is,

$$\dot{e} = \frac{\dot{S}}{C} + \frac{S}{\eta}.$$

- **Kelvin-Voigt Firmoviscous Materials.** Modeled as a spring and a dashpot connected "in parallel" (see Figure 7.8c), the deformation of this type of material is a bit more subtle. In parallel, the dashpot is able to resist the deformation more effectively. When a tensile stress is applied to the apparatus, the spring portion would tend to immediately stretch out; but it can't, and there is no displacement of the system at first, because the dashpot cannot respond instantaneously. As the dashpot plate begins to displace to the right, the spring pulls on it less. This would reduce the stress experienced by the dashpot (and immediately slow its rate of strain). To maintain the stress, as per the conditions of this 'experiment' (Figure 7.8a), the apparatus must be pulled further to the right. The strain of the spring and that of the dashpot 'parallel' each other throughout the experiment.

Given enough time, the dashpot plate will have almost displaced enough to allow the spring to stretch out fully, i.e. to stretch out as much as it would have if the dashpot had not been present, and a new equilibrium is approximately attained. (As illustrated by the graph in Figure 7.8c, the displacement or strain *levels off* to this limiting amount—a contrast to Maxwell solids (especially if Δt is

large), where the strain keeps growing at a constant rate until the impulse ends.)

When the applied stress is removed, the system will strain, but in reverse: as long as the dashpot plate has not returned to its initial position, the spring is still stretched; as the spring tries to contract back to its original length, it will pull the dashpot plate back in. This process slows as the plate gets closer to its initial position because the spring is stretched less, so exerts a smaller restoring force on the dashpot. Curiously, despite possessing a viscous component, Kelvin-Voigt materials end up (after the applied stress has been removed) with no permanent deformation, though it takes an infinite amount of time for them to fully regain their initial state.

Mathematically, deriving the governing equation is fairly straightforward. Because of the 'parallel' nature of this system, it is now the strain e that must be the same for the spring and dashpot at any time. For the spring, the stress corresponding to that strain is given by $S = Ce$. The stress generated for the dashpot is by definition $S = \eta \dot{e}$.

Figure 7.8 (c) Strain (e) versus time is graphed on the left, and the ideal spring/dashpot model of the material type is shown to the right of the graph, for Maxwell, Kelvin-Voigt, and "standard linear solid" classifications. In the standard linear solid, the two springs can have different stiffnesses.

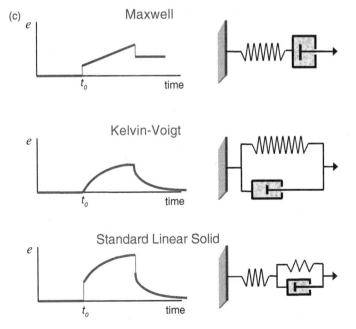

If S now denotes, instead, the total stress on the material, it follows that

$$S = Ce + \eta\dot{e}.$$

By the way, describing viscoelastic types of materials using the "parallel" and "series" terminology of electrical circuits is not just a coincidence. The analysis of electrical circuits using Ohm's Law (voltage = current × resistance) depends on recognizing that the voltage is the same in parallel circuits, whereas the current is the same throughout a series circuit. Wide-ranging analogies exist with 'mechanical circuits,' formed by combinations of springs and dashpots like those considered here, and such analogies can be exploited to mathematically describe the most ornate spring-dashpot combinations; one easy-to-read but rigorous presentation of the subject can be found in Mase (1969).

- **Standard Linear Solid.** The relatively simple mechanical models we have explored to this point—spring, dashpot, spring and dashpot in series, and spring and dashpot in parallel—constitute the four basic *elements* of any mechanical model of material behavior; merely by combining these elements in series or parallel, we can create complex and potentially more realistic material models. One popular example of such models is the standard linear solid, which combines spring and Kelvin-Voigt elements in series (Figure 7.8c); it may be viewed as a Maxwell-type solid but with the dashpot replaced by a Kelvin-Voigt element. The end result of this combination is that a standard linear solid behaves sort of elastically on very short and very long timescales but nonelastically on intermediate timescales. That is, after stresses have been applied for very short times, there is little strain generated in the dashpot, and the overall response is nearly that of the 'solo' spring; if the stress can persist for a long-enough time, the K-V spring can

stretch almost fully, and the overall response approaches that of a sum of the two springs.

How do you think the graph in Figure 7.8c for a standard linear solid would be altered if Δt was very large?

Formulating the governing equation for a standard linear solid combines aspects of the preceding derivations for elastic, Maxwell, and Kelvin-Voigt materials.

- The 'solo' spring element here must be experiencing the same stress, S, as the K-V element it is in series with. That stress will cause the spring to strain by an amount e_1 (with subscript "$_1$" for the solo spring) where $S = C_1 e_1$; equivalently,

$$\dot{S} = C_1 \dot{e}_1,$$

so the spring's rate of strain is equal to $\dot{S}/C_1$.

- As with the Maxwell solid (whose elements are also 'in series'), the total strain e experienced by the standard linear solid must equal the sum of the strains of the elements comprising it:

$$e = e_1 + e_2,$$

where the strain within the K-V element is denoted by e_2. Thus, the total strain *rate* must equal the sum of the elements' strain rates,

$$\dot{e} = \dot{e}_1 + \dot{e}_2,$$

which we can write as

$$\dot{e} = \dot{S}/C_1 + \dot{e}_2$$

after substituting in for the solo spring's strain rate.

- For the K-V element, whose material properties include a spring of stiffness C_2 and a dashpot with viscosity η, the stress-strain relation is

$$S = C_2 e_2 + \eta\dot{e}_2,$$

which in turn implies

$$\dot{e}_2 = (S - C_2 e_2)/\eta.$$

Since

$$e_2 = e - e_1 = e - (S/C_1),$$

the K-V strain rate can be rewritten as

$$\dot{e}_2 = (S - C_2[e - (S/C_1)])/\eta.$$

- Finally, we obtain the governing equation—the stress-strain relation—for the standard linear solid by substituting this last expression into our expression for the total strain rate, and grouping strain terms (e) on the left and stress terms (S) on the right; we find

$$C_2 e + \eta\,\dot{e} = {}^1\!/_{C_1}(\eta\dot{S} + [C_1 + C_2]S).$$

Why do we care how to represent Earth materials? If we can deduce the types of materials that exist in different regions of the Earth, we will be better able to predict their future response to current stresses, or infer past stresses that might have been responsible for their observed deformations. The variety of stresses that act in the Earth System derive mainly from a relatively small number of types of forces: convective, surface loading, tidal, plate tectonic, pressure and wind forcing, all illustrated in earlier chapters of this textbook; and geomagnetic, which will be discussed in a later chapter. Some of these stresses are present throughout much of the Earth, while others act mainly at the surface; some vary on geologic timescales, whereas others vary almost minute to minute.

In response to such a variety of stresses, how well will spring-and-dashpot models describe the behavior of the real Earth? Figure 7.9 shows a number of examples of the real Earth's deformational response to different stresses; *see if you can identify which ideal type of material best represents each situation*.

These examples reinforce a conclusion we could have drawn from our prior geophysical knowledge or even our everyday experience: Earth materials subjected to a set of physical conditions over some timescales might behave like one of our ideal models; but the same materials under different conditions or timescales may exhibit behavior more characteristic of another model. For example, treating the solid earth as purely elastic may be reasonable for short timescales under conditions of low-amplitude stress; but we know that the solid earth can exhibit longer-term fluid-like behavior as well.

Real-World Applications of These Models. Research into Earth's behavior has employed all of the ideal models we have discussed, and more. For studying the viscous flow in the mantle accompanying postglacial rebound, treating the mantle as a Maxwell solid has by far been the most popular approach (e.g., Peltier 1974, Nakiboglu & Lambeck, 1980, Sabadini et al., 1982, Yuen et al., 1982, Wu & Peltier 1984, Nakada & Lambeck 1989, Han & Wahr 1995, Milne & Mitrovica 1998, Mitrovica et al. 2001, Kendall et al. 2006, Spada et al. 2006). In many of those studies, the Earth model was layered, for example with an elastic lithosphere and/or a stratified viscoelastic mantle. And in many of those studies, the true polar wander or change in Earth's rotation rate from postglacial deformation (which we discussed in Chapter 5) was also used to constrain the parameters of the Maxwell solid—especially the viscosity, or viscosity profile, of the mantle.

Researchers have also treated the mantle as a Maxwell solid to estimate the rotational effects of ongoing tectonic activity, including subduction (Spada et al., 1992) and mountain-building (Vermeersen et al., 1994).

Some early studies of postglacial rebound and its effects on Earth's rotation and gravity field also considered more complex constitutive models for the mantle, such as a standard linear solid or a Burgers solid (e.g., Yuen et al., 1986). The latter model, composed of Maxwell and Kelvin-Voigt elements connected in series, was actually developed even earlier to explore the true polar wander hypothetically induced by randomly evolving mass anomalies (Burgers, 1955). Some recent studies of postglacial rebound have also considered complex mantle rheology; for example, Dal Forno et al. (2005) modeled the mantle as a modified Maxwell solid in which the dashpot fluid exhibited

B. Oakland. Prescott School, Ninth and Campbell Streets. A. S. E.

Figure 7.9 Examples of the Earth System's responses to applied stresses.
(a) Bucher Glacier, an outlet glacier of the Juneau Icefield in the Alaska Panhandle; the patterns of rockfall suggest the flow experienced by the glacier. Image from USGS / http://pubs.usgs.gov/of/2004/1216/glaciertypes/glaciertypes.html / last accessed August 22, 2023.
(b) Raised beaches in Bathurst Inlet, Nunavut (formerly part of the Northwest Territories), in the Canadian arctic. Models of Pleistocene deglaciation suggest that this region probably lost its ice load around 10,000–11,000 yr ago. Image from Mike Beauregard / Wikimedia Commons / CC BY 2.0.
(c) Steady creep in Hollister, central CA, along the Hayward Fault causes a distortion of the curb. Stoffer (2008) / USGS.
(d) Scarp created by the 1983 Borah Peak, Idaho earthquake; Photo taken by A. Crone / Crone et al. (1987) / USGS.
(e) Failure of structures in Oakland, California (~20 km from the San Andreas fault) following the 1906 San Francisco earthquake is obvious in this photo; *but what does the integrity of the ground tell you about the type of material constituting the crust?* Lawson et al. (1908) / University of California.
(f) Pahoehoe (i.e., fast-moving, low viscosity) lava flow following an eruption of Kilauea in Hawaii; photo by Mattox (1995) / USGS.

nonlinear behavior (fluids in which the strain rate is proportional to some power of the stress are called *non-Newtonian*).

Taking the mantle to be a Maxwell solid, researchers have studied the stresses generated by postglacial rebound in an elastic (Spada et al., 1991) or a Maxwell viscoelastic (Kaufmann et al., 2005) lithosphere, and investigated the horizontal strains generated during the rebound (James & Morgan, 1990; Mitrovica et al., 1994). With the goal of explaining intraplate earthquakes and extensional features unrelated to postglacial rebound, the bending stresses generated in a lithospheric plate changing curvature as it moves over the surface of an ellipsoidal Earth have been studied using both elastic (Turcotte, 1974) and Maxwell viscoelastic (Dickman & Williams, 1981) models. And, the bending stresses generated in a plate as it enters a trench and begins to subduct have been studied using a viscous lithosphere (De Bremaecker, 1977), a layered elastic-viscous lithosphere (Melosh & Raefsky, 1980), and an elastic-plastic lithosphere (Turcotte et al., 1978, Chapple & Forsyth, 1979).

The stress buildup around a fault, potentially leading to postseismic deformation or recurrent earthquakes, has been studied by Nur and Mavko (1974) and Savage and Prescott (1978) with an elastic lithosphere overlying a Maxwell asthenosphere; Bonafede et al. (1986) extended such analyses by considering a standard linear solid model of the asthenosphere and also a Maxwell model of both the lithosphere and asthenosphere.

A Maxwell mantle has been used to model the global deformation following an earthquake (e.g., Dragoni et al., 1983; Piersanti et al., 1997), the global deformation during true polar wander (Vermeersen et al., 1996), the global deformations accompanying solid earth tides and the wobble of Earth's rotation axis (Nakada & Karato, 2012), and the global deformation caused by changing cryospheric and hydrospheric loads resulting from global warming (e.g., Sabadini et al., 1988, who also considered a Burgers model, and Mitrovica et al., 2001).

Lastly, convection in the mantle or core (or in the oceans or even a magma chamber) is, of course, most directly modeled by treating the Earth, or the portion of it in question, as a viscous fluid. (Incidentally, for lava flow, an elastic-plastic model, in which flow takes place only while stresses exceed a 'yield stress' or yield strength, may be more appropriate; Dragoni et al., 1986.)

Clearly, a perfectly elastic material—whether a simple spring or a system of springs obeying a generalized Hooke's Law—is not sufficient to describe the dynamical behavior of the Earth. But however complex our models become, low-stress elastic behavior is likely to be a critical component in the short term; however many springs and dashpot elements the models have, they all should include a solo spring element. That is, approximating the solid earth as primarily elastic on short timescales is not unreasonable. With such an approximation (which we will make in the remainder of this chapter), our goal will be to infer the elastic properties of Earth's interior. Such inferences will reveal—with much detail—the present structure of the interior, including its layering and the lateral variations in that layering associated with tectonic features and even mantle convection. The inferred properties will also provide clues to the composition and physical conditions of the interior—a windfall we will explore in the next chapter.

7.2.2 Material Properties of an Elastic Medium: *Elastic Parameters*

The spring model of an elastic medium responds to one kind of stress, a tension or compression in the direction along the spring, with one kind of strain, an extension or contraction in the same direction. As described above, the spring is characterized by a single material property, its stiffness, which ultimately determines how much strain is incurred by a given stress. Seismologists treat the solid earth as elastic, so the spring model applies; but—in line with the generalized Hooke's Law discussed earlier—it is necessary to consider a variety of applied stresses, and the variety of strains comprising the solid earth's response.

Imagine, for example, a tennis ball, a 'superball,' a large steel ball bearing, and a large spherical pebble. All of these can be treated as elastic, in that they will respond instantaneously and with no permanent deformation if they are stressed a little. But they are not springs. We perform a "thought experiment" (an actual experiment could be quite damaging!) in which stress is applied by bouncing each of these balls off a very hard surface.

What kind of stress do these balls experience when they hit the surface? Which of the balls will bounce the highest? Why do you think that is so? What material property is reflected in the height of bounce?

It turns out that the steel ball bearing is likely to bounce the highest! However, none of these balls is responding exactly like a spring—it is not their 'stiffness' but rather their resistance to a more 'three-dimensional' *compression* that determines the height of their bounce.

In short, we will need to define a variety of material properties. All of these properties will be represented by *elastic parameters*; as we saw with a spring, they will (mostly) be the proportionality constants between an applied stress and a resulting strain. The easiest way to define them is in the context of basic experiments, illustrated in Figure 7.10. In each case, the surface experiencing the applied stress has an area A.

Something to think about: these properties can be measured under any ambient temperatures and pressures; that is, the experiments can be set up under such conditions before the stress F is applied and measurements are taken.

- *Young's Modulus, E.* Consider a tensional force F acting on both ends of a cylindrical bar of some material. The initial length of the bar is L, and after the bar has been pulled

out its length has increased by an amount ΔL (see Figure 7.10a). The stress on the ends of the bar is F/A, and the resulting longitudinal strain is $\Delta L/L$. In this experiment, **Young's modulus** is defined as

$$E \equiv \frac{tensional\ stress}{longitudinal\ strain} = \frac{F/A}{\Delta L/L}.$$

Young's modulus could be measured from a single such experiment, but for precision and an estimate of measurement uncertainty, it can be useful to perform a series of experiments on the bar over a range of stresses; as shown in Figure 7.10a (right side), the slope of the best-fitting stress-strain line yields a better determination of E.

According to its definition, E would be larger in materials that require a greater stress to produce a given strain, i.e., in materials that *resist* tensional deformation more. For materials constituting the interior of the Earth, we find E is generally in the range $\sim 2\,\mathrm{Mbar}$—$8\,\mathrm{Mbar}$ (thus, using Table 7.1, $\sim 200\,\mathrm{GPa}$—$800\,\mathrm{GPa}$) under interior conditions (e.g., Völgyesi & Moser 1982; see also Stacey & Davis 2008). As an indication of how much resistance to deformation is implied by those values of Young's modulus, consider an E of 5 Mbar (500 GPa); to cause a longitudinal strain of just 0.1%, the definition of E requires that the applied stress be 5,000 bars. In other words, if a material at the Earth's surface had that much resistance to being stretched, it would take a tension pulling on it equivalent to the weight of the top 20 km of the crust to achieve that strain.

- *Incompressibility* or *Bulk Modulus, k.* A compressional force F acting *isotropically*—that is, equally from all directions—on a cube whose initial volume is V causes the cube to *decrease* its volume by an amount ΔV (see Figure 7.10b). The compressive stress on each face of the cube is F/A, and the **volumetric strain**—as with other types of strain, a fractional change in dimensions—is

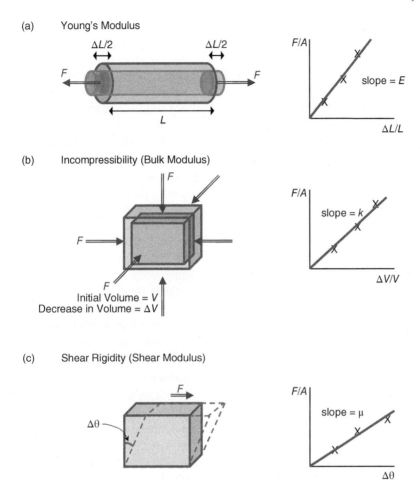

(a) Young's Modulus

(b) Incompressibility (Bulk Modulus)

(c) Shear Rigidity (Shear Modulus)

Figure 7.10 Experiments to measure elastic parameters. In all cases, forces of magnitude *F* are applied to surfaces of area *A*. The graphs to the right illustrate the best-fitting straight line to a series of stress-strain measurements for that experiment; the slope equals the elastic parameter.

$\Delta V / V$. The **bulk modulus** or **incompressibility** is defined as

$$k \equiv \frac{compressive\ stress}{volumetric\ strain} = \frac{F/A}{\Delta V/V}.$$

As before, the bulk modulus is best measured from a set of experiments over a range of stresses; as shown in Figure 7.10b, the slope of the best-fit stress-strain line yields the preferred determination of k.

From its definition, k is larger in materials that require a greater compression to produce a given volumetric strain, i.e. in materials that *resist* compressional deformation more. For materials constituting

the interior of the Earth, we find k is generally in the range $\sim$0.4 Mbar—15 Mbar ($\sim$40 GPa –1,500 GPa). For comparison, the incompressibility of ice at surface pressure is $\sim$0.065 Mbar (Fortes et al., 2005).

All materials resist being compressed, to some extent, even liquids like water and gases like air. Although most people would not intuitively consider fluids to be elastic—and in response to shear stresses, they certainly respond the way we have modeled viscous fluids—they are well described as elastic in situations where they are confined, i.e., when they are compressed. Hydraulic brakes, for example, work very well in transferring the braking action (initiated by stepping on the brakes in your car) to the brake pads, causing the car wheels to slow, because the brake fluid strongly resists being compressed. It thus makes sense to talk about the incompressibility or bulk modulus of water, which we measure to be $\sim$2 GPa = 0.02 Mbar, and air, which has an incompressibility of $\sim$1 bar; both of these values correspond to surface conditions.

The volumetric strain is also called the **dilatation**. It can be related to the 'linear' types of longitudinal strain previously discussed; it turns out that, for small strain,

$$\frac{\Delta V}{V} = \frac{\Delta x}{x} + \frac{\Delta y}{y} + \frac{\Delta z}{z} \equiv e_{xx} + e_{yy} + e_{zz},$$

where the initial volume of the cube, V, equals xyz.

Verify this relationship by calculating ΔV and neglecting very small terms; use the fact that the final volume of the cube is $(x + \Delta x)(y + \Delta y)(z + \Delta z)$.

- *Shear Rigidity* or *Shear Modulus*, μ. A shearing force F acting along the top surface of a cube whose base is held fixed causes the cube to shear out, with an angular distortion $\Delta \theta$. The shear stress on the top face of the cube is F/A, and the shear strain is $\frac{1}{2}\Delta\theta$ (recall that arbitrary-seeming factor of one half in the original definition of shear strain, and watch what happens to it now!). The **shear modulus** or **shear rigidity** is defined as

$$\mu \equiv \frac{\frac{1}{2}shear\ stress}{shear\ strain} = \frac{F/A}{\Delta \theta};$$

see Figure 7.10c.

This definition would also apply to the slightly more complex shearing experiment shown in Figure 7.4b (right-hand side).

From its definition, μ is larger in materials that require a greater shear stress to produce a given angular deformation; that is, μ represents the material's *resistance* to shearing, or bending, or twisting. For solid materials constituting the interior of the Earth, we find μ is generally in the range of $\sim$0.3–3 Mbar ($\sim$30–300 GPa).

It is important to note that fluids do not resist shear stresses at all: if uncontained by any solid boundary, fluids will always flow in response to a shear stress, even an arbitrarily small one, applied to their surface. The shear stress exerted by winds blowing along the surface of a lake, for example, will cause the lake water to circulate. Even if the winds are very weak, so that the fluid moves very slowly, its angular strain will accumulate in time to a large magnitude. With no resistance, and a shear stress differing only marginally from zero able to produce a finite amount of strain, it follows from its definition that $\mu = 0$ *for all fluids*. The shear modulus thus stands in contrast to the bulk modulus, which is nonzero for fluids as well as solids.

- *Poisson's Ratio*, σ. Materials that undergo extension usually thin out as well. Returning to the experiment used to measure Young's modulus, we note that the cylindrical bar experiencing a longitudinal tensional stress should also contract crossways. If the diameter of the end faces was originally D, and the decrease in diameter is ΔD, then the strain in that direction will be $\Delta D/D$. **Poisson's ratio** is defined as

$$\sigma \equiv \frac{\Delta D/D}{\Delta L/L}.$$

(Poisson's ratio is sometimes denoted by the Greek letter ν; to avoid confusion with our symbol for velocity, v, we will break with that tradition.) σ is not an elastic parameter since it does not relate stress to strain; and, unlike elastic parameters, which all have units of bar or Pa (i.e., units of stress, since strain has no units), Poisson's ratio is dimensionless. Nevertheless, it does express a property of any kind of material. For Earth-type materials, σ is generally in the range $\sim$0.15–0.4; seismologists often take it to be approximately $\frac{1}{4}$ (and call such materials *Poisson solids*). It is no surprise that it is less than 1: the greatest deformation will be in the direction of the applied stress.

Curiously, for a few exotic materials—and some common metals and minerals (e.g., pyrite, silica, and zeolites) in particular situations—Poisson's ratio is negative! Such materials are called *auxetic* (Evans et al., 1991; see also Burke, 1997; Baughman et al. 2000 and Grima et al., 2011). Elemental metals with a cubic crystal structure are generally auxetic when stretched in certain directions relative to the crystal structure, especially nonaxial directions (Baughman et al., 1998); for pyrite, single crystals are auxetic (e.g., Alderson & Alderson, 2007; Berardelli, 2010), at least for some orientations (Lakes, 1987); and, aggregate (Kimizuka et al., 2000) and single crystals (Yeganeh-Haeri et al., 1992) of the silica polymorph known as cristobalite are auxetic, as are those of the more familiar quartz (technically, the low-temperature α-quartz), the latter for a wide range of orientations (Ballato, 2010).

- *Lamé's Parameter* or *Lamé's Constant*, λ. In contrast to Poisson's ratio, Lamé's "Constant" *is* an elastic parameter; but there is no physical experiment used to measure it directly in a material. Rather, it arises in the theoretical development of the generalized Hooke's Law relating stress to strain. That theory also finds that it can be connected to the parameters previously discussed; it turns out that

$$\lambda \equiv k - \frac{2}{3}\mu$$

and also

$$\lambda \equiv \frac{\sigma E}{(1 + \sigma)(1 - 2\sigma)}.$$

7.3 Elastic Waves

The elastic properties we have just explored measure in various ways *the resistance of a material to deformation*. That resistance allows elastic waves, more commonly called seismic waves in the case of the Earth, to exist and propagate throughout the material. Consider, for example, a hypothetical cylindrical rod (made of metal wire, or rock, or any other solid) several kilometers long and a few centimeters in diameter. If we shear the rod by twisting one end of it, the shear will *propagate* down the length of the rod: the particles at the end are difficult to twist, because of the molecular bonds that hold them to the adjacent material—it is those bonds that create the rod's resistance to shear (μ, its shear rigidity); in order to succeed in twisting the end of the rod, the adjacent particles must also be twisted; but those particles resist twisting because of their bonds to the molecules next to them, so the twist must involve *those* molecules as well; and so on, with the twist gradually moving down to the other end of the rod.

The greater the resistance of the material to being twisted (i.e., the greater μ is), the stronger the connecting bonds are, the *faster* the twist will propagate down the rod. If it were possible for an infinitely rigid material to exist, the end of such a rod could not be twisted without all of the rod moving together, en masse; in effect the twist will have propagated infinitely fast through the rod. Even a rock type such as granite, whose rigidity is far from infinite, resists shearing to such a degree that twists will move through it at a speed of more than 3 km/sec.

In contrast, if the rod were made of water (we would have to imagine the water was contained within an easily stretched, zero-rigidity membrane), there would be no resistance to shearing ($\mu = 0$), and the end would twist without pulling any adjacent water molecules with it (except for the slight and attenuating effect of viscous drag). In this case, the twist deformation would not propagate through the rod. To put it another way, the speed of propagation through water would be 0.

The propagation of a strain disturbance—the twist—through the rod amounts to a wave of shear deformation propagating through the material. At this point, it is useful to review some basic concepts involving waves.

7.3.1 Descriptions of Waves

There are widely different types of waves that exist in nature: light waves, water waves, air pressure waves, even 'crowd waves' in a sports stadium, to name a few. So what makes something a wave? Most generally, *a **wave** is a disturbance that repeats periodically in space (i.e., over distance) and in time.* (Mathematically, a wave might be defined as a disturbance that can be represented by a function *w* of distance and time, $w = w(x, t)$, in which the function can be written $w = w(x - Vt)$; if this representation is possible, *V* would be the velocity of propagation of the disturbance, and there would be combinations of *x* and *t* that produce the same value of $(x - Vt)$—for example, the value corresponding to the wave crest—and allow *w* to repeat itself.)

The periodicity of the wave in space—the distance over which it repeats, for example the distance from crest to crest or trough to trough—is called its **wavelength**; see Figure 7.11. The symbol usually denoting wavelength is λ (clearly, there are not enough Greek letters to go around…). The periodicity of the wave in time is called its **period**, denoted

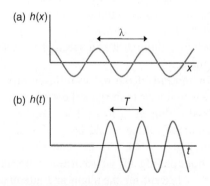

Figure 7.11 Analysis of ripples spreading in a pond, after a rock had been thrown into it. The "disturbance" propagating as a wave is the height $h = h(x, t)$ of the pond surface.
(a) Subsequently, at a fixed moment in time, *h* is measured as a function of distance *x* from the origin; the distance between ripple crests is the wavelength, λ.
(b) At a fixed location in the pond, *h* is measured as a function of time *t*; the time between crests is the wave period, *T*.

T. As an example, imagine a pond into which a rock was thrown; if we photograph the ripples in the pond's surface at a fixed moment in time, the distance from ripple to ripple could be determined, yielding the wavelength of the water waves. Alternatively, if we wade into the pond and remain at a fixed location, watching the water rise and fall as the waves move past us, we can record the time between crests, yielding the wave period.

By considering the propagation of a wave through some medium, and recognizing that the wave peak must cover a distance λ in the time *T* over which it falls and then rises back to a peak, it follows that the speed *V* of propagation must equal λ/*T*. The **frequency** *f* of the wave is defined to be the number of wave cycles (the number of complete wave patterns, from crest to trough back to crest) per time (e.g., per second); it follows that the frequency must be the inverse of the time per cycle: $f = 1/T$. Thus,

$$V = \lambda/T = \lambda f.$$

Waves can be either *traveling waves* or *standing waves*. **Traveling waves**, like the water waves created by throwing a rock into a pond, have a leading edge, a **wavefront**. **Standing waves** are set up by traveling waves that come from opposite directions and combine in such a way as to make it appear that the disturbance is not progressing. For example, if you shake a rope up and down, you can create ripples or waves that travel along the rope away from your hand. But if the rope is fixed at its other end (e.g., attached to a wall), then the ripples traveling out will reflect off the wall; if you shake the rope at just the right frequency, the returning waves will constructively and destructively interfere with the outgoing waves, and it will look like sections of the rope are simply moving up and down—those are the standing waves.

Seiches. One example of naturally occurring standing waves in the Earth System can be seen in lakes or (nearly) land-locked bays. Water waves generated by the winds or pressure differences of a storm, or by the ground shaking from an earthquake, will travel across the lake or bay, and reflect off the distant bordering land. Under the 'right' conditions—a function of the length of the lake, its shallowness, and the nature of the forcing—subsequently generated outgoing waves can constructively interfere with the returning reflected waves; the resulting standing waves are known as **seiches**. The shallowness of the lake is important because, as we noted when discussing tidal friction in Chapter 6, the speed of a water wave—and thus the timing of its reflection—depends on the water depth.

In general, generation of the waves that would constructively interfere to produce a seiche is most effective if the source is very strong or very steady. For a wind source, Darwin (1899) envisioned strong winds causing the water to pile up at one end of the lake; if the winds suddenly weakened, the returning wave could be sufficiently massive to create a large seiche. This scenario plays out in the Adriatic Sea, as strong sirocco winds from the deserts of North Africa, redirected and boosted by the cyclonic winds at the leading edge of a storm system passing eastward over the Mediterranean, blow northwestward into the Adriatic—then suddenly die out after the storm moves further to the east (Leder & Orlić, 2004).

In Europe's Lake Geneva, the site of the first scientifically conducted seiche investigation, the main oscillation has a period of ~74 min; San Francisco Bay experiences seiches—which, incidentally, can persist for hours—with a period of ~45 min (Pugh, 1987). In contrast, in the Adriatic Sea, with Venice at its far end, oscillations are found with periods of 21, 11, and 7 hr (Leder & Orlić, 2004; Pugh, 1987).

In a seiche, the lake appears to tilt back and forth like a seesaw, with one end sinking while the other is rising. The midpoint of the lake is thus a **node** of the oscillation, a location where—thanks in particular to the destructive interference between departing and returning water waves—the water surface exhibits

no rise or fall throughout the entire oscillatory cycle. (Conversely, the ends of the lake are **antinodes**, where the motion is unconstrained and the standing wave amplitude is maximal.) Depending on the constructive and destructive interference, there may *also* be standing waves produced with more than one node. The second mode, with one more node than the "fundamental" mode, has half the wavelength of the fundamental and thus—requiring half the travel time of the waves to produce it—half the period; the third mode has one third the period; and so on. Thus, the three oscillations of the Adriatic Sea are interpreted as fundamental, second, and third modes.

Seiches occur in lakes and bays worldwide, but seem to be relatively frequent—and occasionally deadly—in or nearby the Great Lakes of North America (e.g., Davis, 1890; see also Siegenthaler, 2021).

Seiches generated by earthquakes are relatively rare, but memorable: the great earthquakes of 1755 (in Lisbon, Portugal) and 1950 (in Assam, India), for example, caused seiches in parts of Fennoscandia and Great Britain (Richter, 1958). And the Loma Prieta (California) earthquake in 1989 created a seiche in nearby Monterey Bay (with a period of ~9 min.; Murck et al., 1997). Finally, we note that some of the seiches in Lake Geneva have been attributed to lake bottom debris slides (landslides) triggered by earthquakes (Siegenthaler, 2021).

7.3.2 Elastic Waves: The Wave Equation

In elastic materials, the disturbance that propagates as a wave is some type of *strain*. In the case of that hypothetical cylindrical rod, for instance, a wave of twisting or shear deformation could progress through the rod. As we saw, the ability of that wave to propagate depended fundamentally on the rod's resistance to deformation, that is, on the proportionality between stress and strain for that material.

A Governing Equation for Elastic Waves. If we wanted to characterize such waves mathematically, we would have to begin with Newton's Second Law, $\vec{F} = M\,\vec{a}$, which governs the motion of any particle subjected to a force, for example the twisting force applied to the end of the rod. In this chapter, however, we are not dealing with individual particles but with three-dimensional continua (a **continuum** is simply a medium with a continuous distribution of mass). Rather than individual particles in the medium experiencing an applied stress, it is whole 'clumps' of particles, entire volumes, that will be responding to the stress. We can deal with this situation by considering the forces to be acting on a small 'parcel' of mass within the medium, and replacing Newton's Second Law with an equivalent version 'per unit volume':

$$\vec{F} = \rho\vec{a},$$

where ρ is the material density (mass per unit volume) of the parcel, and where now $\vec{F}$ is the force *per unit volume* acting on it.

The acceleration $\vec{a}$ of the parcel can be written in terms of its velocity $\vec{v}$: by definition, acceleration is the rate at which velocity is changing with time, so $\vec{a} \equiv d\vec{v}/dt$. But the velocity can be expressed further in terms of the parcel's displacement $\vec{u}$. You may be used to thinking of velocity as the rate of change in a mass's position; but if the mass's position is changing over time, it is displacing at the same rate. Thus, $\vec{v} \equiv d\vec{u}/dt$, and we can write our continuum version of Newton's Second Law as

$$\vec{F} = \rho\frac{d\vec{v}}{dt} = \rho\frac{d}{dt}\left(\frac{d\vec{u}}{dt}\right) = \rho\frac{d^2\vec{u}}{dt^2}.$$

Now, the forces (or forces per unit volume) acting on our small material volume are related to the stresses acting on the surface of that volume (or on surfaces within it); the left-hand side of this equation could therefore be rewritten in terms of stress. Invoking generalized Hooke's Law, which tells us that stress is proportional to strain in elastic materials, we could then express the left-hand side in terms of strain; such an expression would also involve various elastic parameters (*which* parameters will depend on the situation, including the type of stress being applied). And, the right-hand side

of our equation can also be written in terms of strain, since—as we found early in this chapter—the strain that a material undergoes can, by definition, be related to the material's displacements.

In sum, then, it is possible to recast the continuum version of Newton's Second Law as an equation involving only the strain. It turns out (e.g., Bullen, 1965) that this equation takes the form

$$\frac{d^2 e}{dx^2} = \frac{1}{V^2}\frac{d^2 e}{dt^2},$$

where e is the particular strain propagating through the material, moving (for example) in the x direction at a speed V. This type of equation is called a **wave equation**. Since the strain wave varies both with position (x) and time as it propagates, that is, e is a function of more than one variable, partial derivatives should be written here; but their use will be deferred until a later chapter.

The Nature of Elastic Waves Derives From the Wave Equation. It turns out that *two* wave equations can be produced by the preceding derivation—both, with the form shown above but involving different kinds of strain. In one version of the wave equation, the strain is a *dilatation*, a volumetric strain; in the other, the strain is a *shear strain*. That is, in the simple, linearly elastic materials we are considering, two kinds of strains can propagate as waves, and—as we will see shortly—they propagate in two quite distinct ways. Elastic waves are therefore an entire 'level' more complex than, say, light waves (electromagnetic radiation), which derive from a single wave equation and always represent the propagation of an electromagnetic disturbance. The greater complexity of elastic waves will need to be accounted for when we discuss the laws—the geophysical equivalents of Snell's Laws (which were formulated for light waves)—governing the passage of elastic waves through the Earth.

Elastic waves in which dilatation propagates are called **P waves**; the "*P*" stands for "primary," because of all the waves to be generated by an earthquake, P waves are the first to arrive at a seismographic station. As a *P* wave passes through a medium, the medium is subject to alternating contraction and expansion (dilation). *P* waves are **longitudinal**: the particle motion associated with this deformation is in the same direction as that in which the wave is propagating. For example, if the long cylindrical rod considered earlier is struck on the end by a hammer (see Figure 7.12a), the compression produced there is transferred to the particles deeper in, whose compression is transferred to the particles still deeper in, and so on. As the wave of compression passes by, and the material then recoils and expands, the particles of the medium will have moved forward, then back—and thus in the same (±) direction as the wave.

One easy way to generate *P* waves in air is simply to clap your hands. Clapping compresses the air, and sends waves of rarefaction alternating with compression radiating out in all directions. And in any direction, the wave is moving in the same direction as the air particles, which move radially out and in. Your ears tell you when the compressed and rarefied air has reached you directly, or (an instant later) traveled out, reflected off walls, and returned to you; that is, *P waves in air are sound waves*. *P* waves within the solid earth are nothing more than sound waves in very incompressible media.

Obviously, the velocity of a *P* wave through any medium depends on the medium's resistance to compression, its incompressibility *k*. When compression occurs at a point, however, the material will end up being sheared as well; see Figure 7.12b. The overall resistance of the medium to the passage of a *P* wave thus includes both its incompressibility and its shear rigidity, μ. Additionally, in order for the material to become deformed and its particles to move, their *inertia* must be overcome; this is reflected in our use of Newton's Second Law as the starting point for the derivation of the wave equation. With our continuum approach to these equations, the inertia is represented not by a particle's mass but by the material density, ρ, within the volume being stressed. In

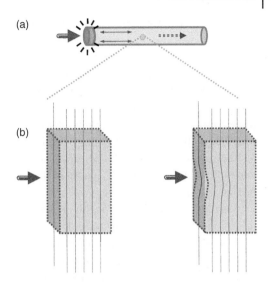

Figure 7.12 Deformation during a *P* wave.
(a) A hammer blow or other compressive stress acting on the end of a cylindrical rod produces a volumetric strain that propagates as a wave through the rod (dashed arrow). The volumetric strain is characterized by contraction then dilation (expansion), with particle motion to the right and left as shown, establishing that this wave is longitudinal.
(b) Close-up of the deformation as the wave passes through the rod. When a localized compressive stress (indicated by the arrow) acts on a portion of a material, it creates shear deformation as well as the change in volume. The vertical lines are imagined to be attached to particles within the material, serving to illustrate the resulting shearing.

sum, we would expect the *P* wave velocity to be proportional in some way to both *k* and μ, and inversely proportional to ρ. Referring one last time to the wave equation, we can guess that the equation will involve all these parameters, but their combination will be identified with V^2 rather than *V*. It turns out finally that the *P* wave velocity is given by

$$V_P \equiv \sqrt{\frac{k + \frac{4}{3}\mu}{\rho}};$$

as expected, the *P* wave travels faster in materials that possess a greater resistance to volumetric or shear deformation.

A common misconception is that *P* waves propagate faster through denser materials than through less dense ones, because of a greater resistance to deformation. In reality, as our formula above more rigorously implies, this will only occur if their greater density amplifies their resistance to deformation by a lot more than the density increase. For example, under surface conditions, iron is typically close to three times as dense as the kinds of silicate rocks found in the crust or mantle. But iron's resistance to compression and shearing is also two to three times as great (Table A6 in Stacey & Davis, 2008), so the velocity of a *P* wave propagating through it is roughly the same in both materials (about 6 km/sec). And, as we will see in the next chapter, conditions in Earth's deep interior actually worsen the resistance of iron to deformation—compared to that of silicates—despite their increases in density.

For a single material, though (as we will also see in the next chapter), it is true that squeezing it—and thereby increasing its density—does indeed make it much more resistant to being deformed, and increases the speed of *P* waves through it.

The propagation of shear strain through a material constitutes an **S wave**, also called a **shear wave**; the shear strain can generally be described as a twisting or *rotational* strain. Historically, the "S" stood for "secondary," because *S* waves travel more slowly than *P* waves and arrive later at seismographic stations. Incidentally, slower does not mean weaker; as a rule, *S* waves generated by earthquakes are larger in amplitude (producing larger ground motion and more strain) than the associated *P* waves.

In contrast to the *P* wave, the shear wave is **transverse**, that is, the particle motion generated by its passage is perpendicular to the direction of the wave. This was demonstrated in our earlier example, the long cylindrical rod whose end was twisted: as the twist propagates down the rod, the material particles 'rotate' *around* its long axis. In general, the twisting can be resolved into two components: one with vertical particle motion (if the wave is moving horizontally), the other with side-to-side particle motion also perpendicular to the direction of travel; see Figure 7.13. The waves in which these components of shear strain propagate are called the **SV** and **SH waves** ("V" for "vertical" and "H" for "horizontal"). Depending on the medium through which these waves propagate, it may be possible for one or the other of these waves to be suppressed at boundaries (this will be discussed later), causing the wave to become polarized; but initially, at least, we can expect to see both *SV* and *SH*, combining to produce a 'complete' *S* wave propagating through the medium.

The speed of *S* waves through a medium will depend on the material's resistance to being sheared, thus on the shear rigidity, μ. Because shearing does not change the volume of the material, however, the material's resistance to compression does not come into play (and because the overall resistance to deformation is less, we can understand why *S* waves travel more slowly than *P* waves). As with *P* waves, initiating the deformation requires that the

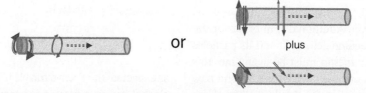

or   plus

Figure 7.13 Propagation of shear strain through a long cylindrical rod. If a twisting stress is applied to the end of the rod, the rod's resistance to shearing will allow the twist, or rotation, to progress down through the rod (dashed arrow), propagating as an *S* wave; the propagation is transverse because the rotation is perpendicular to the direction of propagation. The twist can be resolved into an up-down particle motion plus a side-to-side particle motion; these propagate as *SV* and *SH* waves.

particle inertia be overcome. In sum, it turns out that the *S* wave velocity is given by

$$V_S \equiv \sqrt{\frac{\mu}{\rho}}.$$

As expected, the speed of *S* waves is less than that of *P* waves, that is, $V_S < V_P$ (just compare the formulas for V_S and V_P). As an illustration, for Poisson solids we find $V_P \sim 1.7 V_S$ (Bullen, 1965). And, since $\mu = 0$ in fluids, it follows that $V_S = 0$, confirming that shear waves cannot propagate through a fluid.

P and *S* waves are called **body waves**, because they travel *through* a material, e.g. through the Earth. There are other types of elastic waves, to be discussed later in this chapter, which originate at the earthquake source as *P* and *S* waves but become transformed in the outer regions of the Earth so that they travel along Earth's surface, with their energy concentrated there. They are called **surface waves**.

7.3.3 Elastic Waves: Reflection and Refraction

Materials with different densities and elastic properties will support the passage of elastic waves in different ways, for example with different speeds of propagation. Earth contains diverse materials distributed throughout its interior, plus an overall layering with increasingly dense material at greater depths. It is inevitable that elastic waves generated within the Earth—seismic waves—will encounter *boundaries* as they travel, with differences in material properties on either side of these interfaces. In general, as a wave strikes a boundary, some of the energy will be *transmitted* across the interface, with the remainder of the energy *reflected* back.

To begin with, we consider a two-layer medium. The upper layer consists of material with density ρ_1, incompressibility k_1, and shear rigidity μ_1; as a result of these properties, the *P* wave velocity in this layer is $\sqrt{(k_1 + \frac{4}{3}\mu_1)/\rho_1}$, which we will denote as

V_1. In the lower layer, the corresponding quantities are ρ_2, k_2, μ_2, and $\sqrt{(k_2 + \frac{4}{3}\mu_2)/\rho_2}$ or V_2. For example, our two-layer medium might be the crust and uppermost mantle, with the boundary between them being the Moho; in this case, $V_1 \sim 6$ km/sec (depending on the type of crust) and $V_2 = 8.0$ km/sec.

Seismic waves are generated in nature by earthquakes; they can be produced artificially by subjecting the Earth to an applied stress (e.g., from an explosion, or by striking the Earth's outer surface). Regardless of how the waves have been produced, we start by following a *P* wave as it moves down through the upper layer and strikes the boundary between layers; see Figure 7.14. We measure the orientation of this *incident* wave, and those of the *reflected* and *transmitted* waves resulting from its striking the boundary, by the angles they make measured with respect to the *perpendicular* to the boundary—*not* the angle measured with respect to the boundary itself. These angles will be denoted by θ_i, θ_r, and θ_t, respectively. The laws governing the orientations of the reflected and transmitted waves are

$$\theta_i = \theta_r$$

and

$$\frac{\sin \theta_i}{\sin \theta_t} = \frac{V_1}{V_2}.$$

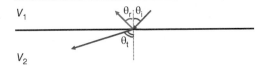

Figure 7.14 Snell's Laws. When a wave encounters a boundary, its energy is partly reflected and partly transmitted. The orientations of the incident, reflected, and transmitted waves—the angles θ_i, θ_r, and θ_t—are measured with respect to the perpendicular (dotted line) to the boundary. The transmitted wave shown here illustrates the situation in which the seismic velocity in the second layer, V_2, is greater than that in the first layer, V_1; in this situation, the law of refraction implies that the transmitted wave is bent *away* from the perpendicular.

Despite the geophysical context, these are called *Snell's Laws*. In optics, such relationships describe the reflection and refraction of light, and are named in honor of a Dutch scientist named Snellius, who in 1621 was (one of) the first in the modern era to quantify those phenomena (note, however, that Berggren (2007; see also Zghal et al. 2007) documents a similar achievement by ibn Sahl, a Persian scientist who in the year 984 correctly deduced the nature of light's refraction—technically, then, *that* law should be called *ibn Sahl's Law*, though we will still call the *set "Snell's" Laws*).

The first of Snell's Laws states that *the angle of incidence equals the angle of reflection*. The second law implies that $\theta_t \neq \theta_i$ when $V_2 \neq V_1$; in other words, the transmitted wave is refracted—changes direction—when encountering a medium of different properties. The transmitted wave is sometimes called the *refracted* wave. In optics, specifically when the wave is initially traveling through a vacuum and V_1 is the speed of light in a vacuum, c, the ratio of velocities is called the **index of refraction**.

As a 'layered earth' example, $\theta_t > \theta_i$ when $V_2 > V_1$; thus, when entering a layer of faster seismic velocity (as shown in Figure 7.14), the P wave refracts *away* from the perpendicular to the boundary, leaving the boundary at a greater angle than that at which the incident wave struck it. In contrast, waves entering slower media are bent toward the perpendicular, i.e. to a more 'vertical' orientation—just consider Snell's Law with $V_2 < V_1$; or, refer to Figure 7.14 but reverse the arrows, with the wave starting in the lower medium then entering the upper medium.

There are limits to the amount of refraction possible. For the case of $V_2 > V_1$, Snell's Law tells us that, as the angle of incidence increases, the angle of transmission increases as well. At some point—at an angle of incidence called the **critical angle**, denoted θ_C—the transmitted angle will equal 90°, and the transmitted wave will travel along the top of the faster layer, just below the interface (see Figure 7.15a). It should be emphasized that, despite this extreme refraction, the transmitted wave has indeed been transmitted into the faster layer, and travels with the velocity V_2 of that layer. This situation, which leads to a useful way to

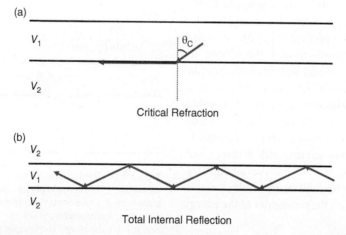

Figure 7.15 (a) Critical refraction. If the seismic velocity in the second layer, V_2, is greater than that in the first layer, V_1, then an incident wave from the first layer striking the boundary at the critical angle (measured with respect to the perpendicular (dotted line) to the boundary) will be transmitted into the faster layer with $\theta_t = 90°$.
(b) Total internal reflection. If the wave velocity in a material is less than that on either side of it, waves traveling through it at angles greater than the critical angle cannot be transmitted across either of the material's boundaries.

explore the crust regionally, is called **critical refraction**; the critically refracted wave is also called the **head wave**. From ibn Sahl's Law (Snell's Law), we have

$$\sin \theta_C = \frac{V_1}{V_2}$$

since $\sin(90°) = 1$.

Taking critical refraction one step further: if the angle of incidence *exceeds* the critical angle, it follows that there should be no transmission into the faster layer—after all, at the critical angle there has already been as much refraction as possible such that the transmitted wave remains in the faster layer. Physically, we might imagine that, at greater than critical incidence, the transmitted wave is refracted by more than 90° and sent right back into the slower layer. Mathematically, $\theta_i > \theta_C$ violates the law: we must always have $\sin\theta_i / \sin\theta_t = V_1/V_2$—which we can write as $\sin\theta_t = (\sin\theta_i) / (V_1/V_2)$—so,

if	$\theta_i > \theta_C$
then	$\sin\theta_i > \sin\theta_C$
or	$\sin\theta_i > V_1/V_2$

—*do you know why that last substitution is valid?*—

leading to

$$\sin \theta_t = (\sin\theta_i)/(V_1/V_2) > 1,$$

which is not possible since no angle has a sine greater than 1.

However, although transmitted waves do not formally propagate through the faster medium when the incident angle is "postcritical," the wave equations in this situation allow for a small amount of strain to 'leak' into the faster medium—as long as it decays away with distance from the boundary. The details of such leakage are beyond the scope of this textbook but may be found in Stein and Wysession (2003).

The situation of (nearly) zero transmission is called **total internal reflection** (see Figure 7.15b), and creates a *waveguide* that keeps the wave energy concentrated in the 'slow' layer. In the case of light waves, total internal reflection allows *optical fibers* to work, for example to transmit visual images around internal organs during surgery in a hospital or, in bundles, to transmit optical signals over long distances for communication or computer networking applications.

Waveguides in the Oceans and the Atmosphere. As mentioned earlier, water and air possess no resistance to shear stresses but do resist compression, allowing P waves—sound waves!—to propagate through the oceans and atmosphere. It turns out that a natural waveguide exists in the world's oceans, in the form of a minimum-velocity layer found typically at depths of around 1 km. The density and incompressibility of seawater vary with depth depending on both the water temperature and the ambient pressure (salinity variations generally have less effect). In the thermocline—also called the "zone of rapid change," an oceanic layer existing below the relatively warm surface waters found in mid- and tropical latitudes, and above the cold waters of the deep ocean—the temperature drops rapidly with depth to just a couple of degrees Celsius. This causes a noticeable increase in density and thus (with $V_P = \sqrt{k/\rho}$ in water) a decrease in the speed of P waves; a minimum is reached at the bottom of the thermocline. Below that, the temperature decreases more, but only slightly, all the way to the seafloor; pressure increases now dominate, and the increase in the water's incompressibility with depth causes the P wave speed to increase. The net result (see Figure 7.16a) is a layer of minimal V_P at the base of the thermocline which is bounded above and below by water with higher V_P.

This layer is known as the **SOFAR** (Sound Fixing And Ranging) **channel**, and its existence allows sound waves—that is, P waves in water—to propagate effectively within it over distances of thousands of miles (see Figure 7.17). It has been speculated that some species of whales communicate long-distance using the SOFAR waveguide; it has also been used by humans to detect remotely the sounds

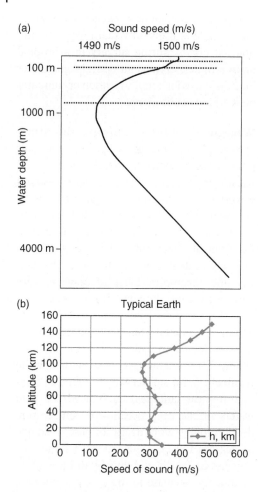

Figure 7.16 (a) The variation of *P* wave (sound) velocity with depth in the ocean, typically like that shown here, produces a layer exhibiting nearly total internal reflection known as the SOFAR channel. As mentioned in the text, the velocity strongly depends on temperature, implying that the depth of the channel will be less at higher latitudes.
Image copyright by University of Rhode Island 2013—downloaded from http://www.dosits.org/science/soundmovement/speedofsound / University of Rhode Island.
(b) The typical variation of *P* wave (sound) velocity with altitude in the atmosphere, excluding the effects of winds and variations in humidity.
Note in these graphs that the ocean depth scale (a) is very small compared to the atmosphere height scale (b). Image from Johnson (2012) / Gary Johnson.

of both submarines and undersea earthquakes (Smith, 2004) and, more recently, the sounds of global warming (e.g., Li & Gavrilov, 2006; MacAyeal et al., 2008).

Strictly speaking, because the boundaries of the SOFAR channel are 'soft,' it operates by 'extreme refraction' within the surrounding fluid rather than by total internal reflection off sharp boundaries. Sound waves exiting the channel upward (for example) encounter a faster medium and refract away from the normal; but that medium is continuously faster away from the channel, so—like a series of layers—it acts to refract the waves more and more. If the 'take-off' angle they start with is great enough, the exiting waves will actually be refracted enough to head back toward the channel (then, as they approach the channel, the continuously slower medium directs them more and more downward back into the channel). They encounter the same situation as they continue on and exit the other side of the channel. The result is a continual up-and-down overshoot superimposed on the main horizontal propagation of the sound wave.

The atmosphere potentially contains three sound waveguides within it (such channels are also called *ducts*). To first approximation, the speed of sound in air depends only on the air temperature (as pointed out by, e.g., Bohn (1988), variations in pressure affect k and ρ proportionally); with the typical variation of temperature with height—cooling as we rise through the troposphere, then warming through the stratosphere (thanks to the absorption of heat by the ozone layer), then cooling again through the mesosphere (which lacks ozone), and finally unlimited warming in the very rarefied thermosphere—we can predict the speed of sound through these layers (see Figure 7.16b). Two low-velocity channels are implied, one between the ground and ~50 km altitude (centered around ~10–30 km above Earth's surface), the other above that, extending to ~110 km altitude (mainly ~80–100 km). On a larger scale, the entire atmosphere between the ground and the

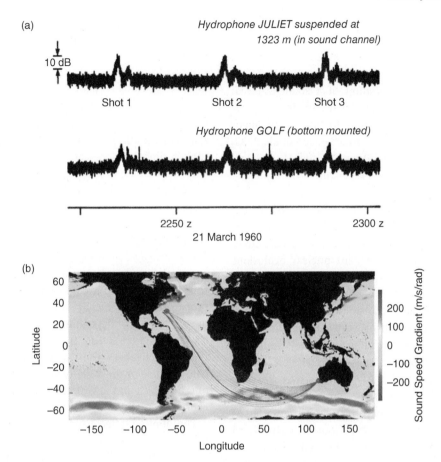

Figure 7.17 Long-distance propagation of sounds through the SOFAR channel.
(a) Hydrophones near Bermuda recorded three shot blasts set off near Perth, Australia, nearly 20,000 km away. Observations originally investigated by Shockley et al. (1982) / AIP Publishing, plotted versus military time; image by Shockley et al. as retouched in Dushaw (2008) / John Wiley & Sons.
(b) The locations of the shot point and receivers are shown here as red-filled circles. The most direct path taken by the sound waves as they traveled through the channel from Perth to Bermuda would have been blocked by southern Africa; the analysis by Dushaw (2008) attempted to determine the actual path they took, accounting also for latitudinal variations in sound speed in the oceans (color scale). The origin of the 'echoes' (in the upper image, about ½ minute after each main signal) recorded by the suspended hydrophone but not the one at the seafloor, 40 km away, has also been investigated.
Image from Dushaw (2008) / John Wiley & Sons.

lower thermosphere (see Figure 7.16b again) can also act as a waveguide, amounting to a third low-velocity sound channel (e.g., de Groot-Hedlin et al., 2011). Like the SOFAR channel, they would operate by 'extreme refraction;' in the atmosphere, though, the ground can act as an additional, 'hard' boundary, enforcing total internal reflection for those sound waves propagating downward which were not already sufficiently refracted higher up.

These channels are much broader and much less sharp than the SOFAR channel (whose width vertically might be up to a kilometer—see Figure 7.16), but it is conceivable that they might nevertheless act as somewhat effective waveguides, allowing the sounds of major events at Earth's surface to propagate to the other side of the globe. Such events would have included the great volcanic explosions of Krakatoa (in Indonesia) in 1883—according to Hedlin et al. (2012), "still considered the loudest sound in recorded human history"—and Mount St. Helens (in the U.S. Pacific Northwest) in 1980, and the mysterious Tunguska blast (probably a mid-air explosive disintegration of a small asteroid; e.g., Phillips, 2008) in Siberia in 1908. A noteworthy recent addition to this set of major events is the "cataclysmic" (Yuen et al., 2022) 2022 eruption in Tonga of the Hunga volcano, likely a rival to Krakatoa (Matoza et al., 2022).

In the earliest days of the Cold War, before seismic monitoring of nuclear tests began, M. Ewing suggested that these atmospheric channels would allow distant nuclear explosions to be detected. A classified military program named "Project Mogul" began, with acoustic monitors (and transmitters for preliminary testing) mounted on very large high-altitude balloons that were adjusted so they would rise to the appropriate heights. It is possible that the crash of one of those balloons over Roswell, New Mexico in 1947—with previous sightings of the balloons already misinterpreted as UFOs, and the rubbery balloon debris from that crash perhaps mistaken for body parts—was the basis for popular 'conspiracy theories' that developed about a government cover-up of aliens crash-landing there (Thomas, 1995; Weaver & McAndrew, 1995).

Unlike low-velocity zones in the oceans and solid earth, the speed of sound in the atmosphere is slow enough, and the atmosphere is mobile enough, that the profile of sound velocity versus height (as in Figure 7.16) can be significantly modified by the existence of winds—and thus, for example, by the global circulation of the troposphere. The net sound velocity at any height will be the vector sum of the sound velocity in a motionless atmosphere plus the wind velocity at that height; in practice, the modification to the profile in Figure 7.16 is determined by the component of the horizontal wind in the same direction that the sound is propagating in.

Strong horizontal winds in the stratosphere (sometimes called the 'super-rotation' of the stratosphere) can also distort the sound velocity profile. Furthermore, seasonal variations in atmospheric temperatures, winds, and the full height of the troposphere (which ideally defines the lower channel's minimum) can modify the profile still more. (In contrast, ocean temperatures have a much reduced seasonal variation, as discussed in Chapter 9, though global differences in ocean structure, such as the lack of a warm surface zone and even a thermocline at high latitudes, lead to global differences in the depth of the SOFAR channel.) The end result for the tropospheric sound channel is that it is only effective as a surface waveguide (trapping sound waves that can be detected at the surface) in certain seasons, at some latitudes, and for sound waves propagating specifically only east-to-west or west-to-east, depending on the season (Diamond, 1963). The tropospheric waveguide's effectiveness also varies diurnally, as a consequence of the daily heating and cooling of the surface and lowest atmosphere (Hedlin et al., 2012).

The limited effectiveness of these atmospheric sound channels casts some doubt on them as the sole explanation for pressure waves observed around the globe following the Krakatoa and Mount St. Helens eruptions. Such observations have instead been attributed (e.g., Paul 1884, LeConte 1884, Harkrider & Press, 1967, Bolt & Tanimoto 1981) primarily to the passage of low-frequency atmospheric "gravity waves" generated by the distant volcanic explosions—explosive equivalents to throwing a rock into the atmospheric 'pond' (Paul, 1884). The ability of such waves to propagate long distances with almost negligible attenuation is a consequence of very limited absorption by the atmosphere at those low frequencies (ReVelle, 2004; Whitaker & Norris, 2008; cf. ReVelle, 1976). Gravity waves as a fluid dynamic phenomenon will be discussed in Chapter 9.

The Basis for Snell's Laws. Snell's Laws may be derived by requiring the particles at the boundary between the two media to behave

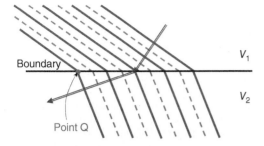

Boundary

V_1

V_2

Point Q

Figure 7.18 Snell's Laws can be viewed as a consequence of a 'sensibility' requirement. This figure illustrates the cross-section of a *P* wave striking a boundary. Successive compressions and expansions are represented by heavy solid and dashed lines. Points on the boundary, including point Q shown, should experience compressions (or expansions) at the same times from both the incident and transmitted waves. Depending on the angle at which the incident wave approaches the boundary and the speeds of the wave on either side of the boundary, such simultaneity can only happen if the transmitted wave is refracted according to the law of refraction.
A similar analysis involving the reflected wave leads to the law of reflection.

'sensibly' during the passage of the wave. To see this, consider the cross-section of a wave being transmitted across a boundary; in the case, for example, of *P* waves as shown in Figure 7.18, there are successive episodes of compression and expansion. We choose a point "Q" on the boundary and note that it experiences successive waves of compression and expansion due both to the incident *P* wave approaching the boundary and to the transmitted wave departing from it. Depending on the orientation and velocity, V_1, of the incoming *P* wave, point Q will experience compression every so often; and depending on its angle of refraction and velocity V_2, the transmitted wave will also induce compressions at point Q. With different angles of refraction, the compressions from the transmitted wave could be induced more, or less, frequently. *The angle of refraction is determined by requiring point Q to behave sensibly, that is, to experience the same frequency of compression from the waves on both sides of the boundary.* This requirement, in turn, leads to the law for refraction; a similar

analysis considering the reflected wave leads to the law for reflection.

The speed at which the compression propagates along the boundary is called the **apparent velocity** (—just imagine if you were standing on the boundary, experiencing the wave's compressions and expansions as it passed by; with a small angle of incidence, those deformations would appear to move quite rapidly along the boundary!). Another way of expressing the 'sensibility' requirement is that *the apparent velocity must remain constant as the wave crosses the boundary*.

Twenty-nine years after Snell formulated the laws of reflection and refraction, the French mathematician Fermat deduced a basic characteristic of nature as a consequence of which those laws could be explained. Those laws dealt originally with light waves, and Fermat's deduction, called the **Principle of Least Time**, said that out of all possible paths which light could take to get from one place to another, the one it *actually* takes is the one it can cover in the least possible time. Note that the least-time path is not necessarily the shortest, distance-wise. For example (see Figure 7.19), in the case of light traveling from point *A* in air to point *C* in water (through which it propagates more slowly), the shortest distance would be the straight line from *A* to *C*; but light following such a path would spend 'too much' time in the water, lengthening its overall travel time. Instead, paths that include a longer segment in air (through which the light travels more quickly) and a shorter segment in water have a net shorter travel time. It turns out that the shortest-time path requires a refraction of the light going from air to water which precisely matches Snell's Law.

A more vivid description of the least-time path was provided by the celebrity physicist Feynman in his *Lectures* series (Feynman et al. 1963). Imagine a 'damsel in distress' in a leaky boat located just offshore in a beach area; the lifeguard running to save her has to decide how much of the run should be on the sand versus how much should be in the water. It hardly

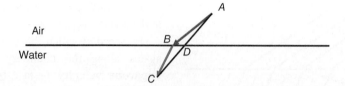

Figure 7.19 Fermat's Principle of Least Time. Whether considering a light wave, as illustrated here, or Feynman's lifeguard running on the beach to rescue someone in the water, the fact that the wave (or runner) travels faster in the air than in the water means that the longer path *ABC* will take less time than the straight-line path *ADC*. The refraction from *AB* to *BC* as predicted by Snell's Law is equivalent to *ABC* being the shortest-time path.

(Technically, as discussed in Feynman et al. (1963) and elsewhere, Fermat's principle is one of *least change*. For example, in the situation illustrated above, the lifeguard entering the water at *B* will indeed take much less time to get to *C* than a lifeguard entering the water at *D* (or any other distant point); but as long as the entry point is close to *B*, the extra time will be very little. That is, as the point of entry approaches *B*, the total travel time shortens significantly, then more modestly, then only slightly, then negligibly. Such a 'leveling off' of the total travel time also applies when approaching *B* from the other side.

That leveling off—which the astute reader will conclude describes a 'slope' of zero in the travel-time curve at *B*—does not happen around *D* (for example): the travel time changes 'steeply' as *D* is approached. Curiously, just as a slope of zero on a curve may indicate either a local minimum or a local maximum, Fermat's principle actually allows for situations in which light encountering a boundary as it travels from one spot to another will take the *greatest* amount of time, out of a range of 'local' possibilities!)

needs to be mentioned that even lifeguards do not swim as fast as they can run on sand . . . .

Just as our sensibility requirement applies equally well to mechanical situations (like elastic waves encountering boundaries) and optical situations, it turns out that Fermat's Principle of Least Time also applies to elastic waves, and can serve as the basis for Snell's Laws in the case of elastic waves. Another consequence of Fermat's principle is the *principle of reciprocity*: if a given path is the least-time path for a wave traveling from point *A* to point *C*, then it is also the least-time path for waves traveling from point *C* to point *A*.

Wave Amplitudes. We have seen that the way waves are transmitted depends on the density and elastic properties of the media, and on the orientations of the waves with respect to boundaries. These same factors determine the relative amounts of energy that are imparted to the reflected and transmitted waves. As an extreme example, when the incident wave hits a boundary at greater than the critical angle, all of its energy is reflected.

In general, to determine the relative amplitudes of the reflected and transmitted waves,

the wave equation must be solved in both layers, using the appropriate media properties and additional "boundary conditions." The latter are physically reasonable constraints:

- that the particle displacement must be the same in either medium right *at* the boundary (i.e., the displacement must be *continuous across* the boundary)—if the boundary was not 'welded' tight, a 'gap' could develop there; and

- that the stress must be continuous across the boundary—if the stress at the boundary were twice as great, say, on one side as on the other, where did that extra stress go?

At some of Earth's boundaries these constraints can be made more specific. For example, at solid-liquid boundaries like the seafloor or core-mantle boundary, the shear stress across the boundary would have to be zero if the viscosity of seawater or core fluid were considered to be negligible; and, with an inviscid fluid, the particle displacement along the boundary can be unconstrained. The normal stress and the displacement perpendicular to the boundary would still, however, have to be continuous across the boundary.

Finally, the outer surface of the solid earth is commonly treated as an even more special type of boundary: a **free surface**, defined as a boundary with no outside stress acting on it. In this case (by continuity of stress), all stresses vanish, and particle displacements associated with the seismic wave are unconstrained, at the surface. Clearly, this is an approximation, since it neglects the normal stresses exerted by the atmosphere on the land and by the oceans on the seafloor.

As an illustration of the kinds of solutions, consider a *P* wave at **normal incidence**, that is, it strikes the boundary perpendicularly (thus, its angle of incidence is zero). Let A_i denote the amplitude of the incident wave, and A_r and A_t the amplitudes of the reflected and transmitted waves. The incident wave has traveled through a medium of density ρ_1 at a speed of V_1; the density and seismic velocity of the other medium are ρ_2 and V_2. Both media are solid. In this case, we would find that

$$\frac{A_r}{A_i} = \frac{\rho_2 V_2 - \rho_1 V_1}{\rho_2 V_2 + \rho_1 V_1} \quad \text{and}$$

$$\frac{A_t}{A_i} = \frac{2\rho_1 V_1}{\rho_2 V_2 + \rho_1 V_1}$$

if $V_2 > V_1$. To account for the opposite directions of the incident and reflected waves, the formula on the left is sometimes written with a reversed numerator, corresponding to $-A_r$. (Keen-eyed readers will notice that the formulas for A_r and A_t seem quite closely related—and they are indeed: continuity of (normal) displacement at the boundary requires that $A_i - A_r = A_t$, so $A_r + A_t = A_i$ or $A_r/A_i + A_t/A_i = 1$.) The product of a medium's density times its seismic velocity is called its **acoustic impedance**; we see that relatively large reflections can be generated at a boundary only if there is a significant 'mismatch' between their acoustic impedances.

Elastic Waves: Converted Waves. So far, the picture we have been developing of body waves within the Earth envisions *P* and *S* waves initially created by an earthquake or other source, with each wave then being partially reflected

and partially transmitted every time it encounters a material boundary within the Earth. Even if the Earth's interior was simply layered, the net result would be a very 'busy' picture indeed, with multiple reflections and transmissions crisscrossing the Earth (though, as suggested by the preceding discussion, many would have vanishingly small amplitudes).

In reality, that picture is further complicated by the fact that P *waves can be converted to* S *waves and* S *waves can be converted to* P *waves at any boundary*! Even worse, those converted reflected and transmitted waves propagate at different angles than the unconverted waves. This surprising phenomenon—a conversion of longitudinal to transverse waves or vice versa—cannot happen with light waves, by the way, which only exist as transverse waves. Such conversion means that in heterogeneous media (i.e., with lots of boundaries), the energy of an initial seismic wave will be scattered far more than an initial light wave would have been.

Observational evidence of elastic-wave conversion includes underground nuclear explosions; such explosions produce pure, isotropic, compressive stresses at the source, yet distant seismographs will record *S* as well as *P* waves from them. This is interpreted to be the result of *P* wave energy partially converting to *S* waves at major boundaries, especially as that energy reflects off the Earth's free outer surface. In contrast, earthquakes are characterized by sudden slip along a fault plane; that very nonisotropic displacement generates both volumetric and shear strain, thus producing both *P* waves and *S* waves even at the source. During the Cold War, when both the United States and the Soviet Union tested nuclear weapons regularly, one way distant observers could distinguish underground nuclear tests from naturally occurring earthquakes was by the relative amplitudes of *P* to *S* waves.

Can you say which situation would produce relatively larger P *waves?*

(Incidentally, later in this chapter we will learn about a different type of seismic wave called a "surface wave;" it turns out that surface waves also tend to be larger than *P* waves in an earthquake, but smaller in a nuclear explosion.)

The ability of elastic waves to undergo conversion can be viewed as a consequence of *particle motion in common*; see Figure 7.20. For example, as an *SV* wave propagates through a medium, it induces transverse particle motion, i.e. particle motion perpendicular to the wave direction. If that wave strikes a boundary, some of its energy will of course be reflected, and the reflected *SV* wave will, trampoline-like, possess similarly reflected transverse particle motion (see Figure 7.20a, left side). But boundaries are strange creatures: with different densities and elastic properties on either side, they behave *intermediate* to both media. The resistance to deformation in the first medium sustained that transverse particle motion, and a specific velocity of its propagation, as the wave moved through it; but the boundary experiences the propagating stress differently. At the point of incidence, the boundary acts like a source—a disturbance—capable of producing strain with a different orientation and

moving at a different speed than the incident wave; however, our governing equations only allow one *other* way for the strain to propagate through the medium: as a *P* wave. How successful the boundary disturbance is in converting the 'incoming' strain to an 'out-going' compressional strain—how much of the *SV* wave energy feeding the disturbance is converted to a *P* wave reflecting off that boundary—depends on how similar the converted wave's longitudinal particle motion is to the incident *SV* wave particle motion (see Figure 7.20a, right side). The converted wave's amplitude thus depends ultimately on the properties of both media *and* on the orientations of both the incident and the converted reflected waves.

According to this explanation, an incident *SH* wave would *not* produce a converted *P* wave upon reflection, because there is no way for its horizontal transverse particle motion to have anything in common with a reflected *P* wave's longitudinal particle motion; no matter what its angle of incidence, the *SH* wave particle motion will always be at right angles to the

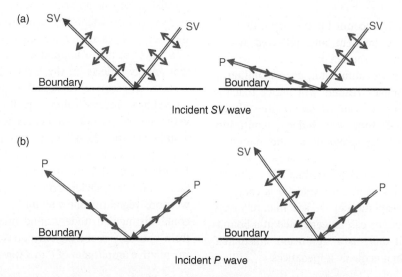

Incident *SV* wave

Incident *P* wave

Figure 7.20 Conversion of reflected elastic waves at a boundary. Illustrations on the left depict unconverted waves; converted waves are on the right. Short solid arrows represent particle motion.
(a) Consider first an incident *SV* wave. The *SV* particle motion reflects off the boundary as if the boundary is a trampoline; but differences in material behavior at the boundary allow a *P* wave whose particle motion is roughly the same as that of the incident *SV* wave to also be created.
(b) Same as (a) but for an incident *P* wave.

P wave's. And, an *SV* wave striking the boundary at normal incidence (i.e., perpendicular to the boundary) would *not* produce a converted reflected *P* wave; that is the only orientation in which their particle motions have nothing in common. By implication, as incident *SV* waves strike a boundary at more and more nearly normal incidence, the amplitude of the converted *P* wave would become vanishingly small.

Our explanation for converted waves works equally well if the incident wave is a *P* wave (see Figure 7.20b); that is, from an incident *P* wave we should, in general, expect to observe a reflected *P* wave *and* a reflected converted *SV* wave (the exception will again be at normal incidence), and the converted wave amplitude will diminish as normal incidence is approached. Furthermore, our entire discussion can apply to the case of transmitted waves, so that an incident *P* or *SV* wave can produce both transmitted *P* and transmitted *SV* waves.

The converted waves must obey Snell's Laws. For the situation shown in Figure 7.21, with θ for *P* wave angles and ϕ for *S* wave angles, we have, altogether,

$$\textit{Reflection} \quad \begin{cases} \theta_i = \theta_r \\ \dfrac{\sin \theta_i}{\sin \phi_r} = \dfrac{(V_P)_1}{(V_S)_1} \end{cases} \quad \text{and}$$

$$\textit{Transmission} \quad \begin{cases} \dfrac{\sin \theta_i}{\sin \theta_t} = \dfrac{(V_P)_1}{(V_P)_2} \\ \dfrac{\sin \theta_i}{\sin \phi_t} = \dfrac{(V_P)_1}{(V_S)_2} \end{cases} ;$$

there are four analogous relations if the incident wave is an *SV* wave. We can view these relations as an *expanded* version of Snell's Laws. They can also be viewed as variations of a single fundamental rule,

$$\frac{\sin(\text{incident wave } \sphericalangle)}{\sin(\text{subsequent wave } \sphericalangle)} = \frac{\text{Velocity of incident wave in incident medium}}{\text{Velocity of subsequent wave in subsequent medium}}.$$

Just as in the case of total internal reflection, there are possible combinations of the incident wave angle and the relative velocities (for example, $(V_P)_1$ and $(V_P)_2$ relative to $(V_S)_1$, if the incident wave is an *S* wave) where the converted waves would violate Snell's Laws (leading to $\sin(\text{subsequent wave } \sphericalangle) > 1$) and are not produced. See the homework exercises for an example of this.

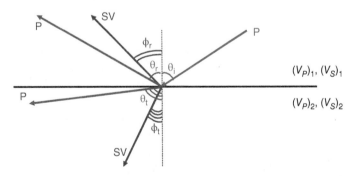

Figure 7.21 Setup for Snell's Laws including converted waves. The incident wave here is a *P* wave. *P* wave orientations are represented by θ angles: angles θ_i, θ_r, and θ_t for the incident, reflected, and transmitted *P* waves; ϕ angles are employed for the *S* waves, ϕ_r and ϕ_t for the reflected and transmitted *S* waves—all measured with respect to the perpendicular to the boundary (dotted line). *P* and *S* wave velocities are $(V_P)_1$ and $(V_S)_1$ in the top layer and $(V_P)_2$ and $(V_S)_2$ in the lower layer. The situation shown here corresponds to $(V_P)_2 > (V_P)_1$.

Note that the fundamental version of Snell's Laws can be rewritten as

$$\frac{\sin(\text{incident wave} \sphericalangle)}{\text{Velocity of incident wave in incident medium}}$$

$$= \frac{\sin(\text{subsequent wave} \sphericalangle)}{\text{Velocity of subsequent wave in subsequent medium}};$$

for example, the situation in Figure 7.21 can be written as

$$\frac{\sin \theta_i}{(V_P)_1} = \frac{\sin \theta_r}{(V_P)_1} = \frac{\sin \phi_r}{(V_S)_1}$$
$$= \frac{\sin \theta_t}{(V_P)_2} = \frac{\sin \phi_t}{(V_S)_2}.$$

That is, the *family* of reflected and refracted, converted and unconverted waves—all the waves generated by the *P* wave initially traveling at velocity $(V_P)_1$ and striking the boundary at angle θ_i—all share the same ratio of $\sin (\sphericalangle)$ to velocity. That ratio,

$$\frac{\sin(\text{wave} \sphericalangle)}{\text{Velocity of wave}},$$

is called the **ray parameter** (often denoted by the symbol *p*). The expanded Snell's Laws can be viewed as a mathematical expression of the requirement that the ray parameter for a given wave remains *constant* as the wave travels through a layered region, crossing or reflecting off the various boundaries, for as long as the waves exist (for as long as $\sin (\sphericalangle)$ is not implied to exceed 1).

On a global scale, the seismic waves propagating within the spherical, layered Earth still must obey Snell's Laws as they encounter each layer boundary. For example, as explained in *Textbox A*, the wave transmitted from layer "one" to layer "two" must obey the law of refraction; but consideration of the geometry of spherical concentric layers reveals that the quantity preserved as the wave propagates through each layer is now

$$\frac{\text{Radius} \times \sin(\text{wave} \sphericalangle)}{\text{Velocity of wave}},$$

defining the ray parameter in the broader situation.

According to this large-scale ray parameter, different 'families' of body waves propagating through the Earth can result from the initial wave leaving the earthquake (or explosion) source with a different 'take-off' angle or from a different source depth. For each family, the sequence of paths they follow will depend on the seismic velocity structure of the interior, and on the kinds of boundaries encountered along the way.

A Global Picture of Seismic Waves. The interior of the Earth is indeed complex in structure, as we will see below; however, most of the boundaries between different layers or structural features actually involve only subtle changes in elastic properties or density. *Three major boundaries* end up causing almost all of the diversity of identifiable seismic waves that we observe at a seismographic station after an earthquake: *the core-mantle boundary* (the **CMB**), *the boundary between crust and mantle* (the Moho), and *the outer surface of the crust* (as noted earlier, often viewed as the Earth's 'free surface').

To keep track of all the body waves that result following an earthquake (or nuclear explosion), we employ a traditional kind of notation (see Figure 7.22). After the initial *P* and *S* waves have been generated, all subsequent reflections off the free surface are denoted simply by concatenating the letters corresponding to the wave types. For example, a wave emerging from the source as a *P* wave and then reflecting without conversion off the outer surface before arriving at the seismographic station would be denoted a *PP* wave. If that same initial *P* wave did convert to an *S* wave at the free surface, it would be denoted a *PS* wave; if it reflected twice off the outer surface before arriving at a station but did not convert again at the second reflection, it would be denoted a *PSS* wave.

Earthquakes occur at various depths within the Earth as great as ~700 km. The *P* and *S* waves produced by an earthquake of course radiate in all directions, leading to some waves

Textbox A: The Ray Parameter in a Spherical Earth

Snell's Laws still apply in a spherical, layered Earth. But it turns out that the ray parameter shared by the pair of waves shown here—and any other 'family' members subsequent to the incidence—must be defined differently from that in a smaller or 'flat' region of the Earth.

In the situation shown here, the incident wave strikes the interface at angle $(\theta_i)_1$ and is transmitted down at angle θ_t; both angles are measured with respect to the perpendicular (dotted line).

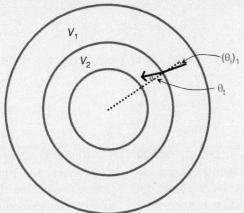

The refraction is still governed by

$$\frac{\sin(\theta_i)_1}{\sin\theta_t} = \frac{V_1}{V_2},$$

but for what follows, we will write this as

$$\frac{\sin(\theta_i)_1}{V_1} = \frac{\sin\theta_t}{V_2}.$$

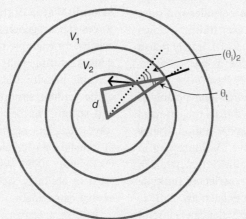

The transmitted wave continues on till it hits the base of the second layer, at an angle $(\theta_i)_2$. The base of the first and second layers is at radius r_1 and r_2, respectively. Thus, in the right triangle marked with bold red, the hypotenuse is of length r_1, and we have

$$\sin(\theta_t) = d/r_1.$$

Textbox A: The Ray Parameter in a Spherical Earth (*Concluded*)

And in the smaller right triangle within (where the hypotenuse is that portion of the dotted line of length r_2),

$$\sin(\theta_i)_2 = d/r_2.$$

Combining these relations yields

$$\sin(\theta_i)_2 = r_1 \sin(\theta_t)/r_2$$

or

$$\sin(\theta_t) = r_2 \sin(\theta_i)_2/r_1.$$

Finally, substituting this into our version of Snell's Law above yields

$$\frac{r_1 \sin(\theta_i)_1}{V_1} = \frac{r_2 \sin(\theta_i)_2}{V_2}.$$

---

that travel more or less straight upward to the free surface; seismologists distinguish these latter with lowercase letters. Thus, for example, an S wave traveling upward from an earthquake and then reflecting off the outer surface as a P wave before arriving at a seismographic station would be called an sP wave. Such arrivals are seen most clearly for deep earthquakes.

To deal with the core-mantle boundary, "c" is used to denote reflection off the boundary and "K" is used for (P) waves traveling through the core fluid. For example, consider an S wave generated by an earthquake, traveling down through the mantle to the CMB. It may reflect back up without conversion as an ScS wave, or with conversion as an ScP wave; or, it may propagate into the outer core—unavoidably as a P wave—so by the time it exits the core and is recorded at a seismographic station, it may be an SKP or an SKS wave. As illustrated in Figure 7.22, another possibility is that, while in the outer core, the wave experiences multiple reflections, leading to waves such as *PKKP* or SKKKS. Such waves in particular can be useful for learning about the outermost core, since a significant fraction of their travel time through the Earth will have been spent in that region.

Finally, waves that remain unconverted P waves as they go 'straight through' the Earth, that is, PKP waves, are sometimes denoted

P'. There is some ambiguity concerning such waves, as—depending on how 'straight through' the core their path is—they may or may not encounter the solid inner core. Waves that travel through the inner core as a P wave are denoted by I; those traveling as an S wave—as a result of conversion at the inner-core boundary—are denoted by J. Thus, for example, a wave traveling through the entire Earth and back up to the surface as a P wave could be denoted PKIKP.

Seismic Waves in the Earth System. Sound waves can be generated naturally within the oceans, for example by iceberg fracturing, 'ice quakes' during collisions between icebergs, volcanic activity, sudden seismic movements of the seafloor, slumping of seafloor sediment, and so on. The oceans also transmit sound waves that had been created elsewhere, e.g., by crustal or even deeper earthquakes (e.g., Okal, 2001, 2008). Whether propagating from crust to ocean or vice versa, the transmitted energy can remain a sound wave throughout or involve a conversion of S wave to sound wave energy. Within the oceans, as implied by our earlier discussion, propagation of the sound waves will take place most effectively (see Okal, 2012) through the SOFAR channel.

To get into (or out of) the SOFAR channel requires a bit of geophysics, however. The

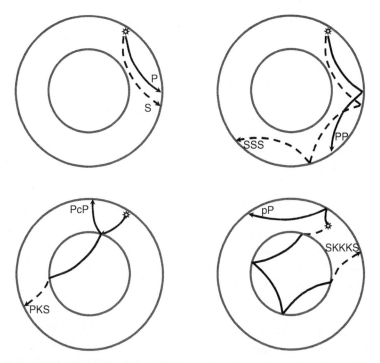

Figure 7.22 A sampling of the variety of body waves resulting from reflection, transmission, and/or conversion after starting out as *P* and *S* waves (solid and dashed curves, respectively) generated by an earthquake. The earthquake source is marked by a star symbol (☼). In these examples, the key boundaries are the core-mantle boundary and the free surface.

velocity contrast between oceans and crust implies that waves coming from land, for example, will be refracted toward the normal, making it unlikely they will be trapped within the essentially horizontal channel. To circumvent that, seismologists have proposed a "downslope conversion" of the sound wave (Okal, 2008), in which it bounces around within the oceans a few times, reflecting off the sea surface and a sloping seafloor that gradually redirects it toward the horizontal.

Sound waves travel through the ocean at speeds of ~1.5 km/sec. When they finally reach a continental seismographic station, even after crossing the crustal boundary and continuing on as *P* waves, they will generally arrive later than the *P* and *S* waves produced by the same earthquake that had followed paths of similar length but avoided the ocean. Arriving third earns them the designation of "*T* wave."

The speed of a *T* wave depends strongly on the ocean temperature, an increase of which will decrease seawater density. As global warming increases temperatures, even at the depths of the SOFAR channel, we would expect that velocity to increase. Because the SOFAR channel extends throughout the breadth of the world's oceans, measurements of such a velocity increase might provide strong confirmation of an actual increase in the global average temperature. A program to conduct such measurements—with the *T* waves generated artificially within the SOFAR channel by strong transducers at specific frequencies, and recorded by hydrophones dispersed ocean-wide—was proposed by Munk and Forbes (1989). The possible impact of such measurements on whales and other marine life, however, eventually raised persistent concerns among environmental groups and scientists, and the program was discontinued. Nevertheless, the ability of the SOFAR channel to act as such an effective waveguide shows that it has the potential to remotely monitor the effects of global warming (e.g., Talandier et al. 2002, Li & Gavrilov, 2006; MacAyeal et al., 2008).

As discussed earlier, sound waves generated by powerful natural and anthropogenic sources can propagate through the atmosphere, helped at least somewhat by the presence of atmospheric waveguides. Propagation over long distances is more likely at low frequencies, for which the atmosphere is less absorptive (e.g., ReVelle, 2004; Whitaker & Norris, 2008). Research focusing on **infrasound**, waves at low enough frequencies—often taken to be frequencies below ~20 Hz—that they are mostly inaudible to humans (e.g., Leventhall, 2007), has been found to be particularly relevant to the Earth System. For example, analysis of the infrasound waves propagating from a meteor explosion above Oregon in 2008 allowed Hedlin et al. (2010) to infer the occurrence at that time of a sudden stratospheric warming event.

We will see in Chapter 8 that analyzing P and S waves can reveal much about the structure and physical state of Earth's interior. In the same way, analysis of infrasound waves allows the structure and state of the atmosphere to be constrained (e.g., Hedlin et al., 2012).

7.4 Surface Waves and Free Oscillations

7.4.1 Surface Waves

After the direct and most of the reflected, refracted, and converted P and S waves produced by an earthquake have passed by us, we generally observe other waves—waves that reach us with larger amplitudes than any recorded earlier. Significantly, rather than there being one or two sinusoidal or up-and-down pulses marking their arrival, for example

as is typical for body waves, they appear as a *train* or sequence of waves,

These waves are also distinguished from body waves by having periods that range from several seconds up to at least several minutes, with a representative value of perhaps ~20 sec; P and S wave periods are usually around 1 sec or less.

Two kinds of these waves are observed. The one that arrives first involves only side-to-side horizontal ground motion, and can therefore be recorded only on horizontal-component seismographs; this is called the **Love wave**, named after the scientist who deduced the conditions under which it occurs (he also wrote the classic reference text on elasticity). The slower wave, arriving soon after the Love wave, is called the **Rayleigh wave**; unlike the Love wave, it involves forward-back horizontal ground motion *and* vertical ground motion.

Love and Rayleigh waves are not simply slow body waves. In addition to being more energetic (larger amplitude) and lower frequency, they have one property that merits them being called "surface waves": their energy is concentrated near the top surface of the earth. More precisely, their amplitude decreases exponentially with depth. If, for example, that 20-sec Rayleigh wave travels along the surface of the earth with an amplitude of 10 cm, then 10 km down its amplitude would be 8.6 cm; 100 km down, its amplitude would be only 0.7 cm; and, at depths well into the mantle (say, 300 km), there would be only microscopic indications that the wave was passing by. Surface waves do not travel *through* the earth; rather, they travel along the top surface, disturbing the interior only slightly.

How do you think the surface waves created by deep and shallow earthquakes might compare?

The particle motions produced by the passage of a surface wave are a clue to its origin. The horizontal ground motion generated by

Love waves is perpendicular to the direction of propagation—just like an *SH* wave; and, indeed, they are created by *SH* waves, though as discussed below, it is not just a simple act of creation. The ground motion generated by a Rayleigh wave as it propagates, horizontally (along the top surface), includes both a longitudinal component and a vertical transverse component (see Figure 7.23). This suggests a combination of *P* and *SV* waves is at play—that is, the existence of the Rayleigh wave is tied to the joint action of *P* and *S* waves.

The velocity of Rayleigh waves is slightly but distinctly less than that of *S* waves. For example, in the case of a free surface overlying a "half-space"—a flat, uniform medium extending indefinitely below the surface, often used as a first approximation to the Earth to study seismic wave propagation—it can be shown theoretically that the Rayleigh wave propagates along the surface at a speed $V_{Rayleigh} \cong 0.92V_S$, where V_S is the velocity of *S* waves within the medium. We might account for $V_{Rayleigh}$ being less than V_S simply as a consequence of the free surface being more deformable than the interior of the medium; with less resistance to deformation, waves propagating along the boundary *should* travel slower than the *P* and *S* waves which travel within the medium.

Somewhat more rigorously, it can be argued (see Knopoff, 2001) that $V_{Rayleigh}$ must be less than V_S because it is traveling along a free surface (or at least a surface experiencing zero shear stress) *and* because its energy must remain concentrated near that surface and attenuate with depth. However, the existence and nature of Rayleigh waves, including their speed, are most thoroughly explained by theoretical considerations, involving solutions to the wave equation constrained by free-surface boundary conditions—considerations that are beyond the scope of this textbook.

So we can view the situation for Rayleigh waves as follows. A *P* or *S* wave strikes the top surface of the earth; reflections take place off the boundary, including conversions of some *P* wave energy to an *SV* wave, or vice versa. But at shallow-enough angles of incidence (~80° off the vertical, as it turns out, so that this process will take place somewhat far from the earthquake source, at a distance ~6 times the source depth; see Stacey & Davis, 2008) there will also be *P* wave-*like* and *S* wave-*like* deformations that couple together and remain near the boundary, propagating along it relatively slowly as an elastic deformation.

In the real, layered Earth, with seismic velocities generally increasing with depth, Rayleigh waves exhibit a phenomenon known as *dispersion*. Longer-wavelength Rayleigh waves, disturbing a greater volume of rock as they propagate along the surface than if they had short wavelengths, penetrate deeper into the Earth than the shorter wavelengths

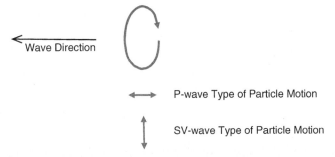

Wave Direction

P-wave Type of Particle Motion

SV-wave Type of Particle Motion

Figure 7.23 Ground motion at the earth's free surface due to the passage of a Rayleigh wave. With a combination of the types of particle motion associated with *P* and *SV* waves, the result is an elliptical particle motion in a *retrograde* sense: at the top of the cycle, each particle moves *away* from the direction the wave is propagating in. This vertical elliptical particle motion creates the "ground roll" experienced away from an earthquake.

(though still decaying away at even greater depths); sensing the higher velocities of those deeper layers, they propagate more quickly and arrive sooner. The result is a 'spread' or extended sequence of Rayleigh wave arrivals at a given station—a **wave train**—with the longer wavelengths showing up first, followed by shorter and shorter wavelengths.

In Chapter 5, we found that the mantle flow generated by lithospheric flexure from surface loads was characterized by a viscosity that depended on the scale of the load (see, e.g., Figure 5.28 of that chapter): the flow generated by spatially limited loads sensed the viscosity of the shallow asthenosphere, whereas the flow from broad loads 'saw' the deeper mantle's viscosity. Why do you think such different phenomena—flexure-generated mantle flow and seismically generated surface waves—exhibit the same kind of spatial-scale sensitivity to depth? Is there a basic physical principle underlying such sensitivity?

In general, **dispersion** is defined as a situation where wave velocity depends on wavelength. Since longer-wavelength waves also have longer periods T, dispersion is sometimes expressed in terms of wave periods: $V = V(T)$. The wave train exhibits increasingly shorter-period arrivals later in the sequence.

Thus, for a two-layer Earth in which S wave velocities are $(V_S)_1$ and $(V_S)_2$, with $(V_S)_1 < (V_S)_2$, the very long-period (= long-wavelength) Rayleigh waves 'see' mainly the lower layer; it's almost as if the upper layer does not exist, and the waves propagate with the Rayleigh wave velocity for the lower layer, $0.92(V_S)_2$. Conversely, very short-period Rayleigh waves hardly 'see' the lower layer, and travel at velocity $0.92(V_S)_1$ (Figure 7.24).

Love waves have a very different origin than Rayleigh waves. As noted above, the ground motion due to a Love wave, as it propagates along the Earth's surface, is pure transverse horizontal particle motion—just like that associated with SH waves. SH waves do not convert to P waves at a boundary, or vice versa, so Love waves do not involve P waves or the coupled deformations of a Rayleigh wave. Instead, Love waves occur when an SH wave is trapped within a layer at Earth's surface, internally reflected up and down at shallow angles as illustrated in Figure 7.15b. For Love waves to develop, however, that top layer must have slower S wave velocities than the medium below it. For the Love waves we see following an earthquake, the Earth's *crust* serves as the top layer. As befits a surface wave, the amplitude of a Love wave decays exponentially with depth—but only below the crust.

The surface layer enabling the Love wave to develop acts like a waveguide for the internal reflections, so they can interfere constructively to create an SH-type disturbance—confined

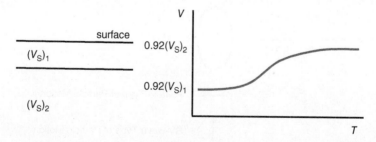

Figure 7.24 Dispersion of Rayleigh waves, expressed in terms of the wave period T. In a two-layer lithosphere, for example (shown on left), with $(V_S)_1 < (V_S)_2$, the wave velocity V depends on the wavelength or period of the wave: shorter-wavelength, shorter-period waves 'see' mainly the upper layer, producing a Rayleigh wave with velocity $0.92(V_S)_1$; very long-wavelength, long-period waves, though, 'see' the deeper half-space, find the effects of the upper layer to be negligible, and travel with velocity $0.92(V_S)_2$. Waves of intermediate period travel at velocities between these two limits.

mainly to the layer—that propagates horizontally through it. The velocity of propagation, V_{Love}, is greater than the S wave velocity of the layer but less than the S wave velocity of the underlying medium. The latter mismatch ($V_{Love} < (V_S)_{medium}$), which is reasonable considering the conditions required to create a waveguide (see Figure 7.15b), ensures that the disturbance stays mainly trapped in the layer and that its energy decays exponentially with depth below the layer.

One Step Further? Intriguingly, our description of the creation of Love waves might be applied to a distinctly non-Love situation: a liquid layer (in which no S waves could propagate) overlying a solid half-space. As studied by Press and Ewing (1950) and noted by Garland (1971), P waves trapped in the liquid layer (from reflections off the lower boundary with a great-enough angle of incidence) could constructively interfere to produce a wave that propagates along the liquid layer with most of its energy similarly concentrated, decaying away with depth—that is, a longitudinal Love wave! And by analogy with Love waves, those surface waves would propagate faster than P waves in the liquid layer!

In the case of the oceans, such waves would therefore travel faster than T waves could propagate through the SOFAR channel. But this scenario fails in the real world: Earth's outer structure is dominated by tectonic processes; as a result, the ocean's liquid-solid interface—the seafloor—is too irregular for reflections to constructively interfere and coherently propagate as a Love-type wave. (However, as we will learn later in this chapter, this type of scenario actually does operate—in the Sun!)

Because the internally reflected SH waves require an angle of incidence greater than critical, the Love wave—like the Rayleigh wave—does not appear in the 'near field' of its earthquake source. For very shallow earthquakes, we would expect surface waves to develop by ~50–100 km away from the source.

Like Rayleigh waves in a layered medium, Love waves are always dispersive, with longer wavelengths sensing more of the faster 'half-space' below and thus propagating faster (similar to what is shown, in terms of period rather than wavelength, in Figure 7.24 for Rayleigh waves). Furthermore, in order for the interference between trapped SH waves in the top layer to succeed in building the Love wave, the 'right' wavelengths must combine, and the constructed wavefront must be traveling at an appropriate speed (other factors, including the thickness of the layer, are also important for favorable constructive interference; see, e.g., Stein & Wysession, 2003 for details). These constraints in effect represent a second way that the wave's wavelengths depend on its velocity, that is, another kind of dispersion. It also 'destructively' eliminates selected wavelengths from the propagation, exacerbating the dispersion.

The dispersion of Love waves, and Rayleigh waves as well, is made still more complicated by the possibility of the waves propagating as higher *modes*. Modes are an important part of the story of free oscillations, the next subject of this chapter; for now, we define the mode of a surface wave simply, according to how its particle motion (e.g., its side-to-side horizontal displacement, in the case of a Love wave) varies with *depth*: the **nth mode** has zero displacement at n depths within the top layer. Those depths are called "**nodes**"; at each nodal depth, the particle motion reverses (we described a 'horizontal' version of nodes and modes when discussing seiches earlier in this chapter). Higher modes of surface waves can thus 'oscillate' a number of times before dying off at greater depths.

These oscillatory modes are still Love and Rayleigh waves: they still satisfy the wave equation and appropriate boundary conditions; their velocities are still bounded by the shear-wave velocities of the half-space below and, if present, the top layer; and they attenuate exponentially with depth in the half-space. In general, the surface wave characteristics we have discussed so far refer to the $n = 0$ mode, which is also called the **fundamental mode**. For any mode, the wave speed and wavelength are greater, the longer the period of the wave. But the higher modes do stand out from the fundamental mode in a few ways (e.g., Stein & Wysession, 2003). Consider, as an example, a region of the earth well approximated by a layer overlying a half-space. Of all the Love waves—all the different modes—that can propagate there with a specific period, the fundamental mode travels the slowest, and the higher the mode, the faster it moves (all within the bounds just mentioned). At long-enough periods, though, the fundamental mode senses only the half-space, so it travels at a speed of $(V_S)_{medium}$—which is already the upper bound, leaving no 'room' for any of the higher modes; that is, at long periods, only the fundamental mode can exist.

As noted at the outset of this discussion, surface wave amplitudes can end up much larger than the body wave arrivals on a seismogram (see Figure 7.25). In fact, beyond the 'near field,' Love and Rayleigh waves inevitably bring more destruction with them than any P and S wave could. As they spread out away from an earthquake source, surface waves weaken *less* than do body waves, because they are confined almost entirely to the Earth's surface; in contrast, body waves spread out through the entire volume of the Earth. Also, the top surface channels a lot of seismic energy into surface waves. And, their wave trains create large, sustained ground motion—in contrast to the 'jolt' of a P wave or strong 'snap' of an S wave—with the range of individual frequencies likely to include some that induce the overlying buildings to sway in resonance.

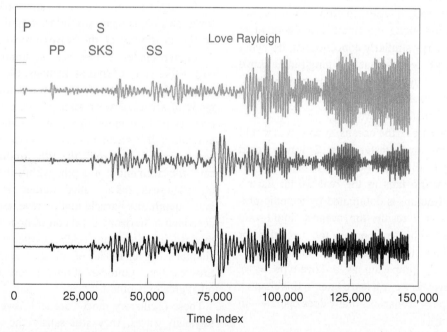

Figure 7.25 Digital seismograms from IRIS station CMB in Columbia, CA, showing the ground motion recorded after a major earthquake off the coast of New Zealand occurred at a depth of 21 km on July 15, 2009. The three component seismometers are set up to measure vertical, N-S, and E-W ground motion (the resulting seismograms are shown, offset, top to bottom). Data graphed every 0.05 sec, so that every 6,000 points represents 5 minutes; the entire span of data shown here is ~90 minutes. Data was downloaded from http://ds.iris.edu/wilber3/find_event. IRIS is the Incorporated Research Institutions for Seismology. Some of the body wave arrivals (as determined by IRIS) and surface wave arrivals are marked above the seismograms.

Because their existence, speed, and dispersive properties depend so strongly on the layer thicknesses and seismic velocities of the crust and upper mantle, Love and Rayleigh waves are a valuable tool for exploring the outermost earth. For instance, surface wave dispersion can be used to infer differences in the properties of oceanic crust versus continental crust (see Figure 7.26).

7.4.2 Free Oscillations

There is a special situation involving surface waves whose ability to shed light on Earth's interior is exceeded only by its comparative rarity of occurrence. If an earthquake of sufficient magnitude takes place, surface waves will be produced (and spread out in all directions away from the source) which are capable of traveling completely around the world, even multiple times. If their wavelengths are *right*, the circuiting waves from opposite directions will combine to produce standing waves, and it will seem as if the whole Earth is vibrating in and out.

These vibrations, called **free oscillations** or **normal modes** or **eigenvibrations** of

the Earth, will mostly be composed of very long wavelengths: shorter-wavelength surface waves go through more cycles (i.e., are higher frequency) as they circuit the globe; the slight loss of wave energy per cycle (from internal friction or other imperfections in the solid earth's elasticity) adds up more, and the waves fade away, leaving only the longer wavelength surface waves to combine into standing waves. As with traveling surface waves, the longer wavelength normal modes penetrate into deeper regions of the Earth; in fact, there are entire *suites* of modes that are sensitive to the deep mantle and the core, potentially yielding key information regarding Earth's deep interior.

Because there are two types of surface waves, there will be two types of free oscillations. Those created from standing Rayleigh waves, which possess vertical as well as horizontal particle motion, are called **spheroidal** modes; such modes are more formally defined as having a component of radial (the global version of vertical) particle motion. For example, if the wavelength of a very, very long-period Rayleigh wave is 20,000 km—half the circumference of the Earth—then, by the time

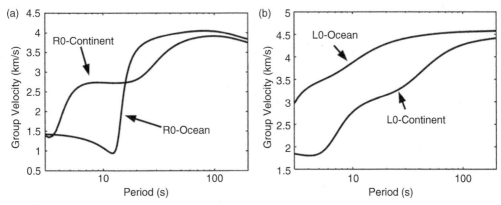

Figure 7.26 Dispersion of surface waves can shed light on the earth's near-surface structure. These graphs show dispersion curves—wave velocity as a function of wave period—for (a) Rayleigh waves and (b) Love waves. The 0 in the label for each wave identifies it as the *fundamental mode* of the wave. Note how different the dispersion is for waves traveling purely continental paths versus purely oceanic (i.e., ocean crust) paths—a result of different crustal thickness, structure, and composition.
Image from Levshin et al. (2018), based on seismic models by Kennett et al. (1995) for continental crust + mantle and by Herrmann (2013) for oceanic crust + mantle.
The velocities shown here are *group velocities*, which reflect the overall speed at which the ensemble of waves propagate, rather than individual "phases" of various wavelengths.

the wave has circled the Earth a few times, from every direction, it will appear as if the Earth is elongating and contracting in a so-called "football mode" of oscillation (see Figure 7.27).

The other type of normal mode derives from standing Love waves; characterized by purely horizontal particle motions, such modes are called **torsional** (as in *twisting*) or **toroidal** (*toroids* are donut-like shapes whose smooth outer surfaces are perfectly 'zonal' or azimuthal, i.e. they have no radial component pushing out or in—at least, for plain donuts!). For example, if Love waves with a 20,000 km wavelength combined to produce a standing wave, we would see half the Earth twisting one way while the other half twisted oppositely (see Figure 7.28).

Because of the constructive and destructive interference leading to standing waves, even in

the simple case of waves produced by shaking one end of a rope attached (at the other end) to a wall, the standing waves exist only at specific, discrete wavelengths. For the rope, the correct wavelengths are achieved by shaking the end at specific frequencies; which frequencies 'work' depends on the velocity at which the shake propagates. Just imagine, for example, two ropes of identical rigidity but one made of material ten times as dense as the other; if both are shaken at a frequency that creates standing waves in the less dense rope, the shake will propagate too slowly in the denser rope to create the 'right' constructive/destructive interference. Similarly, in order for the Earth's surface waves to successfully combine to produce a normal mode, those waves must travel at velocities that produce the 'right' frequencies of deformation.

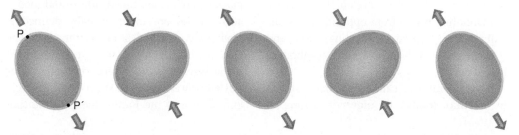

Figure 7.27 Free oscillations, produced when an earthquake causes the Earth to 'ring.' Shown here is the spheroidal oscillation produced from standing Rayleigh waves with a wavelength equal to half of Earth's circumference—the "football mode" oscillation (exaggerated here for visibility). For this mode to derive from Rayleigh waves, we can imagine that the waves were traveling along the great circle between points P and P'.

For clarity, arrows are used here to show the surface displacement at points P and P', but not at points perpendicular to the great circle between P and P' midway between them, that is, not at points on the 'equator' relative to the 'poles' P and P'. Can you describe what those 'equatorial' displacements would look like?

Figure 7.28 Free oscillations, produced when an earthquake causes the Earth to 'ring.' Shown here is the toroidal oscillation produced from standing Love waves with a wavelength equal to half of Earth's circumference. The arrows denote horizontal, side-to-side motion of the Earth; in the left-most picture we see the oscillation initially, with a peak of eastward motion at 45°N 'latitude'—measured relative to P—and a trough of westward motion at 45°S. For this mode to derive from Love waves, we can imagine that the waves were traveling along the great circle between points P and P'.

In sum, the Earth's normal modes will not only exist at specific wavelengths, but at specific frequencies as well; these frequencies are called **eigenfrequencies**. With very large distances to be covered and the velocities of even long-wavelength Love and Rayleigh waves relatively slow, however, the frequencies of the normal modes will be very low; as a result, the modes are typically described according to the period rather than frequency of vibration. Those modes discussed above have some of the longest periods of all free oscillations, including 44 minutes for the toroidal mode of Figure 7.28 and a striking 54 minutes for the football mode! Higher modes—spheroidal and toroidal modes with shorter wavelengths (and thus more cycles of deformation going around the globe)—tend to have periods of only several minutes or less.

A Spherical Harmonic Perspective. The spherical harmonic functions discussed in Chapter 6 can be used to describe the deformations experienced by the Earth during its free oscillations. This is not unexpected, since combinations (i.e., sums) of such functions can be used to describe *any* surface pattern; conveniently, though, it turns out that the particle motions at Earth's surface of a normal mode with a specific period of oscillation can be aptly described using a *single* harmonic. In the spheroidal football mode, for example, where the Earth alternates between oblate and prolate ellipsoidal shapes, the radial particle motion is evidently well characterized by the degree 2 order 0 harmonic—the same harmonic we had used to represent the shape and gravitational effect of Earth's equatorial bulge. Here, incidentally, the 'latitude' and 'longitude' in the harmonic function would be defined relative to the earthquake location (see, e.g., Figure 7.27, also Figure 7.28). This mode is often denoted S_{20}. Higher modes, with shorter wavelengths around the globe, correspond to higher-degree and higher-order spherical harmonics.

Spherical harmonics can also be used to describe the horizontal particle motion of spheroidal and toroidal modes. But it turns out (—a phrase which, by now, should be recognized by the reader as a euphemism for 'it is shown by the solutions to very complicated equations') that, in addition to a description involving an infinite sum of all harmonics, those horizontal particle motions can *alternatively* be described by derivatives of the single harmonic function used for the radial component of the particle motion. For example, for the mode S_{20}, if the degree 2 order 0 harmonic function is denoted by $Y_{20}(\phi, \lambda)$, where ϕ is 'latitude' and λ is 'longitude' or more precisely azimuth relative to the earthquake source, we would find that the 'latitudinal' component of particle displacement for S_{20} (i.e., the component of horizontal particle motion along the 'north'-'south' great circle toward or away from the source) varies spatially according to $dY_{20}(\phi, \lambda)/d\phi$. We would also find that, for the toroidal mode pictured in Figure 7.28, the transverse (i.e., SH-type) particle motion also varies spatially according to $dY_{20}(\phi, \lambda)/d\phi$; for this reason, that normal mode is often denoted T_{20} (—a toroidal mode, related to the 2,0 harmonic).

Spatial derivatives of spherical harmonics arise in the description of free oscillations because, as Stein and Wysession (2003) remind us, the wave equation involves such derivatives of the strains propagating as surface waves. They are part of a bigger mathematical picture—a way to describe vector 'fields,' using a tool called *vector spherical harmonics*—but a discussion of that is beyond the scope of this textbook.

Another complication in the harmonic description of normal modes is that the deformations at the surface penetrate deep into the interior, just as with very long-wavelength surface waves; inevitably, then, a complete description of any mode must include its radial variations. But the spherical harmonics described in Chapter 6—so useful for quantifying surface patterns—are unable by themselves to provide such a description; in fact, those harmonics are more officially called **surface spherical harmonics**. It turns out (...) that those equations which give rise to spherical harmonic solutions can be solved more completely; the net result involves spherical harmonic functions multiplied by various functions of radius. One of these more general types of solutions, in which the functions of radius are found to be powers of either r or $1/r$, is known as **solid spherical harmonics**; solid spherical harmonics find their most frequent use in 3-D descriptions of gravity and magnetic fields. We will discuss them in Chapter 10.

And a related complication—which we already encountered in the context of surface waves—is that the same pattern of surface deformation from a normal mode can be associated with different radial variations. The various possibilities are distinguished by the number of "nodal surfaces," or surfaces where the radial variation goes from positive to negative or vice versa, versus depth within the Earth. If a mode has no nodal surfaces, then, at any latitude and longitude, the deformation of the Earth from the surface down through the interior happens in unison (e.g., all to the left or all to the right, at a given location, for T_{20}); such modes are called **fundamental modes**; the rest are called **higher-order modes**. Given the popular 'musical' view of normal modes—that they result when a great earthquake (defined loosely as one of the very largest earthquakes) makes the Earth 'ring like a bell'—higher-order modes are also called **overtones**. For these modes, particles at a particular latitude and longitude at some instant of time will all be moving to the left (e.g.) above a nodal plane but to the right below it. Furthermore, because overtones 'ring' at different frequencies (a consequence of the elastic properties of the Earth varying with depth), we should expand the notation used to symbolize each mode; as an illustration, an overtone of T_{20} with k nodal surfaces is often denoted $_kT_{20}$ (but if k = 0, $_0T_{20}$ is usually just written as T_{20}).

Earth's normal modes were first widely observed after the great Chilean earthquake of 1960, the largest event of the twentieth century. The confirmation of their existence months later at an international seismological conference—as observations from diverse research groups converged, and were connected to 'fallow' theoretical work stimulated by an earlier great earthquake—was a dramatic (see Bullen, 1965) step up in our scientific understanding of how the Earth behaves. By 1980, more than 1,000 different normal modes had been detected (see, e.g., Gilbert & Dziewonski, 1975), using strain meters, gravity meters (since spheroidal modes change the shape of the Earth, they modify the gravity field), and especially seismometers around the world to measure the deformations following large earthquakes. Such analyses led to the development of the *Preliminary Reference Earth Model* or PREM, a very comprehensive model of Earth's density and elastic properties versus depth (Dziewonski & Anderson, 1981). The use of normal modes to construct Earth models will be discussed in Chapter 8. For now, we note peripherally that normal-mode observations *also* shed light on the dissipative or "anelastic" properties of the interior: it was found necessary to include such slight departures from purely elastic behavior in Earth models in order that surface wave energy would dissipate as they propagate, in turn producing normal modes (from their constructive interference) with smaller amplitude ground motion.

PREM was actually based on a selection of ~900 modes, including 80 fundamental modes (up to $S_{66,0}$ and $T_{36,0}$), as well as additional surface and body wave data. More recent studies, employing higher-quality seismometers and gravimeters with an expanded global distribution and an obviously greater collection of earthquakes to analyze, have been able to increase the variety of both fundamental modes and overtones that are detected. For example, Masters and Widmer (1995) report the detection of more than 600 different normal modes, about 180 of which were fundamental modes (up to $S_{97,0}$ and $T_{85,0}$).

A Closer Look at Normal-Mode Frequencies. There is one final complication regarding the periods or frequencies of Earth's normal modes. (*So many complications, you say? I say: be grateful! The more numerous the complications, the more we can learn about the Earth!*) After a large earthquake, the Earth reverberates with the passage and echoes of numerous body waves, especially early on; minutes later, the shaking of surface waves grows to dominate the ground motion, spreading worldwide; then normal modes ensue, and they may persist for a very long time. But when the ground motion at a site is analyzed to determine the periods or frequencies contributing to the shaking, the normal-mode ground motion is often found to exhibit *several* sinusoidal components with frequencies close to that of a particular normal mode—rather than a single sinusoid—for *each* of the expected normal-mode eigenfrequencies.

This is illustrated indirectly in Figure 7.29a, which shows the crustal strain over time measured at a central California site following the 1960 Chilean earthquake. When the time series of strain is 'filtered' to eliminate sinusoids of all frequencies except those near the expected eigenfrequency—a process called **narrow-band filtering** ("band" refers to a continuous range of frequencies)—we see the type of constructive and destructive interference pattern that is known to result from the superposition of sinusoids of slightly different frequencies, rather than the expected single sinusoid.

By the way, such a pattern is called a "beating phenomenon" (see Figure 7.29b). We can experience that phenomenon in everyday life, for example by listening to two slightly different musical pitches sung or played simultaneously; we will hear a drawn-out wavering, i.e., 'beats' of louder then softer sound, as the combination interferes constructively and then destructively. (Note that the two tones must be very similar in frequency or the beating will

be too rapid for the human ear to distinguish.) This same effect is what piano tuners listen for as they tighten or loosen a piano string while an electronic or mechanical 'tuning fork' vibrates at the desired pitch; when they hear the beats spread out, they know that the string's own pitch is becoming closer and closer to the standard.

A Bit of Math Here Goes a Long Way. Beating phenomena can be appreciated mathematically as well. If two sounds or normal modes can be described by $\sin(\omega_1 t)$ and $\sin(\omega_2 t)$, where ω_1 and ω_2 are the frequencies of the individual sounds, then we hear them together as

$$\sin(\omega_1 t) + \sin(\omega_2 t).$$

Using trigonometric identities for $\sin(A+B)$ and $\sin(A-B)$, and letting $A = \frac{1}{2}(\omega_1 + \omega_2)t$ and $B = \frac{1}{2}(\omega_2 - \omega_1)t$, we can write the total sound as

$$\sin(\omega_1 t) + \sin(\omega_2 t)$$
$$= 2\sin\left(\frac{\omega_1 + \omega_2}{2}t\right)\cos\left(\frac{\omega_2 - \omega_1}{2}t\right).$$

That is, we hear them together as a sinusoid of their average frequency ($\frac{1}{2}[\omega_1+\omega_2]$) multiplied by or *modulated* by the cosine of the difference between their frequencies. If the frequencies are very similar, the average frequency will approximate both frequencies, and (recalling that cosine of an angle nearly zero is close to 1) we will roughly hear a doubling of that sound; but that sinusoid we hear will waver, being gradually amplified and diminished—very gradually, in fact, since the period associated with the 'difference frequency' ($\frac{1}{2}[\omega_2-\omega_1]$) will be very long. Each cycle of amplification and diminishment is one 'beat' of the combined sound.

Can you verify the formula for the total sound? Your derivation should only be a few lines long.

If the analysis of data in Figure 7.29a had employed very wideband filters centered on each expected eigenfrequency, sinusoids with a wide range of frequencies would have

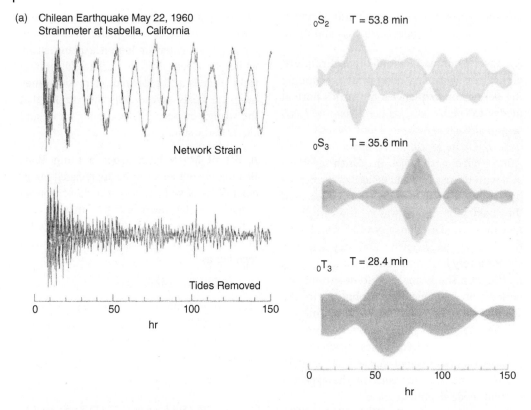

(a) Chilean Earthquake May 22, 1960
Strainmeter at Isabella, California

Network Strain

Tides Removed

$_0S_2$ T = 53.8 min

$_0S_3$ T = 35.6 min

$_0T_3$ T = 28.4 min

Figure 7.29 (a) (Left) Strain measured at a site in central California. Top curve shows the original data, with time 0 the beginning of the great earthquake in Chile. Bottom curve shows the data after the tidally produced strain (as well as all other sinusoidal signals with periods of a few hours or more) has been numerically eliminated.
Note, incidentally, that earth tides can be much larger than free oscillations.
Figure from Geller (1977).
(Right) 'Beating' patterns exhibited by crustal strain at the site, in frequency bands centered on the frequency of the S_{20}, S_{30}, or T_{30} fundamental modes. These time series were obtained by narrow pass filtering the tidally cleaned data shown on the left; they are shown enlarged, not to scale, but all with the same time axis (in hours). Because of the timescale, the oscillations in each time series are too numerous to pick out individually, but the beating pattern—an amplitude growth and shrinking as the individual overtones constructively and destructively interfere—is still clearly seen.
It is also worth noting that these oscillations persisted over *days* (e.g., 150 hours ≡ 6¼ days)! But, as impressive as that is, the record for perpetual shaking might go to the 635-km-deep great earthquake that occurred in 1994 beneath Bolivia: reportedly (see Okal 1996) the ground motion associated with the S_{00} mode (a mode called the *balloon mode*, because the whole earth expands and contracts like a balloon) lasted for at least 116 days!
Figures are taken from Stein and Geller (1978) / Seismological Society of America.

passed through the filter—too wide a range to create a beating phenomenon. Musically, wideband filtering of data results in a whole chorus of sounds, not just a single wavering tone.

Alternatively, with extremely narrow-band filters, we could end up with just a single sinusoid, not a sum, at the frequency of the filter—a single tone, with no wavering. Then, with a consecutive sequence of such filters, we could identify each of the sinusoids comprising the data, and see for ourselves whether the data contains sinusoids of very similar frequencies to any of the eigenfrequencies.

(b)

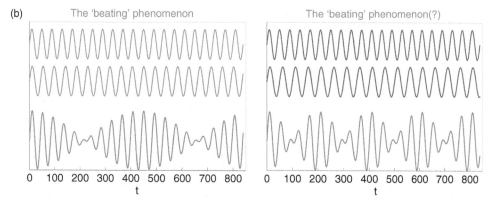

Figure 7.29 (b) A 'beating' phenomenon created by adding two sinusoids of similar frequency. The beating is most apparent if the frequencies are very close (left) but only weakly suggested if the frequencies are not (right). In both illustrations, the two sinusoids are shown in the top portion of the figure (the first sinusoid is the same in both figures), and their sum is shown at the bottom.
With the well-defined beating phenomenon on the left, the sum features a strong modulation, one only vaguely exhibited on the right; in the latter case, there are still minima where the two sinusoids destructively interfere, and maxima where they add, but the cycles are abrupt and brief.

In fact, there is an established mathematical technique for carrying out this filtering procedure and identifying the sinusoids that make up a set of data: **spectral analysis** or **Fourier analysis** (see *Textbox B*, also portions of Chapter Six); in the case of data that is a function of time—a time *series*—the technique is also called **time series analysis**. The results of such an analysis are typically graphed as the amplitude of those sinusoids versus their frequency—that is, the overall amplitude within each very narrow frequency band in that sequence of filters. (Since those bands could be made narrower and narrower, a rule of thumb often used by time series analysts is that the width of those bands is meaningfully never less than $\sim 1/T_{\text{span}}$, where T_{span} is the time span or duration of the data.)

Textbox B: Spectral Analysis

It was mentioned in a textbox in Chapter 6 that sines and cosines can be used to reconstruct any one-dimensional pattern (i.e., any function of a single variable). As an example, with the illustrations from that textbox reproduced below, a triangular pattern was reconstructed to good approximation (shown on the right) with just three sinusoids. Mathematically, that triangular function, which we will denote $TR(t)$, can be well described according to

$$TR(t) \cong 0.86 \cos(\pi t/2) + 0.0955 \cos(3\pi t/2) + 0.0344 \cos(5\pi t/2)$$

(with each term shown below on the left, offset vertically for clarity). For our application to free oscillations, we will use time (t) as the independent variable here. Our interpretation of this reconstruction is similar to what was presented in Chapter 6: if we define 'period' here as the time between zero-crossings, then the 'periods' of the sinusoids building our triangle function are mostly 2, a small amount of 2/3, and a little bit of 2/5.

Textbox B: Spectral Analysis (*Continued*)

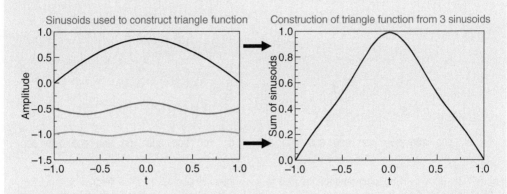

This 'recipe' approximating $TR(t)$ can be simply depicted as

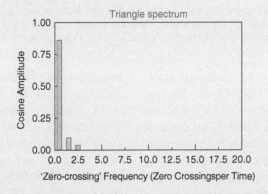

showing the amplitudes of the three cosine functions versus their frequency; following tradition, such depictions—called a **spectrum** or **Fourier amplitude spectrum**—are given in terms of frequency rather than period (or, wavenumber rather than wavelength, if we are dealing with spatial functions).

For example, with our 'zero-crossing' definition of 'period,' a 'period' of two time units for the cosine contributing the most to $TR(t)$ corresponds to a frequency of one half of a zero-crossing per time unit; the other contributing cosines have frequencies of 3/2 and 5/2 zero-crossings per unit time. It should be noted that in this spectrum, we have set the amplitudes of the cosines at all other frequencies to zero, since those other cosine functions contribute nothing to our approximation of $TR(t)$.

In general, we should also specify the sine spectrum, for completeness: cosines alone can only describe functions that are symmetric about the time origin. But for the symmetric triangle function considered here, no sines contribute to it, so their amplitudes would all be set to zero.

For any arbitrary function of time, say $F(t)$, how would we determine its spectrum? There are two equivalent procedures. One option would be to fit $F(t)$ with sines and cosines of every possible frequency; we can then plot their amplitudes versus frequency. For example, one popular 'best-fit' criterion, called "least squares," requires the overall squared error between the candidate sinusoidal function and $F(t)$ to be minimal. The three cosines shown above contributing to $TR(t)$ each represent a least-squared best fit to $TR(t)$ at their frequency. At other frequencies, which contribute only marginally to $TR(t)$, even the best-fitting cosines do not fit well, and their amplitudes are very nearly zero.

Textbox B: Spectral Analysis (*Continued*)

In essence, finding how much of the data best-fits a specific sinusoid is what that narrow-band filtering (discussed in the text) does, if the band is extremely narrow and it's centered on the particular frequency of the sinusoid.

The other, more 'automated' option would be to determine the full sinusoidal expansion that represents $F(t)$. That is, we would write

$$F(t) = \sum_n \left\{ S_n \cdot \sin(2\pi n \Delta t) + C_n \cdot \cos(2\pi n \Delta t) \right\}$$

or

$$F(t) = \{C_0\} + \{S_1 \cdot \sin(2\pi\Delta t) + C_1 \cdot \cos(2\pi\Delta t)\} + \{S_2 \cdot \sin(4\pi\Delta t) + C_2 \cdot \cos(4\pi\Delta t)\}$$
$$+ \{S_3 \cdot \sin(6\pi\Delta t) + C_3 \cdot \cos(6\pi\Delta t)\} + \ldots$$

(noting $\sin(0) = 0$ and $\cos(0) = 1$, and using braces to separate sinusoids of different frequencies) and simultaneously determine all of the sinusoid amplitudes, S_n and C_n. In this expression, the frequencies are (for simplicity) taken to be multiples of a basic frequency interval, Δ cycles per time (with factors of 2π so the sinusoid arguments are in radians), and we henceforth define **period** more rigorously as the time per cycle, i.e. the time between crests or between troughs (not between zero-crossings).

For each frequency the sine and cosine amplitudes S_n and C_n contributing to $F(t)$ yield the recipe of interest. They can be found the same way spherical harmonic coefficients were found in Chapter 6, by orthogonality; orthogonality allows us to 'pick out' each coefficient. For example, if we are interested in the $n = 2$ sine coefficient, then we multiply our expansion by $\sin(2\pi \cdot 2 \cdot \Delta t)$ or $\sin(4\pi\Delta t)$, using color to better visualize the orthogonality; thus,

$$F(t) \cdot \sin(4\pi\Delta t) = C_0 \cdot \sin(4\pi\Delta t) + S_1 \cdot \sin(2\pi\Delta t) \cdot \sin(4\pi\Delta t)$$
$$+ C_1 \cdot \cos(2\pi\Delta t) \cdot \sin(4\pi\Delta t) + S_2 \cdot \sin(4\pi\Delta t) \cdot \sin(4\pi\Delta t)$$
$$+ C_2 \cdot \cos(4\pi\Delta t) \cdot \sin(4\pi\Delta t) + S_3 \cdot \sin(6\pi\Delta t) \cdot \sin(4\pi\Delta t)$$
$$+ C_3 \cdot \cos(6\pi\Delta t) \cdot \sin(4\pi\Delta t) + \ldots,$$

leading to

$$\int F(t) \cdot \sin(4\pi\Delta t) = \int C_0 \cdot \sin(4\pi\Delta t) + \int S_1 \cdot \sin(2\pi\Delta t) \cdot \sin(4\pi\Delta t)$$
$$+ \int C_1 \cdot \cos(2\pi\Delta t) \cdot \sin(4\pi\Delta t) + \int S_2 \cdot \sin(4\pi\Delta t) \cdot \sin(4\pi\Delta t)$$
$$+ \int C_2 \cdot \cos(4\pi\Delta t) \cdot \sin(4\pi\Delta t) + \int S_3 \cdot \sin(6\pi\Delta t) \cdot \sin(4\pi\Delta t)$$
$$+ \int C_3 \cdot \cos(6\pi\Delta t) \cdot \sin(4\pi\Delta t) + \ldots.$$

In this last expression, as with our spherical harmonic analysis, $\int$ symbolizes (without any details) an integration over the entire time span of $F(t)$. The orthogonality condition tells us that all the integrals will be zero *except* the one involving S_2:

$$\int F(t) \cdot \sin(4\pi\Delta t) = 0 + 0 + 0 + S_2 \cdot \int \sin(4\pi\Delta t) \cdot \sin(4\pi\Delta t) + 0 + 0 + 0 + \ldots.$$

Straightforward integrations allow the value of S_2 to be determined—and similarly for any other C_n or S_n (if the multiplier is $\cos(2n\pi\Delta t)$ or $\sin(2n\pi\Delta t)$).

Sinusoidal expansions of the type shown for our second option are called **Fourier series** or Fourier expansions; the option is called **Fourier analysis**, and the depiction of the sine and cosine amplitudes is called a **Fourier amplitude spectrum**.

Textbox B: Spectral Analysis (*Concluded*)

It makes sense to apply spectral analysis to the study of seismology, since seismic waves are periodic phenomena. But spectral analysis is actually a great way to see nearly every aspect of the Earth System. We conclude with an example of Fourier analysis applied to a time series of paleoclimate temperatures inferred from oxygen and deuterium isotopes measured in the Vostok, Antarctica ice core. Technically, those isotope measurements were taken as a function of depth within the core; ideally, carbon dating of bits of organic material trapped within the ice allows the times of the isotope measurements to be specified, but in practice glacial modeling and stratigraphic techniques are needed to obtain the time series (see Petit et al., 1999 for details). The resulting temperature time series, shown in Chapter 5, Figure 7.21b, reveals a repeating sawtooth pattern.

Given the fundamental sawtooth shape (kind of a skewed triangle function …), and its repetition, what would you expect to be the dominant frequencies contributing to the temperature variation?

The answer is found in the spectrum those authors (Petit et al., 1999) construct:

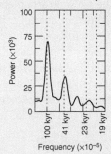

This spectrum is plotted as |amplitude|^2 versus frequency, with frequency in units of cycles per 10^5 yr; the frequencies corresponding to the Milankovitch periods of 100 Kyr, 41 Kyr, 23 Kyr, and 19 Kyr are identified with vertical dashed lines. We see that a wide range of frequencies contribute to that temperature pattern—the 'power' is nonzero over this entire frequency range; but most of the power in the pattern is focused around the two lower Milankovitch frequencies, and (to a lesser extent) the two precessional frequencies.

Incidentally, this particular type of spectrum is actually called a **power spectrum** (or **periodogram**). The amplitude whose squared modulus is being graphed includes both the sine and cosine amplitudes, that is, |amplitude|$^2 = |C_n|^2 + |S_n|^2$.

In effect, that graph, called a **spectrum** because of analogies with light passing through a prism, spreads out a time series into the various tones or 'colors' that comprise it. At a glance (e.g., *Textbox B*), the spectral 'peaks' can indicate which frequencies contribute the most to the time series. Figure 7.30 shows spectra highlighting some of Earth's normal modes. These spectra confirm our inference from the time-domain studies that there are 'clusters' of similar frequencies contributing to various individual types of normal modes, rather than a single frequency.

This clustering is called **mode splitting**, and it has several causes, all readily understood through the equivalence of normal modes and standing surface waves.

- Lateral heterogeneity, that is, the lack of spherical symmetry in the Earth; this formally also includes Earth's ellipticity (equatorial bulge). Lateral heterogeneity

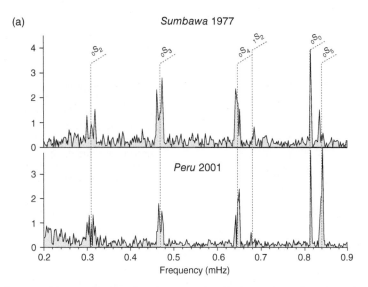

Figure 7.30 (a) Spectra of free oscillations (low-frequency band) observed gravimetrically. Dashed vertical lines indicate the frequencies expected for particular modes; the surrounding peaks show that each mode has been *split* into a cluster of several nearby frequencies.
(Top) At Brasilia following the great Sumbawa, Indonesia earthquake of 1977; based on 150 hours of data from a LaCoste-Romberg gravimeter. (Bottom) At Strasbourg following the great Peruvian earthquake of 2001; based on 167 hours of superconducting gravity data.
Image from Widmer-Schnidrig (2003) / Seismological Society of America.
Frequency units are mHz = 10^{-3} Hertz = 10^{-3} cycles/sec; *can you calculate the period (in sec and then in minutes) corresponding to a frequency of 0.3 mHz?*

can be encountered near the surface, or at the depths penetrated by long-wavelength surface waves as they traverse the globe, and can lead to faster or slower wave velocities. This changes the timing of the constructive and destructive interference that leads to standing waves, thereby modifying the normal-mode frequencies.

The lack of sphericity due to the bulge means that waves traveling in equatorial latitudes will face greater distances and longer travel times, also affecting the wave interference.

• The rotation of the Earth. The connection here can be seen in terms of the Doppler effect. For simplicity, consider the surface waves created by a source on the other side of the Earth. Due to Earth's rotation, that source is moving with the surface waves approaching us from the west but is receding from those waves coming from the east. This results in Doppler shifts in the frequencies

(or wavelengths) and thus speeds of the surface waves; as with heterogeneity, that in turn modifies the normal-mode frequencies.

• The Coriolis force. This force acts to twist the particle motions during free oscillations, in effect converting some toroidal deformation to spheroidal and vice versa. Such *coupling* results in spectral peaks corresponding to the frequencies of specific toroidal modes appearing in the spectra of spheroidal time series, and vice versa. Technically, this is mode 'adding' rather than mode "splitting," but it has the same net result.

Heterogeneity has another consequence for normal modes. With a spherically symmetric Earth, the deformations associated with a normal mode of a particular degree harmonic (ℓ) are the same regardless of where the earthquake causing it was located. This implies (though the full explanation is beyond the scope of this textbook) that the surface

(b)

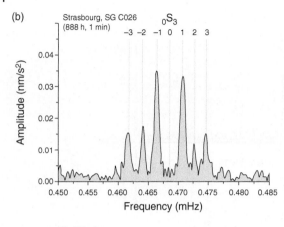

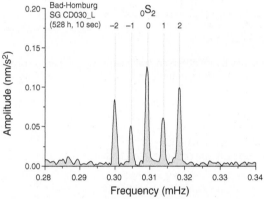

Figure 7.30 (b) Mode splitting as seen in close-ups of free oscillation spectra, observed with superconducting gravimeters following the great Sumatra-Andaman earthquake of 2004. Frequencies predicted according to PREM are indicated by the vertical dotted lines. The "0" marked below the mode designation, above the spectrum, indicates the central, un-split frequency expected for that mode were it not for mode splitting. Locations of gravimeters and duration of observations indicated in upper left for each plot. Images from Rosat et al. (2005) / John Wiley & Sons.

pattern of deformation, described by the harmonic function $Y_{\ell m}(\phi, \lambda)$, has the same eigenfrequency for all choices of the harmonic order m. With a spherically symmetric Earth, fundamental modes and overtones are often symbolized without the order, for example $_kS_2$ instead of $_kS_{20}$. But heterogeneity means that modes of different harmonic orders will have different frequencies—only slightly different, however, at least for low degree or order modes, since in those cases the patterns include broad areas, amounting to broad averages over any heterogeneity.

All of these complications lead to a fundamental challenge in identifying particular eigenfrequencies in normal-mode spectra and interpreting them in terms of Earth's structure and properties: in order to identify and distinguish all the various modes, split or not, we need an earth model capable of accurately predicting surface wave velocities

and normal-mode frequencies; but we cannot construct such a model without knowing those velocities and eigenfrequencies. The solution to this challenge is *bootstrapping* (as in 'pull yourself up by your bootstraps'!), an approach described briefly in the context of satellite geodesy (in Chapter 6) that ultimately enabled high-resolution models of Earth's gravity field to be deduced from satellite observations. For seismic bootstrapping, we can begin with a simple model of seismic velocity versus depth; infer from it the variation of density and elastic properties with depth; and use that to predict the characteristics of some of the more basic normal modes. With those modes now identified, we then refine our 'starting model,' modifying it to predict those eigenfrequencies more accurately and incorporating a greater number of modes. The refinements continue, as the differences between predictions and increasingly well-interpreted observations are minimized.

Refined earth models will be presented in the next chapter and employed to advance our understanding of Earth's interior. Later in this chapter, though, we will see how the most basic of seismic observations can be analyzed to produce a local or regional seismic velocity model—repeated globally, it will become our 'starting model,' allowing the bootstrapping process to proceed!

Normal Modes are Useful as Well as Informative: Some Hints. The extended discussion of free oscillations in this chapter can be justified by their tremendous potential— through their existence, frequencies, and even amplitudes—to reveal the structure and properties of the whole Earth, especially the deep interior. For seismologists, however, free oscillations can also be an invaluable tool in other ways. Like the elements of a stress tensor, the deformations characterizing normal modes serve as fundamental *elements* which, when added correctly and combined with representations of the forces acting at an earthquake source, can predict the global deformations of any earthquake, from 'time zero' onward. In symbolic terms, the displacement $\vec{u}$ corresponding to an earthquake's deformations at some point on or within the Earth can be written in terms of those normal-mode elements as

$$\vec{u}(r, \phi, \lambda; t) = \sum_k \sum_\ell \sum_m$$
$$\times \{ {}_k A_\ell^m \cdot {}_k R_\ell(r) \cdot \vec{H}_\ell^m(\phi, \lambda) \cdot T({}_k \omega_\ell^m t) \}$$

for a spherically symmetric Earth (see, e.g., Stein & Wysession, 2003). This expression predicts the displacement vector $\vec{u}$ at time t at a point with coordinates (r, ϕ, λ); the 'latitude' ϕ and azimuth λ of that location are angles measured relative to the earthquake source location. The corresponding components of the displacement vector are (u_r, u_ϕ, u_λ). The sum is over all fundamental $(k = 0)$ normal modes and their overtones $(k \neq 0)$; for compactness, indices are written vertically $(\omega_\ell^m$ instead of $\omega_{\ell m}$, for example).

This sum involves four types of factors for each mode: a function of the location's radial coordinate (r); a function of the location's surface coordinates (ϕ, λ); a function of time (t); and a constant coefficient. The three functions represent fundamental aspects of the normal-mode deformation, as follows:

- The time dependence $T({}_k \omega_\ell^m t)$ will be a cyclical (sinusoidal) function, with frequency ${}_k \omega_\ell^m$ being one of the normal-mode eigenfrequencies. $T({}_k \omega_\ell^m t)$ may also include a decay over time, if the Earth model dissipates elastic energy.

- The components of the vector $\vec{H}_\ell^m = (H_{r\ell}^m, H_{\phi\ell}^m, H_{\lambda\ell}^m)$ each correspond to a component of the displacement. They are each functions of 'latitude' and azimuth, (ϕ, λ), and, as discussed earlier, will either be a spherical harmonic function of degree ℓ, order m, or a spatial derivative of one.

- As noted earlier, different overtones of a particular normal mode will vary differently with depth below the surface; for the kth overtone, the radial variation is symbolized here by ${}_k R_\ell(r)$.

The coefficients ${}_k A_\ell^m$ are key to the sum, and reflect the proportional contribution by each mode to the total deformation; those proportions depend (among other things) on the relative positions of the site and earthquake source, and on the nature of the source itself. Seismologists often refer to a *moment tensor* to describe the features of the source; the coefficients ${}_k A_\ell^m$ are functions of the elements of that tensor.

Once an Earth model has been constructed and the normal modes have been quantified (thereby yielding expressions for all the R, $\vec{H}$, and T functions), the global deformations can be predicted if the earthquake's moment tensor is known. Alternatively, if we already have observations of the deformation at different sites, we can 'invert' the normal-mode sum to deduce the elements of the moment tensor, quantitatively revealing the conditions operating at the source and the characteristics of the earthquake.

With its abstract generality, our highly symbolic formula for $\bar{u}$ may appear somewhat daunting. It may be comforting to recall that each combination in braces (on the right-hand side) is just one of the many normal-mode sounds made by the Earth as it 'rings like a bell' following the earthquake in question. But in any event, any further discussion of normal-mode 'decompositions' is beyond the scope of this textbook.

It's not the Solid Earth Alone: Part I. Seismologists working at seismographic stations located near a coast have known since the late 1800s that their instruments will record ground motion even when no earthquake has occurred. Such more-or-less continuous seismic activity—amounting to a 'background' of seismic 'noise' in seismograms, and called **microseisms**—was attributed by Wiechert (1904, cited in Hasselmann, 1963) primarily to the pounding of the coast by ocean waves.

Microseisms are also detected in mid-continent, though with smaller amplitude. They also tend to be larger at times of storms over the oceans. Longuet-Higgins (1950) proposed that in deep water, microseisms could be produced by nonlinear interactions between ocean waves, specifically interference between ocean waves (of similar wavelength) coming from opposite directions—such as might be expected near the center of a cyclonic weather system, or perhaps over a broader region if the cyclone is fast-moving. The interference generates pressure waves that persist with depth and couple strongly to the seafloor, generating strong microseisms. The work of Hasselmann (1963) confirmed the plausibility of both ocean wave mechanisms (see also Webb, 2007, Ardhuin et al. 2015).

A spectral analysis of the ground motion recorded during times of no earthquakes reveals that most of the power of microseisms is concentrated at periods between 3 sec and 20 sec. It turns out that the Wiechert mechanism produces microseisms with periods around ~15 sec; the Longuet-Higgins mechanism produces much larger microseisms, with typical periods of ~6 sec.

Given the existence and origin of microseisms, it is perhaps surprising that a global background of *free oscillation* 'noise'—free oscillations occurring independent of any actual earthquakes—was not discovered until 1998 (Nawa et al., 1998; see also references in Fukao et al., 2010). Normal modes excited continuously have been tentatively identified with periods as long as that of the football mode, but are most unequivocal in the range 30–300 sec (Ardhuin et al., 2015); see Figure 7.31. The energy required to sustain the background oscillations—which have been named Earth's **hum** (though perhaps **soft murmur** (Rosat, 2004) is more evocative)—is equivalent to the energy released every day by a moderate magnitude earthquake (5¾ on the Richter scale—Ekstrom, 2001); but actual, daily seismic activity is not sufficient to produce that level of excitation.

Candidates for the source of the hum—which, based on the initial discovery, was presumed to be spheroidal in nature—include atmospheric turbulence (a popular mechanism for the solar hum discussed below); pressure on continental 'walls' and seafloor topography from ocean circulation; and ocean waves. Fukao et al. (2002) argued that atmospheric pressure variations have the potential to excite the observed hum, though they note oceanic sources cannot be ruled out; more recent research by Webb (2007) found atmospheric turbulence to be, overall, too weak an excitation source. Research by Rhie and Romanowicz (2004) determined that the hum at periods between 150 sec and 500 sec during 2000–2001 was generated mainly in the northern hemisphere Pacific Ocean and in the Southern Ocean. There are also indications that the excitation 'peaks' in each hemisphere's winter, giving the overall excitation a semiannual periodicity (Ekstrom, 2001; Rhie & Romanowicz, 2004; see also Nishida et al., 2000; Tanimoto, 2001, and Nishida & Fukao 2007; note contrary results were obtained by Tanimoto et al., 1998).

Intriguingly, Ardhuin et al. (2015) found that the Wiechert mechanism for generating *microseisms*, applied specifically to ocean waves interacting with a sloping seafloor, yields a significant dependence on depth: the greater the depth of that bottom topography, the longer the period of the ocean waves that are able to exert maximum bottom pressure on the seafloor. For coastal locations, microseisms are created at relatively short periods (~15 sec, as noted above). In contrast, 'microseisms' with periods of ~50 sec to ~300 sec are created in deeper water; Ardhuin et al. (2015) pointed to the "shelf break," where the continental shelf drops off to the continental slope, as a prime location for generating them. Excitation will therefore occur around all ocean basins, though in their analysis they found unusually high excitation along the eastern margin of the North Atlantic, associated with a strong storm that had occurred during the year being analyzed.

The work of Ardhuin et al. (2015) implies that Earth's hum is really just the long-period component of a microseism *continuum*, all of it driven by ocean dynamics, with some of the ocean wave activity energized in winter by atmospheric conditions.

The detection of Earth's hum by Nawa et al. (1998) involved a superconducting gravimeter located in Antarctica, and the normal modes discovered were all spheroidal. More recently (as in Figure 7.31), a toroidal hum was detected by Kurrle and Widmer-Schnidrig (2008) using observations of horizontal ground motion—a challenging achievement due to smaller expected amplitudes and greater expected noise levels from non-hum sources. Those authors found that the toroidal observations could not be explained by seismic activity, or by Coriolis coupling of spheroidal to toroidal modes, leaving the oceans as a potential source.

The analysis by Ardhuin et al. (2015) focused only on vertical ground motion and spheroidal modes. But we might imagine that the pressure forces transferring energy from ocean waves into spheroidal modes of the solid earth can also generate shear stresses able to act on the sloping seafloor.

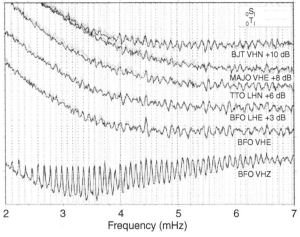

Figure 7.31 Earth hum around the world. Spectra of the horizontal ground motion (north, N, or east, E, component) observed at seismographic stations BFO (Black Forest Observatory, Germany), TTO (Takato, Japan), MAJO (Masushiro, Japan), and BJT (Baijiatuan, China), vertically offset in this figure for clarity. Spectra shown at each station are based on quietest (i.e., nonseismic) days during 1994–2006, added together or "*stacked*" to eliminate *random* noise and increase the signal-to-noise ratio. Vertical lines correspond to frequencies of spheroidal (dotted lines) and toroidal (solid lines) normal modes constituting the hum, predicted from the Earth model denoted 1066A (Gilbert & Dziewonski 1975).
These spectra may be compared with a spectrum at BFO based on vertical (Z) ground motion (bottom curve). Image from Kurrle & Widmer-Schnidrig (2008) / John Wiley & Sons.

This could be accomplished in the absence of bathymetry by lateral differences in the strength of wave-generated pressures acting along the seafloor or—in the presence of bottom topography—by differences in pressure from one side of the topography to the other (a mechanism known as *topographic coupling*); pressure on continental 'walls' from ocean circulation might also be effective (see also Tanimoto, 2008, Saito 2010, Fukao et al. 2010). At present, however, a consensus has not yet been reached on the origin of toroidal hum.

Because of their connection to ocean waves, microseisms and hum (—at least, the spheroidal part) have the potential to tell us about the state of the oceans and the overlying atmosphere. Gutenberg (1947), citing earlier work on the subject as well, demonstrated the use of microseisms, in a project coordinated with the U.S. Navy, for the early detection and location of storms around the world. De Becker (1990) explored the use of microseisms to provide warnings of strong storm surge in the North Sea. Bernard (1990) has tied microseisms to storm activity as a way to explore the connections of the latter to solar forcing. Grevemeyer et al. (2000) found a doubling of North Atlantic storm activity from 1954–1975 to 1974–1998, likely a consequence of global warming, using microseism data. And, Farrell and Munk (2010) have used seafloor observations of microseisms (in both stormy and calm weather) to connect length scales and frequencies of ocean waves to surface wind forcing.

Our discussion has avoided going into details regarding the origin and relevant properties of the oceanic 'gravity waves' and 'infragravity waves' that are responsible for microseisms and Earth hum, and how such waves couple to the seafloor. Those subjects require a basic knowledge of fluid dynamics, which we will postpone till Chapter 9.

It's not the Solid Earth Alone: Part II. Can a fluid body like the Sun 'ring like a bell'?

The Earth exhibits free oscillations because it is capable of sustaining surface waves and because there are sources to excite them. The Sun is certainly energetic enough for us to imagine any number of potential excitation mechanisms. But if shear waves cannot propagate through the Sun, there can be none to become trapped as Love waves or to couple with compressional waves to form Rayleigh waves.

Yet it turns out that the Sun *does* exhibit free oscillations, of a sort; and observing them has proven critical for understanding how the Sun works. The Sun's surface is in constant motion, so discerning the additional motions associated with eigenvibrations is quite challenging; one successful approach is based on the Doppler effect, which causes a slight shift in the wavelengths of light emitted at the Sun's surface depending on whether the oscillating region is moving toward or away from us (see Figure 7.32a).

The Sun's free oscillations are most likely generated by processes in the **convective zone**—the outermost third of the solar interior, where heat transport is dominated by convection rather than radiation. Ulrich (1970) showed that, with the surface of the Sun acting as a reflector to *P* waves traveling up from the interior, and the deeper parts of the convective zone acting as a strong refractor (—almost a waveguide) to waves traveling downward, most of those waves can become trapped within it (see Figure 7.32b). Constructive and destructive interference among the multitude of *P* waves in the zone leads to standing waves of particular wavelengths and frequencies, with 300-second (5-minute) periodicities being the most energetic. The scale of these oscillations is $\sim 10^3$ km or more, both horizontally and vertically.

These trapped standing waves are the free *acoustic* oscillations of the Sun; evidently excited continuously, they are indeed solar hum. One popular excitation mechanism is turbulent disruptions in the convective flow (Goldreich & Keeley, 1977; Goldreich & Kumar, 1990); another is sporadic 'sun-quakes' (Strous

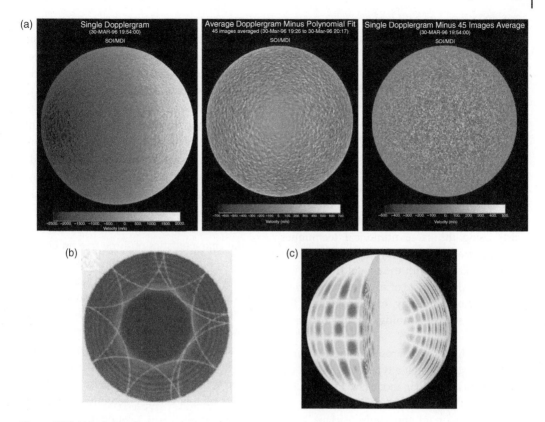

Figure 7.32 Hum in the Sun.
(a) Doppler observations of the Sun's "photosphere" (surface) from the SoHO spacecraft. (Left) 'Dopplergram,' showing relative velocities of parcels on March 30, 1996, at 19:54 GMT: darker shading for parcel motion toward the spacecraft, lighter for motion away, according to the color scale. These relative motions are dominated by the Sun's rotation.
(Middle) The average of 45 dopplergrams (taken within half an hour (±) of the first dopplergram) after the effects of the solar rotation have been removed; this average reveals the patterns of approximately horizontal convective motions near the surface.
(Right) Dopplergram on the left after both solar rotation and that averaged horizontal convection have been removed. This residual dopplergram is dominated by vertical motions, peaking at approximately 500 m/sec, reflecting the spheroidal normal modes of the Sun.
Images from Gough (2010) / Springer Nature.
(b) An example of the paths taken by P waves in the Sun, trapped above by reflections off the free surface and below by refractions within a very fast layer. In this cross-section, the calculated paths are for a normal mode that is the second overtone of a degree 20 harmonic (we could perhaps denote it as $_2A_{20}$, with "A" denoting "acoustic," though these modes are generically called "p modes"). Image from Gough et al. (1996).
(c) Illustration of the kind of surface and internal motions associated with solar free oscillations, in this case a $_{14}A_{20,16}$ mode. Image from https://en.wikipedia.org/wiki/Helioseismology.

et al., 2000, Rimmele et al. 1995), each triggered by localized "catastrophic" cooling and collapse of regions $\sim 10^3$ km in diameter and a few hundred km below the surface—a mechanism which, with perhaps thousands occurring per second globally, the authors estimate to be capable of exciting the entire suite of solar oscillations. Conversely, Ulrich (1970) also demonstrated that the convective zone is naturally 'overstable' (using a term popular among astrophysicists), in that *any* random perturbation (in the flow or just in the state of the fluid), such as those accompanying the passage of a P wave, can grow in an oscillatory manner.

As Doppler techniques and instrumentation have been refined, and as a result of concerted efforts such as GONG (Global Oscillation Network Group) and the SoHO (Solar and Heliospheric Observatory) satellite, the number of acoustic solar modes has skyrocketed; even by 1996, more than 250,000 distinct modes had been identified (Hill et al., 1996). These modes are generally described in a similar way to Earthly normal modes (e.g., as fundamental or overtone modes, all distinguished using spherical harmonics; e.g., Figure 7.32c), and are analyzed by astrophysicists using the same numerical techniques that seismologists employ to find an earth model consistent with the modal frequencies; consequently, this field of study has come to be called **helioseismology**, in the context of the Sun, and **asteroseismology** in application to other stars.

On the subject of asteroseismology, we note that some stars exhibit a type of oscillation we have called the balloon mode (see Figure 7.29a caption)—a purely radial periodic expansion and contraction of the star's 'surface.' The cause of this oscillation is not seismological; most commonly, for example in the case of a type of star called a *Cepheid variable*, it is alternating imbalances in the force balance within a massive star, triggered by fluctuations in its outward heat transport. As the star contracts and expands, it also brightens and dims. In some cases, such as the first such pulsating variable star discovered over 400 yr ago, it may even become too dim to see (e.g., CSIRO, 2015). The amplitude and timescale of the luminosity variation of such stars depend on their mass, composition, and age; the luminosity of a Cepheid variable may fluctuate with a period as short as a day or as long as a few months.

In the remainder of this chapter, we will return to basic seismological observations and learn how they help elucidate the structure and properties of Earth's interior. Helioseismology has brought similar benefits to the study of the Sun (e.g., Bahcall, 1999); we conclude this section by highlighting two of its more profound revelations.

- The fusion reactions within the Sun converting hydrogen to helium and releasing a huge amount of heat also produce neutrinos, nearly massless subatomic particles. Experiments to detect those particles in the second half of the twentieth century suggested that the Sun was producing, roughly, only a third as many neutrinos as we would expect. This **solar neutrino problem** led to concerns that perhaps the fusion reactions in the Sun's core had shut down—with dire consequences for the balance of thermal pressure and gravity within the Sun. But the characteristics of those free oscillations able to penetrate below the convective zone imply that the composition and temperature of solar models are reasonable (e.g., Christensen-Dalsgaard et al., 1996), lending support instead to the somewhat radical idea that previously undetected types of neutrinos are also created during solar fusion reactions. Subsequent experiments by the start of this century actually detected the latter, resolving the solar neutrino problem.

- In Chapter 10, we will see that as the Earth rotates, the convecting fluid in Earth's core must exhibit what is called a *differential rotation* of the core fluid, in order to conserve angular momentum, with deeper layers rotating faster. Differential rotation plays a key role in generating and maintaining Earth's magnetic field. Those same conditions imply that the interior of the Sun should also exhibit differential rotation—at least, within the convective zone. In the case of the Sun, the differential rotation may be more complex, given that it is superimposed on a rotation that is itself not uniform: the surface of the Sun rotates most rapidly at the equator, once every 25 days, decreasing steadily toward the poles, for example once per 28 days in midlatitudes and once per 34 days near the poles (Williams, 2016). (Note that solar scientists refer to such latitudinal nonuniformity as a differential rotation also.)

Spectral analyses of solar free oscillations have revealed mode splitting, which is generally attributed to the Sun's rotation. Using modes of different spherical harmonic degree and order has enabled the latitude dependence of the rotational splitting to be inferred; and with different modes penetrating to different depths, any depth dependence in the splitting can also be deduced. The net result, remarkably, is a rotational *profile* of the Sun—thanks to helioseismology. As an illustration, Thompson et al. (1996; see also Gizon & Birch, 2005) find that the same latitude dependence of the Sun's surface rotation persists more or less to the base of the convective zone, plus there is a kind of equatorial west-to-east shear zone or 'jet' concentrated near the Sun's surface. In contrast, the region below the convective zone (called the **radiative zone**, because heat transport is primarily by radiation) exhibits approximately uniform rotation, latitudinally—essentially a 'rigid-body' rotation—with a speed corresponding to the upper zone's midlatitude angular velocity.

7.5 Seismic Waves and Exploration of the Shallow Earth

An earthquake or explosion will radiate *P* and *S* waves in all directions throughout the Earth's interior. At any seismographic station set up on the Earth's surface, a variety of direct, reflected, and refracted waves produced by that source will arrive over time. Their arrivals can be detected by seismometers recording the ground's displacement (or motion). There are usually three seismometers set up to record all three components of ground motion (e.g., north-south and east-west for the two horizontal components, plus the vertical component); the reconstructed, three-dimensional motion allows us to distinguish the arrival of a *P* wave from an *S* wave, or a Love wave from a Rayleigh wave.

A first analysis of the seismograms generally focuses on the *times* at which the various waves arrive at a station; depending on the application, subsequent analyses may involve the wave *amplitudes* or the pattern of the whole *waveform*. Once the time of occurrence—the 'origin time'—of the earthquake or explosion is known, subtraction from the arrival time yields the time it took each wave to travel from source to receiver. Such **travel times** can be used to learn about the seismic structure and properties of the region through which the waves propagated. These applications will be discussed from a global perspective in the next section and the next chapter; in this section, we focus on local investigations.

Earth's density and elastic properties vary throughout its interior. In the shallow subsurface, variations in material properties resulting from differences in rock type, the region's geological history, recent surface processes, and even water content in subsurface rocks may be significant enough to affect seismic waves. Such variations in material properties are likely to manifest three-dimensionally; but gravity, surface effects, and broad-scale plate structure usually dominate, conspiring to force a mainly vertical variation in properties—a layering, even on local scales.

Seismology provides a number of ways to identify and quantify variations in material properties. In this section, we will focus on techniques that involve the travel times of *critically refracted* waves, an ideal approach for nearly horizontal, layered structures. It should be noted, however, that seismic *reflections* are quite useful in elucidating variations in structure and properties—much as sonar can be used to track variations in ocean-floor bathymetry—whether in a layered or three-dimensional situation, and often with great detail; reflection techniques will not, however, be covered in this textbook.

7.5.1 Refraction Surveys; or, First Arrivals on a Flat Earth

One of the most popular and straightforward seismic exploration techniques is called a *refraction survey*. Refraction methods take advantage of the fact that layering in which seismic velocities *increase* with depth allows waves originating at the surface (e.g., created for the survey) to be critically refracted at and travel along layer boundaries. As those waves—head waves—propagate along a boundary, traveling at the speed of the lower layer, any irregularities or imperfections of the boundary itself serve as sources that send some of the critically refracted energy back up toward the surface; see Figure 7.33a. Considering the principle of reciprocity (or Fermat's Principle of Least Time), it makes sense that those upward-moving waves must leave the boundary at the same critical angle that had turned the incident wave into a head wave. Thus, once observers on Earth's surface are far enough away from the source that downward-moving waves can strike the boundary at the critical angle and create a head wave—and far enough away to allow for the returning wave energy to come upward at the same critical angle—those observers can expect to detect the head wave's return.

Any shallow source of seismic waves, whether an earthquake, explosion, or even just an impulse of stress (e.g., from a truck bouncing up and down in place, a shotgun blast directed into a hole in the ground, or a sledgehammer striking a plate on the ground), will generate waves that radiate in all directions. We focus initially on the P waves created by that source that travel straight out, along

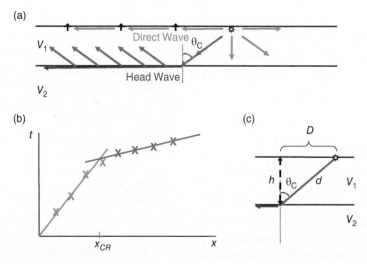

Figure 7.33 Refraction surveys. (a) Geophones (receivers) are set up along the surface away from the source, at locations marked with a ↑. The source radiates seismic waves in all directions. Assuming the seismic velocity in the second layer, V_2, is greater than that in the first layer, V_1, critically refracted waves (head waves, with θ_C denoting the critical angle) can propagate along the boundary between the two layers. With material irregularities characterizing any real-world interface, head wave energy is continually sent back up through the upper layer. This energy, returning at the critical angle, can be detected all along the surface if the receivers are distant enough.
(b) And, at sufficiently great distances, those returning head waves will arrive sooner than any other wave. This is illustrated by the graph of "first arrival" travel times (t), recorded by geophones at various distances (x) from the source: geophones beyond the "cross-over" distance x_{CR} detect the head wave before the direct wave arrives. Each first-arrival data point is marked as an X on the graph, and the best-fitting straight lines to the collection of direct-wave and head-wave data points are also shown. As discussed in the text, the slopes of those lines are $1/V_1$ and $1/V_2$, respectively; the change in slope marks the cross-over.
(c) Geometry of critical refraction. h is the thickness of the upper layer; distances d and D are discussed in the text. A similar diagram (flipped left↔right) would illustrate the final upward stage of the head wave's travel.

the top of the first layer; these waves are called **direct P waves**. Out of all the waves that travel through the upper layer only, including any reflected (or multiply-reflected) P waves and all the S waves, the direct P wave has the shortest travel path (shared by the direct S wave) and will arrive *first* at any point on the surface close to the source.

In a simple refraction survey, we set out an array of geophones—essentially, miniature seismometers, usually designed to detect vertical ground motion—on the surface along a line away from the source. At near distances, the first wave to arrive at each geophone will be the direct P wave; if t is the time this wave takes to reach a geophone at distance x from the source, then $x = V_1 t$ if the velocity of the upper layer is V_1 (the *distance* traveled by the wave *equals* its *rate* of travel *times* the *time* it took). This can be written

$$t = x/V_1,$$

implying that if arrival times are plotted versus distance from the source (as in Figure 7.33b), we will find that—for these near geophones—the graph is a *straight line*. The line passes through the origin (its y intercept is 0), and the slope of the line is $1/V_1$.

For more distant geophones, however, the P wave head wave may arrive first, beating out the direct wave. The head wave clearly covers a much greater distance, as it travels both down to and up from the interface, but for distant geophones most of its trip is spent moving along the top of the second layer, where it propagates more rapidly. Factors that determine its travel time include the thickness of the layer and the critical angle (that angle is in turn a function of V_1/V_2), which together determine its time spent to and from the interface (see Figure 7.33c); and the difference between V_1 and V_2, which determines how long it takes the head wave to significantly overtake the direct wave as they both travel horizontally. Depending on these factors, there will be some distance along the surface—called the **cross-over distance**—beyond which the head wave becomes the first arrival (Figure 7.33b).

Beyond the cross-over distance, any increase (along the surface) in the geophone's distance from the source will correspond to a head wave that traveled (along the top of the second layer) the same extra distance, before arriving first at the geophone; since that distance is covered by the head wave at velocity V_2, it follows that this portion of the first-arrival t versus x graph will also be a straight line, but with a slope of $1/V_2$.

The immediate benefit to graphing first-arrival travel times versus distance in a refraction survey is thus the ability to easily determine the seismic velocities of the two layers (see Figure 7.33b). With a bit of mathematical effort, we can also deduce the thickness, h, of the upper layer, i.e. the depth to the interface. For this analysis, we assume not only that the layers are flat-lying but also that the source is right at the surface of the Earth. Denoting by d the distance traveled by the head wave from the source down to the boundary, or the distance it travels up from the boundary to the geophone, and by D the lateral distance away from the source covered by the head wave as it travels down to or up from the boundary (see Figure 7.33c), basic trigonometry tells us

$$\tan \theta_C = \frac{D}{h},$$

so

$$D = h \, \tan \theta_C = h \frac{\sin \theta_C}{\cos \theta_C}.$$

Meanwhile, Snell's Laws require that the critical angle obeys

$$\sin \theta_C = \frac{V_1}{V_2},$$

which together with the trig identity $\sin^2 + \cos^2 = 1$ implies

$$\cos \theta_C = \sqrt{1 - (\sin \theta_C)^2}$$

$$= \sqrt{1 - (V_1/V_2)^2}$$

$$= \frac{\sqrt{V_2^2 - V_1^2}}{V_2}.$$

Using this relation, we can also write

$$D = h\frac{\sin\theta_C}{\cos\theta_C} = h\frac{V_1}{\sqrt{V_2^2 - V_1^2}}.$$

Recalling once again that distance equals rate times time, these relations allow us to determine how long it took the head wave to travel through each segment of its path. The time spent by the head wave traveling down to or up from the interface is d/V_1, or, since $\cos\theta_C = h/d$ (see Figure 7.33c),

$$\frac{d}{V_1} = \frac{h/\cos\theta_C}{V_1} = h\frac{V_2/V_1}{\sqrt{V_2^2 - V_1^2}}.$$

If the geophone is a distance x from the source, then the distance the head wave covers traveling along the top of the lower layer is $x - 2D$; moving at a speed of V_2, the time it takes for this is

$$\frac{x - 2D}{V_2} = \frac{x}{V_2} - 2h\frac{V_1/V_2}{\sqrt{V_2^2 - V_1^2}}.$$

In sum, the total travel time of the head wave must be

$$t = \left(\frac{h(V_2/V_1)}{\sqrt{V_2^2 - V_1^2}}\right)_{down}$$
$$+ \left(\frac{x}{V_2} - 2h\frac{(V_1/V_2)}{\sqrt{V_2^2 - V_1^2}}\right)_{along}$$
$$+ \left(\frac{h(V_2/V_1)}{\sqrt{V_2^2 - V_1^2}}\right)_{up},$$

which can be written

$$t = \left(\frac{x}{V_2} - 2h\frac{(V_1/V_2)}{\sqrt{V_2^2 - V_1^2}}\right)$$
$$+ \left(\frac{2h(V_2/V_1)}{\sqrt{V_2^2 - V_1^2}}\right)$$

or

$$t = \frac{x}{V_2} + 2h\frac{\sqrt{V_2^2 - V_1^2}}{V_1 V_2}.$$

This confirms our earlier conjecture that the t versus x graph for the head wave will be a straight line with a slope of $1/V_2$. The second term on the right-hand side of the equation is the y intercept of the line. With the straight line extrapolated back to the y axis, an estimate of the y intercept will allow the thickness h of the layer to be estimated as well.

Setting this equation and that for the direct wave equal to each other, the cross-over distance—the distance at which both waves arrive at the same time—can be predicted, in terms of h and the velocities; we find

$$x_{CR} = 2h\sqrt{\frac{(V_2 + V_1)}{(V_2 - V_1)}}.$$

Employing this relation, graphical estimates of the cross-over distance could also be used to estimate h. However, identifying the cross-over distance using t-x graphs tends to be somewhat ambiguous, so this approach to determining h is not preferred.

7.5.2 Refraction Surveys: An Illustration With Possible Hydrogeological Implications

A park at the edge of the Binghamton University campus, cut by a seasonal creek, provides an ideal location to illustrate the technique of seismic refraction. Exposures along the far side of the creek reveal slightly dipping, shaley bedrock overlain by soil and scattered glacial till; but along the near side, the park itself is engineered, and the soil cover is fairly homogeneous. As we will see, the results from our seismic refraction survey are consistent with the park being layered; our interpretation of those results, however, hints at some complications.

Previous refraction surveys as well as the adjacent outcrops implied that a layer interface would be located ~1–2½ m below the surface; this corresponds very roughly to a cross-over distance of ~2–5 m, that is, at roughly that distance away from the source the head wave would become the first arrival. In the present survey, then, to capture a sufficient number of arrivals from the head wave, we could place geophones at a range of distances 3–18 m from the source, at intervals of 3 m; for the direct wave we chose intervals of ~30 cm, up through a distance of 1.5 m.

Incidentally, our rough estimate of the cross-over distance being twice the layer thickness—or the reverse, that the layer thickness is roughly half the cross-over distance—would be exactly true if $V_2 \gg V_1$. Such an approximation is a reasonable first guess in the case of a soil-bedrock interface, because P wave velocities for bedrock are typically much greater than for soil. (If V_1 is not that small relative to V_2, then the cross-over distance would be more than twice the layer thickness, and the layer thickness would be less than half the cross-over distance.)

With such short distances involved, our survey did not require a high-energy source of elastic waves, so we chose to generate the waves using a sledgehammer to strike a metal plate on the ground; the plate allowed the stress generated by the strike to couple more effectively to the ground. The direct and critically

refracted P waves produced by our source were recorded by an array of geophones at distances of 0.3 m, 0.6 m, 0.9 m, 1.2 m, 1.5 m, 3 m, 6 m, 9 m, 12 m, 15 m, and 18 m from the plate. The times of the first arrivals were estimated visually, from the real-time 'seismograms' of particle motion generated when the sledge-hammer struck the plate and displayed on the instrument panel; these times are listed in Table 7.2.

To allow for the possibility that the interface of interest is dipping, we also performed a survey with the source and receivers reversed; first-arrival times for the reversed profile are listed in Table 7.2 as well. We found that differences in the first-arrival times of the two profiles were slight, mostly within the measurement uncertainty of a few tenths of a millisecond; however, the exposed bedrock did reveal a definite though small slope.

There are straightforward techniques in which the data from both profiles can be analyzed to yield the layer thickness (at either end of the survey) *and* the dip of the interface, as well as V_1 and V_2 (one classic reference for this is Dobrin & Savit, 1988); but such techniques are beyond the purview of this textbook. Instead, we simply *averaged* the t versus x data

Table 7.2 Data for seismic refraction survey in a park near Binghamton University, NY.

Distance from source (m)[a]	Time (msec)		Time (msec)
	Forward profile	Reverse profile	Averaged profile
0	0	0	0
0.3	1.0	1.0	1.0
0.6	1.8	1.9	1.85
0.9	2.6	2.7	2.65
1.2	3.7	4.0	3.85
1.5	4.5	4.6	4.55
3	9.1	9.1	9.10
6	12.5	13.0	12.75
9	14.2	14.4	14.30
12	15.6	15.0	15.30
15	16.2	16.4	16.30
18	16.6	17.9	17.25

[a]*Distances shown are rounded off.*

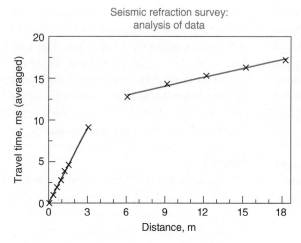

Seismic refraction survey:
analysis of data

y-axis: Travel time, ms (averaged)

x-axis: Distance, m

Figure 7.34 Refraction survey at a park near the Binghamton University campus. First-arrival times are graphed versus distance; they were averaged from forward and reversed profiles to simplify the analysis. Straight lines have been fit by least squares separately to the direct-wave and head-wave sets of data points. The good quality of the fit provides some assurance that the survey area is well described by a layer structure. The fitted lines imply a cross-over distance of nearly 4 m.

from both profiles, in effect averaging out the dip, and continued the analysis to determine the average layer thickness and V_1 and V_2.

From the averaged data in Table 7.2, the best-fitting lines (see Figure 7.34) are

(for the direct wave) $t = 2.9784x + 0.0435$

(for the head wave) $t = 0.3609x + 10.7800$

where time is in milliseconds and distance is in meters. The y intercept of the line for the direct wave, 0.04, is zero within the uncertainties of the data (which exceed 0.1 msec), so within those uncertainties we can consider the line to be passing through the origin—as required for the direct wave.

The slope of the direct-wave line implies a P wave velocity V_1 of 336 m/sec in the upper layer; from the slope of the line for the head wave, we infer $V_2 = 2,771$ m/sec for the P wave velocity in the lower layer. The average depth h of the interface implied by this survey is $\sim$1.8 m.

Verify these values, using the best-fitting lines we have determined.

For a simple interpretation of these results, we note that our value of V_1 is consistent with the upper layer being unconsolidated soil; in such a medium, P wave velocities are typically in the range of 120–500 m/sec (e.g., Haeni, 1988). But what does the second layer correspond to? During the survey, it was noted that the adjacent stream bed—which is underlain

by shaley bedrock—is about 1.5 m below the bank and the study area; perhaps the interface detected by the head wave, at a depth inferred to be $\sim$1.8 m not far from the stream, corresponds to the top of the bedrock. However, another possibility must also be considered.

The Binghamton area experiences almost a meter of rain yearly, and there is often ponding, seepage, and other evidence of a shallow water table throughout the region. At times, even during extended periods of fair weather, the study area is too muddy to conduct a refraction survey. *Could the interface be the water table rather than the top of bedrock?* After all, the adjacent stream is $\sim$1.5 m below the study area.

Theoretically, the **water table** can be defined as the top of the water-saturated zone (in practical situations other, nonequivalent definitions are often used; see Holzer & Bennett, 2003). Now, unconsolidated sediments tend to have relatively high porosities, and if those pore spaces are air-filled, the formations will be relatively easy to compress. Below the water table, however, the voids will by definition be saturated with water. Since water is less compressible than air, the incompressibility of the water-sediment body will be higher, and we can expect to measure higher P wave velocities below the water table. In contrast, incidentally, shear waves are essentially unaffected by the transition from air-filled to water-filled pores: water and air are both fluids, and both have zero rigidity.

The effect of saturation on P wave velocity can be striking. For example, Allen et al. (1980) found from laboratory measurements involving a sand body that a decrease from saturation by 0.2–0.3% could slow P wave propagation by a factor of two. With such sensitivity, even unsophisticated refraction surveys such as ours would easily be able to detect the water table. In fact, during one year, our survey of this area was conducted several days after a bout of rainy weather had ended, yet the effects of moisture content on P wave velocity were still measurable: we found the upper layer velocity V_1 to be ~488 m/sec—about 50% greater than our other, 'fair weather' estimate, and plausibly interpreted as the consequence of some residual moisture not yet drained from the soil layer.

There are a variety of situations in which seismic refraction techniques have yielded key hydrologic information; see e.g. Haeni (1986, 1988) for examples and pitfalls. For brevity, we note here one study by Holzer and Bennett (2003), who compared water tables identified by seismic refraction to well water levels in boreholes, in a central California region; the comparison implied significant unsaturation below the wells' water levels, which the authors speculated was left over from drought conditions a decade earlier. The implication that groundwater recovery from a drought will take many years beyond the apparent recovery of surface reservoirs—what McNutt (2014) called an *underground drought*—is something officials in California, Texas, and other places suffering from ongoing or recent droughts will need to confront; it will also exacerbate the impacts of global warming elsewhere, as the frequency of droughts increases around the world.

In our survey, however (at least with prolonged rain-free conditions), hydrologic interpretations ended up being unnecessary. The velocity we inferred for the lower layer, ~2,800 m/sec, is fully consistent with the interface being the top of the bedrock. P wave velocities in sandstone are typically in the range 1,500–5,500 m/sec; in

limestone, in the range 2,100–7,000 m/sec; and in shale—the most relevant candidate, given the visible geology—in the range 2,700 m/sec—6,100 m/sec (Haeni, 1988).

7.5.3 Refraction Surveys: Thoughts About Multilayered Situations

Refraction surveys can successfully reveal information about a two-layer region as long as P waves travel faster in the lower layer than in the upper, and as long as geophones are set out distantly enough for head waves traveling along the top of the lower layer to be recorded at the surface. But if the region is characterized by more than two layers, and those successively deeper layers are faster than the layers above, head waves can also exist in those deeper layers. As those head waves travel along the deeper interfaces, they also continually generate P waves that travel back up to the surface. If geophones are arrayed far enough from the source, they will eventually pick up some of the deeper head wave returns as first arrivals.

In this situation, what do you think the graph of first-arrival travel times versus distance from the source will look like?

All the head waves generated along a particular deep interface and then returning to the surface will have traveled the same downward and upward paths, if the interface is horizontal; the increase in travel times for those arriving farther out from the source will be in proportion to the increased distance they have traveled along the top of that critical refraction layer. That is, just as in the two-layer case, the t versus x graph for the head wave arrivals associated with a particular layer should be a straight line, and the slope of that line should equal 1 over the velocity of that layer.

Consequently, for a multilayer situation (in which seismic velocity increases with depth from layer to layer), we can expect the t versus x graph to be a sequence of *straight-line segments*, one for each layer, with the segments increasingly horizontal because their

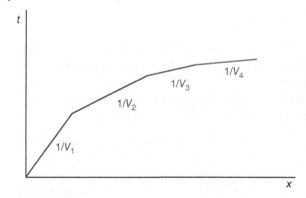

slopes are $1/V_1$, $1/V_2$, $1/V_3$, $1/V_4$, etc., and $V_1 < V_2 < V_3 < V_4$ .... An example of this is shown in Figure 7.35.

Finally, if the number of layers increased indefinitely while their thickness decreased substantially, how would you expect the t versus x graph to look? We will see the answer to that question in the next section.

7.6 Seismic Waves and Exploration of the Whole Earth: Preliminaries

In this section, we return to a global perspective and expand our discussion to include the full variety of *P* and *S* waves generated by large earthquakes and recorded around the world. These waves have sampled some or all of the Earth's depths, with the paths they traveled and thus their travel times to any seismographic station depending on the structure and properties they encountered along the way. As earthquake activity has continued globally, and the records from stations worldwide analyzed, seismologists have built up a collection of travel time data with huge potential—overall, providing a *standard* that can be used

- to locate an earthquake source;
- to quickly identify the various waves (even some unusual ones) on a seismogram; and
- to define anomalous regions on and within the Earth.

Additionally, because travel times depend fundamentally on the wave velocity at each point of the wave path, that standard equivalently yields a model of how those velocities vary with depth—a standard velocity model (to be discussed in the next chapter) that defines our view of the structure of Earth's interior.

7.6.1 Travel Times: Lateral Homogeneity Within the Earth

One of the reasons travel times are so informative is because of a fortunate *seismic symmetry* that exists, approximately, within the Earth—a radial symmetry in elastic properties. That is, no matter where an earthquake occurs, the time it takes for a specific seismic wave to reach seismographic stations around the world depends, to a good approximation, only on the *distance* between earthquake and station. Consider, for example, earthquakes that occur precisely opposite stations located on the other side of the world: a PKIKP wave originating in Hawaii and traveling straight through the core, emerging at the opposite point (on the other side of the Earth) in Zambia, will take about 20 minutes; and, a PKIKP wave starting near Lake Baikal, Russia, traveling straight through the Earth, will arrive at Tierra del Fuego, the southern tip of South America (and the opposite point on Earth's surface from Lake Baikal) *also* in 20 minutes. Despite the completely different regions of the crust, mantle, and core through which those PKIKP waves have traveled, they both covered the same amount of distance—and took the same time to complete their travels.

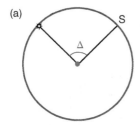 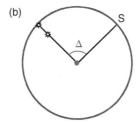

Figure 7.36 (a) For distant earthquakes, their distance from a seismographic station (S) is measured by the central angle, Δ. The earthquake source is marked by a star symbol (☼).
(b) Deep and shallow earthquakes may be the same angular distance from a seismographic station. *In this illustration, which quake is closer to the station?*

For earthquakes occurring far from a seismographic station, their distance to the station is measured angularly, by the central angle, Δ (delta), between their radii; see Figure 7.36a. When we say that travel times are independent of location, we mean that travel times depend only on Δ—*not* on Δ and whether the quake occurred in Hawaii or Lake Baikal, *just* on Δ. For both of those events, by the way, the distance between source and receiver was Δ ~180°.

Travel-time spherical symmetry is a good approximation; even so, observed travel times for a given Δ can vary slightly from place to place (typically by just a few seconds; Fowler, 2005). In practice, the location-independent travel times we use to approximate all observations for each Δ are determined by laterally averaging all the observations for that Δ. A seismic Earth model derived from such radially symmetric travel times will therefore yield a reasonably accurate representation of Earth's interior, overall. However, as we will see in the next chapter, anomalous travel times observed in some regions, though small, point to the presence of important structures within the Earth.

There is one major exception to our inference of travel-time spherical symmetry. Earthquakes have been found to occur at depths up to ~700 km; typically, these deep events are associated with subduction zones (see Frohlich, 1989). It is certainly possible for deep and shallow earthquakes to be the same Δ from a seismographic station; but though their angular distances from that station may be the same, their direct paths to the station are likely to be very different (see Figure 7.36b). Except when Δ is very small, the waves radiated from the deeper source will travel a shorter path to the station, resulting in shorter travel times. Additionally, deeper mantle material resists deformation more than shallower stuff, so seismic waves travel faster when propagating through deeper portions of the Earth; consequently, seismic waves from the deeper source arrive sooner also because they are traveling faster.

The differences in travel times for shallow versus deep events can be substantial; for Δ ~70°, for example, P waves from very deep quakes can arrive more than a minute earlier than near-surface quakes. In practice, then, travel time information is compiled separately for shallow (and intermediate) and deep earthquakes; it is within each category that we find seismic spherical symmetry to be a good approximation.

Seismologists recognize that the Earth is ellipsoidal, and can easily correct travel time information for its effect. For waves traveling straight up from the interior, the extra distance created by Earth's equatorial bulge is at most ~21 km (comparing an equatorial station to a polar one); for P waves the extra travel time through crust is therefore at most ~21 km/5.8 km/sec, or ~3.6 sec, whereas for the slower S waves it is at most ~21 km/3.2 km/sec, or ~6.6 sec.

From abundant travel time measurements, seismologists have compiled tables of travel time versus distance, or (in graphical

(a) (b)

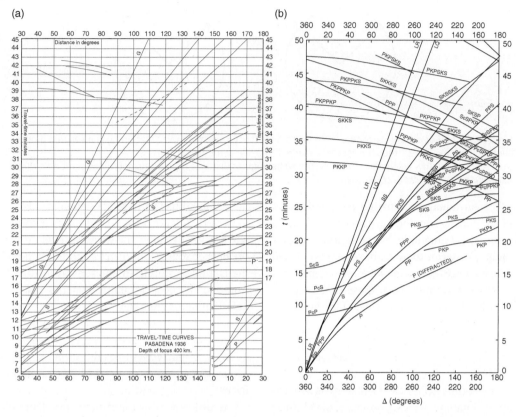

Figure 7.37 Examples of early travel-time curves.
(a) From Gutenberg & Richter (1936), Seismological Society of America, for earthquakes 400 km deep; range of x-axis is 30°–180°, with inset in lower-right from 0°–30°. Gutenberg & Richter (1936) / Seismological Society of America. Travel times (y-axis) are in minutes.
(b) From Jeffreys and Bullen (1940), for shallow earthquakes (surface depth, i.e., 0 km). Image downloaded from the SEG (Society of Exploration Geophysicists) dictionary. Jeffreys & Bullen (1940) / Society of Exploration Geophysicists / CCBY 3.0 / Public domain.

form) *travel-time curves* (also known as *t-Δ curves*). The earliest tabulations were from collaborations involving the most celebrated seismologists of the early to mid-twentieth century: Jeffreys and Bullen (circa 1940) and Gutenberg and Richter (1936); examples of their *t-Δ* curves are shown in Figure 7.37. More recent sets of travel-time curves are shown in Figure 7.38. As noted earlier, these travel-time curves provide a *standard*, a *reference seismic Earth model* to which actual travel-time observations can be compared—in much the same way that the International Gravity Formula allowed comparisons between actual gravity observations and a standard—with

any differences indicating anomalous seismic properties somewhere along the wave path.

We will begin the next chapter with global applications of travel-time curves, using them as a standard for comparison. We end this chapter with two very practical uses for *t-Δ* curves.

7.6.2 Travel Times: Locating Earthquakes

Do the *P* and *S* waves recorded at a seismographic station correspond to a smaller, local tremor or a larger, more distant event? If we compared their wave amplitudes, the results would be ambiguous; but with travel times,

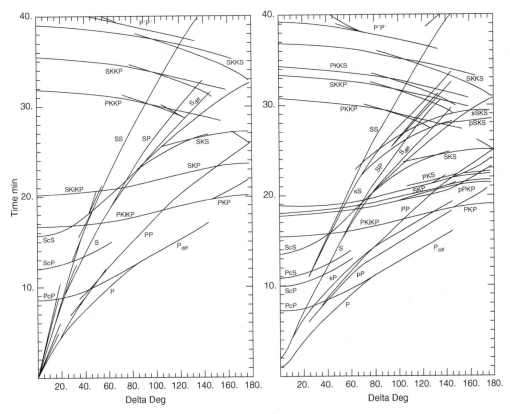

Figure 7.38 Examples of modern travel-time curves, model iasp91 from Kennett & Engdahl (1991) / Oxford University Press, for shallow earthquakes (0 km depth) on left, deep earthquakes (600 km depth) on right. Actually, as discussed in Kennett and Engdahl, such reference curves can be numerically constructed easily for earthquakes of any depth.

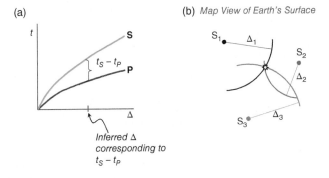

Figure 7.39 Locating earthquakes using travel-time curves.

(a) Travel time (t) curves versus angular distance (Δ) for P and S waves are shown. Note that as the distance Δ from an earthquake increases, it takes even longer for the S wave to arrive after the P wave has arrived. If that delay, t_S-t_P, determined from the seismogram is matched to these curves, the station's distance from the source can be estimated.

(b) Arc method for locating an earthquake. The distances (Δ_1 through Δ_3) from three seismographic stations (S_1 through S_3) are determined using the travel-time curves as in part (a); only the point marked with a star symbol (☼) is the correct distance from all three stations.

we can 'triangulate' earthquake events and unequivocally answer that question.

Because they travel slower than *P* waves, *S* waves will of course arrive at a seismographic station after the *P* waves. More to the point, however, *the farther the seismographic station is from the earthquake, the greater the delay in the S wave arrivals.* Very simply, for example, *P* and *S* waves traveling along the surface at speeds V_P km/sec and V_S km/sec will cover 1 km distance in $1/V_P$ and $1/V_S$ seconds, respectively (and since $V_P > V_S$, the *P* wave indeed covers the distance in less time); so, for each kilometer covered, the *P* wave will get ahead of the *S* wave by that many more seconds—the more kilometers, the farther ahead it is.

On regional and global scales, body waves follow more complex paths—and, as we will see in the next chapter, at varying speeds—through Earth's interior, rather than straight paths along the surface. Nevertheless, the principle of greater delays at greater distances still holds. For example, according to the travel-time curve shown in Figure 7.38 (left side), the *S* wave traveling to a station $\Delta = 30°$ from an earthquake arrives ~5½ minutes later than the *P* wave; but for a station $\Delta = 60°$ away from the earthquake, it arrives ~9 minutes later.

Can you verify these results? First use the figure to determine the travel times for both waves, at each distance.

(a)   IRIS data Station PAB 2013 Great Okhotsk Event

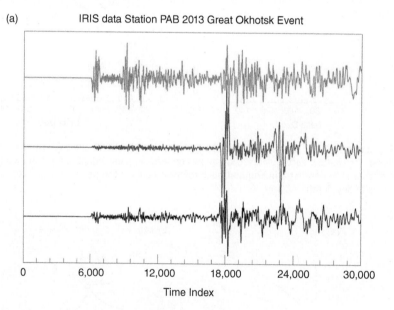

Time Index

Figure 7.40 Use of travel-time curves to interpret seismograms.
(a) Digital seismograms (offset, with the vertical component placed above both horizontal components) from IRIS station PAB in San Pablo, Spain, illustrating the ground motion recorded after a great earthquake in the Sea of Okhotsk (Δ ~84°) on May 24, 2013. Data graphed every 0.05 sec, so that every 6,000 points (large tick marks on the *x* axis) represents 5 minutes.
The first arrival must be a *P* wave; the largest amplitude body wave, as can be seen here at an index of about 18,000, is the *S* wave. *What do you think is the significance of the P wave arriving at PAB primarily with vertical ground motion? Would your explanation also account for the S wave's appearance in all three components of ground motion?*
With a depth of 600 km, this earthquake did not produce distinctive surface waves; even if it had, though, those waves would have traveled too slowly to appear at PAB this early in the seismogram. Consequently, any other features seen here besides the first *P* and *S* waves, especially if they are seen simultaneously on more than one component, might actually represent later body wave arrivals.
Which 'phases' might these arrivals correspond to? (See part b of this figure!)
Data downloaded from http://ds.iris.edu/wilber3/find_stations/4218658

(b)

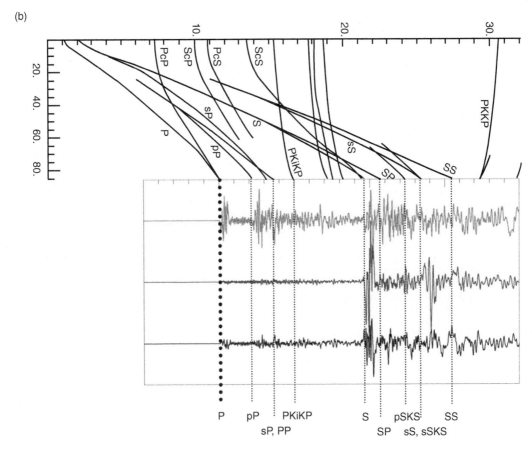

Figure 7.40 Use of travel-time curves to interpret seismograms.
(b) The seismograms in part (a) are overlain with t-Δ curves (from Figure 7.38) at $\Delta = 84°$; the t-Δ curves are 'stretched' so the seismogram and t-Δ timescales (both in minutes) match—for example, the P-to-S interval must be 10 minutes, according to the seismogram in part (a)—and shifted so the P wave arrivals on the seismogram and curve are aligned (heavy dotted line).
The dotted vertical lines are placed at the times of the various arrivals (for $\Delta = 84°$) implied by the t-Δ curves, extended down, and labelled, suggesting where they should occur on the seismograms. (See Figure 7.38 to identify any t-Δ curves whose labels were cut off in the section of the graph used here.)

To locate an earthquake, then, we let t_P and t_S denote the travel times of the P and S waves recorded at a seismographic station following the earthquake. The (angular) distance from the earthquake will be denoted by Δ. Now, if Δ is small, the S wave will have followed closely on the heels of the P wave, and ($t_S - t_P$) will have a small value; but the larger Δ is—as we have argued—and as is evident from Figure 7.39a—the greater will be the value of ($t_S - t_P$). Consulting the t-Δ curves (see Figure 7.39a), we see that the observed ($t_S - t_P$) value can correspond to only one value

of Δ, thereby determining the actual distance of the source from the station.

The basis for that determination, by the way, is the same 'delay time' concept that underlies our estimate of the distance from a lightning storm, with the lightning flash taking the place of the P wave and the sound of thunder analogous to the slower S wave. After we see the lightning, we count the number of additional seconds before the thunder is heard; for every 5 seconds, the source of the lightning and thunder (i.e., the storm) must be an additional mile away (with the lightning having

(a)

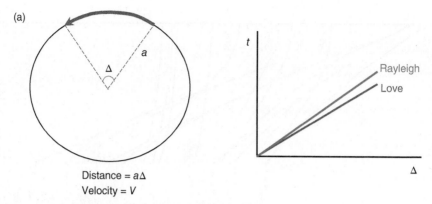

Distance = $a\Delta$
Velocity = V

Figure 7.41 Travel times of surface waves.
(a) Basically, the time-distance relation is a straight line.
(Left) Traveling along the surface of the Earth (whose radius is a), these waves cover a distance ($a\Delta$) in a time t, corresponding to a velocity V where $V = (a\Delta) / t$.
(Right) For either surface wave, the resulting form of the travel time (t-Δ) curves is a straight line, since $t = (a\Delta) / V$ or $t = (a/V) \Delta$; the lines have different slopes because Love waves travel at different velocities than Rayleigh waves.

As shown here, which is the faster wave, Love or Rayleigh, and which arrives first at a seismographic station?

(b)

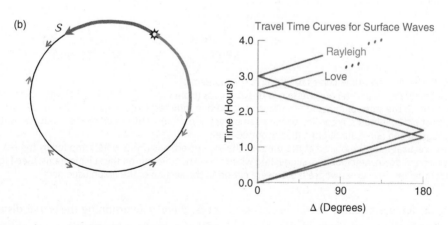

Figure 7.41 Travel times of surface waves.
(b) However, the surface waves generated by a large earthquake (☼) spread out in all directions; the wave traveling along the great circle *away* from our seismographic station (S) will continue on, and eventually arrive back at S. Furthermore, both of the waves shown here will keep circling around, causing repeated arrivals of those waves on the station's seismogram.
The travel time 'curves' for both the 'direct' and 'opposing' surface wave arrivals, whether Love or Rayleigh waves, are indicated on the right (see Stein & Wysession 2003 for a more complete picture). Lines tilted up to the right (increasing travel time with distance) correspond to the direct waves; lines tilting up to the left (increasing travel time for the waves going the 'long way around' to closer seismographic stations) correspond to the opposing waves. The lines continue up the graph indefinitely (as indicated by "...").

We have not yet accounted for the dispersion of the surface waves, i.e., that different period waves travel at (slightly) different speeds. How might that modify our travel-time graph?

traveled at the speed of light—essentially instantaneously—the distance from the storm reflects primarily the speed of sound in air, ~340 m/sec).

Our analysis so far has yielded only the distance of the earthquake from one point, our seismographic station. To actually locate the quake, we can add similar information (Δ) from *two* other stations; in general, there will be only one point on Earth's surface that is simultaneously three specified distances from three specific stations (see Figure 7.39b). This location can be pinpointed using a map of the Earth (or a globe, if we are considering teleseismic distances); we 'swing arcs' of specified distance from each of the three stations, looking for the one point of intersection. This approach, which historically was sometimes called the "arc method" for locating earthquakes, can also be carried out numerically. Numerical analyses can more easily deal with uncertainties in the data or in the travel-time curves (a consequence of regional variations in *P* and *S* wave velocities).

Occasionally, the arc method will appear to fail because the three arcs overshoot, rather than intersecting at a point. The uncertainties just noted could cause an overshoot; however, it may alternatively be the result of the earthquake **focus**—the actual location of the earthquake, in contrast to the **epicenter**, which is just its latitude and longitude coordinates on the surface—being at some depth *below* the surface. The three arcs may be viewed as planar slices of three spheres, each centered on one of the seismographic stations with a radius given by the Δ for that station; if the three spheres intersect at a point below the surface, that point may be a good approximation to the earthquake focus.

7.6.3 Travel Times: Identifying Phases on a Seismogram

Once the distance from the station to the source is known, we can use the *t*-Δ curves

for *other* body waves (pS, PcP, SKP, etc.) to tell us when on the seismogram to expect to see their arrivals. The procedure is illustrated in Figure 7.40.

Those other travel-time curves were constructed from a seismic velocity model based on observed, globally averaged *P* wave and *S* wave travel times (a velocity model based on free oscillations could also be employed). Using techniques outlined in Chapter 8, the averaged travel times were each mathematically 'inverted' to determine the profiles of V_P and V_S within the Earth. This in turn allowed the paths of PP, sSS, PcP, SKKP, and so on, to be delineated as they reflected, refracted, and were converted during their travels through the Earth; with their paths as well as velocities known, their implied travel times were easily calculated (see, e.g., Kennett & Engdahl, 1991).

Including surface waves in travel-time curves is problematic for two reasons. First, surface waves are particularly sensitive to lithospheric properties; since the latter vary significantly—perhaps the most significant lateral variation of all is simply oceanic versus continental crust—their travel times vary from region to region. Second, surface waves are dispersive, leading to multiple travel-time curves, one for each wavelength or period.

Nevertheless, for convenience, some travel-time curves (such as those in Figure 7.37) do include surface waves. The period chosen, 20 sec, corresponds to surface waves that tend to arrive at the beginning of the wave train, and the travel times will either represent a continental path or some type of globally averaged path (cf. Bullen, 1965). The 'curve' for surface wave travel times is a straight line, if path effects are neglected, as explained in Figure 7.41a (see also Figure 7.41b).

8

Seismology and the Interior of the Earth

8.0   Motivation

In this chapter, we expand our view of seismology: instead of focusing specifically on phenomena like body waves or free oscillations, we will use those phenomena as tools to discover the structure and physical state of Earth's interior.

Seismology provides us with an unparalleled, detailed picture of Earth's interior. The velocities of seismic waves and the eigenfrequencies of Earth's normal modes both depend on the material properties—densities and elastic moduli—of the interior. The eigenfrequencies can be 'inverted' by purely mathematical techniques to yield those properties; seismic velocities, calculated from travel-time data, can be combined with straightforward physical relations to infer them. In this chapter, we will explore the latter, physically based approach; but we will use the results from both approaches to learn about the Earth.

The material properties we infer provide clues to the composition of the interior. More importantly, though, the variation of material properties within the Earth, both as a function of depth and laterally, is fundamental to our understanding of how the Earth works. The variations determined by seismology reveal the basic layered structure of the interior and also point to those regions which depart from that structure. Combined with Geoid observations, such seismically determined, three-dimensional departures from lateral symmetry suggest how the Earth operates convectively.

Our inferences of density also allow us to predict the pressure and internal gravity experienced by rock at different depths—our first estimates of the forces acting throughout the interior. The key concept here is that of hydrostatic equilibrium, which was mentioned in Chapter 5 as a term equivalent to Airy isostasy; in this chapter, hydrostatic equilibrium will be introduced quantitatively, and independent of the subject of isostasy. This concept plays a role in all components of the Earth System, and in later chapters we will recast it in a more general form, and see how it underlies the dynamical behavior of the core, as well as the oceans and atmosphere.

In order to understand how and why the density varies with depth, it will be necessary to consider the variation of temperature within the Earth, and that will lead us to another key concept, the adiabatic temperature gradient. We will find this concept important in later chapters as a 'standard' allowing us to predict the instability and 'convective potential' of any fluid, including the core, oceans, and atmosphere, and to some extent, the Earth's mantle as well.

As we progress through this chapter, and then the remaining chapters of this textbook, the topics we study will depend on increasingly complex mathematics for their explication. Readers may benefit from frequently reminding themselves that, no matter how imposing, that math is just a tool to facilitate our understanding of how the Earth and the Earth System work. Furthermore, the symbols of this math, whether (for example) d/dx, $\partial/\partial x$, or ∇, are just shorthand for ideas that—when translated into words (and even the most mathematical of scientists do such translation, automatically though perhaps subconsciously, in every encounter with symbology!)—are perfectly understandable and relatable to the real world.

As a way of easing into both our wider seismological perspective and the mathematical and physical complexities of the solid earth, we will begin this chapter by studying its outermost one or two hundred kilometers (the lithosphere) and then, briefly at first, the mantle beneath (the fluid-like asthenosphere), exploring the seismological basis for plate tectonics and mantle convection. Though plate tectonics is obviously an important subject in itself, in this textbook we will also view it as providing boundary conditions on the dynamics of the interior.

8.1 Seismology and the Dynamic Earth

8.1.1 Defining Plate Tectonics

It is a cardinal rule of seismology that earthquakes can occur anywhere, any time. For instance, in the United States they tend to occur on the West Coast; but in 1811–1812, a devastating series of earthquakes—including three main shocks and several large aftershocks—occurred over a two-month period near St. Louis (centered in a place called "New Madrid"). At least one of the New Madrid shocks was comparable in strength to the famous 1906 San Francisco temblor; and at least one was large enough to ring church bells along the Atlantic coast, in Charleston, South Carolina (Hough et al., 2000; see also Johnston & Schweig, 1996)!

The **seismicity** of a region is the seismic activity within it—the frequency of occurrence of earthquakes and their spatial distribution. In

Chapter 7, we used travel-time curves to locate earthquakes. Cardinal rule notwithstanding, when we plot earthquake locations around the world over a span of years, we find their spatial distribution is actually quite nonrandom; see Figure 8.1. Earthquakes tend to occur in specific regions: the circum-Pacific belt (also called the Ring of Fire), young mountainous areas (such as the Alps, Himalayas, and Andes), and mid-ocean ridges; and they are comparatively rare in oceanic abyssal plains and continental shields (cratons).

Global seismicity thus reveals the surface of the Earth to be divided into distinct 'units' we call **plates** or **lithospheric plates**, each ultimately defined as a section of lithosphere (crust and uppermost mantle), all of whose particles move together, as an ensemble, on geologic time scales; equivalently, a section of lithosphere with nonzero overall motion relative to adjacent sections. This relative motion is accommodated, little by little, by the earthquakes occurring at the plate boundaries.

The relative motion determines the type of plate boundary and the tectonic results. **Convergent** plate boundaries exist where adjacent plates are colliding; the convergence causes plate thickening or subduction, leading to mountain-building or volcanic activity and island arcs. Where adjacent oceanic plates pull apart, at **divergent** boundaries, such rifting enables a more passive magmatism to 'fill in the gap,' creating new oceanic lithosphere as the seafloor continues to spread apart, and even producing mid-ocean ridges. Where plates move past each other, the boundaries are called **transform** boundaries. Depending on the plates' relative motion (which may be skewed with respect to the boundary), the boundary may be a mix of these three types. But at *all* boundaries, seismicity is relatively high.

The type of fault found at a plate boundary is generally consistent with the type of boundary and the relative motion. In simple terms, let's say that an earthquake involves the sudden displacement or *slip* of one block of mass relative to another, adjacent block. As sketched in Figure 8.2, if the relative motion between

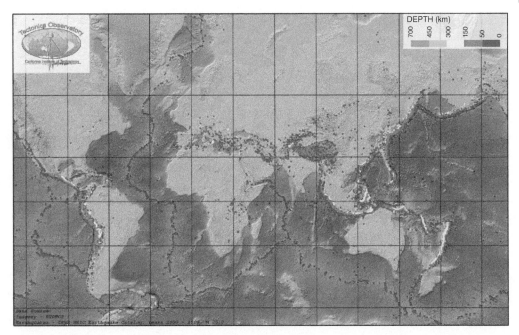

DEPTH (km)

700 450 300 150 50 0

Figure 8.1 Map of global seismicity for 2000–2008. This is a map of epicenters (depths indicated by color—see color code at top right) for earthquakes of at least moderate size (Richter magnitude $\geq$ 5.0); offsets of deeper quakes reflect the tilt of the subducting lithospheric slab experiencing the quakes. In most regions, the seismically active zones at plate boundaries are relatively narrow, and the plate extent sharply marked; in regions like the Himalayas, however, the seismically active zone is diffuse. Superimposed on all of the boundary seismicity, occasional intra-plate events are also seen to occur.
Based on this map, how many plates would you say characterize the Earth?
Courtesy of National Science Foundation produced in 2008 by scientists at Caltech's Tectonics Observatory using USGS epicentral locations.

the blocks is purely horizontal, the fault is called **strike-slip**; if there is a component of vertical relative motion, it is called **dip-slip**. Depending on the direction of relative motion between fault blocks, dip-slip faults, as shown in the figure, are classified as either **normal** or **thrust**.

Convergent and divergent boundaries are primarily dip-slip fault zones, the former, thrust, and the latter, normal; transform plate boundaries are primarily strike-slip. For strike-slip faults, the fault motion is **left lateral** (**right lateral**) if, standing on one block and facing the other, you see the other block move to the left (right).

Given that the net effect of North American and Pacific plate motions is to bring Los Angeles—which is located on the Pacific plate— up to the latitude of San Francisco (over the *next ~20 Myr), would you say the **San Andreas Fault** (the boundary between those plates) is left lateral or right lateral?*

The fault type is also reflected in the orientation of the **fault plane**, defined as the boundary surface between the fault blocks. In strike-slip faults, the fault plane may not be vertical (90° dip) but usually is nearly so, whereas in dip-slip faults, the fault plane is almost *never* vertical. And, in general, the fault plane is shallower for thrust faults than for normal faults (see, e.g., Stacey & Davis, 2008).

Of course, there are complications to the picture we have developed here. The stresses at plate boundaries are often great enough to fracture the lithosphere adjacent to the primary fault plane, creating other faults that can accommodate some of the plate motion. The San Andreas Fault, for example, is actually

(a) *(View from above, at an angle)*

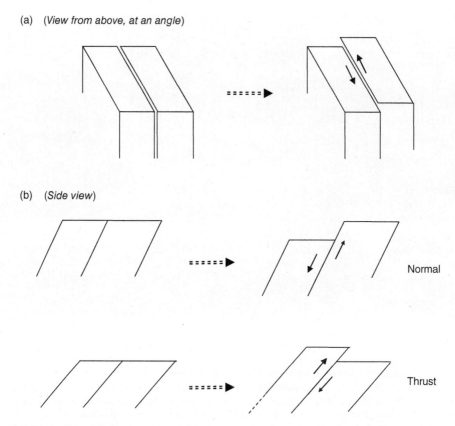

(b) *(Side view)*

Normal

Thrust

Figure 8.2 Ideal types of faults.
(a) In strike-slip faults, the sudden displacement between two blocks, characterizing the earthquake event, is horizontal.
(b) Dip-slip faults exhibit vertical displacement of one fault block relative to the other. Normal faulting corresponds to an extension or rifting of the region; thrust faults (also called *overthrust* or *reverse* faults) facilitate crustal shortening or compression.
These types of faults indicate the regional stresses responsible for the earthquake: normal faulting, typified by "Horst-and-Graben" structures like the alternating mountain ranges and valleys of the Basin and Range province of the western United States, is an expression of *tensional* stress; thrust faulting is associated with *compressional* stress.

a fault *system*; its northern California section includes the Hayward Fault and the Calaveras Fault, both somewhat east of the main fault but also trending southeast–northwest. Significant earthquakes have occurred on those faults as well.

From a global perspective, the San Andreas Fault system can reasonably be viewed as a narrow, almost one-dimensional boundary snaking along a portion of the Earth's surface; but, as Figure 8.1 reminds us, in some regions—the Himalayas being one obvious example—the boundary between plates (the Indian and Eurasian in this case) is a diffuse,

wide zone. Gordon (1998) estimates that the fraction of Earth's surface covered by such zones is ~15%.

For adjacent lithospheric plates moving relative to each other, the stresses building up along a 'locked' segment of the boundary can be relieved by the sudden slip of an earthquake; but a slow persistent microscopic 'creep' of one block past the other can achieve the same result. Such "aseismic" creep can occur episodically or be ongoing; Figure 7.9 in Chapter 7 shows a part of the San Andreas Fault system in central California experiencing ongoing creep. The stress relief from slip or creep does not extend indefinitely along the entire boundary, however, so other segments of the fault will have their own seismic or aseismic activity. The end result is a torturous struggle by the plate to move—torturous, at least, near the plate boundaries, and as seen on human time scales.

8.1.2 Quantifying Plate Motions

Thanks to planetary-scale forces, plates *do* succeed in moving. We can see the net effect in the separation of features on opposite sides of divergent plate boundaries such as the Mid-Atlantic Ridge—features including continental margins, like the west coast of Africa and the east coast of South America, and mountain chains, like the Caledonians of Scotland and the Canadian Appalachians in North America, which (with the plates that carry them) were once joined. As we will shortly learn, rifting began, and the Atlantic basins were created, around 180 Myr ago; since the Atlantic is now very roughly 60° wide in longitude (see Figure 8.1), relative plate motions across the Ridge must be, on average, of the order of 60° per 180 Myr or $1/3°$/Myr, thus nearly 40 km/Myr. On human timescales, if the plate motions were continuous, this would imply the east-west span of the Atlantic increased by ~4 cm every year.

On geologic time scales, plate velocities are determined most directly from spreading rates at mid-ocean ridges. At these divergent boundaries, magma wells up to form new oceanic lithosphere, only to be carried away from the rift as seafloor spreading continues. Oceanic lithosphere contains iron-bearing minerals and, as the new lithosphere cools to a critical point, that iron retains, compass-like, a faint memory of the Earth's magnetic field at the time of cooling. Such 'trace' or *remanent* magnetic fields can be detected by instruments up at the sea surface, carried onboard ships or towed behind them, or at altitude *via* aeromagnetic surveys (e.g., Müller et al., 2008; see also Seton et al., 2014); satellite measurements can also detect some remanent

patterns in the seafloor, though with lower spatial resolution (cf. Sabaka et al., 2004). In Chapter 10, we will learn that the Earth's magnetic field has occasionally and randomly reversed its polarity over time; as oceanic lithosphere cools, the remanent magnetism it retains will accordingly exhibit the prevailing polarity at the time of cooling. Ship-board magnetic surveys moving out, away from the mid-ocean ridge and thus positioned over older and older seafloor, will consequently measure alternating 'normal' and reversed polarities—seafloor magnetic 'stripes.' Matching the pattern of stripes to the known and very distinctive chronology of magnetic field reversals will then reveal the ages of those sections of the seafloor, and allows the rate of seafloor spreading away from the ridge to be estimated. This type of analysis is illustrated in Figure 8.3a; results worldwide (with spreading rates supplemented by data on transform fault orientations and on the amount of slip produced by plate boundary earthquakes) are shown in Figures 8.3b and 8.3c.

Rates of seafloor spreading range between 1 cm/yr and 18 cm/yr, the latter across the East Pacific Rise separating the Pacific and Nazca / Pacific and Antarctic plates; convergence rates at collisional boundaries range from 2 cm/yr to 11 cm/yr, the latter at the New Hebrides Trench, as the Australian plate subducts beneath the Pacific plate, and at the Chilean coast of South America (see also Gordon, 1998).

As plates displace along the Earth's spherical surface, their motion really constitutes an angular velocity (e.g., Morgan, 1968)—though it would take many millions of years for any plate to circuit the Earth once. The *relative* angular velocity of a plate will be measured by the number of degrees covered by the plate per million years relative to its neighbor as it inches over the surface, 'rotating' around some axis; such magnitudes are found to be typically ~0.6°/Myr, and do not exceed 2°/Myr (Gordon, 1998). The direction of the relative angular velocity, treated as a vector, is (consistent with the right-hand rule) the direction of

(a)

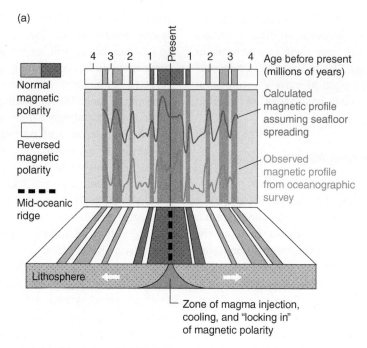

Figure 8.3 Determination of relative plate motions.

(a) Inference of spreading rates from magnetic 'stripes.'

Step One. Magnetic field measurements (blue curve), shown here across the East Pacific Rise, reveal positive peaks and negative troughs due to the magnetization of oceanic lithosphere, as it cooled, by the Earth's field.

Earth's magnetic field has reversed polarity over time. Once the section of lithosphere nearest to the ridge axis (dashed black line) has cooled enough, it will acquire a magnetization in the presence of Earth's field consistent with the field's current "normal" polarity (shown in dark brown color); sections farther out from the ridge had cooled during times of earlier polarities (reversed, white or normal, brown), then were carried away from the ridge by seafloor spreading.

Step Two. Assuming a specific, possibly constant rate of seafloor spreading, the known chronology of magnetic field polarity over time (see the top horizontal bar graph) can be expressed as peaks and troughs versus distance from the ridge axis (red curve). That curve will be more stretched out or more compressed if the seafloor spreading rate is faster or slower than assumed; matching the predicted and observed curves yields the actual spreading rate.

Figure, courtesy of the US Geological Survey, from Kious and Tilling (1996), and downloaded from http://pubs.usgs.gov/gip/dynamic/stripes.html.

that axis; equivalently, we can instead specify the coordinates of the axis' 'pole' on Earth's surface.

Such angular velocity vectors turn out to be a convenient way to describe the relative motion between plates, even if the plate motion is skewed relative to the boundary. And, to predict the relative velocities between plates, along the plate boundaries or elsewhere, the math is straightforward. Using the vector cross product relations developed in Chapter 4 (toward the end of section 4.4.4), the linear

velocity $\vec{v}$ of any point of the plate (located, say, at position $\vec{R}$), relative to another plate, can be found from $\vec{v} = \vec{\omega} \times \vec{R}$, where $\vec{\omega}$ is the relative angular velocity vector for that plate.

Improvements in the chronology of magnetic field polarity reversals (e.g., by Wilson (1993), who used Milankovitch cycles displayed by oxygen isotopes in seafloor sediments to help date the reversals) have caused models of plate motions to be revised. Based on such improvements, for example, DeMets et al. (1994) developed a newer version of their earlier,

(b)

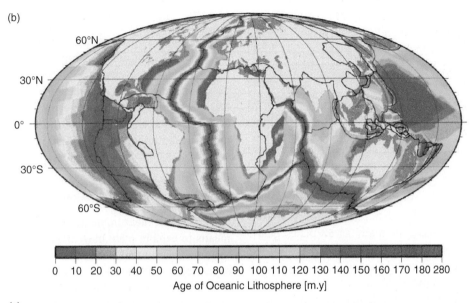

Age of Oceanic Lithosphere [m.y]

(c)

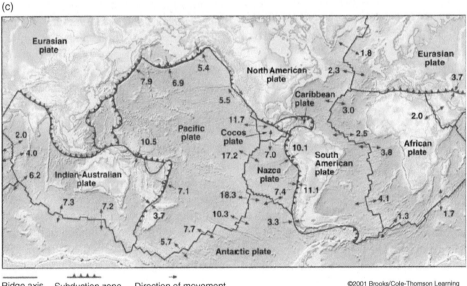

Ridge axis   Subduction zone   Direction of movement   ©2001 Brooks/Cole-Thomson Learning

Figure 8.3 Determination of relative plate motions.

(b) In each region, once the predicted and observed magnetic profiles are matched, as outlined in (a), the polarity chronology indicates the times of the polarity transitions, thus yielding bounds for the age of the lithosphere within each polarity interval. Image from Müller et al. (2008) / John Wiley & Sons.

(c) A plate's spreading rate, together with directional information such as transform boundary orientations, can be converted to an angular velocity representing the relative motion of the entire plate; that angular velocity may be a 'best-fit' approximation if based on multiple estimates plate-wide that are not fully consistent. As outlined in the text, this angular velocity can be used to predict the plate's relative velocity at any boundary. In this image, relative motion at plate boundaries, in cm/yr, is indicated by arrows at the boundary. Image downloaded from http://plcmets.pbworks.com/f/Tectonics+rate+map.jpg/Brooks.

popular plate model NUVEL-1 (DeMets et al., 1990), which described plate motions over the past 3 Myr; their model NUVEL-1A was characterized by plate velocities typically 4%–5% slower than in NUVEL-1. Plate models have also been impacted when deformable zones within large lithospheric plates have been reinterpreted as delineating distinct plates, with their own relative velocities, rather than simply being active portions of a larger plate; for instance, the African plate, which contains the East African rift zone, may more properly be described as distinct 'Nubian' and 'Somali' African plates separated by an active rifting boundary (Jestin et al., 1994). More recent models envision as many as 52 plates globally, several of which are better described as "microplates;" see Figure 8.4.

At the same time, a very different kind of plate motion model has been developed based on space geodetic observations of site displacements between plates. Such observations began in the 1970s with SLR inferences of relative motion between the Pacific and North American plates in California (Smith et al., 1979), then were expanded to include VLBI measurements (discussed below) and eventually GPS (Sengoku, 1998 provides a nice historical summary). As each of these techniques improved in precision and spatial coverage, plate velocities could be estimated with statistical significance, and global models of plate motions could be constructed geodetically.

Space geodetic inferences of plate motion are different than those based on magnetic stripes and other tectonic information because

- many permanent space geodetic sites are far from plate boundaries, where local and regional deformation may obscure the true, overall plate motion ('mobile' sites may, however, include fault zones);
- they are direct measurements (in itself a hugely impressive technological feat!), so they are independent of magnetic polarity chronology; and

- they are, geologically speaking, *instantaneous* rather than an average over hundreds of thousands to millions of years.

Space geodetic measurements thus present an amazing opportunity: being able to confirm that plate movement, a process implied to be taking place incrementally in our lifetimes—implied, because we only see the evidence of its end results (continental drift, mountain-building, and so on) after millions of years of activity—actually is happening *now*.

8.1.3 Plate Motions Through the Ages

Space geodetic measurements also offer us the means to explore plate movements from that incremental, human perspective. Should we expect the instantaneous plate motion to match the geologic long term? And, if a discrepancy is found, with current plate speeds falling short, for example, should we infer that further plate motion is 'overdue,' and will be brought about by large earthquakes yet to occur? These questions assume that the long-term rate is truly representative of the plate's motion. Thus, the more fundamental question is: *have plate motions been constant over time?*

A Geologic Perspective. Plate motions may be agonizingly slow, but that does not mean they remain unchanged for all time. Oceanic plates are created at mid-ocean ridges but consumed at subduction zones; unless an oceanic plate has both types of boundaries, and both are equally active, and they are positioned oppositionally just so, we should expect to see the plate change in size. For example, the North American plate has continued to grow, as seafloor spreading expands the Atlantic Ocean—and its growth has basically been at the expense of the Pacific Ocean; the Pacific plate, mostly surrounded by subduction zones, is shrinking, and will continue to do so for millions of years more. Such long-term plate evolution, which was first documented globally by Garfunkel (1975), also implies that plate boundaries will migrate.

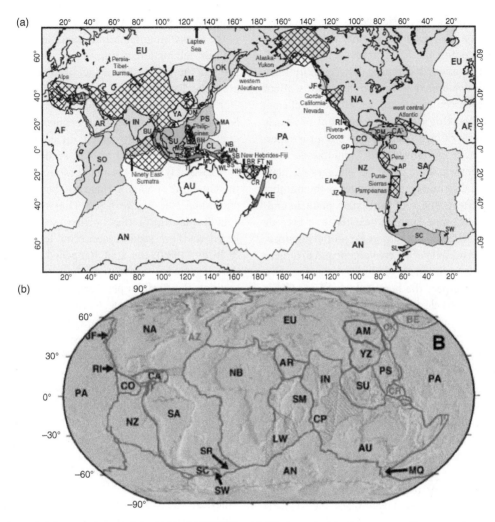

Figure 8.4 Recent plate models. (a) Model PB2002, with 52 plates; hatched areas are active orogenic areas. Bird (2003) / John Wiley & Sons. (b) Model MORVEL, using 25 of the 29 plates shown here; hatched areas are diffuse boundary regions. DeMets et al. (2010) / Oxford University Press. See the cited references for the abbreviations of the plate names.

Note that a merger of these two plate models would lead to even more than 52 plates. For example, in MORVEL the Somalian plate SM (labeled SO in model PB2002) is further divided, with the identification of a portion of it as the "Lwandle" plate (LW).

Plate evolution means that the geometry, dynamics (i.e., force balance), and kinematics (i.e., plate motions) of plate tectonics must all vary on geologically long time scales. An example of this is the sporadic existence of supercontinents. In the past 2 Byr, there have been at least three times when most land mass was aggregated onto a single plate: ~1.8 Byr to ~1.5 Byr ago, as Columbia; ~1 Byr to ~700 Myr ago, as Rodinia; and ~320 Myr to 180 Myr ago, as Pangea (Meert, 2012; Bradley, 2011, with the older time spans very approximate and still subject to active debate). Some researchers argue for additional supercontinents, such as Pannotia or Gondwana (~600 Myr and ~550 Myr ago; see Bradley, 2011), but those can alternatively be considered as less extensive, pre- or post-versions of the three primary supercontinents. The process of supercontinent creation is complex, but certainly must require subduction, to eliminate the seafloor between continents and foster their collision; and, breaking up a supercontinent obviously requires rifting. Thus, as the 'era' of a supercontinent is approached, maintained, and abandoned, there must be changes in plate size and geometry, and changes in the number and effectiveness of particular plate boundaries, as some come into existence or become more active and others weaken or even terminate. And, as specific convergent and divergent boundaries intensify or diminish their activity, we can expect the bounded plates to change their speed and direction of motion. These kinds of variations are illustrated in the context of the past 400 Myr by Matthews et al. (2016).

Such changeability also characterizes plate tectonics in the absence of supercontinents. For example, nearly half of all subduction zones currently operating began their existence in the past 55 Myr; and as those plate boundaries were created and became active, with some even becoming self-sustaining, major changes in the force balance between plates would have taken place (see Gurnis et al., 2004).

The Farallon Plate: A Case Study. One dramatic illustration of plate changeability is provided by the evolution of the Farallon plate, an ancient, mostly oceanic plate once located between the Americas and the Pacific plate and extending more than 80° north-to-south along the coast of the Americas (see Figure 8.5a). The Farallon began subducting to the east, under the Americas, as far back as ~180 Myr ago; meanwhile, its western boundary, with the Pacific plate, was divergent, a mid-ocean ridge as long as the Mid-Atlantic Ridge is today. Over the next ~130 Myr, dominated by collision and subduction, the Farallon–America boundary saw island arcs; the gradual accretion of continental crust resulting in the building up of

(e.g.) Western North America; and the raising of the Rocky Mountains and the Andes.

With subduction continuing to dominate, the Farallon plate shrunk, width-wise. However, the Farallon plate's velocity relative to North America was faster than relative to the South American plate, so the ongoing subduction preferentially ended up consuming most of the Farallon's northern section. And, by ~25–28 Myr ago (Atwater & Molnar, 1973), the Farallon's spreading ridge also began to be consumed (see Figure 8.5b)! This effectively split the plate into a small northern section and larger southern section (according to Gordon & Jurdy (1986), the split happened prior to the ridge being consumed). In-between the sections, the Farallon plate was overridden by the North American plate; North America now made direct contact with the Pacific plate—at a new, transform boundary we would today refer to (in part) as the San Andreas Fault.

Then, as subduction in Central versus South America stressed the Farallon's southern section in different directions, that section also split apart, creating a new spreading ridge 23 Myr ago (Lonsdale, 2005) to facilitate subduction under both Americas (Figure 8.5a). Today, we see the remnants of the Farallon as the Nazca plate off the coast of South America; the Cocos plate off the Central American coast; and the very small Juan de Fuca plate off the North American coast from Vancouver to northern California (plus two even smaller plates)—all bounded on the west by spreading centers and on the east by subduction zones. The southern boundary of the Nazca plate, with Antarctica, is also a spreading center; with the Cocos and Nazca plates separated by that newer spreading ridge, three of Nazca's four boundaries end up being divergent.

It should be noted that Sigloch and Mihalynuk (2013; but see also Liu, 2014) argue for a significantly *more* complex evolution of this region, especially before subduction beneath the Americas began.

The Farallon plate and its descendants exhibited a variety of unsteady behavior associated

(a)

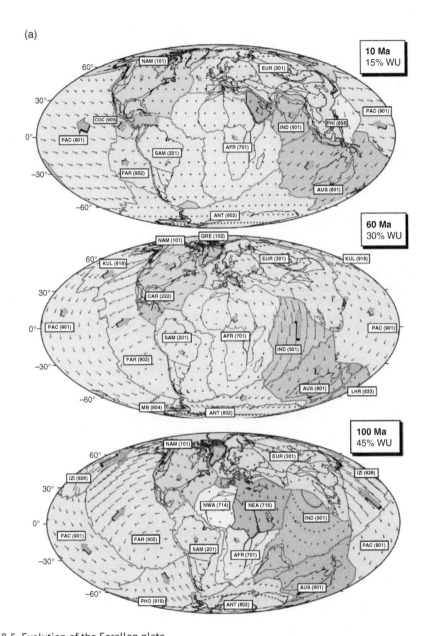

Figure 8.5 Evolution of the Farallon plate.
(a) Overall evolution of the plate (labeled FAR) 100–10 Myr ago.
Note the decrease in size as it continued to subduct beneath the western edge of the North American, Caribbean, and South American plates (NAM, CAR, SAM). Once its western spreading ridge had subducted beneath North America (creating the San Andreas Fault transform boundary between the Pacific (PAC) and North American plates), the remaining more southern section—stressed by subduction under Central and South America in different directions—further split into the Cocos (COC) and Nazca (here still labeled as FAR) plates.
WU percentages are a measure of the uncertainty in plate reconstructions due to the amount of oceanic lithosphere since subducted (and unavailable for direct measurements). Arrows indicate plate motions relative to a 'hot spot' reference frame, discussed later in this chapter. Image from Torsvik et al. (2010) / Elsevier.

(b)

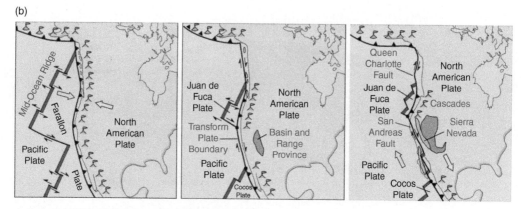

Figure 8.5 Evolution of the Farallon plate.
(b) The Farallon split.
(Left) 40 Myr ago, as the Farallon plate continued to be subducted under the North American plate, its western boundary (a spreading ridge) gradually approached the 'California' coast.
(Middle) By 25–28 Myr ago, the boundary had been overridden by North America, effectively splitting the Farallon plate into northern and southern sections. As of 20 Myr ago (as shown here), the northern section, with continued subduction, remained mostly as the small Juan de Fuca plate; the southern section was comprised mainly of the Cocos and (to the south, out of the picture) Nazca plates.
As subduction progressed, the split broadened, creating a transform boundary between the North American and Pacific plates.
(Right) Today, we call the main part of that transform boundary the San Andreas.
From https://www.nps.gov/subjects/geology/plate-tectonics-transform-plate-boundaries.htm (original images by S. Marshak) / U.S. Department of the Interior.

with such plate evolution. Menard (1978) summarizes several changes in their boundaries, including ridge jumps and abandoned ridge crests. Additionally, for more than 1 Myr preceding its split into the Nazca and Cocos plates, the Farallon's plate motion changed direction, from northeastward to eastward, a direction subsequently retained by the Nazca remnant but not the Cocos (Lonsdale, 2005). Tebbens et al. (1997) focused on the Nazca plate and its neighbors following the split, finding evidence for rift propagation, ridge capture, and triple-junction jumps (triple junctions are points where three plates meet, e.g., the Nazca/Cocos/Pacific plates as seen in Figure 8.4). Tebbens et al. (1997) were also able to correlate slowdowns in spreading rates after the split with ridge-trench collisions. Finally, in the early days of the San Andreas, ~20 Myr ago to ~10 Myr ago, relative motion between the North American and Pacific plates was much slower than today: three times slower than its speed between 10 Myr ago and 4½ Myr ago, and four times

slower than its speed since then (Atwater & Molnar, 1973).

The Farallon and its neighboring plates were not the only ones exhibiting accelerations and decelerations. Using a higher resolution, recalibrated chronology of Earth's magnetic field polarity reversals for the past 80 Myr, Cande and Kent (1992) confirmed that there was a 50% drop in spreading rate across the South Atlantic from 80 Myr ago to 60 Myr ago, an acceleration till 30 Myr ago, then a deceleration to the present time.

Plate motions can be unsteady on short geological time scales as well. A short-term relative plate model called MORVEL (DeMets et al., 2010), essentially based on spreading rates over the past 0.78 Myr only (going back only to the time of the first major reversal of Earth's magnetic field), shows that Nazca plate motion is now slower even compared just to the past 3 Myr (as represented by NUVEL-1A); this includes a more than 10% slowdown in spreading between Nazca and the Antarctic plate and a slowdown of 3% between Nazca and the Pacific plate (together those slowdowns produce a decreased eastward motion of the Nazca plate).

In Our Lifetime. Evidently, then, plates can behave unsteadily on long and short geological time scales; *does such unsteadiness also extend down to human time scales?* The perspective we have just gained on the long term provides a context for the discussion of 'instantaneous' plate motion.

VLBI and the Cosmic Reference Frame. Before that discussion, however, we take a moment to describe a very large-scale space geodetic measurement technique which has proven to be particularly useful for quantifying instantaneous plate motions. This technique, known as **very long baseline interferometry** (**VLBI**), records the radio waves from quasars to determine the distance—the "baseline" length—between receivers, for example if the radio antennas are located on different tectonic plates. **Quasars** are visibly faint, distant objects that actually broadcast intensely powerful radio waves; such 'radio beacons in the sky' are thought most likely to be active galactic nuclei, 'supermassive' black holes at the centers of extremely distant, young galaxies. They are so far from us that their apparent motion, e.g., as the Universe continues to expand, is negligible: for all practical purposes, they are fixed in space, and can serve as a fixed (*quasi*-inertial, perhaps) reference frame.

When a radio wave (technically, wavefront) from a quasar reaches Earth, radio telescopes at different locations around the globe will receive that signal at slightly different times. A location at the 'top' of the globe, relative to the quasar (that is, a location we can picture as being directly 'underneath' the quasar), would record the signal's arrival first, whereas locations away from the top, more distant from the quasar by virtue of the Earth's curvature, would receive the signal after some delay. By matching or *correlating* the signals, that time delay can be deduced; then, using the time-honored relation "distance equals rate times time," we can specify just how much

more distant from the quasar those other sites are. Because the telescopes are located at different positions around the globe, their lines-of-sight—always pointing to the same quasar—are angled differently relative to the curving Earth. With those angles measured, and the time-delayed distances already determined, inferring the baseline distance between each pair of telescopes reduces to a straightforward geometry calculation (see, e.g., Teke et al., 2012; Schuh & Behrend, 2012).

Repeated measurements over time allow changes in baseline lengths to be inferred. Using sites in the Eastern United States (Massachusetts) and Sweden, data collected between 1979 and 1984 led to the first successful detection of instantaneous plate motion by VLBI (Herring et al., 1986; see also Ryan & Ma, 1998, for a personal history of VLBI's early years).

Comparisons of plate velocities from geologic and space geodetic data initially tended to find strong agreement, excluding plate boundaries (e.g., Sengoku, 1998; DeMets et al., 1994). For example, DeMets et al. (1994) concluded that their improved geologic plate model NUVEL-1A was, overall, consistent with (though typically faster than) SLR and VLBI measurements, with an average discrepancy of less than 2%.

Other studies indicated general agreement but with one or two anomalous plates. Larson et al. (1997) concluded from GPS observations that the Pacific plate currently exhibits an angular velocity faster by 10% (citing VLBI measurements, DeMets et al., 1994 had speculated about just that possibility). Norabuena et al. (1999), using mainly GPS but other space geodetic techniques as well, found slower convergence between Nazca and South America, with a nearly 15% drop in relative angular velocity and an even greater drop in convergence rates along the Peru-Chile trench; using GPS only, Angermann et al. (1999) had previously obtained essentially the same results. These studies were in comparison with NUVEL-1A.

Global space geodetic models have suggested that such disparities are not just 'flukes.' Such models include:

- REVEL-2000 (Sella et al., 2002), based on GPS data (and most of that, continuous GPS data) for all but a handful of sites;
- GSRM-1 (Kreemer et al., 2003), based mostly on GPS but supplemented with geologically recent seismological data as well as VLBI and DORIS data;
- an expanded, REVEL-type model based on continuous GPS data by Prawirodirdjo and Bock (2004);
- a series of ITRF models including ITRF2005 (Altamimi et al., 2007), using mainly GPS and VLBI observations (plus a variety of techniques to define a dynamic "international terrestrial reference frame"); and
- GEODVEL (Argus et al., 2010), based on GPS, VLBI, SLR, and DORIS observations.

REVEL-2000 quantified relative velocities among 19 plates and continental blocks; two thirds of the plate pairs exhibited relative motions consistent with NUVEL-1A, within uncertainties, but most of the remaining combinations indicated slower spreading rates now than over the past few million years. Examples of the latter include Arabia-Eurasia (slower by up to 40%), Arabia-Nubia, Eurasia-India, Nazca-Pacific (~7% slower), Nazca-South America, and Nubia-South America (≥ 10% slower). Using 25 plates and continental blocks, GSRM-1 also saw significantly slower motion of several plates relative to Eurasia, including India-Eurasia (30% slower) and Nubia-Eurasia (50% slower), compared to NUVEL-1A.

ITRF2005 determined angular velocities of 15 plates; with reference to their earlier model ITRF2000 (Altamimi et al., 2002), we can infer that the discrepancies with NUVEL-1A persist: major plates are moving faster relative to the Pacific plate, compared to NUVEL-1A (though in all cases by less than ~10%); and the Eurasia–North America separation is faster (by more than ~15%) than in NUVEL-1A.

GEODVEL estimated the angular velocities of 11 plates, and the differences in angular velocity between both neighboring and distant pairs of plates; their broad base of geodetic observations yielded, for most plate pairs, small but statistically significantly different relative angular velocities than other space geodetic models. For example, discrepancies with respect to REVEL-2000 were statistically significant for nearly all plate pairs; their magnitudes ranged as high as ~30% but were typically just several percent. Discrepancies with respect to ITRF2005 were roughly half as large, and were significant for only a third of the plate pairs. Argus et al. (2010) attributed the discrepancies involving space geodetic models mostly to how the various researchers represented the center of the reference frame (see also Kogan & Steblov, 2008), though the choice of data was also a factor.

GEODVEL disparities were much greater with respect to NUVEL-1A and other geologically based plate models, typically more than 10% but as much as ~130%; the greatest disparities involved the Somalia and/or Nazca plates. The majority of relative plate velocities appear to be slower now than over the past few million years, including most spreading rates (which, not surprisingly, have resulted in the Nazca plate exhibiting the most noticeable slowdown).

All of these comparisons are somewhat frustrating. The consensus of most of these studies—including but not limited to the REVEL-2000 and GEODVEL models—is that, in our lifetime, plate motions have been slower than their 3 Myr average velocities (as represented by NUVEL-1A); only the ITRF2005 model finds major plates to be faster than NUVEL-1A at present. But the rigorous analysis by Argus et al. (2010) found significant disparities between GEODVEL and REVEL-2000 and relatively few such between GEODVEL and ITRF2005. The most cautious conclusion we can make is that the uncertainties in either the geologic or geodetic data, or in both, are greater than have been acknowledged. With somewhat less caution, we might instead conclude that, regardless of their uncertainties, space geodetic techniques show plate motions are apparently characterized by unsteadiness (whether accelerations, decelerations, or changes in direction) on human time scales.

8.1.4 A Last Look at Plates and Plate Motions

We defined a lithospheric plate as a portion of the lithosphere all of whose particles share the same velocity, distinct from adjacent plates; the relative motion between adjacent plates is associated with a specific type of plate boundary, and a specific kind of tectonic activity at that boundary. Taken literally, our definition of shared velocity is equivalent to requiring that each plate behave perfectly *rigidly*, with no relative motion or deformation allowed in its interior. On several levels, of course, such a requirement is unrealistic—we know that shear waves travel through the lithosphere, and with less than infinite velocity; that plates can flex in response to surface loads (as discussed in Chapter 5), a response that occasionally even leads to intraplate earthquakes; and that tectonic stresses can cause a plate's interior to strain (consider, for example, the extension occurring in the Basin-and-Range province of the North American plate).

As long as a plate maintains its integrity, and does not break apart (or accrete to another plate), its *identity*, at least, will be preserved—regardless of the deformation from a surface load or passing *S* wave. Practically speaking, then, our definition of a lithospheric plate just needs a little more … flexibility. We can allow the common velocity of all particles of a plate to be supplemented by additional, nonuniform velocities, and the rigid bodily motion of the plate to be supplemented by internal deformations—as long as the plate's boundaries remain well identified.

However, if we want to model plate *velocities*, the question of plate rigidity is unavoidable. The ability of nonrigidity to affect plate velocity is easily imagined for plates on the verge of

rifting apart, for example, or plates with diffuse boundary zones, and could be significant: in the case of several plates with diffuse boundaries studied by Gordon (1998), the relative velocity just across those zones was typically around 1 cm/yr!

Less obviously, perhaps, plate flexure—such as the continental rebound following the melting away of Pleistocene ice loads—can also result in lateral plate movement. As it flexes upward, supported by an underlying asthenospheric inflow beneath each center of uplift, the lithosphere 'domes' or arches regionally, expanding up and out over the rebounding area, thus producing mainly uplift but also outward motion. At the same time, the asthenospheric inflow will drag the lithosphere above, attempting to force it inward toward the center of rebound (James & Morgan, 1990; but see also James & Lambert, 1993 and Mitrovica et al., 1994). (Compounding things further, as discussed in Chapter 5, the rebound and asthenospheric inflow will cause peripheral bulges—created earlier by lithospheric subsidence and asthenospheric outflow away from the loads—to 'deflate,' imparting additional horizontal motions to the lithosphere, though farther out and dispersed.)

The outward motion associated with rebound in Hudson Bay, at rates of ~0.1 cm/yr, has been captured by GPS measurements (Sella et al., 2007; Calais et al., 2006; see Figure 8.6); those same measurements also detected the accompanying vertical rebound of the region, with rates an order of magnitude faster. Argus et al. (2010) obtained similar results, though with a slightly slower uplift, in Fennoscandia (see also Milne et al., 2001). The pattern of horizontal velocities around Hudson Bay makes their connection to rebound unequivocal; but it might be noted that corroboration of their direction and magnitude by theoretical models has been less definitive. Part of the reason for such difficulty is that theoretical predictions of horizontal motion associated with rebound are sensitive to a number of factors, including the thickness of the lithosphere, viscosities of the asthenosphere and lower mantle, and

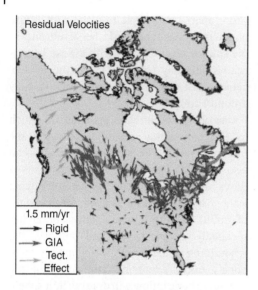

Figure 8.6 Residual horizontal motion around Hudson Bay detected by both continuous and episodic GPS measurements taken between 1993 and 2006. Rigid-body motion of the North American plate was first estimated (from two hundred GPS sites not near this or other deforming regions) and subtracted from the data. Each arrow represents the direction and magnitude of the horizontal motion inferred at that GPS site. Residual motion was interpreted by the authors to be one of three types: present-day glacial isostatic adjustment (red arrows); tectonic effects (blue); or residuals (black arrows) in the estimated rigid-body plate motion. Image from Sella et al. (2007) / John Wiley & Sons.

possible lateral variations in those parameters (e.g., Sella et al., 2007; also James & Morgan, 1990; Mitrovica et al., 1994; Peltier, 1998), as well as the ice history (e.g., Peltier, 1995) and the lithosphere's incompressibility (James & Lambert, 1993).

The GPS measurements analyzed by Sella et al. (2007) suggest that any *remaining* non-rigidity (beyond tectonically active, diffuse zones and the flexure caused by changing surface loads) may well be minor: on the North American plate, at least, its effect on horizontal plate motion at present is everywhere less than 0.1 cm/yr, and typically only ~0.05 cm/yr (see Figure 8.6, black arrows). Argus et al. (2010) drew the same conclusions globally based on an assessment of their global plate model GEODVEL; and, the results of Prawirodirdjo

and Bock (2004) are consistent with these conclusions as well. For most portions of most major plates, then, it appears that plate rigidity is a decent approximation.

Closure. In a tectonically ideal world, with all plates completely rigid and their boundaries sharp, plate velocities obey a simple mathematical rule: the net relative velocity of all the plates, summed as you circle the globe, must be zero (e.g., Morgan, 1968; LePichon, 1968; Chase, 1972; Minster et al., 1974; see also McKenzie & Parker, 1967 and Gordon et al., 1987, who discuss how this criterion may be applied to triple junctions). In other words, as you travel that 'circuit' and arrive back at your original location, the relative velocity between your starting and ending locations is zero.

Symbolically, in the case where three plates, A, B, and C, are traversed in the circuit, and if we denote the linear velocity of (e.g.) plate B relative to plate A by $_B\vec{v}_A$ and the relative angular velocity by $_B\vec{\omega}_A$, this rule amounts to

$$_B\vec{v}_A + {_C\vec{v}_B} + {_A\vec{v}_C} = 0.$$

We can write this as

$$_B\vec{\omega}_A \times \vec{r} + {_C\vec{\omega}_B} \times \vec{r} + {_A\vec{\omega}_C} \times \vec{r} = 0$$

(with $\vec{r}$ a somewhat abstract position vector) or

$$\left(_B\vec{\omega}_A + {_C\vec{\omega}_B} + {_A\vec{\omega}_C}\right) \times \vec{r} = 0,$$

which requires

$$_B\vec{\omega}_A + {_C\vec{\omega}_B} + {_A\vec{\omega}_C} = 0$$

(McKenzie & Parker, 1967). Thus, the criterion can be expressed in terms of either linear or angular relative velocity.

Does the angular velocity version of this criterion make sense to you? Angular velocity may be less intuitive than linear velocity, but you should nevertheless be able to readily confirm this version's validity in the simple case of three plates all moving around the Earth's geographic (N-S) axis. You can take the plates to be north-south sectors; their boundaries, which may be divergent or convergent, are meridians stretching from pole to pole.

This criterion has found two kinds of uses in plate tectonics:

- To develop a globally complete plate model (e.g., Gordon et al., 1990; also, DeMets et al., 1994 and Horner-Johnson et al., 2007 as cited in Argus et al., 2010). If the relative velocity between two plates at some location along a circuit is unknown, the 'missing' velocity—say, $_C\vec{v}_B$—can be estimated by writing the criterion as

$$_C\vec{v}_B = -_B\vec{v}_A - _A\vec{v}_C$$

or

$$_C\vec{v}_B = _A\vec{v}_B + _C\vec{v}_A.$$

This approach could also be used to estimate an unknown or poorly determined angular velocity.

- To assess a plate model (e.g., DeMets et al., 1990; Horner-Johnson et al., 2007). If all relative velocities (linear or angular) are known or have been estimated, they can be summed around the circuit to verify that the criterion is satisfied. If it is not—if the plates fail to achieve "closure"—that may be a consequence of inaccurate plate velocities, perhaps indicating greater data uncertainties than expected. Alternatively (e.g., Gordon et al., 1987; DeMets et al., 2010), failure to achieve closure may be a consequence of plate nonrigidity. A variation of the latter possibility is that plates are rigid except for a deformable zone, such as a diffuse plate boundary (Gordon, 1998). To help rule out some of these alternatives, circuits can be designed excluding known postglacially or tectonically active sites, or with the velocities across deforming zones specified separately, by additional data (e.g., Kreemer et al., 2003; DeMets et al., 2010; Argus et al., 2010).

Anticipating Convection. From the perspective of this textbook, the most fundamental question of plate tectonics is, "Why do plates move?" Of course, the simple answer is "mantle convection"; but evidence from travel times and other seismological data, which we will examine beginning in the next section, will quickly reveal both the mantle's structure and the convection itself to be far from simple. First, however, we briefly consider some elementary though provocative plate-tectonic clues.

- One remarkable clue is apparent in Figure 8.3b: almost all oceanic lithosphere is younger than 200 Myr, and there is none older than 300 Myr—in stark contrast to continental lithosphere, which can be billions of years old! Oceanic lithosphere is consumed at subduction zones—evidently quite effectively—but continental lithosphere is not. One popular explanation for this is the slightly lower density of the latter, which allows it to remain buoyant, withstanding the forces dragging down the former; in cratonic regions, the lower density may be enhanced by depletion of basalt from deeper portions of the lithosphere (Jordan, 1978). However, research by Lenardic and Moresi (1999) and others has shown continental cratons to be, inherently, insufficiently buoyant to prevent subduction. Instead, their long-term stability may possibly be attributed to the shallow cratonic lithosphere possessing a more failure-resistant rheology, or to the deeper lithosphere (or the mantle just below that) being either dehydrated or colder, thus more viscous.

Such research notwithstanding, continental buoyancy cannot be completely ruled out away from the craton, where the continent may be tectonically more active as well as younger, so warmer and conceivably less dense. But the numbers argue against it: as will be detailed later in this chapter, a uniformly granitic continental crust would be just ~7% less dense than the basaltic oceanic crust; more importantly, continental *lithosphere* is, overall, only a few percent less dense than oceanic lithosphere—both types of lithosphere are mostly just uppermost mantle.

A final mechanism for continental stability is the greater thickness of continental lithosphere—two to three times the thickness of oceanic lithosphere at the point of subduction—which makes it much more difficult to bend and begin its descent into the mantle (see, e.g., De Bremaecker, 1977, 1980; Davaille & Limare, 2007).

But regardless of whether these mechanisms are dynamically reasonable, and sufficient or incomplete, the fact that there is a disparity in lithospheric age more generally tells us that *the lithosphere affects its own subduction*. That is, if (or when) we conclude that mantle convection drives plate motions, we will have to reject the simplest model of that convection, in which plates are simply carried about passively by a flowing mantle; forces associated with the plates themselves—whether buoyancy, bending, or other—must provide a significant supplement to the traditional forces of convection (see Davies, 1988a).

- Volcanism at both divergent and convergent plate boundaries seems a rather obvious clue as to what drives plate motions, and we can happily conclude that it is evidence of the surplus of heat energy in Earth's interior that ultimately powers plate activity (and the underlying mantle convection). However, this clue is as much a reflection of surficial, or near-surficial, lithospheric processes as is the preceding clue. As plates are pulled apart at a mid-ocean ridge, magma fills in the 'gap' and solidifies as new oceanic lithosphere; but that magma can have been created by *pressure-release melting*, a consequence of the lower melting point when the overlying pressure suddenly drops (see, e.g., McKenzie, 1984), and does not need to have originated as a plume far below the surface.

Similarly, the volcanic activity we associate with subduction zones is just a by-product of shallow upper-mantle magmatism accompanying the subduction of oceanic lithosphere. In some subduction zones, the magma could be generated by friction at the boundary between the descending slab and the mantle 'wedge' right above it (as suggested in Figure 2.3 of

Chapter 2)—at pressure and temperature conditions found typically just 100–150 km or so below Earth's surface (e.g., Marsh & Carmichael, 1974). Magma generation in the subduction zone may be aided by the presence of water, released from hydrated minerals in the slab and lowering the temperature needed for partial melting; as envisioned by Kelley et al. (2006) and called 'flux melting' (see also Spandler & Pirard, 2013), a stream of rising partial melt is continuously produced as water 'fluxes' from the dehydrating slab through the mantle wedge. In other subduction zones, the mantle wedge itself—hydrated by the subducting slab at those same depths—has been identified as the magma source (see Spandler & Pirard, 2013 for a more complete discussion of these mechanisms). In some cases, partial melting in the wedge might benefit from heat brought to it by laterally extensive horizontal convection, operating within the asthenosphere (Ida, 1983) (horizontal convection is discussed in the next chapter).

Subduction zones correspond to the colder and denser fluid downwelling of mantle convection cells, so it should not be surprising that subduction-zone volcanism requires some way to bring additional heat to the cold slab or to lower the temperature for partial melting to begin.

From a geometric viewpoint, it is reasonable that the source region of subduction-zone volcanism is in the upper mantle: after all, how deep could the source region be, when its surface expression—island arcs, for example, when the overriding plate is oceanic—is just not located that far from the oceanic trench where the subduction began? (To visualize the source region/island arc/ocean trench right triangle, it might be helpful to see Figure 2.3 of Chapter 2.)

- Our final clue to understanding mantle dynamics is perhaps the most remarkable of the three, and historically proved the most provocative: with rare exception, earthquakes are not observed to occur at depths exceeding ~700 km. In fact the strongest deep quakes ever recorded struck beneath the Sea of Okhotsk in 2013 at a depth of 600 km and beneath Bolivia in 1994 at a depth of 635 km; according to Frohlich (1989; also Estabrook, 2004; Gan et al., 2015), the deepest reliably determined earthquakes of any magnitude also occurred at depths above 700 km. (A recent analysis by Kiser et al., 2021 of the strong 2015 Bonin Islands earthquake, which occurred beneath the Philippine Sea south of Japan, revealed that the main, ~660-km-deep shock was followed just minutes later by several 'remote' aftershocks, and some of those were as deep as ~750 km. Though unusual in a number of respects, however, the fact that they were triggered by the main shock makes their relevance to mantle convection indirect.)

In the early days of plate tectonics, the absence of deeper earthquakes was generally interpreted as evidence that subduction—and the convection driving it—did not persist below the upper mantle, or at least that, if there was convection taking place in the lower mantle, it was essentially uncoupled from upper-mantle convection. However, these interpretations were never inescapable. One argument against separate upper- and lower-mantle convection derives from the broad horizontal scale of convection, as inferred from the distance between plate boundaries; upper-mantle circulation would have to have an unusually large 'aspect ratio' (i.e., convection cells would have a high width:depth ratio). Although such dimensions do not rule out convection, they may require unusual or extreme conditions, such as a large lateral temperature gradient, to drive the flow (see, e.g., Houseman, 1983 and references therein, which document some of the intense back and forth regarding layered-mantle convection). Another argument recognizes the high viscosity of the mantle; it is difficult to imagine that flow in opposite directions, at the

boundary between upper- and lower-mantle convection cells, could persist in the face of the expected huge viscous drag forces. Both of these arguments also apply to upper-mantle convection on top of a static lower mantle. And, of course, a lack of earthquakes at great depth does not mean relative motion is not occurring, merely that any such motion is accomplished silently, through some type of steady or episodic creep mechanisms.

From those early days onward, the question of whole-mantle convection versus its alternatives received intense scrutiny. One key issue, as indicated above, was whether the viscosity of the lower mantle was as low as that of the upper mantle, or higher, or even impossibly high. A turning point in the debate was the paper by Kenyon and Turcotte (1983), which showed that two-layer convection would lead to a very hot lower mantle, resulting in a lower-mantle viscosity orders of magnitude less than allowed by deglaciation studies (as discussed in Chapter 5). (The drama played out at that time at a national meeting of the American Geophysical Union, with a talk by Turcotte presenting those results. A major contributor to the theoretical foundations of convection within the Earth (e.g., Turcotte & Oxburgh, 1967, 1972), he had previously argued the merits of both upper-mantle and whole-mantle convection; but at this meeting, he conceded that recent confirmation of the similarity in lower-mantle and upper-mantle viscosities from deglaciation analyses rendered the possibility of whole-mantle convection much more likely.)

In recent years, seismic imaging has found what we infer to be remnants of subducted lithospheric plates in the deepest mantle (!), allowing the fact of mantle-wide convection to gain more general acceptance. This imaging technique, as we will discover later in this chapter, also reveals Earth's interior to be a bizarre realm of material contrasts and convective behavior.

To develop a 'big picture' of Earth's interior, in the hope of understanding the nature of mantle convection, we turn now to travel times.

8.1.5 Travel Times and the Interior of the Earth

With millions of arrival times recorded around the world from numerous earthquakes, processed to determine travel times versus distance for about two dozen types of body waves (e.g., Kennett & Engdahl, 1991), travel-time curves and travel-time tables represent seismic information averaged worldwide; the effects of lateral inhomogeneities in density and elastic properties essentially cancel out. Travel times thus serve as a laterally symmetric *reference*, a *worldwide standard* to which travel times in any region can be compared; discrepancies relative to that standard will imply that regional inhomogeneities exist, superimposed on the global average.

Travel Times and Plumes: Easing Into the Big Picture. For example, the arrival times of *P* waves passing beneath the Yellowstone Plateau are observed to be slightly delayed, compared to worldwide averages, for waves covering the same source-to-receiver distances. The slightly longer travel times imply slower wave speeds, not only right beneath the caldera—which might be expected as a result of the higher temperatures and magma associated with the Yellowstone hotspot—but also deeper, extending all the way down to the lower mantle (Smith et al., 2009; see Figure 8.7). Those slower wave velocities can be interpreted as delineating a deep plume, ~100 km in width, responsible for feeding the hotspot.

The mantle plume below Yellowstone exists within the overall broad-scale convection of the mantle; the upper limb of that convection should include eastward flow beneath the western U.S. continental lithosphere, corresponding to the earlier, eastward subduction of the Farallon/Juan de Fuca plate. If that flow has interacted with the rising plume, it could skew it eastward—producing the distinct

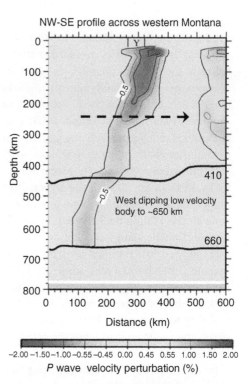

NW-SE profile across western Montana

P wave velocity perturbation (%)

Figure 8.7 *P* wave travel times were compared by Smith et al. (2009) / Elsevier to a standard (they used the iasp91 model of Kennett & Engdahl (1991), displayed in Figure 7.38 of Chapter 7) to infer slower *P* wave velocities beneath the Yellowstone plateau ("Y"). Contours are −0.5%, −0.75%, −1.0%, and −1.5% of *P* wave velocity. (A high-resolution analysis by Huang et al. (2015) revealed even greater anomalies in the top 50 km, which they interpreted as an upper-crustal magma chamber overlying a geochemically distinct lower-crustal magma chamber.) The low-velocity body extends to a depth of at least 660 km; no conclusive results were obtained by Smith et al. at greater depths.

The low velocities within this body are interpreted by Smith et al. to be the result of heat and the partial melting of mantle material, and thus delineate a plume of lower-density mantle material, probably rising from the deep mantle. As discussed in the text, broader-scale mantle convection, including an eastward flow (dashed arrow) beneath the western U.S. continental lithosphere, may have tilted the rising plume, skewing it to the east. Image from Smith et al. (2009) / Elsevier.

tilt (upward to the east) seen in the plume's profile (Figure 8.7). A similar tilt of the mantle plume beneath the Hawaiian hotspot has also been attributed to the effect of superimposed mantle convection (Steinberger, 2000; Nolet et al., 2007; but see also Lithgow-Bertelloni et al., 2001).

Incidentally, the distribution of hotspots around the world has been found by Weinstein and Olson (1989) to be quite nonrandom: they are very likely to be located near divergent boundaries, and unlikely to be located close to convergent boundaries. They interpreted such nonrandomness as another consequence of convective influences—convective upwellings draw the associated mantle plumes toward the surface, whereas downwellings force the plumes downward, away from the surface (see also Weinstein, 1995).

Some studies suggest the Yellowstone plume may penetrate to 800–1,000 km depth (Montelli et al., 2004; Burdick et al., 2008; cf. Sigloch et al., 2008); thus, the plume probably does not extend to the base of the mantle. In our discussion of convection in Chapter 9, we will come to recognize the base of a convection cell as a "thermal boundary layer," with the ability to promote convective upwellings. With a fully convecting mantle, that bottom layer—often referred to as **D double prime** or **D″** (though "Bullen layer" might be more appropriate; Chao, 2000) and perhaps 200–250 km thick—provides physical mechanisms for generating deep-mantle plumes. If the mantle supported two-layer convection, then the base of the upper-layer convection—generally assumed to be at depths of ~650 km, for reasons we will learn about later in this chapter—would also serve to promote plume formation; but the Yellowstone plume is significantly deeper (see Tackley, 2000; also Tackley, 2008, for alternative mechanisms). A variation of the two-layer mechanism, postulating a low-viscosity zone below that upper-layer base as the source of 'mid-mantle' (rather than upper-mantle) plumes, has *also* been proposed (Cserepes & Yuen, 2000); but according to that study, mid-mantle plumes should have very wide diameters, unlike the Yellowstone plume. We conclude that, with no mechanism to produce it except boundary layer processes

at D″, the Yellowstone plume must have had its source (at some time) in the lower mantle.

We might crudely estimate that plumes ascend through the mantle at speeds comparable to that of broad-scale mantle convection (since both reflect thermally driven flow through the same medium), but somewhat faster (since, as seen in Figure 8.7, the plume's ascent is affected but not 'swamped' by the convection it rises through). Velocities of a few to several cm/yr have also been estimated based on experimental and numerical models of plume dynamics (e.g., Olson & Nam, 1986; Liu & Chase, 1989). It might then be noted that, at a speed of ~10 cm/yr, a Yellowstone plume originating at the base of the mantle would have taken ~20 Myr to rise to that depth of 1,000 km; the last of the plume material will reach the surface ~10 Myr from now. Whether such estimates are accurate or not, the circumstances of the Yellowstone plume certainly allow us to conclude that *mantle plumes have a finite lifetime.*

Furthermore, it is believed that the Yellowstone hotspot first became active at Earth's surface as the Snake River Plain volcanic field (to the west of the present hotspot), close to ~20 Myr ago (Smith et al., 2009). Rising from the base of the mantle at ~10 cm/yr, the plume would have had to form ~30 Myr earlier, that is, ~50 Myr ago. Thus, if the plume comes to an end in another ~10 Myr, its lifetime will be roughly ~60 Myr—possibly even less, given that the rate of ascent of plumes may be significantly faster than 10 cm/yr (Olson et al., 1993). In sum, plume lifetimes can evidently be *much* shorter than that of mantle-wide convection cells.

As another example of a hotspot—plume system, we consider briefly the Hawaiian hotspot, perhaps the most well known of all. Its mantle plume has fed the volcanoes building the visible chain of Hawaiian Islands for the past 7 Myr; including older islands, in the Northwest Hawaii Ridge, clearly part of the chain but now eroded to fragments or seamounts, this process extends another 25 Myr into the past (e.g., Garcia et al., 2015). As implied by Figure 8.8,

Figure 8.8 The Hawaiian–Emperor seamount chain.
2006 digital global relief map, from the ETOPOv2 database, centered in the North Pacific. The bathymetry in ETOPOv2 was based mainly on data from the ERS-1 altimetric Geoid, which was discussed in Chapter 6 (see Figure 6.25a). The Geoid reflects seafloor bathymetry, allowing seamounts as well as subduction zones and fracture zones to be seen. Notice the change in trend of the chain, from northwest to north—a bend of about 60°. Image constructed by National Geophysical Data Center / USGS / Wikimedia Commans / Public Domain.

those older seamounts transition to another chain, the Emperor seamounts, the oldest of which, almost at the edge of the Aleutian trench, is 85 Myr old (Regelous et al., 2003). For both chains, the age progresses monotonically (Sharp & Clague, 2006) along the chain, suggesting the islands and seamounts were produced sequentially as the Pacific plate rode over the magma source of the Hawaiian hotspot (Wilson, 1965); the resulting *hotspot track* contains at least 129 current or extinct volcanoes (Sharp & Clague, 2006).

The Hawaiian hotspot is clearly a very active one. As a midplate hotspot, certainly unconnected to the upwellings and downwellings of broad-scale mantle convection, a variety of explanations for its origin have been proposed, some involving a mantle plume (with rates of ascent as high as ~50 cm/yr, estimated from the volume of magma produced in eruptions; DePaolo & Manga, 2003) but others focusing on shallow causes; one of the most intriguing is that the island chain results from a slowly propagating crack in the oceanic lithosphere, with the magma simply welling up from the asthenosphere (Turcotte & Oxburgh, 1973). However, as discussed in the next subsection, actual images of the plume beneath

this hotspot, based on travel-time anomalies relative to the standard, more or less rule out shallow alternatives.

The change in trend from the Hawaiian to Emperor chain (Figure 8.8)—most directly explained as a change in direction of the Pacific plate and occurring 50 Myr ago to 42 Myr ago as dated by Sharp and Clague (2006)—provides a striking, additional example of unsteady tectonics. Such dates allow those authors to connect that change in direction (the Hawaiian-Emperor 'bend') to contemporaneous changes in boundary situations involving the Pacific plate, including the initiation of subduction in the western Pacific (the Izu-Bonin-Mariana trench system, also called the IBM island-arc system, and the Tonga-Kermadec trench; but see also Garcia et al., 2015)—which may in turn have been initiated by the collision of the Indian and Eurasian plates around ~50 Myr ago (as discussed in Chapter 2).

Changes in the direction of Pacific-Farallon spreading (see Figure 8.5a) and in the rate of Farallon subduction beneath North America (Atwater, 1989) may also have contributed to the Hawaiian-Emperor bend.

Finally, our discussion here has so far avoided the question of whether hotspots—Yellowstone, Hawaii, and others around the globe—*and* the underlying mantle plumes creating them are *stationary*, in effect providing a fixed, *absolute* reference frame useful for measuring plate motions with respect to (rather than just measuring plate speeds relative to each other). This issue has proved almost as contentious as the question of upper-mantle versus whole-mantle convection. In the present context, we note that a drift of the Hawaiian hotspot has been proposed as an alternative explanation of the Hawaiian-Emperor bend; but the analysis by Torsvik et al. (2017) considering various plate and true-polar-wander models finds such an explanation to be unlikely: the Pacific plate really did change direction 50–42 Myr ago. On the other hand, Torsvik et al. (2017) note that a southward drift of that hotspot (and the underlying plume) can explain why the Emperor seamount chain is 'stretched out,' compared to the Hawaiian chain.

Travel Times: Global Variations and a Call for Acoustic Tomography. Our preceding discussion may have implied that delineating the Yellowstone plume required only a straightforward comparison of observed and reference travel times. In reality, however, a travel time is the integrated result of the velocities *all along* the path taken by the wave, not just the velocity in a particular place; so, even if the observed and reference seismic waves covered the same total distance (Δ) and followed the same exact paths, the comparison would yield an accurate picture of the plume only if the plume were the only anomalous feature over the entire travel path.

In general, then, the use of travel times will require recognizing that there are heterogeneities or anomalies scattered three-dimensionally throughout the breadth and depth of the Earth. This in turn implies that the waves observed at a seismographic station may have taken different paths, three-dimensionally, and sampled different heterogeneities, compared to the reference waves for that Δ. Every heterogeneity would need to be quantified, and every path delineated.

Fully mapping the interior in three dimensions can be achieved—conceptually, in one fell swoop—by 'flooding the Earth' with seismic waves coming from all different directions. This approach bears some similarity to a technique developed in the 1960s and 1970s by the medical community for imaging our internal organs. That technique came to be called *computer assisted tomography* or *computerized axial tomography*, and the resulting image was called a CAT scan or CT scan. It differs from our geophysical approach, using X-rays rather than sound waves or shear waves, and detecting 'anomalies' according to their amplitude (i.e., how much the X-rays were absorbed by our organs) rather than delays in arrival. Nevertheless, given the similarities, three-dimensional imaging of Earth's interior from travel times is called **tomography**, **acoustic tomography**, or **seismic tomography**.

One way to implement this mapping is to subdivide the Earth into a large number of small volumes, perhaps cubes (or, more generally, volume elements or *voxels*), and simultaneously find the set of seismic velocities of all the cubes that best produces the observed travel times—or, equivalently, the set of velocity anomalies that best produces the travel-time anomalies. Mathematically, the requirement for the 'best' solution is typically achieved in a *least-squares* sense; that is, the differences between predicted and observed travel times, when squared and summed over the whole Earth, should be as small as possible. Optimistically, with the hope that the best solution will yield predicted travel times that are very close to the observed, the observed minus predicted differences are called **travel-time residuals**.

During its travels, the reflection or refraction of each wave as it encounters the next layer or anomalous mass depends on the seismic velocity contrast. To predict the wave's path and thus travel time, we need a *reference* velocity model to begin with, supplying seismic velocities as a function of depth or radius; trial 'perturbations' to the velocity in each cubic volume will then allow us to minimize the prediction error. A reference seismic velocity model is also needed to interpret the travel-time delays or advances in terms of anomalous velocities, as in Figure 8.7.

The construction of a reference seismic velocity model, from travel-time observations, is discussed in the following section. Of fundamental importance in and of itself, that model will facilitate a thorough description of the basic one-dimensional, radial structure of Earth's interior. It will also, finally, serve as the background upon which those three-dimensional heterogeneities of great interest to us are superimposed.

8.2 Seismology and the Large-Scale Structure of the Earth

In this section, we focus on the use of travel times, rather than travel-time residuals, to deduce the overall variation of seismic velocities with depth. That variation will establish the gross (i.e., large-scale) layered structure of the interior, defining what we call the crust, mantle, and core, along with the boundaries that divide them. The use of seismic velocity and its variation with depth to reveal the physical state of the interior (even hint at its material composition) will, along with some tomographic applications, be reserved for the next section.

8.2.1 Travel Times: The Shadow Zone and the Core

Scientists tabulating body wave arrivals around the world long ago realized (Oldham, 1906) that there is a zone of distances Δ, in the approximate range Δ = 98° to Δ = 144° away from any earthquake (Stein & Wysession, 2003 and Kennett & Engdahl, 1991, respectively), in which direct P waves are never recorded (or are recorded with very small amplitude).

This was interpreted as proof that a "core" exists, a central region where P wave velocities are lower than in the overlying mantle. The reason for slower P waves in the core follows from Snell's Law: waves entering the slower core will be deflected downward (i.e., toward the perpendicular to the core boundary), away from Earth's surface, to subsequently reappear at the surface outside that zone. It is as if seismographic stations within the zone are standing 'in the shadow of the core,' leading it to be named the **shadow zone**.

As shown in Figure 8.9, the P wave radiating from the source which arrives at a station 98° distant is the one that had grazed the core

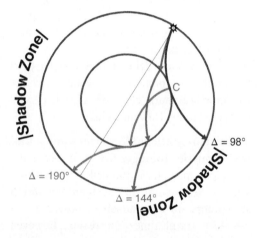

Figure 8.9 The effect of a core with lower seismic velocities than the mantle on P wave transmission through Earth's interior. The earthquake source is marked by a star symbol (✹). The wave just grazing the core surface arrives at distance Δ = 98° from the source. The next deeper wave from the source enters the core, whose slowness deflects it down so that ultimately it reaches the surface at Δ ≈ 190°. Successively deeper waves arrive progressively farther back as much as Δ = 144°; but waves radiated from the source nearly straight down travel through mantle, core, and mantle with only slight refraction. Ultimately, the P wave radiated straight down toward Earth's center travels straight through mantle, core, and mantle (gray dotted line) and arrives at Δ = 180°.
Note that earlier analyses estimated the bounds of the shadow zone to be 103°–140°.
The point marked "C" (and a corresponding one on the other side of the gray dotted line) is discussed in the text.

boundary. The wave starting out pointing just a bit deeper will actually strike and enter the core-mantle boundary (CMB) (at a large incident angle); with the core's lower *P* wave velocity, the transmitted *P* wave will be deflected somewhat downward toward the center of the core. As that wave emerges from the core, the reverse situation applies, and it enters the mantle at a large angle. As suggested in Figure 8.9, this wave reaches the Earth's surface at a distance $\Delta \approx 190°$ (by symmetry, we could as well measure this distance as $\approx 170°$ from the quake).

Succeeding waves enter the core more perpendicularly and are thus refracted by smaller amounts; upon leaving the core, these waves are also refracted by that same smaller amount, and they reach the Earth's surface at Δ less than 190°. In short, waves radiated more and more nearly straight down from the earthquake arrive at distances progressing from $\Delta \approx 190°$ back to $\Delta = 144°$. Then, waves radiated even more straight down travel more or less straight through the Earth, and Δ increases from $\Delta = 144°$ to $\Delta = 180°$.

The bounds of the shadow zone depend on the size of the core as well as on the core-mantle *P* wave velocity contrast; conceivably, the core radius could be estimated from those bounds. However, using reflections off the CMB (i.e., *PcP* and *ScS* waves) proves to be more accurate. Published estimates for the core radius range from 3,479.5 km (Kennett et al., 1995) to 3,485 km (Rubie et al., 2015; Williams, 2019), including also those from free oscillation models, and are encompassed by a value of 3,482 km ± 3 km.

Of course, seismographic stations within the shadow zone can still receive *PP*, *pP*, and other reflected *P* waves. But even for direct *P* waves, the shadow zone is not absolutely 'dark.' For example, the point on the CMB where the *P* wave above just grazes the core and the next *P* wave directed downward from the source penetrates into the core—that is, the point marked C in Figure 8.9—is seen by those *P* waves as the core's 'edge': to one side of it, waves pass by, continuing through the mantle; but waves striking the other side of it must deal with the core. *P* waves striking the edge of the core are **diffracted**—*scattered*, in the same way that light striking the edge or corner of an obstacle can scatter in all directions, even brightening the shadow behind the obstacle a little, or in the same way that sound waves striking the edge of a wall will scatter in all directions so that a person standing around the corner can hear that sound faintly (in addition to hearing sounds transmitted through that wall or reflected off other walls).

Some of the scattered *P* waves travel along the boundary; as they propagate, they send energy back up—some, into the *P* wave shadow zone. These diffracted *P* waves are not head waves (critical refraction is not possible because the incident medium's velocity exceeds the transmitted medium's velocity, i.e., $V_1 > V_2$); but, like head waves, they generate sources of wave energy as they propagate along the interface, and that energy radiates away from the interface.

As might be expected, the arrivals of 'Pdiff' are relatively weak.

A more consequential 'leak' of seismic radiation into the shadow zone, with weak direct *P* waves observed at Δ exceeding 110°, was interpreted by Lehmann in 1936 as proof of a small, *solid inner core* within the outer-core liquid. *P* waves entering the inner core travel faster there; their refraction at the inner-core–outer-core boundary shoots them back up to the surface, where they emerge within the shadow zone. There are actually two *P* wave arrivals observed, both of small amplitude: the faster one, first of the pair to arrive, has traveled as a *P* wave throughout the Earth, even within the inner core; the slower one, arriving a bit later, has traveled as an *S* wave within the inner core—something possible because of the inner core's solidity. These two waves are designated I and J, respectively, in the alphabet-soup dialect of observational seismologists; the two types of arrivals recognized by Lehmann would have been PKIKP and PKJKP in their travels through mantle, outer core, inner core, outer core, and mantle.

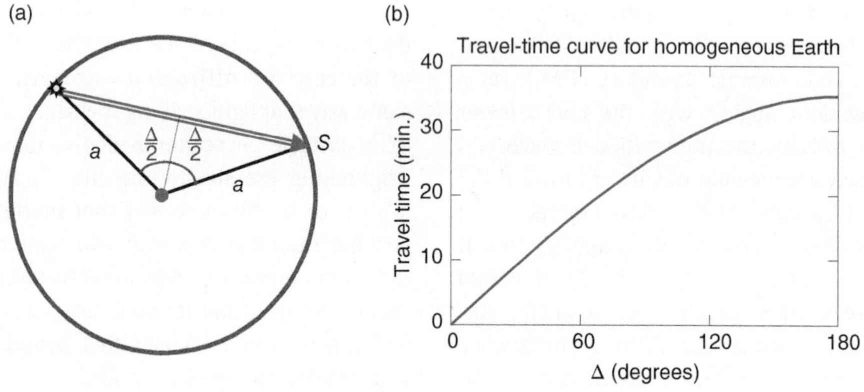

Figure 8.10 Travel-time analysis for a hypothetical homogeneous spherical Earth.
(a) The direct path (double arrow) taken by a body wave from an earthquake source (✵) to a seismographic station (**S**) a distance Δ away is a straight line, since there are no variations in material properties to cause any refraction. The distance traveled along that straight-line path can be determined noting, with reference to each right triangle, that the sine of Δ/2 equals the half-distance divided by *a* (thus the half-distance = *a* sin(Δ/2)).
(b) The predicted *t*-Δ curve for a wave with constant velocity throughout its travels (chosen for this illustration to be 6.0 km/sec) is a pure sinusoid.

Finally, there is in effect an extreme shadow zone for direct *S* waves, since they cannot enter the fluid core without converting to a *P* wave. *With reference to Figure 8.9, how (i.e., over what range) would you draw the* S *wave shadow zone?*

8.2.2 Travel Times: Determining Seismic Velocities Within the Earth

From the point of view of general geophysics, the most important use for travel-time curves is to determine the seismic velocity structure of the Earth's interior.

The travel paths and travel times of body waves propagating through the Earth are best appreciated in contrast to those propagating through a simplified, completely homogeneous Earth, one in which there is no layering and no radial (or lateral) variation in density or elastic properties. In this hypothetical Earth, the waves generated by an earthquake or explosion would radiate away from the source without any refraction; and reflections would occur only as those waves strike the Earth's outer surface. A seismographic station on the lookout for the body waves direct from that source would end up detecting waves that had traveled a straight line to the station; see Figure 8.10a.

As suggested by Figure 8.10a, if the station is a distance Δ away from the source, the direct waves arriving there will have traveled a distance $2a\sin(\Delta/2)$, if a is Earth's radius. So, if the velocity of a *P* wave throughout its trip is V_P—a single quantity because the velocity will remain the same in this homogeneous Earth—then (once again using the relation *distance* = *rate* × *time*) the time it takes the *P* wave to complete its travels will be $2a\sin(\Delta/2)$ / V_P; that is,

$$t = 2a\ \sin(\Delta/2)/V_P.$$

From this relationship, the travel time recorded at a seismographic station a known distance from the source could be used to infer the *P* wave velocity within such a homogeneous Earth (that is, with t, a, and Δ known, we could solve for V_P). Alternatively, the analysis could proceed graphically: *t*-Δ curves could be constructed for a range of possible values of V_P (see Figure 8.10b for one such curve; with several possible values for V_P, there would be a sequence of 'parallel' curves, all starting at the origin); from numerous observations at different stations, the best-fitting curve and thus most representative value for V_P could be chosen.

The preceding equation establishes that the travel-time curves for body waves in a homogeneous Earth would be sinusoidal. But actual travel-time curves for the real Earth (see Chapter 7) do *not* take on a sinusoidal shape; this tells us that the Earth's interior is *not* seismically homogeneous. Furthermore, when the t-Δ curve in Figure 8.10b, using a value for V_P representative of the near-surface, is compared with an actual travel-time curve, we find that travel times for the real Earth are *shorter* than for such a homogeneous Earth, at distances closer than the shadow zone (see Figure 8.11). We conclude that, to produce shorter travel times, *P wave velocity must, overall, increase with depth within the mantle of the real Earth.* (As it turns out, the *core* also exhibits a velocity increase, though the wave paths are too complex to demonstrate this using a Figure 8.11 type of analysis.)

These results hold true for *S* waves as well as *P* waves: in a homogeneous Earth, the *S* wave travel-time curve is sinusoidal, and predicts travel times that (if reasonable for near-surface *S* waves) are too long for *S* waves that have traveled through the mantle, implying that the latter waves must have traveled faster at depth. These results, whether for *P* or *S* waves, are not surprising, given our expectation that material resistance to deformation should increase with depth within the Earth (but more on that, later).

The t-Δ curve for *P* waves in the real Earth for $\Delta < 98°$ is suggestive of the t-x graph we constructed in Chapter 7 for first arrivals in an *n*-layered Earth, predicted assuming velocities increase with depth (shown in Figure 7.35 of that chapter for a four-layer region—now imagine an even greater number of layers!). This reinforces our conclusion that velocities should increase with depth in the real Earth.

The increase in seismic velocity with depth inside the Earth implies in turn that *body waves in the real Earth must follow curved paths away from their source.* This is a consequence of Snell's Laws, which tell us that waves encountering faster regions (as they will when traveling deeper into the real Earth) will refract away from the perpendicular to the boundary; and that waves encountering slower regions (as they will when emerging from the interior) will refract toward the perpendicular to the boundary. The resulting 'concave' curvature of body wave paths was already incorporated into Figure 8.9 (also Figure 7.22 in Chapter 7).

The amount of curvature of the wave paths depends on the velocities of the layers (if Earth's interior is viewed as a succession of layers) or the velocity increase with depth (if the interior is a continuously varying region). It follows that the path of the wave from source to receiver depends on the velocity structure of the interior. This dependence was implied by our discussion of seismic refraction in Chapter 7: the paths followed by waves leaving the source—in particular, *which* downward-moving wave became critically refracted (see Figure 7.33a)—depended on the velocity contrast between the two layers.

In seismic refraction surveys, we found that observations of travel times versus distance can yield the velocities of the layers through which the direct and critically refracted *P* waves have traveled. Ultimately, global observations of *P* and *S* wave travel times can also yield

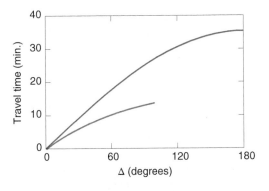

Figure 8.11 Comparison of *P* wave travel times for a homogeneous spherical Earth (red curve, from Figure 8.10b) with those for the real Earth (blue curve, adapted from Kennett and Engdahl (1991) but rescaled to match the scales used for the x- and y-axes here). Only the mantle portion of the real-Earth t-Δ curve—involving values of Δ this side of the shadow zone ($\Delta \leq 98°$)—is shown.

the wave velocities within the Earth. The *t*-Δ observations can be averaged over all latitudes and longitudes if we are interested in just the one-dimensional, radial velocity structure of the interior. Even in this case, however, we are faced with a 'circular challenge': we need to know the distance traversed by a wave—as well as its travel time—in order to infer its velocity (using *distance = rate × time* for each increment of travel); but we need to know the velocities within the Earth to determine its path.

This challenge leads precisely to the same need for bootstrapping that was mentioned in Chapter 7 in the context of free oscillations. And all of this will be encountered later in this chapter in our discussion of acoustic tomography. One mathematically straightforward approach we might take is to postulate a *starting model*—a first guess—of the body wave velocities versus depth; use that model to calculate the paths followed by *P* and *S* waves as they spread out from the source; and build a table of their travel times versus the angular distance (Δ) they end up covering from source to receiver. These estimates are not likely to match travel times observed for the real Earth; with the discrepancies (residuals) in mind, the model can be modified until the residuals are as small as possible. An example of the *P* and *S* wave velocity profiles inferred within the Earth from travel-time data is shown in Figure 8.12a.

As an alternative, that same 'trial-and-error' approach can be applied to observations of the Earth's normal modes. In this case the observations are of the frequencies or periods of various normal modes. These frequencies depend on the density and elastic properties of the interior layers, just as the tones of a bell being struck depend on the materials the bell is made of. That is, as discussed in Chapter 7, the speed of a surface wave depends on the density and elastic properties of the subsurface; to combine 'constructively' and produce the standing waves we call normal modes, those surface waves must be traveling at certain velocities—which are possible only with specific densities and elastic properties of the Earth's interior. Those velocities determine the periods of the normal modes we observe, thereby tying our observations to the

material properties of the interior. Once those properties—k, μ, and ρ—have been inferred, versus depth, they can be combined to yield the *P* and *S* wave velocities of each layer or at each depth. The PREM velocity model (mentioned in Chapter 7), which Dziewonski and Anderson (1981) derived from ~900 spheroidal and toroidal eigenperiods, as well as numerous travel times from various *P* and *S* waves, plus additional types of data, is shown in Figure 8.12b.

The Seismic Velocity Structure of the Earth. Whether determined from travel times or normal modes, the inferred *P* and *S* wave velocity profiles reveal a similar large-scale layering of Earth's interior. There are three categories of this structure, depending on the type of boundary:

- major boundaries, including, of course, Earth's surface (marked with ❶ between Figures 8.12a and 8.12b) and the CMB (marked with ❷);

- secondary boundaries, including two small jumps in V_P and V_S (marked with ❸ and with arrows) in the upper mantle and one (marked with ❹) at the inner-core/outer-core boundary; and

- zones where the velocities exhibit unusual behavior, including a *low-velocity zone* below the lithosphere (marked with ❺) and the base of the mantle, denoted D″ (marked with ❻).

Despite our closeness to it and consequent detailed knowledge of it, and despite its importance for isostasy and exploration seismology, the crust-mantle boundary (marked with ❼) really only amounts to another secondary boundary.

This 6large-scale, vertical seismic velocity structure allows us to divide the interior into several regions (crust, mantle, and core, overall; and, within the core, liquid outer core and solid inner core). Figure 8.12 reveals, however, that the most interesting structure occurs within the mantle: the **upper mantle**, from ❼ to ❸, comprised of the uppermost mantle (beginning right below the Moho and thus including much of the lithosphere), the

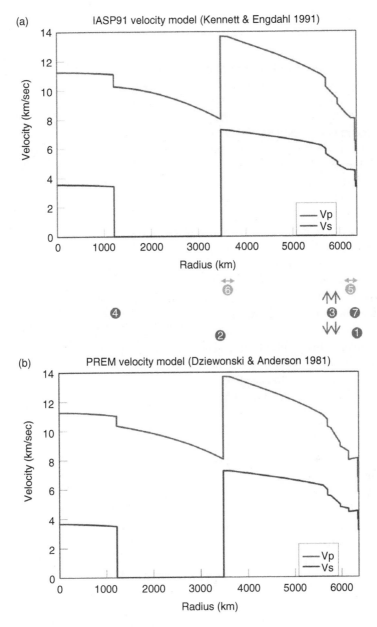

Figure 8.12 Body-wave velocities versus depth.
(a) Model iasp91, adapted from Kennett and Engdahl (1991) based on the travel-time data shown in Chapter 7, Figure 7.38. This is a simplified model, parametrically smoothed in segments and averaged to represent a laterally homogeneous Earth.
(b) The PREM velocity model, based mainly on normal mode data as described in the text, determined by Dziewonski and Anderson (1981) / Elsevier.

low-velocity zone or **LVZ** (i.e., low velocities in contrast to those above and below), the bulk of the upper mantle, and the **transition zone** (beginning and ending at those jumps in velocity ❸); and the **lower mantle**, beginning right below ❺ and ending at ❷, and including

D″ (where the velocity flattens out) in its lowermost region. A few comments about this structure follow.

- First, the LVZ, which begins (somewhat indistinctly) below the lithosphere and, according to the analysis summarized in

Figure 8.12b, terminates at 220 km depth. In general, there are three or four reasons for seismic velocities to change as we penetrate deeper into the Earth: changes in composition or phase, changes in pressure, and changes in temperature. The pressure increase with depth can be expected to increase the material resistance to deformation, resulting in higher velocities. Temperature increases with depth will act to soften the material, leading to slower velocities; as discussed earlier in this chapter, this is mostly caused by temperature's effect on shear rigidity, since even under extreme temperature effects, e.g. partial melting, the material's incompressibility and density would probably not be very different. Thus, outside of the possibility of a compositional change (which is revisited toward the end of this chapter), the LVZ could be interpreted to be mainly the result of increased temperatures within the uppermost mantle.

In Chapter 5, we saw the need for a low-*viscosity* zone below the lithosphere, in order to explain observations of isostatic rebound and subsidence. That weak zone—an asthenosphere (or the top of it)—facilitates the motion of lithospheric plates; in fact, without a low-viscosity zone, i.e. a well-defined asthenosphere, large-scale mantle convection would be sluggish at best. A low-viscosity zone due for example to higher temperatures is consistent with a seismic low-velocity zone.

- However, there are some complications associated with the seismic LVZ. Figure 8.13 presents some results from an analysis of surface waves crossing the Pacific basin (Nishimura & Forsyth, 1989); they show that an LVZ does indeed exist, at least beneath the oceanic lithosphere. But its depth, and how much the *S* wave speed drops within the zone, evidently depend on the age of the overlying lithosphere: as the oceanic lithosphere ages and cools, the LVZ deepens and weakens.

Similarly, then, being even older and colder, continental lithosphere—especially in continental shield areas—would be

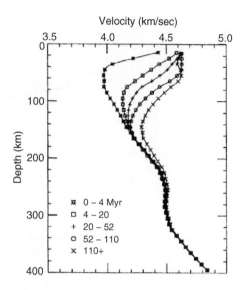

Figure 8.13 Velocity of *S* waves (actually *SV* waves) in the upper mantle beneath the Pacific, as a function of the age of the overlying oceanic lithosphere, based on analysis of surface waves. Image from Nishimura and Forsyth (1989) / Oxford University Press.

expected to overlie a weak LVZ at best. And the relative coolness of the upper mantle there would imply higher viscosities (and thus less vigorous mantle convection) in those regions. Such conditions are also suggested by some of the tomographic analyses discussed later.

If the goal, then, is to produce a laterally averaged velocity model, as in Figure 8.12, the result may not exhibit much of an LVZ, since oceanic LVZs and any continental LVZs would exist at very different depths. For that matter, even a laterally averaged velocity profile beneath oceanic lithosphere alone would feature a reduced LVZ (as one could infer from Figure 8.13), given the range of ages involved.

Finally, we note that the lack of an LVZ in the velocity profiles of Figure 8.12a, especially for *S* waves, is additionally a consequence of how Kennett and Engdahl (1991) mathematically represented the variation of V_P and V_S with depth.

- As shown in Figure 8.12, the transition zone includes two sharp increases in *P* and *S* wave velocities, one at depths around 400 km or 410 km, the other a near-discontinuous jump at about 660 km or 670 km (depending on the model); these are known as the **first and second transitions**. Between them, the velocities increase smoothly, as they do throughout the bulk of the mantle. The transition zone turns out to be crucial for our understanding of the composition and possibly the dynamics of the mantle; its significance will be explored later in this chapter.

Although models iasp91 and PREM clearly reveal the first and second transitions, neither has the resolution to show just how sharp the transitions are. Kennett and Engdahl (1991) recommend using regional travel-time curves and velocity models to determine their sharpness; from a prior study by Bina and Wood (1987) taking that approach we can conclude that the first transition most likely takes place within a depth range of 10 km.

- Compared to the upper mantle, the lower mantle in our laterally averaged, spherically symmetric Earth turns out to be especially unremarkable, displaying (in Figure 8.12) nothing more than an overall increase in velocities. The exception to this blandness is the lowermost mantle—the home of D″, with the core just below—which stands out as a relatively interesting region even when laterally averaged; within it, overall, velocities no longer increase significantly with depth, and may level off or even decrease.

This contrast will make more sense when we see the results of tomographic studies later in this chapter. Seismic tomography will reveal numerous fast and slow structures throughout the entire mantle. In the mid-lower mantle, the seismic velocity anomalies turn out to be small in magnitude and scale, and somewhat random (laterally if not vertically), so lateral averaging will mostly cancel them out. However, in the lowermost mantle, the velocity anomalies are larger magnitude, broader scale, and more coherent, especially the slower-velocity structures; when averaged laterally, there will be less overall cancellation, leading to a velocity variation versus depth with a distinctly reduced trend.

As important as the velocity profiles in Figure 8.12 are, D″ is a reminder that there's more to the story – that a three-dimensional picture of the interior will be needed to truly understand how the Earth works.

- As we had anticipated from the shadow zone, seismic velocities entering the core are substantially lower than in the lower mantle. It is interesting to note that, within the core, seismic velocities increase with depth—not just as the solid inner core is encountered, but throughout the core from the CMB all the way to its center.

Even the velocity structure depicted in Figure 8.12, despite representing the seismic properties of a laterally averaged Earth, contains within it critically important information about the nature of the interior. In the next section, we will discover how to bring such information to light, and then, with that velocity profile as reference, how seismic tomography can add marvelously to that picture.

8.3 Seismic Velocities and the State of Earth's Interior

Our goal here is to use the seismic velocities illustrated in Figure 8.12 to infer the density, elastic properties, physical state (pressure, temperature), and even composition of the interior of the Earth, as a function of depth. We will discuss two approaches to achieve that goal, one empirically based and the other based in theory, then follow that with a third, computationally based exploration of *departures* from that reference state—three-dimensional heterogeneities found tomographically—and the clues they provide regarding mantle convection. Our discussion relating velocities to the interior, though not in chronological order, will span geophysical research from its earliest days, when a theoretical approach was unavoidable, to a time when high-pressure experiments were able to strengthen our picture of the interior, to the present-day era of massive data assimilation. But we will find that all three aspects of our discussion significantly improve our understanding of the state of Earth's interior.

8.3.1 Birch's Rule

As a start, we know that the density of crustal rocks is ~2.7 gm/cm^3 (for granitic rock types)—2.9 gm/cm^3 (for basalt); but the mass of the Earth implies that its mean density is 5.5 gm/cm^3. It follows that the density, ρ, of Earth material must increase, overall, with depth—a conclusion that, as discussed in Chapter 1, is reinforced by Earth's moment of inertia. But we have found that, overall, seismic velocities *also* increase with depth within the interior; exceptions to this rule include only two thin layers (the LVZ and D″) and the CMB, the latter (and possibly D″) a reflection of compositional changes. The similarity in overall trends between ρ and V_P, V_S suggests that it might be possible to express the behavior of seismic velocities within the Earth as primarily a proportional dependence on density (i.e., as one increases, so does the other), plus a dependence on composition.

It should be noted that having velocity directly proportional to density, though reasonable, does *not* simply follow from the explicit, *inverse* dependence of seismic velocities on density ($V_P = \sqrt{(k + 4/3\mu)/\rho}$ and $V_S = \sqrt{\mu/\rho}$).

A test of that proportional dependence is illustrated in Table 8.1, which lists density and P wave velocity measured in the lab for various Earth materials. In each case, an increase in pressure (e.g., from 0.1 GPa to 1 GPa) causes P waves to travel faster through the rock—confirming our expectation that increasing the density of a rock by compressing it causes it to resist compression much more (so that k increases more than ρ does), thus leading to a higher V_P.

The table also suggests that at any pressure, denser compositions tend to sustain faster P waves than less dense compositions. This may seem reasonable at first thought, but it cannot be true in all cases; for example, P waves travel slower through the denser core than through the less-dense mantle above. Evidently, the effect of composition on k does not always parallel its effect on ρ; in fact, it may not be composition itself but only an aspect of composition (such as the strength of interatomic bonds) that is influencing k and thus V_P. Closer atomic spacing and thus greater density do not compensate for the intrinsically weaker interatomic bonds (and thus lower resistance to deformation) of some materials.

In 1961, Birch published results from an experimental study relating P wave velocity to density in a wide range of materials. Using an approach popular in rock mechanics research at the time, he was able to formulate a simple relation between V_P and density that successfully reduced the effects of composition on V_P to a single term involving only the substance's mean atomic weight. The **mean atomic weight** of a mineral is calculated by

Table 8.1 P wave velocities measured in a sample[a] of rock types.

Rock type	Density (gm/cm^3)	At 0.1 GPa pressure	At 1 GPa pressure[b]
		P wave velocity (km/sec)	
Granite	2.64	6.13	6.45
Quartz diorite	2.85	6.44	6.71
Amphibolite	3.12	7.17	7.35
Dunite	3.28	7.87	8.15

[a] Excerpted from Bott (1982) / Table 2.3 / with permission of Elsevier.
[b] Corresponds approximately to a depth of 35 km.

adding up the atomic weights of all atoms in each molecule, then dividing by the number of atoms per molecule. For example, quartz (silica), with a chemical formula SiO_2, would have a mean atomic weight of $(28 + 16 + 16)/3$ or 20.

The relation Birch discovered can be written

$$V_P = a_m + b\rho,$$

where the density ρ of the mineral or rock is in units of gm/cm^3, and m is its mean atomic weight. This relation, called **Birch's rule** or **Birch's law**, is a linear relation between V_P and ρ—about as simple as one could hope for!—with slope b, and y-intercept a_m. a_m is a fixed constant for all substances sharing the same mean atomic weight, m; that is, in general the value of a_m depends on m. The slope b depends only weakly on m and may be treated as a true constant, whose value Birch (1961a, 1964) determined to be about 3.05 (km/sec)/(gm/cm³) if V_P is in units of km/sec. If V_P and ρ were measured and then graphed for a variety of minerals at a variety of pressures, Birch's rule predicts that minerals with the same mean atomic weight would group together along one line; the set of all lines (each for a different m) would be parallel (all with the same slope b). These results are schematically illustrated in Figure 8.14.

For the substances evaluated by Birch, the y-intercepts were all negative ($a_m < 0$). And Birch found that lines for greater m should be positioned farther to the right (more negative a_m, as shown in Figure 8.14). That is, *minerals*

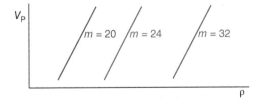

V_P

$m = 20$ $m = 24$ $m = 32$

ρ

Figure 8.14 Schematic depiction of Birch's Rule, after Birch (1961a, b). *P* wave velocity is plotted versus density. Linear relations between V_P and ρ are possible when the data are grouped according to mean atomic weight, m. Data points (not shown) are best fit by lines with the same slope, 3.05 km/sec per gm/cm³ in the original rule. Note that all lines will have negative y-intercepts.

with higher mean atomic weight, though denser, tend to resist deformation less well, and propagate P waves more slowly (—think of iron versus rock!).

Birch's rule was originally derived at a single pressure of 1 GPa, serving as a way to convert measurements of V_P to ρ at the top of the mantle, or vice versa. Holding true at greater pressures (and temperatures), however, it becomes a more profound expression, indicating fundamentally how the physical state of the interior affects seismic velocities: Birch's rule tells us that the net effects of pressure and temperature on V_P are realized through their effects on density. A thorough discussion of Birch's rule may be found in Poirier (2000, also 1994).

Additional experiments have confirmed the validity of Birch's rule; that is, they have demonstrated that the (V_P, ρ) data points are well fit by straight lines when grouped according to mean atomic weight. Other experiments have found a good linear fit between V_P and ρ when the data points are grouped according to atomic number (e.g., Knopoff, 1967). Experiments have also shown that a potentially better fit to the data is achieved by plotting density versus a somewhat different quantity, $\sqrt{\phi}$, where ϕ, defined as

$$\phi \equiv V_P^2 - \frac{4}{3}V_S^2,$$

is known as the **seismic parameter**. The modified version of Birch's rule (McQueen et al., 1964; Wang, 1968; see also Wang, 1970) is

$$\sqrt{\phi} = a'_m + b'\rho,$$

where $b' \cong 2.36$ according to Wang (1970). Note that $\sqrt{\phi}$ involves both *P* wave and *S* wave velocities, implying (at least, on first thought) that V_S is affected by pressure and temperature in similar ways to V_P. Birch's rule (both versions) has been verified at pressures up to ~100 GPa and beyond (Wang, 1970; see Poirier, 2000; Fiquet et al., 2001; Badro et al., 2007 and Vočadlo, 2007; see also Roy & Sarkar, 2017) and for a wide range of substances. On the other hand, it does not work well for rocks under pressures less than a few tenths of a GPa, the pressure (roughly) great enough to close cracks and pore spaces within the rock (Birch, 1961a; Brace, 1965; Poirier, 2000); for surface and near-surface rocks (0.3 Gpa corresponds to a depth of ~10 km), other relations such as the *Nafe-Drake curve* are preferable (that relation and a variety of others are specified in Brocher, 2005).

Given the definitions of V_P and V_S in terms of k, μ, and ρ, it follows that

$$\phi = k/\rho$$

so that

$$\sqrt{\phi} = \sqrt{k/\rho}.$$

$\sqrt{\phi}$ has units of velocity, will always be less than or equal to V_P, and is often known as the **bulk sound velocity**. Because sound waves are waves of compression (P waves) and because $V_S = 0$ in fluids, $\sqrt{\phi}$ is the velocity of sound waves in air or water. For example, in the oceans, k ~22 kbar, yielding a speed of sound ~1.5 km/sec there; in air at the bottom of the troposphere, k ~1.4 bar, implying (with air's density of ~1.23 mg/cm³) that sound travels at ~337 m/sec (1 mile every 5 sec) near Earth's surface.

Table 7.1 in Chapter 7 will allow you to convert the 'meteorological' units for pressure stated here, and used traditionally in ocean and atmosphere situations, to standard cgs or mks units; using mks units, you should verify that the sound speeds highlighted here are correct.

Finally, stimulated by the publication and usefulness of Birch's rule, alternative, power-law velocity–density relations have been developed and evaluated. They can be written in the form

$$\sqrt{\phi} = \hat{a}_m \rho^\nu$$

for some constant $\hat{a}_m$ that depends on mean atomic weight, and a nearly independent exponent ν; for example, a representative value for ν might be ~1.3–1.8 (Chung, 1972; Anderson, 1973; Anderson, 1967). As noted by the latter researchers, this form can be theoretically derived, in contrast to Birch's rule, by considering the behavior of the crystal lattice when minerals are compressed (and

using $k = \rho\phi$ to phrase the results in terms of ϕ); that derivation is beyond the scope of this textbook, but see also *Textbox A*. As shown by Wang (1970) and Chung (1972), the power law predicts $\sqrt{\phi}$ or V_P close in magnitude to the predictions from Birch's rule, for values of ρ and m typical for Earth's mantle (this is reminiscent of the similarity between linear and power-law versions of the Titius-Bode Law, discussed in Chapter 1). Thus, Birch's rule may be viewed as the linear approximation to a theoretically based velocity–density relation.

Birch's Rule: Applications to Earth's Interior. Birch's rule can tell us much about density and composition within the Earth. For simplicity, we begin with the original version of the rule, $V_P = a_m + b\rho$. We already know the density of the uppermost mantle, ~3.3 gm/cm³, from isostasy, and seismic velocities at the top of the mantle are ~8.0 km/sec for the iasp91 velocity model or 8.1 km/sec for the PREM model for V_P, and ~4.5 km/sec for V_S. Substituting into Birch's rule, we can deduce that a_m ~−2.03 or −1.95. For compactness, when presenting results based on iasp91 versus PREM in the rest of this chapter, we will reserve { } for the PREM-based results; for example, we have just deduced that a_m ~−2.03 {−1.95}. As per our discussion of mantle density models and composition later in this chapter, we also consider an uppermost mantle density of 3.4 gm/cm³ (which is also a reasonable choice, as argued, for example, by Stacey & Davis, 2008); this corresponds instead to a_m ~−2.33 {−2.26}. The experiments that led to Birch's rule found such values of the y-intercept (see Birch, 1961a) corresponded to a mean atomic weight m of 21–22.

As we will see shortly, quite a variety of commonly found minerals and rocks have mean atomic weights in the range 20–22; one property they all share, in contrast with many heavier substances (those with m ~24 or greater), is that they all lack appreciable amounts of iron. This supports our reasonable expectation that the mantle is iron-poor—that most of Earth's iron ended up in the core as a consequence of core formation.

Textbox A: Incompressibility and a Power-Law Form of Birch's Rule

---

With a little bit of calculus, we can turn a physically reasonable, experimentally observed property of rocks into a power-law relation between bulk sound velocity and density.

The observed property is (more or less) that *the more a rock is squeezed, the harder it is to squeeze it more.* In other words, the greater the pressure the rock is subjected to, the more incompressible it becomes; as pressure P increases, k increases too.

Furthermore, the experiments suggest that the increases are proportional, that is, that

$$\frac{\mathrm{d}k}{\mathrm{d}P}$$

is roughly constant.

Such a proportionality means that k and P are related linearly, thus

$$k = k'P + k_0,$$

where the slope $\mathrm{d}k/\mathrm{d}P$ is denoted k' for compactness, and the intercept k_0 represents the surface-pressure ($P = 0$) value of k.

Applying the chain rule for derivatives to k, we have

$$\frac{\mathrm{d}k}{\mathrm{d}P} = \frac{\mathrm{d}k}{\mathrm{d}\rho}\frac{\mathrm{d}\rho}{\mathrm{d}P};$$

but the left-hand side is k' and, by definition, we know

$$k = \frac{\mathrm{d}P}{\mathrm{d}\rho/\rho} \quad \text{or} \quad \frac{\rho}{k} = \frac{\mathrm{d}\rho}{\mathrm{d}P},$$

which upon substitution yields

$$k' = \frac{\mathrm{d}k}{\mathrm{d}\rho}\frac{\rho}{k} \quad \text{or} \quad \frac{\mathrm{d}k}{k} = k'\frac{\mathrm{d}\rho}{\rho}.$$

Finally, we know that, for any function y,

$$\frac{\mathrm{d}}{\mathrm{d}y}(\ln y) = \frac{1}{y} \quad \text{or} \quad \mathrm{d}(\ln y) = \frac{\mathrm{d}y}{y}.$$

This allows us to write our incompressibility relation as

$$\mathrm{d}(\ln k) = k'\mathrm{d}(\ln \rho)$$
$$= \mathrm{d}(k' \ln \rho) = \mathrm{d}(\ln \rho^{k'})$$

thus

$$\ln k = \ln \rho^{k'} + \varepsilon'' \quad \text{or} \quad \ln k = \ln(\varepsilon' \cdot \rho^{k'})$$

for some constant of integration ε'' (with $\ln(\varepsilon') = \varepsilon''$), leading to

$$k = \varepsilon' \cdot \rho^{k'}.$$

This implies

$$\phi = \varepsilon' \cdot \rho^{(k'-1)}$$

or

$$\sqrt{\phi} = \varepsilon \cdot \rho^{(k'-1)/2}.$$

Once we have determined densities within the Earth, we can estimate both the elastic moduli (as implied by V_S and V_P together with ρ) and the pressure versus depth, allowing k'

Textbox A: Incompressibility and a Power-Law Form of Birch's Rule (*Concluded*)

---

to be calculated from the change in k and change in P; according to the PREM Earth model (as tabulated in Stacey & Davis, 2008), for example, $k' \sim 3.61$ at ~ 1000 km depth, and ~ 3.15 at $\sim 2,200$ km depth. Later in this chapter we will learn how to estimate k' from ϕ only; using that approach, Birch (1952) found $k' \sim 3 - 4$. Thus, the exponent in our velocity–density relation is in the range 1.0–1.5.

Our derivation leaves ε as yet unspecified, and with the challenge of relating ε to composition or mean atomic weight. Furthermore, if the relation is to apply to a body like the Earth, it must remain applicable as composition changes with depth (but mean atomic weight stays constant, à la Birch). Our derivation may be overly simplistic in other ways as well; for example, the value of k' might be different for different substances, or might change at high pressures (see, e.g., Anderson, 1973, also Poirier, 2000 and Stacey & Davis, 2008, for a more complex perspective.) Nevertheless, it suggests that the relation between seismic velocities and density may be essentially of a power-law nature.

But the mean atomic weight implied by Birch's rule for the uppermost mantle is not merely reasonable; it is specifically consistent with what we believe the composition of the uppermost mantle to be—thanks to various processes that have brought actual pieces of the uppermost mantle to the surface. Those processes include such exotic phenomena as *kimberlite pipes*, 'shot' through the crust (bringing diamonds within our reach!); and, most importantly, *ophiolite sequences*, with oceanic lithosphere—oceanic crust and uppermost mantle—thrust onto land (for our viewing pleasure). Consequently, we are confident that the uppermost mantle consists primarily of peridotite (whose *surface* density, incidentally, is ~ 3.3 gm/cm³). Smaller amounts of eclogite (density ~ 3.4 gm/cm³), created as a high-pressure phase of subducting oceanic crust, should also be present. The specific composition of these ferromagnesian silicates will be discussed shortly; for present purposes, we need to know only that such rocks have mean atomic weights around 21–22—when they contain only small amounts of iron.

To apply Birch's rule to the rest of the Earth, we make the assumption—with some justification (Birch, 1961b; Wang, 1970)—that the rest of the mantle has a mean atomic weight around 21–22, the same as the uppermost mantle. In this case, a_m is known throughout the mantle, and Birch's rule (recalling that b is always 3.05) can be combined with our knowledge of V_P at

all depths in the mantle (from iasp91 or PREM) to easily generate estimates of density all the way down to the CMB. With this approach we find that densities increase from 3.3 gm/cm³ to a maximum of 5.14 gm/cm³ or so at the base of the mantle (where $V_P \approx 13.7$ km/sec). Alternatively, with $a_m \sim -2.33$ {−2.26}, the density range is 3.4 gm/cm³ to 5.24 gm/cm³ or so.

An easy exercise for you would be to calculate the actual densities (not "or so"!) at the base of the mantle implied by Birch's rule, using the four possible values of a_m.

As a bonus, the density of each layer in the mantle yields the mass of that layer, thus (altogether) the mass of the mantle. For example, if a particular layer, of density ρ, extends from radius r_{upper} down to radius r_{lower}, then the volume of that layer equals

$$\frac{4}{3}\pi r_{upper}^3 - \frac{4}{3}\pi r_{lower}^3,$$

so the mass of that layer equals

$$\left(\frac{4}{3}\pi r_{upper}^3 - \frac{4}{3}\pi r_{lower}^3\right)\rho.$$

The mantle's net mass will be the sum of all these individual layer masses. In the discussion that follows, for compactness, we will let ρ_0 symbolize the density of the mantle's topmost layer and denote the mantle's net mass by M_M; the value of M_M turns out to be $\approx 3.82 \times 10^{24}$ kg {3.83×10^{24} kg} using the density profile we have inferred from Birch's rule with $\rho_0 = 3.3$ gm/cm^3, or $\approx 3.91 \times 10^{24}$ kg {3.92 × 10^{24} kg} with $\rho_0 = 3.4$ gm/cm^3.

Birch's Rule and the Core. Our determination of mantle densities and mass has implications for the core. First of all, if the Earth's mean density is ~5.5 gm/cm^3, but the range of densities for the mantle (from Birch's Rule) is only 3.3 gm/cm^3 or 3.4 gm/cm^3 to at most 5.3 gm/cm^3 (*or so...*), then the density within the core must be much greater than 5.5 gm/cm^3. In fact, since the Earth's total mass is known, we can subtract our estimate of M_M to yield a prediction of the core's mass, M_C. For greater accuracy we also subtract the mass of Earth's crust, $\approx 0.028 \times 10^{24}$ kg according to Peterson and dePaolo, 2007 (see also, e.g., Reguzzoni et al., 2013; also, for our estimates based on the older PREM velocities, we use total Earth and crustal masses based on PREM densities, 5.974 × 10^{24} kg and $\approx 0.031 \times 10^{24}$ kg). Our core mass estimates therefore turn out to be $M_C \approx 2.03 \times 10^{24}$ kg using V_P from iasp91 {2.12 × 10^{24} kg from PREM}. Given the size of the core, its average density must therefore be ~11.5 gm/cm^3 {12 gm/cm^3}.

We can go one step further, and consider the possibility that the core has the same composition as the mantle, or a different composition but the same mean atomic weight. As discussed earlier, m ~21–22 generally corresponds to iron-poor silicates and oxides, so from the outset, we expect such a possibility to be unlikely (but it is in line with ideas that were popular at one time; see, e.g., Birch, 1968 and comments in Birch, 1961b (end) or Birch, 1952, and in Dickman, 1983). With this possibility, $V_P = a_m + b\rho$ would hold true in the core as well as the mantle, with the same value of a_m. In this case, however, the greater densities of the core would make $a_m + b\rho$ and thus V_P larger in the core than in the mantle. That is impossible, because there would be no shadow zone observed at Earth's surface if P waves traveled faster in the core than in the mantle. We are forced to conclude that a_m must be different in the core than the mantle; that is, *the core's mean atomic weight must differ from the mantle's.* This is a strong conclusion,

as it requires the *composition* of the core and mantle to be different as well.

Even if a_m is not constrained to have the same value as in the mantle, the applicability of Birch's rule to the core might be questioned; after all, the experiments that led to the rule focused primarily on silicates and oxides, not metals, and not on liquids. However, analyses by Vočadlo (2007), Badro et al. (2007), and Fiquet et al. (2001) based on high-pressure experiments and theoretical modeling have demonstrated that iron, and iron compounds, obey Birch's rule at core pressures (there may be exceptions, at least in the outermost core, if the temperatures are sufficiently hot; Lin et al., 2005; but see also Badro et al., 2007). To add confidence to any inferences we draw now regarding the core's density variation, we will also consider a second, very distinct approach later in this chapter. But first, Birch's rule has more to tell us.

At a minimum, if Birch's rule is relevant in the core, then Figure 8.14 tells us that the simultaneous need for high densities and low seismic velocities in the core requires that it have a (much) higher mean atomic weight than 21–22.

Consequently, we could try different values of a_m in Birch's rule, and for each value use the set of P wave velocities within the core (Figure 8.12) to predict the density throughout the core; and, for each density profile thus produced, we could calculate the total core mass implied by that profile, just as we did for the mantle. The density profile yielding the correct mass, M_C, of the core turns out to be that profile based on a value of a_m of −27.5 (if $\rho_0 = 3.3$ gm/cm^3) or −26.0 (if $\rho_0 = 3.4$ gm/cm^3). With these solutions, the core's density ranges from ~11.6 gm/cm^3 at the top of the core to ~12.7 gm/cm^3 at the center, or 11.1 gm/cm^3 to 12.2 gm/cm^3, with a step up at the inner-core boundary reflecting the small jump in V_P there.

Using the modified Birch's rule instead, the preceding analysis applied to mantle and core would find a'_m ~ −1.6 at the top of the mantle (if $\rho_0 = 3.3$ gm/cm^3) or a'_m ~ −1.9 {−1.8} (if $\rho_0 = 3.4$ gm/cm^3); a mantle density profile ranging from 3.3 gm/cm^3 to 5.26 gm/cm^3 or

3.4 gm/cm^3 to 5.36 gm/cm^3 assuming the same m throughout the mantle; $M_M \approx 3.85 \times 10^{24}$ kg or $\approx 3.94 \times 10^{24}$ kg; $M_C \approx 2.1 \times 10^{24}$ kg or $\approx 2.0 \times 10^{24}$ kg; a value of -18.9 or -17.7 for a'_m in the core; and a core density profile ranging from 11.4 gm/cm^3 to 12.4 gm/cm^3 or 10.9 gm/cm^3 to 11.9 gm/cm^3.

Anderson (1967) collected a large amount of velocity–density data, mostly from high-pressure experiments and including some metals. His data, graphed as ϕ versus ρ, is shown in Figure 8.15a with one of our solutions from the modified Birch's rule overlain. We can see that our solution, with $a'_m = -18.9$,

is positioned close to but to the left of the data for iron, suggesting that our inferred value of a'_m corresponds to a mean atomic weight for the core slightly less than that of iron (which is ~56). With $a'_m = -17.7$, based on $\rho_0 = 3.4$ gm/cm^3, the lower core densities would position the line further to the left, implying a core mean atomic weight even a bit lower. Because silicate minerals are likely to have *much* lower mean atomic weights, we can interpret this as evidence that the core is primarily composed of iron; however, there must be another substance—often called the **light alloy**—present, mixed in with the iron

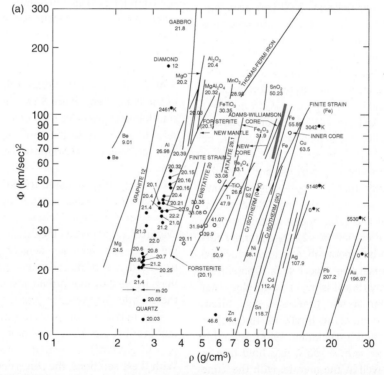

Figure 8.15 (a) Mean atomic weight of the core, inferred from high-pressure experiments and the modified Birch's rule. This figure, from Anderson (1967) / Elsevier, plots on log scales the seismic parameter ϕ, in units of km^2/sec^2, versus density ρ, in gm/cm^3, measured experimentally under high-pressure conditions for a variety of minerals and metals; their mean atomic weights are also shown. (Lines labeled other than with mineral names, e.g., "Thomas-Fermi iron" or "Adams-Williamson Core," are based on various theories and not relevant to our use of the experimental data here.)

Overlain on this plot as a heavy solid red line is the density profile for the core inferred from the modified Birch's rule, as described in the text, based on $\rho_0 = 3.3$ gm/cm^3 and using $a'_m \sim -18.9$ and ϕ from Kennett and Engdahl (1991).

Given the placement of this line, near but to the left of the line for iron (Fe, 55.85), we can infer that our value of a'_m corresponds to a mean atomic weight for the core close to but slightly less than that of iron.

but with a lower mean atomic weight, causing the overall m for the core to be less than 56.

This result must be considered tentative, at this point in our investigation of densities within the Earth; for one thing, the modified Birch's rule solution, as shown in Figure 8.15a (as a heavy red line), does not have quite the same ϕ versus ρ slope as the metals in Anderson's graph. In fact, experiments on various possible light alloy compounds (e.g., Vočadlo, 2007 and Badro et al., 2007; see also Fiquet et al., 2008) confirm the linearity of the velocity versus density relation for each mineral but also find the slopes differ. However, we will see later in this chapter that there is a variety of other evidence strongly supporting the idea of a light alloy in the core.

The critical importance of this light alloy for the dynamics of the outer core will become clear in later chapters. In those chapters we will also consider the possibility that the solid inner core does not contain much light alloy.

In that case, the inner and outer core will have different values of a'_m, and our estimates of core densities would have to be modified.

Back to the Mantle. The compositional implications of Birch's rule for the mantle are less dramatic. As Table 8.2 illustrates, there are a wide variety of minerals found on or in the Earth that possess mean atomic weights around 20–22; iron compounds, however, tend to have mean atomic weights beyond that range. We conclude that a ferromagnesian silicate composition for the entire mantle is *consistent* with Birch's rule, as long as the abundance of iron in those silicates is relatively minor.

For example, a mantle that is predominantly peridotite, plus eclogite, as we believe the uppermost mantle to be, can be consistent with m ~21–22. Peridotite consists of olivines and pyroxenes: olivines are solid solutions of $(Mg, Fe)_2SiO_4$—pure Mg_2SiO_4 is called **forsterite**, and pure Fe_2SiO_4 is called **fayalite**—and pyroxenes are solid solutions of

Table 8.2 Mean atomic weights of various minerals.[a]

Mineral	Chemical formula	M	n	m
Quartz (or stishovite[b])	SiO_2	60	3	20
Periclase	MgO	40	2	20
Enstatite (pyroxene)	$MgSiO_3$	100	5	20
Forsterite (olivine)	Mg_2SiO_4	140	7	20
Albite	$NaAlSi_3O_8$	262	13	20.1
Jadeite (pyroxene)	$NaAlSi_2O_6$	202	10	20.2
Pyrope (Mg-garnet)	$Mg_3Al_2Si_3O_{12}$	403	20	20.2
Corundum	Al_2O_3	102	5	20.4
Anorthite	$CaAl_2Si_2O_8$	276	13	21.4
Diopside (pyroxene)	$CaMgSi_2O_6$	217	10	21.7
—	—	—	—	—
Aegirine (pyroxene)	$NaFeSi_2O_6$	231	10	23.1
Hedenbergite (pyroxene)	$CaFeSi_2O_6$	248	10	24.8
Ferrosilite (pyroxene)	$FeSiO_3$	132	5	26.4
Fayalite (olivine)	Fe_2SiO_4	204	7	29.1
Hematite	Fe_2O_3	160	5	32

[a]M is molecular weight, n is the number of atoms per molecule, and m is the mean atomic weight, calculated as M/n.
[b]Stishovite is a high-pressure form of quartz.

(Mg, Fe)SiO$_3$ or Ca(Mg, Fe)(SiO$_3$)$_2$. Eclogite is composed of pyroxenes like omphacite, (NaAl + CaMg)(SiO$_3$)$_2$, and magnesium-rich garnet, like Mg$_3$Al$_2$(SiO$_4$)$_3$. Peridotite can also include some garnet and other minerals. So, as indicated in Table 8.2, a mantle that is mostly forsterite, with a small amount of fayalite, plus garnet and pyroxenes which are mostly magnesium-rich (or at least iron-poor), has a mean atomic weight in the range implied by Birch's rule.

These inferences are supported by high-pressure experiments on candidate mantle minerals. Figure 8.15b shows the results of such experiments, establishing that rocks consisting mainly of olivine, for example, yield densities and bulk sound velocities consistent with the mantle if they have little fayalite (the line marked (11.2) in the figure); but 50% fayalite (marked (14.3) in the figure) causes the samples to have densities too high and velocities too low for the mantle.

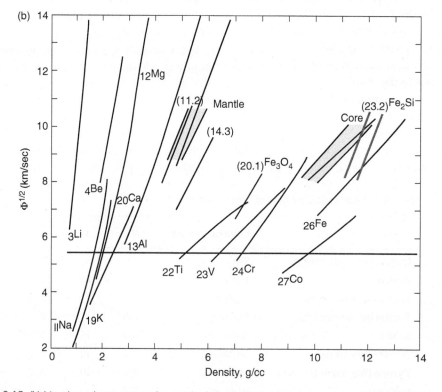

Figure 8.15 (b) Mantle and core properties can be inferred from high-pressure experiments. This figure, from Birch (1968) / NASA / Public Domain, plots the bulk sound velocity $\sqrt{\phi}$, in units of km/sec, versus density ρ, in gm/cm^3, measured experimentally under high-pressure conditions for a variety of elements and some minerals. Numbers adjacent to each line are the element's atomic number or mineral's mean atomic number. The gray-shaded areas denote allowable ranges of density versus $\sqrt{\phi}$ for the mantle and core, based on various contemporaneous density models. Finally, the upper end of the $_{26}$Fe line has been deleted for clarity.

The line marked (11.2) refers to a dunite comprised of > 90% olivine, with the olivine 90% forsterite (Birch, 1968; see also Wilson, 2001); in the line marked (14.3) the olivine is 50% forsterite, 50% fayalite. The latter falls outside the range of acceptable mantle densities and velocities, ruling out a mantle with that much iron.

The line marked (23.2) Fe$_2$Si refers to experiments on an iron–silicon alloy, showing the combination to be an acceptable composition for the core. This combination has a mean atomic weight of ~49. The two bold red lines are the density solutions from the modified Birch's rule, $a'_m = -18.9$ on the right and $a'_m = -17.7$ on the left, plotted versus $\sqrt{\phi}$.

Ringwood (1962 and beyond; see 1991 for a summary of this work) has pointed out that a particular combination of olivines and pyroxenes—which he called **pyrolite** (for pyroxene and olivine)—would have the additional property of producing a basaltic magma upon partial melting (in short, a mid-ocean ridge basalt or **MORB**) yet leave a peridotitic residue. He envisioned the bulk of the (upper) mantle to consist of such an original or 'primitive' mineralogy, which through tectonic processes would end up as the basaltic oceanic crust and uppermost mantle peridotite samples we recognize. For simplicity in our discussion of Earth's composition, however, we will use "peridotite" to refer both to the composition of uppermost mantle samples *and* Ringwood's postulated undepleted mantle.

Figure 8.15b also includes high-pressure experiments involving iron and an iron alloy. Those experiments show that pure iron (the line labeled $_{26}$Fe) yields densities too great and velocities too slow for the core; in contrast, including some amount of light alloy can lower the density and increase the velocities sufficiently to make the samples acceptable. The particular iron-silicon alloy considered in Figure 8.15b has a mean atomic number of 23.2, thus a mean atomic weight of 49.3—lower than that of pure iron (as we also concluded from Figure 8.15a) but much higher than the mantle. Our modified Birch's rule solutions, as seen in Figure 8.15b, are also consistent with the presence of a light alloy in the core, though the same caution noted with respect to Figure 8.15a applies here as well.

Birch's Rule: A Tomographic Application. Our earlier discussion of the Yellowstone plume revealed regions of faster and slower seismic velocities, which were reasonably attributed mainly to temperature effects. Similarly, our presentation of tomographic images later in this chapter will reveal numerous structures in the mantle, in most cases probably associated with—conceivably even driven by—temperature differences. Birch's rule provides a convenient, ultimately experimentally based, way to quantify such effects (to explore

the pressure and temperature dependence of seismic velocity underlying Birch's rule, see, e.g., Anderson, 1967; Wang, 1970).

The 'rate' at which temperature changes cause variations in P wave velocity is

$$\frac{dV_P}{dT};$$

substituting Birch's rule, we have

$$\frac{dV_P}{dT} = \frac{d}{dT}(a_m + b\rho) = b\frac{d\rho}{dT}$$

since a_m is not a function of temperature (for simplicity, we will assume b is not a function of temperature, either; cf. Birch, 1961b). How the density of a material changes when it is heated ($d\rho/dT$) relates to a property called its *thermal expansion*. The **coefficient of thermal expansion**, denoted α, can be defined as

$$\alpha \equiv \frac{-d\rho/\rho}{dT},$$

that is, the fractional decrease in density per increase in temperature. A more popular but equivalent version of this definition is

$$\alpha \equiv \frac{dVOL/VOL}{dT},$$

where VOL is volume here; this is equivalent because the fractional decrease in a sample's density ($-d\rho/\rho$) equals the fractional increase in its volume ($dVOL/VOL$) (for example, if the sample's density decreases by 0.1%, then its volume will have increased by 0.1%). Note that the units for α are 1/degree.

It follows that

$$\frac{dV_P}{dT} = b\frac{d\rho}{dT} = -b\alpha\rho$$

or

$$\Delta V_P = \left(\frac{dV_P}{dT}\right)\Delta T = -b\alpha\rho\Delta T.$$

As an illustration, we might choose $\alpha \sim 0.5 \times 10^{-5}/°C$ (Chopelas & Boehler, 1989) and $\rho \sim 5.1$ gm/cm^3 for the lowermost mantle; $\alpha \sim 1.7 \times 10^{-5}/°C$ (Stacey & Davis, 2008) and $\rho \sim 4.8$ gm/cm^3 for the mid mantle; or $\alpha \sim 2.7 \times 10^{-5}/°C$ (Stacey & Davis, 2008) and $\rho \sim 3.5$ gm/cm^3 for the upper mantle (see also Tackley, 2012). In that case, a region in which P waves travel, say, 0.1 km/sec slower—which is roughly 1% slower—would be ~1,300°C warmer at the base of the mantle or ~400°C warmer in mid mantle or ~350°C warmer in the upper mantle—if all its velocity contrast can be attributed to temperature effects.

Applying these estimates to Figure 8.19 shown later in the chapter, how would you describe the temperature variation within the South Pacific superplume? Is that variation physically reasonable?

8.3.2 The Adams-Williamson Equations

Long before Birch proposed his empirical relation between density and seismic wave velocity, geophysicists were applying physical concepts to infer theoretically the density distribution within the Earth. In 1923, a very useful set of such equations was worked out jointly by Adams and Williamson (Williamson & Adams, 1923; see also Adams & Williamson, 1923). Although most researchers today take a 'black box' approach to determining interior density and seismic properties, for example constructing numerical models of the interior based on free oscillation data, our exploration of the Adams-Williamson equations will show us the power of geophysical theory, tell us things about the state of the interior that numerical values of ρ, k, and μ only obscure, and provide a more constrained and unequivocal view of the interior.

Hydrostatic Equilibrium: A Key Part of Adams-Williamson. From an Earth System Science perspective, however, perhaps the greatest value to be derived from studying the Adams-Williamson equations is that they are based on the concept of *hydrostatic equilibrium*. Hydrostatic equilibrium plays a role in the state and circulation of any fluid—including the atmosphere, the oceans, the core, and the mantle, too—and even in the dynamics of aquifers. As the name implies, **hydrostatic equilibrium** involves a balance (the "equilibrium") between the forces acting on a fluid (the "hydro") at rest (the "static").

As a simple example, we consider a tank, at the surface of the Earth, filled with water; the water is not in motion. The tank exists within the Earth's gravity field, so the water—whose density is ρ_w—experiences a downward gravitational acceleration $\vec{g}$; that gravity imposes a weight, a downward gravitational force, equal to $\rho_w\vec{g}$ per unit volume (ρ_w is mass per unit volume so $\rho_w g$ is (mass × gravity) per unit volume).

As a result of the weight of overlying water, the water within the tank feels a pressure P. Let z denote depth within the tank, with $z = 0$ at the water's surface; see Figure 8.16. As we go deeper into the tank (and z increases), there is more and more water above us, so P increases. Hydrostatic equilibrium tells us just how much of an increase in pressure there must be if the water is to remain at rest. Referring to Figure 8.16, we focus on a 'slice' of water at depth z, of cross-sectional area A and very tiny thickness dz. Let P now specifically denote the pressure at depth z, thus the pressure acting on the top face of the slice. And, let $P + dP$ denote the pressure at depth $z + dz$, thus the pressure

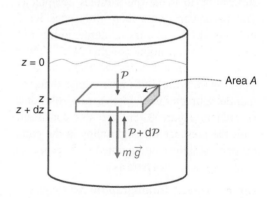

Figure 8.16 Scenario for understanding hydrostatic equilibrium. The tank of water shown here sits at Earth's surface, where the gravitational acceleration is $\vec{g}$. We consider the force balance on a rectangular slice of water of thickness dz and cross-sectional area A, whose top face is at depth z and whose bottom face is at depth $(z+dz)$. The pressure on the slice's top face, due to the weight of overlying water, is P; the slice exerts a downward pressure of magnitude $(P + dP)$ on the water below, which by Newton's Third Law exerts an equal upward pressure on the bottom face of the slice.

exerted by the slice (and everything above it) on the area below the slice; the increase in pressure, dP, is tiny because the increase in the water column, dz, is also tiny. By Newton's Third Law—for every action there is an equal and opposite reaction—the water below the slice must be pushing back (pushing *up*) on the slice, acting on its bottom face, with the same pressure $P + dP$.

The force balance on the slice can now be easily formulated. Recalling that pressure, like any type of stress, has units of force per unit area, the downward **pressure force** on the slice from above it is

$$P A,$$

and the upward force from the pressure of the water below is

$$(P + dP)A.$$

The volume of the slice itself is Adz, so that it contains an amount $\rho_w(Adz)$ of mass. The gravitational force acting on the slice is therefore

$$[\rho_w(Adz)]\vec{g};$$

the direction of this force is, of course, downward. If the slice is not moving, there must be a balance of all the forces acting on it, that is, the net downward force must equal the net upward force:

$$P A + [\rho_w(Adz)]g = (P + dP)A,$$

where g is the magnitude of gravity at the depth where the slice is located. We can divide both sides of this equation by A (our analysis could have been phrased, from the start, in terms of a unit cross-sectional area for the slice), and subtract the common P term from both sides. We then find

$$\rho_w g dz = dP.$$

This is the fundamental statement of hydrostatic equilibrium, quantifying how much the fluid pressure must increase (when the depth increases by dz) in order to keep the volume of water from moving up or down. If the actual increase in fluid pressure is less than this prescribed amount, the upward fluid pressure will not be enough to hold up the

slice, and it will fall deeper into the tank (until it reaches a level where the pressure is great enough); if the actual pressure increase is more than the prescribed amount, that will send the slice upward to higher levels within the tank, until the ambient pressure has decreased sufficiently. In a state of hydrostatic balance, the slice is at precisely the right level to stay put.

It is customary to express the equation above in terms of the 'rate' at which pressure must increase with depth—the pressure *gradient*—to maintain equilibrium; thus,

$$\frac{dP}{dz} = \rho_w g$$

is the official statement of hydrostatic equilibrium. The pressure, P, appearing in this equation is called the *hydrostatic pressure*. If the fluid is in motion, hydrostatic pressure still exists—after all, the weight of overlying fluid does not disappear—but other pressure, associated with the fluid motion, will be acting; such pressure may be called *dynamic pressure*.

As an example of hydrostatic equilibrium, we can predict how fast pressure must increase with depth within the oceans. At sea level, the acceleration of gravity is ~980 cm/sec^2, and the density of seawater is typically 1.03 gm/cm^3. For every kilometer of depth ($\Delta z = 10^5$ cm), then,

$$\Delta P \cong \rho_w g \Delta z \approx 101 \times 10^6 \text{ dyne/cm}^2$$

$$= 101 \text{ bar or 0.1 kbar,}$$

so the hydrostatic pressure increases by 0.1 kbar (10 MPa). Within the troposphere (or, beginning with a tank of air and repeating the same type of derivation as above), the value of g is nearly the same, but the density is smaller by a factor of ~10^3, so the hydrostatic pressure increase moving down through the air is only 0.1 bar (0.01 MPa) per km. (But that pressure gradient exists only near Earth's surface; high up in the troposphere, the air density is much less, as is g, so the pressure gradient is correspondingly smaller.)

Philosophical Guidance From Birch's Rule. The approach we will take to develop the Adams-Williamson equations follows from the simple yet profound implication of Birch's rule noted earlier: we can model the dependence of V_P on 'state' variables such as pressure and temperature through their effects on density; from the modified version of Birch's rule, we can cautiously extend this conclusion to V_S as well. Consistent with our interpretation of velocity anomalies earlier in this chapter, then, we can say that increases in pressure should cause the seismic velocity to increase; since temperature increases have the opposite effect on density, they should cause the seismic velocity to decrease.

From travel-time observations and normal-mode data, geophysicists have deduced that (as shown in Figure 8.12) V_P and V_S generally increase with depth within the Earth. We conclude that *pressure effects dominate over temperature effects throughout most of Earth's interior* (of course, at the core-mantle boundary—the largest change in seismic velocities anywhere inside the Earth—the cause of the changes in V_P and V_S is neither pressure nor temperature). And, as important as those tomographic structures to be discussed later may be for mantle convection, even if we interpret them as primarily thermal anomalies, in terms of seismic velocity they are departures of only a few percent. That is, from a seismic perspective, laterally as well as radially, the interior of the Earth is not 'too' hot.

A Simple Derivation of the Adams-Williamson Equations. The goal of the Adams-Williamson equations is to predict how the density varies with depth within Earth's interior. Letting Earth's density be denoted by $\rho = \rho(z)$ where z is the depth below the surface (we can let $z = 0$ represent the top of the mantle), the goal is to develop a formula for

$$\frac{d\rho}{dz}.$$

$d\rho/dz$ will tell us how much the density changes from layer to layer; once we have determined that change, it can simply be added to the density of a layer to predict the density of the next layer down. Of course, in order for this procedure to work, we will

need to specify a 'starting' density, that is, the density at the top of the mantle. As in our discussion of Birch's rule, that density will be denoted by ρ_0. A reasonable choice for ρ_0 might be the value of 3.3 gm/cm^3 considered in our discussion of isostasy in Chapter 5.

In general, we can envision four factors that affect the density of material within the Earth: *composition*, which determines the basic density of the material; the *phase* (e.g., solid or liquid) of the material; *pressure, P*, an increase in which tends to make any material denser; and *temperature, T*, which tends to have the opposite effect. Symbolically, we could write

$$\rho = \rho(P, T; \text{composition, phase}).$$

As a start, we will apply the Adams-Williamson equations to the mantle only. If we assume as a first approximation that there are no variations in composition or phase within the mantle, then the only factors that can cause changes in the mantle's density will be pressure and temperature:

$$\rho = \rho(P, T).$$

In our discussion henceforth, it will occasionally be important to distinguish between **chemical homogeneity**, in which there are no variations in composition, and what we might call "**material homogeneity**," a stronger restriction in which there are no variations in composition *or* phase. Thus, for now we are assuming that the mantle is materially homogeneous.

For the moment, we will also (!) neglect the dependence of density on temperature—after all, as Birch's rule suggests, pressure effects dominate over temperature effects within the Earth; thus,

$$\rho = \rho(P)$$

for now. Using the "chain rule" from calculus, it follows that *the change in density with depth equals the way density changes with pressure times the change in pressure with depth*:

$$\frac{d\rho}{dz} = \frac{d\rho}{dP}\frac{dP}{dz}.$$

This approach is useful because we can approximate the pressure gradient (the change in pressure with depth) by the hydrostatic pressure gradient:

$$\frac{dP}{dz} \cong \frac{d\mathcal{P}}{dz} = \rho g,$$

with ρ now the density of a 'slice' of mantle at depth z; the gravitational acceleration at that depth is g. An exercise in Chapter 9 will confirm that the hydrostatic approximation is actually an excellent one, despite the fact that the mantle is not static.

The other derivative in our chain-rule expansion, $d\rho/dP$, refers to the way substances respond to an increase in pressure—a material property. It should not be too surprising to find that the material property in question is a quantity we have employed throughout this chapter: the incompressibility, k! k was originally defined as the ratio of the compressive isotropic stress acting on a cube of material to the resulting fractional decrease in the cube's volume. It has previously been noted that, for any given sample of a material being deformed, the fractional decrease in its volume is the same as the fractional increase in its density. In the present discussion, the isotropic compressive stress causing changes in density is dP, the amount by which the pressure increases when moving a bit deeper into the Earth; thus,

$$k \equiv \frac{dP}{d\rho/\rho} \quad \text{or} \quad k \equiv \rho \frac{dP}{d\rho}.$$

Note, by the way, that the speed of sound in a fluid (like air or water), which equals the bulk sound velocity $\sqrt{\phi}$ or $\sqrt{k/\rho}$, can also be expressed as $\sqrt{dP/d\rho}$. Finally, it follows that our other derivative of interest can be written as

$$\frac{d\rho}{dP} = \frac{\rho}{k} = \frac{1}{\phi}.$$

Putting these together yields the Adams-Williamson goal:

$$\frac{d\rho}{dz} = \frac{d\rho}{dP}\frac{dP}{dz} = (\rho g)/\phi.$$

With seismic velocities, and thus ϕ, known in all layers of the mantle, this equation can be solved numerically, 'from the top, down,' to yield the density of each layer. The 'recipe,' or *algorithm* in computational terminology, is as follows.

- Assume a starting density ρ_0 for the top layer; substitute it and the value for ϕ in the top layer into the right-hand side of the equation, which then predicts the 'rate' $d\rho/dz$ at which the density increases from the top layer to the next;

- multiply that rate times the layer thickness to obtain the actual increase in density; then

- add that increase to ρ_0 to find the density of the second layer.

- With the density of that layer—we could call it ρ_1—now known, we can substitute it and the value for ϕ in that layer into the right-hand side of the equation, yielding the rate of density increase;

- multiply that rate by the second layer's thickness to obtain the density increase to the next layer; then

- add that increase to ρ_1 to find the density of the third layer.

This process can be *iterated* (repeated) until the densities of all the mantle layers have been determined.

Mass Implications. The density profile thus deduced can be used to tell us the mass of the mantle. As noted in our discussion of Birch's rule, the general expression for the mass of a uniform spherical layer of density ρ is

$$\left(\frac{4}{3}\pi r_{upper}^3 - \frac{4}{3}\pi r_{lower}^3\right)\rho$$

if that layer extends from radius r_{upper} down to radius r_{lower}. Adding these layer masses then yields the net mass of the mantle; for now, we write this symbolically as

$$M_M = \sum_i \frac{4}{3}\pi(r_i^3 - r_{i+1}^3)\rho_i$$

with i an index to represent each layer, starting at the top of the mantle. *Textbox B* presents such formulas more rigorously, and develops the algorithm to solve Adams-Williamson.

Incidentally, if the solution is fine (as opposed to coarse), so that the number of layers is large and the layer thicknesses Δz_i (where $\Delta z_i = r_{upper} - r_{lower}$ or $r_i - r_{i+1}$) are very small, then the volume of the ith layer is, to good approximation, $4\pi r_i^2 \Delta z_i$—that is, the surface area of the layer times its thickness—and its mass is

$$4\pi r_i^2 \Delta z_i \cdot \rho_i.$$

You could derive that approximation starting with

$$\frac{4}{3}\pi(r_i^3 - r_{i+1}^3)$$

for the volume of a layer; substituting in $r_{i+1} = r_i - \Delta z_i$; and expanding $(r_{i+1})^3$ out. That expansion will have four terms, each smaller than the one before; a key cancellation, and neglect of very small terms, then yields our volume (or mass) approximation. Can you say which of those terms exactly cancels another, and which terms are reasonably neglected?

With this simplification, the mass of the mantle can be well approximated by

$$M_M = \sum_i 4\pi r_i^2 \rho_i \Delta z_i,$$

which, of course, is nothing more than a discrete approximation to

$$\int 4\pi r^2 \rho \, dr.$$

Unfortunately, with Δz_i typically ~ 100 km, neither the PREM nor the iasp91 velocity model—which we will use to specify the value of ϕ in each layer—has a fine enough resolution to warrant the use of this approximation.

Gravity Cannot be Ignored. The astute reader will have noticed that the Adams-Williamson equation ($d\rho/dz = \rho g /\phi$) provides

an algorithm for determining density only if *all* the terms on the right-hand side are known; but we have not yet said anything about g, the acceleration of gravity at the depth where Adams-Williamson is being applied. In mathematical terms, our derivation led to an equation with two unknowns, ρ and g; for a complete solution, we need another equation, one involving gravity. Clues for how to do this come from Chapter 4, where we saw that the inverse-square property of gravity means that the gravity outside a spherically symmetric ball is the same as if all its mass were at a point at its center; and that the gravity within a spherically symmetric shell has to be zero. Putting these properties together, with the help of Figure 8.17, will very directly lead to our second equation.

An observer at depth z below the surface experiences a net gravity due to all of Earth's mass. That mass can be divided into the mass above and the mass below depth z, that is, the mass in the spherical shell extending from the surface down to depth z and the mass in the sphere extending from depth z down to Earth's center (see Figure 8.17); thus

$$g(z) = g_{\text{SHELL ABOVE } z} + g_{\text{SPHERE BELOW } z}.$$

Earth's mean radius will be denoted by a. For our derivation, we will make the simplifying assumption that the Earth is spherically symmetric; although this is only an approximation, we have seen that it is an excellent one, e.g. good to 99.5% in terms of the worldwide variation of gravity. With this assumption, we know

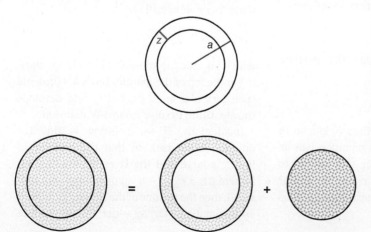

Figure 8.17 Gravity within the Earth. The Earth is modeled as spherically symmetric, with radius *a*. We are interested in the value *g(z)* of gravity at depth *z* below the surface. By dividing the mass of the Earth into the shell above depth *z* plus the sphere below depth *z* (the sphere, therefore, of radius *a−z*), *g(z)* can be easily formulated.

(see Chapter 4) that

$$g_{\text{SHELL ABOVE } z} \equiv 0.$$

We also know that the gravity due to a spherically symmetric ball is the same as if all its mass were concentrated at a point at its center; at depth z, which is a distance $(a-z)$ from the center, it follows that

$$g_{\text{SPHERE BELOW } z} = \frac{GM(z)}{(a-z)^2}$$

if $M(z)$ denotes the mass of the Earth below depth z. In short,

$$g(z) = \frac{GM(z)}{(a-z)^2},$$

where in light of our earlier discussions

$$M(z) = \int_{r=0}^{r=a-z} 4\pi r^2 \rho \, dr.$$

For reference, we note that for a planet or other object that is not only spherically symmetric but uniform in density, our equation for g could be used to demonstrate that gravity must decrease linearly with depth within it: as depth increases, both $M(z)$ and $(a-z)^2$ decrease; but $M(z)$ depends on volume ($\sim(a-z)^3$), so the net effect is that g is proportional to $(a-z)$. Furthermore, at the Earth's center, g is zero—which makes sense because g depends on the mass below, and there is no longer any mass below; or, because at Earth's center we would be pulled upward equally in all directions (due to spherical symmetry), canceling out the gravitational force there.

Finally, our two equations in two unknowns are

$$\frac{d\rho}{dz} = (\rho g)/\phi \text{ and } g = \frac{G}{(a-z)^2} \int_{r=0}^{r=a-z} 4\pi r^2 \rho \, dr,$$

the **Simplified Adams-Williamson Equations**. Since these equations must be solved together, it is now clear that the algorithm presented earlier for solving the Adams-Williamson equations was incomplete. Each time $d\rho/dz$ is to be calculated, leading to an estimate of the density of the next layer, g must first be specified; then, once the mass of the current layer has been subtracted from $M(z)$,

yielding the mass below *that* layer, the value of gravity at the next depth (i.e., at the top of the next layer down) must be determined for the next iteration. We can use the (precise) summation formula written above to calculate layer and total masses, and thus gravity; as before, we need a starting mass, $M(z=0)$, which would be the mass of the whole Earth (or the whole Earth minus the crust, if we focus on the mantle), or equivalently a starting gravity, the gravity at the top of the Earth. See *Textbox B* for more details.

Accounting for Temperature in the Adams-Williamson Equations. It may be true that temperature effects on elastic properties and density are minor in the Earth compared to pressure effects; but they are not completely negligible, either. After all, the temperature is likely to rise by several thousand degrees within the Earth. In this section, we will see that a symbolic use of calculus allows us to account for temperature effects on density very neatly, and also provides an introduction to concepts we will find fundamentally important later in this textbook.

If density is a function of both pressure and temperature, then it can vary when either of the latter changes. Ultimately, the changes we consider will be those associated with an increase in depth. But in general, the change in density when pressure increases by an amount dP can be written in symbolic terms as

$$\left(\frac{d\rho}{dP}\right)_T dP;$$

and the change in density due to a temperature increase of dT can be written

$$\left(\frac{d\rho}{dT}\right)_P dT.$$

The quantities in parentheses are 'rates,' that is, the rate at which density increases when P or T does. For example, if in some situation the density increases by 0.1 gm/cm^3 for every MPa of pressure increase, then the 'rate' $(d\rho/dP)_T$ at which density increases would be 0.1 gm/cm^3 / MPa; and in that situation a pressure increase dP of 5 MPa would lead to a density larger by $(d\rho/dP)_T \cdot dP$ or 0.5 gm/cm^3.

We have already seen that these rates can be related to material properties, such as incompressibility. They are written here with subscripts to instruct us to hold the subscripted variables constant when calculating the rates; that is, the experiments measuring how density changes should be performed by varying only one variable at a time. For example, when measuring the change in density when pressure is varied, the experiment should be carried out with the temperature held constant. In short, those subscripts allow us to 'keep things straight' when assessing the dependence of density on both pressure and temperature.

Mathematically, when a quantity is a function of more than one variable and its derivative is being calculated explicitly with respect to only one of those variables, the derivative is called a *partial derivative*. For example, the rate of change of density with pressure, holding temperature constant, is written as

$$\left(\frac{\partial \rho}{\partial P}\right)_T;$$

or, when there is no doubt about which variables are being held constant, simply

$$\frac{\partial \rho}{\partial P}.$$

The use of ∂ rather than "d" serves to emphasize that some variables are not involved in the differentiation operation. Although this notation may seem intimidating, those faced with the need for it should convince themselves that partial derivatives should be viewed with relief, and even eagerness! For example, let's say f is a function of both x and y, and is defined as

$$f(x, y) = x^x \sin(x^2 - \sin(x \ln x) - e^x) + 2y - 5.$$

and we have been asked to find the partial derivative of f with respect to y. That would be less challenging than the complexity of f might suggest: such a partial derivative means that x will be treated as unchanging; and we know that the derivative of a constant is zero.

Consequently, we have

$$\frac{\partial f}{\partial y} = \frac{\partial \left[x^x \sin(x^2 - \sin(x \ln x) - e^x) + 2y - 5\right]}{\partial y}$$

$$= \frac{\partial \left[x^x \sin(x^2 - \sin(x \ln x) - e^x)\right]}{\partial y}$$

$$+ \frac{\partial [2y]}{\partial y} - \frac{\partial [5]}{\partial y}$$

$$= 0 + 2 - 0$$

$$= 2$$

very easily! A second, more useful example with applications to the atmosphere will be presented below.

If density is a function of both pressure and temperature, then the *total* change in it when pressure is increased by an amount dP and temperature is increased by an amount dT—both, for example, occurring when the depth is increased by a small amount dz—will be

$$d\rho = \underbrace{\left(\frac{\partial \rho}{\partial P}\right)_T dP}_{\text{pressure effect}} + \underbrace{\left(\frac{\partial \rho}{\partial T}\right)_P dT}_{\text{temperature effect}}.$$

If we wished, we could express these changes relative to the increase in depth that caused them in the first place, simply by dividing both sides of this equation by dz; the result is

$$\frac{d\rho}{dz} = \underbrace{\left(\frac{\partial \rho}{\partial P}\right)_T \frac{dP}{dz}}_{\text{pressure effect}} + \underbrace{\left(\frac{\partial \rho}{\partial T}\right)_P \frac{dT}{dz}}_{\text{temperature effect}}.$$

The pressure contribution depends on how the pressure increases with depth, that is, on the *pressure gradient*, dP/dz, within the Earth; and the temperature contribution depends likewise on the *temperature gradient*, dT/dz, within the Earth.

At this point, it is worth digressing briefly to focus on the pressure effect. When a material is compressed, it heats up as well as becoming denser. It follows that pressure experiments measuring the rate at which density changes can be conducted under different conditions. An **adiabatic** experiment is one in which the heat generated by compression is not allowed to escape until after the change in density is measured. In an *isothermal* experiment, the heat of compression is allowed to escape; the change in its density is measured once the sample has cooled back to its original temperature. These two experiments will measure different changes in density: in the isothermal situation, the sample cools and contracts somewhat before its density is recorded, and that contraction makes the sample even denser.

It follows that different types of incompressibility can be defined,

$$k_T \equiv \rho \left(\frac{\partial P}{\partial \rho} \right)_T$$

under isothermal conditions, and

$$k_S \equiv \rho \left(\frac{\partial P}{\partial \rho} \right)_S$$

under adiabatic conditions; the subscript for adiabatic is usually taken to be "S" for entropy rather than "A" because typically (i.e., for reversible thermodynamic processes) constant heat—*adiabaticity*—is equivalent to constant entropy (constant entropy is also called **isentropic**). These expressions use partial derivatives; but no matter how they are written, each k represents a ratio of the applied stress to the resulting volumetric strain. Our description of adiabatic versus isothermal experiments suggests that k_T and k_S will not be equal in magnitude, for any given sample; *can you predict which should be larger—that is, which should imply a greater resistance to compression?*

From our earlier derivation of the 'simplified' Adams-Williamson equations, we know that seismic velocities (or ϕ) will be used to determine densities within the Earth. Which resistance to compression, adiabatic or isothermal, do seismic waves sense as they pass through Earth's interior? The periods of P waves generated by earthquakes tend to be ~1–10 sec (Stein & Wysession, 2003), corresponding to wavelengths of ~10–10^2 km at speeds of ~10 km/sec. Since rocks within the mantle conduct heat poorly, it follows that the heat of compression produced by a P wave has no time to escape before the subsequent dilatation occurs. The resistance to compression experienced by the P wave is thus an adiabatic

resistance, and the P wave velocity should have been explicitly defined as

$$V_P \equiv \sqrt{\frac{k_S + \frac{4}{3}\mu}{\rho}} \, ;$$

the seismic parameter is actually given by

$$\phi = \frac{k_S}{\rho}.$$

Returning to the problem of incorporating temperature effects into the Adams-Williamson equations: it is certainly true that temperature increases steadily throughout Earth's interior; however, the actual temperature increase, dT/dz, is still not well known at different depths. This is partly because there are a number of factors—including various heat sources and dynamic processes—that affect (or have affected in the past) interior temperatures (we will discuss them in the next chapter), and the extent of their contributions is uncertain. We are in the dark concerning dT/dz in part also because it is difficult to predict how temperature variations evolve over time in bodies like the Earth.

But there is no doubt that the interior has experienced compression—as the proto-Earth grew and internal pressures rose to their current values—and that such compression would have caused the interior to heat up, adiabatically, with greater compression at deeper locations producing greater amounts of heat. This compressively driven type of temperature gradient is called the **adiabatic temperature gradient**, and it turns out—as we will discover in the next chapter—that it approximates conditions throughout the Earth System reasonably well, overall. We denote this quantity by

$$\left(\frac{\partial T}{\partial z} \right)_S.$$

(Incidentally, it should be evident both symbolically and physically that we can write the adiabatic temperature gradient as

$$\left(\frac{\partial T}{\partial P} \right)_S \frac{dP}{dz},$$

the first term representing adiabatic heating and the second term—well approximated as ρg—the compression.)

In short, it is important that temperature effects be included in the Adams-Williamson equations; but many of those effects are not well determined. As a compromise, *we can include just that portion of the temperature effects due to adiabatic compression* in computing the changes in density within the Earth. Symbolically, we can write

$$\frac{d\rho}{dz} = \underbrace{\left(\frac{\partial\rho}{\partial P}\right)_T \frac{dP}{dz}}_{\text{pressure effect}} + \underbrace{\left(\frac{\partial\rho}{\partial T}\right)_P \left(\frac{\partial T}{\partial z}\right)_S}_{\text{adiabatic temperature effect}} .$$

But including pressure effects plus the thermal effects of compression is the *same* as dealing with adiabatic compression; it follows that

$$\frac{d\rho}{dz} = \underbrace{\left(\frac{\partial\rho}{\partial P}\right)_S \frac{dP}{dz}}_{\text{adiabatic pressure effect}} .$$

With this compromise, then, our earlier simplistic derivation was not so far wrong; we merely needed to stipulate that the equation referred to adiabatic conditions. Since $k_S = \rho(\partial P/\partial\rho)_S$ and $\phi = k_S/\rho$, we have once again

$$\left(\frac{\partial\rho}{\partial P}\right)_S = \frac{1}{\phi};$$

assuming hydrostatic equilibrium, and—for the calculation of gravity within the Earth—a spherically symmetric Earth, the net results are

$$\frac{d\rho}{dz} = (\rho g)/\phi \text{ and } g = \frac{G}{(a-z)^2}\int_{r=0}^{r=a-z} 4\pi r^2\rho\, dr.$$

These are *the Adams-Williamson equations under the assumption of an adiabatic temperature gradient within the Earth.*

A Full Accounting for Temperature in the Adams-Williamson Equations. We will find it useful to be able to quantify *nonadiabatic* effects on density; this calls for a different approach to deriving the Adams-Williamson equations. Our approach will require four brief steps.

1. **Entropy Is the Key.** Our derivations above were ultimately based on viewing density as

a function of either pressure or pressure and temperature:

$$\rho = \rho(P, T).$$

But, as we have just seen, *entropy* can potentially play a role in determining density variations within the Earth; for example, assuming constant entropy (or, equivalently, constant heat) allows us to formulate reasonable equations for $d\rho/dz$, including even some temperature effects. Following Bullen (1965), we reformulate the Adams-Williamson equations by treating density as a function of pressure (P) and entropy (S) rather than pressure and temperature:

$$\rho = \rho(P, S).$$

With this approach, other thermodynamic variables (such as temperature) should be treated as functions of P and S as well.

As before, we assume no change in composition. The total change in density, for example as we move deeper into the mantle, will then result from either the increase dP in pressure or an increase dS in entropy:

$$d\rho = \underbrace{\left(\frac{\partial\rho}{\partial P}\right)_S dP}_{\substack{\text{pressure effect, with}\\\text{entropy held constant}}} + \underbrace{\left(\frac{\partial\rho}{\partial S}\right)_P dS}_{\substack{\text{entropy effect, with}\\\text{pressure held constant}}} .$$

And, we can express these changes relative to the increase in depth that incurred them in the first place, simply by dividing both sides of this equation by dz; the result is

$$\frac{d\rho}{dz} = \underbrace{\left(\frac{\partial\rho}{\partial P}\right)_S \frac{dP}{dz}}_{\substack{\text{pressure effect, with}\\\text{entropy held constant}}} + \underbrace{\left(\frac{\partial\rho}{\partial S}\right)_P \frac{dS}{dz}}_{\substack{\text{entropy effect, with}\\\text{pressure held constant}}} .$$

In this approach, the pressure contribution derives from the pressure gradient *and* the heat generated by adiabatic compression; and there is an entropy contribution, which depends on the magnitude of the *entropy gradient* within the Earth. This entropy contribution will include all temperature effects beyond that of adiabatic compression.

2. **How Does Temperature Vary Within the Earth?** Bullen considered the total increase in temperature, dT, produced by an increase in depth dz. Since all variables are viewed now as functions of pressure and entropy, we can write

$$dT = \underbrace{\left(\frac{\partial T}{\partial P}\right)_S dP}_{\substack{\text{pressure effect on } T, \\ \text{with entropy held} \\ \text{constant}}} + \underbrace{\left(\frac{\partial T}{\partial S}\right)_P dS}_{\substack{\text{entropy effect on } T, \\ \text{with pressure held} \\ \text{constant}}} ,$$

or

$$\frac{dT}{dz} = \left(\frac{\partial T}{\partial P}\right)_S \frac{dP}{dz} + \left(\frac{\partial T}{\partial S}\right)_P \frac{dS}{dz},$$

similar to the way we wrote the change in density earlier. The temperature gradient on the left-hand side of this equation represents the *actual* temperature gradient within the Earth. As noted parenthetically above, the first term on the right-hand side equals the adiabatic temperature gradient, which we had also denoted by $(\partial T/\partial z)_S$, thus

$$\frac{dT}{dz} = \left(\frac{\partial T}{\partial z}\right)_S + \left(\frac{\partial T}{\partial S}\right)_P \frac{dS}{dz}.$$

The other term on the right-hand side of the equation is therefore the *difference* between the actual temperature gradient and the adiabatic temperature gradient within the Earth; we will denote this difference, which reflects nonadiabatic effects on temperature, by τ:

$$\tau \equiv \left(\frac{\partial T}{\partial S}\right)_P \frac{dS}{dz} = \frac{dT}{dz} - \left(\frac{\partial T}{\partial z}\right)_S.$$

When $\tau = 0$, the actual temperature gradient is adiabatic. If τ is positive, the actual temperature gradient is called **superadiabatic**; if τ is negative, the gradient is **subadiabatic**. We will see later in this textbook that τ is a fundamental parameter that determines the kinds of motion possible within a fluid.

It should be noted that our expansion of the temperature variation (dT) into pres-

sure and entropy components implicitly assumed that there were no changes in phase; a change in phase, e.g. when a solid melts, occurs at a single temperature ($dT = 0$) and pressure ($dP = 0$), but typically involves a change in entropy ($dS \neq 0$). Thus, our entropy-based derivation has the same set of assumptions—besides the possibility of nonadiabatic temperatures—as the earlier derivations: no changes in composition; no changes in phase; hydrostatic equilibrium; and spherical symmetry.

3. **The Entropy Gradient, Briefly.** From the definition of τ,

$$\tau \equiv \left(\frac{\partial T}{\partial S}\right)_P \frac{dS}{dz},$$

it follows that the entropy gradient can be written in terms of τ as

$$\frac{dS}{dz} = \tau \left(\frac{\partial S}{\partial T}\right)_P.$$

Substituting this into the equation for $d\rho/dz$ yields

$$\frac{d\rho}{dz} = \underbrace{\left(\frac{\partial \rho}{\partial P}\right)_S \frac{dP}{dz}}_{\substack{\text{pressure effect, with} \\ \text{entropy held constant}}} + \underbrace{\tau \left(\frac{\partial \rho}{\partial S}\right)_P \left(\frac{\partial S}{\partial T}\right)_P}_{\substack{\text{entropy effect, with} \\ \text{pressure held constant}}}$$

$$= \underbrace{\left(\frac{\partial \rho}{\partial P}\right)_S \frac{dP}{dz}}_{\text{adiabatic pressure effect}} + \underbrace{\tau \left(\frac{\partial \rho}{\partial T}\right)_P}_{\substack{\text{non-adiabatic} \\ \text{temperature effect,} \\ \text{with pressure} \\ \text{held constant}}} .$$

4. **How *Does* Temperature Affect Density?** Finally, we recall the definition of the coefficient of thermal expansion, α,

$$\alpha \equiv \frac{-d\rho/\rho}{dT} = -\frac{1}{\rho}\frac{d\rho}{dT};$$

or, recognizing explicitly here that experiments measuring α are conducted under constant-pressure conditions (so that pressure and thermal effects are not mixed together),

$$\alpha = -\frac{1}{\rho}\left(\frac{\partial \rho}{\partial T}\right)_P.$$

This allows our equation for $d\rho/dz$ to be written

$$\frac{d\rho}{dz} = \underbrace{\left(\frac{\partial\rho}{\partial P}\right)_S \frac{dP}{dz}}_{\text{adiabatic pressure effect}} + \underbrace{-\alpha\rho\tau}_{\substack{\text{non-adiabatic} \\ \text{temperature effect}}} ;$$

and, as we have already seen that the first term on the right-hand side reduces to $(\rho g)/\phi$, we end up with

$$\frac{d\rho}{dz} = (\rho g)/\phi - \alpha\rho\tau \text{ and } g = \frac{G}{(a-z)^2} \int\limits_{r=0}^{r=a-z} 4\pi r^2 \rho\, dr.$$

These are **the full Adams-Williamson equations**. They were actually derived first by Birch (1952), who however used a different approach than shown here (his approach focused on the difference between k_T and k_S).

In retrospect, it should not be surprising that the additional term in these equations takes the form that we found. If α is defined as the fractional decrease in density per change in temperature, then $\rho\alpha$ is the decrease in density per change in temperature; multiplying times τ, which is a temperature change per change in depth, then yields the decrease in density per change in depth. The subtle point here is that, because τ represents the *nonadiabatic* temperature gradient, its multiplication times $\rho\alpha$ yields the density gradient associated with nonadiabatic temperature effects.

It might be noted that the idea of a "total" change in a variable, which we have applied in steps 1 and 2 above to density and temperature (and before that, once again, to density), proves to be exceedingly useful in understanding fluid flow; in Chapter 9, we will formally introduce the related concept of a *total derivative*.

Solutions to the Adams-Williamson Equations: A Simple Application to the Atmosphere. Although the Adams-Williamson equations were developed in order to determine the densities within the Earth, their underlying principles are general enough that they can be employed in other contexts. We consider here the lowest layer of the atmosphere, the troposphere, which—for reasons to be discussed in Chapter 9—can be modeled to good approximation as having an adiabatic temperature gradient. In that case, the first of the Adams-Williamson equations is

$$\frac{d\rho}{dz} = (\rho g)/\phi.$$

Using the calculus relation employed in *Textbox A*,

$$\frac{d(\ln x)}{dx} = \frac{1}{x} \quad \text{or} \quad d(\ln x) = \frac{dx}{x}$$

(which says that small changes in the natural logarithm of a variable are equivalent to the fractional change in the variable itself), we can replace $d\rho/\rho$ with $d(\ln \rho)$, allowing the Adams-Williamson equation to be written

$$\frac{d}{dz}\ln(\rho) = g/\phi.$$

Instead of determining density versus depth within the troposphere, it makes more sense to determine it versus height. In terms of height h, hydrostatic equilibrium would have been written $dP/dh = -\rho g$, with the minus sign reflecting the fact that pressure decreases with height, and leading to

$$\frac{d}{dh}\ln(\rho) = -g/\phi$$

for the first Adams-Williamson equation.

Since the troposphere is only $\sim 12\,$km thick, we can treat gravity as being constant; that eliminates the need for the second Adams-Williamson equation. And, measurements find that the speed of sound ($\sqrt{\phi}$) decreases very nearly linearly through the troposphere, from $\sim 340\,$m/sec at Earth's surface ($h = 0$) to $\sim 300\,$m/sec at $h = 10\,$km (Dubin et al., 1962); thus,

$$\phi = (0.340 - 0.004h)^2$$

where h is in km and ϕ is km^2/sec^2. In this situation, we can solve the Adams-Williamson equation analytically, by simple integration. We have

$$\frac{d}{dh}\ln(\rho) = \frac{-0.01}{(0.340 - 0.004h)^2},$$

where $g \approx 0.01\,km/sec^2$, and this equation is in units of 1/km. The quantity in need of integration is of the form $(A - Bh)^{-2}$. You should be able to verify (e.g., by differentiating the following) that

$$\ln(\rho) = \frac{-25}{3.40 - 0.04h} + c,$$

where c is a constant of integration. Our final result is

$$\rho = \hat{c} e^{-25/(3.40 - 0.04h)},$$

with $\hat{c}$ a related constant of integration. As an example, the density at $h = 10$ km above sea level is smaller than at sea level in the ratio $\exp(-25/3) : \exp(-25/3.4)$ or $0.375 : 1$.

The **scale height** is defined as the distance over which atmospheric density decreases by a factor of e (or over which $\ln(\rho)$ is reduced by 1). With our inferred density profile, the implied scale height is 10.2 km.

Other Approaches Are Possible. Philosophically, our application of Adams-Williamson to the atmosphere might seem incongruous because our analysis did not explicitly deal with temperature—perhaps conveying an impression that temperature is not important for determining the variation in air density. In contrast to the solid earth, though, where we declared that temperature effects were mostly secondary to those of pressure, temperature affects atmospheric density quite strongly. This is illustrated by the *ideal gas law*, which may be most familiar in the form $PV = nR'T$ but which we will write as $P = \rho RT$ or

$$\rho = P/RT.$$

Such a law is typically invoked for individual gases (with R' being the proportionality constant between P, V, and T for that gas), but (as detailed in Haltiner & Martin, 1957) it can also apply to the mix of gases comprising the atmosphere if R is a weighted combination of the R' constants for the various component gases.

To determine the variation of air density with height such that temperature is explicitly accounted for, we might proceed by differentiating the atmospheric ideal gas law:

$$\frac{d}{dh}(\rho) = \frac{d}{dh}(P/RT);$$

or, applying the "product rule" from calculus,

$$\frac{d}{dh}(\rho) = \frac{1}{RT}\frac{d}{dh}(P) + P\frac{d}{dh}(1/RT)$$
$$= \frac{1}{RT}\frac{d}{dh}(P) - \frac{P}{RT^2}\frac{d}{dh}(T)$$
$$= \frac{-\rho g}{RT} - \frac{\rho}{T}\frac{d}{dh}(T).$$

In this last line, we incorporated hydrostatic equilibrium into the first term on the right, as we did earlier, and simplified the second term by using the definition of the ideal gas law. Dividing both sides by ρ, we find

$$\frac{d}{dh}\ln(\rho) = \frac{-g}{RT} - \frac{d}{dh}\ln(T).$$

This is reminiscent of our derivation of the full Adams-Williamson equations, with one term accounting for hydrostatic pressure effects on density and the other accounting for density changes caused by temperature gradients. The present derivation reveals an explicit dependence on T (in the first term on the right); but it also shows (in the second term) a contribution to density which requires specifying the temperature gradient (adiabatic gradient, *if* we assume an adiabatic atmosphere). Such a gradient is formulated in Chapter 9.

Incidentally, the ideal gas law provides a useful illustration of partial differentiation. The thermal expansion of the atmosphere is, by definition,

$$\alpha = -\frac{1}{\rho}\left(\frac{\partial \rho}{\partial T}\right)_P.$$

With P treated as constant and $\rho = P/RT$ (so ρ equals a constant, P/R, times $1/T$), the derivative with respect to T of ρ is simply

$$\left(\frac{\partial \rho}{\partial T}\right)_P = (P/R)\cdot\frac{d}{dT}(T^{-1}) = (P/R)\cdot(-T^{-2})$$
$$= \frac{-P}{RT}\frac{1}{T} = -\rho/T.$$

In short, for the atmosphere,

$$\alpha = 1/T$$

(for actual numerical values of α, remember that—as with any thermodynamic relation—T must be specified in absolute terms, in degrees Kelvin; also, neither the ideal gas law nor this relation holds true when T approaches absolute zero). Interestingly, this result (which is related to *Charles' Law*) implies that the troposphere's thermal expansion coefficient increases with height.

Solutions to the Adams-Williamson Equations in the Interior of the Earth. Obtaining a solution in this case will be more challenging than our application of Adams-Williamson was to the troposphere. Rather than a thin layer, in which gravity could be treated as a constant, the interior is deep enough that g can be expected to vary (and must indeed decrease to zero at Earth's center); so, the second of the Adams-Williamson equations must be solved simultaneously with the first. Additionally, given the complexities of the seismic velocity structure of the interior (Figure 8.12), a single formula for ϕ—linear, quadratic, or otherwise—would not accurately describe its variations with depth, so a simple analytical integration of the equations will not be possible.

Bullen (1936) was the first to employ more-or-less modern seismic velocity models as he attempted to solve the Adams-Williamson equations and determine the densities within the Earth. He focused first on the mantle—after all, the core likely has a different composition than the mantle, and Adams-Williamson was derived assuming that density changes were not the result of changes in composition. Bullen specified a starting density for the top of the mantle and used the values of ϕ, known from seismic velocity models, to solve the simplified version of the equations.

As a way to assess his density solution, Bullen calculated the mass and moment of inertia it implied for the mantle; subtracting these from the known mass and moment of inertia of the Earth, he then deduced a mass, M_C, and moment of inertia, I_C, for the Earth's core. The numbers he found for I_C and for $M_C(R_C)^2$, the core mass times the core radius (R_C) squared, were in the ratio 0.57; that is, his solution implied that

$$I_C = 0.57 M_C (R_C)^2.$$

But such a core is not physically plausible! As we saw in Chapter 1, in a sphere whose moment of inertia exceeds 0.4 times its MR^2 the density must *decrease* with depth. Bullen's solution implied that the core is top-heavy. This is unlikely for several reasons:

- Under the effect of pressure, the core's density should increase with depth.
- P wave velocity does in fact increase with depth inside the core.
- The outer core is fluid, and fluids cannot persist with denser material overlying lighter stuff (they will simply overturn).

Bullen was forced to conclude that his application of the Adams-Williamson equations was faulty—that at least one of the assumptions underlying the equations does not apply to the mantle.

Of all the assumptions, hydrostatic equilibrium is the most obvious to question; after all, the mantle is convecting, and thus not static. But the convective motions are slow, and the pressures driving them are weak. As we will discover from a homework exercise following Chapter 9, such dynamic pressures are orders of magnitude smaller than the hydrostatic pressure; thus hydrostatic equilibrium turns out to be an excellent approximation for the mantle.

An assumption made questionable by our ignorance of actual temperatures within the Earth is that the temperature gradient within the mantle is adiabatic. Could nonadiabatic temperatures change the mantle density profile enough to avoid a top-heavy core? We can use the full Adams-Williamson equations to answer this question. In those equations, the effect of nonadiabatic temperatures on the density gradient ($d\rho/dz$) is given by the term

$$-\alpha\rho\tau.$$

The nonadiabatic change in density incurred over a depth range Δz is then approximately

$$-\alpha\rho\tau\,\Delta z.$$

Laboratory measurements and theoretical modeling for mantle-type rocks imply that the thermal expansion coefficient α decreases with depth through the mantle, from $\sim 3 \times 10^{-5}$

per °C at the top to ~1 × 10^{-5} per °C at the base (these bounds, from e.g., Skinner, 1966 as reported in Robertson, 1988, and Stacey & Davis, 2008, are larger than the values cited earlier). At the same time, mantle density increases from 3.3 gm/cm^3 to 5.2 gm/cm^3. Thus $\alpha\rho$ ranges from ~9.9 × 10^{-5} to ~5.2 × 10^{-5} (in cgs units); for a conservative, maximum effect, we will use the larger value for $\alpha\rho$ throughout the mantle. Consequently, the most we might expect nonadiabatic temperatures to modify our mantle density estimates, from the top to the bottom of the mantle ($\Delta z \cong$ 2890 km), would be

$$(0.286\,\tau)\ \text{gm/cm}^3$$

if τ is in units of °C/km.

The adiabatic gradient will be discussed in the next chapter. We will discover that, in most of the mantle, its magnitude would be estimated as ~0.5 °C/km or less; however, because the actual mantle temperatures are poorly known, their departure from adiabaticity is, as well. Continuing our conservative upper bound estimation, then, we will consider a departure of the actual temperature gradient from adiabatic as large as 100%—an extreme possibility! In this case τ = 0.5 °C/km and, from (0.286 τ), the maximum possible contribution to mantle densities would be ~0.14 gm/cm^3. This is a small modification, and the modifications to densities at shallower depths in the mantle would be even less. Bullen (1965) used more rigorous arguments than ours to find the upper bound on nonadiabatic effects was ~0.07 gm/cm^3. We conclude that, although temperatures within the mantle may differ significantly from adiabatic values, the effects of those temperatures on density are not appreciable—α is just too small. Thus, it is unlikely that relaxing the adiabatic assumption would eliminate an implied top-heavy core.

It seems that the assumptions behind Adams-Williamson most likely responsible for implying the core to be top-heavy are uniform composition and uniform phase within the mantle. That is, *the mantle is either not uniform in composition or exhibits a change in*

phase, or both; the resulting effects on density somehow produce acceptable mantle and core moments of inertia. This strong conclusion serves to remind us that one reason theoretical approaches such as the Adams-Williamson equations are powerful is their ability to test the validity of their underlying assumptions.

Salvaging the Adams-Williamson Equations. It turns out that the Adams-Williamson equations are poorly applied to the mantle only within specific layers. A glance at the mantle's seismic velocity profiles (Figure 8.12) suggests where those layers might be: though V_P and V_S vary throughout the mantle, they vary quite differently in the transition zone (and, to a lesser extent, in the LVZ and D″)—as if they are reflecting more than just the effects of pressure and temperature.

Those layers that depart from material homogeneity (compositional and phase uniformity) can be identified more rigorously with the help of an 'indicator' of material *in*homogeneity: the quantity

$$\frac{dk_S}{dP}$$

(Birch, 1952), which measures the increase in adiabatic incompressibility as we descend deeper into the mantle and pressure increases.

- As mantle rock experiences greater pressures, we expect that it will resist deformation more—its incompressibility will increase—in an approximately proportional manner; that is, dk_S/dP, the rate of increase of k_S, should be roughly constant. This possibility was considered earlier, in *Textbox A*, as a simple basis for deriving a power-law version of Birch's rule.

- Evidently, however, as we descend through the mantle, the failure of Adams-Williamson implies that we will encounter either material of a different composition or the same material but in a different phase. In either case, that material might have a fundamentally different kind of resistance to deformation than the shallower material. In those layers, dk_S/dP might well exhibit a different value than in the shallower layers.

The value of the indicator can be calculated approximately from our knowledge of seismic velocities. First of all, from the "chain rule" of calculus, we have

$$\frac{dk_S}{dP} = \frac{dk_S}{dz}\frac{dz}{dP} = \frac{1}{\rho g}\frac{dk_S}{dz},$$

with hydrostatic equilibrium also applied. By definition, $k_S = \rho\phi$, where ϕ is the seismic parameter; using the "product rule" for taking derivatives,

$$\frac{dk_S}{dz} = \frac{d(\rho\phi)}{dz} = \rho\frac{d\phi}{dz} + \phi\frac{d\rho}{dz}.$$

But the full Adams-Williamson equations tell us that

$$\frac{d\rho}{dz} = (\rho g)/\phi - \alpha\rho\tau;$$

substituting all of this in yields

$$\begin{aligned}
\frac{dk_S}{dP} &= \frac{1}{\rho g}\frac{dk_S}{dz} = \frac{1}{\rho g}\left(\rho\frac{d\phi}{dz} + \phi\frac{d\rho}{dz}\right) \\
&= \frac{1}{g}\frac{d\phi}{dz} + \frac{\phi}{\rho g}\left(\frac{\rho g}{\phi} - \alpha\rho\tau\right) \\
&= \frac{1}{g}\frac{d\phi}{dz} + 1 - \phi\alpha\tau/g.
\end{aligned}$$

Finally, we know that nonadiabatic effects on density will be small, so the τ term can be neglected, leading to

$$\frac{dk_S}{dP} \cong \frac{1}{g}\frac{d\phi}{dz} + 1.$$

As we will discover, gravity is roughly constant throughout the mantle ($g \sim 10\,\text{m/sec}^2$), so our indicator can be calculated solely from observable quantities: seismic velocities and their variation with depth.

Using actual seismic velocities within the mantle from Kennett and Engdahl (1991) or Dziewonski and Anderson (1981), we find that $d\phi/dz$ and thus dk_S/dP are indeed very nearly constant through much of the mantle. Given the velocity profiles in Figure 8.12, such a conclusion is not surprising. Regions of implied material homogeneity (constant dk_S/dP) include the upper mantle right above the transition zone, the mantle between the transitions, and the bulk of the lower mantle (i.e., down to just above D″).

However, the inferred constant value of dk_S/dP differs from region to region; see Table 8.3. Thus, the mantle within each region may be materially homogeneous, but the regions differ from each other by either phase or composition. By implication, processes must be taking place within the transition zone that produces the different composition or phase we infer to exist above, below, and in-between the zone. *It is the transition zone (more specifically, the transitions themselves) where the assumptions of Adams-Williamson significantly fail.*

For a vivid appreciation of the significance of the transition zone, imagine plotting dk_S/dP throughout the whole mantle. The regions noted in Table 8.3 would exhibit nearly constant dk_S/dP values; in contrast, the sharp jumps in V_P and V_S at the first and second transitions (cf. Figure 8.12) would lead to huge values of $d\phi/dz$ (—infinite values, if the transitions are true discontinuities!) and thus highly unrealistic values of dk_S/dP. Such values imply that our equation connecting

Table 8.3 Tests for material homogeneity in the mantle.[a]

Region	Based on iasp91		Based on PREM	
	Depth range[b]	$\dfrac{dk_S}{dP}$ [c]	Depth range[b]	$\dfrac{dk_S}{dP}$ [c]
Upper mantle (below LVZ)	~170–370 km	~5.1	~220–350 km	~3.3
Between first and second transitions	~410–660 km	~4.6	~450–600 km	~6.9
Lower mantle	~770–2670 km	~3.3	~770–2740 km	~3.3

[a]Based on constancy of slope of ϕ versus z curve; dk_S/dP calculated from that slope according to the text. Seismic velocity model iasp91 is from Kennett & Engdahl (1991); PREM is from Dziewonski & Anderson (1981).
[b]Depth ranges chosen for this analysis; within each range, dk_S/dP is nearly constant.
[c]Disparities between the two velocity models in their values of dk_S/dP above and within the transition zone may partly result from their different treatments of the LVZ.

dk_S/dP to $d\phi/dz$—that is, our use of the Adams-Williamson equations—must be inapplicable *at* the first and second transitions. As already discussed, the reason is a lack of material homogeneity, which we can now see occurs at the transitions.

The arguments presented here revealing the first and second transitions to be regions of material inhomogeneity could also be applied to the LVZ and D″, suggesting that they are materially inhomogeneous as well (though, on average, to a (much) lesser extent than the transition zone, according to their seismic velocity profiles or $d\phi/dz$).

Physically, the limitations of applying the Adams-Williamson equations to the mantle can now be understood. Those equations predicted how density should increase with depth but failed to account for additional, 'inhomogeneous' density jumps within the transition zone; they failed to predict extra increases in density because they related density only to seismic velocity (i.e., incompressibility) variations and not to composition or phase. As a result, the solution to the Adams-Williamson equations yielded densities that were erroneously small, thus a mantle whose mass was too small, thus implying the core had 'extra' mass, as high up and close to the lower mantle (where it belonged, according to Earth's moment of inertia) as possible.

With the 'faulty' region now identified, we could return to the Adams-Williamson equations, solving them throughout the mantle except for the transition zone; in the latter region, we could simply 'fudge' a correction (as Bullen did, using a quadratic curve to model the transition-zone density), adjusting it until it produced a reasonable mantle mass—or, even better, simply incorporate another density predictor there, such as Birch's rule (see *Textbox B*).

Textbox B: A Numerical Approach to Adams-Williamson

We will be solving the Adams-Williamson equations for ρ and g using values of the seismic parameter ϕ, i.e., the seismic velocities V_P and V_S, which are specified in the PREM and iasp91 models at various depths. The first step in solving the equations computationally is to set up the Earth model in discrete form—this will allow us to turn the d's of our equations (as in $d\rho/dz$) into delta's (as in $\Delta\rho/\Delta z$).

The Setup

So, let's say there are $N + 1$ values of the seismic parameter—that is, $N + 1$ layers within the Earth where its magnitude is known; we will denote those values by ϕ_i, $i = 0, 1, \ldots, N$:

$$\phi_0, \phi_1, \ldots, \phi_N,$$

with $i = 0$ corresponding to the top layer, $i = 1$ the next layer down, and so on, down to the central layer $i = N$. In other words, the top layer is "layer 0," the next layer is "layer 1," ..., and the central layer is "layer N." And, let's say that "layer i," bounded by upper and lower radii r_i and r_{i+1}, has a thickness Δz_i:

$$\Delta z_0 = r_0 - r_1$$
$$\Delta z_1 = r_1 - r_2,$$
$$\Delta z_2 = r_2 - r_3,$$
$$\ldots$$
$$\Delta z_N = r_N - r_{N+1} \;(= r_N, \text{ since } r_{N+1} = 0).$$

In general, we can write

$$\Delta z_i = r_i - r_{i+1};$$

Textbox B: A Numerical Approach to Adams-Williamson (*Continued*)

see the figure below. These $N + 1$ layer thicknesses are also specified in the seismic model.

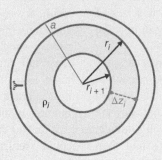

View of "layer i," exaggerated for visibility, defined by outer and inner radii r_i and r_{i+1}, and thickness Δz_i. Earth's outer radius is a. The density within the layer is ρ_i, and the gravity experienced by the weightlifter (in red) standing at the top of the layer is g_i.

If we let ρ_i, $i = 0, 1, ..., N$ denote the density of "layer i" and g_i, $i = 0, 1, ..., N$ denote the gravity at the *top* of that layer, then our goal is to determine the values of ρ_i and g_i for $i = 1, ... , N$. We already know the density ρ_0 of Earth's outermost layer and the gravity g_0 at Earth's surface—which is why we numbered the layers beginning with $i = 0$ rather than $i = 1$. ρ_0 and g_0 are the "starting values" we will use to obtain our solution.

Now let's recast the Adams-Williamson equations in discrete form. These equations were originally

$$\frac{d\rho}{dz} = (\rho g)/\phi \quad and \quad g = \frac{GM(z)}{(a - z)^2},$$

where a is Earth's outer radius and

$$M(z) = \int_{r=0}^{r=a-z} 4\pi r^2 \rho dr$$

is the mass of the Earth below depth z. The first of these equations could also be written

$$\Delta\rho = \frac{d\rho}{dz}\Delta z = [(\rho g)/\phi]\Delta z.$$

Discretized, the equations become

$$\Delta\rho_i = (\rho_i g_i/\phi_i)\Delta z_i \quad and \quad g_i = GM(z_i)/r_i^2,$$

where $\Delta\rho_i = \rho_{i+1} - \rho_i$ is the density increment from "layer i" to "layer $i+1$" (thus $\rho_{i+1} = \rho_i + \Delta\rho_i$), and where $(a - z_i) = r_i$. Note that z_i (= depth) is not the same as Δz_i (= thickness). The role of the indices will become clearer as we develop the algorithm below.

We also need a discrete expression for $M(z_i)$, the total mass below depth z_i. As noted in the text, the relatively low resolution of the PREM and iasp91 models means the mass of each layer should not be approximated; thus, denoting the mass of "layer i" by LM_i, we specify

$$LM_i = \frac{4}{3}\pi (r_i^3 - r_{i+1}^3)\rho_i.$$

The combined mass of all the layers must equal the mass of the model Earth, which we will denote by M_E whether it's the mass of the entire Earth or the Earth minus the crust:

$$M_E = LM_0 + LM_1 + LM_2 + \ ... \ + LM_N.$$

Textbox B: A Numerical Approach to Adams-Williamson (*Continued*)

The mass $M(z_i)$ below depth z_i—which we can equivalently write as $M(r_i)$, the mass within radius r_i—is the mass of the Earth excluding the layers above that depth or radius. Since we will be solving the equations for density and gravity from the top down, the mass of those upper layers will have already been determined. In descending order, then, the mass below Earth's surface (or below the Moho, if we are excluding the crust) is

$$M(r_0) = M_E$$

with r_0 the same as a; the mass below "layer 0," i.e. within radius r_1, is

$$M(r_1) = M_E - LM_0 = M(r_0) - LM_0;$$

the mass below "layer 1," i.e. within radius r_2, is

$$M(r_2) = M(r_1) - LM_1;$$

the mass below "layer 2," i.e. within radius r_3, is

$$M(r_3) = M(r_2) - LM_2;$$

and so on. In general, we have

$$M(r_i) = M(r_{i-1}) - LM_{i-1}$$

—that is, the total mass below any layer equals the mass below the previous, higher layer minus the mass of that layer itself. We will write this general expression as

$$M(r_i) = M(r_{i-1}) - \frac{4}{3}\pi (r_{i-1}^3 - r_i^3)\rho_{i-1}.$$

The Solution, Step-by-Step

To develop the algorithm outlined in the text, we first generate the solution for the top few layers.

$i = 0$

"Layer 0," the top layer ($i = 0$), is set: ρ_0 and g_0 are already specified, as our starting values, with $g_0 = GM_E/a^2$ to insure that our value of surface gravity is consistent with the mass of the Earth. Using the first discretized Adams-Williamson equation, the values of ρ_0 and g_0 imply

$$\Delta\rho_0 = (\rho_0 g_0/\phi_0)\Delta z_0$$

thus

$$\rho_1 = \rho_0 + (\rho_0 g_0/\phi_0)\Delta z_0,$$

yielding the density of "layer 1," the second layer.

For later purposes, we point out that g_0 can also be written

$$g_0 = GM(r_0)/r_0^2.$$

Setting the stage for the second layer, we note that the value of ρ_0 implies that the mass of "layer 0" is

$$LM_0 = \frac{4}{3}\pi (r_0^3 - r_1^3)\rho_0,$$

so the mass $M(r_1)$ below "layer 0" is

$$M(r_1) = M_E - LM_0 = M(r_0) - LM_0.$$

Textbox B: A Numerical Approach to Adams-Williamson (*Continued*)

$i = 1$

From the preceding, and in line with the second discretized Adams-Williamson equation, the value of gravity at the top of "layer 1" ($i = 1$) is

$$g_1 = GM(r_1) / r_1^2.$$

Using the values for ρ_1 and g_1, Adams-Williamson now leads to

$$\Delta\rho_1 = (\rho_1 g_1/\phi_1)\Delta z_1$$

or

$$\rho_2 = \rho_1 + (\rho_1 g_1/\phi_1)\Delta z_1,$$

for the density of "layer 2" ($i = 2$), the third layer.

Setting the stage for that third layer, we note that the value of ρ_1 implies that the mass of "layer 1" is

$$LM_1 = \frac{4}{3}\pi(r_1^3 - r_2^3)\rho_1,$$

which in turn implies that the mass $M(r_2)$ below "layer 1" is

$$M(r_2) = M(r_1) - LM_1.$$

$i = 2$

The value of gravity at the top of "layer 2" ($i = 2$) is

$$g_2 = GM(r_2)/r_2^2.$$

For this layer, then,

$$\Delta\rho_2 = (\rho_2 g_2/\phi_2)\Delta z_2,$$

yielding

$$\rho_3 = \rho_2 + (\rho_2 g_2/\phi_2)\Delta z_2$$

for the density of "layer 3" ($i = 3$). Also, the mass of "layer 2" must be

$$LM_2 = \frac{4}{3}\pi(r_2^3 - r_3^3)\rho_2,$$

implying that the mass $M(r_3)$ below "layer 2" is

$$M(r_3) = M(r_2) - LM_2.$$

"layer i"

The pattern underlying the algorithm should now be apparent. We phrase the algorithm in terms of two sequential layers, labeled "layer $i-1$" above and "layer i" below. The upper and lower radii of "layer i" are r_i and r_{i+1}. The properties of the upper layer—its gravity, density, and mass—have just been computed; we denote these by g_{i-1}, ρ_{i-1}, and LM_{i-1}. From those quantities, we compute the density increment $\Delta\rho_{i-1}$ from "layer $i-1$" to "layer i":

$$\Delta\rho_{i-1} = (\rho_{i-1} g_{i-1}/\phi_{i-1})\Delta z_{i-1}$$

yielding the density ρ_i of "layer i,"

$$\rho_i = \rho_{i-1} + \Delta\rho_{i-1} = \rho_{i-1} + (\rho_{i-1} g_{i-1}/\phi_{i-1})\Delta z_{i-1}.$$

Textbox B: A Numerical Approach to Adams-Williamson (*Concluded*)

The algorithm will guide us in using these computed quantities to calculate the corresponding quantities of the next layer. Very concisely, we have

$$M(r_i) = M(r_{i-1}) - LM_{i-1},$$

$$g_i = GM(r_i)/(r_i)^2,$$

thus

$$\Delta\rho_i = (\rho_i g_i / \phi_i)\Delta z_i$$

so

$$\rho_{i+1} = \rho_i + (\rho_i g_i / \phi_i)\Delta z_i,$$

allowing us to continue the process down to the next layer (beginning with the calculation of LM_i and $M(r_{i+1})$).

Final Comments

Though straightforward, this would be a tedious and intimidating set of calculations, indeed, if the number of layers is large and we have to do the calculations by hand; programmed for a computer, though, the process is quite routine and extremely quick. Given our interpretation of the results of using Adams-Williamson in the mantle, however, it would be advisable to interrupt the computations for the depths within the transition zone, and replace them with other determinations of density and gravity (such as from Birch's rule).

The process we have just described can be called a "forward" calculation: we begin with values of density and gravity for $i = 0$ and successively compute such values for $i = 1, 2, \ldots,$ and finally N. The values we determine are likely to be reasonable, that is, not excessively large (e.g., 10^6 gm/cm^3) or small (e.g., 10^{-6} gm/cm^3); but how do we know if the solution is accurate?

If we are interested only in the mantle's density, then Bullen's experience (as discussed in the text) provides an important guide for our course of action; we will return to this point quantitatively in the text below. However—with all the effort we have devoted in this textbox and this chapter to Adams-Williamson—we can expand our vision and make our goal the determination of both mantle *and* core density. The assumptions underlying the Adams-Williamson equations (hydrostatic equilibrium, adiabatic temperatures, and no changes in phase) are likely somewhat worse approximations for the core than for the mantle, so we will have to treat this expanded view as a tentative exploration.

However, there are clear differences in composition between core and mantle, so the equations will have to be applied separately to each region. That means we will need a starting value for the core's density—the density of the top layer of the core, thus a third starting value to specify. On the other hand, we do not need to specify gravity at the top of the core: since that gravity depends on the mass within the core, its value is already determined by how much mass is left over from the mantle's solution.

Returning to the question of accuracy, we note that our 'whole-Earth' solution will have to obey two global constraints: that it conserves mass; and, guided by Bullen, that it is consistent with Earth's moment of inertia. Given the way the algorithm works, with M_E specified (presumably accurately) from the start and mass removed layer by layer, the test of mass conservation simply amounts to insuring that, at the end, the central layer is left with a reasonable amount of mass, i.e., that M_N is neither negative nor excessively positive. A similar test will apply for the inertia constraint.

The Physical State of Earth's Interior: Models and Modeling. Figure 8.18a (left-hand image) shows the PREM model (Dziewonski & Anderson, 1981) inference of density within Earth's interior, determined by a data 'inversion' consistent with observed normal-mode frequencies and other seismic data. The inversion process simultaneously yielded the elastic moduli k and μ, shown in Figure 8.18b. With density known, gravity within the interior can also be calculated, using the expression for g ($= GM(z)/(a-z)^2$) we derived as part of the Adams-Williamson equations; and, lastly, hydrostatic equilibrium then yields the pressure within the Earth (see Figure 8.18c for P and g).

We have also obtained a density profile for Earth's interior by solving the Adams-Williamson equations using the approach presented in *Textbox B*. Our solution (shown in Figure 8.18a, right-hand image) excluded the crust, so the starting value of 'mass below.' M_E, was set to be smaller than Earth's total mass by the amount specified in Peterson and dePaolo (2007) for the crust. With this solution involving both the mantle and core, whose compositions evidently differ, the assumption of chemical homogeneity obliged us to solve the Adams-Williamson equations separately for those regions. This in turn meant three starting values had to be specified. To generate the mantle solution, we began with the density at the top of the mantle, ρ_0, and the net mass M_E of the mantle + core, or equivalently the gravity at the top of the mantle. We obtained the mantle solution first so that, by subtraction, we would know how much mass the core is required to have (or equivalently, what gravity at the top of the core would be). Then, for the core solution, we had to specify a starting density, ρ_{C0}, the density at the top of the core.

As the solution proceeded downward and the density of each layer was determined, we could subtract off the mass of that layer, allowing the mass below it to be computed. Since the procedure started with M_E, we know that mass will have been conserved if the remaining mass left for the final, central layer of the core was neither negative (resulting from the model having too much mass above the final layer) nor excessively positive (if the mass above that layer was too little).

At the same time, the moment of inertia of each layer could be computed—the inertia of a

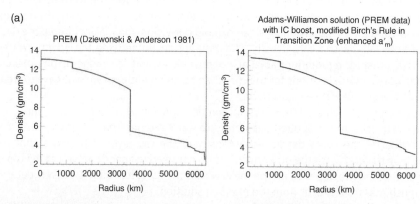

Figure 8.18 Physical state of Earth's interior.
(a) Density profile in gm/cm³:
(Left) based mainly on normal mode data, from PREM (Dziewonski & Anderson, 1981);
(Right) solution to Adams-Williamson equations, with ϕ from PREM. As described in the text, our solution excludes the crust; uses a modified Birch's rule in the transition zone to obtain density; and adds a 0.5 gm/cm³ boost to the density at the inner-core boundary, within the solution process.
Both solutions satisfy the constraints of Earth's mass and moment of inertia, with values as specified in Dziewonski and Anderson (1981) / with permission of Elsevier.

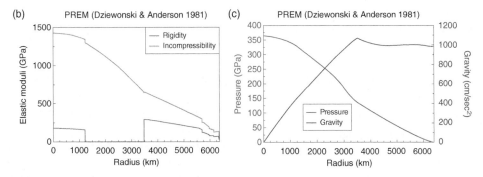

Figure 8.18 Physical state of Earth's interior, based mainly on normal mode data, from PREM. Dziewonski & Anderson (1981) / with permission of Elsevier.
(b) Profile of elastic moduli k, incompressibility, and μ, shear rigidity, in GPa.
(c) Profile of gravity, in cm/sec^2, and hydrostatic pressure, in GPa.

homogeneous sphere is

$$\frac{2}{5}Mr^2 \quad \text{or} \quad \frac{8}{15}\pi\rho r^5,$$

so the inertia of a homogeneous shell of density ρ is

$$\frac{8}{15}\pi\rho(r_{upper}^5 - r_{lower}^5)$$

—and subtracted from the known inertia of the whole Earth (minus crust), which we will denote I_E. We will know that the moment of inertia constraint has been satisfied if the remaining inertia left for the final, central layer of the core was neither negative nor excessively positive. It should be noted that the PREM density model was also constrained to yield the observed mass and moment of inertia of the Earth.

With Adams-Williamson unable to deal with phase transitions, our downward-progressing solution was computed instead from Birch's rule, at all depths within the transition zone. With the possibility that the inner core is different in composition as well as phase from the outer core, however, the density jump at the inner-core boundary was simply represented by an arbitrary 0.5 gm/cm^3 boost in density (comparable to that in PREM; but see also Masters & Gubbins, 2003) at that point in the algorithm.

Our 'hybrid' Adams-Williamson/Birch solution is visually quite similar to the full PREM solution, with a more subdued transition zone

(especially the second transition) being its most obvious difference. This result held true whether we used the original or the modified version of Birch's rule. Looking closer, we see that our hybrid solution also exhibits a slightly greater rate of increase of density, and thus higher actual densities, throughout the transition zone. With a more massive transition zone, it is no wonder that our solution—constrained by M_E—requires that second transition to be muted.

Overall, the density profiles of both models show the same structure as the seismic velocities (Figure 8.12). The elastic moduli more or less show that structure as well; interestingly—though it's likely pure coincidence—the incompressibility of mantle silicates near the base of the mantle has a magnitude essentially equal to that of the fluid iron alloy at the top of the core.

Even more intriguing, gravity at all depths within the mantle—a function both of the mass below the given depth and, inversely, the squared distance from that depth to Earth's center, as discussed earlier—remains roughly constant; throughout the mantle, those competing effects just happen to cancel out, to decent approximation. Such a canceling out requires a particular (and amazingly simple!) density distribution versus depth through the mantle (this is explored in a homework problem for this chapter). Finally, we note that the profiles of gravity and pressure are much smoother than those of density and elastic moduli—a consequence in part of those quantities being the *integrated* effects of density (for gravity) or of density and incompressibility (for pressure).

8.3.3 Seismic Tomography

The technique of acoustic tomography was born over four decades ago (see Rawlinson et al., 2010 for a comprehensive review), with Bois et al. (1971) and Aki and Lee (1976) among the pioneers for local applications and Dziewonski et al. (1977) helping to pave the way for global studies. The technique incorporates millions of travel-time observations around the globe, based on seismic wave arrivals from earthquakes worldwide, to identify anomalous regions. Initially, the focus was on P waves, whose travel times are generally the most reliably determined of all seismic waves (*can you explain why this is so?*) and thus also the most abundantly recorded. However, as first arrivals (*!*), P waves are likely to have bypassed slower-velocity regions along their path; consequently, a good global distribution of P waves 'flooding' the interior is needed in order to delineate the slow regions—or, alternatively, other types of waves should be included in the analysis. Lei and Zhao (2006), for example, explored the use of pP, PP, PcP, and Pdiff, and even core phases such as PKP; with this approach, they were able to identify the seismically slow structure of the Hawaiian plume extending clearly and continuously down to the core-mantle boundary (see Figure 8.19).

Other innovations over the years have included using S wave and surface wave arrival times; using the full waveform of the phase rather than its arrival time; and focusing on the wave amplitude. Because the energy of a seismic wave can be especially attenuated in hot regions, amplitude data might be particularly valuable for establishing the presence of hot and cold regions in the mantle, such as those associated with mantle convection.

Rawlinson et al. (2010) discuss these and other, more exotic data types that have been employed tomographically, such as microseisms and other 'ambient seismic noise' produced by the oceans and atmosphere; ambient noise tomography has proved useful for resolving fine crustal and subcrustal features (e.g., Shapiro et al., 2005).

Innovations involving the method of analysis include alternatives to least squares and alternatives to volumetric cubes—for example, using spherical harmonics to describe the variations in seismic velocity rather than a simple cube-to-cube description. But all these innovations are *tomographic* techniques because they delineate three-dimensional seismic structures within the Earth.

In fact, a multifaceted approach to tomography—whether using multiple data types or multiple methodologies—is probably advisable. Consider, for example, the use of least squares to fit a straight line to a set of data points: we know that the resulting line is likely to fit only a small portion of the data exactly; the remaining data will typically divide into similar portions above and below the line (which is an attractive feature of least squares, in most contexts!). To some extent, seismic tomography might analogously be expected to delineate roughly equal numbers of voxels (e.g.,) with faster and slower velocities than the reference model, regardless of their physical significance. Any correlations between the fast or slow regions and geological features (such as hotspots) or geophysical features (such as geoidal highs and lows, e.g., Hager et al., 1985; Lees & VanDecar, 1991; Li & Romanowicz, 1996) will therefore add confidence to the tomographic results—as will the use of multiple data (several types of P *and* S waves, for example) or multiple analyses.

Below the lithosphere, the structures revealed by global tomographic studies are typically characterized by seismic velocities that differ by $\sim\pm1\%$ to $\pm2\%$ from the initial model (some researchers find greater differences, especially for S waves and especially near the top and bottom of the mantle); such a

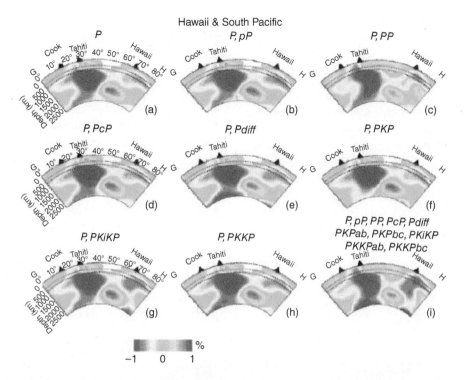

Figure 8.19 Tomographic delineation of the mantle plume beneath Hawaii; cross-sections shown are south-to-north, through the central Pacific basin.

Note that the existence of the Hawaiian plume between 650 km depth and the CMB becomes much more certain when numerous waves in addition to P are included. That multiwave analysis also shows that the greatest velocity anomaly of the Hawaiian plume (note the color code) is at the base of the mantle; interpreting the slowness to be the result of high temperature, the driving force of the plume is evidently in the thermal boundary layer D''.

This image, from Lei and Zhao (2006) / Elsevier, also shows the South Pacific "superplume," and implies that the superplume may interact to a limited extent, for example thermally in mid mantle, with the Hawaiian plume.

As a further complication, we note that Montelli et al. (2006), employing a high-resolution tomographic analysis, find that the South Pacific superplume actually consists of several distinct but nearby plumes, including the Tahiti/Cook Island, Samoa, Solomon Island, and sub-Coral Sea plumes.

magnitude for velocity anomalies is not unreasonable, since those differences stemmed from travel-time differences of several seconds, compared to travel times (used to produce the reference velocity model) of several minutes. With such muted velocity contrasts, the widespread use of bright or intense colors to depict tomographic structures (evident in Figure 8.19 and some other illustrations in this chapter) might not appear justified; but, as we saw in the case of the Yellowstone plume, in an application of Birch's rule, the temperature contrasts inferred for those structures may nevertheless be quite dramatic!

As noted previously, the existence of structures with nonzero velocity contrasts implies that waves traveling through the Earth will actually follow slightly different paths, corresponding to slightly different travel times from source to receiver, than predicted by the initial, one-dimensional velocity model. It turns out that the 'best' solution may be one that also allows for slightly different source locations than originally used. Tomographic studies generally include relocation of the earthquakes, and, since relocation changes the travel times and thus the travel-time residuals, the velocity anomalies must be recalculated as well. By the time the iterations are done, earthquake locations have typically changed by several kilometers (e.g., Vasco et al., 1994).

Focus on and Near the Upper Mantle. One result that emerged from early global tomographic studies was that differences in continental versus oceanic structure persist well below the lithosphere. Evidence supporting a lithospheric influence on the upper mantle includes anomalously fast seismic velocities under continental shields, to depths of perhaps ~400 km or so; and regions of anomalously slow velocity associated with back-arc basins and island-arc volcanism persisting down to ~200 km depth, ringing the Pacific and elsewhere (Figure 8.20a) (e.g., Inoue et al., 1990; Pulliam et al., 1993; Vasco et al., 1994; Su et al., 1994; Zhao, 2004).

On first thought, an influence of continental shields on the mantle immediately below should not be unexpected. Continental shields, being older and less tectonically active, have lost much of their heat; as we will learn in Chapter 9, it is reasonable to expect that coldness to have permeated down a couple of hundred kilometers below the lithosphere over the lifetime of the shield, leaving that sublithospheric layer denser, stiffer, and seismically faster. However, as noted by Jordan (1975) from an analysis of seismic but not tomographic data, those seismically faster regions are able to keep pace, more or less, with the drift of the lithospheric plates on Earth's surface. This is also implied by the tomographic images in Figure 8.20a. But the cooling effect of the shields proceeds by conduction of heat,

(a)

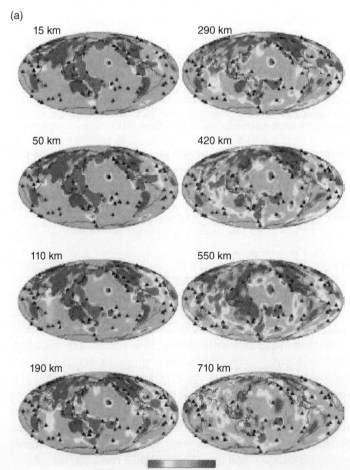

15 km

290 km

50 km

420 km

110 km

550 km

190 km

710 km

−1%   0%   1%

Figure 8.20 Global tomographic analysis, based on travel times of five types of *P* waves, reveals the influence of tectonics on the upper mantle.
(a) Maps of *P* wave velocity variations (see color code) at various depths, relative to global average at that depth. Seismically fast (blue) regions exist below continental shields, persisting to depths of at least 400 km (perhaps greater, though subducting slabs, also with higher *P* wave velocities, may obscure such shield influence at greater depths).
Note also the 'Ring of Warmth,' or at least 'ring of slow seismic velocity' (red), associated with the Ring of Fire around the Pacific basin, underlying back-arc basins and persisting down to ~200 km depth.
Images from Zhao (2004) / with permission of Elsevier.

(b)

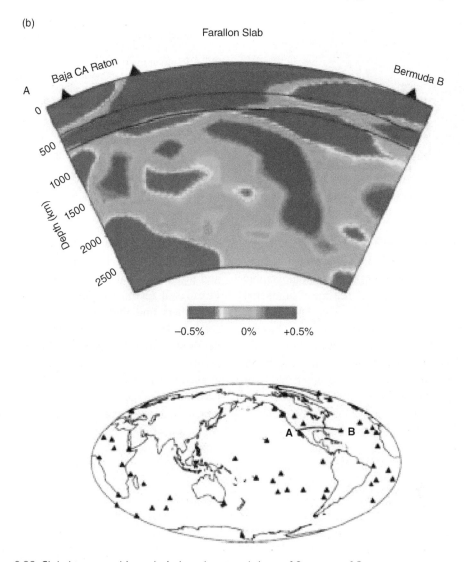

Figure 8.20 Global tomographic analysis, based on travel times of five types of *P* waves.
(b) Tomographic image of the Farallon lithospheric slab, roughly 100 Myr after subduction beneath western North America. Existence of the relict slab in mid to lower mantle based on relatively fast (blue) seismic velocities (see color code below cross section), presumably a reflection of cooler temperatures and a greater stiffness of the slab. Note continental shield effects on upper mantle (faster velocities, in blue) down to 660 km depth.
This cross-section is through mid-North America from west to east (A-B shown on world map below).
Image from Zhao (2004) / with permission of Elsevier.

a process we will discover in Chapter 9 to be quite slow: to effectively reach an additional 100 km depth, say, would take ~300 Myr—a rate two orders of magnitude slower than plate motions. Thus, the seismically faster regions are able to keep up thermally with the shields above them only because there are forces (e.g., viscous drag) that keep them moving *together*.

In effect the result is simply that old continental lithosphere, including now that colder and denser layer below the shield, is much thicker than otherwise thought. We have already inferred that continental plates are

(c)

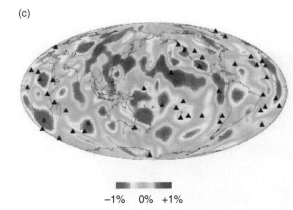

−1%   0%   +1%

difficult to subduct; double their thickness and their subduction becomes even more unlikely.

Another achievement of global tomographic studies is to shed light on the fate of subducted (oceanic) lithospheric slabs. Those studies have revealed that, even as they descend through the mantle and heat up, most slabs retain their identity to some extent into the lower mantle. For example, tomographic analyses delineate an extended region of fast velocities beneath eastern North America that can be identified as the subducted Farallon plate; according to some, the slab may penetrate nearly to the CMB (Grand et al., 1997; van der Hilst et al., 1997), though other studies find its detectable remnant terminates higher in the lower mantle (~2,300 km depth, Zhao, 2004 and Figure 8.20b; > 2,000 km depth, Pulliam et al., 1993; 1,600 km depth, Sigloch et al., 2008; > 1,000 km depth, Inoue et al., 1990). The Farallon slab alone does not tell us how convection works throughout the mantle; but it is one example among many that demonstrates (on a broader scale than plumes) that the lower mantle is not static (see also Silver et al., 1988; van der Hilst et al., 1997).

On closer inspection, however, we see that things are not so straightforward even for subducting oceanic lithosphere. The deep remnant Farallon slab appears to be decoupled from the upper mantle at ~650 km depth (also Megnin & Romanowicz, 2000 and Fukao et al., 2001; cf. Sigloch et al., 2008). Fukao et al. (2001) have found tomographic evidence of detachment of

the Tethys slab (below the Himalayas, a remnant of Indo-Australian subduction) as well, and just below that 650 km depth. Moreover, Fukao et al. (2001), van der Hilst et al. (1991), and others find that many subducting slabs seem to 'level off' as they approach 650 km depth; still other slabs penetrate into the lower mantle, and at quite steep an angle (e.g., Grand et al., 1997; see Goes et al., 2017). There may be some dependence of all these results on the type of tomographic analysis and the types of waves used; but (as noted in Li et al., 2008 and Fukao & Obayasgi, 2013) there is also some consistency in which slabs level off versus reach the deeper mantle.

Explanations independent of a 650-km deep 'divide' are possible: for example, leveling off might be caused simply by the trench migrating backward away from the subduction (e.g., van der Hilst & Seno, 1993; also Kincaid & Olson, 1987); and, slab detachment from a subducting lithospheric plate might result from that lithosphere being too thick to subduct (Bercovici et al., 2015). Nevertheless, the three features together produce an impression that subducted slabs get 'stuck' at ~650 km depth, before 'cascading' down through the lower mantle (e.g., Bijwaard et al., 1998). Such behavior implies that the ~650-km 'boundary' may have some material and fluid dynamic significance. (Aspects of this conclusion are disputed by some researchers, including Masters et al., 1996 and van der Hilst et al., 1997.) Alternatively, Goes et al. (2017)

find that the combination of trench retreat and a dynamically important 650-km boundary can explain the slab behavior. The dynamic implications of that boundary will be further discussed toward the end of this chapter.

The Lower Mantle, and More. As noted earlier, the velocity anomalies revealed by tomography are found by some to be much stronger near the base of the mantle than in mid mantle; this appears to be especially so for shear-wave velocity, and especially over greater length scales (e.g., Lay & Guarnero, 2004; see also Dziewonski, 1984; Tanimoto, 1990). Those stronger anomalies allow us to explore one of the most critical boundary regions of the Earth System: the core–mantle boundary and the boundary layer D″ just above it, potentially associated with (among other things) both broad-scale and plume-style mantle convection. Imaging of and near D″ (e.g., Figures 8.20c, 8.21), providing us with a glimpse of deep-mantle conditions (in terms of velocity anomalies), may thus represent acoustic tomography's most impressive achievement of all.

We have already acknowledged low-velocity regions originating at the base of the mantle, evidently associated with whole-mantle plumes. The primary low-velocity region lies below the Pacific Ocean and includes the Hawaiian plume and the South Pacific 'super-plume' (see Figure 8.19); as seen in Figure 8.21, however—and especially as delineated by shear-wave tomography—this low-velocity region encompasses essentially all of the lower-mantle Pacific. This region, extending upward all the way to the lithosphere, is called the **Pacific Superplume** or **Pacific Megaplume**, although given its connection to several hotspots (see Figure 8.20c), the results of higher-resolution shear-wave analyses, the likelihood of deep-mantle plumes being narrow, and other fluid dynamic considerations, what we see as a superplume may instead be a *cluster* of plumes (e.g., Schubert et al., 2004; Montelli et al., 2006).

A second broad low-velocity region, also better detected by shear-wave tomography, can be inferred from Figures 8.20c and 8.21 to exist in the lower mantle beneath Africa. Both regions have become known as **large**

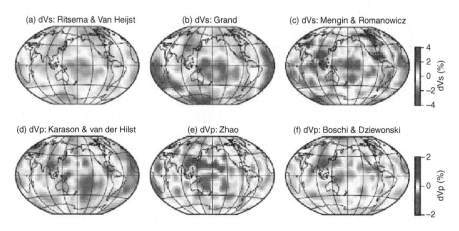

Figure 8.21 Comparison of global tomographic analyses in the lowermost 250 km of the mantle. Faster velocity anomalies are shown in blue, slower in red; but note the different scales (on the right) for *S* wave ((a) through (c)) versus *P* wave ((d) through (f)) velocity anomalies.
Can you think of a reason why anomalous regions based on S wave travel-time residuals would not match the regions inferred from P wave residuals?
Tomographic results shown here are from (a) Ritsema and van Heijst (2000), (b) Grand (2002), (c) Megnin and Romanowicz (2000), (d) Karason and van der Hilst (2001), (e) Zhao (2001), and (f) Boschi and Dziewonski (2000).
Image from Lay and Guarnero (2004) / John Wiley & Sons.

low shear velocity provinces or **LLSVPs** (see McNamara, 2019 for a very extensive review). Koelemeijer et al. (2016) find that these regions also exhibit slower *P* wave velocities, and propose calling them more generally **LLVPs**. Seismic data also suggests that there may be 'cores' of substantially lower velocity (*S* wave and also *P* wave) *within* or at the edges of the LLSVPs; though the LLSVPs are broad and may extend hundreds of kilometers (or even more; Ritsema et al., 1998) out of D″ up into the lower mantle, these cores—called **ultra-low-velocity zones** or **ULVZs**—are typically up to only a few 10^2 km laterally and, vertically, are confined to D″ (see McNamara, 2019 for references).

D″ also contains some high seismic velocity regions, mainly beneath the Americas and Asia (or equivalently, roughly, around the Pacific Ocean basin) (e.g., Dziewonski, 1984; Woodhouse & Dziewonski, 1989; Tanimoto, 1990; Grand, 1994; Li & Romanowicz, 1996; Su et al., 1994; Lay, 1989). The tomographic studies suggest that these anomalies extend up through to at least the mid and even upper mantle, and we may loosely identify them as connecting to subducted lithosphere, probably the Farallon and Tethys slabs. Such a connection is strengthened by the work of Lithgow-Bertelloni and Richards (1998) using a simplified model of slab subduction through the mantle. With an apparent connection from top to bottom, the description of D″ as a "graveyard for subducted lithosphere" (Wysession, 1996) may be particularly apt.

Tomography thus implies that a particular, 'dichotomous' type of convection takes place in the bulk of the mantle: broad-scale downwelling, associated with subducting oceanic lithosphere; and upwelling, in regions overlying LLSVPs, potentially also broad scale in the deepest mantle—corresponding to the breadth of the LLSVPs—but eventually concentrating into narrow plumes. Such a dichotomy (e.g., Figure 8.22; see McNamara, 2019 for other illustrations) has been described by several researchers; however, in detail and over time the actual flow within the mantle is likely to be more complex, with multiple subduction zones, evolving tectonic plates, mantle-wide heterogeneities, and a convective past that is still unfolding. A physical basis for the dichotomy—the strong dependence of mantle viscosity on temperature—will be discussed later in this chapter.

Is Tomography Just Another Word for Temperature? At this point, near the conclusion of our brief investigation into seismic tomography, it might be advisable to step back and confront a general question about its interpretation: *what makes a tomographic feature faster or slower?* Even at the base of the mantle, where we identify features as descended, relict slabs or as LLSVPs, why are they anomalously fast or slow? That is, *how confident are we that (only) temperature differences are responsible?*

Despite the tradition of depicting tomographic images using 'cool' bluish colors for faster media and 'warm' reddish colors for slower, temperature is not the only factor that can cause tomographic velocity anomalies; composition and phase changes (the latter including but not limited to partial melting) can also affect the rigidity, incompressibility, and density of the medium, and thus V_P and V_S.

Since at least some tomographic features are likely to be associated with mantle convection, we can also approach this question from a convective perspective. Convection always involves denser material sinking and lighter material rising; but in the Earth System, those convective density differences inevitably involve compositional as well as temperature effects. We have already seen this in the atmosphere, where water vapor is an important factor affecting the global circulation, and (via phase changes) acts as a huge source of power for regional circulation. In the oceans, salinity helps drive the thermohaline flow. We will find in Chapter 10 that compositional effects may be crucial in powering the core dynamo. It would not, then, be unexpected if composition or phase turns out to be a significant factor driving mantle convection—if the convection is *thermochemical* rather than purely thermal.

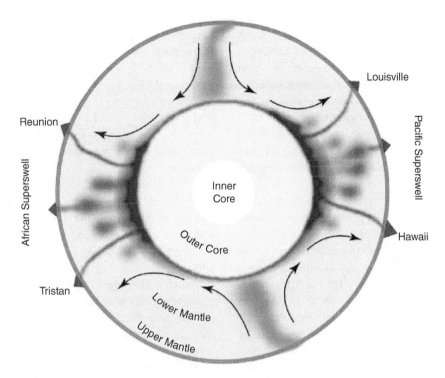

Figure 8.22 Idealized conception of mantle convection, illustrating the possible role of structures inferred tomographically. Dark blue shading: lithospheric slabs, penetrating the ~650-km 'divide' and descending through the lower mantle to D″ (thin orange shading surrounding the outer core). Brown shading: large low shear velocity provinces (LLSVPs), which serve as the base feeding deep- and whole-mantle plumes (fuzzy orange shading, and upward orange conduits connecting to labeled surface hotspots). Ultra-low velocity zones (ULVZs), not shown, would be small patches within the LLSVPs.
Image from Jellinek and Manga (2004), which was modified from Gonnermann et al. (2004).
If you were going to describe this pattern of mantle up- and downwellings—either the pattern of flow or the associated density differences, say, at the CMB—in terms of spherical harmonics, what single harmonic would work best?

In this case, at least some of the tomographic features would reflect compositional or phase differences.

Thermochemical Possibilities. As a somewhat provocative illustration of the interplay between thermal and chemical heterogeneity in the mantle, we briefly reconsider the fate of subducted oceanic lithosphere. Later in this chapter, we will discuss the likelihood that, at increasing depth, the minerals comprising the mantle undergo successive changes in phase—solid-to-solid phase transitions, chemically recombining into different minerals—that are denser and more resistant to deformation, thus more stable under the conditions of greater pressure. As the

lithosphere subducts into the mantle, it will also experience increasing pressures, and undergo such changes in phase; because of its colder temperature, though, it will find itself denser than the surrounding mantle, and continue to subduct, potentially descending all the way into D″.

The lithosphere is not, however, entirely mantle material. In a scenario presented by Christensen and Hofmann (1994), following earlier work by (e.g.) Davies and Gurnis (1986), Ringwood and Irifune (1988), Olson and Kincaid (1991), and Christensen (1989), and further investigated by others, including Jones et al. (2020; see also references therein), the oceanic crust comprising its top layer experiences several phase changes as well, most of which are to denser minerals—except for one occurring near the base of the upper mantle, where for a depth range of about 100 km the formerly crustal material becomes *less* dense! The end result is a more-or-less neutrally buoyant lithosphere (i.e., tending to neither rise nor sink) as it begins to penetrate the 660-km deep transition. This may explain why some subducting slabs (—some, depending on crustal versus total lithospheric mass, compositional variations, temperature contrasts, and the greater convective flow—) seem to level off as they approach that transition.

The slabs (or their remnants) that break through into the lower mantle may (now, once more, denser and colder than the surrounding mantle) descend all the way to D″. Within that bottom layer, in this scenario, high temperatures eventually allow the transformed crust to *separate* from the rest of the slab; convective currents may 'sweep' it to upwelling regions, where it may even become partially entrained into rising mantle plumes. (This scenario helps explain geochemical differences observed between basalts in hotspots—the surface expression of plumes—and mid-ocean ridge basalts; but that discussion is beyond the scope of this textbook.)

A first step in distinguishing compositional from thermal anomalies might be to consider the seismic velocity differences determined by the tomography. Recalling that $V_P = \sqrt{(k + 4/3\mu)/\rho}$ and $V_S = \sqrt{\mu/\rho}$, and that, in general, the incompressibility k is larger than the shear rigidity μ, the fact that the anomalies in V_S in the lower mantle are larger than the anomalies in V_P—roughly twice as large in the deep mantle and up to three times as large near the CMB, according to tomographic imaging (e.g., Woodhouse & Dziewonski, 1989; Grand et al., 1997; Lay & Guarnero, 2004 and our Figure 8.21; also Trampert et al., 2001, who conclude shear anomalies are dominant but not quite as large)—suggests that such anomalies result mainly from changes in μ rather than k. Not surprisingly, we can then attribute much of

the anomalously low shear velocities to hotter temperatures, and also to partial melting in the regions where V_S is very low (partially melted solids, and even fluids, are still fairly incompressible!).

A closer look at tomographic images reveals that anomalous V_P and V_S regions in the deep mantle do not always match, spatially. To the extent that this is not simply an artifact of poor resolution, we can infer that there must also be differences in k between those regions, uncorrelated to those temperature changes that are affecting μ; in this case, composition and phase changes (beyond melting or partial melting) probably have contributed to the anomalies.

Applying these types of relations, Kennett et al. (1998) concluded that in the upper mantle the tomographic anomalies are mostly thermal in origin; but in some orogenic zones, for example, western North America, chemical heterogeneity is present. Thermal effects dominate the anomalies in mid mantle (Koelemeijer et al., 2016) and in most of the lower mantle (Trampert et al., 2001). Koelemeijer et al. (2016) found that significant chemical heterogeneity is present around the top of D″ (including the LLVPs), and according to Trampert et al. (2001), isolated spots of ULVZs may, within uncertainties, be chemically heterogeneous. Finally, near the CMB things may be more uncertain, with Koelemeijer et al. (2016) determining that thermal effects dominate but, e.g., Kennett et al. (1998) and others (see below) having inferred widespread chemical heterogeneity.

Using experimentally derived mineral properties to convert tomographic velocity anomalies to temperature anomalies, Yuen et al. (1993) were able to identify superplumes in the lower mantle, but found the estimated temperature in the plume 'cores' was too high, implying that the anomalous velocities must be partly attributed to compositional differences. Along these lines, Wen (2001), Wen et al. (2001), Tackley (2002) and others inferred the presence of sharp borders for portions of ULVZs (or LLVPs), and argued that temperature effects (including partial melting) could not produce that sharpness (cf. Schuberth et al., 2009), so they must be chemical in nature.

Outside of the shallow mantle, then, the most likely region for chemical heterogeneity appears to be the bottom few hundred kilometers of the mantle, encompassing D″ and perhaps the mantle slightly above. But this conclusion is far from certain; for example, in contrast to the provocative scenario described above, the slab 'graveyard' may be mainly just a thermal resting place. This is implied by the precise analysis of tomographic data by Hutko et al. (2006), who identified what is likely an ancient remnant of the subducted Farallon slab in D″, far below the present-day Cocos plate, resting on the top of the core. They found it to be a colder, denser, and seismically faster region (from their analysis, they also concluded the remnant is folded over!). They suggested compositional variations have contributed to the seismic anomaly, but that thermal effects dominate, with the remnant an estimated 700°F colder than its surroundings. In that condition, any delamination would presumably have been inhibited.

As another example, we note that high temperatures within the ULVZs, rather than chemical heterogeneity, could instead produce the sharp borders inferred to exist at their edges, if the mantle's viscosity was strongly temperature-dependent (see also Ritsema et al., 1998).

Regardless, however, the question of chemical heterogeneity in D″ is and has always been hard to dismiss. There are several mechanisms *in addition* to remnant slabs and delaminated, phase-changed oceanic crust for introducing such heterogeneity into D″; three involve the core:

- A leaky core—infusion of iron into the mantle caused by chemical reactions between mantle and core at the CMB (Knittle & Jeanloz, 1989, but see also Hirose & Lay, 2008);

- A leaky core—fluid iron percolating up from the outer core, then entrained (slowly) by upwelling plumes into the ULVZs (Kellogg & King, 1993; Jellinek & Manga, 2004, and references therein); and

- A 'rough' core surface—undulations of the core boundary, known more familiarly as "bumps on the core," with raised bumps in effect injecting iron into D″.

Core bumps, which (as discussed later in this textbook) may be responsible both for fostering plume development in the deep mantle and for locally stifling the flow in the core that maintains our geomagnetic field, have been inferred from tomography (e.g., Morelli & Dziewonski, 1987; Creager & Jordan, 1986), other types of seismic observations, and Geoid and geomagnetic observations (see, e.g., Loper & Lay, 1995). Conceivably such bumps could be a static feature, an artifact of the way the core and mantle had differentiated; this would require such features to be able to withstand eons of mantle convection. Alternatively, those bumps might be understood as the core boundary's *dynamic topography*, analogous (to some extent, as noted in Chapter 6) to the ocean's dynamic topography but produced and supported instead by convective currents in the mantle—currents that in turn are driven by density anomalies inferred from tomography (Hager et al., 1985; Forte & Peltier, 1989, 1991).

Another possible scenario for D″ is that it is comprised of a very dense phase of mantle material (e.g., Hutko et al., 2006), stable at the high pressures and temperatures of the lowermost mantle, but—perhaps due to lateral variations in temperature or composition, or to the dynamic effects of the overlying mantle convection—with a variable thickness, or even just a 'scattered' presence. We will save discussion of such a phase for later in this chapter, but note here that high-pressure experiments (Murakami et al., 2004) confirm at least the possibility of its existence.

Whether some tomographic anomalies are actually chemical heterogeneities or not, it

Figure 8.23 Tomographic structure of the mantle, from a model by Masters et al. (2000). Top and bottom of this view correspond to Earth's free surface and CMB; view looks northward. Blue denotes regions where V_S is 0.6% faster, whereas red denotes regions where V_S is 1.0% slower. CMB-to-surface red structure is the South Pacific "superplume." Image from Tackley (2000) / American Association for the Advancement of Science.

appears that mantle convection can be understood primarily as being thermally driven. That is, upwelling plumes are buoyant mainly from large temperature differences; compositional buoyancy may aid their rise. Oceanic lithosphere subducts because it is cooler and denser; the phase changes it experiences as it sinks may foster or delay its descent, but ultimately (and perhaps catastrophically, at specific boundaries) its cold-induced negative buoyancy prevails. This view is supported by, for example, Weinstein (1993), Lithgow-Bertelloni and Silver (1998), Forte & Mitrovica (2001), Tackley (2002), Schuberth et al. (2009), and other works cited above.

We will return to this question later in this chapter; for now, you might want to think about how thermochemical convection differs, say, in the atmosphere (beyond what was mentioned earlier).

An issue tangentially related to the question of convective driving forces is whether chemical heterogeneities, passively 'pushed around' by the convecting mantle's flow, can survive over time as compositionally distinct, convectively isolated, reservoirs within the mantle (see, e.g., Kellogg et al., 1999; van der Hilst & Karason, 1999; Tackley, 2000; Li et al., 2014). That issue is beyond the scope of this textbook.

Seismic tomography is a means to investigate the interior three-dimensionally; but

Figures 8.19–8.22 show cross-sections or contour maps that only hint at Earth's full 3-D interior structure. Figure 8.23 provides as our final illustration of tomographic research a more three-dimensional picture of the interior. (It might be noted that the low-velocity anomaly contrast chosen for that figure may imply a more chaotic interior structure than actually prevails.)

8.4   Using Earth Models to Learn About the Composition of the Interior

The question of Earth's composition has loomed in the background throughout this chapter—not just constraining our interpretation and understanding of tomographic structures, but also underlying the reference velocity models used as background for those structures; and, of course, playing a somewhat explicit role in Birch's rule. In this section, we will discuss two ways that the densities determined from Birch's rule, Adams-Williamson, and free oscillation data can be used to elucidate the composition of Earth's interior: equations of state and high-pressure experiments.

8.4.1   Equations of State

Equations of state are relations between density, pressure, and possibly temperature that allow density estimates—corresponding to the P and T of a given depth—to be extrapolated up to surface conditions; the extrapolated densities can then be compared with the STP densities of different candidate minerals.

We have already encountered an equation of state: the ideal gas law. If we knew the pressure and temperature at some altitude within the troposphere, we could use that relation to estimate the density of the atmosphere at that height. Or, if we are given the density at that height, we can extrapolate it down to STP conditions; if its extrapolated density differs from what we know the surface density of air to be, we might infer that its composition (probably its humidity) differs from that at the surface.

Equations of state for the solid earth have been derived from experiments, from theory, and even postulated on an empirical basis. An example of a theoretical relation often used in geophysical applications (see, e.g., Birch, 1952; Christensen & Hoffman, 1994; Suzuki et al., 1998; Poirier; 2000) is the Birch-Murnaghan equation of state, given by

$$P - P_0 = \frac{3}{2}(k_T)_0 \left\{ \left(\frac{\rho}{\rho_0} \right)^{7/3} - \left(\frac{\rho}{\rho_0} \right)^{5/3} \right\}$$
$$\times \left[1 - \frac{3}{4}(4 - k_T') \left\{ \left(\frac{\rho}{\rho_0} \right)^{2/3} - 1 \right\} \right],$$

where $(k_T)_0$ is the isothermal incompressibility at surface pressure and $k_T' = \partial(k_T)/\partial P$. This equation is based on general thermodynamic relations (between pressure and the change in energy of the material as it contracts), combined with expressions describing the strain experienced by the material as its density increases under the high pressures of the interior. In many situations, $k_T' \approx 4$ to good approximation, and the formula simplifies to

$$P - P_0 = \frac{3}{2}(k_T)_0 \left\{ \left(\frac{\rho}{\rho_0} \right)^{7/3} - \left(\frac{\rho}{\rho_0} \right)^{5/3} \right\}.$$

With ρ determined from Birch's rule, Adams-Williamson, or normal modes, and P determined from hydrostatic equilibrium, this equation could be solved for the surface density ρ_0. That is, if material of density ρ exists at a depth within the Earth where the pressure is P, and that material could be brought up to the surface, where the pressure is P_0, its density would (according to this equation) become ρ_0.

As a much simpler exploration, however, we will instead follow the development by Stacey (1977) and consider the definition of (adiabatic) incompressibility:

$$k_S \equiv \rho \left(\frac{\partial P}{\partial \rho} \right)_S.$$

For simplicity, we will write this relation using ordinary derivatives, but we must keep in mind that any 'solution' to it is only a partial solution restricted to adiabatic conditions. Rearranging terms, we can write it as

$$\frac{dP}{k_S} = \frac{d\rho}{\rho} = d(\ln \rho),$$

with the changes in pressure and density measured under adiabatic conditions. For our extrapolation, we add (integrate) all the increments of P/k_S and $\ln \rho$, from the surface, where the pressure is P_0 ($= 1$ bar $= 0.1$ MPa) and the density is ρ_0, down to the depth where the pressure is P and the density is ρ:

$$\int_{P_0}^{P} \frac{dP}{k_S} = \int_{\rho_0}^{\rho} d(\ln \rho);$$

because $\int dx = x$, for any x, this equation becomes

$$\int_{P_0}^{P} \frac{dP}{k_S} = \ln \rho/_{\rho_0} = \ln \rho - \ln \rho_0 = \ln(\rho/\rho_0).$$

To illustrate the use of this equation, we imagine first a situation where the resistance of materials to compression does not grow as the pressure on them increases; this is a poor assumption, but it allows us to treat k_S as a constant. Our equation then simplifies to

$$\frac{1}{k_S} \int_{P_0}^{P} dP = \ln(\rho/\rho_0),$$

thus

$$\frac{1}{k_S}(P - P_0) = \ln(\rho/\rho_0)$$

or

$$P - P_0 = \ln(\rho/\rho_0)^{k_S}.$$

Again, with ρ determined at a given depth from Birch's rule, Adams-Williamson, or normal modes, and P determined from hydrostatic equilibrium, the equation could be solved for the surface density ρ_0. As with any other equation of state, the 'catch' here is that a value of the material property, k_S, representing the minerals existing at that depth, must be specified prior to the extrapolation. And, of course, this would yield an accurate estimate of surface density only to the extent that other temperature effects, in addition to cooling from adiabatic decompression (from the depth of P, ρ up to the surface), would not seriously modify the density.

For greater accuracy we acknowledge that incompressibility is not constant but rather increases with pressure. Following our earlier discussion of 'indicators,' in which dk_S/dP is viewed as roughly constant excluding the effects of material inhomogeneity, we could model k_S as a linear function of pressure:

$$k_S = k_0 + k'(P - P_0);$$

here k_0 is the value of the adiabatic incompressibility at surface pressure ($P = P_0$), and k' is shorthand for the slope, dk_S/dP. We could substitute this expression for k_S into our equation of state,

$$\int_{P_0}^{P} \frac{dP}{k_S} = \ln(\rho/\rho_0),$$

that is,

$$\ln(\rho/\rho_0) = \int_{P_0}^{P} \frac{dP}{k_0 + k'(P - P_0)}.$$

The end result of this integration is

$$\ln(\rho/\rho_0) = \frac{1}{k'}\ln\frac{k_0 + k'(P - P_0)}{k_0} = \frac{1}{k'}\ln(k_S/k_0);$$

in this derivation, we have used the fact that $k_S = k_0 + k'(P - P_0)$ equals k_0 when evaluated at $P = P_0$. Finally, then,

$$\frac{\rho}{\rho_0} = \left(\frac{k_S}{k_0}\right)^{1/k'}.$$

The hard part of this derivation was integrating

$$[k_0 + k'(P - P_0)]^{-1}$$

in general, before evaluating it as a definite integral. If you are unsure about that integration, whose result was

$$\frac{1}{k'}\ln[k_0 + k'(P - P_0)],$$

you can at least verify the reverse—that the derivative of this result is indeed $[k_0 + k'(P - P_0)]^{-1}$.

Since the PREM solution for k_S in the mantle (see Figure 8.18b) does indeed appear fairly linear, in segments, when plotted versus depth (or pressure), we can use this equation of state to learn about the constitution of the mantle. Using PREM, a slope and y-intercept can be calculated for different portions of the mantle (see also Table 8.3); we will use $k' = 3.3$ and $k_0 \cong 134.6$ GPa for the upper mantle and $k' = 3.3$ and $k_0 \cong 200$ GPa for the lower mantle. For the upper mantle, we consider three depths:

- at $z = 265$ km, where $k_S = 157.9$ GPa, the formula allows us to extrapolate the density there, $\rho = 3.46$ gm/cm^3, to a value $\rho_0 = 3.30$ gm/cm^3 at surface pressure ($P = P_0$).

- at $z = 310$ km, where $k_S = 163.0$ GPa, the density there, $\rho = 3.49$ gm/cm^3, can be extrapolated to $\rho_0 = 3.29$ gm/cm^3.

- at $z = 355$ km, where $k_S = 168.2$ GPa, the density there, $\rho = 3.52$ gm/cm^3, can be extrapolated to $\rho_0 = 3.29$ gm/cm^3.

We conclude that over this depth range, the extrapolated surface density appears essentially constant, implying a chemical (and phase) homogeneity *versus depth* through that region of the upper mantle (versus depth, because the underlying seismic model is laterally averaged). Additionally, that extrapolated $\sim$3.3 gm/cm^3 surface density is consistent with a peridotite-type composition.

However, the results of our equation of state extrapolation are not quite the same for the lower mantle. Again using PREM values for k_S and ρ (Figure 8.18), our data now is

- at $z = 871$ km, $k_S = 330.3$ GPa, and $\rho = 4.51$ gm/cm^3;

- at $z = 1,371$ km, $k_S = 412.8$ GPa, and $\rho = 4.79$ gm/cm^3;

- at $z = 1{,}871$ km, $k_S = 492.5$ GPa, and $\rho = 5.05$ gm/cm^3; and

- at $z = 2{,}371$ km, $k_S = 574.4$ GPa, and $\rho = 5.31$ gm/cm^3.

(the depths chosen here correspond to 'round numbers' for the radius). The D″ region is excluded from this analysis. The extrapolated densities are now 3.87 gm/cm^3, 3.85 gm/cm^3, 3.84 gm/cm^3, and 3.86 gm/cm^3, respectively. These values of ρ_0 are approximately constant, implying material homogeneity versus depth, as before; but the value for ρ_0 is definitely *not* consistent with peridotite!

In a pre-PREM version of this analysis, Stacey (1977) derived similar results to ours (his estimate of the ρ_0 based on lower-mantle data was a few percent larger). And, as noted by Stacey, temperature effects only compound the disparity between all these predictions and the STP density of peridotite. That is, because this analysis began with k_S, and k_S was integrated under adiabatic conditions, ρ_0 represents the density of material reduced by *adiabatic* decompression to surface pressure. If the mantle is convecting, as we will discuss in the next chapter, the temperature gradient near its convective boundaries must be superadiabatic; but our extrapolation includes only adiabatic cooling. For extrapolations from the upper mantle, the additional effects of superadiabatic cooling are likely to be relatively small (simply because the distance to the surface is relatively small), but we cannot assume the same for extrapolations from the lower mantle. On the other hand, additional cooling will make ρ_0 larger, exacerbating its discrepancy in comparison to the STP density of peridotite.

This discrepancy highlights a material difference between the upper and lower mantle, and confirms what we could already infer from the seismic velocity profiles of Figure 8.12: something is going on in the transition zone. One possibility is that one or both of the transitions represents a change in composition, to something fundamentally denser below the transition zone than peridotite would be. However, those transitions are not quite as sharp as the Earth's known compositional boundaries (the CMB and the Moho) are. And, a compositional change would leave us with a lower mantle that has a ~20% greater STP density than peridotite (i.e., ~3.85 versus 3.3) but, from Birch's rule, the same mean atomic weight—a fortuitous change in composition, indeed.

The other possibility is that the lower-mantle bulk chemistry is essentially unchanged and the transitions mainly represent changes in the phases of the peridotitic minerals that make up the upper mantle. This possibility, which is explored extensively in the next subsection, avoids the problems faced by a lower mantle of unrelated composition (for example, phase transitions can be gradual or sharp). In this case, the reason our equation of state predicted a high STP density for the lower mantle is simply that it had failed to account for the drop in density as the material decompressed to lower-pressure phases.

The Core. Applying equations of state to the core is challenging. For one thing, the assumption of constant dk_S/dP which underlies our equation of state does not exactly hold true under core conditions (the plot of k_S versus depth in Figure 8.18b in fact shows a clear nonlinearity in the core; and, with that nonlinearity ignored, the predicted surface densities are unrealistically low). To avoid that assumption, Stacey (1972) started with the (integral) definition of incompressibility,

$$\int_{P_0}^{P} \frac{dP}{k_S} = \ln(\rho/\rho_0),$$

and carried out the integration (numerically) from core to surface using laboratory data and extrapolations for k_S rather than a linear approximation. For the top of the core (with a density in his model of 9.9–10.2 gm/cm^3), he found that $\rho_0 = 6.3$–6.5 gm/cm^3. Such extrapolated densities are significantly lower than the density of solid iron at STP (7.9 gm/cm^3), or of solid or even liquid iron at iron's melting point (7.3 gm/cm^3, 7.0 gm/cm^3)—reinforcing our deduction from Birch's rule that the core must contain a light alloy.

Which comparison do you consider to be the most meaningful—ρ_0 versus solid iron at STP, versus solid iron at its melting point, or versus liquid iron at its melting point?

For a more accurate estimate of the density deficit in the core caused by a light alloy, and its implications for the nature of the inner-core–outer-core boundary, more sophisticated equations of state are called for, preferably accounting for temperature effects explicitly (even though the outer core temperature gradient is likely adiabatic); see, for example, Anderson and Isaak (2002) and also Vočadlo et al. (2003).

8.4.2 High-Pressure Experiments

Equations of state allow us to extrapolate the Earth's interior densities *up* to surface conditions, and—by comparing the extrapolated densities with the STP densities of different candidate minerals—we can confirm or rule out those candidates as a primary constituent of the upper mantle, lower mantle, or core. High-pressure experiments are 'the other side of the coin': they take candidates for the composition at some depth *down* to the conditions of that depth; if their high-P,T densities (or other properties) are too dissimilar from what our Earth models tell us prevails there, then they are unlikely to represent the composition of that region. But such experiments can tell us much more—constraining temperatures at depth, revealing the existence of phase changes, and so on—once we are confident of the composition. In short, we will find in this section that high-pressure experiments provide us with a good picture of what is happening in the transition zone; and they substantiate the need for a light alloy in the core. Equally satisfying, the information they provide turns out to be important for our understanding of how the mantle and core convect.

Experimental Verification of Inhomogeneity Within the Transition Zone. High-pressure, high-temperature experiments verify our deductions of material inhomogeneity within the transition zone. The uppermost mantle is primarily peridotite, thus mainly olivines and pyroxenes. Experiments performed on olivines reveal not only a 'normal' increase in density as pressure is first increased, but also—at great enough pressures, ultimately corresponding to conditions within the transition zone—a phase change, from the orthorhombic crystal structure exhibited by olivine at surface conditions to a cubic structure called the **spinel** phase (see Figure 8.24), now known more commonly as **ringwoodite**. Clearly, the denser spinel structure is better suited than the more 'open' orthorhombic configuration to the higher pressures encountered hundreds of kilometers below the surface.

Because olivine is a solid solution of iron and magnesium silicate, the phase transition to ringwoodite will depend on the particular

(a) (b)

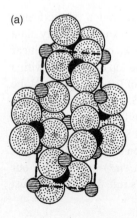 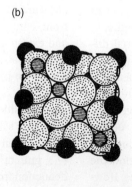

Figure 8.24 Diagram illustrating crystal structure of olivine both in the orthorhombic form found at surface and near-surface conditions (a) and in the more compact cubic packing configuration found in the higher-pressure spinel phase (b). Large spheres represent oxygen ions; small spheres are silicon (black) and magnesium or iron (gray). The higher-pressure spinel structure is also called "ringwoodite."
Figure modified from Garland (1971) / with permission of Elsevier.

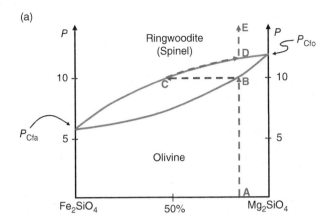

Figure 8.25 Phase diagrams for the olivine→ringwoodite (spinel) phase transition, based on high-pressure experiments performed at high temperature. The vertical axis is pressure, in units of GPa; composition is shown on the horizontal axis.

(a) Idealized phase diagram, based on early experiments at 1,000°C. Composition is expressed here in terms of percentage of forsterite (Mg_2SiO_4) versus fayalite (Fe_2SiO_4). Critical pressures for the phase change of pure fayalite and pure forsterite, P_{Cfa} and P_{Cfo} respectively, are indicated on the vertical axes. The dashed line A-B-C-D-E illustrates a possible phase path for the transition, assuming a mostly forsterite composition. The text explains how this diagram can be interpreted.

composition; Figure 8.25a shows how the phase transition might work under simplified conditions, both for the pure end members (pure fayalite, Fe_2SiO_4, and pure forsterite, Mg_2SiO_4) and for an intermediate composition. With the pure end members, we have a single component system, and the olivine converts 'instantaneously' to the spinel structure when a critical pressure, P_C, is reached. Forsterite resists the phase change more than fayalite does, so a higher critical pressure is required to produce the phase change in forsterite (in this illustration, ~12 GPa or 120 kbar for P_{Cfo} compared to ~6 GPa or 60 kbar for P_{Cfa}).

However, for a mix, the phase change is not 'instantaneous;' instead, the change proceeds over a range of pressures. Between the limits of P_{Cfa} and P_{Cfo}, the upper and lower curves of the phase diagram (in blue) define the pressure bounds within which the phase change has begun but is not yet complete.

In the case of the Earth's mantle, we can assume a mostly forsterite composition. As the pressure is increased, we follow the dashed line in Figure 8.25a upward.

- At the start (point A), under surface pressure conditions, the sample has the orthorhombic structure of olivine.

- As the pressure increases from A to B, we can expect the density (not shown on the graph) to increase; but the sample maintains its orthorhombic crystal structure.

- When the pressure corresponding to point B is reached, the phase change begins, with the fayalite converting first to the spinel structure (point C).

- As the pressure is further increased, more and more of the forsterite converts to the spinel structure; at the pressure corresponding to point D, all of the forsterite has completed the change. The ringwoodite now has the same proportions of Mg and Fe as the original olivine.

- As pressures are increased more (from D to E and beyond), the sample's density will continue to increase, but its crystal structure will remain spinel.

This picture of the olivine→ringwoodite phase change has been complicated by the discovery of an intermediate, "modified" spinel structure, which exists at pressures not quite high enough to complete the change to spinel—but only for compositions dominated by forsterite (Ringwood & Major, 1970, and references therein). Given that intermediate structure, it has been useful (as noted, e.g., in Burnley, 2011) to distinguish the various phases using an α, β, or γ designation: olivine, referred to as α-olivine; the intermediate phase, as β-olivine, β-spinel, or just the β-phase; and ringwoodite (spinel), as γ-olivine or γ-spinel. Although the β-phase is now more commonly called **wadsleyite**, this notation is still occasionally used. A phase diagram based on early experiments exploring the conditions for the appearance of wadsleyite is shown in Figure 8.25b.

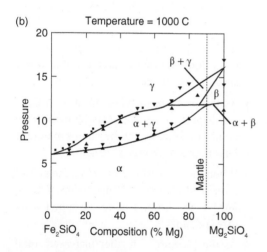

Figure 8.25 Phase diagrams for the olivine→ringwoodite (spinel) phase transition, based on high-pressure experiments performed at high temperature. The vertical axis is pressure, in units of GPa; composition is shown on the horizontal axis.
(b) Phase diagram adapted from Bina and Wood (1987) / John Wiley & Sons, based on experiments at 1,000°C following the discovery of wadsleyite (β-spinel). Sample composition is displayed in terms of its percentage of forsterite; the dashed vertical line corresponds to a dominantly forsterite composition, as is likely for the mantle. As discussed in the text, α signifies olivine; β, wadsleyite; and γ, ringwoodite.
In this phase diagram, P_{Cfa} ~6 GPa (60 kbar), whereas P_{Cfo} ~12 GPa (120 kbar) for the α→β change and ~16 GPa (160 kbar) for the β→γ change.

Wadsleyite is more incompressible than olivine, so seismic waves travel faster through it. But the range of pressures over which the phase change from olivine to wadsleyite is achieved (the region marked α + β in Figure 8.25b) was found to be quite narrow; Bina and Wood (1987) estimated this range to be ~0.2 GPa (2 kbar), corresponding to an increase in depth in the mantle of ~6 km from start to finish (note Katsura & Ito, 1989 found the range to be slightly higher). The effect of this phase change on seismic velocities would thus be to produce a sharp jump, rather than just a gradually higher rate of increase, in the velocity profiles. In contrast, the wadsleyite to ringwoodite phase change (through the region marked β + γ in Figure 8.25b) required ~1.5 GPa for completion in an iron-poor mantle—that is, a depth range of ~45 km—implying that it is gradual enough that it would not produce a jump in seismic velocities.

Bernal (1936) was the first to propose that the jump in seismic velocities at a depth of ~400 km was the consequence of an olivine→ spinel phase change. Much experimental work has shown that those phase changes occur under mantle conditions—broadly speaking, pressures of order 10 GPa (10^2 kbar) and temperatures of order 10^3 °C. And, since the olivine→wadsleyite change produces a nearly discontinuous jump in seismic velocities, it in particular is certainly a good candidate for the cause of the first transition. But does it occur when it's supposed to, that is, at the right depth?

The first transition occurs at a depth of ~400–410 km (based on analysis of seismic waves reflecting off that jump in seismic velocity, we might narrow the range down to 405–410 km; see Katsura et al., 2004 and references within). From hydrostatic equilibrium, which approximates pressure conditions well in the mantle, and densities in the upper mantle determined from normal modes, Adams-Williamson or Birch's rule, that depth corresponds to a pressure of ~13.5 GPa (135 kbar). But the experiments on which the phase diagram in Figure 8.25b is based found the olivine→wadsleyite change to occur at a pressure of ~12 GPa.

Can you think of a way to reconcile the different depths at which this phase change might take place?

In general, phase changes depend on temperature as well as pressure. The experiments in Bina and Wood (1987) were conducted at a nominal 1,000°C; but there is no reason that the temperature at the first transition had to be that.

What would it take to delay the change to a denser phase until higher pressures (by an additional 1.5 GPa) are reached: temperatures that are hotter than 1,000°C, or colder?

Rather than repeat the experiments at a range of temperatures, we can answer that question theoretically using a thermodynamic relation called the *Clausius-Clapeyron equation*. For any kind of phase change, let ΔV denote the associated change in the volume of the sample and ΔS the change in its entropy; if T_{eq} is the temperature at which the two phases are in equilibrium (as an example, for water and ice at sea level, $T_{eq} = 0°C = 273K$), then that relation specifies

$$\frac{dT_{eq}}{dP} = \frac{\Delta V}{\Delta S}.$$

For water/ice, this relation could be used to estimate the melting temperature of ice atop a mountain or beneath the sea; in other words, it can be used to find how much the melting temperature changes (dT_{eq}) when the ambient pressure is different than that at sea level by an amount dP or ΔP (that is, $\Delta T_{eq} = (dT_{eq}/dP) \cdot \Delta P = (\Delta V / \Delta S) \cdot \Delta P$).

The ratio $\Delta V / \Delta S$ is often called the **Clapeyron slope** (though in some contexts, the slope is defined as the reciprocal, $\Delta S / \Delta V$). In general, ΔV and ΔS can each be either positive or negative. The phase changes that take place in the mantle's transition zone result in denser phases, so ΔV will be negative. ΔS is less

clear. The change in the entropy of a 'system' is proportional to the heat gained or lost by the system—for example, as we noted when discussing adiabatic versus isothermal incompressibility, if conditions are adiabatic (no heat in or out), the entropy remains constant. Thus, the sign of ΔS will depend on whether the phase change is exothermic (releases heat, thus negative ΔS) or endothermic (absorbs heat, thus positive ΔS). Experiments show that the olivine→spinel phase transitions are exothermic—in accord with our intuition, because the denser spinel is a more stable configuration at higher pressure—so they correspond to negative ΔS.

To complicate matters further, the magnitudes of ΔV and ΔS may depend on temperature and pressure, requiring laboratory measurements of them to be either conducted at high-P,T conditions or extrapolated from lower-P,T or even STP measurements. In short, with the changes in volume (or density) and entropy estimated under mantle conditions, the Clausius-Clapeyron relation implies that a temperature of ~*1,500°C* would produce olivine→wadsleyite phase transitions that are complete at 13.5 GPa or 405 km depth (Bina & Wood, 1987). This is a reasonable temperature for the upper mantle, lending support to the idea that the first transition corresponds to an olivine-to-wadsleyite phase change, and also demonstrating the value of that phase change as a **geothermometer** for measuring temperatures in Earth's interior.

Actual experiments at such temperatures support our thermodynamic extrapolation. Figure 8.25c shows the results of one experiment, carried out by Fei and Bertka (1999) at 1,600°C. For olivine that is primarily forsterite (as expected for an iron-poor mantle), the transition to wadsleyite begins at ~13.5 GPa or higher; with a composition of 85% (90%) forsterite, that transition begins at ~13.7 GPa (14.0 GPa) and persists for ~½ GPa ~15 km. This phase change therefore satisfies both the depth and depth range constraints for the first seismic transition fairly well.

(c)

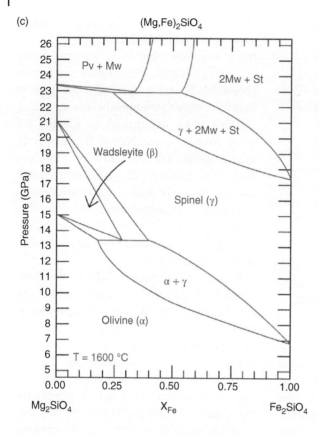

Figure 8.25 Phase diagrams for the olivine→ringwoodite (spinel) phase transition, based on high-pressure experiments performed at high temperature. The vertical axis is pressure, in units of GPa; composition is shown on the horizontal axis.
(c) Phase diagram in Burnley (2011) based on work by Fei and Bertka (1999) at 1,600°C; composition is expressed as a fraction of iron rather than a percentage of magnesium (thus, a scale reversed left to right, compared to (b)). Note that the higher temperature in these experiments led to slightly higher critical pressures for the transition zone phase changes.

Getting to the Second Transition. The experimental results of Fei and Bertka featured in Figure 8.25c also imply that, at 1,600°C and for a composition of 85% (90%) forsterite, the wadsleyite→ringwoodite phase change begins at ~17.0 GPa (18.0 GPa) and reaches completion at ~18.3 GPa (19.2 GPa); completion is thus at depths of ~530 km (553 km). The results of Katsura and Ito (1989), e.g. for ~90% forsterite, were quite similar. For either composition, the range over which this transition takes place is evidently at least ~1 GPa in pressure, ~30 km in depth (as seen in Figure 8.25b, the work of Bina and Wood, 1987 indicates this transition may be still wider, whether temperature effects are accounted for or not).

For conciseness, the wadsleyite→ringwoodite phase change is often referred to as the 520-km transition (e.g., Burnley, 2011; Frost, 2008; Bass & Parise, 2008). It is not detected globally (Frost, 2008; Stacey & Davis, 2008),

which is not surprising given its relatively large depth range (and implied correspondingly gradual velocity increase—leaving no discontinuous 'jump' to bounce seismic waves off). On the other hand, it is likely not totally invisible, seismically: it can enhance the seismic velocity slope within the transition zone (cf. Bina & Wood, 1987; also our Table 8.3). In any case, its width and its depth of completion rule it out as the cause of the 660-km seismic velocity jump.

The Second Transition. Experiments on olivine at even higher pressures (Liu 1975, 1976a and others) reveal the existence of an extreme phase change—really, more of a chemical breakdown—that does correspond to the 660-km discontinuity. At that depth, where the pressure is ~23.5 GPa, ringwoodite transforms into (Mg, Fe)SiO$_3$ *perovskite*, plus a dense (halite structure) version of periclase (MgO) and a dense iron oxide (FeO) called *wüstite*.

Perovskite is an unusual family of minerals, in part because its basic dense cubic structure can be perturbed, by temperature or stress, into other structures (Pradhan & Roy, 2013); see Figure 8.26a. (Mg, Fe)SiO₃ perovskite also stands out in contrast to the phase-change minerals of the first transition because the phase change creating it is endothermic (i.e., energy must be added to the mineral to produce perovskite). Navrotsky (1980) points out that the endothermic nature of perovskite may be more the rule than the exception, for minerals under lower-mantle conditions.

In recent years, (Mg, Fe)SiO₃ perovskite, also called silicate perovskite, was formally renamed **bridgmanite**, honoring a scientist whose instrumental advances a century ago set the stage for modern-day high-pressure experiments (see Sharp, 2014)—an honor indeed, since the large volume of the lower mantle makes bridgmanite Earth's most abundant mineral. The other products of this phase change, the densely packed oxides MgO and FeO, are frequently referred to by a single mineralogical name, either *ferropericlase* or *magnesiowüstite*, to reflect the relative abundance of Mg or Fe.

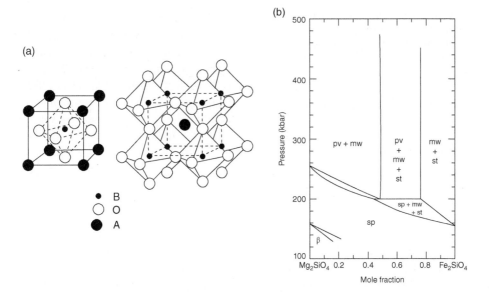

Figure 8.26 (a) Crystal structure of perovskite minerals. Two of the possible structures are shown, cubic (on the left) and orthorhombic (on the right). These minerals all have a chemical formula of the form ABO₃, where O is oxygen, and A and B can include elements such as Mg or Fe (for A) and Si (for B). Figure from Pradhan and Roy (2013) / Pradhan & Roy.
(b) Olivine and its high-pressure phases at 1,000°C, from Yagi et al. (1979b) / Carnegie Science. Mineral abbreviations are β = β-spinel = wadsleyite, sp = spinel = γ-spinel = ringwoodite, pv = perovskite = bridgmanite, mw = magnesiowüstite (or ferropericlase), st = stishovite. In the mantle, the mole fraction of iron olivine is ~0.1 (i.e., ~10% Fe₂SiO₄). Note that an iron-rich mantle (mole fraction exceeding ~0.8) would not support the production of perovskite at high pressure.
The experimental results at high pressures were achieved primarily using a diamond anvil, a compact apparatus based on the simple principle that pressure = force/area; with finely cut diamonds on either side of the sample (their 'culets' featuring a tiny contact area), high pressures—the highest pressures of any static-compression device, in the words of Bass et al. (2008)—can be exerted without the diamond 'anvils' buckling. For example, in the experiments discussed by Ohta et al. (2008) and Tateno et al. (2009), the culets were no bigger than 0.2 mm on a side. And, the transparency of the diamonds allows for controlled heating by lasers, plus easy visual inspection. See, e.g., Bassett (2009), Bass et al. (2008), Boehler (2000), and Block and Piermarini (1976) for more information on this technique.

(c)

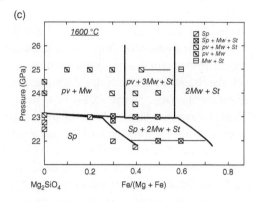

Figure 8.26 (c) Phase diagram for spinel→perovskite (ringwoodite→bridgmanite) at 1,600°C, from high P, T experiments performed by Ito and Takahashi (1989) / John Wiley & Sons. Mineral abbreviations are as in (b).
These results, in contrast to those of Yagi et al. (1979b), show a nearly flat Sp-Pv boundary for an iron-poor mantle, so that the phase transition occurs at essentially the same pressure (depth) no matter what the proportion of iron (up to ~25%, or even ~55%).

High-pressure experiments at 1,000°C by Yagi et al. (1979b), summarized in Figure 8.26b, indicated that the transition to perovskite in an iron-poor mantle would be complete at pressures less than ~25.5 GPa (255 kbar in the figure), i.e. at depths not exceeding 705 km; uncertainties in their pressure estimates were ±1 GPa (±10 kbar). Based on the olivine→wadsleyite geothermometer, we know that the temperature must be significantly higher than 1,000°C throughout the transition zone. Those higher temperatures would *facilitate* the endothermic ringwoodite→perovskite phase change, allowing it to be complete at slightly shallower depths (this inference also follows from the Clausius-Clapeyron equation: since ΔS is positive for endothermic reactions, dT_{eq}/dP will be negative). Yagi et al. (1979b) concluded that the transition is sharp and that it corresponds to the 660-km jump in seismic velocities.

Jeanloz and Thompson (1983) considered more complex mineralogies, and argued that there was too much uncertainty to conclude that a phase change to perovskite was as sharp as observed seismically at 660 km; in contrast,

if the 660-km discontinuity represented a change in chemical composition, there would be no difficulty explaining its sharpness. However, this alternative would imply that the lower mantle was chemically distinct from the upper mantle—thus, either the lower mantle does not convect, or it convects separately from the upper mantle; whole-mantle convection would be ruled out.

The results obtained by Yagi et al. (1979b) were refined by the subsequent work of Ito and Takahashi (1989). Conducting experiments on olivine at 1.600°C (see Figure 8.26c), they found that the phase change to perovskite was even sharper, occurring over a depth range of at most a few kilometers; and that the inferred depth or pressure, 23.1 (± 0.08) GPa, was the same, regardless of the amount of Fe present, as long as the iron fraction did not exceed ~25%. Their findings, supported by the later experiments of Shim et al. (2001), Chudinovskikh and Boehler (2001), and others (see also Figure 8.25c), establish reasonably strongly that the 660-km jump in seismic velocity corresponds to the ringwoodite→bridgmanite phase change, and does not mark a chemical boundary between upper and lower mantle.

Furthermore, these experiments yielded estimates of the temperature at 660-km depth in the range ~1,525°C–1,725°C (implying, though with much uncertainty, a temperature gradient of ~0.3°C/km–0.5°C/km between the first and second transitions). We will see below that such temperatures also hint at the need for whole-mantle convection.

Our explanation of the 660-km discontinuity as a phase change of ringwoodite to bridgmanite depends strongly on the upper mantle being relatively iron-poor. Earlier in this chapter, we concluded repeatedly that this was the case (likely, in fact, for the entire mantle). Can you recall the basis for such a conclusion?

Pyroxene 'Flavors' the Transition Zone.
High-pressure, high-temperature experiments have also focused on pyroxene, the other major component of the upper mantle. For example,

Liu (1976b) found that pyroxene's magnesium end member, MgSiO$_3$ (enstatite), undergoes a series of changes, ultimately transforming from MgSiO$_3$ (now with an ilmenite structure) to bridgmanite at pressures of ~26–25 GPa and temperatures of ~1,000°C–1,400°C. Subsequent experiments by Ono et al. (2001), Ito and Takahashi (1989), and Fei and Bertka (1999, as presented in Burnley, 2011) find the Mg-ilmenite→bridgmanite change takes place at ~24.6 GPa, ~23.6 GPa, ~23.8 GPa, respectively, for a temperature of 1,300°C, and at ~23.8 GPa, ~22.8 GPa, ~23.0 GPa for a temperature of 1,600°C. As detailed below, the intermediate products of that series include spinel (β and γ), so it may not be surprising that the series leads to the same end result, bridgmanite, as those beginning with olivine, at similar pressures and temperatures. And, our comments above concerning the effect of higher temperatures on the transition depth apply here as well (Ito & Takahashi, 1989). The bottom line is that we can view the 660-km depth seismic velocity jump as a consequence of phase changes originating from *both* olivine and pyroxene that end up as bridgmanite.

Based on the research cited in this section, we can summarize transition-zone mineralogy simply as follows:

but these pathways are only valid in an iron-poor mantle. For example, in a dominantly iron mantle, as we could infer from Figures 8.25c and 8.26c, phase changes to wadsleyite and to bridgmanite would not take place; and ilmenite (which is mainly Mg-ilmenite) will only exist if the abundance of iron is less than ~20% (e.g., Jeanloz & Thompson, 1983).

We mentioned earlier that the dense oxides MgO and FeO are sometimes expressed as a solid solution (Mg, Fe)O called magnesiowüstite or ferropericlase. Also noted earlier, SiO$_2$ under mid- and lower-mantle conditions is a dense phase of silica known as **stishovite**. In an iron-rich mantle, the second transition would lead to a lower mantle of stishovite and magnesiowüstite (see Yagi et al., 1979a,b; Jeanloz & Thompson, 1983).

In this summary, phase changes that correspond to one of the seismic discontinuities of the transition zone are indicated with red double arrows. For pyroxene, the various intermediate products of the pyroxene phase reactions occur *within* the transition zone (according to Liu, 1976b, for example, the transition to wadsleyite + stishovite occurs at 19 GPa and that to ringwoodite + stishovite at 20 GPa, at temperatures of ~1,000°C–1,400°C; see also Fei & Bertka, 1999; Burnley, 2011). These phase changes would further enhance the seismic velocity slope within the transition

---

olivine

$$(Mg, Fe)_2SiO_4 \Rightarrow (Mg, Fe)_2SiO_4 \rightarrow (Mg, Fe)_2SiO_4 \Rightarrow (Mg, Fe)SiO_3 + (Mg, Fe)O$$

olivine   wadsleyite   ringwoodite   bridgmanite   periclase
(end-members are   (perovskite)   +wustite
forsterite, fayalite)

pyroxene

$$2(Mg, Fe)SiO_3 \rightarrow (Mg, Fe)_2SiO_4 + SiO_2 \rightarrow (Mg, Fe)_2SiO_4 + SiO_2$$

pyroxene   wadsleyite   stishovite   ringwoodite   stishovite
(end-members are
enstatite, ferrosilite)

$$\rightarrow 2(Mg, Fe)SiO_3 \Rightarrow 2(Mg, Fe)SiO_3$$

ilmenite   bridgmanite
(perovskite)

zone (Table 8.3); but it is only the final phase change to bridgmanite at the (bottom) boundary of the zone that contributes to the 660-km depth velocity jump.

Additional changes in mineralogy not considered in our discussion, resulting from the presence of Al or Ca, may also occur within or just below the transition zone (see, e.g., Frost, 2008; Bass & Parise, 2008).

The Lower Mantle. Experiments by Knittle and Jeanloz (1987) on (Mg, Fe)SiO$_3$ at pressures up to ~110 GPa (corresponding to depths approaching 400 km above the CMB) and temperatures exceeding 2,000°C confirmed that bridgmanite is stable through at least most of the lower mantle. That stability and the phase reactions summarized above suggest that—though the individual minerals are different in the lower mantle than in the upper—the bulk chemistry and overall mean atomic weight appear to be essentially unchanged through the depths of the mantle. Over the years, intense research has challenged this conclusion, exploring the possibilities of the second transition as a chemical discontinuity (as noted earlier); of a different lower-mantle mineralogy in general; and of increased iron content toward the base of the mantle in particular (see, e.g., Matas et al., 2007). The latter possibility, which would have implications for geomagnetism, is briefly discussed in Chapter 10. Models of lower-mantle mineralogy could also benefit from better constraints on lower-mantle temperatures (e.g., Fiquet et al., 2008).

We actually have samples that very likely originated in the lower mantle. These samples, obtained from kimberlites in Orroroo, Australia and Koffiefontein, South Africa, contain diamonds with inclusions of magnesiowüstite (or ferropericlase) and/or enstatite; since bridgmanite is not a stable phase at low pressure, it would have gradually reverted to a phase, like enstatite, which is stable under surface conditions, as mantle convection carried it up toward the surface. High-pressure, high-temperature experiments and consideration of trace mineral partitioning (Kesson &

FitzGerald, 1991; see also Scott-Smith et al., 1984, referenced therein) established that those inclusions existed at pressures of ~30–50 GPa (depths of ~800–1,200 km, well into the lower mantle). Additionally, the mineralogy of the Australian and South African inclusions was similar enough that Kesson and FitzGerald were able to conclude that lower-mantle chemistry must be fairly uniform laterally.

D″. Our story does not end with bridgmanite, however. Mg-rich bridgmanite has been found to undergo a phase change at deep-mantle pressures: called **post-perovskite** (but not post-bridgmanite!), that phase is denser, and with a rutile-like, octahedral sheet lattice structure; it was first observed experimentally at pressures in the range of ~126–134 GPa for pure Mg-bridgmanite, and ~116–126 GPa for Mg-rich bridgmanite with other lower-mantle minerals present, at temperatures ~2,200–2,600K (Murakami et al., 2004, 2005; see also, e.g., Oganov & Ono, 2004). From the Murakami et al. (2004, 2005) experiments, the (P, T) conditions at the phase change itself were estimated to include (125 GPa, 2,500K) for pure bridgmanite and (113 GPa, 2,500K) for the mix. Estimates of the Clapeyron slope allow T_{eq} to be predicted for a given composition at other pressures; in general (since the slope is positive), higher pressures require higher temperatures.

The strong compositional dependence of the pressures and temperatures at which this phase change occurs was discussed by Hutko et al. (2008); we focus first on pure Mg-bridgmanite. Additional (P, T) values defining its phase change were measured experimentally by, e.g., Oganov and Ono (2004) to be (120 GPa, 2250K); Hirose et al. (2006), (113–119 GPa, 2400K); and Tateno et al. (2009), (122 GPa, 2500K). All of these pressures correspond to depths within a few hundred kilometers or so of the CMB, suggesting at first thought that post-perovskite is likely to be a major constituent of the lowermost mantle, including the lowest part, the region we know as D″; D″ is distinguished by pressures ranging *from* ~118–127 GPa (depending on how D″ is defined) up *to* ~136 GPa, the pressure at the core boundary.

However, D″ is thermally complex, and that ends up resulting in a somewhat tentative role for post-perovskite there. Tateno et al. (2009) found that the phase change to post-perovskite took place at 3,520K when the pressure is 136 GPa. But, as we will see in the next chapter, temperatures at the core boundary are still higher than that—they are more like ~3,900K, according to van der Hilst et al. (2007); that is, the bottom of the mantle is too hot for post-perovskite to exist there. If it turned out that all of D″ was too hot, post-perovskite would be irrelevant to Earth's interior.

Within a fully convecting mantle, D″ acts as a thermal boundary layer (this was mentioned earlier in relation to the Yellowstone plume). In Chapter 9, we will learn that thermal boundary layers are distinguished by a steep temperature gradient, in contrast to the convection cell interior (where the temperature rise versus depth is quite gradual). That means temperatures in D″ go from being very high near the CMB to much cooler at the top of D″. A short distance above the core, where the pressures are still high (i.e., still in the realm of bridgmanite→post-perovskite), it might be cool enough that post-perovskite can exist stably. On the other hand, by the top of D″, the pressure and/or temperature might be too low to create post-perovskite, so rock there would have remained bridgmanite.

Add to this scenario the likelihood that temperatures vary laterally (e.g., Kawai & Tsuchiya, 2009; Schuberth et al., 2009), whether at the top of the core or anywhere within D″, and we can envision how D″ might be structured: post-perovskite might be present in some regions, but not everywhere around the globe; where it exists, post-perovskite will not extend to the depth and height of D″—instead, 'melting away' as the CMB is approached, for example—but will take the form of *lenses* (van der Hilst et al., 2007; see also Tateno et al., 2009 and references therein). Some lenses would be higher up or lower down than others, and some locations would have no post-perovskite lenses at all. In short, D″ would be structurally complex, and not a simple and globally uniform layer.

Now expanding our focus, we consider the effects of other mantle minerals on the transition of bridgmanite to post-perovskite. With plate tectonics as a guide, we look to the top of the mantle. In order to end up with a basaltic oceanic crust (MORB) at mid-ocean ridges, Ringwood (as noted earlier in this chapter) had proposed that the mantle composition be pyrolytic. This suggests two rock types may be especially relevant to D″: pyrolite, as a general constituent of the mantle, starting out as that particular combination of olivine and pyroxene, and subject to the various phase changes versus depth we have already discussed; and—as a consequence of ongoing lithospheric subduction, potentially reaching the lowermost mantle—MORB.

High-pressure high-temperature experiments conducted by Ohta et al. (2008) found that pyrolite undergoes a transition to post-perovskite at 2,500K beginning at ~116 GPa, completing the transition at ~121 GPa (thus, a transition over a depth range of 90 km in the lower mantle).

Can you explain why pyrolite's phase change requires a range of depths for its completion?

The products of this phase change actually include ~72% post-perovskite and 21% ferropericlase (Hirose & Lay, 2008). Experiments on MORB at the same temperature yield transition pressures of 112–118 GPa, implying a first appearance for post-perovskite about 70 km shallower than in the case of pyrolite at that temperature. Ohta et al. (2008) also found that, in contrast to bridgmanite or pyrolite, the transitions from MORB lead to decreases in shear-wave velocity.

Post-perovskite is denser than bridgmanite, suggesting that lens boundaries should exhibit jumps in seismic wave velocity. High-pressure experiments have in fact determined that post-perovskite is characterized by a measurable increase in μ, the shear modulus, compared to bridgmanite, while its adiabatic incompressibility k_S is hardly changed (Oganov & Ono, 2004; Hutko et al., 2008; see also Tsuchiya et al., 2004). The net effect is a predicted ~1%–2% change in V_S, a jump up at the change to post-perovskite and a jump down at the base of the lens, where post-perovskite reverts back to bridgmanite. Careful analysis of *S* wave reflections from the lowermost mantle (generically denoted *SdS* waves by some) should allow the lenses to be located (Lay & Garnero, 2007). (Lenses may also be detected seismically from the relatively high anisotropy of post-perovskite, which leads to shear-wave splitting; both of those concepts are discussed later, in the context of the upper mantle, but see also Hirose and Lay, 2008.)

The first detection of anomalously fast *S* wave velocities in the lower mantle was by Lay and Helmberger (1983); they found a jump in V_S at or near the top of D″, located at depths that varied laterally by up to 40 km, with the 'jump' taking perhaps as much as 50 km for completion. Subsequent analyses of seismic data (see, e.g., Lay & Garnero, 2007 and Hirose & Lay, 2008 for reviews) have confirmed the existence of such a "D″ discontinuity" in V_S; it is not global in extent but, where it is present, it has roughly that 1%–2% magnitude. A thickness of ~25 km (Reasoner & Revenaugh, 1999; see also the reviews just cited) for this phase transition (i.e., 'discontinuity') has been inferred from the data.

Thomas et al. (2004a) also identified regions where a second D″ discontinuity exists below the first—a jump down to a slower shear-wave velocity; for example, beneath Eurasia the upper discontinuity was found to be ~300 km above the CMB, while the lower discontinuity was ~70 km above the CMB. In effect, such 'paired discontinuities' mark the upper and lower boundaries of post-perovskite lenses (Lay et al., 2006; see also Thomas et al., 2004b, Hernlund et al., 2005; Tateno et al., 2009).

These phase transitions, predicted to exist in the lowermost mantle, are deep, probably gradual, and nonglobal—a sharp contrast with, say, the 405-km and 660-km discontinuities and thus a significantly greater challenge to detect seismically. Interpreting them is even more challenging because of trade-offs between composition, phase, and temperature; a change to lower *S* wave velocities with depth, for example, can be interpreted to result from any or all of those factors. As a quantitative example, a 10% increase in the amount of iron silicate in post-perovskite will decrease V_S by the same amount as a 1,000K increase in its temperature (Lay et al., 2006). The wide range of temperatures expected in thermal boundary layers adds even more possibilities to temperature's impact on D″ phase changes. And the nonuniqueness is not necessarily reduced even when multiple features (such as paired discontinuities) are analyzed together.

Hutko et al. (2008) were able to clearly locate D″ discontinuities below the Cocos plate at ~300 km and ~50 km above the CMB. They concluded that the volume enclosed within those jumps in seismic velocity is largely post-perovskite, and unlikely to possess the kind of variations in bulk composition some others have envisioned, namely an increase in Fe or Al. In combination with the results from high-*P*, *T* experiments, they estimated the temperature at 300 km above the core boundary to be roughly 2,500K (—thus, post-perovskite as a geothermometer).

In addition to D″ discontinuities, some lower-mantle structures detected by seismic tomography have been interpreted in terms of the post-perovskite phase change. In one portion of the LLSVP beneath the Pacific (in an area, incidentally, not at all close to the region beneath the Cocos plate studied by Hutko et al., 2008, though this LLSVP is adjacent to the sub-Cocos region), Ohta et al. (2008; see also Lay et al., 2006) identified phase changes involving post-perovskite they could ascribe to a mixture of pyrolite and MORB, plus a phase change due to partially melting MORB (thereby creating an ULVZ). They concluded that it was evidence for deep subduction of oceanic crust (ultimately separated from descending oceanic lithosphere).

There is no doubt that post-perovskite exists within D″; but it is easy to imagine that the overall high temperature and lateral variability limit how widespread it is. That is, post-perovskite is likely to occur only in limited, somewhat colder regions of D″ (Schuberth et al., 2009)—perhaps (we might speculate) in association with relict lithospheric slabs; the rest of D″ is just too hot. Our view of the thermal and compositional effects of mantle plumes and lithospheric slabs in D″, discussed earlier in this chapter, may be only marginally modified by post-perovskite.

Convection Connections in the Mantle. Our interpretation of the first and second transitions as the result of phase changes in olivine and pyroxene, together with the experimentally verified stability of bridgmanite—as well as our inferences from Birch's rule, equations of state, and lower-mantle diamond inclusions—means that the major chemistry throughout the mantle can be described (to a first approximation) as wholly peridotitic in origin. Such chemical homogeneity supports our view that the mantle is dynamically well mixed, that is, that it undergoes whole-mantle convection rather than containing regions (e.g., upper versus lower mantle) that remain both convectively and chemically isolated.

In a static mantle, the transition-zone phase changes would exist at depths prescribed by the ambient pressures and temperatures. In a convecting mantle, in particular where the mantle flows vertically, however, the temperature conditions will be perturbed—warmer than the surroundings within upwelling plumes, and colder within downwelling material—and this should distort the transition zone in those regions. To predict the effects, we can recast the Clausius-Clapeyron equation in terms of depth z rather than pressure, using hydrostatic equilibrium:

$$\frac{dT_{eq}}{dz} = \frac{dT_{eq}}{dP}\frac{dP}{dz} = \frac{\Delta V}{\Delta S}\frac{dP}{dz},$$

or

$$\frac{dT_{eq}}{dz} = \frac{\Delta V}{\Delta S}\rho g,$$

where ρ and g are the density and gravity at the depth in question. As noted earlier, hydrostatic equilibrium is a good approximation, despite the obvious lack of 'static' in a convecting mantle.

For an order of magnitude estimate of the distortions, we apply this equation to *up*welling at the first transition, where the olivine density is very roughly 3.6 gm/cm³ = 3,600 kg/m³; for the wadsleyite→olivine phase change, Bina and Wood (1987) estimated $\Delta S \sim 2$ cal/K, which equals 8.374 J/K, and $\Delta V \sim 3.1$ cm³ or 3.1×10^{-6} m³. Taking $g \sim 10$ m/sec², we find $dT_{eq}/dz \sim 0.0133$ K kg m/(sec² J) = 0.0133 K/m. Finally, then, we can rewrite this as $dz/dT_{eq} \sim 75$ m/K. For the Yellowstone plume shown in Figure 8.7, and using the velocity-temperature relation we inferred from Birch's rule, we might estimate the change in temperature at the first transition (where the velocity contrast in that plume is evidently ~0.5%) to be $\Delta T \sim 160$K; this yields an increase in the transition's depth of $(dz/dT_{eq}) \cdot \Delta T \sim 12$ km. For the South Pacific superplume (Figure 8.19i or 8.20a), the velocity contrast, temperature anomaly, and depth increase at the first transition might be double that of the Yellowstone plume.

In a cold, descending lithospheric slab, with ΔT now negative, olivine within the slab would change to wadsleyite at a depth ~10–20 km *shallower* than normal for the first transition. And, because the second transition is endothermic rather than exothermic, the perturbations to its depth are the opposite—and thus *anticorrelated* with the perturbations of the first transition, in both upwelling and downwelling regions!

The topography of the transition zone can be directly inferred from seismic travel-time observations rather than thermodynamic measurements and theory; see Figure 8.27a. Unfortunately, the observations have yielded mixed results, in terms of their convective implications. The results of one of perhaps the most supportive studies, a global analysis by Flanagan and Shearer (1998), are shown in Figure 8.27b. They corrected the travel-time

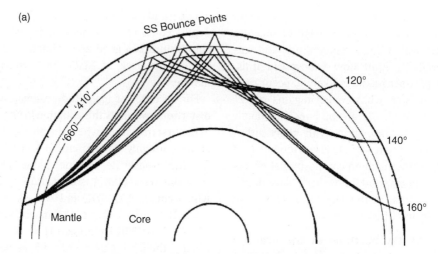

(a)

SS Bounce Points

120°

140°

160°

'410'

'660'

Mantle   Core

Figure 8.27 Transition zone topography as determined by Flanagan and Shearer (1998) / John Wiley & Sons. (a) They compared travel times of *SS* waves from the PREM model with times of waves that bounced off the undersides of the first and second transitions (at depths of 410 km and 660 km); the latter reflected waves, which arrive slightly earlier than SS, are denoted S410S and S660S in their work. Since those 'precursors' follow very similar paths to SS except near their bounce points, differences in their travel times indicate the depth of those discontinuities beneath the SS bounce points.

data for the influence of heterogeneity above the transition zone using a seismic refraction model of the crust by Mooney et al. (1998) and a tomographic mantle-wide model by Masters et al. (1996). They found typical magnitudes of the topography to be ±19 km for the first transition (the range of depths was 398 km to 436 km) and ~±25 km for the second (with a depth range from 633 km to 682 km); in the globally smoothed (spherical harmonic) depiction of their model shown in Figure 8.27b, these magnitudes were ±11 km and ±19 km, respectively.

Our rough predictions of the *magnitude* of the transition-zone topography are generally consistent with these observations. More comprehensive predictions would have accounted for the dependence of ΔV and ΔS measurements on temperature and composition (see, e.g., Bina & Helffrich, 1994), as noted earlier, and would determine temperature perturbations from seismic velocity anomalies by employing more experimentally based, explicit relationships than simply what is implied by Birch's rule.

Regarding our *qualitative* predictions of the transition topography, however, consistency

with the seismic observations is not so evident. The topography of the second transition (in Figure 8.27b) follows our predictions qualitatively, being, for example, deeper beneath subduction zones; but the first transition does not. In earlier research, Bina and Helffrich (1994) had argued that the topography of the first transition may be harder to track in downwelling regions; in that case, perhaps a slightly more robust test of our predictions might be the thickness of the transition zone, measured as the distance between transitions. Figure 8.27b shows that where the second transition is deeper, the transition zone tends to be thicker; and (to some extent) where the second transition is shallower, the transition zone is thinner—thus hinting at the anticorrelation in transition topographies we would expect from Clausius-Clapeyron.

However, other studies (see the review by Shearer, 2000) have found little support for an anticorrelation between first and second transition topographies. Improved corrections for heterogeneity in the crust and uppermost mantle could yield more accurate depths for the transitions, but would have less effect on the transition-zone thickness (see Shearer,

(b)

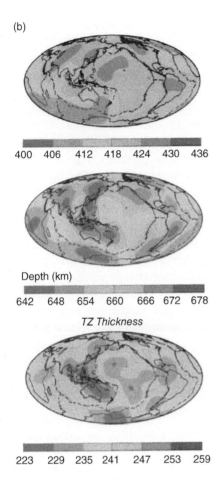

400	406	412	418	424	430	436

Depth (km)

642	648	654	660	666	672	678

TZ Thickness

223	229	235	241	247	253	259

Figure 8.27 Transition zone topography as determined by Flanagan and Shearer (1998) / John Wiley & Sons. (b) Inferred depths to the 410-km discontinuity (upper) and 660-km discontinuity (middle), contoured using spherical harmonics with the color scales shown in kilometer. The difference between those depths yields the thickness of the transition zone, also spherical harmonically contoured in kilometers (bottom).

2000). Deuss et al. (2006) showed that fluctuations in the 660-km transition topography are complicated by compositional as well as temperature perturbations at that depth, with effects that will vary from region to region; these should obscure any correlations (or anti-correlations) with the 410-km transition topography. In contrast, there is more general agreement for there being a convective connection between the 660-km topography and proximity to subduction zones, with a deeper transition when a cold lithospheric slab is nearby.

There is another convection connection worth discussing, also a consequence of the second transition being endothermic. As we will learn in the next chapter, the minerals constituting the mantle transfer heat very slowly; thus, a lithospheric slab subducting at even the slowest rate is able to reach a depth of 660 km without warming up appreciably. The situation then becomes quite curious: just above that depth, the descending slab is colder and therefore denser than its surroundings, with a tendency to continue sinking; but *right at* that depth, the surroundings (which have achieved the ringwoodite→bridgmanite phase change) are denser than the slab (which has not, because it is not warm enough to provide the energy needed to cause the phase change). Depending on a variety of factors—including the slab's temperature and composition, the thermodynamics of the phase change (e.g., the magnitude of the Clapeyron slope), and the length of slab acting as a coherent unit—that density contrast may be great enough to prevent the slab from entering the lower mantle, or sufficient only to slow its descent.

Thus, the second transition may to some extent inhibit whole-mantle convection. In fact, we saw possible evidence for this from seismic tomography, as discussed earlier in this chapter: the Farallon slab and other subducting plates appear to undergo decoupling or detachment at roughly the depth of the second transition (e.g., Figure 8.20b).

Given the support we have found for whole-mantle convection, however, it is unlikely that inhibition by the second phase transition will be completely effective; instead, as suggested by the leveling off of slabs above the second transition and their steepening below, the delayed transition to bridgmanite more likely causes a subducting slab to get stuck, temporarily, after which it 'cascades' down through the lower mantle. Such a situation may lead to times of intermittent layered convection preceding "avalanche instabilities;" alternatively, while the downwelling is intermittent, plume-related upwellings may continue throughout even the episodes of layered convection (see, e.g., Machetel & Weber, 1991; Honda et al., 1993; Weinstein, 1993; Tackley et al., 1993, 1994; Solheim & Peltier, 1994; Goes et al., 2017; also Olson & Yuen, 1982; Christensen & Yuen, 1984, 1985; Ringwood, 1991; Peltier & Solheim, 1992; Bunge et al., 1997). In the latter case, modes of pure whole-mantle convection would be alternating with modes of mixed whole-mantle/layered convection. Either way, the endothermic second phase transition clearly adds a significant unsteadiness to the mantle flow—an unsteadiness already implied from our earlier discussion of plate tectonics.

The downwelling cascade may help explain how remnants of lithospheric slabs make it all the way down to D''. With any further descent hindered by the 660-km phase transition, the slabs pile up above the lower mantle. The newer additions to the pile, energized by the exothermic 410-km transition (e.g., Tackley et al., 1994) but not yet having reached 660 km, are still pushing downward. Speculatively, at some point, as the slabs accumulate, the downward push is sufficient to bring the older slab material—which has meanwhile warmed up somewhat—down to P,T conditions where it can transform into the denser bridgmanite phase. Then, as the pile tumbles further into the lower mantle, the combination of heat and pressure causes successively younger slabs to also change phase, increasing the overall density excess and powering the descent downward through the lower mantle. This ongoing cascade could provide the slab pile with enough momentum to make it to the CMB—with remnants surviving the heat there for a long time.

The ability of the endothermic phase transition to inhibit whole-mantle convection also limits Earth's efficiency as a 'heat engine,' both by slowing the convective cycle and by rationing cold subducting slabs into the lower mantle.

In general, we should not find it surprising that phase transitions can affect mantle convection; in fact, we saw in Chapter 3 that atmospheric convection is strongly affected by the condensation of water vapor as warm humid air rises into the sky. That phase transition is exothermic, and ends up intensifying the upwelling. Interestingly, although the atmosphere is more chaotic than Earth's mantle (and the time scale for its convection is perhaps 11 orders of magnitude shorter!), there are even times when that condensation phase change produces a group of clouds with the same base level—marking that phase boundary in the sky.

In this discussion, we have mainly focused on the inhibition of convective downwelling by the endothermic ringwoodite→bridgmanite phase change. However, upwellings see the reversed phase change when they encounter the transition zone from below. The reversed phase change is exothermic, making the analogy with atmospheric convection even more apt, and analogously enhancing the upwelling. Many of the convective models cited above in fact show mantle plumes persisting throughout both whole-mantle and layered-mantle convection.

How Can Upwelling and Downwelling Be So Different? Such a curiously asymmetric type of convection, with downwelling and upwelling so different in structure, speed, and behavior, is to be expected in part because the viscosity of mantle material is temperature-dependent: the viscosity is lower at higher temperatures. The heat of an upwelling plume lowers the viscosity of the mantle it is rising through, making it easier to ascend, perhaps even creating a conduit to facilitate its ongoing rise. In contrast, the cold of the descending slab viscously thickens the surrounding mantle, somewhat, so that as it descends, the slab brings more of the surroundings along with it; extensive sheet-like structures, instead of plumes, are the result.

Conceivably, mass conservation plays an important role here as well. Those broad, downwelling 'sheets' have the potential to transfer much mass to the depths of the mantle; scattered, upwelling plumes—even a couple of superplumes—have narrower structures, area-wise, and are incapable of returning equivalent amounts of mass to the upper mantle unless their rates of transfer are much faster. The second transition, being endothermic, helps, by inhibiting downwelling (and encouraging any upwelling); but its effectiveness in counteracting the mass imbalance depends on how long the subducting slabs are hung up at the transition.

We know downwelling is accomplished via subducting slabs and upwelling via plumes; if it proves necessary, can you suggest another means of returning mass to the upper mantle—perhaps one that is broader scale than plumes…?

Finally, evidence that the second transition does not completely eliminate whole-mantle convection also comes from the estimates of temperatures at ~405 km depth and ~660 km depth. As noted above, those temperatures imply a temperature increase with depth not exceeding $\frac{1}{2}$°C per km within the transition zone—a rate that is approximately equal to the adiabatic gradient within the mantle. In the next chapter, we will learn more about the physics of convecting fluids, and we will find that the temperature gradient must be much greater than adiabatic at the upper and lower boundaries of a convecting fluid (as we saw with the thermal boundary layer known as D″); such 'superadiabatic' gradients make the fluid unstable, promoting downwelling at the upper boundary and upwelling at the lower boundary. Thus, though the 660-km depth discontinuity may exhibit an inhibiting effect on mantle flow, the temperature gradient inferred from transition-zone phase changes implies that it is not a convective boundary.

High-Pressure Experiments on Iron, and—at Last—Confronting the Need for a Light Alloy in the Core. Hydrostatic pressure at the top of the core is 136 GPa; by the inner-core boundary it has increased to 329 GPa. High-pressure, high-temperature experiments have been carried out on metals at pressures as great as ~500 GPa and temperatures of thousands of degrees; for pure iron,

they yield densities under core conditions that are close to what we believe the density of the core to be, according to Earth models such as PREM (Figure 8.18a). This supports our expectation that iron is most likely the primary constituent of the outer core.

However, such experiments (e.g., Al'tshuler et al., 1958; McQueen & Marsh, 1960, 1966; Jeanloz, 1979; Mao et al., 1990; Boehler, 2000; Uchida et al., 2001; see also Komabayashi, 2014; Komabayashi & Fei, 2010; and Birch, 1964) generally find the high-P,T iron to be about 8%–10% (or ~1.0 gm/cm^3) denser than the outer core; see, for example, Figure 8.28a. Like equations of state and Birch's rule, high-pressure experiments thus point to the need for a lighter substance alloyed with iron in the core. High-P,T experiments measuring other properties, such as incompressibility or bulk sound velocity, also find pure iron to yield different values than Earth models allow for the outer core (e.g., Jeanloz, 1979; Brown & McQueen, 1986).

Historically, candidates for the light alloy have included *oxygen* (a very abundant and active element), *silicon* (a nonvolatile element that could persist in the early Earth even if the Earth formed hot), *carbon* (iron meteorites often contain small amounts of graphite), *sulfur* (iron meteorites often contain small amounts of sulfur compounds, in minerals called *troilite*), and even *hydrogen* (the most abundant element). As argued by Stevenson (1981) and Poirier (1994) it is certainly possible that *most* of these lighter elements are present in the outer core in at least small amounts, as long as (Jeanloz, 1979) they are well mixed throughout the outer core; which ones are significant would depend on how the Earth accreted (e.g., in a hot or cool solar nebula environment) and when the core formed (early or late relative to accretion). An illustration of their effectiveness in reducing the density of the outer core is shown in Figure 8.28b.

Most research has focused on sulfur and silicon. High-pressure experiments have found that an iron-sulfur alloy with ~7%–15% (wt.) sulfur (King & Ahrens, 1973; Alder

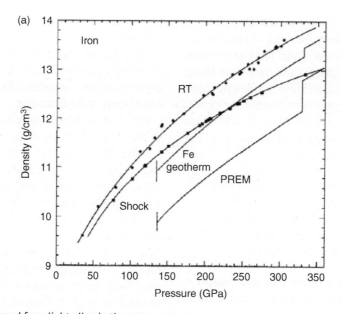

Figure 8.28 The need for a light alloy in the outer core.
(a) High-pressure experiments on iron yield its density, in gm/cm³, shown here versus pressure, in GPa,
rather than depth; the figure is from Boehler (2000) / John Wiley & Sons. For comparison, the densities from
PREM are also shown.
Results are shown here for two types of high-pressure experiments.

1. Densities from the diamond anvil experiments of Mao et al. (1990) were conducted at room temperature
   (RT); the curve labeled "Fe geotherm" takes this density profile and adds in the effects of thermal expansion
   to core temperatures. We see that, under core conditions, pure iron has a density ~1 gm/cm³ or ~10%
   greater than actual core densities throughout the outer core. Such a discrepancy apparently persists into
   the inner core, though to a lesser degree.
2. In contrast to diamond anvils and presses, which create high pressures by 'statically' loading the sample,
   even higher pressures can be achieved 'dynamically' by explosively shooting projectiles into the sample
   (this and other 'dynamic' approaches are described by Ahrens, 1980; see also Ahrens, 1979). The impact
   produces a shock wave, which propagates through the sample and creates very high pressures temporarily.
   Conservation of mass within the sample on either side of the shock 'front,' as the shock wave passes
   through it, allows the shock density to be determined (see also Stacey & Davis, 2008).
   The shock compression is neither adiabatic nor isothermal. Stacey & Davis also discuss how conservation
   of momentum and conservation of energy allow the pressure and temperature of the shocked material to
   be inferred; the equation of state relating these variables, called a *Hugoniot*, then allows the density to be
   extrapolated to core *P,T*.
   Incidentally, the production of a shock wave is conceptually similar to the way a cresting and breaking
   wave is produced at the coast, the only difference being the dependence of wave speed here is on density
   rather than water depth. As the projectile continues to impact the target medium, the density of the
   medium increases, allowing compressional waves to propagate faster; those faster waves catch up with
   the earlier waves, successively slamming into them and building energy until—moments after impact—a
   shock front is created and propagates through the sample.
   The shock-wave data shown here is from Brown and McQueen (1986).

& Trigueros, 1977; Ahrens, 1979) or an
iron-silicon alloy with ~7%–20% silicon
(Balchan & Cowan, 1966; Alder & Trigueros,
1977; Lin et al., 2003) would yield the observed
outer core densities.

Quantitatively, all of these high-*P,T* estimates, whether for iron or its alloys, carry a

number of uncertainties. Major uncertainties derive from poorly known temperatures
within the core (and therefore, for example,
the degree of thermal expansion), a subject
that will be touched on in Chapter 9, and from
difficulties in accounting for phase changes
in (solid) iron. The presence of nickel in the

core also adds to these uncertainties, though to a lesser extent. Nickel is the second-most abundant heavy element in the Universe, after iron; and, it is generally found in iron meteorites (which, as discussed in Chapters 1 and 2, originated as the cores of differentiated proto-planets in the Asteroid Belt). Thus, it is likely that Ni is present in Earth's core as well. At the kind of abundances found in meteorites (~5%–10%, e.g., Ringwood, 1977), an Fe-Ni alloy would have a slightly higher density than pure Fe—up to a few tenths of a gm/cm^3 more at the inner-core boundary, for example, according to high-pressure experiments (Mao et al., 1990; Sakai et al., 2014; also McQueen & Marsh, 1966)—as well as a slightly higher mean atomic weight. To counteract these effects, the fraction of light alloy would also have to be slightly greater.

Among the various candidates for light alloy, sulfur may well be the strongest. The abundance of sulfur in the Sun, in meteorites, and elsewhere throughout the Solar System is much greater than we find in the crust or infer for the mantle; if the light alloy is not sulfur, if the core contained no significant amounts of sulfur, an explanation of Earth's sulfur deficiency would be required. For example, spectral analysis of the solar photosphere reveals an abundance (#atoms) of sulfur relative to silicon in the Sun of ~0.617 (Palme & Jones, 2003), thus a relative abundance by mass of ~0.705; in Earth's crust + mantle, though, that ratio is ~0.00127 (see Newsom, 1995) or ~0.00240 (Stacey & Davis, 2008). That ratio (sulfur relative to silicon in Earth's crust + mantle) is similarly two to three orders of magnitude smaller than in chondrite meteorites (Murthy & Hall, 1972).

In fact, Murthy and Hall (1970, 1972) demonstrated that, although many inert gases and halogens are more volatile than sulfur (see, e.g., Palme & Jones, 2003), Earth's crust and mantle were evidently able to retain a greater percentage of them (relative, e.g., to meteorite abundances) than sulfur, from the time of Earth's accretion and meteorite formation to the present; see Figure 8.29. This is curious, indeed—unless the Earth did manage to retain more sulfur, just not in its crust or mantle.

If sulfur *is* the light alloy in the core, it would have to amount to ~14% of the core's mass, according to Figure 8.28b, in order to yield an Fe-Ni-S alloy under core conditions with a density 1.0 gm/cm^3 less than that of pure iron, as required by high-P,T experiments. For this estimate, the effect of Ni is taken to be an additional ~0.25 gm/cm^3 (Mao et al., 1990; Sakai et al., 2014; but cf. Anderson & Isaak, 2002) (so, for Figure 8.28b we envision an Fe-Ni alloy denser by 1.25 gm/cm^3 than the core). This abundance of sulfur in the core implies that its abundance relative to silicon

Figure 8.28 The need for a light alloy in the outer core.
(b) Fractional abundance of the most popular candidates for the light alloy, plotted versus the density deficit (*as a fraction*) each would create. Figure from Poirier (1994; see also Anderson & Isaak, 2002).
Silicon is less effective in reducing the core's density than sulfur is, despite its lower atomic weight (28 versus 32), because it occupies a smaller volume when mixed with iron than sulfur does, thus counteracting the decrease in density somewhat (see Poirier, 1994).

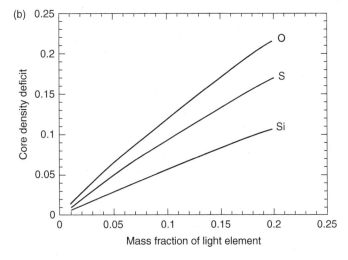

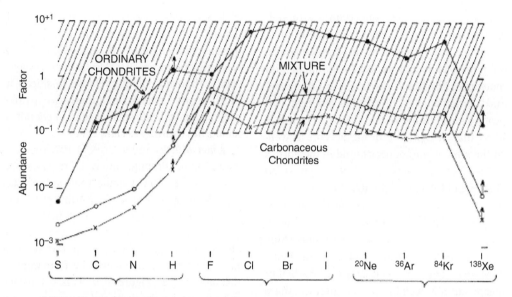

Figure 8.29 One reason sulfur might be a light alloy in the outer core.
This graph compares the abundance of various elements within the Earth, calculated relative to silicon, to their relative abundance in meteorites. For each element, its relative abundance is defined as the number of atoms per 10^6 atoms of silicon. The abundance *factor*, plotted on the y-axis, is the ratio of the element's relative abundance estimated in Earth's crust + mantle (but not core) to the relative abundance in meteorites. Estimates marked with upward arrows are likely underestimates. y-axis is a log scale, with the shaded region correctly marked here as extending from 10^{-1} to 1 to 10^{+1}.
The comparison is shown for three different kinds of meteorites. Carbonaceous chondrites are the most primitive, originating in undifferentiated parent bodies, and may be considered fairly representative of the composition of the primordial solar nebula. Ordinary chondrites, the most commonly found on Earth, may have originated from specific near-Earth parent bodies. "Mixture" refers to a hypothetical combination of both kinds of chondrites plus iron meteorites.
The elements displayed here include halogens and noble gases, almost all of which are more volatile than sulfur. However, despite its lesser volatility, the relative abundance of sulfur is extremely low.
Figure from Murthy and Hall (1972) / NASA / Public Domain.

(S/Si) for the whole Earth would be ~0.32 by mass, still lower than solar and meteoritic abundances but by much less than an order of magnitude, and now close enough that it can more reasonably be ascribed to natural variations in nebula composition at the time of planetary accretion. And, depending on the amount of Ni, this abundance of sulfur would give the core a mean atomic weight of 50.7—

(Can you verify this number? Hint: you will have to convert abundance by mass into abundance by #atoms, for sulfur, iron, and the whole core.)

—a number close to (though slightly higher than) what was implied by our earlier use of Birch's rule.

The amount of sulfur this analysis predicts for the core is thus consistent with a wide variety of constraints, including:

- the density of iron inferred from high-pressure experiments;

- the primordial abundance of sulfur in our Solar System: and

- the mean atomic weight of the core implied by Birch's rule.

Such broad consistency makes sulfur very believable as a light alloy in the core. In Chapter 9, we will discover an additional, dynamic benefit of sulfur being the light alloy in the outer core. In short, our explanation of the deficiency of sulfur in the crust and mantle is simply that, as the core formed during and after Earth's accretion, with molten iron globs sinking through the mantle, sulfur followed the iron down into the core. Sulfur is **siderophilic**—that is, it has a chemical affinity for iron (as evidenced by its tendency to appear as troilite in iron meteorites)—so its association with iron during this process is reasonable.

Densities inferred from high-pressure data also require a light alloy in the solid inner core (e.g., Uchida et al., 2001; Fiquet et al., 2001). But the presence of that inner core complicates the story. The inner core is able to remain solid, despite high temperatures within the core, because of the great central pressures it experiences. As the core cools, the outer core fluid becomes cool enough to solidify at progressively lower pressures—that is, over time, the inner core grows. From the PREM model, the density jump at the inner-core boundary is 0.6 gm/cm^3, or 0.8 gm/cm^3 with the reinterpretation of seismic data by Masters and Gubbins (2003). Some of this is due to the phase change from liquid to solid—about 0.3 gm/cm^3 according to the analysis by Anderson and Duba (1997)—leaving ∼0.5 gm/cm^3 to be explained. Since the presence of a light alloy in the outer core was invoked to reduce the core density by ∼1 gm/cm^3 (∼1.25 gm/cm^3 including the effect of Ni), this in turn suggests that the inner core has just under half the light alloy abundance of the outer core, i.e. ∼6% (that is, to reduce the density by 0.5 gm/cm^3, which is 40% of 1.25 gm/cm^3, we need 40% of the outer core's light alloy abundance of 14%) if the alloy is sulfur.

This conclusion must be handled cautiously: once we allow the possibility of inner and outer cores being compositionally different, the inner-core boundary is no longer required to represent the freezing point of the outer core fluid—just as, off the Antarctic coast, salty bottom-water-to-be remains liquid while the adjacent water ice is frozen. The inner-core boundary can be at a temperature lower than its own freezing point, as long as it is not lower than that of the outer core fluid. Lower inner

core temperatures, it might be noted, would amplify the need for a light alloy.

With such challenges in mind, high-*P,T* data has been interpreted as evidence of a variety of inner core compositions, ranging from pure Fe(/Ni) with no light alloy—Jeanloz (1979) and Brown and McQueen (1986) found seismic velocities observed in the inner core to be consistent with those determined for pure iron at high-*P,T*—to as much as half the light alloy abundance of the outer core (e.g., Jephcoat & Olson, 1987; Anderson & Duba, 1997; and Boehler, 2000). These latter estimates agree well with our own deduction from the density jump at the inner-core boundary and, if the light alloy is sulfur, yield an abundance in the inner core of ∼6%, as noted above. The low-end estimate, a pure Fe(/Ni) inner core within an alloyed outer core, has been a popular view ever since Birch (1952) first proposed it.

Whatever the fraction of light alloy in the inner core, the fact that it is less abundant than in the outer core implies that, as the inner core grows, *the solidification of outer core fluid at the inner-core boundary expels much or all of the light alloy from the solid iron (nickel) matrix, leaving it in the outer core fluid*. This would be analogous (but in an upside-down way) to the process of Antarctic Bottom Water formation (discussed in Chapter 3) resulting from the coastal temperature conditions just referred to: there, freezing of ocean water expelled salt, leaving denser saltwater outside the ice to sink down to the seafloor; here, the freezing of outer core fluid expels light alloy, producing even lighter outer core fluid that will tend to rise up away from the inner core. The result is a *chemical convection*—we could think of it as a *thermosulfine* (rather than thermohaline, if the light alloy is sulfur) *circulation*—and we will see in later chapters that it contributes to the process of geomagnetic field generation (also called the *geodynamo*).

Finally, the relative scarcity of light alloy in the inner core may provide constraints on the number of alloys in the core.

- If the inner core is alloy-free (except for Ni), then most likely only one light alloy is present in the outer core: if two or more were being expelled by solidification, the inner-core boundary would be more like a transition zone (unless, somehow, they were expelled at the same pressures and temperatures). With a single light alloy, it is also reasonable to maintain that it is sulfur, given that element's deficiency in the crust and mantle.

- If the inner core has half the outer core's abundance of light alloy, the situation is ambiguous. One possibility, which we considered above, is that it is the same light alloy in inner and outer core; this requires that the alloy is only partially expelled as the inner core grows. Another possibility is that there are two (or more) light alloys in the outer core, but only one gets expelled upon solidification.

According to a geochemical analysis by Alfe et al. (2002), sulfur should be only slightly less likely to exist in the inner core than in the outer core (i.e., it partitions nearly equally into solid and liquid iron, and its expulsion upon crystallization is slight). In that case, for the inner core to have half the alloy of the outer core, a second, equally abundant light alloy must be present in the outer core, one that was effectively expelled from the crystallizing inner core. Oxygen has in recent years become a popular candidate to consider as the key light alloy (Poirier, 1994). And, the analysis by Alfe et al. (2002) predicts that oxygen will indeed partition into the outer core liquid during solidification. The core model by (e.g.) Stacey and Davis (2008) accordingly postulates an outer core with ~9% (by mass) sulfur and ~5% oxygen, and an inner core with 8% sulfur and essentially no oxygen. (From Figure 8.28b, the effect of 5% oxygen on core density is the same as ~7%–8% sulfur, so their model does in effect correspond to an inner core with half the light alloy of the outer core. Their model also includes ~6%–7% nickel.) Lastly, the analysis by Badro et al. (2007) concluded that oxygen and silicon are the major light alloys in the core (though small amounts of sulfur could

not be ruled out); mostly expelled from the inner core, oxygen would be the primary light alloy of the outer core, leaving silicon as the primary inner core light alloy.

Unfortunately, despite the plausibility of all these candidates, identifying the inner core's actual light alloy(s) with confidence is out of our reach. As pointed out by Poirier (1994), the likelihood of a particular light alloy ending up in the core would have depended first of all on the conditions of core formation and how the early Earth evolved. Uncertainties in the interpretation of high-P,T experiments add to the challenge; for example, some research cited by Poirier (1994) finds that oxygen will fail to be expelled during crystallization and will actually partition into the solid inner core. A comprehensive exploration of this subject is beyond the scope of our textbook. Nevertheless, it is probable that the inner core contains less light alloy than the outer core does, so we can still say that the process of inner core growth supports some level of thermochemical convection in the outer core.

Implications for the Viscosity of the Earth's Interior. With some knowledge of the composition within the Earth, it is possible to better understand important physical properties of the interior. One material property critical to the operation of the entire Earth System is its viscosity. Though traditionally considered a property of fluids, we have also seen it to be useful for describing the deformation of the solid earth. Defined as the ratio of applied stress to resulting strain rate, as discussed in Chapter 7 (see also Chapter 5), the viscosity of olivine, or any other solid mineral, can be measured from its deformation.

Laboratory experiments on olivine at a range of pressures show that its deformation depends on composition (including the percentage of fayalite versus forsterite, and the water content) and grain size (e.g., Karato et al., 1986; Hansen et al., 2011); the experiments additionally suggest that the mechanism of deformation may undergo a transition from creep to grain-boundary sliding, as the applied stress increases through the range of upper-mantle pressures (Ohuchi et al., 2015). These kinds of deformation (see, e.g., Lowrie, 2007) require an 'activation' energy to work effectively (e.g., Hansen et al., 2011), with a higher strain rate and greater deformation possible when more energy (heat) is available. It follows that viscosity—which is described by Tozer (1972) as "creep resistance"—is strongly temperature-dependent, in that the viscosity goes way down at higher temperatures and way up at lower temperatures: at higher temperatures, more of the lattice vacancies, dislocations, and other defects per volume attain the "activation" level needed for creep to occur—what constitutes 'flow' of the material. This can be expressed mathematically, as

$$\eta = \eta_0 \exp(f \cdot T_M / T) = \eta_0 e^{(f \frac{T_M}{T})},$$

where η is the material's viscosity, T_M is its melting temperature, and f is a material constant; $f \cdot T_M$ relates to the activation energy for the creep or grain-boundary sliding mechanism of its deformation.

This expression gives us a *reference point* for temperature effects: the melting point, T_M. The hotter a solid medium is, that is, the closer it is to its melting temperature (so, mathematically, the smaller the exponent $f \cdot T_M / T$ is), the more easily 'flow' can occur. As a function of depth within the mantle, then, temperature competes with pressure to determine how viscosity varies: at greater depths, under greater pressure, the melting temperature T_M should be higher, so more energy ($f \cdot T_M$) is required for extensive creep to take place, and the viscosity will increase—unless the temperature has increased correspondingly.

This kind of temperature dependence is relevant to olivine and pyroxene, and should also apply to other 'polycrystalline' minerals, including perovskite and even periclase (cf. Ammann et al., 2010); in short, the bulk of the mantle should be characterized by a temperature-dependent viscosity. This characterization has several consequences for mantle convection.

- **Convective Regulation.** Overall, that is, globally and over long timescales, we expect that mantle convection—which acts to bring heat out of the interior—should be *self-stabilizing*: if the convection becomes too vigorous, the mantle will cool and the viscosity will increase significantly, causing the convection to slow; if the convection is too sluggish, the heat within the mantle will build up, eventually causing the viscosity to drop and the convection to accelerate. In effect, temperature-dependent viscosity acts as a "thermal regulator" (Tozer, 1972), allowing the mantle to maintain a moderate—well, moderately hot—interior 'climate.'

  This process implies that mantle convection and its surface expression—plate motions—must be unsteady. In the next chapter, we will learn how to estimate the time scales for such convective regulation.

- **Asymmetric Heat Budget.** We have already mentioned that asymmetric mantle convection can be explained as a consequence of temperature-dependent viscosity. Mantle flow will be 'exponentially' easier where the mantle is hotter and correspondingly more difficult to achieve where the mantle is colder; since mantle minerals do not conduct heat to their surroundings very effectively, this implies in turn that hot mantle plumes should be relatively narrow 'conduits' exhibiting rapid flow upward (hence, "plumes"!) (see also, e.g., Loper & Stacey, 1983). The temperature dependence also implies that descending, cold lithosphere will have a high viscosity, leading to broad 'fronts' of slowly sinking lithospheric slabs. Both types of structures are consistent with what we had inferred earlier in this chapter from seismic tomography.

  The narrowness of plumes limits their ability to transport heat up to the surface, and raises the question of how effectively they can cool the mantle. For example, a quick estimate by Davies (1988b) found that the amount of heat brought up by plumes roughly equals the amount of heat currently emanating from the outer core (see also, e.g., Mittelstaedt & Tackley, 2006 for more refined calculations); but the total heat output from the mantle, as we will learn in the next chapter, is an order of magnitude greater! This imbalance would have been even more problematic when the Earth was younger and a greater amount of heat needed to get out.

One possible solution to this disparity is concisely summarized by the phrase "plumes cool the core while slabs cool the mantle" (Mittelstaedt & Tackley, 2006, who provide a further but unverified attribution to this quote). And slab cooling would operate throughout the mantle, thanks to slab penetration into the lower mantle (as we have seen tomographically).

In the next chapter, we will learn more about the basic concepts of heat flow and convection that ultimately determine the mantle's heat budget. For now, we just note that the complexity of the mantle's convective asymmetry, and any associated heat flow imbalance, may be compounded by a much less dramatic but potentially significant component of mantle convection: a background, broad-scale return flow that is generated in response to the convective downwelling driven by (or driving) lithospheric subduction. That is, whether convective currents in the upper mantle are carrying lithospheric plates with them as they flow or are pushed by those plates as the latter move and subduct, those currents eventually head downward (it's not just the slabs themselves that descend). Such 'background' flow also requires a return flow—a sluggish but pervasive upwelling—in order to conserve mass (e.g., Figure 8.22; see also Steinberger & O'Connell, 1998). We have already seen evidence of this background, mantle-wide circulation, in the tilting of the Yellowstone (Figure 8.7) and Hawaiian plumes. This background circulation has the potential to contribute significantly to the mantle's heat budget.

- **Nature of the Asthenosphere.** The mechanical connections between plate motions and the underlying mantle convection have long been debated, and may be complex; for example, the mid-ocean ridges where oceanic lithosphere forms are not necessarily located above deep convective upwellings. That is, lithospheric motion and the mantle's convective flow may be imperfectly coupled, even somewhat independent. How strongly they are coupled depends primarily on the properties of the weak layer comprising the top of the convective flow, and its contrast with the overlying lithosphere. That weak layer—which we associate with the LVZ, the low-(seismic) velocity zone or low-viscosity zone—is, of course, commonly called *the* asthenosphere; we will follow that custom but take this opportunity to remind the reader that the *entire* mantle below the lithosphere is very likely a weak layer, capable of flow.

Given the temperature-dependent viscosity of peridotite, it is hard to avoid concluding that temperature plays a role in the mechanical differences between lithosphere and asthenosphere. At the very least, we might say that, as the coolest part of the mantle (versus depth), the lithosphere will be the most rigid, with the highest viscosity (e.g., De Bremaecker, 1977, 1980).

Correspondingly, the weakness of the asthenosphere has traditionally been attributed to the effects of temperature: at the depths of the LVZ, despite the effects of pressure (which by itself would increase the viscosity, e.g. by increasing T_M), it has become hot enough to lower the rock's viscosity, and hot enough to lower the rock's seismic velocities through decreases in k and especially μ. Maintaining such a thermal perspective, we might also say that the limited depth range of the LVZ is a consequence of pressure effects ultimately dominating—pressure continues to increase significantly with depth, while the rock heats up only slightly more as we descend below the zone into the 'core' of the convective cells (a region where, as we will see in the next chapter, the temperature gradient is at most only ~adiabatic).

However, the thermal perspective is problematic: experiments have found that reasonable predictions of the temperature increase into the asthenosphere do not produce as much of (or as sharp) a drop in viscosity or seismic velocities as are required by isostatic or seismic studies (see Fischer et al., 2010).

Incidentally, a purely thermal explanation of the lithosphere's strength may be problematic as well. Models developed by Solomatov and Moresi (1997) suggested, initially with reference to the planet Venus (Solomatov & Moresi, 1996), that the temperature dependence of peridotite might be so severe that, without any mitigating factors, the cold lithosphere would be too stiff to ever rift apart; this would lead to a single lithospheric layer covering the globe—a *stagnant lid*, prohibiting plate tectonics and severely limiting cooling of the interior. Factors that might lower the lithosphere's effective rigidity, brittle strength, and viscosity enough to allow rifting and prevent a stagnant lid from ever occurring on Earth include compositional variations, preexisting fractures, or some type of rheological weakening within the lithosphere (see, e.g., Karato, 2010 and references therein).

To produce an asthenosphere as weak as seismic and isostatic analyses require, we might speculate that temperatures within the layer are great enough for partial melting to occur; thin 'pockets' of melt, distributed throughout the zone, could effectively decrease both its overall viscosity and seismic velocities. However, experiments summarized in Hirth and Kohlstedt (1996) show that, for 'dry' peridotite (i.e., peridotite with no water present), temperatures in the LVZ are *not* likely to be high enough to produce any melt (Figure 8.30a). It appears that other factors must also play a role in creating the asthenosphere.

Do you agree? According to Figure 8.30a, how deep in the upper mantle below a mid-ocean ridge could the expected temperature be hot enough for partial melting to occur, under dry conditions?

Water Is Crucial Here. Building on earlier work by Kushiro et al. (1968) and others, researchers have established the ability of *water* to substantially weaken the rocks of the asthenosphere (see also, e.g., Karato et al., 1986; Nimmo & McKenzie, 1998; Bolfan-Casanova, 2007; and the reviews by Fischer et al., 2010 and Karato, 2010). We would expect the mantle to contain at least small amounts of water, because the Earth may not yet have fully degassed its primordial share and because water is continually recycled back into the (upper) mantle as oceanic lithosphere subducts. Water in the asthenosphere would have the effect of lowering the viscosity there because it lowers the melting temperature (T_M) and thus activation energy of olivine (Kushiro et al., 1968; also, e.g., Nimmo & McKenzie, 1998). As suggested by the experimentally based results depicted in Figure 8.30a, T_M could be shifted by ~10^2 °C or more, depending on the layer's water content. Water would also increase the number of lattice defects in the minerals (e.g., Bolfan-Casanova, 2007), reducing f (in the expression for η) and thus, again, reducing the activation energy. These effects can also lower the seismic velocities of the layer (as discussed in Hirth & Kohlstedt, 1996).

The effects on viscosity can be quite significant; Hirth and Kohlstedt (1996), for example, find η to be at least two orders of magnitude smaller than η for dry olivine under mid-ocean ridge, upper-mantle conditions.

In lowering the melting temperature, asthenospheric water might also lead to more partial melting. The direct consequence of that would be to lower the overall viscosity and seismic velocities of the layer; however, Karato (1986) has pointed out that partial melting would draw water out from the rock, reducing the effects of water on activation energy and thus, indirectly, hardening the layer. As

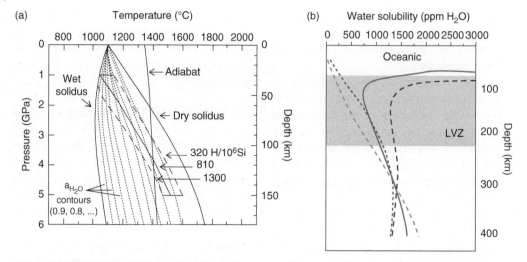

Figure 8.30 The existence of the asthenosphere is more than a temperature *versus* pressure phenomenon. (a) Experimentally based illustrations of the effect of water content on the solidus temperature (i.e., the temperature at which partial melting begins) for peridotite, as a function of depth within the mantle. Olivine water content ranged from none (dry solidus) to saturated (wet solidus). The postulated temperature gradient within the mantle and below a mid-ocean ridge is labeled "adiabat." Note first that in a dry mantle, the LVZ—existing at depths between ~80 km and 220 km, depending on location—will not be hot enough for any melting to occur.
The solid curve labeled 810 represents the solidus for a typical water content (in units equivalent to 50 ppm water by weight) in olivine derived from peridotite with a composition that would also yield mid-ocean ridge basalts. This and the adjacent (dashed) curves suggest that the liquidus, T_M, could also differ by ~10^2 °C at depths corresponding to the LVZ, as a result of modest variations in the amount of water present. Figure from Hirth and Kohlstedt (1996) / with permission of Elsevier.
(b) Experimentally based models of the solubility of water, in ppm by weight, versus depth for pure olivine (dashed green line), pure enstatite (dotted black curve), Al-saturated enstatite (heavy dashed black curve), and a peridotite-type mix with 60% olivine and 40% Al-saturated enstatite (solid red curve). Temperatures are assumed to be typical for oceanic areas, with the depth of the LVZ indicated by heavy gray shading. As discussed in the text, the minimum in solubility suggests that the LVZ is relatively water-rich, thereby lowering the minerals' melting temperature and promoting partial melting.
Figure from Mierdel et al. (2007) / American Association for the Advancement of Science.

implied by Hirth and Kohlstedt (1996), the direct effect would begin to eclipse indirect effects only if the layer is more than 8% partially melted, though other research indicates much smaller amounts of partial melt can cause a net decrease in viscosity (e.g., Takei & Holtzman, 2009; see also Fischer et al., 2010).

The lower boundary of the LVZ, at least for young oceanic regions, appears to be much less sharp than the top boundary (see Figure 8.13). Why this is so deserves an explanation as well: our understanding of the nature of the asthenosphere must be judged incomplete until we can explain why the layer of low viscosity and low seismic velocities ends that

way and neither ends sharply nor extends all the way down through (e.g.) the upper mantle.

One 'fortuitous' possibility is simply that water just happens to become increasingly scarce at depths below ~200 km—perhaps as a trade-off between whole-mantle convection, which would eventually ventilate the mantle below, and plate tectonics, which recycles water back into the upper mantle. However, the discovery by Mierdel et al. (2007) of a minimum in *the solubility of water in peridotite* at the depths of the LVZ may provide a stronger answer. In their experiments, it is not olivine but pyroxene (enstatite) whose ability to store water has the most relevance—especially pyroxene intermixed with aluminum (the latter conceivably derived from garnet). Mierdel et al. (2007) found that the presence of aluminum greatly increases the solubility of water in enstatite, and that this solubility has a strong depth dependence, with a *minimum* at depths that could correspond to the LVZ; see Figure 8.30b. For example, a mix of olivine and Al-saturated enstatite has a solubility minimum at depths ~80– 220 km beneath oceanic plates. At such depths, the expulsion of water from those minerals would raise their activation energy, but that water could also produce abundant melt, leading to a partial melting dominated LVZ. And, at greater depths the expulsion would lessen (as would the availability of water), all these effects would weaken, and the zone would fade away—leaving, in the bulk of the mantle below, a viscosity whose variation with depth is controlled mainly by its temperature dependence.

No one said mantle convection would be simple!

- **Other material properties shed light on convection and the asthenosphere.** Finally, one tangential consequence of the polycrystalline nature of mantle silicates concerns the possible anisotropy of their material properties. Individual crystals of olivine, perovskite, and other silicates are **anisotropic**, meaning their elastic properties are different along one crystal axis than another; for example, the difference in *P* wave velocity through olivine can be as much as 9.9 km/sec versus 7.7 km/sec (Stein & Wysession, 2003). However, their random orientation within the mantle guarantees that the mantle medium is, overall, **isotropic** (i.e., its properties are the same from any direction).

But as microscopic creep and grain-boundary sliding progress during convection, the crystalline fabric of the convecting minerals can become aligned with respect to the direction of 'flow,' producing an anisotropy such that *P* waves and *S* waves propagate at different speeds in the flow direction than in other directions. Furthermore, for a particular direction of propagation, the 'twisting' motion of the shear wave, which we can resolve into two orthogonal components (e.g., *SV* and *SH*, or S_{fast} and S_{slow}), will sense different resistance to deformation in one direction than the other; as those components continue to propagate, at different speeds, a slight but measurable "shear-wave splitting" is produced as one wave advances ahead of the other. This phenomenon, along with other measures of anisotropy, has the potential to reveal the flow directions of the convecting mantle.

The review by Savage (1999) illustrates several regions around the world where flow directions have been inferred. Most studies find anisotropic source regions to be in the upper mantle (or transition zone or, interestingly, D″). As it turns out, the bulk of the lower mantle exhibits only minor anisotropy; experiments suggest this is because perovskite tends to recrystallize randomly and with smaller grain size (thus, isotropically) after deformation or heating, obscuring any flow-generated crystal alignment (Karato & Li, 1992; Meade et al., 1995).

The upper-mantle anisotropy often seems to be connected to tectonic deformations in the overlying crust or lithosphere. Perhaps the strongest anisotropy is found in regions like New Zealand, an obliquely convergent plate boundary area where the tectonics have produced significant shear and compressional deformations (Klosko et al., 1999); although significant shear-wave splitting is inferred at depths as great as perhaps 400 km, the anisotropy clearly correlates with surficial structural features.

Anisotropy has been detected in the asthenosphere, a not-unexpected outcome given the relatively extreme deformability of that layer. The results of Nettles and Dziewonski (2008), shown in Figure 8.30c, delineate the LVZ below the Pacific basin. We see that, although its thickness varies, the depth of its top boundary appears to be

(c)

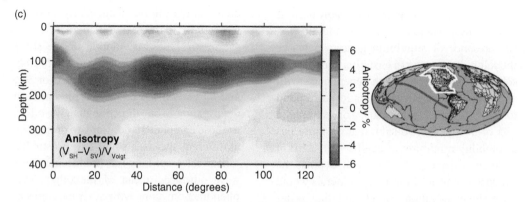

Figure 8.30 The existence of the asthenosphere is more than a temperature *versus* pressure phenomenon. (c) Figure from Nettles and Dziewonski (2008) / John Wiley & Sons showing the anisotropy (measured as differences in SV versus SH velocity, relative to a globally averaged velocity, V_{Voigt}) in the upper mantle beneath the Pacific plate, inferred from extensive surface wave observations and contoured as a function of depth. The location of this cross-section is indicated by a red line on the world map.

As a weak zone, the asthenosphere can be expected to exhibit a higher degree of anisotropy than the lithosphere above or mantle below, thus the inferred region of high anisotropy effectively delineates the LVZ. The fact that its top boundary remains nearly horizontal over the extent of the plate implies (Fischer et al., 2010) that temperature effects—which would cause all thermal boundaries to deepen as the plate cools—must be supplemented by additional factors. As discussed in the text, the effects of water solubility and content may be especially important in the upper portion of the asthenosphere; deeper in the asthenosphere, water may be less dominant (and note that the bottom of the LVZ does indeed deepen away from the East Pacific Rise).

constant regardless of the age or temperature of any portion of the oceanic lithosphere. As we will learn in the next chapter, cooling of the lithosphere as it moves away from the mid-ocean ridge where it was created causes it to subside. The lack of such subsidence in the lithosphere-asthenosphere boundary has been interpreted (Fischer et al., 2010) as still further evidence that nonthermal factors are dominating that boundary, and influencing the character of the entire asthenosphere.

It might be noted that anisotropy is not necessarily limited to the elastic (e.g., seismic) properties of the mantle. The possibility of anisotropic viscosity, and its implications for mantle convection, have been studied since the 1980s. For example, Honda (1986) explored the hypothesis that the LVZ includes horizontal layers of melt, producing a lamination of hard and soft layers, and found that the resulting anisotropy could explain the very broad (lateral) scale of upper-mantle convection. Christensen (1987) suggested that anisotropic viscosity could enhance the formation of mantle plumes. Such studies are, however, beyond the scope of this textbook.

In Chapter 5, we saw that the magnitude of the mantle's viscosity is extraordinarily high, in fact orders of magnitude greater than almost any other physical property found in nature. As a liquid, the outer core should be expected to have a lower viscosity than the mantle; as it turns out, though, the outer core's viscosity is incredibly low—the same viscosity as water, or only slightly higher!

More precisely, the outer core's *molecular* viscosity is, most likely, quite low. An extensive tabulation of its estimates may be found in Secco (1995). Based on theoretical and experimental studies on iron, other liquid metals, and metal alloys at various temperatures and pressures (though most experiments were conducted at low pressure), the viscosity of the outer core fluid is generally predicted to be in the range 10^{-3}–10^{+1} Pa-sec. Some additional studies are summarized in Assael et al. (2006) with values in the same range. For comparison, the viscosity of water at STP is ~10^{-3} Pa-sec (10^{-2} Poise).

As documented by Secco, this range for outer core viscosity (which is based on a number of works including a widely quoted estimate by Gans, 1972) appears to be fairly robust in terms of temperature and type of alloy. That is, the viscosity of molten iron—already at its melting point—will decrease further with an increase in temperature, but only modestly rather than by orders of magnitude. And, higher alloy content (whether Ni, Si, S, or O) produces only minor variations in the fluid's viscosity. Theoretical analysis by Poirier (1988) suggests that the estimates should be somewhat independent of pressure as well. In short, these estimates can be assumed to apply, to a fair approximation, under core conditions. Such robustness in core viscosity is also supported by the work of Dobson (2002), who used a combination of experimental data and equations of state to predict the viscosity of liquid iron at the P,T extremes of the top and bottom of the outer core; he found those viscosities to be well within the range of earlier estimates.

But there is more to the story. A viscous outer core can exert frictional forces on the mantle above, if the core and mantle are in relative motion or are misaligned. A viscous core will also dissipate the elastic energy associated with its deformation and the mechanical energy associated with flow within it. Seismic and dynamic processes associated with the outer core can thus yield observational estimates of its viscosity, potentially allowing us to test whether the actual viscosity is indeed so low.

As tabulated in Secco (1995), however, such observational tests are unfortunately inconclusive. Inferred viscosities span too great a range, generally (but excluding some 'outliers') lying between 10^7 and 10^{12} Pa-sec based on the decay of seismic waves; between 10^3 and 10^{10} Pa-sec based on the decay of periodic rotational phenomena, including the wobble of the

Earth and tidally induced variations in spin rate; and between 10^4 and 10^7 Pa-sec based on theoretical models of geomagnetic phenomena taking place in the outer core. Also, a few of these analyses have been disputed (e.g., Rosat et al., 2004); and with others, the existence of additional energy 'sinks'—such as the core's electrical resistivity, converting to heat the energy of electrical currents generated by the iron's motion through the core's magnetic field—clouds the viscosity estimates.

Perhaps the best we can do is simply recognize that all these estimates indicate a high viscosity (compared to the molecular viscosity of iron), as if the various processes 'see' a much higher viscosity than do laboratory measurements. One way this can be consistent with the lab estimates is if the dynamic processes are measuring a *turbulent* viscosity in the outer core. As we saw in Chapter 6 when discussing tidal friction, the flow of a turbulent fluid experiences greater resistance than would result from its intrinsic molecular friction, because of the chaotic behavior of the fluid. If the core is turbulent, the viscosity we infer from observed core phenomena should be much larger than the experimentally measured values.

In fact—based on concepts that will be explained in the remaining chapters of this textbook—it is likely that the outer core *is* in a state of turbulence. This fact will influence our understanding of how the core convects and how the "dynamo" mechanism manages to sustain the Earth's geomagnetic field. So, it may well be that the core fluid exhibits a turbulent viscosity in the range 10^3–10^{12} Pa-sec.

Getting Back to the 'Big Picture.' Much of what we have learned in this chapter has been interpreted as features of a convecting mantle or core. However, we have not yet established that the mantle *can* convect, or that conditions in the mantle or core—or atmosphere or oceans, for that matter—are such that convection *should* be happening. Establishing the grounds for the onset of convection requires that we first formulate the governing equations of a convecting fluid or fluid-like body.

Our convective inferences for the mantle and core bear some similarities to the situation in the atmosphere or oceans. Secondary constituents can significantly influence the convection: water in the mantle; a light alloy in the core; salt in the oceans; and humidity in the air. Phase changes, whether endothermic or exothermic, are influential as well: crystallization of the inner core; the transition zone; freezing of ice at the ocean's edge; condensation of water vapor. Nevertheless, we expect convection in all those regions to differ markedly, in terms of dominant forces, flow patterns, and timescales. Again, an understanding of the basic governing equations is a necessary first step.

In short, the subject of our next chapter will be *heat flow in the Earth System*—how it works, both physically and mathematically; what the sources of heat are; and what we can learn about the Earth from it.

9

Heat From Earth's Interior

9.0 Motivation

In Chapter 8 we saw images of exotic structures in the mantle, many of which (according to their seismic signatures) are primarily thermal in nature and very likely related to mantle convection. As slow as that convection must be, it nevertheless takes a great amount of energy to drive it. In this chapter, we will learn how to measure the heat that exits Earth's interior, and discover that, from a geologic perspective—that is, compared to the energy involved in plate motions, mountain-building, and other geological activities that we associate with mantle convection—the heat flow out of Earth's interior is indeed huge. The sources of such heat, as we will see, are reasonably identified (though their relative contributions are still debated), and leave little doubt that internal heat sources are prolific enough to power that convection.

In contrast, from the perspective of the Earth System—for example, compared to the solar constant—the magnitude of this heat flow is actually quite small, too small to noticeably alter the daily energetics of the atmosphere or oceans. Examples of heat flow's more subtle influence on the Earth System will be explored briefly toward the end of this chapter. But the impact of Earth's internal heat and mantle convection on the atmosphere and ocean, long or short term, is achieved for the most part indirectly, as we have already seen in earlier chapters of this textbook: through plate activities that change the atmospheric chemistry (via outgassing and lithospheric weathering), the surface topography, and the configuration of ocean basins.

In learning how to measure heat flow, we will first focus on the transmission of heat by conduction; our investigation will lead to the phenomenon of diffusion, which we will find both conceptually and mathematically relevant to a wide variety of processes in the Earth System. And, with the equations governing such processes somewhat complex, we will introduce a simple technique, known as *scaling* the equation, for obtaining approximate solutions; that technique will also allow us to understand key aspects of those processes. For example, one important characteristic of diffusive processes is the *timescale* over which diffusion effectively operates; scaling will easily yield an estimate of that quantity. In the case of heat flow by conduction, scaling will tell us the time it takes for heat to effectively diffuse out of an object, thus (e.g.) the timescale for Earth to cool by conduction. Significantly, that timescale turns out to be dauntingly long.

All of this is a warm-up for our study of convection, which will require us to consider not only how heat is transferred, e.g., via upwellings and downwellings, but also what conditions promote the onset of convection, how convection develops, and what force balances would allow convective motions to persist. For the latter, we will begin to revisit $\vec{F} = M\vec{a}$—but do so gradually, since for fluid

Earth System Geophysics, Advanced Textbook 6, First Edition. Steven R. Dickman.
© 2025 American Geophysical Union. Published 2025 by John Wiley & Sons, Inc.
Companion Website: www.wiley.com/go/Dickman

dynamic applications both the "$\vec{F}$" and the "$\vec{a}$" will be more complicated than we have previously encountered. Scaling the equation will be particularly useful here, allowing us to understand which forces are most important within the mantle, or core, or any other component of the Earth System, as that component convects. And, scaling analyses will allow us to appreciate the importance of *dimensionless ratios* for characterizing the dynamics of the situation.

In sum, some of what we will learn in this chapter will be specific to the mantle, and some of it will be generally applicable to all components of the Earth System. This reinforces a fundamental truth about convection: convection operates in some way throughout the Earth System, but its nature—its timescale, intensity, and ultimate impact—is unique to each component. In Chapter 10, for example, we will conclude our study of the thermal behavior of the 'solid' earth with an exploration of the core, the source of the Earth's geomagnetic field; we will find that the need to maintain that field imposes specific energy requirements on core convection, and also places singular demands on the pattern of core flow.

9.1 Measuring Heat Flow

9.1.1 Basic Ideas and Practical Challenges

Most of us are familiar with the idea that *heat flows from hot to cold*. More precisely, the flow of heat from one place to another depends on the temperature difference between the two places: the greater that difference, the greater the amount of heat flowing; but the greater the distance between the two locations, for the same temperature difference, the weaker the heat flow will be.

Symbolically, we can let dT/dz denote the temperature difference per unit length—the temperature gradient—in the z-direction in some material; and let q_z denote the heat flow through the material in the z-direction. The units of q_z, by the way, are heat energy per unit time per unit area—per unit time because the heat is flowing, and per unit area because it flows across a surface (e.g., the surface of the Earth). Because of the latter property, q_z is also called the heat **flux**. In cgs units q_z might be cal/sec-cm^2; in mks units q_z might be Joules/sec-m^2 or Watts/m^2. The properties of heat flow just noted can be written

$$q_z \propto -\frac{dT}{dz}.$$

The minus sign in this expression ensures that heat flows from hot to cold. For instance (see Figure 9.1), if z is depth within the Earth and the temperature is 30°C at 1 km depth and 60°C at 2 km depth, then $dT/dz = 30$°C/km; the proportionality therefore implies q_z is negative, or in other words heat flows upward, i.e. in the negative z-direction, from 2 km depth to 1 km depth (thus, from hotter to colder). Alternatively, if the temperature was 60°C at 1 km depth and 120°C at 2 km, then $dT/dz = 60$°C/km, and we would expect twice

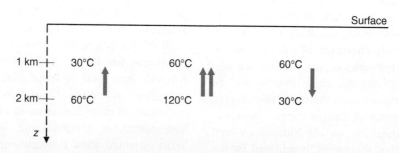

Figure 9.1 Heat flow in the solid earth under three different situations. At depths (z) of 1 km and 2 km below Earth's surface, the temperature is measured to be 30°C and 60°C (left), 60°C and 120°C (middle), or 60°C and 30°C (right). The resulting heat flow is indicated by arrows.

as much heat to flow upward. In contrast, if the temperature was 60°C at 1 km depth and 30°C at 2 km, then $dT/dz = -30°C/km$ (so, $q_z > 0$), and heat would be flowing downward, in the positive z-direction, again from hot to cold.

The actual flow of heat through the material depends on its thermal properties. For a given material, there is generally one constant of proportionality, typically denoted K, allowing us to change the proportionality into an equality:

$$q_z = -K\frac{dT}{dz}.$$

K is called the **thermal conductivity** of the material (to distinguish this symbol K from the symbol for Kelvins, i.e., the absolute temperature scale, in this textbook we will henceforth write °K instead of K for the latter!). Thus, the higher the conductivity, the more heat can flow in response to a temperature gradient.

This relation, which is known as **Fourier's law**, is analogous to Ohm's law in the field of electricity and magnetism (a relation particularly useful concerning electrical circuits), which will be discussed in Chapter 10. It is also analogous to Darcy's law, which governs the flow of groundwater through an aquifer, and Fick's law, which governs the transport of chemical species from regions of high concentration to low. Nature, it seems, is driven by differences....

Thermal conductivities can be measured in the lab or in situ; typical values at STP are 2.5 W/m°C (we could also write 2.5 W/m°K) for igneous rocks, 75 W/m°C for iron (but see later), and 0.60 W/m°C for seawater (e.g., Schatz & Simmons, 1972; Stacey & Davis, 2008; also Nayar et al., 2016); in cgs units, these conductivities are, respectively, 6.0×10^{-3}, 0.18, and 1.4×10^{-3} cal/sec-cm-°C.

So which conducts heat better, rock or water?

The flow of heat governed by Fourier's law is called *conduction*, and the equation itself is a one-dimensional *conduction equation*. In **conduction**, heat flows from one material to another, or from one part of a material to another part, by actual physical contact; the physical contact allows vibrations of heat energy to propagate from one material or portion thereof to the other. For example, when you sit on a hot boulder, your body is heated by conduction. Alternatives to conduction include *convection*, in which the heat is carried by hot materials that are moving, and *radiation*, in which the heat is carried by electromagnetic waves radiating from a hot object. Radiation can travel through a vacuum—no physical contact is needed—and is how the Sun warms the Earth (or that boulder you sat on!); radiation (back into space) is also how Earth gets rid of its excess energy.

Fourier's law provides a straightforward way to estimate the heat coming out of the Earth's interior: measure the temperature gradient, i.e. the temperature at two depths, at or near the Earth's surface. However, doing so on land requires that boreholes be drilled into crustal rock; drilling heats up the rock, and that heat must be allowed to dissipate, or the measured temperature differences will reflect heat flowing away from the borehole rather than heat flowing out of the interior. Later in this chapter, we will learn how to estimate the timescale for an object to cool, and discover that it is quite long; even for a shallow borehole, it might take a few years (e.g., Lachenbruch & Marshall, 1986) for that drilling heat to dissipate.

Compounding the problem is the need for the borehole to be drilled deep enough to avoid contamination by surface temperature fluctuations. Daily fluctuations (with the days typically hotter and the nights cooler) penetrate effectively to depths of ~1 m; seasonal fluctuations, to depths of ~20 m; and fluctuations with a timescale of 10^4 yr, such as the end of the last glacial period (or the end of the ice age), to depths of ~200 m. Incidentally, the strong frequency dependence of the penetration depth is called the **skin effect**, in that higher-frequency fluctuations in surface conditions penetrate only through the 'skin' of the body experiencing the fluctuation.

In contrast, such problems are expected to be minimal when it comes to measuring temperature gradients at the seafloor (and thus inferring the heat flowing out of the largest percentage of Earth's surface). Temperature conditions at the bottom of the ocean should be relatively stable, with the passage from day to night in the abyss imperceptible and the change of seasons almost so. And with the old seafloor there covered by a thick sediment layer, drilling should not be necessary—a long thermal 'spear' or probe dropped vertically into the ocean, piercing the sediment layer, could easily measure the temperature at two points along the probe, yielding the temperature gradient within the layer.

Of course, to determine the heat flow from such temperature gradients, the thermal conductivity of the medium must be known or estimated. Given the lab measurements noted above, can you provide a 'ballpark' estimate of what the sediment layer's conductivity would be?

Though thermal stability at the ocean bottom is generally assumed, it has been questioned (see Davis et al., 2003). One way to test the assumption is with 'multiprobes,' allowing the inferred temperature gradients to be determined nearly simultaneously and compared at multiple adjacent locations; another is with deep boreholes, allowing individual gradient measurements to be taken below any possible 'skin depth.' Alternatively, the bottom water temperature can be monitored for variations. Results (Davis et al., 2003; Chen et al. 2016; Martin et al., 2022) show such variations—seen on timescales of both weeks and years—to be generally small, and their effects on the gradient to be mostly minor. Interestingly, the largest variations are found (Davis et al., 2003; Meinen et al., 2020) in the western portions of the Atlantic and South Pacific Oceans, conceivably due to a dynamic influence of western boundary currents on the deep and bottom waters below.

The first extensive geophysical analyses of land data (from mines, with measurements a mile or so down) date back to 1939 (e.g., Bullard, 1939), whereas the first seafloor measurements were in 1950 (Revelle & Maxwell,

1952). One illustration of the challenges in obtaining land data is that, by 1970, there were only ~550 reliable land values—but by then, nearly 2,500 seafloor values had been reliably determined (Lee, 1970 as cited in Stacey, 1977). In subsequent years (see Pollack et al., 1993), that disparity has been erased.

9.1.2 Heat Flow Data

The kinds of temperature gradients revealed by all these measurements range from ~10°C/km to 50°C/km on land (i.e., in mines and boreholes) and 30°C/km to 100°C/km in marine sediments—thus, roughly 30°C/km on land and perhaps two to three times that at the seafloor. Since heat flow is the product of conductivity and temperature gradient, and the thermal conductivity of seafloor sediments is less than that of rock, because of the presence of water, the heat flow (per unit area) is rarely twice as great through the ocean floor as it is on land (with the exception of young oceanic lithosphere). But even that much oceanic heat flow—a flux merely comparable to continental heat flow—was a surprise to those early scientists: they knew that continental crust is much more radioactive than oceanic crust is; its heat flow should be correspondingly greater, by the heat generated from radioactive decay.

In response to the findings of Revelle and Maxwell (1952), Bullard (1952; also Bullard et al., 1956) suggested that the mantle below the oceanic crust was probably more radioactive than the crust itself, perhaps due to fractionation when the Earth first differentiated; or, alternatively, the large oceanic heat flow observed might be the result of heat brought up from the interior by mantle convection. As we will see later in this chapter, neither explanation is satisfactory.

As the number of seafloor measurements of heat flow grew, and at the same time the tenets of plate tectonic theory became more widely accepted, a fundamental problem with seafloor measurements became apparent: though the heat flow measured at mid-ocean ridges was the highest of any location around the globe, it fell far short of what would be expected for a structure representing the molten 'birthplace' of oceanic lithosphere. Resolution of this problem came from the recognition, by Lister (1972) and others, of the importance of hydrothermally circulating seawater (at the crest of mid-ocean ridges) and pore water (throughout the young, fractured crust). Driven by the extremely hot temperatures there and fostered by the scarce sediment cover and relatively high crustal permeability, that hydrothermal circulation tends to cool the hot lithosphere, *carrying away its heat* and *reducing the conductively measured heat flow*. (That circulation was also responsible for the 'carbonatization' of the upper layers of the oceanic crust, as discussed in Chapter 2, locking up carbon dioxide as part of the Urey Cycle.)

As we understand the situation today, then, calculating the heat flowing out of the seafloor from temperature measurements there (using Fourier's law) leads to *underestimates* of the heat given off by oceanic lithosphere (Lister, 1972); oceanic heat flow is in reality *much* larger than continental heat flow! The average heat flow in various regions, taken from Pollack et al. (1993), is listed in Table 9.1, both including and excluding the effects of hydrothermal circulation experienced by young oceanic lithosphere; heat flow excluding those effects was estimated from a thermal model developed by Stein and Stein (1992) which was based on measurements in locations with enough sediment cover to have suppressed any such circulation. Heat flow is listed in mks units (milliWatts per m^2); these are now generally used rather than the 'classic' cgs units of microcalories per cm^2 per sec, which were known as "heat flow units" or **hfu** (for example, 61 mW/m^2 = 1.46 hfu).

As Table 9.1 illustrates, the effects of hydrothermal circulation are quite substantial for young oceanic lithosphere; excluding those effects, the mid-ocean ridges are revealed to be giving off heat at extremely high rates, as is appropriate for sites of ongoing magmatic activity.

And without the effects of hydrothermal circulation, the cause of high oceanic heat flow, at least near mid-ocean ridges, should now be evident (—*what do you think?*); but we will postpone stating that 'officially' until the end of the next section of this chapter, after we have looked more closely at the data and then formally and quantitatively assessed the various heat sources likely to be important for the mantle.

Returning to Table 9.1, we note that, despite the large number of measurements, the uncertainties shown in this table are not particularly low. But the error bars for measurements in very *young* oceanic regions are much higher still—due in part to the lack of sediment cover there, making measurements difficult to obtain, and in part to the shallowness of the seafloor, making the measurements subject to larger surface temperature fluctuations.

The heat flow estimates in Table 9.1 are regional averages. Noteworthy smaller-scale outliers include those associated with hotspots (Hawaii, Iceland, Yellowstone, etc.), which have heat fluxes ranging from ~60 mW/m^2 to three times that (Plesa et al., 2016 and references therein). And, as reported by Smith et al. (2009), *parts* of the Yellowstone Plateau exhibit heat flux values from the hundreds to tens of thousands of mW/m^2, the latter presumably due to nearby magma bodies. Davies and Davies (2010, but see also Jaupart et al., 2015) estimate that the total effect of hotspots on young oceanic crust is to add 1 Terawatt (1 TW $\equiv 10^{12}$ W) to the global total.

The values listed in Table 9.1 imply an average heat flow of ~100 mW/m^2 out of oceanic lithosphere and 65 mW/m^2 out of continental lithosphere; the global average is 87 mW/m^2 (Pollack et al., 1993). Multiplying by the Earth's surface area yields a total heat loss of 4.4×10^{13} W = 44 TW; 70% of that global heat loss comes from the seafloor. Subsequent work by Davies and Davies (2010; see also Jaupart et al., 2015; Wei & Sandwell, 2006a, 2006b), based on 50% *more* values than Pollack et al. (1993) had considered and employing a refined methodology, yields nearly 47 TW

Table 9.1 Average surface heat flow in different regions of the Earth.[a]

		Heat flow		
	Era or mean age[b]	mW/m²	± st. dev.	Without hydrothermal effects[c], mW/m²
Continental				
	Cenozoic sedimentary, metamorphic	63.9	27.5	–
	Cenozoic igneous	97.0	66.9	–
	Mesozoic sedimentary, metamorphic	63.7	28.2	–
	Mesozoic igneous	64.2	28.8	–
	Paleozoic sedimentary, metamorphic	61.0	30.2	–
	Paleozoic igneous	57.7	20.5	–
	Proterozoic	58.3	23.6	–
	Archean	51.5	25.6	–
Oceanic				
	0.8 Myr	139.5	93.4	806.4
	3.5 Myr	109.1	81.2	286.0
	14.5 Myr	81.9	55.5	142.2
	30.2 Myr	62.3	39.9	93.4
	47.2 Myr	61.7	29.3	75.7
	62.1 Myr	65.1	34.3	65.1
	75.2 Myr	61.5	31.5	61.5
	101.5 Myr	56.3	21.8	56.3
	131.5 Myr	53.0	21.6	53.0
	153.5 Myr	51.3	16.9	51.3

[a] Based on measurements at more than 20,000 locations representing about two-thirds of Earth's surface. Adapted from Pollack et al. (1993).
[b] Geologic era for continental regions; mean age of ocean floor for oceanic areas.
[c] That is, if no heat was carried away by hydrothermal circulation; these values in Pollack et al. were based on the observationally constrained plate cooling model of Stein and Stein (1992).

for the total (this includes the contribution from hotspots), with 69% of the loss worldwide coming from the seafloor, and a global average heat flow of nearly 92 mW/m².

This is a huge amount of energy. For comparison, the seismic wave energy radiated by earthquakes globally has been estimated as anywhere from ~0.01 TW to as much as 0.3 TW (Xu & Chao 2017; Stacey & Davis 2008; Lowrie, 2007; Kanamori 1977), depending on the time span considered and the earthquake threshold (e.g., including only great earthquakes or both

large and great events, spanning the twentieth century or just the past 40 yr). These estimates vary in part because seismic activity, especially involving great earthquakes, has not been uniform over the years.

Additional comparisons are similarly illustrative:

- The 'gravitational' energy associated with raising fault blocks during earthquake events—essentially, the energy of mountain-building—has been estimated as not quite

~7 TW over the past 40 yr (Xu & Chao, 2017; see also Chao et al., 1995).

- The energy of volcanic activity can be estimated as ~0.2–3.2 TW (Verhoogen et al., 1970 then Deligne & Sigurdsson, 2015, but see also Mason et al., 2004 and Pyle, 2015).

- The dissipation of Earth's rotational energy by tidal friction is ~3 TW, as shown later in this chapter, although most of that energy is probably dissipated in the oceans, not the solid earth (see Chapter 6).

Impressively (quantitatively speaking, that is, but not implying a value judgment, positive or negative), our civilization attained 'geophysical' levels of energy consumption during this past century: the rate at which fossil fuels were consumed by human activity around 1970 was estimated at ~8.4 TW (Williams & Von Herzen, 1974), and ~9.5 TW in the year 2000 (Clauser, 2011).

The heat output from Earth's interior dwarfs most of these processes, and exceeds them all. It would seem reasonable to conclude that interior heat is the energy source driving mantle convection and plate tectonics. From the perspective of Earth System Geophysics, then, we could go one step further and assert that all surface processes—atmospheric and ocean circulation, river flow, erosion and weathering, mountain-building, volcanic activity, earthquakes, and so on—have either heat from the Sun or heat from Earth's interior (or both) as their ultimate energy source.

But such a conclusion is premature, at this point in our discussion. First of all, we need to establish that the huge amount of heat passing across Earth's outer surface is physically plausible, i.e. that the sources of heat likely to exist within the Earth are sufficiently prolific. That investigation is the subject of the next section of this chapter. More fundamentally, though, we have yet to demonstrate that convection (as opposed to, for example, conduction) is the mantle's preferred means of transmitting heat out from the interior. That demonstration is

one of the main goals of this chapter, and will be gradually developed throughout.

A Closer Look at the Data. Before exploring the origin and nature of the heat sources, however, it might be helpful to know whether the surface heat flow that we measure originates deep within the mantle or is primarily a shallow, e.g. crustal, phenomenon. As it happens, there are some clues contained in the heat flow measurements themselves. These clues also help us better understand modern global heat flow data sets like the one summarized in Table 9.1.

- Continental heat flow is related to the radioactivity in surface rocks, more precisely to the heat—called **radiogenic heat**—generated by the process of radioactive decay in those rocks. Early researchers found that, on land, locations exhibiting greater radiogenic heat production displayed higher heat flow; surprisingly, the relation turns out to be quite *linear* (see Figure 9.2). The basis for a linear connection has been attributed to different, competing models of the distribution of radioactivity versus depth (e.g., Lachenbruch, 1968; Singh & Negi, 1979), and remains uncertain (see Stacey & Davis, 2008); but the fact of a connection is undisputed.

- Oceanic heat flow exhibits a strong time dependence, in that older regions are colder (i.e. have less heat flow). Continental heat flow, for example the regional averages listed in Table 9.1, hints at a similar time dependence, but much weaker. Perhaps more importantly, the time dependence is roughly linear for continental but very definitely *nonlinear* for oceanic heat flow.

We plot the oceanic heat flow data from Table 9.1, both original and excluding hydrothermal effects, versus age in Figure 9.3, along with a curve of the form

$$q_z = \frac{C}{\sqrt{t}}$$

where t is the age (time since formation at a mid-ocean ridge) of a section of seafloor

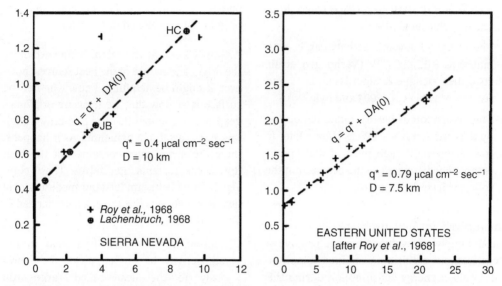

Figure 9.2 Evidence of a radiogenic source for continental heat flow in early measurements. The data (various symbols) is plotted as heat flow, on the *y*-axis and in units of hfu, versus radiogenic heat in surface rocks, on the *x*-axis and in units of 10^{-13} cal of heat produced per cm^3 per sec. Data is for the western United States (Sierra Nevada region) on the left and eastern United States on the right. An apparent linear relationship in both regions is suggested by the dashed lines. Image from Lachenbruch (1970) based on earlier data in Roy et al. (1968) and Lachenbruch (1968).

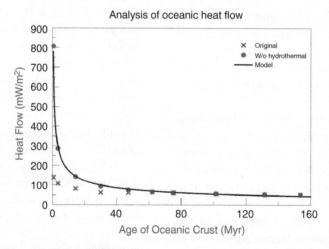

Figure 9.3 Implications of oceanic heat flow.

Regional surface heat flow measurements from Table 9.1 are plotted here versus the age of that portion of the seafloor; both the original values (x) and modeled values excluding (●) the effects of hydrothermal circulation are shown. The black curve is a plot of $C/\sqrt{t}$ versus age *t*, with *C* a constant whose magnitude was chosen here for a reasonable visual fit to the data.

As discussed in the text, the 'one-over-the-square-root-of-age' curve describes the pattern of oceanic heat flow we would expect within a plate tectonics environment; the discrepancy for young oceanic lithosphere between the original measurements and that curve illustrates the significance of hydrothermal corrections. The slight 'leveling off' of the data for older seafloor, compared to the curve, has led some researchers (e.g., see references in Stein & Stein 1992) to postulate additional heat sources, for example heat from underlying mantle convection finally making an appearance. However, in the lithospheric plate model developed by Stein & Stein, the leveling off is much less significant.

whose surface heat flow is q_z, and C is a constant. The curve—which could be described as a 'one-over-the-square-root-of-age' curve—fits most of the heat flow data well; this is significant because, as we will see later in this chapter, *it corresponds to the surface heat flow of an object that is cooling over time.*

Of course, this is a reasonable conclusion, given our understanding of how plate tectonics works: as oceanic plates move away from the mid-ocean ridges where they formed, they cool by giving off heat to the surroundings; in fact, the discovery of such a 'one-over-the-square-root-of-age' pattern of oceanic lithosphere cooling was cited as early proof that plate tectonic theory is correct.

Curiously, our two clues imply that most of the heat flow observed to cross the Earth's surface globally has a shallow—crustal (in the case of continental heat flow) or lithospheric (for oceanic heat flow)—origin. The latter is still tied to mantle convection and interior dynamics, so ~30 TW (the oceanic portion of the global heat output, i.e. 70% of the total) of global heat loss should be connected to Earth's interior.

Our understanding that the primary source of continental heat flow is radiogenic ties each measurement to the mineralogy of that location. This in turn provides a way to estimate the heat flow in locations where no measurements have been taken but the geology is known or can be inferred. Similarly, our understanding that oceanic heat flow is mainly a consequence of lithospheric cooling allows the heat flow to be predicted in oceanic regions too inaccessible for measurements, or regions where contamination by hydrothermal circulation has occurred. Both of these techniques were employed by Pollack et al. (1993), Davies and Davies (2010), and Jaupart et al. (2015) to produce heat flow data sets that are (mostly) globally complete (at some level of resolution). However, as discussed by Davies and Davies (2010), and not surprisingly, Antarctica remains problematic.

A recent depiction by Davies (2013) of global heat flow, based mainly on the global data set compiled by Davies and Davies (2010), is presented in Figure 9.4.

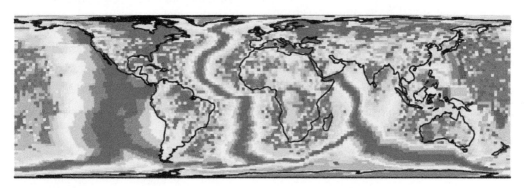

Final Estimate of Heat Flow (mW m^{-2})

4–50	55–58	61–63	70–80	99–129
51–54	59–60	64–69	81–98	130–919

Figure 9.4 State-of-the-art map of global heat flow, constructed by Davies (2013) / John Wiley & Sons, based on over 38,000 measurements. Davies' "final estimate" shown here includes three components: for young oceanic crust, heat flow predictions derive from a 'one-over-square-root-of-age' curve, like that shown in Figure 9.3; in all other regions where data is available, the heat flow estimates are area-weighted averages of the actual data; and in unmeasured locations, estimates are based on correlations of geology to areas with heat flow already determined.

9.1.3 Strengthening Our Theoretical Foundation of Heat Flow: An Introduction to Del

Finally, before moving on to the next section, we take this opportunity to briefly generalize Fourier's law, the basis of the measurements we've been looking at.

The version of Fourier's law we have used describes heat flow in the z-direction. With z referring to depth below the surface, that expression describes well the bulk of Earth's heat flow. But in general, and even for the Earth, heat can flow in any direction. That means heat flow is a vector quantity,

$$\vec{q} \equiv (q_x, q_y, q_z)$$

where we denote the total heat flow by $\vec{q}$ and resolve it into x- and y- as well as z-components. The heat flow in any direction will depend on the temperature differences in that direction; for each direction, we can write a Fourier-type law,

$$q_x = -K\frac{\partial T}{\partial x}, \quad q_y = -K\frac{\partial T}{\partial y},$$

$$q_z = -K\frac{\partial T}{\partial z}.$$

In the general situation, we must make use of partial derivatives: if heat can flow in more than one direction, it must be that temperature varies in more than one direction; that is, temperature is a function of more than one spatial variable.

These three 'component' equations can be combined into a single vector equation, namely,

$$\vec{q} \equiv (q_x, q_y, q_z)$$

$$= \left(-K\frac{\partial T}{\partial x}, -K\frac{\partial T}{\partial y}, -K\frac{\partial T}{\partial z}\right)$$

$$= -K\left(\frac{\partial T}{\partial x}, \frac{\partial T}{\partial y}, \frac{\partial T}{\partial z}\right).$$

The collective term on the right in parentheses is a single quantity with x-, y-, and z-components—that is, it is a vector! Its components are the temperature gradients in the x-, y-, and z-directions; for compactness, it is symbolized as

$$\vec{\nabla} T \equiv \left(\frac{\partial T}{\partial x}, \frac{\partial T}{\partial y}, \frac{\partial T}{\partial z}\right)$$

and called simply "the" temperature gradient (vector), or—even more compactly—"grad T."

By the way, later in this chapter we will go one step further and single out the **gradient operator**, which is known informally as **del**,

$$\vec{\nabla} \equiv \left(\frac{\partial}{\partial x}, \frac{\partial}{\partial y}, \frac{\partial}{\partial z}\right),$$

essentially a three-dimensional gradient 'waiting to happen.' For those who find this too 'unfulfilled,' a more completely specified but still general way to present it is to let it 'operate' on any function f:

$$\vec{\nabla}f \equiv \left(\frac{\partial}{\partial x}, \frac{\partial}{\partial y}, \frac{\partial}{\partial z}\right)f$$

or

$$\vec{\nabla}f \equiv \left(\frac{\partial f}{\partial x}, \frac{\partial f}{\partial y}, \frac{\partial f}{\partial z}\right).$$

In short, our vector heat flow equation can be concisely written as

$$\vec{q} = -K\,\vec{\nabla}T;$$

this three-dimensional version of Fourier's law is called the **conduction equation**. If we consider it as a single vector equation, rather than as a compact way to write the three component equations, we can understand it as follows:

- Heat flows according to temperature gradients, i.e., temperature differences.

- The minus sign ensures that heat flows from hot to cold.

- The magnitude of heat flowing is proportional to the magnitude of the temperature gradient; the proportionality constant, a material property, is the thermal conductivity.

- The net direction of heat flow is the same as but opposite to the direction of the temperature gradient.

Incidentally, there are materials for which a temperature difference in the x-direction can also induce heat flow in the y-direction (imagine, for example, a rock being heated in one direction that possesses a sliver of metal oriented in a cross-direction…).

Such materials would be classified as anisotropic. A simpler kind of thermal anisotropy would be rocks that conduct heat differently in one direction versus another. Very simply, which rock type would you expect is most likely to be anisotropic—sedimentary, igneous, or metamorphic?

9.2 Heat Sources

There are several sources of heat that contribute dramatically to the flux crossing the Earth's surface. These sources are 'four-dimensional'—some originating in Earth's past, some operating throughout time but at various depths—so eventually we will have to address how that heat evolved over time or made it up to the surface.

9.2.1 Radioactivity

Radioactive isotopes decay by ejecting α or β particles from their nuclei. Those particles are energetic and travel large distances, up to centimeters in some cases, until they finally collide with the surrounding medium (see Figure 9.5); upon impact the particles' kinetic energy is converted to heat—radiogenic heat. Within the solid earth, four radioactive isotopes are thought to be responsible for most of the heat currently produced by radioactive decay. These isotopes are ^{238}U, ^{235}U, ^{232}Th, and ^{40}K; the decay of these 'parent' isotopes to final 'daughter' isotopes (i.e., neglecting any intermediate daughter products) can be symbolically written

Figure 9.5 Pleochroic haloes. These rings or 'shells' of darkening and discoloration are the result of damage to the medium—in this case, large crystals of biotite mica—by α particles ejected from uranium nuclei contained in inclusions of small zircon crystals. The radius of the haloes in the biotite, typically ~10–30 microns (Henderson & Bateson, 1934), indicates how far the α particles can fly before impacting the biotite crystal lattice. Image by Alex Strekeisen / http://www .alexstrekeisen.it/english/pluto/pleochroichalo .php / last accessed August 22, 2023. Field of view is 2 mm.

(Dye, 2012; Jaupart et al., 2015), where the heat production is per kg of the parent isotope. Note that there are two possible β-decay paths for ^{40}K; the decay to argon (via an *electron capture* by the nucleus) occurs about 11% of the time, to calcium 89% of the time.

These decay equations do not explicitly include the production of "geoneutrinos" (antineutrinos) associated with β-decay; but the specified heat energy does represent the net result of the α- and β-particles' kinetic energy (converted to heat upon impact) minus the energy carried out of the mantle by the geoneutrinos (Dye, 2012). (Dye also reviews the promising use of geoneutrino detection to quantify the mantle's radiogenic heat; see Jaupart et al., 2015 as well.)

$$^{238}U \rightarrow \ldots \rightarrow {}^{206}Pb + 8\alpha + 6\beta + 0.095 \text{ mW/kg}$$
$$^{235}U \rightarrow \ldots \rightarrow {}^{207}Pb + 7\alpha + 4\beta + 0.568 \text{ mW/kg}$$
$$^{232}Th \rightarrow \ldots \rightarrow {}^{208}Pb + 6\alpha + 4\beta + 0.026 \text{ mW/kg}$$

$$^{40}K \begin{cases} \ldots \rightarrow {}^{40}Ar - \beta \\ \\ \ldots \rightarrow {}^{40}Ca + \beta \end{cases} + 0.028 \text{ mW/kg}$$

Other radioactive isotopes of major interest in the Earth System include $^{87}Rb \to {}^{87}Sr$ and $^{14}C \to {}^{14}N$, neither of which, however, creates significant amounts of heat by their decay; and short-lived isotopes like $^{26}Al \to {}^{26}Mg$, which, as discussed later in this section, would have generated consequential amounts of heat in the early days of the Solar System.

Uranium, thorium, and potassium are found in varying amounts throughout the crust and mantle. Potassium is thousands of times more abundant than uranium; but only one atom of potassium in ~8,500 is ^{40}K, limiting its contribution to Earth's radiogenic heat. And only one atom of uranium in ~140 is ^{235}U, so its contribution is minor. With these differences in abundance, as well as differences in their intrinsic heat production, Th and ^{238}U end up being the primary contributors of radiogenic heat in the crust and mantle, with K comparable but less. Table 9.2 lists typical radiogenic heat production for rock types representative of the crust and mantle.

To assess the contribution of radiogenic heat to the observed heat flow in continental regions

Table 9.2 Typical radiogenic heat production in different rock types.[a]

Rock type	Heat production ($\times 10^{-9}$ mW/kg)
Granite ($\leftrightarrow$ continental crust)	1050
Basalt ($\leftrightarrow$ oceanic crust)	27–180
Peridotite ($\leftrightarrow$ mantle)	1.5–8.4[b]
Chondrite meteorites ($\leftrightarrow$ mantle, sort of[c])	5.2–5.8[d]
Iron meteorites ($\leftrightarrow$ core)	$<3 \times 10^{-4}$

[a] Data from Stacey and Davis (2008), except for peridotite.
[b] Discussed later in the text.
[c] See Chapter 2.
[d] In contrast to the other mantle and crustal rock types shown here, ^{40}K is the primary radiogenic heat source in chondrite meteorites.

(~65 mW/m^2 or a bit more, as discussed earlier), we consider a column of granite 24 km deep with cross-sectional area 1 m^2; since the density of granite is ~2.7 gm/cm^3 or 2,700 kg/m^3, that column would contain a mass of ~6.5×10^7 kg. According to Table 9.2, then, a granite crust 24 km thick would generate 68 mW of radiogenic heat in each 1 m^2 column; after sufficient time had passed for that heat to make it up to the surface, we would measure 68 mW/m^2 heat flow.

How would these numbers change if we were considering Archean crust (whose observed surface heat flow is ~51 mW/m^2, according to Table 9.1) rather than 'average' crust; that is, would we need the thickness of an Archean granitic layer to be more or less than 24 km to produce the observed heat flow? Assume the Table 9.2 values of heat production are still valid.

Continental crust is typically more than 24 km thick, but its composition is likely to be varied; it may have more than 24 km worth of granite in it, or less, and its granite may be more radiogenic or less than the value listed in Table 9.2. In short, given crustal variations in composition and thickness, radiogenic heat may not explain *all* of the heat flow in continental areas; but it is clear that the surface heat flow observed in continental regions is *primarily* the result of radiogenic heat produced within the crust.

However, oceanic crust is thinner than continental crust—and an order of magnitude less radioactive (see Table 9.2). It is clear that radiogenic heat can be only a minor contributor to the observed oceanic heat flow. This is not unexpected, since we have already concluded that such heat flow is consistent with the cooling of the oceanic lithosphere after its creation at mid-ocean ridges. Thus, global surface heat flow observations (i.e., both oceanic and continental) are well accounted for as consequences of surficial—lithospheric or crustal—processes. And, as noted in Chapter 8, the magma appearing at mid-ocean ridges can be explained as the result of shallow, pressure-release melting as oceanic lithosphere rifts apart.

These surface and near-surface processes still leave us with a need to explain where the energy for plate motions and mantle convection comes from.

9.2.2 Gravitational Energy, Part One

Everyone knows that it takes work to oppose gravity, for example to lift up an object; the energy involved is the object's gravitational potential energy. When the lifted object is dropped, that gravitational energy is converted to kinetic energy as the object falls; and, when the object hits the ground, the kinetic energy is, in turn, converted to heat and (if deformation results) strain energy.

4.6 Byr ago, the Earth accreted by the mutual gravitational attraction of small planetesimals and the gas, dust, and ice particles of the primitive solar nebula. The conversion of gravitational energy as all that mass fell toward and then impacted the growing Earth produced an overwhelming amount of heat. We can estimate its magnitude by considering the Earth at a moment in its accretion, already a sphere of some radius, exerting a gravitational force on everything around it; we can then calculate the energy released when a particle—or, more picturesquely, a thin spherical blanket of particles—falls from far away through that gravity field, down to the surface. The total amount of gravitational energy released will be the sum (integral) of these incremental amounts of energy; it ends up being a very large amount of energy in part because the mass of the Earth figures in doubly in this process, once for all the particles falling in and once for the gravity field they are falling through.

The energy of Earth's accretion is thereby estimated at more than 2×10^{32} J. If such an amount of heat was completely retained by the Earth, it would raise the internal temperature on average by more than 30,000°C (cf. Stacey & Davis, 2008; also Jaupart et al., 2015). Or, it's

enough heat to vaporize the Earth twice over. In still other terms, if that heat was released steadily over the life of the Earth, it would be almost 1,400 TW, producing a surface heat flux nearly 30 times the observed amount. We are forced to conclude that *most* of that accretional energy was not retained by the Earth but instead was reradiated back into space as the Earth built up, layer by layer. This, in turn, can be interpreted as placing a limit on how quickly accretion could have transpired—too quickly, and all the heat would be buried by subsequent layers. Alternatively, it limits how 'orderly' accretion was: instead of a steady accretion, layer by blanketing layer, it was accomplished by a decreasing number of increasingly massive impacts (e.g., Rubie et al., 2015), as the primitive nebula evolved into discrete planetesimals, then planetary embryos, then proto-planets (see Chapters 1 and 2); most of the accretional energy was produced in the later, more massive impacts (see Walter & Tronnes, 2004), and any 'blanketing' of outgoing radiation was temporary and incomplete.

9.2.3 Heat of Compression

As the Earth grew by accretion, layer by layer or just particle by particle, the layers below would be increasingly compressed. The compression would heat those layers, and—because rocks conduct heat so poorly—this heating would be adiabatic. We can use the adiabatic temperature gradient to estimate how much temperatures inside the Earth would rise as a result of this heat.

We derive an expression for the adiabatic temperature gradient in *Textbox A*. Our derivation is completely general, so the resulting formula,

$$\left(\frac{\partial T}{\partial z} \right)_S = \frac{\alpha_P T g}{C_P}$$

(letting z denote depth), applies to any component of the Earth System. For the mantle, we saw in Chapter 8 that g is nearly constant. It turns out that C_P is approximately

Textbox A: The Adiabatic Temperature Gradient

In our derivation of the Adams-Williamson equations in Chapter 8, we wrote the adiabatic temperature gradient as

$$\left(\frac{\partial T}{\partial z}\right)_S = \left(\frac{\partial T}{\partial P}\right)_S \frac{dP}{dz};$$

the adiabatic increase in temperature with depth equals the amount of adiabatic heating 'per unit pressure increase' (or, the rate of temperature increase with pressure) times how much the pressure actually increases with depth. In most situations, the pressure increase is well approximated as hydrostatic, so we can write

$$\left(\frac{\partial T}{\partial z}\right)_S = \left(\frac{\partial T}{\partial P}\right)_S \rho g.$$

To proceed further, we use a thermodynamic relation derived by Maxwell (based on the definition of enthalpy, which under some conditions represents the 'heat content' of the system):

$$\left(\frac{\partial T}{\partial P}\right)_S = \left(\frac{\partial V}{\partial S}\right)_P;$$

in this textbox, we will denote volume, as a thermodynamic variable, by V rather than VOL. This expression is actually one of four known as *Maxwell's relations*, not to be confused with the four Maxwell's equations, which involve electromagnetic fields; the latter will be discussed in Chapter 10.

Maxwell's 'enthalpy' relation will lead to a very clear derivation of the adiabatic gradient; ironically, it does so by allowing us to consider the nonadiabatic situations of varying heat and entropy. We first expand this relation as follows:

$$\left(\frac{\partial T}{\partial P}\right)_S = \left(\frac{\partial V}{\partial S}\right)_P = \left(\frac{\partial V}{\partial T}\right)_P \left(\frac{\partial T}{\partial S}\right)_P.$$

Using the definition of the coefficient of thermal expansion, α_P, as in Chapter 8,

$$\alpha_P = \frac{1}{V}\left(\frac{\partial V}{\partial T}\right)_P$$

(with the subscript for α here reminding us that thermal expansion is measured under conditions of constant pressure), we can write our expanded enthalpy relation as

$$\left(\frac{\partial T}{\partial P}\right)_S = \left(\frac{\partial V}{\partial T}\right)_P \left(\frac{\partial T}{\partial S}\right)_P = \alpha_P V \left(\frac{\partial T}{\partial S}\right)_P.$$

This last derivative can be expressed in terms of the *specific heat* or *heat capacity per unit mass* of the medium, defined as the amount of additional heat stored in each kilogram of the material when its temperature is raised by 1°C. We will denote the specific heat, which will be measured at constant pressure, by C_P, and say that the amount of mass in the material is M. Then, if the material's temperature increases by an amount dT, the amount of heat stored within it increases by $C_P \cdot dT$ per kilogram, or $MC_P \cdot dT$ in total.

Let Q symbolize the amount of heat stored in this mass. The increase in Q when the temperature increases by dT is

$$dQ = MC_P \cdot dT;$$

if the increase in T is infinitesimal, then so is the increase in Q. Finally, the second law of thermodynamics tells us that (for reversible reactions) the material's entropy S increases when

Textbox A: The Adiabatic Temperature Gradient (*Concluded*)

its internal heat increases:

$$dS = \frac{dQ}{T};$$

it follows that

$$dS = \frac{MC_P}{T}dT$$

so

$$\left(\frac{\partial T}{\partial S}\right)_P = \frac{T}{MC_P}.$$

Putting it all together yields

$$\left(\frac{\partial T}{\partial P}\right)_S = \alpha_P V\left(\frac{\partial T}{\partial S}\right)_P = \alpha_P V\cdot\frac{T}{MC_P} = \alpha_P\frac{T}{\rho C_P},$$

where $\rho \equiv M/V$ is the density of the medium, thus

$$\left(\frac{\partial T}{\partial z}\right)_S = \left(\frac{\partial T}{\partial P}\right)_S\rho g = \frac{\alpha_P T g}{C_P}.$$

independent of pressure and temperature at mantle conditions (Katsura et al., 2010). Thermal models cited in Chapter 8 suggest that α_P decreases with depth in the mantle; in contrast, T will increase with depth. Consequently, we can expect the adiabatic temperature gradient not to vary too much throughout the mantle.

Numerically, we have $g \sim 10$ m/sec^2 and, according to the thermal model of Stacey and Davis (2008), $C_P \sim 1.2 \times 10^3$ J/°C-kg, throughout the mantle. From Chapter 8, $\alpha_P \sim 2.7 \times 10^{-5}$/°C in the upper mantle and $\sim 1.7 \times 10^{-5}$/°C in mid-mantle. Thus, $\alpha_P \cdot g/C_P \sim 2.3 \times 10^{-7}$/m = $\sim 2.3 \times 10^{-4}$/km in the upper mantle; in mid mantle, $\alpha_P g/C_P \sim 1.4 \times 10^{-4}$/km.

But the magnitude of the adiabatic gradient also depends on how hot the mantle is. The olivine→wadsleyite geothermometer discussed in Chapter 8 implied a temperature of $\sim 1,550$°C at the first transition (Katsura et al., 2010); with the adiabatic gradient derived using the second law of thermodynamics, though, T must be specified in Kelvins, so we use $T \sim 1,800$°K for that depth. Our formula consequently yields a value of ~ 0.4°K/km (or 0.4°C/km) for the adiabatic gradient at ~ 400 km depth.

Based on that rate of increase, the temperature 1,000 km deeper would be ~ 400°K hotter (more, if the actual gradient is superadiabatic). In that case our formula, using a mid mantle value for $\alpha_P \cdot g/C_P$, predicts the magnitude for the adiabatic gradient at that depth to be ~ 0.3°C/km.

More precise estimates of the adiabatic gradient throughout the mantle, based in part on a reanalysis of high-pressure experiments to determine the thermal expansion versus depth, were constructed by Katsura et al. (2010). They show that the adiabat decreases from nearly 0.5°C/km at the top of the *lower* mantle to ~ 0.25°C/km near the base of the mantle; in the upper mantle, there is also a monotonic decrease with depth, except for jump-ups associated with phase transitions, which keep the gradient within the range 0.65°C/km–0.35°C/km.

To assess the impact of adiabatic compression on mantle temperatures, we will (conservatively) take the adiabatic gradient to be ~ 0.4°C/km in the transition zone and ~ 0.33°C/km in the lower mantle. Over the thickness of the transition zone, then, the temperature increases adiabatically by ~ 100°C; through the depth of the lower mantle, it increases by ~740°C. The net effect of compressional heating from 400 km to 2,900 km depth is thus around 840°C.

For comparison we note that, using the post-perovskite phase transition as a geothermometer, the temperature near the base of the mantle has been extrapolated to be ~4,000°C or more (Jaupart et al., 2015; see also Tateno et al., 2009); however, other estimates for the core-mantle boundary, discussed later in this chapter, are centered around ~3,600°C but could be as low as ~3,000°C. Either way, adiabatic gradients alone cannot produce the hot temperatures believed to prevail at the base of the mantle; their contribution to the actual temperature increase is significant, though, ~30%–50% of the roughly ~1,500°C–2,500°C increase from the first transition to the base of the mantle.

In terms of heat rather than temperature, the contribution from adiabatic compression is similarly important but not overwhelming. For the mantle, for simplicity, we will treat that adiabatic temperature increase of 840°C from 400 km to 2,900 km depth as equivalent to an average increase ΔT of $\frac{1}{2} \cdot 840$°C (or roughly 400°C) through that entire depth range. Using the definition of heat capacity in *Textbox A* as a guide, we would estimate the increase in the mantle's heat from compression to be $M C_P \cdot \Delta T \sim 1.6 \times 10^{30}$ J, if an average density of 4.7 gm/cm³ (= 4,700 kg/m³) is used to determine the mass M of the region.

As discussed later in this chapter, the adiabatic gradient in the outer core is, overall, distinctly greater than in the mantle and leads to an adiabatic temperature increase through the outer core of roughly ~1,500°C (more precisely, between 1,300°C and 1,800°C). As an even rougher approximation, we can assume an adiabatic gradient exists through the inner core, yielding a net adiabatic temperature increase from the CMB to the center of the Earth of ~2,200°C. Following the same approach as with the mantle, but using $C_P \sim 0.8 \times 10^3$ J/°C-kg in the core (Stacey & Davis, 2008), we would find the compressional heat added to the core to be ~1.8×10^{30} J.

The total heat generated by Earth's internal compression is then ~3.4×10^{30} J—only a percent or two of the huge accretional energy from Earth's formation, but all of it trapped within the Earth, thus, again, a respectable (if minor) source.

Can you verify these numerical results for the mantle and core? You can use an estimate of the mass of the core from Chapter 8 .

Whether in terms of temperature change or heat production, these estimates illustrate the potential significance of compression as a heat source, but they should be viewed as quite approximate. Compressional heat was first generated during accretion, when Earth's interior was probably quite hot; in addition to T, the parameters α_P and g, and possibly even C_P, may have been very different, yielding a very different magnitude adiabatic gradient than considered here.

Briefly: The Atmosphere. For the atmosphere, our interest is mainly in the magnitude of the gradient. The formula from *Textbox A* still applies, but we can take it one step farther by recalling from Chapter 8 that for a gas obeying the ideal gas law, the coefficient of thermal expansion is simply $1/T$. Our formula then simplifies to

$$\left(\frac{\partial T}{\partial z} \right)_S = \frac{g}{C_P}.$$

For a dry troposphere, $C_P \sim 1,000$ J/kg-°C at surface conditions, yielding an adiabatic temperature gradient of ~9.8°C/km in the lower troposphere. In meteorology, the **lapse rate** is defined as the decrease in temperature with height (but here z denotes depth, not height), so the *dry adiabatic lapse rate* also equals 9.8°C/km. As we saw in Chapter 3, the condensation of water vapor in a rising parcel of air releases heat; this offsets the dry lapse rate, but by an amount that depends on the humidity of the air and thus the temperature as well. Typically, we find the *saturated adiabatic lapse rate* to be ~5°C/km.

The actual temperature gradient in any component of the Earth System may differ significantly from adiabatic. But, as we will see later in this chapter, a fluid that is undergoing convection, or has recently done so, will exhibit a temperature gradient that—from top to bottom, on average—is more or less adiabatic. For the atmosphere, then, the actual temperature gradient will be roughly adiabatic, thus between 5°C/km and 10°C/km; a representative value of the actual lapse rate globally might be ~6½°C/km.

9.2.4 Gravitational Energy, Part Two

The formation of the core during or following accretion would have involved gravitational energy as well. This process differed from accretion in three respects: the particles involved were starting out inside the Earth, not from 'infinitely' far away; as mantle and core differentiated, some particles would rise (to form the mantle), requiring energy, while others fell (to form the core), releasing energy; and the energy released would be trapped within the Earth, unable to radiate away into space. The first two factors had the effect of reducing the amount of energy released, compared to accretion; the energy of differentiation is estimated to be ~10^{31} J—twenty times less than the energy of accretion.

The third factor means that, unlike accretional energy, we cannot dismiss this heat source: it was retained by the Earth, and must have played a role (and probably still does) in the dynamics of the interior. If none of it went into accelerating the process of differentiation, this energy—being almost three times the energy of compression—implies an increase in the *average* interior temperature by ~4,400°C.

9.2.5 Moon-Forming Impact

As discussed in Chapter 6, one very plausible mechanism for creating our Moon is a giant impact with Earth early in its history. If the impact was head-on, by a body as massive as Earth was, then both objects would have had roughly half of Earth's present mass when they collided.

- So, if the Earth had not yet accreted the outer half of its mass, our first step in calculating the energy of the Moon-forming event is to remove the energy of accretion corresponding to that outer half of the Earth. As noted in our discussion of gravitational energy, the Earth's mass figures in doubly in the process of accretion; with only half the Earth accreted, then, the energy generated would be only one fourth of the previously calculated value, that is, ~5×10^{31} J. (Actually, it will be less than that, because the proto-Earth's gravity would have been weaker, and both the acceleration of the infalling particles and their energy of impact would have been less, in the accretion of the inner half of the Earth.)

- As a simple estimate of the impact energy, we treat the impactor as a single (large) particle of mass M. If its velocity at impact was V_{imp}, then its kinetic energy would have been $\frac{1}{2}M(V_{imp})^2$. However, Canup (2012) expresses V_{imp} in terms of V_∞, the impactor's 'initial' velocity, far away from the proto-Earth and well before impact; Canup's model suggests $(V_{imp})^2 \sim 2 \times (V_\infty)^2$, very approximately. According to Canup (2012), V_∞ could have been as large as 11 km/sec, which yields an energy of impact as high as ~3.6×10^{32} J. With preferred values of ~4–5 km/sec for V_∞, corresponding to a low-velocity collision, the collisional energy would have been ~6×10^{31} J.

Can you verify these estimates of the impact energy? Maybe start by rewriting the kinetic energy in terms of V_∞.

In the head-on impactor model, then, the net result is a combined accretional and collisional energy of around 10^{32} J, half our earlier whole-Earth accretional energy estimate.

The results differ somewhat for the glancing blow impactor model. In this case the Earth was already 90% accreted, so our reduction in the accretional energy would be only by

20% (since $0.9 \times 0.9 = 0.81$). The impactor was Mars-sized, so M would be five times smaller than in the head-on collision model. According to Canup (2004), $V_{imp} \sim 9$ km/sec, implying with our simplified approach that the kinetic energy of this impactor was $\sim 2\frac{1}{2} \times 10^{31}$ J (cf. references cited in Clauser, 2011), or more than $\sim 10\%$ of the original accretional energy. However, with a glancing blow, little of that energy would be transferred to the Earth; the net result would be a combined accretional and collisional energy of between 1.6 and 1.8×10^{32} J, that is, a slight drop to 80%–90% of the original, whole-Earth energy of accretion.

9.2.6   Tidal Friction

Tidal friction has significantly slowed the rate of Earth's rotation over the eons, resulting in a major loss of Earth's rotational energy. The kinetic energy of rotation of a body can be calculated from $\frac{1}{2}I\omega^2$, where I is the body's moment of inertia and ω is its angular velocity; this expression is an angular analog to $\frac{1}{2}Mv^2$, the formula for the 'linear' kinetic energy of the body (we used angular analogs in Chapter 6 when discussing torques and conservation of angular momentum). For example, if Earth's day was 22 hours long in Devonian times, its angular velocity has decreased from $2\pi/22$ rad/hr to $2\pi/24$ rad/hr at present; this implies a decrease in Earth's rotational energy from 2.52×10^{29} J to 2.12×10^{29} J over the past 400 Myr.

Throughout the eons in which tidal friction operated, the rotation of the Earth would have been constrained by the need for the Earth-Moon system to conserve angular momentum. With an origin by giant impact (or by fission, for that matter), the Moon would have started out near the Roche limit, a distance just less than three times Earth's radius (Canup & Asphaug, 2001); extrapolating back to that situation, conservation of angular momentum of the system (combined with Kepler's Third Law, to relate the Moon's orbital angular velocity to its orbital radius) implies Earth's daily spin was much faster,

with the day only ~ 5 hours long. From then till now, the decrease in rotational energy has been from 4.66×10^{30} J to 0.21×10^{30} J—a drop of almost 4.5×10^{30} J. (This estimate is four times the frequently cited value given by Stacey and Davis (2008); the disparity is perhaps a consequence of their view that the Moon's evolution was determined by interactions with a hypothetical second moon, and the Moon was never closer to Earth than around eight times the Roche limit.)

Averaged over the lifetime of the Earth, the amount of rotational energy lost per sec has been nearly 32 TW, with a loss per year of 10^{21} J. Over the past 400 Myr, in contrast, the rate of loss per sec and the loss in one year are both lower by a factor of 10. This is not unexpected, since the torque causing tidal friction depends inversely on distance raised to the sixth power, so it was more effective when the Moon was closer (the drastically different averages over time also remind us that taking averages may oversimplify how a phenomenon is represented).

So what happened to all that rotational energy? As we discovered in Chapter 6, most of the energy dissipation by tidal friction takes place in shallow seas; currently, then, tidal friction is only marginally relevant to our investigation of heat sources for the solid earth. In the deep past, however, conditions would have been different. As mentioned in Chapters 1 and 2, there might have been times during the Heavy Bombardment when the oceans were briefly evaporated. More to the point, the young Earth was hotter, thus more anelastic and dissipative, making tidal friction within the solid earth that much more effective.

9.2.7   Another Look at Radioactivity

The Mantle, Present Day. Table 9.2 shows that the mantle produces less radiogenic heat per kg, by at least one or two orders of magnitude, than continental and oceanic crust. In part, this is a consequence of the partial melting that takes place at mid-ocean ridges and at subduction zones (island-arc volcanism); that partial melting preferentially enriches the crust with radiogenic isotopes and leaves the mantle correspondingly depleted. However, the extent of depletion mantle-wide is not well resolved.

There have been efforts to model radiogenic heat in the upper(most) mantle based on its connections with seismic velocity (see, e.g., the thorough analysis by Hasterok & Webb, 2017). Unfortunately, minor variations in composition and material properties can affect heat production, and tend to become obscured when the overall correlations are determined. And regardless of the approach, disparities between estimates of mantle heat per kg become amplified when multiplied by the relatively large magnitude of mantle mass, adding uncertainty to inferences of the mantle's total contribution. On the other hand, we can be fairly certain, because of that large amount of mass (compared to the crust), that the mantle's contribution to Earth's heat budget will not be negligible.

There have been a variety of estimates of the mantle's radiogenic heat.

- For peridotite in the uppermost (lithospheric) mantle, Rudnick et al. (1998) infer 8.37×10^{-9} mW/kg in cratonic settings, 5.38×10^{-9} mW/kg in off-cratonic regions; they argue that larger values would be inconsistent with mantle dynamics.

- Jaupart et al. (2015) quote the latter value of Rudnick et al. (1998) and independently estimate 1.5×10^{-9} mW/kg. (Hasterok & Webb, 2017 deduce much larger values, 75.7×10^{-9} mW/kg for peridotite in the lithospheric mantle and 38.9×10^{-9} mW/kg in the sublithospheric mantle, but they caution for the need to consider larger data sets.)

- The 'lower-mantle reservoir' model by Kellogg et al. (1999; but see also Gale et al., 2013) implies that *below* the depleted mantle, which they infer ends at ~1,600 km depth, the heat production is roughly 7.1×10^{-9} mW/kg.

- For the mantle as a whole, we can also point to research by Davies (1999), Lyubetskaya

and Korenaga (2007), Stacey and Davis (2008), and Jaupart et al. (2015), whose estimates for the mantle's contribution to Earth's heat budget imply the radiogenic heat production by peridotite is 3.1—7.2, 3.1, 5.1, and 2.8, respectively, all in units of 10^{-9} mW/kg.

We will condense all of these values into a very approximate mantle-wide 5×10^{-9} mW/kg.

Using that representative radiogenic heat production and a mass of $\sim 3.9 \times 10^{24}$ kg (Chapter 8), we estimate that the mantle contributes ~20 TW through its intrinsic radioactivity—roughly 40% of the Earth's total surface heat flow. For comparison, estimates by the four research groups just cited range from ~10 TW to ~30 TW; the model postulated by Kellogg et al. (1999) also fits within that range, with a mantle contribution of 14 TW, as does a similar model by Arevalo et al. (2009) with a slightly larger depleted zone and a total mantle contribution of 13 TW. It is premature to say we have just identified the source of 40% of the observed heat flow—we have not yet demonstrated how that heat will have made it up to the surface—but it is hard to avoid concluding that radiogenic heat in the mantle, like the energy of core formation, must play an important role in internal dynamics.

The Mantle, Future and Past. Radiogenic heat in the mantle (as well as the continental crust) will remain an important source of heat far into Earth's future—potentially, tens of billions of years, because it will take that long for radioactive decay to reduce the abundances of those isotopes to negligible amounts. The **half-lives** (the time for half the remaining nuclei of an isotope to decay away) of the 'hottest' isotopes are

> 14.0 Byr for ^{232}Th;
>
> 4.47 Byr for ^{238}U;
>
> 1.25 Byr for ^{40}K,

the latter accounting for both decay paths of ^{40}K; and

> 0.704 Byr for ^{235}U.

With its shorter half-life, so much ^{235}U has already decayed away that its contribution to Earth's radiogenic heat is already minor (by now it is ~140 times less abundant than ^{238}U, despite the fact that the nucleosynthesis (see Chapter 1) that created both isotopes of uranium would have produced them in roughly similar amounts; but see also Tissot et al., 2016 and Arculus, 2021).

Currently, thorium is four times as abundant as uranium; given the radiogenic heat production per kg for ^{232}Th and ^{238}U, roughly estimate how much of those 20 TW of radiogenic heat now produced in the mantle can be attributed to thorium (for the sake of this thought-exercise, neglect the contribution by ^{40}K).

The half-lives of these isotopes imply that ~20 Byr from now, if Earth still exists, radiogenic heat in the mantle will still be producing at least a few TW of heat, thanks mainly to thorium's longevity.

As important as mantle radioactivity is now, and as vital as it may be in the far future for keeping Earth's convective activities energized, it would have been that much more important in the past. All four of the 'hottest' isotopes would have been much more abundant, and all four would have contributed substantially to the mantle's radiogenic heat; additionally, crustal enrichment by partial melting had not progressed for as long, and the mantle had not yet been as depleted in these isotopes (Stacey & Davis, 2008). 4½ Byr ago, for example, those four isotopes would have produced (combined) four times as much heat as now, according to the thermal model of Stacey and Davis.

Additionally, *short-lived isotopes* (many of which are now *extinct isotopes*) would have been present, and some of them—including ^{26}Al, ^{36}Cl, ^{60}Fe, ^{236}U, and ^{247}Cm—would have produced a lot of radiogenic heat. For example, the first three of these decay according to

(Wadhwa et al., 2007; note that ^{36}Cl also decays to ^{36}S, but only 2% of the time). The radiogenic heat production listed here (from McCord & Sotin, 2005 for ^{26}Al and ^{60}Fe; the range specified for ^{36}Cl is an extrapolation based on Fish et al., 1960, with McCord & Sotin, 2005 as a guide) is per kg of the parent isotope.

Created in the nucleosynthesis that 'seeded' our primitive solar nebula with all types of isotopes, these isotopes have all but decayed away by now because their half-lives are so short. In fact—with half-lives such as 0.72 Myr for ^{26}Al, 0.3 Myr for ^{36}Cl, and 2.6 Myr for ^{60}Fe (see Wadhwa et al., 2007; see Rugel et al., 2009 for ^{60}Fe)—none of these isotopes would have lasted in any significant way for more than 10–100 Myr.

The differentiation of some planetesimals during the collapse of the nebula, as briefly discussed in Chapter 2, led to the growth of asteroids whose fragments are today's achondrites and iron meteorites. But differentiation within such small objects was not easily achieved, with the gravitational energy of accretion as well as the heat of compression insufficient; instead, differentiation has been attributed to the heat generated by decay of the ^{60}Fe (but see Rubie et al., 2015) and especially the ^{26}Al incorporated into those planetesimals (Urey, 1955; Fish et al. 1960; Herndon & Herndon 1977; Ghosh & McSween 1998; McCord & Sotin 2005; and Neumann et al. 2012; see also references in Wadhwa et al., 2007). Depending on how long after nucleosynthesis accretion began, and how quickly the asteroids grew, differentiation—with the radiogenic clock ticking—might have been complete or only partially successful; in the latter case, subsequent fragmentation of the asteroid could have produced both achondrites and chondrites (Neumann et al., 2012). On the other hand, asteroids growing after a few million years would have found ^{26}Al and

$$^{26}\text{Al} \rightarrow ... \rightarrow \, ^{26}\text{Mg} - \beta + 138 \text{ mW/kg}$$
$$^{36}\text{Cl} \rightarrow ... \rightarrow \, ^{36}\text{Ar} + \beta + 4-16 \text{ mW/kg}$$
$$^{60}\text{Fe} \rightarrow ... \rightarrow \, ^{60}\text{Ni} + 2\beta + 74.7^{+} \text{ mW/kg}$$

[60]Fe already scarce and would have remained undifferentiated (Wadhwa et al., 2007).

Similarly, the ability of the proto-Earth to incorporate enough [26]Al and [60]Fe for them to boost interior temperatures and really promote differentiation depended on how quickly Earth's accretion progressed. As an illustration, Rubie et al. (2015) found that, on average, Earth's internal temperatures could have risen by 1,000°C or more from [26]Al alone if accretion began less than 1 Myr after nucleosynthesis but only by ~100°C if accretion was delayed for a few million years.

It was estimated by MacDonald (1959) that the net effect of all short-lived isotopes in the very young Earth (including some isotopes not discussed here) might have been an increase in internal temperature of ~2,000°C–3,000°C. And, as another illustration, the work of Rubie et al. (2015) implies that the total amount of energy released in the first million years of Earth's existence by [26]Al and [60]Fe might have been as much as 10^{31} J. In sum, the role of radioactivity in heating the early Earth's interior and promoting its differentiation is uncertain; but its impact had the potential to be quite significant. And its role in facilitating Earth's differentiation indirectly—by differentiating some of the planetesimals Earth would accrete as it grew—was also potentially quite significant.

The Core. Judging by the lack of radiogenic heat measured in iron meteorites (see Table 9.2), we might expect radioactivity to be irrelevant to the heat budget of the core. However (setting aside, for the moment, considerations of chemical convection), the core must be hot enough to convect, in order to generate the geomagnetic field, and—because the field has existed for billions of years and is still going strong—the heat sources must be long term and ongoing. Radioactivity is an obvious choice.

Current models of the accretion and differentiation of the Earth suggest that deep magma oceans within the primitive Earth were geochemical and geophysical incubators for the growth of the core. Under conditions prevailing near the base of these oceans, at pressures and temperatures corresponding approximately to the top of the lower mantle, iron separates out easily from the iron-silicate magma and descends rapidly through the lower mantle to become incorporated into a growing liquid core (e.g., Rubie et al., 2015; Righter 2003). Under the right conditions, sulfur, with its chemical affinity for iron (sulfur's *siderophilic* behavior was discussed in Chapter 8) at low as well as high *P-T* conditions (Gessmann & Wood, 2002), would follow the iron down to become one of the primary light alloys of the core fluid.

It was first suggested by Lewis (1971) and Hall and Murthy (1971) that elements such as potassium might, under appropriate conditions, have a tendency to bond with *sulfur*—such a chemical affinity would make potassium **chalcophilic**—and follow the iron and sulfur into the core.

However, we know that the melting or partial melting of Earth's primordial, chondritic-like minerals during accretion and differentiation, and the partial melting of mantle peridotite or subducting oceanic crust during tectonic activities, would have concentrated major heat-producing isotopes like uranium, thorium, and potassium increasingly upward: in the mantle instead of the core; in the crust instead of the mantle; in the continental crust instead of the oceanic crust. Thus, the core would seem to be the least likely region of the Earth in which to find those isotopes in significant amounts.

But the odds change in the deep magma ocean incubators. Theoretical studies (Bukowinski, 1976, and references in Parker et al., 1996) predict that, at pressures corresponding to the top or interior of the lower mantle, potassium undergoes changes in its electron orbital structure that make it more metal-like, so that it becomes chalcophilic and siderophilic (it also becomes denser). High-pressure experiments have supported the predictions, confirming its change in bonding with respect to iron (Lee & Jeanloz, 2003), nickel (Parker et al., 1996), and sulfur (in an iron-oxygen-sulfur alloy; Gessmann & Wood, 2002). Thus, potassium will alloy with iron, nickel, oxygen, and sulfur near the base of a deep magma ocean (it could happen higher up in the case of sulfur—but not too high; Chabot & Drake, 1999), and then rapidly descend through the mantle, ultimately becoming incorporated into the growing core.

Would the result be enough potassium in the core to affect its energy budget?

- As a preliminary example, it is worth pointing out that a huge amount of K is not required in order to power the geodynamo. With a heat production of 0.028 mW/kg, a mere ~$1\frac{2}{3} \times 10^{18}$ kg of ^{40}K would generate (as an *extreme* example) as much heat in the core as is currently observed to exit the entire Earth, nearly 47 TW; but, as estimated by Roberts et al. (2003), the energy required to maintain the Earth's magnetic field is probably no more than ~2 TW. Given the meager abundance of ^{40}K relative to all potassium, we are talking *in the extreme* about a little more than 10^{22} kg of all K, amounting to just 0.7 percent ($\equiv$ 7,000 ppm) of the core by mass.

- According to high-pressure experiments, potassium's chalcophile behavior suggests its core abundance would be ~100 ppm (Murthy et al., 2003) or up to ~250 ppm (Gessmann & Wood, 2002); its affinity for iron could lead to an abundance of up to 7,000 ppm (Lee & Jeanloz, 2003). Its affinity for nickel (Parker et al., 1996) has not been quantified.

- We know the core has, overall, been cooling because its central region has solidified to become the inner core; and, as cooling has continued, the inner core has grown. Coupled convective models of the core and mantle by Nimmo et al. (2004) show that it is possible for the combination of core heat sources and heat leaving the core to power the geodynamo over the past 3 or even $3\frac{1}{2}$ Byr *and* to produce an inner core that today is of the correct size—and

also to yield the observed surface heat flux—with a core abundance of ~400 ppm potassium.

Given the multiple chemical affinities that could have drawn potassium into the core, 400 ppm might be viewed as a conservative estimate of its abundance; but the results of Nimmo et al. (2004) make it clear that, even at this level, potassium can play an important role in core dynamics.

At 400 ppm, potassium's heat production at present would be ~$2\frac{1}{2}$ TW. Three half-lives ago, or in other words $3\frac{3}{4}$ Byr ago, ^{40}K in the core would have been $2 \times 2 \times 2$ or 8 times as abundant (one doubling for each half-life), and would have been producing 20 TW of heat (8 times $2\frac{1}{2}$ TW) as it decayed. Back then, the inner core very likely did not yet exist—and perhaps would not for another 2 Byr, as noted below—but the geomagnetic field has probably existed, with roughly the same intensity, at least as far back as ~$3\frac{1}{2}$ Byr ago, as discussed in Chapter 2. During this era—without an inner core, and thus without the heat produced by its growth (discussed below)—radiogenic heat in the core would have helped sustain the geomagnetic field, by thermal convection.

9.2.8 Growth of the Inner Core

Though the inner core is growing now, enlarging eon by eon as more outer-core fluid solidifies ('freezes') onto it, it could only have come into existence after the heat of accretion, heat from core formation, and early radiogenic heat pervading the core in its youth had begun to flow out of the core—and into and through the mantle—and the core had cooled substantially.

Given the low thermal conductivity of mantle rock, how successful do you think the core's cooling would have been if the whole mantle was not convecting during this time?

According to the coupled core-mantle convective model of Nimmo et al. (2004), the inner core is probably not more than ~$1\frac{1}{2}$

Byr old; research by Labrosse (2003), Buffett (2003), Roberts et al. (2003), and others cited in Nimmo et al. (2004) yield similar ages (but see also Stacey & Loper, 2007; Gubbins et al. 2004, references cited in Roberts et al., 2003, and an alternate view briefly discussed in McDonough, 2003).

The growth of the inner core provides two additional sources of heat to the outer-core fluid: the heat of fusion, released from each small mass of the iron-nickel alloy as it solidifies; and the release of gravitational energy, thanks to the light alloy expelled during solidification, as the core continues to differentiate into an increasingly lighter outer core overlying a denser inner core. According to Verhoogen (1961), the heat of fusion of pure iron is $\sim\frac{1}{2} \times 10^6$ J/kg at conditions corresponding to the inner-core boundary; Gubbins et al. (2003) use $\frac{3}{4} \times 10^6$ J/kg, whereas Roberts et al. (2003) estimate $\sim1\frac{1}{2} \times 10^6$ J/kg. Given its current mass, $\sim10^{23}$ kg, the amount of heat released over the life of the inner core by its crystallization has therefore been as much as $1\frac{1}{2} \times 10^{29}$ J; if produced at a uniform rate (though in reality its heat production would have increased as its surface area grew), we are talking about a rate of heat generation of ~3 TW. The energy of light alloy separation is $\sim8.8 \times 10^{28}$ J if all the expelled alloy rises to the top of the core (about half as great if it spreads uniformly throughout the outer core) (Stacey & Loper, 2007); this would amount to an average rate of almost 2 TW since the inner core came into existence. Both of these energy sources carry some uncertainty not only depending on how they are modeled but also because the age of the inner core is not well determined.

By itself, these contributions to the core's energy budget represent a modest amount of heat production; but by the time the inner core came into existence, the core's primordial heat had faded. Radiogenic heat, though already diminished, together with the growth of the inner core, may have been critically important in keeping the geodynamo going, maintaining the geomagnetic field at its present strength throughout.

9.2.9 Some Reflections, and What Must Come Next

A summary of the various heat sources we have considered is presented in Table 9.3. But it should be kept in mind that our discussion of heat sources has been incomplete, with the amount of radiogenic heat throughout the interior highly uncertain and, worse, the amount of accretional heat retained by the proto-Earth ultimately not even knowable. Our discussion has also been somewhat scattered, in terms of what the various heat sources are meant to explain; for example, is a given heat source supposed to be keeping the geodynamo going now, or contributing to surface heat flow, or compensating for the early absence of an inner core? More fundamentally, we have invoked some heat sources to keep the interior convecting; but too much heat would prevent the Earth from accreting, or would melt the mantle, or would prevent an inner core from growing. Such trade-offs are unavoidable because we are dealing with a coupled system (mantle + core, at a minimum).

However, there are a number of factors that can help constrain the thermal state of the Earth and its thermal evolution:

- the rapid (< 50 Myr) formation of the core but delayed appearance of the inner core;

- the existence of a geomagnetic field of consistent intensity throughout the past $3\frac{1}{2}$ Byr;

- the mantle being solid but able to convect; and,

- 'boundary conditions,' including the surface heat flow and the temperature at specific depths within the Earth. In Chapter 8, we used phase changes at the first and second transitions to estimate the temperatures at those depths; we now add to that the phase change represented by the inner-core boundary (liquid→solid) (as desirable as it would be to also know the temperature at the CMB, that boundary is compositional, a change that cannot be tied to a specific temperature). Sinmyo et al. (2019) used high-pressure experiments on iron to infer

Table 9.3 Summary of heat sources discussed in the text. Heat 'events' are listed first, followed by ongoing heat-generating processes.

Heat source	Magnitude	Comments
Accretion	2×10^{32} J	Little retained
Accretion plus Moon-forming impact	$(1 - 1.8) \times 10^{32}$ J	Early head-on impact or late glancing blow; little retained
Core formation	1×10^{31} J	All retained
Compression	3.4×10^{30} J	All retained
Mantle radioactivity (short-lived isotopes)	$\leq 1 \times 10^{31}$ J	In Earth's infancy[a]
Tidal friction	32 TW	Averaged over Earth's lifetime
	3.2 TW	Over the past 400 Myr
Mantle radioactivity (excluding short-lived isotopes)	80 TW	In Earth's infancy
	20 TW	At present
Core radioactivity	20 TW[b]	In Earth's infancy
	2½ TW[b]	At present
Inner-core growth	3 TW	Averaged over inner-core lifetime[c]
Light alloy differentiation	2 TW	

[a]That is, in the first million years of the mantle's existence. This time period can be viewed as an 'event' because of the brief presence of very highly radiogenic short-lived isotopes (see the text for details). Note that the core is not included here: assuming that the only short-lived isotope in the core was ^{40}K, which would have produced heat at a much lower rate than those mantle isotopes, the newborn core would have experienced no comparable heat event.
[b]The abundance of ^{40}K in the core is highly uncertain.
[c]Approximated as 1.5 Byr.

the temperature at the inner-core boundary; accounting for the presence of a light alloy, they concluded the temperature is ~4,850°C (± 390°C) there.

Like our heat sources, all these constraints differ as to when they operated and which regions of the Earth they apply to. They can help us learn about Earth's thermal history—*if* we are able to connect Earth's thermal state at different places and at different times. In sum, we need to model heat flow, with equations that explicitly account for the dependence of heat flow on time and space.

As we saw in the work by Nimmo et al. (2004), such an approach can also help us reasonably constrain the magnitudes of heat sources whose uncertainties were previously too great to be useful.

Our conduction equation will be a good place to start, whether our interest is in conduction or convection (—even convecting fluids cannot avoid conducting heat!); but, lacking an explicit time dependence, that equation is insufficient by itself. In the rest of this chapter, we will develop more comprehensive equations to describe the flow of heat within the Earth. These equations can potentially serve as the basis for determining Earth's thermal history, allowing the impact of all heat sources to be gauged. Obtaining actual, complete solutions for mantle flow (and, in Chapter 10, core flow) is beyond the scope of this textbook; nevertheless, our study of these equations, and some simplified solutions they lead to, will allow us to understand the basic concepts and mathematics underlying heat flow by conduction and convection.

9.3 Transmission of Heat in Solids

Conduction of heat always takes place when there are temperature differences within a body. Since it is always operating, conduction is a crucial factor in determining whether convection will occur in any fluid or fluid-like body—whether the heat represented by the higher temperatures can be effectively conducted out of the body or must be carried out by material flow.

Conduction also strongly influences the way convection would proceed; for example, the low thermal conductivity of mantle material is one of the reasons why (as we saw in Chapter 8) the mantle exhibits convective asymmetries. Finally, as we will see in this section, our interpretation of surface heat flow observations requires that we understand how conduction works.

9.3.1 Conduction Plus Conservation Equals Diffusion

We consider here a body that cannot convect, so that heat is only transmitted by conduction. According to the conduction equation, any heat flow depends on the prevailing temperature gradient:

$$\vec{q} = -K \, \vec{\nabla} T.$$

As noted earlier, this equation involves spatial variations in temperature, but says nothing about (thus does not preclude) variations over time.

One-Dimensional Heat Flow; Introducing the Divergence. For simplicity, we will assume for now that temperature differences exist only in one direction, the z-direction; with no variation in the x- or y-direction, there will be no heat flow in those directions. That is, with $T = T(z, t)$, we have

$$\vec{q} \equiv (q_x, q_y, q_z)$$

$$= \left(-K \frac{\partial T}{\partial x}, -K \frac{\partial T}{\partial y}, -K \frac{\partial T}{\partial z} \right)$$

$$= \left(0, 0, -K \frac{\partial T}{\partial z} \right)$$

so $q_x = 0$ and $q_y = 0$,

$$\vec{q} = (0, 0, q_z) = q_z \hat{z}$$

and the conduction equation reverts back to

$$q_z = -K \frac{\partial T}{\partial z}.$$

Now imagine heat flowing vertically through a tiny rectangular volume, the sides of which are aligned with the x, y, and z axes (Figure 9.6). The heat flow q_z may or may not vary with z; if it does not—which, by the way, we could write mathematically as

$$\frac{\partial}{\partial z}(q_z) = 0$$

(the change of q_z with z is zero)—then as much heat is flowing into the volume as leaves it. For instance, if the heat flow is upward ($q_z > 0$), then $\partial(q_z)/\partial z = 0$ means that

Amount of heat entering the volume through the bottom face $= q_z =$ *Amount of heat leaving the volume through the top face.*

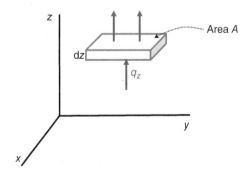

Figure 9.6 Heat flow through a volume ("parcel"). The volume is aligned with the x, y, and z axes; its cross-sectional area is A and its infinitesimal thickness in the z-direction is dz. In this illustration, the heat flow is only in the z-direction, and is upward.

Would you say the heat flow situation shown here corresponds to $\partial(q_z)/\partial z$ being negative, zero, or positive?

On the other hand, if $\partial(q_z)/\partial z \neq 0$ then more heat flows into the volume than out of it, or vice versa. $\partial(q_z)/\partial z$ is the 'rate' at which q_z changes with height z; over the distance dz from bottom to top face, q_z changes by the amount $[\partial(q_z)/\partial z]\cdot dz$.

In short, q_z represents the heat flux into the bottom face of the volume, if $q_z > 0$, and

$$q_z + \frac{\partial q_z}{\partial z}dz$$

represents the heat flux out of the top face. Recalling that q_z is the heat flow per sec per unit area, the total heat flow into the volume per sec is $q_z\cdot A$, and the total heat flow out per sec is

$$\left(q_z + \frac{\partial q_z}{\partial z}dz\right)\cdot A.$$

The difference between the inflow and outflow is

$$\frac{\partial q_z}{\partial z}A dz;$$

this difference is *the amount of heat leaving the volume per sec contributed by the volume itself*.

For example, if this quantity is zero, the amount of heat flowing out of the volume equals the amount flowing in; the volume has not added to or taken from the heat flow passing through it. On the other hand, if the quantity is positive, then the volume must have added to the heat flow out.

Recognizing that the magnitude of the volume whose heat budget we are evaluating is $A dz$, we can pare the mathematics down even further and say that

$$\frac{\partial q_z}{\partial z}$$

is the amount of heat per unit volume leaving the volume, per sec, contributed by the volume itself. Perhaps a clearer way to express this would be to call our small volume a *parcel*; then, $\partial q_z/\partial z$ is *the amount of heat per unit volume leaving the parcel every sec contributed by the parcel itself*.

$\partial q_z/\partial z$ is called the **divergence** of the heat flux, in the case of one-dimensional heat flow. If the divergence of heat flux through a parcel is nonzero—for example, if it is positive, implying that additional heat is leaving the parcel—we will need to explain why that is so, i.e., how our volume manages to contribute some heat of its own.

Three-Dimensional Heat Flow; Invoking Conservation of Energy. It is straightforward to generalize our derivation to the case of three-dimensional heat flow. We would start by replacing our thin volume with a truly infinitesimal one, with dimensions dx, dy, and dz. Considering the 'rate' of heat flow in each direction, for example $\partial(q_x)/\partial x$ in the x-direction, we can then write the corresponding increment in heat flow in that direction, e.g., $[\partial(q_x)/\partial x]\cdot dx$; in terms of the total heat (rather than heat per unit area) per sec in that direction, we have $([\partial(q_x)/\partial x]\cdot dx)\cdot area = ([\partial(q_x)/\partial x]\cdot dx)\cdot dydz$, since the area across which heat is flowing in the x-direction equals $dydz$.

The net result is that, for $T = T(x, y, z; t)$, *the amount of heat per sec leaving the parcel contributed by the parcel itself* is

$$\left(\frac{\partial q_x}{\partial x}dx\right)dydz + \left(\frac{\partial q_y}{\partial y}dy\right)dxdz$$

$$+ \left(\frac{\partial q_z}{\partial z}dz\right)dxdy$$

or

$$\left(\frac{\partial q_x}{\partial x} + \frac{\partial q_y}{\partial y} + \frac{\partial q_z}{\partial z}\right)dxdydz.$$

Dividing this expression by the parcel's volume, we obtain the **divergence of heat flux**, *the amount of heat per unit volume leaving the parcel every sec contributed by the parcel itself*, in this three-dimensional heat flow situation:

$$\left(\frac{\partial q_x}{\partial x} + \frac{\partial q_y}{\partial y} + \frac{\partial q_z}{\partial z}\right).$$

There is a concise way to write the divergence of $\vec{q}$. Earlier we introduced the gradient operator "del," a set of derivatives 'waiting to happen,' defined as

$$\vec{\nabla} \equiv \left(\frac{\partial}{\partial x}, \frac{\partial}{\partial y}, \frac{\partial}{\partial z}\right).$$

Del is a vector, which means we can perform vector operations like dot products and cross products with it. In Chapter 6, we defined the dot product between two vectors in terms of the angle between their directions. An exactly equivalent definition, for the dot product between vector $\vec{A} = (A_x, A_y, A_z)$ and vector $\vec{B} = (B_x, B_y, B_z)$, is

$$\vec{A} \cdot \vec{B} = A_x \cdot B_x + A_y \cdot B_y + A_z \cdot B_z$$

(we could also obtain this result by writing out each vector in terms of unit vectors, $\vec{A} = A_x \hat{x} + A_y \hat{y} + A_z \hat{z}$ and $\vec{B} = B_x \hat{x} + B_y \hat{y} + B_z \hat{z}$; multiplying their product out term by term; and recognizing—according to the original definition of dot product—that dot products like $\hat{x} \cdot \hat{y}$ must be zero). The dot product between the del vector and the heat flux vector would similarly involve multiplying the vectors' components:

$$\vec{\nabla} \cdot \vec{q} = \frac{\partial}{\partial x} \cdot q_x + \frac{\partial}{\partial y} \cdot q_y + \frac{\partial}{\partial z} \cdot q_z$$

or

$$\vec{\nabla} \cdot \vec{q} = \frac{\partial q_x}{\partial x} + \frac{\partial q_y}{\partial y} + \frac{\partial q_z}{\partial z};$$

in this case, the multiplication becomes a differentiation. The divergence of the heat flux can therefore be concisely written as $\vec{\nabla} \cdot \vec{q}$.

The concept of divergence is meaningful in other contexts. How would you interpret $\vec{\nabla} \cdot \vec{v}$, where $\vec{v}$ is the velocity of a fluid through a medium (and into and out of a small volume)? How about $\vec{\nabla} \cdot m\vec{v}$, where $m\vec{v}$ is the fluid's momentum?

In general, it is possible that our parcel is indeed contributing to the heat flowing through the medium, i.e., that $\vec{\nabla} \cdot \vec{q}$ is nonzero. We invoke *conservation of energy* to explain how this is so: any heat leaving the volume which had not entered it must be due either to heat sources within the volume or to cooling of the volume. (This statement specifically applies to a positive divergence of heat; but implicitly we also allow for a negative $\vec{\nabla} \cdot \vec{q}$, that is, a *convergence* of heat, resulting from heat *sinks* within the volume or from the volume *heating up*. Also, if there is motion within the

volume, frictional heating is possible; to keep the math simple, we will just include that in the category of heat sources.)

We will take a symbolic approach to formulating conservation, letting ε denote the heat generated per unit volume per second by heat sources; in most situations, ε will at least include the heat produced by radioactive decay. Thus, $\varepsilon \cdot dxdydz$ will equal the total amount of heat per sec generated within the volume by heat sources. And, as in *Textbox A*, we will let C_P denote the specific heat capacity and ρ the density of material within the volume. Since the amount of mass contained within the volume is $\rho \cdot dxdydz$, the amount of heat it absorbs when it heats up by 1°C will be $\rho dxdydz \cdot C_P$. So, if the volume is heating up by $\partial T / \partial t$ degrees every second, it must be absorbing

$$\rho C_P dxdydz \cdot \frac{\partial T}{\partial t}$$

Joules (or whatever units of energy C_P is measured in) per sec. Putting it all together, conservation of energy tells us that the parcel's contribution to the heat flowing through it is, in units of J/sec,

$$\left(\frac{\partial q_x}{\partial x} + \frac{\partial q_y}{\partial y} + \frac{\partial q_z}{\partial z} \right) dxdydz$$

$$= \varepsilon dxdydz - \rho C_P \frac{\partial T}{\partial t} dxdydz;$$

the minus sign recognizes that the parcel takes heat away from the amount flowing through it if it is heating up.

How does the minus sign work if the parcel is cooling off?

Our expression can be written per unit volume as

$$\left(\frac{\partial q_x}{\partial x} + \frac{\partial q_y}{\partial y} + \frac{\partial q_z}{\partial z} \right) = \varepsilon - \rho C_P \frac{\partial T}{\partial t}$$

or

$$\vec{\nabla} \cdot \vec{q} = \varepsilon - \rho C_P \frac{\partial T}{\partial t}.$$

Finally, we use the conduction equation to write this in terms of one variable only, the temperature:

$$\vec{\nabla} \cdot (-K \vec{\nabla} T) = \varepsilon - \rho C_P \frac{\partial T}{\partial t};$$

in this form, our expression is known as the **diffusion equation**.

This equation is as fundamentally important as it is intimidating! We will learn shortly what it says about the nature of diffusion; but, first, some preliminary comments and mathematical rewriting. This diffusion equation will allow us to determine how the temperature of a material changes with time (the $\partial T/\partial t$ term) as a function of how the temperature is distributed throughout the material (the $\vec{\nabla}$ term). Situations of interest include $\partial T/\partial t \equiv 0$, a **steady state**, in which the temperature *at each point* does not change with time (but T may still vary from place to place—remember, specifying the partial derivative $\partial T/\partial t$ says nothing about T versus the other variables); and $\partial T/\partial t \neq 0$, an *unsteady* situation in which T also depends on time. Obtaining solutions to the diffusion equation in the latter case is that much more challenging, so we will resort to a powerful (and easy!) technique called "dimensional analysis" that yields satisfyingly approximate and still very useful solutions.

It is often assumed that the thermal conductivity K is spatially constant; in that case we can remove it (along with the minus sign) from the set of derivatives constituting the divergence, and the left-hand side of the diffusion equation becomes

$$-K\vec{\nabla}\bullet(\vec{\nabla}T).$$

Using the definition of divergence (e.g., for $\vec{\nabla}\cdot\vec{q}$ it involved the x-derivative of the x-component of $\vec{q}$, and so on), it follows directly that

$$\vec{\nabla}\bullet(\vec{\nabla}T) = \frac{\partial}{\partial x}\frac{\partial T}{\partial x} + \frac{\partial}{\partial y}\frac{\partial T}{\partial y} + \frac{\partial}{\partial z}\frac{\partial T}{\partial z}$$
$$= \frac{\partial^2 T}{\partial x^2} + \frac{\partial^2 T}{\partial y^2} + \frac{\partial^2 T}{\partial z^2}.$$

The left-hand side of this equality is often written compactly as $\nabla^2 T$, that is,

$$\nabla^2 T = \frac{\partial^2 T}{\partial x^2} + \frac{\partial^2 T}{\partial y^2} + \frac{\partial^2 T}{\partial z^2}.$$

This sum of second derivatives is called the **Laplacian** of T.

Our diffusion equation with constant K can therefore be written

$$-K\nabla^2 T = \varepsilon - \rho C_P \frac{\partial T}{\partial t};$$

finally, solving for $\partial T/\partial t$ yields

$$\frac{\partial T}{\partial t} = k\nabla^2 T + \frac{\varepsilon}{\rho C_P}, \quad \text{where } k = \frac{K}{\rho C_P}.$$

The coefficient of the ∇^2 term, k, is called the **thermal diffusivity** or **diffusion coefficient**. A combination of material properties, its units are nevertheless basic: m^2/sec (or length squared per time, in general), hinting that the preceding derivation has resulted in an equation of fundamental character. Using values of K representative of the mantle, as discussed below, as well as experiments to determine k at both upper- and lower-mantle conditions (Chai et al., 1996; Fujisawa et al., 1968), the diffusivity will be in the range $\sim 0.5 \times 10^{-6}$ to 2×10^{-6} m^2/sec; as a typical value, we will take $k \sim 1 \times 10^{-6}$ m^2/sec.

9.3.2 The Nature of Diffusion

At this point, we may seem to be teetering on the brink of becoming an advanced mathematics text. Vector calculus (as in $\vec{\nabla}T$ and $\vec{\nabla}\cdot\vec{\nabla}T$) can appear overwhelming—and will unavoidably become more so in Chapter 10! The analysis below explaining how diffusion works (which I had the pleasure of learning from classes taught by J. Verhoogen) is amazing in part because it turns imposing calculus into simple and easily understood graphical features.

To keep things as basic as possible, we will consider here the simplest kind of diffusion equation, one in which the thermal conductivity is constant and there is no heat generation (e.g., no radioactivity; but heat can still be flowing if there are temperature differences within the body—for example if it is cooling off, from the surface down to the interior). In this case, $\varepsilon = 0$ and the equation reduces to

$$\frac{\partial T}{\partial t} = k\nabla^2 T.$$

We can simplify things further by considering only one-dimensional heat flow, say in the *x*-direction; then

$$\frac{\partial T}{\partial t} = k\frac{\partial^2 T}{\partial x^2}.$$

Time. We ease into the analysis by first noting that our simplified diffusion equation is similar to another equation we already met. In Chapter 7, we discussed the governing equation for seismic waves, the wave equation, which we wrote as

$$\frac{d^2 e}{dx^2} = \frac{1}{V^2}\frac{d^2 e}{dt^2}$$

to describe the propagation of an elastic strain *e* through a material, moving in the *x*-direction at a speed *V*. We can rearrange it as

$$\frac{\partial^2 e}{\partial t^2} = V^2\frac{\partial^2 e}{\partial x^2},$$

also correctly using partial derivatives now. The right-hand side of this equation looks very much like the right-hand side of our simplified diffusion equation, though with different material properties (V^2 instead of k). (And a wave equation that allows for elastic wave propagation in *x*-, *y*-, and *z*-directions will even involve a $V^2 \nabla^2 e$ term instead of just a $V^2\partial^2 e/\partial x^2$ term, at least if the medium is isotropic.)

However, there is a fundamental difference between the wave and diffusion equations in the way they treat *time*. Because of the $\partial^2 e/\partial t^2$ term, which we can also write as

$$\frac{\partial}{\partial t}\frac{\partial e}{\partial t},$$

any solution to the wave equation will also be the solution to a wave equation in which *t* has been replaced by –*t*: the equation is unchanged, since the two minus signs in the time-derivative denominators cancel out; so any expression for *e* that is a solution in 'forward' time will also be a solution in 'backward' time. As an example, if someone records a video of you shaking one end of a rope attached at the other end to a wall, the waves you create will look 'normal' even if the video is played back in reverse. The wave equation does not care about the directionality of time.

But the direction of time is crucial for diffusion processes. If you have a solution to the diffusion equation, that solution will no longer work if you replace *t* with –*t*; the right-hand side of the equation would be unchanged but the left-hand side would have a new minus sign. That is, *diffusion processes are one way in time*. As an example, if you briefly heated the end of a metal rod, you know that the heat will soon spread through the rod; but reversing the process in time—so that a uniformly warm rod becomes hotter and hotter at one end and increasingly cool at the other—is completely unphysical and would yield an impossible result!

Space. Diffusion can also be described as an *averaging* process, in that it occurs until a *uniform level* (in some sense, to be discussed below) has been achieved throughout the medium. We again consider the one-dimensional diffusion of heat, governed by

$$\frac{\partial T}{\partial t} = k\frac{\partial^2 T}{\partial x^2}.$$

If we graphed the temperature as a function of position *x*, at some time *t*, $\partial T/\partial x$ would of course be the slope of the curve. As we change our position, that slope is likely to vary as well; the *change in the slope* with position—which is called the **curvature** of the curve—is measured by

$$\frac{\partial}{\partial x}\frac{\partial T}{\partial x}$$

or, for short, $\partial^2 T/\partial x^2$. Figure 9.7 illustrates the situations of positive and negative curvature.

Can you say what the situation of zero curvature corresponds to?

The diffusion equation says that, since $k > 0$, if the curvature is negative then so is $\partial T/\partial t$ (which equals $k\,\partial^2 T/\partial x^2$). But $\partial T/\partial t < 0$ means that the temperature will decrease with time; this type of scenario is illustrated in Figure 9.8a. Or, if the curvature is positive then $\partial T/\partial t > 0$, so *T* must increase with time (see Figure 9.8b). Note how the temperature profile evolves: early on, while the curvature still has a large magnitude (positive or negative), the

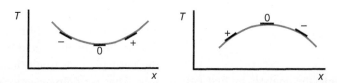

Figure 9.7 Illustration of positive and negative curvature. The curve shown here is of temperature (T) versus position (x). Curvature is defined as the change in the slope of a curve per change in position. The curve on the left has a negative slope at some x-locations; as we move to the right, the slope becomes less negative, then zero, then positive. Since the slope is increasing with x, the curvature is positive. For the curve on the right, the situation is reversed, and the curvature is negative.
Sometimes, positive and negative curvatures are described, respectively, as 'concave-up' and 'concave-down.'

temperature will change at a fast rate; but the more the temperature changes, the lower the curvature (in magnitude), so the slower the change will be. T will only stop evolving (not shown in Figure 9.8) when the curvature has dropped to zero—and a uniform state has been reached—but it will take an infinite amount of time to reach that state.

The temperature profiles illustrated in Figure 9.8 have T 'pinned' to the same value at either end. This might apply to a metal rod cooling down from some initially hot state (Figure 9.8a), with the ends of the rod being, for example, held at room temperature. Or, perhaps the rod is being heated from the middle (Figure 9.8b). But still another possibility is that the temperature is 'pinned' unequally, e.g. to a higher value at the right end. In that case, the diffusion equation still applies, and the rate of cooling or heating up is still proportional to the curvature of the temperature profile; as before, the final state would be a straight line connecting the pinned values—but now that line would no longer be

horizontal. Thus, diffusion operates till a state of zero curvature has been achieved; in the final state, all temperature variations relative to that line (whether horizontal or tilted) have diffused away.

There is one final characteristic of diffusion worth pointing out. In any realistic temperature profile, T is likely to vary irregularly through the medium (e.g., Figure 9.9). Locations where the variations are the most irregular and least smooth—we could call these *high-frequency* (spatially) or *short-wavelength* fluctuations—have the highest curvature, because the slope of the profile changes the most per change in position. Because $\partial T/\partial t$ is proportional to $\partial^2 T/\partial x^2$, such locations will have the highest rates of cooling or heating up: diffusion will quickly even out those irregularities. Of course, heat is flowing, and diffusion is operating everywhere in the medium, so—at the same time that those high-frequency variations are being averaged out—diffusion is also more gradually taking the medium to the long-term averaged state we envisioned in

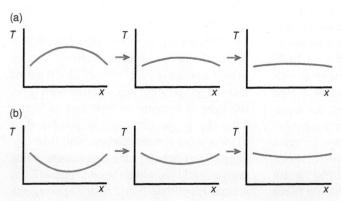

(a)

(b)

Figure 9.8 Depiction of how temperature evolves over time. For simplicity, imagine that the temperature values must remain fixed at the ends.
(a) When the curvature is negative to begin with, the diffusion equation requires T to decrease as time passes (red arrows).
(b) When the curvature is positive to begin with, the diffusion equation requires T to increase over time (red arrows).

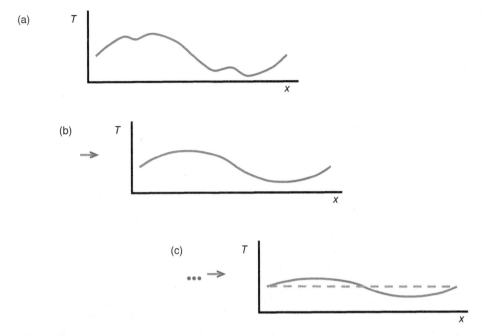

Figure 9.9 Depiction of how a naturally irregular temperature profile evolves over time. For simplicity, imagine that the temperature remains fixed to the same value at both ends. The passage of time is indicated by the red arrows.
(a) The initial temperature profile, which includes 'high-frequency' variations;
(b) the temperature profile after a *short* time has passed;
(c) the temperature profile after a *long* time has passed; and (dashed gray line) after an infinite amount of time has passed.

Figure 9.8. Both processes are shown in the brief transition from Figure 9.9a to 9.9b, and the eventual transition to Figure 9.9c.

Recalling our earlier comments about the one-way nature of diffusion, we can say that the averaged-out states shown in Figures 9.8 and 9.9 will not reverse; as diffusion continues, the local concentrations and deficits of heat suggested by the irregularities in Figure 9.9—and even the mild ones implied by the slight temperature highs and lows in Figure 9.8 (and Figure 9.9)—will smooth out and never reappear in the future. As a final example, we imagine a glass of water into which we spill a drop of vivid red ink. After a few seconds, the ink spot will have diffused somewhat—its edges will no longer be sharp, but we can still identify the small mass of ink; after several more seconds, the ink will have spread out more, turning a small volume of

the water somewhat reddish. Eventually, the ink will have diffused evenly throughout the water, our only 'clue' to its past being a pale, uniform discoloration. This diffusion, like that of heat and other quantities governed by a diffusion equation, is one way: the ink will never recollect in one spot.

9.3.3 Some Solutions to the Diffusion Equation

Preliminary Applications. Solutions to the diffusion equation are easily obtained in some situations, and may help us feel comfortable with the underlying mathematics. We begin with a fictitious Earth that is in a steady state and that contains no heat sources. These conditions mean

$$\frac{\partial T}{\partial t} \equiv 0 \quad \text{and} \quad \varepsilon \equiv 0,$$

respectively, and allow us to reduce the diffusion equation to

$$\nabla^2 T = 0.$$

This equation, often encountered in physics and geophysics, is known as *Laplace's equation*; in some situations, its solutions are the spherical harmonic functions we learned about in Chapter 6.

For additional simplicity, we will consider only a 'local' portion of the Earth (in effect treating the Earth as 'flat') and use Cartesian coordinates (x, y, z) to measure position; z will denote depth below Earth's surface. And we will assume that heat flows vertically only, so $T = T(z)$. In this case the Laplacian reduces to a single term, the second derivative of T with respect to z, and our Laplace's equation becomes

$$\frac{d^2 T}{dz^2} = 0$$

(we can write ordinary derivatives here because, in this example, T is a function of only one variable).

We could solve this by integrating twice. Alternatively, we can 'work backward' by noting that only the derivative of a constant will equal zero; since $d^2 T/dz^2$ is the same as $d/dz(dT/dz)$, that means dT/dz equals a constant, which we will denote by A. With $dT/dz = A$, we then have only to figure out what T could be, such that its derivative is a constant. It turns out the general solution is

$$T = Az + B,$$

where A and B are constants whose values are yet to be determined.

To relate our general solution to the Earth, we impose two boundary conditions: at the surface of the Earth, $z = 0$, we should have

$$T = T_0 \quad \text{and} \quad q_z = q_0,$$

where T_0 is the surface temperature and q_0 is the surface heat flow. Because our solution, $T = Az + B$, requires T to equal B at the surface, it must be that $B = T_0$. And, the conduction equation (Fourier's law) requires that $q_z = -K \cdot dT/dz$, where K is the Earth's thermal

conductivity, whereas our solution requires $dT/dz = A$, so we must have $q_0 = -K \cdot A$, or $A = -q_0/K$. Finally, if z is depth then $q_0 < 0$ (heat flow is upward); thus, A is positive and our solution with Earthly boundary conditions obeyed is

$$T = \left| \frac{q_0}{K} \right| z + T_0$$

(see Figure 9.10).

For slightly greater realism, we can allow for heat sources within the Earth, but assume they are uniformly distributed throughout the interior (so ε = constant). With the Earth still assumed to be in a steady state, the diffusion equation reduces now to

$$\frac{d^2 T}{dz^2} = -\frac{\varepsilon}{K}.$$

This type of equation, incidentally, in which the left-hand side is the Laplacian and the right-hand side is nonzero, is called *Poisson's equation*; like Laplace's equation, it is encountered on occasion in various physics and geophysics disciplines. Working backward *or* just integrating twice, the general solution to this equation is

$$T = -\frac{1}{2} \frac{\varepsilon}{K} z^2 + Az + B,$$

where A and B are arbitrary constants of integration. Then, using the same boundary

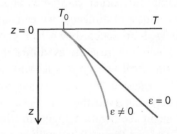

Figure 9.10 Comparison of solutions to the diffusion equation for a flat steady-state Earth. Brown straight line: solution for temperature with no internal heat sources ($\varepsilon = 0$). Blue curve: solution for temperature with uniformly distributed heat sources ($\varepsilon \neq 0$). Both solutions are required to obey boundary conditions at Earth's surface ($z = 0$), with the result that both temperature profiles converge to the same value and the same slope at the surface.

conditions as above, we find

$$T = -\frac{1}{2}\frac{\varepsilon}{K}z^2 + \left|\frac{q_0}{K}\right|z + T_0$$

(see Figure 9.10).

As a somewhat nontrivial exercise: if you begin with this equation, can you show that it satisfies both boundary conditions at z = 0?

As we see in Figure 9.10, the temperature is predicted to be cooler within an Earth that has heat sources than one that does not. This might be explained as a consequence of the boundary conditions: in both cases, the amount of prescribed surface heat flow is the same; but without heat sources, the only way for that heat to get out of the interior is with a higher temperature gradient at depth.

A Quantitative Application to the Earth. The preceding analysis can be made more relevant if we use spherical rather than Cartesian coordinates. This is more than a symbolic change, as it recognizes that Earth is an enclosed body (with a "center," at $r = 0$), not a flat body that conceivably goes on forever (—a situation that could be problematic if Earth has heat sources within it, as suggested by an extrapolation of Figure 9.10!). With heat sources, the diffusion equation in steady state is

$$\nabla^2 T = -\varepsilon/K.$$

In order to express the Laplacian $\nabla^2 T = \vec{\nabla}\cdot(\vec{\nabla}T)$ in spherical coordinates (r, θ, λ), we would start by writing the gradient $\vec{\nabla}T$ in spherical coordinates, then likewise for the divergence. The details are beyond the scope of this chapter; also, we will simplify things somewhat by assuming Earth to be radially symmetric, so that

$$T = T(r).$$

In that case, our diffusion equation ends up becoming

$$\frac{1}{r^2}\frac{d}{dr}\left(r^2\frac{dT}{dr}\right) = -\varepsilon/K.$$

As with a flat Earth, we assume for this example that heat sources are distributed uniformly throughout the interior (so ε is constant); the solution, which can be verified by substituting it into the diffusion equation, is

$$T = \frac{\varepsilon}{6K}(a^2 - r^2) + T_0,$$

where a is the Earth's radius and T_0 is the temperature at $r = a$. Like the flat Earth case, the solution for T is a quadratic function of depth or radius. However, this solution is fundamentally different in other ways: physically, because (as already noted) the solution does not extend infinitely down—if you descend below the center, you start coming out the other side; and mathematically, because the surface heat flow cannot be prescribed independently, as a second boundary condition. The latter situation is mainly a consequence of the steady state condition we've imposed on the Earth, which requires that the total heat flow out of the Earth must exactly equal the total amount of heat produced internally per second (an amount that was determined by Earth's size and its heat production per unit volume, ε). These considerations lead to

$$q_0 = \frac{1}{3}\,\varepsilon a,$$

a relation that can also be obtained directly by invoking Fourier's law,

$$q_r = -K\frac{dT}{dr},$$

differentiating our solution and evaluating it at $r = a$.

To quantify our solution, we can substitute in the value for K, 2.5 W/m°C, noted at the outset of this chapter for igneous rocks, and a value for ε consistent with our discussion of radiogenic heat sources, for example 2×10^{-5} mW/m^3 (corresponding to our representative mantle-wide heat production of 5×10^{-9} mW/kg). Predicted temperatures range from T_0 at the Earth's surface to $T_0 + \varepsilon a^2/6K$ at the center of the Earth, an increase of $\varepsilon a^2/6K$ or 54,120°C. As a 'check' on this solution, we note it predicts a surface heat flow of $\frac{1}{3}\,\varepsilon a$ or $42\frac{1}{2}$ mW/m^2, which is comparable to but

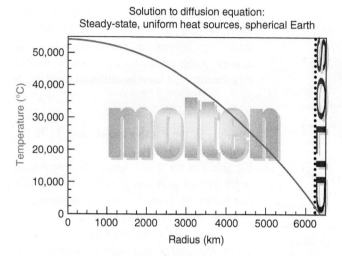

Figure 9.11 Predicted temperatures within an Earth in steady state that contains uniformly distributed heat sources. The heat production is taken to be radiogenic with a magnitude equal to the mantle-wide average value inferred earlier in the text. In this solution (in which T_0 is set to zero, for simplicity), the entire Earth below a thin 'lithosphere' is completely molten, if not vaporized.

somewhat less than the observed heat flow in continental regions as listed in Table 9.1.

This solution, which is depicted in Figure 9.11, is, of course, completely unrealistic for the Earth! Peridotite in the mantle begins to melt at temperatures ~1,400°C, depending on ambient pressure and other factors (e.g., Sarafian et al., 2017); according to our solution, such temperatures would be reached before a depth of 100 km. This solution would predict an Earth that is nearly completely molten (if not substantially vaporized).

The diffusion equation will be very useful in topics covered later in this chapter and in the next. But, as we will now see, all the mathematics we have dealt with till this point was not in vain: the diffusion equation has more to tell us about heat flow.

9.3.4 Learning From Failure: A Deeper Look Into Heat Flow by Conduction

Our first lesson from the solution shown in Figure 9.11 is that *heat flow by conduction is inefficient*: with conduction, heat is transmitted solely through temperature gradients; to successfully get even a reasonable amount of heat out of Earth's interior (think of the value of q_0 implied by this solution), interior temperatures must be extraordinarily high. This should be contrasted with convection, for example as it takes place in a pot of boiling

water: the hot water on the bottom rises as soon as the temperature there reaches a critical point, so the water on the bottom never has a chance to get *too* hot; the water that has risen to the top cools off—by evaporation and conduction—at least somewhat before it cycles back down to the bottom. The result is a much more efficient situation, with much less of a temperature difference from bottom to top.

The fact that the Earth is not vaporized or predominantly molten (the mantle convects as a solid) suggests that at least one of the assumptions our solution was based on must not apply to the Earth.

Constant Heat Sources? We had assumed that heat sources are uniformly distributed throughout the interior.

Earlier in this chapter we investigated the possibility of potassium-40 in the core; we found it to be likely, and possibly in quite impactful amounts. If, instead, the core somehow has no sources of heat (radiogenic or otherwise), the interior would be significantly cooler; in a steady state, for example, temperatures would have to increase more gradually with depth through the mantle and would level off as the core is approached (the core temperature would have to be constant, to maintain that steady state). But even in this case (as a homework exercise for this chapter will help you show), interior temperatures would still be extremely high, reaching tens of thousands of degrees, and the Earth would still be nearly entirely molten.

And on the subject of heat sources: Earth's radiogenic heat has decayed over time. If our hypothetical steady-state solution is based on a single value of ε for all time, it might be argued that a larger value, reflecting its greater magnitude in the past, would provide a fairer picture of the interior temperatures produced when heat is transmitted only by conduction. With a larger ε, of course, our solution would yield an even hotter Earth.

Unchanging Thermal Conductivity? We had assumed that the thermal conductivity of Earth material is constant, i.e., independent of temperature and pressures within the Earth. If it so happened that K increased with depth, then (since $q_r = -K \, dT/dr$), a smaller temperature gradient would be able to conduct the same heat flow (or possibly even more) out of the interior.

It turns out that there is more than one way to conduct heat through materials inside the Earth.

- The 'usual' kind of thermal conductivity proceeds by the propagation of lattice vibrations—called **phonons**—through the medium; we will designate this conductivity by K_{ph}. At high temperatures, the energized phonons increasingly interfere with each other, preventing their effective propagation; the result is that $K_{ph} \propto 1/T$ (e.g., Durham et al., 1987 and references therein). On the other hand, at higher pressures the mineral lattice is compressed and—just like with seismic waves—the waves of vibrational energy propagate faster (e.g., Fujisawa et al., 1968).

  Based on high-pressure experiments carried out over the years (e.g., Schatz & Simmons 1972; Beck et al. 1978; Manga & Jeanloz; 1997; Hofmeister 2005), though, it appears that neither the temperature effect nor the pressure effect dominates within the mantle; the magnitude of K_{ph}

measured under uppermost-, upper-, and lower-mantle conditions, and overall, shows little variation, with values confined to between ~2 W/m°C and ~12 W/m°C. Such a modest variation in K_{ph}—less than an order of magnitude difference, compared to the value we used to construct Figure 9.11—would not, by itself, yield an Earth much less molten than our Figure 9.11 prediction.

- Another way to convey heat energy through a material is by the motion of free electrons. In metals, where there is a 'sea' of mobile electrons, thermal conductivity is essentially conduction by electrons. Of course, those electrons also carry electrical charge, which is why metals are good electrical conductors. In fact, their ability to conduct both heat and charge means their thermal and electrical conductivities 'parallel' each other; it turns out that, at moderate to high temperatures, the ratio of those conductivities in a metal depends only on the temperature:

$$K_{el}/\sigma = LT,$$

where σ is the electrical conductivity (defined analogously to thermal conductivity, as discussed in Chapter 10), T is in units of °K, and L, often called the *Lorenz number*, is the proportionality constant. Based on theory and high-P,T experiments on iron and iron alloys, the value of L is around 2.4×10^{-8} W ohm/(°K)^2, with some differences depending on composition and the model approach (see Yong et al., 2019 and references therein, also Yin et al., 2022). With σ determined from laboratory experiments on iron alloys or from observed variations in the geomagnetic field (discussed in Chapter 10 also), this relation—known as the **Wiedemann-Franz-Lorenz law**—can be applied to the core, allowing $(K_{el})_{core}$ to be estimated.

  Those estimates of $(K_{el})_{core}$ exhibit a near-dichotomy, historically (Olson, 2016): earlier values averaged ~40 W/m°C, whereas those over the past decade or so are typically at least ~100 W/m°C, consistently more than twice as high (—and curiously similar in magnitude to the value of K for pure iron at STP). Such a magnitude for the core's thermal conductivity has potentially significant consequences for (inner) core evolution (see Olson, 2016 and references therein). In any case it is distinctly higher than the mantle's K_{ph}. If we employed this value for the core, we would find the increase in T within the core to be less than what is shown in Figure 9.11; but it would still leave the Earth nearly entirely molten.

- In contrast to metals, silicates (like those comprising the Earth's mantle) are electrical *semiconductors*; free, mobile electrons are much more scarce, restricting the ability of silicates to carry heat as well as electrical charge and relegating electron thermal conductivity to insignificance in the mantle. But there are complications. In a semiconductor, free electrons are created by the absorption of energy, which 'excites' them enough to become mobile and conduct electricity weakly. When one becomes mobile, it leaves behind a positively charged 'hole'; the hole may be mobile, too, changing place as first one nearby electron then another fills the current hole but creates another. Because of their opposite charge, though, the electron and its hole may be electrically bound to each other; such a pair is called an **exciton**. Excitons are electrically neutral, and thus play no role in electrical conductivity; but they carry energy—*the energy that caused them to polarize in the first place*—which can be released during recombination. Excitons thus aid thermal conductivity: imagine the excitons 'popping' as *a wave of polarization* passes through the material, carrying the energy as it propagates.

  For a brief time, exciton thermal conductivity was more than just an esoteric curiosity. As noted by Clark (1957), more excitons could be created in a semiconductor at higher temperatures; this could lead to a high thermal conductivity (we could denote this K_{ex}) in the hot lower mantle.

Subsequent experiments found that K_{ex} could increase to ~4 W/ m°C in the upper mantle (Lubimova, 1967) and to as much as ~400 W/ m°C in the lowermost mantle (Lawson & Jamieson, 1958). However, the interpretation of these quite dramatic results as a manifestation of exciton conductivity was questioned by Mao (1973). Mao concluded that exciton conductivity was probably not important (though he explicitly left open the possibility that it is, at high pressures); and this conclusion appears to have been accepted, without further exploration, to the present day.

- Finally, any mass that is heated will *radiate* some of that heat away. Within the mantle, which is fairly opaque (!), such radiation, for example at infrared wavelengths, will not get very far before it is absorbed by the surrounding mantle. With enough radiation, that surrounding mantle will eventually heat up enough to radiate as well—conveying the heat another centimeter or two (Clark, 1957; Schatz & Simmons, 1972) or more (Hofmeister, 2005) before absorption. Centimeter by centimeter, this process of absorption, heating, and reradiation will continue, providing a very incidental mechanism for transmitting heat away from the hot interior and, ultimately, up to the surface. The heat energy is carried by electromagnetic radiation, i.e. photons, but the net transmission is much slower than the speed of light; in part, this is because of rock's opacity. In the mantle, the efficiency of transmission is further reduced (and the opacity effectively increased) by scattering at grain boundaries (e.g., Shankland et al., 1979; Hofmeister 2005).

  The possibility that 'radiative' thermal conductivity, K_r, contributes significantly to the heat flow out of the mantle depends on whether radiation dominates over opacity. These factors both depend on temperature: rocks radiate *much* more at high temperatures but also become more opaque at high temperatures (cf. Goncharov et al., 2006; Beck et al. 1978; Schatz & Simmons, 1972).

Opacity also increases with higher iron content—as noted in Chapter 8 (and briefly discussed from a different perspective in Chapter 10), some have considered that a possibility in the lowermost mantle—and its dependence on grain size is nonlinear (Goncharov et al., 2006; Hofmeister, 2005; Schatz & Simmons, 1972).

High-pressure experiments have shown that radiative thermal conductivity is at least as important as lattice conductivity in both the upper (Schatz & Simmons, 1972) and lower (Hofmeister, 2005) mantle. Values for the uppermost, upper, and whole mantle from Shankland et al. (1979), Schatz & Simmons (1972), and Hofmeister (2005) are somewhere in the range ~1 W/m°C to ~5 W/m°C, though by implication from the latter work and Manga and Jeanloz (1997) K_r might be as much as 12 W/m°C; in contrast, if the strong dependence on Fe demonstrated by Goncharov et al. (2006) dominates, K_r in the lower mantle might be no more than 5 W/m°C, or even less (cf. Goncharov et al., 2008, 2015).

The thermal conductivity at any depth within the Earth will be the sum of the conductivities K_{ph}, K_{el}, K_{ex}, and K_r representing the four mechanisms of thermal conduction. Because each mechanism has (as pointed out by Mao, 1973) its own kind of dependence on temperature, pressure, and mineralogy, and because there is some uncertainty in the values of each K (judging by their range of estimates), it is difficult to assess their net effect on our prediction in Figure 9.11 of a molten Earth. For the mantle, and excluding K_{ex} as too uncertain, we might simply add the ranges of K_{ph} and K_r (2–12 W/m°C and 1–12 W/m°C, respectively, though more conservatively the latter would be 1–5 W/m°C); we might then conclude that the mantle's thermal conductivity is between 3 and 24 W/m°C (or, more probably, between 3 and 17 W/m°C). It is possible that the lower mantle is characterized by the high end of those ranges, but the depth at which K increases is still unclear.

The effect of exciton conductivity, however, is still an unknown factor. That uncertainty, along with the high end of the preceding estimates, makes it conceivable that the mantle's total thermal conductivity is great enough to get Earth's heat out conductively without requiring the molten interior shown in Figure 9.11. But note that the conductivity increase would have to include the upper half of the mantle—if K is larger only in the lower mantle, temperatures would already have increased too much.

A Steady State? Finally—and most important from a teaching perspective—we have assumed that conduction of heat within the Earth is steady state; if this is not true, perhaps we have not yet reached a point where the internal temperatures are as extreme as what we see in Figure 9.11.

As noted earlier, the full diffusion equation, with $T = T(x, y, z; t)$ and $\partial T/\partial t \neq 0$, can be quite challenging to solve (in fact, such challenges are what led Fourier to apply what we now call *Fourier analysis* to the solution of heat flow problems). One of the simplest *unsteady* diffusion equations is for the case of one-dimensional heat flow in a 'flat Earth' with no heat sources,

$$\frac{\partial T}{\partial t} = k \frac{\partial^2 T}{\partial z^2}.$$

As a mathematical caution, we note first that there are an infinite number of possible solutions to this seemingly basic equation. For example, we can take the derivative of both sides of this equation with respect to z:

$$\frac{\partial}{\partial z}\left(\frac{\partial T}{\partial t}\right) = \frac{\partial}{\partial z}\left(k \frac{\partial^2 T}{\partial z^2}\right),$$

and since the order of partial differentiation should not matter, we can write this as

$$\frac{\partial}{\partial t}\frac{\partial T}{\partial z} = k \frac{\partial^2}{\partial z^2}\frac{\partial T}{\partial z}.$$

That is, if T is a solution to our diffusion equation, then $\partial T/\partial z$ will also be, and by implication $\partial^2 T/\partial z^2$ and $\partial^3 T/\partial z^3$ and so on. Similarly, the integral of our solution for T with respect to z, and the integral of that, and so on, will also be solutions to the same equation. We can add to that derivatives and integrals with respect to t as well.

What will end up forcing us to choose one solution over another is the physical situation we are applying the diffusion equation to—what the 'boundary' conditions are. For unsteady diffusion, with $T = T(z, t)$, this will include restrictions on T or $\partial T/\partial z$ (i.e., q_z) at the Earth's actual boundary at $z = 0$ but also an *initial condition*, at $t = 0$ (the latter, in effect, a 'temporal boundary').

So, one example of a solution to the unsteady one-dimensional diffusion equation is

$$T = \frac{C}{\sqrt{t}}e^{-z^2/4kt}$$

where C is some constant. This expression (which actually can be derived from the next solution presented below) is somewhat typical of unsteady solutions both in its complexity and in its 'nonseparability;' that is, the solution cannot be written as a product of a purely time-dependent term times a completely separate, purely spatial term.

It can be verified by substitution that this is indeed a possible solution to our equation. And, since $T \to 0$ when $t \to 0$, the initial condition it satisfies is $T(z) = 0$ at $t = 0$. However, the boundary condition to which it corresponds at

$z = 0$, namely $T = C/\sqrt{t}$, is physically unlikely, so this solution is *not* directly relevant to the Earth.

Can you explain why that solution is physically unrealistic?

An example of a more relevant solution is

$$T = A \cdot \text{ERF}\left(\frac{z}{2\sqrt{kt}}\right) + B$$

for some constants A and B, where

$$\text{ERF}(u) \equiv \int_{-u}^{u} \frac{1}{\sqrt{\pi}} e^{-v^2} dv.$$

$\text{ERF}(u)$ is known as the **error function**, defined as the area between $-u$ and u under the 'bell-shaped curve,' the so-called *Gaussian* or *"normal"* probability density distribution (more precisely, it is the normal distribution with a mean of 0 and a variance of $\frac{1}{2}$); see Figure 9.12a. Note that $\text{ERF}(0) = 0$ and $\text{ERF}(u) \to 1$ as $u \to \infty$ (that is, the total area under the probability curve is 1). It follows that, according to this solution (with $u = z/2\sqrt{kt}$), $T = B$ at $z = 0$; also, $T = A + B$ at $t = 0$, except at $z = 0$. That is, T always equals B at the surface, and T initially equals $A + B$ in the interior.

For example, if A is positive, then this kind of solution describes a situation where a relatively

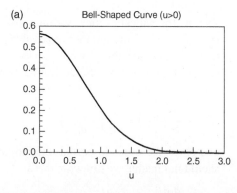
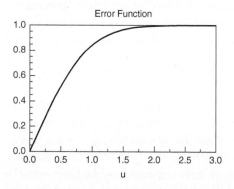

Figure 9.12 Solution to the unsteady diffusion equation versus time, for a 'flat Earth' with no heat sources. (a) The elements of a general solution include a bell-shaped (Gaussian) curve and the error function. As detailed in the text, the former is represented by the formula $\exp(-u^2)/\sqrt{\pi}$; shown on the left is the half of the bell shape corresponding to $u \geq 0$. For $u < 0$, the curve would be the mirror image of $u > 0$ (thus, it is symmetric about $u = 0$; and it would look like a bell!). The latter, shown on the right, corresponds to the area under the bell-shaped curve between $-u$ and $+u$ (or, equivalently, twice the area under the bell-shaped curve between 0 and $+u$).

Figure 9.12 Solution to the unsteady diffusion equation versus time, for a 'flat Earth' with no heat sources.
(b) Error function (ERF) solution, written as $A \cdot \text{ERF}(u) + B$ in the text, with $u = z/\sqrt{(4kt)}$, is shown as a function of time at a depth of 11 km. We used $k = 1 \times 10^{-6}$ m²/sec; for purposes of illustration, we set $B = 0$ and $A = 1$ (yielding a temperature range of only 0 to 1).
Note that, at this depth, by the time $t_C \equiv z^2/k \sim 3.8$ Myr (dotted arrow) has been reached, the temperature has dropped by nearly 50%.

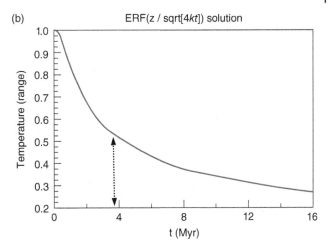

(b)

ERF(z / sqrt[4kt]) solution

cool temperature is imposed at the surface of a mass that was initially warmer within. The boundary condition creates temperature differences, driving thermal conduction that gradually cools the interior.

Our error function solution is quite relevant to the Earth. The heat flow it implies can be obtained from Fourier's law, $q_z = -K \, \partial T/\partial z$, and turns out to equal

$$q_z = -\frac{C}{\sqrt{t}} e^{-z^2/4kt}$$

where the constant $C = AK/\sqrt{\pi k}$. (Yes, we rejected this expression as a solution for T; but it is quite acceptable for q_z!) At Earth's surface ($z = 0$), the predicted heat flow is simply

$$q_z = -\frac{C}{\sqrt{t}};$$

that is, the surface heat flow for a cooling flat 'slab' of Earth material is inversely proportional to the *square root of the age* of the slab.

An illustration of the way temperature varies with time in an error function solution is provided in Figure 9.12b, where we have arbitrarily chosen a depth $z = 11$ km for the case $B = 0°C$ and, for simplicity, $A = 1°C$.

Note that we could equally well have chosen to illustrate T as a function of depth at a particular time. In that case, what do you think the temperature profile would look like (e.g., would T be increasing or decreasing with depth)? How do you think that profile would evolve over time?

We see in Figure 9.12b that the temperature at this depth is initially 1°C, but it decreases over time as that depth 'gives up its heat' to the cool surface. At early times the cooling at depth is rapid; but as cooling proceeds and the temperature difference between that depth and the surface diminishes, the rate of cooling slows.

On the other hand, in many situations we will be more interested in the overall picture—for example, how long it will take for a 'substantial' amount of cooling to happen (or, with $A < 0$, how long for a lot of heating)—rather than details such as how the rate of cooling evolves with time. We can see in Figure 9.12b that the temperature has changed by about 50% after the passage of an amount of time marked by the arrow. We can calculate that time directly by noting that ERF($\frac{1}{2}$) = 0.52. For our diffusion equation, this time—which we will denote by t_C—is determined from

$$\frac{z}{2\sqrt{kt_C}} = \frac{1}{2}$$

or

$$t_C = \frac{z^2}{k}.$$

Can you verify the value for t_C stated in the caption for Figure 9.12b? It is mainly a question of units....

This formula for t_C will be quite informative for understanding thermal diffusion (and, in fact, any diffusive process). In the next subsection we will present a more formal technique for estimating t_C; our approach will be rigorous and illuminating, though it will end up yielding the same formula. We will then return to our discussion of the Figure 9.11 failed solution and, finally, evaluate the steady-state assumption.

Characteristic Times. If we are not interested in the details of how T varies with position or time but just want to know the typical magnitude of that variation, there is a simple but quite powerful technique we can apply to produce such an estimate. This technique, known as *scaling* (in this case, of the diffusion equation), is useful not only because it yields those estimates relatively easily and reasonably accurately, but because the technique is applicable in so many other fields, and to all components of the Earth System. Indeed, scaling and its results appear throughout much of fluid dynamics—as we will see in the rest of this textbook—in part because those order-of-magnitude numbers also facilitate comparisons of the various factors driving fluid flow.

Scaling, consequently, helps us understand the fundamental physics that all our differential equations are meant to describe. And it does that while managing to end up outside the framework of calculus (though it remains consistent with that framework).

Let's say that we want to get an overall idea of how the temperature changes by diffusion in some arbitrary mass; the mass could be the entire Earth, a region thereof, a lithospheric slab, or a magma chamber, to name just a few possibilities, and it could be heating up, cooling off, or experiencing a change in boundary conditions. We begin our scaling analysis by defining the **characteristic time**, which will be denoted by t_C, as a representative time for 'significant' thermal diffusion to occur in that mass. This is a more general (and also more vague) definition of t_C than earlier, but it is not tied to the error function or any other specific solution of the diffusion equation.

We write the time variable as

$$t = t_C \cdot t^*,$$

where t^* is a dimensionless time variable. We can mark the passage of time with either t or t^*, even though they have different magnitudes; for example, in the solution illustrated in Figure 9.12b (where $t_C = 3.8$ Myr), t might be 0 Myr, 1 Myr, 2 Myr, 3 Myr, 4 Myr, and so on, whereas the corresponding t^* would be 0, 0.263, 0.526, 0.789, 1.053, and so on.

t^* is a dimensionless variable because we obtain it by dividing t by 3.8 Myr. If t_C has been well chosen, then $t^* = 1$ will correspond to a time ($t = t_C$) when significant changes in the object's temperature have taken place. In that case, for an overall sense of the solution, we are unlikely to need to consider values of t^* much greater than 1; though it will take an infinite amount of time to achieve the final thermal state, variations of the temperature or heat flow when $t^* = 10$, or 100, or more, can be considered minor details. That is, for most situations of interest, the magnitude of t^* will be of order unity.

For variables like temperature and size that are naturally finite (in contrast to a never-ending variable like time), creating dimensionless variables that are of order unity is solely a matter of choosing scales wisely. So, we denote by T_C the typical or *characteristic* temperature of the body and by L its typical dimension; we then define a dimensionless temperature function and dimensionless spatial variables by

$$T = T_C \cdot T^*,$$
$$x = L \cdot x^*, \quad y = L \cdot y^*, \quad z = L \cdot z^*.$$

In some situations, different scales might be required for different spatial dimensions (thus, L_x, L_y, L_z), but for the present development we will restrict ourselves to one-dimensional heat flow, in the z-direction. If we have chosen our scales well, the resulting dimensionless variables will all be of order unity.

For the lithosphere, an example of well-chosen scales might be (depending on the problem we are investigating) $T_C \sim 1{,}000°C$

and $L \sim 100$ km; those choices would lead to values of T^* between 0 and perhaps 2, and similarly for z^*. Given the breadth of lithospheric plates, we could have chosen a much larger value for L; however, for problems involving vertical heat flow, it makes more sense to specify a vertical dimension.

These scaled variables are substituted into our one-dimensional unsteady heat flow equation (where, for simplicity, we will continue to assume no heat sources). The left-hand side becomes

$$\frac{\partial T}{\partial t} = \frac{T_C}{t_C} \frac{\partial T^*}{\partial t^*}$$

whereas the second derivative on the right-hand side becomes

$$\frac{\partial^2 T}{\partial z^2} = \frac{\partial}{\partial z} \frac{\partial T}{\partial z} = \frac{1}{L} \frac{\partial}{\partial z^*} \left(\frac{T_C}{L} \frac{\partial T^*}{\partial z^*} \right)$$
$$= \frac{T_C}{L^2} \frac{\partial^2 T^*}{\partial z^{*2}}.$$

Our diffusion equation can then be written

$$\frac{T_C}{t_C} \frac{\partial T^*}{\partial t^*} = k \frac{T_C}{L^2} \frac{\partial^2 T^*}{\partial z^{*2}}.$$

Dividing out the common term T_C,

$$\frac{1}{t_C} \frac{\partial T^*}{\partial t^*} = k \frac{1}{L^2} \frac{\partial^2 T^*}{\partial z^{*2}};$$

incidentally, our ability to eliminate T_C reminds us that diffusion is a relative process, driven by temperature differences rather than the absolute, actual temperature values. Finally, we assume that we have scaled all variables reasonably well, so that all asterisked quantities are of order 1, and divide them out to yield our order-of-magnitude ultimate relation:

$$t_C \sim \frac{L^2}{k},$$

which we already found to be true for the ERF solution discussed earlier.

For those not convinced that derivatives of our dimensionless variables will be of order unity just because the variables themselves are (—a valid concern!), we note that diffusive processes are by nature smoothing processes, so that any sharp irregularities—at which times or locations their derivatives may actually be extreme—will fairly quickly be smoothed out, leaving relatively modest derivatives.

Our result for t_C is physically reasonable in implying that the greater the thermal conductivity or diffusivity of an object is, the faster it will heat up or cool off. Our result also implies that the larger the object is, the longer it will take to heat up or cool off—by a factor of four when its size doubles. It is not surprising that t_C depends on the square of the object's size; after all, heat flows across surfaces, i.e., areas. So, if the object is cooling off (for example) and needs to get some amount of heat out, then the heat flux out is less if it is spread over a greater area, and the mass will take longer to cool. More precisely, when the object is larger, all else being the same, gradients are smaller; thus, both heat flow (a reflection of temperature gradients) and its divergence are reduced—doubly reducing the rate at which heat will be given off by the mass as it conserves energy.

The consequences of this simple formula for t_C are enormous. Numerically, we can set $k \sim 1 \times 10^{-6}$ m^2/sec; for L, we consider two situations.

- First, we specify $L = 6.371 \times 10^6$ m, to estimate the time to cool (e.g.) the entire Earth by conduction. In this case,

$$t_C \sim \frac{(6.371 \times 10^6)^2}{1 \times 10^{-6}} \sim 4 \times 10^{19} \text{ sec}$$

—more than a trillion years! That is, even in an Earth without heat sources, it would take almost 300 of Earth's lifetimes to cool its interior, if heat is transmitted only by conduction. Clearly, assuming a conductive steady state was a poor assumption to make when we tried to determine temperatures within the Earth; the solution we obtained, shown in Figure 9.11, shows a temperature profile that has never described the Earth.

This timescale also tells us how *slow* conduction is, as a means of transmitting heat. We have already found conduction to be very inefficient; in contrast, convection is both efficient and potentially rapid, depending on how quickly the fluid can carry the hot and cold material. In the early days of the Earth, when it was much hotter and even the mantle flowed easily, heat transmission through the interior was certain to be dominated by convection; and any situations in which convection was impeded would have been temporary, geologically speaking. Even today, with a cooler Earth and a relatively high mantle viscosity, its efficiency and quicker timescale argue for ongoing mantle-wide convection.

However, such conclusions are premature, since we have not yet studied the physics of convection; that will be our focus in the next section of this chapter.

One last point about the timescale for conduction: its dependence on L^2 implies that conduction will be much slower, and convection much more likely, the larger L is. As we turn from small moons in our Solar System to our Moon, to the Earth, to the giant planets, to the Sun and other stars, the dominance of convection should be increasingly unavoidable.

- Second, we consider $L = 1 \times 10^5$ m, and estimate roughly the time it would take for a thermal event (such as a 'burst of heat' from below) to be conducted through the lithosphere. In this case, $t_C \sim L^2/k = 1 \times 10^{16}$ sec ~317 Myr. Once again, we see how slow conduction is—even for transmission through a distance of only 100 km. Furthermore, since no oceanic lithosphere is older than 200 Myr or so, it is unlikely that any significant fraction of the heat flow observed to cross the seafloor surface had originated below the lithosphere.

What about young oceanic lithosphere? Unlike its heat flow, the thickness of oceanic lithosphere is directly proportional to the square root of its age, so young lithosphere is much thinner. Consider 50-Myr-old lithosphere; estimate its thickness and show that our conclusion still holds true for that lithosphere.

How would the temperature dependence of thermal conductivity further affect our conclusion?

Our conclusion that most oceanic heat flow cannot be attributed to deep heat sources because conduction is *too slow* to have brought their heat up to the surface yet is consistent with the tectonic origin of oceanic lithosphere (hot formation at a mid-ocean ridge) that we inferred earlier from the "one-over-the-square-root-of-age" pattern of oceanic heat flow.

In sum, the observed oceanic heat flow must largely originate *within* the lithosphere itself. Bullard's 'alternative hypothesis' to explain oceanic heat flow observations (that they were the result of heat brought up convectively to the base of the oceanic crust) might never have been proposed had scientists at the time been aware of the existence and properties of the lithosphere.

9.4 Transmission of Heat in Fluids

At last!

In any body exhibiting temperature differences, conduction of heat must happen, unavoidably; but if the body is able to flow, heat can also be transmitted by the motion of hot material. In this section, we will explore the conditions under which convection will take place, and what the thermal implications of that convection might be.

Is the Mantle Relevant Here? Convection is always a possibility, if not an inevitability, in the fluid components of the Earth System; but what about the mantle—are we justified in applying our discussion below to the mantle? In Chapter 8, we interpreted numerous seismic structures as the manifestation of mantle-wide convection; was that premature? Let's review what is already established.

- First of all, we know the mantle can deform. We have already discussed evidence such as the finite speed of elastic waves, the existence of solid earth tides, and most fundamentally the existence of an equatorial bulge.

Another line of evidence, beyond the scope of this textbook (though a great example of the interconnections among components of the Earth System), is the 14-month period of the wobble discussed in Chapter 5; the period is predicted to be 10 months for a perfectly rigid Earth, but significant modifications by the fluidity of the oceans (e.g., Dickman, 1993) and the core, and especially by the elastic flexibility of the mantle (e.g., Smith & Dahlen, 1981), in response to the change in centrifugal force during wobble, end up lengthening each cycle of wobble by 40%.

- We have also discussed evidence that the mantle can flow. The classic evidence for the occurrence of mantle flow is, of course, isostatic rebound and subsidence, in response to past and current changes in surface loads.

  The alignment of Earth's equatorial bulge perpendicular to the spin axis, maintained through time despite ongoing true polar wander, shows that bulge migration must accompany the wander; as explained in Chapter 5, the deformation's broad spatial scale requires that the entire mantle flow to support the bulge's adjustment.

  All told, modeling isostatic and rotational phenomena has yielded a reasonable picture of the mantle's viscosity and its variations with depth (see Chapter 5).

Convection is traditionally discussed in the context of pure fluids; but, for the mantle, its extreme viscosity suggests that that would be an oversimplification. Mantle convection is often described as *solid state*, in which the motion progresses by creep or other types of slow deformation (see Chapter 8); such deformation requires a 'soft' mantle if it is to be effective, that is, the temperature of mantle material should be close to the **solidus**, the temperature at which partial melting begins. We will follow the traditional approach for

now, investigating fluid convection in the mantle and other components of the Earth System; we will then consider how the requirement for a soft mantle modifies our conclusions.

9.4.1 Fluid Stability

Very simply, a fluid is considered to be **stable** if a slight disturbance—a *perturbation*—'settles down' rather than leading to 'catastrophe.' Since convection usually involves vertical motion, we envision perturbations in which a small parcel of fluid is given a tiny push upward, for example. One of three things will happen; which one depends on the fact that the rising parcel encounters lower ambient pressures, decompresses adiabatically, and thus cools somewhat.

- If the magnitude of the fluid's temperature gradient is less than adiabatic—*subadiabatic* conditions—the adiabatically cooled parcel will find itself cooler than the surrounding fluid at its new height, thus denser, and it will sink back down. This is a stable state, keeping perturbations from getting out of hand.

- If the magnitude of the fluid's temperature gradient is greater than adiabatic—*superadiabatic* conditions—the ambient fluid will cool with height at a greater rate than the parcel, so the parcel will, in effect, become warmer and warmer *relative to its surroundings* as it rises. Thus, the parcel will become lighter and lighter, relative to its surroundings, and will rise faster and faster. This is an unstable state, with an exponentially growing displacement (but more on this below).

- If the actual temperature gradient of the fluid is precisely adiabatic, the perturbed parcel will find itself at the same temperature as the surroundings, so—neither more nor less dense than the ambient fluid—it will remain at its new location. This is called a state of **neutral stability**.

These possibilities are illustrated with simple numbers in Figure 9.13a; a real-life 'cartoon quiz' is presented in Figure 9.13b.

In stable fluids, perturbations may produce *gravity waves* (*buoyancy waves*). When the perturbed parcel sinks back down under stable conditions, its momentum may cause it to overshoot its initial level. As it sinks, it is warming adiabatically. If it overshoots, it will find itself warmer than the surroundings, because the surrounding fluid heats up with depth at a slower, less-than-adiabatic rate. The parcel is then less dense than the surroundings, and will rise again; with another overshoot past the initial level, the oscillation will continue.

In a stable fluid that is stratified, with less dense layers overlying denser layers, buoyancy waves are more likely at layer boundaries. If a perturbation to a parcel just below such a boundary pushes it upward above the boundary, it will find itself denser not only because it has cooled adiabatically but also because of its intrinsically greater density—thus resulting in a stronger restoring force. And, when the parcel sinks below its initial level, the lighter fluid just above it (and moving downward with it) will find itself too light, both because of adiabatic warming and because of its intrinsic lightness. In this situation, the waves that result are often called *internal waves*.

9.4.2 How Convection Works in a Fluid

Convection requires that the fluid be initially in an unstable state. The perturbation that sets the fluid into convective motion can either arise naturally, like when a parcel is displaced upward or downward, or it can be forced, for example by heating the fluid from below or cooling it from above. Whether displaced upward or heated, our parcel is **buoyant**: less dense than the surrounding fluid, gravity pulls down less on it than on the surrounding material, so it begins to rise. As it rises, it cools adiabatically, and will continue to rise only if the ambient temperature gradient is superadiabatic.

However, the parcel's rise will be slowed by viscous drag from the surrounding fluid; and if it rises *too* slowly, its heat will be conducted away, leaving it cooler and too dense to keep rising.

Thus, *convection is a struggle between buoyancy, viscous drag, and thermal conduction, in which the forces of buoyancy win*. The importance of each factor may differ from situation to situation, with some factors dominating in one situation but not others. To determine who wins the struggle in a particular situation, we will construct various parameters—dimensionless numbers—that enable us to evaluate different aspects of the struggle. This in turn requires that we quantify the two forces involved in the struggle.

Viscous Drag. Imagine a horizontally extensive body of fluid initially at rest. At some moment we begin to force its top layer to move at a constant velocity, and maintain that forcing for all time. Because the fluid is viscous, that top layer will drag the underlying fluid; that is, the velocity (or, equally well, the momentum) of the top layer will gradually be transferred to the layer just beneath, then the layer below that. If the fluid is otherwise unconstrained, for example if the fluid's bottom boundary is frictionless, then viscous drag will eventually cause the entire fluid body to move uniformly. *Velocity (and momentum) will have diffused to a uniform state.*

In short, the viscous drag force represents *diffusion of momentum* (or *velocity*) of the fluid. Using the diffusion of heat as an analogy, this process can consequently be expressed in the form

$$\eta' \nabla^2 (\rho \vec{v}) \quad \text{or} \quad \eta \nabla^2 \vec{v}$$

where $\vec{v}$ is the fluid velocity, ρ is its density, and $\rho \vec{v}$ its momentum (= mass × velocity) per unit volume. The material property (proportionality constant) reflecting how effective the diffusion will be—analogous to the thermal diffusivity k representing how effective diffusion of heat ($k \nabla^2 T$) will be—is the viscosity η

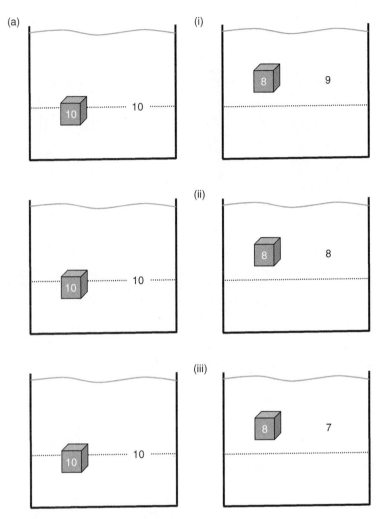

Figure 9.13 (a) Numerical example of fluids that are stable, neutral, and unstable. For numerical simplicity, the temperature gradients used in these hypothetical illustrations are whole numbers. In all cases, the tank of fluid extends up to the wavy blue line.

At the height indicated by the dotted line, both the parcel and the surrounding fluid are at a temperature of 10°C (left-hand drawings). We take the adiabatic gradient to be 2°C per meter (increasing downward). When the parcel is pushed upward (perturbed) a distance of 1 m (right-hand drawings), it cools adiabatically to a temperature of 8°C.

(i) The surrounding fluid is stable, with a temperature gradient of 1°C per meter. At a height of 1 m above the dotted line, its temperature is 9°C; thus, the parcel is cooler, and denser, than its surroundings.

(ii) The surrounding fluid is neutral, with a temperature gradient of 2°C per meter. At a height of 1 m above the dotted line, its temperature is 8°C. In this case, the parcel is the same temperature and density as its surroundings.

(iii) The surrounding fluid is unstable, with a temperature gradient of 3°C per meter. At a height of 1 m above the dotted line, its temperature is 7°C. Here the parcel is warmer and thus less dense than its surroundings.

A really worthwhile exercise would be for you to re-do this illustration, but with the parcel displaced downward *by 1 m.*

(b)

Breeze

(The puffs of smoke here are 'floating in the breeze')

Figure 9.13 (b) Cartoon illustration of stable, neutral, and unstable atmospheric regions (but not in that order!). Smoky air emitted from the chimney is hotter and, overall, slightly lighter than the surrounding air. For simplicity, there is a light breeze blowing, so the smoke does not pile up at the chimney vent.

Can you tell which case corresponds to stable atmospheric conditions? Unstable conditions? A neutral state?

(also called the molecular viscosity or dynamic viscosity), in units of Poise (gm/cm·sec) or Pa-sec (kg/ m·sec), for diffusion of velocity and the **kinematic viscosity** $\eta' \equiv \eta/\rho$, in units of cm^2/sec or m^2/sec, for diffusion of momentum. Clearly (as with k versus K), the units for η' are much more fundamental (length squared per time) and much less 'material property-like' than η, suggesting the equation involving it will be more fundamental as well.

It is certainly reasonable to talk about the diffusion of quantities like $\vec{v}$ or $\rho\vec{v}$; the fact that they are vectors simply means each of their components will be diffusing. Mathematically, this suggests that $\eta' \nabla^2 (\rho\vec{v})$ and $\eta \nabla^2 \vec{v}$ are themselves vectors, each component of which represents the diffusion of that component of $\vec{v}$ or $\rho\vec{v}$. For example,

$$\nabla^2 \rho\vec{v} \equiv (\nabla^2 \rho v_x, \nabla^2 \rho v_y, \nabla^2 \rho v_z)$$

for diffusion of momentum, and similarly for velocity. Conceivably, each component of $\eta' \nabla^2 (\rho\vec{v})$ or $\eta \nabla^2 \vec{v}$ can diffuse in all three dimensions.

The diffusion equation we derived for heat flow was

$$\frac{\partial T}{\partial t} = k\nabla^2 T + \frac{\varepsilon}{\rho C_P},$$

or, in the absence of heat sources,

$$\frac{\partial T}{\partial t} = k\nabla^2 T.$$

Pursuing the analogy with thermal diffusion, we might expect the diffusion of momentum, say, to be governed by a similar type of equation,

$$\frac{\partial}{\partial t}(\rho\vec{v}) = \eta' \nabla^2 (\rho\vec{v}),$$

in the absence of 'momentum sources.' 'Sources' of momentum, by the way, are, of course, nothing more than … *forces*(!), so the more complete version of a momentum diffusion equation would be

$$\frac{\partial}{\partial t}(\rho\vec{v}) = \eta' \nabla^2 (\rho\vec{v}) + \vec{F},$$

with those other forces $\vec{F}$ as yet unspecified. Finally, we note that, for constant density,

$$\frac{\partial}{\partial t}(\rho\vec{v}) = \rho\frac{\partial \vec{v}}{\partial t} = \rho\vec{a},$$

$\vec{a}$ being the fluid's acceleration, yielding

$$\rho\vec{a} = \eta'\nabla^2(\rho\vec{v}) + \vec{F}$$

or

$$\vec{a} = \eta'\nabla^2\vec{v} + \vec{f}.$$

In short, our treatment of viscous drag as a diffusion process has led to a governing equation we recognize as Newton's Second Law! In the equation with $\rho\vec{a}$ on the left-hand side, it is Newton's Law but in units of force per unit volume (since $\rho\vec{a}$ = [mass/vol] × acceleration = [mass × acceleration]/vol), with $\vec{F}$ per unit volume as well; in the equation with $\vec{a}$ on the left-hand side, it is Newton's Law but with ρ divided out, leaving it in units of force per unit mass (and $\vec{f} = \vec{F}/\rho$).

Our identification of viscous diffusion with the Second Law implies that the viscous drag *force* is

$$\eta'\nabla^2(\rho\vec{v})$$

per unit volume. We will shortly use this formulation to evaluate the struggle for convection.

From the diffusion of momentum equation with no sources,

$$\frac{\partial}{\partial t}(\rho\vec{v}) = \eta'\nabla^2(\rho\vec{v}),$$

we can determine the **characteristic time** for momentum diffusion—which we will denote by t_V, representing the time it takes for viscous drag forces to cause appreciable changes in momentum—by scaling the equation, just as we did for diffusion of heat. The result is

$$t_V \sim \frac{L^2}{\eta'},$$

if L is the characteristic dimension of the fluid body.

For example, in the Earth's mantle (with viscosity estimates guided by Chapter 5), we can take $\eta' \sim 10^{21}$ cm^2/sec = 10^{17} m^2/sec for the upper mantle and $L \sim 1,000$ km, leading to

$$t_V \sim 10^{-5}\text{ sec};$$

for the lower mantle or the mantle as a whole η' could be an order of magnitude (or more)

greater, and L might be larger by a factor of 3, yielding the same value for t_V. Evidently, the mantle is so viscous that momentum changes are nearly instantaneous throughout. In contrast, the kinematic viscosity of water is 0.01 cm^2/sec = 10^{-6} m^2/sec, so for the oceans (4 km deep), $t_V \sim \frac{1}{2}$ Myr; viscous drag by molecular viscosity should have negligible impact on ocean dynamics.

Can you verify this estimate for t_V? Thinking of our discussion of tidal friction in Chapter 6, how would this estimate change if we used the turbulent viscosity of the oceans instead?

For any fluid, we might be interested in knowing whether diffusive processes are dominated by heat or momentum (though, in some ways, this is contrasting 'apples' and 'oranges'). For this purpose, we define the **Prandtl number** P_R as the ratio of times for such diffusion:

$$P_R \equiv \frac{t_C}{t_V} = \frac{\text{Time for heat to flow}}{\text{Time for momentum to 'flow'}};$$

P_R is small if thermal diffusion dominates ($t_C \ll t_V$), large if momentum diffusion is more effective ($t_V \ll t_C$). Based on our formulation of these timescales, we find

$$P_R \equiv \frac{\eta'}{k}.$$

For the mantle, $P_R \sim 10^{23}$ or more, making it starkly clear that momentum diffusion is much more effective in the mantle than thermal diffusion is. In the oceans, where $k \sim 1.4 \times 10^{-7}$ m^2/sec (Nayar et al., 2016), $P_R \sim 7$, implying (at least, under nonturbulent conditions) that the two diffusion processes are roughly comparable in effectiveness.

A parameter more commonly used to characterize the state of a fluid is the **Reynolds number** R_E, defined as the ratio of the typical magnitudes of the viscous and 'inertial' forces experienced by the fluid. The **inertial force** is the part of Newton's Second Law, $\vec{F} = M\vec{a}$, that we do not explicitly identify as a force but has the same units as one: $M\vec{a}$. In terms of forces per unit volume—the units we are using for the viscous drag force—this force is $\rho\vec{a}$, since ρ is mass per unit volume. The ratio of these forces is

$$\frac{\text{Inertial force}}{\text{Viscous force}} = \frac{\rho\vec{a}}{\eta'\nabla^2(\rho\vec{v})} = \frac{\rho\partial\vec{v}/\partial t}{\eta'\nabla^2(\rho\vec{v})}$$

and the ratio of their typical magnitudes is

$$\text{RE} = \frac{\text{Inertial force}}{\text{Viscous force}} = \frac{\rho V/t}{\eta'(\rho V)/L^2}$$

$$= \frac{L^2/t}{\eta'} = \frac{LV}{\eta'},$$

after first canceling the common factor ρV and then substituting V in, where $V \sim L/t$ is a characteristic flow velocity of the fluid. According to the definition, a small Reynolds number (for example, if η' is very large) means that changes in velocity—accelerations or velocity gradients—are suppressed by viscous drag; in such situations, with neighboring parcels all moving with similar velocities, the flow is called **laminar**. If the Reynolds number is high (for example, if η' is very small), velocity perturbations are unchecked, and the flow can change drastically from one particle to another or from one instant to the next; the flow is 'scattered,' chaotic, **turbulent**.

For example, in the mantle we might specify $V \sim 3$ cm/yr $\sim 10^{-9}$ m/sec, as a rough estimate from plate movements; this leads to RE $\sim 10^{-20}$–10^{-21}, implying mantle flow is exceedingly laminar. Even within mantle plumes, though η' is much smaller and V may be much higher, the flow is still likely to be laminar.

In the mantle, the extreme value of the Reynolds number makes its interpretation unequivocal. The oceans provide another example equally unequivocal—in some situations. Lateral flow in the oceans is characterized by $V \sim 0.1$ m/sec and $L \sim 1,000$ km for the gyres, and $V \sim 2$ m/sec and $L \sim 100$ km within the western boundary currents; these phenomena yield RE ~ 1–2×10^{11}, clearly implying such lateral circulation will be turbulent. However, vertical flow is less clear. The massive downwelling that produces ABW, for instance, might correspond to $V \sim \frac{1}{2} \times 10^{-4}$ m/sec, assuming a production rate of 4 Sv (Broecker, 1991) over an area of 300 km $\times$ 300 km; this suggests RE $\sim 2 \times 10^5$. But for the upwelling, at a rate of perhaps 1 cm/day or 1×10^{-7} m/sec (see Chapter 3), we have RE ~ 400. Do these values of the Reynolds number imply a state of turbulence? Our intuition, especially as regards the upwelling, would tell us no; but we will save questions about the interpretation of dimensionless numbers for later.

Buoyancy Force. The buoyancy force is a differential gravitational force acting on a parcel of fluid due to its difference in mass compared to the adjacent fluid: if the parcel is relatively deficient in mass, gravity pulls down less on it than on the adjacent fluid, making it 'positively' buoyant; if the parcel has an excess of mass, gravity pulls down more on it, making it 'negatively' buoyant. Symbolically, we can write this as

$$\Delta m\vec{g},$$

where $\vec{g}$ is the acceleration of gravity at the height or depth where the mass difference is Δm. In terms of force per unit volume, buoyancy is associated with a density deficit or density excess $\Delta\rho$ ($=\Delta m/\text{vol}$), and we can write the force as

$$\Delta\rho\vec{g}.$$

Our vector expression ensures that lighter material ($\Delta\rho < 0$) will be buoyant, that is, it will experience a force opposite to $\vec{g}$, and that denser material ($\Delta\rho > 0$) will be negatively buoyant, experiencing a force in the direction of $\vec{g}$.

Buoyancy, and the convection that may result, can be chemical or thermal in origin; in this chapter, we will focus on a purely thermal origin, in which the density deficit (e.g.) is produced by a temperature excess, ΔT. That higher temperature causes the parcel to expand; from the definition of the thermal expansion coefficient, α_P, in Chapter 8, we can write the resulting decrease in density as

$$-\rho\alpha_P\Delta T.$$

If we are dealing with a density excess produced by a decrease ΔT in temperature, that density excess is also expressed as $-\rho \alpha_p \Delta T$ (which is now positive since in this case $\Delta T < 0$). In general, then, the buoyancy force created by ΔT is

$$-(\rho \alpha_p \Delta T)\vec{g}.$$

Thermal convection can be *forced*, if a temperature difference ΔT is imposed, for example on the bottom or top of the fluid body, or *natural*, if the convection arises *spontaneously* from perturbations to our fluid parcel. For a better introduction to the buoyancy force, we will explore the latter situation, and consider a perturbation in which the parcel is nudged vertically through a distance Δz. Before that perturbation, the parcel is at the same temperature as the adjacent fluid. After, the parcel will change its temperature adiabatically, by an amount

$$\left(\frac{\partial T}{\partial z} \right)_S \Delta z.$$

At the displaced location the surrounding fluid's temperature is different by an amount

$$\frac{dT}{dz} \Delta z.$$

As demonstrated in Figure 9.13a, whether the perturbed parcel will keep rising (or keep sinking) will depend not so much on its adiabatically modified temperature as on the difference between that temperature and the ambient temperature. The temperature difference created by the perturbation is

$$\Delta T = \left(\left(\frac{\partial T}{\partial z} \right)_S - \frac{dT}{dz} \right) \Delta z.$$

Defining τ as the temperature gradient of the fluid in excess of adiabatic,

$$\tau \equiv \frac{dT}{dz} - \left(\frac{\partial T}{\partial z} \right)_S,$$

as in Chapter 8 (in the derivation of the full Adams-Williamson equations), we have

$$\Delta T = -\tau \Delta z.$$

To retain consistency with this definition of τ, and with earlier discussions in this chapter,

and to avoid ambiguity (the sign of dT/dz, for example, is the opposite if z is height or depth), we will specify z is depth; so, z increases downward, both temperature gradients (dT/dz and $(\partial T/\partial z)_S$) are positive within the solid earth, $\vec{g}$ is in the positive z-direction (thus, $\vec{g} = g\hat{z}$), and a positive Δz corresponds to a downward displacement of the parcel.

In short, the buoyancy force created when a parcel is displaced by an amount Δz can be written

$$+(\rho \alpha_p \tau \Delta z)\vec{g},$$

or

$$+(\rho \alpha_p \tau \Delta z)\, g\hat{z}.$$

Newton's Second Law and the Buoyancy Force. If the parcel experiences no other forces besides the buoyancy force, then Newton's Second Law—expressed in terms of forces per unit volume—becomes

$$\rho \frac{\partial \vec{v}}{\partial t} = +(\rho \alpha_p \tau \Delta z) g\, \hat{z}$$

or

$$\frac{\partial \vec{v}}{\partial t} = +(\alpha_p \tau \Delta z)\, g\, \hat{z}.$$

Since $\vec{v} \equiv (v_x, v_y, v_z)$ and the x- and y-components of the right-hand side are zero, v_x and v_y stay constant (or zero) over time. The z-component of our equation is

$$\frac{\partial v_z}{\partial t} = +\alpha_p \tau g \Delta z.$$

The vertical velocity v_z of the parcel equals $\partial z/\partial t$, where z is the parcel's position (depth) at all times. For simplicity, we will say the parcel was located at depth $z = 0$ prior to the perturbation; in that case, Δz is just z. Thus,

$$\frac{\partial v_z}{\partial t} = \frac{\partial}{\partial t} \frac{\partial z}{\partial t} = \frac{\partial^2 z}{\partial t^2}$$

and our equation reduces to

$$\frac{\partial^2 z}{\partial t^2} = +(\alpha_p \tau g)z.$$

We can solve this equation to determine how the position z of our perturbed parcel varies with time. There are three types of solutions, depending on the sign of τ.

- If $\tau < 0$, that is, if the fluid is stable, the equation is actually a classic harmonic oscillator equation, which we can write as

$$\frac{\partial^2 z}{\partial t^2} = -(\alpha_P |\tau| g)z \equiv -N^2 z.$$

Its solutions will be sinusoidal functions, such as

$$z = \cos(Nt)$$

where $N = \sqrt{\alpha_P |\tau| g}$ is called the *Brunt-Väisälä frequency* (cf. Tritton, 1988). You can verify this solution by substitution. The solution represents an oscillation—a buoyancy wave!—with a period of $2\pi/N$.

- If $\tau = 0$, the equation reduces to

$$\frac{\partial^2 z}{\partial t^2} = 0,$$

with solutions like

$$z = c_1 t + c_2$$

for some constants c_1 and c_2. These solutions describe a parcel that remains at its perturbed location (c_2), or gradually drifts downward or upward (at a rate c_1), till a later perturbation redirects it. Such 'listless' behavior is just what we would expect in a neutrally stable fluid.

- If $\tau > 0$, that is, if the fluid is unstable, the equation is

$$\frac{\partial^2 z}{\partial t^2} = +(\alpha_P \tau g)z.$$

One solution is

$$z = e^{\sqrt{\alpha_P \tau g}\, t},$$

which describes 'runaway' vertical descent of the parcel—an exponential increase in the parcel's displacement downward. Though in reality such a catastrophic action is unlikely—for example, viscous drag would tend to slow it down—this solution illustrates the potential for extreme motion possible in unstable fluids.

The preceding solutions were a bit imprecise in dealing with the initial perturbation: at time $t = 0$, the first and third solutions imply a displacement of 1, and in the second, the initial

displacement is c_2. How would you make all three solutions consistent, if the perturbation at $t = 0$ is denoted z_0?

From the above solution for $\tau > 0$ (or, from scaling Newton's Second Law with buoyancy forces), we could define a characteristic time for buoyancy to operate:

$$t_B \sim \frac{1}{\sqrt{\alpha_P \tau g}};$$

in this amount of time, the position z increases by a factor of e. For the Earth's mantle, $\alpha_P \sim 2 \times 10^{-5}/°C$, and $g \sim 10 \ m/sec^2$; if, just for illustration, we specify $\tau \sim 0.2 \times 10^{-2} \ °C/km$, we would find

$$t_B \sim \frac{1}{2} \times 10^5 \ sec \sim \frac{1}{2} \ day.$$

This is quite a rapid timescale for the mantle—and it prevails when the mantle's temperature gradient exceeds adiabatic by only half a percent!

For comparison, in the (dry) atmosphere, $\alpha_P \sim 3.48 \times 10^{-3}/°C$ (Stacey & Davis, 2008) and $g \sim 10 \ m/sec^2$; if τ is $\sim 0.05°C/km$, half a percent of the adiabat, we would find

$$t_B \sim 760 \ sec \sim 0.2 \ hours.$$

Clearly, buoyancy is easier to achieve in the atmosphere (even a dry atmosphere) than in the mantle; but the fact that it is easier by less than two orders of magnitude, compared to a mantle that is barely superadiabatic, suggests that buoyancy should also be a consequential force in the mantle, as long as $\tau > 0$.

We can illustrate this point further by representing τ in terms of a typical superadiabatic temperature difference ΔT and a length scale L characterizing the mantle:

$$\tau \sim \frac{\Delta T}{L};$$

in these terms, the characteristic time for buoyancy to operate is

$$t_B \sim \sqrt{\frac{L}{\alpha_P \Delta T g}}.$$

Now, if the mantle sustains a temperature increase beyond adiabatic of just 1°C ($\Delta T = 1°C$) from the top to the bottom of the mantle (for which we specify $L \sim 3{,}000$ km), we find t_B is less than $1\frac{1}{2}$ days!

The Struggle for Convection, and the Rayleigh Number. We are finally ready to consider the feasibility of convection within a fluid. The approach we will take first is to combine the timescales characterizing the processes involved in the struggle for convection, in a way that represents the struggle fairly. We therefore define a dimensionless (i.e., unit-less) number called the **Rayleigh number, R_A**:

$$R_A = \frac{t_C t_V}{t_B^2}.$$

There are two factors of t_B in the denominator because buoyancy has to overcome both viscous drag and heat conduction for convection to take place. In situations where the processes opposing convection take a long time to work, compared with the time for buoyancy, their opposition is ineffective and convection will occur; in this case, R_A is large. If the opposition operates relatively effectively, as indicated by a small R_A, buoyant flow and thus convection will be suppressed.

How large should R_A be for convection to be occurring? In general, we can postulate a **critical Rayleigh number, $(R_A)_{crit}$**, such that the fluid can convect if, and only if, the actual Rayleigh number is greater than the critical value:

$$R_A > (R_A)_{crit}$$
$$\Longleftrightarrow \text{CONVECTION TAKES PLACE.}$$

The value of the critical Rayleigh number depends on the situation: what the geometry and boundary conditions are; whether α_p, k, and η' are allowed to vary (e.g., with depth or temperature); whether the body is rotating, self-gravitating, or neither; and how negligible other forces are. Determination of the critical Rayleigh number requires that the equations governing fluid flow—including equations conserving momentum (i.e., Newton's Second Law), mass, and energy—be solved for a given situation. A number of these situations are evaluated by Schubert et al. (2001); they

deduce $(R_A)_{crit}$ to be in the range ~ 400–$\sim 1{,}700$, with the most extreme values relating to whether the mantle's upper and lower boundaries are fixed or stress-free. For simplicity, we will assume

$$(R_A)_{crit} \sim 10^3$$

for the mantle; since we take mainly an order-of-magnitude approach here, however, our convection criterion should be viewed cautiously (i.e., R_A should exceed $(R_A)_{crit}$ not just barely but, say, by at least an order of magnitude).

From the definition of the Rayleigh number in terms of characteristic times, we have

$$R_A = \frac{g \alpha_p \tau L^4}{\eta' k} \quad \text{or} \quad R_A = \frac{g \alpha_p \Delta T L^3}{\eta' k}.$$

From these expressions, we see that the fundamental physical parameters determining the feasibility of convection are:

- the acceleration of gravity, which provides the basis for buoyancy;

- the coefficient of thermal expansion, which controls how much of a mass deficit is created by a given temperature difference;

- the *superadiabatic* temperature gradient (or difference), which establishes the underlying instability of the fluid;

- the kinematic viscosity, which governs the strength of the viscous drag force; and

- the thermal diffusivity, which determines how quickly any temperature differences can be conductively evened out.

However, even more important is the dependence of R_A on the dimensions of the body—a dependence on L^4 or L^3; double the size of the body, and the chance of convection increases eightfold, given the same temperature difference, or sixteenfold, if the gradient τ can be maintained. The reason for this makes sense: the larger the body, the more inefficient diffusion of heat and momentum are (t_C and t_V become much greater). This dependence on size is the basis for the pervasiveness of convection in the universe—in planets and stars characterized by large L.

For the Earth's mantle, $g \sim 10$ m/sec^2, $\alpha_P \sim 2 \times 10^{-5}/°C$, and $k \sim 1 \times 10^{-6}$ m^2/sec. With a simple approach, we could consider the mantle to be isoviscous, and specify $L \sim 3,000$ km and $\eta' \sim 10^{18}$ m^2/sec (note that this is ten times the value usually quoted for the upper mantle). In that case,

$$\text{R}_\text{A} \sim 1.6 \times 10^{10} \, \tau \quad (\tau \text{ in } °C/m) \quad \text{or}$$

$$\text{R}_\text{A} \sim 1.6 \times 10^{7} \, \tau \quad (\tau \text{ in } °C/km).$$

Before proceeding further, let's pause for a moment and consider what we are hoping to establish. *Can a body of fluid whose viscosity is 24 orders of magnitude greater than that of water actually undergo convection?*

Using our expressions for the Rayleigh number, it is possible to get a sense of the mantle's convective potential.

- With an adiabatic gradient of $\sim\frac{1}{3}°C/km$ in the mantle (though it is actually slightly larger in the upper mantle), the Rayleigh number will be 25 times the critical value—implying that the mantle is certain to be convecting—if the mantle is just $\frac{1}{2}\%$ superadiabatic.

  In terms of temperature differences, the implications are even more dramatic. That adiabatic gradient implies the increase in temperature from the top of the transition zone to the top of D″, a distance of roughly 2,400 km, would be about 800°C under adiabatic conditions; to achieve a gradient half a percent higher would require the temperature to increase by only an additional 4°C!

- If the lower mantle has a higher viscosity, its ability to convect will be challenged but still not out of the realm of the possible. For example, if the lower mantle's viscosity is 100 times that of the upper mantle, i.e., $\eta' \sim 10^{19}$ m^2/sec, then with $L \sim 2,000$ km, we find

$$\text{R}_\text{A} \sim 3 \times 10^{5} \, \tau \quad (\tau \text{ in } °C/km).$$

With this possibility, the lower mantle would need to be more than 1% superadiabatic just to exceed the critical value. 10% superadiabatic would make its convection probable.

If we are going to treat the upper and lower mantle separately, how likely is convection to be in the upper mantle? Use $L \sim 700$ km and $\eta' \sim 10^{17}$ m^2/sec.

For a somewhat more definite conclusion, we can use actual temperatures that were previously estimated from mineral phase changes (see Chapter 8). From the olivine→wadsleyite geothermometer, $T \sim 1,550°C$ at the first transition; from the post-perovskite phase change, T is likely to be at least $\sim 3,500°C$ near the base of the mantle. The implied temperature gradient of $\sim 0.8°C/km$ or more suggests τ will be around 0.5°C/km through most of the mantle, and leads to a Rayleigh number perhaps as great as 10^7. Even with more accurate estimates, accounting for (e.g.) higher adiabatic gradients in the transition zone and D″, the conclusion is unavoidable: in the mantle, the Rayleigh number is way above critical.

Even if, as discussed earlier, exciton thermal conductivity is substantially greater in the lower mantle, so that k is an order of magnitude greater than we have been considering, our conclusion regarding whole-mantle convection would still stand; but if lower-mantle viscosity was higher as well, the likelihood of lower-mantle convection would be less certain.

With Heat Sources, the Struggle is Easier. In a fluid with heat sources, the struggle for convection takes a slightly different form. According to Tritton and Zarraga (1967), the corresponding Rayleigh number is

$$\text{R}_\text{A} = \frac{g\alpha_P L^4}{\eta' k} \left(\frac{\hat{\varepsilon} L}{k C_P} \right),$$

where $\hat{\varepsilon} \equiv \varepsilon/\rho$ is the heat production per unit mass, taken to be uniform throughout the fluid. The quantity in brackets replaces the superadiabatic temperature gradient τ that appears in the traditional definition of R$_\text{A}$. Given the definition of k, that quantity can also be written $\varepsilon L/K$.

In the mantle, $\varepsilon \sim 2 \times 10^{-5}$ mW/m^3 and $K \sim 5$ W/m°C (based on the range 2.5–8 W/m°C discussed earlier); with $L \sim 3,000$ km we find $\varepsilon L/K \sim 12°\text{C/km}$ (—equivalent to a huge superadiabatic gradient), leading to

$$\text{Ra} \sim 1\frac{1}{2} \times 10^{10} \times 12 \sim 2 \times 10^{11}.$$

With a critical Rayleigh number of $\sim 2,800$ when heat sources are present (Roberts, 1967), it appears that convection is unavoidable. Even with $\eta' \sim 10^{19}$ m^2/sec and $L \sim 2,000$ km, corresponding to an extremely viscous lower mantle, we have $\text{Ra} \sim 2\frac{1}{2} \times 10^9$, and the lower mantle easily convects.

What About the Ocean? The thermohaline circulation will be discussed later in this chapter. For now, we assess the ocean's tendency to undergo convection by treating it simply, as a nonturbulent, initially static fluid layer. Using parameter values specified earlier in this chapter, we take $\eta' \sim 10^{-6}$ m^2/sec and $k \sim 1.4 \times 10^{-7}$ m^2/sec; according to Vallis (2005), we have $\alpha_P = 1.7 \times 10^{-4}$/°C. With $L = 4$ km, its Rayleigh number turns out to be $\sim 3 \times 10^{21}\ \tau$ (if τ is in °C/km). η' and k are so small in magnitude that convective instabilities are not inhibited at all.

9.4.3 Heat Transmission in a Convecting Fluid

Even as parcels of hot fluid rise during convection, they are transmitting heat by conduction to their surroundings. Although such transmission must obey conservation of energy, the diffusion equation we derived from it, $\partial T/\partial t = k\,\nabla^2 T$, is not quite applicable.

We can best see this by analogy. Consider a river flowing in a channel that eventually narrows (see Figure 9.14a). Conservation of mass requires that the same amount of mass per second must flow past any point on the bank as any other point; that is, if we set up screens at any two locations (such as the locations in Figure 9.14a marked by dotted lines), with each screen spanning the breadth and depth of the channel, so that all of the river's water flows through each screen, then that volume of water per second must be the same through both screens. Conservation thus mandates that *the river flows faster where the channel is narrower.*

If it so happened that conditions (such as the river's source of water) remained unchanged over time, then at any location the river velocity $\vec{v}$ would remain unchanged as well; that is, $\partial\vec{v}/\partial t = 0$ (where we use a partial derivative to remind us that this rate of change is being measured at a fixed x, y, and z). Despite this steady state, however, the velocity of a small parcel of water can and will change with time, as it flows downstream and encounters the narrows. We say that there has been **advection** of velocity, as the parcel is *carried to* a location where it experiences different conditions, for example where it acquires a different velocity. This terminology is meant to be distinct from convection, a specific process in which heat is *carried with* a parcel as it moves.

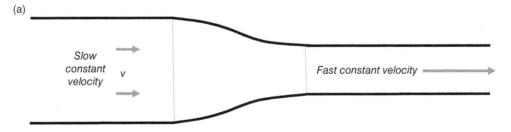

Figure 9.14 Advection in action.
(a) View from above of a river flowing to the right. The river channel narrows from a constant wide width on the left to a constant narrow width on the right. To conserve mass, the velocity v of the river must increase through the narrows (middle area).
For simplicity, we are assuming here that the river water is inviscid; a nonzero viscosity would cause v to decrease close to the banks of the river, in addition to its speeding up downstream.

Thus, in general, if $\vec{v}$ is the velocity of a parcel of fluid in the river, it can change with time whether $\partial \vec{v}/\partial t$ equals 0 or not. With fluids, or fluid-like materials, we will observe changes in quantities over time either when conditions in the spatial environment are changing with time or when (even if conditions are steady, and there are no changes in the spatial environment with time) the environment is nonuniform and the fluid parcel moves to a region experiencing different conditions. Symbolically, we can expect the latter situation to involve both $\vec{v}$ (the parcel will be in motion) and the gradient operator $\vec{\nabla}$ (conditions are varying with position).

As a first step in expressing all this quantitatively, we write

$$\vec{v} = \vec{v}(x, y, z; t),$$

recognizing that, in general, the velocity of the river can be a function of position as well as time. In Figure 9.14a, it varies in the downstream direction; but in more realistic situations, the viscosity of the water will force the velocity to decrease as the river banks and the river bed are approached. We will write the *total* possible change in the parcel's velocity—resulting from changes dx, dy, and dz in the parcel's position *as we track the parcel*, plus the change resulting from the passage dt of time—as

$$\mathrm{D}\vec{v} = \frac{\partial \vec{v}}{\partial x}\mathrm{d}x + \frac{\partial \vec{v}}{\partial y}\mathrm{d}y + \frac{\partial \vec{v}}{\partial z}\mathrm{d}z + \frac{\partial \vec{v}}{\partial t}\mathrm{d}t.$$

Our symbol for the total change, $\mathrm{D}\vec{v}$, is a departure from standard calculus notation (d$\vec{v}$) to emphasize that it is a change measured as we move with the parcel.

Dividing both sides of this expression by the time increment dt, we obtain the total rate at which $\vec{v}$ changes,

$$\frac{\mathrm{D}\vec{v}}{\mathrm{d}t} = \frac{\partial \vec{v}}{\partial t} + \frac{\mathrm{d}x}{\mathrm{d}t}\frac{\partial \vec{v}}{\partial x} + \frac{\mathrm{d}y}{\mathrm{d}t}\frac{\partial \vec{v}}{\partial y} + \frac{\mathrm{d}z}{\mathrm{d}t}\frac{\partial \vec{v}}{\partial z}.$$

But if we are tracking the parcel, i.e., moving with it, then the parcel's change in x-position, dx, per time dt is simply the x-component of its velocity, v_x. The same holds true in the y- and z-directions. Our total change in $\vec{v}$ per time can

thus be written as

$$\frac{\mathrm{D}\vec{v}}{\mathrm{d}t} = \frac{\partial \vec{v}}{\partial t} + v_x\frac{\partial \vec{v}}{\partial x} + v_y\frac{\partial \vec{v}}{\partial y} + v_z\frac{\partial \vec{v}}{\partial z},$$

or, more concisely, given the definitions of vector dot product and the gradient operator $\vec{\nabla}$,

$$\frac{\mathrm{D}\vec{v}}{\mathrm{d}t} = \frac{\partial \vec{v}}{\partial t} + (\vec{v}\bullet\vec{\nabla})\vec{v}.$$

$\mathrm{D}\vec{v}/\mathrm{d}t$ is called a **total derivative**, sometimes denoted $\mathrm{D}\vec{v}/\mathrm{D}t$ (and frequently just denoted $\mathrm{d}\vec{v}/\mathrm{d}t$), because it includes all the possible ways the velocity $\vec{v}$ of a fluid parcel can vary with time as the river flows. As a reminder that the change in parcel velocity we are measuring is obtained by tracking it—*moving with the parcel*—this derivative is also called a **material derivative**. (Actually, it has still another name: *Lagrangian* derivative, in contrast to 'Eulerian' derivatives like $\partial \vec{v}/\partial t$—all of which serves to emphasize that total derivatives must be fairly important, to be given so many names!)

Similarly, though perhaps less intimidating mathematically (since T is not a vector), the temperature $T = T(x, y, z; t)$ of a parcel can change with time because temperatures are changing at specific locations ($\partial T/\partial t \neq 0$) or because the parcel is advected to where it acquires a different temperature. The total change in temperature is

$$\mathrm{D}T = \frac{\partial T}{\partial t}\mathrm{d}t + \frac{\partial T}{\partial x}\mathrm{d}x + \frac{\partial T}{\partial y}\mathrm{d}y + \frac{\partial T}{\partial z}\mathrm{d}z,$$

and the total change with respect to time is

$$\frac{\mathrm{D}T}{\mathrm{d}t} = \frac{\partial T}{\partial t} + (\vec{v}\bullet\vec{\nabla})T$$
$$= \frac{\partial T}{\partial t} + \vec{v}\bullet\vec{\nabla}T.$$

The first term on the right represents temperature changes due to the fluid heating up or cooling down over time; the second measures advective temperature changes as the parcel is carried (at velocity $\vec{v}$) to a region where the temperature is different ($\vec{\nabla}T \neq 0$).

Finally, then, we see that the problem with applying the diffusion equation, as originally written, is that it limits the time variation of T to variations at specific locations, and fails to account for the possible movements of the parcel—an especially critical omission if convection might be occurring.

So, as the parcel is moving, it is the heat that diffuses from *it* to the surroundings, not the heat diffusing from a particular location the parcel had momentarily passed through, which we should be concerned with. We need to conserve the parcel's energy, not the location's; that is, we need to move *with* the parcel as we ensure conservation. The solution is to replace the time derivative we originally employed, $\partial T/\partial t$, with a total time derivative, thus

$$\frac{\mathrm{D}T}{\mathrm{d}t} = k\nabla^2 T$$

(in the absence of heat sources).

However, although we have deduced the correct expression for the diffusion equation for fluids in motion, it is instructive to recast it as close to the original formulation as possible, that is, with the temperature variation *at a specific location* on the left-hand side:

$$\frac{\partial T}{\partial t} = k\nabla^2 T - \vec{v}\cdot\vec{\nabla}T$$

(still without heat sources, for simplicity). We see first of all that, depending on the direction of the fluid's motion, advection may amplify the temperature changes resulting from diffusion, or oppose them. Figure 9.14b illustrates how one of these possibilities may happen. In this sketch, $\vec{v}$ is exactly antiparallel to $\vec{\nabla}T$, so their dot product simplifies, and the equation can be written

$$\frac{\partial T}{\partial t} = k\nabla^2 T + v\nabla T$$

where v and ∇T are the magnitudes of those vectors. Focusing on a location rather than a parcel, for example the fixed point P in the figure, the equation predicts advection will cause the temperature to increase (i.e., it predicts $\partial T/\partial t > 0$). Indeed, in the figure, the flow toward P brings hot fluid to that location.

The sketch in Figure 9.14b depicts advection, but is unlikely to represent a convective flow because it appears skewed with respect to the vertical. Nevertheless, to connect these concepts to convective situations, can you say whether $\vec{v}$ being

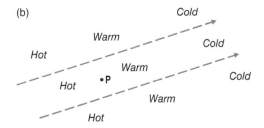

Figure 9.14 Advection in action.
(b) Advection of temperature within a fluid in motion. In this section of the fluid, the temperature gradient is down toward the left ($\vec{\nabla}T$ is always in the direction of greatest temperature increase). For simplicity, we consider a fluid that is moving (dashed arrows) in the exact opposite direction from the temperature gradient. Point P is a fixed location.

antiparallel to $\vec{\nabla}T$ better describes the upwelling limb of a convection pattern or the downwelling limb? How about when they are parallel?

From our recast equation, we see that, in general, the change in temperature with time at a fixed location can result from diffusion of heat or from advection of heat. To determine the relative importance of these mechanisms, we can scale the equation. We let L, T_C, and V_C denote the characteristic dimension, temperature, and velocity of our fluid. t_{CH} will denote the characteristic time for a change in the thermal state of the fluid, that is, a time for it to cool off or heat up; but for generality, we will not necessarily equate this time to t_B or t_C. As before, we will use * to denote dimensionless quantities, which will be of order unity if our characteristic values are truly representative of the situation. After dividing both sides of the equation by T_C and cross-multiplying by t_{CH}, our recast and scaled equation becomes

$$\frac{\partial T^*}{\partial t^*} = k\frac{t_{CH}}{L^2}\nabla^{*2}T^* - \frac{V_C\, t_{CH}}{L}\vec{v}^*\cdot\vec{\nabla}^*T^*$$

In general, we would have to specify V_C according to whatever situation is being evaluated. Given the overarching goal of this textbook, we will proceed by estimating V_C in convective situations. If convection is about to occur, we can expect there to be an approximate balance between the forces promoting and opposing convection:

$$\rho \alpha_p \tau g \vec{z} \cong \eta' \, \nabla^2 (\rho \vec{v}).$$

Scaling this equation yields

$$V_C \sim \frac{\alpha_p g \tau L^3}{\eta'}.$$

For example, in the mantle, we would find $V_C \sim 1$–27 cm/yr for the material property values used earlier, if $L \sim 1{,}000$–3,000 km. Such velocities are plausible in comparison to observed plate speeds (or are at least in the 'right ballpark'), suggesting that estimating V_C from a convective force balance is reasonable and appropriate.

Substituting our expression for V_C into our scaled diffusion-advection equation yields

$$\frac{\partial T^*}{\partial t^*} = \frac{k \, t_{CH}}{L^2} \nabla^{*2} T^*$$
$$- \frac{k \, t_{CH}}{L^2} R_A \, \vec{v}^* \cdot \vec{\nabla}^* T^*.$$

Thus, the Rayleigh number as we have defined it is what determines the relative importance of conduction versus convection, loosely speaking, or more precisely the relative importance of diffusion versus advection in transmitting heat. At high Rayleigh numbers, advection dominates, and contributes much more to the change in temperature at any point. These conclusions hold no matter what the timescale of heat transmission actually is.

Still another way to understand the Rayleigh number focuses separately on the forces and the heat transmission involved in convection. We define a dimensionless number called the **Peclet number**, P$_E$, as the typical ratio of advection to diffusion; that is,

$$\frac{\text{Advection}}{\text{Diffusion}} = \frac{\vec{v} \cdot \vec{\nabla} T}{k \nabla^2 T}$$

so

$$P_E \equiv \frac{\text{Advection}}{\text{Diffusion}} = \frac{L V_C}{k}.$$

When P$_E$ is large, heat transmission in the fluid body is achieved primarily by advection. In a way, the Peclet number is like a thermal Reynolds number (when R$_E$ is large, diffusion

(of momentum) is not very effective either). We might then define the Rayleigh number as

$$R_A \equiv \frac{\text{Buoyancy force}}{\text{Viscous drag force}} \cdot P_E,$$

casting the struggle for convection slightly differently—it is a struggle of buoyancy versus viscous drag, moderated by how effectively advection is in transmitting heat. As before, convection is likely when both aspects of those struggles are won, that is, when R$_A \gg 1$. And, as before, when the forces involved are correctly scaled, we end up with

$$R_A = \frac{g \alpha_p \tau L^4}{\eta' k} \quad \text{or} \quad R_A = \frac{g \alpha_p \Delta T L^3}{\eta' k}.$$

9.4.4 Horizontal Convection

In this textbook we have generally envisioned convection as involving vertical temperature differences, and focused on the vertical components—the upwellings and downwellings—of the resulting convective flow. The "horizontal convection" (called "sideways convection" by Vallis, 2005) discussed in this subsection refers instead to convection driven by lateral temperature differences. Such differences might arise naturally, within the fluid; be imposed at the fluid's side boundaries; or be imposed along the horizontal top or bottom boundary of the fluid. For example, we might encounter lateral temperature differences in the outer core near the inner-core boundary, as the fluid core freezes sporadically and exothermically onto the surface of the inner core; in the core or mantle, in regions that possess irregularly distributed heat sources; in the ocean near a mid-ocean ridge, with the ridge providing a 'sidewall' of contrasting temperature; or at the bottom of the troposphere near the coast, where the land and ocean may heat the overlying air differently.

However, it is the role of horizontal convection in driving the oceans' thermohaline circulation (meridionally overturning circulation, discussed in Chapter 3) that has drawn the greatest interest and the most debate;

research on the subject extends contentiously back more than 100 yr (see, e.g., Mullarney et al., 2004; Hughes & Griffiths 2008; and Scotti & White, 2011). In this case we are dealing with lateral temperature differences imposed along the horizontal top boundary of the oceans (e.g., in the North Atlantic) by a heat exchange with the atmosphere; that is, the boundary condition leading to convection involves heat flux rather than specific temperatures.

We will approach this topic by revisiting the concept of hydrostatic equilibrium. As discussed in Chapter 8, hydrostatic equilibrium represented a balance of vertical forces in a fluid at rest—the pressure force and gravity; expressed in terms of force per unit volume, we found

$$\frac{\mathrm{d}P}{\mathrm{d}z} = \rho g$$

where P is the hydrostatic pressure (if the fluid is not static, dynamic pressures will also exist) and ρ is the fluid density. Our first task is to recast this equation more generally.

Consider a tiny cubic volume within the fluid (Figure 9.15); with sides parallel to the coordinate axes and the lengths of these sides being $\mathrm{d}x$, $\mathrm{d}y$, and $\mathrm{d}z$, the amount of volume contained is $\mathrm{d}\mathrm{VOL} = \mathrm{d}x{\cdot}\mathrm{d}y{\cdot}\mathrm{d}z$. Let P denote the pressure (= force per unit area, of course) *on* the left-hand (shaded) cube face. We can write the pressure on the opposite face as

$$P + \frac{\partial P}{\partial x}\mathrm{d}x,$$

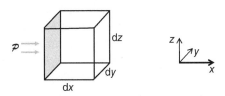

P

$\mathrm{d}z$

$\mathrm{d}y$

$\mathrm{d}x$

z

y

x

Figure 9.15 Derivation of the pressure force. An infinitesimal cubic volume dVOL within a fluid body is shown here. The faces of the cube are perpendicular to the coordinate axes, and the dimensions of the cube are dx, dy, and dz. The pressure on the left (shaded) face is P, and it acts to the right; the pressure on the opposite face will act to the left.

since the increase in pressure equals the rate at which pressure increases with x (that is, $\partial P/\partial x$) times the distance covered from one face to the other, $\mathrm{d}x$. Thus, the net force on the cube in the x-direction is

$$P(\mathrm{d}y{\cdot}\mathrm{d}z) - \left(P + \frac{\partial P}{\partial x}\mathrm{d}x\right)(\mathrm{d}y{\cdot}\mathrm{d}z),$$

recognizing that the area of each face is $(\mathrm{d}y{\cdot}\mathrm{d}z)$ and that the x-direction is to the right in the figure. We can write this as

$$-\left(\frac{\partial P}{\partial x}\mathrm{d}x\right)(\mathrm{d}y{\cdot}\mathrm{d}z)$$
$$= -\frac{\partial P}{\partial x}(\mathrm{d}x{\cdot}\mathrm{d}y{\cdot}\mathrm{d}z)$$
$$= -\frac{\partial P}{\partial x}(\mathrm{d}\mathrm{VOL}).$$

Similarly, the net force on the cube in the y- and z-directions is

$$-\frac{\partial P}{\partial y}(\mathrm{d}\mathrm{VOL}) \quad \text{and} \quad -\frac{\partial P}{\partial z}(\mathrm{d}\mathrm{VOL});$$

so, the net force per unit volume is

$$\left(-\frac{\partial P}{\partial x}, -\frac{\partial P}{\partial y}, -\frac{\partial P}{\partial z}\right) = -\vec{\nabla}P.$$

This is, officially, the **pressure force**, per unit volume.

In its most general form, hydrostatic equilibrium states that there is a balance between the two forces that act on a fluid at rest, the pressure force and the gravitational force; expressed per unit volume, this is

$$-\vec{\nabla}P + \rho\,\vec{g} = 0$$

or

$$\vec{\nabla}P = \rho\,\vec{g}.$$

This expression does not restrict gravity to be in the z-direction, or aligned with one coordinate axis; whatever its direction, the pressure gradient will parallel it—pressure will increase in the direction of gravity.

Now consider an isobaric surface within the fluid body, a surface along which the pressure is constant. If $\vec{\nabla}P$ was not perpendicular to that surface, that would mean there was a component of the gradient acting along the surface; but in that case, P would vary along it. In short, $\vec{\nabla}P$ must be perpendicular to isobaric surfaces within the fluid.

In Chapter 6, we defined level surfaces (equipotential surfaces) as always being perpendicular to the direction of gravity. It follows from our general expression for hydrostatic equilibrium that, in equilibrium, isobars and level surfaces must coincide.

Hydrostatic equilibrium does not require that the fluid have uniform density. However, if ρ varies laterally, then in equilibrium ∇P (the magnitude of $\vec{\nabla} P$) would also vary laterally; that's because $\nabla P = \rho g$ (where g is the magnitude of $\vec{g}$). But, if the pressure gradient—that is, the increase in P downward—varied laterally, then greater depths would no longer share the same value of pressure; for example, if we start at some level surface within the fluid and descend by an amount Δz, then the pressure will increase by $(\partial P/\partial z) \cdot \Delta z = \rho g \Delta z$—thus, by a different amount below locations on that level surface where ρ is different. In other words, isobars will no longer coincide with level surfaces if density varies laterally. *For a fluid in hydrostatic equilibrium, surfaces of constant density* (called **isopycnal** surfaces) *must coincide with surfaces of constant pressure and level surfaces.*

Incidentally, fluids in which surfaces of constant pressure and density coincide are called **barotropic**. The opposite situation, where those surfaces intersect, is called **baroclinic**. Barotropic fluids may or may not be in hydrostatic equilibrium; baroclinic fluids cannot be.

Lastly, if the fluid is chemically homogeneous, then temperature within the fluid is a function only of pressure and density (this is equivalent to saying that the density is a function only of pressure and temperature). Altogether, we have just shown how strong a requirement hydrostatic equilibrium places on a chemically homogeneous fluid: *surfaces of constant temperature, pressure, and density all coincide, and are level surfaces* (e.g., Jeffreys, 1925).

Such a requirement can be stated inversely as well. If temperature variations exist along a horizontal (level) surface within a homogeneous fluid, that fluid cannot be in hydrostatic equilibrium. This disequilibrium applies no matter what the source of the lateral temperature differences is. In all cases, those differences drive fluid motions, and—unlike the 'traditional' kind of convection we have discussed before—it does not matter whether the temperature differences are superadiabatic or not.

Tritton (1988) explains why flow *will* take place: with the hot and cold temperatures laterally offset (i.e., not vertically lined up), their buoyancy forces (with the hotter fluid rising and the colder sinking) create a torque that is unbalanced, causing the fluid to attempt to rotate around a horizontal axis like a 'vortex.' We can contrast that with traditional convection situations, where the vertical temperature differences and thus buoyancy forces are initially aligned, avoiding an intrinsic disequilibrium and requiring additional forcing (e.g., superadiabatic temperatures) to begin convective motions.

That torque and the *vorticity* it creates imply that the resulting horizontal convection will necessarily include vertical motion, resulting from the buoyancy forces, as well as lateral flow. But in contrast to traditional convection, where the lateral flow mainly serves to conserve mass by connecting upwelling and downwelling, the lateral flow in horizontal convection is also driven by the lateral pressure differences caused by the changes in density (as noted above, they forced isobaric surfaces out of alignment with level surfaces). As a consequence, the flow pattern should be much more *asymmetric* than in traditional situations, with the convection cell horizontally elongated and the downwelling (e.g., in the case of thermohaline circulation in the North Atlantic) relatively narrow.

As pointed out by Paparella and Young (2002), Hughes and Griffiths (2008), and Vila et al. (2016), the critical Rayleigh number in horizontal convection is zero. This is a mathematical reminder that horizontal convection does not require superadiabatic conditions for flow to take place; it also serves to emphasize the underlying instability of horizontal convection.

However, the actual flow pattern does depend strongly on the Rayleigh number—in particular, the horizontal Rayleigh number, in which the characteristic length L appropriately (as argued by Hughes & Griffiths, 2008; also, Somerville, 1967) represents the horizontal dimension of the fluid body—as well as on other fluid properties and the geometry of the fluid body (see, e.g., Tritton, 1988). As summarized concisely by Vila et al. (2016), this dependence is thought to lead to three different possible regimes for horizontal convection. With low or moderate Rayleigh numbers (less than ~10^8, when the fluid is water), the flow is laminar and steady; for moderate to high Rayleigh numbers (>10^8, though the cutoff can be as low as ~10^6, as shown in Paparella & Young, 2002 and Hughes & Griffiths 2008), the circulation can 'shake up' the interior and—especially if the Prandtl number is low (<10^2, giving us the third regime)—the flow can be unsteady and turbulent.

Examples of experiments conducted to investigate the nature of horizontal convection under oceanic conditions—hopefully modeling the thermohaline circulation at least partially—are shown in Figure 9.16. The experiments confirm the asymmetric flow pattern we anticipated; but the degree of turbulence (especially in the plume) and its effectiveness in driving the flow and mechanically mixing the interior have been questioned (e.g., Paparella & Young, 2002; but see also Hughes & Griffiths, 2008). Experiments by Mullarney et al. (2004) and Vila et al. (2016) establish the plume's turbulent character; experiments by Mullarney et al. and the analysis by Scotti and White (2011) demonstrate that the broad-scale flow is essentially, though weakly, turbulent.

Interestingly, with low viscosity—a condition under which turbulent flow typically becomes extremely chaotic—horizontal convection is found to dissipate a "surprisingly small" amount of energy (see Paparella & Young (2002) and then Scotti & White (2011)).

In short, this asymmetric flow is too weak to supply much energy to the circulation, and researchers on both sides of the debate about the nature of the convection agree that, in the case of the actual thermohaline circulation, there is a need for additional forcing. To that end, several mechanisms have been proposed, some of which—including surface wind forcing, the preferred mechanism—were mentioned in Chapter 3; as there, we conclude that the observed convective flow is best understood as a *wind-assisted* thermohaline circulation.

The other main driving force for the thermohaline circulation is changes in salinity in the Southern Ocean. In this situation, the key boundary condition involves compositional ('solutal') buoyancy rather than thermal buoyancy—salinity flux instead of heat flux—and one aspect of the circulation becomes how successfully salt rather than (or as well as) cold diffuses from the downwelling plume into the interior. This has been studied as a process of horizontal convection by Pierce and Rhines (1996), also Mullarney et al. (2007).

Such studies have also been useful for understanding the impact, on the underlying thermohaline circulation, of climate events that add a pulse of fresh water to the sea surface (e.g., Belkin et al., 1998; Dickson et al., 1996; Hakkinen, 1995, and references in Mullarney et al., 2007). These events include the "Great Salinity Anomalies" of the Labrador Sea and other portions of the North Atlantic, which appear to have repeated on decadal timescales in recent years; and longer timescale events associated with greater fluctuations in climate. Conceivably the layer of fresh water floating on the ocean surface might interfere sufficiently with the dense saline downwelling plume to not only interfere with the production of deep water but shut it down, for some period of time.

Other situations in which horizontal convection may be operating within the Earth System come to mind.

- Some atmospheric phenomena can be viewed as a consequence of horizontal convection.
  - As discussed in Chapter 3, the Walker circulation is created at sea level (—the

classic horizontal boundary) with positive and negative heat and humidity influxes on one side of the equatorial Pacific versus on the other side.

– Sea breezes (noted by Somerville, 1967) occur in coastal regions when the air is heated more by the land than by the adjacent ocean, or vice versa. During the day, for example, the land typically heats up more than the sea, because of water's greater capacity to store heat, causing air to rise over the land and descend over the sea; the resulting convection cell includes an *offshore* breeze (as a meteorologist would call it), a lateral return flow that replenishes the rising air. At night, the ocean supplies more heat to the air above than the land does, reversing the circulation and creating an *onshore* breeze. These breezes can be enhanced in different seasons.

– **Monsoons** are seasonally alternating *regional*-scale sea breezes whose offshore-breeze mode includes times of heavy rain. The wind patterns are depicted in Figure 9.17 for the Indian and Southeast Asian Monsoons, regions famous for their especially heavy monsoonal rains. During northern hemisphere summer, for example, the Indian and Asian subcontinents warm a lot more than do the Indian and western Pacific Oceans, so the air above those land masses rises; over the relatively cooler oceans, the air sinks. In the winter the flow pattern will reverse.

This circulation will depart somewhat from the idealized patterns of horizontal convection shown in Figure 9.16, since the land there does not remain flat and at sea level, moving away from the coast—in fact, in that region it becomes extremely mountainous—and the thermal forcing provided by the land-ocean differences is irregularly distributed along the lower boundary of the circulating atmosphere (i.e., the coastline is not linear). Nevertheless, once the monsoonal pattern of flow is established, it should have the overall characteristics of a horizontal convection cell, including the latter's

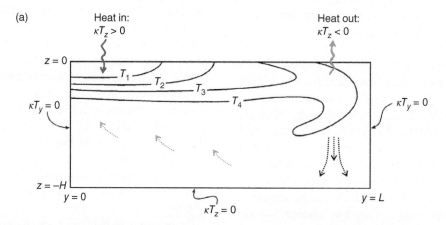

Figure 9.16 Depictions of experimental simulations of horizontal convection as analogs of the thermohaline circulation.

(a) Schematic model of ocean circulation resulting from lateral heating of its top surface. z coordinate increases upward in a basin of depth H (so $T_1 > T_2 > T_3 > T_4$). Calculus shorthand used here denotes derivatives: T_z is $\partial T/\partial z$, and T_y is $\partial T/\partial y$; the side and bottom boundaries of this basin are thus insulating, while the top surface is heated on the left and cooled on the right.

Contours show colder temperatures at greater depths especially on the right, suggesting a flow pattern that includes weak sinking in the region cooled at the surface and implying an even weaker, diffuse upwelling on the left side, away from that region. Figure adapted from Paparella & Young (2002); see also Vallis (2005).

(b)

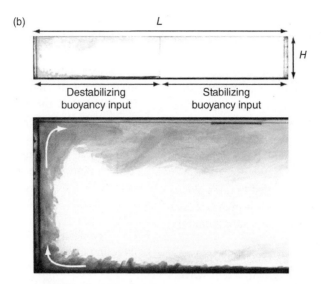

Figure 9.16 Depictions of experimental simulations of horizontal convection as analogues of the thermohaline circulation.
(b) The tank of water shown above was subjected to a uniform cool temperature at its base on the right half-side and uniform heat flux at its base on the left half-side; all other walls were insulating. Dye injected into the fluid allowed the flow pattern to be easily visible; a close-up of the tank's left-most section is shown in the lower image.
This experiment is upside down, compared to the boundary conditions thought to drive the thermohaline circulation of the oceans (not including additional wind forcing); according to Hughes & Griffiths (2008), though, the dynamics seen here—including the narrowness of the major plume, the breadth and sluggishness of the asymmetric return flow, weak mixing through the interior, and the turbulent-like eddies and instabilities—would be the same either way. Figure from Hughes & Griffiths (2008), based on earlier work in Mullarney et al. (2004).

asymmetry, with a strong upwelling plume.

In the summer, that plume will be located over land. The air rising within the plume, drawn in from over the oceans and therefore humid, will cool and condense as it ascends; with the plume intense, we can expect its production of rain to be correspondingly heavy—monsoonal, even. In India, typical rainfall at the peak of the summer monsoon is nearly 3 m per month (e.g., Aguado & Burt, 2007)!

What about in winter, when the circulation reverses and the plume of rising air will be located over the oceans—do you expect there to be similarly intense rains generated? The key word here is "winter"....

- The top of the asthenosphere or low-viscosity zone (or low-velocity zone) represents specific conditions at which lithospheric material has softened or partially melted. Since this boundary is not at the same depth everywhere, the temperature creating it will not be exactly the same all along it (due to pressure effects); and, as noted in Chapter 8, compositional effects may also affect its depth. Nevertheless, as a first approximation, we can treat the boundary as roughly isothermal.

Continental lithosphere is typically at least ~150 km thick; oceanic lithosphere even at its oldest and coldest is no more than ~100 km in thickness. Approaching continental margins, then, the top of the asthenosphere deepens significantly. Despite the roughness of our approximation, it follows that the isotherms marking the top of the asthenosphere change their depth significantly, plunging below the continent; that is, isotherms and level surfaces fail to coincide in the asthenosphere. As arguably the weakest part of the mantle—the zone most likely to be able to flow—it seems unavoidable that convection is occurring within the asthenosphere. (Idea formalized *by C. Chapman.*)

Knopoff (1963) speculated that the shear stresses associated with this asthenospheric circulation cause strike-slip movements of overlying crustal faults.

Hughes and Griffiths (2008) discuss the differences in horizontal convection when, as in the case of the mantle, the Prandtl number is very high.

- We saw in Chapter 8 that the deep mantle, including D'', as inferred from seismic tomography and other seismic analyses, reveals anomalous structures that may be interpreted as relatively hot or cold: low-velocity (or ultra-low-velocity) provinces, serving in part as the base of plumes or superplumes; and high-velocity regions connected to subducted lithospheric slabs. These structures suggest that the core-mantle boundary itself experiences significant differences in temperature.

The fluid at the top of the outer core would share the temperature differences of the CMB; but the flattening of the core-mantle boundary is nearly hydrostatic, as noted in Chapter 6, implying that it is more or less a level surface. Thus, near the top of the outer core, level surfaces will not coincide with isotherms; that part of the core, at least, cannot be in hydrostatic equilibrium.

Olson (2016) has developed a comprehensive model of the CMB's thermal variations constrained by tomography (and other seismic evidence) combined with high-P,T mineralogy and inferred variations in lower-mantle conductivity; expressed in terms of the lateral variations in heat flux across the CMB, he uses those variations to help drive coupled core and mantle convection. Although neither of those components can be described as a 'wide but thin' fluid (in contrast to the situations illustrated in Figure 9.16, and even Figure 9.17), it turns out that the top layer of the fluid core shows some behavior similar to what we've discussed characterizing horizontal convection—including features that dynamically resemble the tropospheric Walker circulation mentioned above!

9.4.5 Temperatures Within a Convecting Fluid; Fluid *Versus* Solid-State Convection

In traditional, 'heated-from-below' or 'cooled-from-above' convection, also known as **Bénard convection** if it takes place in a thin (and usually horizontal) layer, the fluid typically exhibits a simple pattern of alternating upwelling and downwelling, at least away from any side boundaries. The structure of these convection cells is straightforward, easy to understand, and instructive; and despite the fact that they look nothing like the convection patterns we have seen characterizing the oceans or mantle, they are *the* paradigm for idealized convection in general.

The Onset of Convection and Development of Boundary Regions. When a thin fluid layer, that is, a layer with a much greater length or width than height, is heated from below—think "pan of water being heated on a stove"—that heat immediately begins to conduct upward, toward the cooler top boundary. We know that eventually, the entire layer of fluid will be unstable—and the water will begin to convect—once the vertical temperature gradient through the layer has become superadiabatic. But thermal conduction is, in most situations, an inefficient process; the fluid near the bottom absorbs the brunt of the heat, especially in the early stages of heating, and more quickly reaches an unstable (superadiabatic) state, while the fluid higher up has barely begun to warm. Fluid at the bottom

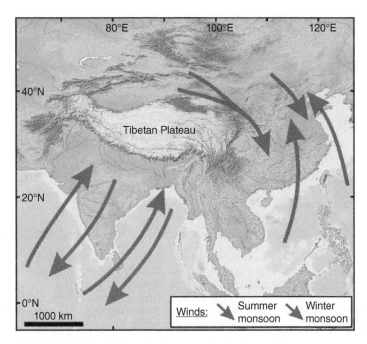

Figure 9.17 Alternating (summer and winter) monsoonal wind patterns, viewed as components of horizontal convection; only the lateral flow at the base of the troposphere is shown here (red and blue arrows).

As discussed in the text, each pattern will be completed by an intense upwelling—over land during the summer, over the ocean during the winter—and a broad, diffuse downwelling.

Two sets of flow patterns are indicated here, representing the Indian and Southeast Asia monsoons (together known as the Southwest Monsoon). Figure by Alexis Licht / https://alexislicht.files.wordpress .com/2014/10/asia-monsoons.jpg / last accessed August 22, 2023.

expands from the heat, becoming increasingly buoyant; hot, buoyant parcels begin to rise as upwelling plumes (Figure 9.18a).

To conserve mass, of course, there must be a replenishment of the upwelled fluid: a flow along the bottom of the layer, in toward the upwellings. But the material flowing in also absorbs heat from below, so plume generation continues. Thus, the bottom of the fluid has transformed into a 'plume factory:' the **lower thermal boundary layer**, characterized by a strongly superadiabatic temperature gradient, and acting as a source region of upwelling plumes.

Meanwhile, the rising plumes carry their heat upward into the fluid interior. Eventually, plumes developing in the bottom thermal boundary layer will be buoyant enough to reach the top surface, where they give off their heat to the air above, by conduction

and evaporation. To the extent that the top boundary maintains a cool temperature, the arriving plumes create a superadiabatic vertical temperature gradient (i.e., hot interior temperatures decreasing sharply upward to the cool top surface). As the plumes arrive at the top, the fluid makes room for the additional material by pushing mass out to the sides; as the plumes cool, some of the mass becomes negatively buoyant enough to sink. In short, as the early stages of convection progress, the top of the fluid develops into an **upper thermal boundary layer**, characterized by a strongly superadiabatic temperature gradient and incipient downwelling.

The thickness of these boundary layers depends on a variety of factors involving the fluid properties and the layer's geometry, as well as the boundary conditions; but it can be roughly estimated through a scaling analysis.

Figure 9.18 Bénard convection: idealized convection in a flat pan heated from below.
(a) The onset of convection envisioned. As the heat forces the bottom boundary layer to become superadiabatic and unstable, buoyant plumes begin to rise ('mushroom').
Sparrow et al. (1970) / Cambridge University Press, actually shows a later stage of convection which, at very high Rayleigh number, has become highly turbulent. Nevertheless, the image is very suggestive of plumes struggling initially to push up through the fluid and reach the top surface.

First, however, it is worth mentioning another kind of boundary layer, one constrained by the transmission of *momentum* by the fluid: a *viscous* boundary layer. Low-viscosity fluids may appear to flow freely; but if the fluid is bounded by any fixed walls, viscous drag—no matter how weak—must slow the fluid down near those boundaries, leading to the fluid's immobility *at* the walls. Ultimately, in the real world, every boundary is a 'no-slip' boundary, for the fluid close enough to it. We can describe the fluid closest to the wall, the **viscous boundary layer**, as a region where viscous drag, i.e., diffusion of momentum, is as important as advection of momentum; in contrast, away from the wall, in the *interior* of the fluid, drag forces may be neglected. (In high-viscosity fluids, distinguishing a viscous boundary layer from the interior may not be useful; in the mantle, for example, it's *all* a viscous boundary layer!)

Similarly, in the interior of a convecting fluid, heat transmission by conduction (diffusion of heat) is minor compared to advection; but in the thermal boundary layer they are both significant. That is,

$$|k\nabla^2 T| \sim |\vec{v} \cdot \vec{\nabla} T|$$

or, after scaling,

$$\frac{k\,T_C}{\delta^2}\nabla^{*2}T^* \sim \frac{V_C\,T_C}{\delta}\vec{v}^* \cdot \vec{\nabla}^* T^*,$$

where the 'length' scale of the boundary layer, in this case its thickness, is denoted, according to tradition, as δ. This implies

$$\frac{k\,T_C}{\delta^2} \sim \frac{V_C\,T_C}{\delta}$$

or

$$\delta \sim \frac{k}{V_C}.$$

To estimate a characteristic velocity within the boundary layer, we can assume that buoyancy and viscous drag forces balance within that layer. Starting from

$$\rho\alpha_P\tau g\,\vec{z} \cong \eta'\nabla^2(\rho\,\vec{v}),$$

we would find

$$V_C \sim \frac{\alpha_P g\,\Delta T\,L^2}{\eta'},$$

equivalent to what we wrote earlier; but in the present context the length scale should be δ

rather than L, so that

$$V_C \sim \frac{\alpha_P g \, \Delta T \, \delta^2}{\eta'}$$

for the boundary-layer velocity. Substituting this expression into the previous scale relation ($\delta \sim k/V_C$) leads to

$$\delta \sim \frac{k\eta'}{\alpha_P g \, \Delta T \delta^2} = \frac{k\eta'}{\alpha_P g \, \Delta T L^3} \frac{L^3}{\delta^2}$$
$$= \frac{1}{\text{RA}} \frac{L^3}{\delta^2},$$

so

$$\delta \sim L \, (\text{RA})^{-1/3}.$$

Note that, even though V_C is scaled by the boundary-layer dimension δ, we retain the length scale L in the definition of RA because the Rayleigh number represents the entire fluid (not just the boundary layer).

This result was also obtained by Tritton (1988) and Bercovici (2011). In short, we find that the thickness of the thermal boundary layer decreases as the Rayleigh number increases and the convection becomes more vigorous.

As an example, we consider the mantle, minus heat sources. We found earlier that a modest superadiabatic temperature gradient implied RA ~25,000; this suggests $\delta \sim L/30$, which for whole-mantle convection yields $\delta \sim 100$ km. For a high-viscosity body like the mantle, the thermal boundary layers are visible as distinct structures: the lithosphere and D″. As an order-of-magnitude estimate, our prediction is surprisingly good (of course, with higher Rayleigh numbers, for example as a result of heat sources, the agreement will not be as perfect).

These thermal boundary layers must be the source region of downwelling and upwelling; can you specify what downwells, and where the upwelling takes place?

Idealized Convection. As the heated fluid body begins to fully participate in the convection, and upwellings and downwellings proliferate, another struggle takes place:

a struggle for territory. Overall, to enable complete circulation, the upwellings and downwellings must alternate; but if they are spaced too closely, their opposing temperatures and directions of motion will cancel each other out, and if they are too far apart, the boundary-layer flow replenishing them will fail, welling up or sinking down before reaching them (Bercovici, 2011).

Once the appropriate adjustments in spacing have taken place, the circulation will emerge as a regular set of convection cells, exhibiting upwellings and downwellings on alternating sides of each cell. Ideally, when the fluid body is thin and the Rayleigh number is not too large, the width of each cell (the distance between upwellings and downwellings) will be more or less comparable to its height.

Convection cell regularity notwithstanding, in 'plan view' (i.e., looking down from above at the top surface of the fluid) there are numerous patterns of motion we might see, reflecting the upwellings and downwellings in the fluid below the surface. The most well-known pattern for Bénard convection is a set of hexagonal cells (see Figure 9.18b, c),

(b)

Figure 9.18 Bénard convection: idealized convection in a flat pan heated from below. (b) Surface pattern after convection has been established in the fluid. Looking down from above at a fluid filled with aluminum flakes (to help visualize the fluid motions), with the shutter open for 10 sec, we see a well-known hexagonal pattern: upwelling in the center of each hexagon, downwelling at the edges. Photo by Van Dyke (1982) / NOAA.

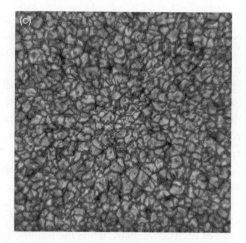

Figure 9.18 (c) Bénard convection in the Sun? This high-definition image, from the Inouye Solar Telescope, shows solar granulation at the top of the Sun's outer convective zone. This image is roughly twice as wide as Earth (for a less close-up view of the solar surface, see Figure 7.32). Each granule represents a convection cell; the darker areas are (relatively) cooler than the brighter ones, and identify sinking masses.
Presumably magnetic forces acting on the plasma create an equivalent surface tension...; so, are we seeing hexagonal Bénard cells, modified by extreme Rayleigh number turbulence?

opposite edges; and longitudinal 'rolls,' a configuration in which the upwelling and downwelling occur in 'sheets' along opposite long edges (see, e.g., Tritton, 1988).

In all these examples of idealized convection, the overall circulation pattern—parcels rising on one side, descending on the other, and flow in-between at the top and bottom to complete the pattern—results in a classic variation of temperature versus height within each cell. This circulation takes heat added to the fluid at the base and brings it up to the top surface, where it can be transferred out of the layer. If the convection is efficient, the central 'core' of each cell will be unaffected by the heat flowing in or out of those boundaries, remaining unchanged thermally and essentially motionless as well. The core of each convection cell will therefore be *adiabatic*; however, in a shallow fluid (like the Bénard pan on a stove) or a very incompressible fluid, where sinking and rising parcels undergo negligible compression or decompression, this would barely differ from temperature remaining constant with depth through the core (see Bercovici, 2011). Either way, of course, the boundary layers are superadiabatic; see Figure 9.18d.

with an upwelling plume in the middle of each hexagon and downwelling at the edges; however, this particular pattern has been attributed to the effects of surface tension (see Bercovici, 2011), which acts to pull all particles along the surface equally toward each other. Fluids in which the viscosity is temperature-dependent can also yield a hexagonal pattern (Tritton, 1988); given how easily upwelling plumes can develop in such fluids, and how narrow those plumes can be, such a plume-centered pattern is understandable if we want the heat transfer from top to bottom (managed at the cell edges) to be as efficient as that from bottom to top.

Beyond hexagons, surface patterns commonly observed (if the Rayleigh number is not too large) include various 'square' formations, reflecting upwelling and downwelling at

(d)

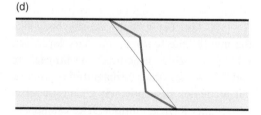

Figure 9.18 Bénard convection: idealized convection in a flat layer heated from below. (d) Schematic temperature profile of the fluid layer, averaged laterally, once convection is established. Shading indicates thermal boundary layers; boundary layer thicknesses are highly exaggerated for clarity. Temperature in the 'core' of the convection cell increases with depth approximately adiabatically (brown line); temperature in the boundary layers increases strongly superadiabatically with depth (red lines). The overall profile, from top to bottom (faint black line), is also superadiabatic—not unexpected, given that the fluid is convecting.

At very high Rayleigh numbers, the convection is no longer ideal: the boundary layers and core all become turbulent, and the core is no longer isolated (see Tritton, 1988).

Implications of an Ideal Convective Structure. The core, mantle, and atmosphere can be expected to possess the vertical convective structure we have just described, at least as a first approximation. This structure has some important consequences.

- Because distinctly superadiabatic temperatures are required near any convective boundary, we can conclude, as noted in Chapter 8, that the mantle's second transition is probably not a thermal boundary layer. Our conclusion does not, however, mean that this transition cannot *influence* the mantle's convective behavior.

- As pointed out by Stein and Wysession (2003), the mantle's convective structure leads us to expect that the lower mantle should be more viscous than the upper mantle: temperature increases only slightly from the asthenosphere down to D″—the adiabatic gradient is weak—producing only a mild decrease in viscosity, which can be more than offset by the effects of a substantial increase in pressure (see also Chapter 8, and the subsection below on the solidus).

- With adiabatic gradients playing such a key role in the behavior of different components of the Earth System, it is not surprising that researchers have developed an alternative, essentially nonthermodynamic, way to calculate adiabatic temperatures within the Earth. The basis for this alternative is a quantity called the **Grüneisen parameter** or **Grüneisen ratio**, γ, defined as

$$\gamma \equiv \frac{\alpha_P k_S}{\rho\, C_P},$$

where k_S is the adiabatic incompressibility. We can write the adiabatic gradient in terms of γ:

$$\left(\frac{\partial T}{\partial z}\right)_S = \frac{\alpha_P\, T g}{C_P} = \frac{\rho \gamma\, T g}{k_S}.$$

Bringing in hydrostatic equilibrium, that is, dividing the left-hand side by $\partial P/\partial z$ and the right-hand side by ρg, this relation becomes

$$\left(\frac{\partial T}{\partial P}\right)_S = \frac{\gamma}{k_S}\, T.$$

Use of this formula and our final result below will require T to be specified in Kelvins. For simplicity, we will no longer write the derivative with a subscript, but keep in mind that this derivation assumes adiabatic conditions:

$$\frac{\partial T}{\partial P} = \frac{\gamma}{k_S}\, T$$

or

$$\frac{\partial T}{T} = \frac{\gamma}{k_S}\, \partial P.$$

Incorporating the definition of k_S, from Chapter 8,

$$k_S = \rho \left(\frac{\partial P}{\partial \rho}\right)_S = \rho \frac{\partial P}{\partial \rho},$$

we have

$$\frac{\partial T}{T} = \gamma \frac{\partial \rho}{\rho}.$$

Finally, integrating this versus depth in the convection cell's core, from the top where density and temperature are ρ_0 and T_0 to some greater depth where they are ρ and T, we find

$$\ln(T/T_0) = \gamma \ln(\rho/\rho_0);$$

thus

$$T = T_0 \left(\frac{\rho}{\rho_0}\right)^\gamma$$

(the mathematical steps in this derivation are similar to what was presented in *Textbox A*, Chapter 8). Within the solid earth, this formula allows us to estimate adiabatic temperatures from a model of the interior density.

Our derivation is accurate only if γ is constant. Amazingly enough, given the combination of physical parameters that produce the Grüneisen parameter, it does turn out to be roughly constant, according to experiments on a variety of materials at high pressures and temperatures; for example, values tabulated in Anderson (2002) range from 1.3 to 1.6 in the outer core and 1.1 to 1.3 in the mantle. Theoretical calculations by Vočadlo et al. (2003) determined a similar range for γ in the outer core, with preferred values between 1.53 and 1.50. Thus, our estimates of the adiabatic temperature deep within the Earth may be reasonably accurate even if the individual thermodynamic parameters comprising γ (especially α_P; Stacey & Davis, 2004) are uncertain at those depths.

This seismological alternative for predicting temperatures within the Earth allows our confidence in those estimates to be assessed. As an application, we take $\gamma \sim 1.5$ from Vočadlo et al. (2003) for the outer core and combine it with the estimate of 5120°K (~4,850°C) for T by Sinmyo et al. (2019) for the temperature at the inner-core boundary; using PREM density estimates of $\rho = 12.2$ gm/cm^3 *just above* the inner-core boundary and $\rho_0 = 9.9$ gm/cm^3 at the top of the core, we predict a temperature $T_0 \sim 3,743°$K (~3,470°C) at the CMB. Or, using the density model we computed in Chapter 8, with $\rho = 12.5$ gm/cm^3 and $\rho_0 = 10.1$ gm/cm^3, we predict $T_0 \sim 3,720°$K (~3,450°C) at the CMB. For comparison, Sinmyo et al. (2019) estimated $T_0 \sim 3,760°$K (~3,490°C) at the CMB using the same value of γ and densities from a model by Vočadlo et al. (2003).

On the other hand, if we use estimates of the adiabatic temperature gradient in the outer core, we can extrapolate temperatures from the inner-core boundary (still using $T \sim 4,850°$C for the temperature there) up to the core-mantle boundary. For example, taking the gradient to roughly equal 0.8°C/km (e.g., Vočadlo et al., 2003, Lowrie 2007), we would find $T \sim 3,050°$C near the core-mantle boundary. Or, using the results of an analysis by Stacey and Davis (2004) estimating temperatures throughout the core—but just using their values from

the middle of the outer core as the basis for still another inference of the adiabatic gradient—we would take the gradient to be almost 0.54°C/km, leading to an estimate of $T \sim 3,630°$C near the core-mantle boundary.

Combining both types of estimates, we might conclude that the CMB temperature is between ~3,000°C and ~3,600°C, though probably closer to the higher value. Lay et al. (2008) suggest a range of ~3,000°C–~4,000°C from similar considerations.

- We can follow the same approaches in the mantle. As discussed in Chapter 8, the olivine→wadsleyite geothermometer implies a temperature of ~1,550°C at the 400 km transition. Earlier in this chapter, we approximated the adiabatic gradient as ~0.4°C/km through the transition zone, suggesting the temperature T_0 must be about 1,650°C (1,920°K) just below the 660 km transition. With PREM densities of $\rho_0 = 4.38$ gm/cm^3 at that depth and $\rho = 5.49$ gm/cm^3 at the top of D″, a Grüneisen parameter of 1.2 then yields a temperature T of ~2,250°C at the top of D″. On the other hand, using an adiabatic gradient of ~0.33°C/km in the lower mantle, we can directly estimate the increase in temperature through the rest of the mantle, finding, for example, that the temperature at the top of D″ is ~2,340°C. Similarly, but with updated values and a more variable adiabatic gradient, Katsura et al. (2010) estimated a temperature of ~2,460°C at the top of D″.

Combining both types of estimates, it appears that the temperature at the top of D″ is between ~2,250°C and ~2,450°C. We have thus demonstrated a mismatch, amounting to roughly 600°C–1,300°C, in the temperature predicted at the top of the core versus the bottom—or near-bottom—of the mantle. More precisely, this difference in temperature exists between the bottom of D″ (whose temperature should be that of the CMB) and the top—and thereby reveals the lower thermal boundary layer of the mantle! As a layer roughly 150 km thick (in the PREM model), it therefore sustains a temperature gradient of ~4°C/km–9°C/km; such a gradient is higher than the lower mantle's adiabatic gradient—as is appropriate for a thermal boundary layer—by an order of magnitude.

In contrast, the mantle's upper thermal boundary layer, the lithosphere, is characterized by a temperature difference of ~1,300°C from top to bottom (e.g., Turcotte & Schubert, 2002), which for oceanic plates implies a gradient of ~13°C/km.

Can you suggest a reason why the outer thermal boundary layer of the mantle has a larger gradient than its 'inside' thermal boundary layer?

The astute reader may have noticed that, when we extrapolated temperatures from the inner-core boundary up to the CMB (using either a Grüneisen parameter approach or the temperature gradient), we had assumed adiabatic temperatures throughout the core, thus neglecting the presence of a superadiabatic thermal boundary layer at the top of the core (we also neglected to consider the possible existence instead of a stable, *sub*adiabatic layer at the top of the core, a topic of recent renewed interest among geophysicists, though beyond the scope of this textbook). Our estimates of the CMB temperature were higher than they would have been had we accounted for that boundary layer (extrapolating up through that layer, the temperature should have decreased superadiabatically), causing a greater D″–CMB mismatch. For example, if we envision such an upper thermal boundary layer to be perhaps ~200 km thick (see, e.g., Bouffard et al., 2019 and references therein, also Stacey & Davis, 2004), with a temperature gradient in that layer twice adiabatic (as in Stacey & Davis), the *additional* cooling up

through the boundary layer would have been ~100°C. In this case, the mismatch would have been close to our earlier estimate, ranging between ~500°C and 1,200°C, so the superadiabaticity of D″ would have been affected only slightly; but other scenarios are possible.

Finally, the temperature ranges we deduced for D″ and for the core-mantle boundary allow for the possible presence in D″ of a post-perovskite phase, but only at the 'low' end of both ranges; more likely, as we concluded in Chapter 8, any post-perovskite is limited to cold pockets within D″, or (if D″ is cold enough) perhaps a few swaths of the upper reaches of D″.

The Mantle: Adiabat Versus Solidus. In ideal convection, the adiabatic temperature gradient plays a crucial role, defining when the fluid becomes unstable and dominating the temperature profile in the 'core' of each convection cell. But in the mantle, convection is solid-state—even for hot upwelling plumes, flow is achieved by ongoing creep; the success of any instability in driving the convection depends more on how soft and 'strain-able' the mantle is than on how much unstable parcels cool when rising.

As we saw in Chapter 8 when discussing viscosity, the ability of a material to deform depends on how hot it is, that is, how near it is to melting. For a body like the mantle, composed of multiple minerals, this more precisely means how near it is to *beginning* to melt. The temperature at which partial melting begins, as noted earlier, is called the *solidus* (so named because the material is completely solid below that temperature). The mantle's ability to deform, creep, and convect at any depth depends on how close its temperature there is to the solidus (e.g., Wang 2016; Stacey & Davis, 2008; Lowrie 2007; Turcotte & Schubert, 2002)—leading some to call the mantle's flow "subsolidus convection."

Like the viscosity, then, the strain rate $\dot{e}$ can be represented as an exponential function of T/T_M, where both variables are in units of Kelvins and where we now explicitly recognize that, for the mantle, T_M represents the solidus. The ratio T/T_M, which plays a more fundamental role in the rheological behavior of solid materials than the temperature itself does, is termed the **homologous temperature**.

For example, at a depth of 100 km, roughly the top of the asthenosphere in oceanic regions, we might have $T \cong 1{,}300°C \approx 1{,}600°K$ and $T_M = 1{,}750°K$, the latter based on the solidus determined by Hirschmann (2000) for dry peridotite; at that depth, then, $T/T_M \sim 0.9$. These numbers, by the way, indicate a situation where the asthenosphere's low viscosity and high deformability are not due primarily to partial melting (since T is less than T_M), but would instead simply be a consequence of relatively hot temperatures. As discussed in Chapter 8, though, the presence of hydrated minerals would lower T_M; depending on how much T_M is lowered, partial melting might end up contributing to the asthenosphere's weakness as well.

According to Stacey and Davis (2008), T/T_M is less than 0.5 (on average) in the lithosphere and in subducted slabs still in the upper mantle; those relatively cold structures clearly behave as highly rigid, elastic-brittle solids. In the upper mantle, low-viscosity and highly deformable plastic behavior take place when temperatures are hot enough for T/T_M to exceed ~ 0.66 (Wang, 2016).

Viscosity actually depends on the *inverse* of the homologous temperature (see Chapter 8), which is why a higher homologous temperature means lower viscosity and higher strain rate. Because the viscosity of the lower mantle is, roughly, one to two orders of magnitude greater than the upper mantle's viscosity (Table 5.3), we can conclude that the homologous temperature *decreases*, somewhat, with depth—that is, T_M increases with depth more than T does. The latter conclusion is consistent with *but stronger than* our statement in Chapter 8 that the increase in pressure with depth causes T_M to increase.

These trends are unlikely, however, to continue all the way to the core-mantle boundary. As the source region for whole-mantle plumes, D″ will have lower viscosity (and thus a higher homologous temperature) than the mantle just above it. And, the homologous temperature should increase even more as the hot lower boundary of D″ (i.e., the CMB) is approached and the viscosity decreases further. Thus, T/T_M increases with depth within D″, and T_M no longer increases more rapidly than T does.

In fact, in either thermal boundary layer the increase in T with depth can be expected to dominate; it is mainly the 'core' of the mantle convection cell where T_M has the greater increase. Another way to say that is, in the core of the mantle convection cell, T_M increases with depth at a greater than adiabatic rate. Such a statement can actually be proved under some conditions as a consequence of the Clausius-Clapeyron equation we met in Chapter 8 (e.g., Lowrie, 2007).

Finally, then, we see that the adiabat controls the temperature within a solid-state convection core, and determines the stability of parcels in the cell; but it is the nearness of the adiabat to the solidus that determines whether flow will ensue, and how effective it will be.

A speculative depiction of the variations in T and T_M through the mantle is provided in Figure 9.19. It should be noted that many other depictions are possible, as there is not widespread agreement on temperatures or temperature gradients in the mantle—and, for that matter, in the core—in part because their estimation may be based on poorly known parameters or not have included the complexities of Earth's mineralogy.

9.5 More on Surface Heat Flow in the Earth System

There are additional connections between the topics discussed in this chapter and the state of the Earth System.

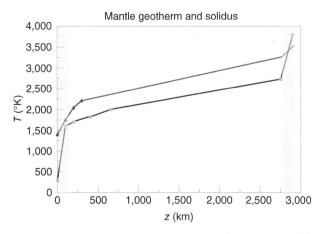

Figure 9.19 Hypothetical scenario of temperatures within the mantle, both the actual temperatures ('geotherm') and the solidus, versus depth *z*. Thermal boundary layers are loosely indicated by shading. Geotherm (black line segments): based on geothermometer experiments and other estimates of temperatures at specific depths (green-filled circles) and adiabatic predictions (line segments), as discussed in the text.
Solidus (purple line segments): for upper mantle depths, based on the work of Hirschmann (2000) (filled triangle symbols); for the lower mantle, guided by the considerations of homologous temperature as discussed in the text. The solidus is drawn arbitrarily intersecting the geotherm within the lower boundary layer (D″) as a 'nod' to ULVZs (see Chapter 8).

What does that intersection imply about the physical state of the deepest part of D″?

9.5.1 Geothermal Heat and the Thermohaline Circulation

Our first application explores the possible connections between heat flowing up and out of the oceanic crust and the slowly circulating oceans above. For conciseness, and despite the more apt alternative names discussed in Chapter 3, we will continue to refer to the oceanic motion here simply as the thermohaline circulation.

Can the Heat Flux From Earth's Interior Really Affect the Thermohaline Circulation? We noted at the outset of this chapter that heat from within the Earth represents a tiny fraction of the ocean's heat budget. Heat flux exiting the solid earth averages ~105 mW/m² in oceanic areas (Davies & Davies, 2010), whereas insolation is on average about 170 W/m² (Chapter 3)—more than three orders of magnitude greater. Not surprisingly, nearly all models of global ocean circulation have neglected forcing by 'geothermal' heat (Scott et al., 2001).

However, the main drivers of the thermohaline circulation are only indirectly related to insolation. And for convective flow, heating from above is less effective than heating from below. For both reasons, the preceding energy comparison is not appropriate, and suggests caution before dismissing the possibility of geothermal heat having an impact on ocean circulation.

Some simple calculations reinforce the need for caution—if we maintain a geologic perspective. Consider the average heat flow through the seafloor, ~60 mW/m² for older seafloor (though more in young regions; see Table 9.1). Even if, just for the sake of argument, there were no mid-ocean ridges or hotspots and the oceanic heat flow was the same 'cold' value of 60 mW/m² everywhere, this heat flux would add more than ~21 TW of heat to the oceans; that translates into 6.8×10^{20} J per year. If this heat remained in the oceans, whose mass is ~1.4×10^{21} kg (Stacey & Davis, 2008) and whose heat capacity is ~4,000 J/°C per kg of seawater (Nayar et al., 2016), it would raise the temperature of all the oceans' seawater by 100°C in less than a million years (cf. Munk, 1966)—an extreme result, taking place over a geologic instant!

Can you verify the value of the total heat per year and the temperature increase quoted here?

Actually, that amount of heat would not quite cause the oceans to boil away every million years, because we have not included the heat of vaporization in our calculations. Nevertheless, the increase in seawater temperature would have catastrophic effects on marine life and (by the increased evaporation) on climate as well, and must be ruled out. That is, the oceans must deal with the heat from Earth's interior, either convectively or diffusively conveying it to the surface where it can be taken up by the atmosphere.

We can estimate the typical time for that bottom heat to diffuse up through the water column from $t_C \equiv L^2/k$; with $k \sim 1.4 \times 10^{-7}$ m²/sec (Nayar et al., 2016), we find the characteristic time for diffusion to be $\sim 3\frac{1}{2}$ Myr. If the ocean interior is turbulent, though, k can be as much as three orders of magnitude greater; for example, models of the downward diffusion of surface heat and upward diffusion of mass, representing the thermohaline return flow, have been interpreted as consistent with $k \sim 1 \times 10^{-4}$ m²/sec (e.g., Munk, 1966, Emile-Geay & Madec 2009, and others; but see also MacKinnon et al., 2013). In that case, turbulent diffusion of the bottom heat up to the surface might happen as quickly as several thousand years. In contrast, the timescale for heat transport by the thermohaline circulation, that is, the time for one half-cycle of that circulation, is (Chapter 3) typically in the range 1–2×10^3 yr. It thus appears that diffusion of heat is too slow a process to transport the bulk of the heat coming out of the seafloor—leaving convection, or at least advection, by the thermohaline circulation as the most effective way to keep the oceans from 'glacially' overheating. This inference is consistent with the conclusions of both

Scott et al. (2001) and Emile-Geay and Madec (2009).

As a final argument, we note that the pattern of vertical heat flow within the bottom waters of the ocean, as inferred (using Fourier's law) from global ocean data by Mashayek et al. (2013; see also Emile-Geay & Madec, 2009) is generally similar to the pattern of geothermal heat flow versus age seen in Figure 9.4. This similarity hints at the ability of the heat emanating from Earth's interior, despite its faintness, to impact the oceans.

Predicted Effects. Results from numerical experiments that model the thermohaline circulation with and without heat flow from the seafloor tend to agree with the preceding speculations on the effect of that heat flow. Numerical experiments conducted by Scott et al. (2001) in idealized (flat, rectangular) ocean basins with a geothermal heat flux of 50 mW/m² show a definite intensification of the thermohaline circulation by a few Sverdrups (thus, by $\sim 20\%$; see Chapter 3), with most of the amplification taking place near the equator or at high latitudes (also found by Mashayek et al., 2013). Similar work by Mullarney et al. (2006) (but with the heat flux implied to be ~ 100 mW/m²) predicts a 10% increase in the strength of the circulation, and also hints at a stronger component of bottom flow along the seafloor (from downwelling to upwelling regions).

Adcroft et al. (2001) carried out numerical modeling with geometrically realistic ocean basins, with a geothermal heat flux of 50 mW/m². They found a 25% increase in the strength of the thermohaline circulation in the Pacific and Indian Ocean basins—where much of that circulation's upwelling takes place (see Figure 3.10)—amounting to almost 2 Sv additional transport. Emile-Geay and Madec (2009) employed realistic ocean basins with both uniform and realistically varying geothermal heat flow; they found the ABW cell intensified by $3\frac{1}{2}$ Sv due to uniform heat flow but by only $2\frac{1}{2}$ Sv with realistic heat flow (with realistic heat flow, the hotter seafloor tended to be elsewhere, i.e. at shallow depths, causing less intensification). Hofmann and Maqueda (2009) also employed realistic ocean basins with both uniform and realistically varying geothermal heat flow; in their model, the intensification of the ABW cell was 3 Sv, and $1\frac{1}{2}$ Sv in the NADW cell.

These models do not agree precisely, but all suggest that heat flow from the ocean bottom acts to strengthen the thermohaline circulation, by something like a few Sverdrups at most.

9.5.2 Subsurface Temperature Variations and Climate Change

Early in this chapter, we discussed how heat flow from Earth's interior could be inferred from measurements of surface or near-surface temperature gradients, and mentioned the importance of taking those measurements deeply enough to avoid 'contamination' from the surface temperature variations imposed at Earth's surface by meteorological and climatological processes. Those surface fluctuations typically include a wide range of timescales—daily, seasonal, decadal, and so on—with the longer timescale fluctuations diffusing to greater depths before attenuating, according to the 'skin effect.'

Of course, that diffusion is not instantaneous; the deeper the measurement, the longer the diffusion takes (as we know from considering t_C). In other words, the deeper our measurement, the farther in the past is the surface temperature fluctuation that contaminated it. Our final application in this chapter explores efforts by researchers (Pollack & Huang, 2000 provide a concise history of those efforts; see also references cited in Roy et al., 2002) to use measurements at depth in order to estimate such past temperature variations.

Paleoclimatologists have long used climate proxy data, such as tree ring thickness and elemental isotope ratios, to infer the temperatures that characterized past climates, especially on regional and global scales. But all proxy data require 'calibrating;' that is,

they must be related to temperature (when we used $^{18}O/^{16}O$ in earlier chapters, it was only to illustrate past temperature *patterns*, so we avoided such calibrations). The calibrating relation is determined by statistical correlations and/or climate modeling, techniques that may limit the relation's usefulness: if the determination is theoretical, it may be too simplified (perhaps ignoring nonlinear or multivariable relationships); if based on comparison with actual temperature data—which is available globally only for the past century and a half—it may have missed long-term trends (e.g., Christiansen & Ljungqvist, 2017; Pollack & Huang, 2000).

Borehole data shares some similarities with climate proxy data, including a need to separate the temperature variations of interest from unrelated variations, and having the potential to reveal both regional and global climatic trends. But it stands apart from proxies because it involves direct measurements of temperature, and does not require calibration.

There are a few challenges in using borehole data. For example, groundwater flow passing by the borehole may advect heat to or away, obscuring the temperature effect we are interested in. And there has been some debate about whether the surface *ground* temperature, inferred from the borehole data, will match the surface *air* temperature, which is what we want to learn about: snow cover will insulate the ground, and soil moisture will 'buffer' the ground, in both cases preventing it from feeling even colder air temperatures (e.g., Smerdon et al., 2006, Mann et al. 2009); according to Pollack and Huang (2000) and others, though, the long-term effect worldwide may be minor. We can interpret the analysis presented in Pollack and Huang (2000) to conclude that the overall (global, long-term) disparity between air and ground temperatures will amount to ~0.03°C or less.

Quantifying the Skin Effect. We consider a shallow region of the Earth in which the temperature $T = T(z, t)$ can be a function of both depth and time. This region is subject to a meteorologically or climatologically forced, periodically varying temperature at its outer boundary, the Earth's surface:

$$T (z = 0, t) = T_0 \sin(\omega t),$$

where T_0 is the amplitude of this fluctuation and ω is its angular frequency (i.e., its period $= 2\pi/\omega$). We want to find $T = T (z, t)$ as a solution to the one-dimensional unsteady diffusion equation we derived earlier, subject to this time-varying boundary condition. Anticipating that the subsurface temperature perturbation produced by this boundary constraint will exhibit a similar periodic variation, but with an amplitude that attenuates with depth (according to the skin effect), and with an increasing lag of its peaks and troughs at greater depths, we attempt a solution of the form

$$T(z, t) = T_0 \, e^{-\hat{\varepsilon} z} \, \sin(\omega t - \tau z).$$

With this formulation, $\hat{\varepsilon}$ is a decay constant representing how rapidly T attenuates with depth, and τz is a phase lag accounting for the perturbation's delay in reaching greater depths. Note that, with the amplitude of this postulated solution set equal to T_0, it already satisfies the boundary condition at $z = 0$.

Substituting this expression into the diffusion equation,

$$\frac{\partial T}{\partial t} = k \frac{\partial^2 T}{\partial z^2},$$

we have (with differentiation with respect to z required twice on the right-hand side)

$$T_0 \, e^{-\hat{\varepsilon} z} \omega \cos(\omega t - \tau z)$$
$$= k \, T_0 \{ (\hat{\varepsilon}^2 - \tau^2) e^{-\hat{\varepsilon} z} \sin(\omega t - \tau z)$$
$$+ 2 \hat{\varepsilon} \tau e^{-\hat{\varepsilon} z} \cos(\omega t - \tau z) \}.$$

Obtaining this is not as difficult as it might appear; for practice with differentiation, try verifying this result.

Canceling out terms common to both sides reduces it to

$$\omega \cos(\omega t - \tau z)$$
$$= k \{ (\hat{\varepsilon}^2 - \tau^2) \sin(\omega t - \tau z)$$
$$+ 2 \hat{\varepsilon} \tau \cos(\omega t - \tau z) \}.$$

However, it is not possible for a cosine function to equal a sine function (or a sine function plus a cosine function) at all times; that is, if $C_1 \sin(\omega t) = C_2 \cos(\omega t)$ for all t, the only way this is possible is if C_1 and C_2 are both zero—otherwise, the oscillations of the left-hand side would always be 90° out of phase with those of the right-hand side. Another way to say this is that sine and cosine functions are *orthogonal*—a concept we explored in Chapters 6 and 7. It follows that

$$(\hat{\varepsilon}^2 - \tau^2) = 0,$$

or

$$\tau = \pm \hat{\varepsilon}.$$

This allows us to simplify the preceding equation to

$$\omega \cos(\omega t - \hat{\varepsilon} z) = k \{ \pm 2 \hat{\varepsilon}^2 \cos(\omega t - \hat{\varepsilon} z) \},$$

which leads to

$$\omega = k \{ \pm 2 \hat{\varepsilon}^2 \}$$

or, finally,

$$\hat{\varepsilon} = \sqrt{\omega / 2k}.$$

Our solution can thus be written

$$T = T_0 \, e^{-\hat{\varepsilon} z} \, \sin(\omega t - \hat{\varepsilon} z)$$

with $\hat{\varepsilon}$ given by the preceding formula. The **skin depth**, now defined officially as the depth over which the temperature decays by a factor of e ($\cong 2.7$), equals $1/\hat{\varepsilon}$.

Can you show that the formula for $\hat{\varepsilon}$ leads to the various skin depths noted earlier in this chapter for daily, annual, and interglacial-scale surface fluctuations?

Characteristics of the skin effect are shown in Figure 9.20a, using this solution. A vivid illustration of the skin effect, including its dependence on the period of the surface temperature fluctuation, is provided in Figure 9.20b, from the work of Smerdon et al. (2006). This same work also illustrates how disparities may arise between the surface air and surface ground temperatures; see Figure 9.20c.

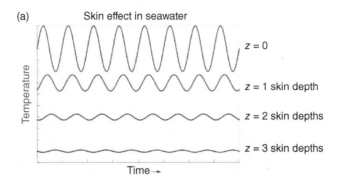

Figure 9.20 (a) Temperature variations over time versus depth in the ocean, illustrating the skin effect. The temperature variation imposed at the sea surface ($z = 0$) is an annual cycle. This annual variation persists at all depths, but with diminishing amplitude and an increasing lag in its peaks and troughs.
These temperature profiles were calculated using the solution derived in the text. As depicted here in general terms, it can apply to any medium; for seawater subjected to an annual variation, the skin depth turns out to be ~1.2 m.

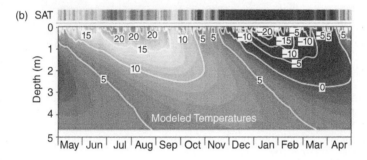

Figure 9.20 (b) Illustration of the skin effect. Observed surface air temperatures (SAT) in Fargo, North Dakota from May 1981 through April 1982, and the temperatures versus depth within the ground, theoretically predicted based on those surface air temperatures.
The surface temperatures vary seasonally from hot (25°C–30°C, orange) to moderate (~5°C, green) to cold (−15°C––20°C, dark blue). Note the attenuation of these temperatures with depth.
The attenuation also exhibits a frequency dependence, in that the hottest summer temperatures are brief and do not penetrate very far into the ground, whereas the cold winter temperatures are long-lasting and penetrate farther. Finally, note the increasing lag with depth in the temperature change: the winter cold, for example, penetrates 3 m down—but that happens in March and April, two months after the peak of winter. Image from Smerdon et al. (2006) / John Wiley & Sons.

Inferring Climate Change From Borehole Data. We have just shown that surface temperature fluctuations will penetrate, albeit diminishingly, to some depth below the surface, potentially allowing them to be detected in borehole measurements. But boreholes exist within the Earth's overall geothermal environment; the surface heat flow we measure reflects the need to get a lot of heat out of the interior. The signals associated with a surface fluctuation will thus be superimposed on a 'background' geothermal temperature gradient.

As a simple approximation, we can describe that gradient as linear, corresponding (as discussed earlier in this chapter) to a flat Earth in a steady state with no heat sources. This approximation is actually fairly reasonable: the borehole's depth will typically be no more than a kilometer or so (e.g., Cohen, 2012; cf. Kukkonen et al., 2011), making the Earth's sphericity irrelevant; and heat sources will turn the linear

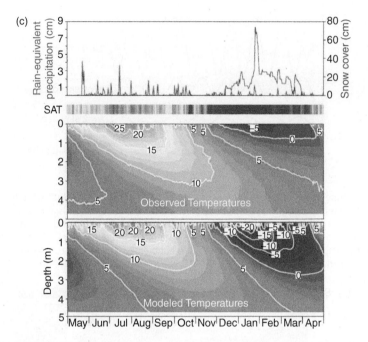

Figure 9.20 (c) Illustration of the challenges in using borehole temperatures to infer surface air temperature fluctuations.

The observed surface air temperatures (SAT) in Fargo, North Dakota from May 1981 through April 1982, and the theoretically predicted temperatures, both shown in Figure 9.20b, are reproduced here together with the actual borehole temperatures versus depth within the ground.

The top graph shows the precipitation (brown) and snow cover (black) during this time span.

Note especially that the winter snow cover significantly reduces both the magnitude of cold penetrating into the ground and how far it penetrates.

What do you think the consequences would be if the observed borehole temperatures were used 'at face value' to infer winter temperatures at the surface?

Image from Smerdon et al. (2006) / John Wiley & Sons.

gradient into a quadratic, but near the surface the two solutions become indistinguishable (see, e.g., Figure 9.10). However, in one sense the approximation does call for caution: in defining the background geotherm as steady state, all transient fluctuations—whether a periodic fluctuation induced by meteorological changes in surface temperatures, a trending change due to global warming or cooling, or other, unrelated fluctuations—are mixed together in the temperature 'perturbation' we end up solving for.

Thus, we will write the background temperature variation as

$$T_B(z) = Az + T_{B0},$$

where T_{B0} is the long-term average surface temperature characterizing the background. As discussed earlier in this chapter, we could relate the coefficient A to the regional heat flux and thermal conductivity, if necessary. The net temperature we observe in the borehole will be the *sum* of T_B and all the transient perturbations, including those caused by meteorological and climatic processes.

A hypothetical example of borehole data is shown as the brown curve in Figure 9.21 (actually constructed from the formula for T_B plus the periodic solution just derived). At greater depths (with surface influence consequently reduced), the data looks increasingly

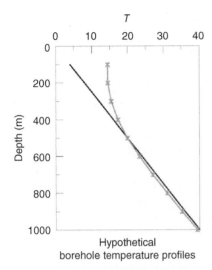

Figure 9.21 Hypothetical example of borehole temperature measurements.
Brown data (×) and curve: 'actual' temperature measurements.
Black line: background geothermal gradient, estimated from lower portion of borehole data (the portion presumably least influenced by surface temperature fluctuations).

like it could be well fit by a straight line—as expected, the linear geothermal gradient.

- On the simplest level, we could extrapolate both the line and the data curve up to the surface ($z = 0$) and compare the values. In this example, we might infer a 15°C difference, and conclude that the net effect of all surface perturbations (meteorological and otherwise) was warming by that amount.

- With slightly more sophistication, we might note that the borehole temperatures exceed the background gradient for shallow depths, but are slightly less than background for greater depths. Since surface temperature variations that occurred farther in the past would have penetrated deeper by now, we can conclude that surface temperatures must have been cooler in the past, then warmer more recently.

- Most rigorously, we could construct a set of 'reduced' data, in which the background temperatures (predicted according to the

linear gradient) are subtracted from the observed temperatures; such data would be mainly, if not entirely, the result of climate and meteorological fluctuations. We will symbolize this data as ΔT:

$$\Delta T = T - T_B = T - (Az + T_{B0})$$

where T is the original borehole data. The goal would be to analyze ΔT in order to deduce the surface temperature fluctuations.

There are different approaches one might take to achieve that goal. In line with our discussion above, we could assume the surface fluctuation can be represented as a single periodicity at frequency ω, and find the amplitude T_0 and skin-depth parameter $\hat{\varepsilon}$ that best fit the data; or, Fourier-like, we might assume multiple periodicities are involved, and solve for multiple amplitudes and skin depths. Or, we could apply the methods of "inverse theory," inverting the data to solve for the surface forcing characteristics (e.g., Shen & Beck 1991; Dahl-Jensen et al., 1998; Pollack & Huang, 2000; Roy et al., 2002; Kukkonen et al. 2011; Roberts et al. 2013; Kneier et al. 2018).

Another approach may be particularly well suited for climate studies: instead of assuming a periodic surface forcing, we could model the temperature variation at the surface as a linear trend— for example, a gradual warming—and determine the characteristics of the trend that best fit the reduced data. Those characteristics would include the slope of the trend, which indicates how many degrees per century or millennium the surface temperature has changed, and the trend's duration; alternatively, the maximum temperature change MAXT, which equals the slope times the duration, could be determined in place of one of those parameters. If the trend has gone on for an amount of time t^*, it turns out that the climatological perturbation $\Delta T(z)$ to the borehole temperature at depth z must theoretically equal

$$\Delta T(z) = 4 \cdot \text{MAX}T \times$$
$$\int_{-u}^{u} \int_{-v}^{v} \text{ERFC}(w) \, dw \, dv$$

at the present time, where

$$u = \frac{z}{2\sqrt{kt^*}}$$

(Roy et al., 2002; see also Lachenbruch & Marshall, 1986 and Chisholm & Chapman 1992). In this expression, v and w are 'dummy' variables of integration, and the result of this double integration is evaluated between $-u$ and $+u$, with u having the value shown. $\text{ERFC}(w) \equiv 1 - \text{ERF}(w)$ is called the *complementary error function* (visually, $\text{ERFC}(u)$ would be an upside-down image of the error function shown in the right-hand graph of Figure 9.12a, starting at $u = 0$ with a value of 1 and descending rapidly to the x-axis as u increases). In hindsight, it is not surprising that the solution involves integrals of the error function, because the boundary condition it is a solution to is an integral of the boundary condition whose solution was the error function.

This solution is nevertheless quite daunting; but the point is we can use it to predict $\Delta T(z)$ at each depth, for various choices of MAXT and t^*, ultimately allowing the parameter values most consistent with the data to be identified. The error analysis of Chisholm and Chapman (1992) suggests that, in practice, t^* will not be as well resolved (or have as low uncertainties) as MAXT, especially if the trend's duration is long; after all, most of the temperature change diffuses through the subsurface early on. They recommend more advanced inversion techniques to determine t^* accurately in such cases.

Once the optimal parameters have been resolved, and depending on how close the best predictions are to the actual data, it might also be possible to judge whether the underlying assumptions—including that the reduced temperatures are entirely the result of boundary forcing by a (climatic) trend in surface temperature—are reasonable.

Roy et al. (2002, also Roy & Chapman, 2012) applied this last technique to 70 borehole sites spread throughout much of the Indian subcontinent, finding an increase of almost 0.9°C in the average surface temperature over the past 150 yr. The rate at which temperatures have been changing per century implied by their work is in overall agreement with meteorological data for the region (and also with the global average; see Chapter 3).

Using a variety of numerical techniques to model borehole data through ice on the West Antarctic Ice Sheet Divide, with a physical model that accounted for movement of ice and consequent advection of heat, Orsi et al. (2012) determined that West Antarctica experienced the cold of the Little Ice Age (LIA) at a level roughly comparable to the northern hemisphere. With respect to the average temperature over the past 1,000 yr, Orsi et al. found the cold period in the southern hemisphere spanned 1300–1800 CE; the coldest portion (between 1500 CE and 1700 CE) was around 0.4°C colder than the prior "medieval warm period" and at least 0.4°C colder than the average since 1800.

Orsi et al. also found that West Antarctic climate warmed during the latter half of the twentieth century, and the warming accelerated toward the end, to a rate of 0.8°C per decade. Based on numerical modeling of borehole data in *East* Antarctica, Roberts et al. (2013) inferred a warming of nearly 0.4°C during those same decades. Although these two estimates differ significantly, both groups of authors cited additional data supporting their results; both concluded they are seeing regional rather than local effects—a situation not completely unexpected since the boreholes are in geographically distinct parts of the continent, with differing local and regional meteorology. It is worth noting that, even if the average rate of Antarctic warming over the past few decades is 'only' 0.4°C/decade, such a rate is still several times faster than the typical warming worldwide during the twentieth century, and faster also than the accelerated warming worldwide at the end of that century (see Chapter 3). Such behavior is consistent with our view that global warming will be amplified in the polar regions of the globe.

To the extent the analysis by Orsi et al. (2012) of Little Ice Age cold is representative of the entire southern hemisphere, their results argue against the idea that the LIA was essentially a North Atlantic, or at least northern hemisphere, event. On the other hand, if regional effects have influenced the West Antarctic borehole data, the actual amount of southern hemisphere cooling may be less than (or greater than) what those authors estimated. But the fact that the time spans of overall and maximum cooling at the West Antarctic site agree with northern hemisphere models suggests that the LIA was in some way a global phenomenon.

Finally, Huang et al. (2008) assembled a massive database, including surface air temperature and borehole heat flow data sets as well as borehole temperature data. The borehole data included data from all continents except Antarctica, and went deep enough for their analysis to extend 20,000 yr into the past. They found Earth's surface warmed from the Last Glacial Maximum till the mid-Holocene by 5°C–6°C, reaching a high up to 2°C warmer than temperatures in the second half of the twentieth century.

The analysis by Huang et al. did not detect the cooling associated with the Younger Dryas about 12,000 yr ago. Given what you know about how diffusion works, can you explain why a Dryas 'signal' was not measured?

Additionally, according to their global data set, the LIA was characterized by temperatures about 1°C lower than the latter half of the twentieth century; and the last decade of the century was another ½°C higher.

10

Geomagnetism and the Dynamics of the Core

10.0   Motivation

The core—in particular, the fluid outer core—may well be the most provocative component of the Earth System. With ten times the density of the oceans, it nevertheless has almost the same 'watery' viscosity. It undergoes convection of the traditional fluid kind, in contrast to the mantle, but its great thickness results in convection patterns very different than we see in thin fluid layers like the oceans and atmosphere. There are pieces of core material all around us—well, cores of asteroidal proto-planets, anyway—yet it is the only component of the Earth System no one has been able to touch. And, despite being the most distant component from us, traces of its intimate behavior are all around us, in the magnetic field it continually recreates and constantly changes.

What we know about the core's behavior mostly derives from observations of Earth's magnetic field, including paleomagnetic measurements, which hint at processes taking place in the core on long timescales. We will begin our discussion of Earth's field with some basics—including an exploration of the simplest kind of magnetic field, that of a bar magnet or 'dipole,' and a description of the different properties of the field that are typically measured—but it will be our discussion of its time variations, on short and long timescales, that is most important for revealing what's going on in the core.

Perhaps the best-known time variations are the reversals in polarity of the dipole portion of Earth's field; a reversal typically occurs roughly $\sim 10^5$–10^6 yr after the previous one, though with some unpredictability. It might be argued, however, that other variations, on both shorter and longer timescales, tell us even more about core dynamics.

Throughout this textbook, we have mostly implied that the magnetic field originates in the core; in this chapter we will confirm the strong connection between field and core. However, the field we observe at Earth's surface also includes contributions from other components of the Earth System. A spherical harmonic description of the field will help us distinguish the various sources of the observed field, and allow us to connect the core with the appropriate portion. As we will see, observations of Earth's field can then be used to provide fundamental constraints on mantle properties, ocean tides, and the behavior of the mid- and upper atmosphere.

A strong connection between core and field exists because of the **dynamo** (or **geodynamo**), the process of core convection that not only gets heat out of the core but also maintains the magnetic field; 'dynamo action' from the core flow induces electrical currents that are ultimately able to bolster the field. But it is only possible for electrical currents to be induced because the core is a good electrical conductor—which *also* means that parcels of core fluid can experience the same forces

Earth System Geophysics, Advanced Textbook 6, First Edition. Steven R. Dickman.
© 2025 American Geophysical Union. Published 2025 by John Wiley & Sons, Inc.
Companion Website: www.wiley.com/go/Dickman

exerted by the field that its motions are helping to create. It follows that the best way to understand core convection *or* the nature of the Earth's magnetic field is to treat them as *coupled* phenomena.

The study of fluid motion in this kind of situation is called *magnetohydrodynamics* or *hydromagnetics*—terms that may induce feelings of intimidation at any level of study. And there is no avoiding it: this chapter will involve some mathematics. We will need a more complex version of $\vec{F} = M\,\vec{a}$, to include magnetic forces; and we will need to simultaneously deal with a coupled equation that describes the generation and decay of magnetic fields. We will even find it worthwhile to discuss some mathematical 'theorems,' which will guide us in understanding the long-term nature of Earth's magnetic field and the constraints on flow in the core. Much of hydromagnetics is beyond the scope of this textbook's (upper-level) introductory approach; but we will find that our earlier forays into math as a way to understand geophysical phenomena have prepared a relatively painless and productive foundation for us to gain some understanding of core convection.

For example, we will discover that magnetic fields weaken by the same process of diffusion we studied in detail in Chapter 9. And our understanding of diffusion will also lead to an awesome way—known as "frozen flux"—for magnetic fields to be strengthened.

In some ways this chapter will represent very specialized topics, and might not appear suitable for the broad perspective on the Earth System adopted for this textbook. But what we study in this chapter will add to our understanding of convection. And, since the existence of our atmosphere (as discussed in Chapter 2)—perhaps even life as we know it—may well depend on the protection afforded by Earth's magnetic field, and since our discussion of the geodynamo relates to dynamo processes in the Sun and in other planets, it might not be such an anomalous way to conclude this textbook after all.

10.1 The Earth's Magnetic Field

There has been some degree of recognition since the twelfth century, if not earlier, that the Earth possesses a magnetic field: at that time, the *compass* was used to aid navigation at sea (see Guarnieri, 2014, also Jonkers et al., 2003; Nabighian et al., 2005; Lowrie, 2007; and Jonkers, 2003). The navigators were taking advantage of the fact that the Earth's field is very similar to that of a simple bar magnet.

10.1.1 Dipole Fields

The field of a bar magnet, illustrated in Figure 10.1a, is a dipole field. The lines drawn

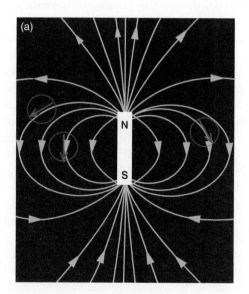

Figure 10.1 Dipole fields.
(a) The field of a bar magnet. Field lines (also called lines of force or flux lines) indicate by their direction how compass needles (red arrows) will line up at any location. The north (N) and south (S) poles of the magnet are the points all field lines disperse from or converge to.
The illustration shown here is a cross-section; altogether, the field lines actually wrap completely around the bar magnet.
Image by R. Russell, downloaded from https://www.windows2universe.org/physical_science/magnetism/magnetic_multipole_fields.html&edu=high and modified.

from end to end are called **lines of force**, **field lines**, or **flux lines**, and have the property that magnetic objects—like compass needles—are continually being forced to align with them. The points from which the field lines radiate are the **poles** of the field, with the **north pole** being where field lines leave the magnet and the **south pole** being where they gather.

This type of field is called a **dipole field** because it has two poles. The line connecting the poles is called the **dipole axis**.

There is an interesting property of all magnetic fields, called the 'no monopole' rule, which we can easily envision with bar magnets. If we take a bar magnet and break it, halfway between the poles, we end up with two smaller bar magnets—two smaller dipoles—not a single north pole and a single south pole. We can repeat this ad infinitum (Figure 10.1b) but will never be able to isolate a single magnetic pole. This property can be expressed mathematically by saying that no volume encloses a different number of flux lines leaving the volume than entering it; that is, the *divergence* of magnetic flux lines is always zero.

Can you verify this in Figure 10.1a? Consider a volume the size of the entire image; try counting the number of flux lines leaving the top half of the image versus those entering in the lower half.

Our assertion is equivalent to saying that the divergence of magnetic *fields* is always zero.

In contrast, electric charges, which produce electric fields, do not only exist as pairs of opposite charges; electric 'poles' (positive electric charges and negative ones) can be isolated. Not surprisingly, you might imagine that the divergence of an electric field will depend on how much net electric charge (positive + negative) is enclosed by a volume.

Incidentally, one way to produce a dipole magnetic field is to run an electric current through a circular loop of wire (see Figure 10.1c). The axis of the dipole field created in this way is perpendicular to the loop. It may be worth pointing out that, when a dipole field is generated by an electrical current, no one would expect to be able to isolate either of its magnetic poles....

Flux lines can tell us more than just the direction of a magnetic field. The *intensity* of the magnetic force exerted on a compass needle or other magnetic materials is proportional to the *concentration* of flux lines. The more the flux lines bunch up, the quicker the compass needle is brought into alignment, the stronger the magnetic force is. Thus, as seen in Figure 10.1a, the field of a bar magnet is strongest at the poles and weakest at the magnetic 'equator,' where the field lines are most spread apart.

The concentration of flux lines could be formally defined as the number of lines penetrating a surface area, per unit area (like heat *flux*, magnetic "flux" indicates that quantities

Figure 10.1 Dipole fields.
(b) Thought experiment with a bar magnet. The magnet is broken in half repeatedly, each time halfway between the poles. But no matter how far we take it, each portion always contains a north pole and a south pole; in other words, it is impossible to isolate a single magnetic pole.

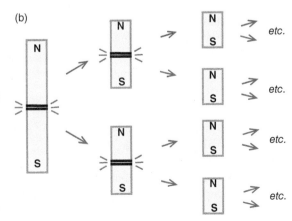

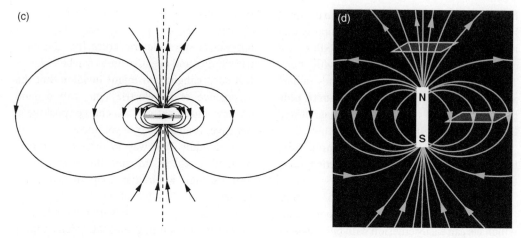

Figure 10.1 Dipole fields.
(c) The magnetic field produced by a circular loop of electrical current (denoted here by *i*). Outside the loop, the field lines strongly resemble those of a bar magnet oriented vertically (i.e., perpendicular to the loop). Despite what's shown here (for clarity), all field lines are closed. Image downloaded from https://www.fossilhunters.xyz/solar-atmosphere/magnetic-field-data.html /fossilhunters.
(d) The strength of a magnetic field depends on the concentration of flux lines, i.e. the number of flux lines penetrating a surface, per unit area. The unit-area surface shown here (with white lines marking its edges, and seen in a nearly side view) experiences a much higher concentration over the poles than around the magnet's equator, indicating a much stronger field near the poles. Image credits as in Figure 10.1a.

are measured per unit area). As we see in Figure 10.1d, a relatively large number of flux lines penetrate the unit-area surface above the north pole of a magnet, as we would expect given the very strong magnetic force near the poles; on the other hand, the same unit-area surface placed to the side of the magnet's equator, where the field lines are spread out the most, will be intersected by fewer flux lines, corresponding to a weaker field there.

10.1.2 An 'Elemental' Description of Magnetic Fields, with Reference to the Earth

The magnetic field experienced at any location on Earth is a vector quantity, possessing both direction (along the flux line) and magnitude (as just noted, reflected in the local concentration of flux lines). To help us visualize its three-dimensional quality, we consider for the moment an imaginary Earth that contains a small but strong bar magnet at its center; for

simplicity, we will say the magnet is positioned with its dipole axis oriented along Earth's geographic axis (Figure 10.2a). Despite the great distance of the magnet from the surface, an observer standing right on the North or South Pole would measure a field that was completely vertical, pointing straight up out of the Earth or straight down into the Earth. An observer standing on the equator would measure a completely horizontal field, pointing straight from north to south along a meridian. And, an observer standing somewhere on this hypothetical Earth between equator and pole would see a field that has both horizontal and vertical components, but still points along a meridian.

For the *real* Earth, the magnetic field may differ from the preceding example in some fundamental ways, as we will shortly discover, but it is still a vector quantity. Like any vector, it can be described either in terms of its vector components or in terms of its magnitude and orientation. We will symbolize the field at

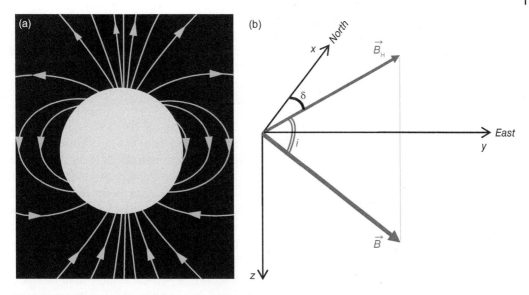

Figure 10.2 The magnetic field as a vector quantity.
(a) The field created by a hypothetical dipole at the Earth's center, as observed on and above the Earth's surface; this dipole is oriented north-south, the same as Earth's geographic poles, so it would be called an *axial* dipole. Note in this cross-section that an observer standing at the surface would say the field is entirely vertical at the poles and entirely horizontal everywhere along the equator; at other latitudes, the observer would measure both horizontal and vertical components of the field. Image credits as in Figure 10.1a.
(b) The field $\vec{B}$ at this location is drawn pointing somewhere between east and north, and points downward; its projection upward to the horizontal plane is $\vec{B}_H$. The orientation of the field is completely specified by the angles δ and *i*, the declination and inclination.
From basic geometry, we can relate the components of $\vec{B}$ to the observed quantities *B*, *i*, and δ. For example, to find the vertical component B_z of $\vec{B}$, we treat $\vec{B}$ as the hypotenuse of a right triangle, and note that $\sin(i) = B_z/B$.
To find the other components, you can focus on the right triangle involving δ, *whose hypotenuse is* $\vec{B}_H$, *and use the fact that* $B_H = B\cos(i)$.

a location of interest by $\vec{B}$, ignoring a further complication beyond the scope of this text-book: that there are two different kinds of magnetic field (this comes into play when considering the field *within* a magnet or other magnetizable material, according to whether the medium's magnetization is or is not included as part of the field; see, e.g., Lowrie, 2007). $\vec{B}$ is sometimes called the *magnetic flux density*.

At our location, we set up a 'local' *x-y-z* coordinate system, with the *x*-axis pointing toward Earth's North (geographic) Pole, which for the real Earth is not necessarily connected to any magnetic pole; the *y*-axis points eastward, and the *z*-axis points vertically *downward*, into the ground (see Figure 10.2b). Relative to that coordinate system, we can then specify the

field $\vec{B}$ according to its components B_x, B_y, and B_z; B_z is positive, for example, when the field at that location points into the ground, and negative if it points out of the ground.

Alternatively, we could focus on those quantities that are directly observable—the field's magnitude and direction. To that end, we combine the *x*- and *y*-components of the field into a single *horizontal* component, the vector $\vec{B}_H$; that component could also be defined (see Figure 10.2b) as the 'projection' of the field $\vec{B}$ onto the local horizontal or *x-y* plane. We then define two observable angles to complete our alternate description of the field. The **inclination *i*** of the field at our location is the 'dip angle,' the angle that $\vec{B}$ makes with the horizontal or, in other words, the angle between

$\vec{B}$ and $\vec{B}_H$. This angle can easily be measured with a compass by holding it vertically. The **declination δ** of the field is the angle between $\vec{B}_H$ and the *x*-axis, the latter pointing to true north, and can be measured while holding the compass horizontally.

In the hypothetical example pictured in Figure 10.2a, what is the declination observed at any location on the surface of that Earth?

Observers commonly measure *i*, δ, and the field strength *B* (i.e., the magnitude of $\vec{B}$). Using Figure 10.2b as a guide, these observables can be used to determine the components of $\vec{B}$, viz.:

$$B_z = B \sin(i)$$

$$B_y = B \cos(i) \sin(\delta)$$

$$B_x = B \cos(i) \cos(\delta).$$

Many observers use a simpler and more 'anglicized' notation not followed here: *D* rather than δ for declination and *I* for inclination (though *I* is also used to denote electrical current); *X*, *Y*, and *Z* (not to be confused with Cartesian coordinates *x*, *y*, and *z*) for B_x, B_y, and B_z; $\vec{H}$ instead of $\vec{B}_H$ (though that 'other' kind of magnetic field is usually symbolized by $\vec{H}$); and, occasionally, *F* in place of *B* for field strength. All seven of these quantities, however they are symbolized, are known as the **elements** of the magnetic field.

Other quantities helpful in describing magnetic fields include the locations of the **magnetic poles** (also called '*dip*' poles), which are places on Earth's surface where the field is vertical, i.e. where $i = \pm 90°$. As implied in our discussion of a hypothetical Earth whose magnetic field is due entirely to an axial dipole at Earth's center (as in Figure 10.2a), the orientation of a dipole source at depth can be inferred from surface measurements by finding the magnetic poles, that is, finding where the field inclination is vertical (if other magnetic fields are superimposed, as with the real Earth, further analysis will be necessary).

A Taste of Things to Come. We began this section with an assertion that the Earth's magnetic field was "very similar" to that of a bar magnet or dipole. Global observations, however, reveal complications on a number of levels.

- Measurements of the field show that the magnetic poles do not coincide with the geographic North and South Poles, unlike the hypothetical example shown in Figure 10.2a. This implies that the declination measured in most places will not be zero (compass needles will point toward the magnetic, not geographic, poles).

- Detailed measurements suggest that the field will be vertical at more than two locations (discussed later)—that there are more than two magnetic poles! Earth's field is not just a simple dipole.

- Measurements demonstrate that the field changes over time. These changes evidently take place on short and long 'human' timescales, and on short and long geological timescales. Unlike Earth's gravity field, which also varies over time but whose variations typically require modern instrumentation to detect, the most basic of the changes in Earth's magnetic field have been known for centuries. For example, some of the earliest systematic measurements of field properties in European cities found that the declination and inclination both vary, on timescales of decades to centuries; see Figure 10.3.

10.1.3 Magnetic Fields: An Overview of the Earth System

We have seen that magnetic fields can be produced by 'rock magnetism' (like bar magnets), and also by electrical currents (such as that current loop in Figure 10.1c). It follows that contributions to Earth's magnetic field will come from *every* component of the Earth System.

The Atmosphere. Different layers of the atmosphere contribute uniquely to Earth's

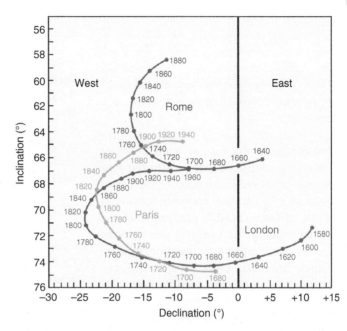

Figure 10.3 Changes in declination (*x*-axis) and inclination (*y*-axis) over time, measured in three European cities. In these locations, the declination appears to be oscillating (to the west then the east) while the inclination has been mostly decreasing. Image from British Sundial Society; adapted from http://sundialsoc.org .uk/old/Glossary/ap09-new.htm

magnetic field. Since we live in the troposphere, the most obvious to us is lightning—sporadic electrical currents—which produce locally strong magnetic fields. If our goal is to measure other components of Earth's magnetic field, this source can be minimized, simply by not taking measurements during lightning storms (!); since the life of a typical thunderstorm cell is less than an hour (e.g., NOAA, 2020), such a restriction is usually not much of an inconvenience.

Lightning can, however, interact with the ionosphere above, resulting in more global—though still transient—fluctuations in Earth's magnetic field. The stronger, and more visible, part of a lightning strike is the **return stroke**, a ground-to-cloud electrical discharge through the 'channel' of heated air that was created by the original cloud-to-ground strike a few moments earlier. The electrical disturbance created at cloud level by the return stroke may be large enough to perturb the ionosphere, creating changing electro-magnetic fields that produce radio-frequency electromagnetic radiation; this can also happen when there is cloud-to-cloud lightning. Guided by the Earth's magnetic field (as first explained by Storey, 1953), those perturbations may travel, as radio

waves, from one hemisphere to the other and back. Following the outer field lines (see the dipole field images in Figures 10.1 and 10.2), the waves can loop around high above before returning to the Earth.

And their detection is somewhat novel. Due to the slower speed of radio waves in the ionosphere, the waves experience dispersion—the same phenomenon we discussed in Chapter 7 in the context of seismic waves—so that different wavelengths travel at slightly different speeds. High above the Earth, they will encounter plasma trapped from the solar wind by our magnetic field (those regions are called the **Van Allen radiation belts**); that plasma also dispersively slows them down. Their spread-out arrival makes them sound to an observer kind of like 'whistling,' and they have been named **whistlers** (see also, e.g., Helliwell, 1965; Urrutia & Stenzel, 2018). Whistlers have been heard by space probes passing near Jupiter, Neptune, and most recently Saturn (Akalin et al., 2006), confirming the existence of their magnetic fields and ionospheres or plasma 'disks,' and providing definitive evidence that lightning occurs on all those planets.

The magnetic field created by a planet stays with that planet, as it orbits the Sun and, in particular, as it rotates on its axis. For a volume of plasma surrounding a planet, can you explain why that volume becomes 'flattened out,' eventually resembling (more or less) a disk shape?

Interestingly, in the case of Saturn, Akalin et al. (2006) concluded that the whistler they detected originated in a region of the planet in almost total darkness (possible because of the tilt of Saturn's axis), implying that the weather creating that lightning was powered by internal heat rather than heat from the Sun.

Other, less-exotic processes involving our atmosphere also produce globally detectable magnetic fields. Daytime heating by the Sun of the middle ionosphere (i.e., the layer at ~100 km altitude) causes it to flow, mostly within the sunlit hemisphere; the motion of its charged particles amounts to an electric current, creating a diurnal magnetic field known as the **solar quiet variation**, **Sq** (see, e.g., Campbell, 1989; Svalgaard, 2016; Yamazaki & Maute, 2017). Because of its ability to create a magnetic field, that flow of charged particles is sometimes called the **ionospheric wind dynamo**; in the thin upper atmosphere, that wind attains speeds of 30 m/s or more (e.g., Richmond, 1995). Measured at different locations on Earth's surface, the amplitude of Sq is typically a tenth of a percent of Earth's total field; that amplitude includes a substantial boost by coupling to thermal tides in the lower atmosphere, and a slight boost from gravitational tides (lunar and solar) in the lower atmosphere. The variation of Sq with latitude is also enhanced near the (magnetic) equator, due to a strong current in the ionosphere known as the *equatorial electrojet*.

Numerically, the magnitude of Sq is in the range ~10–100 nanoTeslas (nT) (Yamazaki & Maute, 2017), where 1 nT $\equiv 10^{-9}$ Teslas, and 1 Tesla $\equiv$ 1 volt-sec/m^2 is the appropriate unit for the magnetic field we are discussing (the equivalent cgs unit is Gauss, where 1 Gauss $\equiv 10^5$ nT). For comparison, Earth's total field

at the surface is $\sim \frac{1}{2} \times 10^5$ nT or $\frac{1}{2}$ Gauss (e.g., Thébault et al., 2015).

Magnetic storms are interactions between Earth's magnetic field (extending well above the Earth) and magnetically charged plasma previously ejected from the Sun (e.g., during a solar flare). Viewed from an astronomical perspective, the magnetic field enveloping our planet is sometimes called the *magnetosphere*; magnetic storms can thus be defined as interactions between the magnetosphere and the solar wind, especially when the latter is intensified by solar flares. Those interactions cause sharp fluctuations in the field measured at Earth's surface; produce the aurora ("northern lights" and "southern lights") as a by-product; and, by the changing electromagnetic fields, can even disrupt flow in the ionosphere (e.g., Huang et al., 2005; Kikuchi et al., 2008; Yamazaki & Maute, 2017). Humboldt, over two hundred years ago, was apparently the first to associate auroras with the perturbations to surface magnetic fields—perturbations he called "magnetic thunderstorms" (Kamide & Maltsev, 2007).

The magnitude of a typical magnetic storm might be several hundred nT or so (Yamazaki & Maute, 2017), much larger than the daily quiet variation but nevertheless only a few percent at most of the total field. Figure 10.4 shows measurements of the horizontal component (B_H) of the total field at the surface of the Earth over the span of one month; B_H is particularly sensitive to magnetic storm events. Superimposed on the 'main' field, whose horizontal-component magnitude is ~10^4 nT at that location, we see both the daily Sq variation and some magnetic storms. The timescale of a storm is a few days or so, with each storm developing within a day and dying off after a couple more days; the Sq signal is most clearly seen on the truly quiet days between storms (which is why it is called the "quiet variation").

To the extent that we can isolate the atmospheric component of Earth's total field, and then separate daily and storm contributions, magnetic field measurements can tell us about

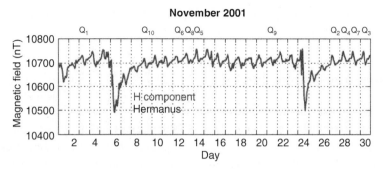

Figure 10.4 Magnetic field measurements in Hermanus, South Africa. The horizontal component shown here is typically of magnitude ~10,700 nT. Two storms were recorded during the month of November 2001—perhaps not a surprising occurrence, since 2001 was a solar maximum year. The first storm shown began and peaked during November 6, but recovery evidently took another day or two.
Days when storm influence was minimal are optimal for seeing the daily solar quiet variation, Sq; the clearest quiet days in this data set are indicated by $Q_1 - Q_{10}$.
Image from Yamazaki and Maute (2017) / Springer Nature / CCBY 4.0 / Public Domain.
Time series like this, but with some modification (one is called the *Disturbance Storm Time Index* or *Dst Index*, another the *Ring Current Index or RC Index*), are also used to monitor and highlight geomagnetic storm activity (see, e.g., Rich et al., 2007; Hamilton et al., 1988; Olsen et al., 2014).

the properties of the atmosphere, and how different layers of the atmosphere interact. For example, the magnitude of Sq depends on the electrical conductivity of the ionosphere, thus the extent to which its daylit side is ionized by solar radiation (a process that takes place mainly at "extreme" ultraviolet wavelengths; Svalgaard, 2016). The magnitude also reflects the ability of *some* tides in the troposphere and the ozone layer to couple more strongly to Sq than others do (Yamazaki & Maute, 2017, and references therein; also Richmond, 1995).

The magnitude of Sq depends on the 11-yr solar cycle as well: when the Sun is more active, it radiates more extreme ultraviolet, which increases the ionosphere's conductivity. Models of solar emission evaluated by Lean et al. (2003) show significant differences, but all roughly indicate doubled emission at solar max. Svalgaard (2016) finds that Sq—or, more precisely, yearly averages of the range of its daily variation—can provide a way to monitor long-term changes in solar ultraviolet radiation and thus sunspot activity; such long-term trends can be assessed to predict possible solar contributions to present-day climate change.

Still another interaction between atmospheric layers involving Sq is worth mentioning.

A **stratospheric sudden warming** is a rapid increase in polar stratospheric temperatures that causes a weakening of the polar vortex (in Chapter 3, we noted that ozone-destroying polar stratospheric clouds grow best in cold, still air—those are the conditions found *within* a polar vortex, away from the strong winds circulating around at its edge). Sudden warming events tend to occur in winter, once every year or two; the event in January 2009, as an example, featured a rise of about 50°C over the span of a week (Yamazaki & Maute, 2017). The warming spreads up through the ionosphere and beyond, and can intensify the tides, and even the equatorial electrojet, which had already amplified Sq.

The Oceans. In physics classes, when studying how 'dynamos' and other electrical generators work, we learn a basic principle of electromagnetism: if a loop of wire—or any other electrically conducting material—is moved through a magnetic field, electrical currents will be produced in the wire. Because seawater is electrically conducting (–think 'sea of ions' rather than 'sea of electrons'), its *motion* relative to Earth's magnetic field generates electrical currents within the oceans; those currents in turn create magnetic fields of their own. Seawater's conductivity is orders of magnitude lower than, say, that of iron or any other metal (though it is also at least an order of magnitude greater than that of the upper mantle); on the other hand, some components of its circulation are quite vigorous. The fields created by ocean currents might therefore be expected to be smaller than Earth's, but not negligible.

The oceans' electrical conductivity is a function of temperature (ions in seawater are more mobile at higher temperatures) as well as salinity, so it varies both spatially and temporally. Tyler et al. (2017) calculated its magnitude oceanwide from temperature and salinity measurements, finding spatial variations associated with freshwater influx at the coasts and with latitudinal variations in insolation; seasonal variations in both of these types of forcing also result in seasonal variations in the oceans' conductivity. Though quite measurable, and physically significant, however, all these variations in conductivity are approximately within an order of magnitude of the oceanwide average—a remarkably restrained variation, as noted by Tyler et al. (2017), compared to the more extreme range of conductivities deduced for other components of the Earth System.

The general circulation of the oceans creates a magnetic field; for a brief history of early efforts to predict and detect that field, or to use observations of it to infer the circulation, see Manoj et al. (2006a) and Minami (2017). The magnitude of this field is ~10 nT, as measured at sea level (e.g., Manoj et al., 2006a; Wardinski et al., 2007; Kuvshinov, 2008) (and, as with other contributions to Earth's total field, this component can add to or oppose the total, depending on its direction). The lumbering Antarctic Circumpolar Current, discussed in Chapter 3 and a part of that general circulation, produces by itself a field of magnitude 4 nT measured at sea level (Vivier et al., 2004). Western boundary currents can also produce measurable fields; Lilley et al. (1993) reported the detection of fields (magnitude 10–25 nT, measured at the sea floor) associated

with transient features of the East Australian Current (see Figure 3.14a, Chapter 3).

The oceans' tidal motions generate magnetic fields as well (see, e.g., Figure 10.5). Those fields can be predicted from tide models; however, given their distinctive frequencies, they can also be directly estimated from field observations (though for some tidal constituents, the contributions from Sq must also be accounted for, or the measurements restricted to nighttime—a procedure followed by Schnepf et al., 2014). At present, magnetic signals of the major semidiurnal and diurnal lunar tides have been identified (e.g., Grayver & Olsen, 2019; also references within Petereit et al., 2019); measured at sea level, those signals

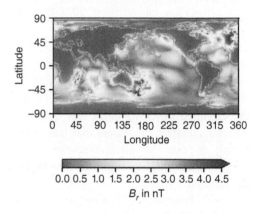

Figure 10.5 Magnetic field generated by the oceanic motions of the principal semidiurnal lunar tide; the field's amplitude is that experienced at sea level.

These predictions employed a model of the tide based on TOPEX altimetry data (Egbert & Erofeeva, 2002), determined ocean conductivity from observations of temperature and salinity, and averaged the predictions over the time span 1990–2016.

The field shown here is the vertical component of the tidally induced field; from a global perspective, the vertical component is better named the *radial* component.

Image by Petereit et al. (2019) / John Wiley & Sons, Inc.

The tidal magnetic field predicted here is weak in several regions, including the equator, and strongest in a few large regions and a number of smaller coastal areas. Can you identify which of those regions owes its characteristics to tidal dynamics? To the process of field generation?

typically amount to several nT (e.g., Kuvshinov, 2008; Petereit et al., 2019). Interestingly, as pointed out by both of the latter authors, the signals are evidently amplified—by an order of magnitude—in shallow-ocean (shelf) regions.

Magnetic fields are also created during tsunamis. Measured at sea level, the vertical component of $\vec{B}$ associated with the 2004 Sumatra tsunami would have been up to ~3 nT (predicted by Tyler, 2005, using sea-surface heights determined from Jason altimetry) or ~6 nT (predicted by Manoj et al., 2006b as reported in Kuvshinov, 2008). Tyler (2005) noted that the field perturbation would have been greater if the tsunami location had not been at the magnetic equator.

Toh et al. (2011) reported the detection of magnetic signals from tsunamis caused by shallow earthquakes in the northwest Pacific (Kuril Trench) in 2006 and 2007; both of the latter were great earthquakes, though not nearly as energetic as the giant Sumatra upheaval. The three components of the field, measured by seafloor magnetometers, were all 'pulses' (but with somewhat distinctive 'signatures') of magnitude roughly one or two nT, for both events. Toh et al. (2011) suggested that the signature of the vertical component, which they described as a 'bipolar' (downward-upward) peak, could be used to provide a slightly advanced warning of the tsunami's arrival—an idea that led to improvements in tsunami monitoring (Minami, 2017).

Minami (2017) reviews the detection of those and other tsunamis by both land-based and ocean-bottom magnetometers. The mechanism by which tsunamis couple to the lower atmosphere and the ionosphere, potentially facilitating that advanced warning, is discussed by Wu et al. (2020).

Finally, the 'dynamo' principle referred to above has a flip side: if a magnetic field is changing—so that its flux lines are moving, with respect to a conducting material—electrical currents will be generated in that conductor (in either version of the principle, there is relative motion between the conductor and the flux lines). Thus, even without any motion taking place within the oceans, changing magnetic fields within the Earth System will induce electrical currents (and therefore magnetic fields) in the oceans. According to Kuvshinov (2008), the oceans amplify the field perturbations of magnetic storms significantly in this way, with the oceanic contribution near the coast as much as ~50%; for Sq, the oceanic contribution is slightly less comparable (see also Svalgaard, 2016), but is also enhanced in coastal regions.

All of these magnetic fields provide opportunities to improve our understanding of the Earth System, spatially and temporally. The strong regional magnetic fields produced by features of the wind-driven ocean circulation like the Antarctic Circumpolar Current and the Kuroshio Current could be used to discriminate between circulation models, by comparing predictions from those models with observations (Manoj et al., 2006a); the comparison would be facilitated by using data from nearby coastal geomagnetic observatories. Discrepancies between observations and predictions of the fields generated by ocean tides have been suggested as a way to constrain spatial variations in ocean conductivity (Wardinski et al., 2009); alternatively, such discrepancies have been studied as a way to improve ocean tide models (Saynisch et al., 2018). Satellite observations may be preferred for such investigations (Tyler et al., 2003), since tidal magnetic fields are global in character; satellite data is discussed later in this chapter.

Magnetic fields produced by tsunamis have been used as tools to delineate their source region; for example, Ichihara et al. (2013) used ocean-bottom magnetometer data to determine that the tsunami generated by the great 2011 Tohoku earthquake off the coast of Japan actually originated about 100 km away from the earthquake source, about 1 minute after the quake began. This and other applications of tsunami magnetic fields are discussed by Minami (2017).

Because of the temperature dependence of the oceans' electrical conductivity, the possibility exists of inferring climate change from variations in oceanic conductivity over time. Much of the work in this regard has focused on the corresponding variations in the vertical component (globally speaking, the radial component) of the magnetic field induced by major ocean tides. Simulations by, for example, Saynisch et al. (2017), Petereit et al. (2019), and Irrgang et al. (2019) have shown that variations stemming from trends in temperature as the oceans evolve are small but detectable, confirming that magnetic field observations could be used as a proxy for ocean heat content to quantify global warming. Irrgang et al. also combined their model with recent field observations to construct tentative predictions, for a 2-yr span beyond the model's end, of the total heat content of the upper layers of the oceans; those predictions turned out to be within ~20% – 33% of what was observed subsequently for the oceans' uppermost 700 m, and within ~20% or better for the oceans between depths of 700 m and 2000 m.

Trossman and Tyler (2019) demonstrated that conductance is potentially even more useful than conductivity for inferring the heat content of the oceans, at least on an annual basis. **Conductance** is defined as the depth-integrated conductivity (or, equivalently, the conductivity averaged throughout the depth of a column of seawater, times the depth of that column). A depth-integrated approach makes sense because, on most timescales of interest, the electric currents induced by oceanic motion permeate the entire ocean column; in other words, the 'skin depth' associated with diffusion of electric currents and magnetic fields through the ocean layer exceeds the ocean's depth.

Later in this chapter, we will formally define and quantify electrical conductivity and electrical diffusivity; you will be able to calculate just how different the skin depths of the oceans and core (for example) are, on various timescales.

Trends in oceanic conductivity due to changes in temperature *and* salinity, caused for example by an influx of freshwater from Greenland ice sheet melting that weakens the thermohaline circulation, may also be detectable in tidal magnetic field observations (Saynisch et al., 2016). The trends predicted in the tidally induced field in this situation are toward greater intensity from warming temperatures but reduced intensity from freshwater influx; however, Petereit et al. (2019) have shown that the nature of those trends differs to a significant extent *regionally*, suggesting that ground-level, regional measurements may allow both kinds of trends to be detected and distinguished.

Regional investigations will certainly benefit from the amplification these induced magnetic fields experience in coastal regions. Interestingly, whether it is warming oceans or Greenland glacial melting, Petereit et al. (2019) and Saynisch et al. (2016) argue that such amplification is mainly a consequence of changes in electrical conductivity, not changes in the tidal dynamics in shallow seas: a given increase in heat, for example, can impact the conductivity of a shallow column of water more than a deep column.

The Crust and Mantle, Part I: Rock Magnetism. The primary contribution made by the *lithosphere* to Earth's magnetic field comes from rock magnetism. Our thought experiment with bar magnets, breaking them in half a number of times in an attempt to isolate a magnetic monopole (Figure 10.1b), could actually be continued down to the subatomic level, and we would find dipoles on the smallest scales. That is, a bar magnet—or any magnetized material—is really a collection of tiny, atomic or subatomic dipoles that have lined up, reinforcing each other's magnetic field.

Very simply, electrons within an atom *spin* on their axis as they orbit the atom's nucleus (a more rigorous view of the concept of electron spin is presented comprehensively by Sebens, 2019; see also Ohanian, 1986). The spin can be either prograde, so if the fingers of your right hand curled around in the direction of spin, your thumb would point *up* (counterclockwise, as viewed from above, like Earth's spin); or retrograde, in which case your thumb would point *down*. The laws of quantum mechanics require that no more than two electrons occupy an orbital region (leading, for almost all elements, to several energy levels or orbitals around each nucleus) *and* that, if an orbital has two electrons in it, their spin must be opposite.

The spin of an electron on its axis is essentially an electric current; like the current loop in Figure 10.1c, that spin will produce a dipole magnetic field. In effect, each electron is its own dipolar bar magnet.

Are you starting to see why monopoles do not exist?

If an orbital is filled—that is, if it contains an electron with an 'up' spin and an electron with a 'down' spin—then the net magnetic field produced by the electrons in that orbital will be zero. Alternatively, if a particular atom contains an unfilled orbital, i.e. that orbital has one electron, the field of that electron will not be canceled out; and, if the atom has more than one orbital with a single electron, the dipole field of each unpaired electron can, potentially, reinforce the others. In effect, each atom will be its own, potentially stronger, dipolar bar magnet.

If those atomic dipoles are strong enough, they can influence nearby atomic dipoles, with the outcome being a whole region—called a **magnetic domain**—of atomic dipoles aligned together, with their dipole fields adding. The alignment may occur spontaneously, or may be helped along by the presence of an ambient magnetic field. The end result is a magnetized material, with an induced magnetization that is very small but measurable. Domains may be the size of a small grain, and can shrink or grow over time in response to an external magnetic field; crystal defects and changes in mineralogy can inhibit their growth, but also prevent their collapse. When a strong magnetic field is applied, the domains can align, effectively merging to create a much more 'macroscopic' and perhaps even permanent magnetic field.

Iron, nickel, and cobalt are relatively common elements that can exhibit strong magnetic behavior. Iron has two unfilled inner orbitals—inner, meaning that even when an atom of iron is chemically combined with other elements, those orbitals will remain unfilled. The fields of the unpaired electrons are usually aligned (i.e., either both 'up' or both 'down'), creating a strong atomic dipole and ultimately producing a stronger, larger, and more long-lasting domain. This type of behavior, in which the atomic dipoles all align, is called **ferromagnetic**.

In compounds, spacing between ferromagnetic atoms can change their response to magnetic fields, reducing the overall magnetism. For example, in NiO, the atomic dipoles alternate in polarity, canceling out their neighbor's field; such **antiferromagnetic** minerals exhibit no permanent magnetic field of their own (Stacey & Davis, 2008). (There is a special subclass of antiferromagnetic minerals, including hematite, Fe_2O_3, in which the alternating dipoles are tilted—"canted"—rather than antiparallel; cancellation is incomplete, allowing a small permanent magnetization.) In **ferrimagnetic** materials, also known as **ferrites**, an alternation in dipole polarity occurs but the antiparallel dipoles are weaker than the parallel dipoles, resulting in a moderately strong permanent magnetization. The classic example of a ferrite mineral is *magnetite*, Fe_3O_4; rocks bearing naturally (spontaneously) magnetized magnetite are called **lodestones**—historically, serving as the earliest compass needles.

In contrast, the atomic dipoles in **paramagnetic** substances are too weak to influence their neighbors, leading to a random set of dipole orientations and no macroscopic magnetization. In the presence of a strong externally applied magnetic field, however, those orientations will become less random, producing a small net magnetic field—which will disappear as soon as the external field is gone.

Still another type of magnetism deserves mention, for conceptual completeness and exotic achievement. The rock magnetism we have been discussing is caused by subatomic electrical currents, as electrons spin on their axes. But the movement of electrons as they orbit the nucleus of an atom also constitutes an electrical current. These currents create dipole fields as well, and in response to an external magnetic field the orbits can be modified, aligning the dipoles and building a macroscopic magnetic field. This mechanism of producing a field is called **diamagnetism**. All substances respond to an external field diamagnetically, though to different extents; but in all cases, diamagnetic fields are weak, so if the constituent atoms contain unpaired electrons, ferromagnetism or ferrimagnetism will dominate.

Diamagnetic fields are opposite in polarity to the field that induces them; their opposition creates, in effect, a slight repulsion from that field—so, when the inducing field is that of a magnet, a repulsion from that magnet. Using this repulsion, it is possible to (weakly) levitate objects that would, at first glance, be considered 'non-magnetic,' against Earth's gravity. This includes small animals, apparently with no adverse health effects beyond a brief period of anxiety (Liu et al., 2010; Berry & Geim, 1997 and references therein), giving new meaning to the term "animal magnetism."

In sum, given the various types of rock magnetism, we can expect crustal (lithospheric) magnetism to be quite variable from place to place, depending on composition and especially iron content; geological factors such as degree of weathering and crystallinity can also be influential. Another crucial factor is *temperature*: at higher temperatures, the atomic dipoles are unable to stay aligned, a consequence of both their greater separation (from thermal expansion; Butler, 1992) and thermal agitation (as phrased by Stacey & Davis, 2008), and the rock's otherwise permanent magnetization disappears.

The temperature at which magnetization is lost—in effect, rendering those rock samples paramagnetic—is called the **Curie temperature**, often denoted T_C. Different substances can have quite different Curie temperatures; for example, for pure iron, $T_C = 1{,}043°$K ($\sim$770°C), whereas for magnetite, $T_C = 858°$K

($\sim$580°C). Other examples include nickel, with $T_C = 631°$K ($\sim$354°C), and hematite, with T_C $\sim$948°K ($\sim$675°C) (Kittel, 1968; Lowrie, 2007).

The existence of a Curie temperature has two consequences. First of all, it means that only the shallowest part of the solid earth can contribute to Earth's total magnetic field by rock magnetism; the rest of the solid earth is simply too hot. Following Chapter 9, we can quantify this from above and below.

- Temperature gradients measured in the crust at Earth's surface allow us to estimate the depth at which the Curie temperature is reached. On land, this depth is perhaps 20 km or so, making rock magnetism a mostly crustal phenomenon—mostly, because in a few places the continental crust may be thinner than that. In oceanic regions away from mid-ocean ridges, T_C is probably reached by a depth of $\sim$12 km below the sea floor, implying that rock magnetism can occur in the topmost oceanic mantle as well as the oceanic crust (the analysis by Dyment & Arkani-Hamed, 1998 supports this, finding that the crustal magnetic fields alone are too weak to explain observations).

- The olivine$\rightarrow$wadsleyite geothermometer yields an estimate of the temperature at 400 km depth; assuming an adiabatic gradient in the convecting mantle above that depth allows us to estimate the temperature just below the thermal boundary layer, that is, at the base of the lithosphere. For either 200-km thick continental or 100-km thick oceanic lithosphere (away from the ridge), that temperature is roughly 1,450°C. A Curie temperature of $\sim$600°C is thus reached at a depth of 40% of the lithospheric thickness, $\sim$80 km for continental and $\sim$40 km for oceanic lithosphere—implying substantial contributions from mantle rock magnetism in either case.

These estimates may be viewed as lower and upper bounds on the actual Curie depths in the mantle. After verifying the estimates presented here (using numbers as needed from Chapter 9), you should consider the basic assumptions underlying both types of estimates; why don't they yield the same estimates, and which do you think is closer to reality?

Clearly, rock magnetism in the uppermost mantle should not be discounted. Nevertheless, because of its fundamental shallowness, this component of Earth's field is often called the **crustal field**.

The second consequence of a Curie temperature is even more wide-ranging. The elimination of a rock's magnetization at temperatures exceeding the Curie point means that a rock heated to such temperatures, and then cooled down, will 'start over' when it has cooled sufficiently, acquiring a new permanent magnetization corresponding to the ambient magnetic field at the time of cooling (for an important example of this, see Figure 10.6). That newly acquired and permanent magnetization is called its **remanent magnetization**, or more specifically its **thermal remanent magnetization**; amazingly enough, that remanence could potentially (e.g. if the rock is not heated again) be retained stably for tens of billions of years (Stacey & Davis, 2008). Our ability to use such rocks to infer the characteristics of Earth's magnetic field in the past—the topic of *paleomagnetism*—will be discussed later.

The magnitude of the crustal field is typically around a few hundred nT, with variations perhaps several times stronger (e.g., Sabaka et al., 2002; cf. Whaler, 1994; Langel & Whaler, 1996; Langel, 1982). However, locally the field can be much stronger; after all, at some locations

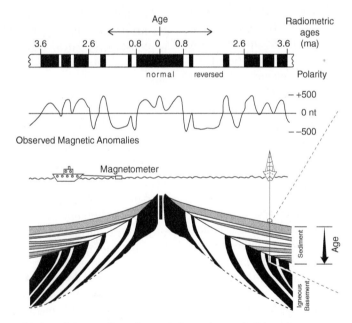

Figure 10.6 Magnetization of the sea floor. As newly formed oceanic lithosphere moves away from the mid-ocean ridge (bottom sketch), it cools (and subsides); when it cools below the Curie temperature, it acquires a permanent ("remanent") magnetization induced by Earth's magnetic field at that time (graph, with values between ±500 nT). As Earth's field varies, for example changing from 'normal' to 'reversed' polarity, and seafloor spreading continues, a record of those variations is preserved in the lithosphere, and can be detected by (e.g.) instruments towed behind a sea-going vessel.
The pattern of variations—also called *magnetic stripes* (top illustration, showing the top of the oceanic crust)—versus time depends on the rate of seafloor spreading as well as the history of reversals of Earth's magnetic field (see also Figure 8.3).
Can you explain how the alternating field (depicted with white or black) might produce the 'skewed' pattern of magnetization in the deeper crust shown in the bottom sketch?
Illustration from Hambach et al. (2008) / Copernicus Gesellschaft / CCBY 3.0 / Public Domain.

we may be quite close to sources of significant rock magnetism.

A couple of locations really stand out, one in Eastern Europe, the other in Central Africa; they are perhaps the largest and strongest crustal anomalies in the world. Curiously, though unrelated, they both evidently date to the Archean.

- The *Kursk Magnetic Anomaly* covers a large area north of the Black Sea, stretching from the Ukrainian Shield northeast toward Moscow. According to Taylor et al. (2014), it is associated with a massive iron ore deposit (mostly magnetite and hematite) including extensive banded iron formations (the most likely age of BIFs was discussed in Chapter 2). As summarized by Taylor et al. (also Taylor & Frawley, 1987; Ravat et al., 1993), the Kursk feature is associated with anomalous amplitudes approaching 2×10^5 nT for the vertical component! Not surprisingly—just think of that ore deposit as a (very) strong bar magnet—observers there find the inclination of the total field to be vertical, defining a third magnetic pole, albeit a more 'local' one (a homework exercise for this chapter explores such a circumstance quantitatively).

- The *Bangui Magnetic Anomaly*, covering a sizeable fraction of the Central African Republic, is several times the area of the Kursk anomaly (Girdler et al., 1992; Taylor, 2007), though comparable in intensity to it (cf. Regan & Marsh, 1982). The broad scale of this anomaly, as observed from aeromagnetic surveys, suggests that it is associated with a source as deep as the lower crust (Green, 1976). The region is also characterized by a strong, −120 mGal gravity low (Boukeke, 1994, cited in Taylor, 2007).

    Regan and Marsh (1982) argued that the Bangui anomaly is due to an iron-rich plutonic intrusion that cooled and isostatically subsided (the subsidence would have produced the sedimentary basin we see there now, and the accompanying gravity low). Alternatively, the suggestion by Green

(1976), developed by Girdler et al. (1992) and Ravat et al. (2002; see also Taylor, 2007), that it is the result of an iron meteorite impact, has received some support from the analysis by Garai et al. (2006) demonstrating that the carbonado diamonds found there probably had an extraterrestrial origin. Carbonado diamonds, also known as "black diamonds," are ceramic-like aggregates of nanometer-scale diamonds containing a variety of metals and metal alloys; they are found only in the Central African Republic and a tectonically related region of Brazil, and were presumably deposited in the same massive impact event. The Bangui carbonados were radiometrically dated to the late Archean (Ozima & Matsumoto, 1997).

The nature of the crustal field—also called the **anomalies field**—is best understood from a global perspective. Questions of a global nature include the degree of randomness in the field versus large-scale patterns; what kinds of correlations exist with global geology or tectonic province; and the extent to which the crustal field is induced (i.e., magnetized by present-day fields that are external to the crust) versus remanent (magnetized by past fields, in rocks that have subsequently been carried around by plate motions). With the help of satellite data, and using the language of spherical harmonics, such questions will be explored later in this chapter.

The Crust and Mantle, Part II: An Exploration of Electrical Conductivity. Although rocks are generally mediocre conductors of electricity—in Chapter 9, mantle silicates were referred to as semiconductors—they are, with few exceptions, not complete electrical insulators; thus, changing external magnetic fields will induce electrical currents in the crust and mantle. In fact, if we falsely treat the lithosphere as an insulator, predictions of the magnetic fields induced by ocean currents will not agree as well with observations (Kuvshinov, 2008).

Just as we saw with heat conduction and temperature variations imposed at Earth's surface, electrical currents and magnetic fields diffusing through an electrically conducting medium will exhibit a skin effect; consequently, the longer the timescale of variation by an external magnetic field, the greater the skin depth: the deeper it will penetrate into an electrical conductor, and the deeper electrical currents will be induced. For example, according to Olsen (1998, 1999), variations on timescales of half a day penetrate ~400 km into the crust and mantle, whereas ~40-day variations can reach depths of ~1,200 km, and annual timescales reach ~1,800 km; see also Kelbert et al. (2009) and Grayver et al. (2017).

Of course, the induced electrical currents—also called **magnetotelluric** currents—create their own magnetic fields. Measurements of those fields, or of the currents themselves (if they are shallow enough for us to sense directly) or their associated electric fields, can be used to determine the electrical conductivity at various depths within the crust and mantle (e.g., Vozoff, 1990; Lizarralde et al., 1995; Olsen, 1998; Jones, 1999; Dobson & Brodholt, 2000; Eaton et al., 2009).

The magnetic fields originating in the oceans, troposphere, ionosphere, and magnetosphere vary on the right timescales to probe the crust and upper mantle; for shallow exploration, artificially created high-frequency fields broadcast from the surface can be employed in "controlled source" experiments (but see, e.g., Constable et al., 1998a and Hoversten et al., 1998).

To enable quantitative comparisons, we formally define electrical conductivity, provide a familiar context for it, and describe its units, in *Textbox A*. Conductivity, which we will denote by σ, is one of those material properties whose magnitude varies over a broad range (e.g., Waff, 1974)—as broadly even as viscosity (see Chapter 5). The electrical conductivities of metals are of course at the high end of that range; for example, $\sigma \sim 1 \times 10^7$ Siemens/m (S/m) for pure iron at STP (Stacey & Davis, 2008). In contrast, the conductivities of granite and basalt range from 10^{-6} S/m to 1 S/m (Lowrie, 2007; Glover, 2015; Chave et al., 1990; cf. Stacey & Davis, 2008), and that of quartz at STP can range as low as 10^{-18} S/m (Halpern & Erlbach, 1998), depending on the presence of impurities (e.g., Jain & Nowick, 1982), earning its reputation as an electrical insulator.

Textbox A: Electrical Conductivity and Ohm's Law

Electrically charged particles can exert forces on other charged particles—repelling forces if the particles have 'like' charges (both positive or both negative), attracting if the charges are 'unlike.' Analogous to gravitational forces, these electrical forces are proportional to the product of the charges and inversely proportional to the squared distance between the charges; this relationship is called Coulomb's law. And, also analogously, we can divide the force by the magnitude of the charge experiencing the force; the result—the electrical force per unit charge—is called the **electric field**. We will denote that field, which is a vector quantity just like any other force or field, by $\vec{E}$.

In response to an applied electric field, charged particles will move, if they can, in effect creating an electrical current. Electrical currents usually represent the flow of electrons but can include ions as well. How big the current is—how much *charge per second* flows past an observation point—depends on the abundance of charge and its mobility. Electrical conductivity is the material property which, according to those factors, determines how much of a current is created by a given electric field acting on a particular substance.

As with heat flow, though, electrical current flows through surfaces, so it makes sense to measure the current in terms of the charge flowing per second per unit area; unlike heat flow, however, the flow of charge per sec per unit area is called the **current density**, not the 'current

Textbox A: Electrical Conductivity and Ohm's Law (*Continued*)

flux' (i.e., the greater the amount of charge flowing across a unit area, the 'denser' the current is). The traditional symbol for current density is $\vec{J}$.

Yes, the current density is a vector too; can you say what its direction should be?

In symbolic terms, we have simply

$$\vec{J} \propto \vec{E}.$$

Introducing a proportionality constant, namely the electrical conductivity σ, to express just how strong a current is produced by $\vec{E}$, we can write this relation as

$$\vec{J} = \sigma \vec{E}.$$

This equation is actually the official version of Ohm's law!

Allowing for some familiarity and some ambiguity, we can develop the units for σ.

- Electrical current, measured by the flow of charge per sec (as noted above), is commonly expressed as *Amperes* or *amps;* the mks units for J are thus *amp/m²*. And, the accepted unit for electrical charge, the **Coulomb**, is defined as the amount of charge carried per second past a location by a 1 amp current (that is, 1 amp = 1 Coulomb/sec).

- With energy defined as the ability to do work, and work defined as force × distance, the units for force could be energy/distance or Joule/m. E must have units of force per charge; it therefore has the units of Joule/(Coulomb-m). Alternatively, if we define a **volt** as 1 Joule per Coulomb—a situation where the strength of the electric field requires 1 Joule of energy to be expended in order to move 1 Coulomb of charge against it through a distance of 1 m—then E has the units of volt/m.

- Putting it all together, the units for σ ($\equiv J/E$, from Ohm's law) must be amp/(volt·m). As we will see shortly, we can define 1 **ohm** as 1 volt/amp; the units for σ are consequently 1/(ohm·m), sometimes written as 1 mho/m. 1/ohm is called 1 **Siemens**; the standard unit for electrical conductivity is, at last, Siemens/m or S/m.

Once again, we see that the units for material properties can be cumbersome.

To obtain the more familiar version of Ohm's law, the version we learn in elementary physics classes, will take a bit of ... work.

If a charged particle is experiencing an electrical force, then we must do work to move the charge against that force; the work done is called the **electrical potential** energy, or—if the particle is a unit charge—the work done is called the electrical potential, and denoted V_{el}. That work reflects the strength of the field; more precisely, the work done per distance the particle is moved reflects the strength of the field; more concisely, the *gradient* of the work done reflects the strength of the field. And, at different locations, V_{el} and its gradient will vary, reflecting variations in field strength.

We have just verbally derived a key underlying relation: the gradient of the electric potential is proportional to (and, as it turns out, equals) the electric field! Of course, there is more to it: accounting for the directions of gradient and field, and the possible need for a minus sign; but—in this chapter especially—being able to avoid some mathematics may be appreciated. Our verbal derivation may sound reminiscent of the discussion in Chapter 6 of gravitational potential, equipotential, and gradients—and with good reason, since the relation between the gravity field and gravitational potential is quite analogous.

Denoting the electric potential by V_{el}, then, we can write Ohm's law as

$$\vec{J} = -\sigma \vec{\nabla} V_{el},$$

Textbox A: Electrical Conductivity and Ohm's Law (*Concluded*)

---

with a minus sign included so that energy is expended (work is done) when moving the charge against the field. This expression looks exactly like our equation for heat flow (Chapter 9)—the thermal conduction equation.

Finally, consider an electrical circuit, where the wire—whose conductivity is σ—has cross-sectional area A and total length L. The current flowing through the wire is I. We define the intrinsic **electrical resistivity** ρ of the wire by

$$\rho = 1/\sigma,$$

and the **resistance** R of the circuit by

$$R = \rho L/A$$

(e.g., Dai & Karato, 2009). As implied by this formula, the resistance of the circuit is effectively increased if its length is greater, that is, if charges need to be forced through a greater distance around the circuit; on the other hand, if the electric field driving the flow acts on a greater number of electrons or ions—due to an increase in A—then more charges will flow, and in effect the resistance is lower.

If the electric field driving the current is associated with a potential V_{el}, it is the *difference* ΔV_{el} in potential going around the circuit that represents the work done moving a unit electrical charge against the field through the distance L; or, moving *with* the field (as the current flows), the energy released is $-\Delta V_{el}$. The **voltage** V of the circuit is defined as that difference $-\Delta V_{el}$ in electric potential (the greater the voltage, the stronger the field driving the current). Now we can rewrite our 'gradient' version of Ohm's law in scalar (nonvector) terms:

$$J = -\sigma\,(\Delta V_{el})/L = \sigma V/L = V/\rho L$$

or, since the current density is I/A,

$$I/A = V/\rho L;$$

this simplifies to

$$V = IR.$$

With our earlier definition of ohm, what must the units for R be?

Based on geomagnetic observations ringing the north Pacific (also electric field measurements from abandoned submarine telecommunications cables stretching across the floor of the Pacific), Utada et al. (2003) inferred the conductivity to increase monotonically from the uppermost mantle through the transition zone, rising from ~10^{-3} S/m to around 1 S/m at and below the second transition; they also found the observations to be consistent with sharp jumps up in σ at the first and second transitions.

Below the transition zone down to depths of 1,000–1,500 km, many researchers, including Utada et al. (2003), also Egbert and Booker (1992), Schultz et al. (1993), Lizarralde et al. (1995), Olsen (1998, 1999) and Kelbert et al. (2009), estimated values of σ clustering around ~1 S/m. Going deeper, the analysis by Constable (1993) showed that σ increases (from ~1 S/m at the base of the transition zone) to ~2 S/m at depths of around 2,000 km; Olsen (1999) had the increase at such depths reaching values between 1 S/m and ~7 S/m.

Both Constable and Olsen cited high-pressure (diamond anvil) experiments by Shankland et al. (1993) on lower-mantle minerals yielding estimates of σ in that range and exhibiting a similar increase versus depth. Other high-pressure experiments, summarized by Dobson and Brodholt (2000), also indicated such magnitudes and trends.

This overall trend in σ from top to bottom of the mantle is reasonable, and may be plausibly accounted for by the increase in temperature with depth; but such an explanation for that trend glosses over variations worth noting—some, for what they might tell us about the Earth, others for what they tell us about the nature of electrical conductivity.

Those variations occur regionally as well as versus depth.

- For the lithosphere beneath the East Pacific Rise, the conductivity inferred by Baba et al. (2006) was an order of magnitude smaller than that by Utada et al. (2003) for the North Pacific; the conductivity determined by Baba et al. (2010) for the western Pacific was in-between. Of course, differences in lithospheric conductivity are not unexpected; for oceanic lithosphere, for example, the older and colder western Pacific lithosphere just east of the Mariana Trench is less conductive by a factor of three than the younger Philippine Plate (technically, the Mariana microplate) under which it subducts—and it remains less conductive as it subducts, down to at least 200 km depth (Baba et al., 2010). At lithospheric depths, the conductivity of the northeast Pacific (Lizarralde et al., 1995) is similar to (Schultz et al., 1993) or even greater than (Eaton et al., 2009) that of the Canadian Shield, but still an order of magnitude lower than the lithospheric conductivity, ~10^{-2} S/m, in the U.S. Southwest (Egbert & Booker, 1992) and in Europe (Olsen, 1998).

- Just below the lithosphere, however, the variations in conductivity challenge our expectations. For reasons we will discuss shortly, the asthenosphere—the seismic low-velocity zone—is often viewed by researchers, especially those focusing on continental regions, as an *electrical asthenosphere*, a layer of enhanced electrical conductivity (see Jones, 1999 for a brief history). But the enhancement is uneven; the preceding authors, for example, find σ below the lithosphere to be similar for the northeast Pacific and U.S. Southwest, ~10^{-1} S/m, but at least an order of magnitude smaller below the Canadian Shield and Europe. And in some cases, for example in the northeast Pacific (Lizarralde et al., 1995), the enhancement, though the right magnitude, is too deep to line up with the zone of low seismic velocities. Our discussion below describing the factors affecting σ should shed light on some of the reasons for these mismatches.

- Regional variations in conductivity extend at least through the transition zone as well. Within the transition zone, the conductivity may vary by an order of magnitude from one part of the Pacific to another (Baba et al., 2010); and, as shown by Kelbert et al. (2009), above the second transition σ varies from region to region around the world by more than an order of magnitude.

Conductivity as a Material Property. In most minerals, electrical conductivity proceeds by the motion of electrons—as they 'hop' from ferrous to ferric ions (electron hopping, which can be described by $Fe^{+2} \rightarrow Fe^{+3} + e^-$ (Xu et al., 1998), is also called *polaron conductivity*)—and by the migration of ions (also called proton conduction) (Karato, 2011). Mathematically, these mechanisms are described by exponential terms where the exponent depends on the appropriate 'activation energy' and on the inverse temperature (similar to the way we represented mantle viscosity in Chapter 8). It is not surprising, then, that electrical conductivity in silicates is relatively sensitive to temperature; to mineralogy, especially iron content (Verhoeven et al., 2009, Li & Jeanloz, 1991a; 1990); to water (hydrogen) content (Li & Jeanloz, 1991b; Karato, 2011); and to how oxidizing the environment is (more precisely, to the oxygen *fugacity*) (e.g., Xu et al., 2000). The presence of partial melt may also be important, either directly (e.g., Baba et al., 2010) or indirectly (see Karato, 2011).

The dependence of σ on these factors helps us appreciate why estimates of conductivity can vary so much from region to region. High-*P,T* experiments and geochemical modeling suggest their potential impact on σ.

- For olivine in the upper mantle, as shown in Karato (2011), the addition of 0.01% (by weight) water increases the electrical conductivity by almost two orders of magnitude, compared to olivine under dry conditions.

- In the lower mantle, changing the temperature gradient from adiabatic (∼0.3°C/km, as discussed in Chapter 9) yields a nearly proportional change in conductivity, according to the analysis by Verhoeven et al. (2009): when the gradient is increased by a factor of 2.3, the conductivity increases by a factor of ∼2.2; when the gradient is reduced by a factor of 2/3, the conductivity decreases by ∼0.8.

- A temperature change may also lead to a change in phase. If the temperature is increased to the liquidus and beyond, so that the rock is completely molten, the mobility of its constituent particles will be greater, and its conductivity will be much higher. The degree of mobility, and which ions carry most of the charge, depend on composition; as an illustration, Gaillard and Marziano (2005) cite research on silicates showing an increase from ∼10^{-10}–10^{-3} S/m for solid crystals to ∼10^{-2}–10 S/m for magma.

  For a partially melted rock, there is still a dependence on composition, including (for a multicomponent rock) the composition of the melt versus that of the 'matrix' (e.g., Roberts & Tyburczy, 1999). Additionally, the ability of electrical currents to flow will depend on how 'connected' the melt inclusions are—thus, the degree of melting, the shapes of the inclusions, and so on; isolated

pockets of melt are not an effective way to raise the conductivity (e.g., Waff, 1974).

- When the lower-mantle iron concentration increases from a nominal 10% to 14%, Verhoeven et al. (2009) find the lower-mantle conductivity increases by a factor of four; but when it decreases to 6%, the conductivity becomes eight times smaller.

  However, the impact of iron content depends on which iron minerals are involved. Li and Jeanloz (1990), exploring iron abundances in (Mg, Fe)O magnesiowüstite (or ferropericlase) experimentally at lower-mantle conditions, found an almost exponential dependence of σ on iron content; their subsequent work (Li & Jeanloz, 1991a) also demonstrated that, at lower-mantle conditions, magnesiowüstite/ferropericlase is intrinsically more conductive than (Mg, Fe)SiO₃ perovskite. Thus, the impact of Fe depends on which minerals it is incorporated into; at lower-mantle conditions, magnesiowüstite/ferropericlase with 10% Fe, perovskite with 25% Fe, or a blend with 20% Fe, all have about the same conductivity, ∼10 S/m.

- Mineralogy indirectly compounds the issue. Dai and Karato (2009) showed that in the upper mantle the presence of pyroxene can lead effectively to an enhanced conductivity, compared to olivine alone, because it allows a higher water content.

  Xu et al. (1998) concluded that, under lower-mantle conditions, a minor amount (less than 3% by weight) of alumina (Al_2O_3) present in perovskite can increase its conductivity by a factor of more than three—presumably because its incorporation would have led to an increase in the fraction of Fe^{+3} in the perovskite.

The dependence of σ on all these factors points to the value of using conductivity to learn about Earth's interior; but it also implies that the interpretation of a particular estimate of σ in terms of these factors will probably be nonunique. For now, we briefly explore two applications that relate to the principal theme of this textbook.

Case Study #1: The Asthenosphere; Convection may be Universal, but is Plate Tectonics? Convection in Earth's mantle is closely tied to plate motions; those motions, in turn, are possible because lithospheric plates ride on top of the weak layer we identify as the asthenosphere. As we saw in Chapter 9, the Rayleigh number tells us that, for a body subjected to significant (i.e., superadiabatic) temperature differences, the possibility of it convecting increases greatly as its size increases; there is little doubt, then, that some moons and many planets have experienced internal convection at some point in their lives. But plate tectonics is a very different story.

In Chapter 8 we discussed possible reasons for the weakness of the asthenosphere, one of which was the presence of small amounts of water; the water would either be stored in hydrated minerals or dissolved in partially melted rock (Hirschmann, 2010). If water is the main reason for the existence of the asthenosphere, then 'dry' worlds would not exhibit plate tectonics. For this case study we explore whether determinations of electrical conductivity can shed light on what makes the asthenosphere weak.

The conductivity of the asthenosphere has been found to be relatively high, ~0.1 S/m, below the East Pacific Rise and extending a few hundred kilometers to the east (Evans et al., 2005; also Baba et al., 2006); the conductivity is also apparently *anisotropic*, with σ measured in the direction along plate motion up to roughly an order of magnitude greater than σ measured perpendicular to that. Naif et al. (2013) obtained similar results for the asthenosphere beneath the Cocos plate, 3,000 km to the north of the site studied by Evans et al., though with somewhat reduced anisotropy (σ is only twice as great in the direction of plate motion).

The higher conductivity of the asthenosphere is not likely to be the result of higher temperatures, even if its low strength is: as noted by Dai and Karato (2009), the variations in σ observed from region to region would imply an unreasonably high temperature variation. Instead, Dai and Karato (2009) (as well as Evans et al., 2005 and Baba et al., 2006, also Hirth et al., 2000) attribute the enhanced conductivity of the asthenosphere primarily to its water content (at least ~0.01% by weight), consistent with our discussion in Chapter 8 pointing to water as a key factor in making the asthenosphere so weak. In this case, the regional variations in asthenospheric σ would derive from variations in water content.

Furthermore, in Chapter 8 we also discussed the anisotropy in seismic velocities and implied anisotropy in crystalline 'fabric' associated with flow away from mid-ocean ridges; the latter would include but not be limited to a *lattice-preferred orientation* of olivine crystals, as they deform under flow-related stresses (e.g., Evans et al., 2005). With water as the basis for the high conductivity, anisotropy in σ would not be unexpected, whether from aligned hydrated minerals or from water (and silicate melt) along grain boundaries. If high temperatures were the direct cause of the high conductivity, it is unclear how conductive anisotropy would result.

The high asthenospheric conductivity has also been attributed to the presence of partial melt (e.g., Shankland & Waff, 1977); in fact, as discussed in Chapter 8, small amounts of water in the asthenosphere (less than 0.01%, according to Hirth & Kohlstedt, 1996) will promote partial melting there. Naif et al. (2013) envisioned what they termed a "damp" asthenosphere (less than 0.03% water), able to produce enough partial melt to account for the enhanced conductivity of the top ~25 km of the asthenosphere beneath the Cocos plate.

To affect the electrical conductivity, the pockets of melt must be connected. The strain-induced anisotropy in mineral fabric, amplified by shear flow within the asthenosphere, could be expected to lead to parallel channels of melt, resulting not only in enhanced conductivity but anisotropy as well (e.g., Caricchi et al., 2011; also Naif et al., 2013).

Perhaps the most intriguing variation on the theme of partial melt in the asthenosphere, advanced by Gaillard et al. (2008), begins with small amounts of carbonate in the mantle; ambient conditions would lead to carbonate melt, called **carbonatite**, dispersed beneath and away from mid-ocean ridges at depths exceeding ~80 km.

Could the asthenosphere maintain some amount of carbonate over the lifetime of the Earth? Check out our discussion of the Urey cycle in Chapter 2, in particular the portion where we implied the cycle might 'leak' into the Earth's interior….

Gaillard et al. (2008) found that carbonatite is highly conductive—two to three orders of magnitude more so than silicate melt, based on measurements at asthenospheric temperatures—and tends to spread out along grain boundaries; the latter 'wetting' property ensures that its connectivity will be high, and also implies it will take on the anisotropy of the mineral fabric seen around mid-ocean ridge environments. They noted that the fraction of carbonatite melt required to explain asthenospheric conductivity was consistent with the CO_2 content of mid-ocean ridge basalts.

Hoping to narrow down the most likely causes of the electrical asthenosphere, and recalling that the asthenosphere is also the seismic low-velocity zone, we can try imposing seismological constraints. Perhaps the simplest would be that the conditions producing partial melt or mineral hydration exist at the depths where seismic velocities actually decrease. By modeling the stability of partial melts under uppermost mantle conditions, Hirschmann (2010) found that carbonatite melt is less likely to exist in the hotter asthenosphere beneath young oceanic plate environments; in other words, the depths where it would occur are not where the major seismic velocity decreases of the LVZ are seen. He attributed the high asthenospheric conductivity instead to silicate melt, with a relatively large fraction concentrated at the top of the asthenosphere.

And, looking at the seismic velocity contrast between the lithosphere and asthenosphere in the western Pacific, Kawakatsu et al. (2009) found it could be explained by partial melt, which would also produce a higher asthenospheric conductivity, but not water content, which in their model yields too small a velocity contrast. Kawakatsu et al. postulated a layered melt structure, so the effect of the melt on velocities would not be too great (cf. Shankland et al., 1981).

A seismic perspective adds further complications. As pointed out by researchers involved in the "Mantle Electromagnetic and Tomography" ("MELT") project (MELT Seismic Team, 1998), melt channels or wet grain boundaries aligned in the direction of plate motion (though producing anisotropically enhanced conductivity) should cause lower seismic velocities in that same direction; but the observed seismic anisotropy is the opposite (Wolfe & Solomon, 1998; Forsyth et al., 1998). This may suggest that the anisotropy is crystalline (lattice-preferred orientation of hydrated minerals; Evans et al., 2005); alternatively, it may be that the 'contrary' anisotropy actually occurs in the lithosphere above (MELT Seismic Team, 1998).

Is water, partial melt, or another factor most responsible for both the enhanced conductivity of the asthenosphere *and* the seismic LVZ? Utada and Baba (2014) conclude that it "is still an open question." At the introductory level of this textbook, perhaps a better answer is, simply and unfortunately, "yes."

Case Study #2: Mid-Mantle Electrical Conductivity and the Scale of Mantle Convection. In Chapter 8 we learned about the debate over whole-mantle versus upper-mantle (or layered) convection; the evidence that we considered included estimates of lower-mantle viscosity, the existence of whole-mantle plumes, and the geochemical properties of the transition zone. Additional evidence was mentioned in Chapter 9, involving the thickness of the second transition. Overall, this evidence implied that whole-mantle convection was likely; but the endothermic nature of the second transition increased the possibility that convective downwelling through that transition is inhibited or even suppressed.

Mantle conductivity provides more fuel for the debate. Dobson and Brodholt (2000) posed the hypothesis that a thermal boundary layer exists at or below the base of the upper mantle—thus, that the upper mantle undergoes its own convection—and tested it through its implied effects on electrical conductivity. The presence of a thermal boundary layer (see Chapter 9) would result in higher temperatures than otherwise as the boundary is approached from above, and thus higher temperatures throughout the lower mantle; for the sake of comparison, Dobson and Brodholt envisioned temperatures higher by 1,000°C.

High-P,T measurements of conductivity in transition-zone minerals were extrapolated to conditions corresponding to whole-mantle versus layered-mantle convection, that is, to smaller and larger temperature gradients. With a larger temperature gradient, the higher temperatures would lead to higher electrical conductivities. Instead of focusing on conductivity, though, Dobson and Brodholt predicted the magnetic fields that would be induced by external fields on various timescales—thus, by fields penetrating to various depths—in a mantle with either conductivity profile. They then compared their predictions with observations from Constable (1993) (more precisely, the quantities compared, called *response functions*, were ratios of the vertical to horizontal components of the total field, i.e. the external forcing field plus the induced internal field, measured at Earth's surface). The results of their comparison were about as unequivocal as we could expect: if the lower mantle is comprised of perovskite, the lower-mantle conductivities require a hot mantle; if the lower mantle is perovskite with small amounts of alumina present, the conductivities require the cooler mantle.

The choice presented by these alternatives is consistent with the work of Xu et al. (1998) cited earlier, showing that the presence of alumina can significantly increase σ, rendering hotter temperatures unnecessary. Dobson and Brodholt argue that the scenario with alumina is the more likely one, pointing to the chemical analysis of diamonds by McCammon et al. (1997) implying that alumina is present, in small amounts, in the lower mantle.

However, other factors capable of affecting σ may also be at work in or below the transition zone—most obviously magnesiowüstite/ferropericlase, which we know the lower mantle should contain. From our extended discussion above regarding water in the asthenosphere, we might also wonder whether water is present lower in the mantle; in fact, research on mantle conductivity suggests there may be up to 1% water in the transition zone (Kelbert et al., 2009; Karato, 2011; Munch et al., 2018). High-P,T solubility measurements support this possibility, showing that significant amounts of water can be stored in lower-mantle (e.g., Murakami et al., 2002) and especially transition-zone (see, e.g., references in Bercovici & Karato, 2003) minerals.

Water in or near the transition zone would also promote the development of partial melt there—with particularly significant consequences, according to the transition-zone model proposed by Bercovici and Karato (2003). In their model the broad-scale, slowly upwelling mantle (to be distinguished from narrow, rapidly upwelling and relatively hotter mantle plumes) *dehydrates* as it rises through the first transition (wadsleyite→olivine, as discussed in Chapter 8); the water accumulating from this transition leads to a thin layer of partial melt at the top of the transition zone. The process of partially melting 'filters' or depletes the upwelling mantle material of some trace elements, thus potentially explaining differences in chemistry found between hotspot basalts (derived from those deep-mantle plumes) and mid-ocean ridge basalts (created by pressure-release melting in the shallow—and, as we now understand, depleted—upper mantle). Such a process removes the need to postulate upper- versus lower-mantle chemical reservoirs, isolated by layered-mantle convection, as a way of explaining the differences in basalt chemistry, and results in one less reason to reject whole-mantle convection (see also Hofmann, 2003).

Dobson and Brodholt (2000) acknowledge the possibility that magnesiowüstite/ferropericlase and small amounts of water might be present in the lower mantle but point out that, if anything, both would increase the conductivity, counteracting the need for higher temperatures and strengthening the conclusion that the lower mantle is not subject to the higher temperatures caused by a mid-mantle thermal boundary layer.

Our Exploration of Electrical Conductivity, Concluded. The extreme variability in σ seen in the crust and mantle extends to the atmosphere as well. The electrical conductivity of the atmosphere is around $\sim 2 \times 10^{-14}$ S/m near Earth's surface under fair-weather conditions (Rycroft et al., 2000; Siingh et al., 2007), but increases up to $\sim 10^{-9}$ S/m at a height of 65 km (Cho & Rycroft, 1998), mainly as a consequence of increased ionization with altitude from cosmic rays (e.g., Siingh et al., 2007). Above that height σ increases at a faster rate, reaching $\sim 10^{-2}$ S/m at a height of 100 km (Cho & Rycroft), thanks to abundant electrons produced by upper-atmosphere activity (including magnetic storms and whistlers) and fewer collisions between them and the surrounding gas in an increasingly 'thin' atmosphere (see also, e.g., Kokorowski et al., 2012; Svalgaard, 2016). A number of other sources of charged particles, and varying physical conditions (e.g. locally amplified magnetic fields), add significant variability to the conductivity.

The lower atmosphere's conductivity leads to some interesting situations. Though extremely low, the tropospheric conductivity nevertheless permits weak, ongoing electrical currents from the ionosphere to reach the ground; those currents are termed 'fair-weather currents,' as they exist remote from any thunderstorm. The currents are part of a so-called *global atmospheric electric circuit* maintained by worldwide thunderstorm activity, which drives electrical currents upward to the ionosphere; currents from ground to thunderclouds (including lightning return strokes) complete the circuit (Rycroft et al., 2000; Siingh et al., 2007).

The low ground-level conductivity also helps to turn the lower atmosphere into a *waveguide* for electromagnetic waves, in much the same way that seismic low velocities in the oceans' SOFAR channel created a waveguide for sound waves (see Chapter 7); the ionosphere serves as the waveguide's upper boundary. The waves are created by lightning. Unlike sound waves in the oceans, however, electromagnetic waves in the atmosphere can circle the globe without encountering any barriers; waves traveling within the earth-ionosphere waveguide in opposite directions can, therefore, constructively and destructively interfere (Chapter 7). The net result is electromagnetic *normal modes*, which are known as **Schumann resonances**; the fundamental mode has a frequency of ~ 8 cycles per second (Siingh et al., 2007; Price, 2016). With 500–1,000 thunderstorms active around the globe each day on average (Rycroft et al., 2000; Williams et al., 2000), numerous modes are excited more or less continuously.

Global warming might be expected to affect both the frequencies and the amplitudes of the Schumann resonances. Warmer tropospheric temperatures, especially in the tropics—where most thunderstorms and most lightning occur—should increase storm intensities and greatly increase lightning 'flash rates,' resulting in larger amplitude resonances (Williams, 1992; Price & Rind, 1994; Romps et al., 2014). Warmer tropical temperatures can expand the size of the lightning source region and shift the latitude of the effective 'center' of that region, modifying the frequencies of the Schumann resonances (Balling & Hildebrandt, 2000). Conceivably, the 'falling sky' effect mentioned in Chapter 3, in which warming of the troposphere is accompanied by cooling of the middle atmosphere and sinking of the ionosphere, would change the depth of the electromagnetic waveguide and thereby modify its frequencies as well. Tracking changes in the Schumann resonances—and accounting for changes in ionospheric height and storm intensity/lightning frequency due to other factors (e.g., Hudson et al., 2016; Sátori & Zieger, 1999; Blakeslee et al., 2014)—may thus be useful for studying global warming (see also Rycroft et al., 2000; Price & Asfur, 2006; Price, 2016).

Rounding out our survey of electrical conductivity in the Earth System: the oceanic conductivity, calculated from observed ocean temperatures and salinities, ranges from 0.1 S/m to 6.5 S/m along the sea surface, with a global, depth-averaged value of 3.3 S/m (Tyler et al., 2017); estimates of the conductivity of the core, based on experiments and theory for iron and iron alloys, have ranged from a low of $\sim 2 \times 10^5$ S/m (Stacey & Loper, 2007) to values more recently as high as $\sim 10 - 20 \times 10^5$ S/m (see, e.g., Yin et al., 2022; Xu et al., 2018 and Gomi & Yoshino, 2018, and references therein). The latter, higher estimates of σ, with implications according to the Wiedemann-Franz-Lorenz law (see Chapter 9) of an unusually high *thermal* conductivity for the core, have sparked a heated debate about core convection and the age of the inner core (e.g., O'Rourke & Stevenson, 2016; Pozzo et al., 2012; Yong et al., 2019; Olson, 2013).

The Core. The largest component of the Earth's magnetic field by far is that originating in the core; also called the **main field** or **geomagnetic field** (though those terms carry some ambiguity, as discussed below), its magnitude at the surface of the Earth is typically almost $\sim$50,000 nT or ½ Gauss. A contour map of its amplitude is shown in Figure 10.7, along with a map of the inclination that would be measured from it at locations around the world. These maps reveal basic similarities to a dipole field: the amplitude is maximum in polar regions, minimum near the 'equator' (as in Figure 10.1d); and the inclination flips from up to down as the 'equator' is crossed (see Figure 10.2a).

Looking more closely, however, we see that the amplitude maxima do not coincide with Earth's geographic poles; for that matter, it is not completely clear from Figure 10.7a which maxima would correspond to the poles of our dipole field. Evidently *the field we see in the figure is more than just a dipole field.* Considering both Figure 10.7a and Figure 10.7b, we can at least say that one of the poles is likely near the coast of Antarctica; but that suggests the other pole should be in northern Canada. And, if we drew a line through the minimum amplitudes around the world, it would be tilted with respect to the geographical equator but consistent with (and roughly perpendicular to) our inferred dipole axis. Evidently, the axis of the dipole field best representing Earth's geomagnetic field is tilted with respect to Earth's geographic axis.

Also, the inclinations in Figure 10.7b are positive in the northern hemisphere and negative in the southern hemisphere; that is, the geomagnetic field lines point downward into the Earth in the northern hemisphere and out of the Earth in the southern hemisphere. This is the opposite of what we see in Figure 10.2. Thus, the Earth's dipole field is upside down as well as tilted—which, to be clear, is contrary to what was illustrated in Figures 10.1 and 10.2 early in this chapter. This polarity, in which the geomagnetic south pole is near the geographic North Pole, is officially called the dipole's **normal** polarity; the alternative is known as the **reversed** polarity.

A mathematical analysis of the geomagnetic field is required to quantify the properties of the dipole field. Based on recent observations, the International Geomagnetic Reference Field for 2020 (Alken et al., 2021), known as *IGRF-13*, reveals that the dipole axis of the best-fitting dipole field is tilted by $\sim 9\frac{1}{3}°$; the intersections of this axis with Earth's surface—called the **geomagnetic poles**—are at (80.7°N, 72.7°W), where the field lines go into the Earth, and (80.7°S, 107.3°E), where they exit the Earth. In contrast, the principal magnetic poles, defined by $i = \pm 90°$ but excluding any 'local' poles, are at (86.5°N, 162.9°E) and (64.1°S, 135.9°E);

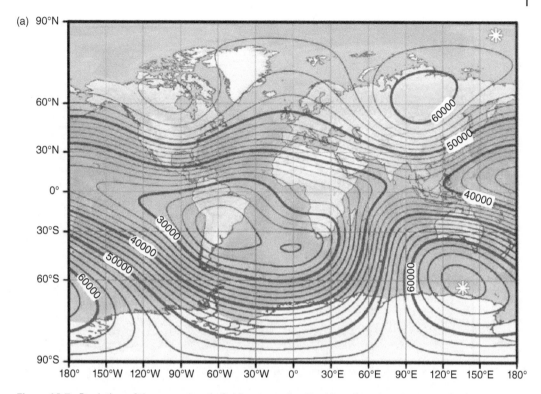

Figure 10.7 Depiction of the geomagnetic field measured at Earth's surface, based on a model of the field called IGRF-13. IGRF-13 is an edition of the *International Geomagnetic Reference Field* appropriate for years 2015–2020, and predictive through 2025; the field changes too much over time for that edition to remain very accurate after that.
(a) The intensity (magnitude) of the field during 2020, contoured at intervals of 2,000 nT.
Image from Alken et al. (2021) / Springer Nature / CCBY 4.0 / Public Domain.

note that, unlike the geomagnetic poles, the magnetic poles are not antipodal.

The dipole field we have just described constitutes the major component of the geomagnetic field; most of what remains, with an amplitude at Earth's surface about one-fourth as strong (e.g., Stacey & Davis, 2008; Langel & Estes, 1982), is called the **non-dipole field**. There are several reasons why we identify the geomagnetic field (dipole *and* non-dipole components) as originating in the core, or at least exclude other possible source regions.

- Distance: the strength of the **external** components of Earth's magnetic field (those components originating in the atmosphere or magnetosphere), and that of the field associated with the oceans, increase as we get closer to those source regions. At coastal

locations, we observe stronger oceanic magnetic fields; at higher altitudes, including satellite altitudes, we measure stronger ionospheric and magnetospheric fields. But the main field gets weaker as we rise above Earth's surface.

- Timescale: the components of Earth's field associated with the atmosphere, magnetosphere, and oceans vary on geologically instantaneous timescales—minutes, hours, and days—associated with external factors such as tidal forces and solar radiation. Variations in the main field that we measure act on timescales of years to decades, as seen in Figure 10.3.

Does our failure to detect variations in the main field on very short timescales imply that such variations do not exist? Can you think of

(b)

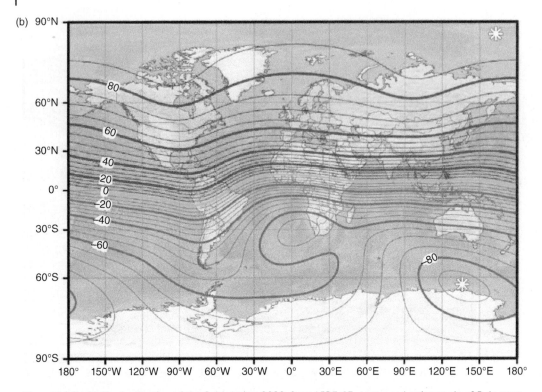

Figure 10.7 (b) The inclination of the field during 2020, from IGRF-13, contoured at intervals of 5 degrees. Asterisks indicate the locations of the principal magnetic poles (defined in the text). Image from Alken et al. (2021) / Springer Nature / CCBY 4.0 / Public Domain.

a way rapid field variations originating in the core might, somehow, get 'screened out' before 'diffusing' up to the surface?

On the other hand, the fact that we see significant variations in the main field on timescales of years tells us that it probably does not originate in the mantle, whose timescale for change is centuries to eons.

- Magnitude: the magnitudes of the dipole field and even the non-dipole field are, overall, much greater than the other components of the Earth's field (the only exceptions are a couple of regions with extreme crustal magnetization, as discussed earlier). Our understanding of how external and oceanic magnetic fields are generated would require unrealistically intense electric currents generated within them in order to produce a 50,000 nT field, and unrealistically intense crust and mantle magnetization worldwide

if those regions are the cause of the main field.

- Spatial scale: the broad scale of the geomagnetic field, where the highs and lows are separated by several thousands of kilometers (see Figure 10.7), is consistent with a deep source, one distant enough below the surface that we can sense only the overall pattern, not individual details. If the geomagnetic field had sources located in the upper mantle, we would expect to see highs and lows of those sources 'poking out' through the contour map.

The underlying idea—more precisely stated as *short-wavelength fluctuations in the field correspond to shallow sources*—was briefly discussed in Chapter 6 as a way to interpret Geoid undulations. Figure 10.8 illustrates how it holds true for the simple case of a dipole source buried below the surface.

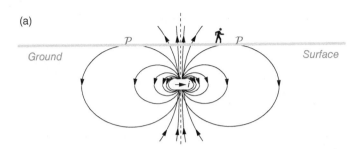

Figure 10.8 The 'spread' or wavelength of a magnetic field due to a buried dipole source reflects its depth of burial.

(a) In preparation for demonstrating that principle (in part (b) of this figure), we highlight the vector nature of the field seen here, due to a buried dipole, that we would experience as we walk along the surface of the Earth (green line). The dipole chosen for this illustration is from Figure 10.1c, and we show its field lines above *and* below the surface for better visualization.

At the surface, the field is vertical when we stand right above the dipole (on the dipole axis); mostly vertical, if we are slightly to the side of the dipole axis; increasingly tilted as we move further away from the axis; and finally, at point $\mathcal{P}$, completely horizontal.

Beyond point $\mathcal{P}$ —*you will have to draw the field lines, 'walk' to where they intersect the ground surface, and then identify their orientation*—the field lines point downward, more and more; of course, by this point we are getting pretty far from the dipole.

And at the same time, the field is getting weaker as we move to the side, as our distance from the dipole increases.

In sum, then, the *vertical component* of the field is maximum over the dipole source (the field is 'all vertical' there, and we are closest to the source); decreases as we move to the side, till it becomes zero (where the field is all horizontal); then starts to become increasingly negative (as the field points more and more downward) but is so far from the source it ultimately just approaches zero.

Do the curves in Figure 10.8b show this pattern?

If we removed the broad-scale dipole component of the main field from Figure 10.7, we would be left with the variations that made Figure 10.7 an imperfect dipole—an extra maximum, an irregular 'equator,' and so on. These features are still broad scale, though not as broad as the dipole field. It appears that the non-dipole field we observe originates in the core as well, but with its sources located in the shallow core.

Identifying the source region of the main field as the core leads to a fundamental clue regarding its origin: it is not produced by rock magnetism. This is true despite the core being mostly iron, and despite the fact (first demonstrated by Gilbert in 1600) that a uniformly magnetized sphere has the same dipole field as a bar magnet. Rather, our rejection of rock magnetism prevails because, at the depths of the core, the temperature most certainly exceeds the Curie temperature of the iron alloy

there. (And, to emphasize: even if permanent magnetization were possible at those depths, the intensity of magnetization required—for a sphere 3,000 km below us to produce a surface field of ~50,000 nT—would be unrealistically high.)

It follows that the geomagnetic field must be produced by electrical currents generated within the core. Given the high electrical conductivity and fluidity of the outer core, plus the likelihood that it is actively convecting, we can reasonably speculate that somehow, in the presence of a preexisting magnetic field, fluid motions in the core induce electrical currents such that the resulting magnetic fields reinforce the initial field. Such a 'self-excited' dynamo process will be explored in the last section of this chapter.

For convenience, though clearly not for realism, the source of the Earth's dipole field is often represented as a bar magnet, placed

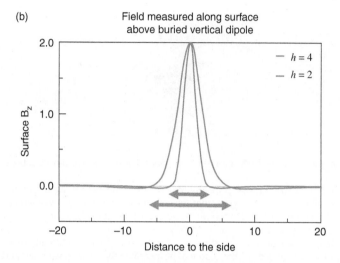

Figure 10.8 The 'spread' or wavelength of a magnetic field due to a buried dipole source reflects its depth of burial.

(b) This graph shows the vertical component B_z of the dipole field measured by an observer moving along the surface from right over the dipole to one side or the other, for dipoles buried at two different depths h. As demonstrated in (a), B_z drops to zero as the observer moves away from the dipole; then 'overshoots,' becoming negative; and approaches zero from below with increasing distance.

To see the overshoot more clearly, the strengths of the two dipoles are different, adjusted such that their fields have the same peak value.

The 'spread' or wavelength of the field can be defined as the lateral distance between zero-crossings; these are indicated in the graph by thick two-sided arrows (which do not represent the dipole orientation).

Evidently *the deeper source has a greater spread.*

The math underlying this illustration is developed in a homework exercise for this chapter.

upside down at the center of the Earth with its axis tilted appropriately relative to Earth's rotation axis. As a way of representing the field at the Earth's surface, Figure 10.7 suggests that this is not a bad first approximation.

Of course, we could replace the central bar magnet with an equivalent current loop—upside down (i.e., with the current traveling clockwise) and slightly tilted. But even with the correct orientation, and choosing a current loop rather than magnetized minerals, we would not be representing conditions in the core accurately at all. It is an oversimplification to imagine that Earth's dipole field is produced by a single current loop; instead, that loop is just a way to represent the total effect, at Earth's surface, of all the electrical current systems operating throughout the core.

Bar magnets—or, rather, current loops—could also be used to represent the non-dipole field's sources. There are two ways this is possible. First, as shown in Figure 10.9a, dipole sources located anywhere besides the center of the Earth—called **excentric** or, more commonly, **eccentric dipoles**—will create a field that lacks the symmetry of a dipole field, as seen from Earth's surface. Beginning with Vestine (1953), researchers investigating correlations between changes in the field and changes in Earth's spin rate have found it useful to approximate the geomagnetic field (dipole + the broadest-scale non-dipole components) by a *single* eccentric dipole. In the analysis by Chulliat et al. (2015), for example, the optimal eccentric dipole for 2015 was parallel to the centered dipole representing the dipole field, i.e. it had the same tilt, but was located almost 600 km away from the center of the Earth.

Earth's non-dipole field is sometimes approximated by a *set* of excentric dipole sources (i.e.,

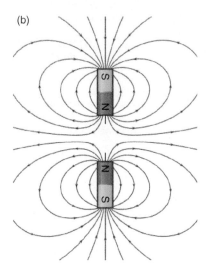

Figure 10.9 Sources of non-dipole magnetic fields.
(a) Current loops or bar magnets located
off-center—also called *eccentric dipoles*—produce
magnetic fields that are not symmetric, as
measured along the surface of the Earth; that is,
they produce fields that would be classified as
non-dipole.
The location of this hypothetical eccentric dipole
within the Earth is indicated by a faint rectangle.
Image credits as in Figure 10.1a.
(b) "Quadrupole" magnetic field created by two
opposite dipole sources (shown here as bar
magnets, but they could equally well be current
loops). These field lines are determined according
to the constraint that lines leave N and return to S.
As in Figure 10.1 and other illustrations of field
lines, this image is actually a cross-section; in
reality, the field lines here wrap around the
quadrupole axis.
Image downloaded from https://www.wikiwand
.com/en/Force_between_magnets.

current loops), dispersed throughout the outer core. The basis for this interpretation is discussed later in this chapter, in the context of the geomagnetic 'spectrum;' but we will also find that it is useful for understanding how the core sustains the geomagnetic field.

Second, combinations of adjacent bar magnets or current loops can produce distinctly non-dipolar fields; Figure 10.9b shows a **quadrupole field**, produced by a pair of bar magnets with opposite polarity that are aligned axially.

Try combining those two magnets instead by placing them side by side; do you get the same quadrupole field?

For even more excitement, try combining the pair of magnets representing a quadrupole with a single magnet next to them. How would you describe the field so produced? (Some would call it an octupole *field, by the way.)*

There is a lot that stands out about the geomagnetic field in addition to its impressive magnitude; one of its most amazing and important characteristics is that it varies in so many ways with time. Taking a mostly global perspective, we will discuss its spatial as well as temporal features in the next two sections.

10.2 Global Descriptions of the Internal Field

By now, we have identified a number of components of Earth's magnetic field (see Figure 10.10). In this section, we will learn how observations of the internal field can be analyzed in order to separate its core and crustal components. This will require a global rather than local approach, and is best achieved with the help of satellite observations and the use of spherical harmonics. The spherical harmonics employed here, a three-dimensional version traditionally called 'solid' spherical harmonics, will allow us to infer the magnitude and spatial patterns of the main field at other heights and depths, even right at the top of the core.

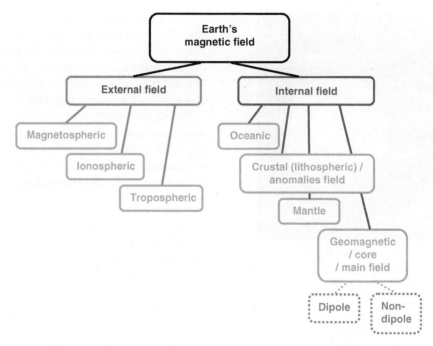

Figure 10.10 Magnetic field 'tree:' classification of the components of the Earth's magnetic field. These components can also be distinguished by their type of source; the source of the anomalies field, for example, is rock magnetism.

As discussed later in the text, it might be argued that distinguishing dipole from non-dipole components of the core field is less meaningful *inside* the core than outside it.

10.2.1 Satellite Missions Dedicated to Observing Earth's Magnetic Fields

Earth's magnetic field has long been observed from land, sea, and even air (global aeromagnetic measurements took off around 1953, thanks to NASA's *Project Magnet*; Langel & Baldwin, 1991). Land- and sea-based observations, because of their proximity to the oceans and to continental and oceanic crust, have been especially useful for recording the oceanic component of the field and for revealing the small spatial variations of the crustal field. In contrast, since the 1960s, satellites have provided truly global, consistent coverage (e.g., Olsen & Stolle, 2012), allowing us to obtain an accurate picture of the broad features of the internal field—and capturing its changes on short timescales. Moreover, their proximity to the magnetosphere and (in some cases) the ionosphere makes their observations critical for understanding Earth's external fields. All in all, satellite observations have led to a prolific increase in research on Earth's magnetic field (Hulot et al., 2010).

Satellite measurements of the Earth's gravitational field, as we saw in Chapter 6, could be described as 'inferential,' in that gravity (or the Geoid) was inferred either from satellite orbit perturbations or from altimetric measurements of sea-surface heights. Satellite measurements of the Earth's magnetic field, however, are direct, achieved simply through the presence of a magnetometer on board the satellite. The magnetometer can be *scalar*, in which case it measures the magnitude (intensity) of the field at its location, or *vector*, in which case three mutually perpendicular components of the field are recorded.

For determining the internal components of the field, satellite measurements during magnetospherically quiet days (see Figure 10.4) are preferred, with daytime measurements excluded to minimize the effects of the daily ionospheric currents (e.g., Finlay et al., 2015;

also Langel, Phillips, & Horner, 1982; Langel, Schnetzler, Phillips, & Horner, 1982). Even at night, though, the ionosphere may be quite active at high latitudes, as electrons drain from the Van Allen radiation belts into the ionosphere below, strengthening polar ionospheric currents (e.g., Maus et al., 2006a; Andersson et al., 2012; see also Aakjaer et al., 2016); consequently, measurements in polar regions are also excluded. By the way, the 'precipitation' of energetic electrons out of the radiation belts is driven by whistlers (Voss et al., 1984)—discussed earlier in this chapter—as well as by magnetic storms.

On the other hand, to determine the external components of the field, the core field—which dominates even at satellite altitudes—must first be removed. At satellite heights, its strength is typically ~40,000 nT; representative magnitudes of the other components at those heights are as follows.

- The range of crustal anomalies is typically ±20 nT (e.g., Purucker et al., 2002; Maus et al., 2002; Maus et al., 2006a; Olsen & Stolle, 2012). The Kursk magnetic anomaly is ~25–40 nT (peak to trough) (see Taylor et al., 2014; Taylor & Frawley, 1987), whereas Bangui is ~800 nT (Olsen et al., 2017). (Note that the values actually detected by the satellite may be inaccurate, especially if the satellite has sampled the field at a low spatial resolution as it orbits, and 'misses' the true high or low; see Maus et al., 2002 for an illustration of this in the case of Bangui.)
- Magnitudes of the various oceanic components of the field are ~1–2 nT for ocean tides and ocean general circulation (Tyler et al., 2003; Vivier et al., 2004; Manoj et al., 2006a; Kuvshinov, 2008; Sabaka et al., 2016; Grayver & Olsen, 2019; Petereit et al., 2019; see also Minami, 2017) but only ~0.2 nT for the field due to a large tsunami.

Earlier in this chapter, we saw that the fields due to all these ocean phenomena had similar magnitude at ground level. Why do you think the tsunami field attenuated significantly more with height than the other fields?

- Magnitudes of the external components of the field include ~10 nT for the ionospheric solar quiet variation Sq (Yamazaki & Maute, 2017) and ~20–50 nT for auroral electrojets (Stolle et al., 2016; Maus et al., 2006a); for the non-stormy magnetosphere, ~25–50 nT (Olsen & Stolle, 2012; Yamazaki & Maute, 2017).

With a solid spherical harmonic description allowing ground-based field models to be extrapolated to satellite heights, the core field can easily be subtracted from satellite data. Figure 10.11 provides an example of the stages of a satellite-based analysis focusing on the ionospheric field; as this figure suggests, removal of an accurate crustal field is also required for external field investigations—which is not unexpected given their comparable magnitudes.

The converse holds true as well. The sources of the external field are spatially and temporally complex, necessitating further refinements to those models (see, e.g., Egbert et al., 2021) in order to resolve features of the internal field. As an illustration, Maus et al. (2002) noted situations where estimates of the crustal field, inferred as the residual from satellite observations after core and external field models had been subtracted, were rendered highly if not completely uncertain due to unmodeled additional sources of the magnetospheric field. Subsequent data processing was able to reduce the effects of the latter somewhat.

Langel (1979) provides a tabulation of the earliest satellites used to measure Earth's magnetic field. Satellite missions that have had a major impact on research concerning Earth's field are listed in Table 10.1; some are illustrated in Figure 10.12. The Polar Orbiting Geophysical Observatories (POGO) series of six satellites included three lower-altitude satellites, preferred for studying the internal field. The Swarm mission is comprised of three satellites, two of which have side-by-side orbits with a maximum separation of ~170 km, while the third is at a slightly higher altitude. All satellites listed in Table 10.1 have near-polar orbits.

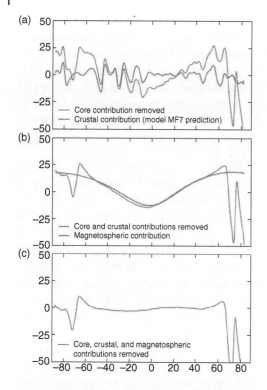

Figure 10.11 Components of Earth's magnetic field measured by the CHAMP satellite nightside during one orbit in 2010. Field intensity (in nT) plotted versus 'dipole latitude,' that is, latitude (in degrees) with respect to the dipole field (for example, a dipole latitude of 0° corresponds to the dipole's magnetic equator).
Analysis and images by Olsen and Stolle (2012) / Annual Reviews.
(a) Gray curve: observed field intensity minus the core field intensity. Blue curve: crustal field intensity according to model MF7, http://www.geomag.us/models/MF7.html, constructed in 2010 by Maus.
(b) Gray curve: top panel's gray curve minus blue curve, that is, observed minus core and crustal fields. Blue curve: modeled effects of magnetosphere.
(c) Gray curve: middle panel's gray curve minus blue curve, that is, observed minus core, crustal, and magnetospheric fields. Note that even during nighttime, there are strong external field contributions at high latitudes.

Table 10.1 is not all-inclusive, omitting a few satellite missions that have focused on the magnetosphere and its interactions with the solar wind. For example, the Magnetospheric Multiscale Mission, a set of four satellites,

was launched by NASA in 2015 (e.g., Burch et al., 2016; see https://sci.esa.int/cluster for still another mission). With an orbital altitude ranging from ~2,500 km to as much as ~153,000 km as it studies the far reaches of the magnetosphere, it is more properly described as a *solar-terrestrial* probe.

As with ground-based observations, satellite measurements can be used to infer the electrical conductivity of the Earth System; because of the skin effect, the results likewise depend on the periods of the signals being analyzed. Using data from the four most recent missions listed in Table 10.1 as well as from magnetic observatories, Püthe et al. (2015) used magnetospheric signals of periods 1 ½ days and longer to determine σ through the entire mantle; they obtained similar results to what we outlined earlier in this chapter for the topmost and lower mantle but found most of the upper mantle, from the asthenosphere through the transition zone, to be more resistive (thereby resulting in a sharp 'return' to higher conductivity at ~900 km depth). In contrast, with the same data sets except Swarm, Grayver et al. (2016) used magnetic signals from semidiurnal ocean tides to constrain conductivity in the upper mantle, finding some evidence for a jump in σ by two orders of magnitude at the ocean lithosphere-asthenosphere boundary. With a timescale smaller by a factor of 3 compared to Püthe et al. (2015), their work was evidently more sensitive to upper-mantle conductivity.

Using Swarm as well as Ørsted, CHAMP and SAC-C data, a simultaneous analysis of tidal *and* magnetospheric signals—thus, involving both semidiurnal and 1½-day timescales—allowed Grayver et al. (2017) to constrain both upper- and mid-mantle conductivities. Their results (see Figure 10.13) confirm the conclusions of Grayver et al. (2016) for the upper mantle and Püthe et al. (2015) for the lower mantle, and also reveal an increase in σ either steadily through the transition zone or at the second transition. They also show that reasonable amounts of water in the upper mantle and transition zone can account for their inferred conductivity profile.

Table 10.1 Satellite missions with a major focus on Earth's magnetic field.[a]

Mission	Active years	Country/agency of origin	Instrumentation[b]	Orbit Inclin.	Orbit Altitude (km)
POGO		NASA	Scalar		
OGO-2	1965–1967			87.4°	414–1,510
OGO-4	1967–1969			86.0°	416–900
OGO-6	1969–1970			82.0°	413–1,077
Magsat	1979–1980	NASA	Scalar, vector; star imager	96.8°	352–561 ($\rightarrow$ 190)[c]
Ørsted	1999–2014	Denmark	Scalar, vector; star imager; GPS	96.5°	640–880
CHAMP	2000–2010	Germany	Scalar, vector; star imager; GPS	87.2°	454 ($\rightarrow$ 300)[c]
SAC-C	2000–2013	Argentina, U.S. (also Denmark, Brazil, Italy, France)	Scalar; GPS	98.2°	702
Swarm[d]	2013–	ESA	Scalar, vector; star imager; GPS		
A				87.4°	465
B				88.0°	520
C				87.4°	465

[a]**POGO**: Jackson & Vette, 1975; Stockmann et al., 2015; cf. Langel, 1979. **Magsat**: Langel et al., 1980. **Ørsted**: Olsen, 2002; Sabaka et al., 2020. **CHAMP**: Olsen et al., 2017; also 2009. **SAC-C**: NASA, 2000; Maus et al., 2006b;. **Swarm**: Olsen et al., 2015; Chulliat et al., 2015.
Note that numerous authors, including some of those cited here, may provide slightly different values for nearly all the entries in this table.

[b]Scalar magnetometers measure B at that location; vector magnetometers measure three mutually perpendicular components of $\vec{B}$, which—with the help of the star camera(s), to orient the magnetometer at that moment in space—can be converted to B_x, B_y, and B_z. GPS receivers allow for precise orbit determination; see Stockmann et al. (2015).

[c]Orbital decay to lower altitude.

[d]Beginning in 2018, a fourth satellite, launched in 2013 by the Canadian Space Agency and originally named Cassiope, was operationally included in the Swarm constellation to aid in the study of external fields (see http://www.esa.int/Applications/Observing_the_Earth/Swarm/Swarm_trio_becomes_a_quartet).
Decay of the Swarm satellite orbits is being addressed for A and C by a sequence of orbital boosts or "raises" that began in 2022 (see https://earth.esa.int/eogateway/documents/20142/37627/Orbit-Evolution-of-the-Swarm-Mission.pdf/).

10.2.2 Spherical Harmonics, Once Again

With the goal of a big-picture view of the geomagnetic field, it is not surprising to find spherical harmonics as the mathematical tool of choice; we have already seen how useful they are for understanding global patterns, for example in Chapter 6, where they helped us interpret variations in the gravitational field and Geoid. For magnetic fields, there is an even more important reason to bring spherical harmonics into the picture: to see those patterns *three-dimensionally*. In Chapter 6, maps

illustrating Earth's gravitational field and Geoid represented variations at the surface of the Earth; in effect, those applications of harmonics (see also Chapter 7) were limited to *surface* spherical harmonics. Although such applications—treating the harmonic expansions as a kind of two-dimensional Fourier series—are perfectly valid, under some conditions harmonics can be used to quantify how fields vary with height as well as with latitude and longitude. In those situations, *solid spherical harmonics* are indispensable.

For geomagnetism, the ability to predict how different a measurement would be at a different

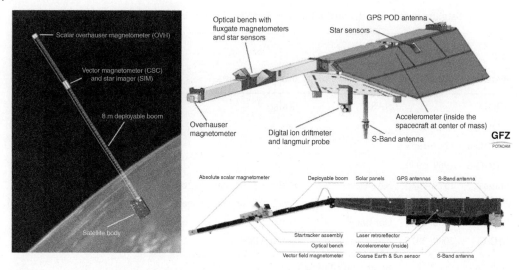

Figure 10.12 Illustrations of the Ørsted, CHAMP, and Swarm satellites (in clockwise order, from above left). The scalar and vector magnetometers are located on the satellite's boom; the length of each satellite including boom is ~9 m. Images from Chulliat et al. 2015 / NOAA.

height or radius is crucial in two respects. Although many observations of Earth's global magnetic field, especially over the past two decades, have been made with instruments onboard satellites, i.e. above Earth's surface, we may be more interested in the field at ground level. And, all observations, whether satellite or ground based, are made thousands of kilometers above the source of the geomagnetic field, the core; it is important to know what the field is like right at the top of the core.

Mathematical Development of Solid Spherical Harmonics for Potential Fields.

So, the difference between surface spherical harmonics and solid spherical harmonics is that the former can be used to describe any pattern on the surface of a sphere, whereas the latter is the basis of a predictive technique for physically real quantities and must, therefore, obey the laws of physics.

To see how our development might proceed, we refer to gravity for guidance; it turns out that, at locations where no mass is present (e.g., outside the solid earth, to good approximation), gravitational potential obeys an equation that reduces to *Laplace's equation*. We have not defined magnetic potential, in order to avoid (or at least postpone) some intimidating

mathematics; however, at locations where no electrical currents are present—or, as in the mantle, crust, and troposphere (excluding thunderstorms), where currents may be weak enough to neglect—the magnetic potential has a well-defined meaning, similar to that of gravitational potential, and ends up also obeying Laplace's equation.

Following tradition, we will denote this magnetic potential by W. As with gravitational potential, we can use the gradient operator to relate the potential to the field:

$$\vec{B} = -\mu_0 \vec{\nabla} W;$$

μ_0 is a 'scale factor,' called the *magnetic permeability*, that converts $\vec{\nabla} W$ to the units we have been dealing with so far for $\vec{B}$.

Wait just a minute, you say—we have never written the gravity field in terms of gravitational potential, like

$$\vec{g} = -\vec{\nabla} U$$

or in any other way! Are you sure? In Chapter 6, we defined surfaces of constant U to be perpendicular to $\vec{g}$—just what we would expect from this gradient expression. We showed that the units for potential are g times distance, consistent with this gradient expression. And we

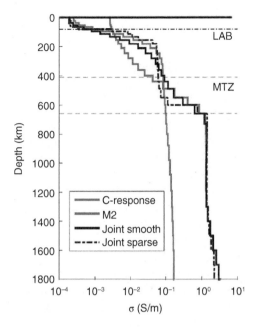

Figure 10.13 Electrical conductivity of the mantle inferred from satellite data. Depths corresponding to the lithosphere-asthenosphere boundary (LAB) and the mantle transition zone (MTZ) are highlighted.

As discussed in the text, analysis based on longer-period signals (red curve) is sensitive to the deeper mantle conductivities; that based on shorter-period signals—in particular, the principal lunar semidiurnal tide (blue curve)—can delineate the uppermost mantle σ well. Solutions based on both longer and shorter periods (black curves), in contrast, are reliable through the entire portion of the mantle shown here.

In this analysis by Grayver et al. (2017) / John Wiley & Sons, solutions were obtained with (solid black curve) and without (dashed black curve) a constraint that the resulting σ have only smooth variations with depth from layer to layer. All solutions allowed for variable conductivity in the overlying oceans.

suggested that the magnitude of the potential's gradient must be proportional to the strength of gravity (we literally said "Places where the potential has steeper gradients ... will mark the locations of stronger gravitational forces").

Incidentally, in Textbox A, we employed a similar relation for the electric force field in terms of electric potential....

As discussed earlier in this chapter, the fact that magnetic monopoles do not exist is equivalent

to saying that the divergence of the magnetic field is zero, that is,

$$\vec{\nabla} \cdot \vec{B} = 0.$$

Substituting in our gradient expression for $\vec{B}$ leads to

$$\vec{\nabla} \cdot (-\mu_0 \vec{\nabla} W) = 0$$

or

$$\vec{\nabla} \cdot (\vec{\nabla} W) = 0;$$

that is,

$$\nabla^2 W = 0,$$

which is Laplace's equation. In Chapter 9, we expanded the Laplacian ∇^2 as a sum of x, y, and z second derivatives,

$$\nabla^2 = \frac{\partial^2}{\partial x^2} + \frac{\partial^2}{\partial y^2} + \frac{\partial^2}{\partial z^2};$$

to quantify variations in height or radius on the global Earth, though, we must express those derivatives in spherical coordinates.

Get ready for some fairly intimidating equations; but don't worry—with a wave of our hand, most of it will quickly disappear.

Our spherical coordinates will be radius r, colatitude θ, and longitude λ, corresponding to radially outward, southward, and eastward directions. Using colatitude θ rather than latitude ϕ ($\equiv 90°-\theta$), which was our preferred variable in Chapter 6, *Textbox B* because of its familiarity, allows us to have a right-handed coordinate system. The gradient operator in spherical coordinates (r, θ, λ) is

$$\vec{\nabla} = \left(\frac{\partial}{\partial r}, \frac{1}{r} \frac{\partial}{\partial \theta}, \frac{1}{r \sin \theta} \frac{\partial}{\partial \lambda} \right).$$

The presence of r and $r\sin\theta$ factors in the denominator means in effect that all three derivatives are taken with respect to length: dr; $r d\theta$ (north-south arc length along the spherical surface of radius r); and $r\sin\theta d\lambda$ (west-east arc length). Using this expression for $\vec{\nabla}$, and an unfortunately more complex equation for divergence ($\vec{\nabla} \cdot$) also in spherical coordinates, we would eventually find that we can write Laplace's equation as

$$0 = \nabla^2 W = \frac{1}{r^2} \frac{\partial}{\partial r} \left(r^2 \frac{\partial W}{\partial r} \right)$$
$$+ \frac{1}{r^2 \sin \theta} \frac{\partial}{\partial \theta} \left(\sin \theta \frac{\partial W}{\partial \theta} \right)$$
$$+ \frac{1}{r^2 \sin^2 \theta} \frac{\partial^2 W}{\partial \lambda^2}.$$

The way we will solve this equation is to use a popular and fairly straightforward technique known as **separation of variables**, in which we assume W depends separately on r, θ, and λ:

$$W = \mathcal{R}(r) \cdot \Theta(\theta) \cdot \Lambda(\lambda).$$

Thus, we will be avoiding nonseparable solutions, a simple (and fictitious) example of which is $\sin(\theta - \lambda)$ (—*try writing that as a function of only θ times a function of only λ!*). Our approach will be to substitute this expression for W into Laplace's equation and solve for each of the unknown functions $\mathcal{R}$, Θ, and Λ; afterward, we can simply multiply them together to yield W.

Because each of these functions is a function of only one variable, each derivative will only involve one of the functions—from the point of view of each variable, it will be like the other two functions are just constants—and we can therefore replace each partial derivative with an ordinary derivative. For example, the r-derivative simplifies according to

$$\frac{\partial W}{\partial r} = \frac{\partial}{\partial r}(\mathcal{R} \cdot \Theta \cdot \Lambda) = \frac{\partial \mathcal{R}}{\partial r} \cdot \Theta \cdot \Lambda$$
$$= \frac{d\mathcal{R}}{dr} \cdot \Theta \cdot \Lambda.$$

Upon substitution, we have

$$0 = \frac{\Theta \cdot \Lambda}{r^2} \frac{d}{dr} \left(r^2 \frac{d\mathcal{R}}{dr} \right)$$
$$+ \frac{\mathcal{R} \cdot \Lambda}{r^2 \sin \theta} \frac{d}{d\theta} \left(\sin \theta \frac{d\Theta}{d\theta} \right)$$
$$+ \frac{\mathcal{R} \cdot \Theta}{r^2 \sin^2 \theta} \frac{d^2 \Lambda}{d\lambda^2},$$

and if we multiply both sides of the equation by r^2 and divide both sides of the equation by $\mathcal{R} \cdot \Theta \cdot \Lambda$, this simplifies to

$$0 = \frac{1}{\mathcal{R}} \frac{d}{dr} \left(r^2 \frac{d\mathcal{R}}{dr} \right)$$
$$+ \frac{1}{\Theta \sin \theta} \frac{d}{d\theta} \left(\sin \theta \frac{d\Theta}{d\theta} \right)$$
$$+ \frac{1}{\Lambda \sin^2 \theta} \frac{d^2 \Lambda}{d\lambda^2}.$$

What makes this approach so appealing is that the first term on the right appears to depend *only* on r (since $\mathcal{R}$ is a function only of r), whereas the second and third terms *cannot* depend on r at all; it follows that, if we change the radius (height) of our location—conceivably changing the magnitude of the first term—but maintain the same colatitude and longitude (so the second and third terms do not change), the terms must still sum to 0! The only way this can hold true is if the first term always equals a constant, i.e. does not change when r is varied, and the second plus third term always equals the negative of that constant.

The equation that will determine $\mathcal{R}$ thus looks something like

$$\frac{1}{\mathcal{R}} \frac{d}{dr} \left(r^2 \frac{d\mathcal{R}}{dr} \right) = \text{constant};$$

its solution will tell us how W varies with r. Applying the same type of argument to the second and third terms—writing the preceding equation such that one term only involves λ, for example, and the other terms do not—would eventually succeed as well, allowing us to obtain solutions for Λ and Θ; but that process (which is especially challenging for the Θ solution) is unnecessary here. That's because we already know what the solution will be: surface spherical harmonic functions, the solution to Laplace's equation (whether $\nabla^2 W = 0$ or $\nabla^2 U = 0$) when there is no radial dependence!

So, for constants ℓ and m associated with the second and third terms, we can write the solution as

$$\Theta(\theta) \Lambda(\lambda) = Y_{\ell m}(\theta, \lambda).$$

What we had written in Chapter 6, *Textbox B* as a single spherical harmonic, $Y_{\ell m}(\theta, \lambda)$, is actually, as we now see, a *product* of two separated quantities: one, a function of latitude or colatitude (such solutions are named *Legendre functions*, after the person who first derived them); the other, a function of longitude (a homework exercise will reveal the latter to be a sinusoidal function). In obtaining these solutions we would discover that they are mathematically possible only if ℓ and m are integers, and if $m \leq \ell$. It turns out (luckily) that when these functions are substituted into our Laplacian equation, the net result is that the second and third terms add up simply to $-\ell \cdot (\ell + 1)$ (which is indeed a constant), implying

$$\frac{1}{R}\frac{d}{dr}\left(r^2\frac{dR}{dr}\right) = \ell(\ell + 1).$$

There are two possible solutions to this equation:

$$R = \frac{1}{r^{\ell+1}} \text{ or } R = r^\ell,$$

which can easily be verified by substituting them in.

Go ahead—try it! The r-dependence we have identified here is critically important, and worth some confidence-building!

Finally, for each choice of constants ℓ and m, the resulting product $R(r)\cdot\Theta(\theta)\cdot\Lambda(\lambda)$ will be a solution for W, i.e., it will make $\nabla^2 W$ equal 0; these solutions will take one of two forms,

$$\frac{1}{r^{\ell+1}}Y_{\ell m}(\theta, \lambda) \quad \text{or} \quad r^\ell \cdot Y_{\ell m}(\theta, \lambda).$$

Furthermore, we can multiply any of these solutions by a constant, and its Laplacian will still equal 0; in fact, we can add *all* these solutions—each corresponding to a particular ℓ and m, and multiplied by some constant ("coefficient")—and the sum (formally called a *linear combination* of the individual solutions) will still have a zero Laplacian. In all of these cases, the resulting W still satisfies $\nabla^2 W = 0$.

In short, we have just demonstrated that the most general solution to Laplace's equation in spherical coordinates—whether representing a potential (gravitational, electric, or magnetic) or, as in Chapter 9, a diffusing quantity (like temperature) in steady state in the absence of sources—will look something like

$$W = \sum_{\ell \geq 0, m} \left\{ \frac{1}{r^{\ell+1}} \cdot \widetilde{C}_{\ell m} \cdot Y_{\ell m}(\theta, \lambda) \right\}$$
$$+ \sum_{\ell \geq 0, m} \{ r^\ell \cdot \widetilde{D}_{\ell m} \cdot Y_{\ell m}(\theta, \lambda) \},$$

a sum of all possible solutions; our concise symbol here for summation is meant to include sums over both ℓ and m, with the restrictions on those indices as stated earlier. The coefficients, $\widetilde{C}_{\ell m}$ for an inverse radial dependence and $\widetilde{D}_{\ell m}$ for a direct radial dependence, determine the proportions of the degree ℓ order m harmonic to be included in the sum for W, in order to represent a particular situation.

What we have written here are *solid spherical harmonic* representations. If (*after all this derivation!*) we decided to apply this expansion at just a single value of r, say $r = a$, then the radial factors in the expansion would remain constant, and we could combine the terms into

$$W = \sum_{\ell \geq 0, m} \{ C'_{\ell m} \cdot Y_{\ell m}(\theta, \lambda) \},$$

where $C'_{\ell m}$ $(= a^{-(\ell+1)} \cdot \widetilde{C}_{\ell m} + a^{\ell} \cdot \widetilde{D}_{\ell m})$ is the combined coefficient of the harmonic function $Y_{\ell m}$. That is, the solid spherical expansion on a particular spherical surface is just the familiar set of (surface) spherical harmonics we have previously used to create surface patterns.

Alternatively, however, with their explicit dependence on r, we can use solid spherical harmonics to extrapolate the magnetic potential—and thus the field—to different heights or depths. Such an extrapolation is called **upward** or **downward continuation**. As implied in our earlier introduction to magnetic potentials, these extrapolations will be accurate as long as (strong) electrical currents are not present.

Using Solid Spherical Harmonics to Learn About Earth's Internal Fields. Our expansion of the magnetic potential into spherical harmonics requires one more step before it can be applied to the Earth: deciding which radial dependence to use in the expansion.

In the course of this chapter, we have learned about a number of external and internal components of Earth's magnetic field. The sources of the external fields are all above the solid surface of the Earth, and as we measure those fields at increasing heights their amplitudes grow because we are getting closer to their sources. Such behavior, whether of the field or its potential, is accurately described by a direct radial dependence (r^ℓ increases as r increases); thus, $\widetilde{D}_{\ell m}$ coefficients would be appropriate to use in order to describe those fields. On the other hand, with increasing height above Earth's surface we are further from the sources of the internal field; this is true even for crustal or mantle components induced by external field variations (the sources of those components are the electrical currents induced in the crust and mantle). For the internal field, then, the potential and the field are best characterized by an inverse radial dependence ($r^{-(\ell+1)}$ decreases as r increases), and we would use observations to infer the various $\widetilde{C}_{\ell m}$ coefficients comprising that field.

Focusing on the internal field, our most general expression for the magnetic potential would seem to be

$$W = \sum_{\ell \geq 0, m} \left\{ \frac{1}{r^{\ell+1}} \cdot \widetilde{C}_{\ell m} \cdot Y_{\ell m}(\theta, \lambda) \right\},$$

where for each ℓ the sum over m is restricted to $m \leq \ell$. Since $Y_{00} \equiv$ constant, however (see Chapter 6, *Textbox B*), the $\ell = 0$ term exhibits no variation with latitude or longitude; for that term, W equals a constant divided by r. The $\ell = 0$ magnetic field, obtained from the gradient of W, is therefore purely radial (only the radial derivative is nonzero), and, depending on the sign of $\widetilde{C}_{00}$, would either be a purely outward 'north' monopole or a purely inward 'south' monopole. Of course, magnetic monopoles do not exist, so we conclude $\widetilde{C}_{00}$ must be 0 for the Earth and our sum should, in effect, begin at $\ell = 1$:

$$W = \sum_{\ell \geq 1, m} \left\{ \frac{1}{r^{\ell+1}} \cdot \widetilde{C}_{\ell m} \cdot Y_{\ell m}(\theta, \lambda) \right\}$$
$$= \frac{1}{r^2} \cdot \widetilde{C}_{10} \cdot Y_{10}(\theta, \lambda) + \frac{1}{r^2} \cdot \widetilde{C}_{11} \cdot Y_{11}(\theta, \lambda)$$
$$+ \frac{1}{r^3} \cdot \widetilde{C}_{20} \cdot Y_{20}(\theta, \lambda) + \frac{1}{r^3} \cdot \widetilde{C}_{21} \cdot Y_{21}(\theta, \lambda)$$
$$+ \frac{1}{r^3} \cdot \widetilde{C}_{22} \cdot Y_{22}(\theta, \lambda) + \dots,$$

where we have written out the $\ell = 1$ terms (with $m = 0$ or 1), then the $\ell = 2$ terms (with $m = 0$ or 1 or 2), and so on.

The $\ell = 1$, $m = 0$ term,

$$\frac{1}{r^2} \cdot \widetilde{C}_{10} \cdot Y_{10}(\theta, \lambda),$$

is, more explicitly,

$$\text{constant} \cdot \frac{\cos \theta}{r^2},$$

using the definition of the degree 1 order 0 spherical harmonic. This term corresponds to a dipole magnetic field! That is, if we take its minus gradient in spherical coordinates, we would find the field it is associated with can be written

$$\text{constant} \cdot \frac{2 \cos \theta}{r^3} \hat{r} + \text{constant} \cdot \frac{\sin \theta}{r^3} \hat{\theta},$$

where $\hat{r}$ is the unit vector pointing radially outward and $\hat{\theta}$ is the unit vector pointing north-to-south along the spherical surface at radius r. If we graphed this vector expression for the field using different values of r and θ, we would find ourselves recreating the various positions of the compass needle shown in Figure 10.1a, in effect tracing out the field lines of an axial or vertical dipole.

Can you verify the above vector expression for the field associated with the $\ell = 1$, $m = 0$ potential? Also, use this expression to confirm that the field is outward at the north magnetic pole ($\theta = 0°$) for a positive constant, inward at the south magnetic pole ($\theta = 180°$), and horizontal at the magnetic equator ($\theta = 90°$).

Incidentally, this last expression serves as a reminder that even the simplest kinds of magnetic fields, dipole fields, are fundamentally different than gravitational fields: they depend inversely on distance cubed rather than distance squared; and they are not central force fields (which would have only a radial ($\hat{r}$) direction for a source at the center of the coordinate system). It is that 'sideways,' twisting force of bar magnets and magnetic fields—for a dipole field, represented by the $\hat{\theta}$ component—that challenges our intuition (and also our math skills, as we will see later in this chapter).

Unit Complications. The explicit dependence on radius in our general, spherical harmonic solution leads to an inconvenient numerical complication. Each time ℓ increases by 1 in the summation, the radial term decreases by another factor of r, $\sim 6 \times 10^6$ m (if we are near the Earth's surface) or $\sim 3 \times 10^6$ m (if we are near the core boundary), going from $1/r^2$ to $1/r^3$ to $1/r^4$, and so on. Beyond the dipole term, for W to maintain a reasonable magnitude, the $\widetilde{C}_{\ell m}$ coefficients would have to increase correspondingly. This would quickly get out of hand, as the values of the coefficients grow astronomically.

An easy solution is to rewrite the spherical harmonic expansion as

$$W = \frac{1}{r} \sum_{\ell \geq 1, m} \left(\frac{R}{r}\right)^{\ell} \cdot \widehat{C}_{\ell m} \cdot Y_{\ell m}(\theta, \lambda)$$

where, if we are describing the Earth's magnetic potential, R is Earth's mean radius, but for other bodies R would be appropriately different. This expression is the same as our previous one as long as

$$(R)^{\ell} \cdot \widehat{C}_{\ell m} \equiv \widetilde{C}_{\ell m},$$

which in turn implies that

$$\widehat{C}_{\ell m} \equiv R^{-\ell} \cdot \widetilde{C}_{\ell m},$$

so, though the $\widetilde{C}_{\ell m}$ coefficients grow without limit, the $\widehat{C}_{\ell m}$ coefficients do not.

A slight variation on this with an additional factor of R^2, namely

$$W = \frac{R^2}{r} \sum_{\ell \geq 1, m} \left(\frac{R}{r}\right)^{\ell} \cdot C_{\ell m} \cdot Y_{\ell m}(\theta, \lambda),$$

with correspondingly modified coefficients $C_{\ell m}$, is more commonly used: this formulation not only keeps the magnitude of the coefficients under control ($R^2 C_{\ell m} \equiv \widehat{C}_{\ell m}$ for all ℓ and m) but—as an added benefit—these new coefficients, $C_{\ell m}$, have the same units as magnetic fields, e.g., nT.

Additional Complications. We know the Earth's dipole field is not quite axial. Thinking of bar magnets and dipoles as vectors, we could create a slightly tilted dipole by adding to the preceding axial dipole a second, smaller, horizontal one.

Mathematically, because dipoles are characterized by a $1/r^2$ potential (or a $1/r^3$ field, as we have just seen), that second dipole would have to come from another degree $\ell = 1$ term in the summation for the potential. Conveniently enough, the degree 1 order 1 term meets our requirements. With $Y_{11} \sim \sin\theta$, the $\ell = 1, m = 1$ potential equals a constant times $\sin\theta/r^2$; this would be the formula for an axial dipole potential ($\sim \cos\theta/r^2$) if the coordinate system was tilted by 90° (since $\sin(\theta+90°)$ equals $\cos\theta$). To say it in other words, the $\ell = 1, m = 1$ potential corresponds to a dipole referenced to an equatorial axis, whereas the $\ell = 1, m = 0$ potential is referenced to a polar axis.

However, focusing on horizontal dipoles is a slippery mathematical slope: once we admit to the need for a horizontal dipole, rigor demands that we specify the orientation of that dipole—whether it points in the direction of 0°E longitude, 90°E longitude, or somewhere else; any of these are possible. By implication, our symbol $Y_{11}(\theta, \lambda)$ for the spherical harmonic function must allow for such differences; that is, there must be two functions, one referenced to 0°E and the other to 90°E. More generally, our symbol $Y_{\ell m}(\theta, \lambda)$ for the spherical harmonic function must allow for such differences when $m \neq 0$; that is, there must be two functions of longitude for each nonzero order m, one referenced to 0°E and the other to 90°E.

Earlier, we mentioned that, as a function of longitude, $Y_{\ell m}(\theta, \lambda)$ is sinusoidal; can you guess what those two functions for each m might be?

Furthermore, there must be a spherical harmonic coefficient for each of these functions; instead of one $C_{\ell m}$, there will be, say, one $g_{\ell m}$ and one $h_{\ell m}$ (using traditional symbols). With certain definitions, and using a version of Legendre functions called *Schmidt functions*, the set of $g_{\ell m}$ and $h_{\ell m}$ are known as **Gauss coefficients**.

These additional complications are critical for an accurate description of the geomagnetic field; though regretting the loss of rigor, we will nevertheless declare them beyond the scope of this textbook.

The degree $\ell = 2$ terms in our spherical harmonic expansion turn out to correspond to quadrupole sources at the center of the Earth. The $\ell = 2$, $m = 0$ term corresponds to an axial quadrupole (perhaps like that pictured in Figure 10.9b); we know it is axial because with $m = 0$, it is a zonal harmonic, i.e. it does not vary with longitude. The $\ell = 2$, $m = 1$ and $\ell = 2$, $m = 2$ terms account for any tilt in the quadrupole source. As one might imagine, the higher-degree terms represent higher multipole sources.

The $1/r^{(\ell + 1)}$ radial dependence of the magnetic potential, and the implied $1/r^{(\ell + 2)}$ radial dependence of the geomagnetic field, have significant implications for our understanding of the Earth's internal field. At the surface of the Earth, the field we measure is dominantly dipolar; just outside the core, at a radius of just over half the radius to the surface, the dipole potential will be ~4 times as big (by the factor $1/r^{(\ell + 1)}$ with $\ell = 1$ and $r \sim \frac{1}{2}$), and the dipole field eight times as big. But the non-dipole terms are magnified even more: the quadrupole ($\ell = 2$) potential and field will be boosted by factors of 8 and 16, respectively, going from the surface to the core; the $\ell = 3$ terms by factors of 16 and 32; the $\ell = 4$ terms by factors of 32 and 64; and so on, with still higher-degree components correspondingly further magnified. Downward continuation, in this case from Earth's surface to the top of the core, clearly has the potential to significantly transform the field. At a minimum, we can say that representing the geomagnetic field as just a dipole will be a less satisfactory approximation at the core-mantle boundary than at the surface.

Before carrying out downward continuation, however, we need to make sure that we can separate out the core and crustal fields, so that only the core field will be extrapolated to the CMB; this will be explored in the next subsection, and the results of the extrapolation will be analyzed in the section to follow. We will find that the non-dipole components of the downward-continued core field do not dramatically overwhelm the downward-continued dipole field. But their greater 'respectability' in magnitude will suggest that distinguishing the dipole component from all the non-dipole components (a perspective we will take repeatedly with the surface field) is perhaps not as physically meaningful from the core's perspective. The question of dipole versus non-dipole will lie behind much of what we do in the balance of this chapter.

10.2.3 Back to the Surface: A Closer Look at the Crustal (and Geomagnetic) Fields

In 1838, based on sparse observations scattered around the world, Gauss estimated the spherical harmonic coefficients of Earth's magnetic field (Garland, 1979), thereby producing the first global picture of the field; with external components and even crustal contributions relatively small, the field he was mapping with spherical harmonics amounted to (more or less) the geomagnetic field—the core field.

In the years following his achievement, measurements began to accumulate worldwide, and by the last third of the twentieth century, global coverage—from land, sea, and air, as well as satellite measurements—allowed high-resolution, high-accuracy surface maps of Earth's internal field to be constructed. Global coverage has ultimately allowed the determination of spherical harmonic coefficients ("Gauss coefficients") through degree 800 (Lesur et al., 2016)—a far cry from Gauss' expansion through degree 4, and no longer a representation of just the core field.

Figure 10.14 shows the field derived from such measurements (Dyment et al., 2016) *with harmonics 1 through 15 omitted*; as justified below, those very low-degree harmonics, most of which are much larger than all the higher harmonics, and which together look like the field shown in Figure 10.7, can be taken to correspond to the core field. Omitting the core field reveals the much smaller crustal (lithospheric) field. In this subsection we will discuss the crustal field in general terms, neglecting its rich variations in favor of a bigger picture, one

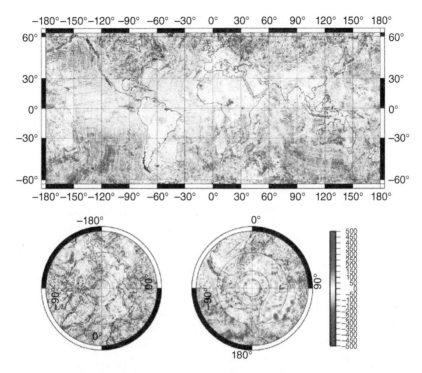

Figure 10.14 The World Digital Magnetic Anomaly Map (WDMAM) depiction of the crustal (lithospheric) anomalies magnetic field. This map is the result of a massive collaboration (see wdmam.org for details) in which nearly 26 million measurements of field intensity were gridded and combined with models of oceanic lithosphere magnetization to produce crustal anomalies with a spatial resolution corresponding to harmonic degree 800. The anomalies are determined relative to the 1990 main field, in units of nT (see color scale). An explanation of their approach can also be found in Lesur et al. (2016) and Dyment et al. (2015). This image, from Dyment et al. (2016) and available at wdmam.org, was downloaded in May 2020. The polar views are centered on the North (left) and South (right) Poles.

with spherical harmonic implications. According to Figure 10.14, then, we find the crustal field on land to be strongest in North America, Eastern Europe, and Western Australia; it is weakest in South America, most of Africa and the Mid-East, and central Asia, though data availability may be a factor (Dyment et al., 2016). Over the oceans it is weakest throughout the tropics. In general, the field is a few times stronger over land than over the oceans (e.g., Maus, 2008). Surprisingly, however, it appears that the field is weaker in Antarctica than over the Arctic Ocean (Maus et al., 2002).

There are some connections—Frey (1982) called them "geographical coincidences"—between crustal anomalies and tectonic features; see Langel (1982) for references and also Dyment and Arkani-Hamed (1998). Using Magsat data, for example, Frey found that

ocean basins and abyssal plains are associated with negative anomalies, mid-ocean ridges and submarine plateaus with positive anomalies, at satellite altitudes; subduction zones also tend to exhibit positive anomalies, including those ringing the Pacific (Arkani-Hamed & Strangway, 1986; but see also Manoj et al., 2006a).

In general, cratons (continental shields) may be expected to have strong signals, simply because, due to their age and tectonic inactivity, they will be colder; a greater Curie depth would give them a thicker layer of magnetization (Maus, 2008; Olsen et al., 2017). A thicker magnetized layer would also be expected in older oceanic lithosphere compared to younger oceanic lithosphere, and expected even in subduction zones; Arkani-Hamed and Strangway (1987) showed that, as a consequence of that thickening plus their cooling effect on the surrounding mantle as they subduct, the magnetic anomalies over oceanic lithosphere 100 Myr old could be twice as strong as that over oceanic lithosphere 40 Myr old.

However, a consensus on tectonics as the underlying explanation for major features of the anomalies field seems uncertain, whether from differences in data analysis (for example, Arkani-Hamed & Strangway, 1986 found no significant correlation between the anomalies field and cratons using a particular range of spherical harmonic degrees) or from differences in interpretation (for example, Frey, 1982 noted that in many cases, tectonic boundaries and structures *limited* a magnetic anomaly rather than defined it).

It is clear from the detail seen in Figure 10.14 that the source of this field is shallow (as in crustal or lithospheric); to the extent that the field correlates with tectonic features, shallow sources are also implied (e.g., Regan et al., 1975). We also know that the high-degree spherical harmonics used to describe the field (up through degree 800) must correspond to shallow sources. But what about the lowest-degree harmonics (degrees 1–15 in the case of that figure)? As suggested by our discussion above and in Chapter 6, a broad scale of fluctuation at Earth's surface is consistent with a deep source *but does not require it.*

For example, the crustal field includes seafloor magnetic stripes (Figure 10.6), corresponding to reversed alternating with normal polarity remanent magnetization of the oceanic lithosphere. During eras when the field reversed frequently, or seafloor spreading was slow, those stripes would be relatively narrow. But for a period of more than 35 Myr in the middle Cretaceous, Earth's dipole field did not reverse (or experienced only very brief reversals; see, e.g., Lowrie, 2007 but then Tarduno, 1990; Granot et al., 2012). During this long **Cretaceous quiet interval** (also called the **Cretaceous Normal Superchron**), the magnetization of new oceanic lithosphere was largely unchanged, creating in effect a broad stripe: a pattern of broad crustal magnetization (see Dyment & Arkani-Hamed, 1998; Dyment

et al., 2015) that could be characterized only with *low* harmonics.

Differences in continental versus oceanic crustal magnetization is also a pattern that would be described mainly with lower-degree harmonics (Counil et al., 1991). In short, it is possible that removing the low harmonics in Figure 10.14 may have eliminated some portion of the anomalies field as well as the main field.

The Spherical Harmonic Spectrum. Such a conclusion is reinforced by examining the strength of the internal magnetic field as a function of harmonic degree; since the degree of a spherical harmonic is like a 'wave number' or spatial frequency, it is not surprising that such a 'breakdown' amounts to a spectral analysis, similar in some ways to the power spectrum introduced at the end of Chapter 7, *Textbox B.* The magnetic field's power for degree ℓ will look something like

$$(\ell + 1) \sum_m |C_{\ell m}|^2$$

according to traditional formulations, where the sum involves all the coefficients for that degree harmonic (all $C_{\ell m}$ for that ℓ), and the factor $(\ell + 1)$ reflects the conversion of potential (W) to field ($\vec{B}$).

Can you explain the factor $(\ell + 1)$ mathematically? Think about the general relation between the potential W and the radial component of the field. Now, what if you wrote W out as a series of spherical harmonics (for simplicity, in terms of $\widetilde{C}_{\ell m}$ rather than $C_{\ell m}$)?

The spectrum of the internal field modeled by Sabaka et al. (2004) is shown in Figure 10.15a. It reveals sharply different spectral characteristics for the low-degree versus higher-degree components of the field: the former exhibit an approximately *exponential decrease* in power with ℓ, from the dipole ($\ell = 1$) term through the increasingly weak non-dipole terms (the decrease in power looks linear because of the graph's log scale); but beyond $\ell = 15$–18, the field *strengthens*, albeit at a relatively low rate. Other researchers have marked the 'elbow' as occurring at a value of ℓ as low as

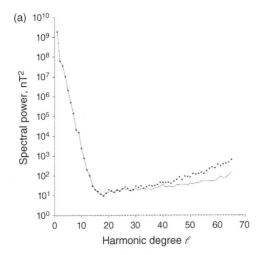

(a)

Figure 10.15 Spherical harmonic spectrum (traditional formulation, formally called the Lowes-Mauersberger spectrum; see, e.g., Maus, 2008) of Earth's internal magnetic field, with the summed power of all harmonics of each degree plotted versus that degree. Spectral values are in units of nT^2, as discussed in the text, and plotted on a log scale.
(a) The internal magnetic field analyzed here was calculated by Sabaka et al. (2004) / John Wiley & Sons based on surface and satellite observations between 1960 and 1985 (filled circles), or based on that data extended to year 2000 (solid line).
The low-degree and high-degree portions of the spectra—which we associate with the geomagnetic field and the anomalies field, respectively—are clearly distinctly different in character.
Image from Sabaka et al. (2004) / John Wiley & Sons.

10–12; Meyer et al. (1983; also, e.g., Voorhies et al., 2002) view those ℓ (13 and 14 in their work) as intermediate 'transition' values (see also Olsen et al., 2014).

With such distinct spectral characteristics, there is no hesitation in identifying the high-degree terms as the crustal (lithospheric) anomalies field; after all, there is no way a core field, with sources thousands of kilometers below the surface, could produce spatial variations with short wavelengths. And, it makes sense to view the low-degree terms—including an overwhelmingly strong dipole, and other terms whose strength declines in a consistent pattern as ℓ increases—as the components of the core field. This is mostly correct; but, as already suggested, there are complications.

First, as we have noted above, although high harmonics require a shallow source, low-degree harmonics can be produced by deep or shallow sources. In fact, it is easy to imagine extending the anomalies spectrum from high degrees down to an ℓ less than 13, perhaps even down to $\ell \sim 1$; such an extension would represent the crustal contribution to the low-degree, broad-scale component of the field.

An example of how that extension of the anomalies spectrum might look is shown in Figure 10.15b (top). In this example (from Jackson, 1990), the direct contributions that broad-scale crustal anomalies make to the observed field are roughly constant over a range of harmonic degrees, though they become relatively insignificant for the lowest-degree harmonics.

Second, the existence of such anomalies also has implications for the *higher*-degree spectrum. To represent a broad-scale feature, whether a Cretaceous Superchron or a magnetized continent, with spherical harmonics generally requires some secondary, high-degree harmonics (in addition to those low-degree harmonics that correspond to the feature's breadth); high harmonics are especially necessary if the feature has a sharp boundary.

Hold on, you say! In Chapter 7, Textbox B shows an excellent approximation to a triangle shape using a sum of only three sinusoids; where are the high harmonics (i.e., where are higher-frequency sinusoids) for that situation? The answer is: just beyond your view! In the range $t = -1$ to $t = +1$, the sum of those sinusoids does indeed nearly equal the triangle function TR, but outside that range—sinusoids having a fundamentally repeatable character—the sum keeps adding up to triangle functions (we know what the sum equals at $t = 0$; how does it add up when $t = 4$? How about when $t = 8$)? This repeatability was irrelevant in that textbox discussion, because we were only interested in representing TR within the range ± 1; but if we had wanted to represent a triangle that was zero everywhere outside of ± 1 (—truly a sharp boundary!), we would have needed lots of other sinusoids to counteract the repeatability.

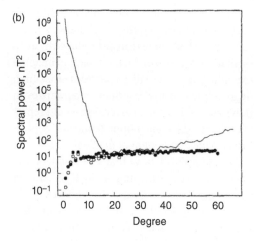

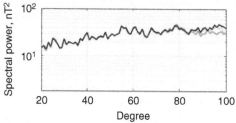

Figure 10.15 Spherical harmonic spectrum (traditional formulation) of Earth's internal magnetic field, with the summed power of all harmonics of each degree plotted versus that degree. Spectral values are in units of nT^2, as discussed in the text, and plotted on a log scale. (b) Other views of the lithospheric field. (Top graph) Spectrum of the internal field (curve), as in (a), and of just the crustal field (symbols), predicting the long-wavelength portion of the latter were it not obscured by the main field. Those predictions were based on models by Hahn et al. (1984) and Jackson (1990); see Jackson for details. Figure from Jackson (1990) / John Wiley & Sons. (Bottom graph) Extension of the crustal spectrum in (a) to higher harmonics, based on more recent satellite observations (1999–2019) as well as surface data, as analyzed by Sabaka et al. (2020). Note that the spectrum 'whitens,' further leveling off, at degrees > 60. Image adapted from Sabaka et al. (2020) / Springer Nature / CCBY 4.0 / Public domain.

Consequently, when the low-degree harmonics of the crustal field are omitted, the high harmonics associated with such broad-scale features remain—leaving us with high harmonics we naively try to interpret as actual features, though they are in reality just an 'edge' effect produced by those broad features' boundaries (e.g., Counil et al., 1991). Such a challenge can be addressed to some extent by 'forward' modeling of crustal sources.

It can be argued that the shape of the high-degree portion of the spectrum in Figure 10.15a relates to whether the crustal field is mainly induced or remanent—whether its magnetization is mostly induced by the present geomagnetic field or left over from past fields frozen in when the rock formed. Induced magnetization would be fairly coherent, as the inducing field is mostly dipolar; for example, the magnetization would be relatively weak near the (geomagnetic) equator and strong near the geomagnetic poles. Remanent magnetizations would be less ordered and more random, as a result of subsequent displacements (from plate motions) and differences between inducing fields at the times of magnetization. The roughly flat spectrum of the crustal field in Figure 10.15a (see also Figure 10.15b, bottom), which implies nearly the same contribution from every harmonic, is often viewed as evidence of such randomness. Along those lines, it might be noted that a more rigorous formulation of the geomagnetic spectrum—developed by Maus (2008) and amounting to an 'averaging' of the traditional spectrum according to

$$\frac{\ell+1}{2\ell+1} \sum_m |C_{\ell m}|^2$$

—does indeed produce a flatter spectrum, for ℓ exceeding ~15.

The distinctive long-wavelength portion of the spectrum has been used to assess the source *region* of the core field. Envisioning the observed non-dipole field to be the result of randomly oriented dipoles in a single layer below the Earth's surface, we know that the shallower the layer, the greater the proportion of shorter wavelength variations we would observe in the field at the surface—so, the less steep the spectrum would be. Conversely,

Figure 10.15 (c) Spherical harmonic spectrum (traditional formulation) of Earth's internal magnetic field, as in (a), compared with the internal magnetic field of Mars. Spectral values are in units of nT², and plotted on a log scale, with filled circles for Earth, plusses for Mars. The Mars spectrum looks similar in form to that of Earth's crustal field, including that it levels off at higher degrees (Langlais et al., 2019), *but* it lacks the sharp rise of Earth's core field at low degrees. See also Figure 14 of Hulot et al. (2010).

Image from Voorhies et al. (2002) / John Wiley & Sons.

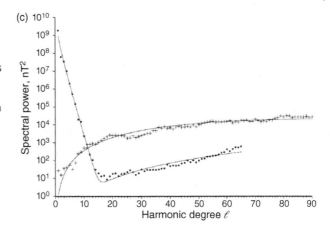

the deeper the layer, the fewer the 'ripples' we would sense at the surface, and the smaller they would be—so the spectrum would be steeper. According to the analysis by Voorhies et al. (2002), consistency with the observed low-degree spectrum (as in Figure 10.15a) implies that such a layer is most likely in the outer core, close to the core-mantle boundary, and perhaps ~30 km thick. (To be clear: there may well be sources deeper within the core; that outermost-core layer relates to the non-dipole field we observe.)

It is instructive to compare the spectrum of Earth's internal magnetic field with that of Mars (Figure 10.15c); measurements of the latter were taken by the Mars Global Surveyor (Purucker et al., 2000), and subsequently by the Mars Atmosphere and Volatile Evolution (MAVEN) orbiter (e.g., Langlais et al., 2019). The Mars spectrum shows the same pattern versus harmonic degree as Earth's crustal field; but a sharp rise at low degree is absent: Mars does not have a core field! There are a couple of additional conclusions we can draw from the Mars spectrum.

- If the crustal field was created 'externally,' for example from impacts, it would be mostly local or regional, thus restricted to intermediate or high harmonics. The presence of a crustal field with global characteristics suggests that it was magnetized globally—by a core field. Thus, Mars' core

(which, we found in Chapter 1, is similar in proportion to Earth's) must have sustained dynamo action at some point in the past. In Chapter 1, we briefly discussed some ways that such an ancient dynamo might have been turned off early in Mars' history.

- Mars' crustal magnetization is spatially complex, even including extensive demagnetized regions (Acuña et al., 1999; Lewis & Simons, 2012). Overall, it is stronger, degree by harmonic degree, than Earth's—about an order of magnitude stronger; that is, Mars' spectrum is two orders of magnitude larger than Earth's (Figure 10.15c), but the spectrum is proportional to the square of the harmonic coefficients.

In part, we could attribute this to a higher iron content in the Red Planet's crust; but it could also be the result of a thicker layer of magnetization. Interpreting the high harmonics of the Earth's spectrum with the same arguments they applied to the low-degree non-dipole portion of the spectrum, Voorhies et al. (2002) concluded that the source region was, not unexpectedly, within 20 km of Earth's surface; but for Mars, they found the source region was nearly 50 km below the Martian surface. If we view these estimates as equivalent to layer thicknesses, then Mars' crustal magnetization 'per unit area' would have been several times greater than Earth's.

Regional spectral analysis by Lewis and Simons (2012) found significant variability in the inferred layer thicknesses of Mars' magnetized crust, but with a typical thickness comparable to that estimated by Voorhies et al. (2002). They were also able to conclude that Mars' crustal magnetization was unlikely to be the result of hydrothermal alteration.

Recent surface measurements of Mars' crustal field by the *InSight* (Interior Exploration using Seismic Investigations, Geodesy and Heat Transport) mission imply an even stronger local magnetization below the Elysium Plains than inferred from satellite data, supporting evidence from Martian meteorites (discussed later in this chapter) that Mars' ancient dynamo was as strong as Earth's is (Johnson et al., 2020).

Incidentally, the spectrum of the lunar crust is quite similar to that of Mars, but about two orders of magnitude smaller than Earth's rather than two orders greater (see Hulot et al., 2010). We leave it to the interested reader to speculate on whether that is consistent with the Moon having a very small core at present.

On a physical basis, the spectra shown in Figure 10.15 cannot continue to rise indefinitely as ℓ increases, or even stay flat; we can argue that the trend must ultimately be downward. For simplicity, imagine that the crustal field is a *product* of the inducing field and a global spatial distribution representing the crust's ability to be magnetized (this property, called its *magnetic susceptibility*, is formally introduced in the next section of this chapter). If Earth's anomalies field is mostly induced magnetization, the crustal spectrum will have to trend downward, beginning at some low-to-intermediate harmonic degree, as the harmonics of the inducing field (the main field) die away. Or, if the anomalies field is mostly remanent, the downward trend will reflect the distribution of magnetized crust—think of that distribution as being described by spherical harmonics—and is likely to occur at much higher harmonic degree.

The lithospheric field model developed by Olsen et al. (2017; see also Maus, 2008) based on CHAMP and Swarm data appears to peak around degree 100 (spatial scale ~400 km), then decline; the model displayed in Figure 10.14 peaks between degrees 100 and ~170 (Lesur et al., 2016). These models imply the spectrum has been influenced more by remanence than by induction.

So, Remanent or Induced? We tend to associate the lithospheric magnetic field with remanent magnetization fundamentally because of the existence—and importance—of seafloor stripes. And we have just found two reasons related to the crustal field's spectrum (its flatness and high 'corner frequency') to prefer a remanent origin for the crust's/ lithosphere's magnetization. But it turns out that induced magnetization plays a role in determining significant features of that field as well, suggesting we should explore this question a bit further; that exploration will also lead us to a fascinating concept.

A vivid illustration of the importance of induced versus remanent magnetizations in Earth's anomalies field is given in Figure 10.16, from the analysis by Purucker et al. (2002). The induced field was predicted from seismic models of crustal thickness, assumed mineralogy, and geotherms, which yielded estimates of the depth to the Curie temperature and ultimately the thickness of magnetizable crust worldwide; the Curie temperature depth varied with lithospheric age, ranging in North America from 29 km to 81 km depth. The remanent field, limited in their analysis to oceanic lithosphere due to uncertainties in continental remanence in some locations, was a recalibrated version of an earlier model by Dyment and Arkani-Hamed (1998). All fields in Figure 10.16 are shown at 400 km altitude for ultimate comparison with satellite data. As we see in Figures 10.16a and 10.16b, induced and remanent fields are somewhat comparable in magnitude in oceanic regions, at least for the parts of the globe shown; not surprisingly, the remanent field exhibits more

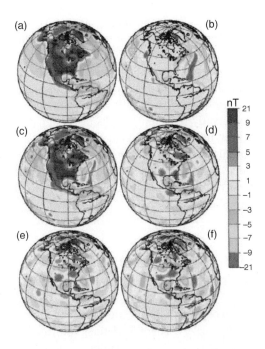

Figure 10.16 Spherical harmonic representation of induced and remanent anomalies fields at a satellite altitude of 400 km above Earth's surface (Purucker et al. (2002) / John Wiley & Sons). Field intensities are in nT as indicated by the color code. Left to right, top to bottom:
(a) Predicted induced field;
(b) predicted remanent field, but for magnetized *oceanic* crust only;
(c) predicted total crustal field, i.e. sum of (a) and (b);
(d) portion of total predicted field of harmonic degrees 15–26, thus with long-wavelength field removed.
Satellite observations, restricted to the same harmonic degrees as the predicted field (d), for comparison:
(e) Magsat and (f) Ørsted.
Images from Purucker et al. (2002) / John Wiley & Sons.

variability. The induced field in continental regions (Figure 10.16a)—largely attributable by Purucker et al. (2002) to the thick magnetizable layer comprising the extensive North American craton (which is consistent with what we see in Figure 10.14)—generally dominates both oceanic fields, induced and remanent.

The total (induced + remanent) crustal field predicted by Purucker et al. (2002) is shown in Figure 10.16c; with low harmonics (degrees 1–14) and high harmonics (degrees > 26) removed, for subsequent comparison with satellite data, we are left with the map in Figure 10.16d. (By the way, the difference between Figures 10.16c and 10.16d is consistent with our assertion that broad-scale harmonics are not necessarily attributable only to deep sources.) The good agreement between Figure 10.16d and the maps from two satellite data sets (Figures 10.16e and 10.16f) suggests that the underlying models are reasonable; thus, the anomalies field is more continental than oceanic, and (though with some uncertainty) more induced than remanent.

From an analysis of more recent satellite data, and with the lithospheric field expanded through harmonic degree 185, Olsen et al. (2017) also concluded that the anomalies field has more power in continental than oceanic regions.

Annihilating Induced Magnetization. Paradoxically, despite the implied dominance of continental magnetization, the anomalies field is predicted (Purucker et al., 2002, as we see in Figure 10.16c) and observed (Lesur et al., 2016, as we see in Figure 10.14) to be relatively weak throughout South America and in parts of Africa and Europe. Although Lesur et al. (2016) attribute the weakness in South America to poor data quality, an additional explanation should be considered: magnetic *annihilators* (Maus & Haak, 2003). We note first that South America and Africa are at low geomagnetic latitudes; the geomagnetic equator intersects those continents at or near their widest extents, guaranteeing a weak, and mostly horizontal, induced magnetization.

In Chapter 6, we discussed the fact that gravitational fields are fundamentally nonunique; for example, any number of spherically symmetric density distributions within a sphere can produce the same gravity field outside the sphere, as long as those density distributions add up to the same total mass: in all cases, that gravity equals $-(GM/r^2)\hat{r}$, where M is the total mass of the sphere. The difference between two such distributions (e.g., a greater density in one internal shell, compensated by a lower density in a second internal shell) must by itself contribute nothing to the external gravity (since applying the difference to the first distribution gives us the second distribution, thus it does not change the gravity at all). That difference in density distributions could be called a gravitational annihilator.

In this chapter, we have written both the electric and magnetic fields in terms of a potential—just as gravitational acceleration can be. It turns out that, because they are 'potential' fields, they also possess a fundamental nonuniqueness: a variety of electric or magnetic source distributions can produce the same electric or magnetic field, external to the body—just like gravity (Blakely, 1995).

For magnetic fields, their nonuniqueness means that **magnetic annihilators**— distributions of crustal magnetization that by themselves contribute nothing to the field outside the crust—can exist. Perhaps the simplest magnetic annihilator is a uniformly magnetized spherical shell, with a magnetization of internal origin (Runcorn, 1975). Most annihilators are symmetric about, and strongest at, the geomagnetic equator (Maus & Haak, 2003). Strikingly, Maus and Haak (2003) also found that, when the inducing field is the geomagnetic field, the two main annihilators are most of South America and most of Africa. Thus, those continents may well possess strong induced magnetizations; but *their* magnetizations will contribute little to the observed anomalies field.

Finally, because the inducing field is primarily dipolar, the nature of the anomalies field can be tested in another way. Lesur et al. (2016) divided the observed anomalies field (Figure 10.14) by a variable factor with the same pattern as our dipole field (see Figure 10.7), thus ~67,000 nT near the geomagnetic poles and ~22,000 nT near the geomagnetic equator. If the anomalies field is induced, dividing by that factor will 'equalize' the field globally; if the field is remanent, it will

not be equalized and might vary even more disparately around the world. Although the anomalies field shown in Figure 10.14 exhibits such a dipole pattern, Lesur et al. (2016) found that the test failed for oceanic lithosphere, implying that the oceanic anomalies field is dominantly remanent; results over land were mixed, indicating that induced and remanent contributions are both important for the continental anomalies field (though it is possible their test would not work for the continental annihilators).

10.3 Snapshots in Time of the Geomagnetic Field

The geomagnetic field varies on a huge range of timescales. These variations provide us with a glimpse into numerous facets of the core's dynamical behavior—and set the stage for our ultimate goal (and the final topic of this chapter), learning how the core is able to maintain its field.

10.3.1 Current and Recent Snapshots

Dipole Versus Non-dipole. We begin by briefly reviewing and highlighting some relevant features of the main field. Approximated with the lowest-degree harmonics (degrees 1 through ~12–15), this field dominates the crustal field, and for that matter all other contributors to the internal and external fields. At Earth's surface we observe the main field to display a mostly dipolar character, though at the surface of the core its non-dipole components are more pronounced. An example of such an extrapolation, by downward continuation of solid spherical harmonics (treating the mantle as nonconducting), is shown in Figure 10.17.

The main field originates in the core, created and maintained by electrical currents generated there by 'dynamo action.' But it is an oversimplification to imagine that there is one centered current loop responsible for the dipole field, or one eccentric current loop

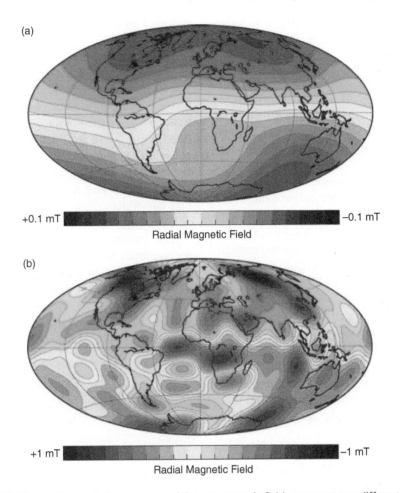

(a)

+0.1 mT   −0.1 mT

Radial Magnetic Field

(b)

+1 mT   −1 mT

Radial Magnetic Field

Figure 10.17 The vertical (radial) component of the geomagnetic field, as seen at two different levels:
(a) at the surface of the Earth, using spherical harmonics through degree 12, contoured with the color code shown and a range of ± 0.1 mT (0.1 mT $= 10^5$ nT);
and
(b) at the surface of the core, with the above surface observations downward-continued, contoured with the color code shown and a range of ± 1 mT (1 mT $= 10^6$ nT).
Yellow/brown colors correspond here to a vertically outward direction; blue colors indicate the opposite. The comparison reveals significantly more mid- and short-wavelength irregularities—'patchiness'—in the core's field, which are attenuated before reaching the Earth's surface.
Images by R. Holme, from Jones (2011) / Annual Reviews.

responsible for the dipole and non-dipole fields; instead, there are a number of electrical current systems dispersed throughout the outer core fluid which give rise to all components of the field, and which together *sum* to a field that looks mostly dipolar from our vantage point at Earth's surface. Analysis of the non-dipole spectrum by Voorhies et al. (2002), as noted earlier, places those current systems in the outer part of the outer core.

Philosophically speaking, the non-dipole part of the field is not just a 'side effect' of an imperfect dipole field. Paleomagnetic data suggests that non-dipoles have always been a permanent part of the Earth's field (Schneider & Kent, 1990, and other references below)—they are simply part of the way the field maintains itself. Furthermore, as we learn about the time dependence of the geomagnetic field, we should not be surprised if our snapshots in time show a different intensity and type of time dependence for components of the non-dipole field than they do for the resultant—that is, the dipole—field.

Present-Day and Historical Time Variations. Changes in the geomagnetic field occur in all ways, and—as far as we can tell—on all timescales. Changes in the magnetic elements (declination, inclination, etc.) have been recorded by magnetic observatories and on voyages across the oceans for hundreds of years (see, e.g., Jackson et al., 2000; Jonkers et al., 2003; Jonkers, 2003). As observations of the field accumulated, it became increasingly evident that those changes called for continual revisions of the global description of the field. The International Geomagnetic Reference Field (**IGRF**) of any particular epoch (see, e.g., Finlay et al., 2010a) consequently includes not only a listing of all spherical harmonic coefficients above noise level, but also the *rates* at which those coefficients have been changing with time—their **secular variation** (also called the **geomagnetic secular variation** or **GSV**).

For example, IGRF-5 (Langel, 1988) not only specified a value of 52 nT for C_{60}, the degree 6 axial coefficient, it listed a prediction for its secular variation to be +1.4 nT/yr between 1985 and 1990. Although small in absolute value, that secular variation implied a rate of change in C_{60} of nearly 3% per year, which—if unchanged—would lead to its doubling in less than four decades. Several of the small coefficients would even change by 100% in a decade or less! Most of the harmonics were predicted to vary at a slower rate, typically perhaps ~10% per decade; nevertheless, they represent major changes in the non-dipole field over a human lifetime.

Overall, the geomagnetic field—as represented by the single largest harmonic coefficient, the axial dipole term C_{10}—was predicted in IGRF-5 to be changing by ~0.08% per year. This is slower than individual non-dipole components by an order of magnitude (~8% per century versus ~10% per decade), yet it still

implies a geologically shocking change: 100% in less than 1300 yr!

However, none of those changes lasts; some amplify over time, while others wane. It turns out, for example, that the axial dipole intensity had been changing at a rate averaging less than 0.06% per year earlier in the twentieth century (see Lowrie, 2007); and, in IGRF-11 (see, e.g., Macmillan & Finlay, 2010) its rate of change between 2010 and 2015 was predicted to be just under 0.04% per year. Apparently, the relatively high rate of change toward the end of the twentieth century, seen in IGRF-5, was temporary.

Such capriciousness pervades the entire geomagnetic spectrum. For example, by 2010, according to IGRF-11, the degree 6 axial coefficient had reached a magnitude of 73 nT—unexpected in comparison to IGRF-5—and was now predicted to *decrease* between 2010 and 2015, though at a small rate of −0.3 nT/yr.

Given the amplitude of C_{60} and its rate of change as specified above for IGRF-5, what amplitude should it have had, 25 yr later? The rate of change of C_{60} went from +1.4 nT/yr in IGRF-5 to −0.3 nT/yr in IGRF-11; is that slowdown in its rate of change qualitatively consistent with the disparity in your prediction?

The changes in magnetic elements at each point on the Earth's surface lead to changes in the locations of magnetic poles, i.e. places where the inclination is ±90°. And changes in the degree 1 coefficients would lead to changes in the locations of the geomagnetic poles, and thus the tilt of the dipole axis. Figure 10.18 shows the changes in the magnetic poles and geomagnetic poles over time—the secular variation of the poles.

Changes in the geomagnetic field's secular variation have come to be called its **geomagnetic secular acceleration** (or **GSA**). Our discussion of the $C_{\ell m}$ clearly indicates a secular acceleration of at least some field coefficients. From Figure 10.18, it is equally clear that there has been a secular acceleration (or deceleration) of the poles as well.

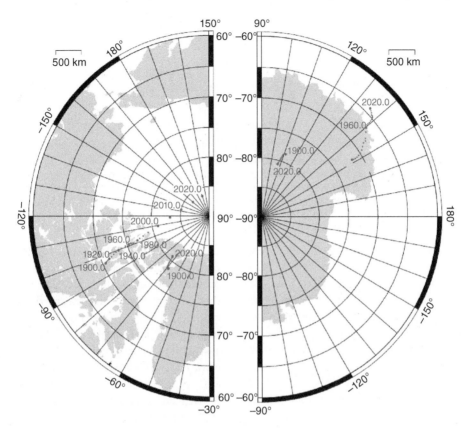

Figure 10.18 The secular motion of the magnetic and geomagnetic poles (defined in the text) according to IGRF-12: from 1900 through 2015, and predicted to 2020, for the northern hemisphere (left) and southern hemisphere (right). Magnetic poles (red) have moved more than the geomagnetic poles (blue), especially in the northern hemisphere.
Image from Thébault et al. (2015) / Springer Nature.

Over the time span considered in Figure 10.18, the shift in the poles associated with the dipole field (that is, the geomagnetic poles) is smaller by a factor of ∼3–5 than the shift in the poles associated with the total field (i.e., the magnetic poles). This is qualitatively consistent with what we saw regarding the secular variation in the magnitude of dipole versus non-dipole spherical harmonic coefficients.

In the hierarchy of geomagnetic secular variation, we must leave room for still another 'level' of variation. Using data from magnetic observatories, scientists began to realize in the 1970s that trends of increasing or decreasing GSV—such trends being another indication of secular acceleration—could themselves suddenly change (as in Figure 10.19), in a coordinated way worldwide (for historical

perspectives see, e.g., Malin et al., 1983; Courtillot & Le Mouël, 1984; Mandea et al., 2010). Just as we might call an abrupt change in the acceleration of a car we are riding in or a carousel that we are sitting on (cf. Chapter 4) a 'jerk,' these abrupt changes in magnetic acceleration are called **geomagnetic jerks** (also, less popularly, **secular variation impulses**).

Jerk events take place over several months to about a year (e.g., Mandea et al., 2010; Courtillot & Le Mouël, 1984), making them "abrupt" from a geomagnetic viewpoint. The first one to be identified (and the most well known) happened in 1969 (Courtillot et al., 1978) but subsequent analyses have revealed a few more recent events, plus numerous possible events going back a couple of centuries (see the historical review in Mandea et al., 2010).

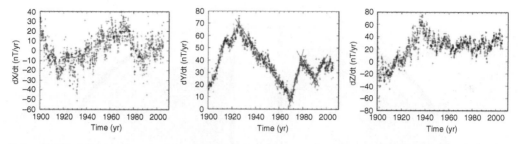

Figure 10.19 Geomagnetic jerk, as seen in observations at magnetic observatory Niemegk, in Germany. Secular variation (rate of change with time, d/dt, in nT/yr) of X, Y, and Z components during the twentieth century. Note the trends in these variations (highlighted for the Y-component by lines, in red, fit to the data); abrupt changes in the trends—jerks—occur roughly around 1920, 1970, and 1990 (and possibly other times as well). But note also that the jerks are not seen equally in all three components. Image from Mandea et al. (2010) / Springer Nature.

Do you think spherical harmonic coefficients should also exhibit geomagnetic jerks over time? Consider the variations in C_{60} from IGRF-5 to IGRF-11 noted earlier; are those trends consistent with a jerk? If so, roughly when might that jerk have occurred?

Check out the secular variation for C_{60} predicted from IGRF-6 through IGRF-10 (http://www.ngdc.noaa.gov/IAGA/vmod/igrf_old_models.html) to see how accurate your estimate was. (Also check out Chulliat & Maus, 2014!)

Using satellite as well as observatory data, Olsen and Mandea (2008) found evidence for a geomagnetic jerk in ~2003 (as did Chulliat & Maus for a jerk in 2011).

Modeling the flow in the core using the observed geomagnetic secular variation as a guide, Chambodut et al. (2007) found that extreme values in the pressure exerted by the flow at the top of the core correlate well with the occurrence of jerks. And, even more intriguingly, dynamical models by Aubert and Finlay (2019) have shown that jerks can be created from a "buoyancy release" during convective upwellings within the outer core; how that 'pulse' of buoyancy leads to a jerk depends on the existence of other phenomena, to be discussed later but including our very next topic.

Westward Drift. The global manifestation of geomagnetic secular variation is not limited to a shift in the poles. Figure 10.20a shows maps of the vertical component of the non-dipole

field in 1835 and in 1965 (TOU, 1971). These maps reveal that some of the highs and lows of the field have shifted westward during this time interval—a **westward drift** of the non-dipole field! For example, the contour high in 1835 labeled "A" has drifted from the east coast of Africa (longitude 45°E) to the west coast (0°E) by 1965, implying a drift rate of ~1/3°/yr. Considering the whole picture, a rate of ~0.2°/yr is generally quoted (e.g., Bullard et al., 1950; Yukutake, 1968).

One can easily imagine that, as a 'bubble' of non-dipole field—such as the one centered east of Greenland seen in Figure 10.20a (illustrating the field's vertical component)—drifts through a region, the elements of the field, including its inclination and declination, will sweep through a cycle of variation, producing the kinds of curves shown in Figure 10.3.

In the map of geomagnetic field intensity B displayed in Figure 10.7a, there is a feature of special interest, located at present over South America: a major low in field intensity, called the **South Atlantic Anomaly (SAA)** or **South Atlantic Magnetic Anomaly**. As that figure shows, it is one of the major contributors to the departure of the geomagnetic field from purely dipolar behavior. Figure 10.20b, from Finlay et al. (2010a; see also Macmillan et al., 2009), suggests how this feature has been evolving over time, as its center drifts westward and the low in intensity deepens.

The westward drift is primarily a feature of the non-dipole field; but the dipole

Figure 10.20 Aspects of westward drift.
(a) The non-dipole field drifts westward over time.
(Upper) Vertical component of the non-dipole field for 1835.
(Lower) Vertical component of the non-dipole field for 1965.
Dashed lines are negative in upper plot and positive in lower plot; contour interval is 20 μT (0.02 mT). At many locations, the field was changing during 1835–1965; at the same time, some of the highs and lows of the field were also drifting westward.
TOU (1971) / Open University Press.

(a)

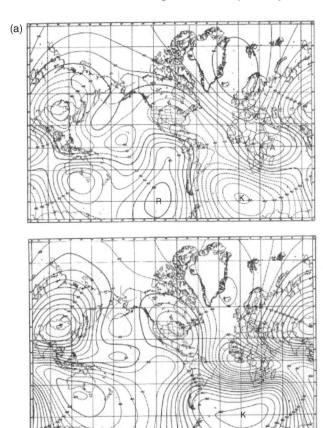

field exhibits a drifting character as well, though—like the change in its intensity, relative to the non-dipole field—at a slower rate. According to McDonald and Gunst (1968) and Barton (1989), and as seen in Figure 10.18, the longitude of the geomagnetic poles shifted westward by ∼4° over the twentieth century.

Another way to document the westward drift is illustrated in Figure 10.20c, which shows the vertical (radial) component of the total field (dipole + non-dipole) predicted, in this case for years 1690 and 1990 by Jackson et al. (2000), at the core boundary. Although the total field is shown, the downward continuation that enabled the prediction also serves to highlight the non-dipole field, as we discussed above. In the figure the westward drift of feature "A" is still seen, but begins farther to the east.

Still another illustration is provided in Figure 10.20d, showing the westward drift of

the GSV—yes, the field *and* the secular variation both drift westward! The rate of westward drift of the secular variation is estimated at ∼0.3°/yr (e.g., Bullard et al., 1950; Yukutake, 1968).

Even more provocatively, the rate of westward drift of the non-dipole field appears to be affected by whatever phenomenon causes geomagnetic jerks: for example, a trend of decreasing drift rate in the decades preceding the 1969 jerk changed to an increasing drift rate afterward (Courtillot & Le Mouël, 1984).

Finally, we note that the eccentric dipole best fitting the geomagnetic field (or, at a minimum, fitting the dipole + quadrupole field) *also* drifts westward over time (e.g., Vestine, 1953; Vestine & Kahle, 1968; Cain et al., 1985, and references therein). Interestingly, irregularities in its motion, on decadal or multidecadal timescales, were found to correlate

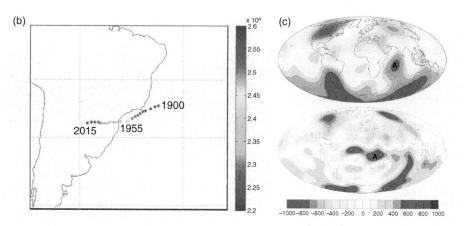

Figure 10.20 Aspects of westward drift.
(b) Center of the South Atlantic Anomaly from 1900 to 2015. Its drift westward over South America is apparent, as is a trend toward lower total field intensity (color code in nT, yielding a red-to-blue range from 26 to 22 µT).
In recent years, there has been some concern about the impact of this lower field intensity on Earth's external field, with potential consequences for orbiting satellites and for air travelers (see, e.g., Olson & Amit, 2006).
Image from Finlay et al. (2010) / Springer Nature.
(c) Drift of the vertical (radial) component of the geomagnetic field, as seen at the core boundary.
(Upper) Radial component of the field for 1690.
(Lower) Radial component of the field for 1990.
Blue contours indicate field lines are into the core; red are out; contour interval is 100 µT.
Note that the contoured high in Figure 10.20a labeled "A" (and which we have labeled "A" here as well) was, in 1690, southwest of India and is, in 1990, in central Africa, as seen at the core boundary.
Images from Jackson et al. (2000) / The Royal Society.

with variations in Earth's spin on the same timescales; a slowing of Earth's rotation rate, for example, corresponds to a slowing of the eccentric dipole's westward drift.

Such a correlation suggests that some type of balance or compensation is at play—perhaps conservation of angular momentum, if we assume that the core fluid is moving *with* the field: the increase in angular momentum from reduced westward drift of the core balances the loss in Earth's rotational angular momentum implied by the increase in the length of day. Both the assumption and the potential role of angular momentum conservation will be explored below.

All of the preceding results are consistent with each other and support the idea that the westward drift is a real and fundamental characteristic of the core field. But, inevitably, reality is more challenging (see Nilsson et al., 2020 for an extensive summary of those

complications). First of all, a westward drift is evident in Figure 10.20 a-d; but clearly not all highs and lows participate in the drift. That has led some researchers to view the non-dipole field as the sum of a 'standing' secular variation, involving places where the field may change over time but there is no drift, plus a drifting secular variation, in which the field changes and drifts (e.g., Yukutake & Tachinaka, 1969; McElhinny & Merrill, 1975). One variation on this theme hypothesizes that the drifting field is confined to one hemisphere, the 'Atlantic' hemisphere, between ±90° longitude, ultimately reflecting the weakness of GSV in the Pacific (e.g., Holme et al., 2011; McElhinny & Merrill, 1975; Bloxham & Gubbins, 1985). Alternatively, the drift may be global but latitude dependent, with a much lower (Yukutake, 1967; see also Lowrie, 2007; Livermore et al., 2017) or higher (see Finlay & Jackson, 2003) drift rate at low latitudes. Still

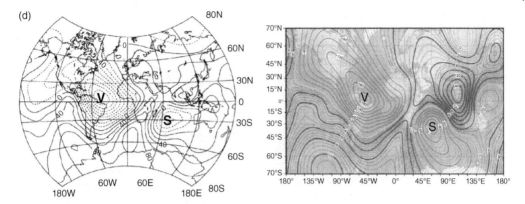

Figure 10.20 Aspects of westward drift.
(d) Westward drift of the geomagnetic secular variation.
(Left) Time derivative of the radial (vertical) component of the field at 1990, observed at Earth's surface.
Contour interval is 20 nT/yr; dashed contours are negative.
Image adapted from Langel (1992) / Society of Geomagnetism and Earth.
(Right) Time derivative of the radial (vertical) component of the field for 2015–2020, at Earth's surface.
Contour interval is 5 nT/yr; blue contours are negative.
Image adaptedfrom Chulliat et al. (2015) / U.S. Department of Defense.
Despite the different map projections, we can note the shift to the west of the two major negative highs:
one (labeled "S"), centered at 30°S, ~67°E in 1990, drifts to ~33°S, 60°E; the other (labeled "V"), just north
of the equator, drifts from ~45°W to ~50°W by 2015.
What drift rates do these motions correspond to?

other research suggests the possibility of eastward drift in an equatorial zone (Yukutake, 1968) or at high latitude (Bloxham & Gubbins, 1985).

The westward drift has played a significant role in our understanding of core dynamics and the generation of the core field. We elaborate here on three major examples.

- Thinking of the non-dipole field as the result of smaller dipole sources (local or regional electrical current loops) distributed throughout the outermost core: if the non-dipole field is drifting westward, then so must those sources. If the sources derive from localized or regional flow patterns, then those flow patterns—that is, the core fluid—must also be moving westward, relative to the mantle and presumably the rest of the core.

  The westward drift thus provides a baseline, order-of-magnitude estimate of flow speeds in the core. In Chapter 4, we retrospectively demonstrated that, if a particle at position $\vec{R}$ is orbiting around an axis with an angular velocity of $\vec{\Omega}$, then its linear velocity as it orbits around is $\vec{\Omega} \times \vec{R}$; the magnitude of that velocity is $\Omega R \sin\theta$, where θ is the angle between $\vec{\Omega}$ and $\vec{R}$. For a parcel of fluid at the equator of the outermost core (thus $\theta = 90°$, $R = 3{,}480$ km), moving around with the non-dipole field, its linear velocity associated with the westward drift is thus ~12 km/yr or ~4 × 10⁻⁴ m/sec, if we take the drift rate to be $\Omega \sim 0.2°$/yr (= 0.0035 rad/yr).

In Chapter 9, we saw that a dimensionless number called the Reynolds number and defined as

$$\mathrm{R_E} \equiv \frac{LV}{\eta'}$$

could be viewed as an indication of the state of a fluid, i.e., whether flow at a typical speed of V is laminar or turbulent. Taking the outer core's representative dimension L to be ~10³ km, the 'classic' estimate of core viscosity η', 10⁻⁶ m²/sec (Gans, 1972, with similar experimental/theoretical results by

Poirier, 1988; Vočadlo et al., 2000; Dobson, 2002; Pozzo et al., 2013; Posner et al., 2017 and references therein, Li et al., 2022, and other references cited in Chapter 8) yields a Reynolds number of $\sim 4 \times 10^8$. Some of those viscosity estimates account for the light alloy in the core (which tends to increase the viscosity slightly—see Posner et al., 2017); and some of them predict an increase with depth, by up to an order of magnitude at the inner-core boundary (Dobson, 2002, or for a radically different view see Smylie et al., 2009 also Cormier, 2009 but then Zharkov, 2009). Such a Reynolds number implies that, amazingly enough, the outer core is in a turbulent state—despite flowing at such a slow drift rate. Thus, when discussing how the core field regenerates itself, we will need to consider turbulent dynamo processes.

Turbulence explains why estimates of the core's viscosity have spanned such a wide range—up to 12 orders of magnitude in pre-1991 estimates (Lumb & Aldridge, 1991), 15 orders of magnitude in the estimates cited in Chapter 8. Higher viscosity values typically correspond to observational analyses (e.g., estimates by Koot et al., 2010 derived from observations of perturbations in Earth's rotation), as opposed to experimental or theoretical research, suggesting that those higher values were in effect the *turbulent* viscosity of the outer core (just as, in Chapter 6, we inferred tidal dissipation in shallow seas could be associated with a turbulent oceanic viscosity at least ~ 9 orders of magnitude higher than the molecular viscosity of water). Molodenskii (2010) estimates that turbulence increases the outer core's effective η' by a factor of 10^6, similar to what is implied by our discussion in Chapter 8.

Turbulence may provide a physical explanation of why many studies have found the westward drift to be so nonuniform. For that matter, analyses of geomagnetic jerks have revealed several candidate jerks to be regional rather than global, or global but with delays from one hemisphere to the

other; such lack of coherence might also be a consequence of significant turbulence throughout the outer core.

- Every 9,000 years, if current rates do not change, the non-dipole field will circle around Earth's axis five times, and the geomagnetic poles—the poles of the dipole field—will circle around once. On those timescales or longer, then, the geomagnetic field may average to an axially symmetric field: an axial dipole + axial quadrupole + ... field (e.g., Schneider & Kent, 1990; Constable et al., 2016). As a good first approximation, then, paleomagnetic studies can treat their data as records of past axial dipole fields. And on long-enough timescales, that dipole axis will approximate Earth's rotation axis as well.

As implied by Figures 10.1a and 10.2a (and as first realized by Gilbert, 1600), the tilt (inclination) of flux lines due to a centered axial dipole has a very specific relation to the latitude (or colatitude) of the site at which the tilt is being measured. Consequently, if the inclination has been determined from paleomagnetic measurements, the 'paleolatitude' of that site—that is, the central angle between the site and the paleo-equator (implying, in turn, the central angle between the site and the paleopole)—can be deduced. Our spherical harmonic development earlier showed that the magnetic field of a centered axial dipole could be written

$$\text{constant} \cdot \frac{2\cos\theta}{r^3}\,\hat{r} + \text{constant} \cdot \frac{\sin\theta}{r^3}\,\hat{\theta}$$

in spherical coordinates. At any site on the Earth's surface, the direction of $\hat{r}$ is locally upward, whereas $\hat{\theta}$ points directly southward; it follows (see Figure 10.21a and caption) that the inclination i of the field at that location is given by

$$\tan i = \frac{\text{radial component}}{\text{horizontal component}}$$

$$= \frac{\text{constant} \cdot \dfrac{2\cos\theta}{r^3}}{\text{constant} \cdot \dfrac{\sin\theta}{r^3}} = \frac{2\cos\theta}{\sin\theta}.$$

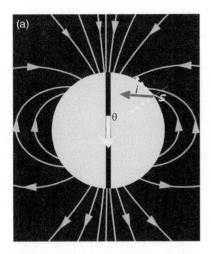

Figure 10.21 The paleomagnetic field as a vector quantity.
(a) At site S on the Earth's surface, at colatitude θ, the field due to an axial dipole at the Earth's center has both horizontal (north-south) and radial (vertically upward or downward) components, as discussed in the text.
At this northern-hemisphere location, the field (bold red arrow), which of course is tangent to a flux line, points into the Earth. Both components of the field are shown as heavy dotted yellow lines. The inclination i is the angle between the field and the horizontal; the tangent of that angle is the ratio of "opposite over adjacent" sides, thus B_r/B_θ. Image credits as in Figure 10.1a.

geographic location of that pole to be predicted (Figure 10.21b); this pole is called the **virtual geomagnetic pole** or **VGP**. The accuracy of the VGP determination (or of the underlying assumptions) can be tested by assessing whether different sites of the same age yield the same pole position.

Furthermore, analysis of paleomagnetic data from sites of different ages reveals that the pole has moved over time; such motion is called an **apparent polar wander path**. Very simply, there are two possible causes of polar wander: either the pole itself has actually been moving, relative to all the sites and to the whole Earth—a **true polar wander**; or the sites have been moving, relative to each other and to the pole. In the first case, the sites would show the same or consistent apparent polar wandering paths; in the second case, different sites would show different and inconsistent polar wander paths.

Of course, continental drift will generally cause the second situation to prevail (and those differing polar wander paths can be used to quantify or constrain past plate motions). Currently, however, actual true polar wander is also taking place, at a rate (~1° of latitude

This can be written

$$\tan i = 2\cot\theta,$$

which we can use to determine the angular distance θ of the paleopole from the site. Technically, θ is the paleo-colatitude of the site; with latitude and colatitude complementary angles, and the paleolatitude $\phi \equiv (90° - \theta)$ measuring the angular distance between the site and the paleo-equator, we could write this expression as

$$\tan i = \frac{2\cos\theta}{\sin\theta} = \frac{2\sin\phi}{\cos\phi} = 2\tan\phi.$$

With the analyzed sample's horizontal orientation at the site also measured, the direction of the ancient field—the direction from the site toward the paleopole (a sort of paleo-declination)—allows the

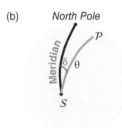

Figure 10.21 The paleomagnetic field as a vector quantity.
(b) A sample at the site labeled point S on the Earth's surface has recorded Earth's ancient magnetic field. 1. Measurements reveal the elements of the ancient field to include inclination i and declination δ (as well as giving an estimate of the field intensity). 2. As discussed in the text, the inclination implies a specific value for the angular distance θ from the site to the paleopole; but in which direction from S? 3. The declination δ gives us the correct direction toward the paleopole, P.

per Myr), which is an order of magnitude faster (e.g., Dickman, 1979) than typical paleomagnetic rates of apparent polar wander. The connection of this true polar wander to Pleistocene deglaciation, though, as discussed in Chapter 5, suggests that fast true polar wander will generally be limited to geologically sudden events, such as times of rapid and massive global climate change.

- The axial symmetry induced in the average field by its westward drift may in turn have consequences of a profound nature for the field. Those consequences emerge from rules derived by Cowling (1933, 1957) concerning the effectiveness of dynamo action under different conditions; despite subsequent verifications, modifications, and generalizations by numerous researchers (including Backus & Chandrasekhar, 1956; Todoeschuck & Rochester, 1980; Hide & Palmer, 1982; Roberts & Soward, 1992; and Núñez, 1996, but see also Ingraham, 1995), all versions are still known as "**Cowling's theorem.**" To begin with, we consider the earliest version of the theorem, which states (more or less) that "it is impossible to maintain a steady axially symmetric magnetic field by fluid motions."

A simple proof of this version of Cowling's theorem, though not rigorous, may be instructive. We consider first the most basic field, a pure dipole field created by a current loop (as in Figure 10.1c). By symmetry, each flux line must be planar (i.e., contained within a meridional (N-S) plane—any plane intersecting the dipole axis) and circle around the current loop (in Figure 10.1c, we see flux lines in the plane of the page, circling the loop). These circles become smaller, closer to the loop, and vanish at the loop (as often explained, the flux lines are in opposite directions on opposite sides of each circle, so they cancel out in the limit of approaching the loop). That is, the field equals zero on the current loop.

As we have seen throughout this chapter, electrical currents and thus magnetic fields can be created by moving an electrical conductor through a magnetic field, or by imposing a changing magnetic field on the conducting material. Both of these types of induction are unnecessary in Figure 10.1c *if* the current producing the dipole field is powered by a battery inside the loop, or a current is generated elsewhere and then fed to the loop by a wire, not shown. *But* if we want the dipole field to be reinforced by dynamo action, induction is the only way to achieve that: either move the conducting material comprising the loop through the preexisting magnetic field, or require that field to vary over time.

Can you see where this is going?

Neither of these mechanisms work under the conditions imposed by the theorem: if the field is steady, i.e. unchanging with time, 'motional' induction is the only way to induce currents in the loop; but because the field is symmetric, it must be zero at the loop (as we saw), so moving the loop generates nothing either.

Subsequent research (cited above) has shown that dynamo failure can also result from other imposed conditions, and can happen even if the magnetic field is unsteady (time dependent); but in all cases, the field can fail to be sustained *only* if it is axially symmetric:

> "*It is impossible to maintain by fluid motions an axially symmetric magnetic field.*"

In general terms, then, we might conclude that the non-dipole field, ever changing, helps keep the dipole field alive by providing an asymmetry that the dipole field can 'feed' off of. And, in this context, we can view the core's inferred turbulence as healthy for the field, providing chaotically varying nonaxisymmetric flow that will help generate an asymmetric distribution of electrical currents and regenerate the field. Of course, there is (much) more to the story, as we will discover in the remainder of this chapter.

On longer timescales, however, we can anticipate things becoming more extreme: averaged over many cycles of westward drift, the total field will more closely approximate an axially symmetric field, making regeneration more of a challenge. Later in this chapter, we will estimate a timescale for the field to weaken in the absence of regeneration. On such timescales—longer than we have discussed so far—we should not find it surprising to see substantial variations in the magnitude of the field, as those capricious fluctuations noted earlier, in the face of axial symmetry, fail to balance each other and maintain the 'status quo,' and instead build up to major variations. And we cannot rule out the grandest secular variation of all, every now and then, when the dipole magnitude decreases drastically: a change in the polarity of the dipole, amounting to a reversal in the character of the overall field.

This last philosophic implication—that reversals are fundamentally a part of the dynamo process—is supported by both familiar and exotic evidence. First, in our discussion of the 11-yr cycle of sunspot activity (Chapter 3), we noted that the underlying periodicity is a 22-yr cycle in which the magnetic field of the Sun has its own 'normal' polarity for 11 yr then a 'reversed' polarity for the next 11 yr (observational aspects of the solar cycle are discussed, e.g., in Hathaway, 2015, and theoretical aspects in Charbonneau, 2010). Certainly there must be significant differences between the solar and terrestrial dynamos, not only because their timescales are so different but because reversals in one are so regular and in the other almost random (see, e.g., Parker, 1971); but the fact that both experience reversals suggests that the occurrence

of reversals is an even more fundamental property.

Our second, more exotic evidence that reversals are to be expected with dynamos comes from a planet that currently has no dynamo. In our brief discussion earlier in this chapter about the magnetic spectrum of Mars, we explained why the observed crustal magnetization was likely remanence from a dynamo that operated within Mars' core early in its history.

Analysis of data from the Mars Global Surveyor led Connerney et al. (1999, but see also Harrison, 2000) to identify coherent magnetic stripes covering a fairly large region of the Martian highlands, suggesting that Earth-like plate tectonics—a 'seafloor' spreading, of sorts—prevailed on the very young Mars. Further analysis by Connerney et al. (2005) allowed the data to be interpreted in terms of familiar tectonic structures (tectonic-type features have also been inferred by Yin (2012) for portions of *Valles Marineris*, the 'Grand Canyon of Mars' seen in Chapter 1, Figure 1.30). Despite earlier work by Sleep (1994) comprehensively arguing that Martian tectonics could be Earth-like, rather than a single immobile plate (a global 'stagnant lid') as previously envisioned, however, the tectonic implication of those stripes has remained controversial (see, e.g., Kerr, 1999; Kobayashi & Sprenke, 2010). One alternative explanation, developed by Kobayashi & Sprenke (2010), postulates a single global plate that can move relative to the asthenosphere below, and explains the stripes in terms of Martian hotspot (plume) tracks resulting from a net lithospheric motion. We note that, if plate tectonics could ever have operated on Mars, the most likely time would have been early in Mars' history, *before* the dynamo was suppressed and mantle convection was inhibited (see Chapter 1).

In any case, the alternating magnetism recorded in those stripes seems unquestionable—and tells us that the Martian dynamo underwent polarity reversals during its brief existence!

A last philosophical point: the equations governing the dynamo, which we will develop toward the end of this chapter, tell us that whatever conditions support a particular magnetic field will *also* support the reversed field; that is, if $\vec{B}$ is a solution to the dynamo equations, then $-\vec{B}$ is also a solution (see also Roberts & King, 2013). (In fact, with the equations so unequivocal, it was commonly thought in the early days of numerical dynamo simulations that such models had to achieve a reversal before they could be considered successful!)

If we want more than a philosophical understanding of reversals, though, we will need to study paleomagnetic data from our own planet (as well as the dynamo governing equations). Such data will shed light on what happens during a reversal, and on what timescales they occur. Paleomagnetic 'snapshots' are presented in the next subsection of this chapter; before getting to that, we complete our discussion of the westward drift.

Westward Drift, Differential Rotation, and Core-Mantle Coupling. The relative motion implied by the westward drift is called a **differential rotation** of the outermost core. We observe it relative to Earth's surface; theoretically, we expect that it reflects relative motions within the core, rather than the entire core rotating rigidly relative to the mantle. The westward drift is achieved by the relevant portion (e.g., the outermost layer) of the core rotating west-to-east, like the rest of the Earth, but slower; thus, all of the Earth has a prograde angular velocity, but that of the outer(most) core is slightly less.

One mechanism by which a differential rotation *within* the core might be created is, at first glance, straightforward: conservation of angular momentum, applied to each parcel of core fluid, as the core convects within a rotating Earth (Bullard et al., 1950). For now,

we will invoke the 'traditional' description of convection; in the last section of this chapter, though, we will have to consider some fairly extreme ways that convection might be occurring. Parcels rise or sink by gravity, whose direction within the core is toward the center; the direction of rise or fall will be radial as well. Sinking parcels get closer to the rotation axis, decreasing their moment of inertia and (like those perennial ice skaters forever pulling their arms in) causing them to speed up to compensate; rising parcels have the opposite experience so their rotation slows as they rise. With this mechanism, the outer layers of the core should be progressively left behind, resulting in a westward drift of the outermost core relative to the deeper core.

However, the westward drift we need to explain—the differential rotation we observe—is a differential rotation between the outer core, where the sources of the non-dipole field reside, and the solid earth above, upon which our observatories sit. The simplest explanation is that conservation of angular momentum, though speeding up the deeper layers and slowing the outer layers, does not change the average rotational speed of the outer core; because the non-dipole field we observe is confined to the outer layers, its westward drift relative to the rest of the core is seen as a westward drift relative to the Earth.

More precisely, even when the convection is radial, the dependence of each parcel's momentum on how near it is to the rotation axis implies that its differential rotation will similarly be a function of distance from that axis. This dependence would result in progressively slower 'cylinders' rather than spherical shells, farther away from the axis (see Figure 10.22a). As a consequence, the outermost-core fluid at the equator would be most left behind, while the outermost-core fluid near the poles would hardly be left behind at all.

Incidentally, either of these explanations of the observed differential rotation relies on the viscosity of the outer core being low enough

(a)

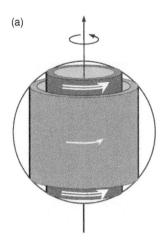

Figure 10.22 Alternative views of secular variation and westward drift.
(a) Cylindrical differential rotation: as a result of convection in the core, cylindrical volumes—rather than spherical shells—farther from the rotation axis must spin progressively slower, to conserve the core's angular momentum.
This type of constraint on the core's flow is discussed later in this chapter.
Image from Holme (2015) / with permission of Elsevier; original illustration by S. Zatman.

that the core fluid can move relative to the bottom of the mantle without resistance—outside of a very small viscous boundary layer. We can estimate the thickness of that layer in two slightly different ways using the mathematics of Chapter 9, combined with the strategy we followed there to calculate the thickness of thermal boundary layers. First, defining the viscous boundary layer as a region where momentum diffusion and advection are comparable (equivalently, a layer in which viscous drag is not negligible), we have

$$\eta' \nabla^2(\rho \vec{v}) \sim (\vec{v} \cdot \vec{\nabla}) \rho \vec{v};$$

here $\vec{v}$ is fluid velocity, and using density ρ allows us to convert the advection of velocity term from Chapter 9 into a momentum advection term. Scaling terms, with characteristic fluid velocity V_C and boundary layer dimension δ, we find

$$\delta \sim \eta'/V_C.$$

Using parameter values specified earlier, we find $\delta \sim \frac{1}{4} \times 10^{-2}$ m; even with a turbulent core, δ is probably only a few kilometers. Alternatively, recognizing the importance of rotation—a fact we will come to appreciate even more, later in this chapter—we can envision the boundary layer (in this situation, generally called the **Ekman boundary layer**) as the only part of the core where viscous drag manages to balance the Coriolis force. In this case, we could write

$$\eta' \nabla^2(\rho \vec{v}) \sim 2\rho \vec{\Omega} \times \vec{v}$$

within that boundary layer, where $\vec{\Omega}$ is the Earth's (or core's) angular velocity. Denoting that layer's thickness by δ_E, we find

$$\delta_E \sim \sqrt{\eta'/2\Omega};$$

an exact solution of the force balance (e.g., Pedlosky, 1982) would yield the more frequently quoted estimate of Ekman boundary layer thickness,

$$\delta_E = \sqrt{\eta'/\Omega},$$

implying in this situation $\delta_E \sim 10^{-1}$ m under laminar conditions and $\sim 10^2$ m under turbulent conditions.

Other explanations of an observed differential rotation have also been proposed. For example, Bullard et al. (1950) argued that electromagnetic coupling between the lower mantle and the core as a whole could lock the mantle to the 'average core,' which as just noted rotates somewhat faster than the outermost layer or shell. This in turn would require that the lowermost mantle be electrically conducting (to support the fields that would lock the core to the mantle); the analysis by Bullard et al. (1950) considered a conductivity of 10 S/m.

A variety of subsequent work lends support for such a value of σ. Frequency-dependent extrapolations of mid-mantle conductivity models to the lowermost mantle by Olsen (1999) find σ to be 10 S/m or greater. Laboratory experiments (e.g., Shankland et al., 1993 and most other experimental work cited therein, and Xu et al., 2000, 2003) extrapolated to high pressure and temperature find

~10 S/m to be a reasonable upper bound on σ in or near D″, though experiments extrapolated by Poirier and Peyronneau (1992) to the bottom of the mantle yielded ~200 S/m. Constraints from geomagnetic jerks imply a value for the average lower-mantle conductivity that in turn allows σ ≥ 10 S/m in or near D″ (Mandea Alexandrescu et al., 1999) or restricts σ to be less than 100 S/m (Ducruix et al., 1980).

Observed decadal changes in the length of day can be accounted for by coupling to the core, with a core flow constrained by the GSV, if the lowermost mantle has an electrical conductance (σ × layer thickness) of ~10^8 S (Holme, 1998; see also Buffett, 2015, 1992). With a layer thickness corresponding to all of D″, we might then infer a conductivity of ~500 S/m or more. However, positing a model where mantle conductivity is low except in a thin layer above the core boundary, Holme and de Viron (2013) found that coupling of the core to that layer could account for time delays inferred between jerks and changes in the length of day if (with a layer conductance of ~10^8 S) that layer was right above the core boundary, and no more than 100 km thick, implying that the layer's conductivity had to exceed ~1,000 S/m.

It may be worth considering a historical perspective on such results.

- Jerks can constrain the conductivity in or near D″ only if, as we have implicitly been assuming, they are a core phenomenon. Simply because we employ the spherical harmonics of the internal field to analyze jerks, however, does not automatically guarantee that they are part of the internal field. Characterized by relatively rapid variations in time, it is conceivable that they instead originate as external field events that have somehow contaminated the harmonic coefficients of the internal field (in which case they should be treated simply as 'noise' in the coefficients). In fact, for several decades the 'rule of thumb' in geomagnetism was that any variations in the surface field shorter than 3–4 yr were probably external in origin (this was also noted by Courtillot & Le Mouël, 1984). But we can point to the work of several researchers (e.g., Malin & Hodder, 1982; Bloxham et al., 2002; see also references cited in Chambodut et al., 2007) as establishing the internal nature of jerks.

- For us to be able to see fluctuations in the core field with periods shorter than 3 or 4 yr—that is, the ability of the mantle to allow approximately yearly variations from the core to reach the surface—has implications for lower-mantle conductivity, as we have just reviewed; but some clarification is in order. We have noted that magnetic fields weaken by diffusion, the process we encountered when discussing heat flow in Chapter 9, and that as a result they also experience a skin effect which attenuates shorter-period variations of the field more than longer-period ones.

But there is one significant difference between thermal and magnetic skin effects. With thermal diffusion, the skin depth depended directly on the thermal conductivity (specifically, skin depth ~$\sqrt{k/\omega}$, as we saw in Chapter 9, where $2\pi/\omega$ is the period of the variation), so that heat flow variations penetrate deeper through materials of higher conductivity. In contrast, as you will prove in a homework exercise for this chapter, the magnetic skin depth depends *inversely* on electrical conductivity (that is, skin depth ~$\sqrt{1/(\sigma\omega)}$): magnetic field fluctuations penetrate more deeply through poorly conducting materials; in good conductors, electrical currents induced in response to the fluctuations are able to effectively cancel them out.

It may, therefore, be more precise to call this phenomenon a **screening effect**, in that highly conducting materials *screen out* (attenuate) magnetic field fluctuations more effectively than do poor conductors. As just stated, though, it is still true that the skin depth is greater for longer-period variations: shorter-period magnetic fluctuations are screened out more completely than longer-period ones.

The fact that there is *any* screening of core fluctuations—even if, according to the old rule of thumb, it was 'only' at periods shorter than 3 or 4 yr—requires that the mantle be treated as a decent electrical conductor, despite its being comprised of semiconducting silicates; most likely, this is the result of high temperatures or a high iron content, in the lower(most) mantle. The subsequent detection and identification of shorter-period features (i.e., jerks) originating in the core—the fact that those 'impulses' were not fully screened out—implied lower temperatures and/or iron content than previously thought. But, in the end, its impact on our understanding of lower-mantle conditions is not so dramatic. Imagine, for example, if the decrease in the period of fluctuations screened out, by a factor of 4 (from ~4 yr for the old rule to ~1 yr for jerks), was matched by a decrease in σ by the same factor of 4; the skin depth and thus the spatial attenuation would be unchanged. The problem is that the range of experimental and observational results for σ in the lower mantle, such as those cited above, allow a factor of 4 adjustment in it without forcing us to modify our inferences of lower-mantle conditions.

What we can say, beyond any consideration of the state of the lower mantle, is that the detection of approximately yearly fluctuations in the core field has widened our window into core dynamics a bit; activities near the top of the core are obscured (screened out) only for timescales less than a year.

According to Stacey and Davis (2008), electromagnetic coupling between core and mantle is, however, problematic for any reasonable value of σ in the lower mantle; in addition to insuring over the long term that the mantle keeps pace with the average core, it is strong enough that it would lock the topmost core to the lowermost mantle in perhaps a century, eliminating the westward drift (were the drift not continually excited).

Alternative mechanisms coupling the mantle to the 'average' core have also been envisioned (e.g., Rochester, 1970; see also Buffett, 1996 and Dumberry & Bloxham, 2006, e.g., who discuss their effectiveness on long timescales). One of the more subtle types of mechanisms is gravitational coupling, acting between a mantle and core in which neither is perfectly spherically symmetric (spherically symmetric masses, like point masses, cannot exert gravitational torques on each other). The mantle's asymmetries would stem from its ellipsoidal shape and from density anomalies distributed throughout; the core's would stem from density anomalies associated with outer core convection (but see also Stevenson, 1987a), the slightly ellipsoidal shape of the inner core, and irregularities in the inner core (see also, e.g., Szeto & Smylie, 1984; Aurnou & Olson, 2000; Dumberry, 2008; Aubert et al., 2008)—perhaps even hemisphere-scale 'lopsidedness' (Alboussière et al., 2010; Monnereau et al., 2010; also Frost et al., 2021 and references therein). Mechanisms involving the inner core would rely on the tight electromagnetic coupling between inner and outer core to insure the latter's participation (e.g., Gubbins, 1981 and Aurnou & Olson, 2000; but see also Aubert & Dumberry, 2011).

Stacey and Davis (2008) point to gravitational coupling as being able to maintain a westward drift where the mantle stays locked to the 'average core;' this is substantiated by Aubert (2013; see Aubert et al., 2013 for more details, including the contribution to this mechanism by core convection, especially at the base of the outer core).

Differential rotation is thought to play an essential role in generating the Earth's magnetic field. This will be explored in the final section of this chapter.

A Truly Alternative View. Our explanation of differential rotation in the core as a consequence of convection subject to angular momentum conservation is a simplified one in several respects; for example, we have neglected turbulence and envisioned idealized convection cells. Such complications are coincidentally avoided with a completely different explanation for the westward drift; this alternative, in its original form, postulated that the flow of outermost-core fluid past topographic irregularities at the core boundary—bumps on the core—generates a certain kind of hydromagnetic wave (discussed later), called an Alfvén wave (Hide, 1966), which ultimately is observed at the surface as a geomagnetic secular variation. In other versions, the source of the waves is not restricted to the core boundary or to bumps, and other types of hydromagnetic waves—termed *slow magnetic Rossby waves* (Hori et al., 2015), *Magnetic-Coriolis (MC) waves* (Finlay et al., 2010b; Gillet et al., 2022), and *Magnetic-Archimedean-Coriolis (MAC) waves* (Buffett et al., 2016), depending on the force balance creating them—are envisioned. ("Archimedean" refers to the buoyancy force; for example, we invoked Archimedes' principle in Chapter 5 as the basis for why things like oceanic and continental crust float.)

These waves, oscillations of both the fluid and the magnetic field, can propagate both eastward and westward. However, their source (e.g., the bump) is carried to the east by the Earth's daily spin; the recession of the source from its westward-moving waves causes a Doppler shift of those waves to lower frequencies (Figure 10.22b, left). The electrically conducting mantle screens all magnetic fluctuations emanating from the core, to some extent; but those lower-frequency variations, with periods of years to centuries, are only mildly attenuated, and make it up to the surface. In contrast, as a source is carried to the east by the rotating Earth, eastward-moving waves will shift to higher frequencies, resulting in periods of the order of days; those fluctuations in the magnetic field get screened out by the overlying mantle much more completely before they can reach the surface.

At the Earth's surface, then, we will be unaware of any eastward-propagating waves but will observe the slowly oscillating westward-propagating waves as a secular variation of the field with an overall sense of westward drift. According to this mechanism, parts or even all of the secular variation and westward drift are actually a partially screened collection of hydromagnetic oscillations superimposed on the main field. Support for this viewpoint comes from a different way to represent the geomagnetic secular variation, one in which the observations are filtered to highlight the temporally and spatially variable part of the field at specific latitudes; as an illustration, Figure 10.22b (right) from Hulot et al. (2010) shows changes over the past 300 yr in the geomagnetic field along the equator. The tilt of the contours shows the ongoing westward drift of the highs and lows of the equatorial field; for example, the high centered near ~80°E in 1650 exhibits an ongoing drift bringing it to ~20°E by 1950, yielding a drift amounting to ~60°, thus a drift rate of ~0.2°/yr. Moreover, at any year, the highs and lows themselves correspond to 4 or 5 wavelengths of east-west oscillation—and they have persisted for over 300 yr!

Still Another Alternative View, With a Bigger Role for the Inner Core. We have already discussed the growth of the inner core as a way to power the geodynamo. In recent years, though, models of the core have been developed in which features of the secular variation also appear as a consequence of the inner core.

- Aubert (2013) showed that gravitational coupling of the inner core to the mantle, acting to maintain the inner core's alignment, causes outer core fluid to circulate around it, gyre-like, producing a westward drift. Such a gyre had already been inferred from observations and field modeling under various conditions (Aubert, 2014; see Pais & Jault, 2008; Gillet et al., 2009; also Barrois et al., 2018; Kloss & Finlay, 2019, and references within).

- Aubert et al. (2013) determined that asymmetric growth of the inner core—that 'lopsidedness' cited above, inferred from analysis of seismic waves (see Buffett, 2010a

(b)

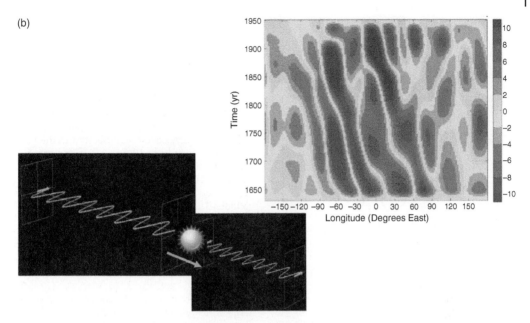

Figure 10.22 Alternative views of secular variation and westward drift.

(b) Hydromagnetic waves in the outer core.

(Left) These waves are transverse oscillations of both the magnetic field and the fluid. In this simplified scenario, their source, symbolized by ☼, is moving in the direction shown by the arrow. Waves are generated in different directions (perhaps, for example, following along the flux lines of the core's magnetic field); waves shown here generated in the same direction as the source's motion are Doppler-shifted to higher frequencies/shorter wavelengths, whereas waves generated in the opposite direction (and 'left behind') are Doppler-shifted to lower frequencies/longer wavelengths.

Image adapted from https://bestanimations.com/gifs/Horizontally-Polarized-Light-Wave.html.

(Right) Observations of Earth's main field suggest the existence of such waves. Here we see the radial component B_r of the main field at the equator, after removal of (i) the time-averaged axisymmetric field (axial dipole + axial quadrupole + etc., averaged over time) and (ii) variations on timescales exceeding 400 yr.

The amplitude of this residual B_r (the color code is in units of 10^4 nT) is plotted at each longitude of the equator versus time. Moving up from 1650 to 1950, each high (red) and low (purple) drifts westward, as shown by their shifts toward the left. And, during any one year, there are four or five sets of alternating highs and lows—that is, four or five 'waves.'

Image from Hulot et al. (2010), after Finlay and Jackson (2003).

and references within) and explained with mineralogical models (see those above references)—would promote asymmetric outer core convection, distorting the gyre (e.g., Figure 10.22c); the resulting geomagnetic secular variation was able to match observations well, including the various 'reality-challenging' latitude- and longitude-dependent features mentioned earlier in our discussion of the westward drift.

We note that these latter results also allow the core field to avoid the pitfalls of Cowling's theorem.

10.3.2 Snapshots Further Back in Time

Archaeomagnetic and Paleomagnetic Time Variations. We have already discussed how hot, magnetizable rocks can record the ambient magnetic field when they cool below the Curie temperature. In fact, any magnetizable material can undergo thermal remanent magnetization, if heated sufficiently; as it turns out, high-enough temperatures would have been achieved in the clay kilns at archaeological sites, leaving both the kiln lining and the artifacts fired there with a remanent magnetization. Remanence in prehistoric and

(c)

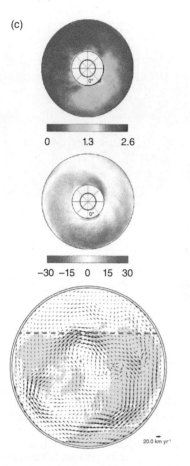

0   1.3   2.6

−30 −15 0 15 30

20.0 km yr⁻¹

Figure 10.22 Alternative views of secular variation and westward drift.
(c) The inner core promotes a westward drift and additional secular variation.
(Top and middle) These cross-sections through the equator of the core illustrate a scenario (Aubert, 2015, but see also Aubert et al., 2013) in which thermochemical convection originating at the inner-core boundary drives the geodynamo.
(Top) The density deficits associated with the convection, in units of 10^{-4} kg/m³, with most of the buoyancy forcing concentrated around longitudes 0°–90°E as the inner core grows lopsidedly.
(Middle) The azimuthal component of the resulting flow, in km/yr, with blue denoting westward flow and red, eastward, featuring a dominantly westward drift, spiraling in from the core-mantle boundary to the inner-core boundary. Images from Aubert (2015) / Oxford University Press.
(Bottom) Another view of the westward, eccentric large-scale gyre in the core, in this case from a flow model inferred from observations by Gillet et al. (2015) and plotted in equatorial cross-section by Dumberry and More (2020). With a view from above the North Pole, the continents of the northern hemisphere are shown in gray and allow us to see the weakness or even absence of the gyre in the Pacific region (marked by a pink dashed line); that near-absence results in the gyre's asymmetry and eccentricity.
Image from Dumberry & More (2020) / Springer Nature.

historical lava flows provides another source of archaeomagnetic (and also paleomagnetic) data. Together with paleomagnetic data, these allow us to develop an understanding of how the Earth's magnetic field varies on timescales all the way from years and centuries to millennia and eons.

Ancient magnetic fields (archaeo- and paleo-) can also be recorded in lake and marine sediments: as magnetizable grains settle through the water column, they can become aligned with the ambient field; the resulting sedimentary deposit can preserve that 'directional' record of the ambient field, yielding a **depositional remanent magnetism**. Accounting for any compaction and subsequent deformation or displacement, that remanent magnetism can tell us the inclination and declination of the field at the time of deposition.

Taking the intensity of the remanent field to be proportional to the intensity of the field that originally induced it, thermal and depositional remanent magnetism have also been used to infer ancient field intensities. The inference is somewhat problematic for depositional remanence, however, because of the clumping of grains during or following their descent (Tauxe et al., 2006). The proportionality constant, a measurable material property of the sample, is called its **magnetic susceptibility**.

Archaeomagnetic data is especially valuable for telling us about the nature of the geomagnetic field, both its magnitude and its orientation, during the past 3,000 yr (e.g., Donadini et al., 2009); paleomagnetic data is our only hope of learning about Earth's field in the distant geologic past. In this subsection, we will review what such data tells us about the nature of that field; we will focus on the dipole field, but begin by briefly highlighting the non-dipole field. It will turn out, though, that our more extensive review of the dipole field also has important implications for the ancient non-dipole field.

A number of researchers have established that the non-dipole field existed in ancient times.

- Schneider and Kent (1990) used a spherical harmonic analysis of deep-sea sediments 0–2.5 Myr in age to determine that quadrupole and octupole components of the field have persisted through that time span.

- Other analyses focusing on the past 5 Myr, summarized by Merrill and McFadden (2003), also find such persistence, for both axial and nonaxial quadrupole and octupole components; but they conclude that, because of nonuniqueness and disparities between models, only the axial quadrupole is statistically significant.

- Coupland and Van der Voo (1980) modeled all available paleomagnetic data for the past 130 Myr in terms of spherical harmonics, concluding that an axial quadrupole component was present for the past 100 Myr; an axial octupole term was important for the past 50 Myr.

- In contrast, Lee and Lilley (1986) found quadrupole and octupole components existed over the past 195 Myr; the axial quadrupole amplitude C_{20} varied but (ignoring its large uncertainties) was never more than ±10% of C_{10} (similar to the results of Livermore et al., 1984), whereas the axial octupole amplitude C_{30} was typically 10% of C_{10} but occasionally even larger. Interestingly, the smallest amplitude of C_{30} has been during the past few million years.

- More recent studies confirm the long-term presence of at least an axial quadrupole component, but question the reality of an axial octupole (see Hulot et al., 2010 and references therein).

The presence of a non-dipole field also manifests itself in routine paleomagnetic data analysis through the **far-sided effect**, in which reconstructions of the paleo-dipole from averaged measurements project the paleopole to be located several degrees beyond the geographic pole—on the 'far side' of the globe, relative to the measurement site—rather than at the pole itself. As noted by Wilson (1972; see also Butler, 1992), this can be explained as the true dipole being slightly offset vertically above Earth's equator. But as we have seen, an eccentric dipole is equivalent to a combination of dipole plus non-dipole (at least quadrupole) sources.

The ancient non-dipole field has exhibited many of the same characteristics that are seen in the current field. Nilsson et al. (2014) concluded from artifacts, sediments, and igneous rocks that there is clear evidence of a westward drift over the past 4 Kyr, at least in high latitudes of the northern hemisphere (see also Nilsson et al., 2020). Using the same types of data, Constable et al. (2016) found geographic similarities between the field or its secular variation now and through the past 10 Kyr: a strong northern hemisphere field; a weak and highly variable field in the southern hemisphere; and a weak field in at least the western Pacific. And looking at Hawaiian lavas that are younger than 1 Myr old, Doell and Cox (1965) found evidence of a weak or absent non-dipole field in the Pacific for at least the past several hundred thousand years.

For the dipole field, a quantity commonly used instead of the spherical harmonic coefficients C_{10} and C_{11} to describe its strength is the **dipole moment**; thinking of dipoles as being produced by current loops, for instance, the dipole moment is defined as the area of the loop times the amount of current (alternatively, in terms of our expansions it turns out that the dipole moment is proportional to $\sqrt{(C_{10})^2 + (C_{11})^2}$. At present, the geomagnetic dipole moment is about 7.8×10^{22} Amp m^2 (Constable, 2005). As we see in Figure 10.23, the dipole field strength has remained more or less the same (roughly 7 to 9×10^{22} Amp m^2) for the past 10 Kyr (e.g., Constable et al., 2016; the work of McElhinny & Senanayake, 1982 shows slightly greater variability, with a range

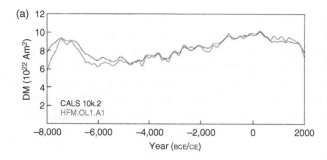

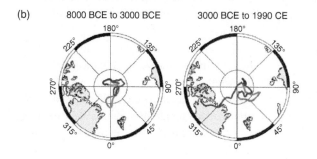

Figure 10.23 The dipole field over the past 10 Kyr.
(a) Strength of the dipole field since 8000 BCE, based on two models combining archaeological, volcanic, and especially sedimentary data. Strength is measured in terms of the dipole moment, defined in the text. Image from Constable et al. (2016) / Elsevier.
(b) Position of the north geomagnetic pole over the past 10 Kyr (the south geomagnetic pole is not shown but is just antipodal to the north geomagnetic pole). The position of the pole is shown from 8000 BCE to 3000 BCE (left) and 3000 BCE to 1990 CE (right), progressing continuously during those intervals along the curves from dark blue to light blue to green to red.
Image from Korte et al. (2011) / Elsevier.

of $\sim$7 to 11×10^{22} Amp m^2 for about the same time period).

But there are some dramatic details in this data. Much interest has focused on the relatively sharp downward trend in the dipole moment during the past millennium (see, e.g., Olson & Amit, 2006). From Figure 10.23, we can estimate that it has dropped 15% over the most recent $\sim$250 yr, according to one of the models shown in that figure, $\sim$30% over the past $\sim$600 yr according to the other; such changes are equivalent to rates of 6% and 5% per 100 yr. These rates are similar to our quick estimate earlier in this chapter (focusing on C_{10}), which corresponded to a drop of $\sim$8% per century. In work by Korte et al. (2011) involving an earlier model than those in Figure 10.23, the dipole moment declined at a somewhat slower rate, $\sim$20% over the past $\sim$750 yr.

This downward trend began by the year $\sim$750 CE (or, some time between $\sim$550 CE and $\sim$750 CE), according to Poletti et al. (2018). Such a decrease in the most important component of the main field, persisting for centuries, raised concerns that it signals an impending polarity reversal of the field; see Olson and Amit (2006), who evaluated that possibility

geophysically and concluded that a reversal was not imminent.

We note here that a similarly sustained decrease in the dipole moment also occurred $\sim$9 Kyr ago (see both models in Figure 10.23); after 1,000 yr of decline, however, the dipole moment 'stabilized' at a moderate value, and since then underwent a slow increase to current levels.

Korte et al. (2011) have shown that during the past few hundred years, the dipole axis has been more tilted than at any other time since at least 8000 BCE (see Figure 10.23)! Close inspection of that figure shows the tilt to have lessened a bit over the past century, and the geomagnetic poles are now heading toward the geographic poles (cf. Figure 10.18).

Clues to understanding the cause of the centuries-long dipole moment decline were uncovered by Olson and Amit (2006; see also Gubbins et al., 2006). They looked at **reversed flux patches** on the core surface, regions where the field lines are directed opposite to the 'normal' flux expected given the current dipole polarity. We can see examples of reversed flux patches in Figure 10.17, blue areas in a sea of yellow and brown in the core's southern hemisphere. According to the color code for that figure, field lines associated with

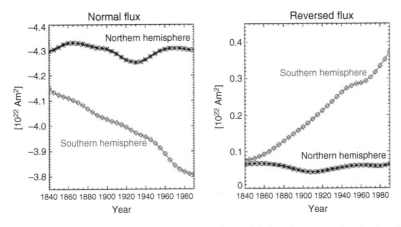

Figure 10.24 The role of the core's southern hemisphere in explaining the recent drop in the dipole moment.

The areas of normal flux and reversed flux patches (defined in the text) on the surface of the core are shown versus time between 1840 and 1990. Although the area of reversed flux patches is much less than the area of normal flux (note the different scales for the *y*-axes), its growth in the southern hemisphere is remarkable.

In these graphs, the field used to estimate flux area or flux patch area is not the radial component, B_r, of the core field but $B_r \cos\theta$ (where θ is colatitude), which is more relevant to the axial dipole moment. Image from Olson and Amit (2006) / Elsevier.

blue regions point down into the interior of the Earth into the core; field lines associated with yellow and brown regions point outward. In the southern hemisphere of the core, then, the field lines of the blue reversed flux patches, pointing into the core, are surrounded by yellow regions where the field lines point outward. In the core's northern hemisphere, the opposite situation prevails—yellow reversed patches, with field lines pointing outward, are surrounded by a blue sea of inward-pointing field lines. However, the total area of reversed patches is much greater in the southern hemisphere.

The analysis by Olson and Amit (2006) (see Figure 10.24) revealed that, during the present-day decline in dipole moment, the total area of normal flux and reversed flux patches in the core's northern hemisphere remained more or less the same, whereas the southern hemisphere was quite active, with the area of normal flux declining as that of reversed flux patches grew. As suggested by our discussion of Figure 10.17, the patchiness of the field is a reflection of the existence and strength of the non-dipole field; in contrast, the 'sea' of normal flux those patches are embedded within reflects the dipole field itself. We might then interpret the growth of reversed flux patches as an indication of ongoing struggles between the dipole and non-dipole components of the field (see also Finlay et al., 2016). The remarkable growth of the reversed patches appears to have come at the expense of the dipole field.

Timescales in the Context of Reversals. Without a doubt the most well-known variation in the field over time is its reversal in dipole polarity, from normal polarity—our current situation, and defined earlier—to the opposite, reversed polarity, or from reversed back to normal polarity. In one sense, it is almost unbelievable that a global-scale feature of the Earth could change to its opposite—it's not like Earth could suddenly start rotating east to west, for example, or all lithospheric plates could suddenly move in the opposite direction. The geomagnetic field exists because of motions in the fluid core; but its reversals do *not* mean that those fluid motions are also reversed!

In the final section of this chapter, we will discuss several magnetohydrodynamic concepts that will help us understand how the dynamo works and how reversals are possible. Before that, though, we will find it instructive to review the *timescales* of reversals, properties that (among other things) provide clues to their origin. Various scenarios can be pictured. One possible reason for a reversal might be changes in the boundary conditions the core fluid is subjected to; most likely, that means a change in the conditions at the core-mantle boundary, i.e. a change in lower-mantle conditions. Such a mechanism would have a very long timescale, characteristic of geological processes. A second possible reason might be changes in the pattern of the core flow itself, for example the growth of relatively small-scale flow into much broader structures, perhaps concentrated in one hemisphere. This type of mechanism would have a relatively short timescale, reflecting the lifetime of regional flow structures. Still another possibility might be changes in the *balance* between different processes; one popular idea is that the field's polarity reflects the core's climate: when cyclones dominate over anticyclones in the outer core, the net result is one dipole polarity, and when the balance shifts in favor of anticyclones, the reverse polarity takes over. Such a 'climate change' would be characterized by intermediate timescales.

Figure 10.25 shows a sequence of reversals revealed by paleo-inclinations between ~23 Myr and ~31 Myr ago, as analyzed by Tauxe and Hartl (1997). The reversals are marked by jumps in the inclination; for example, nearly 24 Myr ago (at a depth of ~59 m in the

figure), the inclination jumped from almost +60° to −60° (note that older values are to the right, younger values are to the left). Small fluctuations—'noise'—before and after that jump (and throughout the span of data) correspond to typical secular variation; large but 'temporary' fluctuations, in which the inclination does not reach or persist at a reversed value, are not true reversals but have been called **excursions**.

Jumps in paleo-declination accompanying a reversal—no matter where the site is, i.e. what meridian it is located on, a paleo-compass would then point to the opposite geomagnetic pole—are even more vivid and have also been used to identify reversals (e.g., Valet & Meynadier, 1993).

There are two relevant timescales underlying Figure 10.25: the duration of reversals, and the frequency at which they occur. A visual inspection of some of the clearest jumps suggests a typical duration of ~25–30 Kyr, if the data sampling is ~6 Kyr; recent high-resolution studies by Korte et al. (2019) reveal even shorter durations for some excursions. And, there are roughly ~25 reversals throughout the span of 8 Myr—though gaps in the data and some unclear or incomplete jumps make that count highly uncertain—implying on average one reversal every ~300,000 yr. But we can see even in this one data set that the rate of reversals is not constant; near the top

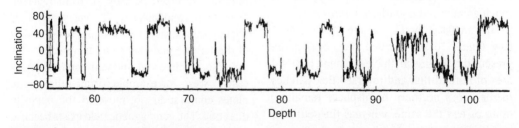

Figure 10.25 Paleomagnetic data reveals that Earth's field has undergone polarity reversals throughout its past. The data shown here of the field inclination was measured at high resolution in marine sediment cores retrieved from the Deep Sea Drilling Project at a location 26°S, 5°W, in the South Atlantic between southern Africa and eastern South America. The inclinations (in degrees) are plotted versus depth (in meters) below the sea floor.

This depth range corresponds to ~23 Myr to ~31 Myr ago, a time span of roughly 8 Myr; the cores were sampled at ~4-cm increments, yielding an average resolution of ~6 Kyr.

Image from Tauxe and Hartl (1997) / Oxford University Press.

of the section, reversals occur quite frequently (see also Constable, Tauxe, & Parker, 1998).

Using the timescale implied in the figure caption, what reversal rate would you estimate characterizes the top 5 m of these sediment cores?

Reversals are, of course, documented in the magnetic stripes of the oceanic crust, thanks to seafloor spreading (see Figure 10.6). As noted earlier, the width of a stripe depends on the rate of spreading, and also on the reversal rate. Since plate motions can be unsteady—in the case of the Farallon plate (Chapter 8), on timescales as short as 1 Myr—variable spreading rates must be accounted for, so ties to magnetostratigraphic data (see Figure 10.25), with consistently determined ages, are important (e.g., Pavlov & Gallet, 2005). Figure 10.26

shows the pattern of geomagnetic reversals associated with plate motions over the past 160 Myr, as determined by Gee & Kent (2007; see also Lowrie & Kent, 2004 and Cande & Kent, 1995); its depiction in black and white is called the **geomagnetic polarity timescale**.

In Figure 10.26, we see that reversals occurred more frequently during 0–30 Myr and 125–155 Myr than during 30–125 Myr. Such long, geologic-type timescales of variation in the reversal frequency suggest an influence of the mantle, conceivably through changes in temperature conditions at the core-mantle boundary, most likely associated with changes in mantle convection patterns (e.g., Jones, 1977; Olson & Christensen, 2002 and references within, Hulot et al., 2010 and references within; see also below).

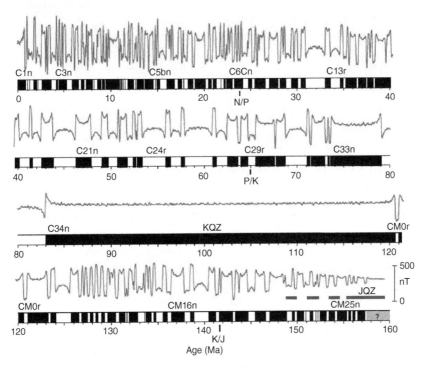

Figure 10.26 Geomagnetic polarity timescale for the past 160 Myr, as inferred from seafloor magnetization; black = normal polarity, white = reversed polarity. The Cretaceous Normal Superchron ("KQZ") and some key 'chrons' (e.g., C1n, C13r), used—whether normal or reversed periods—to calibrate the timescale, are labeled here.
Red curve: magnetic field predicted at the surface due to shallow crustal magnetization with the polarities shown; intensity scale, in nT, is shown at the lower right. The reason the Cretaceous Superchron is also called the *Cretaceous quiet zone* (*KQZ*) is evident from the perspective of the crustal magnetic field seen here. Image from Gee and Kent (2007) / Elsevier.

The nearly 40-Myr-long Cretaceous Super-chron has been attributed to the mantle as well: the growth or collapse of mantle super-plumes, originating in LLSVPs and other D'' structures, changes the heat flow leaving the core and creates conditions that, respectively, either promote or suppress geomagnetic field reversals (Olson & Amit, 2015; Amit & Olson, 2015); in short, the collapse of a superplume would lead to a superchron.

How does the heat flux out of some region of the core change if a superplume develops above that region? Is it
Superplume grows → decreased heat flow out of core → more frequent reversals?
or
Superplume grows → increased heat flow out of core → more frequent reversals?
So, do reversals flourish in a contained or venti-lated hot core? What do you think is the effect of such changes in heat flux on the vigor of the core's convection?

The polarity timescale in Figure 10.26 con-tains about 275 reversals over a period of less than 160 Myr. On average, then, a reversal occurred every ~580 Kyr; excluding the Creta-ceous Superchron, and reducing the time span accordingly to 120 Myr, we should expect a reversal every 440 Kyr. Alternatively, recogniz-ing that reversal rates have varied throughout this time span and focusing only on the most recent 30 Myr or so, we would find typical 'chron' lengths to be ~250 Kyr (Lowrie & Kent, 2004)—closer to the 300 Kyr or less we estimated above from Figure 10.25. However, the *most recent* reversal, from the reversed **Matuyama** chron to the current, normal **Brunhes** chron, is dated at about 780 Kyr (e.g., Cande & Kent, 1995; see also Langereis et al., 1997). Even accounting for variability in reversal frequencies, it does seem we are overdue for another reversal.

Interestingly, we also see from Figure 10.26 that the geomagnetic field has tended to be normally polarized during the past 160 Myr more than reverse-polarized; this holds true even excluding the Cretaceous Superchron. However, as we will learn in the final section of this chapter, the equations governing the geomagnetic field do not have an explicit pref-erence for one polarity over the other. In effect, then, the polarity timescale in Figure 10.26 suggests that further in the past a reverse polar-ity likely dominated for comparable lengths of time.

A glimpse of geomagnetic field variations in the distant past (Pavlov & Gallet, 2005) is provided in Figure 10.27. We see that the geo-magnetic field experienced at least two other superchrons, both with reversed polarity, both during the Paleozoic (see also Hulot et al., 2010 and references cited therein for inferences of reversals and superchrons even farther in the past). The three Phanerozoic superchrons each lasted tens of millions of years (between ~20 or 25 Myr and 50 Myr; see also Hounslow, 2016).

Unexpectedly, the superchron midpoints occur 192 Myr apart, according to Figure 10.27. It would be surprising if such regularity per-sisted through the past (i.e., as a "cycle"); nevertheless, we note that, more recently, evidence for a normal polarity superchron has been found from the late Proterozoic, about 1 Byr ago (Gallet et al., 2012).

(What do you think—how well does that Pro-terozoic superchron fit the pattern?)

In any case, the clear nonrandomness and long timescales involving the three Phanerozoic superchrons both point, again, to a mantle influence (see also Olson et al., 2014).

For at least two of the three Phanerozoic superchrons, Figure 10.27 also suggests that reversal rates were higher ~25 Myr before each superchron, or equivalently that there was a dramatic decrease in reversal rate leading up to each superchron (Pavlov & Gallet, 2005). Some paleomagnetic data even implies that there were periods of time, in the Cambrian and perhaps the mid-Jurassic and Ediacaran as well, when the dynamo was hyperactive and the reversal frequency was extreme—as many as 15 reversals per million years or more

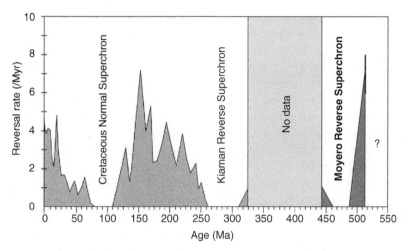

Figure 10.27   Rate of polarity reversals over the past half billion years based on magnetostratigraphic analyses. Data plotted here is the average reversal rate (per million years) within each geologic stage. The three known superchron time periods are labeled; the Kiaman Reverse Superchron is also known as the Permo-Carboniferous Reverse Polarity Superchron (e.g., Olson & Amit, 2015).
Image from Pavlov and Gallet (2005) / International Union of Geological Sciences. Note the *x*-axis increments are 8 ⅓ Myr.

(e.g., Gallet et al., 2019; Tivey et al., 2006; and Bazhenov et al., 2016).

Bringing Intensity Into the Picture. Intensity variations provide another way to explore the timescales of the geomagnetic field; they are also important for understanding reversals. For the full 11 Myr time span of data analyzed by Tauxe and Hartl (1997; also Constable, Tauxe, & Parker, 1998), only part of which was illustrated in Figure 10.25, the authors concluded that the field was somewhat stronger when the reversal frequency was lower. As Figure 10.28 shows, the connection is clearer when intensity is graphed versus polarity interval: longer chrons (due to a lower reversal frequency) correlate with a stronger field.

Figure 10.28   Possible dependence of paleomagnetic field intensity on length of polarity intervals, based on the same data whose inclination variations were shown in Figure 10.25.
(Bottom plot, with left-hand *y*-axis) Length of each polarity interval (triangle symbol) versus its mean age. The curve connecting the symbols highlights their variation.
(Upper plot, with right-hand *y*-axis) Intensity, relative to the average intensity during this time span, plotted (as tiny x symbols) versus age; smoothed data is shown by heavy black line.
Image adapted from Constable et al. (1998) / John Wiley & Sons.

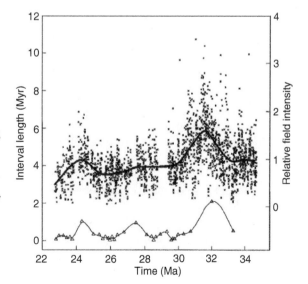

Similarly, extremely high reversal rates appear to correspond to very weak fields (Bono et al., 2019; Thallner et al., 2021; Lloyd et al., 2022 and references therein).

Throwing caution to the paleo-wind, we might go one step farther and ask whether reversals themselves correlate with a correspondingly weak field. To answer that question we turn to intensity data over the past 1 Myr. Figure 10.29a shows estimates by Guyodo and Valet (1999) of the field intensity (as measured by the axial dipole moment) over the past 800 Kyr; this time span includes a number of excursions as well as the most recent (Matuyama-Brunhes) reversal. The excursions and the reversal all evidently occur at times when the axial dipole moment drops to a low level—at most, 4×10^{22} A m^2 (that is, 50% of the current dipole level), perhaps even lower (2.5×10^{22} A m^2, according to Channell et al., 2009). This was interpreted by Langereis et al. (1997; see also Korte et al., 2019) as implying that there is a larger non-dipole:dipole ratio during excursions (and, we might add, during a reversal) than during 'stable' periods.

It appears that reversals are tied to a sharp decrease or even disappearance of the (axial) dipole field. Strong additional evidence in support of such a 'failing dipole' mechanism comes from the striking discovery by Valet et al. (2005) that not only does the dipole intensity drop during a reversal, the change in intensity also follows a consistent pattern; see Figure 10.29b. This pattern is demonstrated by the five most recent reversals, all of which are either to or from the reversed-polarity Matuyama chron; in addition to the Matuyama to Brunhes reversal, at 780 Kyr as noted above, they involve the normal **Jaramillo** and **Olduvai** 'subchrons,' dated by Gee and Kent (2007) as 0.99–1.07 Myr and 1.77–1.95 Myr, respectively. For example, the *lower Olduvai* reversal was a reversed-to-normal event (Matuyama to Olduvai) around 1.95 Myr ago, and the *upper Olduvai* reversal was a normal-to-reversed event (Olduvai back to Matuyama) around 1.77 Myr ago.

Valet et al. (2005) found there was typically a two-thirds decrease in the dipole moment during a reversal. Furthermore, the drop in dipole intensity leading up to each reversal was somewhat gradual, taking place over up to 80 Kyr; the recovery, however, was much quicker, and generally complete within ~10 Kyr (and in one case, as fast as ~5 Kyr)!

The much more gradual decline of the dipole moment compared to its recovery suggests 'philosophically' that creating the conditions for a reversal is somewhat difficult (and also implies why, if conditions are not quite right, we might end up with just an excursion); but ultimately the dynamo is robust, and no matter what the conditions—or whether the reversal is normal-to-reversed or the opposite—dynamo processes reassert themselves and quickly produce a strong field. The nature of reversals will be explored from a more physics-based framework in the next section.

Geomagnetic timescales can be estimated from the kinds of changes in intensity we see in Figure 10.29. We should not expect that such estimates will necessarily be the same as the timescales inferred using changes in direction, i.e. inclination (as in Figure 10.25) or declination, even at comparable resolution (cf. Korte et al., 2019). However, the duration of excursions and reversals appears in Figure 10.29a to be in the range ~10–30 Kyr, which is similar to our estimate of 25–30 Kyr from Figure 10.25.

Also from Figure 10.29a, excursions can occur *at least* as often as every ~40 Kyr (the Laschamp excursion is dated at ~40 Kyr); if the Calabrian and Emperor excursions are indeed distinct, the frequency of occurrence could be as low as every ~20 Kyr. And some evidence exists for the so-called Mono Lake excursion (not shown in the figure) ~33 Kyr ago (e.g., Korte et al., 2019). In Figure 10.29b, the duration of reversals is somewhat ambiguous, especially because their onset is so gradual; if we define their duration by a complete recovery in intensity, the five reversals show durations lasting between 25 Kyr and 50 Kyr or

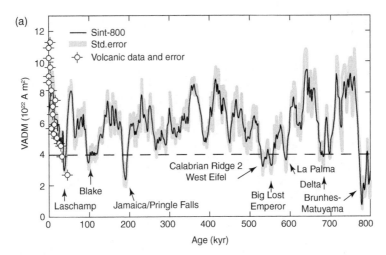

(a)

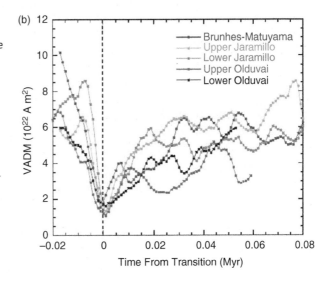

Figure 10.29 (a) Intensity of the dipole field during the past 800 Kyr from a combination of marine sediment data. The intensities are represented by the virtual axial dipole moment (which derives from C_{10} but not C_{11}); these estimates were calibrated by comparison with volcanic measurements of intensity during the most recent 40 Kyr.

Excursions in the magnetic field during this period are identified by name, as is the most recent reversal that occurred in our past: the change from the reversed-polarity epoch known as the Matuyama to our current normal-polarity epoch, the Brunhes. The authors mark a key intensity level (horizontal dashed line; see also Langereis et al., 1997) below which excursions are observed.

Image from Guyodo and Valet (1999).

Figure 10.29 (b) Dipole intensity before, during, and after a reversal. The magnitude of the virtual axial dipole moment, plotted from 0.08 Myr before the reversal to 0.02 Myr after. The five reversals analyzed include the most recent, from the reversed Matuyama chron to the current normal Brunhes chron, and reversals within the Matuyama involving normal subchrons Jaramillo and Olduvai, as discussed in the text.

Image from Valet et al. (2005) / Springer Nature.

longer. But the 'moment' of reversal is much more brief, only ~10–15 Kyr.

What's the world like in the midst of a reversal? An analysis by Clement (2004) suggests the answer. It turns out that the duration of a polarity reversal recorded at a site depends on the latitude of the site. The four reversals

Clement studied are the most recent experienced on Earth, those shown in Figure 10.29b from the Matuyama-Brunhes back to the upper Olduvai; they thus include reverse-to-normal and normal-to-reverse polarity switches. With each reversal marked by a jump in the field's direction, as discussed earlier, the duration of

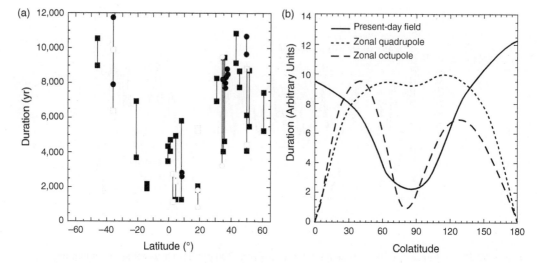

Figure 10.30 Duration of reversals depends on latitude.
(a) Duration (in years) of the four most recent reversals, estimated by two different methods from measurements at various sites, plotted versus the latitude of the site. Solid square symbols are for the Matuyama to Brunhes reversal; open circles are for the 'Upper Jaramillo' (Jaramillo to Matuyama) reversal; open squares are for the 'Lower Jaramillo' (Matuyama to Jaramillo) reversal; and solid circles are for the 'Upper Olduvai' (Olduvai to Matuyama) reversal.
(b) Apparent duration of a reversal predicted at any colatitude when the dipole component drops to zero then re-emerges with opposite polarity. Each point of one of these curves illustrates how long it takes, at a site with a given colatitude, for the field direction (declination or inclination) to complete its 'jump' when the field contains that dipole plus a quadrupole (dotted line); that dipole plus an octupole (dashed line); or that dipole plus all the non-dipole components in our present-day field.
Images from Clement (2004).

the jump was measured magnetostratigraphically in two ways, in effect yielding upper and lower bounds. Clement found (Figure 10.30a) that sites at high latitudes experienced longer reversal durations—by at least a couple of thousand years—than sites at low latitudes.

Clement showed that this latitude dependence could be explained as a consequence of the type of ambient field being recorded (see Figure 10.30b; note that the curves shown there are *not* being plotted versus time). In his analysis, the dipole was envisioned dropping to zero intensity then rebuilding with opposite polarity. If the only field present was the dipole, sites at all latitudes would record the change in direction at the same time; but if other fields (quadrupole, octupole, …) are present during the reversal, the net change in direction would be experienced sooner, or later, depending on the strength and polarity of those other fields at the site location.

For example, consider an ambient field which is mostly dipolar but with a small quadrupole component, with both components axial (see Figure 10.30b). At a site at the equator, the total (dipolar + quadrupolar) field will be nearly horizontal, so the inclination (e.g.) will be nearly 0°. If, during the reversal, the dipole field weakens to zero, the inclination at the site will be that of a quadrupole field (see Figure 10.9b), either +90° or −90° depending on the quadrupole's polarity. The inclination there will remain at ±90° until the dipole has recovered enough to 'strong-arm' the inclination back to nearly 0° (with a sign change due to the polarity reversal, if the site is not quite at the equator). The 'wrong' inclination persists as long as the dipole field is too weak to dominate, resulting in a longer estimated duration.

At the poles, in contrast, because axial quadrupole fields always have a ±90° inclination there, they will (depending on the signs) either reinforce the initial $\pm 90°$ inclination of the axial dipole, as the dipole weakens, delaying the apparent onset of the reversal (and thus shortening its duration), or oppose the initial $\pm 90°$ inclination and thus anticipate the reversed inclination; either way, they have shortened the reversal. At both the equator and the poles, this behavior (Figure 10.30b, dotted curve) is the opposite of what is actually observed (Figure 10.30a).

Like the dipole field, the octupole field is antisymmetric about the equator, so that both should have the same inclination, $0°$, at the equator; sites at very low latitudes, then, should not record any delay in the dipole's directional reversal. Figure 10.30b confirms this conclusion.

Finally, Figure 10.30b also demonstrates that a field like our present-day dipole + non-dipole field should exhibit shorter-duration reversals at low latitudes and longer reversals at high latitudes—which is what is observed.

Most fundamentally, then, Figure 10.30 tells us that the Earth's field does not completely vanish during a reversal: a non-dipole field must still be present. Life on the surface, along with the atmosphere, consequently is shielded from incoming solar wind and cosmic-ray particles (see Chapters 1 and 2), but—for the duration of a reversal, perhaps several thousand years—only by the much weaker non-dipole field.

10.4 Generation of the Geomagnetic Field

How the core is able to maintain its field—the final subject of this chapter and textbook—is arguably the most challenging topic we will encounter, from both conceptual and mathematical points of view. Indeed, even some of the basic governing equations are beyond the scope of this book. Nevertheless, it is worth even a limited exploration: it will expand our understanding of the variety of ways convection can operate; reveal some of the 'inner

workings' of the core; and succeed in explaining (to some extent) why the geomagnetic field has the properties we have been learning about. Any frustration at our vagueness or lack of rigor here should be considered motivation to continue pursuing the topics of this textbook at a higher level.

We have already argued that the geomagnetic field must originate in the core, and that electrical currents generated in the core by dynamo action must be the source of the field. It is worth noting that historically, a number of alternative mechanisms were proposed, and eventually ruled out—a testament both to creativity and scientific thoroughness. Durand-Manterola (2009) reviews those alternatives.

10.4.1 Preliminary Assessments

As a way of 'easing into' the topic, we first review key factors that should figure in to any theory of the geodynamo. With the Sun, nearly all the planets, and even a variety of smaller bodies in the Solar System showing evidence of a current or past dynamo (e.g., Stevenson, 2010; Olson & Christensen, 2006), our discussion may include them as well. For all such bodies, we will for conciseness refer to their dynamo source region as their 'core,' though in at least a few cases the dynamo action takes place mainly in their outer layers.

Ultimate Constraints: Field Properties the Dynamo Will Need to Reproduce. There are several characteristics of the geomagnetic field that might be viewed as ultimate constraints on the process of field generation and maintenance.

● The geomagnetic field has probably existed for as long as the Earth has had a core; that is, the geodynamo has been able to adjust to changing conditions within an evolving Earth and continue to operate. Several researchers have found paleomagnetic evidence for the field at least $\sim 3\frac{1}{2}$ Byr ago (McElhinny & Senanayake, 1980; Hale & Dunlop, 1984; Hale, 1987;

Tanaka et al., 1995; Yoshihara & Hamano, 2004; Tarduno et al., 2010), and there is some evidence for a field at least as old as 4.2 Byr (Tarduno et al., 2020)!

For perspective, we note that the Allan Hills (Antarctica) Martian meteorite ALH84001 discussed in Chapter 1 has been found to contain magnetite and other minerals with a stable remanent magnetism dating back ~4 Byr, implying that Mars also had an active dynamo and internally produced field at that time (Weiss et al., 2002). For that matter, asteroid Vesta is thought to have had a core dynamo nearly 4 Byr ago (Fu et al., 2012).

- The properties of the early geomagnetic field—its intensity, variability, and spectrum—were probably similar to the present-day field.
  - According to Yoshihara and Hamano (2004) (cf. Hale & Dunlop, 1984; Hale, 1987; and Tanaka et al., 1995), the field's dipole moment 3½ Byr ago was at least 2×10^{22} Am²; although weaker than today's field by a factor of 4, that is still comparable to the field at times in the recent past (see Figure 10.29). In contrast, Tarduno et al. (2010) found that the field 3½ Byr ago reached 50% of its present-day magnitude. It seems reasonable to conclude that the geodynamo worked more or less as effectively in the early Earth as it does today (see also Smirnov et al., 2016).

    Incidentally, the strength of the ALH84001 magnetization (—a reflection of the strength of the ancient Martian dynamo) is estimated to have been comparable to the field that we measure today at the *Earth*'s surface (Weiss et al., 2002, 2008).
  - The earliest reversal found (so far) in Earth's field took place in the Archean, 3.2 Byr ago (Layer et al., 1996).
  - Evidence for non-dipole components of the geomagnetic field, as discussed earlier, goes back 'only' ~0.2 Byr (Lee & Lilley, 1986); considering the weak magnitude of the non-dipole field and its tendency

to get 'averaged out' over time, though, its existence farther back in time should probably not be ruled out.

The longevity of the geodynamo also suggests that it may be 'robust' with respect to changes in the patterns of flow in the outer core. Later in this chapter we will find that the inner core has a strong, even restrictive, influence on flow above and below it in the outer core; and as the inner core grew, the volume of outer core fluid it affected would have increased. But if theories of a late inner core (see Chapter 9) are correct, it must be that the inner core's influence is not critical to the creation or maintenance of the field, or to its reversals; as far as the core flow driving the dynamo is concerned, the field does equally well with or without an inner core. (Note that this 'robustness' is a different issue from the dynamo's energy needs—which, as discussed in Chapter 9, may have depended on the growth of the inner core, especially once the heat of accretion, core formation, and/or radioactive decay had waned.)

Ingredients for a Successful Dynamo. Since the magnetic field created by a planetary dynamo results from electrical currents, the most obvious requirement is that its 'core' be highly conducting; equally obviously, that will not be sufficient by itself. A quick survey of our Solar System, noting which bodies possess planetary magnetic fields and which do not, sheds light on what other factors are important in the generation of a field. From Table 1.1 (Chapter 1), we see that the Sun and all eight planets (officially speaking) have magnetic fields—except for Venus. Venus is our 'sister' planet, and there is not much that sets it apart from Earth; but one stark difference is Venus' incredibly slow rotation rate.

Rotation. In fact, we have already seen evidence that rotation is important in generating Earth's magnetic field. The geomagnetic poles are somewhat close to Earth's rotation poles (especially considering *Cowling's theorem*). The largest component of Earth's field is the dipole component, and the largest component of the dipole field is its axial component. The westward drift is roughly zonal, and the long-term average field is axial.

One reason rotation is so important is because it creates a Coriolis force. The source regions of the bodies listed in that table are all fluid, so we can expect the Coriolis force to have a strong, potentially dominant influence on those fluids' motions. The specific role of the Coriolis force will be discussed later; but in general terms, and with Earth's atmospheric circulation in mind, we can say that the Coriolis force tends to *organize* the flow, imposing structure and making it more coherent, thereby helping to produce a large-scale field (Aurnou, 2004; see also Busse, 2002 and Dietrich et al., 2013).

Figure 10.31 shows just how fundamental a planet's spin is to the generation of its magnetic field; planetary rotation rate turns out to be a surprisingly good predictor of the strength of the field's dipole moment. Figure 10.31 suggests at first glance that the absence of a field for Venus could be attributed to its extremely slow rotation rate; on closer inspection, though, it implies only that Venus should have a very weak field, slightly weaker than that of Mercury (which spins 4 times faster) and much weaker than that of Earth (which spins 240 times faster).

... and Convection. Of course, other factors are also fundamentally important in determining the strength of a planet's magnetic field (see, e.g., Christensen & Aubert, 2006; Durand-Manterola, 2009; Jones, 2011). Besides electrical conductivity, for example, the mass of the core is an obvious factor. And, since the dynamo will most likely be driven by convective motions within the core, factors like the fluid buoyancy and the convective heat flux should be quite relevant as well; these would indicate how strongly the dynamo is driven and how well heat is transferred out of the core, respectively.

As a measure of these latter factors, and building on the work of Christensen and Aubert (2006), Olson and Christensen (2006)

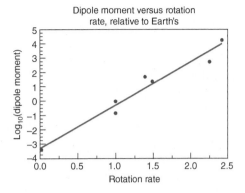

Figure 10.31 Dependence of planetary dipole moment on rotation rate. Both quantities are plotted relative to the values for Earth; the *y*-axis is a log scale (thus, the values for Earth plot at data point (1, 0)). From left to right, the data points are for Mercury, Mars and Earth, Uranus, Neptune, Saturn, and Jupiter.
Dipole moment data is taken from Olson and Christensen (2006).
Venus is excluded from this analysis. Mars' field refers to its ancient, now-extinct dynamo-generated field. The other dipole moments are observationally based. For simplicity, current rotation rates are used, even for Mars. The best-fitting straight line (slope = 3) is also shown; its correlation coefficient of 0.98 suggests that the dependence is very strong. The inferred strength of the ancient Martian dipole based on remanent magnetism in Martian meteorite ALH84001 (Weiss et al., 2002, 2008), as discussed in the text, combined with its probable low-latitude origin (Barlow, 1997), suggests Mars' dipole moment could have been as large as or larger than Earth's; that would bring the data point for Mars up to (1, 0), and improve the straight-line fit even more.

focused on a modified Rayleigh number based on a 'combination' parameter they termed the *convective buoyancy flux*, similar to the buoyancy force ($\rho \alpha_p \Delta T g$, as in Chapter 9, for thermal buoyancy), but per unit mass instead of per unit volume (thus $\alpha_p \Delta T g$) and with the superadiabatic temperature difference ΔT replaced with a term involving the total advected heat flow exiting the core. Evaluating a large variety of numerical dynamos from this perspective, Olson and Christensen (2006) found that the intensity of the dipole moment they produced depended strongly on the convective buoyancy flux parameter (and the modified Rayleigh number). The magnitude of this parameter was also found to be important in determining both the onset of dynamo action and the character of the resulting magnetic field's spectrum.

The importance of convection, and the particular significance of the convective buoyancy flux parameter, can be appreciated by comparing the situations for Earth and Venus within the framework developed by Olson and Christensen (2006). If we assume an Earth-like magnitude of convected heat out of Venus' core, thus a buoyancy flux parameter similar in value for Earth and Venus (see Olson & Christensen, 2006), and then hypothetically apply that parameter value to the dipole moment - modified Rayleigh number relationship they had determined from numerical dynamo studies, we would predict a dipole field of respectable intensity for Venus, even with that planet's sluggish rate of rotation. The lack of an observed field for Venus tells us that the actual value of its buoyancy flux parameter must be much smaller; that is, what is preventing Venus from creating a measurable magnetic field is insufficient advected heat flow out of its core.

Whatever it is that suppresses the heat flowing out of Venus' core represents an additional factor that distinguishes Earth from its 'sister' planet. From our discussion in Chapter 2, one factor that comes to mind is Venus' runaway greenhouse atmosphere. Thanks to that atmosphere, Venus may simply be too dry or its surface too hot to promote Earth-type plate tectonics (e.g., Solomon et al., 1992; Simons et al., 1997); that is, Venus has a 'stagnant lid': its mantle undergoes thermal convection and some tectonic activities, especially volcanism (e.g., Esposito, 1984; Filiberto et al., 2020 and Gülcher et al., 2020), but without ongoing plate motions (e.g., Solomatov & Moresi, 1996; Nimmo, 2002). In the absence of cold lithospheric plates subducting into Venus' mantle, its interior would cool very inefficiently, and heat would quickly build up at the core boundary. And a hot core boundary would eliminate the superadiabatic temperature differences needed to drive core convection and the dynamo (see also Nimmo, 2002).

What if the core convection was chemical, driven by the growth of an inner core—what would the effect of a hot core-mantle boundary be in that case? (Pursuing this further, you may want to see Gastine et al., 2020.)

But that explanation is not absolute, either. Volcanic activity, possible even with a stagnant lid, can bring heat up through the lithosphere to the surface, cooling the mantle below and potentially maintaining core convection. This kind of heat transfer (see Moore & Webb, 2013; also Morgan & Phillips, 1983; Turcotte, 1989) is accomplished via lithospheric conduits, nicknamed 'heat pipes,' which funnel hot magma to the surface. On Earth, such heat transfer is just a small component of that associated with plate tectonics, but on Venus under some conditions, even without tectonic activity, it could lead to a cooler mantle and a functioning dynamo (Driscoll & Bercovici, 2014).

So, in the past, Venus may have had a convecting core and a measurable (though probably weak) magnetic field. And, even if its core was not convecting, a field might have been produced by a very unusual source: a silicate dynamo! This is envisioned as taking place in a deep or 'basal' magma ocean (O'Rourke, 2020; see also Stixrude et al., 2020), whose low viscosity and high conductivity could have facilitated field generation; with a slow-cooling interior, that basal layer might have persisted to the present day (thinning as it cooled, though, it would have become too thin for dynamo action during the past half billion years).

In sum, there are ways that a dynamo could have operated within Venus, depending on thermal conditions, despite the additional challenge of a weak Coriolis force. We might consequently expect to have seen a weak field in Venus' past, though more episodically as Venus evolved to its current state.

10.4.2 Fields Weaken, Fields Strengthen

In this subsection, we will learn what happens when a magnetic field grows weaker or stronger, find ways to express those processes mathematically, and develop a governing equation for the evolution of a magnetic field. Our starting point will be the flux—magnetic flux lines—characterizing the field.

Field Weakening, and its Implications. In our earliest discussion of magnetic fields, we saw how flux lines could be used to describe both the intensity and direction of a magnetic field; where flux lines were spaced closer together, for example, the field was stronger. It follows that, if the field weakens over time, the flux lines must spread farther apart as the weakening occurs; that is, the flux lines *diffuse* through space, or through whatever medium they are penetrating. Field weakening is thus a diffusion process.

We can use the mathematical formalism developed in Chapter 9 and earlier in this chapter to express this diffusion. In general, the field, $\vec{B}$, will diffuse in all three directions, so we write its diffusive weakening as

$$\frac{\partial}{\partial t}(\vec{B}) = k_M \nabla^2 (\vec{B})$$

$$\equiv k_M \left\{ \frac{\partial^2 \vec{B}}{\partial x^2} + \frac{\partial^2 \vec{B}}{\partial y^2} + \frac{\partial^2 \vec{B}}{\partial z^2} \right\};$$

k_M is the coefficient for diffusion, called the **magnetic diffusivity**. This vector equation is actually three equations in one—one equation for each component of $\vec{B}$—since each component of it can weaken, i.e. diffuse, in all three directions; for compactness, we will continue to write it in vector form.

As we have seen, this type of equation can be scaled to determine a timescale t_M for diffusion, i.e. for significant weakening of the magnetic field; we find

$$t_M \sim \frac{L^2}{k_M},$$

where now L is a characteristic dimension of the dynamo's source region, typically the region's thickness or radius. As in other diffusion situations, the larger the 'core,' the longer it takes for significant diffusion; in this case, that means dynamos operating in larger cores are likely to create fields that weaken more

slowly. Anticipating the importance of convection in regenerating the field, and recalling how strongly convection depends on L, it is not surprising that dynamo-generated fields are more likely to survive in astronomical ('large L') environments, whether within our Solar System or beyond.

In the core, or in any electrically conducting substance, field weakening will happen because the electrical currents responsible for the field face resistance as they flow. The resistance converts electrical energy to heat, 'draining' the currents and requiring an energy source to maintain dynamo action and regenerate the currents. The lower the electrical conductivity σ, the greater the resistance (as implied in *Textbox A*) and the faster the weakening. For this to be consistent with our timescale formula, k_M must be inversely proportional to σ. Additionally, the units for k_M must be length²/time—which is the case for any diffusivity (and what we have seen for thermal and momentum diffusivities)—in order for t_M to be in units of time. Both of these issues are resolved if

$$k_M = \frac{1}{\mu_0 \sigma}.$$

The magnetic permeability μ_0, which we encountered earlier as a 'scale factor' that would yield the correct units for $\vec{B}$, has a value of $4\pi \times 10^{-7}$ Henry/m or $4\pi \times 10^{-7}$ Ohm-sec/m.

If σ is in units of Siemens/m, can you verify the units for k_M?

Earlier in this chapter, we took the value of σ to be in the range $\sim 2 \times 10^5$–2×10^6 S/m for Earth's core. Taking the thickness of the outer core to be $\sim 2{,}200$ km, it follows that the timescale for the entire geomagnetic field to appreciably weaken is $\sim 1.22 \times 10^{12}$–1.22×10^{13} sec ($\sim 39{,}000$–$390{,}000$ yr); more simply, if we treat L as a characteristic dimension of order $1{,}000$ km, the timescale does not exceed $100{,}000$ yr (see, e.g., Finlay et al., 2010b for other estimates of timescales). But we know the field has been around for billions of years, leading to an inescapable conclusion: Earth's magnetic field requires continual regeneration.

The low end of the timescales implied by t_M is comparable to the timescale we inferred for the duration of a field reversal, consistent with our inference that reversals are characterized by a significant intensity drop. However, an alternative interpretation of such processes will be suggested later.

The dependence of field weakening and diffusion on σ has a flip side: the greater the conductivity, the slower the diffusion, i.e. the slower the flux lines spread out. Approaching the limit of infinite conductivity, the flux lines exhibit no perceptible spreading; that is, they remain *frozen* into the core material. The flux lines would remain frozen in whether the material is stationary or flowing, and no matter how it might flow.

For example, plasma—the super-heated gas comprising the Sun, and also found across the Solar System—is extremely electrically conductive; whatever the ambient magnetic field is, in direction and magnitude, that field's flux lines must move with the plasma, no matter how chaotic the gas motions might become. As an extreme consequence, if moving 'parcels' (small volumes) of plasma collide in some region, and their frozen-in flux lines happen to have opposite polarity, those flux lines will cancel out as the gas parcels merge. To preserve a zero divergence of flux lines overall—imagine a volume that cuts through a portion of this region including flux lines going in, then vanishing—the flux lines just outside the 'null' region, still oppositely directed on one side versus the other, must avoid entering that region and instead loop around, from one side of the null region to the other, and connect. This phenomenon, called **magnetic reconnection**, can release a huge amount of magnetic energy as the region expands explosively, ejecting plasma at high speed; it is associated with solar flares, and thus magnetic storms and auroras (see Hesse & Cassak, 2020).

Who Said Shear Waves Can't Propagate Through Fluids? Frozen flux can be studied in the lab with the help of highly conducting metals such as mercury, sodium, gallium (or its alloys), and rubidium, all in their liquid state (e.g., Lundquist, 1949; Tigrine et al., 2019; Alboussière et al., 2011; and Stefani et al., 2021, respectively), and even plasma (Gekelman et al., 2003). Despite their high conductivities, though, the small length scale of laboratories (compared to the core, say) drastically limits the time t_M before appreciable flux line diffusion occurs; in the work of Tigrine et al. (2019), for example, the waves they produce take about $\frac{1}{2}$ sec to traverse the apparatus, by which time the disturbances are no longer perceptible as waves.

For simplicity, let's imagine there is a pan of gallium in our lab, and nearby, at the same height, a strong magnet oriented vertically. In this configuration, the flux lines penetrating the liquid—more or less along the dipole's equator—will be essentially vertical (e.g., Figure 10.1a). Or, if we use a stronger magnet placed farther away, the field lines penetrating the mercury will be less curved and more nearly vertical, even if the pan of liquid is deeper (in the next subsection, we will learn another way to produce straight flux lines). If the pan is shaken sideways, once, we would see that the shear created at the top (free) surface propagates quickly, vertically downward, as a shear wave, reflecting off the base of the pan and continuing back upward. With gallium's high electrical conductivity, the frozen-in flux lines must move with the fluid, and magnetometer measurements (or even the needle of a compass adjacent to the pan) will show an oscillation of the magnetic field which propagates in parallel, downward then upward: this is a hydro-magnetic phenomenon, called an **Alfvén wave**!

In fact, if no magnet was present, there would be no coherent oscillations; after its sudden displacement, the fluid would simply shake chaotically and briefly.

Alternatively, if we turned the magnet sideways and positioned it such that essentially horizontal flux lines penetrated the fluid, we could create oscillations that traveled a longer distance, laterally, eventually reflecting off the sides of the pan. In that experiment, though, we would need to distinguish between the Alfvén wave; waves created by the density contrast at the fluid interface (buoyancy or gravity waves); and waves created by the fluid's surface tension. And, to keep the wave from damping out too soon, the magnet would have to be very strong (cf. Alboussière et al., 2011).

In our discussion of seismic waves, in particular traveling versus standing waves (Chapter 7), we used an example of a rope, fixed at one end, being shaken to create transverse waves. In our thought experiment here, the flux lines behave like those ropes, with the magnetic oscillations traveling along the flux lines as a transverse magnetic wave. And the fluid, frozen to the flux lines, joins in with its own transverse wave motion. In sum, our interpretation is that the flux lines have imparted a tension or *rigidity* to the fluid. This is consistent with our understanding of seismic waves, where the propagation of shear waves through a solid medium depends on its rigidity.

The propagation speed of Alfvén waves depends on the strength of the magnetic field—which determines the tension in the flux lines—and on the fluid density (Alfvén, 1942; Finlay et al., 2010b). Using the field intensity shown in Figure 10.17, we could estimate that an Alfvén wave traveling around the core would move at a speed of $< \frac{1}{2}$ cm/sec—definitely greater than zero, but far slower than the km/sec speed of shear waves in Earth's solid interior.

However, the intensities shown in that figure are of the field at the surface of the core, estimated by downward continuation. A better estimate of the Alfvén wave velocity comes from analyses of observed changes in the length of day (i.e., in Earth's rotation rate), which reveal periodic variations on different timescales, including a timescale of roughly 6 yr (e.g., Abarca del Rio et al., 2000; Gross, 2001; see also Gross, 2007). Gillet et al. (2010) confirmed that a 6-yr periodicity has occurred through most of the twentieth century and, using various models of core flow inferred from the secular variation of the geomagnetic field, they argue that the 6-yr variation results

from the propagation of Alfvén waves through the core—more precisely, from the periodic variation in angular momentum associated with those hydromagnetic waves. Those waves travel (outward, for instance, from near the inner-core boundary) through the bulk of the fluid core, covering that distance on a timescale of about 4 yr, implying a typical velocity through the interior of ~1–2 cm/sec. This is a faster speed than our first estimate because the strength of the field is greater, deeper into the outer core; according to Gillet et al. (2010), the typical intensity within the core is a few mT (see also Buffett, 2010b; Jackson, 2010).

Compare this intensity to that shown in Figure 10.17; do you think the difference between surface and interior intensities of the core field is reasonable?

In compressible fluids, frozen flux can lead to changes in the strength of the field (i.e. the concentration of flux lines): if the fluid is squeezed together the flux lines will become closer. For example, as our own T-Tauri proto-sun continued to contract, on its way to becoming our Sun (as discussed in Chapter 1), the flux lines of its T-Tauri magnetic field—essentially frozen into its plasma—also contracted together, amplifying that already strong field (see Thompson et al., 2003).

Even more generally, with frozen flux it is possible for the *character* of the field to change fundamentally. For example, if the flow 'twists around,' so will the flux lines; flux lines that are meridional (north-south) could become azimuthal (east-west).

Can you think of any type of force that has a twisting effect?

Focusing on the other variable determining the magnitude of t_M: our view of the non-dipole field as the product of smaller dipole sources (local or regional electrical current loops) distributed throughout the outermost core implies that those sources will weaken faster than the overall field; with $L \sim 10\text{--}10^2$ km, for example, $t_M \sim 8\text{--}800$ yr at most. This is consistent with the observed timescales for the non-dipole secular variation.

Finally, and as a lead-in to our next discussion, we emphasize that the diffusion equation we wrote above for $\vec{B}$ is incomplete; as written, it is not yet a general governing equation for magnetic fields. Thinking of the diffusion equations we have previously explored, we know immediately what's missing: a *source term*—a way to generate (currents and) magnetic fields.

The Lorentz Force. In an electrically conducting medium, currents flow when the mobile charges within it (electrons or ions) are forced to move by an electric field. As developed in *Textbox A*, the electrical force they experience is proportional to both the field and the amount of charge; symbolically, if the electric field is denoted by $\vec{E}$ and it acts on a charge q, the resulting force is simply $q\vec{E}$.

This force exists whether the charged particle is static or already in motion. But if the charge is moving, it can experience another force—if it is moving through a magnetic field. If the velocity of the charge is $\vec{v}$ and the ambient magnetic field is $\vec{B}$, this force, called the **Lorentz force**, is given by

$$q\vec{v} \times \vec{B}.$$

The cross product here makes sense: if $\vec{v}$ was parallel to $\vec{B}$, the conductor (i.e. the charges within it) would not be cutting across any field lines, so no new current (i.e., new component of the charges' velocity) would be induced.

As outlined in *Textbox A*, the current created by an electric field $\vec{E}$ is proportional to $\vec{E}$, with a current density $\vec{J} = \sigma\vec{E}$. If the Lorentz force acting on q as it moves is $q\vec{v}\times\vec{B}$, the equivalent electric field (electric force per unit charge) created by the motion of q is $\vec{v}\times\vec{B}$, and the resulting current density is $\sigma\vec{v}\times\vec{B}$.

That current is in the direction of $\vec{v}\times\vec{B}$, which is perpendicular to (and thus distinct from) the initial direction $\vec{v}$ of the charge itself. And this new current will produce a magnetic field.

It seems we have found a way to our missing 'source term' for magnetic fields: the motion of electrical charges through a preexisting magnetic field.

This type of source term is the most challenging we have encountered so far. Unlike heat sources (e.g., radioactive decay), which are quite straightforward, and momentum sources, which require only a moment's thought to see their relevance, the source term for magnetic field diffusion is much less direct. Quantifying the connection between $\vec{v}\times\vec{B}$ and $\partial\vec{B}/\partial t$ is beyond the scope of this textbook (as an indication of what would be involved: earlier, in the course of deriving a Laplacian and then a spherical harmonic expansion for the magnetic potential, we encountered the 'dot product' $\vec{\nabla}\cdot\vec{B}$; the kind of mathematics needed to produce a rigorous source term would look more like $\vec{\nabla}\times\vec{B}$, but with even more cross products). Instead, we will denote the source term abstractly as $S(\vec{v}\times\vec{B})$ (meaning S is a function of $\vec{v}$ and $\vec{B}$, but also that S 'operates' on the components of $\vec{v}$ and $\vec{B}$, by differentiating them). Our governing equation for the weakening *or* strengthening of magnetic fields can thus be written

$$\frac{\partial}{\partial t}(\vec{B}) = \underbrace{k_M \nabla^2(\vec{B})}_{weakening} + \underbrace{S(\vec{v}\times\vec{B})}_{strengthening};$$

with nT/sec as the units for each term in this equation (—just look at the left-hand side), S takes the Lorentz factor ($\vec{v}\times\vec{B}$) and divides it by meters (if $\vec{v}$ is in units of m/sec) to yield the same units. If the math symbolized by S was explicit, this equation would be called the **induction equation**.

According to the induction equation, successful dynamo action within Earth's core will depend on fluid motions ($\vec{v}$) within the ambient core field ($\vec{B}$) somehow being able to regenerate ($S(\vec{v}\times\vec{B})$) the field before appreciable diffusion ($\nabla^2(\vec{B})$) can weaken it fatally. We will explore this equation throughout the remainder of this chapter; for now, we note two important conclusions that can be drawn from it about any dynamo.

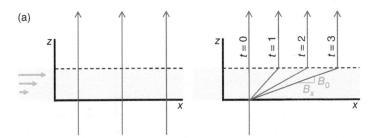

Figure 10.32 Stretching of field lines by shear flow in a highly conducting fluid.
(a) The field lines (a few of which are illustrated here, as red lines) are initially those of a uniform vertical magnetic field, $\vec{B} = B_0\hat{z}$. The fluid layer (shaded) is located between the x-axis and the dashed horizontal line. The shear flow considered here is directed to the right, and velocities increase with z, as shown by the bold blue arrows.
(Left) Before the fluid starts moving, the field lines within the layer are all vertical. Then, at time 0, the layer is set into motion.
(Right) Because of frozen flux, the field lines within the layer move with the fluid, so they are sheared as well; the result is the creation of a new, horizontal component of the field within the layer: an x-component, to add to the original z-component. In this illustration, we focus on *one* flux line at three subsequent moments in time, showing how the x-component grows over time as the flow continues. At time 3, the decomposition of the field line into both the original z-component (we had denoted that field B_0) and the new x-component is indicated in green.

- Every term in the induction equation is proportional to $\vec{B}$; thus, if a particular function $\vec{B}$ is a solution to this equation, so is $10\cdot\vec{B}$ and $100\cdot\vec{B}$ and .... That is, the induction equation does not place a limit on the magnitude of the field generated by the dynamo, no matter how well or poorly the core conducts electricity; that limit must be provided by the one remaining variable: the fluid velocity.

  But it is not just the amplitude of the fluid motion that determines the strength of the magnetic field: to produce a more intense field, the induction equation tells us that $\vec{v} \times \vec{B}$ (or, more precisely, $S(\vec{v} \times \vec{B})$), not just $\vec{v}$, must be bigger. The success of the dynamo, and the strength of the field it generates, thus depend on the vector (and differential) qualities of the flow.

- Because every term in the induction equation is proportional to $\vec{B}$, it follows that $-\vec{B}$ is also a solution to this equation. That is, the governing equation for any planetary (or stellar) magnetic field does not require a specific polarity for the field;

reversed polarities are equally likely—even with exactly the same fluid velocity field!

The expression for the Lorentz force can be verified experimentally. It also can be derived theoretically, based on considerations of special relativity: transforming our frame of reference to that of the charge, as it flies through space, we would find that the charge experiences $\vec{v} \times \vec{B}$ as an electric field—what *we* see as an exotic "Lorentz" force, the charge sees as an everyday electric force (e.g., Feynman, 1963).

Field Strengthening. Though we are ultimately interested in field strengthening as a way of sustaining a dynamo, it is instructive to see how it can take place outside the context of a dynamo. We will imagine that the fluid we are dealing with is a perfect or nearly perfect electrical conductor, so that the flux lines penetrating through it from any ambient magnetic field diffuse negligibly if at all during the time interval of interest—that is, frozen flux will be a reasonable approximation. This guarantees the field will not weaken for a time, though by itself it does not necessarily lead to a stronger field.

For simplicity, we will also assume that the ambient field is a uniform vertical magnetic field and that the fluid is a horizontal layer (see Figure 10.32a). At time $t = 0$, a velocity profile in which the fluid velocity increases linearly with height through the layer is imposed on the fluid. Once the fluid is in motion, the flux lines within the layer also move, staying with the fluid. Because of the shear flow, those flux lines 'shear out'—*stretching* with the flow—progressively as the flow continues (see Figure 10.32a, right). This stretching creates a new component of the field, B_x.

With the creation of this new component, we have succeeded in increasing the magnitude of the total field. And as time increases, and the fluid carries the flux lines further downstream, B_x grows even stronger, as does the total field. Conceivably, B_x could grow arbitrarily large; but in practice, as the field intensifies, the fluid 'stiffens' and finds it increasingly difficult to continue shearing.

In short, magnetic fields can be strengthened if the fluid they are frozen into is flowing *nonuniformly*. Theoretically, this follows from the induction equation, whose source term $S(\vec{v} \times \vec{B})$ involves, among other things, spatial derivatives of $\vec{v}$.

Figure 10.32b illustrates a more complicated situation where the fluid velocity first increases and then decreases, with height; as the shear flow carries the flux lines along, B_x grows, but in opposite directions, reflecting the inversion of the shear gradient.

In Earth's core, we believe a differential rotation is probably responsible for the westward drift observed at Earth's surface. That differential rotation should, accordingly, cause flux lines of the geomagnetic field in the core to be stretched out. In this case, the stretching should be azimuthal, that is, east-west, creating field lines that wrap around the core. Those field lines are described as *toroidal* (as in having the shape of a toroid), meaning they do not include a radial component. This is in contrast to the kinds of fields we have been dealing with throughout this chapter: dipole, quadrupole and other multipole fields all include a radial

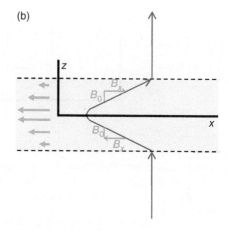

Figure 10.32 Stretching of field lines by shear flow in a highly conducting fluid.
(b) The fluid velocity gradient determines how the field lines stretch.
In this illustration, the fluid layer (shaded) is located between the dashed horizontal lines. As in part (a), the field lines are initially those of a uniform vertical magnetic field, $\vec{B} = B_0\hat{z}$. Here we focus on one field line (in red).
As indicated by the velocity profile on the far left (blue arrows), the flow in this layer—in contrast to (a)—is directed to the left, with velocities increasing in magnitude then decreasing as z increases.
Because of frozen flux, the field lines within the layer move with the fluid, so they are sheared as well. Our field line of interest is shown at some moment not long after the flow has developed.
- At any depth z in the lower half of the layer, an increasingly negative B_x develops over time—just as we would expect from part (a), if the flow in (a) had been to the left.
- At any height z in the upper half of the layer, conditions are reversed, so an increasingly positive B_x develops over time.

At any moment, B_x is strongest in midlayer, and weakens toward the layer's boundaries. If the fluid velocity is zero at the boundaries, the flux line there and outside the layer remains motionless.

(outward or inward) component, and are all described as **poloidal**.

Such a classification was encountered in Chapter 7 when we discussed the Earth's free oscillations and distinguished toroidal from spheroidal modes. However, in contrast to toroidal free oscillations, which we can detect directly, for example by lateral displacements of the solid earth's surface, there is no way to directly detect the toroidal magnetic field here at the surface: the toroidal field lines exist entirely within the core! This is true despite the fact that the toroidal field is thought to be several times stronger (Sreenivasan & Narasimhan, 2017; Hori et al., 2015; Hardy et al., 2020, also Stacey & Davis, 2008; cf. Glatzmaier & Roberts, 1995) than the poloidal field in the core.

Interestingly, toroidal fields are also expected in the oceans. As discussed earlier in this chapter, the movement of the oceans through Earth's magnetic field induces electrical currents and associated (poloidal) magnetic fields that can even be measured at satellite altitudes; but their shear flow will also produce toroidal fields confined to the oceans. Estimates of the toroidal fields created by tidal flow (Tyler et al., 2003) and other types of ocean currents (cf. Tyler & Mysak, 1995) yielded intensities as much as two orders of magnitude greater than the poloidal fields; more recent analysis finds that the two tidally produced fields are instead comparable in magnitude, that the strongest toroidal amplitudes are in shallow-ocean regions (Dostal et al., 2012), and that the toroidal field created by the Antarctic Circumpolar Current is at least a few times larger than the poloidal field induced by the ACC (Velímský et al., 2019).

For illustration, we consider a differential rotation of the outermost core in which westward drift is greatest at the equator and reduced at higher latitudes in both hemispheres; such a scenario is supported by some recent analyses (as discussed earlier in this chapter). The corresponding equator-to-pole velocity gradients are oppositely directed in this case (one decreases to the north and the other decreases to the south), similar to what is illustrated in Figure 10.32b, and yields a toroidal field—created from poloidal flux lines stretched oppositely above and below the equator—that winds oppositely around the core in northern versus southern hemispheres.

Whatever the details of the westward drift, the toroidal flux lines are created by differential rotation from preexisting poloidal flux lines.

The 'other side of the coin' is also possible, with the help of dynamo action: as we will see now, the energy of a toroidal field can be used to create new poloidal energy; under the right conditions, that new poloidal field can even reinforce the preexisting one. Toroidal fields can thus play a pivotal role in the regeneration of planetary magnetic fields by dynamo action.

10.4.3 Examples of Simple Dynamos

The core dynamo is sometimes described as a *self-sustaining* or *self-exciting dynamo*, in that it starts with a preexisting field $\vec{B}$, and—by turning part of it into a toroidal field—is then able to create poloidal field lines which end up strengthening the original $\vec{B}$; it's as if $\vec{B}$ is feeding off itself. In reality, of course, fluid motions are required, to induce the electrical currents responsible for both the toroidal field and the poloidal changes; for the dynamo to continue operating, energy must be continually injected into the system, at a sufficient level to drive those motions. As noted with reference to the Lorentz force, however, if those motions are not the 'right' kind, dynamo action will fail to reinforce the original poloidal field. These key features—an energy source and the 'right' kind of motion—are illustrated by a simple mechanical device called a *disk dynamo*.

In this chapter, we have generally focused on the kind of magnetic field—a dipole!—produced by (circular) current loops. In order to understand the disk dynamo, we need to consider two other geometries besides a single current loop; first, a straight wire; it turns out that the magnetic field associated with an electrical current traveling through a straight wire is a series of concentric, circular flux lines surrounding the wire (Figure 10.33a). This makes sense if we imagine cutting the loop in Figure 10.1c and straightening it out: the oval field lines are no longer repelled by field lines on the other side of the loop, and will align concentrically.

Second, consider a series of current loops all lined up, and connected (in practice, we would use one long wire, wound repeatedly (a)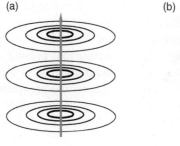

(b)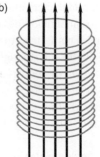

Figure 10.33 The magnetic fields (black) associated with two different configurations of electrical current (orange).
(a) The magnetic field of a straight-line current consists of sets of concentric circular flux lines throughout the length of the current. With the current in the direction shown, the field lines are all counterclockwise, looking down from above (—a right-hand rule!).
Can you verify the direction of the field lines (as seen when the current points toward you) using Figure 10.1c?
(b) The magnetic field within an electromagnetic coil consists of a set of approximately straight lines parallel to the coil. Note that the greater the number of 'windings,' the stronger the field within.

around a cylindrical metal 'core' for added strength—creating a simple *electromagnetic coil* or *solenoid*). From Figure 10.1c, we see that the field line in the center of a single current loop is straight, perpendicular to the loop and aligned with the dipole axis; field lines off-center (but still within the loop) are nearly so. It is not surprising, then, to find that the magnetic field line through the center of the coil (Figure 10.33b) is straight, and the other field lines within the coil are very nearly so. As with Figure 10.1c, and according to the right-hand rule, the field lines will point upward if the current moving through the coils is counterclockwise (as seen from above).

The simplest disk dynamo begins with a disk attached to an axle; both are made of a metal or other conducting material. The disk will be spinning within an ambient magnetic field and, as with any conductor moving in the presence of a magnetic field, its spin will generate an electrical current. We can infer the nature of that current by invoking the Lorentz force.

Let's say that the ambient field is vertically upward, the axle is also vertical, the disk is horizontal, and the disk is spinning in a prograde fashion (counterclockwise as viewed from above); see Figure 10.34a. Any mobile charges within the disk are carried around with the

(a) $\vec{B}$ (b)

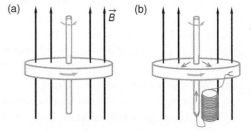

Figure 10.34 How a disk dynamo works.
(a) Setup for the disk dynamo. An electrically conducting disk and axle are spun in a counterclockwise sense (blue arrows) in the presence of a uniform ambient magnetic field $\vec{B}$.
(b) Currents (orange arrows) are induced in the spinning disk; by the Lorentz force, they are directed radially outward toward the edge of the disk. A coil is connected to the disk and axle (brush connections allow both to continue spinning); with the 'circuit' thereby closed, current will flow. If the current flows through the coil in the right sense, it will produce a strong vertical field reinforcing the original, vertically upward field.
Should the current in the coil flow clockwise or counterclockwise to yield a reinforced field?

disk, so their velocity $\vec{v}$ is azimuthal. Based on the definition of vector cross products, it follows that $\vec{v} \times \vec{B}$ is directed radially outward.

Those charges will actually flow if the 'circuit' is complete. Using 'brush' connections, we attach a wire to the outer edge of the disk, and connect the other end of it to the bottom of the axle; the brush connections maintain contact with the disk and axle but do not impede their spin. The current thus enabled flows outward in the disk, from axle to edge; it then runs down the wire, then back up the axle to the disk.

The spinning disk generates an electrical current and thus a magnetic field; but the field associated with a straight connecting wire—see Figure 10.33a—will be concentric circles, and no matter how that wire is oriented, such field lines will not reinforce the initial vertical field. However, if the wire is vertically coiled, with the coil located somewhere between the disk and the axle (see Figure 10.34b), the disk currents can generate a very strong vertical field from the coil that reinforces the initial field!

This dynamo generates a magnetic field as long as it keeps spinning. In effect, the spinning converts mechanical energy to electrical and magnetic energy. But currents lose energy as they travel through imperfect electrical conductors, with their energy converted to heat and dissipated away. To keep these currents going, mechanical energy must be continually supplied. And, the magnetic field created will not be reinforcing unless the currents are made to follow a particular coiled ('helical') path.

Significantly (though perhaps not surprising, at this point), the disk dynamo will equally well reinforce a reversed ambient field—without the sense of spin needing to be changed!

See if you can verify this: start with a vertically downward field; spin the disk in the same prograde manner; then use the Lorentz force to determine how the induced current moves through the system.

More complex laboratory dynamo systems have also been developed, including coupled-disk dynamos (Rikitake, 1958) and cylindrical or 'barrel' dynamos (Lowes & Wilkinson, 1963, 1968); see Figure 10.35. A two-body dynamo generates two interacting magnetic fields, each a reservoir for the magnetic energy required to sustain the other. For the Earth, the corresponding fields are the poloidal and toroidal fields in the core; that is, we envision the toroidal field strengthening by feeding off the poloidal field—as we saw earlier with the help of frozen flux—*and the*

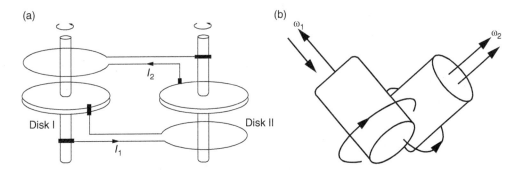

Figure 10.35 Alternatives to a simple disk dynamo.
(a) Coupled disk dynamo. The current from each disk coils around the other disk's axle, allowing the motion of that other disk within the coil's magnetic field to generate additional current.
Image from Rikitake (1958) / American Association for the Advancement of Science.
(b) Cylinder dynamo. Both cylinders spin in the same sense (e.g. both prograde, as seen here); their spin axes are perpendicular to each other. The initial inducing magnetic fields are suggested by the unlabeled arrows. Currents induced by one cylinder's spin create the ambient magnetic field for the other. The currents are able to flow through the entire dynamo system because both cylinders have been placed within cavities of a large block (not shown) of the same material; liquid mercury is used to lubricate the cylinders and electrically connect them and the block.
Image modified from Lowes and Wilkinson (1963) / Springer Nature. Those authors later (1968) found that a 45° angle between cylinder axes created a spontaneously reversing dynamo.

poloidal field regenerating by feeding off the toroidal field (e.g., Moffatt, 1978).

Significantly (—yes, again!), two-disk and two-cylinder dynamos exhibit reversals (e.g., Rikitake, 1958; Lowes & Wilkinson, 1968) and, within a polarity regime, repeated fluctuations in magnitude (Cook & Roberts, 1970).

Actual disk dynamos can be found in nature. In the ocean, they are created from *mesoscale eddies*, which spin electrically conducting seawater around in the presence of Earth's magnetic field. They have been detected in association with meanders of the Australian western boundary current (Lilley et al., 1993), with their dynamo action as a likely explanation for the strong magnitude of the magnetic field observed there (noted earlier in this chapter; see also Tyler & Mysak, 1995). They are also expected as submerged eddies (nicknamed 'Meddies') in the eastern or central North Atlantic that originate as 'salt lenses' in the Mediterranean Sea (Lilley et al., 1993). One such Meddy, first detected in 1984, lasted for at least 2 yr, spinning at a depth of ~1 km with a period of ~6 days, and with an initial diameter of ~50 km (Armi et al., 1989); an earlier one, described by Armi et al. (1989), was twice as large and lasted for several more years.

Finally, we note that an initial applied field is required for the disk dynamo, to feed the growth of the dynamo's field. This was initially true for the geodynamo as well. Conceivably, though, that field could have been quite weak, if the geodynamo was very effective; in the laboratory, for example, the fields produced by successfully operating dynamos are much larger than the initial field. Possible 'feeder' fields available to the young Earth include the Sun's magnetic field and the disk-like galactic magnetic field (e.g., Moffatt, 1978; Roberts & Soward, 1992, and references therein; for a scenario explaining how the very first galactic-scale magnetic fields in the universe came about, see Mtchedlidze et al., 2022).

An 'internal' feeder field is also possible, associated with electric currents produced in the early Earth right at the core-mantle boundary by a thermoelectric (thermocouple) effect.

In this mechanism, two materials with different conductivities will experience a difference in electric potential (i.e., a voltage; see *Textbox A*), and thus a current, if they are in contact and a temperature contrast exists between them. This effect was once hypothesized (Elsasser, 1939) as the source of Earth's observed field itself (see Stevenson, 1987b and Giampieri & Balogh, 2002 for an application to Mercury); though rejected because the required temperature difference would have been unreasonably high, less extreme temperature differences producing a much weaker current cannot be ruled out as the cause of the weak feeder field.

10.4.4 Inescapable Wisdom From Unavoidable Equations

The equations governing dynamo behavior are among the most challenging in geophysics. However, throughout this textbook, and as this chapter has progressed, we have actually been introduced to many of the forces and other quantities that appear in such equations. As we near the end of this textbook, it is worth taking a few painless moments to put it all together; even without attempting any rigorous solutions, there is much we can learn from these equations.

We have already discussed our less-intimidating version of the induction equation, an equation that governs the evolution and behavior of magnetic fields:

$$\frac{\partial}{\partial t}(\vec{B}) = k_{\mathrm{M}} \nabla^2(\vec{B}) + \mathsf{S}(\vec{v} \times \vec{B}).$$

This equation is important but, by itself, is not enough. As noted earlier, successful (re)generation of a field requires that the source region exhibit the 'right' kinds of fluid motion; to see if such motion is possible in the core, we also need an equation governing flow there. Mathematically, of course, a second equation is necessary simply because our equation for $\vec{B}$ involves a second variable, namely $\vec{v}$.

The law governing motion of any kind is conservation of momentum, that is, Newton's Second Law, $\vec{F} = M \, \vec{a}$; expressed in terms of forces per unit volume, and with acceleration $\vec{a}$ expressed in terms of velocity, this law becomes

$$\vec{F}/\text{VOL} = \rho\vec{a}$$

$$= \rho\frac{d\vec{v}}{dt} = \rho\left(\frac{\partial\vec{v}}{\partial t} + (\vec{v}\cdot\vec{\nabla})\vec{v}\right),$$

where ρ is the density of a parcel of fluid whose volume is VOL. In this formulation we have also written out the total derivative of $\vec{v}$ (symbolized here as $d\vec{v}/dt$ rather than $D\vec{v}/dt$; see Chapter 9).

The forces that can act on a parcel of fluid in the core include gravity and centrifugal force (which we will lump together); the Coriolis force; pressure forces; viscous drag (which we saw earlier is mostly negligible); and magnetic forces. We quantified (or at least symbolized) some of these in and prior to Chapter 9; to deal with the magnetic force on a parcel of fluid, we begin with the Lorentz force, which expresses the force on an electrical charge in motion. If the parcel contains an amount of charge q within its volume, and is moving at velocity $\vec{v}$, it follows that the force per unit volume, $q\vec{v}\times\vec{B}/$ VOL (which we can write as $[q\vec{v}/\text{VOL}]\times\vec{B}$), is equivalent to $\vec{J}\times\vec{B}$ where $\vec{J}$ is the current density (charge per cross-sectional area per sec); this equivalence is illustrated in Lowrie (2007).

Using our quantifications of all the various forces, our expression for Newton's Second Law (with all terms per unit volume) can be written

$$\rho\frac{\partial\vec{v}}{\partial t} + \rho(\vec{v}\cdot\vec{\nabla})\vec{v} = \vec{F}/\text{VOL}$$

$$= \rho\vec{g} - 2\rho\vec{\Omega}\times\vec{v} - \vec{\nabla}P$$

$$+ \eta'\nabla^2(\rho\vec{v}) + \vec{J}\times\vec{B}$$

(see also, e.g., Kono & Roberts, 2002). In its full glory, this equation is known as the **Navier-Stokes equation**; if it seems daunting and impenetrable, we should not be surprised: under various conditions, and with suitable accompaniments, this very equation accounts for the flow of the entire Earth System—atmosphere, oceans, mantle, and core!

With Navier-Stokes as the starting point for all fluid dynamic studies, each particular situation—each dynamical 'regime'—will require that some relevant terms be retained while others may be dropped. For example, in the simplest situation possible—a non-conducting fluid that is at rest—nearly all the terms in the equation vanish or are irrelevant, leaving

$$\rho\vec{g} = \vec{\nabla}P,$$

a relation we recognize as hydrostatic equilibrium (see Chapters 8 and 9). In other situations, fewer terms will vanish; in mantle convection studies, for example, the Coriolis term will be negligible, since v is so small, and the Lorentz force is likely to be negligible because electrical currents are weak. And, in some situations a 'hybrid' strategy, where one component of the equations is simplified but not the others, has proven itself quite useful; for example, if we approximate the oceans as a thin horizontal fluid layer, we might assume hydrostatic equilibrium for the vertical component, leaving more complex "shallow-water equations" (which still include various forces) for the horizontal components that allow us to determine the lateral flow (e.g., Vallis, 2017). One version of the shallow-water equations, called the *Laplace tide equations*, has been widely used (by, e.g., Dickman, 1989 among many others) to both predict and understand ocean tides.

For the geodynamo, viscous drag is negligible through most of the core, making it safe to neglect the term involving the viscosity η'. On the other hand, we expect that the core is undergoing convection (either thermal or compositional, or both); thus, not only is gravity ($\rho\vec{g}$) important (requiring that it be balanced overall by a 'background' pressure gradient), the buoyancy force associated with gravity acting on mass excesses or deficits ($\Delta\rho$) is key as well. To deal with convection, we might specify

$$\rho = \rho_0 + \Delta\rho,$$

where ρ_0 is the background density profile and $\Delta\rho$ are the convective excesses and deficits experiencing buoyant forces. Instead of $\rho\vec{g}$ in the Navier-Stokes equation, we would then have $\rho_0\vec{g} + \Delta\rho\vec{g}$. For thermal convection, as discussed in Chapter 9, $\Delta \rho = -\rho_0 \alpha_P \Delta T$ (where α_P is the thermal expansion coefficient).

With this approach, then, one conceptual simplification we can make is to subtract out a background state of hydrostatic equilibrium, defined by $\rho_0 \vec{g} = \vec{\nabla} P$ (see, e.g., Tritton, 1980); the Navier-Stokes equation 'simplifies' to

$$\rho \frac{\partial \vec{v}}{\partial t} + \rho(\vec{v} \cdot \vec{\nabla})\vec{v}$$
$$= \Delta \rho \vec{g} - 2\rho \vec{\Omega} \times \vec{v} - \vec{\nabla} P'$$
$$+ \eta' \nabla^2 (\rho \vec{v}) + \vec{J} \times \vec{B},$$

where P' denotes any remaining non-hydrostatic (dynamic) pressure.

Some Quick Scaling Analyses. In general, scaling the terms of the Navier-Stokes equation will help clarify which forces control the dynamical behavior of the core fluid versus which can be neglected, thereby allowing us to simplify the equations being solved. The 'other side of the coin' is worth pointing out as well: physical and numerical models of a particular phenomenon, say, the core dynamo, will not be relevant to the real Earth unless they scale similarly, placing them in the correct turbulent, nearly inviscid, and high-rotation regimes (see Olson & Christensen, 2006).

- Viscous drag versus inertial forces: the Reynolds number

  We defined this dimensionless number in Chapter 9 as

$$\text{RE} \equiv \frac{\text{Inertial force}}{\text{Viscous force}} = \frac{\rho \vec{a}}{\eta' \nabla^2 (\rho \vec{v})}$$
$$= \dots = \frac{LV}{\eta'}.$$

With reference to the Navier-Stokes equation, the inertial force per unit volume, $\rho \vec{a}$, could be written as either of the terms on the left-hand side of that equation (not just $\rho \partial v / \partial t$, as considered in Chapter 9).

It should be easy for you to verify that the formula for the Reynolds number ends up being the same when the inertial force is represented by $\rho(\vec{v} \cdot \vec{\nabla})\vec{v}$!

For the core, as noted earlier in this chapter, RE might be as large as $\sim 4 \times 10^8$, justifying our decision to neglect viscous drag.

- Coriolis versus inertial forces: the **Rossby number**

  This dimensionless number is

$$\text{Ro} \equiv \frac{\text{Inertial force}}{\text{Coriolis force}} = \frac{\rho \vec{a}}{2\rho(\vec{\Omega} \times \vec{v})}$$
$$= \dots = \frac{V}{2\Omega L}.$$

In the core, using V as inferred from the westward drift and taking the typical core dimension to be 10^6 m, we find Ro $\sim 3 \times 10^{-6}$. It appears that, just as the viscous drag term could be neglected in comparison to inertial forces, the latter can be neglected in comparison to the Coriolis force.

- Coriolis versus buoyancy forces: a modified Rayleigh number

  This dimensionless number is

$$\text{RA}_\text{B} \equiv \frac{\text{Buoyancy force}}{\text{Coriolis force}} = \frac{\Delta \rho \vec{g}}{2\rho(\vec{\Omega} \times \vec{v})}$$
$$= \frac{\rho \alpha_P \Delta T \vec{g}}{2\rho(\vec{\Omega} \times \vec{v})} = \dots = \frac{\alpha_P g \Delta T}{2\Omega V}$$

where thermal buoyancy has been assumed for illustration; this formulation is similar to the "modified Rayleigh number" of Christensen and Aubert (2006) and to the square of the "convective Rossby number" of Aurnou et al. (2020).

In the core, using V as inferred from the westward drift and taking $g \sim 8 \, \text{m/sec}^2$ (see Chapter 8) and $\alpha_P \sim 10^{-5} \, (°\text{C})^{-1}$ (Kono & Roberts, 2002), we have $\text{RA}_\text{B} \sim 1,370 \, \Delta T$. Even with a (superadiabatic) temperature difference of just 1°C vertically, the buoyancy force would be a thousand times as strong as the Coriolis force. Buoyancy (i.e., convection) is likely to play an important role in core dynamics.

- Lorentz versus Coriolis forces: the **Elsasser number**

This dimensionless number is

$$\Lambda \equiv \frac{\text{Lorentz force}}{\text{Coriolis force}} = \frac{\vec{J} \times \vec{B}}{2\rho(\vec{\Omega} \times \vec{v})}$$

(e.g., Kono & Roberts, 2002). Reducing this to a simple formula is beyond the scope of this textbook (it involves the same kind of calculus that underlies $S(\vec{v} \times \vec{B})$ in the induction equation). It turns out that $J \sim B/\mu_0 L$, which leads to

$$\Lambda = \frac{B^2/\mu_0 L}{2\rho\Omega V};$$

other parameterizations and scalings to construct this ratio are possible (e.g., Soderlund et al., 2012; Christensen & Aubert, 2006; Kono & Roberts, 2002; Stevenson, 2010).

For the core, taking its typical dimension to be 10^6 m, its typical flow velocity to be that of the westward drift, and its density to be $\sim 10^4$ kg/m^3 (10 gm/cm^3), we find $\Lambda \sim$ 1,360 B^2, if B is in Tesla; consequently, $\Lambda \sim 1$ if $B \sim 0.025$ T ($= 25$ mT). As we saw in Figure 10.17, an extrapolation of the poloidal field from the surface to the core boundary yields a field typically $\sim \frac{1}{2}$ mT in magnitude; based on Alfvén wave speeds, its intensity in the interior is probably a few mT, implying $\Lambda \sim 10^{-2}$. And, if we took B to be the core's toroidal field, which might be one or two orders of magnitude larger than the poloidal field, Λ could well be ~ 1 or greater; it should be cautioned, however, that even with $\Lambda > 1$, the influence of the Coriolis force on the flow structure (described later in this subsection) would still be dominant (Soderlund et al., 2012; also Calkins, 2018).

- Growth versus decay: the magnetic Reynolds number

This dimensionless parameter often figures in to dynamo assessments. The **magnetic Reynolds number** is defined (e.g., Roberts & Soward, 1992) as the ratio of the strengthening versus weakening terms in

the induction equation, i.e.,

$$\text{RE}_{\text{MAG}} \equiv \frac{\text{Magnetic strengthening}}{\text{Magnetic weakening}}$$

$$= \frac{S(\vec{v} \times \vec{B})}{k_{\text{M}} \nabla^2(\vec{B})} = \dots = \frac{LV}{k_{\text{M}}},$$

where k_{M} is the magnetic diffusivity; we are able to complete the scaling even though S has not been specified because (as noted above) S has units of $1/L$. With this definition we can view RE_{MAG} as a measure of how survivable a body's magnetic field is, i.e., how successful a dynamo is in maintaining its field. By evaluating a large variety of numerical dynamos, Olson and Christensen (2006; see also Stevenson, 2003) established that successful dynamo action in convecting and rotating fluids takes place when RE_{MAG} exceeds a critical value of ~ 40.

For the core, with a characteristic dimension of 10^6 m and other parameters as specified earlier, $\text{RE}_{\text{MAG}} \sim 10^2$–$10^3$.

RE_{MAG} can also be interpreted as a *frozen flux number*, an indicator of the validity of the frozen flux approximation (but see also Love, 1999); such conditions are expected to roughly hold for RE_{MAG} exceeding ~ 10 (Stacey & Davis, 2008).

The 'classification' (in name only) of RE_{MAG} as a Reynolds number derives from the formal definition of the latter but with $\vec{B}$ substituting in for the momentum:

$$\text{RE} \equiv \frac{\text{Inertial force}}{\text{Viscous force}} = \frac{\vec{v} \cdot \vec{\nabla}(\rho\vec{v})}{\eta' \nabla^2(\rho\vec{v})}$$

$$\text{becomes} \frac{\vec{v} \cdot \vec{\nabla}(\vec{B})}{k_{\text{M}} \nabla^2(\vec{B})} = \dots = \frac{LV}{k_{\text{M}}} \equiv \text{RE}_{\text{MAG}}.$$

If RE_{MAG} is much greater than 1, strengthening of the field is likely to be happening faster than diffusing of the field can weaken it. The timescale t_{M} for the latter was estimated earlier at tens of thousands of years. But most of the reversal process happens much more quickly than t_{M}; for example, for at least two of the reversals shown in Figure 10.29b, the decline in dipole moment takes place in (much) less than

20 Kyr. It has thus been argued that reversals are a consequence of active dynamo processes—strengthening of the opposite polarity magnetic field—rather than passive weakening; thanks to frozen flux, that is, it is not so much a 'failing dipole' as a forcibly 'collapsing dipole' leading up to a reversal (Stacey & Davis, 2008; Olson et al., 2009; Hulot et al., 2010; Finlay et al., 2016).

- Coriolis force versus viscous drag, sort of: the Taylor and Ekman numbers

These dimensionless parameters may be particularly useful for our discussion near the end of this chapter. Defined officially as the ratio of the Coriolis force to the viscous drag force, in turbulent situations, and with the ratio squared (e.g., Brandenburg et al., 1990), the **Taylor number** is

$$\text{TA} \equiv \left(\frac{\text{Coriolis force}}{\text{Turbulent viscous force}} \right)^2$$

$$= \left(\frac{2\rho(\vec{\Omega} \times \vec{v})}{\widehat{\eta}' \nabla^2(\rho\vec{v})} \right)^2 = \dots$$

$$= \left(\frac{2\Omega L^2}{\widehat{\eta}'} \right)^2$$

where $\widehat{\eta}'$ denotes the turbulent (kinematic) viscosity. Sometimes (Sakai, 1997) 2Ω is replaced with the Coriolis parameter f ($\equiv 2\Omega\cos\theta$, defined in Chapter 4). Alternatively, we could define the Taylor number as $(\text{RE}_T / \text{RO})^2$, with RE_T a turbulent version of the Reynolds number.

Using the values of RE and RO specified above, and assuming $\widehat{\eta}' \sim 10^6 \eta'$ for illustration (so that $\text{RE}_T \sim 10^{-6} \cdot \text{RE}$), we find $\text{TA} \sim 10^{15}$.

The basis for the **Ekman number**, which is widely used in oceanographic studies, is similar to that for the Taylor number, though not restricted to turbulent flow:

$$\text{EK} \equiv \frac{\text{Viscous force}}{\text{Coriolis force}} = \frac{\eta' \nabla^2(\rho\vec{v})}{2\rho(\vec{\Omega} \times \vec{v})}$$

$$= \dots = \frac{\eta'}{2\Omega L^2}.$$

Earlier in this chapter we employed the force balance implicit in this definition to estimate the thickness δ_E of an Ekman boundary layer.

As with the Taylor number, sometimes 2Ω is replaced with the Coriolis parameter f in the Ekman number (e.g., Vallis, 2017). In turbulent conditions, $\text{EK} \equiv 1/\sqrt{\text{TA}}$, leading to a value of $\sim 3 \times 10^{-8}$ for the core. If nonturbulent conditions apply, $\text{EK} \equiv \text{RO} / \text{RE}$; for the core, then, under nonturbulent conditions, $\text{EK} \sim 3 \times 10^{-14}$.

One term from Navier-Stokes that we have not included in all these scaling comparisons is the pressure gradient. The pressure gradient accommodates the other forces, allowing a force balance to be achieved with whichever other forces are dominating the flow (Tritton, 1980; see also Christensen & Aubert, 2006).

Our scaling analysis is rough, meant to yield orders of magnitude rather than precise numerical values. It has also ignored the vector nature of the forces (i.e., which components are being scaled), a simplification probably more serious for the Elsasser number (which has cross products in both its numerator and its denominator) than the others.

Which $\vec{B}$, poloidal or toroidal, is more relevant to the Elsasser number? That is, the comparison between the Coriolis and Lorentz forces should involve the same component of the $\vec{\Omega} \times \vec{v}$ and $\vec{J} \times \vec{B}$ vectors. Pick a likely direction for $\vec{v}$ in the core, and use it to determine the associated direction of $\vec{\Omega} \times \vec{v}$. $\vec{J} \times \vec{B}$ must also have that latter direction. For $\vec{J} \times \vec{B}$ in that direction, which kind of $\vec{B}$ field—toroidal or poloidal—contributes more effectively to $\vec{J} \times \vec{B}$?

Despite our simplifications, one basic conclusion stands out from our scaling analysis: in Earth's core, the fluid motions that support the dynamo are most influenced by convection in the core (the buoyancy force), Earth's rotation (including the Coriolis force and also conservation of angular momentum), and core magnetism (the Lorentz force). However, depending on the situation, there are different ways these forces (along with the pressure gradient) can balance, leading to different types of flow.

Our formulation of the Navier-Stokes equation makes explicit the conditions that must prevail in those situations. For example, in Chapter 3, we discussed regional atmospheric weather systems in which a balance between the Coriolis and pressure forces resulted in cyclonic (low-pressure) and anti-cyclonic (high-pressure) circulations; such a balance was called *geostrophic* flow. Ideal, pure geostrophic flow must be both steady and smoothly varying (so that both parts of $d\vec{v}/dt$ vanish) with a fluid that is inviscid, poorly conducting (electrically), and not convecting. Under these conditions, the Navier-Stokes equation reduces to

$$0 = -2\rho\vec{\Omega} \times \vec{v} - \vec{\nabla}\mathcal{P}'$$

or

$$2\rho\vec{\Omega} \times \vec{v} = -\vec{\nabla}\mathcal{P}'.$$

It follows from the definition of vector cross products that in geostrophic flow, $\vec{v}$ must be perpendicular to the pressure gradient; that is, the flow is parallel to (i.e., along) isobars.

The geostrophic balance in the atmosphere is mainly a horizontal one. But in cyclonic weather systems, as noted in Chapter 3, friction can oppose the Coriolis force slightly, changing the balance somewhat and allowing the pressure gradient to force a weak spiraling of the winds in or out. The convergence or divergence of winds is accompanied by a slight buoyant updraft or downdraft associated with the central low or high pressure. In short, the resulting *quasi-geostrophic* flow is a three-dimensional spiral, a pattern described as *helical*.

Geostrophic, quasi-geostrophic, and other types of force balances have been envisioned in the core. As with the hydromagnetic waves mentioned earlier, **MAC** and **MC** can refer to conditions in which magnetic, convective (or 'Archimedean'), and Coriolis forces (together with the pressure gradient) determine the flow (e.g., Christensen & Aubert, 2006); the force balance and resulting flow are called

magnetostrophic. Pure magnetostrophic flow also must be steady and smooth, and requires an inviscid fluid. In contrast to geostrophy, however, the fluid must also be a strong electrical conductor, and the buoyancy force may be even stronger; quantitatively, all these conditions reduce the Navier-Stokes equation to

$$2\rho\vec{\Omega} \times \vec{v} = -\vec{\nabla}\mathcal{P}' + \Delta\rho\vec{g} + \vec{J} \times \vec{B}$$

for magnetostrophy.

For the core, one further complication stands out. Most of the dimensionless numbers highlighted above depend explicitly on L, so different force balances can prevail on different length scales. In the core, as a consequence, the magnetic field ends up playing different roles on different scales.

In fact, consideration of length scales in the Elsasser number, as defined above, led Aurnou and King (2017) to argue that the Lorentz force is as important as the Coriolis force—that is, the flow is magnetostrophic—*only* on very small length scales, estimated for Earth's core as ~1 km; on larger scales, and overall, a quasi-geostrophic convective flow dominates.

The quasi-geostrophic nature of the core's large-scale flow, to a first approximation, and a consequent secondary role for magnetic forcing, was also found in numerical dynamo simulations by, for example, Aubert (2019), Schwaiger et al. (2019), and Schaeffer et al. (2017) (with an exception in the work of Schaeffer et al. (2017) for regions above and below the inner core). Like Aurnou and King (2017), those researchers did find the Lorentz force to be at least somewhat significant on smaller scales (except with multipolar, as opposed to dipolar, magnetic fields, according to Schwaiger et al., 2019).

Overall, these studies suggest that—rather than actively producing its own strengthening —the core's large-scale magnetic field is just a passive beneficiary of convective dynamo action; the latter is "only weakly influenced" (Soderlund et al., 2012) by the former. The most significant determinant of large-scale flow structure is the one force not excluded by any of these flow models or scaling analyses: the Coriolis force! Core flow is thus mostly quasi-geostrophic, and (among other things) should exhibit helical flow patterns. Experiments discussed later in this chapter bear these conclusions out. However, it might be noted that the role of the magnetic field is not completely secondary; on small scales, the field provides a quite effective means to (ohmically) dissipate convective energy (Aubert, 2019).

Then There's Turbulence. The high Reynolds number expected for the outer core implies that the flow there is likely to be turbulent. Turbulence compounds the challenges both in obtaining and in interpreting solutions to the Navier-Stokes equation. One approach that has been reasonably successful distinguishes 'higher-order'—small-scale—random turbulent fluctuations from the 'mean' flow; the fluctuations are assumed to average out over appropriately larger spatial domains (e.g., Olson, 1983; Moffatt, 1978). Taking such a 'perturbation' approach we might, for example, write the fluid velocity as

$$\vec{v} = \vec{v}_0 + \vec{v}', \text{ with } <\vec{v}'> = 0;$$

here $\vec{v}_0$ refers to the background 'mean flow,' while $\vec{v}'$ represents the superimposed turbulent fluctuations experienced by the flow and $<\vec{v}>$ symbolizes its *spatial* average, vector component by vector component. Note that $<\vec{v}_0>$ $\equiv \vec{v}_0$ (hence its name, "mean" flow!). Similar expressions apply to the other variables in the equation, for example

$$\vec{B} = \vec{B}_0 + \vec{B}', \text{ with } <\vec{B}'> = 0,$$

so that $\vec{B}_0$ represents the 'mean field' in the core (see, e.g., Kono & Roberts, 2002).

Applying these expressions to the induction equation gives us an idea of the quantitative challenges faced when accounting for turbulent conditions. When our expressions for $\vec{v}$ and $\vec{B}$ are substituted in, we have

$$\frac{\partial}{\partial t}(\vec{B}_0 + \vec{B}') = k_M \nabla^2 (\vec{B}_0 + \vec{B}')$$
$$+ S(\{\vec{v}_0 + \vec{v}'\} \times \{\vec{B}_0 + \vec{B}'\}).$$

We can take the spatial average of this entire equation. Assuming the mathematical operations can be interchanged, for example

$$<\frac{\partial}{\partial t}(\vec{B})> = \frac{\partial}{\partial t}(<\vec{B}>)$$

(the average of the time derivative equals the time derivative of the average), the left-hand side of the induction equation simplifies nicely:

$$<\frac{\partial}{\partial t}(\vec{B}_0 + \vec{B}')> = \frac{\partial}{\partial t} <(\vec{B}_0 + \vec{B}')>$$
$$= \frac{\partial}{\partial t}(<\vec{B}_0> + <\vec{B}'>)$$
$$= \frac{\partial}{\partial t}(\vec{B}_0).$$

However, the right-hand side is a bit more involved, and the averaged induction equation reduces first to

$$\frac{\partial}{\partial t}(\vec{B}_0) = k_M \nabla^2 (\vec{B}_0)$$
$$+ S(<\{\vec{v}_0 + \vec{v}'\} \times \{\vec{B}_0 + \vec{B}'\}>)$$

and then, eventually,

$$\frac{\partial}{\partial t}(\vec{B}_0) = k_M \nabla^2 (\vec{B}_0)$$
$$+ (S(\vec{v}_0 \times \vec{B}_0) + S(<\vec{v}' \times \vec{B}'>)).$$

We obtained this result first by eliminating the $\vec{v}_0 \times \vec{B}'$ and $\vec{v}' \times \vec{B}_0$ terms, each of which averages to zero; but if $\vec{v}'$ and $\vec{B}'$ are spatially correlated (e.g., they are both positive, or both negative, in the same locations), then $\vec{v}' \times \vec{B}'$ does not average to zero, and instead amounts to an additional, *turbulent* source term for $\vec{B}$. That is, under some conditions small-scale turbulence ($\vec{v}'$) can power the mean field ($\vec{B}_0$).

In the (full) Navier-Stokes equation, averaging the advective acceleration $\rho(\vec{v} \cdot \vec{\nabla})\vec{v}$ similarly leads to

$$\rho_0(\vec{v}_0 \cdot \vec{\nabla})\vec{v}_0 + \rho_0 <(\vec{v}' \cdot \vec{\nabla})\vec{v}'> .$$

The second of these terms shows once again that turbulence can contribute to the mean flow solution (see Tritton, 1980, who develops additional expressions for and an interpretation of the turbulent advection term).

Once such time-averaged equations have been derived, it is useful to subtract them from the original 'turbulent' equations. For the induction equation, this yields

$$\frac{\partial}{\partial t}(\vec{B}_0 + \vec{B}') = k_M \nabla^2 (\vec{B}_0 + \vec{B}') + S(\{\vec{v}_0 + \vec{v}'\} \times \{\vec{B}_0 + \vec{B}'\})$$

$$- \frac{\partial}{\partial t}(\vec{B}_0) - k_M \nabla^2 (\vec{B}_0) - (S(\vec{v}_0 \times \vec{B}_0) + S(<\vec{v}' \times \vec{B}'>))$$

or

$$\frac{\partial}{\partial t}(\vec{B}') = k_M \nabla^2 (\vec{B}') + S(\vec{v}_0 \times \vec{B}') + S(\vec{v}' \times \vec{B}_0) + S(\vec{v}' \times \vec{B}') - S(<\vec{v}' \times \vec{B}'>),$$

showing how the smaller-scale part ($\vec{B}'$) of the magnetic field evolves as the turbulence and large-scale (mean) fields feed off each other.

The Rest of the Story. Finally, a fully rigorous fluid dynamic model, turbulent or not, must also conserve energy and mass. Our exploration of energy conservation in Chapter 9 allows us to write

$$\frac{\partial T}{\partial t} + \vec{v} \cdot \vec{\nabla} T = k \nabla^2 T + \frac{\varepsilon}{\rho C_P}$$

where ε symbolizes heat sources. With a fluid in motion, heat is generated by friction as flow continues, so ε must include a term involving $\vec{v}$ (the slower the flow, the less the frictional heat). For the core, Ohmic dissipation (because the electrical conductivity is less than infinite) also takes place, so ε must also include a term involving $\vec{J}$. Additionally, ε may be time dependent, including, in different eras, radioactivity and/or heat of fusion (from inner-core crystallization).

The energy equation is obviously critically important if the dynamo is driven by thermal convection; in that case, the Navier-Stokes equation will explicitly include the temperature variable, in its formulation of the buoyancy force. But it is also crucial for compositionally driven dynamos, as it lets us determine whether temperature conditions promoting the 'nucleation' and growth of an inner core (e.g., Driscoll & Du, 2019) have been achieved.

As suggested by a brief discussion in Chapter 9 regarding the concept of divergence, the quantity $\vec{\nabla} \cdot \vec{v}$ should be relevant o the question of mass conservation. If the divergence of the fluid velocity is negative, for example, then there is more flow into small volumes than out of it; but in order for the fluid to get 'stuffed' into a volume, it must be compressible. The divergence will be zero if and only if the fluid is incompressible. This is frequently taken as the condition of mass conservation:

$$\vec{\nabla} \cdot \vec{v} = 0.$$

More general expressions appropriate for compressible fluids are possible. As you might guess from analogies with thermal diffusion, for example, and with the preceding two equations in mind (also recalling that the $\nabla^2 T$ term in the diffusion equation originated as $-\vec{\nabla} \cdot \vec{q}$), we could try expressing mass conservation something like energy conservation, viz.,

$$\frac{\partial \rho}{\partial t} + \vec{v} \cdot \vec{\nabla} \rho = -\rho(\vec{\nabla} \cdot \vec{v}) + \varepsilon_{mass},$$

where multiplying $\vec{\nabla} \cdot \vec{v}$ times density on the right-hand side gives us the required units of mass (per volume) per time. Based on the thermal diffusion analogy, we have written ε_{mass} as a mass analog to the diffusion equation's heat source term. However, with mass conservation—"mass can neither be created nor destroyed"—such a source term must be zero within the fluid; our general expression for mass conservation (also known as **continuity**) is thus

$$\frac{\partial \rho}{\partial t} + \vec{v} \cdot \vec{\nabla} \rho = -\rho \vec{\nabla} \cdot \vec{v}.$$

Our introduction to the four governing equations—induction, Navier-Stokes, continuity, and energy conservation—is somewhat daunting, even after anticipating simplifications that are possible based on scaling analyses. Their joint solution is the basis for some of the dynamo models presented next.

10.4.5 Dynamo Flow in a Taylor-Proudman World

The idea of a self-exciting dynamo mechanism was first proposed to explain the solar magnetic field (Larmor, 1919 as cited in Moffatt, 1978). Conditions in the Sun relevant to field generation are different than those in our core; for example, solar differential rotation is actually associated with an eastward drift (e.g., Thompson et al., 2003) rather than our own westward drift, and the Sun's angular velocity is significantly (Brandenburg et al., 1990) less than ours. Nevertheless, much of our understanding of how the geodynamo works derives from research on the solar dynamo; the paradigm developed by Parker (1955) has been used for both. In this subsection, we will begin with a brief discussion of that paradigm, and use it to contrast with more recent models of the geodynamo.

The Parker Dynamo. Also known as the **Parker-Levy dynamo**, this paradigm (Parker, 1955; Levy, 1972a) describes one way a solar or planetary magnetic field can be sustained by dynamo action. Its visual simplicity is remarkable, given the complexity and imposing nature of the underlying mathematics used by its creators to substantiate it (—that mathematics, based on the equations we outlined above), and it provides a relatively easygoing context for some basic concepts and terminology.

This paradigm identifies three steps to regenerating the Earth's (or the Sun's) dynamo-driven field.

1. Differential rotation within the core shears and stretches vertical (i.e., meridional) poloidal field lines into azimuthal, toroidal ones.

   Figure 10.32 shows how this could work. By itself, of course, this action does nothing to strengthen the poloidal field, and leaves us instead with a weakening poloidal field feeding what seems like an ever-growing toroidal field.

2. The core convection features cyclonic systems characterized by regional counterclockwise and clockwise circulation around upwellings and downwellings—just like the low- and high-pressure systems in our atmosphere. At the center of each cyclone, upwelling core fluid drags the east-west, toroidal flux lines up along with it, creating flux loops, while the circulation of the fluid around the upwelling twists the loops into a north-south alignment; see Figure 10.36a.

   In the southern hemisphere, that circulation is opposite; however, the direction of the toroidal flux lines is also opposite, so the polarity of the twisted flux loops is the same in both hemispheres. Also, in both hemispheres, the circulation is opposite for the anticyclones ('high-pressure systems') created by convective downdrafts. In all cases, the flux loops are twisted north-south and—now possessing a radial component—those loops have become poloidal in character.

3. Over time those newly poloidal flux loops weaken, and diffuse out, coalescing with the new flux loops from other cyclonic systems, and eventually with the preexisting (dipolar) field, sustaining it; see Figure 10.36b.

Parker (1969) and Levy (1972b) envisioned perhaps a dozen or so convective 'weather systems' spread throughout the core. But all convective systems will not be equally effective in turning toroidal into poloidal field lines, in part because the Coriolis force is not equally strong at all latitudes (e.g., Levy, 1972a).

The vigorousness of each regional circulation—of each 'vortex'—is measured by a quantity proportional to their overall spin rate called the **vorticity**. Vorticity is actually a vector quantity, with its direction reflecting the sense of spin (counterclockwise or clockwise) of the vortex according to the right-hand rule. As a physically meaningful quantity, vorticity must obey the laws of physics; it turns out that the governing equation for vorticity—derived from the Navier-Stokes equation, by the way—looks just like the induction equation for magnetic fields! Such a connection may seem less surprising if you look at Figure 10.33a and imagine the magnetic

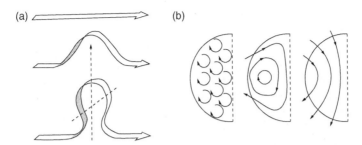

Figure 10.36 The Parker-Levy mechanism for regenerating the magnetic field of the Sun or Earth.
(a) Close-up: an initially east-west toroidal flux line, frozen into the core fluid, is (from top to bottom) lifted with the fluid by the convective buoyancy at the center of a 'low-pressure system' but also twisted by the regional circulation around that 'low' until the flux loop aligns north-south, perpendicular (dashed line) to the plane of the page.
(b) Big picture, side view, north to south pole: the flux loops thus created are poloidal; as they diffuse, they coalesce (from left to right) into a larger poloidal field, eventually adding to the original dipolar field. Image modified from Jacobs (1963).

field lines as representing a vortex encircling the current.

Can you guess what the coefficient for diffusion of vorticity (i.e., the analog to the coefficient k_M representing the diffusion of $\vec{B}$) would be? Hint: what physical parameter has the power to weaken or diffuse an eddy?

In step 2 of the Parker paradigm, we specified the effects on magnetic flux lines of both the vertical motion (e.g., upwelling) and horizontal circulation (e.g., counterclockwise flow) of a vortex in the core. Of course, both types of motion happen simultaneously, as with atmospheric cyclones, with the net effect being a *spiraling* flow as each parcel takes a helical path. We define the **helicity** of this flow pattern as the product (technically the vector dot product) of the two types of motion: the vorticity and the upwelling velocity. In dynamo models, the magnitude and sign of the helicity are seen to be crucial for the regenerative ability of the dynamo (see, e.g., Amit & Olson, 2004).

Steps 1 and 2 of the Parker dynamo depend on frozen flux, and take place on dynamically short timescales; step 3 takes place on a sufficiently long timescale for the frozen flux condition to have relaxed. The creation of toroidal field by differential rotation in step 1 is called the **ω effect**, with ω symbolizing both Earth's rotation (Ω) and the differential

rotation that is a consequence of Ω. Step 2's buoyant lifting and twisting during convection can be called an **α effect**, with α viewed very simply as a graphical symbol of the helical *twisting* that turns a toroidal flux line into a poloidal one, or more physically as a reference to the buoyancy force. Altogether, this type of dynamo is called an **αω dynamo**.

α can also symbolize the source of the poloidal field in more general situations; for example, in turbulent dynamos, α has come to signify the mean intensity of the turbulent source term, $<\vec{v}' \times \vec{B}'>$ (see, e.g., Moffatt, 1978; Olson, 1983). Various types of dynamos, including also an **α² dynamo** (where the turbulence, possibly exhibiting a 'cyclonic' or helical character due to the effects of rotation (Parker, 1970; Levy, 1976), is a source of both poloidal *and* toroidal field), are discussed in, e.g., Roberts (1972), Roberts and Soward (1992), and Jones (2011).

The relevance of the Parker-Levy dynamo is underscored by its ability to produce not only a reinforced poloidal field, but reversals of that field as well (Parker, 1969; Levy, 1972c). The reversals happen when *toroidal* field lines are created in the opposite direction to what is needed for poloidal reinforcement. With some models of the shear accompanying differential rotation, for example, the polarity of the toroidal field is different at different latitudes, resulting in a zone (—patch?) or latitude belt of reversed field (Levy, 1972b, 1972c). In that case, changes in the proportion of high-latitude and low-latitude cyclones, as they migrate into or out of that belt (or as new cyclones develop and old ones die away), can produce a net reversed toroidal field, thus ultimately a reversed poloidal field. In the Parker paradigm, the polarity of the dipole field is thus a question of the core's 'climate;' reversals are a reflection of climate change in the core.

Parker-Levy dynamos can also exhibit *periodic* reversals, through magnetic (not hydromagnetic) oscillations between the poloidal and toroidal field components called Parker waves (e.g., Dietrich et al., 2013); Parker waves can also (Moffatt, 1978) be traveling magnetic waves (see Parker, 1955; Sheyko et al., 2016; Schrinner et al., 2012). However, such waves—despite their potential importance in explaining the solar cycle (as noted by Dietrich et al. (2013); see also Grote & Busse, 2000)—are beyond the scope of this chapter.

The Parker dynamo is an example of a **kinematic dynamo**, one in which the fluid velocities are specified in advance, and then the induction equation is solved (or used as a guide) to determine if the resulting magnetic field reinforces the original poloidal field or not. But its dependence on the 'right' kind of velocity shear to achieve its goal highlights the need for **dynamic dynamo** models (also called **magnetohydrodynamic** or **MHD dynamos**), in which $\vec{v}$ is simultaneously determined as the solution to some version of the Navier-Stokes equations that includes the Lorentz force. This need is especially critical given the ambiguity in our observations of the westward drift (e.g., whether it is latitude dependent, and if so, how), since that drift is the basis for our inferences of differential rotation (see also Amit & Olson, 2004).

We can build on the basic Parker-Levy dynamo paradigm to account for more realistic force balances and scales (spatial or temporal) characterizing the flow and field generation. For example, under rapidly rotating turbulent conditions, scaling analyses and experiments suggest that the quasi-geostrophic flow in Earth's core would be better represented by a Coriolis/inertial/Archimedean (sometimes known by the acronym CIA) force balance than by a MAC balance (e.g., Guervilly et al., 2019; Aurnou et al., 2020, also Schwaiger et al., 2019); this is consistent with our conclusion, noted earlier, of a reduced role for the Lorentz force. Furthermore, rather than that dozen or so convective 'weather systems,' each a regional structure extending laterally over broad distances, a more modern view of the dynamo envisions hundreds of large-scale upwellings and downwellings, but with each surrounded by much-smaller-scale ('local') counterclockwise or clockwise circulation (cf. Grooms et al., 2010; Guervilly et al., 2019; Aurnou et al., 2020; and for the Sun, Featherstone & Hindman, 2016 and Vasil et al., 2021). We will see vivid illustrations of this kind of flow pattern in our final discussions below on this subject.

Those achievements in our understanding of core dynamics derive, at least in part, from a recognition of the importance of another rotational effect—the final one to be considered in this chapter!—which, as we will now see, has the potential to change the nature of core flow profoundly. Our progress in understanding core behavior, incidentally, has been matched by improved models of the core dynamo, both computational and experimental (see reviews by, e.g., Roberts & King, 2013 and Le Bars et al., 2022); however, exploration of that topic is beyond the scope of this chapter.

Taylor Columns and Tangent Cylinders. Till now we have envisioned flow in the core as consisting of two parts: a radial component, associated with convection; and an azimuthal component, associated with the westward drift. Both components are influenced by the Coriolis force, as we have discussed. But in situations where the Coriolis force completely dominates the flow, it acquires a very different character.

Returning to the Navier-Stokes equation, we consider a situation ruled by geostrophy:

$$2\rho\vec{\Omega} \times \vec{v} = \vec{\nabla}P',$$

with all other forces zero or at least small enough to neglect, as noted earlier. *Textbox B* shows how this equation can be manipulated to yield the Taylor-Proundman theorem:

$$(\vec{\Omega}\cdot\vec{\nabla})\vec{v} = 0.$$

In particular, if the Earth is rotating around its z-axis, then $\vec{\Omega} = \Omega_z\hat{z}$, and the theorem reduces to

$$\frac{\partial\vec{v}}{\partial z} = 0;$$

that is, the velocity cannot change with z.

Textbox B: Extreme Geostrophy and the Taylor-Proudman Theorem

We begin with the geostrophic version of the Navier-Stokes equation,

$$2\rho\vec{\Omega}\times\vec{v} = -\vec{\nabla}P'.$$

We will assume that the Earth's angular velocity $\vec{\Omega}$ is constant. For this derivation, we will also assume that the fluid density ρ is constant and rewrite the equation as

$$\vec{\Omega}\times\vec{v} = -\vec{\nabla}(P'/2\rho).$$

This will be a three-step derivation; the first step is to write the equation explicitly in terms of the components of all the vectors. For simplicity, we use x-y-z coordinates throughout and write

$$\vec{v} = (v_x, v_y, v_z) = v_x\hat{x} + v_y\hat{y} + v_z\hat{z}$$

where $\hat{x}, \hat{y}$, and $\hat{z}$ are unit vectors. Similarly,

$$\vec{\Omega} = (\Omega_x, \Omega_y, \Omega_z) = \Omega_x\hat{x} + \Omega_y\hat{y} + \Omega_z\hat{z}$$

(note that at the end, we will simply consider $\vec{\Omega} = \Omega_z\hat{z}$). Substituting these vector expressions into the left-hand side of the equation gives us

$$(\Omega_x\hat{x} + \Omega_y\hat{y} + \Omega_z\hat{z})\times(v_x\hat{x} + v_y\hat{y} + v_z\hat{z}) = -\vec{\nabla}(P'/2\rho).$$

We can multiply out all the various cross products (—a tedious operation but ultimately worth the effort!). Using $\hat{x}\times\hat{y} = \hat{z}, \hat{y}\times\hat{z} = \hat{x}, \hat{z}\times\hat{x} = \hat{y}$ (and also $\hat{y}\times\hat{x} = -\hat{z}, \hat{x}\times\hat{x} = 0$, and so on, all by the rules of vector cross products), we end up with

$$\hat{x}(\Omega_y v_z - \Omega_z v_y) + \hat{y}(\Omega_z v_x - \Omega_x v_z) + \hat{z}(\Omega_x v_y - \Omega_y v_x) = -\vec{\nabla}(P'/2\rho)$$

$$= -\frac{\partial(P'/2\rho)}{\partial x}\hat{x} - \frac{\partial(P'/2\rho)}{\partial y}\hat{y} - \frac{\partial(P'/2\rho)}{\partial z}\hat{z}.$$

Of course, as with any vector equation, this is really three equations in one, for each of the $\hat{x}$-, $\hat{y}$-, and $\hat{z}$-components:

$$\hat{x}(\Omega_y v_z - \Omega_z v_y) = -\frac{\partial(P'/2\rho)}{\partial x}\hat{x} \quad \text{or} \quad (\Omega_y v_z - \Omega_z v_y) = -\frac{\partial(P'/2\rho)}{\partial x},$$

$$\hat{y}(\Omega_z v_x - \Omega_x v_z) = -\frac{\partial(P'/2\rho)}{\partial y}\hat{y} \quad \text{or} \quad (\Omega_z v_x - \Omega_x v_z) = -\frac{\partial(P'/2\rho)}{\partial y},$$

$$\hat{z}(\Omega_x v_y - \Omega_y v_x) = -\frac{\partial(P'/2\rho)}{\partial z}\hat{z} \quad \text{or} \quad (\Omega_x v_y - \Omega_y v_x) = -\frac{\partial(P'/2\rho)}{\partial z}.$$

Textbox B: Extreme Geostrophy and the Taylor-Proudman Theorem (*Concluded*)

The next step is to take the $\hat{x}$-derivative of the $\hat{y}$-component equation, and subtract it from the $\hat{y}$-derivative of the $\hat{x}$-component equation. That is,

$$\frac{\partial}{\partial y}\{(\Omega_y v_z - \Omega_z v_y)\} = \frac{\partial}{\partial y}\left\{-\frac{\partial(P'/2\rho)}{\partial x}\right\}$$

$$-\frac{\partial}{\partial x}\{(\Omega_z v_x - \Omega_x v_z)\} = -\frac{\partial}{\partial x}\left\{-\frac{\partial(P'/2\rho)}{\partial y}\right\}.$$

The right-hand side becomes

$$-\frac{\partial^2(P'/2\rho)}{\partial y\,\partial x} + \frac{\partial^2(P'/2\rho)}{\partial x\partial y},$$

which is zero—and the reason we took those particular derivatives. Since $\vec{\Omega}$ is constant, taking the derivatives of all the terms on the left-hand side is straightforward, and the whole thing reduces to

$$\left(\Omega_y\frac{\partial v_z}{\partial y} - \Omega_z\frac{\partial v_y}{\partial y}\right) - \left(\Omega_z\frac{\partial v_x}{\partial x} - \Omega_x\frac{\partial v_z}{\partial x}\right) = 0.$$

Combining terms, we can write this as

$$\Omega_x\frac{\partial v_z}{\partial x} + \Omega_y\frac{\partial v_z}{\partial y} - \Omega_z\left(\frac{\partial v_x}{\partial x} + \frac{\partial v_y}{\partial y}\right) = 0.$$

The final step is to invoke conservation of mass, which, as discussed earlier for incompressible fluids, takes the form $\vec{\nabla}\cdot\vec{v} = 0$. This can be written

$$\frac{\partial v_x}{\partial x} + \frac{\partial v_y}{\partial y} + \frac{\partial v_z}{\partial z} = 0,$$

which allows us to modify the geostrophic equation to

$$\Omega_x\frac{\partial v_z}{\partial x} + \Omega_y\frac{\partial v_z}{\partial y} + \Omega_z\frac{\partial v_z}{\partial z} = 0,$$

or

$$(\vec{\Omega}\cdot\vec{\nabla})v_z = 0.$$

Similarly, we could take the $\hat{z}$-derivative of the $\hat{x}$-component and subtract it from the $\hat{x}$-derivative of the $\hat{z}$-component, leading to

$$(\vec{\Omega}\cdot\vec{\nabla})v_y = 0;$$

and subtract the $\hat{y}$-derivative of the $\hat{z}$-component from the $\hat{z}$-derivative of the $\hat{y}$-component, yielding

$$(\vec{\Omega}\cdot\vec{\nabla})v_x = 0.$$

All three of these results can be combined into a single vector equation,

$$(\vec{\Omega}\cdot\vec{\nabla})\vec{v} = 0,$$

which is known historically (see Finlay et al., 2010b) as the Taylor-Proudman theorem.

(a)

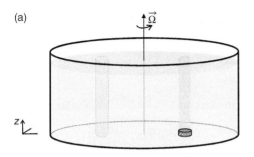

Figure 10.37 Exploration of the Taylor-Proudman theorem.
(a) Experimental demonstration of the theorem. This tank of water is rotating about a central axis aligned with the z-direction; z also denotes height above the base of the tank.
(Left side) Each column of water moves as a unit, since particles throughout its entire height have the same $\vec{v}$.
(Right side) If a 'bump' exists at the bottom of the tank, water cannot move off it without violating the Taylor-Proudman theorem. If colored dye is injected into the water within that column, it will (to first approximation) stay within the column.

Consider first a deep tank of water rotating fast enough about an axis for geostrophic conditions to apply. We will call that axis the z-axis, and let z measure the height above the bottom of the tank (see Figure 10.37a, left). At any location within the tank, according to this theorem, whatever the flow is there, it will be unchanged at points above or below; in effect, the entire column of water must move together. Such a column is called a **Taylor column**, and has the effect of imparting a *rigidity* to the water.

This same experimental set-up can provide an even more dramatic example of the power of the Coriolis force to restrict flow, if there is a 'bump' somewhere along the bottom of the tank (Figure 10.37a, right). The column of water above that bump cannot move off to the side: if it did, some of its water would move downward, some upward, as the column stretched to fill the larger vertical space—violating the theorem. Though other Taylor columns might drift through the water (in areas where the tank bottom is flat), migrating around, that column is in effect

'trapped' by the Coriolis force! And we can demonstrate this by adding some dye to the water in that column; after time has passed, the dye remains in that column (in reality, chemical diffusion and slight departures from geostrophy may allow a relatively tiny amount of dye to leak out).

Based on such a demonstration, we might conclude that long-lived 'spots' visible at the surface of a rotating fluid body might be a surface manifestation of a Taylor column, trapped by a solid irregularity ('bump') at the base of the fluid layer. The prime example that comes to mind is the Great Red Spot of Jupiter, not only because it has been around for hundreds of years, but also because (as noted in Chapter 3) flow within the Spot is nearly stagnant, whereas flow outside it is characterized by fierce winds. However, it is hard to attribute the Spot to an underlying solid 'bump,' given its tropical location plus the small diameter of Jupiter's solid inner core (see Table 1.1 (Chapter 1) footnote).

Jupiter inspires speculation on so many different levels; so how about this: we know many of Jupiter's moons are captured asteroids (e.g., Stacey & Davis, 2008); what if one large asteroid, either through orbital mechanics or tidal friction, impacted Jupiter, sinking down to become a 'bump' for the atmosphere above? (What about the size of the Spot, though?)

We will be relating this phenomenon to a core that is not just rotating strongly but convecting as well. Because convection is, at least in the simplest situations, radial, whereas the constraints of Taylor-Proudman 'rigidify' the flow, inhibiting it parallel to the rotation axis, we may wonder how convection will work in such situations. In Chapter 9, we explored the classic, 'pan of water heated from below' experiment known as Bénard convection. Figure 10.37b shows the extension of Bénard convection to rapid rotation, at high Rayleigh number. Strictly speaking, with buoyancy forces important we no longer

(b)

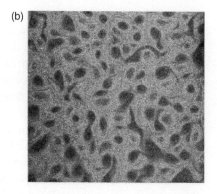

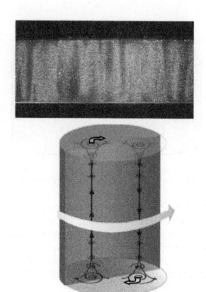

Figure 10.37 Exploration of the Taylor-Proudman theorem.
(b) Bénard convection with Taylor columns.
(Left, upper right, and lower right) Photographs of a tank of water, heated from below, cooled from above, and rapidly rotating about a vertical axis. The tank is also filled with microscopic thermotropic liquid-crystal capsules, which turn reddish when cold and bluish when warm.
(Left) View from above. Columns with ascending flow within are green/blue; with descending flow, red.
(Upper right) Side view. Red and green/blue still identify the two types of columns. It may be useful to imagine this figure as created by a cross-section through the tank with the view from above in mind; columns will not alternate in type, or be equally spaced—an unavoidable situation, given the 'migratory' aspect of the Taylor-Proudman theorem, as discussed in the text (see Figure 10.37a).
Images from Sakai (1997) / Cambridge University Press.
(Lower right) Depiction of flow and vorticity patterns of rising (left) and sinking (right) convective Taylor columns. Water rising off the hot lower boundary has a cyclonic (counterclockwise, or positive) vorticity; water sinking off the cool upper boundary spins with an anticyclonic (clockwise, or negative) vorticity.
As water rises within the initially cyclonic column, it spins with decreasingly positive, then zero, then increasingly negative vorticity; overall, the column's contribution to the angular momentum of the body is zero. In sinking columns, the vorticity variation with height is the opposite, but the overall effect is the same.
Are these patterns of vorticity reasonable to you? Consider a low-pressure system in the atmosphere. Convergence of winds at the surface lead to cyclonic flow around the low, as we know, spiraling in; at the top of the troposphere, the updraft diverges out. How does that spin?
Image modified from King and Aurnou (2012).

have geostrophic flow, the basis for the Taylor-Proudman theorem (nor is the flow exactly incompressible, as was required in the derivation of that theorem). In this nearly geostrophic situation, the primary flow is a set of *convective Taylor columns*, each one exhibiting either upward or downward flow (see Grooms et al., 2010; King & Aurnou, 2012; Sakai, 1997, and references therein). Outflow at the top or bottom boundary can feed other columns. Convectively, these columns serve to pump warm water from the heated lower boundary to the cool upper boundary, and facilitate the reverse for cool water (King & Aurnou, 2012). Within each column, the net flow is spiral, as indicated in Figure 10.37b (lower right). In line with our initial interpretation of the theorem, the columns are able to migrate at will around the tank, 'dancing' (King & Aurnou, 2012) as they swirl around.

Columnar convection is problematic in at least one respect. In these experiments, the vertical flow we see, whether upward or downward, cannot continue indefinitely, because the boundaries of the tank are impermeable; but any change in the flow versus z would violate the Taylor-Proudman theorem. The analysis by King et al. (2009) suggests that those violations are confined to *boundary layers*, where the Coriolis force can be eclipsed by other forces to ensure continuity of flow. Viscous forces are an obvious possibility, at least if we are close enough to the boundaries.

The geophysically relevant viscous boundary layer here should be the Ekman boundary layer, which as we saw earlier defines a region where momentum diffusion and *the Coriolis force* are comparable. From our earlier discussion, the thickness of that boundary layer is

$$\delta_E = \sqrt{\eta'/\Omega}.$$

Our focus on δ_E highlights the importance of the Ekman layer thickness in determining whether convection in a rotating body will be mostly classic (with or without turbulence) or rotationally controlled and dominated by convective Taylor columns (King et al., 2009). The faster the rotation, the thinner the Ekman layer, allowing the Coriolis force to dominate throughout the bulk of the tank, with viscous drag forces competitive only in very thin layers quite close to the boundaries. For the Bénard convection experiments presented in Figure 10.37b, the tank was rotating once every ~1 sec to ~10 sec, depending on the run, yielding Ω ~6 rad/sec to 0.6 rad/sec; in these experiments, then, the combined Ekman layer thickness (top + bottom) was no more than ~0.2 cm, or about a percent of the tank depth (which ranged from 3 cm to 20 cm).

As the core of the Earth convects and rotates, there is a similar tendency for the outer core fluid to align with the rotation axis, as convective Taylor columns; but compared with those tank experiments, there are differences worth mentioning. Because of the inner core, the Taylor-Proudman theorem effectively divides the volume of the outer core into two regions; we can mark the divide by positioning an imaginary cylinder—called the **tangent cylinder**—around the inner core, lined up with the rotation axis (see Figure 10.37c, top). Outside the tangent cylinder, outer core fluid can form 'vertical' columns, just like in our tank, to the extent that other forces (including now magnetic forces) allow nearly geostrophic flow. Some of the earliest experiments on fluids within rotating spherical shells (Busse & Carrigan, 1976; also Busse, 1970), vividly confirming the existence of such columns, generated huge interest among geophysical fluid dynamicists (including a standing ovation at at least one conference).

Within the tangent cylinder, however, the flow may differ dramatically from the convective Taylor columns of the wider core. The inner core acts as a pair of giant 'bumps,' one per hemisphere, at the base of the rotating fluid; the drop-offs at the 'sides' of the inner core are precipitous (see Figure 10.37c, left, top), trapping the outer core fluid above by the Taylor-Proudman theorem. The confinement of the fluid is strong but not complete, however; a slight exchange between fluid inside and outside of the tangent cylinder is observed experimentally due to edge effects, increasingly so if the convection is more intense (e.g., Aurnou et al., 2003; Schaeffer et al., 2017; cf. Aujogue et al., 2018).

Like columns in the rotating tank, convective Taylor columns in the outer core would be expected to sustain 'vertical' flow throughout their height, except right near the core-mantle boundary, where no-flow conditions would apply; presumably those conditions could be achieved, as with the tank, within its Ekman boundary layers. Using parameter values specified earlier for the core, $\delta_E \sim 10^{-1}$ m if the flow is not turbulent, ~10^{+2} m if the core is turbulent. In these thin non-geostrophic layers, outflow from columns with rising flow and inflow to columns with sinking flow would also provide nongeostrophic continuity; the process—called **Ekman pumping**—is described in more detail by Kono and Roberts (2002; see also Jones, 2011).

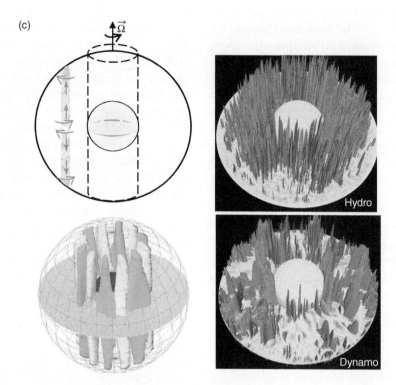

Figure 10.37 Exploration of the Taylor-Proudman theorem.
(c) Taylor-Proudman in the outer core.
(Left, top) The dashed lines encasing the inner core form a tangent cylinder, which in effect divides the outer core into two types of regions: the volume outside the cylinder, where Taylor columns can roam with no interference from the inner core; and the volumes above and below the inner core, where any motion is constrained by its presence.
One Taylor column (in blue) is shown to illustrate some concepts. Ideally, as discussed in the text, flow (straight arrows) within the column is parallel to the rotation axis; however, near the core-mantle boundary, which is a 'no-flow' boundary, the flow exiting the column is forced to diverge. Deflected around (curved arrows) by the Coriolis force, the flow has thereby gained helicity.
At the equator, the opposite situation prevails, with convergent flow into the columns.
(Left, bottom) Set of Taylor columns outside the tangent cylinder in the dynamo model of Kageyama and Sato (1997). Yellow columns are cyclonic; green columns are anticyclonic. The tilt of their top and bottom ends is consistent with no-flow requirements at the CMB. As discussed by Kono and Roberts (2002) / John Wiley & Sons, the widening of anticyclonic columns, for example, in their midsection (at the equator) and narrowing at their ends is a consequence of outflow and inflow at those locations; the pattern is reversed for cyclonic columns, and continuity holds.
In some dynamo models (see Jones, 2011), the convective Taylor columns drift westward over time, and the columns grow and decay over time.
Image from Kono and Roberts (2002) / John Wiley & Sons.
(Right-hand side two images) Taylor columns in a very fast rotating and convecting core without ("Hydro") and with ("Dynamo") Lorentz forces. The high rotation rate in these simulations is reflected in a very high Taylor number, which we estimate at more than 10^5 times greater than for the model shown bottom left. Blue columns are anticyclonic; orange columns are cyclonic.
The Lorentz force evidently stretches the thin Taylor 'tubes' into more sheet-like structures.
Images adapted from Yadav et al. (2016).

In the outer core, a thin equatorial zone—separating upwelling and downwelling columns in one hemisphere from those in the other (see Figure 10.37c, left, top)—acts as another boundary layer to the Taylor columns, facilitating flow between columns to insure continuity.

Outside the tangent cylinder, the curvature of the core-mantle boundary is an impediment to the migration of Taylor columns: they can move around the core, but only azimuthally (i.e., east-west), maintaining their initial latitudes in order to avoid stretching (or compressing) and thereby violate the Taylor-Proudman theorem. This is suggested by the illustration in Figure 10.37c (left, bottom). Alternatively, Amit and Olson (2004) describe one way the columns can stretch or shrink 'proportionally' as they move toward or away from the tangent cylinder, violating the theorem minimally (but also obeying the no-flow condition).

Under more extreme rotation, the Taylor columns proliferate (Figure 10.37c, right-hand side two images), increasing greatly in number but shrinking in girth.

Within the tangent cylinder, rotation is able to suppress convection unless the Rayleigh number is high (Olson et al., 1999). The experiments of Aurnou et al. (2003) and Aujogue et al. (2018) find that lateral temperature differences as the convection progresses generate a 'thermal wind' within the tangent cylinder, which leads to an anticyclonic azimuthal circulation—a vortex—around the rotation axis. In fact, such a polar vortex has been inferred by Olson and Aurnou (1999) and others from observations of the geomagnetic field (see Figure 10.37d). Its anticyclonic sense is consistent with it being an upwelling from deeper within the core, perhaps all the way from the inner-core boundary, then diverging at the top of its rise just beneath the core-mantle boundary.

As exotic as Taylor columns are, they are not restricted to the core. Oceanic plumes and 'chimneys' are essentially convective Taylor

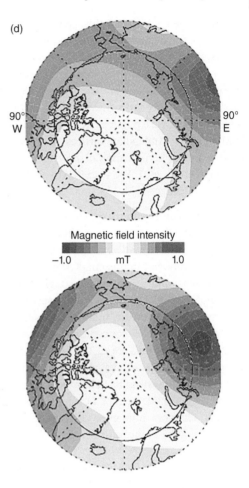

(d)

90° W   90° E

Magnetic field intensity

−1.0   mT   1.0

Figure 10.37 Exploration of the Taylor-Proudman theorem.
(d) Taylor-Proudman in the outer core, continued. Extrapolated geomagnetic field at the top of the core, as seen from its north pole down to 60°N, in 1870 (upper) and 1990 (lower), derived from the model by Bloxham and Jackson (1992). The domain of the tangent cylinder is shown by the black circle. The azimuthal (westward) motion of a prominent feature—a magnetic "low" (i.e. reversed flux patch), in yellow—during this time interval suggests the existence of an anticyclonic vortex around the pole. See also Hulot et al. (2002).
Image from Olson et al. (1999) / Springer Nature.

columns, created as part of the oceans' horizontal convection (Chapter 9; see Jones & Marshall, 1993; also Aurnou et al., 2003). We might also note the similarity of the polar vortex within the core's tangent cylinder to the stratospheric polar vortex in our atmosphere (as mentioned earlier in this chapter, from Chapter 3, the depletion of ozone occurs most effectively in polar stratospheric clouds, which require a cold *and still* stratosphere to form; those cloud conditions are most likely found in winter and early spring *within* a long-lasting polar vortex—see, e.g., Kretschmer et al., 2018; WMO, 2018). And polar vortices have been observed elsewhere in the Solar System (e.g., Svedhem et al., 2007; Yadav & Bloxham, 2020; Cappucci, 2023).

Regaining the Big Picture. Over the course of this textbook, we have gradually increased our use of mathematics, both symbolically and practically; but in this chapter, its use has skyrocketed—and appropriately so, given the nature of the topic! At this point, however, we will step back, with the intention of simply reinforcing some basic concepts about the generation and maintenance of the geomagnetic field. For those interested in learning more about the complexities of field production, or issues we may have deemed beyond the scope of this chapter, numerous comprehensive reviews may be found in the works of Olson et al. (1999), Kono & Roberts (2002), Hulot et al. (2010), Jones (2011), Roberts and King (2013), Moffatt and Dormy (2019), Le Bars et al. (2022), and in the extensive compilation edited by Schubert (2015), to name just a few. Some of those reviews also address one more issue worth noting but which we deem beyond the scope of this chapter: the need to employ unrealistic values of some parameters when modeling the core flow or dynamo, to avoid problems of numerical precision; one example of this is the Ekman number, which we saw earlier to be extremely small under core conditions.

Stepping back, then, we have a fluid outer core whose electrical conductivity is finite but high enough to carry and stretch magnetic flux lines. With the help of frozen flux, our observed poloidal magnetic field is regenerated by feeding off a toroidal magnetic field that exists within the core; that toroidal field was created in the first place from the original poloidal field. Rotation (including the Coriolis force but perhaps also conservation of angular momentum) and convection (possibly turbulent) are vital in promoting the toroidal-poloidal conversions and also in providing important constraints on the flow.

One obvious and reasonable conclusion from our discussions of dynamos and dynamo processes is that the creation and maintenance of the geomagnetic field results from a *combined*, Parker/Taylor-type dynamo mechanism (—a 'Taylored' Parker dynamo, perhaps?): that is, a Parker dynamo, but in the context of a strongly rotating core. For example, with strong rotation, the convective Taylor columns, characterized by helical flow, would have no problem twisting toroidal field into poloidal field, easily achieving step 2 of the Parker paradigm (see Figure 10.38a).

Paradigm step 1 has multiple possibilities for its realization; evidently our hybrid dynamos can be either $\alpha\omega$ or α^2, depending on how the toroidal field is first created. In an α^2 dynamo, as envisioned (see Figure 10.38b) by Olson et al. (1999), production of toroidal field derives from flow taking place in the equatorial plane between columns, to satisfy continuity; that flow also carries and stretches poloidal field lines azimuthally.

In convecting and strongly rotating $\alpha\omega$ dynamos, the toroidal field is created by differential rotation in the manner we discussed earlier, though there are differences from model to model (indeed, as we have seen, inferences of differential rotation from observations vary widely as well). For example, in the convecting, strongly rotating dynamo model developed by Kageyama and Sato (1997), westward drift is confined to the regions of the outer core occupied by Taylor columns, near to the tangent cylinder in their model; but, approaching the outer core boundary, differential rotation near the equator is either nonexistent (Kageyama & Sato, 1997) or, for a very similar model, eastward (Kono & Roberts, 2002). In the rapidly rotating convective dynamo models of Glatzmaier and Roberts (1995) and Schaeffer et al. (2017), differential rotation is confined to the tangent cylinder.

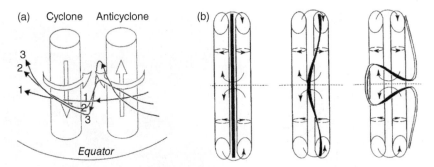

Figure 10.38 Using convective Taylor columns to achieve steps of the Parker paradigm.
(a) Step 2, in an αω Taylor-Parker dynamo: the flow in and around convective Taylor columns turns toroidal field into poloidal (illustrations in Jones (2011) may also prove enlightening). Three stages of this conversion are numbered as shown. Note the northern hemisphere upwelling column is labeled anticyclonic because the outflow at the top is deflected to its right, resulting in clockwise circulation. Image from Kageyama and Sato (1997) *and* from Kono and Roberts (2002).
(b) Step 1, in an α² Taylor-Parker dynamo: upwelling and downwelling (curved arrows) in convective Taylor columns feed adjacent columns at their ends (near the core-mantle boundary) and near the equatorial plane (dotted horizontal line). In particular, the outflow and inflow at the equator push poloidal flux lines (heavy vertical lines between columns) westward, stretching them and producing a toroidal field. See Olson et al. (1999) and Jones (2011), also Roberts and King (2013), for further details.
Image from Olson et al. (1999) / Springer Nature.

For years, the goal of numerical dynamo modeling was to show not only that a self-regenerating field was possible—that a field feeding off itself, with energy supplied by fluid motions, could keep its strength up—but that such a field would also spontaneously reverse its dipole polarity. The first to achieve that goal was the convectively driven, rapidly rotating dynamo model developed by Glatzmaier and Roberts (1995). Their model maintained its strength for almost 40,000 yr (of 'model time') before succumbing to a polarity reversal. During the reversal, which in their model took almost 1000 yr, the dipole moment dropped by an order of magnitude; it recovered right after.

Their model actually appeared to initiate several reversals, though only the one was completed during those 40 Kyr. (Reportedly, even after publication, Glatzmaier and Roberts continued to run their model, for at least another 250 Kyr; two other reversals were recorded.) The failure of most reversal 'attempts' to take hold has been attributed to a stabilizing effect of the inner core—a resistance to reversals (as first noted by Hollerbach & Jones, 1995). Convection in the outer

core causes fluctuations in the core magnetic field at all times, as documented (in part) by the secular variation; some of those fluctuations might conceivably lead to a polarity reversal, if they can grow to dominate the core. As we noted earlier, the timescale for a reversal (see Figure 10.29b) has also been viewed as a 'dynamic' timescale, shorter than the timescale for magnetic field diffusion and thus implying that reversals represent 'active' rather than passive processes. Because of its solidity, though, the field within the inner core can change only by diffusion—giving time for new fluctuations in the outer core to counteract the reversal before it becomes established throughout the entire core.

We started this chapter as most would, thinking of Earth's field as that of a dipole to (very) good approximation. But a single dipole, whether embodied as an intense bar magnet or single current loop, at or near Earth's center, is simply physically unreasonable. With that fact in mind, we vaguely talked about an alternate perspective, in which the dipole field was viewed as the averaged effect of a system of smaller dipole sources situated around the core: the behavior of the "dipole" was really the

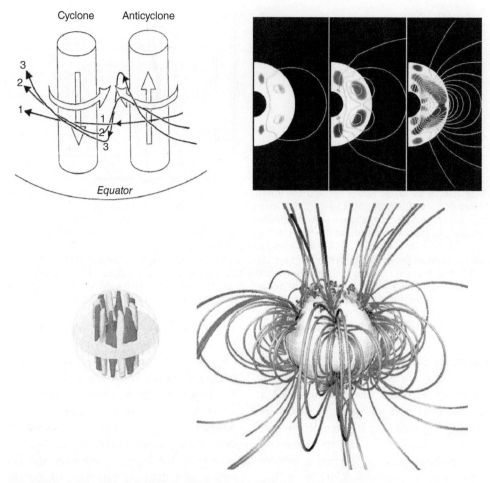

Figure 10.39 Conceptual evolution of a dynamo, clockwise from the lower left.
(Lower left) Predicted flow within the core, outside the tangent cylinder; the inner core, mostly hidden in this view, is in red. The equatorial plane is in pink. Fluid within the various columns shown here alternates between cyclonic (green) and anticyclonic (yellow) vorticities.
(Upper left) Close-up of Taylor columns turning a toroidal flux line into a poloidal one; see also the Figure 10.38 caption.
(Upper right) The success of the Taylor columns: generation of the poloidal field (yellow lines) by convectively driven dynamo action; in this side view, as time passes (from left to right), the development of a dipole-type field is evident. We also see color contours of the toroidal field (created from shear flow within the core, feeding off the original poloidal field), with red for westward-directed field lines, i.e. where the toroidal field points out of the page, and purple for the opposite.
(Lower right) Predicted field outside the core: not a dipole but dipole-like.
Images from Kageyama and Sato (1997); upper left image is *also* from Kono and Roberts (2002).

behavior of the ensemble of sources. Until now, that perspective has remained vague, but, in fact, it is generally implied by dynamo models; in Figure 10.39 (bottom right), for example, we see the evidence of such an ensemble! Thanks to the columnar nature of the convection, the individual sources have added together to produce a much stronger, nearly axial, and mostly dipolar geomagnetic field.

10.4.6 Some Final Thoughts

… Meanwhile, far from the core, and high above Earth's surface, it was *not* a dark and stormy night—or stormy daytime.

And that was a problem, magnetically speaking, of course. Deep in the mantle, the African and Pacific superplumes continued to grow, ensuring convection in the core would remain vigorous and chaotic. It was the time of a geomagnetic reversal, and the dipole field was down.

This reversal was particularly momentous. It began what we would come to call the Olduvai subchron, around 1.95 Myr ago. For several thousand years, the atmosphere would be exposed to the ravages of the solar wind; even the surface would, at least sporadically, feel the intense radiation of an unmasked Sun.

Eroding the atmosphere, the solar wind would have eaten away at the ozone layer, leaving a 'window' for ultraviolet light to pour through the atmosphere all the way down to the surface. Some wavelengths of that light (known as UV-B) would have been the 'right' size to damage strands of DNA; any life exposed to such radiation would potentially mutate,

or produce mutated offspring. As discussed in Chapter 2, the high-energy charged particles we call "cosmic rays" also have the potential to damage DNA; with a magnetic field too weak to deflect those particles toward the poles, life at the surface would, again, be left unprotected.

This was also a time of changes in the climate system, putting stress on living creatures. In parts of Africa, for example, there was a shift from woodlands to open grasslands as the region became more arid; this change was largely driven by tectonic uplift in East Africa (Sepulchre et al., 2006; see Amit & Olson, 2015)—a good time for our ancestors to evolve the ability to walk upright, perhaps, and definitely time for them to develop new hunting (or scavenging) and survival strategies.

Beginning with discoveries in the mid-twentieth century, paleontologists have been able to reconstruct our probable evolution to present-day *Homo sapiens* (see, e.g., Catt & Maslin, 2012). As seen in Figure 10.40, our

Figure 10.40 Evolution of *Homo sapiens* and other hominin species over the past 6 Myr.
Note the dramatic increase in species diversity ('radiation' of hominins) beginning ~2.0 Myr ago, around the time of the lower Olduvai reversal. Image from Catt and Maslin 2012 / Elsevier.
Another burst of diversity is evident in this image around 2 ½ Myr ago, corresponding to the previous, Gauss-to-Matuyama reversal.

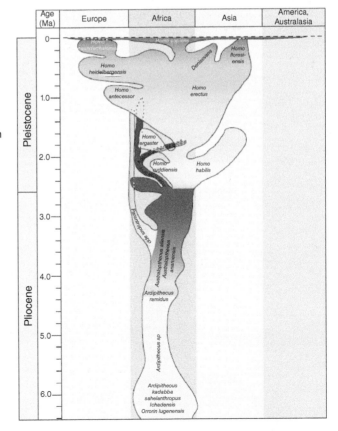

oldest *Homo* ancestor, *Homo habilis*, first appeared around 2½ Myr ago (which was actually the time of the reversal prior to lower Olduvai), according to fossil remains found in the so-called *Cradle of Humanity* of East Africa—specifically in Olduvai Gorge (e.g., Leakey & Leakey, 1964; Leakey et al., 1964). But our most direct ancestor, *Homo erectus*, whose cranial capacity was increased compared to its progenitors, and who spread widely and rapidly to neighboring continents (Catt & Maslin, 2012), appeared at the time of the Olduvai reversal.

Momentous, indeed.

References

Aakjaer, C., Olsen, N., & Finlay, C. (2016). Determining polar ionospheric electrojet currents from *Swarm* satellite constellation magnetic data. *Earth, Planets and Space, 68.* doi:10.1186/s40623-016-0509-y

Abarca del Rio, R., Gambis, D., & Salstein, D. (2000). Interannual signals in length of day and atmospheric angular momentum. *Annales Geophysicae, 18,* 347–364.

Abbot, D., Voigt, A., & Koll, D. (2011). The Jormungand global climate state and implications for Neoproterozoic glaciations. *Journal of Geophysical Research, 116,* D18103. doi:10.1029/2011JD015927

Abe, Y. (1997). Thermal and chemical evolution of the terrestrial magma ocean. *Physics of the Earth and Planetary Interiors, 100,* 27–39.

Abe, Y., Abe-Aouchi, A., Sleep, N., & Zahnle, K. (2011). Habitable zone limits for dry planets. *Astrobiology, 11,* 443–460.

Ablain, M., Cazenave, A., Larnicol, G., Balmaseda, M., Cipollini, P., Faugère, Y., et al. (2015). Improved sea level record over the satellite altimetry era (1993–2010) from the Climate Change Initiative project. *Ocean Science, 11,* 67–82.

Acuña, M., Connerney, J. E., Ness, N. F., Lin, R. P., Mitchell, D., Carlson, C. W. et al. (1999). Global distribution of crustal magnetization discovered by the Mars Global Surveyor MAG/ER experiment. *Science, 284,* 790–793.

Adams, F., Hollenbach, D., Laughlin, G., & Gorti, U. (2004). Photoevaporation of circumstellar disks due to external far-ultraviolet radiation in stellar aggregates. *Astrophysical Journal, 611,* 360–379.

Adams, L., & Williamson, E. (1923). The composition of the Earth's interior. *Smithsonian Institution Annual Report, 18,* 241–260.

Adcroft, A., Scott, J., & Marotzke, J. (2001). Impact of geothermal heating on the global ocean circulation. *Geophysical Research Letters, 28,* 1735–1738.

Adelberger, E., Heckel, B., & Nelson, A. (2003). Tests of the gravitational inverse-square law. *Annual Review of Nuclear and Particle Science, 53,* 77–121.

Adhikari, S., Caron, L., Steinberger, B., Reager, J., Kjeldsen, K., Marzeion, B., et al. (2018). What drives 20th century polar motion? *Earth and Planetary Science Letters, 502,* 126–132.

Agnor, C., Canup, R., & Levinson, H. (1999). On the character and consequences of large impacts in the late stage of terrestrial planet formation. *Icarus, 142,* 219–237.

Aguado, E., & Burt, J. (2007). *Understanding weather and climate* (4th edition). NJ: Pearson Prentice Hall, 562 pp.

A'Hearn, M., Belton, M. J. S., Delamere, W. A., Kissel, J., Klaasen, K. P., McFadden, L. A., et al. (2005). Deep impact: Excavating Comet Tempel 1. *Science, 310,* 258–264.

Aharoni, U. (2007). Visualization of the Coriolis and centrifugal forces. *Published December 29, 2007.* Available at https://www.youtube.com/watch?v=49JwbrXcPjc

Ahrens, T. (1979). Equations of state of iron sulfide and constraints on the sulfur content of the Earth. *Journal of Geophysical Research, 84,* 985–998.

Earth System Geophysics, Advanced Textbook 6, First Edition. Steven R. Dickman.
© 2025 American Geophysical Union. Published 2025 by John Wiley & Sons, Inc.
Companion Website: www.wiley.com/go/Dickman

Ahrens, T. (1980). Dynamic compression of Earth materials. *Science, 207*, 1035–1041.

Ahrens, T. (1990). Earth accretion. In J. Jones & H. Newsom (Eds.), *Origin of the Earth* (pp. 211–227). Houston, TX: Oxford University Press.

Ahrens, T. (1993). Impact erosion of terrestrial planetary atmospheres. *Annual Review of Earth and Planetary Sciences, 21*, 525–555.

Akalin, F., Gurnett, D., Averkamp, T., Persoon, A., Santolik, O., Kurth, W., & Hospodarsky, G. (2006). First whistler observed in the magnetosphere of Saturn. *Geophysical Research Letters, 33*, L20107. doi:10.1029/2006GL027019

Aki, K., & Lee, W. (1976). Determination of the three-dimensional velocity anomalies under a seismic array using first P arrival times from local earthquakes: 1. A homogeneous initial model. *Journal of Geophysical Research, 81*, 4381–4399.

Al'tshuler, L., Krupnikov, K., & Brazhnik, M. (1958). Dynamic compressibility of metals under pressures from 400,000 to 4,000,000 atmospheres. *Soviet Physics JETP, 34*(7), 614–619.

Alboussière, T., Cardin, P., Debray, F., La Rizza, P., Masson, J., Plunian, F., et al. (2011). Experimental evidence of Alfvén wave propagation in a Gallium alloy. *Physics of Fluids, 23*, 096601.

Alboussière, T., Deguen, R., & Melzani, M. (2010). Melting-induced stratification above the Earth's inner core due to convective translation. *Nature, 466*, 744–747.

Alder, B., & Trigueros, M. (1977). Suggestion of a eutectic region between the liquid and solid core of the Earth. *Journal of Geophysical Research, 82*, 2535–2539.

Alderson, A., & Alderson, K. (2007). Auxetic materials. *Journal of Aerospace Engineering, 185*, 565–575. doi:10.1243/09544100 JAERO185

Alfe, D., Gillan, M., & Price, G. (2002). Composition and temperature of the Earth's core constrained by combining ab initio calculations and seismic data. *Earth and Planetary Science Letters, 195*, 91–98.

Alfonsi, L., & Spada, G. (1998). Effect of subductions and trends in seismically induced Earth rotational variations. *Journal of Geophysical Research, 103*, 7351–7362.

Alfvén, H. (1942). Existence of electromagnetic-hydrodynamic waves. *Nature, 150*, 405–406.

Alken, P., Thébault, E., Beggan, C. D., Amit, H., Aubert, J., Baerenzung, J. et al. (2021). International Geomagnetic Reference Field: the thirteenth generation. *Earth Planets Space, 73*. https://doi.org/10.1186/s40623-020-01288-x

Allen, N., Richart, F., & Woods, R. (1980). Fluid wave propagation in saturated and nearly saturated sands. *Journal of Geotechnical and Geoenvironmental Engineering, 106*, 235–254.

Alley, M. (2003). *The craft of scientific presentations*. New York, NY: Springer-Verlag, 241 pp.

Allwood, A., Walter, M., Kamber, B., Marshall, C., & Burch, I. (2006). Stromatolite reef from the Early Archaean era of Australia. *Nature, 441*, 714–718.

ALMA Partnership, Brogan, C. L., Pérez, L. M., Hunter, T. R., Dent, W. R. F., Hales, A. S., et al. (2015). The 2014 ALMA long baseline campaign: First results from high angular resolution observations toward the HL Tau region. *Astrophysical Journal Letters, 808*. doi:10.1088/2041-8205/808/1/L3

Alt, J., & Teagle, D. (1999). The uptake of carbon during alteration of ocean crust. *Geochimica et Cosmochimica Acta, 63*, 1527–1535.

Altamimi, Z., Collilieux, X., Legrand, J., Garayt, B., & Boucher, C. (2007). ITRF2005: A new release of the International Terrestrial Reference Frame based on time series of station positions and Earth Orientation Parameters. *Journal of Geophysical Research, 112*, B09401. doi:10.1029/2007JB004949

Altamimi, Z., Sillard, P., & Boucher, C. (2002). ITRF2000: A new release of the International Terrestrial Reference Frame for earth science applications. *Journal of Geophysical Research, 107*. doi:10.1029/2001JB000561

Alvarez, L., Alvarez, W., Asaro, F., & Michel, H. (1980). Extraterrestrial cause for the

Cretaceous-Tertiary extinction. *Science, 208,* 1095–1108.

Amit, H., & Olson, P. (2004). Helical core flow from geomagnetic secular variation. *Physics of the Earth and Planetary Interiors, 147,* 1–25.

Amit, H., & Olson, P. (2015). Lower mantle superplume growth excites geomagnetic reversals. *Earth and Planetary Science Letters, 414,* 68–76.

Ammann, M., Brodholt, J., & Dobson, D. (2010). Simulating diffusion. *Reviews in Mineralogy and Geochemistry, 71,* 201–224.

Ammon, C., Ji, C., Thio, H., Robinson, D., Ni, S., Hjorleifsdottir, V., et al. (2005). Rupture process of the 2004 Sumatra-Andaman earthquake. *Science, 308,* 1133–1139.

Anbar, A., Duan, Y., Lyons, T., Arnold, G., Kendall, B., Creaser, R., et al. (2007). A whiff of oxygen before the Great Oxidation Event? *Science, 317,* 1903–1906.

Anderson, D. (1967). A seismic equation of state. *Geophysical Journal of the Royal Astronomical Society, 13,* 9–30.

Anderson, J., Campbell, J., Jurgens, R., Lau, E., Newhall, X., Slade, M., III, & Standish, E., Jr. (1992). Recent developments in solar-system tests of general relativity. In H. Sato & T. Nakamura (Eds.), *Proceedings of the Sixth Marcel Grossman Meeting on General Relativity* (pp. 353–355). London: World Scientific.

Anderson, O. (1973). Comments on the power law representation of Birch's Law. *Journal of Geophysical Research, 78,* 4901–4914.

Anderson, O. (2002). The power balance at the core–mantle boundary. *Physics of the Earth and Planetary Interiors, 131,* 1–17.

Anderson, O., & Duba, A. (1997). Experimental melting curve of iron revisited. *Journal of Geophysical Research, 102,* 22,659–22,669.

Anderson, O., & Isaak, D. (2002). Another look at the core density deficit of Earth's outer core. *Physics of the Earth and Planetary Interiors, 131,* 19–27.

Andersson, M., Verronen, P., Wang, S., Rodger, C., Clilverd, M., & Carson, B. (2012). Precipitating radiation belt electrons and enhancements of mesospheric hydroxyl during 2004–2009. *Journal of Geophysical Research, 117,* D09304. doi:10.1029/2011JD017246

Andrews, D. (1982). Could the Earth's core and moon have formed at the same time? *Geophysical Research Letters, 9,* 1259–1262.

Angermann, D., Klotz, J., & Reigber, C. (1999). Space-geodetic estimation of the Nazca-South America Euler vector. *Earth and Planetary Science Letters, 171,* 329–334.

Aoyama, Y., & Naito, I. (2001). Atmospheric excitation of the Chandler wobble, 1983-1998. *Journal of Geophysical Research, 106,* 8941–8954.

Arculus, R., (2021). The cosmic origins of uranium. Accessed Nov. 2022 from https://world-nuclear.org/information-library/nuclear-fuel-cycle/uranium-resources/the-cosmic-origins-of-uranium.aspx, in original form a paper presented at the Uranium Institute Mid-Term Meeting, Adelaide.

Ardalan, A., & Grafarend, E. (2001). Somigliana-Pizzetti gravity: the international gravity formula accurate to the sub-nanoGal level. *Journal of Geodesy, 75,* 424–437.

Ardhuin, F., Gualtieri, L., & Stutzmann, E. (2015). How ocean waves rock the Earth: Two mechanisms explain microseisms with periods 3 to 300 s. *Geophysical Research Letters, 42,* 765–772. doi:10.1002/2014GL062782

Arevalo, R., McDonough, W., & Luong, M. (2009). The K/U ratio of the silicate Earth: Insights into mantle composition, structure and thermal evolution. *Earth and Planetary Science Letters, 278,* 361–369.

Argus, D., Gordon, R., Heflin, M., Ma, C., Eanes, R., Willis, P., et al. (2010). The angular velocities of the plates and the velocity of Earth's centre from space geodesy. *Geophysical Journal International, 180,* 913–960.

Argus, D., & Gross, R. (2004). An estimate of motion between the spin axis and the hotspots over the past century. *Geophysical Research Letters, 31.* doi:10.1029/2004GL019657

Arkani-Hamed, J., & Olson, P. (2010). Giant impact stratification of the Martian core. *Geophysical Research Letters, 37,* L02201. doi:10.1029/2009GL041417

Arkani-Hamed, J., & Strangway, D. (1986). Band-limited global scalar magnetic anomaly map of the Earth derived from Magsat data. *Journal of Geophysical Research, 91*, 8193–8203.

Arkani-Hamed, J., & Strangway, D. (1987). An interpretation of magnetic signatures of subduction zones detected by MAGSAT. *Tectonophysics, 133*, 45–55.

Armi, L., Hebert, D., Oakey, N., Price, J., Richardson, P., Rossby, T., & Ruddick, B. (1989). Two years in the life of a Mediterranean salt lens. *Journal of Physical Oceanography, 19*, 354–370.

Armitage, P. (2014). Lecture notes on the formation and early evolution of planetary systems (an online version of *Astrophysics of Planet Formation*, Armitage, 2010), arXiv:astro-ph/070148. Downloaded from https://arxiv.org/abs/astro-ph/0701485 January 2017.

Arnet, F., Kahle, H., Klingelé, E., Smith, R., Meertens, C., & Dzurisin, D. (1997). Temporal gravity and height changes of the Yellowstone caldera, 1977–1994. *Geophysical Research Letters, 24*, 2741–2744.

Arvidson, R., Seelos, F., IV, Deal, K., Koeppen, W., Snider, N., Kieniewicz, J., et al. (2003). Mantled and exhumed terrains in Terra Meridiani, *Mars. Journal of Geophysical Research, 108*, 8073. doi:10.1029/2002JE001982

Ashok, K., Behera, S., Rao, S., Weng, H., & Yamagata, T. (2007). El Niño Modoki and its possible teleconnection. *Journal of Geophysical Research, 112*. doi:10.1029/2006JC003798

Asphaug, E. (2014). Impact origin of the Moon? *Annual Review of Earth and Planetary Sciences, 42*, 551–578.

Assael, M., Kakosimos, K., Banish, R., Brillo, J., Egry, I., Brooks, R., et al. (2006). Reference data for the density and viscosity of liquid aluminum and liquid iron. *Journal of Physical and Chemical Reference Data, 35*, 285–300.

Atwater, T. (1989). Plate tectonic history of the northeast Pacific and western North America. In E. Winterer, D. Hussong, & R. Decker (Eds.), *The Geology of North America: Vol. N, The eastern Pacific Ocean and Hawaii* (pp. 21–72). Boulder: Geological Society of America.

Atwater, T., & Molnar, P. (1973). Relative motion of the Pacific and North American plates deduced from sea-floor spreading in the Atlantic, Indian and South Pacific Oceans. In R. Kovach & A. Nur (Eds.), *Proceedings of the Conference on Tectonic Problems of the San Andreas Fault* (reprinted as Geological Sciences, v. XIII, Stanford University, pp. 136–148).

Aubert, J. (2013). Flow throughout the Earth's core inverted from geomagnetic observations and numerical dynamo models. *Geophysical Journal International, 192*, 537–556.

Aubert, J. (2014). Earth's core internal dynamics 1840–2010 imaged by inverse geodynamo modelling. *Geophysical Journal International, 197*, 1321–1334.

Aubert, J. (2015). Geomagnetic forecasts driven by thermal wind dynamics in the Earth's core. *Geophysical Journal International, 203*, 1738–1751.

Aubert, J. (2019). Approaching Earth's core conditions in high-resolution geodynamo simulations. *Geophysical Journal International, 219*, S137–S151.

Aubert, J., & Dumberry, M. (2011). Steady and fluctuating inner core rotation in numerical geodynamo models. *Geophysical Journal International, 184*, 162–170.

Aubert, J., & Finlay, C. (2019). Geomagnetic jerks and rapid hydromagnetic waves focusing at Earth's core surface. *Nature Geoscience, 12*, 393–398.

Aubert, J., Finlay, C., & Fournier, A. (2013). Bottom-up control of geomagnetic secular variation by the Earth's inner core. *Nature, 502*, 219–223.

Aujogue, K., Pothérat, A., Sreenivasan, B., & Debray, F. (2018). Experimental study of the convection in a rotating tangent cylinder. *Journal of Fluid Mechanics, 843*, 355–381.

Aurnou, J. (2004). Secrets of the deep (News and Views). *Nature, 428*, 134–135.

Aurnou, J., Andreadis, S., Zhu, L., & Olson, P. (2003). Experiments on convection in Earth's core tangent cylinder. *Earth and Planetary Science Letters, 212,* 119–134.

Aurnou, J., Heimpel, M., Allen, L., King, E., & Wicht, J. (2008). Convective heat transfer and the pattern of thermal emission on the gas giants. *Geophysical Journal International, 173,* 793–801.

Aurnou, J., Horn, S., & Julien, K. (2020). Connections between nonrotating, slowly rotating, and rapidly rotating turbulent convection transport scalings. *Physical Review Research, 2.* doi:10.1103/PhysRevResearch.2.043115

Aurnou, J., & King, E. (2017). The cross-over to magnetostrophic convection in planetary dynamo systems. *Proceedings of the Royal Society A, 473,* 20160731. http://dx.doi.org/10.1098/rspa.2016.0731

Aurnou, J., & Olson, P. (2000). Control of inner core rotation by electromagnetic, gravitational and mechanical torques. *Physics of the Earth and Planetary Interiors, 117,* 111–121.

AVISO (2013). Principle (of Doppler tracking with the DORIS system), from Centre National d'Etudes Spatiales, http://www.aviso.oceanobs.com/en/doris/principle.html

Awramik, S. (2006). Respect for stromatolites. *Nature, 441,* 700–701.

Baba, K., Chave, A., Evans, R., Hirth, G., & Mackie, R. (2006). Mantle dynamics beneath the East Pacific Rise at 17°S: Insights from the Mantle Electromagnetic and Tomography (MELT) experiment. *Journal of Geophysical Research, 111,* B02101. doi:10.1029/2004JB003598

Baba, K., Utada, H., Goto, T., Kasaya, T., Shimizu, H., & Tada, N. (2010). Electrical conductivity imaging of the Philippine Sea upper mantle using seafloor magnetotelluric data. *Physics of the Earth and Planetary Interiors, 183,* 44–62.

Babanin, A., Rogers, W., de Camargo, R., Doble, M., Durrant, T., Filchuk, K., et al. (2019). Waves and swells in high wind and extreme fetches, measurements in the Southern Ocean. *Frontiers in Marine Science, 6.* doi:10.3389/fmars.2019.00361

Backus, G., & Chandrasekhar, S. (1956). On Cowling's theorem on the impossibility of self-maintained axisymmetric homogeneous dynamos. *Proceedings of the National Academy of Sciences, 42,* 105–109.

Badro, J., Fiquet, G., Guyot, F., Gregoryanz, E., Occelli, F., Antonangeli, D., & d'Astuto, M. (2007). Effect of light elements on the sound velocities in solid iron: Implications for the composition of Earth's core. *Earth and Planetary Science Letters, 254,* 233–238.

Bahcall, J. (1999). Ulrich's explanation for the solar five minute oscillations. *Astrophysical Journal, 525,* 1199–1200.

Bailey, E., Batygin, K., & Brown, M. (2016). Solar obliquity induced by Planet Nine. *Astronomical Journal.* Preprint downloaded as arXiv:1607.03963v2 from https://arxiv.org/abs/1607.03963

Balchan, A., & Cowan, G. (1966). Shock compression of two iron-silicon alloys to 2.7 megabars. *Journal of Geophysical Research, 71,* 3577–3588.

Baldwin, R. (2006). Was there ever a Terminal Lunar Cataclysm? *With lunar viscosity arguments, Icarus, 184,* 308–318.

Ballato, A. (2010). Poisson's ratios of auxetic and other technological materials. *IEEE Transactions on Ultrasonics, Ferroelectrics, and Frequency Control, 57,* 7–15.

Ballesteros-Paredes, J., & Hartmann, L. (2007). Remarks on rapid vs. slow star formation. *Revista Mexicana de Astronomia y Astrofisica, 43,* 123–136.

Ballesteros-Paredes, J., Klessen, R., Mac Low, M., & Vázquez-Semadeni, E. (2007). Molecular cloud turbulence and star formation. In B. Reipurth, D. Jewitt, K. Keil (Eds.), *Protostars and Protoplanets V* (pp. 63–80), Tucson: University of Arizona Press.

Balling, R., & Hildebrandt, M. (2000). Evaluation of the linkage between Schumann Resonance peak frequency values and global and regional temperatures. *Climate Research, 16,* 31–36.

Balmaseda, M., Trenberth, K., & Kallen, E. (2013). Distinctive climate signals in reanalysis of global ocean heat content. *Geophysical Research Letters, 40,* 1754–1759.

Bambach, R. (1999). Energetics in the global marine fauna: A connection between terrestrial diversification and change in the marine biosphere. *Geobios, 32,* 131–144.

Bamber, J., Riva, R., Vermeersen, B., & LeBrocq, A. (2009). Reassessment of the potential sea-level rise from a collapse of the West Antarctic Ice Sheet. *Science, 324,* 901–903. doi:10.1126/science.1169335

Barboni, M., Boehnke, P., Keller, B., Kohl, I., Schoene, B., Young, E., & McKeegan, K. (2017). Early formation of the Moon 4.51 billion years ago. *Science Advances, 3,* e1602365. 10.1126/sciadv.1602365

Bannister, M. T., Gladman, B. J., Kavelaars, J. J., Petit, J. M., Volk, K., & Chen, Y. T. (2018). OSSOS. VII. 800+ Trans-Neptunian objects – The complete data release. *Astrophysical Journal, Supplement Series, 236.* https://doi.org/10.3847/1538-4365/aab77a

Bard, E., Hamelin, B., & Fairbanks, R. (1990). U–Th ages obtained by mass spectrometry in corals from Barbados: Sea level during the past 130,000 years. *Nature, 346,* 456–458.

Barge, P., & Sommeria, J. (1995). Did planet formation begin inside persistent gaseous vortices?*Astronomy & Astrophysics, 295,* L1–L4.

Bargo, M. (2001). The ground sloth *Megatherium americanum:* Skull shape, bite forces, and diet. *Acta Palaeontologica Polonica, 46,* 173–192.

Barlow, N. (1997). The search for possible source craters for Martian meteorite ALH84001. *Proceedings of the 28th Annual Lunar and Planetary Science Conference* (abstract).

Bar-On, Y., Phillips, R., & Milo, R. (2018). The biomass distribution on Earth. *Proceedings of the National Academy of Sciences, 115,* 6506–6511. www.pnas.org/cgi/doi/10.1073/pnas.1711842115

Barrois, O., Hammer, M., Finlay, C., Martin, Y., & Gillet, N. (2018). Assimilation of ground and satellite magnetic measurements: Inference of core surface magnetic and

velocity field changes. *Geophysical Journal International, 215,* 695–712.

Barron, E. (1983). A warm, equable Cretaceous: The nature of the problem. *Earth Science Reviews, 19,* 305–338.

Barthelmes, F. (2013). Definition of functionals of the geopotential and their calculation from spherical harmonic models (Scientific Technical Report STR09/02), rev. 2013. Potsdam: Helmholtz Centre, 32 pp. (Downloaded from http://icgem.gfz-potsdam .de/home)

Barton, C. (1989). Geomagnetic secular variation: direction and intensity. In D. James (Ed.), *The Encyclopedia of solid earth sciences.* New York, NY: Van Nostrand Reinhold, 560–577 as cited in Lowrie 2007.

Bass, J., & Parise, J. (2008). Deep Earth and recent developments in mineral physics. *Elements, 4,* 157–163.

Bass, J., Sinogeikin, S., & Li, B. (2008). Elastic properties of minerals: A key for understanding the composition and temperature of Earth's interior. *Elements, 4,* 165–170.

Bassett, W. (2009). Diamond anvil cell, 50th birthday. *High Pressure Research, 29,* 163–186.

Batygin, K., Bodenheimer, P., & Laughlin, G. (2016). In situ formation and dynamical evolution of hot Jupiter systems. *Astrophysical Journal, 829.* doi:10.3847/0004-637X/829/2/114

Baughman, R., Dantas, S., Stafström, S., Zakhidov, A., Mitchell, T., & Dubin, D. (2000). Negative Poisson's ratios for extreme states of matter. *Science, 288,* 2018–2022.

Baughman, R., Shacklette, J., Zakhidov, A., & Stafström, S. (1998). Negative Poisson's ratios as a common feature of cubic metals. *Nature, 392,* 362–365.

Bazhenov, M., Levashova, N., Meert, J., Golovanova, I., Danukalov, K., & Fedorova, N. (2016). Late Ediacaran magnetostratigraphy of Baltica: Evidence for magnetic field hyperactivity? *Earth and Planetary Science Letters, 435,* 124–135.

Beaulieu, J., Bennett, D., Fouqué, P., Williams, A., Dominik, M., Jørgensen, U., et al. (2006).

Discovery of a cool planet of 5.5 Earth masses through gravitational microlensing. *Nature, 439*, 437–440.

Beck, A., Darbha, D., & Schloessin, H. (1978). Lattice conductivities of single-crystal and polycrystalline materials at mantle pressures and temperatures. *Physics of the Earth and Planetary Interiors, 17*, 35–53.

Beckwith, S., Sargent, A., Chini, R., & Güsten, R. (1990). A survey for circumstellar disks around young stellar objects. *Astronomical Journal, 99*, 924–945.

Beerling, D., & Royer, D. (2011). Convergent Cenozoic CO_2 history. *Nature Geoscience (Commentary), 4*, 418–420.

Bekker, A., & Holland, H. (2012). Oxygen overshoot and recovery during the early Paleoproterozoic. *Earth and Planetary Science Letters, 317–318*, 295–304.

Bekker, A., Slack, J., Planavsky, N., Krapež, B., Hofmann, A., Konhauser, K., & Rouxel, O. (2010). Iron formation: The sedimentary product of a complex interplay among mantle, tectonic, oceanic, and biospheric processes. *Economic Geology, 105*, 467–508.

Belcher, C., & McElwain, J. (2008). Limits for combustion in low O_2 redefine paleoatmospheric predictions for the Mesozoic. *Science, 321*, 1197–1200.

Belkin, I., Levitus, S., Antonov, J., & Malmberg, S. (1998). "Great salinity anomalies" in the North Atlantic. *Progress in Oceanography, 41*, 1–68.

Bell, R., Childers, V., Arko, R., Blankenship, D., & Brozena, J. (1999). Airborne gravity and precise positioning for geologic applications. *Journal of Geophysical Research, 104*, 15281–15292.

Bell, T. (2006). Bizarre Lunar Orbits. In T. Phillips (Ed.), *Science @ NASA*. Downloaded from https://science.nasa.gov/science-news/science-at-nasa/2006/06nov_loworbit

Belloche, A., Garrod, R., Müller, H., & Menten, K. (2014). Detection of a branched alkyl molecule in the interstellar medium: *iso*-propyl cyanide. *Science, 345*, 1584–1587.

Belloche, A., Menten, R., Comito, C., Müller, H., Schilke, P., Ott, J., et al. (2008). Detection of amino acetonitrile in Sgr B2(N). *Astronomy & Astrophysics, 482*, 179–196.

Bengtsson, L., Hodges, K., Esch, M., Keenlyside, N., Kornblueh, L., Luo, J., & Yamagata, T. (2007). How may tropical cyclones change in a warmer climate? *Tellus, 59A*, 539–561.

Berardelli, P. (2010). These materials can't be stretched thin. *Science News.* Downloaded June 2022 from https://www.science.org/content/article/these-materials-cant-be-stretched-thin

Bercovici, D. (2003). The generation of plate tectonics from mantle convection. *Earth and Planetary Science Letters, 205*, 107–121.

Bercovici, D. (2011). Mantle convection. In H. Gupta (Ed.), *Encyclopedia of Solid Earth Geophysics* (2nd edition). Dordrecht: Springer, 28 pp.

Bercovici, D., & Karato, S. (2003). Whole-mantle convection and the transition-zone water filter. *Nature, 425*, 39–44.

Bercovici, D., Schubert, G., & Ricard, Y. (2015). Abrupt tectonics and rapid slab detachment with grain damage. *Proceedings of the National Academy of Sciences, 112*, 1287–1291. www.pnas.org/cgi/doi/10.1073/pnas.1415473112

Berger, A. (1988). Milankovitch theory and climate. *Reviews of Geophysics, 26*, 624–657.

Berger, A., & Loutre, M. (1999). Parameters of the Earth's orbit for the last 5 million years in 1 kyr resolution, https://doi.org/10.1594/PANGAEA.56040. Downloaded from PANGEA website Dec. 2017. A *supplement to* Berger, A., & Loutre, M. (1991). Insolation values for the climate of the last 10 million of years, *Quaternary Science Reviews, 10*, 297–317. https://doi.org/10.1016/0277-3791(91)90033-Q

Berger, A., Loutre, M., & Dehant, V. (1989). Influence of the changing lunar orbit on the astronomical frequencies of pre-Quaternary insolation patterns. *Paleoceanography, 4*, 555–564.

Berggren, L. (2007). Ibn Sahl: Abū Sa'd al-'Alā' ibn Sahl. In T. Hockey et al. (Eds.), *Biographical encyclopedia of astronomers (Springer Reference)*. New York, NY: Springer, p. 567.

Bergin, E., & Tafalla, M. (2007). Cold dark clouds: The initial conditions for star formation. *Annual Review of Astronomy and Astrophysics, 45*, 339–396.

Berkner, L., & Marshall, L. (1965). On the origin and rise of oxygen concentration in the Earth's atmosphere. *Journal of the Atmospheric Sciences, 22*, 225–261.

Bernal, J. (1936). Geophysical discussion. *The Observatory, 59*, 268.

Bernard, P. (1990). Historical sketch of microseisms from past to future. *Physics of the Earth and Planetary Interiors, 63*, 145–150.

Berner, R. (1998). The carbon cycle and CO_2 over Phanerozoic time: The role of land plants. *Philosophical Transactions of the Royal Society of London B, 353*, 75–82.

Berner, R. (2006). GEOCARBSULF: A combined model for Phanerozoic atmospheric O_2 and CO_2. *Geochimica et Cosmochimica Acta, 70*, 5653–5664.

Berner, R., & Canfield, D. (1989). A new model for atmospheric oxygen over Phanerozoic time. *American Journal of Science, 289*, 333–361.

Berner, R., & Kothavala, Z. (2001). GEOCARB III: A revised model of atmospheric CO_2 over Phanerozoic time. *American Journal of Science, 301*, 182–204.

Berry, M., & Geim, A. (1997). Of flying frogs and levitrons. *European Journal of Physics, 18*, 307–313.

Bertotti, B., Iess, L., & Tortora, P. (2003). A test of general relativity using radio links with the Cassini spacecraft. *Nature, 425*, 374–376.

Bhatiani, S., Dai, X., & Guerras, E. (2019). Confirmation of planet-mass objects in extragalactic systems. *Astrophysical Journal, 885*. https://doi.org/10.3847/1538-4357/ab46ac

Bicknell, P. (1968). Did Anaxagoras observe a sunspot in 467 B.C.? *Isis, 59*, 87–90.

Biemann, K., Oro, J., Toulmin, P., III, Orgel, L., Nier, A., Anderson, D., et al. (1977). The search for organic substances and inorganic volatile compounds in the surface of Mars. *Journal of Geophysical Research, 82*, 4641–4658.

Bijwaard, H., Spakman, W., & Engdahl, B. (1998). Closing the gap between regional and global travel time tomography. *Journal of Geophysical Research, 103*, 30355–30378.

Bills, B. (2005). Variations in the rotation rate of Venus due to orbital eccentricity modulation of solar tidal torques. *Journal of Geophysical Research, 110*, E11007. doi:10.1029/2003JE002190

Bills, B., Adams, K., & Wesnousky, S. (2007). Viscosity structure of the crust and upper mantle in western Nevada from isostatic rebound patterns of the late Pleistocene Lake Lahontan high shoreline. *Journal of Geophysical Research, 112*, B06405. doi:10.1029/2005JB003941

Bills, B., Currey, D., & Marshall, G. (1994). Viscosity estimates for the crust and upper mantle from patterns of lacustrine shoreline deformation in the Eastern Great Basin. *Journal of Geophysical Research, 99*, 22059–22086.

Bills, B., & May, G. (1987). Lake Bonneville: Constraints on lithospheric thickness and upper mantle viscosity from isostatic warping of Bonneville, Provo, and Gilbert Stage shorelines. *Journal of Geophysical Research, 92*, 11493–11508.

Bina, C., & Helffrich, G. (1994). Phase transition Clapeyron slopes and transition zone seismic discontinuity topography. *Journal of Geophysical Research, 99*, 15,853–15,860.

Bina, C., & Wood, B. (1987). Olivine-spinel transitions: Experimental and thermodynamic constraints and implications for the nature of the 400-km seismic discontinuity. *Journal of Geophysical Research, 92*, 4853–4866.

Birch, F. (1952). Elasticity and constitution of the Earth's interior. *Journal of Geophysical Research, 57*, 227–286.

Birch, F. (1961a). The velocity of compressional waves in rocks to 10 kilobars, Part 2. *Journal of Geophysical Research, 66*, 2199–2224.

Birch, F. (1961b). Composition of the Earth's mantle. *Geophysical Journal of the Royal Astronomical Society, 4*, 295–311.

Birch, F. (1964). Density and composition of mantle and core. *Journal of Geophysical Research, 69*, 4377–4388.

Birch, F. (1965). Energetics of core formation. *Journal of Geophysical Research, 70*, 6217–6221.

Birch, F. (1968). On the possibility of large changes in the Earth's volume. *Physics of the Earth and Planetary Interiors, 1*, 141–147.

Birchfield, G. (1977). A study of the stability of a model continental ice sheet subject to periodic variations in heat input. *Journal of Geophysical Research, 82*, 4909–4913.

Birchfield, G., & Weertman, J. (1978). A note on the spectral response of a model continental ice sheet. *Journal of Geophysical Research, 83*, 4123–4125.

Bird, P. (2003). An updated digital model of plate boundaries. *Geochemistry, Geophysics, Geosystems, 4*, 1027. doi:10.1029/2001GC000252

Biskaborn, B., Smith, S. L., Noetzli, J., Matthes, H., Vieira, G., Streletskiy, D. A., et al. (2019). Permafrost is warming at a global scale. *Nature Communications, 10*. https://doi.org/10.1038/s41467-018-08240-4

Biver, N., Bockelée-Morvan, D., Colom, P., Crovisier, J., Germain, B., Lellouch, E., et al. (1997). Long-term evolution of the outgassing of Comet Hale–Bopp from radio observations. *Earth, Moon & Planets, 78*, 5–11.

Bjerknes, J. (1969). Atmospheric teleconnections from the equatorial Pacific. *Monthly Weather Review, 97*, 163–172.

Blake, G. (1977). The rate of change of G. *Monthly Notices of the Royal Astronomical Society, 178*, 41P–43P.

Blakely, R. (1995). *Potential theory in Gravity and Magnetic applications*. Cambridge, England: Cambridge University Press, 441 pp.

Blakemore, R. (1975). Magnetotactic bacteria. *Science, 190*, 377–379.

Blakeslee, R., Mach, D., Bateman, M., & Bailey, J. (2014). Seasonal variations in the lightning diurnal cycle and implications for the global electric circuit. *Atmospheric Research, 135–136*, 228–243.

Blamey, N., Brand, U., Parnell, J., Spear, N., Lécuyer, C., Benison, K., et al. (2016). Paradigm shift in determining Neoproterozoic atmospheric oxygen. *Geology, 44*, 651–654.

Block, S., & Piermarini, G. (1976). The diamond cell stimulates high-pressure research. *Physics Today, Sept. 1976*, 44–55.

Bloxham, J., & Gubbins, D. (1985). The secular variation of Earth's magnetic field. *Nature, 317*, 777–781.

Bloxham, J., & Jackson, A. (1992). Time-dependent mapping of the magnetic field at the core-mantle boundary. *Journal of Geophysical Research, 97*, 19537–19563.

Bloxham, J., Zatman, S., & Dumberry, M. (2002). The origin of geomagnetic jerks. *Nature, 420*, 65–68.

Blue, C. (2016). Life's first handshake: Chiral molecule detected in interstellar space (press release June 14, 2016). NRAO. Accessed at https://public.nrao.edu/news/pressreleases/2016-chiral-gbt

Bockelee-Morvan, D., Wink, J., Despois, D., Colom, P., Biver, N., Crovisier, J., et al. (1997). A molecular survey of Comet C/1995 O1 (Hale–Bopp) at the IRAM telescopes. *Earth, Moon & Planets, 78*, 67.

Boehler, R. (2000). High-pressure experiments and the phase diagram of lower mantle and core materials. *Reviews of Geophysics, 38*, 221–245.

Boehnke, P., & Harrison, T. (2016). Illusory late heavy bombardments. *Proceedings of the National Academy of Sciences, 113*, 10802–10806.

Bogard, D. (1995). Exposure-age-initiating events for Martian meteorites: Three or four?. *26th Annual Lunar and Planetary Science Conference*, pp. 143–144. Downloaded March 30, 2017 from http://www.lpi.usra.edu/lpi/meteorites/ALH84001_References.html#TRE95

Böhm, E., Lippold, J., Gutjahr, M., Frank, M., Blaser, P., Antz, B., et al. (2015). Strong and deep Atlantic meridional overturning circulation during the last glacial cycle. *Nature, 517*, 73–76.

Bohn, D. (1988). Environmental effects on the speed of sound. *Journal of the Audio Engineering Society*, *36*, 223–231.

Bois, P., La Porte, M., Lavergne, M., & Thomas, G. (1971). Essai de determination automatique des vitesses sismiques par mesures entre puits. *Geophysical Prospecting*, *19*, 42–83.

Bolfan-Casanova, N. (2007). Fuel for plate tectonics (perspectives). *Science*, *315*, 338–339.

Bolt, B., & Tanimoto, T. (1981). Atmospheric oscillations after the May 18, 1980 eruption of Mount St. *Helens. EOS*, *62*, 529–530.

Bonafede, M., Dragoni, M., & Morelli, A. (1986). On the existence of a periodic dislocation cycle in horizontally layered viscoelastic models. *Journal of Geophysical Research*, *91*, 6396–6404.

Bono, R., Tarduno, J., Nimmo, F., & Cottrell, R. (2019). Young inner core inferred from Ediacaran ultra-low geomagnetic field intensity. *Nature Geoscience*, *12*, 143–147.

Bontognali, T., Sessions, A., Allwood, A., Fischer, W., Grotzinger, J., Summons, R., & Eiler, J. (2012). Sulfur isotopes of organic matter preserved in 3.45-billion-year-old stromatolites reveal microbial metabolism. *Proceedings of the National Academy of Sciences*, *109*, 15146–15151.

Borgonie, G., Linage-Alvarez, B., Ojo, A. O., Mundle, S. O. C., Freese, L. B., van Rooyen, C., et al. (2015). Eukaryotic opportunists dominate the deep-subsurface biosphere in South Africa. *Nature Communications*, *6*. doi:10.1038/ncomms9952

Bosak, T., Knoll, A., & Petroff, A. (2013). The meaning of stromatolites. *Annual Review of Earth and Planetary Sciences*, *41*, 21–44.

Bosch, W. (2010). Satellite Altimetry, ESPACE altimetry lesson 1. Downloaded from ftp://ftp.sirgas.org/pub/espace

Boschi, L., & Dziewonski, A. (2000). Whole Earth tomography from delay times of P, PcP, and PKP phases: Lateral heterogeneities in the outer core or radial anisotropy in the mantle? *Journal of Geophysical Research*, *105*, 13,675–13,696.

Boss, A. (2006). Ask Astro. *Astronomy*, *34*, 70–71.

Boss, A., & Peale, S. (1986). Dynamical constraints on the origin of the Moon. In W. Hartmann, R. Phillips, & G. Taylor (Eds.), *Origin of the Moon* (pp. 59–101). Houston, TX: The Lunar & Planetary Institute, 781 pp.

Bott, M. (1982). *The interior of the earth* (2nd edition). London: Elsevier Sci. Pub., 403 pp.

Bottke, W., & Andrews-Hanna, J. (2017). A post-accretionary lull in large impacts on early Mars. *Nature Geoscience*, *10*, 344–348. doi:10.1038/NGEO2937

Bouffard, M., Choblet, G., Labrosse, S., & Wicht, J. (2019). Chemical convection and stratification in the Earth's outer core. *Frontiers of Earth Science*, *7*. doi:10.3389/feart.2019.00099

Boukeke, D. (1994). Structures crustales D'Afrique Centrale déduites des anomalies gravimétriques et magnétiques: Le domaine précambrien de la République Centrafricaine et du Sud-Cameroun. *ORSTOM TDM 129*. (cited in Taylor, 2007)

Boy, J.-P., Hinderer, J., & Gegout, P. (1998). Global atmospheric loading and gravity. *Physics of the Earth and Planetary Interiors*, *109*, 161–177.

Brace, W. (1965). Some new measurements of linear compressibility of rocks. *Journal of Geophysical Research*, *70*, 391–398.

Bradley, D. (2011). Secular trends in the geologic record and the supercontinent cycle, *Earth Science Reviews*, *108*, 16–33. doi:10.1016/j.earscirev.2011.05.003

Bradley, R., & Jones, P. (1993). "Little Ice Age" summer temperature variations: their nature and relevance to recent global warming trends. *The Holocene*, *3*, 367–376.

Brandenburg, A., Moss, D., Rüdiger, D., & Tuominen, I. (1990). The nonlinear solar dynamo and differential rotation: A Taylor number puzzle? *Solar Physics*, *128*, 243–251.

Bray, R., & Loughhead, R. (1964). *Sunspots*. New York, NY: John Wiley & Sons, 303 pp. (reprinted 1979 by Dover Publications, NY).

Bredow, E., Steinberger, B., Gassmöller, R., & Dannberg, J. (2017). How plume-ridge interaction shapes the crustal thickness pattern of the Réunion hotspot track. *Geochemistry, Geophysics, Geosystems*, *18*, 2930–2948.

Brennan, P. (2021). Exoplanet exploration: Planets beyond our Solar System. Accessible from https://exoplanets.nasa.gov/discovery/exoplanet-catalog/

Brenner, A., Fu, R., Evans, D., Smirnov, A., Trubko, R., & Rose, I. (2020). Paleomagnetic evidence for modern-like plate motion velocities at 3.2 Ga. *Science Advances, 6*, eaaz8670.

Brocher, T. (2005). Empirical relations between elastic wavespeeds and density in the Earth's crust. *Bulletin of the Seismological Society of America, 95*, 2081–2092.

Broecker, W. (1987). The biggest chill. *Natural History Magazine, 96*, 74–82.

Broecker, W. (1991). The great ocean conveyor. *Oceanography, 4*, 79–89.

Broecker, W., & van Donk, J. (1970). Insolation changes, ice volumes, and the ^{18}O record in deep-sea cores. *Reviews of Geophysics Space Physics, 8*, 169–198.

Brohan, P., Kennedy, J., Harris, I., Tett, S., & Jones, P. (2006). Uncertainty estimates in regional and global observed temperature changes: A new data set from 1850. *Journal of Geophysical Research, 111*, D12106. doi:10.1029/2005JD006548

Brook, E., White, J., Schilla, A., Bender, M., Barnett, B., Severinghaus, J., et al. (2005). Timing of millennial-scale climate change at Siple Dome, West Antarctica, during the last glacial period. *Quaternary Science Reviews, 24*, 1333–1343.

Brovar, V., & Yurkina, M. (2000). Mikhail Sergeevich Molodensky–Life and Work (1996). In H. Moritz & M. Yurkina (Eds.), *M. S. Molodensky: In Memoriam* (pp. 5–38). Graz: Technical Institute.

Brown, J., & McQueen, R. (1986). Phase transitions, Grüneisen parameter, and elasticity for shocked iron between 77 GPa and 400 GPa. *Journal of Geophysical Research, 91*, 7485–7494.

Brown, M. (2012). The compositions of Kuiper Belt objects. *Annual Review of Earth and Planetary Sciences, 40*, 467–494.

Brownlee, D. (2014). The Stardust Mission: Analyzing samples from the edge of the Solar System. *Annual Review of Earth and Planetary Sciences, 42*, 179–205.

Buffett, B. (1992). Constraints on magnetic energy and mantle conductivity from the forced nutations of the Earth. *Journal of Geophysical Research, 97*, 19581–19597.

Buffett, B. (1996). Gravitational oscillations in the length of day. *Geophysical Research Letters, 23*, 2279–2282.

Buffett, B. (2003). The thermal state of Earth's core. *Science, 299*, 1675 & 1677.

Buffett, B. (2010a). The enigmatic inner core. *Science, 328*, 982–983.

Buffett, B. (2010b). Tidal dissipation and the strength of the Earth's internal magnetic field. *Nature, 468*, 952–955.

Buffett, B. (2015). 8.08 Core–Mantle interactions, In G. Schubert (Ed.), *Treatise on Geophysics* (2nd edition, pp. 213–224). Amsterdam: Elsevier.

Buffett, B., Knezek, N., & Holme, R. (2016). Evidence for MAC waves at the top of Earth's core and implications for variations in length of day. *Geophysical Journal International, 204*, 1789–1800.

Buizert, C., Adrian, B., Ahn, J., Albert, M., Alley, R. B., Baggenstos, D., et al. (WAIS Divide Project Members). (2015). Precise interpolar phasing of abrupt climate change during the last ice age. *Nature, 520*, 661–665.

Bukowinski, M. (1976). The effect of pressure on the physics and chemistry of potassium. *Geophysical Research Letters, 3*, 491–494.

Bullard, E. (1939). Heat flow in South Africa. *Proceedings of the Royal Society London, A, 173*, 474–502.

Bullard, E. (1952). Comment. *Nature, 170*, 200.

Bullard, E., Freedman, C., Gellman, H., & Nixon, J. (1950). The westward drift of the Earth's magnetic field. *Philosophical Transactions of the Royal Society of London, A, 243*, 67–92.

Bullard, E., Maxwell, A., & Revelle, R. (1956). Heat flow through the deep sea floor. *Advances in Geophysics, 3*, 153–181.

Bullen, K. (1936). The variation of density and the ellipticities of strata of equal density within the Earth. *Monthly Notices of the Royal*

Astronomical Society, Geophysical Supplements, *3*, 395–401.

Bullen, K. (1965). *An Introduction to the Theory of Seismology* (3rd edition). Cambridge, England: Cambridge University Press.

Bunge, H., Richards, M., & Baumgardner, J. (1997). A sensitivity study of three-dimensional spherical mantle convection at 10^8 Rayleigh number: Effects of depth-dependent viscosity, heating mode, and an endothermic phase change. *Journal of Geophysical Research, 102*, 11,991–12,007.

Burbank, D., Leland, J., Fielding, E., Anderson, R., Brozovic, N., Reid, M., & Duncan, C. (1996). Bedrock incision, rock uplift and threshold hillslopes in the northwestern Himalayas. *Nature, 379*, 505–510.

Burbine, T., & Greenwood, R. (2020). Exploring the bimodal Solar System via sample return from the Main Asteroid Belt: The case for revisiting Ceres. *Space Science Reviews, 216*. https://doi.org/10.1007/s11214-020-00671-0

Burch, J., Moore, T., Torbert, R., & Giles, B. (2016). Magnetospheric multiscale overview and science objectives. *Space Science Reviews, 199*, 5–21.

Burdick, S., Li, C., Martynov, V., Cox, T., Eakins, J., Mulder, T., et al. (2008). Upper mantle heterogeneity beneath North America from travel time tomography with global and USArray transportable array data. *Seismological Research Letters, 79*, 384–392.

Burgers, J. (1955). Rotational motion of a sphere subject to viscoelastic deformation. *Proceedings of the Koninklijke Nederlandse Akademie van Wetenschappen, 58*, 219–237.

Burke, M. (1997). A stretch of the imagination. *New Scientist, 154*, 36–39.

Burnley, P. (2011). Phase equilibria at high pressure. In *Teaching mineralogy* (from the Science Education Resource Center's *Teach The Earth*). Downloaded August 2022 from https://serc.carleton.edu/NAGTWorkshops/mineralogy/mineral_physics/phase_equilibria.html

Burov, E., & Diament, M. (1995). The effective elastic thickness (T_e) of continental lithosphere: What does it really mean? *Journal of Geophysical Research, 100*, 3905–3927.

Busse, F. (1970). Thermal instabilities in rapidly rotating systems. *Journal of Fluid Mechanics, 44*, 441–460.

Busse, F. (2002). Convective flows in rapidly rotating spheres and their dynamo action. *Physics of Fluids, 14*, 1301–1314.

Busse, F., & Carrigan, C. (1976). Laboratory simulation of thermal convection in rotating planets and stars. *Science, 191*, 81–83.

Butler, R. (2004). *Paleomagnetism: Magnetic domains to geologic terranes*, electronic ed., originally Butler, R. (1992). *Paleomagnetism: Magnetic Domains to Geologic Terranes*. Boston, MA: Blackwell.

Butterfield, N. (2009). Oxygen, animals and oceanic ventilation: An alternative view (editorial). *Geobiology, 7*, 1–7.

Čadek, O., & Fleitout, L. (1999). A global geoid model with imposed plate velocities and partial layering. *Journal of Geophysical Research, 104*, 29,055–29,075.

Čadek, O., & Fleitout, L. (2003). Effect of lateral viscosity variations in the top 300 km on the geoid and dynamic topography. *Geophysical Journal International, 152*, 566–580.

Cain, J., Schmitz, D., & Kluth, C. (1985). Eccentric geomagnetic dipole drift. *Physics of the Earth and Planetary Interiors, 39*, 237–242.

Calais, E., Han, J., DeMets, C., & Nocquet, J. (2006). Deformation of the North American plate interior from a decade of continuous GPS measurements. *Journal of Geophysical Research, 111*, B06402. doi:10.1029/2005JB004253

Calkins, M. (2018). Quasi-geostrophic dynamo theory. *Physics of the Earth and Planetary Interiors, 276*, 182–189.

Callahan, M., Smith, K., Cleaves, H., II, Ruzicka, J., Stern, J., Glavin, D., et al. (2011). Carbonaceous meteorites contain a wide range of extraterrestrial nucleobases. *Proceedings of the National Academy of Sciences, 108*, 13995–13998.

Calmant, S., Francheteau, J., & Cazenave, A. (1990). Elastic layer thickening with age of the oceanic lithosphere: A tool for prediction of the age of volcanoes or oceanic crust. *Geophysical Journal International, 100*, 59–67.

Cambiotti, G., & Sabadini, R. (2013). Gravitational seismology retrieving Centroid-Moment-Tensor solution of the 2011 Tohoku earthquake. *Journal of Geophysical Research, 118*, 183–194.

Cambiotti, G., Douch, K., Cesare, S., Haagmans, R., Sneeuw, N., Anselmi, A., et al. (2020). On earthquake detectability by the Next-Generation Gravity Mission, *Surveys in Geophysics, 41*, 1049–1074.

Cambiotti, G., Wang, X., Sabadini, R., & Yuen, D. (2016). Residual polar motion caused by coseismic and interseismic deformations from 1900 to present. *Geophysical Journal International, 205*, 1165–1179.

Cameron, A. (1975). The origin & evolution of the Solar System. *Scientific American, 233*(3), 32–41.

Cameron, A., & Truran, J. (1977). The supernova trigger for formation of the Solar System. *Icarus, 30*, 447–461.

Campbell, W. (1989). An introduction to quiet daily geomagnetic fields. *Pure and Applied Geophysics, 131*, 315–331.

Cande, S., & Kent, D. (1992). A new geomagnetic polarity time scale for the Late Cretaceous and Cenozoic. *Journal of Geophysical Research, 97*, 13,917–13,951.

Cande, S., & Kent, D. (1995). Revised calibration of the geomagnetic polarity timescale for the Late Cretaceous and Cenozoic. *Journal of Geophysical Research, 100*, 6093–6095.

Cane, M., & Zebiak, S. (1985). A theory for El Niño and the Southern Oscillation. *Science, 228*, 1085–1087.

Canfield, D. (1998). A new model for Proterozoic ocean chemistry. *Nature, 396*, 450–453.

Canfield, D., Poulton, S., Knoll, A., Narbonne, G., Ross, G., Goldberg, T., & Strauss, H. (2008). Ferruginous conditions dominated later Neoproterozoic deep-water chemistry. *Science, 321*, 949–952.

Canfield, D., Poulton, S., & Narbonne, G. (2007). Late-Neoproterozoic deep-ocean oxygenation and the rise of animal life. *Science, 315*, 92–95.

Canfield, D., & Teske, A. (1996). Late Proterozoic rise in atmospheric oxygen concentration inferred from phylogenetic and sulphur-isotope studies. *Nature, 382*, 127–132.

Canfield, D., Zhang, S., Frank, A., Wang, X., Wang, H., Su, J., et al. (2018). Highly fractionated chromium isotopes in Mesoproterozoic-aged shales and atmospheric oxygen. *Nature Communications, 9.* doi:10.1038/s41467-018-05263-9

Cannell, A. (2018). The engineering of the giant dragonflies of the Permian: revised body mass, power, air supply, thermoregulation and the role of air density. *Journal of Experimental Biology, 221*, jeb185405. doi:10.1242/jeb.185405

Canup, R. (2004). Simulations of a late lunar-forming impact. *Icarus, 168*, 433–456.

Canup, R. (2012). Forming a Moon with an Earth-like composition via a giant impact. *Science, 338*, 1052–1055.

Canup, R. (2014). Lunar-forming impacts: processes and alternatives. *Philosophical Transactions of the Royal Society, A, 372*, 20130175.

Canup, R., & Asphaug, E. (2001). Origin of the Moon in a giant impact near the end of the Earth's formation. *Nature, 412*, 708–712.

Capitaine, N., Wallace, P., & Chapront, J. (2003). Expressions for IAU 2000 precession quantities. *Astronomy & Astrophysics, 412*, 567–586.

Cappucci, M. (2023). Scientists marvel at whirlpool of plasma forming polar vortex on sun. *Washington Post*. Downloaded from https://www.washingtonpost.com/weather/2023/02/15/polar-vortex-sun-nasa-/

Caricchi, L., Gaillard, F., Mecklenburgh, J., & Le Trong, E. (2011). Experimental determination of electrical conductivity during deformation of melt-bearing olivine aggregates: Implications for electrical anisotropy in the oceanic low velocity zone. *Earth and Planetary Science Letters, 302*, 81–94.

Carr, M. (1990). Mars. In J. Beatty & A. Chaikin (Eds.), *The new solar system* (3rd edition, pp. 53–64). Cambridge: Sky Publishing.

Carr, M., & Head, J., III. (2010). Geologic history of Mars. *Earth and Planetary Science Letters, 294*, 185–203.

Cassan, A., Kubas, D., Beaulieu, J. P., Dominik, M., Horne, K., Greenhill, J., et al. (2012). One or more bound planets per Milky Way star from microlensing observations. *Nature, 481*, 167–169.

Cassen, P. (1994). Utilitarian models of the Solar Nebula. *Icarus, 112*, 405–429.

Catling, D., & Kasting, J. (2017). *Atmospheric evolution on inhabited and lifeless worlds.* Cambridge, England: Cambridge University Press, 592 pp.

Catling, D., Zahnle, K., & McKay, C. (2001). Biogenic methane, hydrogen escape, and the irreversible oxidation of early Earth. *Science, 293*, 839–843.

Catt, J., & Maslin, M. (2012). The prehistoric human time scale. In F. Gradstein, J. Ogg, M. Schmitz, & G. Ogg (Eds.), *The geologic time scale* (Vol. 2, pp. 1011–1032). Amsterdam: Elsevier, 1176 pp.

Cazenave, A., Dominh, K., Guinehut, S., Berthier, E., Llovel, W., Ramillien, G., Met al. (2009). Sea level budget over 2003–2008: A reevaluation from GRACE space gravimetry, satellite altimetry and Argo. *Global and Planetary Change, 65*, 83–88.

Chabot, N., & Drake, M. (1999). Potassium solubility in metal: The effects of composition at 15 kbar and 1900°C on partitioning between iron alloys and silicate melts. *Earth and Planetary Science Letters, 172*, 323–335.

Chadburn, S., Burke, E., Cox, P., Friedlingstein, P., Hugelius, G., & Westermann, S. (2017). An observation-based constraint on permafrost loss as a function of global warming. *Nature Climate Change, 7*, 340–344.

Chai, M., Brown, J., & Slutsky, L. (1996). Thermal diffusivity of mantle minerals. *Physics and Chemistry of Minerals, 23*, 470–475.

Chaisson, E., & McMillan, S. (2005). *Astronomy today* (5th edition). New Jersey: Pearson Prentice Hall.

Chambers, J. (2001). Making more terrestrial planets. *Icarus, 152*, 205–224.

Chambers, J. (2005). Planet formation. In A. M. Davis (Ed.), *Meteorites, comets & planets* (pp. 461–475) (*Treatise on Geochemistry*, Vol. 1, exec. eds. Holland & Turekian), Amsterdam: Elsevier.

Chambers, J. (2010). Terrestrial planet formation. In S. Seager (Ed.), *Exoplanets.* Tucson: University of Arizona Press.

Chambodut, A., Eymin, C., & Mandea, M. (2007). Geomagnetic jerks from the Earth's surface to the top of the core. *Earth, Planets and Space, 59*, 675–684.

Channell, J., Xuan, C., & Hodell, D. (2009). Stacking paleointensity and oxygen isotope data for the last 1.5 Myr (PISO-1500). *Earth and Planetary Science Letters, 283*, 14–23.

Chao, B. (1988). Correlation of interannual length-of-day variation with El Niño/Southern Oscillation, 1972–1986. *Journal of Geophysical Research, 93*, 7709–7715.

Chao, B. (1995). Anthropogenic impact on global geodynamics due to reservoir water impoundment. *Geophysical Research Letters, 22*, 3529–3532.

Chao, B. (2000). Renaming D Double Prime. *EOS Transactions AGU, 81*, 46.

Chao, B., & Ding, H. (2016). Global geodynamic changes induced by all major earthquakes, 1976–2015. *Journal of Geophysical Research, 121*, 8987–8999.

Chao, B., & Gross, R. (1987). Changes in the Earth's rotation and low-degree gravitational field introduced by earthquakes. *Geophysical Journal of the Royal Astronomical Society, 91*, 569–596.

Chao, B., Gross, R., & Dong, D. (1995). Changes in global gravitational energy induced by earthquakes. *Geophysical Journal International, 122*, 784–789.

Chao, B., & O'Connor, W. (1988). Global surface-water-induced seasonal variations in the Earth's rotation and gravitational field. *Geophysical Journal, 94*, 263–270.

Chao, B., Wu, Y., & Li, Y. (2008). Impact of artificial reservoir water impoundment on global sea level. *Science, 320*, 212–214.

Chapelle, G., & Peck, L. (1999). Polar gigantism dictated by oxygen availability. *Nature, 399*, 114–115.

Chapman, C., Cohen, B., & Grinspoon, D. (2007). What are the real constraints on the existence and magnitude of the late heavy bombardment?. *Icarus, 189*, 233–245.

Chapple, W., & Forsyth, D. (1979). Earthquakes and bending of plates at trenches. *Journal of Geophysical Research, 84*, 6729–6749.

Charbonneau, P. (2010). Dynamo models of the solar cycle. *Living Reviews in Solar Physics, 7*. http://www.livingreviews.org/lrsp-2010-3

Chase, C. (1972). The *N*-plate problem of plate tectonics. *Geophysical Journal of the Royal Astronomical Society, 29*, 117–122.

Chaussidon, M., & Liu, M. (2015). Timing of nebula processes that shaped the precursors of the terrestrial planets. In J. Badro & M. Walter (Eds.), *The early Earth: Accretion and differentiation* (AGU Monograph 212, pp. 1–26). New Jersey: John Wiley & Sons and AGU, 181 pp.

Chave, A., Flosadóttir, A., & Cox, C. (1990). Some comments on seabed propagation of ULF/ELF electromagnetic fields. *Radio Science, 25*, 825–836.

Chelton, D., & Enfield, D. (1986). Ocean signals in tide gauge records. *Journal of Geophysical Research, 91*, 9081–9098.

Chen, G., Wang, X., & Qian, C. (2016). Vertical structure of upper-ocean seasonality: Extratropical spiral versus tropical phase lock. *Journal of Climate, 29*, 4021–4030.

Chen, J., & Wilson, C. (2005). Hydrological excitations of polar motion, 1993–2002. *Geophysical Journal International, 160*, 833–839.

Chen, J., Cazenave, A., Dahle, C., Llovel, W., Panet, I., Pfeffer, J., & Moreira, L. (2022). Applications and challenges of GRACE and GRACE Follow-On satellite gravimetry. *Surveys in Geophysics, 43*, 305–345. https://doi.org/10.1007/s10712-021-09685-x

Chen, J., Wilson, C., Blankenship, D., & Tapley, B. (2009). Accelerated Antarctic ice loss from satellite gravity measurements. *Nature Geoscience, 2*, 859–862.

Chen, J., Wilson, C., Ries, J., & Tapley, B. (2013). Rapid ice melting drives Earth's pole to the east. *Geophysical Research Letters, 40*. doi:10.1002/grl.50552

Cheney, R., & Miller, L. (1988). Mapping the 1986-1987 El Niño with GEOSAT altimeter data. *EOS (Transactions AGU), 69*, 754–755.

Cheng, M., & Tapley, B. (2004). Variations in the Earth's oblateness during the past 28 years. *Journal of Geophysical Research, 109*. doi:10.1029/2004JB003028

Cheng, M., Tapley, B., & Ries, J. (2013). Deceleration in the Earth's oblateness. *Journal of Geophysical Research, 118*, 740–747. doi:10.1002/jgrb.50058

Chisholm, T., & Chapman, D. (1992). Climate change inferred from analysis of borehole temperatures: An example from western Utah. *Journal of Geophysical Research, 97*, 14155–14175.

Cho, M., & Rycroft, M. (1998). Computer simulation of the electric field structure and optical emission from cloud-top to the ionosphere. *Journal of Atmospheric and Solar-Terrestrial Physics, 60*, 871–888.

Chopelas, A., & Boehler, R. (1989). Thermal expansion measurements at very high pressure, systematics, and a case for a chemically homogeneous mantle. *Geophysical Research Letters, 16*, 1347–1350.

Christ, A., Talaia-Murray, M., Elking, N., Domack, E. W., Leventer, A., Lavoie, C., et al. & the LARISSA Group (2015). Late Holocene glacial advance and ice shelf growth in Barilari Bay, Graham Land, west Antarctic Peninsula. *Geological Society of America Bulletin, 127*, 297–315. doi:10.1130/B31035.1

Christensen, U. (1987). Some geodynamical effects of anisotropic viscosity. *Geophysical Journal of the Royal Astronomical Society, 91*, 711–736.

Christensen, U. (1989). Models of mantle convection: One or several layers. *Philosophical Transactions of the Royal Society of London, A, 328*, 417–424.

Christensen, U., & Aubert, J. (2006). Scaling properties of convection-driven dynamos in rotating spherical shells and application to planetary magnetic fields. *Geophysical Journal International, 166,* 97–114.

Christensen, U., & Hofmann, A. (1994). Segregation of subducted oceanic crust in the convecting mantle. *Journal of Geophysical Research, 99,* 19,867–19,884.

Christensen, U., & Yuen, D. (1984). The interaction of a subducting lithospheric slab with a chemical or phase boundary. *Journal of Geophysical Research, 89,* 4389–4402.

Christensen, U., & Yuen, D. (1985). Layered convection induced by phase transitions. *Journal of Geophysical Research, 90,* 10,291–10,300.

Christensen-Dalsgaard, J., Dappen, W., Ajukov, S. V., Anderson, E. R., Antia, H. M., Basu, S., et al. (1996). The current state of solar modeling. *Science, 272,* 1286–1292.

Christiansen, B., & Ljungqvist, F. (2017). Challenges and perspectives for large-scale temperature reconstructions of the past two millennia. *Reviews of Geophysics, 55,* 40–96. doi:10.1002/2016RG000521

Christodoulidis, D., Smith, D., Williamson, R., & Klosko, S. (1988). Observed tidal braking in the Earth/Moon/Sun system. *Journal of Geophysical Research, 93,* 6216–6236.

Chudinovskikh, L., & Boehler, R. (2001). High-pressure polymorphs of olivine and the 660-km seismic discontinuity. *Nature, 411,* 574–577.

Chulliat, A., & Maus, S. (2014). Geomagnetic secular acceleration, jerks, and a localized standing wave at the core surface from 2000 to 2010. *Journal of Geophysical Research, 119,* 1531–1543. doi:10.1002/2013JB010604

Chulliat, A., Macmillan, S., Alken, P., Beggan, C., Nair, M., Hamilton, B., et al. (2015). *The US/UK World Magnetic Model for 2015–2020: Technical Report.* NOAA: National Geophysical Data Center. doi:10.7289/V5TB14V7

Chung, D. (1972). Birch's Law: Why is it so good? *Science, 177,* 261–263.

Church, J., & White, N. (2011). Sea-level rise from the late 19th century to the early 21st century, *Surveys in Geophysics, 32,* 585–602.

Cianetti, S., Giunchi, C., & Spada, G. (2002). Mantle viscosity beneath the Hudson Bay: An inversion based on the Metropolis algorithm. *Journal of Geophysical Research, 107.* doi:10.1029/2001JB000585

Cieza, L., Casassus, S., Tobin, J., Bos, S. P., Williams, J. P., Perez, S., et al. (2016). Imaging the water snow-line during a protostellar outburst. *Nature, 535,* 258–261.

Cimbala, J. (2016). *Some Instabilities.* Downloaded from https://www.me.psu.edu/cimbala/me522/Some_Instabilities.pdf

Ciobanu, M., Burgaud, G., Dufresne, A., Breuker, A., Redou, V., Ben Maamar, S., et al. (2014). Microorganisms persist at record depths in the subseafloor of the Canterbury Basin. *ISME Journal, 8,* 1370–1380.

Ciraci, E., Velicogna, I., & Swenson, S. (2020). Continuity of the mass loss of the world's glaciers and ice caps from the GRACE and GRACE Follow-On missions. *Geophysical Research Letters, 47.* https://doi.org/10.1029/2019GL086926

Clark, P., Alley, R., & Pollard, D. (1999). Northern hemisphere ice-sheet influences on global climate change. *Science, 286,* 1104–1111.

Clark, P., Dyke, A., Shakun, J., Carlson, A., Clark, J., Wohlfarth, B., et al. (2009). The last glacial maximum. *Science, 325,* 710–714.

Clark, P., & Mix, A. (2002). Ice sheets and sea level of the Last Glacial Maximum. *Quaternary Science Reviews, 21,* 1–7.

Clark, S. (1957). Radiative transfer in the Earth's mantle. *Transactions AGU, 38,* 931–938.

Clauser, C. (2011). Radiogenic heat production of rocks. In H. Gupta (Ed.), *Encyclopedia of solid earth geophysics* (2nd edition). Dordrecht: Springer. doi:10.1007/978-90-481-8702-7_74, preprint downloaded from ResearchGate Oct. 7, 2019.

Clette, F., Svalgaard, L., Vaquero, J., & Cliver, E. (2014). Revisiting the sunspot number: A 400-year perspective on the Solar Cycle. *Space Science Reviews, 186,* 35–103.

Clinton, W. (1996). President Clinton statement regarding Mars Meteorite Discovery. White House: Office of the Press Secretary. Accessed at http://www2.jpl.nasa.gov/snc/clinton.html

Cloud, P. (1972). A working model of the primitive Earth. *American Journal of Science*, *272*, 537–548.

Cobb, A., & Pudritz, R. (2014). Nature's starships. I. Observed abundances and relative frequencies of amino acids in meteorites. *Astrophysical Journal*, *783*. doi:10.1088/0004-637X/783/2/140

CODATA (2010). CODATA, International Committee on Data for Science and Technology. http://physics.nist.gov/cuu/Constants/index.html, accessed March 2013.

CODATA (2018). CODATA, International Committee on Data for Science and Technology. http://physics.nist.gov/cuu/Constants/index.html, accessed March 2022.

Cohen, A. (2012). Scientific drilling and biological evolution in ancient lakes: Lessons learned and recommendations for the future. *Hydrobiologia*, *682*, 3–25.

Cohen, J., Screen, J., Furtado, J., Barlow, M., Whittleston, D., Coumou, D., et al. (2014). Recent Arctic amplification and extreme mid-latitude weather. *Nature Geoscience*, *7*, 627–637.

Cohen, K., Finney, S., Gibbard, P., & Fan, J. (2013). The ICS International Chronostratigraphic Chart. *Episodes*, *36*, 199–204.

Cohen, Y., Padan, E., & Shilo, M. (1975). Facultative anoxygenic photosynthesis in the cyanobacterium *Oscillatoria limnetica*. *Journal of Bacteriology*, *123*, 855–861.

Cole, D., Reinhard, C., Wang, X., Gueguen, B., Halverson, G., Gibson, T., et al. (2016). A shale-hosted Cr isotope record of low atmospheric oxygen during the Proterozoic. *Geology*, *44*, 555–558.

Connerney, J., Acuña, M., Ness, N., Kletetschka, G., Mitchell, D., Lin, R., & Reme, H. (2005). Tectonic implications of Mars crustal magnetism. *Proceedings of the National Academy of Sciences*, *102*, 14970–14975.

Connerney, J., Acuña, M., Wasilewski, P., Ness, N., Rème, H., Mazelle, C., et al. (1999). Magnetic lineations in the ancient crust of Mars. *Science*, *284*, 794–798.

Connolly, M., Connolly, R., Soon, W., Velasco Herrera, V., Cionco, R., & Quaranta, N. (2021). Analyzing atmospheric circulation patterns using mass fluxes calculated from weather balloon measurements: North Atlantic region as a case study. *Atmosphere*, *12*. https://doi.org/10.3390/atmos12111439

Constable, C. (2005). Dipole moment variation. In D. Gubbins & E. Herrera-Bervera (Eds.), *Encyclopedia of geomagnetism and paleomagnetism (preprint)*.

Constable, C., Korte, M., & Panovska, S. (2016). Persistent high paleosecular variation activity in southern hemisphere for at least 10000 years. *Earth and Planetary Science Letters*, *453*, 78–86.

Constable, C., Tauxe, L., & Parker, R. (1998). Analysis of 11 Myr of geomagnetic intensity variation. *Journal of Geophysical Research*, *103*, 17735–17748.

Constable, S. (1993). Constraints on mantle electrical conductivity from field and laboratory measurements. *Journal of Geomagnetism and Geoelectricity*, *45*, 707–728.

Constable, S., Orange, A., Hoversten, M., & Morrison, F. (1998). Marine magnetotellurics for petroleum exploration, Part I: A sea-floor equipment system. *Geophysics*, *63*, 816–825.

Cook, A. (1977). The moment of inertia of Mars and the existence of a core. *Geophysical Journal of the Royal Astronomical Society*, *51*, 349–356.

Cook, A., & Roberts, P. (1970). The Rikitake two-disc dynamo system. *Mathematical Proceedings of the Cambridge Philosophical Society*, *68*, 547–569.

Cooper, A., Turney, C. S. M., Palmer, J., Hogg, A., McGlone, M., Wilmshurst, J., et al. (2021a). A global environmental crisis 42,000 years ago. *Science*, *371*, 811–818.

Cooper, A., Turney, C. S. M., Palmer, J., Hogg, A., McGlone, M., Wilmshurst, J., et al. (2021b). Response to comment on "A global environmental crisis 42,000 years ago." *Science*, *374*. doi:10.1126/science.abi9756

Corliss, J., Dymond, J., Gordon, L., Edmond, J., von Herzen, R., Ballard, R., et al. (1979).

Submarine thermal springs on the Galápagos Rift. *Science, 203,* 1073–1083.

Cormier, V. (2009). A glassy lowermost outer core. *Geophysical Journal International, 179,* 372–380.

Correia, A., & Laskar, J. (2001). The four final rotation states of Venus. *Nature, 411,* 767–770.

Cory, R., Crump, B., Dobkowski, J., & Kling, G. (2013). Surface exposure to sunlight stimulates CO_2 release from permafrost soil carbon in the Arctic. *Proceedings of the National Academy of Sciences, 110,* 3429–3434.

Costard, F., Forget, F., Mangold, N., & Peulvast, J. (2002). Formation of recent Martian debris flows by melting of near-surface ground ice at high obliquity. *Science, 295,* 110–113.

Counil, J., Cohen, Y., & Achache, J. (1991). The global continent-ocean magnetization contrast. *Earth and Planetary Science Letters, 103,* 354–364.

Coupland, D., & Van der Voo, R. (1980). Long-term nondipole components in the geomagnetic field during the last 130 m.y. *Journal of Geophysical Research, 85,* 3529–3548.

Courtillot, V., Ducruix, J., & Le Mouël, J. (1978). Sur une accélération récente de la variation séculaire du champ magnétique terrestre. *Comptes Rendus de l'Académie des Sciences Paris, D, 287,* 1095–1098 (cited in Mandea Alexandrescu et al. 1999).

Courtillot, V., & Le Mouël, J. (1984). Geomagnetic secular variation impulses. *Nature, 311,* 709–716.

Courtillot, V., & Olson, P. (2007). Mantle plumes link magnetic superchrons to phanerozoic mass depletion events. *Earth and Planetary Science Letters, 260,* 495–504.

Cowling, T. (1933). The magnetic field of sunspots. *Monthly Notices of the Royal Astronomical Society, 94,* 39–48.

Cowling, T. (1957). The dynamo maintenance of steady magnetic fields. *Quarterly Journal of Mechanics and Applied Mathematics, 10,* 129–136.

Creager, K., & Jordan, T. (1986). Aspherical structure of the core-mantle boundary from PKP travel times. *Geophysical Research Letters, 13,* 1497–1500.

Crittenden, M. (1963a). Effective viscosity of the Earth derived from isostatic loading of Pleistocene Lake Bonneville. *Journal of Geophysical Research, 68,* 5517–5528.

Crittenden, M. (1963b). *New Data on the Isostatic Deformation of Lake Bonneville,* USGS Prof. Paper 454-E. Washington, DC: US Government Printing Office.

Crockford, P., Hayles, J., Bao, H., Planavsky, N., Bekker, A., Fralick, P., et al. (2018). Triple oxygen isotope evidence for limited mid-Proterozoic primary productivity. *Nature, 559,* 613–616.

Crone, A., Machette, M., Bonilla, M., Lienkaemper, J., Pierce, K., Scott, W., & Bucknam, R. (1987). Surface faulting accompanying the Borah Peak earthquake and segmentation of the Lost River fault, Central Idaho. *Bulletin of the Seismological Society of America, 77,* 739–770.

Cronin, J., Cooper, G., & Pizzarello, S. (1995). Characteristics and formation of amino acids and hydroxy acids of the Murchison meteorite. *Advances in Space Research, 15,* 91–97.

Crossley, D., Hinderer, J., & Riccardi, U. (2013). The measurement of surface gravity, *Reports on Progress in Physics, 76,* 47pp. doi:10.1088/0034-4885/76/4/046101

Crossley, D., & Stevens, R. (1976). Expansion of the Earth due to a secular decrease in G – Evidence from Mercury. *Canadian Journal of Earth Sciences, 13,* 1723–1725.

Crossley, D., & Xu, S. (1998). Analysis of superconducting gravity data from Table Mountain, Colorado. *Geophysical Journal International, 135,* 835–844.

Crowe, S., Døssing, L., Beukes, N., Bau, M., Kruger, S., Frei R., & Canfield, D. (2013). Atmospheric oxygenation three billion years ago. *Nature, 501,* 535–539.

Crowley, T. (1992). North Atlantic Deep Water cools the southern hemisphere. *Paleoceanography, 7,* 489–497.

Cserepes, L., & Yuen, D. (2000). On the possibility of a second kind of mantle plume.

Earth and Planetary Science Letters, *183*, 61–71.

CSIRO (2015). Pulsating variable stars. Downloaded from http://www.atnf.csiro.au/outreach/education/senior/astrophysics/variable_pulsating.html. January 27, 2016.

Ćuk, M., & Stewart, S. (2012). Making the Moon from a fast-spinning Earth: A giant impact followed by resonant despinning. *Science*, *338*, 1047–1052.

Currie, R. (1991). Deterministic signals in tree-rings from North America. *International Journal of Climatology*, *11*, 861–876.

Currie, R. (1993). Luni-solar 18.6- and 10–11-year solar cycle signals in South African rainfall. *International Journal of Climatology*, *13*, 237–256.

Currie, R. (1995). Luni-solar 18.6- and solar cycle 10–11-year signals in Chinese dryness-wetness indices. *International Journal of Climatology*, *15*, 497–515.

Currie, R. (1996). Mn and Sc signals in north Atlantic tropical cyclone occurrence. *International Journal of Climatology*, *16*, 427–439.

D'Alessio, P., Cantó, J., Calvet, N., & Lizano, S. (1998). Accretion disks around young objects. I. *The detailed vertical structure. Astrophysical Journal*, *500*, 411–427.

D'Angelo, G., & Podolak, M. (2015). Capture and evolution of planetesimals in circumjovian disks. *Astrophysical Journal*, *806*. doi:10.1088/0004-637X/806/2/203

Dahl, T., Hammarlund, E., Anbar, A., Bond, D., Gill, B., Gordon, G., et al. (2010). Devonian rise in atmospheric oxygen correlated to the radiations of terrestrial plants and large predatory fish. *Proceedings of the National Academy of Sciences*, *107*, 17911–17915.

Dahle, C., Murböck, M., Flechtner, F., Dobslaw, H., Michalak, G., Neumayer, K., et al. (2019). The GFZ GRACE RL06 monthly gravity field time series: Processing details and quality assessment. *Remote Sensing*, *11*, 2116; doi:10.3390/rs11182116

Dahl-Jensen, D., Mosegaard, K., Gundestrup, N., Clow, G., Johnsen, S., Hansen, A., & Balling, N. (1998). Past temperatures directly from the Greenland ice sheet. *Science*, *282*, 268–271.

Dai, A. (2010). Drought under global warming: A review. In *Wiley Interdisciplinary Reviews: Climate Change*, *2*, 45–65. doi:10.1002/wcc.81

Dai, A. (2012). Erratum. *In Wiley Interdisciplinary Reviews: Climate Change*, *3*, 617. Image from *NCAR/UCAR AtmosNews* Oct. 19, 2010 release: "Climate change: Drought may threaten much of globe within decades," with update July 3, 2012.

Dai, L., & Karato, S. (2009). Electrical conductivity of orthopyroxene: Implications for the water content of the asthenosphere. *Proceedings of the Japan Academy, Series B*, *85*, 466–475.

Dai, X., & Guerras, E. (2018). Probing extra-galactic planets using quasar microlensing. *Astrophysical Journal Letters*, *853*. https://doi.org/10.3847/2041-8213/aaa5fb

Dal Forno, G., Gasperini, P., & Boschi, E. (2005). Linear or nonlinear rheology in the mantle: A 3D finite-element approach to postglacial rebound modeling. *Journal of Geodynamics*, *39*, 183–195.

Dangendorf, S., Hay, C., Calafat, F., Marcos, M., Piecuch, C., Berk, K., & Jensen, J. (2019). Persistent acceleration in global sea-level rise since the 1960s. *Nature Climate Change*, *9*, 705–710.

Dangendorf, S., Marcos, M., Wöppelmann, G., Conrad, C., Frederikse, T., & Riva, R. (2017). Reassessment of 20th century global mean sea level rise. *Proceedings of the National Academy of Sciences*, *114*, 5946–5951.

Darwin, G. (1899). *The Tides (and kindred phenomena in the Solar System)*. Boston, MA: Houghton, Mifflin & Co.

Davaille, A., & Limare, A. (2007). Laboratory studies of mantle convection. In G. Schubert (Ed.), *Treatise on Geophysics* (Vol. 7, pp. 89–165). Amsterdam: Elsevier.

David, L. (2016). Remembering a big scoop about a small rock. *SpaceNews Magazine, Sept. 12*, 2016. Accessed at https://www.spacenewsmag.com/feature/remembering-a-big-scoop-about-a-small-rock

David, L., & Alm, E. (2011). Rapid evolutionary innovation during an Archaean genetic expansion. *Nature, 469,* 93–96.

Davies, G. (1988a). Role of the lithosphere in mantle convection. *Journal of Geophysical Research, 93,* 10,451–10,466.

Davies, G. (1988b). Ocean bathymetry and mantle convection 1. Large-scale flow and hotspots. *Journal of Geophysical Research, 93,* 10,467–10,480.

Davies, G. (1999). *Dynamic Earth: Plates, Plumes and Mantle Convection.* Cambridge, England: Cambridge University Press.

Davies, G., & Gurnis, M. (1986). Interaction of mantle dregs with convection: lateral heterogeneity at the core-mantle boundary. *Geophysical Research Letters, 13,* 1517–1520.

Davies, J. (2013). Global map of solid Earth surface heat flow. *Geochemistry, Geophysics, Geosystems, 14,* 4608–4622. doi:10.1002/ggge.20271

Davies, J., & Davies, D. (2010). Earth's surface heat flux. *Solid Earth, 1,* 5–24.

Davies, N., Garwood, R., McMahon, W., Schneider, J., & Shillito, A. (2022). The largest arthropod in Earth history: Insights from newly discovered *Arthropleura* remains (Serpukhovian Stainmore Formation, Northumberland, England). *Journal of the Geological Society, 179.* https//doi.org/10 .1144/jgs2021-115 | Vol. 179 | 2022 | jgs2021-115

Davis, E., Wang, K., Becker, K., Thomson, R., & Yashayaev, I. (2003). Deep-ocean temperature variations and implications for errors in seafloor heat flow determinations. *Journal of Geophysical Research, 108.* doi:10.1029/2001JB001695

Davis, M. (2000). *Late Victorian Holocausts: El Niño Famines and the Making of the Third World.* Brooklyn: Verso.

Davis, R., Hayden, B., Gay, D., Phillips, W., & Jones, G. (1997). The North Atlantic subtropical anticyclone. *Journal of Climate, 10,* 728–744.

Davis, W. (1890). Oscillations of lakes (seiches). *Science, 15,* 117.

De Becker, M. (1990). Continuous monitoring and analysis of microseisms in Belgium to forecast storm surges along the North Sea coast. *Physics of the Earth and Planetary Interiors, 63,* 219–228.

De Bremaecker, J. (1977). Is the oceanic lithosphere elastic or viscous? *Journal of Geophysical Research, 82,* 2001–2004.

De Bremaecker, J. (1980). Correction. *Journal of Geophysical Research, 85,* 3952.

De Groot-Hedlin, C., Hedlin, M., & Walker, K. (2011). Finite difference synthesis of infrasound propagation through a windy, viscous atmosphere: application to a bolide explosion detected by seismic networks. *Geophysical Journal International, 185,* 305–320.

De Jong, T., & van Soldt, W. (1989). The earliest known solar eclipse record redated. *Nature, 338,* 238–240.

de Linage, C., Rivera, L., Hinderer, J., Boy, J.-P., Rogister, Y., Lambotte, S., & Biancale, R. (2009). Separation of coseismic and postseismic gravity changes for the 2004 Sumatra-Andaman earthquake from 4.6 yr of GRACE observations and modelling of the coseismic change by normal-modes summation. *Geophysical Journal International, 176,* 695–714.

De Niem, D., Kührt, E., Morbidelli, A., & Motschmann, U. (2012). Atmospheric erosion and replenishment induced by impacts upon the Earth and Mars during a heavy bombardment. *Icarus, 221,* 495–507.

De Wit, M., & Furnes, H. (2016). 3.5-Ga hydrothermal fields and diamictites in the Barberton Greenstone Belt – Paleoarchean crust in cold environments. *Science Advances, 2,* e1500368.

DeConto, R., & Pollard, D. (2003). Rapid Cenozoic glaciation of Antarctica induced by declining atmospheric CO_2. *Nature, 421,* 245–249.

Deen, S. (2021). 2014 UN271: A possible dwarf planet from the Oort Cloud on a tour through the Solar System. Accessed June 28, 2021 from https://groups.io/g/mpml/topic/83645454# 36493

Deligne, N., & Sigurdsson, H. (2015). Global rates of volcanism and volcanic episodes. In H. Sigurdsson et al. (Eds.), *Encyclopedia of volcanoes* (2nd edition, pp. 265–272). Amsterdam: Elsevier. 10.1016/B978-0-12-385938-9.00014-6

DeMets, C., Gordon, R., & Argus, D. (2010). Geologically current plate motions, *Geophysical Journal International, 181*, 1–80.

DeMets, C., Gordon, R., Argus, D., & Stein, S. (1990). Current plate motions, *Geophysical Journal International, 101*, 425–478.

DeMets, C., Gordon, R., Argus, D., & Stein, S. (1994). Effect of recent revisions to the geomagnetic reversal time scale on estimates of current plate motions. *Geophysical Research Letters, 21*, 2191–2194.

Deming, D., & Knutson, H. (2020). Highlights of exoplanetary science from *Spitzer*. *Nature Astronomy, 4*, 453–466.

DePaolo, D., & Manga, M. (2003). Deep origin of hotspots – The mantle plume model. *Science, 300*, 920–921.

Dermott, S. (Ed.). (1978). *The origin of the Solar System* (NATO Advanced Study Institute on the Origin of the Solar System, University of Newcastle-Upon-Tyne, 1977). Chichester: John Wiley & Sons.

Des Marais, D. (1997). Isotopic evolution of the biogeochemical carbon cycle during the Proterozoic Eon. *Organic Geochemistry, 27*, 185–193.

Deuss, A., Redfern, S., Chambers, K., & Woodhouse, J. (2006). The nature of the 660-kilometer discontinuity in Earth's mantle from global seismic observations of PP precursors. *Science, 311*, 198–201.

Dey, N., & Dickman, S. (2010). Spectral and Geographical Variability in the Oceanic Response to Atmospheric Pressure Fluctuations, as Inferred from "Dynamic Barometer" Green's Functions. *Journal of Geophysical Research, 115*, C09026. doi:10.1029/2009JC005609

Diamond, J. (1997). *Guns, Germs, and Steel: The fates of human societies*. NYC: Norton & Company.

Diamond, M. (1963). Sound channels in the atmosphere. *Journal of Geophysical Research, 68*, 3459–3464.

Dicke, R. (1962). The Earth and cosmology. *Science, 138*, 653–664.

Dicke, R. (1966). The secular acceleration of the Earth's rotation and cosmology. In B. Marsden & A. Cameron (Eds.), *The Earth-Moon system* (pp. 98–164). New York, NY: Plenum Press.

Dickens, G., Castillo, M., & Walker, J. (1997). A blast of gas in the latest Paleocene: Simulating first-order effects of massive dissociation of oceanic methane hydrate. *Geology, 25*, 259–262.

Dickens, G., O'Neil, J., Rea, D., & Owen, R. (1995). Dissociation of oceanic methane hydrate as a cause of the carbon isotope excursion at the end of the Paleocene. *Paleoceanography, 10*, 965–971.

Dickman, S. (1977). Secular trend of the Earth's rotation pole: Consideration of motion of the latitude observatories. *Geophysical Journal of the Royal Astronomical Society, 51*, 229–244.

Dickman, S. (1979). Continental drift and true polar wandering. *Geophysical Journal of the Royal Astronomical Society, 57*, 41–50.

Dickman, S. (1981). Investigation of controversial polar motion features using homogeneous International Latitude Service data. *Journal of Geophysical Research, 86*, 4904–4912.

Dickman, S. (1983a). *The Earth and Its Mountains* by R.A. Lyttleton – Book review. *Physics of the Earth and Planetary Interiors, 33*, 333–336.

Dickman, S. (1983b). The rotation of the ocean-solid earth system. *Journal of Geophysical Research, 88*, 6373–6394.

Dickman, S. (1988). Theoretical investigation of the oceanic inverted barometer hypothesis. *Journal of Geophysical Research, 93*, 14941–14946.

Dickman, S. (1989). A complete spherical harmonic approach to luni-solar tides, *Geophysical Journal International, 99*, 457–468.

Dickman, S. (1993). Dynamic ocean-tide effects on Earth's rotation. *Geophysical Journal International, 112*, 448–470.

Dickman, S. (2010). Rotationally acceptable ocean tide models for determining the response of the oceans to atmospheric pressure fluctuations. *Journal of Geophysical Research, 115.* doi:10.1029/2010JB007556

Dickman, S., & Gross, R. (2010). Rotational evaluation of a long-period spherical harmonic ocean tide model. *Journal of Geodesy, 84,* 457–464. (See also Dickman, S. (2010). Rotationally acceptable ocean tide models for determining the response of the oceans to atmospheric pressure fluctuations. *Journal of Geophysical Research, 115.* doi:10.1029/2010JB007556.)

Dickman, S., & Nam, Y. (1998). Constraints on Q at long periods from Earth's rotation, *Geophysical Research Letters, 25,* 211–214.

Dickman, S., & Williams, D. (1981). Viscoelastic membrane tectonics. *Geophysical Research Letters, 8,* 199–202.

Dickson, R., Lazier, J., Meincke, J., Rhines, P., & Swift, J. (1996). Long-term coordinated changes in the convective activity of the North Atlantic. *Progress in Oceanography, 38,* 241–295.

Dieng, H., Cazenave, A., von Schuckmann, K., Ablain, M., & Meyssignac, B. (2015). Sea level budget over 2005–2013: Missing contributions and data errors. *Ocean Science, 11,* 789–802.

Dietrich, W., Schmitt, D., & Wicht, J. (2013). Hemispherical Parker waves driven by thermal shear in planetary dynamos. *Europhysics Letters, 104.* doi:10.1209/0295-5075/104/49001

Ding, Y., Luo, B., & Feng, Y. (1983). Peak years of various solar cycles. *Chinese Journal of Astronomy and Astrophysics, 7,* 24–30.

Dirac, P. (1937). The cosmological constants (Letters to the Editor). *Nature, 139,* 323.

Dobbs, C., Burkert, A., & Pringle, J. (2011). Why are most molecular clouds not gravitationally bound?. *Monthly Notices of the Royal Astronomical Society, 413,* 2935–2942.

Dobrin, M., & Savit, C. (1988). *Introduction to geophysical prospecting* (4th edition). New York, NY: McGraw-Hill Book Company.

Dobson, D. (2002). Self-diffusion in liquid Fe at high pressure. *Physics of the Earth and Planetary Interiors, 130,* 271–284.

Dobson, D., & Brodholt, J. (2000). The electrical conductivity and thermal profile of the Earth's

mid-mantle. *Geophysical Research Letters, 27,* 2325–2328.

Doell, R., & Cox, A. (1965). Paleomagnetism of Hawaiian lava flows. *Journal of Geophysical Research, 70,* 3377–3405.

Doin, M., Fleitout, L., & McKenzie, D. (1996). Geoid anomalies and the structure of continental and oceanic lithospheres. *Journal of Geophysical Research, 101,* 16119–16135.

Dokka, R. (2011). The role of deep processes in late 20th century subsidence of New Orleans and coastal areas of southern Louisiana and Mississippi. *Journal of Geophysical Research, 116,* B06403. doi:10.1029/2010JB008008

Donadini, F., Korte, M., & Constable, C. (2009). Geomagnetic field for 0–3 ka: 1. New data sets for global modeling. *Geochemistry, Geophysics, Geosystems, 10,* Q06007. doi:10.1029/2008GC002295

Dones, L., Brasser, R., Kaib, N., & Rickman, H. (2015). Origin and evolution of the cometary reservoirs. *Space Science Reviews, 197,* 191–269.

Dostal, J., Martinec, Z., & Thomas, M. (2012). The modelling of the toroidal magnetic field induced by tidal ocean circulation. *Geophysical Journal International, 189,* 782–798.

Douglass, D., & Clader, D. (2002). Climate sensitivity of the Earth to solar irradiance, *Geophysical Research Letters, 29.* 10.1029/2002GL015345

Dragoni, M., Bonafede, M., & Boschi, E. (1986). Downslope flow models of a Bingham liquid: Implications for lava flows. *Journal of Volcanology and Geothermal Research, 30,* 305–325.

Dragoni, M., Yuen, D., & Boschi, E. (1983). Global postseismic deformation in a stratified viscoelastic Earth: Effects on Chandler wobble excitation. *Journal of Geophysical Research, 88,* 2240–2250.

Draine, B. (2003). Interstellar dust grains. *Annual Review of Astronomy and Astrophysics, 41,* 241–289.

Drake, M., & Campins, H. (2006). Origin of water on the terrestrial planets. In D. Lazzaro, S. Ferraz-Mello, J. A. Fernández (Eds.),

Asteroids, Comets, Meteors (Proceedings of the IAU Symposium 229) (pp. 381–394). Cambridge, England: Cambridge University Press.

Driese, S., Jirsa, M., Ren, M., Brantley, S., Sheldon, N., Parker, D., & Schmitz, M. (2011). Neoarchean paleoweathering of tonalite and metabasalt: Implications for reconstructions of 2.69 Ga early terrestrial ecosystems and paleoatmospheric chemistry. *Precambrian Research, 189*, 1–17.

Driscoll, P., & Bercovici, D. (2014). On the thermal and magnetic histories of Earth and Venus: Influences of melting, radioactivity, and conductivity. *Physics of the Earth and Planetary Interiors, 236*, 36–51.

Driscoll, P., & Du, Z. (2019). Geodynamo conductivity limits. *Geophysical Research Letters, 46*, 7982–9789.

Dubin, M., Sissenwine, N., & Wexler, H. (1962). *U.S. Standard Atmosphere, 1962*. Washington, DC: NASA Langley Research Center.

Ducruix, J., Courtillot, V., & Le Mouël, J. (1980). The late 1960s secular variation impulse, the eleven year magnetic variation and the electrical conductivity of the deep mantle. *Geophysical Journal of the Royal Astronomical Society, 61*, 73–94.

Dullemond, C., & Dominik, C. (2004). Flaring vs. self-shadowed disks: The SEDs of Herbig Ae/Be stars. *Astronomy & Astrophysics, 417*, 159–168.

Dullemond, C., & Dominik, C. (2005). Dust coagulation in protoplanetary disks: A rapid depletion of small grains. *Astronomy & Astrophysics, 434*, 971–986.

Dumberry, M. (2008). Gravitational torque on the inner core and decadal polar motion. *Geophysical Journal International, 172*, 903–920.

Dumberry, M., & Bloxham, J. (2006). Azimuthal flows in the Earth's core and changes in length of day at millennial timescales. *Geophysical Journal International, 165*, 32–46.

Dumberry, M., & More, C. (2020). Weak magnetic field changes over the Pacific due to high conductance in lowermost mantle. *Nature Geoscience, 13*, 516–520.

Dunlop, J., Muir, M., Milne, V., & Groves, D. (1978). A new microfossil assemblage from the Archaean of Western Australia. *Nature, 274*, 676–678.

Dunn, P., Torrence, M., Kolenkiewicz, R., & Smith, D. (1999). Earth scale defined by modern satellite ranging observations. *Geophysical Research Letters, 26*, 1489–1492.

Durand-Manterola, H. (2009). Dipolar magnetic moment of the bodies of the solar system and the Hot Jupiters. *Planetary and Space Science, 57*, 1405–1411.

Durham, W., Mirkovich, V., & Heard, H. (1987). Thermal diffusivity of igneous rocks at elevated pressure and temperature. *Journal of Geophysical Research, 92*, 11615–11634.

Dushaw, B. (2008). Another look at the 1960 Perth to Bermuda long-range acoustic propagation experiment. *Geophysical Research Letters, 35*, L08601. doi:10.1029/2008GL 033415

Dye, S. (2012). Geoneutrinos and the radioactive power of the earth. *Reviews of Geophysics, 50*, RG3007. doi:10.1029/2012RG000400

Dyment, J., & Arkani-Hamed, J. (1998). Contribution of lithospheric remanent magnetization to satellite magnetic anomalies over the world's oceans. *Journal of Geophysical Research, 103*, 15423–15441.

Dyment, J., Choi, Y., Hamoudi, M., Lesur, V., & Thébault, E. (2015). Global equivalent magnetization of the oceanic lithosphere. *Earth and Planetary Science Letters, 430*, 54–65.

Dyment, J., Lesur, V., Hamoudi, M., Choi, Y., Thébault, E., Catalan, M., WDMAM Task Force, WDMAM Evaluators & WDMAM Data Providers (2016). World Digital Magnetic Anomaly Map version 2.0. Downloaded from wdmam.org May 2020.

Dziewonski, A. (1984). Mapping the lower mantle: Determination of lateral heterogeneity in P velocity up to degree and order 6. *Journal of Geophysical Research, 89*, 5929–5952.

Dziewonski, A., & Anderson, D. (1981). Preliminary reference Earth model. *Physics of the Earth and Planetary Interiors, 25*, 297–356.

Dziewonski, A., Hager, B., & O'Connell, R. (1977). Large-scale heterogeneities in the lower mantle. *Journal of Geophysical Research, 82*, 239–255.

Eaton, D., Darbyshire, F., Evans, R., Grütter, H., Jones, A., & Yuan, X. (2009). The elusive lithosphere-asthenosphere boundary (LAB) beneath cratons. *Lithos, 109*, 1–22.

Eder, S., Cadiou, H., Muhamad, A., McNaughton, P., Kirschvink, J., & Winklhofer, M. (2012). *Proceedings of the National Academy of Sciences, 109*, 12022–12027.

Egbert, G., Alken, P., Maute, A., & Zhang, H. (2021). Modelling diurnal variation magnetic fields due to ionospheric currents. *Geophysical Journal International, 225*, 1086–1109.

Egbert, G., & Booker, J. (1992). Very long period magnetotellurics at Tucson Observatory: Implications for mantle conductivity. *Journal of Geophysical Research, 97*, 15099–15112.

Egbert, G., & Erofeeva, S. (2002). Efficient inverse modeling of barotropic ocean tides. *Journal of Atmospheric and Oceanic Technology, 19*, 183–204.

Eguchi, J., Seales, J., & Dasgupta, R. (2020). Great Oxidation and Lomagundi events linked by deep cycling and enhanced degassing of carbon. *Nature Geoscience, 13*, 71–76.

Ehrenfreund, P., & Charnley, S. (2000). Organic molecules in the interstellar medium, comets, and meteorites: A voyage from dark clouds to the early Earth. *Annual Review of Astronomy and Astrophysics, 38*, 427–483.

Eickmann, B., Hofmann, A., Wille, M., Bui, T., Wing, B., & Schoenberg, R. (2018). Isotopic evidence for oxygenated Mesoarchaean shallow oceans. *Nature Geoscience, 11*, 133–138.

Ekstrom, G. (2001). Time domain analysis of Earth's long-period background seismic radiation. *Journal of Geophysical Research, 106*, 26,483–26,493.

El Amri, C., Maurel, M., Sagon, G., & Baron, M. (2005). The micro-distribution of carbonaceous matter in the Murchison meteorite as investigated by Raman imaging. *Spectrochimica Acta, A, 61*, 2049–2056.

Elkins-Tanton, L. (2008). Linked magma ocean solidification and atmospheric growth for Earth and Mars. *Earth and Planetary Science Letters, 271*, 181–191.

Elkins-Tanton, L. (2012). Magma oceans in the inner Solar System. *Annual Review of Earth and Planetary Sciences, 40*, 113–139.

Elsasser, W. (1939). On the origin of the Earth's magnetic field. *Physical Review, 55*, 489–498.

Elsila, J., Dworkin, J., Bernstein, M., Martin, M., & Sandford, S. (2007). Mechanisms of amino acid formation in interstellar ice analogs. *Astrophysical Journal, 660*, 911–918.

Elsila, J., Glavin, D., & Dworkin, J. (2009). Cometary glycine detected in samples returned by Stardust. *Meteoritics & Planetary Science, 44*, 1323–1330.

Emanuel, K. (2005). Increasing destructiveness of tropical cyclones over the past 30 years. *Nature, 436*, 686–688. doi:10.1038/nature03906. See also Emanuel, K., (2005). Emanuel replies. *Nature, 438*, E13. doi:10.1038/nature04427

Emanuel, K., Sundararajan, R., & Williams, J. (2008). Hurricanes and Global Warming: Results from downscaling IPCC AR4 simulations. *Bulletin of the American Meteorological Society, 89*, 347–367.

Emile-Geay, J., & Madec, G. (2009). Geothermal heating, diapycnal mixing and the abyssal circulation. *Ocean Science, 5*, 203–217.

Epstein, S., Krishnamurthy, R., Cronin, J., Pizzarello, S., & Yuen, G. (1987). Unusual stable isotope ratios in amino acid and carboxylic acid extracts from the Murchison meteorite. *Nature, 326*, 477–479.

Eriksson, K., & Simpson, E. (2000). Quantifying the oldest tidal record: The 3.2 Ga Moodies Group, Barberton Greenstone Belt, South Africa. *Geology, 28*, 831–834.

Erlykin, A., Harper, D., Sloan, T., & Wolfendale, A. (2017). Mass extinctions over the last 500 myr: An astronomical cause? *Palaeontology, 60*, 159–167.

Erwin, D., Laflamme, M., Tweedt, S., Sperling, E., Pisani, D., & Peterson, K. (2011).

The Cambrian Conundrum: Early divergence and later ecological success in the early history of Animals. *Science, 334,* 1091–1097.

Esposito, L. (1984). Sulfur dioxide: Episodic injection shows evidence for active Venus volcanism. *Science, 223,* 1072–1074.

Estabrook, C. (2004). Seismic constraints on mechanisms of deep earthquake rupture. *Journal of Geophysical Research, 109,* B02306. doi:10.1029/2003JB002449

Eucken, A. (1944a). Physikalisch-chemische Betrachtungen über die früheste Entwicklungsgeschichte der Erde (Physico-chemical observations on the early history of the Earth). *Nachrichten Der Akademie Der Wissenschaften Göttingen, Mathematisch–Physikalische Klasse, 1,* 1–25.

Eucken, A. (1944b). Über den Zustand des Erdinnern (On the state of the Earth's interior). *Naturwissenschaften, 32,* 112–121.

Evans, K., Nkansah, M., Hutchinson, I., & Rogers, S. (1991). Molecular network design. *Nature, 353,* 124.

Evans, R., Hirth, G., Baba, K., Forsyth, D., Chave, A., & Mackie, R. (2005). Geophysical evidence from the MELT area for compositional controls on oceanic plates. *Nature, 437,* 249–252.

Fairbanks, R. (1989). A 17,000-year glacio-eustatic sea level record: Influence of glacial melting dates on the Younger Dryas event and deep ocean circulation. *Nature, 342,* 637–642.

Farquhar, J., Bao, H., & Thiemens, M. (2000). Atmospheric influence of Earth's earliest sulfur cycle. *Science, 289,* 756–758.

Farquhar, J., Savarino, J., Airieau, S., & Thiemens, M. (2001). Observation of wavelength-sensitive mass-independent sulfur isotope effects during SO_2 photolysis: Implications for the early atmosphere. *Journal of Geophysical Research, 106,* 32829–32839.

Farquhar, J., Zerkle, A., & Bekker, A. (2011). Geological constraints on the origin of oxygenic photosynthesis. *Photosynthesis Research, 107,* 11–36.

Farrell, W., & Clark, J. (1976). On postglacial sea level. *Geophysical Journal of the Royal Astronomical Society, 46,* 647–667.

Farrell, W., & Munk, W. (2010). Booms and busts in the deep. *Journal of Physical Oceanography, 40,* 2159–2169.

Featherstone, N., & Hindman, B. (2016). The emergence of solar supergranulation as a natural consequence of rotationally constrained interior convection. *Astrophysical Journal Letters, 830.* doi:10.3847/2041-8205/830/1/L15

Fedorov, A., Hu, S., Lengaigne, M., & Guilyardi, E. (2015). The impact of westerly wind bursts and ocean initial state on the development, and diversity of El Niño events. *Climate Dynamics, 44,* 1381–1401.

Fei, Y., & Bertka, C. (1999). Phase transitions in the Earth's mantle and mantle mineralogy. In Y. Fei, C. Bertka, & B. Mysen (Eds.), *Mantle Petrology: Field observations and high pressure experimentation* (pp. 189–207). Houston, TX: Spec. Pub. 6, Geochemical Society.

Ferrón, H., Martínez-Pérez, C., & Botella, H. (2017). Ecomorphological inferences in early vertebrates: Reconstructing *Dunkleosteus terrelli* (Arthrodira, Placodermi) caudal fin from palaeoecological data. *PeerJ, 5,* e4081. doi:10.7717/peerj.4081

Feynman, R. (1963). Electromagnetism (Chapter 1). *The Feynman lectures on Physics,* II. Boston, MA: Addison-Wesley. Downloaded from https://www.feynmanlectures.caltech .edu/II_01.html

Feynman, R., Leighton, R., & Sands, M. (1963). *The Feynman lectures on Physics: Mainly mechanics, radiation, and heat.* Reading, MA: Addison-Wesley.

Filiberto, J., Trang, D., Treiman, A., & Gilmore, M. (2020). Present-day volcanism on Venus as evidenced from weathering rates of olivine. *Science Advances, 6.* doi:10.1126/sciadv .aax7445

Finlay, C., & Jackson, A. (2003). Equatorially dominated magnetic field change at the surface of Earth's core. *Science, 300,* 2084–2086.

Finlay, C., Aubert, J., & Gillet, N. (2016). Gyre-driven decay of the Earth's magnetic dipole. *Nature Communications, 7.* doi:10.1038/ncomms10422

Finlay, C., Dumberry, M., Chulliat, A., & Pais, M. (2010). Short timescale core dynamics: Theory and observations. *Space Science Reviews, 155*, 177–218.

Finlay, C., Maus, S., Beggan, C. D., Bondar, T. N., Chambodut, A., Chernova, T. A. et al. (International Association of Geomagnetism and Aeronomy, Working Group V-MOD) (2010). International Geomagnetic Reference Field: The eleventh generation. *Geophysical Journal International, 183*, 1216–1230.

Finlay, C., Olsen, N., & Tøffner-Clausen, L. (2015). DTU candidate field models for IGRF-12 and the CHAOS-5 geomagnetic field model. *Earth, Planets and Space, 67*. doi:10.1186/s40623-015-0274-3

Fiquet, G., Badro, J., Guyot, F., Requardt, H., & Krisch, M. (2001). Sound velocities in iron to 110 Gigapascals. *Science, 291*, 468–471.

Fiquet, G., Guyot, F., & Badro, J. (2008). The Earth's lower mantle and core. *Elements, 4*, 177–182.

Fischbach, E., & Talmadge, C. (1999). *The Search for Non-Newtonian Gravity*, New York, NY: Springer, https://doi.org/10.1007/978-1-4612-1438-0_1

Fischer, K., Ford, H., Abt, D., & Rychert, C. (2010). The lithosphere-asthenosphere boundary. *Annual Review of Earth and Planetary Sciences, 38*, 551–575.

Fischer, W., Hemp, J., & Valentine, J. (2016). How did life survive Earth's great oxygenation? *Current Opinion in Chemical Biology, 31*, 166–178.

Fischer-Gödde, M., & Becker, H. (2011). What is the age of the Nectaris basin? New Re-Os constraints for a pre-4.0 Ga bombardment history of the moon. *42nd Lunar and Planetary Science Conference*. Downloaded Feb. 15, 2017 from http://www.lpi.usra.edu/meetings/lpsc2011/pdf/1414.pdf

Fish, R., Goles, G., & Anders, E. (1960). The record in the meteorites. III. On the development of meteorites in asteroidal bodies. *Astrophysical Journal, 132*, 243–258.

Flanagan, M., & Shearer, P. (1998). Global mapping of topography on transition zone velocity discontinuities by stacking SS precursors. *Journal of Geophysical Research, 103*, 2673–2692.

Flasar, F., & Birch, F. (1973). Energetics of core formation: A correction. *Journal of Geophysical Research, 78*, 6101–6103.

Folland, C., Rayner, N., Brown, S., Smith, T., Shen, S., Parker, D., et al. (2001). Global temperature change and its uncertainties since 1861. *Geophysical Research Letters, 28*, 2621–2624.

Forsberg, R. (2016). Theoretical fundamentals of gravity field modelling using airborne, satellite and surface data, presentation at *NOAA Airborne Gravity for Geodesy Summer School*, May 2016. Downloaded April 2022 from https://www.ngs.noaa.gov/grav-d/2016SummerSchool/history-fundamentals.shtml

Förste, C., Bruinsma, S., Abrikosov, O., Lemoine, J., Schaller, T., Götze, H., et al. (2014). EIGEN-6C4: The latest combined global gravity field model including GOCE data up to degree and order 2190 of GFZ Potsdam and GRGS Toulouse. *5th GOCE User Workshop*. Paris. Downloaded from http://icgem.gfz-potsdam.de/Foerste-et-al-EIGEN-6C4.pdf

Forsyth, D., Webb, S., Dorman, L., & Shen, Y. (1998). Phase velocities of Rayleigh waves in the MELT experiment on the East Pacific Rise. *Science, 280*, 1235–1238.

Forte, A., & Mitrovica, J. (2001). Deep-mantle high-viscosity flow and thermochemical structure inferred from seismic and geodynamic data. *Nature, 410*, 1049–1056.

Forte, A., & Peltier, W. (1989). Core-mantle boundary topography and whole-mantle convection. *Geophysical Research Letters, 16*, 621–624.

Forte, A., & Peltier, W. (1991). Mantle convection and core-mantle boundary topography: Explanations and implications. *Tectonophysics, 187*, 91–116.

Fortes, A., Wood, I., Alfredsson, M., Vočadlo, L., & Knight, K. (2005). The incompressibility and thermal expansivity of D_2O ice II determined by powder neutron diffraction.

Journal of Applied Crystallography, *38*, 612–618.

Foukal, P., & Lean, J. (1990). An empirical model of total solar irradiance variation between 1874 and 1988. *Science*, *247*, 556–558.

Fowler, C. (2005). *The Solid Earth: An introduction to Global Geophysics* (2nd edition). Cambridge, England: Cambridge University Press.

Francis, J., & Vavrus, S. (2012). Evidence linking Arctic amplification to extreme weather in mid-latitudes. *Geophysical Research Letters*, *39*, L06801. doi:10.1029/2012GL051000

Franks, P., Royer, D., Beerling, D., Van de Water, P., Cantrill, D., Barbour, M., & Berry, J. (2014). New constraints on atmospheric CO_2 concentration for the Phanerozoic. *Geophysical Research Letters*, *41*, 4685–4694.

Fray, S., Diez, C., Hänsch, T., & Weitz, M. (2004). Atomic interferometer with amplitude gratings of light and its applications to atom based tests of the equivalence principle. *Physical Review Letters*, *93*, 240404 (arXiv:physics/0411052).

Frederikse, T., Jevrejeva, S., Riva, R., & Dangendorf, S. (2018). A consistent sea-level reconstruction and its budget on basin and global scales over 1958-2014. *Journal of Climate*, *31*, 1267–1280.

Frei, R., Gaucher, C., Poulton, S., & Canfield, D. (2009). Fluctuations in Precambrian atmospheric oxygenation recorded by chromium isotopes. *Nature*, *461*, 250–254.

Fretwell, P., Pritchard, H. D., Vaughan, D. G., Bamber, J. L., Barrand, N. E., Bell, R., et al. (2013). Bedmap2: Improved ice bed, surface and thickness datasets for Antarctica. *Cryosphere*, *7*, 375–393.

Frey, H. (1982). MAGSAT scalar anomaly distribution: The global perspective. *Geophysical Research Letters*, *9*, 277–280.

Fricker, H. (2010). Are the ice sheets melting? Response of Antarctica and Greenland to climate change. *Climate and National Security Conference, June 22*, 2010.

Frohlich, C. (1989). The nature of deep-focus earthquakes. *Annual Review of Earth and Planetary Sciences*, *17*, 227–254.

Fröhlich, C. (2009). Evidence of a long-term trend in total solar irradiance. *Astronomy & Astrophysics*, *501*, L27–L30. doi:10.1051/0004-6361/200912318

Frost, D. (2008). The upper mantle and transition zone. *Elements*, *4*, 171–176.

Frost, D., Lasbleis, M., Chandler, B., & Romanowicz, B. (2021). Dynamic history of the inner core constrained by seismic anisotropy. *Nature Geoscience*, *14*, 531–535.

Fu, R., Weiss, B., Shuster, D., Gattacceca, J., Grove, T., Suavet, C., et al. (2012). An ancient core dynamo in asteroid Vesta. *Science*, *338*, 238–241.

Fujisawa, H., Fujii, N., Mizutani, H., Kanamori, H., & Akimoto, S. (1968). Thermal diffusivity of Mg_2SiO_4, Fe_2SiO_4, and NaC1 at high pressures and temperatures. *Journal of Geophysical Research*, *73*, 4727–4733.

Fukao, Y., & Obayashi, M. (2013). Subducted slabs stagnant above, penetrating through, and trapped below the 660 km discontinuity. *Journal of Geophysical Research*, *118*, 5920–5938.

Fukao, Y., Nishida, K., & Kobayashi, N. (2010). Seafloor topography, ocean infragravity waves, and background Love and Rayleigh waves. *Journal of Geophysical Research*, *115*. doi:10.1029/2009JB006678

Fukao, Y., Nishida, K., Suda, N., Nawa, K., & Kobayashi, N. (2002). A theory of the Earth's background free oscillations. *Journal of Geophysical Research*, *107*. doi:10.1029/2001JB000153

Fukao, Y., Widiyantoro, S., & Obayashi, M. (2001). Stagnant slabs in the upper and lower mantle transition region. *Reviews of Geophysics*, *39*, 291–323.

Gaillard, F., & Iacono-Marziano, G. (2005). Electrical conductivity of magma in the course of crystallization controlled by their residual liquid composition. *Journal of Geophysical Research*, *110*, B06204. doi:10.1029/2004JB003282

Gaillard, F., Malki, M., Iacono-Marziano, G., Pichavant, M., & Scaillet, B. (2008). Carbonatite melts and electrical conductivity in the asthenosphere. *Science*, *322*, 1363–1365.

Gale, A., Dalton, C., Langmuir, C., Su, Y., & Schilling, J. (2013). The mean composition of ocean ridge basalts. *Geochemistry, Geophysics, Geosystems, 14*, 489–518.

Gallet, Y., Pavlov, V., & Korovnikov, I. (2019). Extreme geomagnetic reversal frequency during the Middle Cambrian as revealed by the magnetostratigraphy of the Khorbusuonka section (northeastern Siberia). *Earth and Planetary Science Letters, 528*. https://doi.org/10.1016/j.epsl.2019.115823

Gallet, Y., Pavlov, V., Halverson, G., & Hulot, G. (2012). Toward constraining the long-term reversing behavior of the geodynamo: A new "Maya" superchron ~1 billion years ago from the magnetostratigraphy of the Kartochka Formation (southwestern Siberia). *Earth and Planetary Science Letters, 339–340*, 117–126.

Gan, W., Frohlich, C., & Jin, Z. (2015). Origin and significance of deep earthquake clusters surrounding a pronounced seismic gap in northeast China. *Journal of Asian Earth Sciences, 100*, 91–97.

Gans, R. (1972). Viscosity of the Earth's core. *Journal of Geophysical Research, 77*, 360–366.

Gao, C., Robock, A., & Ammann, C. (2008). Volcanic forcing of climate over the past 1500 years: An improved ice core-based index for climate models. *Journal of Geophysical Research, 113*, D23111. doi:10.1029/2008JD010239

Garai, J., Haggerty, S., Rekhi, S., & Chance, M. (2006). Infrared absorption investigations confirm the extraterrestrial origin of carbonado diamonds. *Astrophysical Journal, 653*, L153–L156.

Garaud, P., & Lin, D. (2007). The effect of internal dissipation and surface irradiation on the structure of disks and the location of the snow line around sun-like stars, *Astrophysical Journal, 654*, 606–624.

Garcia, M., Smith, J., Tree, J., Weis, D., Harrison, L., & Jicha, B. (2015). Petrology, geochemistry, and ages of lavas from Northwest Hawaiian Ridge volcanoes. *Geological Society of America Special Paper, 511*, 1–25.

García-Berro, E., Isern, J., & Kubyshin, Y. (2007). Astronomical measurements and constraints on the variability of fundamental constants. *Astronomy and Astrophysics Review, 14*, 113–170.

Garfunkel, Z. (1975). Growth, shrinking, and long-term evolution of plates and their implications for the flow pattern in the mantle. *Journal of Geophysical Research, 80*, 4425–4432.

Garland, G. (1971). *Introduction to Geophysics: Mantle, core, and crust*. Philadelphia: Saunders Publishing.

Garland, G., (1979). The contributions of Carl Friedrich Gauss to geomagnetism. *Historia Mathematica, 6*, 5–29.

Garlick, S., Oren. A., & Padan, E. (1977). Occurrence of facultative anoxygenic photosynthesis among filamentous and unicellular cyanobacteria. *Journal of Bacteriology, 129*, 623–629.

Garrick-Bethell, I., Nimmo, F., & Wieczorek, M. (2012). Structure and formation of the lunar farside highlands. *Science, 330*, 949–951.

Garrison, T. (1993). *Oceanography: An invitation to Marine Science*. Belmont, CA: Wadsworth Pub. Co.

Gaspar, A., & Rieke, G. (2020). New HST data and modeling reveal a massive planetesimal collision around Fomalhaut. *Proceedings of the National Academy of Sciences, 117*, 9712–9722.

Gastine, T., Aubert, J., & Fournier, A. (2020). Dynamo-based limit to the extent of a stable layer atop Earth's core. *Geophysical Journal International, 222*, 1433–1448.

Gee, J., & Kent, D. (2007). 5.12 Source of oceanic magnetic anomalies and the geomagnetic polarity timescale. In M. Kono (Ed.), *Treatise on Geophysics, vol. 5: Geomagnetism* (pp. 455–507). Amsterdam: Elsevier.

Gekelman, W., Van Zeeland, M., Vincena, S., & Pribyl, P. (2003). Laboratory experiments on Alfvén waves caused by rapidly expanding plasmas and their relationship to space phenomena. *Journal of Geophysical Research, 108*. doi:10.1029/2002JA009741

Geller, R. (1977). Amplitudes of rotationally split normal modes for the 1960 Chilean and 1964 Alaskan earthquakes, *Ph.D. Thesis* (Part II), California Institute of Technology.

Genda, H., & Abe, Y. (2003). Survival of a proto-atmosphere through the stage of giant impacts: The mechanical aspects. *Icarus, 164,* 149–162.

Genda, H., & Abe, Y. (2005). Enhanced atmospheric loss on protoplanets at the giant impact phase in the presence of oceans. *Nature, 433,* 842–844.

Gerland, P., Raftery, A., Sevcíková, H., Li, N., Gu, D., Spoorenberg, T., et al. (2014). World population stabilization unlikely this century. *Science, 346,* 234–237.

Gessmann, C., & Wood, B. (2002). Potassium in the Earth's core? *Earth and Planetary Science Letters, 200,* 63–78.

Ghosh, A., & McSween, H., Jr. (1998). A thermal model for the differentiation of asteroid 4 Vesta, based on radiogenic heating. *Icarus, 134,* 187–206.

Giampieri, G., & Balogh, A. (2002). Mercury's thermoelectric dynamo model revisited. *Planetary and Space Science, 50,* 757–762.

Gibney, E. (2014). Force of nature gave life its asymmetry. *Nature.* doi:10.1038/nature. 2014.15995

Gilbert, F., & Dziewonski, A. (1975). An application of normal mode theory to the retrieval of structural parameters and source mechanisms from seismic spectra. *Philosophical Transactions of the Royal Society of London, A, 278,* 187–269.

Gilbert, W. (1600). *De Magnete (On the Magnet, Magnetick).* London: Chiswick Press (1900). Available at https://www.gutenberg.org/files/ 33810/33810-h/33810-h.htm#BV.6

Gillet, N., Jault, D., Canet, E., & Fournier, A. (2010). Fast torsional waves and strong magnetic field within the Earth's core. *Nature, 465,* 74–77.

Gillet, N., Jault, D., & Finlay, C. (2015). Planetary gyre, time-dependent eddies, torsional waves, and equatorial jets at the Earth's core surface. *Journal of Geophysical Research, 120,* 3991–4013.

Gillet, N., Pais, M., & Jault, D. (2009). Ensemble inversion of time-dependent core flow models. *Geochemistry, Geophysics, Geosystems, 10,* Q06004. doi:10.1029/2008GC002290

Gillies, G. (1997). The Newtonian gravitational constant: recent measurements and related studies. *Reports on Progress in Physics, 60,* 151–225.

Gillon, M., Jehin, E., Lederer, S., Delrez, L., de Wit, J., Burdanov, A., et al. (2016). Temperate Earth-sized planets transiting a nearby ultracool dwarf star. *Nature, 533,* 221–224 and Supplemental Methods.

Gillon, M., Triaud, A. H., Demory, B. O., Jehin, E., Agol, E., Deck, K. M., et al. (2017). Seven temperate terrestrial planets around the nearby ultracool dwarf star TRAPPIST-1. *Nature, 542,* 456–460 and Supplemental Methods.

Girdler, R., Taylor, P., & Frawley, J. (1992). A possible impact origin for the Bangui magnetic anomaly (Central Africa). *Tectonophysics, 212,* 45–58.

Gizon, L., & Birch, A. (2005). Local helioseismology. *Living Reviews in Solar Physics, 2,* 131 pp., online article cited January 2016.

Glantz, M. (2001). *Currents of change: Impacts of El Niño and La Niña on climate and society.* Cambridge, England: Cambridge University Press.

Glatzmaier, G., & Roberts, P. (1995). A three-dimensional convective dynamo solution with rotating and finitely conducting inner core and mantle. *Physics of the Earth and Planetary Interiors, 91,* 63–75.

Glavin, D., Callahan, M., Dworkin, J., & Elsila, J. (2011). The effects of parent body processes on amino acids in carbonaceous chondrites. *Meteoritics & Planetary Science, 45,* 1948–1972.

Glover, P. (2015). Geophysical properties of the near surface Earth: Electrical properties. In G. Schubert (Ed.), *Treatise on Geophysics* (2nd edition, pp. 89–137). Amsterdam: Elsevier.

Goes, S., Agrusta, R., van Hunen, J., & Garel, F. (2017). Subduction-transition zone interaction: A review. *Geosphere, 13,* 644–664.

Gold, T. (1992). The deep, hot biosphere. *Proceedings of the National Academy of Sciences, 89,* 6045–6049.

Gold, T., & Soter, S. (1969). Atmospheric tides and the resonant rotation of Venus. *Icarus, 11*, 356–366.

Goldbogen, J., & Madsen, P. (2018). The evolution of foraging capacity and gigantism in cetaceans. *Journal of Experimental Biology, 221*, jeb166033. doi:10.1242/jeb.166033

Goldbogen, J., Cade, D. E., Wisniewska, D. M., Potvin, J., Segre, P. S., Savoca, M. S., et al. (2019). Why whales are big but not bigger: Physiological drivers and ecological limits in the age of ocean giants. *Science, 366*, 1367–1372.

Goldreich, P., & Keeley, D. (1977). Solar seismology. II. The stochastic excitation of the solar *p*-modes by turbulent convection. *Astrophysical Journal, 212*, 243–251.

Goldreich, P., & Kumar, P. (1990). Wave generation by turbulent convection. *Astrophysical Journal, 363*, 694–704.

Goldreich, P., & Tremaine, S. (1980). Disk-satellite interactions. *Astrophysical Journal, 241*, 425–441.

Goldsmith, D. (1976). *The Universe*. Menlo Park: W.A. Benjamin, Inc.

Gomes, R., Deienno, R., & Morbidelli, A. (2016). The inclination of the planetary system relative to the solar equator may be explained by the presence of Planet 9. *Astronomical Journal*. Preprint downloaded as arXiv:1607.05111v1 from https://arxiv.org/abs/1607.05111

Gomes, R., Levison, H., Tsiganis, K., & Morbidelli, A. (2005). Origin of the cataclysmic Late Heavy Bombardment period of the terrestrial planets. *Nature, 435*, 466–469.

Gomi, H., & Yoshino, T. (2018). Impurity resistivity of fcc and hcp Fe-based alloys: Thermal stratification at the top of the core of super-Earths. *Frontiers of Earth Science, 6*. doi:10.3389/feart.2018.00217

Goncharov, A., Haugen, B., Struzhkin, V., Beck, P., & Jacobsen, S. (2008). Radiative conductivity in the Earth's lower mantle. *Nature, 456*, 231–234.

Goncharov, A., Lobanov, S., Tan, X., Hohensee, G., Cahill, D., Lin, J., et al. (2015). Experimental study of thermal conductivity at high pressures: Implications for the deep Earth's interior. *Physics of the Earth and Planetary Interiors, 214*, 11–16.

Goncharov, A., Struzhkin, V., & Jacobsen, S. (2006). Reduced radiative conductivity of low-spin (Mg, Fe)O in the lower mantle. *Science, 312*, 1205–1208.

Gonnermann, H., Jellinek, M., Richards, M., & Manga, M. (2004). Modulation of mantle plumes and heat flow at the core mantle boundary by plate-scale flow: Results from laboratory experiments. *Earth and Planetary Science Letters, 226*, 53–67.

Gonzalez, P., Neilson, R., Lenihan, J., & Drapek, R. (2010). Global patterns in the vulnerability of ecosystems to vegetation shifts due to climate change. *Global Ecology and Biogeography, 19*, 755–768. doi:10.1111/j.1466-8238.2010.00558.x and also Supporting Information, Appendix S4.

Gordon, A. (1986). Interocean exchange of thermocline water. *Journal of Geophysical Research, 91*, 5037–5046.

Gordon, R. (1998). The plate tectonic approximation: Plate nonrigidity, diffuse plate boundaries, and global plate reconstructions. *Annual Review of Earth and Planetary Sciences, 26*, 615–642.

Gordon, R., & Jurdy, D. (1986). Cenozoic global plate motions. *Journal of Geophysical Research, 91*, 12,389–12,406.

Gordon, R., DeMets, C., & Argus, D. (1990). Kinematic constraints on distributed lithospheric deformation in the equatorial Indian ocean from present motion between the Australian and Indian plates. *Tectonics, 9*, 409–422.

Gordon, R., Stein, S., DeMets, C., & Argus, D. (1987). Statistical tests for closure of plate motion circuits. *Geophysical Research Letters, 14*, 587–590.

Gore, A. (2006). *An Inconvenient Truth*. Warner Bros.

Gough, D. (2010). Vainu Bappu Memorial Lecture: What is a sunspot? In S. Hassan & R. Rutten (Eds.), *Magnetic coupling between the interior and atmosphere of the Sun,*

Astrophysics and space science proceedings (pp. 37–66). Berlin: Springer.

Gough, D., Kosovichev, A. G., Toomre, J., Anderson, E., Antia, H. M., Basu, S., et al. (1996). The seismic structure of the Sun. *Science, 272,* 1296–1300.

Gould, J. (2010). Magnetoreception. *Current Biology, 20,* R431–R435.

Gould, W. (1985). Physical oceanography of the Azores front. *Progress in Oceanography, 14,* 167–190.

Graf, H., & Zanchettin, D. (2012). Central Pacific El Niño, the "subtropical bridge," and Eurasian climate. *Journal of Geophysical Research, 117.* D01102. doi:10.1029/2011JD016493

Graham, J., Aguilar, N., Dudley, R., & Gans, C. (1995). Implications of the late Palaeozoic oxygen pulse for physiology and evolution. *Nature, 375,* 117–120.

Grand, S. (1994). Mantle shear structure beneath the Americas and surrounding oceans. *Journal of Geophysical Research, 99,* 11,591–11,621.

Grand, S. (2002). Mantle shear-wave tomography and the fate of subducted slabs. *Philosophical Transactions of the Royal Society of London, A, 360,* 2475–2491.

Grand, S., van der Hilst, R., & Widiyantoro, S. (1997). Global seismic tomography: A snapshot of convection in the Earth. *GSA Today, 7,* 1–7.

Grandstaff, D. (1980). Origin of uriniferous conglomerates at Elliot Lake, Canada and Witwatersrand, South Africa: Implications for oxygen in the Precambrian atmosphere. *Precambrian Research, 13,* 1–26.

Graner, F., & Dubrulle, B. (1994). Titius- Bode laws in the solar system. *Astronomy & Astrophysics, 282,* 262–268.

Granot, R., Dyment, J., & Gallet, Y. (2012). Geomagnetic field variability during the Cretaceous Normal Superchron. *Nature Geoscience, 5,* 220–223. doi:10.1038/NGEO1404

Gray, W. (1984). Atlantic seasonal hurricane frequency. Part I: El Nino and 30 mb Quasi-Biennial Oscillation influences. *Monthly Weather Review, 112,* 1649–1668.

Grayver, A., Munch, F., Kuvshinov, A., Khan, A., Sabaka, T., & Tøffner-Clausen, L. (2017). Joint inversion of satellite-detected tidal and magnetospheric signals constrains electrical conductivity and water content of the upper mantle and transition zone. *Geophysical Research Letters, 44,* 6074–6081. doi:10.1002/2017GL073446

Grayver, A., Schnepf, N., Kuvshinov, A., Sabaka, T., Manoj, C., & Olsen, N. (2016). Satellite tidal magnetic signals constrain oceanic lithosphere-asthenosphere boundary. *Science Advances, 2,* e1600798. doi:10.1126/sciadv.1600798

Grayver, A., & Olsen, N. (2019). The magnetic signatures of the M_2, N_2, and O_1 oceanic tides observed in Swarm and CHAMP satellite magnetic data. *Geophysical Research Letters, 46,* 4230–4238. https://doi.org/10.1029/2019GL082400

Green, A. (1976). Interpretation of Project MAGNET aeromagnetic profiles across Africa. *Geophysical Journal of the Royal Astronomical Society, 44,* 203–228.

Grevemeyer, I., Herber, R., & Essen, H. (2000). Microseismological evidence for a changing wave climate in the northeast Atlantic Ocean. *Nature, 408,* 349–352.

Grijalva, K., Bürgmann, R., & Banerjee, P. (2011). Using geodetic data to understand postseismic processes following the Sumatra-Andaman earthquake. *Berkeley Seismological Laboratory Annual Report July 2009-June 2010.* Accessed Feb. 8, 2018 from https://seismo.berkeley.edu/annual_report/ar09_10/node6.html

Grima, J., Manicaro, E., & Attard, D. (2011). Auxetic behaviour from connected different-sized squares and rectangles. *Proceedings of the Royal Society A, 467,* 439–458.

Grooms, I., Julien, K., Weiss, J., & Knobloch, E. (2010). Model of convective Taylor columns in rotating Rayleigh-Bénard convection. *Physical Review Letters, 104,* 224501.

Gross, R. (2001). A combined length-of-day series spanning 1832–1997: LUNAR97. *Physics of the Earth and Planetary Interiors, 123*, 65–76.

Gross, R. (2007). 3.09 Earth rotation variations – Long period. In T. Herring (Ed.), *Treatise on Geophysics, Vol. 3: Geodesy* (pp. 239–294). Amsterdam: Elsevier.

Gross, R., Fukumori, I., & Menemenlis, D. (2003). Atmospheric and oceanic excitation of the Earth's wobbles during 1980–2000. *Journal of Geophysical Research, 108*, 2370. doi:10.1029/2002JB002143

Gross, R., & Vondrak, J. (1999). Astrometric and space-geodetic observations of polar wander. *Geophysical Research Letters, 26*, 2085–2088.

Grossman, J. (2017). Meteoritical Bulletin Database. Accessed on March 29, 2017 at https://www.lpi.usra.edu/meteor/metbull .php; website maintained by The Meteoritical Society and the Lunar and Planetary Institute.

Grossman, L. (1977). Chemical fractionation in the solar nebula. In J. Pomeroy, & N. Hubbard (Eds.), *The Soviet-American Conference on Cosmochemistry of the Moon and Planets, Part 2* (pp. 787–796). Washington, DC: NASA.

Grote, E., & Busse, F. (2000). Hemispherical dynamos generated by convection in rotating spherical shells. *Physical Review E, 62*, 4457–4460.

Groten, E. (2004). Fundamental parameters and current (2004) best estimates of the parameters of common relevance to astronomy, geodesy, and geodynamics. *Journal of Geodesy, 77*, 724–797.

Grünwaldt, H., Neugebauer, M., Hilchenbach, M., Bochsler, P., Hovestadt, D., Bürgi, A., et al. (1997). Venus tail ray observation near Earth. *Geophysical Research Letters, 24*, 1163–1166.

Guarnieri, M. (2014). Once upon a time…the compass. *IEEE Industrial Electronics Magazine, 8*(2), 60–63.

Gubbins, D. (1981). Rotation of the inner core. *Journal of Geophysical Research, 86*, 11695–11699.

Gubbins, D., Alfè, D., Masters, G., Price, G., & Gillan, M. (2003). Can the Earth's dynamo run on heat alone? *Geophysical Journal International, 155*, 609–622.

Gubbins, D., Alfè, D., Masters, G., Price, G., & Gillan, M. (2004). Gross thermodynamics of two-component core convection. *Geophysical Journal International, 157*, 1407–1414.

Gubbins, D., Jones, A., & Finlay, C. (2006). Fall in Earth's magnetic field is erratic. *Science, 312*, 900–902.

Gudmundsson, G., Krug, J., Durand, G., Favier, L., & Gagliardini, O. (2012). The stability of grounding lines on retrograde slopes. *The Cryosphere, 6*, 1497–1505.

Guervilly, C., Cardin, P., & Schaeffer, N. (2019). Turbulent convective length scale in planetary cores. *Nature, 570*, 368–371.

Gülcher, A., Gerya, T., Montési, L., & Munch, J. (2020). Corona structures driven by plume-lithosphere interactions and evidence for ongoing plume activity on Venus. *Nature Geoscience, 13*, 547–554.

Gurnis, M., Hall, C., & Lavier, L. (2004). Evolving force balance during incipient subduction. *Geochemistry, Geophysics, Geosystems, 5*, Q07001. doi:10.1029/2003GC000681

Gutenberg, B. (1947). Microseisms and weather forecasting. *Journal of Meteorology, 4*, 21–28.

Gutenberg, B., & Richter, C. (1936). Materials for the study of deep-focus earthquakes. *Bulletin of the Seismological Society of America, 26*, 341–390.

Gwinn, C., Herring, T., & Shapiro, I. (1986). Geodesy by radio interferometry: Studies of the forced nutations of the Earth 2. *Interpretation. Journal of Geophysical Research, 91*, 4755–4765.

Gyory, J., Mariano, A., & Ryan, E. (2013). The canary current. In *Surface Currents in the Atlantic Ocean*. Downloaded May 10, 2018 from http://oceancurrents.rsmas.miami.edu/ atlantic/canary.html

Haeni, F. (1986). Application of seismic refraction methods in groundwater modeling studies in New England. *Geophysics, 51*, 236–249.

Haeni, F. (1988). *Techniques of Water-Resources Investigations of the United States Geological Survey, Book 2: Collection of environmental data*. USGS Pub., U.S. Government Printing Office: Washington, DC. Chapter D2,

Application of seismic-refraction techniques to hydrologic studies.

Hager, B., & Clayton, R. (1989). Constraints on the structure of mantle convection using seismic observations, flow models, and the geoid. In W. Peltier (Ed.), *Mantle convection: Plate tectonics and global dynamics* (pp. 657–763). New York, NY: Gordon & Breach Science Publishers.

Hager, B., & Richards, M. (1989). Long-wavelength variations in Earth's geoid: Physical models and dynamical implications. *Philosophical Transactions of the Royal Society of London, A, 328*, 309–327.

Hager, B., Clayton, R., Richards, M., Comer, R., & Dziewonski, A. (1985). Lower mantle heterogeneity, dynamic topography and the geoid. *Nature, 313*, 541–545.

Hahn, A., Ahrendt, H., Meyer, J., & Hufen, J. (1984). A model of magnetic sources within the Earth's crust compatible with the field measured by the satellite Magsat. *Geologisches Jahrbuch der BGR, Hannover, 75*, 125–156. (cited in Jackson, A. (1990). Accounting for crustal magnetization in models of the core magnetic field. *Geophysical Journal International, 103*, 657–673.)

Hakkinen, S. (1995). Simulated interannual variability of the Greenland Sea deep water formation and its connection to surface forcing. *Journal of Geophysical Research, 100*, 4761–4770.

Hale, C. (1987). The intensity of the geomagnetic field at 3.5 Ga: Paleointensity results from the Komati Formation, Barberton Mountain Land, South Africa. *Earth and Planetary Science Letters, 86*, 354–364.

Hale, C., & Dunlop, D. (1984). Evidence for an early Archean geomagnetic field: a paleomagnetic study of the Komati formation, Barberton greenstone belt, South Africa. *Geophysical Research Letters, 11*, 97–100.

Halevy, I., Fischer, W., & Eiler, J. (2011). Carbonates in the Martian meteorite Allan Hills 84001 formed at $18 \pm 4\,°C$ in a near-surface aqueous environment. *Proceedings of the National Academy of Sciences, 108*, 16895–16899.

Hall, H., & Murthy, V. (1971). The early chemical history of the Earth: Some critical elemental fractionations. *Earth and Planetary Science Letters, 11*, 239–244.

Halpern, A., & Erlbach, E. (1998). *Theory and problems of beginning Physics II* (Schaum Outline Series). New York, NY: McGraw-Hill.

Haltiner, G., & Martin, F. (1957). *Dynamical and Physical Meteorology*. New York, NY: McGraw-Hill.

Hambach, U., Rolf, C., & Schnepp, E. (2008). Magnetic dating of Quaternary sediments, volcanites and archaeological materials: An overview. *E&G Quaternary Science Journal, 57*, 25–51.

Hamilton, D., Gloeckler, G., Ipavich, F., Stüdemann, W., Wilken, B., & Kremser, G. (1988). Ring current development during the great geomagnetic storm of February 1986. *Journal of Geophysical Research, 93*, 14343–14355.

Hammer, S. (1939). Terrain corrections for gravimeter stations. *Geophysics, 4*, 184–194.

Han, D., & Wahr, J. (1995). The viscoelastic relaxation of a realistically stratified Earth, and a further analysis of postglacial rebound. *Geophysical Journal International, 120*, 287–311.

Han, T., & Runnegar, B. (1992). Megascopic eukaryotic algae from the 2.1-billion-year-old negaunee iron-formation, Michigan. *Science, 257*, 232–235.

Handler, P. (1984). Possible association of stratospheric aerosols and El Nino type events. *Geophysical Research Letters, 11*, 1121–1124.

Hansen, B. (2009). Formation of the terrestrial planets from a narrow annulus. *Astrophysical Journal, 703*, 1131–1140.

Hansen, J., & Lebedeff, S. (1987). Global trends of measured surface air temperature. *Journal of Geophysical Research, 92*, 13,345–13,372.

Hansen, J., Ruedy, R., Sato, M., & Lo, K. (2010). Global surface temperature change. *Reviews of Geophysics, 48*, RG4004. doi:10.1029/2010 RG000345

Hansen, J., Sato, M., Ruedy, R., Lo, K., Lea, D., & Medina-Elizade, M. (2006). Global temperature change. *Proceedings of the*

National Academy of Sciences, *103*, 14,288–14,293.

Hansen, L., Zimmerman, M., & Kohlstedt, D. (2011). Grain boundary sliding in San Carlos olivine: Flow law parameters and crystallographic-preferred orientation. *Journal of Geophysical Research*, *116*, B08201. doi:10.1029/2011JB008220

Haqq-Misra, J., Domagal-Goldman, S., Kasting, P., & Kasting, J. (2008). A revised, hazy methane greenhouse for the Archean Earth. *Astrobiology*, *8*, 1127–1137.

Hardy, C., Livermore, P., & Niesen, J. (2020). Enhanced magnetic fields within a stratified layer. *Geophysical Journal International*, *222*, 1686–1703.

Hare, J., Furgeson, J., & Brady, J. (2008). The 4D microgravity method for waterflood surveillance: Part IV — Modeling and interpretation of early epoch 4D gravity surveys at Prudhoe Bay, Alaska. *Geophysics*, *73*, WA173–WA180.

Harkrider, D., & Press, F. (1967). The Krakatoa air-sea waves: An example of pulse propagation in coupled systems. *Geophysical Journal of the Royal Astronomical Society*, *13*, 149–159.

Harper, D., Hammarlund, E., & Rasmussen, C. (2014). End Ordovician extinctions: A coincidence of causes. *Gondwana Research*, *25*, 1294–1307.

Harris, N., Hassler, B., Tummon, F., Bodeker, G. E., Hubert, D., Petropavlovskikh, I., et al. (2015). Past changes in the vertical distribution of ozone – Part 3: Analysis and interpretation of trends. *Atmospheric Chemistry and Physics*, *15*, 9965–9982.

Harrison, C. (2000). Questions about magnetic lineations in the ancient crust of Mars (Technical Comment). *Science*, *287*, 547a.

Harrison, D., & Schopf, P. (1984). Kelvin-wave-induced anomalous advection and the onset of surface warming in El Niño events. *Monthly Weather Review*, *112*, 923–933.

Harvey, B., Shaffrey, L., & Woollings, T. (2014). Equator-to-pole temperature differences and the extra-tropical storm track responses of the CMIP5 climate models, *Climate Dynamics*, *43*, 1171–1182.

Hasselmann, K. (1963). A statistical analysis of the generation of microseisms. *Reviews of Geophysics*, *1*, 177–210.

Hasterok, D., & Webb, J. (2017). On the radiogenic heat production of igneous rocks. *Geoscience Frontiers*, *8*, 919–940.

Hathaway, D. (2015). The solar cycle. *Living Reviews in Solar Physics*, *12*. doi:10.1007/lrsp-2015-4

Haug, G., & Tiedemann, R. (1998). Effect of the formation of the Isthmus of Panama on Atlantic Ocean thermohaline circulation. *Nature*, *393*, 673–676.

Hayes, W., & Tremaine, S. (1998). Fitting selected random planetary systems to Titius–Bode laws. *Icarus*, *135*, 549–557.

Hays, J., Imbrie, J., & Shackleton, N. (1976). Variations in the Earth's orbit: Pacemaker of the ice ages. *Science*, *194*, 1121–1132.

Hays, M. (2020). Nyack people & places: Balance rock. *Nyack News & Views*, Sept. 24, 2020. Downloaded from https://nyacknewsandviews.com/2020/09/nyack-people-places-balance-rock/

Head, J., Bloch, J., Hastings, A., Bourque, J., Cadena, E., Herrera, F., et al. (2009). Giant boid snake from the Palaeocene neotropics reveals hotter past equatorial temperatures. *Nature*, *457*, 715–717.

Head, J., Mustard, J., Kreslavsky, M., Milliken, R., & Marchant, J. (2003). Recent ice ages on Mars. *Nature*, *426*, 797–802.

Hearty, P., Kindler, P., Cheng, H., & Edwards, R. (1999). A +20 m middle Pleistocene sea-level highstand (Bermuda and the Bahamas) due to partial collapse of Antarctic ice. *Geology*, *27*, 375–378.

Hedlin, M., Drob, D., Walker, K., & de Groot-Hedlin, C. (2010). A study of acoustic propagation from a large bolide in the atmosphere with a dense seismic network. *Journal of Geophysical Research*, *115*, B11312. doi:10.1029/2010JB007669

Hedlin, M., Walker, K., Drob, D., & de Groot-Hedlin, C. (2012). Infrasound:

Connecting the solid earth, oceans, and atmosphere. *Annual Review of Earth and Planetary Sciences*, *40*, 327–354.

Hegler, F., Posth, N., Jiang, J., & Kappler, A. (2008). Physiology of phototrophic iron(II)-oxidizing bacteria: Implications for modern and ancient environments. *FEMS Microbiology Ecology*, *66*, 250–260.

Heiskanen, W., & Moritz, H. (1967). *Physical Geodesy*. San Francisco, CA: Freeman & Co.

Hellings, R., Adams, P., Anderson, J., Keesey, M., Lau, E., Standish, E., et al. (1983). Experimental test of the variability of *G* using Viking Lander ranging data. *Physical Review Letters*, *51*, 1609–1612.

Helliwell, R. (1965). *Whistlers and related ionospheric phenomena*. Redwood City, CA: Stanford University Press.

Henderson, G., & Bateson, S. (1934). A quantitative study of pleochroic haloes. – I. *Proceedings of the Royal Society London, A*, *145*, 563–581.

Henry, L., McManus, J., Curry, W., Roberts, N., Piotrowski, A., & Keigwin, L. (2019). North Atlantic ocean circulation and abrupt climate change during the last glaciation. *Science*, *353*, 470–474.

Heppenheimer, T. (1980). Secular resonances and the origin of eccentricities of Mars and the Asteroids. *Icarus*, *41*, 76–88.

Herman, G., & Johnson, W. (1978). The sensitivity of the general circulation to Arctic sea ice boundaries: A numerical experiment. *Monthly Weather Review*, *106*, 1649–1664.

Herndon, J., & Herndon, M. (1977). Aluminum-26 as a planetoid heat source in the early solar system. *Meteoritics*, *12*, 459–465.

Hernlund, J., Thomas, C., & Tackley, P. (2005). A doubling of the post-perovskite phase boundary and structure of the Earth's lowermost mantle. *Nature*, *434*, 882–886.

Herring, T., Shapiro, I., Clark, T., Ma, C., Ryan, J., Schupler, B., et al. (1986). Geodesy by radio interferometry: Evidence for contemporary plate motion. *Journal of Geophysical Research*, *91*, 8341–8347.

Herrmann, R. (2013). Computer programs in seismology: An evolving tool for instruction and research. *Seismological Research Letters*, *84*, 1081–1088.

Hesse, M., & Cassak, P. (2020). Magnetic reconnection in the space sciences: Past, present, and future. *Journal of Geophysical Research*, *125*. doi.org/10.1029/2018JA025935

Hessler, A., Lowe, D., Jones, R., & Bird, D. (2004). A lower limit for atmospheric carbon dioxide levels 3.2 billion years ago. *Nature*, *428*, 736–738.

Hester, J., Burstein, D., Blumenthal, G., Greeley, R., Smith, B., Voss, H., & Wegner, G. (2002). *21st Century Astronomy*. New York, NY: W. Norton & Co.

Heyl, P., & Chrzanowski, P. (1942). A new determination of the constant of gravitation. *Journal of Research of the National Bureau of Standards*, *29*, 1–31.

Hide, R. (1966). Free hydromagnetic oscillations of the Earth's core and the theory of the geomagnetic secular variation. *Philosophical Transactions of the Royal Society A*, *259*, 615–650.

Hide, R., & Malin, S. (1970). Novel correlations between global features of the Earth's gravitational and magnetic fields. *Nature*, *225*, 605–609.

Hide, R., & Palmer, T. (1982). Generalization of Cowling's Theorem. *Geophysical & Astrophysical Fluid Dynamics*, *19*, 301–309.

Hiesinger, H., Head, J., III, Wolf, U., Jaumann, R., & Neukum, G. (2011). Ages and stratigraphy of lunar mare basalts: A synthesis. In W. Ambrose & D. Williams (Eds.), *Recent advances and current research issues in lunar stratigraphy* (Geol. Soc. America Spec. Pap. 477, pp. 1–51). doi:10.1130/2011.2477(01)

Hiesinger, H., Jaumann, R., Neukum, G., & Head, J., III. (2000). Ages of mare basalts on the lunar nearside. *Journal of Geophysical Research*, *105*, 29239–29275.

Hildenbrand, T., Berger, B., Jachens, R., & Ludington, S. (2000). Regional crustal structures and their relationship to the distribution of ore deposits in the Western

United States, based on magnetic and gravity data. *Economic Geology, 95*, 1583–1603.

Hill, F., Stark, P. B., Stebbins, R. T., Anderson, E. R., Antia, H. M., Brown, T. M., et al. (1996). The solar acoustic spectrum and eigenmode parameters, *Science, 272*, 1292–1295.

Hillier, J. (2005). *The Bathymetry of the Pacific Ocean Basin and its Tectonic Implications.* Ph.D. Thesis, University of Oxford, 330 pp. Downloaded Nov. 2017 from https://www.researchgate.net/profile/John_Hillier2/publication/33688132_The_bathymetry_of_the_Pacific_Ocean_Basin_and_its_tectonic_implications/links/558003e608ae26eada90729b.pdf

Hines, D., Backman, D. E., Bouwman, J., Hillenbrand, L. A., Carpenter, J. M., Meyer, M. R., et al. (2006). The formation and evolution of planetary systems (FEPS): Discovery of an unusual debris system associated with HD 12039. *Astrophysical Journal, 638*, 1070–1079.

Hinnov, L., & Wilson, C. (1987). An estimate of the water storage contribution to the excitation of polar motion. *Geophysical Journal of the Royal Astronomical Society, 88*, 437–459.

Hirose, K., & Lay, T. (2008). Discovery of post-perovskite and new views on the core-mantle boundary region. *Elements, 4*, 183–189.

Hirose, K., Sinmyo, R., Sata, N., & Ohishi, Y. (2006). Determination of post-perovskite phase transition boundary in $MgSiO_3$ using Au and MgO pressure standards. *Geophysical Research Letters, 33*, L01310. doi:10.1029/2005GL024468

Hirschmann, M. (2000). Mantle solidus: Experimental constraints and the effects of peridotite composition. *Geochemistry, Geophysics, Geosystems, 1*, 2000GC000070.

Hirschmann, M. (2010). Partial melt in the oceanic low velocity zone. *Physics of the Earth and Planetary Interiors, 179*, 60–71.

Hirschmann, M., & Withers, A. (2008). Ventilation of CO2 from a reduced mantle and consequences for the early Martian greenhouse. *Earth and Planetary Science Letters, 270*, 147–155.

Hirth, G., & Kohlstedt, D. (1996). Water in the oceanic upper mantle: implications for rheology, melt extraction and the evolution of the lithosphere. *Earth and Planetary Science Letters, 144*, 93–108.

Hirth, G., Evans, R., & Chave, A. (2000). Comparison of continental and oceanic mantle electrical conductivity: Is the Archean lithosphere dry? *Geochemistry, Geophysics, Geosystems, 1, Paper number* 2000GC000048.

Hodgskiss, M., Crockford, P., Peng, Y., Wing, B., & Horner, T. (2019). A productivity collapse to end Earth's Great Oxidation. *Proceedings of the National Academy of Sciences, 116*, 17207–17212.

Hoerling, M., & Ting, M. (1994). Organization of extratropical transients during El Niño. *Journal of Climate, 7*, 745–766.

Hoffman, P, Kaufman, A., Halverson, G., & Schrag, D. (1998). A Neoproterozoic snowball Earth. *Science, 281*, 1342–1346.

Hoffman, P., Abbot, D. S., Ashkenazy, Y., Benn, D. I., Brocks, J. J., Cohen, P. A., et al. (2017). Snowball Earth climate dynamics and Cryogenian geology-geobiology. *Science Advances, 3*, e1600983.

Hofmann, A. (2003). Just add water. *Nature, 425*, 24–25.

Hofmann, M., & Morales Maqueda, M. (2009). Geothermal heat flux and its influence on the oceanic abyssal circulation and radiocarbon distribution. *Geophysical Research Letters, 36*, L03603. doi:10.1029/2008GL036078

Hofmeister, A. (2005). Dependence of diffusive radiative transfer on grain-size, temperature, and Fe-content: Implications for mantle processes. *Journal of Geodynamics, 40*, 51–72.

Holdahl, S., & Sauber, J. (1994). Coseismic slip in the 1964 Prince William Sound earthquake: A new geodetic inversion. *Pure and Applied Geophysics, 142*, 55–82.

Holland, H. (1973). The Oceans: A possible source of iron in iron-formations. *Economic Geology, 68*, 1169–1172.

Holland, H. (1999). When did the Earth's atmosphere become oxic? A reply. *The Geochemical News, 100*, 20–22.

Holland, H. (2002). Volcanic gases, black smokers, and the Great Oxidation Event. *Geochimica et Cosmochimica Acta, 66,* 3811–3826.

Holland, H. (2006). The oxygenation of the atmosphere and oceans. *Philosophical Transactions of the Royal Society, B, 361,* 903–915.

Hollerbach, R., & Jones, C. (1995). On the magnetically stabilizing role of the Earth's inner core. *Physics of the Earth and Planetary Interiors, 87,* 171–181.

Holme, R. (1998). Electromagnetic core-mantle coupling II: Probing deep mantle conductance. In M. Gurnis, M. Wysession, E. Knittle, & B. Buffett (Eds.), *The core–mantle boundary region* (pp. 139–151). American Geophysical Union.

Holme, R. (2015). Chapter 8.04: Large-scale flow in the core. In G. Schubert (Ed.), *Treatise on Geophysics* (2nd edition, Vol. 8). Amsterdam: Elsevier, http://dx.doi.org/10.1016/B978-0-444-53802-4.00138-X

Holme, R., & de Viron, O. (2013). Characterization and implications of intradecadal variations in length of day. *Nature, 499,* 202–205.

Holme, R., Olsen, N., & Bairstow, F. (2011). Mapping geomagnetic secular variation at the core–mantle boundary. *Geophysical Journal International, 186,* 521–528.

Holzer, T., & Bennett, M. (2003). Technical note: Unsaturation beneath a water table. *Environmental & Engineering Geoscience, 9,* 379–385.

Honda, S. (1986). Strong anisotropic flow in a finely layered asthenosphere. *Geophysical Research Letters, 13,* 1454–1457.

Honda, S., Balachandar, S., Yuen, D., & Reuteler, D. (1993). Three-dimensional mantle dynamics with an endothermic phase transition. *Geophysical Research Letters, 20,* 221–224.

Hoover, R., Maciejewski, A., & Roberts, R. (2009). Eigendecomposition of images correlated on S^1, S^2, and $SO(3)$ using spectral theory. *IEEE Transactions on Image Processing, 18,* 2562–2571.

Hori, K., Jones, C., & Teed, R. (2015). Slow magnetic Rossby waves in the Earth's core. *Geophysical Research Letters, 42,* 6622–6629. doi:10.1002/2015GL064733

Horner-Johnson, B., Gordon, R., & Argus, D. (2007). Plate kinematic evidence for the existence of a distinct plate between the Nubian and Somalian plates along the Southwest Indian Ridge. *Journal of Geophysical Research, 112,* B05418. doi:10.1029/2006JB004519

Hough, S., Armbruster, J., & Seeber, L. (2000). On the Modified Mercalli intensities and magnitudes of the 1811–1812 New Madrid earthquakes. *Journal of Geophysical Research, 105,* 23,839–23,864.

Hounslow, M. (2016). Geomagnetic reversal rates following Palaeozoic superchrons have a fast restart mechanism. *Nature Communications, 7.* doi:10.1038/ncomms12507

House, M. (1995). Orbital forcing timescales: an introduction. *Geological Society of London, Special Publication, 85,* 1–18.

Houseman, G. (1983). Large aspect ratio convection cells in the upper mantle. *Geophysical Journal of the Royal Astronomical Society, 75,* 309–334.

Hoversten, M., Morrison, F., & Constable, S. (1998). Marine magnetotellurics for petroleum exploration, Part II: Numerical analysis of subsalt resolution. *Geophysics, 63,* 826–840.

Howell, E. (2022). Halley's Comet: Facts about history's most famous comet, Jan. 13, 2022. Downloaded from https://www.space.com/19878-halleys-comet.html

Hoyle, F., & Wickramasinghe, C. (1980). The origin of life. In Hoyle, F. (Ed.), *Evolution from space (The Omni Lecture and Other Papers on the Origin of Life)* (pp. 35–52). New Jersey: Enslow Publishers.

Huang, C., Foster, J., & Kelley, M. (2005). Long-duration penetration of the interplanetary electric field to the low-latitude ionosphere during the main phase of magnetic storms. *Journal of Geophysical Research, 110,* A11309. doi:10.1029/2005JA011202

Huang, H., Lin, F., Schmandt, B., Farrell, J., Smith, R., & Tsai, V. (2015). The Yellowstone magmatic system from the mantle plume to the upper crust. *Science, 348*, 773–776.

Huang, J., & McElroy, M. (2014). Contributions of the Hadley and Ferrel circulations to the energetics of the atmosphere over the past 32 years. *Journal of Climate, 27*, 2656–2666.

Huang, S., Pollack, H., & Shen, P. (2008). A late Quaternary climate reconstruction based on borehole heat flux data, borehole temperature data, and the instrumental record. *Geophysical Research Letters, 35*, L13703. doi:10.1029/2008GL034187

Huber, K., Czesla, S., & Schmitt, J. (2017). Discovery of the secondary eclipse of HAT-P-11 b. *Astronomy & Astrophysics, 597*. doi:10.1051/0004-6361/201629699

Hudson, T., Horseman, A., & Sugier, J. (2016). Diurnal, seasonal, and 11-yr solar cycle variation effects on the virtual ionosphere reflection height and implications for the Met Office's lightning detection system, ATDnet. *Journal of Atmospheric and Oceanic Technology, 33*, 1429–1441.

Hughes, D. (1987). The history of Halley's Comet. *Philosophical Transactions of the Royal Society of London A, 323*, 349–367.

Hughes, G., & Griffiths, R. (2008). Horizontal convection. *Annual Review of Fluid Mechanics, 40*, 185–208.

Hughes, T. (1973). Is the West Antarctic Ice Sheet Disintegrating? *Journal of Geophysical Research, 78*, 7884–7910.

Hulot, G., Eymin, C., Langlais, B., Mandea, M., & Olsen, N. (2002). Small-scale structure of the geodynamo inferred from Oersted and Magsat satellite data. *Nature, 416*, 620–623.

Hulot, G., Finlay, C., Constable, C., Olsen, N., & Mandea, M. (2010). The magnetic field of Planet Earth. *Space Science Reviews, 152*, 159–222.

Hunt, H. (2016). How does a bike stay upright? Surprisingly, it's all in the mind. *The Conversation*, available from https://theconversation.com/how-does-a-bike-stay-upright-surprisingly-its-all-in-the-mind-59829

Hutko, A., Lay, T., Garnero, E., & Revenaugh, J. (2006). Seismic detection of folded, subducted

lithosphere at the core–mantle boundary. *Nature, 441*, 333–336.

Hutko, A., Lay, T., Revenaugh, J., & Garnero, E. (2008). Anticorrelated seismic velocity anomalies from post-perovskite in the lowermost mantle. *Science, 320*, 1070–1074.

Huybers, P. (2006). Early Pleistocene glacial cycles and the integrated summer insolation forcing. *Science, 313*, 508–511.

Huybers, P., & Wunsch, C. (2005). Obliquity pacing of the late Pleistocene glacial terminations. *Nature, 434*, 491–494.

Huybrechts, P. (2009). West-side story of Antarctic ice. *Nature, 458*, 295–296.

Ibarra, D., Caves Rugenstein, J., Bachan, A., Baresch, A., Lau, K., Thomas, D., et al. (2019). Modeling the consequences of land plant evolution on silicate weathering. *American Journal of Science, 319*, 1–43.

Ichihara, H., Hamano, Y., Baba, K., & Kasaya, T. (2013). Tsunami source of the 2011 Tohoku earthquake detected by an ocean-bottom magnetometer. *Earth and Planetary Science Letters, 382*, 117–124.

Ida, Y. (1983). Convection in the mantle wedge above the slab and tectonic processes in subduction zones. *Journal of Geophysical Research, 88*, 7449–7456.

IERS Conventions (2010). G. Petit & B. Luzum (Eds.), *IERS Technical Note 36*. Frankfurt am Main: Verlag des Bundesamts für Kartographie und Geodäsie, 2010. ISBN 3-89888-989-6

Iess, L., Militzer, B., Kaspi, Y., Nicholson, P., Durante, D., Racioppa, P., et al. (2019). Measurement and implications of Saturn's gravity field and ring mass. *Science, 364*, eaat2965. doi:10.1126/science.aat2965

Imbrie, J., & Imbrie, J. (1980). Modeling the climatic response to orbital variations. *Science, 207*, 943–953.

Ingersoll, A. (2013). *Planetary Climates*. Princeton, NJ: Princeton University Press.

Ingersoll, A., & Porco, C. (1978). Solar heating and internal heat flow on Jupiter. *Icarus, 35*, 27–43.

Ingersoll, A., Dowling, T., Gierasch, P., Orton, G., Read, P., Sanchez-Lavega, A., et al. (2004).

"Dynamics of Jupiter's Atmosphere." In F. Bagenal, T. Dowling, & W. McKinnon, *Jupiter: The Planet, Satellites and Magnetosphere.* Cambridge, England: Cambridge University Press. Downloaded July 2015.

Ingraham, R. (1995). A note on Cowling's Theorem. *Physica Scripta*, *52*, 60–61.

Inoue, H., Fukao, Y., Tanabe, K., & Ogata, Y. (1990). Whole mantle P-wave travel time tomography. *Physics of the Earth and Planetary Interiors*, *59*, 294–328.

IPCC (2007). Climate Change 2007: The Physical Science Basis. In S. Solomon, D. Qin, M. Manning, Z. Chen, M. Marquis, K.B. Averyt, et al. (Eds.), *Contribution of Working Group I to the Fourth Assessment Report of the Intergovernmental Panel on Climate Change.* Cambridge University Press, Cambridge, United Kingdom and New York, NY, USA.

IPCC (2013). Climate Change 2013: The Physical Science Basis. In T. Stocker, D. Qin, G. Plattner, M. Tignor, S. Allen, & J. Boschung, (Eds.), *Contribution of Working Group I to the Fifth Assessment Report of the Intergovernmental Panel on Climate Change.* Cambridge, England: Cambridge University Press. The whole report is available at https://www.ipcc.ch/report/ar5/wg1/

IPCC (2022). Climate Change 2021: The Physical Science Basis. In V. Masson-Delmotte, P. Zhai, A. Pirani, S. Connors, C. Péan, S. Berger, et al. (Eds.), *Working Group I Contribution to the Sixth Assessment Report of the Intergovernmental Panel on Climate Change.* Cambridge, England: Cambridge University Press.

Irrgang, C., Saynisch, J., & Thomas, M. (2019). Estimating global ocean heat content from tidal magnetic satellite observations. *Nature Scientific Reports*, *9*. https://doi.org/10.1038/s41598-019-44397-8

Irving, M. (2021). Extremely eccentric minor planet to visit inner Solar System this decade. *New Atlas.* Accessed June 28, 2021 from https://newatlas.com/space/2014-un271-comet-solar-system-close-pass/

Isley, A., & Abbott, D. (1999). Plume-related mafic volcanism and the deposition of banded iron formation. *Journal of Geophysical Research*, *104*, 15,461–15,477.

Ito, E., & Takahashi, E. (1989). Postspinel transformations in the system Mg_2SiO_4-Fe_2SiO_4 and some geophysical implications. *Journal of Geophysical Research*, *94*, 10,637–10,646.

Jackson, A. (1990). Accounting for crustal magnetization in models of the core magnetic field. *Geophysical Journal International*, *103*, 657–673.

Jackson, A. (2010). A new turn for Earth's rotation (News & Views). *Nature*, *465*, 39–40.

Jackson, A., Jonkers, A., & Walker, M. (2000). Four centuries of geomagnetic secular variation from historical records. *Philosophical Transactions of the Royal Society of London, A*, *358*, 957–990.

Jackson, J., & Vette, J. (1975). *The Orbiting Geophysical Observatories: OGO Program Summary (NASA SP-7601).* Washington, DC: NASA Scientific & Technical Information Office.

Jacobs, J. (1963). *The Earth's core and geomagnetism.* Oxford: Pergamon Press.

Jacobs, J. (1987). *The earth's core* (2nd edition). London: Academic Press.

Jagniecki, E., Lowenstein, T., Jenkins, D., & Demicco, R. (2015). Eocene atmospheric CO_2 from the nahcolite proxy. *Geology*, *43*, 1075–1078. https://doi.org/10.1130/G36886.1

Jain, H., & Nowick, A. (1982). Electrical conductivity of synthetic and natural quartz crystals. *Journal of Applied Physics*, *53*, 477–484.

Jakosky, B., Brain, D., Chaffin, M., Curry, S., Deighan, J., Grebowsky, J., et al. (2018). Loss of the Martian atmosphere to space: Present-day loss rates determined from MAVEN observations and integrated loss through time. *Icarus*, *315*, 146–157.

Jakosky, B., Grebowsky, J., Luhmann, J., & Brain, D. (2015). Initial results from the MAVEN mission to Mars. *Geophysical Research Letters*, *42*, 8791–8802.

Jakosky, B., Slipski, M., Benna, M., Mahaffy, P., Elrod, M., Yelle, R., et al. (2017). Mars' atmospheric history derived from upper-atmosphere measurements of ^{38}Ar/^{36}Ar. *Science*, *355*, 1408–1410.

Jamero, M., Onuki, M., Esteban, M., Billones-Sensano, X., Tan, N., Nellas, A., et al. (2017). Small-island communities in the Philippines prefer local measures to relocation in response to sea-level rise. *Nature Climate Change*, *7*, 581–586.

James, T., & Ivins, E. (1997). Global geodetic signatures of the Antarctic ice sheet. *Journal of Geophysical Research*, *102*, 605–633.

James, T., & Lambert, A. (1993). A comparison of VLBI data with the ICE-3G glacial rebound model. *Geophysical Research Letters*, *20*, 871–874.

James, T., & Morgan, J. (1990). Horizontal motions due to post-glacial rebound. *Geophysical Research Letters*, *17*, 957–960.

Jannasch, H., & Mottl, M. (1985). Geomicrobiology of deep-sea hydrothermal vents. *Science*, *229*, 717–725.

Janson, M., Carson, J., Lafrenière, D., Spiegel, D., Bent, J., & Wong, P. (2012). Infrared non-detection of Fomalhaut b: Implications for the planet interpretation. *Astrophysical Journal*, *747*. doi:10.1088/0004-637X/747/2/116

Jaramillo, E., Royle, S., Claire, M., Kounaves, S., & Sephton, M. (2019). Indigenous organic-oxidized fluid interactions in the Tissint Mars Meteorite. *Geophysical Research Letters*, *46*, 3090–3098.

Jarrett, R., & Malde, H. (1987). Paleodischarge of the late Pleistocene Bonneville Flood, Snake River, Idaho, computed from new evidence. *Geological Society of America Bulletin*, *99*, 127–134.

Jaupart, C., Labrosse, S., Lucazeau, F., & Mareschal, J. (2015). Temperatures, heat, and energy in the mantle of the Earth. In G. Schubert (Ed.), *Treatise on Geophysics* (2nd edition, Vol. 7, pp. 223–270). Oxford: Elsevier.

Javaux, E., & Lepot, K. (2018). The Paleoproterozoic fossil record: Implications for the evolution of the biosphere during Earth's middle-age. *Earth Science Reviews*, *176*, 68–86.

Jeanloz, R. (1979). Properties of iron at high pressures and the state of the core. *Journal of Geophysical Research*, *84*, 6059–6069.

Jeanloz, R., & Thompson, A. (1983). Phase transitions and mantle discontinuities. *Reviews of Geophysics Space Physics*, *21*, 51–74.

Jefferts, K., Penzias, A., & Wilson, R. (1973). Deuterium in the Orion Nebula. *Astrophysical Journal*, *179*, L57–L59.

Jeffreys, H. (1925). On fluid motions produced by differences of temperature and humidity. *Quarterly Journal of the Royal Meteorological Society*, *51*, 347–356.

Jeffreys, H., & Bullen, K. (1940). *Seismological tables*. London: British Association for the Advancement of Science, Gray Milne Trust, reprinted *1970*, pp. 1–50.

Jekeli, C. (2016). Brief history of airborne gravity and data needs, presentation at NOAA *Airborne Gravity for Geodesy Summer School*, May 2016. Downloaded March 2022 from https://www.ngs.noaa.gov/grav-d/2016SummerSchool/history-fundamentals.shtml

Jellinek, M., & Manga, M. (2004). Links between long-lived hot spots, mantle plumes, D″, and plate tectonics. *Reviews of Geophysics*, *42*, RG3002. doi:10.1029/2003RG000144

Jensen, L., Rietbroek, R., & Kusche, J. (2013). Land water contribution to sea level from GRACE and Jason-1 measurements. *Journal of Geophysical Research*, *118*, 212–226.

Jephcoat, A., & Olson, P. (1987). Is the inner core of the Earth pure iron? *Nature*, *325*, 332–335.

Jestin, F., Huchon, P., & Gaulier, J. (1994). The Somalia plate and the East African Rift System: present-day kinematics. *Geophysical Journal International*, *116*, 637–654.

Jin, F. (1997). An equatorial ocean recharge paradigm for ENSO. Part I: Conceptual model. *Journal of the Atmospheric Sciences*, *54*, 811–829.

Johansen, A., Blum, J., Tanaka, H., Ormel, C., Bizzarro, M., & Rickman, H. (2014). The multifaceted planetesimal formation process. In H. Beuther, R. Klessen, C. Dullemond & T.

Henning (Eds.), *Protostars and Planets VI* (pp. 547–570). Tucson: University of Arizona Press.

Johansson, J., Davis, J., Scherneck, H.-G., Milne, G., Vermeer, M., Mitrovica, J., et al. (2002). Continuous GPS measurements of postglacial adjustment in Fennoscandia, 1. Geodetic results. *Journal of Geophysical Research, 107.* doi:10.1029/2001JB000400

John, B. (2010). Stonehenge and the Ice Age: On postglacial rebound and relative sea-level, June 6, 2010 blog. Downloaded March 2022 from https://brian-mountainman.blogspot .com/2010/06/on-postglacial-rebound.html

Johnson, C., Mittelholz, A., Langlais, B., Russell, C., Ansan, V., Banfield, D., et al. (2020). Crustal and time-varying magnetic fields at the InSight landing site on Mars. *Nature Geoscience, 13,* 199–204.

Johnson, G. (2012). Atmosphere Models for Earth, Mars, and Titan. Downloaded December 10, 2015 from website http:// exrocketman.blogspot.com/2012/06/ atmosphere-models-for-earth-mars-and.html

Johnson, J., Gerpheide, A., Lamb, M., & Fischer, W. (2014). O_2 constraints from Paleoproterozoic detrital pyrite and uraninite. *Geological Society of America Bulletin.* doi:10.1130/B30949.1

Johnston, A., & Schweig, E. (1996). The enigma of the New Madrid earthquakes of 1811–1812. *Annual Review of Earth and Planetary Sciences, 24,* 339–384.

Johnston, D. (2011). Multiple sulfur isotopes and the evolution of Earth's surface sulfur cycle. *Earth-Science Reviews, 106,* 161–183.

Johnston, P., & Lambeck, K. (1999). Postglacial rebound and sea level contributions to changes in the geoid and the Earth's rotation axis. *Geophysical Journal International, 136,* 537–558.

Jones, A. (1999). Imaging the continental upper mantle using electromagnetic methods. *Lithos, 48,* 57–80.

Jones, C. (2011). Planetary magnetic fields and fluid dynamos. *Annual Review of Fluid Mechanics, 43,* 583–614.

Jones, G. (1977). Thermal interaction of the core and the mantle and long-term behavior of the geomagnetic field. *Journal of Geophysical Research, 82,* 1703–1709.

Jones, G., Stott, P., & Christidis, N. (2013). Attribution of observed historical near-surface temperature variations to anthropogenic and natural causes using CMIP5 simulations. *Journal of Geophysical Research, 118,* 4001–4024.

Jones, H., & Marshall, J. (1993). Convection with rotation in a neutral ocean: A study of open-ocean deep convection. *Journal of Physical Oceanography, 23,* 1009–1039.

Jones, P., & Kelly, P. (1983). The spatial and temporal characteristics of Northern Hemisphere surface air temperature variations. *Journal of Climate, 3,* 243–252.

Jones, P., Raper, S., Bradley, R., Diaz, H., Kelly, P., & Wigley, T. (1986). Northern hemisphere surface air temperature variations: 1851-1984. *Journal of Climate and Applied Meteorology, 25,* 161–179.

Jones, T., Maguire, R., van Keken, P., Ritsema, J., & Koelemeijer, P. (2020). Subducted oceanic crust as the origin of seismically slow lower-mantle structures. *Progress in Earth and Planetary Sciences, 7.* https://doi.org/10.1186/ s40645-020-00327-1

Jonkers, A. (2003). *Earth's magnetism in the age of sail* (1st Amer. Ed.). Baltimore: JHU Press.

Jonkers, A., Jackson, A., & Murray, A. (2003). Four centuries of geomagnetic data from historical records. *Reviews of Geophysics, 41,* 1006. doi:10.1029/2002RG000115

Jordan, T. (1975). The continental tectosphere. *Reviews of Geophysics Space Physics, 13,* 1–12.

Jordan, T. (1978). Composition and development of the continental tectosphere. *Nature, 274,* 544–548.

Jørgensen, B., Kuenen, J., & Cohen, Y. (1979). Microbial transformations of sulfur compounds in a stratified lake (Solar Lake, Sinai). *Limnology and Oceanography, 24,* 799–822.

Joughin, I., & Alley, R. (2011). Stability of the West Antarctic ice sheet in a warming world.

Nature Geoscience, 4, 506–513. doi:10.1038/NGEO1194

Jouzel, J., Masson-Delmotte, V., Cattani, O., Dreyfus, G., Falourd, S., Hoffmann, G., et al. (2007). Orbital and millennial Antarctic climate variability over the past 800,000 years, *Science, 317,* 793–796.

Julian, P., & Chervin, R. (1978). A study of the Southern Oscillation and Walker circulation phenomenon. *Monthly Weather Review, 106,* 1433–1451.

Jurkowski, G., Ni, J., & Brown, L. (1984). Modern uparching of the Gulf coastal plain. *Journal of Geophysical Research, 89,* 6247–6255.

Kageyama, A., & Sato, T. (1997). Generation mechanism of a dipole field by a magnetohydrodynamic dynamo. *Physical Review E, 55,* 4617–4626.

Kaiser, A., Klok, C., Socha, J., Lee, W.-K., Quinlan, M., & Harrison, J. (2007). Increase in tracheal investment with beetle size supports hypothesis of oxygen limitation on insect gigantism. *Proceedings of the National Academy of Sciences, 104,* 13198–13203.

Kaiser, R., Stockton, A., Kim, Y., Jensen, E., & Mathies, R. (2013). On the formation of dipeptides in interstellar model ices. *Astrophysical Journal, 765.* doi:10.1088/0004-637X/765/2/111

Kalas, P., Graham, J., Chiang, E., Fitzgerald, M., Clampin, M., Kite, E., et al. (2008). Optical images of an exosolar planet 25 light-years from Earth. *Science, 322,* 1345–1348.

Kallmeyer, J., Pockalny, R., Adhikari, R., Smith, D., & D'Hondt, S. (2012). Global distribution of microbial abundance and biomass in subseafloor sediment. *Proceedings of the National Academy of Sciences, 109,* 16213–16216.

Kamide, Y., & Maltsev, Y. (2007). Geomagnetic storms. In Y. Kamide & A. Chian (Eds.), *Handbook of the solar-terrestrial environment* (pp. 355–374). Berlin: Springer-Verlag.

Kanamori, H. (1977). The energy release in great earthquakes. *Journal of Geophysical Research, 82,* 2981–2987.

Kaneps, A. (1979). Gulf Stream: Velocity fluctuations during the late Cenozoic. *Science, 204,* 297–301.

Kanzaki, Y., & Murakami, T. (2015). Estimates of atmospheric CO_2 in the Neoarchean – Paleoproterozoic from paleosols. *Geochimica et Cosmochimica Acta, 159,* 190–219.

Kappler, A., Pasquero, C., Konhauser, K., & Newman, D. (2005). Deposition of banded iron formations by anoxygenic phototrophic Fe(II)-oxidizing bacteria. *Geology, 33,* 865–868.

Karason, H., & van der Hilst, R. (2001). Tomographic imaging of the lowermost mantle with differential times of refracted and diffracted core phases (PKP, Pdiff). *Journal of Geophysical Research, 106,* 6569–6587.

Karato, S. (2010). Rheology of the Earth's mantle: A historical review. *Gondwana Research, 18,* 17–45.

Karato, S. (2011). Water distribution across the mantle transition zone and its implications for global material circulation. *Earth and Planetary Science Letters, 301,* 413–423.

Karato, S., & Li, P. (1992). Diffusion creep in perovskite: Implications for the rheology of the lower mantle. *Science, 255,* 1238–1240.

Karato, S., Paterson, M., & FitxGerald, J. (1986). Rheology of synthetic olivine aggregates: Influence of grain size and water. *Journal of Geophysical Research, 91,* 8151–8176.

Kargel, J., Cogley, J., Leonard, G., Haritashya, U., & Byers, A. (2011). Himalayan glaciers: The big picture is a montage. *Proceedings of the National Academy of Sciences, 108,* 14709–14710.

Karhu, J., & Holland, H. (1996). Carbon isotopes and the rise of atmospheric oxygen. *Geology, 24,* 867–870.

Kasting, J. (1987). Theoretical constraints on oxygen and carbon dioxide concentrations in the Precambrian atmosphere. *Precambrian Research, 34,* 205–229.

Kasting, J. (1991). CO_2 condensation and the climate of early Mars. *Icarus, 94,* 1–13.

Kasting, J. (1993). Earth's early atmosphere. *Science, 259,* 920–926.

Kasting, J. (2013). What caused the rise of atmospheric O$_2$? *Chemical Geology, 362*, 13–25.

Kasting, J., & Pollack, J. (1983). Loss of water from Venus. I. Hydrodynamic escape of hydrogen. *Icarus, 53*, 479–508.

Kasting, J., & Siefert, J. (2002). Life and the evolution of Earth's atmosphere. *Science, 296*, 1066–1068.

Kasting, J., Whitmire, D., & Reynolds, R. (1993). Habitable zones around main sequence stars. *Icarus, 101*, 108–128.

Katsura, T., & Ito, E. (1989). The system Mg$_2$SiO$_4$-Fe$_2$SiO$_4$ at high pressures and temperatures: Precise determination of stabilities of olivine, modified spinel, and spinel. *Journal of Geophysical Research, 94*, 15,663–15,670.

Katsura, T., Yamada, H., Nishikawa, O., Song, M., Kubo, A., Shinmei, T., et al. (2004). Olivine-wadsleyite transition in the system (Mg,Fe)$_2$SiO$_4$. *Journal of Geophysical Research, 109*, B02209. doi:10.1029/2003JB002438

Katsura, T., Yoneda, A., Yamazaki, D., Yoshino, T., & Ito, E. (2010). Adiabatic temperature profile in the mantle. *Physics of the Earth and Planetary Interiors, 183*, 212–218.

Kaufman, M. (2017). A four-planet system in orbit, directly imaged and remarkable. *Many Worlds*, accessible from https://exoplanets.nasa.gov/news/1404/a-four-planet-system-in-orbit-directly-imaged-and-remarkable/

Kaufmann, G., Wu, P., & Ivins, E. (2005). Lateral viscosity variations beneath Antarctica and their implications on regional rebound motions and seismotectonics. *Journal of Geodynamics, 39*, 165–181.

Kaula, W. (1969). A tectonic classification of the main features of the Earth's gravitational field. *Journal of Geophysical Research, 74*, 4807–4826.

Kaula, W. (1979). The moment of inertia of Mars. *Geophysical Research Letters, 6*, 194–196.

Kaula, W., Schubert, G., Lingenfelter, R., Sjogren, W., & Wollenhaupt, W. (1974). Apollo laser altimetry and inferences as to lunar structure. *5th Lunar and Planetary Science Conference*, pp. 3049–3058. Downloaded from http://adsabs.harvard.edu/abs/1974LPSC....5.3049K

Kawai, K., & Tsuchiya, T. (2009). Temperature profile in the lowermost mantle from seismological and mineral physics joint modeling. *Proceedings of the National Academy of Sciences, 106*, 22119–22123.

Kawakatsu, H., Kumar, P., Takei, Y., Shinohara, M., Kanazawa, T., Araki, E., & Suyehiro, K. (2009). Seismic evidence for sharp lithosphere-asthenosphere boundaries of oceanic plates. *Science, 324*, 499–502.

Ke, J., Luo, J., Shao, C., Tan, Y., Tan, W., & Yang, S. (2021). Combined test of the gravitational inverse-square law at the centimeter range. *Physical Review Letters, 126*. doi:10.1103/PhysRevLett.126.211101

Keeling, C., Piper, S., Bacastow, R., Wahlen, M., Whorf, T., Heimann, M., & Meijer, H. (2001). Exchanges of atmospheric CO2 and 13CO2 with the terrestrial biosphere and oceans from 1978 to 2000. I. Global aspects. SIO Reference Series, No. 01-06. San Diego, CA: Scripps Institution of Oceanography.

Keigwin, L. (1982). Isotopic paleoceanography of the Caribbean and East Pacific: Role of Panama uplift in late Neogene time. *Science, 217*, 350–353.

Kelbert, A., Schultz, A., & Egbert, G. (2009). Global electromagnetic induction constraints on transition-zone water content variations. *Nature, 460*, 1003–1006.

Kelley, K., Plank, T., Grove, T., Stolper, E., Newman, S., & Hauri, E. (2006). Mantle melting as a function of water content beneath back-arc basins. *Journal of Geophysical Research, 111*, B09208. doi:10.1029/2005JB 003732

Kellogg, L., Hager, B., & van der Hilst, R. (1999). Compositional stratification in the deep mantle. *Science, 283*, 1881–1884.

Kellogg, L., & King, S. (1993). Effect of mantle plumes on the growth of D″ by reaction between the core and mantle. *Geophysical Research Letters, 20*, 379–382.

Kelly, D., & Wood, A. (2006). The chemolithotrophic prokaryotes. In M. Dworkin, S. Falkow, E. Rosenberg, K. Schleifer, & E.

Stackebrandt (Eds.), *The Prokaryotes: A Handbook on the Biology of Bacteria, Vol. 2: Ecophysiology and Biochemistry* (3rd edition, pp. 441–456). Berlin: Springer-Verlag.

Kendall, R., Latychev, K., Mitrovica, J., Davis, J., & Tamisiea, M. (2006). Decontaminating tide gauge records for the influence of glacial isostatic adjustment: The potential impact of 3-D Earth structure. *Geophysical Research Letters, 33*. doi:10.1029/2006GL028448

Kennett, B., & Engdahl, E. (1991). Travel times for global earthquake location and phase identification. *Geophysical Journal International, 105*, 429–465.

Kennett, B., Engdahl, E., & Buland, R. (1995). Constraints on seismic velocities in the Earth from travel times. *Geophysical Journal International, 122*, 108–124.

Kennett, B., Widiyantoro, S., & van der Hilst, R. (1998). Joint seismic tomography for bulk sound and shear wave speed in the Earth's mantle. *Journal of Geophysical Research, 103*, 12,469–12,493.

Kennett, J. (1977). Cenozoic evolution of Antarctic glaciation, the Circum-Antarctic Ocean, and their impact on global paleoceanography. *Journal of Geophysical Research, 82*, 3843–3860.

Kennett, J., Houtz, R., Andrews, P., Edwards, A., Gostin, V., Hajos, M., et al. (1975). Cenozoic paleoceanography in the southwest Pacific Ocean, Antarctic glaciation, and the development of the Circum-Antarctic Current. In J. Kennett, R. Houtz, et al., *Initial Reports of the Deep Sea Drilling Project, XXIX* (pp. 1155–1169). Washington, DC: U.S. Government Printing Office.

Kenyon, P., & Turcotte, D. (1983). Convection in a two-layer mantle with a strongly temperature-dependent viscosity. *Journal of Geophysical Research, 88*, 6403–6414.

Kerr, R. (1983). *El Chichón climate effect estimated. Science, 219*, 157.

Kerr, R. (1996). Ancient life on Mars? *Science, 273*, 864–866.

Kerr, R. (1999). Signs of plate tectonics on an infant Mars. *Science, 284*, 719–722.

Kesson, S., & Fitz Gerald, J. (1991). Partitioning of MgO, FeO, NiO, MnO and Cr_2O_3 between magnesian silicate perovskite and magnesiowüstite: Implications for the origin of inclusions in diamond and the composition of the lower mantle. *Earth and Planetary Science Letters, 111*, 229–240.

Khan, A., Liebske, C., Rozel, A., Rivoldini, A., Nimmo, F., Connolly, J., et al. (2018). A geophysical perspective on the bulk composition of Mars. *Journal of Geophysical Research, 123*, 575–611.

Kiehl, J., & Dickinson, R. (1987). A study of the radiative effects of enhanced atmospheric CO2 and CH4 on early Earth surface temperatures. *Journal of Geophysical Research, 92*, 2991–2998.

Kikuchi, T., Hashimoto, K., & Nozaki, K. (2008). Penetration of magnetospheric electric fields to the equator during a geomagnetic storm. *Journal of Geophysical Research, 113*, A06214. doi:10.1029/2007JA012628

Kimizuka, H., Kaburaki, H., & Kogure, Y. (2000). Mechanism for negative Poisson ratios over the α-β transition of Cristobalite, SiO_2: A molecular-dynamics study. *Physical Review Letters, 84*, 5548–5551.

Kincaid, C., & Olson, P. (1987). An experimental study of subduction and slab migration. *Journal of Geophysical Research, 92*, 13,832–13,840.

King, D., & Ahrens, T. (1973). Shock compression of iron sulphide and the possible sulfur content of the Earth's core. *Nature Physical Science, 243*, 82–84.

King, E., & Aurnou, J. (2012). Thermal evidence for Taylor columns in turbulent rotating Rayleigh-Bénard convection. *Physical Review E, 85*. doi:10.1103/PhysRevE.85.01631

King, E., Stellmach, S., Noir, J., Hansen, U., & Aurnou, J. (2009). Boundary layer control of rotating convection systems. *Nature, 457*, 301–304.

King, S. (2002). Geoid and topography over subduction zones: The effect of phase transformations. *Journal of Geophysical Research, 107*. doi:10.1029/2000JB000141

Kious, J., & Tilling, R. (1996). *This Dynamic Earth: the Story of Plate Tectonics*, online edition version 1.20 viewed Feb. 29, 2016 at http://pubs.usgs.gov/gip/dynamic/stripes.html

Kirkby, J., Duplissy, J., Sengupta, K., Frege, C., Gordon, H., Williamson, C., et al. (2016). Ion-induced nucleation of pure biogenic particles. *Nature, 533*, 521–526.

Kirschvink, J. (1992). Late Proterozoic low-latitude glaciation: The snowball Earth. In J. Schopf & C. Klein (Eds.), *The Proterozoic Biosphere* (pp. 51–52). Cambridge, England: Cambridge University Press.

Kirschvink, J., & Kopp, R. (2008). Palaeoproterozoic ice houses and the evolution of oxygen-mediating enzymes: The case for a late origin of photosystem II. *Philosophical Transactions of the Royal Society B, 363*, 2755–2765.

Kiser, E., Kehoe, H., Chen, M., & Hughes, A. (2021). Lower mantle seismicity following the 2015 Mw 7.9 Bonin Islands deep-focus earthquake. *Geophysical Research Letters, 48*. 10.1029/2021GL093111

Kite, E., Williams, J., Lucas, A., & Aharonson, O. (2014). Low palaeopressure of the Martian atmosphere estimated from the size distribution of ancient craters. *Nature Geoscience, 7*, 335–339.

Kittel, C. (1968). *Introduction to solid state physics* (3rd edition). New York, NY: John Wiley & Sons.

Kjeldsen, K., Korsgaard, N. J., Bjørk, A. A., Khan, S. A., Box, J. E., Funder, S., et al. (2015). Spatial and temporal distribution of mass loss from the Greenland Ice Sheet since AD 1900. *Nature, 528*, 396–400.

Kjellsson, J., & Döös, K. (2012). Lagrangian decomposition of the Hadley and Ferrel cells. *Geophysical Research Letters, 39*, L15807. doi:10.1029/2012GL052420

Klein, H. (1977). The Viking biological investigation: General aspects. *Journal of Geophysical Research, 82*, 4677–4680.

Klosko, E., Wu, F., Anderson, H., Eberhart-Phillips, D., McEvilly, T., Audoine, E., et al. (1999). Upper mantle anisotropy in the New Zealand region. *Geophysical Research Letters, 26*, 1497–1500.

Kloss, C., & Finlay, C. (2019). Time-dependent low-latitude core flow and geomagnetic field acceleration pulses. *Geophysical Journal International, 217*, 140–168.

Klotzbach, P., & Landsea, C. (2015). Extremely intense hurricanes: Revisiting Webster et al. (2005) after 10 years. *Journal of Climate, 28*, 7621–7629.

Klotzbach, P., Wood, K., Bell, M., Blake, E., Bowen, S., Caron, L., et al. (2022). A hyperactive end to the Atlantic hurricane season: October-November 2020. *Bulletin of the American Meteorological Society, 103*, E110–E128.

Kneier, F., Overduin, P., Langer, M., Boike, J., & Grigoriev, M. (2018). Borehole temperature reconstructions reveal differences in past surface temperature trends for the permafrost in the Laptev Sea region, Russian Arctic. *arktos, 4*. https://doi.org/10.1007/s41063-018-0041-3

Knittle, E., & Jeanloz, R. (1987). Synthesis and equation of state of (Mg, Fe)SiO$_3$ perovskite to over 100 gigapascals. *Science, 235*, 668–670.

Knittle, E., & Jeanloz, R. (1989). Simulating the core-mantle boundary: An experimental study of high-pressure reactions between silicates and liquid iron. *Geophysical Research Letters, 16*, 609–612.

Kniveton, D. (2017). Questioning inevitable migration (news & views). *Nature Climate Change, 7*, 548–549.

Knoll, A. (2011). The multiple origins of complex multicellularity. *Annual Review of Earth and Planetary Sciences, 39*, 217–239.

Knoll, A., & Barghoorn, E. (1977). Archean microfossils showing cell division from the Swaziland System of South Africa. *Science, 198*, 396–398.

Knopoff, L. (1963). Horizontal convection in the Earth's mantle: A mechanism for strike-slip faulting (abstract). *Science, 140*, 383.

Knopoff, L. (1967). Density-velocity relations for rocks. *Geophysical Journal of the Royal Astronomical Society, 13*, 1–8.

Knopoff, L. (2001). Rayleigh waves without cubic equations. *Computational Seismology and Geodynamics, 7*, 23–27.

Knudsen, P., Bingham, R., Andersen, O., & Rio, M. (2011). Enhanced mean dynamic topography and ocean circulation estimation using GOCE preliminary models. In *Proceedings of 4th International GOCE User Workshop*, Munich. Downloaded from https://earth.esa.int/web/guest/-/goce-4th-international-user-workshop-2011-sessions-8263

Knutson, T., Camargo, S., Chan, J., Emanuel, K., Ho, C., Kossin, J., et al. (2019). Tropical cyclones and climate change assessment, Part I: Detection and attribution. *Bulletin of the American Meteorological Society, 100*, 1987–2007.

Knutson, T., McBride, J., Chan, J., Emanuel, K., Holland, G., Landsea, C., et al. (2010). Tropical cyclones and climate change. *Nature Geoscience, 3*, 157–163.

Kobayashi, D., & Sprenke, K. (2010). Lithospheric drift on early Mars: Evidence in the magnetic field. *Icarus, 210*, 37–42.

Koelemeijer, P., Ritsema, J., Deuss, A., & van Heijst, H. (2016). SP12RTS: A degree-12 model of shear- and compressional-wave velocity for Earth's mantle. *Geophysical Journal International, 204*, 1024–1039.

Koerner, D. (1997). Analogs of the early Solar System. *Origins of Life and Evolution of Biospheres, 27*, 157–184.

Kogan, M., & Steblov, G. (2008). Current global plate kinematics from GPS (1995–2007) with the plate-consistent reference frame. *Journal of Geophysical Research, 113*, B04416. doi:10.1029/2007JB005353

Koike, M., Nakada, R., Kajitani, I., Usui, T., Tamenori, Y., Sugahara, H., & Kobayashi, A. (2020). In-situ preservation of nitrogen-bearing organics in Noachian Martian carbonates. *Nature Communications, 11*. doi.org/10.1038/s41467-020-15931-4

Kokorowski, M., Seppälä, A., Sample, J., Holzworth, R., McCarthy, M., Bering, E., & Turunen, E. (2012). Atmosphere-ionosphere conductivity enhancements during a hard solar energetic particle event. *Journal of Geophysical Research, 117*, A05319. doi:10.1029/2011JA017363

Kolaczek, B., Nuzhdina, M., Nastula, J., & Kosek, W. (2000). El Niño impact on atmospheric polar motion excitation. *Journal of Geophysical Research, 105*, 3081–3087.

Kolodny, Y., Kerridge, J., & Kaplan, I. (1980). Deuterium in carbonaceous chondrites. *Earth and Planetary Science Letters, 46*, 149–158.

Komabayashi, T. (2014). Thermodynamics of melting relations in the system Fe-FeO at high pressure: Implications for oxygen in the Earth's core. *Journal of Geophysical Research, 119*, 4164–4177.

Komabayashi, T., & Fei, Y. (2010). Internally consistent thermodynamic database for iron to the Earth's core conditions. *Journal of Geophysical Research, 115*, B03202. doi:10.1029/2009JB006442

Konhauser, K., Hamade, T., Raiswell, R., Morris, R., Ferris, F., Southam, G., & Canfield, D. (2002). Could bacteria have formed the Precambrian banded iron formations? *Geology, 30*, 1079–1082.

Kono, M., & Roberts, P. (2002). Recent geodynamo simulations and observations of the geomagnetic field. *Reviews of Geophysics, 40*, 1013. doi:10.1029/2000RG000102

Konopliv, A., Asmar, S., Folkner, W., Karatekin, Ö., Nunes, D., Smrekar, S., et al. (2011). Mars high resolution gravity fields from MRO, Mars seasonal gravity, and other dynamical parameters. *Icarus, 211*, 401–428.

Konopliv, A., Yoder, C., Standish, E., Yuan, D., & Sjogren, W. (2006). A global solution for the Mars static and seasonal gravity, Mars orientation, Phobos and Deimos masses, and Mars ephemeris. *Icarus, 182*, 23–50.

Koot, L., Dumberry, M., Rivoldini, A., de Viron, O., & Dehant, V. (2010). Constraints on the coupling at the core-mantle and inner core boundaries inferred from nutation observations. *Geophysical Journal International, 182*, 1279–1294.

Kopp, G. (2016). Magnitudes and timescales of total solar irradiance variability. *Journal of*

Space Weather and Space Climate, 6, A30. doi:10.1051/swsc/2016025

Kopp, G., & Lean, J. (2011). A new, lower value of total solar irradiance: Evidence and climate significance. *Geophysical Research Letters, 38*, L01706. doi:10.1029/2010GL045777

Kopparapu, R., Ramirez, R., Kasting, J., Eymet, V., Robinson, T., Mahadevan, S., et al. (2013). Habitable zones around main-sequence stars: New estimates. *Astrophysical Journal, 765*. doi:10.1088/0004-637X/765/2/131

Korotaev, A., Benken, K., & Sabaneyeva, E. (2020). "Candidatus Mystax nordicus" aggregates with mitochondria of its host, the Ciliate Paramecium nephridiatum. *Diversity, 12*, 251. doi:10.3390/d12060251

Korte, M., Brown, M., Panovska, S., & Wardinski, I. (2019). Robust characteristics of the Laschamp and Mono Lake geomagnetic excursions: Results from global field models. *Frontiers of Earth Science, 7*. doi:10.3389/feart.2019.00086

Korte, M., Constable, C., Donadini, F., & Holme, R. (2011). Reconstructing the Holocene geomagnetic field. *Earth and Planetary Science Letters, 312*, 497–505.

Kossin, J. (2017). Hurricane intensification along United States coast suppressed during active hurricane periods. *Nature, 541*, 390–393.

Kossin, J. (2018). A global slowdown of tropical-cyclone translation speed. *Nature, 558*, 104–107.

Kossin, J., Knapp, K., Olander, T., & Velden, C. (2020). Global increase in major tropical cyclone exceedance probability over the past four decades. *Proceedings of the National Academy of Sciences, 117*, 11975–11980, also Correction, *117*, 29990.

Koster van Groos, A. (1988). Weathering, the carbon cycle, and the differentiation of the continental crust and mantle. *Journal of Geophysical Research, 93*, 8952–8958.

Krause, A., Mills, B., Zhang, S., Planavsky, N., Lenton, T., & Poulton, S. (2018). Stepwise oxygenation of the Paleozoic atmosphere. *Nature Communications, 9*. doi:10.1038/s41467-018-06383-y

Kreemer, C., Holt, W., & Haines, J. (2003). An integrated global model of present-day plate motions and plate boundary deformation. *Geophysical Journal International, 154*, 8–34.

Kretschmer, M., Coumou, D., Agel, L., Barlow, M., Tziperman, E., & Cohen, J. (2018). More-persistent weak stratospheric polar vortex states linked to cold extremes. *Bulletin of the American Meteorological Society, 99*, 49–60.

Kring, D., & Cohen, B. (2002). Cataclysmic bombardment throughout the inner solar system 3.9-4.0 Ga. *Journal of Geophysical Research, 107*. 10.1029/2001JE001529

Kuang, W., Jiang, W., & Wang, T. (2008). Sudden termination of Martian dynamo?: Implications from subcritical dynamo simulations. *Geophysical Research Letters, 35*, L14204. doi:10.1029/2008GL034183

Kukkonen, I., Rath, V., Kivekäs, L., Šafanda, J., & Čermak, V. (2011). Geothermal studies of the Outokumpu Deep Drill Hole, Finland: Vertical variation in heat flow and palaeoclimatic implications. *Physics of the Earth and Planetary Interiors, 188*, 9–25.

Kulikov, Y., Lammer, H., Lichtenegger, H., Penz, T., Breuer, D., Spohn, T., et al. (2007). A comparative study of the influence of the active young Sun on the early atmospheres of Earth, Venus, and Mars. *Space Science Reviews, 129*, 207–243.

Kumar, A., Bhambri, R., Kumar Tiwari, S., Verma, A., Kumar Gupta, A., & Kawishwar, P. (2019). Evolution of debris flow and moraine failure in the Gangotri Glacier region, Garhwal Himalaya: Hydro-geomorphological aspects. *Geomorphology, 333*, 152–166.

Kump, L., Kasting, J., & Crane, R. (2010). *The Earth System* (3rd edition). New York, NY: Prentice Hall / Pearson.

Kurrle, D., & Widmer-Schnidrig, R. (2008). The horizontal hum of the Earth: A global background of spheroidal and toroidal modes. *Geophysical Research Letters, 35*. doi:10.1029/2007GL033125

Kushiro, I., Syono, Y., & Akimoto, S. (1968). Melting of a peridotite nodule at high pressures and high water pressures. *Journal of Geophysical Research, 73*, 6023–6029.

Kuvshinov, A. (2008). 3-D global induction in the oceans and solid earth: Recent progress in modeling magnetic and electric fields from sources of magnetospheric, ionospheric and oceanic origin. *Surveys in Geophysics, 29*, 139–186.

Kvenvolden, K., Lawless, J., Pering, K., Peterson, E., Flores, J., Ponnamperuma, C., et al. (1970). Evidence for extraterrestrial amino-acids and hydrocarbons in the Murchison meteorite. *Nature, 228*, 923–926.

Laakso, T., & Schrag, D. (2017). A theory of atmospheric oxygen. *Geobiology, 15*, 366–384.

Labrosse, S. (2003). Thermal and magnetic evolution of the Earth's core. *Physics of the Earth and Planetary Interiors, 140*, 127–143.

Labrosse, S., Hernlund, J., & Coltice, N. (2007). A crystallizing dense magma ocean at the base of the Earth's mantle. *Nature, 450*, 866–869.

Lachenbruch, A. (1968). Preliminary geothermal model of the Sierra Nevada. *Journal of Geophysical Research, 73*, 6977–6989.

Lachenbruch, A. (1970). Crustal temperature and heat production: Implications of the linear heat-flow relation. *Journal of Geophysical Research, 75*, 3291–3300.

Lachenbruch, A., & Marshall, B. (1986). Changing climate: Geothermal evidence from permafrost in the Alaskan arctic. *Science, 234*, 689–696.

Lachmy, O., & Kaspi, Y. (2020). The role of diabatic heating in Ferrel cell dynamics, *Geophysical Research Letters, 47*. 10.1029/2020GL090619

Lada, C., & Lada, E. (2003). Embedded clusters in molecular clouds. *Annual Review of Astronomy and Astrophysics, 41*, 57–115.

Lagabrielle, Y., Goddéris, Y., Donnadieu, Y., Malavieille, J., & Suarez, M. (2009). *Earth and Planetary Science Letters, 279*, 197–211.

Lainey, V., Arlot, J., Karatekin, O., & Van Hoolst, T. (2009). Strong tidal dissipation in Io and Jupiter from astrometric observations. *Nature, 459*, 957–959.

Lakes, R. (1987). Foam structures with a negative Poisson's ratio. *Science, 235*, 1038–1040. (Also see Downs & Palmer 1994. *American Mineralogist, 79*, 9–14.)

Lambeck, K. (1980). *The Earth's variable rotation*. Cambridge, England: Cambridge University Press.

Lambeck, K. (1988). *Geophysical geodesy*. Oxford: Oxford University Press.

Lambeck, K. (2002). Sea level change from mid Holocene to recent time: An Australian example with global implication. In J. Mitrovica & B. Vermeersen (Eds.), *Ice sheets, sea level and the dynamic earth (Geodynamics Series 29)*. American Geophysical Union.

Lambeck, K., Hide, R., & Moffatt, H. (1982). Bumps on the core-mantle boundary: Geomagnetic and gravitational evidence revisited: Discussion. *Philosophical Transactions of the Royal Society of London, A, 306*, 287–289.

Lambeck, K., Smither, C., & Johnson, P. (1998). Sea-level change, glacial rebound and mantle viscosity for northern Europe. *Geophysical Journal International, 134*, 102–144.

Lambert, A., Coutier, N., & James, T. (2006). Long-term monitoring by absolute gravimetry: Tides to postglacial rebound. *Journal of Geodynamics, 41*, 307–317.

Lambert, W. (1945). The international gravity formula. *American Journal of Science, 347 (Daly Volume)*, 360–392.

Lambrechts, M., & Johansen, A. (2012). Rapid growth of gas-giant cores by pebble accretion. *Astronomy & Astrophysics, 544*. doi:10.1051/0004-6361/201219127

Landerer, F., Jungclaus, J., & Marotzke, J. (2009). Long-term polar motion excited by ocean thermal expansion. *Geophysical Research Letters, 36*, L17603. doi:10.1029/2009GL039692

Lane, N., & Martin, W. (2010). The energetics of genome complexity. *Nature, 467*, 929–934.

Langel, R. (1979). Near-Earth satellite magnetic field measurements: A prelude to Magsat. *Eos, Transactions, American Geophysical Union, 60*, 667–668.

Langel, R. (1982). The magnetic Earth as seen from MAGSAT, initial results. *Geophysical Research Letters, 9*, 239–242.

Langel, R. (1988). International Geomagnetic Reference Field revision 1987. *Eos, Transactions, American Geophysical Union, 69*, 557–558.

Langel, R. (1992). International Geomagnetic Reference Field: The sixth generation. *Journal of Geomagnetism and Geoelectricity, 44*, 679–707.

Langel, R., & Baldwin, R. (1991). *Geodynamics branch data base for main magnetic field analysis.* Greenbelt MD: NASA Technical Memorandum 104542.

Langel, R., & Estes, R. (1982). A geomagnetic field spectrum. *Geophysical Research Letters, 9*, 250–253.

Langel, R., Estes, R., Mead, G., Fabiano, E., & Lancaster, E. (1980). Initial geomagnetic field model from Magsat vector data. *Geophysical Research Letters, 7*, 793–796.

Langel, R., Phillips, J., & Horner, R. (1982). Initial scalar magnetic anomaly map from MAGSAT. *Geophysical Research Letters, 9*, 269–272.

Langel, R., Schnetzler, C., Phillips, J., & Horner, R. (1982). Initial vector magnetic anomaly map from MAGSAT. *Geophysical Research Letters, 9*, 273–276.

Langel, R., & Whaler, K. (1996). Maps of the magnetic anomaly field at Earth's surface from scalar satellite data. *Geophysical Research Letters, 23*, 41–44.

Langereis, C., Dekkers, M., de Lange, G., Paterne, M., & van Santvoort, P. (1997). Magnetostratigraphy and astronomical calibration of the last 1.1 Myr from an eastern Mediterranean piston core and dating of short events in the Brunhes. *Geophysical Journal International, 129*, 75–94.

Langlais, B., Thébault, E., Houliez, A., Purucker, M., & Lillis, R. (2019). A new model of the crustal magnetic field of Mars using MGS and MAVEN. *Journal of Geophysical Research, 124*, 1542–1569.

Larsen, C., Motyka, R., Freymueller, J., Echelmeyer, K., & Ivins, E. (2005). Rapid viscoelastic uplift in southeast Alaska caused by post-Little Ice Age glacial retreat. *Earth and Planetary Science Letters, 237*, 548–560.

Larson, K., Freymueller, J., & Philipsen, S. (1997). Global plate velocities from the Global Positioning System. *Journal of Geophysical Research, 102*, 9961–9981.

Larson, R. (1991). Latest pulse of Earth: Evidence for a mid-Cretaceous superplume. *Geology, 19*, 547–550.

Laskar, J. (1989). A numerical experiment on the chaotic behaviour of the Solar System. *Nature, 338*, 237–238.

Laskar, J. (2000). On the spacing of planetary systems. *Physical Review Letters, 84*, 3240–3243.

Laughlin, G., & Bodenheimer, P. (1994). Nonaxisymmetric evolution in protostellar disks. *Astrophysical Journal, 436*, 335–354.

Lawler, A. (1996). Finding puts Mars exploration on front burner. *Science, 273*, 865.

Lawson, A., & Jamieson, J. (1958). Energy transfer in the Earth's mantle. *Journal of Geology, 66*, 540–551.

Lawson, A., Leuschner, A. O., Gilbert, G. K., Davidson, G., Reid, H. F., Burkhalter, C., et al. (1908). California Earthquake of April 18, 1906, Vol. I: Report of the State Earthquake Investigation Commission, Vol. I. Washington, DC: Carnegie Institution Publication No. 87.

Lawver, L., & Gahagan, L. (2003). Evolution of Cenozoic seaways in the circum-Antarctic region. *Palaeogeography, Palaeoclimatology, Palaeoecology, 198*, 11–37.

Lay, T. (1989). Structure of the core-mantle transition zone: A chemical and thermal boundary layer. *EOS, Transactions of the American Geophysical Union, 70*, 49–59.

Lay, T., & Guarnero, E. (2004). Core-Mantle boundary structures and processes. In R. Sparks & C. Hawkesworth (Eds.), *The state of the Planet: Frontiers and challenges in Geophysics* (Vol. 19, pp. 25–41). Geophysical Monograph Series *150*, IUGG.

Lay, T., & Guarnero, E. (2007). Reconciling the post-perovskite phase with seismological observations of lowermost mantle structure. In K. Hirose, J. Brodholt, T. Lay & D. Yuen (Eds.), *Post-Perovskite: The last mantle phase transition* (pp. 129–153). Geophysical Monograph Series *174*, AGU.

Lay, T., & Helmberger, D. (1983). A lower mantle S-wave triplication and the shear velocity structure of D". *Geophysical Journal of the Royal Astronomical Society, 75,* 799–837.

Lay, T., Hernlund, J., & Buffett, B. (2008). Core–mantle boundary heat flow. *Nature Geoscience, 1,* 25–32.

Lay, T., Hernlund, J., Garnero, E., & Thorne, M. (2006). A post-perovskite lens and D′ heat flux beneath the central Pacific. *Science, 314,* 1272–1276.

Lay, T., Kanamori, H., Ammon, C. J., Nettles, M., Ward, S. N., Aster, R. C., et al. (2005). The Great Sumatra-Andaman Earthquake of 26 December 2004. *Science, 308,* 1127–1133. The website http://www.tectonics.caltech.edu/outreach/highlights/sumatra/why.html was also useful for presenting the basic properties of this earthquake.

Layer, P., Kröner, A., & McWilliams, M. (1996). An Archean geomagnetic reversal in the Kaap Valley Pluton, South Africa. *Science, 273,* 943–946.

Le Bars, M., Barik, A., Burmann, F., Lathrop, D., Noir, J., Schaeffer, N., & Triana, S. (2022). Fluid dynamics experiments for planetary interiors. *Surveys in Geophysics, 43,* 229–261.

Le Pichon, X. (1968). Sea-floor spreading and continental drift. *Journal of Geophysical Research, 73,* 3661–3697.

Le Traon, P. (2007). *Satellite Altimetry* (ESA 2nd Advanced Training Course in Ocean Remote Sensing, Hangzhou).

Leakey, L., & Leakey, M. (1964). Recent discoveries of fossil hominids in Tanganyika: at Olduvai and near Lake Natron. *Nature, 202,* 5–7.

Leakey, L., Tobias, P., & Napier, J. (1964). A new species of the genus *Homo* from Olduvai Gorge. *Nature, 202,* 7–9.

Lean, J., & Rind, D. (2008). How natural and anthropogenic influences alter global and regional surface temperatures: 1889 to 2006. *Geophysical Research Letters, 35,* L18701. doi:10.1029/2008GL034864

Lean, J., Warren, H., Mariska, J., & Bishop, J. (2003). A new model of solar EUV irradiance variability 2. Comparisons with empirical models and observations and implications for space weather. *Journal of Geophysical Research, 108, A2,* 1059. doi:10.1029/2001JA 009238

Leblanc, F., Modolo, R., Curry, S., Luhmann, J., Lillis, R., Chaufray, J., et al. (2015). Mars heavy ion precipitating flux as measured by Mars Atmosphere and Volatile Evolution. *Geophysical Research Letters, 42,* 9135–9141.

Lecar, M., Podolak, M., Sasselov, D., & Chiang, E. (2006). On the location of the snow line in a protoplanetary disk. *Astrophysical Journal, 640,* 1115–1118.

LeConte, J. (1884). Atmospheric waves from Krakatoa, *Science, 3*(71), 701–702.

Leder, N., & Orlić, M. (2004). Fundamental Adriatic seiche recorded by current meters. *Annales Geophysicae, 22,* 1449–1464.

Lee, K., & Jeanloz, R. (2003). High-pressure alloying of potassium and iron: Radioactivity in the Earth's core? *Geophysical Research Letters, 30.* doi:10.1029/2003GL018515

Lee, S., & Kim, H. (2003). The dynamical relationship between subtropical and eddy-driven jets. *Journal of the Atmospheric Sciences, 60,* 1490–1503.

Lee, S., & Lilley, F. (1986). On paleomagnetic data and dynamo theory. *Journal of Geomagnetism and Geoelectricity, 38,* 797–806.

Lee, T., & McPhaden, M. (2010). Increasing intensity of El Niño in the central-equatorial Pacific. *Geophysical Research Letters, 37,* L14603. doi:10.1029/2010GL044007

Lee, W. (1970). On the global variations of terrestrial heat-flow. *Physics of the Earth and Planetary Interiors, 2,* 332–341.

Lees, J., & VanDecar, J. (1991). Seismic tomography constrained by bouguer gravity anomalies: Applications in western Washington. *Pure and Applied Geophysics, 135,* 31–52.

Lei, J., & Zhao, D. (2006). Global P-wave tomography: On the effect of various mantle and core phases. *Physics of the Earth and Planetary Interiors, 154,* 44–69.

Leinert, C., van Boekel, R., Waters, L. B. F. M., Chesneau, O., Malbet, F., Köhler, R. et al. (2004). Mid-infrared sizes of circumstellar disks around Herbig Ae/Be stars measured with MIDI on the VLTI. *Astronomy & Astrophysics, 423*, 537–548.

Lemoine, F., Kenyon, S., Factor, J., Trimmer, R., Pavlis, N., Chinn, D., et al. (1998). *The Development of the Joint NASA GSFC and the National Imagery and Mapping Agency (NIMA) Geopotential Model EGM96, NASA/TP—1998-206861*. Greenbelt, MD: NASA.

Lemoine, F., Klosko, S., Chinn, D., & Cox, C. (2002). The development of NASA gravity models and their dependence on SLR. *13th International Workshop on Laser Ranging: Proceedings from the Science Session*. http://cddis.gsfc.nasa.gov/lw13/

Lenardic, A., & Moresi, L. (1999). Some thoughts on the stability of cratonic lithosphere: Effects of buoyancy and viscosity. *Journal of Geophysical Research, 104*, 12,747–12,758.

Lenton, T., Dahl, T., Daines, S., Mills, B., Ozaki, K., Saltzman, M., & Porada, P. (2016). Earliest land plants created modern levels of atmospheric oxygen. *Proceedings of the National Academy of Sciences, 113*, 9704–9709.

Lesur, V., Hamoudi, M., Choi, Y., Dyment, J., & Thébault, E. (2016). Building the second version of the World Digital Magnetic Anomaly Map (WDMAM). *Earth, Planets and Space, 68*. doi:10.1186/s40623-016-0404-6

Leventhall, G. (2007). What is infrasound? *Progress in Biophysics and Molecular Biology, 93*, 130–137.

Levin, G., & Straat, P. (1976). Viking labeled release biology experiment: Interim results. *Science, 194*, 1322–1329.

Levin, G., & Straat, P. (2016). The case for extant life on Mars and its possible detection by the Viking labeled release experiment. *Astrobiology, 16*, 798–810.

Levison, H., Dones, L., Chapman, C., Stern, S., Duncan, M., & Zahnle, K. (2001). Could the lunar "Late Heavy Bombardment" have been triggered by the formation of Uranus and Neptune? *Icarus, 151*, 286–306.

Levison, H., Kretke, K., Walsh, K., & Bottke, W. (2015). Growing the terrestrial planets from the gradual accumulation of submeter-sized

objects. *Proceedings of the National Academy of Sciences, 112*, 14180–14185.

Levitus, S., Antonov, J., & Boyer, T. (2005). Warming of the world ocean, 1955-2003. *Geophysical Research Letters, 32*, L02604. doi:10.1029/2004GL021592

Levitus, S., Antonov, J., Boyer, T., Locarnini, R., Garcia, H., & Mishonov, A. (2009). Global ocean heat content 1955–2008 in light of recently revealed instrumentation problems. *Geophysical Research Letters, 36*, L07608. doi:10.1029/2008GL037155

Levshin, A., Barmin, M., & Ritzwoller, M. (2018). Tutorial review of seismic surface waves' phenomenology. *Journal of Seismology, 22*, 519–537.

Levy, E. (1972a). Effectiveness of cyclonic convection for producing the geomagnetic field. *Astrophysical Journal, 171*, 621–633.

Levy, E. (1972b). Kinematic reversal schemes for the geomagnetic dipole. *Astrophysical Journal, 171*, 635–642.

Levy, E. (1972c). On the state of the geomagnetic field and its reversals. *Astrophysical Journal, 175*, 573–581.

Levy, E. (1976). Generation of planetary magnetic fields. *Annual Review of Earth and Planetary Sciences, 4*, 159–185.

Lewis, J. (1971). Consequences of the presence of sulfur in the core of the Earth. *Earth and Planetary Science Letters, 11*, 130–134.

Lewis, K., & Simons, F. (2012). Local spectral variability and the origin of the Martian crustal magnetic field. *Geophysical Research Letters, 39*, L18201. doi:10.1029/2012GL052708

Lewis, K., Aharonson, O., Grotzinger, J., Kirk, R., McEwen, A., & Suer, T. (2008). Quasi-periodic bedding in the sedimentary rock record of Mars. *Science, 322*, 1532–1535.

Lewkowicz, A., & Way, R. (2019). Extremes of summer climate trigger thousands of thermokarst landslides in a High Arctic environment. *Nature Communications, 10*. https://doi.org/10.1038/s41467-019-09314-7

Li, B., & Gavrilov, A. (2006). Hydroacoustic observation of Antarctic ice disintegration events in the Indian Ocean. *Proceedings of Acoustics 2006*, 479–484.

Li, C., van der Hilst, R., Engdahl, R., & Burdick, S. (2008). A new global model for P wave speed variations in Earth's mantle. *Geochemistry, Geophysics, Geosystems, 9*, Q05018. doi:10.1029/2007GC001806

Li, Q., Xue, C., Liu, J. P., Wu, J. F., Yang, S. Q., Shao, C. G. et al. (2018). Measurements of the gravitational constant using two independent methods. *Nature, 560*, 582–588.

Li, W., Li, Z., Ma, Z., Zhang, P., Lu, Y., Wang, C., et al. (2022). Ab initio determination on diffusion coefficient and viscosity of FeNi fluid under Earth's core condition. *Nature Scientific Reports, 12*. https://doi.org/10.1038/s41598-022-24594-8

Li, X., & Jeanloz, R. (1990). High pressure-temperature electrical conductivity of magnesiowüstite as a function of iron oxide concentration. *Journal of Geophysical Research, 95*, 21609–21612.

Li, X., & Jeanloz, R. (1991a). Effect of iron content on the electrical conductivity of perovskite and magnesiowüstite assemblages at lower mantle conditions. *Journal of Geophysical Research, 96*, 6113–6120.

Li, X., & Jeanloz, R. (1991b). Phases and electrical conductivity of a hydrous silicate assemblage at lower-mantle conditions. *Nature, 350*, 332–334.

Li, X., & Romanowicz, B. (1996). Global mantle shear velocity model developed using nonlinear asymptotic coupling theory. *Journal of Geophysical Research, 101*, 22,245–22,272.

Li, Y., Deschamps, F., & Tackley, P. (2014). The stability and structure of primordial reservoirs in the lower mantle: Insights from models of thermochemical convection in three-dimensional spherical geometry. *Geophysical Journal International, 199*, 914–930.

Liang, Z., & Gao, S. (2021). Study of the formation of the Arctic cell associated with the two-wave middle-high latitude circulation. *Atmospheric Research, 258*. doi.org/10.1016/j.atmosres.2021.105616

Liard, J., Palinkas, V., & Jiang, Z. (2011). The self-attraction effect in absolute gravimeters and its influence on the CIPM key comparisons during the ICAG2009. *Report BIPM-2012-01.*

Lide, D. (editor-in-chief) (1996). *CRC handbook of chemistry and physics* (77th edition). Boca Raton: CRC Press.

Lilley, F., Filloux, J., Mulhearn, P., & Ferguson, I. (1993). Magnetic signals from an ocean eddy. *Journal of Geomagnetism and Geoelectricity, 45*, 403–422.

Lillis, R., Frey, H., & Manga, M. (2008). Rapid decrease in Martian crustal magnetization in the Noachian era: Implications for the dynamo and climate of early Mars. *Geophysical Research Letters, 35*, L14203. doi:10.1029/2008GL034338

Lin, J., Campbell, A., Heinz, D., & Shen, G. (2003). Static compression of iron-silicon alloys: Implications for silicon in the Earth's core. *Journal of Geophysical Research, 108*, 2045. doi:10.1029/2002JB001978

Lin, J., Sturhahn, W., Zhao, J., Shen, G., Mao, H., & Hemley, R. (2005). Sound velocities of hot dense iron: Birch's law revisited. *Science, 308*, 1892–1894.

Lira, C. (2013). "Biography of James Watt: a summary," supplement to *Introductory Chemical Engineering Thermodynamics*, by R. Elliott & C. Lipa (updated 5/21/13). Downloaded from http://www.egr.msu.edu/~lira/supp/steam/wattbio.html\ignorespacesSept.\ignorespaces2017

Lis, D., Mehringer, D., Benford, D., Gardner, M., Phillips, T., Bockelée-Morvan, D., et al. (1997). New molecular species in Comet C/1995 O1 (Hale–Bopp) observed with the Caltech submillimeter observatory. *Earth, Moon and Planets, 78*, 13–20.

Lissauer, J., & de Pater, I. (2013). Chapter 1, Introduction (pp. 1–23) and Chapter 15, Planet Formation (pp. 413–451). *Fundamental planetary science: Physics, chemistry and habitability*, Cambridge, England: Cambridge University Press.

Lister, C. (1972). On the thermal balance of a mid-ocean ridge. *Geophysical Journal of the Royal Astronomical Society, 26*, 515–535.

Lithgow-Bertelloni, C., & Richards, M. (1998). The dynamics of Cenozoic and Mesozoic plate motions. *Reviews of Geophysics, 36*, 27–78.

Lithgow-Bertelloni, C., Richards, M., Conrad, C., & Griffiths, R. (2001). Plume generation in natural thermal convection at high Rayleigh and Prandtl numbers. *Journal of Fluid Mechanics, 434*, 1–21.

Lithgow-Bertelloni, C., & Silver, P. (1998). Dynamic topography, plate driving forces and the African superswell. *Nature, 395*, 269–272.

Liu, F., Li, J., Wang, B., Liu, J., Li, T., Huang, G., & Wang, Z. (2017). Divergent El Niño responses to volcanic eruptions at different latitudes over the past millennium. *Climate Dynamics, 50*, 3799–3812. doi:10.1007/s00382-017-3846-z

Liu, J., Yuan, X., Rind, D., & Martinson, D. (2002). Mechanism study of the ENSO and southern high latitude climate teleconnections. *Geophysical Research Letters, 29*. doi:10.1029/2002GL015143

Liu, L. (1975). Post-oxide phases of forsterite and enstatite. *Geophysical Research Letters, 2*, 417–419.

Liu, L. (1976a). Orthorhombic perovskite phases observed in olivine, pyroxene and garnet at high pressures and temperatures. *Physics of the Earth and Planetary Interiors, 11*, 289–298.

Liu, L. (1976b). The high-pressure phases of $MgSiO_3$. *Earth and Planetary Science Letters, 31*, 200–208.

Liu, L. (2014). Constraining Cretaceous subduction polarity in eastern Pacific from seismic tomography and geodynamic modeling. *Geophysical Research Letters, 41*, 8029–8036.

Liu, M., & Chase, C. (1989). Evolution of midplate hotspot swells: Numerical solutions. *Journal of Geophysical Research, 94*, 5571–5584.

Liu, Y., Zhu, D., Strayer, D., & Israelsson, U. (2010). Magnetic levitation of large water droplets and mice. *Advances in Space Research, 45*, 208–213.

Livermore, P., Hollerbach, R., & Finlay, C. (2017). An accelerating high-latitude jet in Earth's core. *Nature Geoscience, 10*, 62–69. doi:10.1038/NGEO2859

Livermore, R., Vine, F., & Smith, A. (1984). Plate motions and the geomagnetic field – II. Jurassic to Tertiary. *Geophysical Journal of the Royal Astronomical Society, 79*, 939–961.

Lizarralde, D., Chave, A., Hirth, G., & Schultz, A. (1995). Northeastern Pacific mantle conductivity profile from long-period magnetotelluric sounding using Hawaii-to-California submarine cable data. *Journal of Geophysical Research, 100*, 17837–17854.

Lloyd, S., Biggin, A., Paterson, G., & McCausland, P. (2022). Extremely weak early Cambrian dipole moment similar to Ediacaran: Evidence for long-term trends in geomagnetic field behaviour? *Earth and Planetary Science Letters, 595*. https://doi.org/10.1016/j.epsl.2022.117757

Lodders, K. (2003). Solar System abundances and condensation temperatures of the elements. *Astrophysical Journal, 591*, 1220–1247.

Lognonné, P., Banerdt, W. B., Pike, W. T., Giardini, D., Banfield, D., Christensen, U., et al. and the SEIS Science Commissioning Team (2019). SEIS: Overview, deployment and first science on the ground. *50th Lunar and Planetary Science Conference*. Downloaded from https://www.hou.usra.edu/meetings/lpsc2019/pdf/2246.pdf

Lohmann, K., Lohmann, C., & Putnam, N. (2007). Magnetic maps in animals: Nature's GPS (Commentary). *Journal of Experimental Biology, 210*, 3697–3705.

Longuet-Higgins, M. (1950). A theory of the origin of microseisms. *Philosophical Transactions of the Royal Society of London, A, 243*, 1–35.

Longwell, C., Flint, R., & Sanders, J. (1969). *Physical geology*. New York, NY: John Wiley & Sons.

Lonsdale, P. (2005). Creation of the Cocos and Nazca plates by fission of the Farallon plate. *Tectonophysics, 404*, 237–264.

Loper, D., & Lay, T. (1995). The core-mantle boundary region. *Journal of Geophysical Research, 100,* 6397–6420.

Loper, D., & Stacey, F. (1983). The dynamical and thermal structure of deep mantle plumes. *Physics of the Earth and Planetary Interiors, 33,* 304–317.

Lopez, P., Chevallier, P., Favier, V., Pouyaud, B., Ordenes, F., & Oerlemans, J. (2010). A regional view of fluctuations in glacier length in southern South America. *Global and Planetary Change, 71,* 85–108.

Love, J. (1999). A critique of frozen-flux inverse modelling of a nearly steady geodynamo. *Geophysical Journal International, 138,* 353–365.

Lovejoy, T. (1997). Quoted in *Scientists say Amazon rain forests ready to burn*, article summarizing Woods Hole press release in *Binghamton Press & Sun-Bulletin,* December 4, 1997.

Lovis, C., Segransan, D., Mayor, M., Udry, S., Benz, W., Bertaux, J., et al. (2011). The HARPS search for southern extra-solar planets XXVII. Up to seven planets orbiting HD 10180: probing the architecture of low-mass planetary systems. *Astronomy & Astrophysics, 528.* doi:10.1051/0004-6361/201015577

Lowenstein, T., & Demicco, R. (2006). Elevated Eocene atmospheric CO_2 and its subsequent decline. *Science, 313,* 1928.

Lowes, F., & Wilkinson, I. (1963). Geomagnetic dynamo: A laboratory model. *Nature, 198,* 1158–1160.

Lowes, F., & Wilkinson, I. (1968). Geomagnetic dynamo: An improved laboratory model. *Nature, 219,* 717–718.

Lowrie, W. (2007). *Fundamentals of geophysics* (2nd edition). Cambridge, England: Cambridge University Press.

Lowrie, W., & Kent, D. (2004). Geomagnetic polarity timescales and reversal frequency regimes. In J. Channell, D. Kent, W. Lowrie & J. Meert (Eds.), *Timescales of the Paleomagnetic Field* (pp. 117–129). AGU Geophysical Monograph *145.* Washington, DC: American Geophysical Union.

Lozovsky, M., Helled, R., Rosenberg, E., & Bodenheimer, P. (2017). Jupiter's formation and its primordial internal structure. *Astrophysical Journal, 836.* doi:10.3847/1538-4357/836/2/227

Lubimova, E. (1967). Theory of thermal state of the Earth's mantle. In T. Gaskell (Ed.), *The Earth's mantle* (pp. 231–323). New York, NY: Academic Press, as cited in Mao (1973).

Lumb, I., & Aldridge, K. (1991). On viscosity estimates for the Earth's fluid outer core and core-mantle coupling. *Journal of Geomagnetism and Geoelectricity, 43,* 93–110.

Lundquist, S. (1949). Experimental investigations of magneto-hydrodynamic waves. *Physical Review, 76,* 1805–1809.

Lynden-Bell, D., & Pringle, J. (1974). The evolution of viscous discs and the origin of the nebular variables. *Monthly Notices of the Royal Astronomical Society, 168,* 603–637.

Lyons, T., Reinhard, C., & Planavsky, N. (2014). The rise of oxygen in Earth's early ocean and atmosphere. *Nature, 506,* 307–315.

MacAyeal, D., Okal, E., Aster, R., & Bassis, J. (2008). Seismic and hydroacoustic tremor generated by colliding icebergs. *Journal of Geophysical Research, 113,* F03011. doi:10.1029/2008JF001005

MacDonald, G. (1959). Calculations on the thermal history of the Earth. *Journal of Geophysical Research, 64,* 1967–2000.

MacDonald, G. (1964). Tidal friction. *Reviews of Geophysics, 2,* 467–541.

Macdougall, D. (2004). *Frozen Earth: The once and future story of ice ages.* Berkeley: University of California Press, 267 pp.

Machetel, P., & Weber, P. (1991). Intermittent layered convection in a model mantle with an endothermic phase change at 670 km. *Nature, 350,* 55–57.

MacKinnon, J., St. Laurent, L., & Naveira Garabato, A. (2013). Diapycnal mixing processes in the ocean interior (draft), published as Chapter 7 – Diapycnal mixing processes in the ocean interior. *International Geophysics, 103,* 159–183 (part of *Ocean Circulation and Climate: A 21st Century Perspective,* G. Siedler, S. Griffies, J. Gould & J.

Church (Eds.). Amsterdam: Elsevier Publishing.)

Macmillan, S., & Finlay, C. (2010). The International Geomagnetic Reference Field. In M. Mandea & M. Korte (Eds.), *Geomagnetic observations and models: IAGA Special Sopron Book Series, 5* (pp. 265–276). New York, NY: Springer (preprint of 2011 publication).

Macmillan, S., Turbitt, C., & Thomson, A. (2009). Ascension and Port Stanley geomagnetic observatories and monitoring the South Atlantic Anomaly. *Annales Geophysicae, 52*, 83–95.

Maher, K., & Stevenson, D. (1988). Impact frustration of the origin of life. *Nature, 331*, 612–614.

Maier-Reimer, E., Mikolajewic, U., & Crowley, T. (1990). Ocean general circulation model sensitivity experiment with an open Central American isthmus. *Paleoceanography, 5*, 349–366.

Malin, S., & Hodder, B. (1982). Was the 1970 geomagnetic jerk of internal or external origin? *Nature, 296*, 726–728.

Malin, S., Hodder, B., & Barraclough, D. (1983). Geomagnetic secular variation: A jerk in 1970. In J. Cardus (Ed.), *75th Anniversary Volume of Ebro Observatory* (pp. 239–256). Tarragona: Ebro Observatory (cited in Mandea et al., 2010).

Manabe, S., & Stouffer, R. (1988). Two stable equilibria of a coupled ocean-atmosphere model. *Journal of Climate, 1*, 841–866.

Mandea Alexandrescu, M., Gibert, D., Le Mouël, J., Hulot, G., & Saracco, G. (1999). An estimate of average lower mantle conductivity by wavelet analysis of geomagnetic jerks. *Journal of Geophysical Research, 104*, 17735–17745.

Mandea, M., Holme, R., Pais, A., Pinheiro, K., Jackson, A., & Verbanac, G. (2010). Geomagnetic jerks: Rapid core field variations and core dynamics. *Space Science Reviews, 155*, 147–175.

Manga, M., & Jeanloz, R. (1997). Thermal conductivity of corundum and periclase and implications for the lower mantle. *Journal of Geophysical Research, 102*, 2999–3008.

Manger, G. (1963). Porosity and bulk density of sedimentary rocks. *Geological Survey Bulletin, 1144-E*, 60 pp.

Mann, M., Schmidt, G., Miller, S., & LeGrande, A. (2009). Potential biases in inferring Holocene temperature trends from long-term borehole information. *Geophysical Research Letters, 36*, L05708. doi:10.1029/2008GL036354

Manoj, C., Kuvshinov, A., Maus, S., & Lühr, H. (2006). Ocean circulation generated magnetic signals. *Earth, Planets and Space, 58*, 429–437.

Manoj, C., Neetu, S., Kuvshinov, A., & Harinarayana, T. (2006). Magnetic fields, generated by the Indian Ocean tsunami. In *Proceedings of the First Swarm International Science Meeting*, Nant, France (cited in Kuvshinov, 2008).

Mao, H. (1973). Thermal and electrical properties of the Earth's mantle. In *Carnegie Institution of Washington, Year Book 72, 1972–1973* (pp. 557–564).

Mao, H., Wu, Y., Chen, L., & Shu, J. (1990). Static compression of iron to 300 GPa and $Fe_{0.8}Ni_{0.2}$ alloy to 260 GPa: Implications for composition of the core. *Journal of Geophysical Research, 95*, 21,737–21,742.

Marchi, S., Chapman, C., Fassett, C., Head, J., Bottke, W., & Strom, R. (2013). Global resurfacing of Mercury 4.0-4.1 billion years ago by heavy bombardment and volcanism. *Nature, 499*, 59–61.

Marhold, S., Wiltschko, W., & Burda, H. (1997). A magnetic polarity compass for direction finding in a subterranean mammal. *Naturwissenschaften, 84*, 421–423.

Marion, J. (1980). *Physics and the physical universe* (3rd edition). New Jersey: John Wiley & Sons.

Markwick, P. (1998). Fossil crocodilians as indicators of Late Cretaceous and Cenozoic climates: Implications for using palaeontological data in reconstructing palaeoclimate. *Palaeogeography, Palaeoclimatology, Palaeoecology, 137*, 205–271.

Marois, C., Macintosh, B., Barman, T., Zuckerman, B., Song, I., Patience, J., et al. (2008). Direct imaging of multiple planets orbiting the star HR 8799. *Science, 322,* 1348–1352.

Marone, C., Scholtz, C., & Bilham, R. (1991). On the mechanics of earthquake afterslip. *Journal of Geophysical Research, 96,* 8441–8452.

Marotta, A., Restelli, F., Bollino, A., Regorda, A., & Sabadini, R. (2020). The static and time-dependent signature of ocean–continent and ocean–ocean subduction: the case studies of Sumatra and Mariana complexes. *Geophysical Journal International, 221,* 788–825.

Marsden, B. (1993). Periodic comet Shoemaker-Levy 9, in IAU Circular 5800: 1993e. http://www.cbat.eps.harvard.edu/iauc/05800/05800.html

Marsh, B., & Carmichael, I. (1974). Benioff Zone magmatism. *Journal of Geophysical Research, 79,* 1196–1206.

Marshall, C., Emry, J., & Olcott Marshall, A. (2011). Haematite pseudomicrofossils present in the 3.5-billion-year-old Apex Chert. *Nature Geoscience, 4,* 240–243.

Marshall, S., James, T., & Clarke, G. (2002). North American Ice Sheet reconstructions at the Last Glacial Maximum. *Quaternary Science Reviews, 21,* 175–192.

Marsiat, I., Berger, A., Gallee, H., Fichefet, T., & Tricot, C. (1988). Modelling the long-term variations of a coupled-climate model over the past 125,000 years: A test of the astronomical theory. *Chemical Geology (abstracts), 71,* 368.

Martin, M., Venkatesan, R., Weller, R., Tandon, A., & Joseph, K. (2022). Seasonal temperature variability observed at abyssal depths in the Arabian Sea. *Nature Scientific Reports, 12.* https://doi.org/10.1038/s41598-022-19869-z

Martin, R., & Livio, M. (2012). On the evolution of the snow line in protoplanetary discs. *Monthly Notices of the Royal Astronomical Society, 425,* L6–L9.

Martins, Z., Botta, O., Fogel, M., Sephton, M., Glavin, D., Watson, J., et al. (2008). Extraterrestrial nucleobases in the Murchison meteorite. *Earth and Planetary Science Letters, 270,* 130–136.

Marzeion, B., Leclercq, P., Cogley, J., & Jarosch, A. (2015). Brief Communication: Global reconstructions of glacier mass change during the 20th century are consistent. *The Cryosphere, 9,* 2399–2404.

Mase, G. (1969). *Continuum mechanics (Schaum Outline Series).* New York, NY: McGraw-Hill Publishers.

Mashayek, A., Ferrari, R., Vettoretti, G., & Peltier, W. (2013). The role of the geothermal heat flux in driving the abyssal ocean circulation. *Geophysical Research Letters, 40,* 3144–3149. doi:10.1002/grl.50640

Mason, B., Pyle, D., & Oppenheimer, C. (2004). The size and frequency of the largest explosive eruptions on Earth. *Bulletin of Volcanology, 66,* 735–748.

Masters, G., & Gubbins, D. (2003). On the resolution of density within the Earth. *Physics of the Earth and Planetary Interiors, 140,* 159–167.

Masters, G., Johnson, S., Laske, G., & Bolton, H. (1996). A shear-velocity model of the mantle. *Philosophical Transactions of the Royal Society of London A, 354,* 1385–1411.

Masters, G., Laske, G., Bolton, H., & Dziewonski, A. (2000). The relative behavior of shear velocity, bulk sound speed, and compressional velocity in the mantle: Implications for chemical and thermal structure. In S. Karato, et al. (Ed.), *Earth's Deep Interior: Mineral Physics and Tomography From the Atomic to the Global Scale* (pp. 63–87). AGU Geophysical Monograph *117.* Washington, DC: American Geophysical Union.

Masters, T., & Widmer, R. (1995). Free oscillations: Frequencies and attenuations. In T. Ahrens (Ed.), *Global Earth Physics: A handbook of physical constants.* Washington, DC: American Geophysical Union.

Matas, J., Bass, J., Ricard, Y., Mattern, E., & Bukowinski, M. (2007). On the bulk composition of the lower mantle: Predictions and limitations from generalized inversion of radial seismic profiles. *Geophysical Journal International, 170,* 764–780.

Mathews, P., Herring, T., & Buffet, B. (2002). Modeling of nutation and precession: New nutation series for nonrigid Earth and insights into the Earth's interior. *Journal of Geophysical Research, 107*. 10.1029/2001JB000390

Matoza, R., Fee, D., Assink, J. D., Iezzi, A. M., Green, D. N., Kim, K., et al. (2022). Atmospheric waves and global seismoacoustic observations of the January 2022 Hunga eruption, Tonga. *Science, 377*, 95–100. 10.1126/science.abo7063 (first release).

Matthews, K., Maloney, K., Zahirovic, S., Williams, S., Seton, M., & Müller, D. (2016). Global plate boundary evolution and kinematics since the late Paleozoic. *Global and Planetary Change, 146*, 226–250.

Mattila, K., Lemke, D., Haikala, L., Laureijs, R., Léger, A., Lehtinen, K., et al. (1996). Spectrophotometry of UIR bands in the diffuse emission of the galactic disk. *Astronomy & Astrophysics, 315*, L353–L356.

Mattox, T. (1995). Photo glossary of volcano terms: Ropy pahoehoe, in USGS website *Volcano Hazards Program – Reducing volcanic risk*, at http://web.archive.org/web/20070102035046/http://volcanoes.usgs.gov/Products/Pglossary/pahoehoe_ropy.html (downloaded August 27, 2014).

Matzel, J., Ishii, H., Joswiak, D., Hutcheon, I., Bradley, J., Brownlee, D., et al. (2010). Constraints on the formation age of cometary material from the NASA Stardust mission. *Science, 328*, 483–486.

Maus, S. (2008). The geomagnetic power spectrum. *Geophysical Journal International, 174*, 135–142.

Maus, S., & Haak, V. (2003). Magnetic field annihilators: Invisible magnetization at the magnetic equator. *Geophysical Journal International, 155*, 509–513.

Maus, S., Rother, M., Hemant, K., Stolle, C., Lühr, H., Kuvshinov, A., & Olsen, N. (2006). Earth's lithospheric magnetic field determined to spherical harmonic degree 90 from CHAMP satellite measurements. *Geophysical Journal International, 164*, 319–330.

Maus, S., Rother, M., Holme, R., Lühr, H., Olsen, N., & Haak, V. (2002). First scalar magnetic anomaly map from CHAMP satellite data indicates weak lithospheric field. *Geophysical Research Letters, 29*. doi:10.1029/2001GL013685

Maus, S., Rother, M., Stolle, C., Mai, W., Choi, S., Lühr, H., et al. (2006). Third generation of the Potsdam Magnetic Model of the Earth (POMME). *Geochemistry, Geophysics, Geosystems, 7*, Q07008. doi:10.1029/2006GC001269

Maycock, A., Randel, W. J., Steiner, A. K., Karpechko, A. Y., Christy, J., Saunders, R., et al. (2018). Revisiting the mystery of recent stratospheric temperature trends. *Geophysical Research Letters, 45*, 9919–9933.

Mayor, M., & Queloz, D. (1995). A Jupiter-mass companion to a solar-type star. *Nature, 378*, 355–359.

McCammon, C., Hutchison, M., & Harris, J. (1997). Ferric iron content of mineral inclusions in diamonds from São Luiz: A view into the lower mantle. *Science, 278*, 434–436.

McCarthy, D., & Luzum, B. (1996). Path of the mean rotational pole from 1899 to 1994. *Geophysical Journal International, 125*, 623–629.

McCord, T., & Sotin, C. (2005). Ceres: Evolution and current state. *Journal of Geophysical Research, 110*, E05009. doi:10.1029/2004JE002244

McCord, T., Carlson, R., Smythe, W., Hansen, G., Clark, R., Hibbitts, C., et al. (1997). Organics and other molecules in the surfaces of Callisto and Ganymede. *Science, 278*, 271–275.

McCreary, J. (1976). Eastern tropical ocean response to changing wind systems: With application to El Niño. *Journal of Physical Oceanography, 6*, 632–645.

McDonald, K., & Gunst, R. (1968). Recent trends in the Earth's magnetic field. *Journal of Geophysical Research, 73*, 2057–2067.

McDonough, W. (2003). Compositional model for the Earth's core. *In Treatise on Geochemistry, 2*, 547–568.

McElhinny, M., & Merrill, R. (1975). Geomagnetic secular variation over the past 5 M.y. *Reviews of Geophysics and Space Physics, 13*, 687–708.

McElhinny, M., & Senanayake, W. (1980). Paleomagnetic evidence for the existence of the geomagnetic field 3.5 Ga ago. *Journal of Geophysical Research*, *85*, 3523–3528.

McElhinny, M., & Senanayake, W. (1982). Variations in the geomagnetic dipole 1: The past 50,000 years. *Journal of Geomagnetism and Geoelectricity*, *34*, 39–51.

McElhinny, M., Taylor, S., & Stevenson, D. (1978). Limits to the expansion of Earth, Moon, Mars and Mercury and to changes in the gravitational constant. *Nature*, *271*, 316–321.

McGuire, B., Carroll, B., Loomis, R., Finneran, I., Jewell, P., Remijan, A., & Blake, G. (2016). Discovery of the interstellar chiral molecule propylene oxide (CH_3CHCH_2O). *Science*, *352*, 1449–1452.

McKay, A., & Roth, N. (2021). Organic matter in cometary environments. *Life*, *11*. http://doi.org/10.3390/life11010037

McKay, D., Gibson, E., Jr., Thomas-Keprta, K., Vali, H., Romanek, C., Clemett, S., et al. (1996). Search for past life on Mars: Possible relic biogenic activity in Martian meteorite ALH84001. *Science*, *273*, 924–930.

McKee, D., Yuan, X., Gordon, A., Huber, B., & Dong, Z. (2011). Climate impact on interannual variability of Weddell Sea Bottom Water. *Journal of Geophysical Research*, *116*, C05020. doi:10.1029/2010JC006484

McKenzie, D. (1984). The generation and compaction of partially molten rock. *Journal of Petrology*, *25*, 713–765.

McKenzie, D., & Parker, R. (1967). The North Pacific: An example of tectonics on a sphere. *Nature*, *216*, 1276–1280.

McMahon, S., & Parnell, J. (2014). Weighing the deep continental biosphere. *FEMS Microbiology Ecology*, *87*, 113–120.

McManus, J., Francois, R., Gherardi, J., Keigwin, L., & Brown-Leger, S. (2004). Collapse and rapid resumption of Atlantic meridional circulation linked to deglacial climate changes. *Nature*, *428*, 834–837.

McMichael, C., Dasgupta, S., Ayeb-Karlsson, S., & Kelman, I. (2020). A review of estimating population exposure to sea-level rise and the relevance for migration. *Environmental Research Letters*, *15*, 123005.

McNamara, A. (2019). A review of large low shear velocity provinces and ultra low velocity zones. *Tectonophysics*, *760*, 199–220.

McNutt, M. (1980). Implications of regional gravity for state of stress in the Earth's crust and upper mantle. *Journal of Geophysical Research*, *85*, 6377–6396.

McNutt , M. (2014). The drought you can't see (Editorial). *Science*, *345*, 1542.

McPhaden, M. (2016). The 2015-2016 El Niño. Accessed at https://www.youtube.com/watch?v=sVkJk3R9G2s

McPhaden, M., & Picaut, J. (1990). El Niño–Southern Oscillation displacements of the western equatorial Pacific warm pool. *Science*, *250*, 1385–1388.

McQueen, R., & Marsh, S. (1960). Equation of state for nineteen metallic elements from shock-wave measurements to two megabars. *Journal of Applied Physics*, *31*, 1253–1269.

McQueen, R., & Marsh, S. (1966). Shock-wave compression of iron-nickel alloys and the Earth's core. *Journal of Geophysical Research*, *71*, 1751–1756.

McQueen, R., Fritz, J., & Marsh, S. (1964). On the composition of the Earth's interior. *Journal of Geophysical Research*, *69*, 2947–2965.

McSweeney, R. (2019). Extreme Weather Q&A: How is Arctic warming linked to the 'polar vortex' and other extreme weather? Blog dated January 31, 2019. Downloaded from https://www.carbonbrief.org/qa-how-is-arctic-warming-linked-to-polar-vortext-other-extreme-weather

Meade, C., Silver, P., & Kaneshima, S. (1995). Laboratory and seismological observations of lower mantle isotropy. *Geophysical Research Letters*, *22*, 1293–1296.

Medvedev, M., & Melott, A. (2007). Do extragalactic cosmic rays induce cycles in fossil diversity? *Astrophysical Journal*, *664*, 879–889.

Meehl, G., Arblaster, J., Fasullo, J., Hu, A., & Trenberth, K. (2011). Model-based evidence of deep-ocean heat uptake during surface-temperature hiatus periods. *Nature Climate Change*, *1*, 360–364.

Meert, J. (2012). What's in a name? The Columbia (Paleopangaea/Nuna) super-continent. *Gondwana Research*, *21*, 987–993.

Megnin, C., & Romanowicz, B. (2000). The three-dimensional shear velocity structure of the mantle from the inversion of body, surface and higher-mode waveforms. *Geophysical Journal International*, *143*, 709–728.

Meijaard, J., Papadopoulos, J., Ruina, A., & Schwab, A. (2007). Linearized dynamics equations for the balance and steer of a bicycle: A benchmark and review. *Proceedings of the Royal Society A*, *463*, 1955–1982.

Meinen, C., Perez, R., Dong, S., Piola, A., & Campos, E. (2020). Observed ocean bottom temperature variability at four sites in the northwestern Argentine Basin: Evidence of decadal deep/abyssal warming amidst hourly to interannual variability during 2009–2019. *Geophysical Research Letters*, *47*. 10.1029/2020GL089093

Meissner, K. (2007). Younger Dryas: A data to model comparison to constrain the strength of the overturning circulation. *Geophysical Research Letters*, *34*, L21705. doi:10.1029/2007GL031304

Melosh, H., & Raefsky, A. (1980). The dynamical origin of subduction zone topography. *Geophysical Journal of the Royal Astronomical Society*, *60*, 333–354.

Melosh, H., Freed, A., Johnson, B., Blair, D., Andrews-Hanna, J., Neumann, G., et al. (2013). The origin of Lunar Mascon Basins. *Science*, *340*, 1552–1555.

MELT Seismic Team (Forsyth, D., Scheirer, D., Webb, S., Dorman, L., Orcutt, J., Harding, A., et al.) (1998). Imaging the deep seismic structure beneath a mid-ocean ridge: The MELT experiment. *Science*, *280*, 1215–1218.

Menard, H. (1978). Fragmentation of the Farallon plate by pivoting subduction. *Journal of Geology*, *86*, 99–110.

Mercer, J. (1968). Antarctic ice and Sangamon sea level. *International Association of Scientific Hydrology Symposium*, *79*, 217–225.

Mercer, J. (1978). West Antarctic ice sheet and CO_2 greenhouse effect: A threat of disaster. *Nature*, *271*, 321–325.

Merrill, R., & McFadden, P. (2003). The geomagnetic axial dipole field assumption. *Physics of the Earth and Planetary Interiors*, *139*, 171–185.

Meyer, J., Hufen, J., Siebert, M., & Hahn, A. (1983). Investigations of the internal geomagnetic field by means of a global model of the Earth's crust. *Journal of Geophysics*, *52*, 71–84.

Mierdel, K., Keppler, H., Smyth, J., & Langenhorst, F. (2007). Water solubility in aluminous orthopyroxene and the origin of Earth's asthenosphere. *Science*, *315*, 364–368.

Miles, B., Stokes, C., & Jamieson, S. (2015). Pan–ice-sheet glacier terminus change in East Antarctica reveals sensitivity of Wilkes Land to sea-ice changes. *Science Advances*, *2*, e1501350.

Miller, G., Geirsdottir, A., Zhong, Y., Larsen, D. J., Otto-Bliesner, B. L., Holland, M. M., et al. (2012). Abrupt onset of the Little Ice Age triggered by volcanism and sustained by sea-ice/ocean feedbacks. *Geophysical Research Letters*, *39*, L02708. doi:10.1029/2011GL050168

Milne, G., Davis, J., Mitrovica, J., Scherneck, H.-G., Johansson, J., Vermeer, M., & Koivula, H. (2001). Space-geodetic constraints on glacial isostatic adjustment in Fennoscandia. *Science*, *291*, 2381–2385.

Milne, G., & Mitrovica, J. (1998). Postglacial sea-level change on a rotating Earth. *Geophysical Journal International*, *133*, 1–19.

Minami, T. (2017). Motional induction by tsunamis and ocean tides: 10 years of progress. *Surveys in Geophysics*, *38*, 1097–1132.

Minster, B., Jordan, T., Molnar, P., & Haines, E. (1974). Numerical modelling of instantaneous plate tectonics. *Geophysical Journal of the Royal Astronomical Society*, *36*, 541–576.

Minton, D., & Malhotra, R. (2010). Dynamical erosion of the asteroid belt and implications for large impacts in the inner Solar System. *Icarus*, *207*, 744–757.

Mitchell, J., Beebe, R., Ingersoll, A., & Garneau, G. (1981). Flow Fields within Jupiter's Great Red Spot and White Oval BC. *Journal of Geophysical Research*, *86*, 8751–8757.

Mitrovica, J., Davis, J., & Shapiro, I. (1994). A spectral formalism for computing three-dimensional deformations due to surface loads 2. Present-day glacial isostatic adjustment. *Journal of Geophysical Research*, *99*, 7075–7101.

Mitrovica, J., & Forte, A. (2004). A new inference of mantle viscosity based upon joint inversion of convection and glacial isostatic adjustment data. *Earth and Planetary Science Letters*, *225*, 177–189.

Mitrovica, J., Gomez, N., & Clark, P. (2009). The sea-level fingerprint of West Antarctic collapse. *Science*, *323*, 753. doi:10.1126/science.1166510

Mitrovica, J., Hay, C., Morrow, E., Kopp, R., Dumberry, M., & Stanley, S. (2015). Reconciling past changes in Earth's rotation with 20th century global sea-level rise: Resolving Munk's enigma. *Science Advances*, *1*, doi:10.1126/sciadv.1500679

Mitrovica, J., Milne, G., & Davis, J. (2001). Glacial isostatic adjustment on a rotating earth. *Geophysical Journal International*, *147*, 562–578.

Mitrovica, J., Wahr, J., Matsuyama, I., & Paulson, A. (2005). The rotational stability of an ice-age earth. *Geophysical Journal International*, *161*, 491–506.

Mittlefehldt, D. (1994). ALH84001, a cumulate orthopyroxenite member of the Martian meteorite clan. *Meteoritics*, *29*, 214–221.

Mix, A., & Ruddiman, W. (1984). Oxygen-isotope analyses and Pleistocene ice volumes, *Quaternary Research*, *21*, 1–20.

Mizuno, H. (1980). Formation of the giant planets. *Progress of Theoretical Physics*, *64*, 544–557.

Mizuno, H., Nakazawa, K., & Hayashi, C. (1978). Instability of a gaseous envelope surrounding a planetary core and formation of giant planets. *Progress of Theoretical Physics*, *60*, 699–710.

Moffatt, H. (1978). *Magnetic field generation in electrically conducting fluids*. Cambridge, England: Cambridge University Press. See also Moffatt, H. & Dormy, E. (2019).

Moffatt, H., & Dormy, E. (2019). *Self-exciting fluid dynamos*. Cambridge, England: Cambridge University Press.

Mohajerani, Y. (2020). Record Greenland mass loss. *Nature Climate Change (News & Views)*, *10*, 803–804.

Mojzsis, S., Harrison, T., & Pidgeon, R. (2001). Oxygen-isotope evidence from ancient zircons for liquid water at the Earth's surface 4,300 Myr ago. *Nature*, *409*, 178–181.

Molodenskii, S. (2010). Correctives to the scheme of the Earth's structure inferred from new data on nutation, tides, and free oscillations. *Izvestiya, Physics of the Solid Earth*, *46*, 555–579.

Monnereau, M., Calvet, M., Margerin, L., & Souriau, A. (2010). Lopsided growth of Earth's inner core. *Science*, *328*, 1014–1017.

Montelli, R., Nolet, G., Dahlen, T., & Masters, G. (2006). A catalogue of deep mantle plumes: New results from finite-frequency tomography. *Geochemistry, Geophysics, Geosystems*, *7*, Q11007. doi:10.1029/2006GC001248

Montelli, R., Nolet, G., Dahlen, T., Masters, G., Engdahl, E., & Hung, S. (2004). Finite-frequency tomography reveals a variety of plumes in the mantle. *Science*, *303*, 338–343. Also *Supplemental Online Material* (downloaded from www.sciencemag.org/cgi/content/full/1092485/DC1).

Montmerle, T., Augereau, J., Chaussidon, M., Gounelle, M., Marty, B., & Morbidelli, A. (2006). 3. Solar System formation and early evolution: The first 100 million years. *Earth, Moon, Planets*, *98*, 39–95.

Mooney, W., Laske, G., & Masters, G. (1998). CRUST 5.1: A global crustal model at 5°×5°. *Journal of Geophysical Research*, *103*, 727–747. (The actual citation in Flanagan & Shearer (1998) is to their 1995 AGU talk on CRUST 5.0, an unpublished model.)

Moore, J., & Sharma, M. (2013). The K-Pg impactor was likely a high velocity comet. *44th Lunar and Planetary Science Conference*. Downloaded Feb. 27, 2017 from http://www.lpi.usra.edu/meetings/lpsc2013/pdf/2431.pdf

Moore, W., & Webb, A. (2013). Heat-pipe Earth. *Nature, 501*, 501–505.

Morbidelli, A. (2008). Origin and dynamical evolution of comets and their reservoirs. Downloaded from https://arxiv.org/abs/astro-ph/0512256 (submitted 9 Dec 2005).

Morbidelli, A., & Nesvorny, D. (2012). Dynamics of pebbles in the vicinity of a growing planetary embryo: Hydro-dynamical simulations. *Astronomy & Astrophysics, 546.* doi:10.1051/0004-6361/201219824

Morbidelli, A., Tsiganis, K., Crida, A., Levison, H., & Gomes, R. (2007). Dynamics of the giant planets of the Solar System in the gaseous protoplanetary disk and their relationship to the current orbital architecture. *Astronomical Journal, 134*, 1790–1798.

Morelli, A., & Dziewonski, A. (1987). Topography of the core-mantle boundary and lateral homogeneity of the liquid core. *Nature, 325*, 678–683.

Morgan, J. (1968). Rises, trenches, great faults, and crustal blocks. *Journal of Geophysical Research, 73*, 1959–1982.

Morgan, P., & Phillips, R. (1983). Hot spot heat transfer: Its application to Venus and implications to Venus and Earth. *Journal of Geophysical Research, 88*, 8305–8317.

Moritz, H. (2000). Geodetic reference system 2000. *Journal of Geodesy, 74*, 128–162.

Morrison, R. (1991). Quaternary stratigraphic, hydrologic, and climatic history of the Great Basin, with emphasis on Lakes Lahonton, Bonneville, and Tecopa. In R. Morrison (Ed.), *Quaternary nonglacial geology: Conterminous US* (pp. 283–320). Boulder, CO: Geological Society of America.

Moucha, R., Forte, A., Mitrovica, J., & Daradich, A. (2007). Lateral variations in mantle rheology: Implications for convection related surface observables and inferred viscosity models. *Geophysical Journal International, 169*, 113–135.

Mtchedlidze, S., Domínguez-Fernández, P., Du, X., Brandenburg, A., Kahniashvili, T., O'Sullivan, S., et al. (2022). Evolution of primordial magnetic fields during large-scale structure formation. *Astrophysical Journal, 929.* https://doi.org/10.3847/1538-4357/ac5960

Muir, M., & Hall, D. (1974). Diverse microfossils in Precambrian Onverwacht group rocks of South Africa. *Nature, 252*, 376–378.

Mullarney, J., Griffiths, R., & Hughes, G. (2004). Convection driven by differential heating at a horizontal boundary. *Journal of Fluid Mechanics, 516*, 181–209.

Mullarney, J., Griffiths, R., & Hughes, G. (2006). The effects of geothermal heating on the ocean overturning circulation. *Geophysical Research Letters, 33*, L02607. doi:10.1029/2005GL024956

Mullarney, J., Griffiths, R., & Hughes, G. (2007). The role of freshwater fluxes in the thermohaline circulation: Insights from a laboratory analogue. *Deep Sea Research I, 54*, 1–21.

Müller, J., & Biskupek, L. (2007). "Variations of the gravitational constant from lunar laser ranging data." *Classical and Quantum Gravity, 24*, 4533–4538.

Müller, D., Sdrolias, M., Gaina, C., & Roest, W. (2008). Age, spreading rates, and spreading asymmetry of the world's ocean crust. *Geochemistry, Geophysics, Geosystems, 9*, Q04006. doi:10.1029/2007GC001743

Müller, H., Endres, C., Stutzki, J., & Schlemmer, S. (2016). Molecules in Space. Accessed 11/20/16 at http://www.astro.uni-koeln.de/cdms/molecules

Muller, P., & Sjogren, W. (1968). Mascons: Lunar mass concentrations. *Science, 161*, 680–684.

Muller, R., & MacDonald, G. (1997). Spectrum of 100-kyr glacial cycle: Orbital inclination, not eccentricity. *Proceedings of the National Academy of Sciences, 94*, 8329–8334.

Munch, F., Grayver, A., Kuvshinov, A., & Khan, A. (2018). Stochastic inversion of geomagnetic observatory data including rigorous treatment of the ocean induction effect with implications for transition zone water content and thermal structure. *Journal of Geophysical Research, 123*, 31–51.

Munk, W. (1950). On the wind-driven ocean circulation. *Journal of Meteorology, 7*, 79–93.

Munk, W. (1966). Abyssal recipes. *Deep Sea Research*, *13*, 707–730.

Munk, W. (1997). Once again: Once again – Tidal friction. *Progress in Oceanography*, *40*, 7–35.

Munk, W., & Forbes, A. (1989). Global ocean warming: An acoustic measure? *Journal of Physical Oceanography*, *19*, 1765–1778.

Munk, W., & Wunsch, C. (1998). Abyssal recipes II: Energetics of tidal and wind mixing, *Deep-Sea Research I*, *45*, 1977–2010.

Murakami, H., Levin, E., Delworth, T., Gudgel, R., & Hsu, P. (2018). Dominant effect of relative tropical Atlantic warming on major hurricane occurrence. *Science*, *362*, 794–799.

Murakami, M., Hirose, K., Kawamura, K., Sata, N., & Ohishi, Y. (2004). Post-perovskite phase transition in $MgSiO_3$. *Science*, *304*, 855–858.

Murakami, M., Hirose, K., Sata, N., & Ohishi, Y. (2005). Post-perovskite phase transition and mineral chemistry in the pyrolytic lowermost mantle. *Geophysical Research Letters*, *32*, L03304. doi:10.1029/2004GL021956

Murakami, M., Hirose, K., Yurimoto, H., Nakashima, S., & Takafuji, N. (2002). Water in Earth's lower mantle. *Science*, *295*, 1885–1887.

Murck, B., Skinner, B., & Porter, S. (1997). *Dangerous Earth: An introduction to geologic hazards*. New York, NY: John Wiley & Sons.

Murthy, R., & Hall, H. (1970). The chemical composition of the Earth's core: Possibility of sulphur in the core. *Physics of the Earth and Planetary Interiors*, *2*, 276–282.

Murthy, R., & Hall, H. (1972). The origin and chemical composition of the Earth's core. *Physics of the Earth and Planetary Interiors*, *6*, 123–130.

Murthy, V., van Westrenen, W., & Fei, Y. (2003). Experimental evidence that potassium is a substantial radioactive heat source in planetary cores. *Nature*, *423*, 163–165.

Nabighian, M., Grauch, V., Hansen, R., LaFehr, T., Li, Y., Peirce, J., et al. (2005). The historical development of the magnetic method in exploration. *Geophysics*, *70*, 33ND–61ND.

Nadiga, B., & Aurnou, J. (2008). A tabletop demonstration of atmospheric dynamics: Baroclinic instability. *Oceanography*, *21*, 196–201.

Naghibi, S., Jalali, M., Karabasov, S., & Alam, M.-R. (2017). Excitation of the Earth's Chandler wobble by a turbulent oceanic double-gyre. *Geophysical Journal International*, *209*, 509–516.

Naif, S., Key, K., Constable, S., & Evans, R. (2013). Melt-rich channel observed at the lithosphere-asthenosphere boundary. *Nature*, *495*, 356–359.

Nakada, M., & Karato, S. (2012). Low viscosity of the bottom of the Earth's mantle inferred from the analysis of Chandler wobble and tidal deformation. *Physics of the Earth and Planetary Interiors*, *192–193*, 68–80.

Nakada, M., & Lambeck, K. (1989). Late Pleistocene and Holocene sea-level change in the Australian region and mantle rheology. *Geophysical Journal*, *96*, 497–517.

Nakada, M., & Okuno, J. (2003). Perturbations of the Earth's rotation and their implications for the present-day mass balance of both polar ice caps. *Geophysical Journal International*, *152*, 124–138.

Nakamura, K., & Kato, Y. (2004). Carbonatization of oceanic crust by the seafloor hydrothermal activity and its significance as a CO_2 sink in the Early Archean. *Geochimica et Cosmochimica Acta*, *68*, 4595–4618.

Nakiboglu, S., & Lambeck, K. (1980). Deglaciation effects on the rotation of the Earth. *Geophysical Journal of the Royal Astronomical Society*, *62*, 49–58.

Nakiboglu, S., & Lambeck, K. (1983). A reevaluation of the isostatic rebound of Lake Bonneville. *Journal of Geophysical Research*, *88*, 10439–10447.

Namiki, N., Iwata, T., Matsumoto, K., Hanada, H., Noda, H., Goossens, S., et al. (2009). Farside gravity field of the Moon from four-way Doppler measurements of SELENE (Kaguya). *Science*, *323*, 900–905.

NASA (2000). *NASA facts: SAC-C satellite.* FS-2000-11-012-GSFC. Greenbelt: Goddard Space Flight Center.

NASA Science News (2001). Planet gobbling dust storms. http://science1.nasa.gov/science-news/science-at-nasa/2001/ast16jul_1/

Nastula, J., & Ponte, R. (1999). Further evidence for oceanic excitation of polar motion. *Geophysical Journal International, 139,* 123–130.

Navarro-González, R., Vargas, E., de la Rosa, J., Raga, A., & McKay, C. (2010). Reanalysis of the Viking results suggests perchlorate and organics at midlatitudes on Mars. *Journal of Geophysical Research, 115,* E12010. doi:10.1029/2010JE003599

Navarro-González, R., Vargas, E., de la Rosa, J., Raga, A., & McKay, C. (2011). Correction to "Reanalysis of the Viking results suggests perchlorate and organics at midlatitudes on Mars." *Journal of Geophysical Research, 116,* E08011. doi:10.1029/2011JE003854

Navrotsky, A. (1980). Lower mantle phase transitions may generally have negative pressure-temperature slopes. *Geophysical Research Letters, 7,* 709–711.

Nawa, K., Suda, N., Fukao, Y., Sato, T., Aoyama, Y., & Shibuya, K. (1998). Incessant excitation of the Earth's free oscillations. *Earth, Planets and Space, 50,* 3–8.

Nayar, K., Sharqawy, M., & Lienhard V, J. (2016). Seawater thermophysical properties library. Downloaded Sept. 23, 2019 from http://web.mit.edu/seawater
See also Nayar, K., Sharqawy, M., Banchik, L., & Lienhard V, J. (2016). Thermophysical properties of seawater: A review and new correlations that include pressure dependence. *Desalination, 390,* 1–24. doi: 10.1016/j.desal.2016.02.024

Nemchin, A., Timms, N., Pidgeon, R., Geisler, T., Reddy, S., & Meyer, C. (2009). Timing of crystallization of the lunar magma ocean constrained by the oldest zircon. *Nature Geoscience Letters, 2,* 133–136.

Nepstad, D., Lefebvre, P., Lopes da Silva, U., Tomasella, J., Schlesinger, P., Solorzano, L., et al. (2004). Amazon drought and its implications for forest flammability and tree growth: A basin-wide analysis. *Global Change Biology, 10,* 704–717. doi:10.1111/j.1529-8817.2003.00772.x

Nerem, R., Beckley, B., Fasullo, J., Hamlington, B., Masters, D., & Mitchum, G. (2018). Climate-change–driven accelerated sea-level rise detected in the altimeter era. *Proceedings of the National Academy of Sciences, 115,* 2022–2025. https://doi.org/10.1073/pnas.1717312115

Nerem, R., Chambers, D., Choe, C., & Mitchum, G. (2010). Estimating mean sea level change from the TOPEX and Jason altimeter missions. *Marine Geodesy, 33*(S1), 435–446. Doi:10.1080/01490419.2010.491031

Nerem, R., Tapley, B., & Shum, C. (1990). Determination of the ocean circulation using Geosat Altimetry. *Journal of Geophysical Research, 95,* 3163–3179 and 3439–3441.

Nettles, M., & Dziewonski, A. (2008). Radially anisotropic shear velocity structure of the upper mantle globally and beneath North America. *Journal of Geophysical Research, 113,* B02303. doi:10.1029/2006JB004819

Nettleton, L. (1939). Determination of density for reduction of gravimeter observations. *Geophysics, 4,* 176–183.

Neumann, W., Breuer, D., & Spohn, T. (2012). Differentiation and core formation in accreting planetesimals. *Astronomy & Astrophysics, 543,* A141. doi:10.1051/0004-6361/201219157

Newsom, H. (1995). Composition of the Solar System, planets, meteorites, and major terrestrial reservoirs. In T. Ahrens (Ed.), *Global Earth Physics: A handbook of physical constants* (pp. 159–189). Washington, DC: American Geophysical Union.

Newton, R. (1968). Experimental evidences for a secular decrease in the gravitational constant G. *Journal of Geophysical Research, 73,* 3765–3772.

Nieto, M. (1972). *The Titius-Bode law of planetary distances: Its history and theory.* New York, NY: Pergamon Press.

Nilsson, A., Holme, R., Korte, M., Suttie, N., & Hill, M. (2014). Reconstructing Holocene geomagnetic field variation: New methods, models and implications. *Geophysical Journal International, 198*, 229–248.

Nilsson, A., Suttie, N., Korte, M., Holme, R., & Hill, M. (2020). Persistent westward drift of the geomagnetic field at the core–mantle boundary linked to recurrent high-latitude weak/reverse flux patches. *Geophysical Journal International, 222*, 1423–1432.

Nimmo, F. (2002). Why does Venus lack a magnetic field? *Geology, 30*, 987–990.

Nimmo, F., & McKenzie, D. (1998). Volcanism and tectonics on Venus. *Annual Review of Earth and Planetary Sciences, 26*, 23–51.

Nimmo, F., & Stevenson, D. (2000). Influence of early plate tectonics on the thermal evolution and magnetic field of Mars. *Journal of Geophysical Research, 105*, 11969–11979.

Nimmo, F., Price, G., Brodholt, J., & Gubbins, D. (2004). The influence of potassium on core and geodynamo evolution. *Geophysical Journal International, 156*, 363–376.

Nishida, K., & Fukao, Y. (2007). Source distribution of Earth's background free oscillations. *Journal of Geophysical Research, 112*. doi:10.1029/2006JB004720

Nishida, K., Kobayashi, N., & Fukao, Y. (2000). Resonant Oscillations Between the Solid Earth and the Atmosphere. *Science, 287*, 2244–2246.

Nishimura, C., & Forsyth, D. (1989). The anisotropic structure of the upper mantle in the Pacific. *Geophysical Journal, 96*, 203–229.

NOAA (2018). Webpage "Global surface temperature anomalies." *National Centers for Environmental Information*. Downloaded from https://www.ncdc.noaa.gov/monitoring-references/faq/anomalies.php

NOAA (2020). Severe Weather 101 – Thunderstorms. National Severe Storms Laboratory. Downloaded from https://www.nssl.noaa.gov/education/svrwx101/thunderstorms/types/

NOAA (2021). Webpage "Heinrich and Dansgaard-Oeschger events." *National Centers for Environmental Information*, dated Oct. 2021. Downloaded from https://www.ncei.noaa.gov/products/paleoclimatology/paleo-perspectives/abrupt-climate-change

Noffke, N., Eriksson, K., Hazen, R., & Simpson, E. (2006). A new window into Early Archean life: Microbial mats in Earth's oldest siliciclastic tidal deposits (3.2 Ga Moodies Group, South Africa). *Geology, 34*, 253–256.

Nolet, G., Allen, R., & Zhao, D. (2007). Mantle plume tomography. *Chemical Geology, 241*, 248–263.

Norabuena, E., Dixon, T., Stein, S., & Harrison, C. (1999). Decelerating Nazca-South America and Nazca-Pacific plate motions. *Geophysical Research Letters, 26*, 3405–3408.

Norman, M., & Nemchin, A. (2014). A 4.2 billion year old impact basin on the Moon: U–Pb dating of zirconolite and apatite in lunar melt rock 67955. *Earth and Planetary Science Letters, 388*, 387–398.

Núñez, M. (1996). The decay of axisymmetric magnetic fields: A review of Cowling's theorem. *SIAM Review, 38*, 553–564.

Nur, A., & Mavko, G. (1974). Postseismic viscoelastic rebound. *Science, 183*, 204–206.

O'Dea, A., H. A. Lessios, A. G. Coates, R. I. Eytan, S. A. Restrepo-Moreno, Cione, A. L., et al. (2016). Formation of the Isthmus of Panama. *Science Advances, 2*, e1600883. doi:10.1126/sciadv.1600883

O'Keefe, J., Eckels, A., & Squires, R. (1959). Vanguard measurements give pear-shaped component of Earth's figure. *Science, 129*, 565–566.

O'Rourke, J., & Stevenson, D. (2016). Powering Earth's dynamo with magnesium precipitation from the core. *Nature, 529*, 387–389.

Och, L., & Shields-Zhou, G. (2012). The Neoproterozoic oxygenation event: Environmental perturbations and biogeochemical cycling. *Earth-Science Reviews, 110*, 26–57.

Oganov, A., & Ono, S. (2004). Theoretical and experimental evidence for a post-perovskite phase of $MgSiO_3$ in Earth's D″ layer. *Nature, 430*, 445–448.

Ogawa, R., & Heki, K. (2007). Slow postseismic recovery of geoid depression formed by the

2004 Sumatra-Andaman Earthquake by mantle water diffusion. *Geophysical Research Letters, 34*, L06313. doi:10.1029/2007GL029340

Ohanian, H. (1986). What is spin? *American Journal of Physics, 54*, 500–505.

Ohishi, M. (2008). Molecular spectral line surveys and the organic molecules in the interstellar molecular clouds. In S. Kwok & S. Sandford (Eds.), *Organic Matter in Space: Proceedings of the IAU Symposium, 251*, pp. 17–25. doi:10.1017/S174392130802108X

Ohta, K., Hirose, K., Lay, T., Sata, N., & Ohishi, Y. (2008). Phase transitions in pyrolite and MORB at lowermost mantle conditions: Implications for a MORB-rich pile above the core–mantle boundary. *Earth and Planetary Science Letters, 267*, 107–117.

Ohuchi, T., Kawazoe, T., Higo, Y., Funakoshi, K., Suzuki, A., Kikegawa, T., & Irifune, T. (2015). Dislocation-accommodated grain boundary sliding as the major deformation mechanism of olivine in the Earth's upper mantle. *Science Advances, 1*, e1500360.

Okal, E. (1996). Radial modes from the great 1994 Bolivian earthquake: No evidence for an isotropic component to the source. *Geophysical Research Letters, 23*, 431–434.

Okal, E. (2001). "Detached" deep earthquakes: Are they really? *Physics of the Earth and Planetary Interiors, 127*, 109–143.

Okal, E. (2008). The generation of *T* waves by earthquakes. *Advances in Geophysics, 49*, 1–65. doi:10.1016/S0065-2687(07)49001-X

Okal, E. (2012). T waves. In H. Gupta (Ed.), *Encyclopedia of solid earth geophysics*. Dordrecht: Springer. doi:10.1007/978-90-481-8702-7

Oldham, R. (1906). The constitution of the interior of the Earth, as revealed by earthquakes. *Quarterly Journal of the Geological Society of London, 62*, 456–475.

Olsen, N. (1998). The electrical conductivity of the mantle beneath Europe derived from *C*-responses from 3 to 720 hr. *Geophysical Journal International, 133*, 298–308.

Olsen, N. (1999). Long-period (30 days–1 year) electromagnetic sounding and the electrical conductivity of the lower mantle beneath Europe. *Geophysical Journal International, 138*, 179–187.

Olsen, N. (2002). A model of the geomagnetic field and its secular variation for epoch 2000 estimated from Ørsted data. *Geophysical Journal International, 149*, 454–462.

Olsen, N., & Mandea, M. (2008). Rapidly changing flows in the Earth's core. *Nature Geoscience, 1*, 390–394.

Olsen, N., & Stolle, C. (2012). Satellite geomagnetism. *Annual Review of Earth and Planetary Sciences, 40*, 441–465.

Olsen, N., Hulot, G., Lesur, V., Finlay, C., Beggan, C., Chulliat, A., et al. (2015). The Swarm Initial Field Model for the 2014 geomagnetic field. *Geophysical Research Letters, 42*, 1092–1098.

Olsen, N., Lühr, H., Finlay, C., Sabaka, T., Michaelis, N., Rauberg, J., & Tøffner-Clausen, L. (2014). The CHAOS-4 geomagnetic field model. *Geophysical Journal International, 197*, 815–827.

Olsen, N., Mandea, M., Sabaka, T., & Tøffner-Clausen, L. (2009). CHAOS-2 – A geomagnetic field model derived from one decade of continuous satellite data. *Geophysical Journal International, 179*, 1477–1487.

Olsen, N., Ravat, D., Finlay, C., & Kother, L. (2017). LCS-1: A high-resolution global model of the lithospheric magnetic field derived from CHAMP and *Swarm* satellite observations. *Geophysical Journal International, 211*, 1461–1477.

Olson, P. (1983). Geomagnetic polarity reversals in a turbulent core. *Physics of the Earth and Planetary Interiors, 33*, 260–274.

Olson, P. (2013). The new core paradox. *Science, 342*, 431–432.

Olson, P. (2016). Mantle control of the geodynamo: Consequences of top-down regulation. *Geochemistry, Geophysics, Geosystems, 17*, 1935–1956.

Olson, P., & Amit, H. (2006). Changes in Earth's dipole. *Naturwissenschaften, 93*, 519–542.

Olson, P., & Amit, H. (2015). Mantle superplumes induce geomagnetic superchrons. *Frontiers of Earth Science, 3.* doi:10.3389/feart.2015.00038

Olson, P., & Aurnou, J. (1999). A polar vortex in the Earth's core. *Nature, 402,* 170–173.

Olson, P., & Christensen, U. (2002). The time-averaged magnetic field in numerical dynamos with non-uniform boundary heat flow. *Geophysical Journal International, 151,* 809–823.

Olson, P., & Christensen, U. (2006). Dipole moment scaling for convection-driven planetary dynamos. *Earth and Planetary Science Letters, 250,* 561–571.

Olson, P., Christensen, U., & Glatzmaier, G. (1999). Numerical modeling of the geodynamo: Mechanisms of field generation and equilibration. *Journal of Geophysical Research, 104,* 10383–10404.

Olson, P., Driscoll, P., & Amit, H. (2009). Dipole collapse and reversal precursors in a numerical dynamo. *Physics of the Earth and Planetary Interiors, 173,* 121–140.

Olson, P., Hinnov, L., & Driscoll, P. (2014). Nonrandom geomagnetic reversal times and geodynamo evolution. *Earth and Planetary Science Letters, 388,* 9–17.

Olson, P., & Kincaid, C. (1991). Experiments on the interaction of thermal convection and compositional layering at the base of the mantle. *Journal of Geophysical Research, 96,* 4347–4354.

Olson, P., & Nam, I. (1986). Formation of sea-floor swells by mantle plumes. *Journal of Geophysical Research, 91,* 7181–7191.

Olson, P., Schubert, G., & Anderson, C. (1993). Structure of axisymmetric mantle plumes. *Journal of Geophysical Research, 98,* 6829–6844.

Olson, P., & Yuen, D. (1982). Thermochemical plumes and mantle phase transitions. *Journal of Geophysical Research, 87,* 3993–4002.

Olson, S., & Hearty, P. (2009). A sustained +21 m sea-level highstand during MIS 11 (400 ka): Direct fossil and sedimentary evidence from Bermuda. *Quaternary Science Reviews, 28,* 271–285.

Olson, S., Kump, L., & Kasting, J. (2013). Quantifying the areal extent and dissolved oxygen concentrations of Archean oxygen oases. *Chemical Geology, 362,* 35–43.

Ono, S., Katsura, T., Ito, E., Kanzaki, M., Yoneda, A., Walter, M., et al. (2001). *In situ* observation of ilmenite-perovskite phase transition in $MgSiO_3$ using synchrotron radiation. *Geophysical Research Letters, 28,* 835–838.

Onstott, T., Balkwill, D., Boone, D., Colwell, F., Griffin, T., Kieft, T., et al. (Taylorsville Basin Working Group) (1994). D.O.E. seeks origin of deep subsurface bacteria. *EOS Transactions, American Geophysical Union, 75,* 385, 395–396.

Oppenheimer, M. (1998). Global warming and the stability of the West Antarctic Ice Sheet. *Nature, 393,* 325–332.

Oro, J. (1961). Comets and the formation of biochemical compounds on the primitive Earth. *Nature, 190,* 389–390.

Orsi, A., Cornuelle, B., & Severinghaus, J. (2012). Little Ice Age cold interval in West Antarctica: Evidence from borehole temperature at the West Antarctic Ice Sheet (WAIS) Divide. *Geophysical Research Letters, 39,* L09710. doi:10.1029/2012GL051260

Ossa Ossa, F., Spangenberg, J. E., Bekker, A., König, S., Stüeken, E. E., Hofmann, A., et al. (2022). Moderate levels of oxygenation during the late stage of Earth's Great Oxidation Event. *Earth and Planetary Science Letters, 594.* https://doi.org/10.1016/j.epsl.2022.117716

Ostro, S. (1993). Planetary radar astronomy. *Reviews of Modern Physics, 65,* 1235–1279.

Owen, T., Cess, R., & Ramanathan, V. (1979). Enhanced CO2 greenhouse to compensate for reduced solar luminosity on early Earth. *Nature, 277,* 640–642.

Ozima, M., & Tatsumoto, M. (1997). Radiation-induced diamond crystallization: Origin of carbonados and its implications on meteorite nano-diamonds. *Geochimica et Cosmochimica Acta, 61,* 369–376.

Pagani, M., Caldeira, K., Archer, D., & Zachos, J. (2006). An ancient carbon mystery. *Science, 314,* 1556–1557.

Pahlevan, K., & Stevenson, D. (2007). Equilibration in the aftermath of the lunar-forming giant impact. *Earth and Planetary Science Letters, 262,* 438–449.

Paillard, D. (1998). The timing of Pleistocene glaciations from a simple multiple-state climate model. *Nature, 391,* 378–381.

Pais, M., & Jault, D. (2008). Quasi-geostrophic flows responsible for the secular variation of the Earth's magnetic field. *Geophysical Journal International, 173,* 421–443.

Pall, P., Aina, T., Stone, D. A., Stott, P. A., Nozawa T., & Hilberts A. G. J. (2011). Anthropogenic greenhouse gas contribution to flood risk in England and Wales in autumn 2000. *Nature, 470,* 382–385. doi:10.1038/nature09762

Palme, H. (2000). Are there chemical gradients in the inner Solar System? *Space Science Reviews, 192,* 237–262.

Palme, H., & Jones, A. (2003). Solar System abundances of the elements. *In Treatise on Geochemistry, 1,* 41–61.

Pannella, G. (1972). Paleontological evidence on the Earth's rotational history since early Precambrian. *Astrophysics and Space Science, 16,* 212–237.

Pannella, G., MacClintock, C., & Thompson, M. (1968). Paleontological evidence of variations in length of synodic month since late Cambrian. *Science, 162,* 792–796.

Paparella, F., & Young, W. (2002). Horizontal convection is non-turbulent. *Journal of Fluid Mechanics, 466,* 205–214.

Parker, E. (1955). Hydromagnetic dynamo models. *Astrophysical Journal, 122,* 293–314.

Parker, E. (1969). The occasional reversal of the geomagnetic field. *Astrophysical Journal, 158,* 815–827.

Parker, E. (1970). The generation of magnetic fields in astrophysical bodies. I. The dynamo equations. *Astrophysical Journal, 162,* 665–673.

Parker, E. (1971). The generation of magnetic fields in astrophysical bodies. IV. The solar and terrestrial dynamos. *Astrophysical Journal, 164,* 491–509.

Parker, L., Atou, T., & Badding, J. V. (1996). Transition element-like chemistry for

potassium under pressure. *Science, 273,* 95–97.

Partin, C., Bekker, A., Planavsky, N., Scott, C., Gill, B., Li, C., et al. (2013b). Large-scale fluctuations in Precambrian atmospheric and oceanic oxygen levels from the record of U in shales. *Earth and Planetary Science Letters, 369–370,* 284–293.

Partin, C., Lalonde, S., Planavsky, N., Bekker, A., Rouxel, O., Lyons, T., & Konhauser, K. (2013a). Uranium in iron formations and the rise of atmospheric oxygen. *Chemical Geology, 362,* 82–90.

Paul, H. (1884). Atmospheric waves from Krakatoa. *Science, 3*(65), 531–532.

Paulson, A., Zhong, S., & Wahr, J. (2007). Inference of mantle viscosity from GRACE and relative sea level data. *Geophysical Journal International, 171,* 497–508.

Pavlis, N., Holmes, S., Kenyon, S., & Factor, J. (2008). An Earth gravitational model to degree 2160: EGM2008. 2008 European Geoscience Union Meeting, Vienna.

Pavlis, N., Holmes, S., Kenyon, S., & Factor, J. (2012). The development and evaluation of the Earth Gravitational Model 2008 (EGM2008). *Journal of Geophysical Research, 117,* B04406. doi:10.1029/2011JB008916

Pavlis, N., Holmes, S., Kenyon, S., & Factor, J. (2013). Correction to "The development and evaluation of the Earth Gravitational Model 2008 (EGM2008)." *Journal of Geophysical Research, 118,* 2633. doi:10.1002/jgrb.50167

Pavlov, A., & Kasting, J. (2002). Mass-independent fractionation of sulfur isotopes in Archean sediments: Strong evidence for an anoxic Archean atmosphere. *Astrobiology, 2,* 27–41.

Pavlov, A., Brown, L., & Kasting, J. (2001). UV shielding of NH_3 and O_2 by organic hazes in the Archean atmosphere. *Journal of Geophysical Research, 106,* 23267–23287.

Pavlov, A., Hurtgen, M., Kasting, J., & Arthur, M. (2003). Methane-rich Proterozoic atmosphere? *Geology, 31,* 87–90.

Pavlov, A., Kasting, J., Brown, L., Rages, K., & Freedman, R. (2000). Greenhouse warming by

CH$_4$ in the atmosphere of early Earth. *Journal of Geophysical Research*, *105*, 11981–11990.

Pavlov, V., & Gallet, Y. (2005). A third superchron during the Early Paleozoic. *Episodes*, *28*, 78–84.

Peale, S., Cassen, P., & Reynolds, R. (1979). Melting of Io by tidal dissipation. *Science*, *203*, 892–894.

Pearson, P., Ditchfield, P., Singano, J., Harcourt-Brown, K., Nicholas, C., Olsson, R., et al. (2001). Warm tropical sea surface temperatures in the Late Cretaceous and Eocene epochs. *Nature*, *413*, 481–488.

Pedlosky, J. (1982). *Geophysical fluid dynamics*. New York, NY: Springer-Verlag.

Peixoto, J., & Oort, A. (1992). *Physics of Climate*. New York, NY: American Institute of Physics.

Peltier, W. (1974). The impulse response of a Maxwell Earth. *Reviews of Geophysics Space Physics*, *12*, 649–669.

Peltier, W. (1995). VLBI baseline variations from the ICE-4G model of postglacial rebound. *Geophysical Research Letters*, *22*, 465–468.

Peltier, W. (1998a). A space geodetic target for mantle viscosity discrimination: Horizontal motions induced by glacial isostatic adjustment. *Geophysical Research Letters*, *25*, 543–546.

Peltier, W. (1998b). Postglacial variations in the level of the sea: Implications for climate dynamics and solid-earth geophysics. *Reviews of Geophysics*, *36*, 603–689.

Peltier, W., & Jiang, X. (1996). Glacial isostatic adjustment and Earth rotation: Refined constraints on the viscosity of the deepest mantle. *Journal of Geophysical Research*, *101*, 3269–3290.

Peltier, W., & Solheim, L. (1992). Mantle phase transitions and layered chaotic convection. *Geophysical Research Letters*, *19*, 321–324.

Peltier, W., & Wu, P. (1982). Mantle phase transitions and the free air gravity anomalies over Fennoscandia and Laurentia. *Geophysical Research Letters*, *9*, 731–734.

Peltier, W., Forte, A., Mitrovica, J., & Dziewonski, A. (1992). Earth's gravitational field: Seismic tomography resolves the enigma of the Laurentian anomaly. *Geophysical Research Letters*, *19*, 1555–1558.

Peltzer, G., Rosen, P., Rogez, F., & Hudnut, K. (1998). Poroelastic rebound along the Landers 1992 earthquake surface rupture. *Journal of Geophysical Research*, *103*, 30131–30145.

Penland, S., & Ramsey, L. (1990). Relative sea-level rise in Louisiana and the Gulf of Mexico: 1908-1988. *Journal of Coastal Research*, *6*, 323–342.

Perryman, M. (2018). *The Exoplanet Handbook* (2nd edition). Cambridge, England: Cambridge University Press.

Persson, M., Futaana, Y., Ramstad, R., Masunaga, K., Nilsson, H., Fedorov, A., & Barabash, S. (2020). Global Venus-solar wind coupling and oxygen ion escape, *Europlanet Science Congress 2020*, online, September 21–October 9, 2020, EPSC2020-344. https://doi.org/10.5194/epsc2020-344

Persson, M., Futaana, Y., Ramstad, R., Schillings, A., Masunaga, K., Nilsson, H., et al. (2021). Global Venus – Solar wind coupling and oxygen ion escape. *Geophysical Research Letters*, *48*, e2020GL091213. https://doi.org/10.1029/2020GL091213

Petereit, J., Saynisch-Wagner, J., Irrgang, C., & Thomas, M. (2019). Analysis of ocean tide-induced magnetic fields derived from oceanic in situ observations: Climate trends and the remarkable sensitivity of shelf regions. *Journal of Geophysical Research*, *124*, 8257–8270. https://doi.org/10.1029/2018JC014768

Peterson, B., & dePaolo, D. (2007). Mass and composition of the continental crust estimated using the CRUST2.0 model. *Eos, Transactions American Geophysical Union. Fall Meeting*, Abstract V33A-1161.

Peterson, T., & Owen, T. (2005). Urban heat island assessment: Metadata are important. *Journal of Climate*, *18*, 2637–2646.

Petigura, E., Howard, A., & Marcy, G. (2013). Prevalence of Earth-size planets orbiting Sun-like stars. *Proceedings of the National Academy of Sciences*, *110*, 19273–19278.

Petit, J., Jouzel, J., Raynaud, D., Barkov, N. I., Barnola, J. -M., Basile, I., et al. (1999). Climate and atmospheric history of the past 420,000 years from the Vostok ice core, Antarctica. *Nature, 399*, 429–436.

Pfennig, N., & Widdel, F. (1982). The bacteria of the sulphur cycle. *Philosophical Transactions of the Royal Society of London B, 298*, 433–441.

Pham, L., Karatekin, Ö., & Dehant, V. (2011). Effects of impacts on the atmospheric evolution: Comparison between Mars, Earth, and Venus. *Planetary and Space Science, 59*, 1087–1092.

Philander, S. (1981). The response of equatorial oceans to a relaxation of the Trade Winds. *Journal of Physical Oceanography, 11*, 176–189.

Phillips, R., Zuber, M., Solomon, S., Golombek, M., Jakosky, B., Banerdt, W., et al. (2001). Ancient geodynamics and global-scale hydrology on Mars. *Science, 291*, 2587–2591.

Phillips, T. (2008). The Tunguska impact – 100 years later. *NASA Science: Science News.* Accessed Dec. 16, 2015 from http://science .nasa.gov/science-news/science-at-nasa/ 2008/30jun_tunguska/

Picin, A., Benazzi, S., Blasco, R., Hajdinjak, M., Helgen, K., Hublin, J., et al. (2021). Comment on "A global environmental crisis 42,000 years ago." *Science, 374*, doi:10.1126/science. abi8330

Pickrell, J. (2006). Top 10: Controversial pieces of evidence for extraterrestrial life. *New Scientist*, online article dated September 4, 2006, retrieved March 8, 2017.

Pierce, D., & Rhines, P. (1996). Convective building of a pycnocline: Laboratory experiments. *Journal of Physical Oceanography, 26*, 176–190.

Piersanti, A., Spada, G., & Sabadini, R. (1997). Global postseismic rebound of a viscoelastic Earth: Theory for finite faults and application to the 1964 Alaska earthquake. *Journal of Geophysical Research, 102*, 477–492.

Pizzarello, S., Krishnamurthy, R., Epstein, S., & Cronin, J. (1991). Isotopic analyses of amino acids from the Murchison meteorite. *Geochimica et Cosmochimica Acta, 55*, 905–910.

Planavsky, N., Asael, D., Hofmann, A., Reinhard, C., Lalonde, S., Knudsen, A., et al. (2014). Evidence for oxygenic photosynthesis half a billion years before the Great Oxidation Event. *Nature Geoscience, 7*, 283–286.

Planavsky, N., Cole, D., Reinhard, C., Diamond, C., Love, G., Luo, G., et al. (2016). No evidence for high atmospheric oxygen levels 1,400 million years ago. *Proceedings of the National Academy of Sciences, 113*, E2550–E2551.

Planavsky, N., Reinhard, C., Wang, X., Thomson, D., McGoldrick, P., Rainbird, R., et al. (2014). Low mid-Proterozoic atmospheric oxygen levels and the delayed rise of animals. *Science, 346*, 635–638.

Plesa, A., Grott, M., Tosi, N., Breuer, D., Spohn, T., & Wieczorek, M. (2016). How large are present-day heat flux variations across the surface of Mars? *Journal of Geophysical Research, 121*, 2386–2403. doi:10.1002/ 2016JE005126

Plescia, J., & Saunders, R. (1979). The chronology of the Martian volcanoes. *Proceedings of the Tenth Lunar and Planetary Science Conference*, 2841–2859.

Poirier, J. (1988). Transport properties of liquid metals and viscosity of the Earth's core. *Geophysical Journal, 92*, 99–105.

Poirier, J. (1994). Light elements in the Earth's outer core: A critical review. *Physics of the Earth and Planetary Interiors, 85*, 319–337.

Poirier, J. (2000). *Introduction to the physics of the earth's interior* (2nd edition). Cambridge, England: Cambridge University Press.

Poirier, J., & Peyronneau, J. (1992). Experimental determination of the electrical conductivity of the material of the Earth's lower mantle. In Y. Syono & M. Manghnani (Eds.), *High-pressure research: Application to Earth and planetary sciences* (pp. 77–87). Tokyo: AGU TERRAPUB.

Poletti, W., Biggin, A., Trindade, R., Hartmann, G., & Terra-Nova, F. (2018). Continuous millennial decrease of the Earth's magnetic axial dipole. *Physics of the Earth and Planetary Interiors, 274*, 72–86.

Pollack, H., & Huang, S. (2000). Climate reconstruction from subsurface temperatures.

Annual Review of Earth and Planetary Sciences, 28, 339–365.

Pollack, H., Hurter, S., & Johnson, J. (1993). Heat flow from the earth's interior: Analysis of the global data set. *Reviews of Geophysics, 31*, 267–280.

Pollack, J. (1990). Atmospheres of the terrestrial planets. In J. Beatty & A. Chaikin (Eds.), *The new solar system* (3rd edition, pp. 91–106). Cambridge: Sky Publishing.

Pollack, J., Hubickyj, O., Bodenheimer, P., Lissauer, J., Podolak, M., & Greenzweig, Y. (1996). Formation of the giant planets by concurrent accretion of solids and gas. *Icarus, 124*, 62–85.

Pollard, D., & DeConto, R. (2009). Modelling West Antarctic ice sheet growth and collapse through the past five million years. *Nature, 458*, 329–333. doi:10.1038/nature07809

Pollitz, F., Burgmann, R., & Banerjee, P. (2006). Post-seismic relaxation following the great 2004 Sumatra-Andaman earthquake on a compressible self-gravitating Earth. *Geophysical Journal International, 167*, 397–420.

Porter, M. (2013). Iron 2013: Major iron deposits of Australia, downloaded Aug. 10, 2017 from Porter GeoConsultancy at http://www .portergeo.com.au/tours/iron2013/ iron2013deposits.asp

Porter, S., & Knoll, A. (2000). Testate amoebae in the Neoproterozoic era: Evidence from vase-shaped microfossils in the Chuar Group, Grand Canyon. *Paleobiology, 26*, 360–385.

Posner, E., Rubie, D., Frost, D., Vlček, V., & Steinle-Neumann, G. (2017). High P–T experiments and first principles calculations of the diffusion of Si and Cr in liquid iron. *Geochimica et Cosmochimica Acta, 203*, 323–342.

Potts, L., & von Frese, R. (2003). Comprehensive mass modeling of the Moon from spectrally correlated free-air and terrain gravity data. *Journal of Geophysical Research, 108*. doi:10.1029/2000JE001440

Poulton, S., Fralick, P., & Canfield, D. (2004). The transition to a sulphidic ocean ~1.84 billion years ago. *Nature, 431*, 173–177.

Pozzo, M., Davies, C., Gubbins, D., & Alfè, D. (2012). Thermal and electrical conductivity of iron at Earth's core conditions. *Nature, 485*, 355–358.

Pozzo, M., Davies, C., Gubbins, D., & Alfè, D. (2013). Transport properties for liquid silicon-oxygen-iron mixtures at Earth's core conditions. *Physical Review B, 87*. doi:10.1103/PhysRevB.87.014110

Pradhan, S., & Roy, G. (2013). Study the crystal structure and phase transition of $BaTiO_3$ – A perovskite. *Researcher, 5*, 63–67.

Prawirodirdjo, L., & Bock, Y. (2004). Instantaneous global plate motion model from 12 years of continuous GPS observations. *Journal of Geophysical Research, 109*, B08405. doi:10.1029/2003JB002944

Prentice, A. (1979). Formation of the Solar System. *Proceedings of the Astronomical Society of Australia, 3*, 300–308.

Press, F., & Ewing, M. (1950). Propagation of explosive sound in a liquid layer overlying a semi-infinite elastic solid. *Geophysics, 15*, 426–466.

Price, C. (2016). ELF electromagneticwaves from lightning: The Schumann resonances. *Atmosphere, 7*, 116. doi:10.3390/atmos7090116

Price, C., & Asfur, M. (2006). Can lightning observations be used as an indicator of upper-tropospheric water vapor variability? *Bulletin of the American Meteorological Society, 34*, 291–298.

Price, C., & Rind, D. (1994). Possible implications of global climate change on global lightning distributions and frequencies. *Journal of Geophysical Research, 99*, 10823–10831.

Pugh, D. (1987). *Tides, surges and mean sea-level: A handbook for engineers and scientists.* Chichester: John Wiley & Sons.

Pulliam, R., Vasco, D., & Johnson, L. (1993). Tomographic inversions for mantle P wave velocity structure based on the minimization of l^2 and l^1 norms of International Seismological Centre travel time residuals. *Journal of Geophysical Research, 98*, 699–734.

Purucker, M., Langlais, B., Olsen, N., Hulot, G., & Mandea, M. (2002). The southern edge of cratonic North America: Evidence from new satellite magnetometer observations. *Geophysical Research Letters, 29*. doi:10.1029/2001GL013645

Purucker, M., Ravat, D., Frey, H., Voorhies, C., Sabaka, T., & Acuña, M. (2000). An altitude-normalized magnetic map of Mars and its interpretation. *Geophysical Research Letters, 27*, 2449–2452.

Püthe, C., Kuvshinov, A., Khan, A., & Olsen, N. (2015). A new model of Earth's radial conductivity structure derived from over 10 yr of satellite and observatory magnetic data. *Geophysical Journal International, 203*, 1864–1872.

Pyle, D. (2015). Sizes of volcanic eruptions. In H. Sigurdsson et al. (Eds.), *Encyclopedia of volcanoes* (2nd edition, pp. 257–264). Amsterdam: Elsevier. http://dx.doi.org/10.1016/B978-0-12-385938-9.00013-4

Qian, W., Wu, K., & Chen, D. (2015). The Arctic and Polar cells act on the Arctic sea ice variation. *Tellus A, 67*, 27692. http://dx.doi.org/10.3402/tellusa.v67.27692

Qian, W., Wu, K., & Liang, H. (2016). Arctic and Antarctic cells in the troposphere. *Theoretical and Applied Climatology, 125*, 1–12.

Qian, W., Wu, K., Leung, J., & Shi, J. (2016). Long-term trends of the Polar and Arctic cells influencing the Arctic climate since 1989. *Journal of Geophysical Research, 121*, 2679–2690.

Queloz, D. (2006). Light through a gravitational lens. *Nature, 439*, 400–401.

Quinn, W. (1974). Monitoring and predicting El Niño invasions. *Journal of Applied Meteorology, 13*, 825–830.

Radkov, A., & Moe, L. (2014). Bacterial synthesis of D-amino acids. *Applied Microbiology and Biotechnology, 98*, 5363–5374.

Rahmstorf, S. (2013). Sea level in the 5th IPCC report. *RealClimate (blog).* Downloaded September 2017 from http://www.realclimate.org/index.php/archives/2013/10/sea-level-in-the-5th-ipcc-report/

Raj Pant, S., Goswami, A., & Finarelli, J. (2014). Complex body size trends in the evolution of sloths (Xenarthra: Pilosa). *BMC Ecology and Evolution, 14*. http://www.biomedcentral.com/1471-2148/14/184

Ramanathan, V. (1988). The greenhouse theory of climate change: A test by an inadvertent global experiment. *Science, 240*, 293–299.

Ramirez, R., Kopparapu, R., Zugger, M., Robinson, T., Freedman, R., & Kasting, J. (2013). Warming early Mars with CO_2 and H_2. *Nature Geoscience, 7*, 59–63.

Randel, W., Covey, C., & Polvani, L. (2021). Global Climate: Stratospheric temperature and winds. In J. Blunden & T. Boyer (Eds.), *State of the Climate in 2020* (pp. S37 –S38). Special Supplement to *Bulletin of the American Meteorological Society, 102*.

Raphael, R. (2021). America's worst winter ever. *HistoryNet.* Downloaded March 2021 from https://www.historynet.com/americas-worst-winter-ever.htm

Rasmusson, E. (1985). El Niño and variations in climate. *American Scientist, 73*, 168–177.

Rasmusson, E., & Carpenter, T. (1982). Variations in tropical sea surface temperature and surface wind fields associated with the Southern Oscillation / El Niño. *Monthly Weather Review, 110*, 354–384.

Rasmusson, E., & Carpenter, T. (1983). The relationship between eastern equatorial Pacific sea surface temperatures and rainfall over India and Sri Lanka. *Monthly Weather Review, 111*, 517–528.

Rasmusson, E., & Wallace, J. (1983). Meteorological aspects of the El Niño / Southern Oscillation. *Science, 222*, 1195–1202.

Ravat, D., Hinze, W., & Taylor, P. (1993). European tectonic features observed by Magsat. *Tectonophysics, 220*, 157–173.

Ravat, D., Wang, B., Wildermuth, E., & Taylor, P. (2002). Gradients in the interpretation of satellite-altitude magnetic data: An example from central Africa. *Journal of Geodynamics, 33*, 131–142.

Rawlinson, N., Pozgay, S., & Fishwick, S. (2010). Seismic tomography: A window into deep

Earth. *Physics of the Earth and Planetary Interiors*, *178*, 101–135.

Raymo, M. (1994). The initiation of northern hemisphere glaciation. *Annual Review of Earth and Planetary Sciences*, *22*, 353–383.

Raymo, M. (1997). The timing of major climate terminations. *Paleoceanography*, *12*, 577–585.

Raymo, M., & Mitrovica, J. (2012). Collapse of polar ice sheets during the stage 11 interglacial. *Nature*, *483*, 453–456. doi:10.1038/nature10891

Raymo, M., Ruddiman, W., & Froelich, P. (1988). Influence of late Cenozoic mountain building on ocean geochemical cycles. *Geology*, *16*, 649–653.

Rayner, N., Parker, D., Horton, E., Folland, C., Alexander, L., Rowell, D., et al. (2003). Global analyses of sea surface temperature, sea ice, and night marine air temperature since the late nineteenth century. *Journal of Geophysical Research*, *108*, 4407. doi:10.1029/2002JD002670

Reager, J., Gardner, A., Famiglietti, J., Wiese, D., Eicker, A., & Lo, M. (2016). A decade of sea level rise slowed by climate-driven hydrology. *Science*, *351*, 699–703.

Reasenberg, R., Shapiro, I., MacNeil, P., Goldstein, R., Breidenthal, J., Brenkle, J., et al. (1979). *Viking* relativity experiment: Verification of signal retardation by solar gravity. *Astrophysical Journal*, *234*, L219–L221.

Reasoner, C., & Revenaugh, J. (1999). Short-period P wave constraints on D″ reflectivity. *Journal of Geophysical Research*, *104*, 955–961.

Rees, B., & Okal, E. (1987). The depth of the deepest historical earthquakes. *Pure and Applied Geophysics*, *125*, 699–715.

Reesink, A., & Bridge, J. (2007). Influence of superimposed bedforms and flow unsteadiness on formation of cross strata in dunes and unit bars. *Sedimentary Geology*, *202*, 281–296.

Regan, R., Cain, J., & Davis, W. (1975). A global magnetic anomaly map. *Journal of Geophysical Research*, *80*, 794–802.

Regan, R., & Marsh, B. (1982). The Bangui Magnetic Anomaly: Its geological origin. *Journal of Geophysical Research*, *87*, 1107–1120.

Regelous, M., Hofmann, A., Abouchami, W., & Galer, S. (2003). Geochemistry of lavas from the Emperor seamounts, and the geochemical evolution of Hawaiian magmatism from 85 to 42 Ma. *Journal of Petrology*, *44*, 113–140.

Reguzzoni, M., Sampietro, D., & Sanso, F. (2013). Global Moho from the combination of the CRUST2.0 model and GOCE data. *Geophysical Journal International*, *195*, 222–237.

Reid, G. (1987). Influence of solar variability on global sea surface temperatures. *Nature*, *329*, 142–143.

Reid, J. (1981). On the mid-depth circulation of the world ocean. In B. Warren & C. Wunsch (Eds.), *Evolution of physical oceanography, scientific surveys in honor of Henry Stommel* (pp. 70–111), Cambridge, MA: MIT Press.

Reipurth, B., Chini, R., Krügel, E., Kreysa, E., & Sievers, A. (1993). Cold dust around Herbig-Haro energy sources – A 1300-μm survey. *Astronomy & Astrophysics*, *273*, 221–238.

Renaud, J., & Henning, W. (2018). Increased tidal dissipation using advanced rheological models: Implications for Io and tidally active exoplanets. *Astrophysical Journal*, *87*. https://doi.org/10.3847/1538-4357/aab784

Renne, P., Deino, A., Hilgen, F., Kuiper, K., Mark, D., Mitchell, W., III, et al. (2013). Time scales of critical events around the Cretaceous-Paleogene boundary. *Science*, *339*, 684–687.

ReVelle, D. (1976). On meteor-generated infrasound. *Journal of Geophysical Research*, *81*, 1217–1230.

ReVelle, D. (2004). Recent advances in bolide entry modeling: A bolide potpourri. *Earth, Moon, and Planets*, *95*, 441–476.

Revelle, R., & Maxwell, A. (1952). Heat flow through the floor of the eastern North Pacific Ocean. *Nature*, *170*, 199–200.

Reynolds, J. (1960). Isotopic composition of primordial xenon. *Physical Review Letters*, *4*, 351–354.

Rhie, J., & Romanowicz, B. (2004). Excitation of Earth's continuous free oscillations by atmosphere-ocean-seafloor coupling. *Nature, 431*, 552–556. doi:10.1038/nature02942

Rice, W., & Armitage, P. (2003). On the formation timescale and core masses of gas giant planets. *Astrophysical Journal, 598*, L55–L58.

Rich, F., Bono, J., Burke, W., & Gentile, L. (2007). A space-based proxy for the Dst index. *Journal of Geophysical Research, 112*, A05211. doi:10.1029/2005JA011586

Rich, P., Rich, T., Wagstaff, B., McEwen-Mason, J., Douthitt, C., Gregory, R., & Felton, E. (1988). Evidence for low temperatures and biologic diversity in Cretaceous high latitudes of Australia. *Science, 242*, 1403–1406.

Richards, M., Bunge, H., Ricard, Y., & Baumgardner, J. (1999). Polar wandering in mantle convection models. *Geophysical Research Letters, 26*, 1777–1780.

Richards, M., Duncan, R., & Courtillot, V. (1989). Flood basalts and hot-spot tracks: Plume heads and tails. *Science, 246*, 103–107.

Richards, M., & Hager, B. (1984). Geoid anomalies in a dynamic Earth. *Journal of Geophysical Research, 89*, 5987–6002.

Richmond, A. (1995). The ionospheric wind dynamo: Effects of its coupling with different atmospheric regions. In R. Johnson & T. Killeen (Eds.), *The upper mesosphere and lower thermosphere: A review of experiment and theory* (pp. 49–65). Washington: American Geophysical Union.

Richter, C. (1958). *Elementary seismology*. San Francisco: Freeman & Co.

Ridgwell, A., Watson, A., & Raymo, M. (1999). Is the spectral signature of the 100 kyr glacial cycle consistent with a Milankovitch origin? *Paleoceanography, 14*, 437–440.

Riding, R., Fralick, P., & Liang, L. (2014). Identification of an Archean marine oxygen oasis. *Precambrian Research, 251, 232–237. Also Supplementary Data.*

Righter, K. (2003). Metal-silicate partitioning of siderophile elements and core formation in the early Earth. *Annual Review of Earth and Planetary Sciences, 31*, 135–174.

Rignot, E., Velicogna, I., van den Broeke, M., Monaghan, A., & Lenaerts, J. (2011). Acceleration of the contribution of the Greenland and Antarctic ice sheets to sea level rise. *Geophysical Research Letters, 38*. doi:10.1029/2011GL046583

Rikitake, T. (1958). Oscillations of a system of disk dynamos. *Mathematical Proceedings of the Cambridge Philosophical Society, 54*, 89–105.

Rimmele, T., Goode, P., Strous, L., & Stebbins, R. (1995). Dark lanes in granulation and the excitation of solar oscillations. In J. Hoeksema, V. Domingo, B. Fleck, & B. Battrick (Eds.), *Helioseismology: Proceedings of 4th SOHO Workshop*, 329–334.

Rind, D., Russell, G., Schmidt, G., Sheth, S., Collins, D., deMenocal, P., & Teller, J. (2001). Effects of glacial meltwater in the GISS coupled atmosphere-ocean model 2. A bipolar seesaw in Atlantic Deep Water production. *Journal of Geophysical Research, 106*, 27355–27365.

Ringwood, A. (1962). A model for the upper mantle. *Journal of Geophysical Research, 67*, 857–867.

Ringwood, A. (1977). Composition of the core and implications for origin of the earth. *Geochemical Journal, 11*, 111–135.

Ringwood, A. (1979). Composition and origin of the Earth. In M. McElhinny (Ed.), *The Earth: Its origin, structure and evolution* (pp. 1–58). New York, NY: Academic Press.

Ringwood, A. (1991). Phase transformations and their bearing on the constitution and dynamics of the mantle (inaugural Ingerson lecture). *Geochimica et Cosmochimica Acta, 55*, 2083–2110.

Ringwood, A., & Irifune, T. (1988). Nature of the 650–km seismic discontinuity: Implications for mantle dynamics and differentiation. *Nature, 331*, 131–136.

Ringwood, A., & Major, A. (1970). The system Mg_2SiO_4 Fe_2SiO_4 at high pressures and temperatures. *Physics of the Earth and Planetary Interiors, 3*, 89–108.

Riser, S., Freeland, H. J., Roemmich, D., Wijffels, S., Troisi, A., Belbéoch, M., et al. (2016).

Fifteen years of ocean observations with the global Argo array. *Nature Climate Change, 6,* 145–153.

Ritsema, J., Ni, S., Helmberger, D., & Crotwell, P. (1998). Evidence for strong shear velocity reductions and velocity gradients in the lower mantle beneath Africa. *Geophysical Research Letters, 25,* 4245–4248.

Ritsema, J., & van Heijst, H. (2000). Seismic imaging of structural heterogeneity in Earth's mantle: Evidence for large-scale mantle flow. *Science Progress, 83,* 243–259.

Rivoldini, A., Van Hoolst, T., Verhoeven, O., Mocquet, A., & Dehant, V. (2011). Geodesy constraints on the interior structure and composition of Mars. *Icarus, 213,* 451–472.

Robbins, S., & Keller, G., Jr. (1992). Complete Bouguer and isostatic residual gravity maps of the Anadarko Basin, Wichita Mountains, and surrounding areas, Oklahoma, Kansas, Texas, and Colorado. *U.S. Geological Survey Bulletin 1866 (Chapter G),* pp. G1–G11.

Roberts, J., Lillis, R., & Manga, M. (2009). Giant impacts on early Mars and the cessation of the Martian dynamo. *Journal of Geophysical Research, 114,* E04009. doi:10.1029/2008JE003287

Roberts, J., & Tyburczy, J. (1999). Partial-melt electrical conductivity: Influence of melt composition. *Journal of Geophysical Research, 104,* 7055–7065.

Roberts, J., Moy, A., van Ommen, T., Curran, M., Worby, A., Goodwin, I., & Inoue, M. (2013). Borehole temperatures reveal a changed energy budget at Mill Island, East Antarctica, over recent decades. *The Cryosphere, 7,* 263–273.

Roberts, P. (1967). Convection in horizontal layers with internal heat generation. *Theory. Journal of Fluid Mechanics, 30,* 33–49.

Roberts, P. (1972). Kinematic dynamo models. *Philosophical Transactions of the Royal Society of London, A, 272,* 663–698.

Roberts, P., & King, E. (2013). On the genesis of the Earth's magnetism. *Reports on Progress in Physics, 76.* doi:10.1088/0034-4885/76/9/096801

Roberts, P., & Soward, A. (1992). Dynamo theory. *Annual Review of Fluid Mechanics, 24,* 459–512.

Roberts, P., Jones, C., & Calderwood, A. (2003). Energy fluxes and ohmic dissipation in the Earth's core. In C. Jones, A. Soward & K. Zhang (Eds.), *Earth's core and lower mantle* (pp. 100–129). London: Taylor & Francis.

Robertson, E. (1988). *Thermal properties of rocks* (US Geol. Surv. Open-File Rep. 88–441). Downloaded July 2016 from https//pubs.usgs.gov

Robine, J., Cheung, S. L. K., Le Roy, S., Van Oyen, H., Griffiths, C., Michel, J.-P., & Herrmann, F. R. (2008). Death toll exceeded 70,000 in Europe during the summer of 2003. *Comptes Rendus Biologies, 331,* 171–178.

Rochester, M. (1970). Core-mantle interactions: Geophysical and astronomical consequences. In L. Mansinha, D. Smylie, & A. Beck (Eds.), *Earthquake displacement fields and the rotation of the earth* (pp. 136–148). Amsterdam: Springer.

Rodgers, S., & Charnley, S. (2001). Organic synthesis in the coma of Comet Hale-Bopp? *Monthly Notices of the Royal Astronomical Society, 320,* L61–L64.

Rogers, J. (1995). *The giant planet Jupiter.* Cambridge, England: Cambridge University Press.

Rohli, R., & Vega, A. (2015). *Climatology* (3rd edition). Burlington, MA: Jones & Bartlett Learning.

Romps, D., Seeley, J., Vollaro, D., & Molinari, J. (2014). Projected increase in lightning strikes in the United States due to global warming. *Science, 346,* 851–854.

Rosat, S. (2004). *Variations temporelles de la gravite en relation avec la dynamique interne de la Terre: Apport des gravimètres supraconducteurs.* PhD Thesis, University of Strasbourg. Downloaded from http://www.srosat.com/PRO/These/HTML/these.htm

Rosat, S., Hinderer, J., Crossley, D., & Boy, J. (2004). Performance of superconducting gravimeters from long-period seismology to tides. *Journal of Geodynamics, 38,* 461–476.

Rosat, S., Sato, T., Imanishi, Y., Hinderer, J., Tamura, Y., McQueen, H., & Ohashi, M.

(2005). High-resolution analysis of the gravest seismic normal modes after the 2004 Mw = 9 Sumatra earthquake using superconducting gravimeter data. *Geophysical Research Letters, 32*. doi:10.1029/2005GL023128

Roscoe, S. (1969). Huronian rocks and uraniferous conglomerates. *Geological Survey of Canada Paper 68–40*. Ottawa: Dept. of Energy, Mines & Resources.

Rosmorduc, V., Benveniste, J., Bronner, E., Dinardo, S., Lauret, O., Maheu, C., et al. (2011). J. Benveniste & N. Picot (Eds.), *Radar altimetry tutorial*, http://www.altimetry.info

Ros-Tonen, M., & van Boxel, J. (1999). El Niño in Latin America: The case of Peruvian fishermen and north-east Brazilian peasants. *European Review of Latin American and Caribbean Studies, 67*, 5–20.

Roy, J., Lapointe, P., & Anderson, P. (1975). Paleomagnetism of the oldest red beds and the direction of the late Aphebian polar wander relative to Laurentia. *Geophysical Research Letters, 2*, 537–540.

Roy, K., & Peltier, W. (2011). GRACE era secular trends in Earth rotation parameters: A global scale impact of the global warming process? *Geophysical Research Letters, 38*. doi:10.1029/2011GL047282

Roy, R., Blackwell, D., & Birch, F. (1968). Heat generation of plutonic rocks and continental heat flow provinces. *Earth and Planetary Science Letters, 5*, 1–12.

Roy, S., & Chapman, D. (2012). Borehole temperatures and climate change: Ground temperature change in south India over the past two centuries. *Journal of Geophysical Research, 117*, D11105. doi:10.1029/2011JD017224

Roy, S., Harris, R., Rao, R., & Chapman, D. (2002). Climate change in India inferred from geothermal observations. *Journal of Geophysical Research, 107*, 2138. 10.1029/2001JB000536

Roy, U., & Sarkar, S. (2017). Large dataset test of Birch's law for sound propagation at high pressure. *Journal of Applied Physics, 121*, 225901, https://doi.org/10.1063/1.4984793

Rubie, D., Nimmo, F., & Melosh, H. (2015). Formation of the Earth's core. In G. Schubert (Ed.), *Treatise on Geophysics* (Vol. 9, pp. 43–76). Amsterdam: Elsevier.

Ruddiman, W. (2003). The Anthropogenic Greenhouse Era began thousands of years ago. *Climatic Change, 61*, 261–293.

Ruddiman, W. (2008). *Earth's climate: Past and future* (2nd edition). New York, NY: W.H. Freeman & Co.

Ruddiman, W. (2014). *Earth's climate: Past and future* (3rd edition). New York, NY: W.H. Freeman & Co.

Ruddiman, W., & Kutzbach, J. (1991). Plateau uplift and climatic change. *Scientific American, 264*, 66–75.

Rudnick, R., McDonough, W., & O'Connell, R. (1998). Thermal structure, thickness and composition of continental lithosphere. *Chemical Geology, 145*, 395–411.

Rugel, G., Faestermann, T., Knie, K., Korschinek, G., Poutivtsev, M., Schumann, D., et al. (2009). New measurement of the ^{60}Fe half-life. *Physical Review Letters, 103*. doi:10.1103/PhysRevLett.103.072502

Runcorn, K. (1975). On the interpretation of lunar magnetism. *Physics of the Earth and Planetary Interiors, 10*, 327–335.

Ryan, J., & Ma, C. (1998). NASA-GSFC's geodetic VLBI program: A twenty-year retrospective. *Physics and Chemistry of the Earth, 23*, 1041–1052.

Rycroft, M., Israelsson, S., & Price, C. (2000). The global atmospheric electric circuit, solar activity and climate change. *Journal of Atmospheric and Solar-Terrestrial Physics, 62*, 1563–1576.

Rye, R., & Holland, H. (1998). Paleosols and the evolution of atmospheric oxygen: A critical review. *American Journal of Science, 298*, 621–672.

Rye, R., Kuo, P., & Holland, H. (1995). Atmospheric carbon dioxide concentrations before 2.2 billion years ago. *Nature, 378*, 603–605.

Sabadini, R., Di Donato, G., Vermeersen, L., Devoti, R., Luceri, V., & Bianco, G. (2002). Ice mass loss in Antarctica and stiff lower mantle viscosity inferred from the long wavelength time dependent gravity field. *Geophysical Research Letters, 29.* doi:10.1029/2001GL014016

Sabadini, R., Yuen, D., & Boschi, E. (1982). Polar wandering and the forced responses of a rotating, multilayered, viscoelastic planet. *Journal of Geophysical Research, 87,* 2885–2903.

Sabadini, R., Yuen, D., & Gasperini, P. (1988). Mantle rheology and satellite signatures from present-day glacial forcings. *Journal of Geophysical Research, 93,* 437–447.

Sabaka, T., Olsen, N., & Langel, R. (2002). A comprehensive model of the quiet-time, near-Earth magnetic field: Phase 3. *Geophysical Journal International, 151,* 32–68.

Sabaka, T., Olsen, N., & Purucker, M. (2004). Extending comprehensive models of the Earth's magnetic field with Ørsted and CHAMP data. *Geophysical Journal International, 159,* 521–547.

Sabaka, T., Tøffner-Clausen, L., Olsen, N., & Finlay, C. (2020). CM6: A comprehensive geomagnetic field model derived from both CHAMP and Swarm satellite observations. *Earth, Planets and Space, 72.* https://doi.org/10.1186/s40623-020-01210-5

Sabaka, T., Tyler, R., & Olsen, N. (2016). Extracting ocean-generated tidal magnetic signals from *Swarm* data through satellite gradiometry. *Geophysical Research Letters, 43,* 3237–3245. doi:10.1002/2016GL068180

Sagan, C., & Khare, B. (1979). Tholins: Organic chemistry of interstellar grains and gas. *Nature, 277,* 102–107.

Sagan, C., & Mullen, G. (1972). Earth and Mars: Evolution of atmospheres and surface temperatures. *Science, 177,* 52–56.

Sahney, S., Benton, M., & Falcon-Lang, H. (2010). Rainforest collapse triggered Carboniferous tetrapod diversification in Euramerica. *Geology, 38,* 1079–1082.

Sahoo, S., Planavsky, N., Kendall, B., Wang, X., Shi, X., Scott, C., et al. (2012). Ocean oxygenation in the wake of the Marinoan glaciation. *Nature, 489,* 546–549.

Sailor, R., & Dziewonski, A. (1978). Measurements and interpretation of normal mode attenuation. *Geophysical Journal of the Royal Astronomical Society, 53,* 559–582.

Saito, T. (2010). Love-wave excitation due to the interaction between a propagating ocean wave and the sea-bottom topography. *Geophysical Journal International, 182,* 1515–1523.

Sakai, S. (1997). The horizontal scale of rotating convection in the geostrophic regime. *Journal of Fluid Mechanics, 333,* 85–95.

Sakai, T., Takahashi, S., Nishitani, N., Mashino, I., Ohtani, E., & Hirao, N. (2014). Equation of state of pure iron and $Fe_{0.9}Ni_{0.1}$ alloy up to 3 Mbar. *Physics of the Earth and Planetary Interiors, 228,* 114–126.

Salcedo-Castro, J., de Camargo, R., Marone, E., & Sepúlveda, H. (2015). Using the mean pressure gradient and NCEP/NCAR reanalysis to estimate the strength of the South Atlantic Anticyclone. *Dynamics of Atmospheres and Oceans, 71,* 83–90.

Sandu, C., & Kiefer, W. (2012). Degassing history of Mars and the lifespan of its magnetic dynamo. *Geophysical Research Letters, 39,* L03201. doi:10.1029/2011GL050225

Sandwell, D., & Smith, W. (1997). Marine gravity anomaly from Geosat and ERS 1 satellite altimetry. *Journal of Geophysical Research, 102,* 10039–10054.

Sandwell, D., Müller, R., Smith, W., Garcia, E., & Francis, R. (2014). New global marine gravity model from CryoSat-2 and Jason-1 reveals buried tectonic structure. *Science, 346,* 65–67.

Sarafian, E., Gaetani, G., Hauri, E., & Sarafian, A. (2017). Experimental constraints on the damp peridotite solidus and oceanic mantle potential temperature. *Science, 355,* 942–945.

Sasgen, I., Wouters, B., Gardner, A., King, M., Tedesco, M., Landerer, F., et al. (2020). Return to rapid ice loss in Greenland and record loss in 2019 detected by the GRACE-FO satellites. *Nature Communications, 1.* https://doi.org/10.1038/s43247-020-0010-1

Sato, T., Miura, S., Sun, W., Sugano, T., Freymueller, J., Larsen, C., et al. (2012). Gravity and uplift rates observed in southeast Alaska and their comparison with GIA model predictions. *Journal of Geophysical Research, 117.* doi:10.1029/2011JB008485

Sátori, G., & Zieger, B. (1999). El Niňo related meridional oscillation of global lightning activity. *Geophysical Research Letters*, *26*, 1365–1368.

Savage, J., & Prescott, W. (1978). Asthenosphere readjustment and the earthquake cycle. *Journal of Geophysical Research*, *83*, 3369–3376.

Savage, J., Prescott, W., Lisowski, M., & King, N. (1981). Strain accumulation in Southern California, 1973-1980. *Journal of Geophysical Research*, *86*, 6991–7001.

Savage, M. (1999). Seismic anisotropy and mantle deformation: What have we learned from shear wave splitting? *Reviews of Geophysics*, *37*, 65–106.

Saynisch, J., Irrgang, C., & Thomas, M. (2018). Estimating ocean tide model uncertainties for electromagnetic inversion studies. *Annales Geophysicae*, *36*, 1009–1014.

Saynisch, J., Petereit, J., Irrgang, C., Kuvshinov, A., & Thomas, M. (2016). Impact of climate variability on the tidal oceanic magnetic signal – A model-based sensitivity study. *Journal of Geophysical Research*, *121*, 5931–5941. doi:10.1002/2016JC012027

Saynisch, J., Petereit, J., Irrgang, C., & Thomas, M. (2017). Impact of oceanic warming on electromagnetic oceanic tidal signals: A CMIP5 climate model-based sensitivity study. *Geophysical Research Letters*, *44*, 4994–5000. doi:10.1002/2017GL073683

Scanlon, B., Zhang, Z., Save, H., Sun, A. Y., Müller Schmied, H., Van Beek, L. P. H. et al. (2018). Global models underestimate large decadal declining and rising water storage trends relative to GRACE satellite data. *Proceedings of the National Academy of Sciences*, *115*, E1080–E1089. www.pnas.org/cgi/doi/10.1073/pnas.1704665115

Schaeffer, N., Jault, D., Nataf, H., & Fournier, A. (2017). Turbulent geodynamo simulations: A leap towards Earth's core. *Geophysical Journal International*, *211*, 1–29.

Schatz, J., & Simmons, G. (1972). Thermal conductivity of Earth materials at high temperature. *Journal of Geophysical Research*, *77*, 6966–6983.

Scherer, R., Aldahan, A., Tulaczyk, S., Possnert, G., Engelhardt, H., & Kamb, B. (1998). Pleistocene collapse of the West Antarctic Ice Sheet. *Science*, *281*, 82–85.

Scherstén, A., Elliott, T., Hawkesworth, C., Russell, S., & Masarik, J. (2006). Hf–W evidence for rapid differentiation of iron meteorite parent bodies. *Earth and Planetary Science Letters*, *241*, 530–542.

Schlamminger, S., Wagner, T., Choi, K., Gundlach, J., & Adelberger, E. (2007). Tests of the equivalence principle. Presentation at AAPT, Seattle.

Schmidt, B., Blankenship, D., Patterson, G., & Schenk, P. (2011). Active formation of "chaos terrain" over shallow subsurface water on Europa. *Nature*, *479*, 502–505.

Schmitt-Kopplin, P., Gabelica, Z., Gougeon, R., Fekete, A., Kanawati, B., Harir, M., et al. (2010). High molecular diversity of extraterrestrial organic matter in Murchison meteorite revealed 40 years after its fall. *Proceedings of the National Academy of Sciences*, *107*, 2763–2768. doi:10.1073/pnas.0912157107

Schmitz, W. (1995). On the interbasin-scale thermohaline circulation. *Reviews of Geophysics*, *33*, 151–173.

Schmitz, W. (1996). On the World Ocean Circulation: Volume II, ONR: *Woods Hole Oceanographic Institution Technical Report, WHOI-96-08.*

Schneider, D., Bickford, M., Cannon, W., Schulz, K., & Hamilton, M. (2002). Age of volcanic rocks and syndepositional iron formations, Marquette Range Supergroup: Implications for the tectonic setting of Paleoproterozoic iron formations of the Lake Superior region. *Canadian Journal of Earth Sciences*, *39*, 999–1012.

Schneider, D., & Kent, D. (1990). The time-averaged paleomagnetic field. *Reviews of Geophysics*, *28*, 71–96.

Schneider, J., Lucas, S., Werneburg, R., & Rößler, R. (2010). Euramerican late Pennsylvanian / early Permian arthropleurid/ tetrapod associations – Implications for the habitat and paleobiology of the largest terrestrial arthropod. *New Mexico Museum of Natural History and Science Bulletin*, *49*, 49–70.

Schnepf, N., Manoj, C., Kuvshinov, A., Toh, H., & Maus, S. (2014). Tidal signals in ocean-bottom magnetic measurements of the Northwestern Pacific: Observation versus prediction. *Geophysical Journal International*, *198*, 1096–1110.

Schoof, C. (2007). Ice sheet grounding line dynamics: Steady states, stability, and hysteresis. *Journal of Geophysical Research*, *112*, F03S28. doi:10.1029/2006JF000664

Schopf, J. (1993). Microfossils of the Early Archean Apex Chert: New evidence of the antiquity of life. *Science*, *260*, 640–646.

Schopf, J., & Packer, B. (1987). Early Archean (3.3-billion to 3.5-billion-year-old) microfossils from Warrawoona Group, *Australia. Science*, *237*, 70–73.

Schrenk, M., Brazelton, W., & Lang, S. (2013). Serpentinization, carbon, and deep life. *Reviews in Mineralogy and Geochemistry*, *75*, 575–606.

Schrinner, M., Petitdemange, L., & Dormy, E. (2012). Dipole collapse and dynamo waves in global direct numerical simulations. *Astrophysical Journal*, *752*. doi:10.1088/0004-637X/752/2/121

Schubert, G. (2009). Io's escape (News & Views). *Nature*, *459*, 920–921.

Schubert, G. (2015). *Treatise on geophysics* (2nd edition, Vol. 8). London: Elsevier.

Schubert, G., Masters, G., Olson, P., & Tackley, P. (2004). Superplumes or plume clusters? *Physics of the Earth and Planetary Interiors*, *146*, 147–162.

Schubert, G., Russell, C., & Moore, W. (2000). Timing of the Martian dynamo. *Nature*, *408*, 666–667.

Schubert, G., Turcotte D., & Olson, P. (2001). *Mantle convection in the Earth and Planets*. Cambridge, England: Cambridge University Press.

Schuberth, B., Bunge, H., & Ritsema, J. (2009). Tomographic filtering of high-resolution mantle circulation models: Can seismic heterogeneity be explained by temperature alone? *Geochemistry, Geophysics, Geosystems*, *10*, Q05W03. doi:10.1029/2009GC002401

Schuh, H., & Behrend, D. (2012). VLBI: A fascinating technique for geodesy and astrometry. *Journal of Geodynamics*, *61*, 68–80.

Schultz, A., Kurtz, R., Chave, A., & Jones, A. (1993). Conductivity discontinuities in the upper mantle beneath a stable craton. *Geophysical Research Letters*, *20*, 2941–2944.

Schwaiger, T., Gastine, T., & Aubert, J. (2019). Force balance in numerical geodynamo simulations: A systematic study. *Geophysical Journal International*, *219*, S101–S114.

Scott, A., & Glasspool, I. (2006). The diversification of Paleozoic fire systems and fluctuations in atmospheric oxygen concentration. *Proceedings of the National Academy of Sciences*, *103*, 10861–10865.

Scott, C., Lyons, T., Bekker, A., Shen, Y., Poulton, S., Chu, X., & Anbar, A. (2008). Tracing the stepwise oxygenation of the Proterozoic ocean. *Nature*, *452*, 456–460.

Scott, J., Marotzke, J., & Adcroft, A. (2001). Geothermal heating and its influence on the meridional overturning circulation. *Journal of Geophysical Research*, *106*, 31141–31154.

Scotti, A., & White, B. (2011). Is horizontal convection really "non-turbulent?" *Geophysical Research Letters*, *38*, L21609. doi:10.1029/2011GL049701

Scrutton, C. (1964). Periodicity in Devonian coral growth. *Palaeontology*, *7*, 552–558.

Sebens, C. (2019). How electrons spin. *Studies in History and Philosophy of Modern Physics*, *68*, 40–50.

Secco, R. (1995). Viscosity of the outer core. *A handbook of physical constants: Mineral Physics and Crystallography* (pp. 218–226). Washington, DC: American Geophysical Union.
See also Wagner, T., Schlamminger, S., Gundlach, J., & Adelberger, E. (2012). Torsion-balance tests of the weak equivalence principle. *Classical and Quantum Gravity*, *29*, 184002 (arXiv:1207.2442v1).

Seed, R., Bea, R. G., Abdelmalak, R. I., Athanasopoulos, A. G., Boutwell, G. P., Bray,

J. D., et al. (2006). *Investigation of the Performance of the New Orleans Flood Protection Systems in Hurricane Katrina on August 29, 2005*: Chapter Three: Geology of the New Orleans region, 50 pp. Downloaded from http://www.ce.berkeley.edu/projects/neworleans/report/CH_3.pdf

Seeds, M., & Backman, D. (2013). *Foundations of astronomy* (12th edition). Boston, MA: Brooks/Cole.

Sella, G., Dixon, T., & Mao, A. (2002). REVEL: A model for recent plate velocities from space geodesy. *Journal of Geophysical Research*, *107*. doi:10.1029/2000JB000033

Sella, G., Stein, S., Dixon, T., Craymer, M., James, T., Mazzotti, S., & Dokka, R. (2007). Observation of glacial isostatic adjustment in "stable" North America with GPS. *Geophysical Research Letters*, *34*, L02306. doi:10.1029/2006GL027081

Sengoku, A. (1998). A plate motion study using Ajisai SLR data. *Earth Planets Space, 50*, 611–627.

Sephton, M. (2002). Organic compounds in carbonaceous meteorites. *Natural Product Reports*, *19*, 292–311.

Sepulchre, P., Ramstein, G., Fluteau, F., Schuster, M., Tiercelin, J., & Brunet, M. (2006). Tectonic uplift and Eastern Africa aridification. *Science*, *313*, 1419–1423.

SETI (2017). SETI@home Classic: In *Memoriam*, at http://setiathome.berkeley.edu/classic.php and SETI@home's transition to BOINC, at http://setiathome.berkeley.edu/transition.php

Seton, M., Whittaker, J., Wessel, P., Müller, D., DeMets, C., Merkouriev, S., et al. (2014). Community infrastructure and repository for marine magnetic identifications. *Geochemistry, Geophysics, Geosystems*, *15*, 1629–1641.

Shackleton, N. (1967). Oxygen isotope analyses and Pleistocene temperatures re-assessed. *Nature*, *215*, 15–17.

Shackleton, N. (2000). The 100,000-year ice-age cycle identified and found to lag temperature, carbon dioxide, and orbital eccentricity. *Science*, *289*, 1897–1902.

Shackleton, N., & Opdyke, N. (1973). Oxygen isotope and palaeomagnetic stratigraphy of equatorial Pacific core V28-238: Oxygen isotope temperatures and ice volumes on a 10^5 year and 10^6 year scale. *Quaternary Research*, *3*, 39–55.

Shankland, T., & Waff, H. (1977). Partial melting and electrical conductivity anomalies in the upper mantle. *Journal of Geophysical Research*, *82*, 5409–5417.

Shankland, T., Nitsan, U., & Duba, A. (1979). Optical absorption and radiative heat transport in olivine at high temperature. *Journal of Geophysical Research*, *84*, 1603–1610.

Shankland, T., O'Connell, R., & Waff, H. (1981). Geophysical constraints on partial melt in the upper mantle. *Reviews of Geophysics and Space Physics*, *19*, 394–406.

Shankland, T., Peyronneau, J., & Poirier, J. (1993). Electrical conductivity of the Earth's lower mantle. *Nature*, *366*, 453–455.

Shannon, A., Jackson, A., Veras, D., & Wyatt, M. (2015). Eight billion asteroids in the Oort cloud. *Monthly Notices of the Royal Astronomical Society*, *446*, 2059–2064.

Shapiro, I., Smith, W., Ash, M., Ingalls, R., & Pettengill, G. (1971). Gravitational constant: Experimental bound on its time variation. *Physical Review Letters*, *26*, 27–30.

Shapiro, N., Campillo, M., Stehly, L., & Ritzwoller, M. (2005). High-resolution surface-wave tomography from ambient seismic noise. *Science*, *307*, 1615–1618.

Sharp, T. (2014). Bridgmanite–named at last. *Science*, *346*, 1057–1058.

Sharp, W., & Clague, D. (2006). 50-Ma initiation of Hawaiian-Emperor Bend records major change in Pacific plate motion. *Science*, *313*, 1281–1284.

Sharqawy, M., Lienhard V, J., & Zubair, S. (2010). Thermophysical properties of seawater: A review of existing correlations and data. *Desalination and Water Treatment*, *16*, 354–380.

Shaw, H., & Moore, J. (1988). Magmatic heat and the El Niño cycle. *EOS (Transactions AGU)*, *69*, 1553, 1564–1565.

Shearer, P. (2000). Upper mantle seismic discontinuities. In S. Karato, A. Forte, R. Liebermann, G. Masters, & L. Stixrude (Eds.), *Earth's deep interior: Mineral Physics and Tomography from the atomic to the global scale*, (pp. 115–131). AGU Geophysical Monograph *117*. Washington: American Geophysical Union.

Sheldon, N. (2006). Precambrian paleosols and atmospheric CO_2 levels. *Precambrian Research*, *147*, 148–155.

Shen, P., & Beck, A. (1991). Least squares inversion of borehole temperature measurements in functional space. *Journal of Geophysical Research*, *96*, 19965–19979.

Shepherd, A., Ivins, E., Geruo, A., Barletta, V. R., Bentley, M. J., Bettadpur, S. et al. (2012). A reconciled estimate of ice-sheet mass balance. *Science*, *338*, 1183–1189.

Sheth, H. (2006). The Deccan beyond the plume hypothesis, http://www.mantleplumes.org/Deccan.html; August 29, 2006 version. Downloaded August 12, 2013.

Sheyko, A., Finlay, C., & Jackson, A. (2016). Magnetic reversals from planetary dynamo waves. *Nature*, *539*, 551–557.

Shim, S., Duffy, T., & Shen, G. (2001). The post-spinel transformation in Mg_2SiO_4 and its relation to the 660-km seismic discontinuity. *Nature*, *411*, 571–574.

Shimoyama, A., & Ogasawara, R. (2002). Dipeptidesand diketopiperazines in the Yamato-791198 and Murchison carbonaceous chondrites. *Origins of Life and Evolution of Biospheres*, *32*, 165–179.

Shindell, D., Lamarque, J.-F., Schulz, M., Flanner, M., Jiao, C., Chin, M., et al. (2013). Radiative forcing in the ACCMIP historical and future climate simulations. *Atmospheric Chemistry and Physics*, *13*, 2939–2974.

Shockley, R., Northrop, J., & Hansen, P. (1982). SOFAR propagation paths from Australia to Bermuda: Comparison of signal speed algorithms and experiments. *Journal of the Acoustical Society of America*, *71*, 51–60.

Siegenthaler, C. (2021). Seiches and the slide/seiche dynamics; subcritical and supercritical subaqueous mass flows and their deposits. Examples from Swiss lakes, *Swiss Journal of Geosciences*, *114*. https://doi.org/10.1186/s00015-021-00394-6

Sigloch, K., McQuarrie, N., & Nolet, G. (2008). Two-stage subduction history under North America inferred from multiple-frequency tomography. *Nature Geoscience*, *1*, 458–462.

Sigloch, K., & Mihalynuk, M. (2013). Intra-oceanic subduction shaped the assembly of Cordilleran North America. *Nature*, *496*, 50–57.

Sigman, D., & Boyle, E. (2000). Glacial/interglacial variations in atmospheric carbon dioxide. *Nature*, *407*, 859–869.

Siingh, D., Gopalakrishnan, V., Singh, R., Kamra, A., Singh, S., Pant, V., et al. (2007). The atmospheric global electric circuit: An overview. *Atmospheric Research*, *84*, 91–110.

Silver, P., Carlson, R., & Olson, P. (1988). Deep slabs, geochemical heterogeneity, and the large-scale structure of mantle convection: Investigation of an enduring paradox. *Annual Review of Earth and Planetary Sciences*, *16*, 477–541.

Simons, M., & Hager, B. (1997). Localization of the gravity field and the signature of glacial rebound. *Nature*, *390*, 500–504.

Simons, M., Solomon, S., & Hager, B. (1997). Localization of gravity and topography: Constraints on the tectonics and mantle dynamics of Venus. *Geophysical Journal International*, *131*, 24–44.

Simpson, R., Jachens, R., Blakely, R., & Saltus, R. (1986). A new isostatic residual gravity map of the conterminous United States, with a discussion on the significance of isostatic residual anomalies. *Journal of Geophysical Research*, *91*, 8348–8372.

Singh, R., & Negi, J. (1979). A reinterpretation of the linear heat flow and heat production relationship for the exponential model of the heat production in the crust. *Geophysical Journal of the Royal Astronomical Society*, *57*, 741–744.

Sinmyo, R., Hirose, K., & Ohishi, Y. (2019). Melting curve of iron to 290 GPa determined in a resistance-heated diamond-anvil cell. *Earth and Planetary Science Letters*, *510*, 45–52.

Skinner, B. (1966). Thermal expansion. In S. Clark, Jr. (Ed.), *Handbook of Physical Constants* (Memoir 97 Section 6), (pp. 75–96). New York, NY: Geological Society of America.

Skinner, B., & Porter, S. (1995). *The dynamic earth* (3rd edition). New York, NY: John Wiley & Sons.

Slack, J., & Cannon, W. (2009). Extraterrestrial demise of banded iron formations 1.85 billion years ago. *Geology, 37*, 1011–1014.

Slater, G., Goldbogen, J., & Pyenson, N. (2017). Independent evolution of baleen whale gigantism linked to Plio-Pleistocene ocean dynamics. *Proceedings of the Royal Society B, 284*, 20170546. http://dx.doi.org/10.1098/rspb .2017.0546

Sleep, N. (1994). Martian plate tectonics. *Journal of Geophysical Research, 99*, 5639–5655.

Sleep, N. (2000). Evolution of the mode of convection within terrestrial planets. *Journal of Geophysical Research, 105*, 17563–17578.

Sleep, N. (2005). Evolution of the continental lithosphere. *Annual Review of Earth and Planetary Sciences, 33*, 369–393.

Sleep, N., & Zahnle, K. (2001). Carbon dioxide cycling and implications for climate on ancient Earth. *Journal of Geophysical Research, 106*, 1373–1399.

Sleep, N., Zahnle, K., & Lupu, R. (2014). Terrestrial aftermath of the Moon-forming impact. *Philosophical Transactions of the Royal Society, A, 372*, 20130172. http://dx.doi.org/10 .1098/rsta.2013.0172

Sleep, N., Zahnle, K., Kasting, J., & Morowitz, H. (1989). Annihilation of ecosystems by large asteroid impacts on the early Earth. *Nature, 342*, 139–142.

Smerdon, J., Pollack, H., Cermak, V., Enz, J., Kresl, M., Safanda, J., & Wehmiller, J. (2006). Daily, seasonal, and annual relationships between air and subsurface temperatures. *Journal of Geophysical Research, 111*, D07101. doi:10.1029/2004JD005578

Smirnov, A., Tarduno, J., Kulakov, E., McEnroe, S., & Bono, R. (2016). Palaeointensity, core thermal conductivity and the unknown age of the inner core. *Geophysical Journal International, 205*, 1190–1195.

Smith, B., Fricker, H. A., Gardner, A. S., Medley, B., Nilsson, J., Paolo, F. S., et al. (2020). Pervasive ice sheet mass loss reflects competing ocean and atmosphere processes. *Science, 368*, 1239–1242.

Smith, D. (2004). Ears in the ocean. *Oceanus, 42*, 1–3.

Smith, D., Kolenkiewicz, R., Dunn, P., & Torrence, M. (1979). The measurement of fault motion by satellite laser ranging. *Tectonophysics, 52*, 59–67.

Smith, D., Zuber, M., Neumann, G., & Lemoine, F. (1997). Topography of the Moon from the Clementine lidar. *Journal of Geophysical Research, 102*, 1591–1611.

Smith, J., Andersen, T. J., Shortt, M., Gaffney, A. M., Truffer, M., Stanton, T. P., et al. (2017). Sub-ice-shelf sediments record history of twentieth-century retreat of Pine Island Glacier. *Nature, 541*, 77–80.

Smith, M., & Dahlen, F. (1981). The period and Q of the Chandler wobble. *Geophysical Journal of the Royal Astronomical Society, 64*, 223–281.

Smith, R., Jordan, M., Steinberger, B., Puskas, C., Farrell, J., Waite, G., et al. (2009). Geodynamics of the Yellowstone hotspot and mantle plume: Seismic and GPS imaging, kinematics, and mantle flow. *Journal of Volcanology and Geothermal Research, 188*, 26–56.

Smylie, D., Brazhkin, V., & Palmer, A. (2009). Direct observations of the viscosity of the Earth's outer core and extrapolation of measurements of the viscosity of liquid iron. *Physics Uspekhi Fizicheskikh Nauk, 52*, 79–92.

Soderlund, K., King, E., & Aurnou, J. (2012). The influence of magnetic fields in planetary dynamo models. *Earth and Planetary Science Letters, 333–334*, 9–20.

Soffen, G. (1976). Scientific results of the Viking missions. *Science, 194*, 1274–1276.

Solheim, L., & Peltier, W. (1994). Avalanche effects in phase transition modulated thermal convection: A model of Earth's mantle. *Journal of Geophysical Research, 99*, 6997–7018.

Solomatov, V. (2007). Magma oceans and primordial mantle differentiation. *In Treatise on Geophysics, 9,* 91–119.

Solomatov, V., & Moresi, L. (1996). Stagnant lid convection on Venus. *Journal of Geophysical Research, 101,* 4737–4753.

Solomatov, V., & Moresi, L. (1997). Three regimes of mantle convection with non-Newtonian viscosity and stagnant lid convection on the terrestrial planets. *Geophysical Research Letters, 24,* 1907–1910.

Solomon, S., Smrekar, S., Bindschadler, D., Grimm, R., Kaula, W., McGill, G., et al. (1992). Venus tectonics: An overview of Magellan observations. *Journal of Geophysical Research, 97,* 13,199–13,255.

Somerville, R. (1967). A nonlinear spectral model of convection in a fluid unevenly heated from below. *Journal of the Atmospheric Sciences, 24,* 665–676.

Spada, G., Antonioli, A., Cianetti, S., & Giunchi, C. (2006). Glacial isostatic adjustment and relative sea-level changes: the role of lithospheric and upper mantle heterogeneities in a 3-D spherical Earth. *Geophysical Journal International, 165,* 692–702.

Spada, G., Ricard, Y., & Sabadini, R. (1992). Excitation of true polar wander by subduction. *Nature, 360,* 452–454.

Spandler, C., & Pirard, C. (2013). Element recycling from subducting slabs to arc crust: A review. *Lithos, 170–171,* 208–223.

Sparrow, E., Husar, R., & Goldstein, R. (1970). Observations and other characteristics of thermals. *Journal of Fluid Mechanics, 41,* 793–800.

Sperling, E., Wolock, C., Morgan, A., Gill, B., Kunzmann, M., Halverson, G., et al. (2015). Statistical analysis of iron geochemical data suggests limited late Proterozoic oxygenation. *Nature, 523,* 451–454.

Squyres, S., Grotzinger, J. P., Arvidson, R. E., Bell, J. F., III, Calvin, W., Christensen, P. R. et al. (2004). In situ evidence for an ancient aqueous environment at Meridiani Planum, Mars. *Science, 306,* 1709–1714.

Sreenivasan, B., & Narasimhan, G. (2017). Damping of magnetohydrodynamic waves in a rotating fluid. *Journal of Fluid Mechanics, 828,* 867–905.

Stacey, F. (1972). Physical properties of the Earth's core. *Geophysical Surveys, 1,* 99–119.

Stacey, F. (1977). *Physics of the earth* (2nd edition). New York, NY: John Wiley & Sons.

Stacey, F., & Davis, P. (2004). High pressure equations of state with applications to the lower mantle and core. *Physics of the Earth and Planetary Interiors, 142,* 137–184.

Stacey, F., & Davis, P. (2008). *Physics of the Earth* (4th edition). Cambridge, England: Cambridge University Press.

Stacey, F., & Loper, D. (2007). A revised estimate of the conductivity of iron alloy at high pressure and implications for the core energy balance. *Physics of the Earth and Planetary Interiors, 161,* 13–18.

Stähler, S., Savas, C., Duran, A. C., García, R., Giardini, D., Huang, Q., et al. (2021). Seismic detection of the Martian core by InSight. In *52nd Lunar and Planetary Science Conference.* Downloaded from https://www.hou.usra.edu/meetings/lpsc2021/pdf/1545.pdf

Stanley, S., & Bloxham, J. (2006). Numerical dynamo models of Uranus' and Neptune's magnetic fields. *Icarus, 184,* 556–572.

Stefani, F., Forbriger, J., Gundrum, T., Herrmannsdörfer, T., & Wosnitza, J. (2021). Mode conversion and period doubling in a liquid rubidium Alfvén-wave experiment with coinciding sound and Alfvén speeds. *Physical Review Letters, 127,* 275001.

Steffen, H., & Wu, P. (2011). Glacial isostatic adjustment in Fennoscandia – A review of data and modeling. *Journal of Geodynamics, 52,* 169–204.

Stein, C., & Stein, S. (1992). A model for the global variation in oceanic depth and heat flow with lithospheric age. *Nature, 359,* 123–129.

Stein, C., Stein, S., Merino, M., Keller, G., Flesch, L., & Jurdy, D. (2014). Was the Midcontinent Rift part of a successful seafloor-spreading episode? *Geophysical Research Letters, 41,* 1465–1470. doi:10.1002/2013GL059176

Stein, J. (1910). A newly discovered document on Halley's Comet in 1066. *Observatory, 33,* 234–238.

Stein, S., & Geller, R. (1978). Time-domain observation and synthesis of split spheroidal and torsional free oscillations of the 1960 Chilean earthquake: Preliminary results. *Bulletin of the Seismological Society of America, 68,* 325–332.

Stein, S., & Wysession, M. (2003). *An introduction to seismology, earthquakes, and earth structure.* Oxford: Blackwell Publishing.

Steinberger, B. (2000). Plumes in a convecting mantle: Models and observations for individual hotspots. *Journal of Geophysical Research, 105,* 11,127–11,152.

Steinberger, B., & Holme, R. (2002). An explanation for the shape of Earth's gravity spectrum based on viscous mantle flow models. *Geophysical Research Letters, 29,* 2019. doi:10.1029/2002GL015476

Steinberger, B., & Holme, R. (2008). Mantle flow models with core-mantle boundary constraints and chemical heterogeneities in the lowermost mantle. *Journal of Geophysical Research, 113,* B05403. doi:10.1029/2007JB005080

Steinberger, B., & O'Connell, R. (1998). Advection of plumes in mantle flow: Implications for hotspot motion, mantle viscosity and plume distribution. *Geophysical Journal International, 132,* 412–434.

Steinberger, B., & O'Connell, R. (2002). The convective mantle flow signal in rates of true polar wander. In J. Mitrovica & L. Vermeersen (Eds.), *Ice sheets, sea level, and the dynamic earth* (pp. 233–256). AGU Geodynamics series, *29.* Washington, DC: American Geophysical Union.

Stephenson, F. (2003). Historical eclipses and Earth's rotation (Harold Jeffreys Lecture 2002). *Astronomy & Geophysics, 44,* 2.22–2.27. doi:10.1046/j.1468-4004.2003.44222.x. See also Morrison, L., & Stephenson, F. (2002). Ancient eclipses and the Earth's rotation. In H. Rickman (Ed.), *Highlights of Astronomy* (Vol. 12, pp. 338–341). International Astronomical Union.

Stevenson, D. (1981). Models of the Earth's core. *Science, 214,* 611–619.

Stevenson, D. (1987a). Limits on lateral density and velocity variations in the Earth's outer core. *Geophysical Journal of the Royal Astronomical Society, 88,* 311–319.

Stevenson, D. (1987b). Mercury's magnetic field: A thermoelectric dynamo? *Earth and Planetary Science Letters, 82,* 114–120.

Stevenson, D. (2003). Planetary magnetic fields. *Earth and Planetary Science Letters, 208,* 1–11.

Stevenson, D. (2010). Planetary magnetic fields: Achievements and prospects. *Space Science Reviews, 152,* 651–664.

Stixrude, L., Scipioni, R., & Desjarlais, M. (2020). A silicate dynamo in the early Earth. *Nature Communications, 11.* https://doi.org/10.1038/s41467-020-14773-4

Stocker, T., & Johnsen, S. (2003). A minimum thermodynamic model for the bipolar seesaw. *Paleoceanography, 18,* 1087. doi:10.1029/2003PA000920

Stockmann, R., Christiansen, F., Olsen, N., & Jackson, A. (2015). POGO satellite orbit corrections: An opportunity to improve the quality of the geomagnetic field measurements? *Earth, Planets and Space, 67.* doi:10.1186/s40623-015-0254-7

Stoffer, P. (2008). *Where's the Hayward fault? A green guide to the fault. USGS Open-File Report* 2008-1135.

Stoker, C., Boston, P., Mancinelli, R., Segal, W., Khare, B., & Sagan, C. (1990). Microbial metabolism of tholin. *Icarus, 85,* 241–256.

Stoks, P., & Schwartz, A. (1981). Nitrogen-heterocyclic compounds in meteorites: Significance and mechanisms of formation. *Geochimica et Cosmochimica Acta, 45,* 563–569.

Stolle, C., Michaelis, I., & Rauberg, J. (2016). The role of high-resolution geomagnetic field models for investigating ionospheric currents at low Earth orbit satellites. *Earth, Plan Space, 68.* doi:10.1186/s40623-016-0494-1

Stolper, D., & Keller, C. (2018). A record of deep-ocean dissolved O_2 from the oxidation state of iron in submarine basalts. *Nature, 553,* 323–327.

Stommel, H., & Arons, A. (1960). On the abyssal circulation of the world ocean- II. An idealized model of the circulation pattern and amplitude in oceanic basins. *Deep-Sea Research, 6*, 217–233.

Stone, P. (1973). The dynamics of the atmospheres of the major planets. *Space Science Reviews, 14*, 444–459.

Storey, L. (1953). An investigation of whistling atmospherics. *Philosophical Transactions of the Royal Society of London, A, 246*, 113–141.

Stramma, L. (1984). Geostrophic transport in the Warm Water Sphere of the eastern subtropical North Atlantic. *Journal of Materials Research, 42*, 537–558.

Strauss, B., Kulp, S., Rasmussen, D., & Levermann, A. (2021). Unprecedented threats to cities from multi-century sea level rise. *Environmental Research Letters, 16*, 114015.

Strigari, L., Barnabè, M., Marshall, P., & Blandford, R. (2012). Nomads of the galaxy. *Monthly Notices of the Royal Astronomical Society, 423*, 1856–1865.

Strom, K., Strom, S., Edwards, S., Cabrit, S., & Skrutskie, M. (1989). Circumstellar material associated with solar-type pre-main-sequence stars: A possible constraint on the timescale for planet building. *Astronomical Journal, 97*, 1451–1470.

Strom, R., Banks, M., Chapman, C., Fassett, C., Forde, J., Head J., III, et al. (2011). Mercury crater statistics from MESSENGER flybys: Implications for stratigraphy and resurfacing history. *Planetary and Space Science, 59*, 1960–1967.

Strom, S., & Edwards, S. (1993). Physical properties and evolutionary time scales of disks around solar-type and intermediate mass stars. In J. Phillips, J. Thorsett, S. Kulkarni (Eds.), *Planets around Pulsars* (Vol. 36, pp. 235–256). Downloaded from http://articles.adsabs.harvard.edu/pdf/1993ASPC...36..235S

Strom, S., Edwards, S., & Skrutskie, M. (1993). Evolutionary time scales for circumstellar disks associated with intermediate- and solar-type stars. In E. Levy & J. Lunine (Eds.), *Protostars and Planets III* (pp. 837–866). Tucson: University of Arizona Press.

Downloaded from http://articles.adsabs.harvard.edu/pdf/1993prpl.conf..837S

Strous, L., Goode, P., & Rimmele, T. (2000). The dynamics of the excitation of solar oscillations. *Astrophysical Journal, 535*, 1000–1013.

Stuiver, M., & Daddario, J. (1963). Submergence of the New Jersey coast. *Science, 142*, 951.

Su, W., Woodward, R., & Dziewonski, A. (1994). Degree 12 model of shear velocity heterogeneity in the mantle. *Journal of Geophysical Research, 99*, 6945–6980.

Sumi, T., Kamiya, K., Bennett, D., Bond, I., Abe, F., Botzler, C., et al. in the *Microlensing Observations in Astrophysics Collaboration* and Udalski, A., Szymański, M., Kubiak, M., Pietrzyński, G., Poleski, R., Soszyński, I., et al. in the *Optical Gravitational Lensing Experiment Collaboration* (2011). Unbound or distant planetary mass population detected by gravitational microlensing. *Nature, 473*, 349–352.

Sumino, H., Burgess, R., Mizukami, T., Wallis, S., Holland, G., & Ballentine, C. (2010). Seawater-derived noble gases and halogens preserved in exhumed mantle wedge peridotite. *Earth and Planetary Science Letters, 294*, 163–172.

Suter, I., Zech, R., Anet, J., & Peter, T. (2014). Impact of geomagnetic excursions on atmospheric chemistry and dynamics. *Climate of the Past, 10*, 1183–1194.

Suzuki, A., Ohtani, E., & Kato, T. (1998). Density and thermal expansion of a peridotite melt at high pressure. *Physics of the Earth and Planetary Interiors, 107*, 53–61.

Svalgaard, L. (2016). Reconstruction of solar extreme ultraviolet flux 1740–2015. *Solar Physics, 291*, 2981–3010. doi:10.1007/s11207-016-0921-2

Svedhem, H., Titov, D., Taylor, F., & Witasse, O. (2007). Venus as a more Earth-like planet. *Nature, 450*, 629–632.

Swanson, R. (2003). Evidence of possible sea-ice influence on Microwave Sounding Unit tropospheric temperature trends in polar regions. *Geophysical Research Letters, 30*, 2040. doi:10.1029/2003GL017938

Szeto, A., & Smylie, D. (1984). Coupled motions of the inner core and possible geomagnetic implications. *Physics of the Earth and Planetary Interiors, 36,* 27–42.

Tackley, P. (2000). Mantle convection and plate tectonics: Toward an integrated physical and chemical theory. *Science, 288,* 2002–2007.

Tackley, P. (2002). Strong heterogeneity caused by deep mantle layering. *Geochemistry, Geophysics, Geosystems, 3.* doi:10.1029/2001GC000167

Tackley, P. (2008). Layer cake or plum pudding? *Nature (News & Views), 1,* 157–158.

Tackley, P. (2012). Dynamics and evolution of the deep mantle resulting from thermal, chemical, phase and melting effects. *Earth Science Reviews, 110,* 1–25.

Tackley, P., Stevenson, D., Glatzmaier, G., & Schubert, G. (1993). Effects of an endothermic phase transition at 670 km depth in a spherical model of convection in the Earth's mantle. *Nature, 361,* 699–704.

Tackley, P., Stevenson, D., Glatzmaier, G., & Schubert, G. (1994). Effects of multiple phase transitions in a three-dimensional spherical model of convection in Earth's mantle. *Journal of Geophysical Research, 99,* 15,877–15,901.

Takei, Y., & Holtzman, B. (2009). Viscous constitutive relations of solid-liquid composites in terms of grain boundary contiguity: 1. Grain boundary diffusion control model. *Journal of Geophysical Research, 114,* B06205. doi:10.1029/2008JB 005850

Talandier, J., Hyvernaud, O., Okal, E., & Piserchia, P. (2002). Long-range detection of hydroacoustic signals from large icebergs in the Ross Sea, Antarctica. *Earth and Planetary Science Letters, 203,* 519–534.

Talmadge, C., Berthias, J., Hellings, R., & Standish, E. (1988). Model-independent constraints on possible modifications of Newtonian gravity. *Physical Review Letters, 61,* 1159–1162.

Tanaka, H., Kono, M., & Uchimura, H. (1995). Some global features of palaeointensity in geological time. *Geophysical Journal International, 120,* 97–102.

Tanimoto, T. (1990). Long-wavelength S-wave velocity structure throughout the mantle. *Geophysical Journal International, 100,* 327–336.

Tanimoto, T. (2001). Continuous free oscillations: Atmosphere-solid earth coupling. *Annual Review of Earth and Planetary Sciences, 29,* 563–584.

Tanimoto, T. (2008). Humming a different tune. *Nature, 452,* 539–541.

Tanimoto, T., Um, J., Nishida, K., & Kobayashi, N. (1998). Earth's continuous oscillations observed on seismically quiet days. *Geophysical Research Letters, 25,* 1553–1556.

Tapley, B., Bettadpur, S., Ries, J., Thompson, P., & Watkins, M. (2004). GRACE measurements of mass variability in the Earth System. *Science, 305,* 503–505.

Tapley, B., Bettadpur, S., Watkins, M., & Reigber, C. (2004). The gravity recovery and climate experiment: Mission overview and early results. *Geophysical Research Letters, 31.* doi:10.1029/2004GL019920

Tapley, B., Ries, J., Bettadpur, S., Chambers, D., Cheng, M., Condi, F., et al. (2005). GGM02 – An improved Earth gravity field model from GRACE. *Journal of Geodesy, 79,* 467–478.

Tapley, B., Ries, J., Bettadpur, S., Chambers, D., Cheng, M., Condi, F., & Poole, S. (2007). The GGM03 mean Earth gravity model from GRACE. 2007 Fall AGU Meeting, San Francisco.

Tapley, B., Watkins, M. M., Flechtner, F., Reigber, C., Bettadpur, S., Rodell, M., et al. (2019). Contributions of GRACE to understanding climate change. *Nature Climate Change, 9,* 358–369.

Tarduno, J. (1990). Brief reversed polarity interval during the Cretaceous Normal Polarity Superchron. *Geology, 18,* 683–686.

Tarduno, J., Cottrell, R., Bono, R., Oda, H., Davis, W., Fayek, M., et al. (2020). Paleomagnetism indicates that primary magnetite in zircon records a strong Hadean geodynamo. *Proceedings of the National Academy of Sciences, 117,* 2309–2318. www.pnas.org/cgi/doi/10.1073/pnas.1916553117

Tarduno, J., Cottrell, R., Davis, W., Nimmo, F., & Bono, R. (2015). A Hadean to Paleoarchean geodynamo recorded by single zircon crystals. *Science, 349*, 521–524.

Tarduno, J., Cottrell, R., Watkeys, M., & Bauch, D. (2007). Geomagnetic field strength 3.2 billion years ago recorded by single silicate crystals. *Nature, 446*, 657–660.

Tarduno, J., Cottrell, R., Watkeys, M., Hofmann, A., Doubrovine, P., Mamajek, E., et al. (2010). Geodynamo, solar wind, and magnetopause 3.4 to 3.45 billion years ago. *Science, 327*, 1238–1240.

Tartèse, R., Anand, M., Gattacceca, J., Joy, K., Mortimer, J., Pernet-Fisher, J., et al. (2019). Constraining the evolutionary history of the Moon and the inner Solar System: A case for new returned lunar samples. *Space Science Reviews, 215*. https://doi.org/10.1007/s11214-019-0622-x

Tateno, S., Hirose, K., Sata, N., & Ohishi, Y. (2009). Determination of post-perovskite phase transition boundary up to 4400 K and implications for thermal structure in D″ layer. *Earth and Planetary Science Letters, 277*, 130–136.

Tauxe, L., & Hartl, P. (1997). 11 million years of Oligocene geomagnetic field behavior. *Geophysical Journal International, 128*, 217–229.

Tauxe, L., Steindorf, J., & Harris, A. (2006). Depositional remanent magnetization: Toward an improved theoretical and experimental foundation. *Earth and Planetary Science Letters, 244*, 515–529.

Tavares, F. (2020). Arrokoth revealed: A first in-depth look at a pristine world. Downloaded April 2021 from https://www.nasa.gov/feature/ames/arrokoth-first-look

Taylor, L., Leake, J., Quirk, J., Hardy, K., Banwart, S., & Beerling, D. (2009). Biological weathering and the long-term carbon cycle: Integrating mycorrhizal evolution and function into the current paradigm. *Geobiology, 7*, 171–191.

Taylor, P. (2007). Bangui Anomaly. In D. Gubbins & E. Herrero-Bervera (Eds.), *Encyclopedia of geomagnetism and paleomagnetism*. Dordrecht: Springer. https://doi.org/10.1007/978-1-4020-4423-6

Taylor, P., & Frawley, J. (1987). Magsat anomaly data over the Kursk region, U.S.S.R. *Physics of the Earth and Planetary Interiors, 45*, 255–265.

Taylor, P., Kis, K., & Wittmann, G. (2014). Satellite-altitude horizontal magnetic gradient anomalies used to define the Kursk Magnetic Anomaly. *Journal of Applied Geophysics, 109*, 133–139.

Taylor, T., & Taylor, E. (1990). *Antarctic Paleobiology: Its role in the reconstruction of Gondwana*. New York, NY: Springer-Verlag.

Tebbens, S., Cande, S., Kovacs, L., Parra, J., LaBrecque, J., & Vergara, H. (1997). The Chile ridge: A tectonic framework. *Journal of Geophysical Research, 102*, 12,035–12,059.

Teke, K., Tanir Kayikçe, E., Böhm, J., & Schuh, H. (2012). Modelling very long baseline interferometry (VLBI) observations. *Journal of Geodesy and Geoinformation, 1*, 17–26.

Tera, F., Papanastassiou, D., & Wasserburg, G. (1973). A lunar cataclysm at ∼ 3.95 AE and the structure of the lunar crust. *Lunar Science Conference IV*, pp. 723–725.

Tera, F., Papanastassiou, D., & Wasserburg, G. (1974). Isotopic evidence for a terminal lunar cataclysm. *Earth and Planetary Science Letters, 22*, 1–21.

Thallner, D., Biggin, A., & Halls, H. (2021). An extended period of extremely weak geomagnetic field suggested by palaeointensities from the Ediacaran Grenville dykes (SE Canada). *Earth and Planetary Science Letters, 568*. https://doi.org/10.1016/j.epsl.2021.117025

Thébault, E., Finlay, C. C., Beggan, C. D., Alken, P., Aubert, J., Barrois, O., et al. (2015). International Geomagnetic Reference Field: The 12th generation. *Earth Planets Space, 67*. doi:10.1186/s40623-015-0228-9

Thiagarajan, S. (2015). Simulating the motion of a self-stable bicycle. COMSOL Blog, available from Simulating the Motion of a Self-Stable Bicycle | COMSOL Blog.

Thoma, M., Jenkins, A., Holland, D., & Jacobs, S. (2008). Modelling circumpolar deep water intrusions on the Amundsen Sea continental shelf, *Antarctica. Geophysical Research Letters, 35*, L18602. doi:10.1029/2008GL034939

Thomas, C., Garnero, E., & Lay, T. (2004). High-resolution imaging of lowermost mantle structure under the Cocos plate. *Journal of Geophysical Research, 109*, B08307. doi:10.1029/2004JB003013

Thomas, C., Kendall, J., & Lowman, J. (2004). Lower-mantle seismic discontinuities and the thermal morphology of subducted slabs. *Earth and Planetary Science Letters, 225*, 105–113.

Thomas, D. (1995). The Roswell Incident and Project Mogul. Skeptical Inquirer, 19. Downloaded December 10, 2015 from http://www.csicop.org/si/show/roswell_incident_and_project_mogul

Thompson, M., Christensen-Dalsgaard, J., Miesch, M., & Toomre, J. (2003). The internal rotation of the sun. *Annual Review of Astronomy and Astrophysics, 41*, 599–643.

Thompson, M., Toomre, J., Anderson, E. R., Antia, H. M., Berthomieu, G., Burtonclay, D., et al. (1996). Differential rotation and dynamics of the solar interior. *Science, 272*, 1300–1305.

Thorne, C., Harmar, O., Watson, C., Clifford, N., Biedenharn, D., & Measures, R. (2008). *Current and historical sediment loads in the lower Mississippi river. Report to the U.S. Army*. London.

Thrasher, B., & Sloan, L. (2009). Carbon dioxide and the early Eocene climate of western North America. *Geology, 37*, 807–810.

Tice, M., & Lowe, D. (2004). Photosynthetic microbial mats in the 3,416-Myr-old ocean. *Nature, 431*, 549–552.

Tigrine, Z., Nataf, H., Schaeffer, N., Cardin, P., & Plunian, F. (2019). Torsional Alfvén waves in a dipolar magnetic field: Experiments and simulations. *Geophysical Journal International, 219*, S83–S100.

Tissot, F., Dauphas, N., & Grossman, L. (2016). Origin of uranium isotope variations in early solar nebula condensates. *Science Advances, 2*, e1501400 (corrected version).

Tivey, M., Sager, W., Lee, S., & Tominaga, M. (2006). Origin of the Pacific Jurassic quiet zone. *Geology, 34*, 789–792.

Todoeschuck, J., & Rochester, M. (1980). The effect of compressible flow on anti-dynamo theorems. *Nature, 284*, 250–251.

Toh, H., Satake, K., Hamano, Y., Fujii, Y., & Goto, T. (2011). Tsunami signals from the 2006 and 2007 Kuril earthquakes detected at a seafloor geomagnetic observatory. *Journal of Geophysical Research, 116*, B02104. doi:10.1029/2010JB007873

Toon, O., Pollack, J., Ward, W., Burns, J., & Bilski, K. (1980). The astronomical theory of climatic change on Mars. *Icarus, 44*, 552–607.

Torsvik, T., Doubrovine, P., Steinberger, B., Gaina, C., Spakman, W., & Domeier, M. (2017). Pacific plate motion change caused the Hawaiian-Emperor Bend. *Nature Communications, 8*. doi:10.1038/ncomms15660

Torsvik, T., Steinberger, B., Gurnis, M., & Gaina, C. (2010). Plate tectonics and net lithosphere rotation over the past 150 My. *Earth and Planetary Science Letters, 291*, 106–112.

Tosca, N., & McLennan, S. (2006). Chemical divides and evaporite assemblages on Mars. *Earth and Planetary Science Letters, 241*, 21–31.

Tosi, N., Čadek, O., & Martinec, Z. (2009). Subducted slabs and lateral viscosity variations: Effects on the long-wavelength geoid. *Geophysical Journal International, 179*, 813–826.

Tosi, N., Sabadini, R., Marotta, A., & Vermeersen, L. (2005). Simultaneous inversion for the Earth's mantle viscosity and ice mass imbalance in Antarctica and Greenland. *Journal of Geophysical Research, 110*. doi:10.1029/2004JB003236

TOU (1971). *The Earth, its shape, internal structure and composition / The Earth's magnetic field*, M. Pentz et al. (Eds.), Science Foundation Course Units 22 / 23. Portsmouth: The Open University Press.

Touboul, M., Kleine, T., Bourdon, B., Palme, H., & Wieler, R. (2007). Late formation and prolonged differentiation of the Moon inferred from W isotopes in lunar metals. *Nature, 450,* 1206–1209.

Tozer, D. (1972). The present thermal state of the terrestrial planets. *Physics of the Earth and Planetary Interiors, 6,* 182–197.

Trampert, J., Vacher, P., & Vlaar, N. (2001). Sensitivities of seismic velocities to temperature, pressure and composition in the lower mantle. *Physics of the Earth and Planetary Interiors, 124,* 255–267.

Treiman, A. (1995). S≠NC: Multiple source areas for Martian meteorites. *Journal of Geophysical Research, 100,* 5329–5340.

Trenberth, K., & Caron, J. (2001). Estimates of meridional atmosphere and ocean heat transports. *Journal of Climate, 14,* 3433–3443.

Trenberth, K., & Fasullo, J. (2013). An apparent hiatus in global warming? *Earth's Future, 1,* 19–32. doi:10.1002/2013EF000165

Tritton, D. (1980). *Physical fluid dynamics.* Berkshire (England): Van Nostrand Reinhold.

Tritton, D. (1988). *Physical fluid dynamics* (2nd edition). New York, NY: Oxford University Press.

Tritton, D., & Zarraga, M. (1967). Convection in horizontal layers with internal heat generation. *Experiments. Journal of Fluid Mechanics, 30,* 21–31.

Trossman, D., & Tyler, R. (2019). Predictability of ocean heat content from electrical conductance. *Journal of Geophysical Research, 124,* 667–679.

Trouet, V., Scourse, J., & Raible, C. (2012). North Atlantic storminess and Atlantic meridional overturning circulation during the last millennium: Reconciling contradictory proxy records of NAO variability. *Global and Planetary Change, 84–85,* 48–55.

Tsuchiya, T., Tsuchiya, J., Umemoto, K., & Wentzcovitch, R. (2004). Phase transition in MgSiO$_3$ perovskite in the Earth's lower mantle. *Earth and Planetary Science Letters, 224,* 241–248.

Tucker, J., & Mukhopadhyay, S. (2014). Evidence for multiple magma ocean outgassing and atmospheric loss episodes from mantle noble gases. *Earth and Planetary Science Letters, 393,* 254–265.

Turbet, M., Bolmont, E., Chaverot, G., Ehrenreich, D., Leconte, J., & Marcq, E. (2021). Day-night cloud asymmetry prevents early oceans on Venus but not on Earth. *Nature, 598,* 276–280.

Turcotte, D. (1974). Membrane tectonics. *Geophysical Journal of the Royal Astronomical Society, 36,* 33–42.

Turcotte, D. (1989). A heat pipe mechanism for volcanism and tectonics on Venus. *Journal of Geophysical Research, 94,* 2779–2785.

Turcotte, D., & Oxburgh, E. (1967). Finite amplitude convective cells and continental drift. *Journal of Fluid Mechanics, 28,* 29–42.

Turcotte, D., & Oxburgh, E. (1972). Mantle convection and the new global tectonics. *Annual Review of Fluid Mechanics, 4,* 33–66.

Turcotte, D., & Oxburgh, E. (1973). Mid-plate tectonics. *Nature, 244,* 337–339.

Turcotte, D., & Schubert, G. (1982). *Geodynamics: Applications of continuum physics to geological problems.* New York, NY: John Wiley & Sons.

Turcotte, D., & Schubert, G. (2002). *Geodynamics* (2nd edition). Cambridge, England: Cambridge University Press.

Turcotte, D., McAdoo, D., & Caldwell, J. (1978). An elastic-perfectly plastic analysis of the bending of the lithosphere at a trench. *Tectonophysics, 47,* 193–205.

Turekian, K., & Clark, S., Jr. (1969). Inhomogeneous accumulation of the Earth from the primitive solar nebula. *Earth and Planetary Science Letters, 6,* 346–348.

Turner, G., Cadogan, P., & Yonge, C. (1973). Argon selenochronology. *Proceedings of the Fourth Lunar and Planetary Science Conference, 2,* 1889–1914.

Tyler, R. (2005). A simple formula for estimating the magnetic fields generated by tsunami flow. *Geophysical Research Letters, 32,* L09608. doi:10.1029/2005GL022429

Tyler, R., Boyer, T., Minami, T., Zweng, M., & Reagan, J. (2017). Electrical conductivity of the global ocean. *Earth, Planets and Space, 69.* doi:10.1186/s40623-017-0739-7

Tyler, R., Maus, S., & Lühr, H. (2003). Satellite observations of magnetic fields due to ocean tidal flow. *Science, 299*, 239–241.

Tyler, R., & Mysak, L. (1995). Motionally-induced electromagnetic fields generated by idealized ocean currents. *Geophysical & Astrophysical Fluid Dynamics, 80*, 167–204.

Uchida, T., Wang, Y., Rivers, M., & Sutton, S. (2001). Stability field and thermal equation of state of ε-iron determined by synchrotron X-ray diffraction in a multianvil apparatus. *Journal of Geophysical Research, 106*, 21799–21810.

Ulrich, R. (1970). The five-minute oscillations on the solar surface. *Astrophysical Journal, 162*, 993–1002.

Unnikrishnan, C., Mohapatra, A., & Gillies, G. (2001). Anomalous gravity data during the 1997 total solar eclipse do not support the hypothesis of gravitational shielding. *Physical Review D, 63*. doi:10.1103/PhysRevD.63.062002

Urey, H. (1951). The origin and development of the earth and other terrestrial planets. *Geochimica et Cosmochimica Acta, 1*, 209–277.

Urey, H. (1952). *The planets: Their origin and development*. New Haven, CT: Yale University Press.

Urey, H. (1955). The cosmic abundances of potassium, uranium, and thorium and the heat balances of the Earth, the Moon, and Mars. *Proceedings of the National Academy of Sciences, 41*, 127–144.

Urrutia, J., & Stenzel, R. (2018). Whistler modes in highly nonuniform magnetic fields. I. Propagation in two-dimensions. *Physics of Plasmas, 25*, 082108. doi.org/10.1063/1.5030703

USGS (1997). Introduction to Potential Fields: Gravity (FS-239-95). Downloaded April, 2022 from https://pubs.usgs.gov/fs/fs-0239-95/

Utada, H., & Baba, K. (2014). Estimating the electrical conductivity of the melt phase of a partially molten asthenosphere from seafloor magnetotelluric sounding data. *Physics of the Earth and Planetary Interiors, 227*, 41–47.

Utada, H., Koyama, T., Shimizu, H., & Chave, A. (2003). A semi-global reference model for electrical conductivity in the mid-mantle beneath the north Pacific region. *Geophysical Research Letters, 30*, 1194. doi:10.1029/2002GL016092

Uzan, J.-P. (2003). The fundamental constants and their variation: Observational and theoretical status. *Reviews of Modern Physics, 75*, 403–455.

Uzan, J.-P. (2011). Varying constants, gravitation and cosmology. *Living Reviews in Relativity, 14*, 155 pp (arXiv:1009.5514v1).

Valet, J., & Meynadier, L. (1993). Geomagnetic field intensity and reversals during the past four million years. *Nature, 366*, 234–238.

Valet, J., Meynadier, L., & Guyodo, Y. (2005). Geomagnetic dipole strength and reversal rate over the past two million years. *Nature, 435*, 802–805.

Valet, J., & Valladas, H. (2010). The Laschamp–Mono Lake geomagnetic events and the extinction of Neanderthal: A causal link or a coincidence? *Quaternary Science Reviews, 29*, 3887–3893.

Vallis, G. (1986). El Niño: A chaotic dynamical system? *Science, 232*, 243–245.

Vallis, G. (2005). *Atmospheric and oceanic fluid dynamics*. Cambridge, England: Cambridge University Press. Available from www.princeton.edu/~gkv/aofd

Vallis, G. (2017). *Atmospheric and oceanic fluid dynamics* (2nd edition). Cambridge, England: Cambridge University Press.

van der Hilst, R., de Hoop, M., Wang, P., Shim, S., Ma, P., & Tenorio, L. (2007). Seismostratigraphy and thermal structure of Earth's core-mantle boundary region. *Science, 315*, 1813–1817.

van der Hilst, R., Engdahl, R., Spakman, W., & Nolet, G. (1991). Tomographic imaging of subducted lithosphere below northwest Pacific island arcs. *Nature, 353*, 37–43.

van der Hilst, R., & Karason, H. (1999). Compositional heterogeneity in the bottom 1000 kilometers of Earth's mantle: Toward a hybrid convection model. *Science, 283*, 1885–1888.

van der Hilst, R., & Seno, T. (1993). Effects of relative plate motion on the deep structure and penetration depth of slabs below the Izu-Bonin and Mariana island arcs. *Earth and Planetary Science Letters, 120*, 395–407.

van der Hilst, R., Widiyantoro, S., & Engdahl, R. (1997). Evidence for deep mantle circulation from global tomography. *Nature, 386*, 578–584.

Van Dyke, M. (1982). *An album of fluid motion.* Stanford: Parabolic Press, 176 pp., as cited in Cimbala (2016), with credit for the photo in our Fig. 18b ascribed to M. Velarde, M. Yuste & J. Salan.

Van Hengstum, P., Scott, D., & Javaux, E. (2009). Foraminifera in elevated Bermudian caves provide further evidence for +21 m eustatic sea level during Marine Isotope Stage 11. *Quaternary Science Reviews, 28*, 1850–1860.

Van Valen, L. (1971). The history and stability of atmospheric oxygen. *Science, 171*, 439–443.

Vance, D. (1994). Iron – The environmental impact of a universal element. *National Environmental Journal*, 4, 24–25. Available at http://2the4.net/iron.htm, copyright 2008; downloaded July 2014.

Varadi, F., Runnegar, B., & Ghil, M. (2003). Successive refinements in long-term integrations of planetary orbits. *Astrophysical Journal, 592*, 620–630.

Vasavada, A., & Showman, A. (2005). Jovian atmospheric dynamics: An update after *Galileo* and *Cassini*. *Reports on Progress in Physics, 68*, 1935–1996.

Vasco, D., Johnson, L., Pulliam, R., & Earle, P. (1994). Robust inversion of IASP91 travel time residuals for mantle P and S velocity structure, earthquake mislocations, and station corrections. *Journal of Geophysical Research, 99*, 13,727–13,755.

Vasil, G., Julien, K., & Featherstone, N. (2021). Rotation suppresses giant-scale solar convection. *Proceedings of the National Academy of Sciences, 118*. https://doi.org/10.1073/pnas.2022518118

Velicogna, I. (2009). Increasing rates of ice mass loss from the Greenland and Antarctic ice sheets revealed by GRACE. *Geophysical Research Letters, 36*. doi:10.1029/2009GL040222

Velicogna, I., Mohajerani, Y., Landerer, G. A, F., Mouginot, J., Noel, B., Rignot, E., et al. (2020). Continuity of ice sheet mass loss in Greenland and Antarctica from the GRACE and GRACE Follow-On missions. *Geophysical Research Letters, 47*. https://doi.org/10.1029/2020GL087291

Velímský, J., Šachl, L., & Martinec, Z. (2019). The global toroidal magnetic field generated in the Earth's oceans. *Earth and Planetary Science Letters, 509*, 47–54.

Verdun, J., Klingelé, E., Bayer, R., Cocard, M., Geiger, A., & Kahle, H. (2003). The alpine Swiss-French airborne gravity survey. *Geophysical Journal International, 152*, 8–19.

Verhoeven, O., Mocquet, A., Vacher, P., Rivoldini, A., Menvielle, M., Arrial, P., et al. (2009). Constraints on thermal state and composition of the Earth's lower mantle from electromagnetic impedances and seismic data. *Journal of Geophysical Research, 114*, B03302. doi:10.1029/2008JB005678

Verhoogen, J. (1961). Heat balance of the Earth's core. *Geophysical Journal of the Royal Astronomical Society, 4*, 276–281.

Verhoogen, J., Turner, F., Weiss, L., Wahrhaftig, C., & Fyfe, W. (1970). *The Earth: An introduction to physical geology.* New York, NY: Holt, Rinehart & Winston.

Vermeersen, L., Sabadini, R., & Spada, G. (1996). Compressible rotational deformation. *Geophysical Journal International, 126*, 735–761.

Vermeersen, L., Sabadini, R., Spada, G., & Vlaar, N. (1994). Mountain building and Earth rotation. *Geophysical Journal International, 117*, 610–624.

Vermeij, G. (2016). Gigantism and its implications for the history of life. *PLoS ONE, 11*, e0146092. doi:10.1371/journal.pone.0146092

Vestine, E. (1953). On variations of the geomagnetic field, fluid motions, and the rate of the Earth's rotation. *Journal of Geophysical Research (Terrestrial Magnetism and Atmospheric Electricity), 58*, 127–145.

Vestine, E., & Kahle, A. (1968). The westward drift and geomagnetic secular change. *Geophysical Journal of the Royal Astronomical Society*, *15*, 29–37.

Vickers-Rich, P., & Rich, T. (1993). Australia's polar dinosaurs. *Scientific American*, *269*, 50–55.

Vickery, A., & Melosh, H. (1990). Atmospheric erosion and impactor retention in large impacts, with application to mass extinctions. *Geological Society of America Special Paper*, *247*, 289–300.

Vigny, C., Simons, W., Abu, S., Bamphenyu, R., Satirapod, C., Choosakul, N., et al. (2005). Insight into the 2004 Sumatra–Andaman earthquake from GPS measurements in southeast Asia. *Nature*, *436*, 201–206.

Vila, C., Discetti, S., Carlomagno, G., Astarita, T., & Ianiro, A. (2016). On the onset of horizontal convection. *International Journal of Thermal Sciences*, *110*, 96–108.

Vimont, I., Hall, B., Montzka, S., Dutton, G., Siso, C., Crotwell, M., & Gentry, M. (2021). Global Climate: Ozone-depleting Substances. In J. Blunden & T. Boyer (Eds.), *State of the Climate in 2020* (pp. S87–S89). Special Supplement to *Bulletin of the American Meteorological Society*, 102.

Vivier, F., Maier-Reimer, E., & Tyler, R. (2004). Simulations of magnetic fields generated by the Antarctic Circumpolar Current at satellite altitude: Can geomagnetic measurements be used to monitor the flow? *Geophysical Research Letters*, *31*, L10306. doi:10.1029/2004GL019804

Vočadlo, L. (2007). Ab initio calculations of the elasticity of iron and iron alloys at inner core conditions: Evidence for a partially molten inner core? *Earth and Planetary Science Letters*, *254*, 227–232.

Vočadlo, L., Alfè, D., Gillan, M., & Price, G. (2003). The properties of iron under core conditions from first principles calculations. *Physics of the Earth and Planetary Interiors*, *140*, 101–125.

Vočadlo, L., Alfè, D., Price, G., & Gillan, M. (2000). First principles calculations on the diffusivity and viscosity of liquid Fe-S at experimentally accessible conditions. *Physics of the Earth and Planetary Interiors*, *120*, 145–152.

Völgyesi, L., & Moser, M. (1982). The inner structure of the Earth, *Periodica Polytechnica Chemical Engineering*, *26*, 155–204.

von Schuckmann, K., & le Traon, P. (2011). How well can we derive Global Ocean Indicators from Argo data? *Ocean Science Discussions*, *8*, 999–1024.

Voorhies, C., Sabaka, T., & Purucker, M. (2002). On magnetic spectra of Earth and Mars. *Journal of Geophysical Research*, *107*, E6, 5034. doi:10.1029/2001JE001534

Voosen, P. (2021). Kauri trees mark magnetic flip 42,000 years ago. *Science*, *371*, 766.

Voss, H., Imhof, W., Walt, M., Mobilia, J., Gaines, E., Reagan, J., et al. (1984). Lightning-induced electron precipitation. *Nature*, *312*, 740–742.

Vozoff, K. (1990). Magnetotellurics: Principles and practice. *Proceedings of the Indian Academy of Sciences (Earth & Planetary Sciences)*, *99*, 441–471.

Wacey, D., Kilburn, M., Saunders, M., Cliff, J., & Brasier, M. (2011). Microfossils of sulphur-metabolizing cells in 3.4-billion-year-old rocks of Western Australia. *Nature Geoscience*, *4*, 698–702. Also *Supplement*, 18 pp.

Wacey, D., McLoughlin, N., & Brasier, M. (2008). Looking through windows onto the earliest history of life on Earth and Mars. In J. Seckbach & M. Walsh, (Eds.), *From Fossils to Astrobiology: Cellular origin, life in extreme habitats and Astrobiology* (Vol. 11, pp. 39–68). Dordrecht: Springer.

Wacey, D., McLoughlin, N., Whitehouse, M., & Kilburn, M. (2010). Two coexisting sulfur metabolisms in a ca. 3400 Ma sandstone. *Geology*, *38*, 1115–1118.

Wada, Y., van Beek, L., Sperna Weiland, F., Chao, B., Wu, Y., & Bierkens, M. (2012). Past and future contribution of global groundwater depletion to sea-level rise. *Geophysical Research Letters*, *39*, L09402. doi:10.1029/2012GL051230

Wadhwa, M., Amelin, Y., Davis, A., Lugmair, G., Meyer, B., Gounelle, M., & Desch, S. (2007).

From dust to planetesimals: Implications for the solar protoplanetary disk from short-lived radionuclides. In B. Reipurth, D. Jewitt, & K. Keil (Eds.), *Protostars and Planets, V* (pp. 835–848). Tucson: University of Arizona Press.

Waff, H. (1974). Theoretical considerations of electrical conductivity in a partially molten mantle and implications for geothermometry. *Journal of Geophysical Research, 79,* 4003–4010.

Wagner, C., & McAdoo, D. (2004). Time variations in the GRACE gravity field: Applications to global hydrologic mass flux. In C. Reigber & B. Tapley (Eds.), *Proceedings of the Joint CHAMP/GRACE Science Meeting,* GFZ Potsdam.

Wahl, S., Hubbard, W. B., Militzer, B., Guillot, T., Miguel, Y., Movshovitz, N., et al. (2017). Comparing Jupiter interior structure models to *Juno* gravity measurements and the role of a dilute core. *Geophysical Research Letters, 44,* 4649–4659. doi:10.1002/2017GL073160

Wahl, S., Wilson, H., & Militzer, B. (2013). Solubility of iron in metallic hydrogen and stability of dense cores in giant planets. *Astrophysical Journal, 773.* doi:10.1088/0004-637X/773/2/95

Wahr, J. (1981). The forced nutations of an elliptical, rotating, elastic and oceanless earth. *Geophysical Journal of the Royal Astronomical Society, 64,* 705–727.

Walcott, R. (1972a). Late Quaternary vertical movements in eastern North America: Quantitative evidence of glacio-isostatic rebound. *Reviews of Geophysics Space Physics, 10,* 849–884.

Walcott, R. (1972b). Past sea levels, eustasy and deformation of the Earth. *Quaternary Research, 2,* 1–14.

Walcott, R. (1973). Structure of the earth from glacio-isostatic rebound. *Annual Review of Earth and Planetary Sciences, 1,* 15–37.

Walker, M., Head, M., Berkelhammer, M., Björck, S., Cheng, H., Cwynar, L., et al. (2018). Formal ratification of the subdivision of the Holocene Series/Epoch (Quaternary System/Period): Two new Global Boundary Stratotype Sections and Points (GSSPs) and three new stages/subseries. *Episodes, 41,* 213–223.

Walsh, M., & Lowe, D. (1985). Filamentous microfossils from the 3,500-Myr-old Onverwacht Group, Barberton Mountain Land, South Africa. *Nature, 314,* 530–532.

Walter, F., Brown, A., Mathieu, R., Myers, P., & Vrba, F. (1988). X-ray sources in region of star formation III. Naked T Tauri stars associated with the Taurus-Auriga complex, *Astronomical Journal, 96,* 297–325.

Walter, M., & Tronnes, R. (2004). Early Earth differentiation. *Earth and Planetary Science Letters, 225,* 253–269.

Walter, M., Buick, R., & Dunlop, J. (1980). Stromatolites 3,400-3,500 Myr old from the North Pole area, Western Australia. *Nature, 284,* 443–445.

Wang, C. (1968). Constitution of the lower mantle as evidenced from shock wave data for some rocks. *Journal of Geophysical Research, 73,* 6459–6476.

Wang, C. (1970). Density and constitution of the mantle. *Journal of Geophysical Research, 75,* 3264–3284.

Wang, C. (2014). Ocean circulation links to global sea surface temperature biases in climate models. *AOML Program Review, NOAA.* Downloaded March 12, 2021 from https://www.aoml.noaa.gov/wp-content/uploads/2019/06/Climate_Models_Wang.pdf

Wang, Q. (2016). Homologous temperature of olivine: Implications for creep of the upper mantle and fabric transitions in olivine. *Science China Earth Sciences, 59,* 1138–1156.

Wang, W., Wang, Y., & Wu, R. (2005). A new view on the Ferrel cell. *Chinese Journal of Geophysics, 48,* 539–545.

Ward, W. (1979). Present obliquity oscillations of mars: Fourth-order accuracy in orbital *e* and *I. Journal of Geophysical Research, 84,* 237–241.

Ward, W. (1981). Solar nebula dispersal and the stability of the planetary system I. *Scanning secular resonance theory. Icarus, 47,* 234–264.

Ward, W. (1997). Protoplanet migration by nebula tides. *Icarus, 126,* 261–281.

Wardinski, I., Dostal, J., Mandea, M., & Thomas, M. (2009). Toward monitoring magnetic signals generated by ocean tides (abstract), *Geophysical Research Abstracts, 11*, 2009 EGU General Assembly, abstract id EGU2009-2818.

Wardinski, I., Lesur, V., Kuvshinov, A., & Mandea, M. (2007). Ocean circulation generated magnetic signals: Application to geomagnetic field modeling (abstract), *EOS Transactions AGU*, Fall Meeting abstract id GP33C-1441.

Warren, B. (1983). Why is no deep water formed in the North Pacific? *Journal of Materials Research, 41*, 327–347.

Wasson, J., & Warren, P. (1980). Contribution of the mantle to the lunar asymmetry. *Icarus, 44*, 752–771.

Watts, A. (2001). *Isostasy and flexure of the lithosphere*. Cambridge, England: Cambridge University Press.

Watts, A., & Ribe, N. (1984). On geoid heights and flexure of the lithosphere at seamounts. *Journal of Geophysical Research, 89*, 11,152–11,170.

Watts, A., & Zhong, S. (2000). Observations of flexure and the rheology of oceanic lithosphere. *Geophysical Journal International, 142*, 855–875.

Way, M., Del Genio, A., Kiang, N., Sohl, L., Grinspoon, D., Aleinov, I., et al. (2016). Was Venus the first habitable world of our solar system? *Geophysical Research Letters, 43*, 8376–8383.

WCRP (2018) (Cazenave, A., Meyssignac, B., Ablain, M., Balmaseda, M., Bamber, J., Barletta, V., et al.). Global sea-level budget 1993–present. *Earth System Science Data, 10*, 1551–1590.

Weaver, R., & McAndrew, J. (1995). *The Roswell report: Fact versus Fiction in the New Mexico Desert*. Washington, DC: Headquarters United States Air Force.

Webb, S. (2007). The Earth's "hum" is driven by ocean waves over the continental shelves. *Nature, 445*, 754–756 *plus* supplementary information.

Weber, K., Achenbach, L., & Coates, J. (2006). Microorganisms pumping iron: Anaerobic microbial iron oxidation and reduction. *Nature Reviews, 4*, 752–764.

Weber, R., Lin, P., Garnero, E., Williams, Q., & Lognonné, P. (2011). Seismic detection of the lunar core. *Science, 331*, 309–312.

Webster, P., Holland, G., Curry, J., & Chang, H. (2005). Changes in tropical cyclone number, duration, and intensity in a warming environment. *Science, 309*, 1844–1846.

Weertman, J. (1974). Stability of the junction of an ice sheet and an ice shelf. *Journal of Glaciology, 13*, 3–11.

Wei, M., & Sandwell, D. (2006a). Estimates of heat flow from Cenozoic seafloor using global depth and age data. *Tectonophysics, 417*, 325–335.

Wei, M., & Sandwell, D. (2006b). Reply to comment on: "Estimates of heat flow from Cenozoic seafloor using global depth and age data." *Tectonophysics, 428*, 101–103.

Wei, Y., Fraenz, M., Dubinin, E., Woch, J., Lühr, H., Wan, W., et al. (2012). Enhanced atmospheric oxygen outflow on Earth and Mars driven by a corotating interaction region. *Journal of Geophysical Research, 117*, A03208. doi:10.1029/2011JA017340

Weigelt, M., van Dam, T., Jäggi, A., Prange, L., Tourian, M., Keller, W., & Sneeuw, N. (2013). Time-variable gravity signal in Greenland revealed by high-low satellite-to-satellite tracking. *Journal of Geophysical Research, 118*, 3848–3859.

Weinstein, S. (1993). Catastrophic overturn of the earth's mantle driven by multiple phase changes and internal heat generation. *Geophysical Research Letters, 20*, 101–104.

Weinstein, S. (1995). The effects of a deep mantle endothermic phase change on the structure of thermal convection in silicate planets. *Journal of Geophysical Research, 100*, 11,719–11,728.

Weinstein, S., & Olson, P. (1989). The proximity of hotspots to convergent and divergent plate boundaries. *Geophysical Research Letters, 16*, 433–436.

Weiss, B., Fong, L., Vali, H., Lima, E., & Baudenbacher, F. (2008). Paleointensity of the ancient Martian magnetic field. *Geophysical*

Research Letters, *35*, L23207. doi:10.1029/2008GL035585

Weiss, B., Vali, H., Baudenbacher, F., Kirschvink, J., Stewart, S., & Shuster, D. (2002). Records of an ancient Martian magnetic field in ALH84001. *Earth and Planetary Science Letters*, *201*, 449–463.

Wells, J. (1963). Coral growth and geochronometry. *Nature*, *197*, 948–950.

Wen, L. (2001). Seismic evidence for a rapidly varying compositional anomaly at the base of the Earth's mantle beneath the Indian Ocean. *Earth and Planetary Science Letters*, *194*, 83–95.

Wen, L., Silver, P., James, D., & Kuehnel, R. (2001). Seismic evidence for a thermo-chemical boundary at the base of the Earth's mantle. *Earth and Planetary Science Letters*, *189*, 141–153.

Westerling, A. (2016). Increasing western US forest wildfire activity: Sensitivity to changes in the timing of spring. *Philosophical Transactions of the Royal Society B*, *371*, 20150178.

Westerling, A., Hidalgo, H., Cayan, D., & Swetnam, T. (2006). Warming and earlier spring increase western U.S. forest wildfire activity. *Science*, *313*, 940–943. doi:10.1126/science.1128834

Wetherill, G. (1985). Occurrence of giant impacts during the growth of the terrestrial planets. *Science*, *228*, 877–879.

Whalen, A., & Baalke, R. (2003). Stardust: NASA's comet sample return mission. Downloaded March 13, 2017 from http://stardust.jpl.nasa.gov/science/overview.html

Whaler, K. (1994). Downward continuation of Magsat lithospheric anomalies to the Earth's surface. *Geophysical Journal International*, *116*, 267–278.

Whipple, F. (1950). A comet model. I. The acceleration of Comet Encke. *Astrophysical Journal*, *111*, 375–394.

Whipple, F. (1951). A comet model. II. Physical relations for comets and meteors. *Astrophysical Journal*, *113*, 464–474.

Whitaker, R., & Norris, D. (2008). Infrasound propagation. pp. 1497–1519, In D. Havelock, S.

Kuwano & M. Vorländer (Eds.), *Handbook of signal processing in acoustics*. New York, NY: Springer.

Whitman, W., Coleman, D., & Wiebe, W. (1998). Prokaryotes: The unseen majority. *Proceedings of the National Academy of Sciences*, *95*, 6578–6583.

Widmer-Schnidrig, R. (2003). What can superconducting gravimeters contribute to normal-mode seismology? *Bulletin of the Seismological Society of America*, *93*, 1370–1380.

Wiechert, U., Halliday, A., Lee, D., Snyder, G., Taylor, L., & Rumble, D. (2001). Oxygen isotopes and the Moon-forming giant impact. *Science*, *294*, 345–348.

Wijffels, S., Willis, J., Domingues, C., Barker, P., White, N., Gronell, A., et al. (2008). Changing expendable bathythermograph fall rates and their impact on estimates of thermosteric sea level rise. *Journal of Climate*, *21*, 5657–5672.

Will, C. (1981). *Theory and experiment in gravitational physics*. Cambridge, England: Cambridge University Press.

Will, C. (2006). The confrontation between general relativity and experiment. *Living Reviews in Relativity*, *9*, 3, online version cited March 2013. www.Livingreviews.org/lrr-2006-3

Williams, D. (2008). The Apollo 15 Hammer-Feather Drop (video). http://nssdc.gsfc.nasa.gov/planetary/lunar/apollo_15_feather_drop.html

Williams, D. (2013a). Mars Fact Sheet, *updated* 01 July 2013. http://nssdc.gsfc.nasa.gov/planetary/factsheet/marsfact.html

Williams, D. (2013b). Venus Fact Sheet. NASA website http://nssdc.gsfc.nasa.gov/planetary/factsheet/venusfact.html, downloaded 1/7/14.

Williams, D. (2016a). *Planetary Fact Sheets*. Greenbelt, MD: NASA Goddard Space Flight Center. Downloaded from https://nssdc.gsfc.nasa.gov/planetary/planetfact.html

Williams, D. (2016b). Sun Fact Sheet. Downloaded from http://nssdc.gsfc.nasa.gov/planetary/factsheet/sunfact.html

Williams, D. (2019). *Planetary Fact Sheets: Earth Fact Sheet*. Greenbelt, MD: NASA Goddard Space Flight Center. Available at https://nssdc.gsfc.nasa.gov/planetary/planetfact.html

Williams, D., & Von Herzen, R. (1974). Heat loss from the Earth: New estimate. *Geology, 2*, 327–328.

Williams, E. (1992). The Schumann resonance: A global tropical thermometer. *Science, 256*, 1184–1187.

Williams, E., Rothkin, K., Stevenson, D., & Boccippio, D. (2000). Global lightning variations caused by changes in thunderstorm flash rate and by changes in the number of thunderstorms. *Journal of Applied Meteorology, 39*, 2223–2230.

Williams, G. (1989). Late Precambrian tidal rhythmites in South Australia and the history of the Earth's rotation. *Journal of the Geological Society of London, 146*, 97–111.

Williams, G. (1997). Precambrian length of day and the validity of tidal rhythmite paleotidal values. *Geophysical Research Letters, 24*, 421–424.

Williams, G. (2000). Geological constraints on the Precambrian history of Earth's rotation and the Moon's orbit. *Reviews of Geophysics, 38*, 37–59.

Williams, J., Blitz, L., & McKee, C. (1999). The structure and evolution of molecular clouds: From clumps to cores to the IMF. In V. Mannings, A. Boss, S. Russell (Eds.), *Protostars and Planets IV*. Tucson: University of Arizona Press. Downloaded from arXiv:astro-ph/9902246

Williams, J., & Cieza, L. (2011). Protoplanetary disks and their evolution. *Annual Review of Astronomy and Astrophysics, 49*, 67–117.

Williams, J., & Dickey, J. (2002). Lunar geophysics, geodesy, and dynamics. *13th International Workshop on Laser Ranging*, Washington, DC. Downloaded Sept. 1, 2013 from http://ilrs.gsfc.nasa.gov/docs/williams_lw13.pdf

Williams, J., & Dickey, J. (2002). Lunar geophysics, geodesy, and dynamics. *13th International Workshop on Laser Ranging*, Washington, DC. Downloaded Sept. 1, 2013 from http://ilrs.gsfc.nasa.gov/docs/williams_lw13.pdf

Williams, J., Newhall, X., & Dickey, J. (1996). Relativity parameters determined from lunar laser ranging. *Physical Review D, 53*, 6730–6739.

Williams, J., Turyshev, S., & Boggs, D. (2004). Progress in Lunar Laser Ranging tests of relativistic gravity. *Physical Review Letters, 93*. doi:10.1103/PhysRevLett.93.261101

Williams, R., Grotzinger, J. P., Dietrich, W. E., Gupta, S., Sumner, D. Y., Wiens, R. C., et al. (2013). Martian fluvial conglomerates at Gale Crater. *Science, 340*, 1068–1072.

Williamson, E., & Adams, L. (1923). Density distribution of the Earth. *Journal of the Washington Academy of Sciences, 13*, 413–428.

Willson, R. (1982). Solar irradiance variations and solar activity. *Journal of Geophysical Research, 87*, 4319–4326.

Wilson, D. (1993). Confirmation of the astronomical calibration of the magnetic polarity timescale from sea-floor spreading rates. *Nature, 364*, 788–790.

Wilson, J. (1965). Convection currents and continental drift: XIII. Evidence from ocean islands suggesting movement in the Earth. *Philosophical Transactions of the Royal Society of London, A, 258*, 145–167.

Wilson, R. (1972). Palaeomagnetic differences between normal and reversed field sources, and the problem of far-sided and right-handed pole positions. *Geophysical Journal of the Royal Astronomical Society, 28*, 295–304.

Wilson, R., Tudhope, A., Brohan, P., Briffa, K., Osborn, T., & Tett, S. (2006). Two-hundred-fifty years of reconstructed and modeled tropical temperatures. *Journal of Geophysical Research, 111*, C10007. doi:10.1029/2005JC003188

Wilson, S. (2001). *United States Geological Survey Certificate of Analysis*. Dunite, Twin Sisters Mountain DTS-2B. Downloaded 8/3/16 from http://crustal.usgs.gov/geochemical_reference_standards/pdfs/dunitedts2.pdf

Wiltschko, W., & Wiltschko, R. (2005). Magnetic orientation and magnetoreception in birds and other animals. *Journal of Comparative Physiology, A, 191*, 675–693.

Wittmann, A., & Xu, Z. (1987). A catalogue of sunspot observations from 165 BC to AD 1984. *Astronomy and Astrophysics, Supplement Series, 70*, 83–94.

WMO (World Meteorological Organization) (2018). *Scientific assessment of ozone depletion.* Global Ozone Research and Monitoring Project – Report No. 58, Geneva.

Wolfe, C., & Solomon, S. (1998). Shear-wave splitting and implications for mantle flow beneath the MELT region of the East Pacific Rise. *Science, 280*, 1230–1232.

Wolszczan, A., & Frail, D. (1992). A planetary system around the millisecond pulsar PSR1257 + 12. *Nature, 355*, 145–147.

Wolter, K. (2011). Extended Multivariate ENSO Index (MEI.ext). *NOAA Earth System Research Laboratory: Physical Sciences Division.* Downloaded Sept. 2015 from http://www.esrl.noaa.gov/psd/enso/mei.ext/index.html

Wolter, K., & Timlin, M. (2011). El Niño/Southern Oscillation behaviour since 1871 as diagnosed in an extended multivariate ENSO index (MEI.ext). *International Journal of Climatology, 31*, 1074–1087.

Woodhouse, J., & Dziewonski, A. (1989). Seismic modelling of the Earth's large-scale three-dimensional structure. *Philosophical Transactions of the Royal Society of London, A, 328*, 291–308.

Woolum, D., & Cassen, P. (1999). Astronomical constraints on nebular temperatures: Implications for planetesimal formation. *Meteoritics & Planetary Science, 34*, 897–907.

Woon, D. (2017). Interstellar and circumstellar molecules. *The Astrochymist* (online compilation). http://www.astrochymist.org/astrochymist_ism.html

Wordsworth, R. (2016). The climate of early Mars. *Annual Review of Earth and Planetary Sciences, 44*, 381–408.

Wordsworth, R., Forget, F., Millour, E., Head, J., Madeleine, J., & Charnay, B. (2013). Global modelling of the early Martian climate under a denser CO_2 atmosphere: Water cycle and ice evolution. *Icarus, 222*, 1–19.

Worley, S., Woodruff, S., Reynolds, R., Lubker, S., & Lott, N. (2005). ICOADS Release 2.1 data and products. *International Journal of Climatology, 25*, 823–842.

Wright, J., Fakhouri, O., Marcy, G., Han, E., Feng, Y., Johnson, J., et al. (2011). The exoplanet orbit database. *Publications of the Astronomical Society of the Pacific, 123*, 412–422.

Wu, P., & Peltier, W. (1984). Pleistocene deglaciation and the Earth's rotation: A new analysis. *Geophysical Journal of the Royal Astronomical Society, 76*, 753–791.

Wu, Y., Llewellyn Smith, S., Rottman, J., Broutman, D., & Minster, J. (2020). Time-dependent propagation of tsunami-generated acoustic–gravity waves in the atmosphere. *Journal of the Atmospheric Sciences, 77*, 1233–1244.

Wunsch, C. & Heimbach, P. (2013). Two decades of the Atlantic meridional overturning circulation: anatomy, variations, extremes, prediction, and overcoming its limitations. *Journal of Climatology, 26*, 7167–7186.

Wunsch, C. (2002). What is the thermohaline circulation? *Science, 298*, 1179–1181.

Wynn, J., Kucks, R., & Grybeck, D. (1999). An Interpretation of the Aeromagnetic and Gravity Data and Derivative Maps of the Craig and Dixon Entrance 1°x3° Quadrangles and the Western Edges of the Ketchikan and Prince Rupert Quadrangles, Southeastern Alaska. USGS Open-File Report 99-316. Downloaded April 2022 from https://pubs.er.usgs.gov/publication/ofr99316

Wyrtki, K. (1975). El Niño – The dynamic response of the equatorial Pacific Ocean to atmospheric forcing. *Journal of Physical Oceanography, 5*, 572–584.

Wysession, M. (1996). Large-scale structure at the core-mantle boundary from diffracted waves. *Nature, 382*, 244–248.

Xie, S. (1998). Ocean-atmosphere interaction in the making of the Walker circulation and equatorial cold tongue. *Journal of Climate, 11*, 189–201.

Xiong, J. (2006). Photosynthesis: What color was its origin? *Genome Biology, 7.* doi:10.1186/gb-2006-7-12-245

Xu, C., & Chao, B. (2017). Coseismic changes of gravitational potential energy induced by global earthquakes based on spherical-Earth elastic dislocation theory. *Journal of Geophysical Research, 122,* 4053–4063. doi:10.1002/2017JB014204

Xu, C., Liu, J., Song, C., Jiang, W., & Shi, C. (2000). GPS measurements of present-day uplift in the Southern Tibet. *Earth Planets Space, 52,* 735–739.

Xu, J., Zhang, P., Haule, K., Minar, J., Wimmer, S., Ebert, H., & Cohen, R. (2018). Thermal conductivity and electrical resistivity of solid iron at Earth's core conditions from first principles. *Physical Review Letters, 121.* doi:10.1103/PhysRevLett.121.096601

Xu, Y., McCammon, C., & Poe, B. (1998). The effect of alumina on the electrical conductivity of silicate perovskite. *Science, 282,* 922–924.

Xu, Y., Shankland, T., & Poe, B. (2000). Laboratory-based electrical conductivity in the Earth's mantle. *Journal of Geophysical Research, 105,* 27,865–27,875.

Xu, Y., Shankland, T., & Poe, B. (2003). Correction to "Laboratory-based electrical conductivity in the Earth's mantle." *Journal of Geophysical Research, 108,* 2314. doi:10.1029/2003JB002552

Xue, C., Liu, J., Li, Q., Wu, J., Yang, S, Liu, Q, Shao, C., Tu, L., Hu, Z., & Luo, J. (2020). Precision measurement of the Newtonian gravitational constant. *National Science Review, 7,* 1803–1817.

Yadav, R., & Bloxham, J. (2020). Deep rotating convection generates the polar hexagon on Saturn. *Proceedings of the National Academy of Sciences, 117,* 13991–13996.

Yadav, R., Gastine, T., Christensen, U., Wolka, S., & Poppenhaeger, K. (2016). Approaching a realistic force balance in geodynamo simulations. *Proceedings of the National Academy of Sciences, 113,* 12065–12070.

Yagi, T., Bell, P., & Mao, H. (1979). Phase relations in the system MgO-FeO-SiO$_2$ between 150 and 700 kbar at1000°C. *Carnegie Institution Yearbook 1978–1979, 78,* 614–618.

Yagi, T., Mao, H., & Bell, P. (1979). Lattice parameters and specific volume for the perovskite phase of orthopyroxene composition, (Mg,Fe)SiO$_3$. *Carnegie Institution Yearbook 1978–1979, 78,* 612–613.

Yamazaki, Y., & Maute, A. (2017). Sq and EEJ – A review on the daily variation of the geomagnetic field caused by ionospheric dynamo currents. *Space Science Reviews,* 299–405.

Yang, X., & Wang, Q. (2002). Gravity anomaly during the Mohe total solar eclipse and new constraint on gravitational shielding parameter. *Astrophysics and Space Science, 282,* 245–253.

Yao, T., Thompson, L., Yang, W., Yu, W., Gao, Y., Guo, X., et al. (2012). Different glacier status with atmospheric circulations in Tibetan Plateau and surroundings. *Nature Climate Change, 2,* 663–667. doi:10.1038/NCLIMATE1580

Yau, K., & Stephenson, F. (1988). A revised catalogue of Far Eastern observations of sunspots (165 B.C.to AD 1918). *Quarterly Journal of the Royal Astronomical Society, 29,* 175–197.

Yeganeh-Haeri, A., Weidner, D., & Parise, J. (1992). Elasticity of alpha-cristobalite: A silicon dioxide with a negative Poisson's ratio. *Science, 257,* 650–652.

Yeomans, D., Rahe, J., & Freitag, R. (1986). The history of Comet Halley. *Journal of the Royal Astronomical Society of Canada, 80,* 62–86.

Yin, A. (2012). Structural analysis of the Valles Marineris fault zone: Possible evidence for large-scale strike-slip faulting on Mars. *Lithosphere, 4,* 286–330.

Yin, Y., Zhang, Q., Zhang, Y., Zhai, S., & Liu, Y. (2022). Electrical and thermal conductivity of Earth's core and its thermal evolution – A review. *Acta Geochimica, 41,* 665–688.

Yoder, C. (1995). Astrometric and geodetic properties of Earth and the Solar System. In T. Ahrens (Ed.), *Global Earth Physics: A handbook of physical constants* (pp. 1–31).

Washington, DC: American Geophysical Union. doi:10.1029/RF001p0001

Yoder, C., Williams, J., Dickey, J., Schutz, B., Eanes, R., & Tapley, B. (1983). Secular variation of Earth's gravitational harmonic J2 coefficient from Lageos and non-tidal acceleration of Earth rotation. *Nature, 303,* 757–762.

Yong, W., Secco, R., Littleton, J., & Silber, R. (2019). The iron invariance: Implications for thermal convection in Earth's core. *Geophysical Research Letters, 46,* 11065–11070.

Yorke, H., & Bodenheimer, P. (1999). The Formation of protostellar disks. III. The influence of gravitationally induced angular momentum transport on disk structure and appearance. *Astrophysical Journal, 525,* 330–342.

Yoshihara, A., & Hamano, Y. (2004). Paleomagnetic constraints on the Archean geomagnetic field intensity obtained from komatiites of the Barberton and Belingwe greenstone belts, South Africa and Zimbabwe. *Precambrian Research, 131,* 111–142.

Youdin, A., & Goodman, J. (2005). Streaming instabilities in protoplanetary disks. *Astrophysical Journal, 620,* 459–469.

Young, G., von Brunn, V., Gold, D., & Minter, W. (1998). Earth's oldest reported glaciation: Physical and chemical evidence from the Archean Mozaan Group (~2.9 Ga) of South Africa. *Journal of Geology, 106,* 523–538.

Young, I., & Ribal, A. (2019). Multiplatform evaluation of global trends in wind speed and wave height. *Science, 364,* 548–552.

Young, N., Schweinsberg, A., Briner, J., & Schaefer, J. (2015). Glacier maxima in Baffin Bay during the Medieval Warm Period coeval with Norse settlement. *Science Advances, 1,* e1500806. doi:10.1126/sciadv.1500806

Yu, S., Tornqvist, T., & Hu, P. (2012). Quantifying Holocene lithospheric subsidence rates underneath the Mississippi Delta. *Earth and Planetary Science Letters, 331–332,* 21–30. doi:10.1016/j.epsl.2012.02.021

Yuen, D., Cadek, O., Chopelas, A., & Matyska, C. (1993). Geophysical inferences of thermal-chemical structures in the lower mantle. *Geophysical Research Letters, 20,* 899–902.

Yuen, D., Sabadini, R., & Boschi, E. (1982). Viscosity of the lower mantle as inferred from rotational data. *Journal of Geophysical Research, 87,* 10,745–10,762.

Yuen, D., Sabadini, R., Gasperini, P., & Boschi, E. (1986). On transient rheology and glacial isostasy. *Journal of Geophysical Research, 91,* 11,420–11,438.

Yuen, D., Scruggs, M. A., Spera, F. J., Zheng, Y., Hu, H., McNutt, S. R., et al. (2022). Under the surface: Pressure-induced planetary-scale waves, volcanic lightning, and gaseous clouds caused by the submarine eruption of Hunga Tonga-hunga ha'apai volcano provide an excellent research opportunity. *Earthquake Research Advances,* preprint.

Yukutake, T. (1967). The westward drift of the Earth's magnetic field in historic times. *Journal of Geomagnetism and Geoelectricity, 19,* 103–116.

Yukutake, T. (1968). The drift velocity of the geomagnetic secular variation. *Journal of Geomagnetism and Geoelectricity, 20,* 403–414.

Yukutake, T., & Tachinaka, H. (1969). Separation of the Earth's magnetic field into drifting and standing parts. *Bulletin of the Earthquake Research Institute Tokyo University, 47,* 65 as cited in McElhinny & Merrill 1975.

Zachos, J., Röhl, U., Schellenberg, S., Sluijs, A., Hodell, D., Kelly, D., et al. (2005). Rapid acidification of the ocean during the Paleocene-Eocene Thermal Maximum. *Science, 308,* 1611–1615.

Zahel, W. (1978). The influence of solid earth deformations on semidiurnal and diurnal oceanic tides. In P. Brosche & J. Sündermann (Eds.), *Tidal friction and the Earth's rotation* (pp. 98–124). Berlin: Springer-Verlag.

Zahnle, K. (2001). Decline and fall of the Martian empire. *Nature, 412,* 209–213.

Zahnle, K. (2005). Being there (News and views). *Nature, 433,* 814–815.

Zahnle, K., & Sleep, N. (1997). Impacts and the early evolution of life. In P. Thomas, C. Chyba,

& C. McKay (Eds.), *Comets and the origin and evolution of life* (pp. 175–208). New York, NY: Springer-Verlag.

Zaranek, S., & Parmentier, E. (2004). Convective cooling of an initially stably stratified fluid with temperature-dependent viscosity: Implications for the role of solid-state convection in planetary evolution. *Journal of Geophysical Research, 109*, B03409. doi:10.1029/2003JB002462

Zbinden, E., Holland, H., Feakes, C., & Dobos, S. (1988). The Sturgeon Falls paleosol and the composition of the atmosphere 1.1 Ga BP. *Precambrian Research, 42*, 141–163.

Zeitlin, C., Hassler, D. M., Cucinotta, F. A., Ehresmann, B., Wimmer-Schweingruber, R. F., Brinza, D. E. et al. (2013). Measurements of energetic particle radiation in transit to Mars on the Mars Science Laboratory. *Science, 340*, 1080–1084.

Zghal, M., Bouali, H., Ben Lakhdar, Z., & Hamam, H. (2007). The first steps for learning optics: Ibn Sahl's, Al-Haytham's and Young's works on refraction as typical examples. *Proceedings of SPIE, 9665*. doi:10.1117/12.2207465

Zhang, K., & Schubert, G. (2000). Teleconvection: Remotely driven thermal convection in rotating stratified spherical layers. *Science, 290*, 1944–1947.

Zhang, P., Wang, B., & Wu, Z. (2019). Weak El Niño and winter climate in the mid- to high latitudes of Eurasia. *Journal of Climate, 32*, 405–421.

Zhang, S., Wang, X., Wang, H., Bjerrum, C., Hammarlund, E., Costa, M., et al. (2016a). Sufficient oxygen for animal respiration 1,400 million years ago. *Proceedings of the National Academy of Sciences, 113*, 1731–1736.

Zhang, S., Wang, X., Wang, H., Bjerrum, C., Hammarlund, E., Dahl, T., & Canfield, D. (2016b). Reply to Planavsky et al.: Strong evidence for high atmospheric oxygen levels 1,400 million years ago. *Proceedings of the*

National Academy of Sciences, 113, E2552–E2553.

Zhang, W., Wang, L., Xiang, B., Qi, L., & He, J. (2015). Impacts of two types of La Niña on the NAO during boreal winter. *Climate Dynamics, 44*, 1351–1366.

Zhang, Y., & Zindler, A. (1993). Distribution and evolution of carbon and nitrogen in Earth. *Earth and Planetary Science Letters, 117*, 331–345.

Zhao, D. (2001). Seismic structure and origin of hotspots and mantle plumes. *Earth and Planetary Science Letters, 192*, 251–265.

Zhao, D. (2004). Global tomographic images of mantle plumes and subducting slabs: Insight into deep Earth dynamics. *Physics of the Earth and Planetary Interiors, 146*, 3–34.

Zharkov, V. (2009). On estimating the molecular viscosity of the Earth's outer core: Comment on the paper by D E Smylie et al. *Physics Uspekhi Fizicheskikh Nauk, 52*, 93–95.

Zhong, S., & Davies, G. (1999). Effects of plate and slab viscosities on the geoid. *Earth and Planetary Science Letters, 170*, 487–496.

Zhong, Y., Miller, G., Otto-Bliesner, B., Holland, M., Bailey, D., Schneider, D., & Geirsdottir, A. (2011). Centennial-scale climate change from decadally-paced explosive volcanism: A coupled sea ice-ocean mechanism. *Climate Dynamics, 37*, 2373–2387.

Zhou, M., Paduan, J., & Niiler, P. (2000). Surface currents in the Canary Basin from drifter observations. *Journal of Geophysical Research, 105*, 21,893–21,911.

Zimorski, V., Mentel, M., Tielens, A., & Martin, W. (2019). Energy metabolism in anaerobic eukaryotes and Earth's late oxygenation. *Free Radical Biology and Medicine, 140*, 279–294.

Zuckerman, B., Turner, B., Johnson, D., Clark, F., Lovas, F., Fourikis, N., et al. (1975). Detection of interstellar trans-ethyl alcohol. *Astrophysical Journal, 196*, L99–L102.

Index

a

Absolute gravimeters 270, 272
Acceleration of gravity 235
Accretion 9, 15–16, 613, 620
  Earth's 74, 82, 609
  pebble accretion 17, 20
  runaway gas accretion 24
Accretionary disks 8
Acoustic impedance 445
Acoustic tomography 513
Adams-Williamson equations
  532, 610
  assumptions to question
    545
  derivation of 534–535
  Earth's interior 544–545,
    552–554
  full equations 542
  goal of 534
  gravity 536–537
  hydrostatic equilibrium
    532–533
  indicator of inhomogeneity
    545–547
  mass implications 535–536
  numerical approach
    547–551
  simplified equations 537
  solutions 542–545
  temperature 537–542
Adiabatic (isentropic)
    experiment 538
Adiabatic temperature gradient
  225, 539, 609–611,
  663–665
Adriatic Sea 433
Advection 649, 651
Agricultural Revolution 169
Airy isostasy 290–296
  depth of compensation
    289–292

hydrostatic equilibrium 291
the Moho 293–294
mountain roots 291–293
oceanic response to
  atmospheric pressure
  variations 294–296
regional. *See* Flexural isostasy
Alaska Panhandle 312–313,
  317, 319
Albedo 121
Alfvén waves 742, 760–761
ALH84001 Martian meteorite
  63–65, 756
Aliens 54–55
Allan Hills ALH84001 63–65,
  756
  Roswell, New Mexico 442
α effect 777
αω dynamo 777
α² dynamo 777
Altimetry, satellite 368–379
Alumina 700
Amazing theorem 229, 236,
  243, 389
Amino acids, in Murchison
  meteorite 61
Ampere 694
Ancient magnetic fields 744
Anelasticity 403
Angle of refraction 443
Angular momentum 14–15,
  39, 341–342, 399
  conservation 6, 11, 14, 39,
    321, 738, 741–742
  planetary spacings 38–40
  problem 14
  torque and 341–342
Angular velocity 253, 350, 506
Animal magnetism 690
Anisotropy 593–595
Anomalies magnetic field 692

Anomalistic year 350
Anomalous potential 382
Anoxic conditions 108
Antarctica ice core 307, 308
Antarctic Bottom Water (ABW)
  138
Antarctic Circumpolar Current
  (ACC) 143, 148–149,
  765
Antarctic ice sheets 205–209
Antarctic ozone hole 215–216
Antarctic warming 141, 674
Anthropogenic climate change
  163, 173
  direct consequences
    193–201
  indirect consequences
    201–222
Anticyclones 129, 130
Antiferromagnetism 689
Antinode 433
Antioxidants 114
Aphelion 349
Apollo Moon 48
Apparent polar wander path
  735
Apparent velocity 443
Archaean 94, 100
Archimedes' principle 290,
  291, 297
ARGO floats 210–211
Arrokoth 13
Asteroid Belt 6, 7, 10, 22, 23,
  26–28, 31, 41, 60–62
Asteroid Ida 27
Asteroids 616
Asteroseismology 474
Asthenosphere 590–595, 657,
  698–699
  anisotropy 593–595
  conductivity of 696–698

Earth System Geophysics, Advanced Textbook 6, First Edition. Steven R. Dickman.
© 2025 American Geophysical Union. Published 2025 by John Wiley & Sons, Inc.
Companion Website: www.wiley.com/go/Dickman

Asthenosphere (*contd.*)
  convection within   657–658
  damp   698
  electrical   696–699
  nature of   590–591
  water in   591–593
Astronomical unit (AU)   13,
    348
Asymmetric mantle convection
    589
2005 Atlantic hurricane season
    220–221
Atmosphere
  Adams-Williamson
      application   542–543
  carbon dioxide   94–99
  Earth   70, 71, 82–84
  magnetic field   682–684
  other terrestrial planets
      84–85
  oxygen   99–118
  runaway greenhouse   92
  scale height   543
  temperature gradient (lapse
      rate)   612–613
  waveguides   439–442
Atmospheric circulation
  Jupiter's   131–133
  nonrotating Earth
      123–124
  role of Sun   120–122
  rotating Earth   124–125
  three-cell   123, 125–131
  troposphere   122–123
Atomic dipoles   689
Aurora   684
Australia   96
Autumnal equinox   346
Auxetic materials   431
Axial precession   340, 342,
    350, 357
Axial strain   415

b

Backward spin   4, 5
Bacteria   56, 57
Baltic Sea   317
Banded iron formations
    (BIFs)   105–108
Bangui magnetic anomaly
    692, 709
Bar   410
Bar magnets   678–680,
    705–706

produced by current loop
    679–680
Baroclinic fluids   654
Barotropic fluids   654
Beating phenomenon   461–463
Bénard convection   658–662,
    781–782
BIFROST (Baseline Inferences
    for Fennoscandian
    Rebound Observations,
    Sea-level, and Tectonics)
    program   307
Big Bang   9
Big burp   81–82
"Big G"   230–232
Binghamton, NY, flood of June
    2006   220
Binghamton University
    campus, seismic
    refraction   478–481
Biosphere   85–86
Bipolar jets   19
Bipolar seesaw   142
Birch's rule   522–532
  applications to Earth's
      interior   524–527
  and core   527–529
  and mantle   529–531
  mean atomic weight
      522–524
  modified   523, 527
  philosophical guidance from
      533–534
  power-law form of   525–526
  schematic depiction of   523
  tomographic application
      531–532
Black diamonds   692
Bode's Law. *See* Titius-Bode Law
Body forces   412
Body tides   329
Body waves   437, 451
Bombardment. *See* Heavy
    Bombardment
Bonneville Flood   312
Bootstrapping   468
Borehole data   599–600, 669,
    671–675
  and climate change   669,
      671–675
Bouguer anomaly   276–277,
    280, 293–294, 302–304
Bouguer approximation
    277–278, 301

Bouguer correction   276–277
Bouguer's discovery   287–289
Boundary conditions   619
Boundary layers   658–662
  Ekman   739
  temperatures within
      662–666
  thermal   659–660, 662
  viscous   660
Bradbury, Ray   62
Bridgmanite   573
Brown dwarf   49
Brunhes chron   750, 753
Bruns formula   382
Brunt-Väisälä frequency
    646
Bucher Glacier   426
Bulk modulus   428–430
Bulk sound velocity   524
Bumps on the core   563
Buoyancy   227, 640, 769–770
Buoyancy force   644–647
Buoyancy waves   640

c

California Current   146
Cambrian explosion   109
Canary Current   146
Capture hypothesis   405
Carbonado diamonds   692
Carbonate reservoir   90
Carbonate-silicate cycle. *See*
    Urey cycle
Carbonatite   699
Carbon dioxide (CO_2)   88, 90
  abundance   170, 171
  estimates   97–98
  Ice House climate constraints
      on   94–97
  lower levels of   96–97
Carbonic acid   90
Carina Nebula   20
Carousel force   12
Cavendish experiment
    239–242
Celestial pole   343
Central force   228
Centrifugal acceleration   250,
    256, 259–261
Centrifugal force   11, 249–256,
    344
  gravity *vs.*   258–263
  indirect effects on gravity
      261–265

Kepler's Third Law 266–267
   magnitude of 254–256
Centripetal force 249
Cepheid variable 474
Chagos-Laccadive ridge 387
Chalcophile element 617
CHAMP (Challenging
      Minisatellite Payload)
      391, 710–712
Characteristic time
   diffusion of heat 636–638
   diffusion of momentum 643
Chemical convection 587
Chemical homogeneity 534,
      654
Chemical weathering 90–91
Chemosynthetic organisms 99
Chilean earthquake (1960)
      460–462
Chirality 61
Chlorofluorocarbons (CFCs)
      171
Chromium 103
Circumstellar disk 8, 11–15
Clairaut's theorem 263–264,
      366
Clapeyron slope 571, 576
Clathrates 170, 172
Clausius-Clapeyron equation
      571, 579
   Clapeyron slope 571
Climate
   impact of precession
      345–356
   refugees 176
   reversal 93
   Urey cycle and 93
Climate change 142, 669–675.
      See also Anthropogenic
      climate change; Global
      warming
Climate computer models
      197–198, 200
Climate seesaw 141–142
Climatic precession 351
Clumps 15–18
CO_2 partial pressure (pCO_2)
      98
Coal forests 111
Coastal upwelling 147
Coefficient of thermal
      expansion 531
Cold Jupiters 27
Columnar convection 783

Comet(s) 12–13, 25
   Churyumov-Gerasimenko
      59
   Hale-Bopp 57, 58
   Halley 12, 13, 26
   Shoemaker-Levy 9 337
   Tempel 1 59
   Wild 2 12, 58–59
Comet tails 25
Compass 678
Complementary error function
      674
Compressional heat 609–613
Computer assisted tomography
      (CAT) 513
Condensation temperature 6
Conductance 688
Conduction of electricity
      696–698
   electrical conductivity
      692–696, 701–702, 710,
      713
   Ohm's Law 424, 693–695
Conduction of heat 599, 621,
      630–638
   alternatives to conduction
      599
   conduction equation
      (Fourier's Law) 599,
      606
   thermal conductivity 599,
      631–633
Conservation of angular
      momentum 6, 11, 14,
      39, 321, 738, 741–742
Conservation of energy
      622–623, 775
Conservation of mass
      (continuity) 649, 775
Continental crust 608
Continental heat flow
      603–605
Continental lithosphere 297,
      300, 306, 309, 310,
      507–508, 520
Continental shields 556
Continuity 775
Continuum 43, 243, 247,
      434–435, 471
Convection 71–72, 560, 561,
      599, 757–758
   Bénard 658
   chemical 588
   columnar 783

core 71, 72
   double-diffusive 140
   horizontal 652–658
   idealized 661–663
   mantle 76, 81, 93, 325,
      507–510, 560, 561,
      589–590, 593, 638–639
   thermal 75–76
   upper-and lower-mantle
      509
Convective buoyancy flux 757,
      758
Convective regulation 589
Convective Taylor columns
      782, 787
Convective zone 472–475
Convergent plate boundaries
      492, 494
Cooling of upper atmosphere
      180, 213–215
Core 23, 344–345
   Birch's rule 527–529
   bumps 563
   condensation 72–74
   convection 71, 72
   differentiation 74–75
   Earth's magnetic field
      702–707
   equations of state 567–568
   leaky 563
   light alloys 583–588
   mean atomic weight 528
   radiogenic heat 617–619
Core formation 75–82, 85–86
   biosphere consequences
      85–86
   core consequences 75–76
   mantle consequences 76–80
   ocean-atmosphere
      consequences 80–82
Core-mantle boundary (CMB)
      448, 515, 658, 664–666,
      785
Core-mantle coupling
      739–741
Coriolis deflection 123–126,
      128–131, 136, 145–148,
      158, 258
Coriolis force 123–125,
      129–131, 134, 145, 183,
      249, 251, 256–258, 378,
      402, 467, 757, 773, 778,
      783
Coriolis parameter 258, 378,
      772

Corona 4
CoRoT (Convection, Rotation,
    and planetary Transits)
    49
Cosmic rays 789
Coulomb (unit) 694
Coulomb's law 693
Cowling's theorem 736, 756
Cratons 719
Creep 494–495, 509, 589, 639
    creep resistance 589
Cretaceous Normal Superchron
    720, 749–751
Cretaceous quiet interval 720
Crichton, Michael, *Jurassic Park*
    62
Critical angle 438
Critical Rayleigh number
    647
Critical refraction 439
Cross-over distance 477–479
Cross product. *See* Vector, cross
    product
Crustal field 691, 718–726
Crustal subsidence/rebound
    304–307, 312–314
Cryogenian 95, 96
Curie temperature 690–691
Current density 693
Cyclones 129, 130, 379

d
Dalton Minimum 167
Damped harmonic oscillator
    395
Dansgaard-Oeschger events
    141
Darcy's law 599
Darwin, *The Tides* 401
D double prime (D″) 511, 519,
    521, 560–563, 576–579,
    665
Debris disks 28
Deccan Traps 387
Declination 682, 683
Deep-ocean oxygenation 110
Del (gradient operator) 606,
    622–624
Density
    differences 138–140
    and seismic velocities 522
    temperature and 541–542
Deoxyribonucleic acid (DNA)
    55

Depositional remanent
    magnetism 744
Depth of compensation 288,
    290–292
Detrital uraninite 104
Devonian jawed fish 114
Diamagnetism 690
Differential force 331
Differential gravity 332
Differential rotation 474,
    738–741, 764–765,
    776–778, 786
Differentiation 73–75, 80, 85
Diffraction 515
Diffusion 624–627, 740
    as an averaging process
        625–626
    of momentum 640, 642, 660
    one-way in time 625
    temperature profile
        625–626
Diffusion coefficient 624
Diffusion equation 623 762
    derivation 1-D, 621–622
    derivation 3-D, 622–624
    nature of diffusion 624–627
    solutions 627–638
Dilatation 430, 434
Dimensional analysis 624
Dimensionless numbers 640
    Ekman number 772
    Elsasser number 770–771
    magnetic Reynolds number
        771
    Peclet number 652
    Prandtl number 643
    Rayleigh number 647, 654,
        770
    Reynolds number 643, 770
    Rossby number 770
    Taylor number 772
Dipole 702, 787
    eccentric 706, 707
    excentric 706
Dipole axis 679
Dipole field 678–680, 726–728,
    745, 746
Dipole moment 745
Dipole moment decline 746,
    747, 752
Dip-slip faults 493, 494
Direct imaging of exoplanets
    51, 53
Direct *P* wave 477

Disk dynamo 765–768
    initial "feeder" field 768
    mesoscale eddies 768
Dispersion 453–457, 683
Displacement gradient 418
Disturbing potential 382
Diurnal inequality 334
Diurnal tide 334
Divergence 244–246, 622–624,
    713, 760, 775
Divergence of heat flux
    622–624
Divergent plate boundaries
    492, 494
Doldrums 129
Domains 689
Doppler technique 50–51
Dot product 369, 623
Double-diffusive convection
    140
Doubling of planetary spacings
    40–41
Downslope conversion 451
Downward continuation 715
Dragonfly 115
Droughts 185, 186, 217
Dualistic theories 3–5
Ductile materials 419
Dynamical flattening 344
Dynamic ellipticity 344
Dynamic pressure 533
Dynamic topography
    377–380, 388–389, 563
Dynamic viscosity 315
Dynamo 677, 755–756
    conceptual evolution of
        788
    disk 765–768
    equations 738
    kinematic 778
    magnetohydrodynamic
        778
    Parker 776–778
    preliminary requirements
        from geomagnetic field
        755–756
        from solar system
        perspective 69–70,
        756–758
    reversals 737
    self-exciting 765
    spontaneously reversing
        787
Dynes 410

e

Early Eocene Climate Optimum (EECO) 93
Earth
  accretion 74, 82, 609
  albedo 121
  atmosphere 70, 71, 82–83, 121
  creation of 71
  curvature 122
  differences in environment 54
  diffusion equation 629–630
  equatorial bulge 320, 321, 323, 325, 332–333, 335, 344
  flattening 248–249, 262, 263
  free oscillations 472
  Gravitational Model 2008 (EGM2008) 384
  gravity field 236–248, 366–368, 383, 390–392
  gravity *vs.* centrifugal force 258–261
  Heavy Bombardment 34–37
  hum 271, 470–472
  idealized 263
  incoming solar radiation 120–122
  iron core 7
  isostatic response to surface loads
    anomalous regions 300–313
    changing surface loads 296–300
    mantle rheology 313–319
  moment of inertia 45
  normal modes of 271
  obliquity 35, 333, 347
  orbit 403
  orientation 35
  pear shape 359, 367
  perturbations to 347–348
  precession
    axial 340, 342
    core, geomagnetic field and 344–345
    impact on climate 345–356
    prograde 342
    rate of 343–344
    retrograde 342
  precessional torque 338–339

pressure 410–412
radius 238–239, 248, 263–264, 287–288
responses to applied stresses 425–427
rotation
  length of day 322, 325–328
  true polar wander 323–328, 735–736
  wobble 322–324, 639
rotation reference frame 249–252, 467
shape 120, 248–249, 262
strain 416
structure 72, 75
tidal friction
  consequences 401–403
  effect on spin 401
tides 329
Urey cycle 92
weighing the Earth 239–242
Earthquake 271, 322, 416, 445, 446, 687
  afterslip 393
  arc method 489
  Chilean earthquake (1960) 460–462
  depth of 508–509
  epicentre 489
  focus 489
  Loma Prieta earthquake (1989) 433
  P and *S* waves 450
  seiches 433
  seismic waves 437
  Sumatra-Andaman earthquake (2004) 392–393, 468
  travel times 484–489
Earth's interior
  Adams-Williamson equations 544–545, 552–554
  application of Birch's rule 524–527
  heat flux 667–668
  physical state of 552–554
  travel times and 510–514
  viscosity of 588–589
Earth's magnetic field 85–86, 682–707. *See also* Dynamo
  atmosphere 682–684, 701–702

buried dipole source and 704–706
components of 707, 708
core 702–707
crust/mantle 688–701
dipole *vs.* non-dipole 726–728
distance 703
external components of 703
importance of
  convection 757–758
  dynamo 755–756
  rotation 756–757
magnetic storms 684–685
magnitude 704
oceans 685–688, 702
poloidal 764, 765
rock magnetism 688–692
satellite missions 708–711
scaling analysis 770–772
spatial scale 704
spherical harmonics 711–718
strengthening 763–765
temporal variations 703, 728–730
  archaeomagnetic and paleomagnetic 743–755
  intensity 751–755
  polarity reversals 747–751
toroidal 764–765
unit complications 712, 717
weakening 759–760
East Pacific Rise 160
Eccentric dipole 706
Eccentricity of Earth's orbit 348, 349
Eclipse 231, 242–243, 326, 398–399
Ecliptic 4, 5
Eddy-driven jet 134
Eddy viscosity 394
Eigenfrequencies 459, 491
Eigenvibrations 457
Ekman boundary layer 739, 783
Ekman number 772, 786
Ekman pumping 783
Elastic-brittle materials 419
Elastic constant 420
Elastic limit 419
Elastic materials 419–421

Elastic modulus   420
Elastic parameter   420,
    427–431
Elastic-plastic (ductile)
    materials   419
Elastic waves   431–457
    converted waves   445–448
    reflection and refraction
        437–452
    wave equation   433–437
Electrical asthenosphere   696
Electrical conductivity
        692–696, 701–702, 710,
        713
Electrical current   693, 694
Electrical potential energy   694
Electrical resistivity   695
Electrical resistance   695
Electric field   693
Electron spin   688, 689
Elements of magnetic field
        680–682
Elevation correction   274–275
Ellipticity of Earth's orbit   348
El Niño   150–151. *See also* La
    Niña
El Niño-Southern Oscillation
        (ENSO) events   154,
        163
    chronology   152–153
    global warming   185,
        191–193
    return to normalcy   159–160
    strong   155–158
    weak   158–159
    westerly wind bursts   160,
        161
Elsasser number   770
Empire State Building, NYC
        297
Energy conservation   622–624
Energy, definition of   56, 370,
        694
Enstatite   90
Enthalpy   610
Entropy   540–541
Entropy gradient   541
Envisat   375
Eötvös correction   286–287
Eötvös effect   286
Equations of state   564–568
Equatorial bulge   320, 321,
        323–325, 332–333, 335,
        344

Equatorial countercurrent
        147
Equatorial electrojet   684
Equilibrium ocean tide   333
Equinox
    autumnal   346
    precession of   346, 360
    vernal   346
Equipotential surface
        368–371, 379
Eratosthenes' method (Earth's
        radius)   238–239
Error function   634, 674
ERS-1   375
ERS-2   375, 376
Eukaryotes   112–113
Eurasian flooding of 2010
        221
Europa (moon)   60
European heat wave (2003)
        219, 220
Eustasy   179
Evidence for life
    in Solar System   57–70
    in Universe   54–57
Excentric dipole   706
Excitons   632
Excursions   748
Exoplanet detection   49–54
    direct imaging   51, 53
    Doppler technique   50–51
    microlensing   51, 52
    transit method   49–50
Exoplanets   49
External field   703
Extrasolar asteroid belts   27
Extreme geostrophy   779–780
Extreme weather effects
        181–185, 216–222

f

Faculae   164
Faint young Sun paradox   88
Falling sky effect   180, 701
Farallon plate   500–503
Farallon slab   558, 560
Far-sided effect   745
Fault plane   493
Faults   493, 494
Fault zones   419
Fayalite   529
Feedback loops   121
Fermat's Principle   443–444
Ferrel cell   125, 127, 128

Ferrimagnetic materials
        (ferrites)   689
    lodestone   689
Ferrites   689
Ferromagnetism   689
Ferropericlase   575, 700
Fick's law   599
Field lines   679
Field strengthening   763–765
Field weakening   759–760
Finite strength   419
Finland–Scandinavia region,
        post-glacial uplift
        311
First Oxygenation Event   103,
        104, 108
First transition   521, 570–571
Fish   114
Fission hypotheses   405–406
Flattening   248
Flexural isostasy   296–300
    imperfect compensation
        299–300
    Vening-Meinesz-Turcotte
        mechanism   298
Flexural rigidity   298
Flexure   297
Flood   217, 220, 221, 312
Fluid convection   639
    buoyancy force   644–647
    fluid stability   639–642
    heat sources   648–649
    heat transmission   649–658
    mechanism   640–649
    ocean   649
    solid-state *vs.*   658–666
    as a struggle   640, 647–649
Fluids
    non-Newtonian fluids   427
    stress *vs.* strain   420–422
Flux   245–246
Flux lines   679
Flux-melting   508
Focus, earthquake   489
Forebulge   305, 312
Forseismic bootstrapping   468
Forsterite   529, 569
Fossils
    charcoal   109
    eukaryotic   113
    microbial   100–102
Fourier analysis   463, 633
Fourier's law   599, 606
Fractionation   104

Free-air anomaly 276, 303–304
Free-air correction 276–277
Free electrons 631
Free oscillations (normal modes) 271, 457, 469–470
 eigenfrequencies 459–462
 fundamental modes 460
 higher-order modes (overtones) 460
 hum 470–472
 mode-splitting 466
 noise 470
 spectra of 466, 467
 spherical harmonics 459–460
 spheroidal 457
 Sun 472–475
 toroidal 458
Free surface 445
Frequency
 Brunt-Väisälä 646
 wave 432
Friction as shear stress 412
Frozen flux 678, 760–764
Full Adams-Williamson equations 542
Fundamental mode 456, 460

g
Gal 269
Gauss coefficients 717
Gauss' Law 245–246
Generalized Hooke's Law 421, 427, 434
General precession 351
Geodetic surveys 304
GEODVEL 504
Geodynamo 677
Geographical coincidences 719
Geoid 329
 definition of 371, 377
 global 383–394
 and satellite altimetry 368–379
 slope of 383
 spheroid *vs.* 379–383
 undulations of 387–389
Geoidal height 380–382, 388
 Bruns formula 382
 disturbing potential 382

Geomagnetic field 702, 703. *See also* Earth's magnetic field
Geomagnetic jerks 729–731
Geomagnetic polarity timescale 749–750
Geomagnetic poles 702–703, 729
Geomagnetic secular acceleration (GSA) 728
Geomagnetic secular variation (GSV) 728, 729
Geometric progression 39
Geoneutrinos 607
Geopotential 370
Geostrophic currents 379
Geostrophic flow 130, 773, 778
Geothermometer 571–572
Gerstenkorn event 400, 405
Giant impact hypothesis 406–407
Giant planets 22
Gigantism 115–116
Glacial isostatic adjustments 209
Glacial melting 201–208
Gleissberg Cycle 169
Global atmospheric electric circuit 701
Global average temperature 193–197
Global conveyor-belt circulation 139, 142–144
Global oxygenation 103–105
 BIF perspective 105–108
Global Positioning System (GPS) 383, 503
Global warming 119, 141, 173–175, 701
 indirect consequences of 174–185, 201–222
 cooling of upper atmosphere 180, 213–215
 ENSO 185, 191–193
 extreme weather effects 181–185, 216–222
 interactions with one another 185, 188
 irregularities 188–191
 jet streams 183–184
 ozone holes 180–181, 215–216
 sea-level rise 174–180, 208–214

 shifts in natural vegetation 185, 188
GOCE (Gravity field and steady-state Ocean Circulation Explorer) 384
GONG (Global Oscillation Network Group) 474
GRACE (Gravity Recovery and Climate Experiment) 205–209, 211, 212, 383, 384, 386, 391–393
Gradient operator (del) 606
Grand Canyon of Mars 737
Gravitational coupling 741
Gravitational energy 18, 609, 613
Gravitational potential 370
Gravitational potential energy 189, 370
Gravitational tides 329
Gravity 227–228
 acceleration of 235
 Adams-Williamson equations 536–537
 attracting force 228
 centrifugal force *vs.* 258–261
 differential 332
 indirect effects of centrifugal force 261–265
 International Gravity Formula 264–265, 274, 276
 inverse-square law force 228–230, 243–246
 Law of Gravitation 228–235
 observed 261, 370
 superposition principle 242–248
Gravity field 236
 Earth's 236–242
 of three-dimensional Earth 242, 244–247
Gravity measurements and corrections
 absolute measurements 270
 base station 270, 275, 279–283
 Bouguer anomaly 276–277, 280, 293–294, 302–304
 Bouguer correction 276–277, 279
 drift correction 270, 281–282, 287

Gravity measurements and corrections (*contd.*)
  elevation correction 274–275
  elevation effect 274
  Eötvös correction 286–287
  Eötvös effect 286
  example of 276
  free-air anomaly 276, 303–304
  free-air correction 276–277
  gravimeters 270–271
  instrumental drift 281
  isostatic anomaly 301–302
  isostatic correction 301
  latitude correction 280–281
  mass correction 276–279
  mass effect 276–277
  moving platforms 285–287
  Nettleton's method 284–285
  relative measurements 270
  satellite measurments. *See* Satellite
  spatial anomalies 274
  superconducting gravimeters 270–272, 467–471
  Susquehanna River valley 281–283
  topographic effects 278
  topographic (terrain) correction 279–280
Gravity on the Spheroid 264
Gravity waves 442, 640
Great Oxidation Event 103, 105
Great Red Spot of Jupiter 132, 781
Greenhouse effect 88, 89
Greenhouse gas abundance 170–173. *See also* Global warming
Greenhouse gases 94, 355
Greenland ice sheets 205–209, 688
Grounding line 178
Growth cores 23–24
Grüneisen parameter (Grüneisen ratio) 663
Gulf Stream 145, 146
Gyre 143, 145–147

h

Habitable zone 72, 92
Hadley cell 125–127

Half-life 615
Hawaiian–Emperor seamount chain 512
Hawaiian hotspot 511–513
Head wave 439
Heat capacity (per unit mass) 610
Heat flow
  by conduction 599, 630–638
  continental 602–605
  by convection 599
  data 600–605
  from hot to cold 598
  heat flux 598
  huge amount of energy 601–603
  magnitude, *vs.* other global processes 601–603
  oceanic 602–605
  one-dimensional 621–622
  by radiation 599
  theoretical foundation of 606–607
  thermal conductivity 599
  three-dimensional 622–624
Heat flow units (hfu) 601
Heat of compression 609–613
Heat of fusion 619
Heat sources
  compressional heat 609–613
  core radioactivity 617–618
  fluid convection 648–649
  gravitational energy 609, 613
  inner-core growth 618–619
  mantle radioactivity 614–617
  Moon-forming impact 613–614
  radioactivity 607–609
  summary of 619–621
  tidal friction 614
Heat transmission 649–658
Heavy Bombardment 29–37, 84, 91
  Earth System 34–37
  late 30
  Moon 33–34
  terminal phase 32
Heinrich events 140
Helicity 777
Helioseismology 474
Helium 9

Herbig-Haro (HH) phase 19, 29
Higher-order modes 460
High-pressure experiments 568, 587, 617–618, 631, 633, 696
  convection connections in mantle 579–583
  on iron and core's light alloy 583–588
  lower mantle 576
  material inhomogeneity within transition zone 568–571
  post-perovskite 576–579
  pyroxene 574–576
  second transition 572–574, 581, 582
High-pressure system 130–131
Himalayas 305
Hole 632
Homo erectus 790
Homo habilis 790
Homologous temperature 666
Homo sapiens 789
Hooke's Law 420–421, 427. *See also* Generalized Hooke's Law
Horizontal convection 652–658
Horizontal pendulum 239
Horse latitudes 129, 131
Hot House climate 93, 305–306
Hot Jupiters 27
Hudson Bay 317, 322–324, 326, 368
Hum
  Earth 470–472
  Sun 472, 473
Humboldt Current 146
Huronian 94, 107
Hurricane 136, 182, 217–218, 220–221, 294–296
Hurricane breeding grounds 136, 137
Hurricane Katrina 296, 305
Hydrogen 9
Hydrologic cycle 211
Hydromagnetics 678
Hydrostatic equilibrium 532–533, 653–654, 663, 769

Hydrostatic pressure   533
Hydrothermal circulation
     601–602

i

Ibn Sahl's Law   438
Ice Age   306–313
Ice House climate   93–97,
     305–306
Icequakes   450
Ice-rafted debris   140
ICESat (Ice, Cloud, and land
     Elevation Satellite)
     206, 375
Ice-sheet growth   354
Ice sheets   306–310
Ice tectonics   60
Ideal Earth   263
Ideal gas law   543, 565
Ideal materials
     models of   420–425
     types   419–420
Igneous rocks   77
IGRF-13   702
Impact basins   30
Impact degassing   82
Inclination   681–683
Incompressibility   429–430,
     525–526, 562, 565, 566
Index of refraction   438
Induced magnetization
     724–725
Induction equation   762–763
Inertia   233–235. *See also*
     Moment of inertia
Inertial coupling   345
Inertial force   643–644
Inertial path   249–252
Inertial reference frame
     249–250
Infrasound   452
Inner core   742–743, 756
Inner-core growth   618–619
Inner planets   6, 12, 21–22, 45
Inner Solar System   26–28
InSAR (Interferometric
     Synthetic Aperture
     Radar)   206, 207
Insects   115
InSight (Interior Exploration
     using Seismic
     Investigations, Geodesy
     and Heat Transport)
     724

Insolation   121, 354–356
Insolation difference   122
Intergovernmental Panel on
     Climate Change (IPCC)
     119
Intermolecular collisions   18
Internal heat   189, 190
Internal waves   640
International Commission on
     Stratigraphy (ICS)   36
International Geomagnetic
     Reference Field (IGRF)
     728
International Gravity Formula
     264–265, 274, 276
Inter-tropical convergence zone
     (ITCZ)   127, 129, 131
Inverse-square nature of gravity
     228–230, 243–246
Inverse theory   673
Inverted barometer   295
Io (moon)   59, 337
Ionosphere   180, 330, 683–685
Ionospheric wind dynamo   684
Iron   689
Iron enrichment   110
Isentropic 539
Isopycnal surfaces   654
Isostasy   269, 287–328
     Airy   290–296
     Bouguer's discovery
         287–288
     flexural (regional)   298–300
     Pratt   289–290
     principle of   288
Isostatic anomaly   301–302
Isostatic correction   300–301
Isostatic response to surface
     loads
     anomalous regions   300–313
     changing surface loads
         296–300
     mantle rheology   313–319
Isothermal experiment
     538–539
Isotropy   593–595
ITRF2005   504

j

J_2 262
Jack Hills   89
Jaramillo sub-chron   752
Jason-1   375
Jason-2   375

Jawed fish   114
Jerks   729–730, 740
Jet streams   133–136, 183–184
Jupiter
     atmospheric circulation
         patterns   131–133
     cold   27
     gravity   27
     Great Red Spot   132, 781
     hot   27
     mass   23
     moons of   59–60
     perturbations to Earth's orbit
         347–348

k

Kelvin-Voigt firmoviscous
     materials   422–424
Kepler-90 planetary system   42
Kepler's Laws   265–267
Kepler space telescope   49
Kepler's Second Law   265–266,
     337
Kepler's Third Law   40, 41, 51,
     227, 238, 242, 265–267,
     330–332, 361, 399, 614
Kinematic dynamo   778
Kinematic viscosity   315, 642
Kinetic energy   189, 607, 609
Kuiper Belt   6, 13, 24, 32, 57–59
Kuroshio   145
Kursk magnetic anomaly   692,
     709

l

LAGEOS (LAser GEOdynamics
     Satellite)   360
Lake Bonneville   312, 317, 319,
     391
Lake Geneva   433
Lamé's parameter (Lamé's
     constant)   431
Laminar flow   644
Land plants   110–111
La Niña   161–163
Laplace's equation   366, 628,
     713–715
Laplace tide equations   769
Laplacian   624, 629
Lapse rate   612
Large low shear velocity
     provinces (LLSVPs)
     559–561

Late heavy bombardment 30
Latent heat 189, 190
Latitude
  correction 280–281
Laurentide ice sheet 320
Law of Gravitation 261, 267.
      See also Newton's Law of
      Gravitation
  direction and magnitude
      228
  gravity formula 230–235
  superposition principle
      242–248
  vector description of
      232–235
  verbal formulation 228–230
Laws of quantum mechanics
      689
Leaky core 563
Least squares 464, 480, 513,
      554
Left lateral strike-slip fault 493
Legendre functions 714–715,
      717
Length of day (LOD) 321
Level surface 368–371, 654
Libration 337
Life
  alien 54–55
  earliest 100–105
  requirements for 55–57
  in Solar System 57–70
  in Universe 54
Light alloy 80, 528, 529,
      583–588
Lightning 683, 701
  return stroke 683
Lines of force (flux lines) 679
Lithosphere 297, 319
  continental 297, 300, 306,
      309, 310, 507–508, 520
  flexural rigidity 298
  oceanic 76, 77, 91, 297–300,
      310, 495, 507, 564
  relict (remnant) slabs 500,
      509, 557–563, 579, 582
  rheology 300
Lithospheric plates 492–508
Little Ice Age (LIA) 165–167,
      674, 675
Load tide 330
Local Airy isostasy. *See* Airy
      isostasy
Lodestones 689

Lomagundi excursion 104
Longitudinal strain 415–416
Longitudinal wave 435, 445
Long-period tides 335
Lorentz force 762–763, 766,
      769, 773
Lorenz number 631
Loss of ice 201–208
Love wave 452–457. *See also*
      Surface waves
Lower Jaramillo reversal 753,
      754
Lower mantle 519, 521, 546,
      559–560, 573, 576
  conductivity 697
  convection 319
Lower Olduvai reversal 752,
      753
Lower thermal boundary layer
      659
Lowes-Mauersberger spectrum
      721
Low-pressure system 130
Low-velocity zone (LVZ)
      518–520, 699
Low-viscosity channel 319
Low-viscosity zone (LVZ) 590
Lunar laser ranging (LLR)
      230, 400
Lunar torque 396–397

m

Magma 508
Magma ocean 22, 36, 74, 82
Magnesiowüstite 575, 700
Magnetic annihilators 725,
      726
Magnetic-Archimedean-Coriolis
      (MAC) waves 742
Magnetic-Coriolis (MC) waves
      742
Magnetic diffusivity 759
Magnetic domain 689
Magnetic field 48. *See also*
      Earth's magnetic field
  crustal 718–726
  divergence of fields 679,
      713
  elemental description of
      680–682
  field strength from flux line
      spacing 679–680
  how fields strengthen
      761–765, 786

  how fields weaken
      758–760
  frozen flux 678, 760–764
  measurements 682
  measurements in Hermanus,
      South Africa 685
  as vector quantity 681
Magnetic flux density 681
Magnetic permeability 712,
      759
Magnetic poles 682, 729
Magnetic reconnection 760
Magnetic Reynolds number
      771
  as a frozen flux number 771
Magnetic storms 684–685
Magnetic susceptibility 724,
      744
Magnetic tide 330
Magnetite 689
Magnetohydrodynamic (MHD)
      dynamos 778
Magnetohydrodynamics 678
Magnetometer 708
Magnetosphere 684
Magnetospheric Multiscale
      Mission 710
Magnetostrophic flow 773
Magnetotelluric currents
      693
Main field 702, 703
Malcolm, Ian 62
Mantle
  adiabat *vs.* solidus 665–666
  asymmetries 741
  Birch's rule 529–531
  convection 76, 81, 93, 325,
      507–510, 561, 579–583,
      590, 593, 638–639
  core-mantle coupling
      739–741
  electrical conductivity 713
  lower 519, 521, 559–560,
      576
  material inhomogeneity in
      545–547
  as Maxwell solid 425, 427
  plumes 107, 387, 510–513,
      554–555, 559–561, 590
  radiogenic heat 614–620
  rheology 313–319
  temperatures 667
  tomographic structure 564
  upper 518, 521, 556–559

upwelling and downwelling differ 582–583
viscosity 314, 316–319, 326, 589
Mantle Electromagnetic and Tomography (MELT) project 699
Maria (lunar) 29–30
Marine gravity anomalies 373–374
Marine sediment cores, oxygen isotope analysis in 352, 355, 356
Mars
  atmosphere 62, 65
  crustal magnetization 723–724
  differentiation 85
  dust storm on 357, 358
  evidence of life 62–69
  hemispherical view of 66
  loss of surface water 85
  magnetic field 69, 85
  meteorites 63–65, 756
  Milankovitch and 357–358
  moment of inertia 48
  plate tectonics 737
  precession 357
  presence of water 62, 65, 67–68
  sedimentary rocks on 68
  south polar region 358
  Viking Lander 62–63
  volcanoes 65, 66
Martian ice age 358
Martian meteorite ALH 84001 63–65, 756
Mascons 33
Mass conservation 582, 775
Mass correction 276–277
Material derivative 650
Material homogeneity 534
Matuyama chron 750
Maunder Minimum 165
Maxwell's relations 610
Maxwell viscoelastic material (Maxwell solid) 422, 423
Mean atomic weight 522–524, 526–530
Mean sea level 371, 377
Mean sea surface 377
Mean solar day 259

Medieval warm period 166, 674
Mercury 30, 31, 350
Meridional overturning circulation (MOC) 142
Meta-sedimentary rocks 100
Meteorites, Martian 63–64, 756
Meteorological equator 126, 127, 129
Methane (CH_4) 94, 100, 107, 108, 170, 172
Methane hydrates 170
Mg-rich bridgmanite 576
Microbial fossils 100–102
Microgravity studies 271–274
Microlensing 51, 52
Microseisms 470–472
Mid-cretaceous hot house 93
Mid-latitude cell 125
Mid-latitude westerlies 124
Mid-mantle electrical conductivity 699–701
Mid-ocean ridge basalt (MORB) 531
Midwest U.S. tornado activity 219
Milankovitch 351
  hypothesis 351–356
  and Mars 357–358
Mineral enrichment 110, 117
Mississippi River 305, 317
Mitochondria 112
Mode splitting 466, 468
Modified Rayleigh number 770
Mohorovičic discontinuity (Moho) 293–294, 297, 298
Molecular cloud 8
Molecular cloud core 8
Molecular viscosity 315
Molybdenum enrichment 110
Moment of inertia 42–45, 262, 405
  continuous mass 43
  definition 43
  Earth 45
  homogeneous and non-homogeneous spheres 43–45
  Mars 48
  Moon 48, 405
Moment tensor 469

Monistic theories 3, 6–7
Mono Lake excursion 752
Monsoons 656
Moon 48–49
  creation of 74
  far side of 29
  Heavy Bombardment 33–34
  libration of 337
  Man in the Moon 30, 34
  maria 29–30
  mascons 33
  moment of inertia 48, 405
  near side of 29, 30
  orbit, tidal friction effects 399–401
  origin of 405–407
    capture hypothesis 405
    fission hypotheses 405–406
    giant impact hypothesis 406–407
    same-face moons 34
    tidal locking 404
Moon-forming impact 613–614
Moons of Jupiter 59–60
MORVEL plate model 502
Mountain root 290
Multilevel global conveyor belt 143, 144
Murchison meteorite 61

n
Nafe-Drake curve 524
Nanofossils 64
Narrow-band filtering 461
NASA
  Deep Impact mission 59
  Exoplanet Exploration website 52
  MAVEN mission 85
Navier-Stokes equation 769, 770, 773
Neap tide 335
Nebula
  collapse 9–10
  primitive 8
  theory 6, 13–14
Negative feedback 121
Neoproterozoic Oxygenation Event 109–110
Neptune 45, 48
Nettleton's method 284–285
Neutral stability 639, 641–642

Newton's First Law   249
Newton's Law of Gravitation
        228–235, 261, 267
Newtons (N)   410
Newton's Second Law
        235–242, 316, 341, 434,
        643, 645–647, 768–769
Newton's Third Law   249, 533
Nickel   585
'No monopole' rule   679
Non-dipole field   703, 706–707,
        726–728, 745
Nontidal acceleration   321
Nonuniqueness
    gravity field   389, 392, 725
    magnetic field   726, 745
Normal incidence   445
Normal faults   493
Normal modes. *See* Free
        oscillations
Normal polarity   702
Normal stress   410
North Atlantic Deep Water
        (NADW)   138, 140, 141
North Atlantic Ocean   149–150
Northern hemisphere ice sheets
        309
North pole   679
*n*th mode   455
Nuclear winter   26
Nucleosynthesis   9, 616, 617
NUVEL-1A   498, 504
NUVEL-1 Plate models   498

o
^{16}O   307–309
^{18}O   307–309, 352, 353
Oblate spheroid   248
Obliquity   35, 333, 347
Observed gravity   260, 370
Ocean circulation   136
    plate tectonics and   149
    thermohaline circulation
        137–143
    wind-driven circulation
        143–150
        Antarctic Circumpolar
            Current   143
        equatorial countercurrent
            147
        gyre   143, 145–147
        heat transfer   147–150
        undercurrents   147
Oceanic crust   608

Oceanic heat flow   602–605
Oceanic lithosphere   76, 77,
        91, 297–300, 310, 495,
        507, 564
Oceanic plates   498
Ocean loading   320
Oceans   83–84, 88
    convection   649
    currents   378–379
    electrical conductivity   686
    magnetic field   685–688
    response to tidal force
        335–338
    tidal friction without
        403–404
    viscosity   394
    waveguides   439–442
Ocean temperatures   201, 202
Offshore breeze   656
Ohm   694
Ohm's Law   424, 693–695
Olduvai
    Gorge   790
    reversals   752–754, 789–790
    subchron   752, 789
Olivine   575, 697
    fayalite, forsterite   529–531
    to ringwoodite phase
        transition   569–570, 572
    structure of   568
    to wadsleyite phase transition
        569–572
Olympus Mons   65, 66
ω effect   777
One-dimensional heat flow
        621–622
Oort Cloud   24–25, 57–59
Orbit
    aphelion   349
    eccentricity   348, 349
    ellipticity   348
    geometry   348–351
    Jupiter's perturbations
        347–350
    perihelion   349
Orbital period   39
Orbital resonances   27
Organic chemicals   55, 56
Organic compounds   56
Orion Nebula   21
Ørsted   710, 711
Outer core viscosity   595
Outer planets   22–24, 45
Outer Solar System   24–26

Outgassing   73, 80–82, 97
Overtones   460
Oxygen (O_2)   588
    chronology   116–118
    estimates of   116–118
    and evolution   112–116
    history of   99
    oasis   103
    stages of development
        99–112
        III stage: oxygenic dark
            ages   108–109
        II stage: global oxygenation
            103–108
        I stage: earliest life
            100–105
        IV stage: oxygen-rich
            109–112
Ozone depletion   786
Ozone holes   180–181, 215–216

p
Pacific superplume
        (megaplume)   559, 789
Paleocene-eocene thermal
        maximum (PETM)   93
Paleolatitude   734
Paleomagnetic data   231, 727,
        744, 745, 748
    far-sided effect   745
Paleomagnetic field   735
Paleomagnetism   691
Paleosols   97, 103, 105
Panspermia   26
Paramagnetic materials   689
Parker (Parker-Levy) dynamo
        776–778
Parker waves   778
Partial derivative   538
Partial melting   591, 614–615,
        617, 665, 698
Pascal-sec   316
Pascals (Pa)   410
Pebble accretion   17, 20
Peclet number   652
Perihelion   229, 349. *See also*
        Precession of the
        perihelion
Period
    orbital   39
    waves   432
Peripheral bulge   305, 311–312
Perovskite (Bridgmanite)
        572–575

Peru Current 146

Phanerozoic 36, 37, 97, 98, 109, 114, 115

Phanerozoic superchrons 750

Phonons 631

Photosynthesis 99, 100, 103, 108, 113

Pizzetti-Somigliana latitude formula 272

Plages 164

Planetary properties 45–48

Planetary spacings 37–42. *See also* Titius-Bode Law

Planetesimals 17, 82, 616

Plasma 760

Plate mechanism 298

Plate models 498, 499, 502, 504, 507

Plate motions
  Farallon plate 500–503
  geologic perspective 498–500
  quantifying 495–498, 507
  time scales 498–505
  VLBI 503

Plates 492

Plate tectonics 71, 76–79, 93, 96, 149, 297, 323, 492–509, 698–699, 737

Plate velocities 495–498, 505–507
  closure 507
  instantaneous 503

Pleistocene 312

Pleistocene blitzkrieg 141

Pleistocene deglaciation 208, 304, 310, 312, 320, 322, 324, 327

Pleochroic haloes 607

Plumes 510–513, 554–555, 559–561, 590

Pluto's spin 5

Poincaré coupling 345

Poincaré force 250–251

Poise 316

Poisson's equation 628

Poisson solids 430

Poisson's ratio 430–431

Polar convection cell 125

Polar drift 326–328

Polar easterlies 125

Polar Front 129

Polar front jet 134

Polaris 343

Polar jet 133–134

Polaron conductivity 696

Polar Orbiting Geophysical Observatories (POGO) 709, 711

Polar stratospheric clouds (PSCs) 181, 182

Polar vortex 685, 785

Poles 679, 682

Poloidal field 764, 765

Population growth 167, 173, 174

Poroelastic rebound 393–394

Positive feedback 121, 171, 178

Post-earthquake fault creep 393

Post-glacial rebound 209

Post-perovskite 576–579

Potassium 607–608, 615–618

Power spectrum 466

Prandtl number 643

Pratt isostasy 289–290

Precambrian 109, 111

Precession 338, 340
  climatic 351
  Earth's
    axial 340, 342, 350
    core and geomagnetic field 344–345
    impact on climate 345–356
  general 351
  moving bicycle 339–340
  orbital 350, 351
  prograde 342
  rate of 343–344, 359
  retrograde 342
  satellite orbital 359–368
  spinning top 340

Precessional torque 338–339

Precession of the ecliptic 351

Precession of the equator 351

Precession of the equinoxes 329, 346, 351, 360

Precession of the perihelion 350

Preliminary Reference Earth Model (PREM) 460, 521, 552–554, 566, 567

Present atmospheric level (PAL) 109

Pressure 410, 411

Pressure force 533, 653

Pressure gradient 533, 538

Pressure-release melting 508

Primitive nebula 8–11, 28–29

Principle of equivalence 237–238

Principle of isostasy 288, 294

Principle of least time 443–444

Principle of reciprocity 444

Prograde precession 342

Project Mogul 442

Prokaryotes 112

Proportional limit 419

Protoplanetary disk 8, 12, 15–17, 19

Proto-star 18–19

P wave 434–437
  conversion 445
  diffraction 515
  incident 445–447
  sound wave equivalence 435
  travel times 510, 554, 556–559
  velocity 435, 436, 440, 517–523, 556

Pyrolite 531, 577–578

Pyroxene 574–576, 697

q

Quadrupole field 707

Quasars 503

Quasi-geostrophic flow 773

r

Radial drift 17

Radiant heat 189

Radiation 599

Radiational tides 330

Radiative thermal conductivity 632, 633

Radiative zone 475

Radioactivity 607–609, 614–618

Radiogenic heat 74, 603, 607
  core 617–618
  mantle 614–617

Radio waves 683

Rain, acidic 90

Rate of uplift 316–317

Rayleigh number 647–648, 652

Rayleigh wave 452–457. *See also* Surface waves

Ray parameter 448–450

Reference ellipsoid 264

Reference seismic velocity
    model   514
Reflection   437–452
Refraction   437–452
Refraction survey   475–476
    Binghamton University
        campus   478–481
    cross-over distance   477–479
    direct *P* wave   477
    first-arrival travel times
        475–478
    geophones   476, 477
    head wave   476–477
    multilayered situations
        481–482
Regional Airy isostasy
    297–298
Relative gravimeters   270
Relative plate motions
    495–497
Remanent magnetization   691,
    722, 724–725, 744
Resonant gaps   27, 28
Response function   700
Retrograde precession   342
Return stroke   683
REVEL-2000   504
Reversals   737–738, 747–751,
    787, 789
Reversed flux patches   746–747
Reversed polarity   702
Reynolds number   643, 733,
    734
Rheology
    lithosphere   300
    mantle   313–319
Ribonucleic acid (RNA)   55
Right-hand rule   253–255
Right lateral strike-slip fault
    493
Rigid lid   76
Rings of Saturn   336
Ringwoodite (spinel)   568–569
    olivine phase transition
        569–570, 572
    wadsleyite phase transition
        572
Roche limit   336, 337
Rock-dwelling bacteria   56, 57
Rock magnetism   682, 688–692
Rossby number   770
Rossby wave   135
Rotation and the dynamo
    756–757

Rotational forces   251–252
Rotation as reality   249–252
Rule of thumb   280, 740, 741
Runaway gas accretion   24
Runaway greenhouse
    atmosphere   92
Runaway growth   20
Russian heat wave (2010)
    221

S
SAC-C   710, 711
Sagittarius B2 molecular cloud
    56
Same-face moons   34
San Andreas Fault   493–494
Satellite
    altimetry   212–214, 368–379
    measurements of Earth's
        gravity field   366–368,
        390–392
    measurements of Earth's
        magnetic field   708–711
    oceanography   375–379
    orbital precession   359–368
    range   360–361
    range rate   361
Satellite tracking
    doppler tracking   361
    satellite laser ranging (SLR)
        361
Saturn
    rings   336
    whistler   684
Sawtooth pattern   352–356
Sawtooth-pattern timescales
    306–309
Scalar   235
Scale height   543
Scaling   636
Schmidt functions   717
Schumann resonances   701
Screening effect   740
Sea breeze   656
Sea floor magnetization   691
Sea level drop   310
Sea level rise   174–180,
    208–214, 320
Search for Extra-Terrestrial
    Intelligence (SETI)
    program   55
Seasonal average   126
Seasonal markers in Earth's
    orbit   346

Sea surface temperature (SST)
    151, 155–158, 162, 191
Secondary eclipse   50
Second Law of thermodynamics
    610
Second Oxygenation Event
    109
Second transition   521,
    572–575
Secular variation   728, 730,
    739, 742–745
Secular variation impulses   729
Sediment cores   309
Sediment deposition   305
Seiches   432–433
    node, antinode   433
Seismicity   492, 493
Seismic parameter   523
Seismic refraction surveys
    475–476, 517
    Binghamton University
        campus   478–481
    cross-over distance   477–479
    first-arrival travel times
        475–478
    geophones   476, 477
    multilayered situations
        481–482
Seismic (acoustic) tomography
    513, 554–564
Seismic velocities   516–521
    interior velocity structure
        518–521
Seismic velocity–density
    relation   522
Seismic waves   48, 409,
    431–438, 445–457. *See
    also* Elastic waves
    body waves *vs.* surface waves
        437
    reflection and refraction
        437–438
Seismogram   486, 487, 489
Seismology   409, 491
Self-exciting dynamo   765
Self-gravitation   296
Semiconductors   632, 692
Semidiurnal tide   334
Separation of variables   714
Settling toward the mid-plane
    16
Shadow zone   514–516
Shear modulus (shear rigidity)
    430, 431

Shear strain   415–416, 434, 436
Shear stress   314–317, 410, 412
Shear wave   436, 760–761. *See also* S wave
Shear-wave splitting   593
Short-lived isotopes   616, 617
SH wave   436
Sidereal day   259, 260
Sidereal year   347
Siderophilic   587
Siemens   694
Silicates   632
Silicon   583
Skin depth   670, 671, 740, 741
Skin effect   599, 669–671, 693, 740
Slow magnetic Rossby waves   742
Snell's Laws   434, 437–439, 447–448, 477, 517
   basis   442–444
   ray parameter   448–450
Snowball Earth   96, 108
Snow line   21
SOFAR (Sound Fixing and Ranging) channel   439–442, 450–452
Soft murmur   470
SoHO (Solar and Heliospheric Observatory)   474
Solar constant   120
Solar day   259, 260
Solar hum   472, 473
Solar irradiance   120
Solar maximum   164
Solar neutrino problem   474
Solar quiet variation (Sq)   684
Solar System   5
   dualistic theories   3–5
   evidence for life   57–70
   formation
      stage 0   4 1/2 billion years ago, 8–9
      stage 1, nebula's collapse   9–10
      stage 2, circumstellar disk formation   11–15
      stage 3, clumping   15–18
      stage 4, heat generation   18
      stage 5, proto-star   18–19
      stage 6, protoplanetary disk   19–28
      stage 7, Sun ignited   28–29

stage 8, bombardment   29–37
      inner   26–28
      monistic theories   3, 6–7
      outer   24–26
      properties of   37–49
Solar wind   4
Solid earth response to tidal forces   335–338
Solid-earth tides   329
Solid spherical harmonics   460, 711–715
Solid-state convection   658–666
Solidus   639, 665–666
Sound waves   435, 439–442, 450–452
   infrasound   452
   T waves   451
South Atlantic Anomaly (SAA)   730
Southern Ocean   148
Southern Oscillation   151–155
Southern Oscillation Index (SOI)   153–154
South pole   679
Space geodetic techniques   326
Specific heat   610, 623
Spectral analysis   352–353, 463–466, 474
   narrow-band filtering   461
Spectrum   464–466
Spherical harmonics   361–366
   Earth's internal magnetic field   715–716
   free oscillations   457–461
   harmonic degree and order   363, 367
   high-degree   720
   solid   460, 711–715
   surface   460
   vector   459
   wavelength *vs.* source depth   373, 389–390, 704–706, 720–725
Spherical harmonic geomagnetic spectrum   720–724
Spheroid   264, 379–383
Spheroidal oscillation   457
Spinel phase (Ringwoodite)   568–569
Spin-orbit coupling   34, 337, 404

Spörer Minimum   167
Spring tide   335
Stability
   fluid   639–642
   neutral   639
Stagnant lid   76
Standardization of temperature data   193–194
Standard linear solid   423–425
Standing waves   432
Star   18
Starlette   360
Static ocean tide   333
Stationary orbits   229
Steady state   624, 629–630, 633–636
Steric rise in sea level   210, 211
Stishovite   575
Stokes coefficients   366
Strain   414
   longitudinal   415–416
   in real-life situations   416–417, 419–420
   rigorous description of   417–418
   rotational   436
   shear   415–416, 434, 436
   stress-strain relations of various materials   419–431
   tensor   418
   tidal   417
   volumetric   430
Stratosphere   180
Stratospheric cooling   213–215
Stratospheric sudden warming   685
Streaming instabilities   17
Stress   410
   decomposition   413
   Earth system's response to   425–427
   normal   410
   rigorous description of   412–414
   sedimentological setting   411
   shear   410, 412
   stress-strain relations of various materials   419–431
   surface orientation   413
   vector   412–414

Stress tensor 414
Strike-slip faults 493, 494
Stromatolites 101
Strong ENSO event 155–158
Subadiabatic conditions 639
Subadiabatic temperature
    gradient 541
Subduction-zone volcanism
    508
Subsolidus mantle convection
    665
Subtropical jet 134
Sulfate aerosols 198, 199
Sulfur
    in Earth's core 583–588
    siderophilic 587
Sumatra-Andaman earthquake
    (2004) 392–393, 468
Summer solstice 346
Sun
    atmospheric circulation
        120–122
    changing luminosity 87–88
    convective zone 472
    corona 4
    evolution of 87
    faint young Sun paradox 88
    free oscillations (solar hum)
        472–473
    fusion reaction 474
    Kepler's Laws 266
    rotation 164
    rotational profile 475
    solar neutrino problem 474
Sun-quakes 472
Sunspot cycles 164–165
Sunspots 163–164
    Earth's climate variations
        165–169
    Gleissberg Cycle 169
Superadiabatic conditions 639
Superadiabatic temperature
    gradient 541
Supernova 9–10
Superposition principle
    242–248
Surface currents 147, 148
Surface spherical harmonics
    460
Surface temperatures, changes
    in 200–201
Surface waves 437, 446,
    452–457
    dispersion 453–457
    fundamental mode 456

modes 455
nodes 455
travel times 488, 489
Susquehanna River valley,
    gravity data 281–283
Sverdrups (Sv) 143
SV wave 436
Swarm 709–711
S wave 436–437
    conversion 445
    SV and *SH* components 436
    travel times 483–485
    velocity 437
Symmetric strain tensor 418
Synchronous rotation 337,
    404

t

Tangent cylinder 783
Taylor column 781–788
    in the oceans and atmosphere
        785–786
Taylor number 772
Taylor-Proudman theorem
    779–782
Technology 54–55
Teleconnections 155
Teleseisms 416
Temperature
    Adams-Williamson equations
        537–542
    anomalies 193–195
    and density 541–542
    rock magnetism 690
    trends 198–201
    variations 541
Temperature gradient 538,
    539, 599–600, 606
Terminal phase heavy
    bombardment 32
Thermal conductivity 599,
    631–633
Thermal convection 75–76
Thermal diffusivity 624
Thermal expansion coefficient
    210, 531
Thermal inertia 126
Thermal remanent
    magnetization 691
Thermocline 162
Thermohaline circulation
    137–143, 655, 667–669
Thermonuclear reaction 28
Thermosphere 180

Thermosteric rise in sea level
    210, 211
Thermosulfine 587
Tholins 57
Thorium 607–608, 615–616
Three-cell circulation 123,
    125, 218
    climate implications
        128–129
    complications of 125–128
    global warming 182
    regional circulation
        129–131
Thrust faults 493
Tidal bulge 332–335
Tidal force 330–332
    as a differential gravity
        331–332
    response of oceans 332–335
    response of solid Earth
        335–338
Tidal friction 5, 329, 394, 397,
    614, 620
    consequences for Earth
        system 401–403
    effect on Earth's spin
        397–399, 401
    effect on Moon's orbit
        399–401
    sources of friction 394–396
    without oceans 403–404
Tidal heating 336–338
Tidal imbalance 331
Tidal locking 34, 337, 404
Tidal phase lag 396, 403, 404
Tidal rhythmites 397–398
Tidal strain 417
Tides 329–338, 417
    body 329
    diurnal 334
    Earth 329
    gravitational 329
    load 330
    long-period 335
    magnetic 330
    neap 335
    radiational 330
    secondary 330
    semidiurnal 334
    solid-earth 329
    spring 335
    tidal phase lag 396
Timescales of Ice Age 306–313
Time series analysis 463

Tissint 64
Titius-Bode Law 399
  angular momentum 38–40
  astronomical confirmation of
    41–42
  astrophysical implications of
    41
  doubling expression 40–41
  planetary spacings 37–38
  predictions *vs.* reality 38
  tests of 42
2011 Tohoku earthquake and
    tsunami 687
Tomography 513, 554–564
TOPEX-Poseidon 375
Topographic coupling 472
Topographic (terrain) correction
    279–280
Tornado 183, 219
Toroidal field 764–765, 777,
    786
Toroidal oscillation 458
Torque 240, 359
  and angular momentum
    341–342
  evidence of 397–399
  lunar 396–397
  precessional 338–339
Torsion constant 239
Total derivative 650
Total internal reflection 439
Trade Winds 124, 126, 136,
    143, 150, 155, 160
Transform boundaries 492
Transition zone 519–521, 546,
    567–576, 579–582,
    699–701
Transition-zone topography
    580–581
Transit method 49–50
Transverse wave 436, 445
Trappist planetary system
    42
TRAPPIST (Transiting Planets
    and Planetesimals Small
    Telescope) 50
Traveling waves 432
Travel-time residuals 513
Travel times 475, 510
  t-Δ curves 483–485
  and Earth's interior
    510–514
    seismic velocity structure
    516–521

shadow zone and core
    514–516
  first-arrival 475–478. *See
    also* Refraction survey
  homogeneous spherical Earth
    516–517
  locating earthquakes
    484–489
  location-independent 483
  and plumes 510–513
  real Earth 517
  spherical symmetry 483
  surface wave 488, 489
Tropical convection cell 125
Tropical easterlies 124, 126
Tropical year 347
Troposphere 96, 122–123
True polar wander 323,
    735–736
Tsunami 392, 687
T-Tauri phase 7, 19, 82
Turbulence 16, 734, 774–775
Turbulent flow 15–16, 132,
    142, 395, 472, 644, 734
  mean flow 774
Turbulent (eddy) viscosity 394
T wave 451
Two-cell circulation 128

u
Ultimate strength 419
Ultra-low-velocity zones
    (ULVZs) 560, 561
Undercurrents 147
Underground drought 481
Undulations of the Geoid
    387–389
Unit vector 232
Universal Constant of
    Gravitation 230–232
Unsteady diffusion equation
    633–635
Uplift 305, 307, 311–313,
    316–318, 335
Upper atmosphere cooling
    180
Upper Jaramillo reversal 753,
    754
Upper mantle 518–521,
    556–559
Upper Olduvai reversal 752,
    753
Upper thermal boundary layer
    659

Upward continuation 715
Uraninite 103–104
Uranium 607–608, 615–616
Uranus 45, 48
Urban heat islands 196
Urey cycle 90–91
  implications of 91–93
  influence on climate 93
  limitations of 94

v
Vanadium enrichment 110
Van Allen radiation belts 683
Vector 228–229, 232, 233
  addition 233–234
  cross product 253–254
  spherical harmonics
    459–460
Velocity
  advection of 649
  anomalies 559, 560
  apparent 443
  ocean current 378–379
Velocity gradient 314, 316
Velocity profile 314–316
Vening-Meinesz–Turcotte
    mechanism 298
Venus 756
  atmosphere 73, 85
  core and mantle 73
  gravity 350
  iron core 7
  loss of surface water 85
  spin 5
  stagnant lid 758
Vernal equinox 346
Very long baseline
    interferometry (VLBI)
    503
Virtual geomagnetic pole (VGP)
    735
Viscoelastic materials 420,
    422–425
  analogy with Ohm's Law
    424
Viscosity 227, 314, 666
  convective regulation 589
  dynamic 315
  Earth's interior 588–589
  eddy 394
  estimates 317, 318
  fluid 314–316
  kinematic 315, 642

Viscosity (*contd.*)
  mantle 316–321, 326–327, 509
  molecular 315
  oceans 394
  outer core 595
  turbulent 394
Viscous boundary layer 660
Viscous drag 640, 642–644
  momentum diffusion 640
Viscous materials 421–422
Volcanic dust 168
Volcanic eruptions 160–161, 167–169
Volcanoes 65, 66, 80–81
Volt 694
Voltage 695
Volume forces 412
Volume transport 143, 146
Volumetric strain 430
Vorticity 134, 654, 776

W

Wadsleyite 570–575
Walker circulation 158–160, 655–656
Water 90, 99, 591–593
  requirement for life 56
  transition zone 700
Water cycle 211
Water table 480
Wave equation 433–437
Waveguide 439–442, 701
Wave of polarization 632
Waves 432–433, 742. *See also* Elastic waves

amplitudes 444–445
body 437, 452
frequency 432
infrasound 452
longitudinal 435, 445
period 432
P wave 434–438, 445–448
speed of 395
standing 432
surface 437, 446, 452–457
S wave 436
transverse 436, 445
traveling 432
T wave 451
wavefront 432
wavelength 432
Wave train 454
Weak ENSO event 158–160
Weighing the Earth 239–242
Weight 235
West Antarctic Ice Sheet (WAIS) 176–180
Westerly wind bursts (WWBs) 160, 161
Western boundary current 145
Westward drift 730–738, 743, 744, 786
  from hydromagnetic waves 742–743
  from the inner core 742–744
West Wind Drift 143
Whistlers 683–684
Whole-mantle convection 581–583

Wiedemann-Franz-Lorenz law 631
Wildfire 109, 185, 187, 217, 221–222
Wind-driven circulation 143–150
  Antarctic Circumpolar Current 143
  equatorial countercurrent 147
  gyre 143, 145–147
  heat transfer 147–150
  undercurrents 147
Wind shear 182
Winter solstice 346
Wobble 322–324, 639
Wolf Minimum 167
Wollastonite 90
Work, definition of 368–369
Work done 369
World Digital Magnetic Anomaly Map (WDMAM) 719
Wüstite 572

Y

Yellowstone plume 511, 531
Younger Dryas 140–141
Young's modulus 428–429

Z

Zenith angle 238, 287
Zircon 89
Zonal harmonics 363, 366
Zone of rapid change 439